T5-AEW-821

Reservoir Characterization III

Proceedings
Third International
Reservoir Characterization Technical Conference
Tulsa, Oklahoma
November 3-5, 1991

Edited by

Bill Linville
IIT Research Institute
National Institute for Petroleum and Energy Research
Bartlesville, Oklahoma

Conference Co-Chairmen

Thomas E. Burchfield
IIT Research Institute
National Institute for Petroleum and Energy Research
Bartlesville, Oklahoma

Thomas C. Wesson
United States Department of Energy
Bartlesville, Oklahoma

PennWell Books
A Division of
PennWell Publishing Company
Tulsa, Oklahoma

1421 South Sheridan/P.O. Box 1260
Tulsa, Oklahoma 74101

International Reservoir Characterization Technical Conference (3rd: 1991: Tulsa, Okla.)
Reservoir characterization III: proceedings, Third International Reservoir Characterization Technical Conference, Tulsa, Oklahoma, November 3–5, 1991/edited by Bill Linville; conference co-chairmen, Thomas E. Burchfield, Thomas C. Wesson.
p. cm.
"Doe conf—911125."
NIPER–558."
Includes bibliographical references.
ISBN 0-87814-392-0
1. Oil fields—Congresses. 2. Gas reservoirs—Congresses. 3. Petroleum—Geology—Congresses. 4. Natural gas—Geology—Congresses. I. Linville, Bill. II. Title. III. Title: Reservoir characterization three. IV. Title: Reservoir characterization 3.
TN870.5.I53 1991
622'.3382—dc20 92–43076
CIP

Printed in the United States of America

1 2 3 4 5 97 96 95 94 93

Table of Contents

Session 3
Modeling/Description of Interwell Region

Session 4
Optimization of Reservoir Management

Session 5
Poster Presentations

Contributing Authors

Yngve Aasum
Saad M. Al-Haddad
William Almon
A. Aly
U. G. Araktingi
Paul Armitage
Khalid Aziz
Mary L. Barrett
W. M. Bashore
Steve Begg
R. A. Beier
R. Bibby
Peter Blumling
Rodney R. Boade
R. E. Bretz
Alan A. Burzlaff
Ming-Ming Chang
Mark A. Chapin
Ioannis Chatzis
H. Y. Chen
A. K. Chopra
William M. Colleary
Patrick W. M. Corbett
Dwight Dauben
Roussos Dimitrakopoulos
Christine Doughty
K. H. Duerre
Thomas L. Dunn
James M. Eagan
M. A. Ferris
Robert J. Finley
R. M. Flores
G. E. Fogg
Mike Fowler
John A. French
Thierry Gallouët
G. E. Gould
D. W. Green
Jeffry D. Grigsby
Dominique Guérillot
H. H. Hardy
Bob R. Harris, Jr.
K. J. Heffer
Adolfo Henriquez
Kevin Hestir
T. A. Hewett
Leif Hinderaker
William E. Howard
John R. Hulme
Neil Humphreys
Marios A. Ioannidis
Susan Jackson
Jerry L. Jensen
Hou Jiagen
Alistair Jones
Apostolos Kantzas
Ekrem Kasap
H. Kazemi
C. Wm. Keighin
Mohan Kelkar
Charles Kerans
Dennis R. Kerr
Peter King
Zsolt Komlosi
David C. Kopaska-Merkel
Tibor Kuhn
Richard P. Langford
J. C. Lean
Raymond A. Levey
Jane C. S. Long
F. J. Lucia
G. F. Luger
Ernest Majer
Ernest A. Mancini
Steven D. Mann
A. J. Mansure
Stephen Martel
Yves Mathieu
David F. Mayer
Colin McGill
Carl A. Mendoza
Robert M. Mink

Marco A. S. Moraes
Larry Myer
Jim Myers
Richard J. Norris
Mark A. Nozaki
Henning Omre
F. M. Orr, Jr.
J. J. Osowski
W. J. Parkinson
Godofredo Perez
John Peterson
R. A. Phares
Edward D. Pittman
S. W. Poston
Yuanchang Qi
Xin Quanlin
Andrew R. Scott
R. K. Senger
A. A. Shinta
D. Scott Singdahlsen
Mark A. Sippel
Paul G. Soustek
Eve S. Sprunt
Rudy Strobl
Edward A. Sudicky
Ronald C. Surdam
Berry H. Tew
J. M. D. Thomas
Roderick W. Tillman
Håkon Tjelmeland
Liviu Tomutsa
T. T. B. Tran
Kelly Tyler
Francois Verdier
Jose' M. Vidal
Joseph P. Vogt
J. R. Waggoner
A. W. Walton
W. Lynn Watney
R. C. Wattenbarger
G. P. Willhite
John Williams
D. S. Wolcott
Li-Ping Yuan
Liu Zerong

Preface

This book contains the Proceedings of the Third International Reservoir Characterization Technical Conference held November 3-5, 1991, at the Westin Hotel in Tulsa, Oklahoma. The IIT Research Institute National Institute for Petroleum and Energy Research (IITRI/NIPER) and the U.S. Department of Energy Bartlesville Project Office (DOE/BPO)were co-sponsors. The theme of the conference was *Reservoir Characterization Requirements for Different Stages of Development.* Thomas E. Burchfield of NIPER and Thomas C. Wesson, DOE/BPO, were Co-Chairmen. Edith Allison, DOE/BPO, was Project Manager, and Bill Linville of NIPER was Conference Coordinator. The opening address was given by Dr. Donald A. Juckett, Director of the DOE Office of Geoscience Research. Mrs. Denise Bode, President of the Independent Petroleum Association of America (IPAA) gave the luncheon address, Monday, Nov. 4, and Farouk Al-Kasim, consultant, Stavanger, Norway, gave the Tuesday luncheon address. Those three speeches are reproduced in these Proceedings.

The first conference on reservoir characterization was held in Dallas in the summer of 1985, with registration of about 175, and the second conference was held in June 1989, also in Dallas, with registration of 233. Total registration for the third conference was 247, and the following countries were represented: England, Norway, Canada, France, Japan, Scotland, China, Hungary, Germany, Sweden, Venezuela, Australia, and the United States.

Four tutorials were presented on Sunday, November 3, the opening day of the conference. They were as follows:

Opportunities and Strategies for Incremental Recovery in Mature Oil and Gas Reservoirs--The Texas Experience by Noel Tyler and Robert Finley, Bureau of Economic Geology, University of Texas at Austin;

Introduction to Geostatistics by Mohan Kelkar, University of Tulsa;

Seismic Waves: An Engineer's Friend by Jack Caldwell, Schlumberger;

Geomechanics for Reservoir Management by Lawrence W. Teufel, Sandia National Laboratories.

Tutorial presentations are not included in these Proceedings but may be available from the authors.

The conference consisted of four half-day sessions. Each session consisted of six 30-minute presentations and one alternate. A Poster Session was presented Monday evening, November 4. Session topics and co-chairmen were as follows:

Heterogeneities/Anisotropies—Co-Chairmen: William Almon, Texaco E&P, and Rodney R. Boade, Phillips Petroleum Company
Field Studies and Data Needs—Co-Chairmen: Neil Humphreys, Mobil E&P Services, and Charles Kerans, Bureau of Economic Geology, University of Texas at Austin
Modeling/Description of Interwell Region—Co-Chairmen: Steve Begg, BP Exploration-Alaska, and Mohan Kelkar, University of Tulsa
Optimization of Reservoir Management—Co-Chairmen: Mike Fowler, Exxon Company USA, and Susan Jackson, National Institute for Petroleum and Energy Research (NIPER)
Poster Session—Chairman: Dwight Dauben, K&A Energy Consultants

Each session contained presentations based on prepared manuscripts. Following the conference, manuscripts were sent to selected professional scientists and engineers for peer reviews and returned to authors for final revisions based on reviews. Completed manuscripts appear in these Proceedings.

Reservoir characterization is a procedure for generating the input for numerical reservoir simulation and optimizing reservoir management strategies. To improve our understanding of oil and gas production processes, reservoir characterization must:

- accurately reflect the internal and external geology of the reservoir,
- be as quantitative and precise as possible, and
- achieve a degree of detail that accommodates the limitations of simulation.

Reservoir characterization emphasizes both geology and simulation and requires collaboration among geoscientists and engineers. The Third International Reservoir Characterization Technical Conference provided a forum to present research results relevant to recovering the large amount of oil-in-place remaining after primary production.

We thank Dr. Mike Madden, Dr. Min Tham, and Mrs. Susan Jackson, all of NIPER, for their help in organizing the conference, and Mr. Ron Kendall, Mrs. Edna Hatcher, and Miss Beverly Fox, also all of NIPER, for their help with registration and other support services, and we sincerely thank the session chairmen, authors, and reviewers for their efforts and cooperation and Halliburton Services for sponsoring the Poster Session reception. We especially thank the members of the Steering Committee for their help in organizing the program and other work on the conference:

Edith Allison
U. S. Department of Energy
Lee Allison
Utah Geological Survey
Bill Almon
Texaco Exploration & Production Co.
Steve Begg
BP Exploration

Rod Boade
Phillips Petroleum Co.
Thomas E. Burchfield
NIPER
Jack Caldwell
Schlumberger
Frank Carlson
Amoco Production Co.
Dwight Dauben
K & A Energy Consultants
Howard Dennis
McKenzie Petroleum Co.
Mike Fowler
Exxon USA

Neil Humphreys
Mobil Exploration & Producing Services
Susan Jackson
NIPER
Mohan Kelkar
University of Tulsa
Charles Kerans
Bureau of Economic Geology, University of Texas
Bob Lemmon
U.S. Department of Energy
Bill Linville
NIPER
Mike Madden
NIPER
James Russell
Russell Petroleum Co.
Min Tham
NIPER
Thomas C. Wesson
U. S. Department of Energy

Bill Linville
National Institute for Petroleum and Energy Research

Thomas E. Burchfield
National Institute for Petroleum and Energy Research

Thomas C. Wesson
U. S. Department of Energy

KEYNOTE SPEAKER
Monday, November 4, 1991

Dr. Donald A. Juckett
***Director*, Office of Geoscience Research**
Office of Fossil Energy
U.S. Department of Energy

On November 21, 1988, Dr. Juckett began his duties as Director of the Office of Geoscience Research which develops and administers strategic programs for geoscience research in Fossil Energy/DOE. He is responsible for supporting geoscience technology transfer among federal, state, academic, and private sector organizations to ensure that information is available to all interested parties. Prior to his present position, he served as Director, Hydrocarbon Systems and Surface Detection, Research and Services Division for Phillips Petroleum Company. He received his doctorate in physical chemistry from the State University of New York.

Remarks of Dr. Donald A. Juckett
***Director*, Office of Geoscience Research**
Office of Fossil Energy
U. S. Department of Energy

On behalf of Admiral Watkins and the Department of Energy, Tom Wesson and the Bartlesville Project Office, and the National Institute for Petroleum and Energy Research, I welcome you to the Third International Reservoir Characterization Technical Conference. We have an exciting program planned which focuses on a concept critical to the future of oil and gas development.

International gatherings like this are important because they allow us to solidify contacts, develop better understanding of energy issues and concerns in our respective countries, and most importantly, advance the cause of cooperation and friendship throughout the world. In times of peace, it is easy to call one another "friend," but as the Gulf War proved, it is the strength of our friendship in times of crisis that truly matters.

The lessons of that experience are certainly a useful background for understanding the importance and significance of this conference as we focus on improving ultimate recovery. Some of the lessons we learned from the Gulf crisis were that supplies were adequate, even during a serious loss of supply because producers around the world immediately increased production to maximum levels; and that there were ample strategic stocks and coordinated planning for their drawdown.

As a result, the world economy weathered this latest "shock," the largest to date, with surprising ease. But it did reinforce the critical importance of building production capacity outside the Persian Gulf. Diversified supply sources will encourage greater market stability and ease geopolitical pressures in the Middle East.

Many areas of the world offer potential for increased production. While the Middle East will remain the dominant player in world oil markets, there are also opportunities in the former Soviet Republics, in the Western Hemisphere, and in Southeast Asia.

The Department of Energy has just spent 2 years analyzing present and future energy issues as we developed the National Energy Strategy. For the first time ever, we addressed the issues of energy development and use within their full context of energy security, economic growth, environmental quality, and an interdependent world community.

To quote remarks made by Deputy Secretary of Energy Henson Moore at last month's meeting of the Society of Petroleum Engineers, "The Strategy recognizes that oil will remain essential to the world economy for the foreseeable future, and that the best way to ensure adequate supplies and stable prices is through free markets and free trade. For the first time, stated U.S. energy policy is now to actively support U.S. industry in participating in world oil development. To further this goal, some 18 months ago the Department of Energy established a Deputy Assistant Secretary for Export Assistance, whose sole job is to facilitate energy trade and support U.S. companies in investing and working in energy projects around the world.

Specific proposals in the National Energy Strategy call for increased production in all producing regions around the world, including the U.S., increased strategic stock and international cooperation to avoid panic responses in the event of an expected supply disruption, and increased commercial development and trade in non-petroleum fuels and technologies, particularly in the transportation, manufacturing, and electric generation sectors."

National Energy Strategy

In the NES, as we call it, we outline a strategy for achieving a balance between our increasing need for energy at reasonable prices, our commitment to a safer, healthier environment, our determination to maintain an economy second to none, and our goal to reduce dependence by ourselves and our friends and allies on potentially unreliable energy suppliers.

The NES includes both legislative and Administrative activity. Implementation of certain initiatives requires the passing of a comprehensive energy bill, and the Administration will continue to work with Congress to achieve this. However, despite current setbacks on the legislative front, we will continue implementation of the Administrative initiatives. An important item on the list is our need for energy at reasonable prices. This need involves both oil and gas.

NPC Gas Study

Secretary Watkins has asked the National Petroleum Council to conduct a comprehensive analysis of the potential for natural gas to make a larger contribution. This study is in progress and will be completed next year. The NPC believes that a key requirement of the study will be the integration of all of the elements: demand, local distribution capabilities, transportation and storage requirements, and supply, including imports. As an advisory group reporting to the Secretary of Energy, the work of the NPC is very important to the Department of Energy. But it is interesting to note that the industry is learning more from itself as a result of this assignment than is the Government.

Gas Research Plan

The Office of Fossil Energy has been working with the gas industry to develop a new R&D Strategy which takes a systems approach. In the supply area, it will focus on reservoir characterization and more efficient drilling and recovery technologies. But it will give new and added focus to utilization, with high efficiency gas turbines, gas to liquids technology, improved gas clean-up, and improved combustion technology for reduced nitrogen oxide emissions. Work in delivery and storage will include areas such as rock cavern storage and storage well production technologies.

Advanced Oil Recovery Technology

While the 22 conservation initiatives contained in the NES are expected to save up to 3.4 MM BOPD by 2010, the fact remains that we are dependent on oil and will remain so for some time. Furthermore, since oil and gas are finite resources, it is important that we focus our efforts on increasing recovery efficiency. And understanding the reservoir environment is critical to our ability to do so.

We have refocused our R&D programs to reflect that need. There are 18 initiatives in the NES which relate to the oil and gas industry, but the most important initiative addressing improved recovery efficiency is the Advanced Oil Recovery Technology initiative.

This initiative will be implemented through the DOE's *Oil Research Program Implementation Plan,* published in April 1990. The overall approach to the program is to identify those types of oil deposits that have both the greatest potential for improved oil recovery and the greatest risk of abandonment within the next 5 to 10 years.

The program began with the development of a reservoir classification system based on the theory of similarity which states that reservoirs existing in geological formations deposited under similar environments should have similar characteristics. Once classified, the reservoir classes were prioritized based on volume of remaining resource and urgency in preventing abandonment and loss of access to the resource. The first class we will focus on is fluvial dominated deltaic reservoirs. Some of the papers presented today will address this reservoir type.

Program Opportunity Notices

On October 15, 1991, we initiated the joint government-industry program with the release of two competitive Program Opportunity Notices. Let me emphasize that this is an industry-driven program, and we appreciate the presence of the Independent Petroleum Association of America today and their interest and support of this program. The first solicitation will focus on achieving near-term results and the second on achieving mid-term results.

The near-term program calls for projects that will demonstrate improved oil recovery techniques that can reduce abandonment rates and that can be applied commercially within the next 5 years. Many of these techniques, in fact, may already be in use. If that is the case, we want to replicate those technologies in different areas of the country, and then disseminate the results to operators in geologically similar fields.

The mid-term activity should produce technologies that would be widely applicable in the latter half of this decade. Here we are moving beyond simply maintaining access to a reservoir, we are looking for techniques that increase economically recoverable oil.

In both cases, we're asking for industry's best proposals and envision giving the companies selected a great degree of latitude in carrying out a field project. In return, we ask that they invest their own dollars for up to at least 50% of the project's total cost. And we will require an aggressive technology transfer effort. Proposals are due January 15th. We hope to carry out as many as 10 competitions over the next several years.

We have high hopes for this program. We have seen many good projects experience technical or economic failure because of poor understanding of the reservoir. The more we learn about the geology of a specific reservoir or reservoir type, the more there seems we do not know. This program will attempt to build on all existing knowledge and apply it to as many reservoirs as can derive benefit from it.

Closing

In closing, I must reemphasize the importance and significance of your contributions. Reservoir characterization is critical to our ability to meet the goals set by President Bush to satisfy our increasing need for energy at reasonable prices. As I've discussed, reservoir characterization is a critical component of the National Energy Strategy. It is a critical component of the mission of the Office of Geoscience Research. Reservoir characterization is at the heart of the new oil and gas research programs. Reservoir characterization is one ofthe toughest jobs we have in oil and gas development. Thank you for the work you are doing. Thank you for the work you will do. And thank you for joining us here at the Third International Reservoir Characterization Technical Conference.

LUNCHEON SPEAKER
Monday, November 4, 1991

Denise A. Bode, *President*
Independent Petroleum Association of America

A native of Bartlesville, Oklahoma, Mrs. Bode became the first woman president of the IPAA on February 1, 1991. She was a founding partner of the Washington, DC lobbying firm of Gold and Liebengood, Inc. She began working for Senator David L. Boren (D-Okla.) in 1976, when he was Governor of Oklahoma, and followed him to Washington in 1978 when he was elected to the Senate. She received a law degree from George Mason University in 1982 and a master of law degree from Georgetown University in 1984. In 1976, she received an undergraduate degree in political science from the University of Oklahoma. Mrs. Bode resides in Alexandria, Virginia with her husband John and their son, Sean.

Remarks of Denise A. Bode
President
Independent Petroleum Association of America

It is a pleasure to be here to visit with you about challenges facing independent producers and how we at the new IPAA are setting about to meet those challenges.

I guess you all know that I am the first woman president of IPAA in our 61-year history as an organization. My leadership at the association has changed the image of the independent in one tangible way. In fact, as you all could guess, I get teased a lot by some of the men who head other energy associations in Washington. But I just tell them (as Will Rogers used to say) "Wait, because the time isn't far off when a woman won't know any more than a man."

I grew up in the oil patch with an intense interest in our government, its strengths and its inequities. My career has been devoted to the development of public policy both in the public and private sector. I first began working for Governor David Boren of Oklahoma as his Assistant for Energy and Natural Resources. Every morning outside my window, I could see the oil wells still producing oil from underneath our state capitol. Later when he was elected to the Senate, I moved to Washington and served as his Tax and Legal Counsel, handling the intricate energy and tax issues that arose from his position on the Senate Finance Committee. I learned about running a business by starting my own firm with other former Hill staff and members of Congress, which we were fortunate to sell last year to the largest public relations firm--Burson Marstellar. And I learned from Senator Boren that government's role in dealing with business should be to limit barriers, not to create them.

I took on the leadership role for the domestic producers because of my great respect for the industry and the people in it. The folks still in business are the survivors, the best and the brightest--the daring wildcatters and the entrepreneurial dreamers who are brilliant, efficient, effective business men and women.

You too are survivors who have made it through the last decade of hard times. But for the sake of country, you both must do more than just survive. You must prosper. With my apologies to Winston Churchill, in my opinion it is the independent producer who has the lion's heart.

As I said earlier, I have been asked to report to you on activities in Washington. But before I begin, I have to tell you this story one of my independents told me. He told me the prices he was getting for his natural gas were at all time lows, he couldn't attract investors, his bank wouldn't loan him money, and the wells he drilled were all dry holes. At his wits end, he sat down and wrote a letter to God asking for $10,000 to drill a well. He mailed the letter and somehow it ended up on George Bush's desk. Being a former independent producer, the President felt a great deal of sympathy for him. But being back on a government salary, he could only afford to put $100 in the envelope and send it back. When the independent received the anonymous letter postmarked Washington, he was pleased. He sat down and wrote back to God, saying "Thank you, God, for the money, but next time don't send it through Washington."

TECHNOLOGY TRANSFER PROGRAM

The major integrated companies are abandoning the domestic resource base at an accelerating rate. Exploration and production research, traditionally performed by the majors, is being severely curtailed or refocused on foreign resources. Independents face the same economic and regulatory problems as the majors. But capital constraints deny them the opportunity of seeking alternative foreign exploration and production opportunities. Rather, independents must choose between learning to produce better and smarter in the domestic environment or abandoning the industry altogether. While some independents have been forced to accept the latter choice, I am here to tell you that most independents want to stay the course. Our province is the lower-48 states. We stand prepared to do what is necessary to produce the resource efficiently in order to meet the needs of the nation.

With the departure of the major integrated companies, thousands of oil and gas producing leases and properties are being sold or scheduled for abandonment. Many of these properties are strong candidates for improved recovery processes. With leaner operating budgets and reduced overhead costs, independents can economically apply improved recovery technologies where the majors cannot. Low, unstable prices and technical uncertainty, however, make the risks of such investments seem formidable to some independents. Achievement of this enormous potential depends squarely on reducing the perceived risks through the effective demonstration and transfer of R&D to the producing community.

IPAA has followed with great interest the efforts of Department of Energy(DOE) in order to refocus the federal oil and gas research programs to address the nation's highest priority technology problems. We encourage DOE to continue its efforts toward developing a "borehole to burnertip" plan for the natural gas research program that is equally as well focused and well-reasoned as the oil program plan.

DOE has begun implementation of the Oil Research Program for the first of several prioritized classes of reservoirs--fluvial dominated deltas through a Program Opportunity Notice (PON). Although we have had real concerns about the PON mechanism used to implement the program, IPAA nonetheless played an active role in providing input to DOE as to the structure of the PON's. Of the 35 candidate fields that were proposed by operators for near-term or mid-term PON activities, sixty percent of the responses came from independent oil and gas companies. While the PON process continues cautiously, one reservoir class at a time, independents need access to technology, information, and analytical tools now.

DOE's oil program called for active technology transfer but did not spell out what it meant. IPAA has taken the responsibility during the past year to think through what is needed to affect technology transfer to independents. Our Improved Oil and Gas Recovery Task Force, under the leadership of Jim Russell of Abilene, Texas, has determined that one of the most important roles IPAA can play is to serve as a strong national facilitator of technology transfer to domestic oil and gas producers.

By serving as an "umbrella organization" for technology transfer, IPAA and its state and regional cooperating associations have the unique capability to bring together all of the organizations and institutions that are critical to achieving the common goals of the Department of Energy, the oil and gas research programs, and the producing community. Through the coordinated approach we have developed, IPAA wants to work with the R&D community, universities, professional societies, geologic surveys, and the states to intensify technology transfer efforts.

IPAA's proposed program would focus on holding workshops around the country to both identify problems and provide technological solutions. But the main component would be the establishment of regional resource centers throughout the main producing areas of the country. They would provide operators with access to technical information through a library, expert referral services, and access to computer workstations that contain project histories, field and reservoir data, and analytical software for reservoir and project analyses. Whereas major companies already have research centers and data resources that their personnel routinely use to plan the development of a project, most independents lack physical and economic access to such resources and personnel.

IPAA is currently working with DOE to try establish this type of technology transfer program. With the current depressed state of the industry, many independents may not be around to benefit from the mid-term and long-term phases of DOE's program. The rotary rig count is at an all-time low. Crude oil production in the lower 48 states is at its lowest level in 40 years, and all other industry indicators are equally as pessimistic. The time to act is now, and IPAA's proposed technology transfer program provides a very important tool for the survival of independents and the preservation of the domestic resource base.

WASHINGTON REPORT

As I said, I am here to report to you on the progress I believe the domestic industry is making in Washington. When I took this job in February, I had a number of meetings with policy-makers, key energy industry officials and leaders of the state independent producers associations. Those discussions helped me to advise our leadership on the direction we must go as advocates for independent producers. The challenges confronting us:

--Uniting the independent producers, both the state and national association, behind a short, focused legislative and regulatory agenda which we have done and are aggressively pursuing (I want to elaborate on this a little later);

--Building coalitions both with other segments of the energy industry and beyond the energy industry with other business groups, we are clearly taking a leadership role in pursuing coalitions, particularly on environmental and natural gas issues;

--Working with both sides of the political aisle because we were long thought of as a one-party organization and we need to work with both the Democratic Congress and the Republican administration (we have started the first IPAA PAC);

--And finally, we desperately need to nurture a different image, an image that accurately depicts the independent producer of today--the small business man and woman.

You know J.R. Ewing is dead, Southfork is now a flea market and the domestic energy industry couldn't be more delighted. What really makes me happy is that it wasn't Bobby that killed J.R., or CBS, or even an angry independent producer fed up with the bad rap, but the American people, who have grown weary of what J.R. represented. Now is the time for us to change our image. The domestic oil industry must be seen as we are--composed of a cross section of the American public--young and old, rich and poor, male and female--chasing after the American dream like every other American. The domestic industry must be seen as serving a national security interest, an interest which sent over a half-million troops into a desert half way around the world at enormous cost. I for one don't want to have to worry that in ten years when my son is eighteen, he may have to fight in another war to protect our foreign energy supply. It is up to all of us to tell our story. We are talking with the national press about the plight of the independent producer today. IPAA conducted a survey of independents to obtain a more accurate profile. We will use this information, which is quite extensive, to dispel old images. For example, did you know independents are established business people with an average of 26 years in business, the lion's share have less than 20 employees and most have less than 10, the majority are under the age of 60 years, 93 percent are college graduates and more than 15 percent have held elected or appointed government office, a demonstration of your dedication to the community. What you see is a profile of a small business operation. And as a small business operating here in the United States, don't we deserve to be treated like any other small business by our government? We don't ask for special treatment, just not to be singled out.

You don't realize how ignorant the American public is about my vital industry until you raise a drilling rig as part of a week of educating sessions about the domestic petroleum industry. We had thirteen members of Congress kicking off the week with Bennett Johnston, Chairman of the Energy and Natural Resources Committee, taking the lead. Johnston claimed that it was ironic that the rig was in the shadow of the Smithsonian because it may be the only place to see a drilling rig in the time to come. It is a good thing we erected it during this busy tourist season in Washington. We had a number of visitors, including one cute couple from Indiana who walked around the rig looking very perplexed until they saw a member of the rig crew demonstrating the changing of pipe. The lady said, "Oh, so that's where they jump from." The owner of the rig from West Virginia has considered its conversion to a bungi jumping device as it has been idle for over six months.

The 1600 or more unemployed rigs and the hundreds of thousands of unemployed workers need to be put back to work. These are critical times for the domestic petroleum industry and thus for producing states--that demand bold and dramatic action. Nationwide, producers are organizing like never before behind a common agenda because their survival is at stake. They include over thirty-three state producer organizations.

We have been aggressive in our public relations effort to demand attention from policy makers and the public. Our testimony before the Republican and Democratic platforms is extremely important because without the understanding that there is a crisis, we have no impetus for action by policy makers in Washington, or at the state level.

It is also important to separate myths from facts. People may ask, "Why should we act if the U.S. is in fact 'drilled up?". John Kennedy said, "The greatest enemy of the truth is very often not the lie--deliberate, contrived, and dishonest--but the myth--persistent, persuasive and realistic." The myth that the U.S. is "drilled up" is just that--a myth. The U.S. has vast reserves of oil and natural gas out there. There are at least 60 billion barrels of oil and 74 billion barrels equivalent of natural gas still out there. We have barely scratched the surface. They may not be in "elephant" fields which a large company needs for it to be economical, but in an infinite number of smaller "jack rabbit" fields well suited to the independents. There are also 460,000 stripper wells with 2.6 billion barrels of reserves this nation sorely needs to preserve. They are the true strategic petroleum reserve--those wells that supply 20% of oil and 13% of natural gas. With many companies leaving the country there are billions of dollars of these prospects on the market for independents. We are the future of the domestic industry. We won't be forced out of the U.S. We will stay on to find, develop and maintain these other untapped domestic resources. But, the government regulation must be sensitive as smaller businesses must take the lead in developing these resources.

We have started an educational effort with a voluntary yet comprehensive public relations plan to talk about the domestic producer of the future. Our Wildcatter's Week was the first such effort which we hope will become an annual event. With the potential of one-hundred forty new members of Congress coming to Washington, next year is another huge window of opportunity to change the image of the producer. We have also established a financial relations committee to promote reinvestment in our industry, and appointed a resource base task force with leading geologists to dispel the myths about the resource base.

Understanding provides the basis for the agreement between the state associations and IPAA on our united domestic energy agenda. A year ago in May, we jointly developed this strategy which includes:

--Elimination of tax barriers to capital formation in our industry by allowing producers to deduct their ordinary necessary business expenses. Drilling costs and percentage depletion is of paramount importance to domestic producers. This is our clear first priority for any new tax legislation. As you are aware, under existing law, drilling costs and depletion are considered preference items in the alternative minimum tax. The lion's share of producers find that under the alternative minimum tax, they cannot deduct their everyday business expenses, unlike many other small businesses. The AMT, contrary to traditional principles of taxation, taxes directly the capital invested to produce new jobs here at home in America, rather than income generated by these investments. This influences the producers decision to drill--in essence putting a cap on drilling. The rig count as it sits at 656, a modern low, bears evidence of the impact of this governmental barrier. More producers have pointed out to us that every major drilling expense that the independent producer incurs to find new oil and natural gas resources, to comply with environmental laws, or to improve production from existing reserves is penalized under the AMT. At a time when both the state and federal government is focused on piling on additional environmental requirements and costs, we cannot write off our environmental expenditures as quickly as a multinational oil will operate. Domestic production will soon be up to the independent producers. And according to my members, it is a challenge they are prepared to meet.

Tax-writers support our cause, because we are not asking for special treatment like new tax credits, but to just be allowed to use our ordinary business deductions like any other business. The bottom line is that current federal tax policies contribute to the decline in U.S. crude oil production. Since passage of the Tax Reform Act of 1986, domestic crude oil production has declined by more than 2 million barrels per day, despite interim price increases of nearly 100 percent. At the same time, the U.S. is increasingly reliant on foreign sources of oil. A decade ago, oil imports averaged approximately 37 percent of demand. Current forecasts are that oil imports will rise to more than 55 percent of demand by 1995, adding 470 billion dollars to the 1995 trade deficit. If this huge outflow of purchasing power was redirected to our own economy, the result would increase industrial activity and create more than one million new jobs.

--The need for oil price stability. Our government must become an advocate for the domestic oil industry. Consuming state members need to convey their recognition that it is in America's best interests to keep jobs here in America, that domestic oil produced here is better than oil shipped in on a tanker, that the right price encourages conservation as well as switching to clean burning natural gas. Reaching out to both parties can make a difference.

--Expansion of natural gas markets through natural gas pipeline rate overhaul, encouraging longer term sales contracts, creating a level playing field for domestic natural gas producers and lifting of the antitrust restriction on independent producer cooperatives to market natural gas. Winston Churchill once said it is better to be making the news than taking it; to be an actor rather than a critic. Independents have not been actors in Washington for some time. The natural gas issues being debated today both on the Hill in the National Energy Strategy and at FERC have given us that opportunity of which we have made the most. If we had been players many of our agenda items would have been included in the National Energy Strategy, or at least in the Energy Committee bills when introduced. But the perception of many policy makers was that we did not have to be dealt with because we had lost the ability to turn out our grassroots membership to influence opinion.

We have provided policy makers with a short, focused agenda which if implemented will address the crisis and begin to lift the barriers to our industry.

-- Expand natural gas markets and address regulatory barriers.

-- Eliminate the alternative minimum tax on drilling and depletion.

-- Maintain state based regulation for by-products of production as the Resource Conservation Recovery Act is considering.

With focus we are turning the tide.

It is critical that both the Congress and the new people we may replace them with understand this crisis and act in the ways we have suggested to address this crisis.

Last month, the Federal Energy Regulatory Commission adopted one of the major elements of our independent agenda by overhauling natural gas pipeline rate design and adopting the straight fixed variable rate design in Order 636. This new design will equalize the rates between Canada and the United States, so that imports no longer have that advantage over domestic produced natural gas. That way, as natural gas demand increases, we will have a shot at that new demand. The natural gas industry has formed a Natural Gas Council composed of IPAA, Inter-state Natural Gas Association of America (INGAA), American Gas Association (AGA) and Natural Gas Supply Association (NGSA). All of us are working together to increase demand by marketing natural gas.

Despite the positive developments at FERC, we had some very bad news as the House included in its version of the National Energy Strategy Legislation a provision which would effectively repeal state conservation laws in as many as 39 states. The House provision specifically prohibits regulation of natural gas production that has the purpose or effect of generally restricting gas production or gas prices. It also changes the venue for cases on these issues from the states to the District of Columbia. Senate Energy Committee Chairman Bennett Johnston again committed at a hearing last Thursday in his committee to strike this provision from the bill, as has Congressman Dingell, Chairman of the House Energy and Commerce Committee.

But there is some good news for producers in this House Energy Package. Policy makers have gone a long way to lift one of the principal government barricades to drilling--the alternative minimum tax penalty on drilling costs and percentage depletion--yes, percentage depletion, which is the lifeblood of the marginal producer, has been exempted from AMT-- the ordinary and necessary expenses of a producer. Right now, if a producer hasn't stopped drilling, he or she can end up with effective tax rates in excess of 100%, like George Yates of New Mexico. The average tax rate of an independent who reinvests and drills under AMT is 70% - yes 70% plus. We now have one of the highest effective tax rates among business. Whenever the government needs money, they turn to us and we are forced to belly up to the bar - we pay FICA taxes, state and local income, federal income, severance taxes, fees for permits, superfund taxes, oil spill liability taxes, and the windfall profits tax. Now when we are in serious trouble - when we are all in a state of free-fall - we are asking just for equal treatment.

When independents started talking about AMT relief for our industry last year, most folks laughed and said it will never happen. Well, we made it happen. We broke the barrier. Thanks to senators Breaux, Bentsen, and Boren, some relief was included in the growth package that passed in Congress. Although the President vetoed the bill, it was not because of AMT relief, which the President came out and endorsed. Congressman Bill Archer of Houston led the House Ways and Means Committee, with the advice and support of the Democratic leadership, included this in a special tax title of their energy bill. It is the first time in decades that the Ways and Means Committee has initiated reforms that would provide tax relief for our industry. The Senate Finance Committee went further and made permanent the changes passed in the House. Now we are hoping that the Senate can retain those changes passed on the floor and in conference. It is worth a billion dollars to our hard hit industry.

The final barrier is probably the most difficult, but also where we find the most consensus. The environmental barrier prevents a producer from getting to the ground to drill.

Everyone I know who is an independent producer considers him or herself an environmentalist. That is why many chose to be a geologist, petroleum engineer or whatever field led them to join our industry. Our members have embraced the cause for a cleaner world. In October, at our annual meeting, independents wanted to further demonstrate their responsibility as good environmental stewards by approving, for the first time, official environmental standards for our members. It is also important to note that in Washington, policy makers also believe that natural gas is part of the solution to cleaner air. In fact, many policy makers are working hard to eliminate barriers to its use, and are also looking for incentives.

Yes, we all believe we are good environmental stewards, but frankly we are not considered reliable authorities. In fact, the agencies that regulate the environment are not even considered good enough for the job. Only the environmental lobby seems to be qualified. For example, in September when I testified before the House and Senate on behalf of the domestic producers of oil and natural gas, I peered over a high stack of books which included the federal and state environmental regulations, 30 hours of IPAA video tapes which educate our members on environmental rules, the EPA study, and the IOGCC's report on oil field wastes. I wanted to demonstrate visually our existing environmental commitment, which is why we have helped sponsor numerous educational trips for members of Congress and their staffs to show first hand how responsibly we operate and to let them talk to state officials. During the hearing on the reauthorization of the Resource Conservation and Recovery Act known as RCRA, (which imposes federal rules on hazardous waste and its disposal) the big environmental lobby attempted to consider <u>all</u> our by-products of production such as mud and saltwater as hazardous. I was horrified by the "environmentalists'" presentation due to <u>their</u> lack of science and facts. In this case what P.J. O'Rourke said about our government is so true, "The mystery of government is not how Washington works but how to <u>make it stop</u>." As advocates for independents we must make our case, because the potential consequences to our industry would be disastrous. A federal RCRA-based program would shut down practically all stripper wells and a sizable number of additional low-volume wells in the first year. Stripper production represents 20 percent of the oil production and 13 percent of the natural gas production in the country. That is more than the U.S. imports from Saudi Arabia, our largest foreign oil supplier. That would force many more domestic independent producers and almost all small independents out of business. I already feel like an advocate for the homeless--when you consider two-thirds of the domestic producers have gone out of business in the last decade. This change will impact state and local government and business unrelated to energy. This dramatic loss of production would also impact consumer prices. And all these changes would be for naught because oil and gas wastes are safely managed under current state and federal requirements, such as Safe Drinking Water Act, Clean Water Act. RCRA would be an additional layer of federal regulation where none is needed. The EPA has done a two year extensive study that determined that no new regulation is needed.

I was appalled at the so-called "environmentalists'" complete disregard for all the work done by the IOGCC Council on Regulatory Needs which included environmental concerns. They had a fellow come up from the Louisiana Attorney General's office--Mr. Willy Fountenot--who offered no empirical data to eliminate our exemption--but simply showed slides of oil field horror stories in Louisiana. I guess what bothered me was these horror stories were shown to demonstrate the need for legislation, but the examples he presented were already a violation of <u>existing</u> laws--federal and state. Congressman Tauzin asked Mr. Fountenot why he wasn't back in Louisiana enforcing the law rather than asking for another layer of federal rules where existing law already regulates. The state's presentation was out of sync with the other producing states' regulators. In fact, Bob Krueger of the Texas State Railroad Commission and Tim Dowd of the IOGCC testified on the panel with me in opposition to federal intervention on oil field wastes because they said that the states do a better job regulating the industry in their state.

But the big environmental lobby asks for legislation by anecdote, not with science. But anecdotes do not justify a new overlay of regulation on the majority of businesses who are law abiding. Particularly when the EPA has studied whether the federal law on oil field wastes should be changed and has found that no new legislation is needed. Especially when you consider the loss of production that would result from a loss of our exemption from being classified as hazardous--20% of the U.S. oil production and 13% of the natural gas in this country would be shut down. We are talking about 1.2 million sites in the U.S. This is the critical issue that could mean life or death for this industry. But the brilliance of the environmental lobby's strategy on this and other environmental issues is that they discredit science and government, and that they persuade the public habitually to place its trust in their proclamation. And that is what is happening: only 15% of the American public trusts what government scientists say, only 6% trust scientists seen as representing industry, 68% implicitly believe the political activists, and 67% agree with the statement, "threats to the environment are only as serious as "environmental groups say they are."

Now consider the environmental lobby which IPAA and the other energy trade associations face in Washington everyday. In its September edition, Outside magazine reported on 21 of these organizations: total annual budget $535,607,650; total membership 7,769,000; with a total staff of 3806. Let me tell you money is an effective weapon. 90% of these funds go to support political activities and litigation, not environmental improvements. Anymore, only a small portion of their funds go to conservation education. In fact, my husband's law firm rents space in the new $15 million headquarters of the National Wildlife Federation in Washington. I doubt if the 975,000 members are unaware that the $87.2 million budget is going to pay for a staff of 860 lobbyists, lawyers, public relations staff in Washington, new buildings, and chauffeured limo's for the staff director among other things. Please don't talk to me about the money and influence of the oil lobby.

We need strong environmental laws, but they must be responsible and must not impede economic growth. The independents have taken on the challenge of being good environmental stewards, which is a big challenge for any small business. But it is a challenge we are determined to meet. Much of our efforts will go to assist state producers in the states in publicizing the measures they are undertaking and have undertaken to be good environmental citizens. We have stopped fighting a defensive battle. We are about to go on the offensive because as good environmental stewards, we as an industry have a lot to be proud of.

While other energy trade groups in Washington are downsizing their domestic exploration & production and federal lands lobbying shops, I have moved to improve our presence in this area. We will also be taking more of an active role in lobbying on offshore drilling issues. After all, independents control 35 percent of offshore leases, and the number is growing fast. You will also see us more active in projecting the rights of producers in the courts.

I believe with the conservative Supreme Court that we have now, there are real opportunities. I believe that we should take our cues from our friends in the environmental lobby, and tie in a litigation arm to take advantage of the opportunity that may be there in the courts.

Let me close by sharing with you a conversation I had last month with Dan Yergin, the author of The Prize (who just won the Pulitzer for his work.) He said that he had learned a lesson from his eight years of research in the petroleum industry, that the possession of oil has determined the outcome of every armed conflict in the twentieth century. He also pointed out again that dependence on oil in the Middle East has hurt the U.S. no less than six times since World War II. The 1973 Arab oil embargo reduced real GNP by 2.7 percent, and the Iranian revolution triggered a 3.6 percent drop in real GNP in 1979. And the war in Kuwait and Iraq resulted in the greatest harm--the loss of American lives. Yesterday, we learned Iraq is back in business. And the instability still exists. Instead of relying on American G.I.'s for our oil, we should be enacting legislation that puts U.S. natural gas and oil field workers back to work in the United States.

But achieving our agenda is not just about industry jobs or even financial security for our states, but about our country's future in competing with a new world economic order.

For almost all of this century, the oil and gas industry made this country the greatest in all of the world. As producers of this precious resource, we are looking beyond the success of today to the future of generations still to come. That has always been the hope of America, and it can be in the next century if you and I work together to make it so.

What differentiates us from other small businesses is that without our product--energy-- there would be no other American business. We are truly America's business. And America should be in the business of protecting such an important contributor. To put it in terms that any environmentalist can understand, America can't afford for us to become an endangered species.

But our leadership team of members and I can't do the job alone. We need you to be active now, especially now, when times are not so good. This is when we can have the most impact on a company's bottom line. We need every voice, speaking in concert, with a focused agenda, working with both the Republican administration and Democratic Congress, and stating proudly and accurately who domestic producers are, and what it means to be a part of this great industry. We need your help to get the legal barriers we independents have identified out of the way to keep a domestic industry alive. As I stated earlier, it is the independent producer who has the lion's heart. It is my great honor to lead the team who gives voice and teeth to the lion in Washington.

LUNCHEON SPEAKER
Tuesday, November 5, 1991

Farouk Al-Kasim, ***President***
PETROTEAM a.s., Stavanger, Norway

Farouk Al-Kasim is an independent consultant on resource management and heads his own company PETROTEAM. He is also Chairman of the Government-sponsored research program on improved recovery SPOR and of the newly established consultant company on petroleum engineering PETEC. He was one of the founders of the Norwegian Petroleum Directorate where he was the Director of Resource Management from 1973 until January 1991. From 1968 to 1973, he was advisor to the Norwegian Ministry of Industry in Oslo on matters related to petroleum activities. Until 1968, he worked with Iraq Petroleum Company in various positions within geology, petroleum engineering, and management. He graduated from Imperial College, London University as a petroleum geologist in 1957.

RESERVOIR DESCRIPTION IN NORTH SEA PERSPECTIVE

Farouk Al-Kasim

Petroteam a. s.
Nedre Tastasjoen 26
4007 Stavanger, Norway

It is a great pleasure and an honor to be invited to speak to you here today. I would like to thank the organizers of this conference for availing me of this splendid opportunity to talk to you about a favorite subject; namely, *"Reservoir Description in North Sea Perspective."*

The North Sea experience as a whole can be simply summarized in the age old saying that "need is the mother of invention." Let me briefly set the scene for you at the beginning of oil activities in the early sixties so that you may appreciate the challenges and the incentives of early operations in the North Sea.

Firstly, the North Sea represented a hope for indigenous energy at a time when Western Europe was becoming painfully aware of its chronic dependence on external sources.

Secondly, the climatic and marine conditions of the North Sea were much more severe than hitherto encountered in the Gulf of Mexico, Venezuela, or offshore California.

Thirdly, the North Sea as a vital waterway and fishing ground made strict demands on the standards of safety and environmental protection.

Because of these factors, the North Sea development not only required the upgrading of current offshore technology, but it also entailed a much higher cost. Fortunately, the large size of the initial discoveries made it still possible to meet these challenges.

Looking back at early field-development in the North Sea, the majority of fields seem to have faced some or all of the following challenges:

1. Large gravity-base structures played a major role in early development by providing a central production unit. A great deal of consideration was therefore given to the optimal location of these platforms. This obviously put emphasis on reservoir description.

2. Since the number of well slots on the platform was ultimately limited by cost considerations, well spacing had to be wide at 1 Km or more. The location of wells

was therefore critical. Also, this underlined the importance of reliable reservoir description.

3. Not all parts of the reservoir could be drained by extended drilling. In elongated structures, therefore, careful consideration had to be given to the feasibility of additional platforms or sub-sea completion units to drain the extremities of the field. This again required careful assessment of the reservoir.

4. In most cases, the limited free space on deck made it exceedingly difficult, and at any rate costly, to introduce later modifications to accommodate improved or enhanced recovery schemes.

5. In many cases, investment considerations ruled out the gradual approach to field-development in favor of full-scale development from the very beginning.

In the difficult process of adapting to these operational constraints, there had to be occasions for afterthoughts and some regrets. It was these somewhat negative experiences which prompted the emergence of a special North Sea approach to reservoir management in general, and to reservoir description in particular.

From here onwards, I will be talking about the Norwegian experience in reservoir description. Although there are few points of difference, the Norwegian experience, I believe, is fairly representative of the North Sea approach as a whole.

Very early in the field development experience on the Norwegian Shelf, it was realized that optimal plans for improved recovery had to be considered in the early stages of field development planning. Unless recovery-optimization strategies were adopted in the field development plan, the chances were rather slim for introducing such strategies at a later stage of depletion.

Without a reliable and fairly detailed understanding of the dynamic properties of the reservoir, however, the optimization of depletion strategies, including improved recovery, would be at best a risky proposition. This consideration placed added demands on reservoir description in the delineation phase.

The Norwegian Authorities contributed to this trend by insisting that the licensees thoroughly consider improved recovery schemes early in the field development plans. The virtual prohibition of gas-flaring made it in effect imperative to consider gas-injection. On the other hand, the marine environment made water injection an obvious alternative to consider. Should the licensees decide against improved recovery schemes, they must document in detail the basis for their conclusion. Moreover, the authorities, in their ongoing consultations with the licensees prior to the approval of field development plans, have consistently required the licensees to do the following:

a) Continue to improve reservoir description both during and after delineation, and

b) Continue to investigate, and maintain flexibility for additional improved recovery schemes as and when required.

Fortunately, the geophysical and geological community was quick in rising to the challenge of improved reservoir characterization. The spectacular improvements in seismic acquisition and processing techniques, together with parallel improvements in interpretation, have been invaluable to the delineation effort. Here again, we owe much of these achievements to the marine environment which allows the rapid testing and promotion of improved seismic methods at reasonable cost.

The North Sea approach from the mid-seventies to the eighties also put great emphasis on delineation drilling in large and complicated fields, numerous delineation wells, as many as 33 wells in one case, have been drilled prior to decisions on field depletion strategy. Appraisal wells are, of course, necessary in almost all cases, but their high cost in the North Sea requires that they be carefully located in order to resolve the most critical reservoir uncertainties. For this reason, we have seen a gratifying trend towards high-density seismic grids which today extend all the way to 3-D surveys.

Although the value of delineation drilling was indisputable, there soon emerged a realization that delineation wells should preferably be augmented by test data on the dynamic properties of the reservoir. What made this realization particularly acute was the increase in sedimentation heterogeneity and tectonic complexities in some of the more recent fields.

Whereas delineation wells may provide some evidence on pressure communication in geological times, the role played by the various heterogeneity in a production regime can only be evaluated through suitable field testing.

The Norwegian authorities have been encouraging the trend towards prolonged production testing, test production, or, if necessary, early production schemes. The objective is to maximize data on the dynamic properties of reservoirs before commitment to a full-scale field development plan. If the licensees can demonstrate that they are unable to utilize the associated gas that is produced in these tests, the authorities have gone to the unusual step of allowing some gas-flaring in order to secure dynamic reservoir data. For those of you who are unfamiliar with Norwegian operations, let me assure you that this is an outstanding concession for the authorities to make.

For one category of resources, namely the thin oil-zones, the concept of pilot production testing has been extremely important. The total in-place potential in these thin-layered reservoirs of around 7.0 billion barrels of oil makes it imperative to try and find ways of commercially extracting these resources. The recovery of oil from these thin layers is made difficult either by gas catering or water coning or both. The challenge is to achieve a sufficiently high production rate, with tolerable gas and water cuts, in order to allow commercial exploitation.

The successful testing of the thin-oil zone below the Troll gas-field is perhaps the first positive step that will open the way for the commercial exploitation of these thin oil zones. Two tests were conducted by horizontal drilling from a mobile test-production vessel "Petro Jarl" in sandstone layers of 22 and 12 meters, respectively. The results indicate that horizontal barriers can play a major role in restraining coning and catering effects. The challenge is to make heterogeneity work for you. For this, an accurate reservoir description is obviously imperative.

Petro Jarl was also employed in other fields such as the Oseberg oil-field. Here, the purpose of test-production was to gather information on the dynamic behavior of the reservoir prior to the final decision on depletion strategy. After 7 months of production, valuable data were gathered that led to suitable adjustments in the plans.

But what if reservoir uncertainty remains too big even after test production has been tried? For example, what if the field were criss-crossed with different faulting systems whose transmiscibility could not be reliably predicted on the basis of testing? Or, what if the size and properties of the reservoir or aquifer could only be established after considerable depletion history?

Experience from the North Sea indicates that the phased development approach may be the best strategy in such cases. Not only does it reduce investment exposure, but the experience gained from the first phase will ensure a tailor-made approach in the subsequent phases. The drawback is, of course, delayed revenue. In cases of high reservoir risk, however, we must learn to accept that a secure revenue later is much better than a shaky revenue sooner.

It must be emphasized, however, that the objective of any early production scheme must be clearly defined. There is always the risk that once the necessary approvals are obtained and production is well on its way, someone will fall for the temptation of sacrificing information for the sake of maximizing production. The Norwegian authorities have been extremely careful in making sure that the licensees understand and respect that the objective of test-production is to facilitate the collection of data and not the other way round.

But what role will reservoir description play in future production in the North Sea? Let us first look at fields which are already on production.

At the beginning of the eighties, there was a growing realization that peak production was fast approaching in a number of major fields, particularly on the UK side. This prompted companies and governments alike to embark on extensive research and development efforts on various forms of improved recovery including EOR techniques. As you would expect, reservoir characterization soon became a vital part of almost all approaches, and was thus given a prominent place in all research and development efforts. These efforts are still going strong today.

On the Norwegian Shelf, only 17% of the total oil and gas reserves have been produced. Moreover, there is a reasonable hope that improved recovery can be achieved beyond the originally estimated reserves, if only accurate and reliable reservoir description could eliminate some of the risks. The injection of chemical and WAG schemes in particularly well suited fields, or segments of fields, hold an interesting reward if prudently applied.

In the final analysis, however, field trials may have to be conducted to verify the most critical reservoir response parameters. These include not only reservoir characterization but also the interaction between fluids and rock surfaces.

In both field trials and full-scale field applications, the need for accurate monitoring of the flow patterns can be crucial. Intensified efforts to improve the mapping of inter-well heterogeneity and fluid fronts both before and during injection are of the utmost importance. Norwegian research and trials on the use of tracers for these and other production purposes can provide a useful technique to reservoir management. In this and other respects, there is a need for an optimal number of observation wells.

The positive impact of horizontal wells on production from thin-layered zones and peripheral resources of large deposits can be extremely important in improving total recovery. An important side-effect, however, would be the new perspective of sampling reservoir beds along their bedding planes. Although this may present operational challenges, its contribution to a three-dimensional description of heterogeneity is fascinating.

Let us now turn to fields that are yet to be developed.

The experiences that will be gained in improving recovery from existing fields will also have great value in the development and successful depletion of future fields. Most of the remaining fields on the Norwegian Shelf are of the medium to small size compared to the giant fields of the seventies and early eighties. Undeveloped proven resources are of the order of 16 billion barrels of oil equivalents spread over some 60 structures. Most of these structures are less than 150 million barrels oil equivalents in size, and may, therefore, be dependent on existing or planned infrastructures. The timing of their development is, therefore, also restricted by the operational constraints of the infrastructure.

Fortunately, however, there is today a variety of field development concepts that can be tailor-made to suit the individual challenges in small petroleum accumulations. These have, without doubt, expanded the scope for economic recovery from small discoveries and residual pools, but only if reservoir risk is manageable. The need for accuracy and confidence is obvious. One could even envisage a transition into a situation where the distinction between the delineation, testing and production phases is virtually non-existent.

In short, ladies and gentlemen, the development and depletion of the remaining resources will require a higher degree of accuracy in reservoir description. Also, in anticipation of this requirement, the thrust of current research and development, whether by companies or Government, individually or collectively, is very clearly in the direction of improved reservoir description. The approach is to gather and analyze field data from exposures in order to establish a quantitative three-dimensional description of sedimentation environments. By relating reservoir data from cores and logs to these sedimentation models, it is hoped to provide meaningful realizations of the salient parameters to reservoir simulation. This requires a thorough knowledge of how the various heterogeneity influence fluid flow, a subject that is obviously receiving great attention, particularly with regard to chemicals and polymer injection. On the monitoring side, efforts have been concentrated on seismic methods for interwell mapping, and on the use of tracers in verifying flow between wells.

To summarize this brief review, I am tempted to coin a couple of phrases which I hope will remind you of the North Sea Experiences.

1. On the North Sea's technical achievements in meeting the challenging operational environment, I am tempted to leave you with the following thought:

   "If need is the mother of invention, a sure reward should be the father."

2. On the importance of reservoir description to field-development, I would like you to remember that:

   "To avoid the perils of reservoir uncertainty, please test before you invest."

SESSION 1

Heterogeneities/Anisotropies

Co-Chairmen

William Almon, Texaco E&P

Rodney R. Boade, Phillips Petroleum Company

A STOCHASTIC MODEL FOR COMPARING PROBE PERMEAMETER AND CORE PLUG MEASUREMENTS

Jerry L. Jensen
Patrick W. M. Corbett

Department of Petroleum Engineering
Heriot-Watt University
Edinburgh, Scotland EH14 4AS

I. INTRODUCTION

The past few years has seen a growing awareness of the influence of small-scale (i.e., mm to dm size) heterogeneities upon reservoir behavior. At this scale, lamination and bedform are the prevalent geologic elements of the reservoir. While there are often significant visual similarities (e.g., geometry and contrast) between the many laminae and bedforms which make up the reservoir, the corresponding permeability contrasts are only recently being established by detailed mapping (e.g., van Veen, 1975; Corbett and Jensen, 1992a). As shown by Kortekaas (1985), Ringrose et al. (1991), Kelkar and Gupta (1991), and others, the impact of these permeability heterogeneities can be to significantly impair the recovery of oil for a variety of recovery processes.

For a given reservoir and recovery process, the impact of small-scale variations can only be assessed once their presence is recognized and their correlation and variability quantified. In the past, both core flooding and core plug data have been used for these purposes. Each technique has its own particular advantages and shortcomings. Core floods can provide flow-based information on how the formation heterogeneity affects a process, but they are expensive and must be undertaken on cores which represent well (i.e., at the appropriate scale, magnitude, and structure) the variability in the reservoir. In contrast, core plugs are less expensive and can be used to more extensively sample the reservoir. Consequently, core plugs have become the industry standard method for conveying small-scale variability.

Plug data, however, require a flow model in order to translate the observed variability into an impact on performance. Furthermore, because of the cost and mechanics of taking core plugs, selection depths may be biased towards the thicker, more permeable strata, giving a false impression of the permeability variation and its structure. For these and other reasons, permeability data from core plugs have been found in some cases (e.g., van Veen, 1975; Weber, 1986; Martin and Evans, 1988) to poorly capture small-scale variability.

An alternative to core plug permeability measurements is to use the probe (or mini) permeameter. Although not new in concept (e.g., Dykstra and Parsons, 1950), the probe permeameter has only recently seen widespread use to assess small-scale permeability variation. It has the advantages of being a relatively inexpensive, non-destructive measurement which can be made almost anywhere on the rock surface. A number of studies (e.g., Weber et. al., 1972; Goggin et. al., 1988a; Dreyer et. al., 1990; Corbett and Jensen, 1992a) have used the probe permeameter to successfully explore small-scale variability on outcrop surfaces as well as subsurface cores. Similar to core plug permeabilities, probe data must be interpreted with a flow model in order to translate variability into effect on process performance. Probe measurements are often made under less favorable conditions than core plugs. Factors such as residual fluids, poor seal integrity, and non-Darcy effects can degrade the measurement. Hence, probe measurements are not necessarily a substitute for core plug data.

So when can probe data augment or replace core plug measurements? When does the ability to measure frequently--if less accurately--give better, more reliable variability assessments? In this study, we propose a measurement model which attempts to address these questions.

The model we propose and subsequently describe is a computer model which uses the Monte Carlo method. It enables us to assess and compare different sampling methods and strategies. The model also provides a way of assessing the sensitivities of sampling strategies to changes in formation features. After describing the measurement model, we investigate three applications: correlations of permeability with wireline density data; permeability variation; and medium-scale permeability prediction. The results

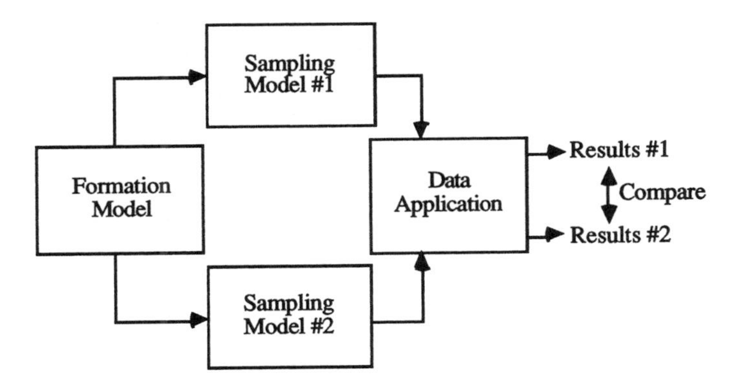

Fig. 1. The three measurement model components being used to compare two different sampling strategies. Modified from Jensen (1990).

are compared to observations using field data. We find that the model, despite having some simplified features, gives results which agree with measurements from a North Sea field.

II. MEASUREMENT MODEL DESCRIPTION

The model consists of three components: the formation model, the sampling model, and the data application (Fig. 1). Together, the three components emulate any process which collects formation data and applies them to make some decision or estimate concerning the formation. We first describe each component in general terms, then we discuss their interrelationships and how they are used together. A specific example will be presented in Section III.

The formation model (Fig. 2) consists of a series of discrete rock "elements" which collectively represent an interval of the formation. The properties (thickness, lithology, porosity, permeability, etc.) are assigned either in a deterministic or a probabilistic manner to each element. Depending on the sophistication of the rock property assignments, any element property may be influenced by the properties of neighboring elements. In this study, each element has constant properties and we assume that a vertical section is being emulated. More elaborate formation models could, for example, incorporate variation within each element to reflect local lateral variation in cores. For the homogeneous elements used here, the element size physically represents a compromise between the true scale of permeability variation (mm's) and the resolution (cm's) of the measurements modeled.

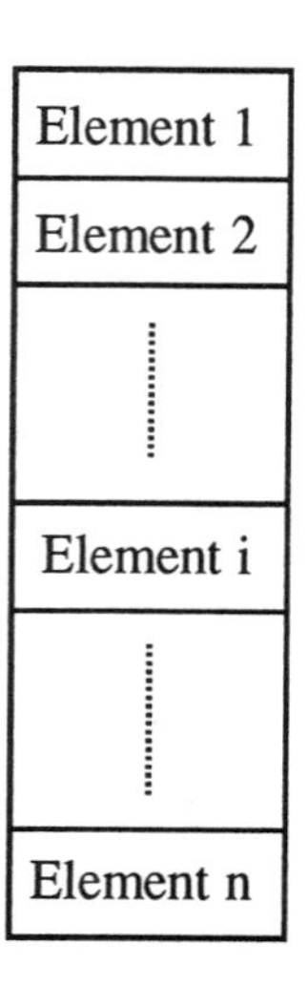

Fig. 2. The basic discretized formation model. Modified from Jensen (1990).

The sampling model captures how the properties of the rock elements are measured. The sampling model encompasses three features of any measurement: the measurement locations, the volume of investigation, and the measurement precision. Measurement locations depend on the measurement being modeled. For example, core plug samples might be taken at discrete points 30cm apart. Additional features involving location can be included, such as skipping a measurement if the element lithology at the chosen location is a shale. The investigation volume concerns the number of rock elements whose properties are being measured at a given location and how that measurement weights the contribution of each element. A 3cm diameter horizontal core plug, for example, reflects the properties of only one element if the elements are 5cm thick. The measurement precision depends upon any further disturbing influences which might be included such as statistical errors, variations in the environment (e.g., mudcake), sample treatment, or operator error.

The third component, the data application, uses the measurements obtained from the sampling model. It emulates the way the data will be analysed and used. For this model, we require a quantitative application involving a well defined procedure. For example, computing the average of permeability measurements can be easily modeled. Zonations could be more difficult to systematize.

Together, the formation and sampling models represent the data acquisition process. They represent both the vagaries and physics of the formation and the data acquisition process. The sampling and formation models produce a series of measurements, precise or otherwise, which reflect properties of the formation and which are used in the application. The application model is necessary because only after the application is considered can an assessment be made of the impact of measurement errors and imprecision on the ultimate outcome. To fully compare the performance of two measurement techniques (Fig. 1), the measurements from both methods need to be used in the intended application and the results compared.

By using the measurement model with the Monte Carlo method, the merits of measurements and procedures can be explored. The Monte Carlo method generates hundreds of "formations" using the formation model, makes "measurements" of the formations, and uses the measurements in the data application. The results can then be examined for sensitivities and behavior which reflect the appropriateness of a particular measurement procedure and data application. Clearly, the more realistic are the formation, sampling, and data application models, the better the results will reflect the observed measurement behavior. We have found, however, that even simple models can give useful results.

III. EXAMPLE APPLICATION

The procedure to be modeled is the correlation of the wireline bulk densities, ρ_b, with permeability measurements. The aim of the procedure is to develop a reliable predictor of permeability for use in uncored wells. In the modelling, we will use both core plug and probe measurements and see which method gives a better result. We assume that a relationship between porosity (ϕ) and permeability (k) exists at the formation element scale and that core plug measurements are "correct". The differences in vertical investigation between the wireline, core plug, and probe measurements are addressed. We assess the quality of the resulting k--ρ_b relationships using the coefficients of correlation and the slopes of the regression lines. The role of sample spacing is also investigated.

A. The Formation Model

Figure 3 shows the formation model geometry, which represent a vertical section through a hypothetical clastic reservoir 12m thick with 3cm elements. The section thickness was chosen to give a large number of elements while still being computationally tractable. The reservoir is flanked by shoulder beds of shale each over 50cm thick, so that measurements towards the top and bottom of the section may be influenced by these shoulder beds.

For element i, $1 \leq i \leq 400$, the formation model assigns lithology (l_i), porosity (ϕ_i), permeability (k_i), and density (ρ_i) values by the following steps.

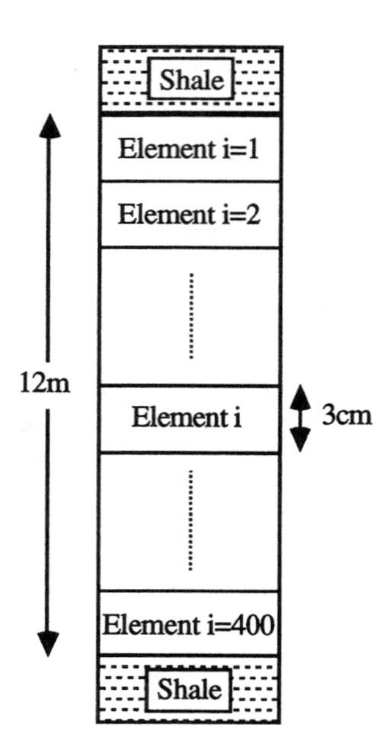

Fig. 3. The formation model. Each element can be either sand or shale except for the shoulder elements, which are shale.

1. The lithology of either sandstone ($l_i = 1$) or shale ($l_i = 0$) is randomly assigned, assuming the average amount of shale is V (%) for the reservoir and no correlation exists between the lithologies of neighboring elements.
2. ϕ_i is assigned according to l_i and l_{i-1}. If $l_i = 0$ (shale), no porosity is defined. If $l_i = 1$ (sand) and $l_{i-1} = 0$, ϕ_i is assigned a random value from a normally distributed population with mean porosity $\bar{\phi}$ and standard deviation σ_ϕ [abbreviated as $\phi_i \sim N(\bar{\phi}, \sigma_\phi^2)$]. If $l_{i-1} = l_i = 1$, then

$$\phi_i = \bar{\phi} + r_\phi(\phi_{i-1} - \bar{\phi}) + t_i\sigma_\phi\sqrt{1 - r^2}, \tag{1}$$

where r_ϕ is the correlation coefficient of porosity between adjacent sandstone elements and $t_i \sim N(0, 1)$. Eq. (1) is a first order Markov model (Haan, 1977, p. 294) which uses a nearest-neighbor approach to introduce interelement correlation.
3. The permeability of each sandstone element is given by $k_i = \exp(a\phi_i + b + \varepsilon_i)$ md, where $\varepsilon_i \sim N(0, \sigma_\varepsilon^2)$, while the permeability of each shale element is set at $k_i = 0$. The exponential relationship is used so that the sandstone permeability distribution is log normal, i.e., $\ln(k_i) \sim N(a\bar{\phi} + b, a^2\sigma_\phi^2 + \sigma_\varepsilon^2)$. ε_i is a "noise" element which causes the ϕ--log(k) correlation to be imperfect. It accounts, for example, for any textural and diagenetic influences which the porosity does not reflect, but which do affect the permeability. The values of a and b are selected to give a realistic variation in permeability values:

$$a = \frac{\sqrt{\ln(C_v^2 + 1) - \sigma_\varepsilon^2}}{\sigma_\phi} \tag{2}$$

and

$$b = \ln(250) - \bar{\phi}a, \tag{3}$$

where C_v is the coefficient of variation, a common measure of permeability heterogeneity (Lake and Jensen, 1991), and the constant 250 chosen to set the median (i.e., 50th percentile) permeability at 250 md. Eq. (2) was derived from the fact that

$$C_v = \sqrt{\exp\left(a^2\sigma_\phi^2 + \sigma_\varepsilon^2\right) - 1}$$

because the k_i are log-normally distributed.
4. Element densities are either $\rho_i = 2.45$ g/cc for $l_i = 0$ or $\rho_i = 2.68(1 - \phi_i) + \phi_i$ g/cc for $l_i = 1$, assuming a grain density of 2.68 g/cc and a pore fluid density of 1.0 g/cc.

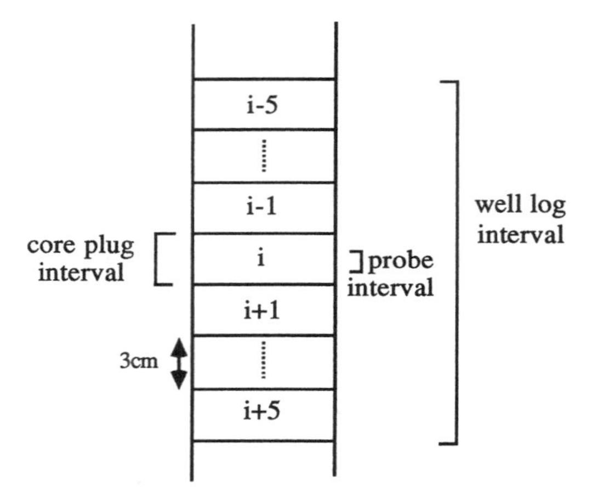

Fig. 4. The sampling relationships of the three measurements modeled.

The shale shoulder beds also have permeability 0 and a density of 2.45 g/cc, giving $\rho_i = 2.45$ g/cc and $k_i = 0$ for i<1 and i>400. Thus, each element has a permeability and a density assigned to it for the sampling models to operate on.

B. The Measurement Models

Three sampling models are considered because three measurements are involved. The wireline density log sampling model (Fig. 4) is described first. Typically, wireline density data, $(\rho_b)_i$, are recorded every 15cm of tool travel and the vertical investigation is about 30cm. So that, at the measurement depth, the tool is responding to the formation densities within approximately ±15cm from that depth. We modeled this volumetric averaging as an arithmetic average of the element densities within ±5 units of the measure point:

$$(\rho_b)_i = \frac{1}{11} \sum_{j=i-5}^{j=i+5} \rho_j .$$

We did not include noise on the density log data since corruption would not affect the core plug--probe comparison; it would simply degrade both correlations.

The core plug sampling model was straightforward, given that the elements are sufficiently large and sized so that a plug never has to come from more than one element. The measurement location of the plug, however, involves an appreciation that plugs are often not taken in shalier lithologies. Thus, while plug sampling points have a nominal spacing of 30cm (10 elements), that spacing may vary if a plug point lands on a shale. In this model of measurement location, if the 30cm plug point fell on a shale element, the program searched for the nearest sandstone element and, provided that it was within 6cm, the permeability of that sandstone element was assigned to the plug. The measure point which was ascribed to the plug was not changed, however. If the program could find no sandstone element within 6cm of the plug point, the plug was assigned a 0 md permeability. Hence, the core plug permeability at location i, $(k_p)_i = k_j$ where $i-2 \leq j \leq i+2$. This model of core plug permeability measurement is somewhat idealized but, since we are using core plugs as the "standard", we only need to be careful about how we characterize the probe permeameter measurements in comparison with core plug measurements.

The probe sampling model consisted of a measurement taken every element, i.e., every 3cm. The probe measurement at location i, $(k_m)_i$, included a multiplicative noise component δ_i: $(k_m)_i = k_i \bullet \exp(\delta_i)$, where $\delta_i \sim N(0, 0.35)$. This model is based on Fig. 5, a comparison of core plug and probe permeameter measurements for Rannoch Formation samples. For each plug, four probe measurements were made on each end, giving eight probe data per plug value.

The "noise" factor, δ, includes surface preparation, invasion, surface roughness, and sample heterogeneity effects as they existed for the samples measured. Since this model comes from a laboratory one-to-one comparison of core plug and probe measurements, it accounts for all differences observed between core plug and probe permeabilities taken on the same rock sample, including the differences in resolution. It does not, however, compensate for differences in measurement when different samples are used. For example, core plug permeability measurements are usually made on cleaned rock, whereas probe measurements are often made on uncleaned samples (e.g., Corbett and Jensen, 1992a). A separate factor may need to be included if this

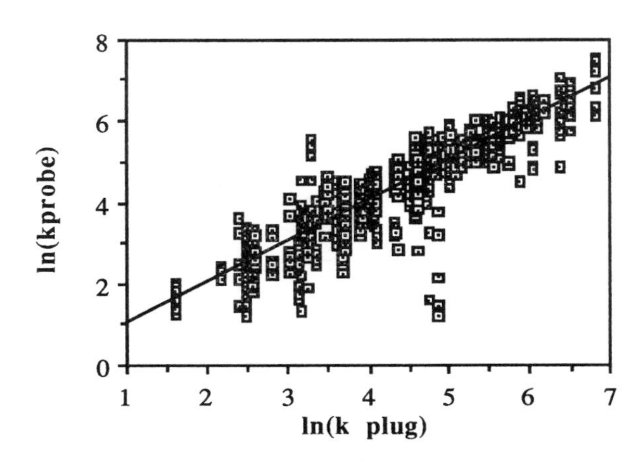

Fig. 5. Core plug--probe permeability relationship. Each plug measurement has eight probe measurements, four on each face of the plug.

were the case. Another element which can significantly change δ is the pore structure of the rock. Figure 8 of Goggin et al. (1988), obtained using their probe permeameter on a number of sandstone plugs, gives a surprisingly similar model to the one for the probe we have used (Jensen, 1990). Goggin (1991), however, shows a much larger variability with probe measurements and carbonate plugs.

The final part of the probe sampling model includes a running average (arithmetic average) taken of every p^{th} probe measurement ($p = 1, 2, 5,$ or 10) to form a permeability "log" with approximately the same measurement resolution as that of the wireline density log (Fig. 4). The permeability log measure point was taken to be the centre of the 11-element span of the running average:

$$(k_{log})_i = \frac{p}{10+p}\left[(k_m)_{i-5} + (k_m)_{i-5+p} + \ldots + (k_m)_{i+5}\right].$$

C. The Data Application

The data application, correlation of density log data with permeability measurements, conceptually results in the production of two plots for each formation generated: ρ_b versus $\log(k_p)$ and ρ_b versus $\log(k_{log})$. Each plot has 40 points on it, corresponding to the 40 depths (i.e., every 30 cm) at which core plug, permeability log, and density log data were calculated. Because the Monte Carlo process generates many realisations of the formation, we chose to analyse two statistics from each plot rather than try to study individually the many correlations generated. The statistics chosen were the coefficient of correlation, r, and the slope of the least-square regression line, m. The coefficients of correlation obtained from the ρ_b versus $\log(k_p)$ and ρ_b versus $\log(k_{log})$ correlations are termed r_p and r_{log}, respectively. Similarly, the slopes of the regression lines (i.e., $\log(k)$ on ρ_b) are m_p and m_{log}.

D. Model Results

Table I gives the results for the case of V = 10% shale and $r_\phi = 0.95$. All results for the simulations, m_p, m_{log}, r_p, r_{log}, and $r_{log} \div r_p$ are listed in terms of their median (50%) values (rather than arithmetic averages) for the 500 iterations performed per case. The slope inter-quartile ranges (i.e., the differences between the 75% and 25% points) are listed in brackets to indicate the variability of the results over all the simulations. Expressing the results in quartiles (i.e., the 25, 50, and 75% points) has two advantages over the arithmetic average and standard deviation. Firstly, median values are invariant under monotonic transformations so that the median values of other statistics (e.g., the coefficient of determination, R^2) are simply related to the median values of the statistics listed. Secondly, for robustness since we did not analyse the results from each realisation. That is, one or two wild results could substantially influence the average or standard deviation, whereas the quartiles would not budge. See, for example, Lewis and Orav (1989, Ch. 6) for a discussion of robust estimators.

The values of r_p in Table I are realistic for ρ_b--log(k_p) correlations. The ρ_b--log(k_{log}) relation shows a stronger linear element (as measured by r) than the ρ_b--log(k_p) correlation. The coefficients of correlation have a weak sensitivity to C_v. The median $r_{log} \div r_p$ increases with C_v. This behavior reflects the fact that, as C_v increases, the relative influence of the measurement corruption, δ, diminishes. Increasing σ_ε degrades the strength of both relationships; r_p is more sensitive to σ_ε than r_{log} and, hence, the ratio $r_{log} \div r_p$ usually increases as σ_ε increases.

The value $-a \div 1.68$, also listed on Table I, represents the slope one might expect for the ρ_b--log(k) plots because this is the slope of the ρ_b--log(k) relation for the sandstone elements. Neither the core plug nor the permeability log plots consistently give this slope, although the probe data come closest. If the measurement model for the core plugs had not had the shale avoidance scheme, m_p would be significantly altered from the listed values; m_p increases with V. Both m_p and m_{log} are about equally affected by C_v. The variability of m_p is always greater than the variability of m_{log}. σ_ε does not appear to exert a substantial effect upon either m_p or m_{log} for the two higher C_v cases.

The porosity--log permeability relationship for the sandstone elements is linear in this model. The ρ_b--log(k) correlations, however, may not necessarily reflect that relationship because of several corrupting factors: the noise ε; the density log response is a combination of the porosity of several elements; and shales, which have a different density-permeability relationship, are also present. If V is large (e.g., 40% or more), the latter two factors will predominate, giving a particularly weak relationship for the plug data. Two other factors for the permeability "log" which may impair correlation are the noise δ on the probe measurements and the nonlinear behavior of the logarithmic function. Hence, r may not be 1 for either correlation.

Table I. Measurement Model Regression and Correlation Performance for the Bulk Density--Permeability Application. Variabilities are in parenthesis.

C_v	σ_ε	$-a \div 1.68$	m_p	m_{log}	r_p	r_{log}	$r_{log} \div r_p$
0.50	0.10	-5.50	-6.36 (0.95)	-6.01 (0.84)	-0.84	-0.87	1.04
	0.25	-4.77	-5.42 (1.22)	-5.27 (0.97)	-0.73	-0.82	1.12
	0.40	-2.99	-3.46 (1.62)	-3.31 (0.96)	-0.45	-0.64	1.40
1.00	0.10	-9.84	-11.1 (1.44)	-10.4 (0.94)	-0.85	-0.93	1.10
	0.25	-9.45	-10.8 (1.93)	-9.99 (1.05)	-0.82	-0.92	1.12
	0.40	-8.69	-9.96 (2.07)	-9.26 (1.13)	-0.76	-0.90	1.18
1.50	0.10	-12.9	-14.7 (2.05)	-13.5 (1.30)	-0.85	-0.93	1.09
	0.25	-12.6	-14.4 (2.22)	-13.2 (1.20)	-0.84	-0.93	1.11
	0.40	-12.0	-13.9 (2.57)	-12.7 (1.54)	-0.80	-0.92	1.15

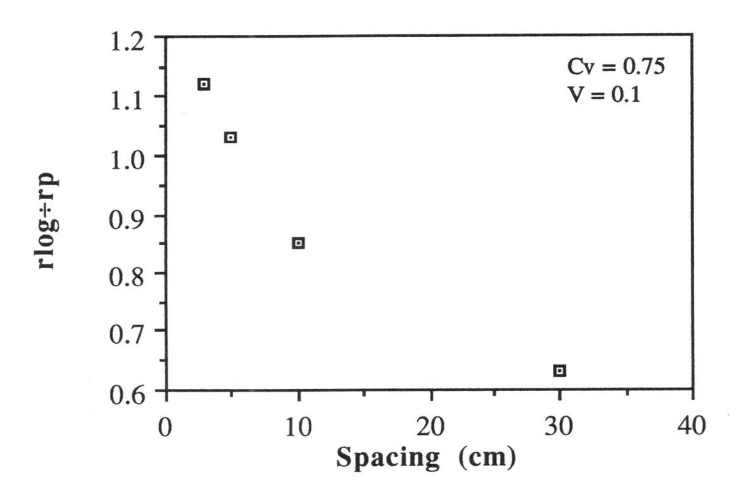

Fig. 6. Degradation of the median probe-based correlation as sample spacing increases.

Figure 6 is a plot of the median of $r_{log} \div r_p$ versus p for the case V = 0.1, C_v = 0.75 and σ_ε = 0.25. The median ratio declines rapidly when fewer samples are used to produce the permeability log. The steepness of the decline suggests that, for this application, closely spaced samples quickly lose their utility as the spacing increases. Thus, simply halving the sample spacing from 30cm to 15cm would not lead to an improved correlation.

Thus, the model results suggest that probe data, suitably averaged, may perform as well or better than core plug data for correlations with wireline density measurements. Since no measurement errors were included in the core plug permeability measurement model, we can expect the model results to be on the conservative side. The two measurement methods, as modeled, do not produce exactly the same product. k_{log} represents the permeability of the total formation, shale as well as sand. k_p, on the other hand, pertains only to the sand if the shale avoidance scheme always finds sand within two elements of the intended sampling depth. Hence, as V increases, k_p - k_{log} will increase. Depending on the needs of the user, probe measurements show promise for providing a more reliable predictor of formation permeability for uncored wells.

E. Comparsion of Model Results and Field Measurements

The preceding results were compared to those obtained from measurements taken on Rannoch Formation core taken from the Statfjord Field. The Rannoch is a very laminated, micaceous, fine-grained sandstone whose permeability variation is predominantly controlled by the amount of mica. The laminae thicknesses are highly variable, ranging from a few mm to a few cm. The study of Corbett and Jensen (1992a), however, suggests that some formation parameters can be estimated without sampling every laminae. Thus, it seems likely that the measurement model, having 3cm element thickness, can still give results which are useful for comparison with Rannoch data.

The permeabilities of two four-meter intervals of Rannoch slabbed core were measured at 1cm spacing along four lines with the Statoil probe permeameter. Core plug porosity and permeability data were also available on 30cm spacing. See Corbett and Jensen (1992a) for further details of the sampling program.

At the core plug scale, the porosity-permeability relation was very strong (Fig. 7). After depth adjustment, the density-core plug permeability association was also good (Fig. 8 left). The probe data were averaged similar to the previously described model prior to correlating with the wireline density. At each core plug depth, 30 probe measurements (i.e., 15 1cm measurements above and below the plug depth) were averaged to obtain a "permeability log". Figure 8 (right) shows the resulting correlation with the bulk density. The cloud of data about the least-squares regression line is clearly reduced and the influence of the high density point (ρ_b = 2.36g/cc) is diminished for the latter plot. The slope of the line is also slightly reduced, indicating that the probe-based predictor would give smaller predictions. This reflects the fact that, in these intervals, the core plugs missed some of the low permeability regions (Corbett and Jensen, 1992a). The variability of the correlation strengths with sample spacing was also examined for both permeability measurements.

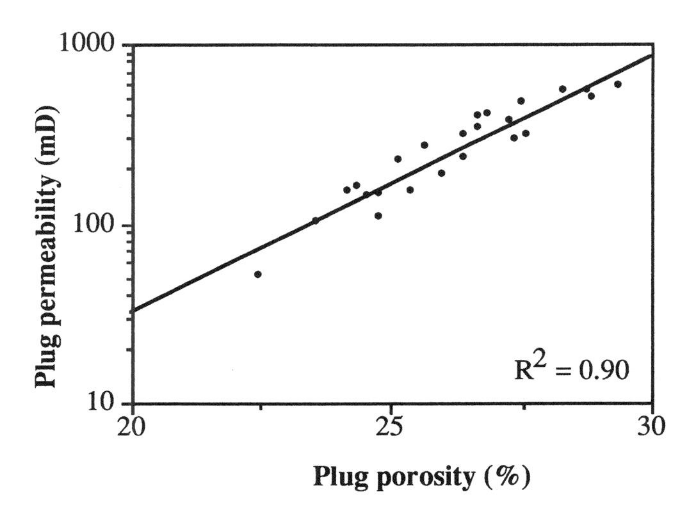

Fig. 7. Rannoch Formation core plug porosity--permeability correlation.

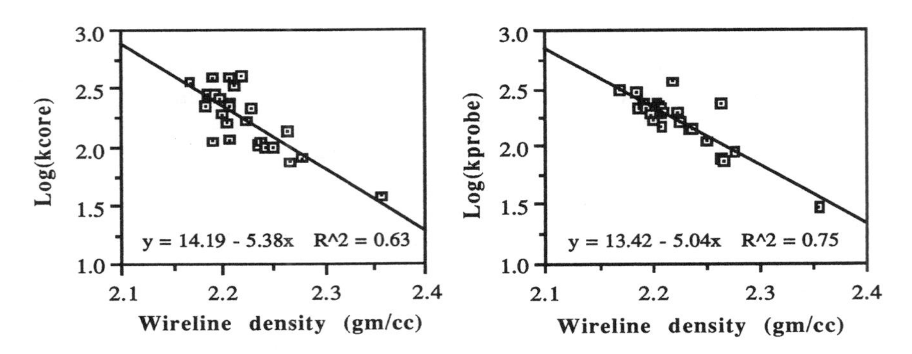

Fig. 8. Density--permeability relations for the idealized plug and averaged probe data.

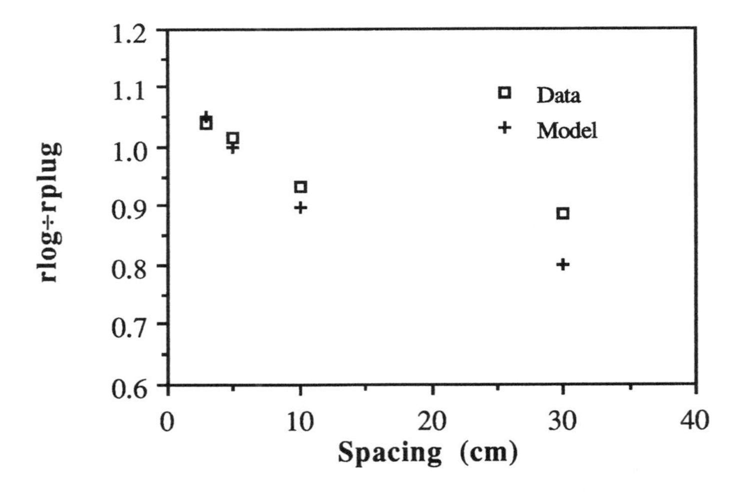

Fig. 9. Comparison of model and Rannoch Formation results for four sample spacings.

Figure 9 shows the median ratio $r_{log} \div r_{plug}$ versus spacing for the model and field data. The parameter values chosen for the model (i.e., C_v, ϕ, etc.) closely duplicated the actual formation and probe figures. The value of r_ϕ = 0.95 was based on correlograms of core plug porosity giving a first-lag (30cm) correlation of about 0.6 so that, at 3cm, the correlation is $(0.6)^{0.1}$ = 0.95. The shale percentage, V = 0, was chosen because no permeability measurement was less than 2mD. While the short spacing results are close, the longer spacing model results are more pessimistic than the field performance. This discrepancy could be caused by a too simplistic porosity correlation model, an insufficiently discretized formation model, or assuming perfect plug measurements. The match at the extremes could have been improved by increasing r_ϕ to 0.96 or 0.97, but we had no justification for the change. In all, the similarity of results is very encouraging and suggests that the model is capturing the important features of this data application.

IV. FURTHER MEASUREMENT MODEL APPLICATIONS

We now consider two other applications of the model: prediction of average permeability and quantitative estimates of permeability variability. Again, we investigate the precision and relative variability of estimates derived from probe and core plug measurements.

A. Predicting Average Permeability

During the scaling-up process of reservoir properties, small-scale permeability measurements are used to predict the aggregate permeability of portions of the reservoir. The arithmetic or harmonic averages are often used to combine small-scale measurements to reflect bed-parallel or bed-transverse flow, respectively. The behavior of both core plug and probe data for both averages were examined using the model previously described. We did not examine the merits of either average to represent the bulk permeability in this study.

The formation model consisted of a sand-only ($V = 0$) formation with uncorrelated, 3cm thick elements of 250 md median permeability. The core plug sampling model, as in the previous application, was an uncorrupted permeability measurement, $(k_p)_i$, every 10 elements (30cm). The probe measurement model differed from the previous application in that no permeability "log" was produced; only the probe data $(k_m)_i$ were used.

The true harmonic and arithmetic means of the formation, $(k_t)_h$ and $(k_t)_a$, are given by

$$(k_t)_h = \exp(\mu - \tfrac{1}{2}\sigma^2) \qquad \text{(4a)}$$

and $$(k_t)_a = \exp(\mu + \tfrac{1}{2}\sigma^2) \qquad \text{(4b)}$$

where $\ln(k) \sim N(\mu, \sigma^2)$, as previously described in Step 3 of the formation model. $\sqrt{(k_t)_a(k_t)_h}$ is also the median permeability of the formation (250 md) as determined by Eq. (3). From Eqs. (2) and (4) it follows that

$$(k_t)_h = \exp\left\{ \mu - \tfrac{1}{2}\left[\ln(1+C_v^2)\right]^2 \right\} \tag{5a}$$

and $$(k_t)_a = \exp\left\{ \mu + \tfrac{1}{2}\left[\ln(1+C_v^2)\right]^2 \right\}. \tag{5b}$$

The data applications were the harmonic and arithmetic averages of the core plug ($(k_p)_h$ and $(k_p)_a$) and probe data ($(k_m)_h$ and $(k_m)_a$):

$$(k_x)_h = \left\{ \frac{1}{n} \sum_{i=1}^{n} [(k_x)_i]^{-1} \right\}^{-1} \tag{6a}$$

and $$(k_x)_a = \left\{ \frac{1}{n} \sum_{i=1}^{n} (k_x)_i \right\}, \tag{6b}$$

where x = p or m and the summations are over n = 400 data for the probe or n = 40 data for the core plug measurements.

The model is examining the competing effects of a few precise samples (ideal core plug data) and many less precise samples (probe data). Table II, which shows the averages for 500 iterations along with the standard deviations in parenthesis, indicates two distinct differences in behavior for estimates based on the two sampling methods. First, the increased number of measurements reduces the variability of the estimates for all cases. The probe averages have about one-half to one-third the plug variabilities. Only for the two higher C_v cases, however, does the plug variability exceed 15% of the true average value. Second, the probe averages are biased compared to the plug averages at all heterogeneity levels. The probe harmonic averages are about 12% low and the arithmetic averages are about 12% too large. This level of bias is still probably acceptable for most reservoir description purposes.

Table II. Model results for harmonic and arithmetic permeability averages.

C_v	$(k_t)_h$	$(\overline{k}_p)_h$	$(\overline{k}_m)_h$	$(k_t)_a$	$(\overline{k}_p)_a$	$(\overline{k}_m)_a$
0.50	224	225 (±18)	197 (±7)	280	280 (±21)	316 (±12)
1.00	177	181 (±28)	157 (±10)	354	353 (±59)	401 (±26)
1.50	139	147 (±32)	124 (±11)	451	451 (±103)	513 (±46)

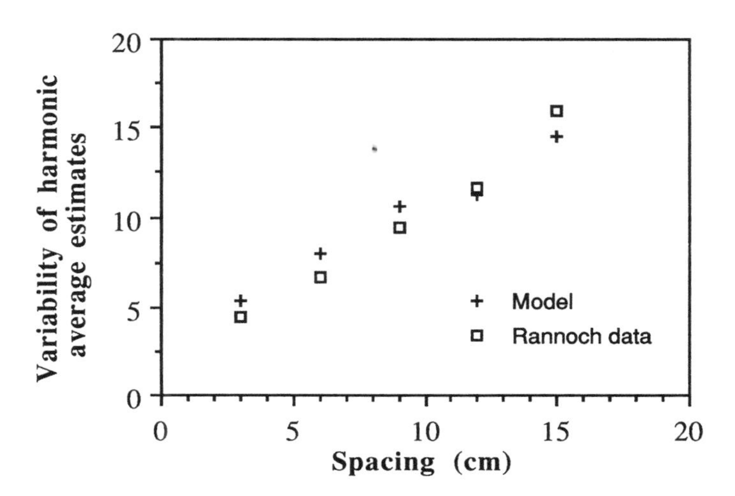

Fig. 10. Comparison of model and Rannoch Formation harmonic average performance.

Figure 10 shows model results for $(k_m)_h$ variability (in md) as the sample distance increases. For these results, n = 130 elements were used to permit comparison with Rannoch Formation data from one of the four-meter intervals. The satisfactory match suggests that the model is emulating the sampling behavior well.

Consequently, the model results suggest that, to estimate the aggregate permeability of a region using the harmonic average, the corrupted probe measurements can give consistently better results than the few, idealized plug measurements. However, when the arithmetic average is to be used, more careful selection may be required, depending upon the formation heterogeneity. Considering the errors of actual plug measurements, the previous constraints on probe utility could probably be relaxed somewhat.

B. Estimation of Permeability Variation

Formation heterogeneity can be assessed in a variety of ways. For this study, we chose C_v as the measure to be estimated:

$$(C_v^x)_{est} = \frac{\sqrt{\left\{ \frac{1}{n-1} \sum_{i=1}^{n} \left[(k_x)_i - (k_x)_a \right]^2 \right\}}}{(k_x)_a}, \qquad (7)$$

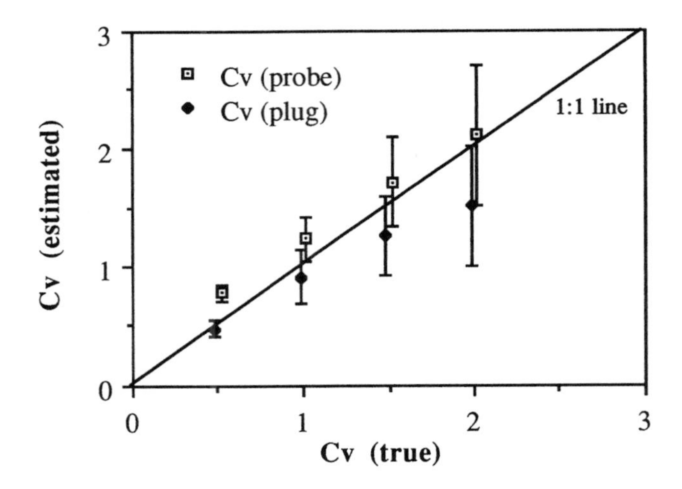

Fig. 11. Comparison of averaged heterogeneity estimates based on probe and idealized plug measurements.

where x = p or m. Eq. 7 is the standard deviation divided by the arithmetic average. The model was used to give average values and standard deviations of $(C_v)_{est}$ for various C_v for the two measurement methods.

The model results (Fig. 11) indicated that the variability of $(C_v^m)_{est}$ was similar to that of $(C_v^p)_{est}$ for all heterogeneity levels. The model results, however, show that the estimated values differed significantly for the two sampling methods. The variability of the probe measurement (δ) appears as a formation with a heterogeneity of about $C_v = 0.6$ which, when combined with the intrinsic formation variation for $C_v = 0.5$, produces a formation with an apparent $C_v = 0.8$. This result agrees with theory that uncorrelated variations add according to the variances:

$$\sigma^2_{measurement} + \sigma^2_{formation} = \sigma^2_{total}$$

where $\sigma = \ln(1 + C_v^2)$. The effect, however, diminishes rapidly for formations where $C_v > 1$, since $\sigma^2_{measurement} < \frac{1}{2}\sigma^2_{formation}$. At these higher C_v levels, the insufficient sampling of the plugs becomes evident.

Hence, statistical heterogeneity assessments on relatively homogeneous sediments may not benefit by taking many, less precise measurements instead of a few precise measurements. The advantages gained using one method instead of another will depend on the errors associated with each measurement along with the intrinsic formation variability to be quantified. For very heterogeneous ($C_V > 1$) materials, the probe measurement is preferred because of the low cost and non-destructive benefits; for homogeneous materials ($C_V < 0.5$) plugs, being more accurate and because few samples are required, are to be preferred; and for heterogeneous materials ($0.5 < C_V < 1.0$) both measurements have their merits. Spacing of the required number of samples (e.g., see Corbett and Jensen, 1992b) will determine which method is to be preferred. For example, in an aeolian sandstone described by Weber (1987), the crossbed sets are clearly laminated and heterogeneous ($C_V \sim 1.5$). Hence, probe assessment of heterogeneity in this case is appropriate and may be usefully used with correlations (e.g., Kelkar and Gupta, 1991) to predict the recovery performance of this facies.

V. FURTHER COMMENTS

The general approach of measurement modelling used here builds on the concepts of Jensen (1990). The model components, however, have been improved and the results tested against field data. The model shows that having many more data does not always bring advantages.

More sophisticated models for plugs (e.g., noise and systematic errors), formation (e.g., greater variety and correlation of lithologies and lateral permeability variations) and probe (e.g., residual fluids and core damage) can be developed. These can be used to investigate systematic differences between plug and probe performance. Differences between plugs and probe measurement are to be expected in all comparative studies. This model highlights differences that are due solely to the statistics of sampling and provides a means of quantifying them. Hence, comparisons of measurements with this knowledge will shed light on the reservoir significance of differences.

The harmonic average was investigated here because of its relevance to bed-series flow and, hence, vertical permeability estimation. Our experience shows that low permeability laminae that are present in many reservoirs are below the resolution of many probes (i.e., < 5mm) when horizontal measurements are taken on slabbed surfaces. This model can be used to explore how the distribution of these laminae affects both vertical plugs and horizontal probe measurements, leading to a better understanding of the meaning and performance of the harmonic average as an estimator of vertical permeability.

VI. CONCLUSIONS

We have proposed a measurement model which permits examination of sampling schemes and comparison of measurement methods. We used the measurement model for three different formation evaluation purposes and found that it gave results in agreement with field measurements. The model results showed that more data than the usual one-foot (30cm) sampling schemes may achieve significantly improved formation permeability estimates. The benefits, however, depend on the measurement type and the data application.

The model, when used to compare probe permeameter and idealized core plug measurements for correlations with the wireline bulk density log, showed that:

1. Averaged probe measurements can give stronger correlations than core plugs.
2. The probe--bulk density relationship degrades as fewer probe measurements are made.
3. Simply halving the spacing in a sampling scheme may not produce a perceptible benefit.
4. The degradation behavior of the model is slightly pessimistic compared to field data.

Estimation of the harmonic and arithmetic average permeability generally benefitted from the more frequent probe measurements. Probe-based estimates showed lower variability but more bias than estimates from idealized plug data. Heterogeneity estimates (C_v) are positively biased at low heterogeneity levels ($C_v \leq 0.5$) but give equal or better performance at higher levels ($C_v \geq 1$).

The model gives clear guidance on the appropriateness of plug and probe measurements and their respective sampling schemes. Many of the results accord with intuition and the model quantifies these assessments. Some results, however, are counter intuitive and, hence, lead to new insights into the data collection and application processes.

ACKNOWLEDGEMENTS

The authors would like to thank Greg Geehan, Andy Hurst, and Christian Halvorsen for their help in the acquisition of the Rannoch Formation data. Statoil and BP are also thanked for technical support and their partners for allowing use of these data.

REFERENCES

1. Corbett, P. W. M., and Jensen, J. L.: "Variation of Reservoir Statistics According to Sample Spacing and Measurement Type for Some Intervals in the Lower Brent Group," *The Log Analyst* (1992a) **33**, No. 1, 22-41.

2. Corbett, P. W. M., and Jensen, J. L.: "Estimating the Mean Permeability: How Many Measurements Do You Need?" *First Break* (1992b) **10**, No. 3, 89-94.
3. Dreyer, T., Scheie, Å., and Walderhaug, O.: "Minipermeameter-Based Study of Permeability Trends in Channel Sand Bodies," *AAPG Bulletin* (April 1990) **74**, No. 4, 359-374.
4. Dykstra, H., and Parsons, R. L.: "The Prediction of Oil Recovery by Water Flood," *API Secondary Recovery of Oil in the United States*, API, Washington, DC (1950) 160-174.
5. Goggin D. J.: "Minipermeametry: Is It Worth the Effort?," *Proc.*, Petroleum Science and Technology Institute Minipermeametry in Reservoir Studies Conference, Edinburgh (1991), June 27, 21p.
6. Goggin, D. J., Chandler, M. A., Kocurek, G., and Lake, L. W.: "Patterns of Permeability in Eolian Deposits: Page Sandstone (Jurassic), Northeastern Arizona," *SPEFE* (June 1988a) 297-306.
7. Goggin, D. J., Thrasher, R. L., and Lake, L. W.: "A Theoretical and Experimental Analysis of Minipermeameter Response Including Gas Slippage and High Velocity Flow Effects," *In Situ* (1988b) **12**, No. 1&2, 79-116.
8. Haan, C. T.: *Statistical Methods in Hydrology*, Iowa State Univ. Press, Ames (1977), 378p.
9. Jensen, J. L.: "A Model for Small-Scale Permeability Measurement With Applications to Reservoir Characterization," paper SPE/DOE 20265 presented at the 1990 Enhanced Oil Recovery Symposium, Tulsa, Apr. 22-25, 10p.
10. Kelkar, M. G., and Gupta, S. P.: "A Numerical Study of Viscous Instabilities: Effect of Controlling Parameters and Scaling Considerations," *SPERE* (Feb. 1991) 121-128.
11. Kortekaas, T. F. M.: "Water/Oil Displacement Characteristics in Crossbedded Reservoir Zones," *SPEJ* (Dec. 1985) 917-926.
12. Lake, L. W., and Jensen, J. L.: "A Review of Heterogeneity Measures Used in Reservoir Characterization," *In Situ* (1991) **15**, No. 4, 409-439.
13. Lewis, P. A. W., and Orav, E. J.: *Simulation Methodology for Statisticians, Operations Analysts, and Engineers*, **1**, Wadsworth and Brooks/Cole Inc., Belmont, Ca. (1989), 416p.
14. Martin, J. H., and Evans, P. F.: "Reservoir Modeling of Marginal Aeolian/Sabkha Sequences, Southern North Sea (UK Sector)," paper SPE 18155 presented at the 1988 Annual Technical Conference and Exhibition of the SPE, Houston, Oct. 2-5.
15. Ringrose, P. S., Sorbie, K. S., Feghi, F., Pickup, G. E., and Jensen, J. L.: "Relevant Reservoir Characterisation: Recovery Process, Geometry and Scale," *Proc.*, Sixth European IOR Symposium, Stavanger (1991), May 21-23, 15-25.
16. van Veen, F. R.: "Geology of the Leman Gas-Field," *Petroleum and the Continental Shelf of North-West Europe*, A. W. Woodland (ed.), Applied Science Pubs. Ltd., Essex (1975) **1**, 223-231.
17. Weber, K. J., Eijpe, R., Leijnse, D., and Moens, C.: "Permeability

Distribution in a Holocene Distributary Channel-Fill Near Leerdam (The Netherlands)," *Geologie En Mijnbouw* (1972) **51**, No. 1, 53-62.

18. Weber, K. J.: "How Heterogeneity Affects Oil Recovery," *Reservoir Characterization,* L. W. Lake and H. B. Carroll, Jr. (eds.), Academic Press, Orlando (1986), 487-544.
19. Weber, K. J.: "Computation of Initial Well Productivities in Aeolian Sandstone on the Basis of a Geological Model, Leman Gas Field, U. K.," *Reservoir Sedimentology,* R. W. Tillman and K. J. Weber (eds.), Society of Economic Paleontologists and Mineralogists, Tulsa (1987) **40**, 333-354.

PERMEABILITY HETEROGENEITIES OF CLASTIC DIAGENESIS

Thomas L. Dunn
Ronald C. Surdam

Department of Geology and Geophysics
University of Wyoming
Laramie, Wyoming 82071

ABSTRACT

Within clastic sediments, post-depositional processes (e.g., infiltration, compaction, and diagenesis) produce and alter permeability heterogeneity on a range of scales from microscopic fabrics, through bedding and laminations, up to basin-wide features. Diagenetic alterations have been shown to be complexly but systematically related to the burial and thermal histories of the sediments. Cementation patterns alter the dimensions of macroscopic depositional features. The recently recognized need for dimensional data on reservoir heterogeneity related to depositional systems has renewed emphasis on outcrop studies. Because the burial histories and diagenetic alterations of subsurface reservoirs and their outcropping depositional analogs may be significantly different, it is important to understand what changes in clastic reservoir heterogeneity can be attributed solely to post-depostional processes.

Within fluvial volcanogenic sandstones, variations of 0.10 to 0.15 in the Dykstra-Parsons coefficient of permeability heterogeneity can be attributed to diagenetic events related to meteoric invasion and uplift. Similar variations can be ascribed to the presence of infiltrated materials. Where their effects can be isolated, compactional processes are observed to decrease permeability heterogeneity.

I. INTRODUCTION

Post-depositional processes such as diagenesis, compaction, and infiltration of fines affect permeability heterogeneity in sandstones. Prior work has concentrated on the sedimentologic controls on permeability heterogeneity, and few studies have examined the diagenetic aspects. Given an understanding of how rocks are progressively altered during burial, it is difficult to imagine that burial diagenesis does not in some manner alter the permeability distribution or impose changes in the anisotropy or heterogeneity of a sandstone.

This is a study of the petrology, diagenesis, and permeability heterogeneity of the sandstones of the Comodoro Rivadavia Formation, San Jorge Basin, southern Argentina. The Comodoro Rivadavia Formation (CRF) is a late Cretaceous fluvial-alluvial sequence deposited along the northern flank of a rifted basin. The CRF varies greatly in composition and contains abundant volcanogenic material. The CRF has been only shallowly buried.

The term *reservoir heterogeneity* refers to the spatial variation of those rock properties that affect the production of commercial quantities of hydrocarbons: porosity, permeability, wettability, and pore aperture distribution. Lithology and rock fabric, determined by depositional processes, govern the initial values of these rock properties and, to some extent, their first-order variations. Derived parameters such as relative permeability and sweep efficiency are also, in part, a function of lithology and texture (see, Morgan and Gordon, 1970). Changes in lithology during the burial of clastic rocks hopefully do not occur; barring some form of alchemy, changes in texture and mineralogy are the only observed effects of burial diagenesis. Hence, variations in the derived parameters, relative permeability and sweep efficiency, also occur during burial diagenesis. Chemical pathways for the diagenetic evolution of the mineralogy of sandstones undergoing burial are presented by Surdam et al (1989a,b,c).

II. GEOLOGY OF THE COMODORO RIVADAVIA FORMATION

The San Jorge Basin (Figure 1) is the oldest and one of the most prolific oil provinces of Argentina. The basin produces approximately 200,000 BOPD (Pucci, 1987) the majority of which comes from the north flank of the basin, from Upper Cretaceous fluvial sandstones of the Chubutiano Group. The oil is principally lacustrine-sourced and has an API gravity ranging between 21 and 25°. The formation waters are fresh to brackish (approximately 1000 to 15,000 ppm TDS; Khatchikian and Lesta, 1973). Production comes from many thin sands with highly variable water saturations. The reservoirs are volcanogenic sandstones and contain an abundance of authigenic cements including clays, zeolites, carbonates, quartz and feldspars. Hence, evaluation of oil production from electric logs is difficult due to the combination of electrically

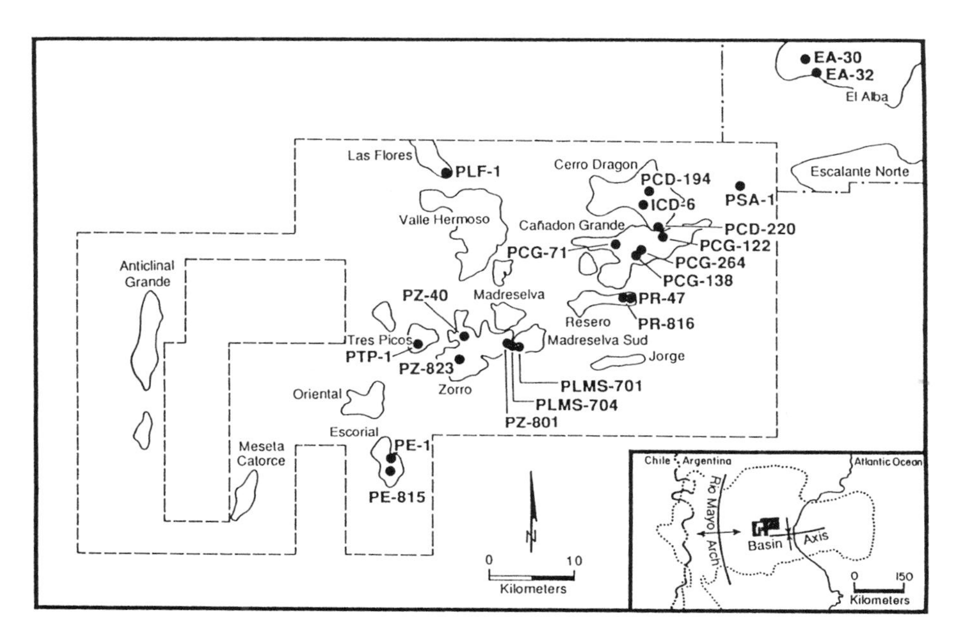

Figure 1. Location map of fields and sampled wells within the outlined study area (the Amoco Argentina Oil Company Contract Area). Inset at lower right is a map showing the location of the study area within the San Jorge Basin and gross structural features. After Dunn (1992).

resistive water and the somewhat electrically conductive reservoir rock.

The Upper Cretaceous forms a wide apron of sediment spread throughout Argentina along the flanks of the western (Pacific) convergent margin. Within the northern portion of the basin, the fraction of sediment which is sandstone varies from 0.29 along the basin axis to 0.49 on the rift margin (Figure 2). Isopach maps of a portion of the study area (Figure 3) illustrate the meanderbelt channel sand development. Overall, the alluvial architecture in any given area does not change significantly. The coarse deposits are predominantly point bar sequences and abandoned channel deposits (Figures 4 and 5).

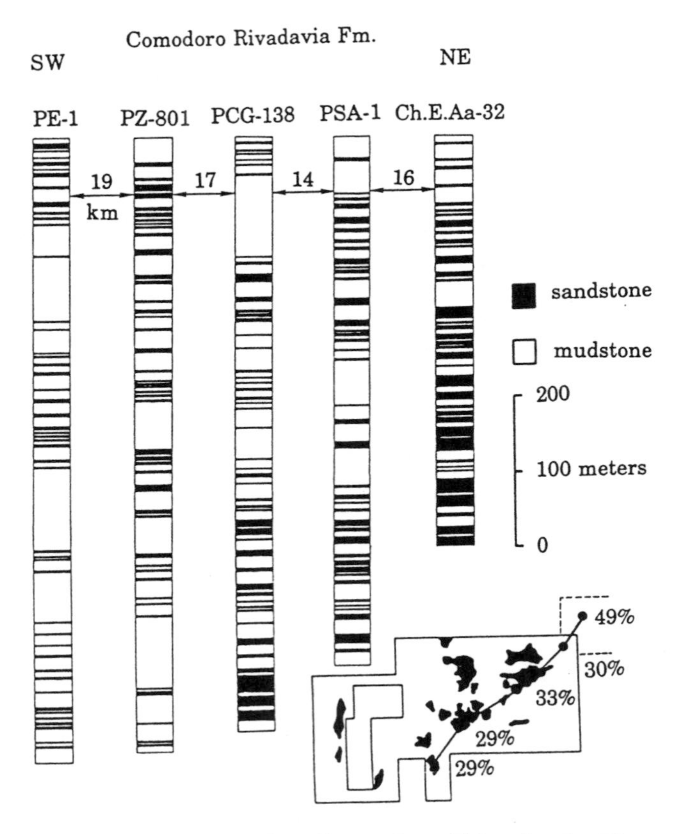

Figure 2. Cross section of the Comodoro Rivadavia Formation, hung on the top of the formation. The positions of the five wells are shown in the index map in the lower right corner. After Dunn (1992).

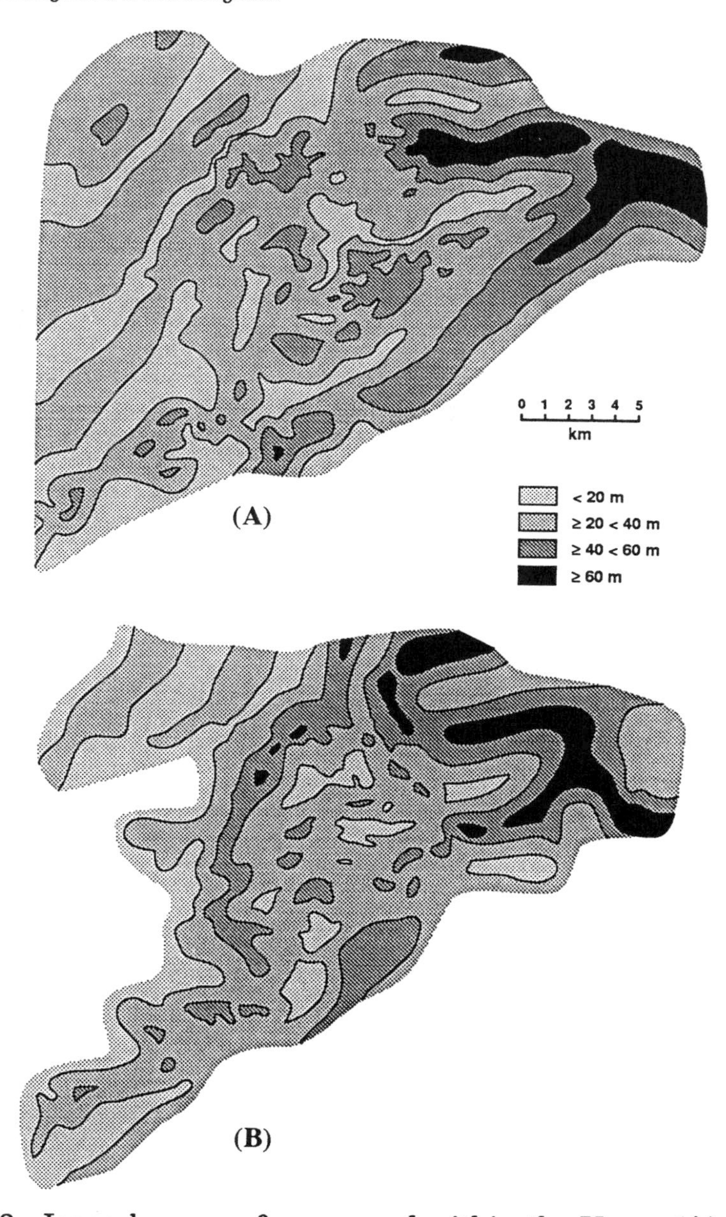

Figure 3. Isopach maps of gross sand within the Upper (**A**) and Lower (**B**) Comodoro Rivadavia Formation. The maps show the series of channels entering the basin from the northeast carrying principally the San Jorge petrofacies. After unpublished maps by C. Manzolillo and A. Silveira; courtesy of Amoco Argentina Oil Company.

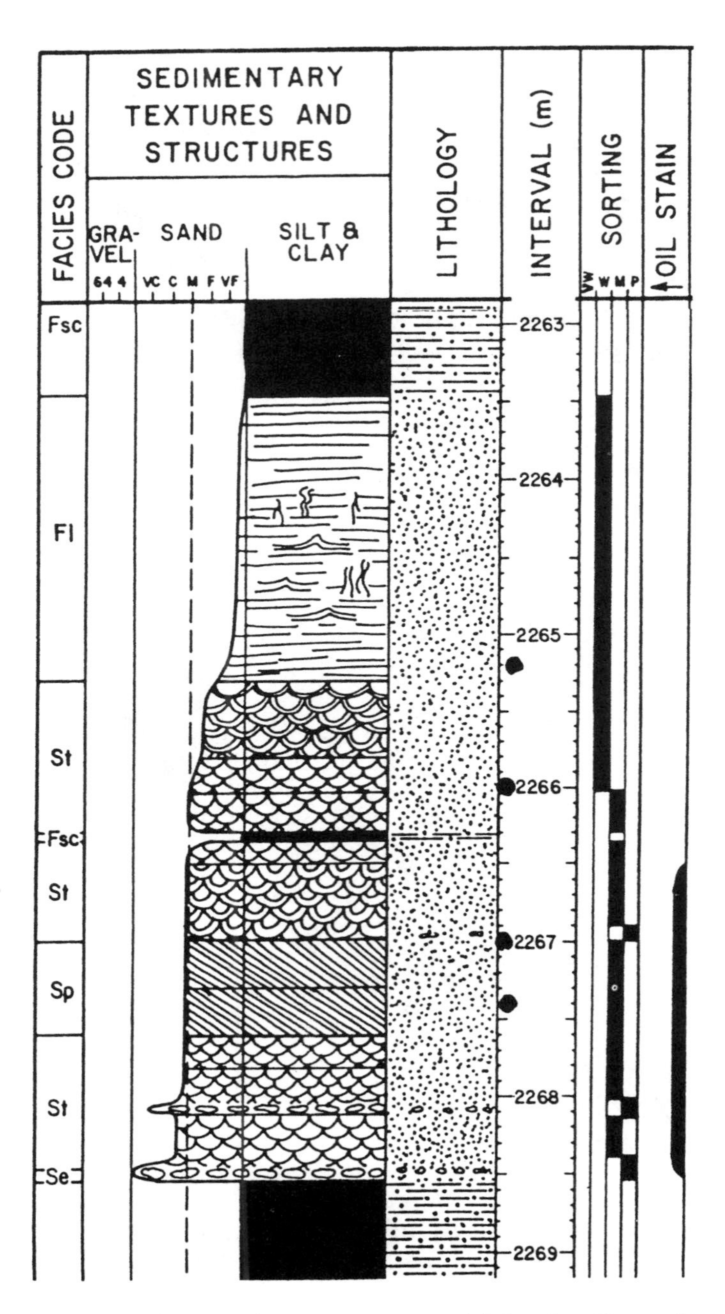

Figure 4. Typical point bar sequence within the Comodoro Rivadavia Formation fluvial deposits. After Dunn (1992).

The sandstones vary greatly in composition (Figure 6). The great range represents variations in the amount of volcanogenic material (principally plagioclase and volcanic lithic fragments). These variations in composition occur in close proximity, indicating mixed provenances within the drainage basin. Petrologic analysis of sandstones reported by Teruggi and Rosetto (1963) suggest that the strongly lithic petrofacies (referred to herein as the Zorro petrofacies) was derived from west. The isopach maps indicate a northeasterly source entering the basin carrying the predominant petrofacies (referred to herein the San Jorge petrofacies). The Cañadon Grande petrofacies is a diagentically altered

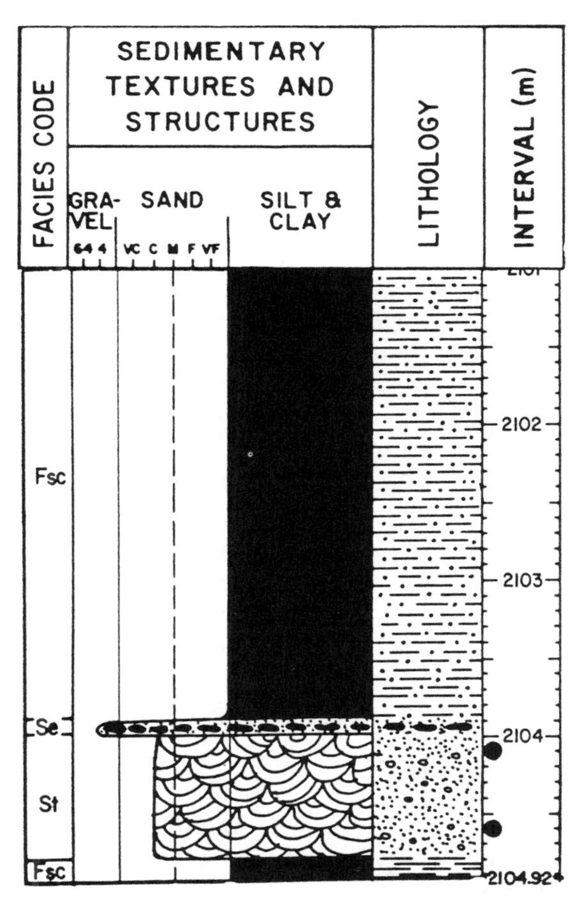

Figure 5. Abandoned channel deposit in the Comodoro Rivadavia Formation. After Dunn (1992).

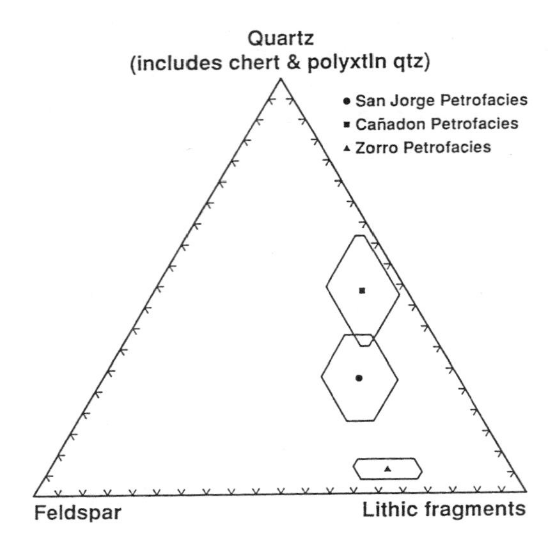

Figure 6. Ternary Q-F-L diagram showing the average and one standard deviation compositions of the Comodoro Rivadavia Formation petrofacies. After Dunn (1992).

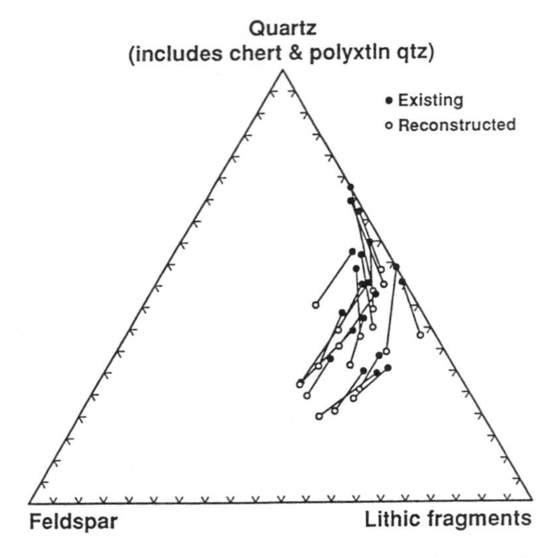

Figure 7. Ternary Q-F-L diagram showing the change in compostion accompanying the extensive leaching of the San Jorge petrofacies producing the Cañadon Grande petrofacies. The leaching produced abundant kaolinite cements. After Dunn (1992).

daughter product of the San Jorge petrofacies: extensive leaching of labile grains such as the volcanic lithic fragments and feldspar depleted the sandstones in those components and produced a much more quartzose composition (Figure 7).

The porosity types (as distinguished petrographically) within each petrofacies are distinct (Figure 8), reflecting the dominant authigenic assemblage and porosity loss due to compaction within each group. The Zorro petrofacies, characterized by the abundance of porefilling smectite and zeolite has pore systems which are dominated by microporosity. The Cañadon Grande petrofacies, characterized by leached grains and extensive kaolinite formation, has pore systems dominated by moldic and micropores, and only minor amounts of clear intergranular primary pores are present. The predominant petrofacies within the study area, the San Jorge petrofacies, contains on average roughly equal proportions of microporous aggregates, moldic pores, and clear, unobstructed primary intergranular pores. Where these petrofacies are juxtaposed, their different pore systems cause macroscopic and megascopic heterogeneities in permeability and fluid saturation.

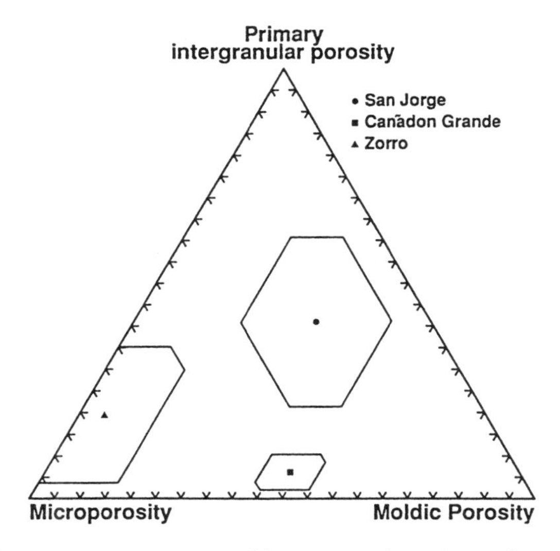

Figure 8. Ternary pore-type diagram showing the average and one standard deviation of the pore types for each petrofacies recognized in the study area. After Dunn (1992).

III. DIAGENESIS AND PERMEABILITY HETEROGENITY

A clear indication that diagenesis is controlling the permeability distribution in a sandstone is the degradation of the relationship between grain size or sorting, and permeability. The lack of a strong correlation between these parameters is characteristic of volcaniclastic sandstones. For example, within the Comodoro Rivadavia Formation there is no visible correlation between sorting or grain size with permeability, despite the well developed porosity-versus-log-permeability relationship (Figure 9). Grain size and sorting are frequently cited (cf. Beard and Weyl, 1973) as the dominant controls on porosity and permeability in sandstones; yet it is important to keep in mind that strong relationships between porosity and permeability are found in diverse lithologies–even basalts, where grain size and sorting have no meaning. Clearly, within the Comodoro Rivadavia Fromation, other features control porosity and permeability. As mentioned above, the pore-type distributions of the Comodoro Rivadavia Formation sandstones clearly reflect their diagenetic alteration.

A. Compactional Porosity Loss

Sandstones lose porosity after deposition through the compactional processes of grain packing rearrangement and ductile material deformation. This is particularly true for the first 3 km of burial. Excluding the effect of early cements, the change in slope on a depth-versus-porosity plot for both sandstones and shales is always maximized during early burial. The original composition of the sandstone plays a significant role in determining the magnitude of this effect, since detrital phyllosilicates and microcrystalline masses, along with aggregate lithic grains containing them, tend to behave plastically during compaction.

Poor sorting, abundant matrix, and early carbonate, zeolite, and clay cements create low permeability. Hence, the range of permeabilities of any given sandstone body (including both cemented and uncemented regions) during early burial is likely very great but highly variable. In environments not conducive to the formation of these early cements, the range of permeabilities (and therefore of heterogeneity) should not be as

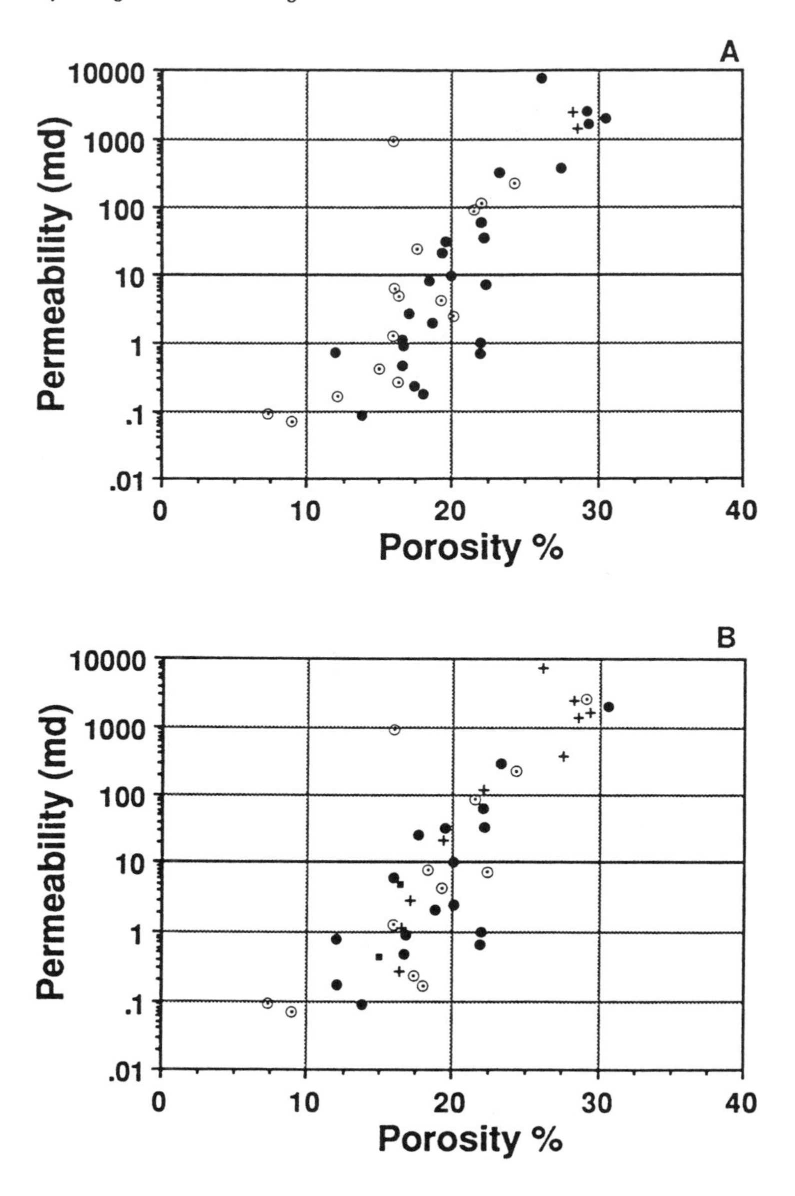

Figure 9. Porosity vs. permeabilty with grain size (**A**) and sorting (**B**) for the PCG-264 core illustrating the lack of control of initial textural controls on the reservoir quality of these sandstones. The symbols in **A** are ʘ = coarse-grained, • = medium-grained, and + = fine-grained. The symbols in **B** are + = very-well-sorted, • = well-sorted, ʘ = moderately-well-sorted, and ▪ = poorly-sorted. After Dunn (1992).

great. As porosity and permeability decline with increasing compaction and cementation, the range of permeabilities decreases as the maximum permeability drops. This trend is shown schematically in Figure 10.

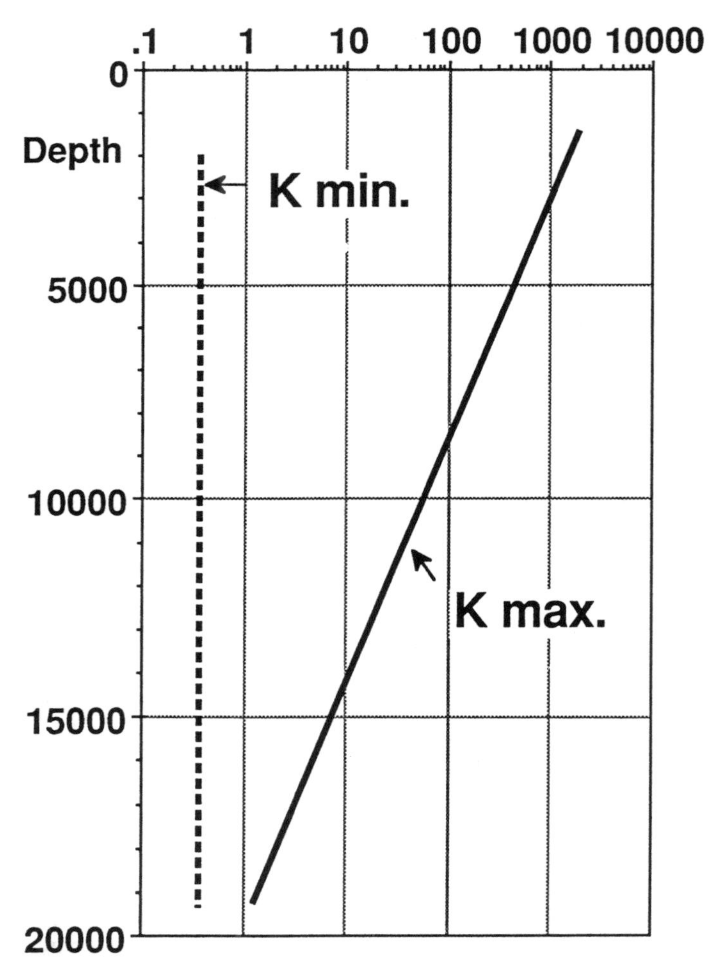

Figure 10. Schematic decrease in permeabilty variation simply due to compaction of sediment. With the decreasing range of permeabilities, there should be a commensurate decrease in permeability heterogeneity.

An estimate of the compactional porosity loss can be obtained from the variation in the intergranular volume (IGV) with burial. The IGV is a summation of open intergranular pore space and intergranular pore space occupied by cements. Within the Comodoro Rivadavia Formation, the mean IGV decreases by approximately 5% (phi units) per 500m increase in burial depth (Figure 11). This change is commensurate with a decrease in the Dykstra-Parsons coefficient (Figure 12).

Hence, compactional processes alone tend to decrease both mean permeability and permeability heterogeneity. These variations will be greatest within the first 3 km of burial and will be a significant component of permeability heterogeneity within shallow buried sandstones that contain abundant ductile lithic fragments.

B. Kaolinite Formation

Within the northeast portion of the study area, there is an interval of sandstone (the Cañadon Grande petrofacies) which has been extensively leached of its soluble detrital grains and cements. Only minor amounts of feldspar remain. Volcanic lithic fragments are altered to siliceous relicts. The moldic pores and the surrounding intergranular pore space contains abundant kaolinite. The interval is relatively thick (≈ 200m) and relatively isolated. Its updip position and extent of alteration suggest that meteoric fluids were likely responsible for the leaching. Uplifts during the Tertiary are known to have occurred, but their magnitude and timing are poorly known. Commercial kaolin deposits are present along the river valleys in the basin to the north (Murray, 1988).

The leached zone,containing abundant kaolinite, is characterized by a large increase in the range of conventional core analysis gas permeabilites (Figure 12). In the Comodoro Rivadavia Formation sandstones, above and below the leached zone, permeabilites range three orders of magnitude. Within the leached zone, the permeabilites range over six orders of magnitude. This variation is not evident from the well-behaved porosity-versus-log permeability relationship (Figure 13). The variation in permeability is reflected in the Dykstra-Parsons coefficient calculated for these intervals, which increases from 0.84 to 0.99 going into the leached zone and reduces to 0.76 below it (Figure 12). These values above and below the leached zone compare with other San Jorge petrofacies heterogeneity calculations such as that for the PZ-40 core, where the Dykstra-Parsons coefficient is 0.7, (Figure 14). The PZ-40 core contains small amounts of clay rims, infiltration structures and laumontite pore-filling cements. The presence of laumontite was not found to have substantial impact on the permeability heterogeneity of the Comodoro Rivadavia Formation sandstones.

The variation is due to the heterogenous degree of diagenetic

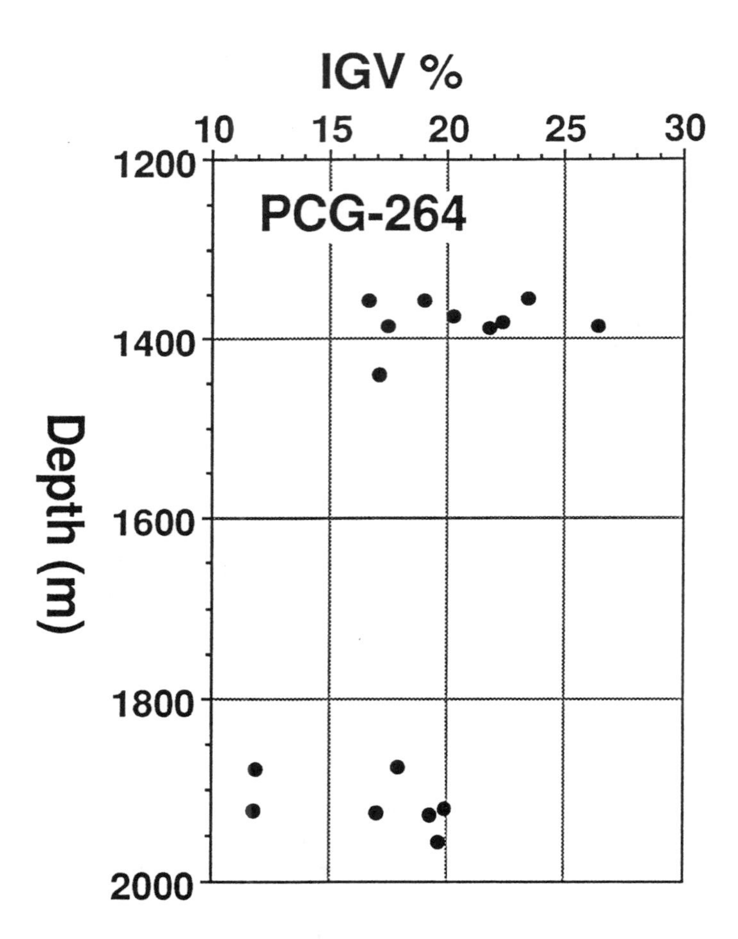

Figure 11. Intergranular volume (IGV %) for San Jorge petrofacies sandstones from the PCG-264 core, showing an average decrease of approximately 5%. A decrease in the Dykstra-Parsons coefficient accompanies this porosity loss due to compaction. After Dunn (1992).

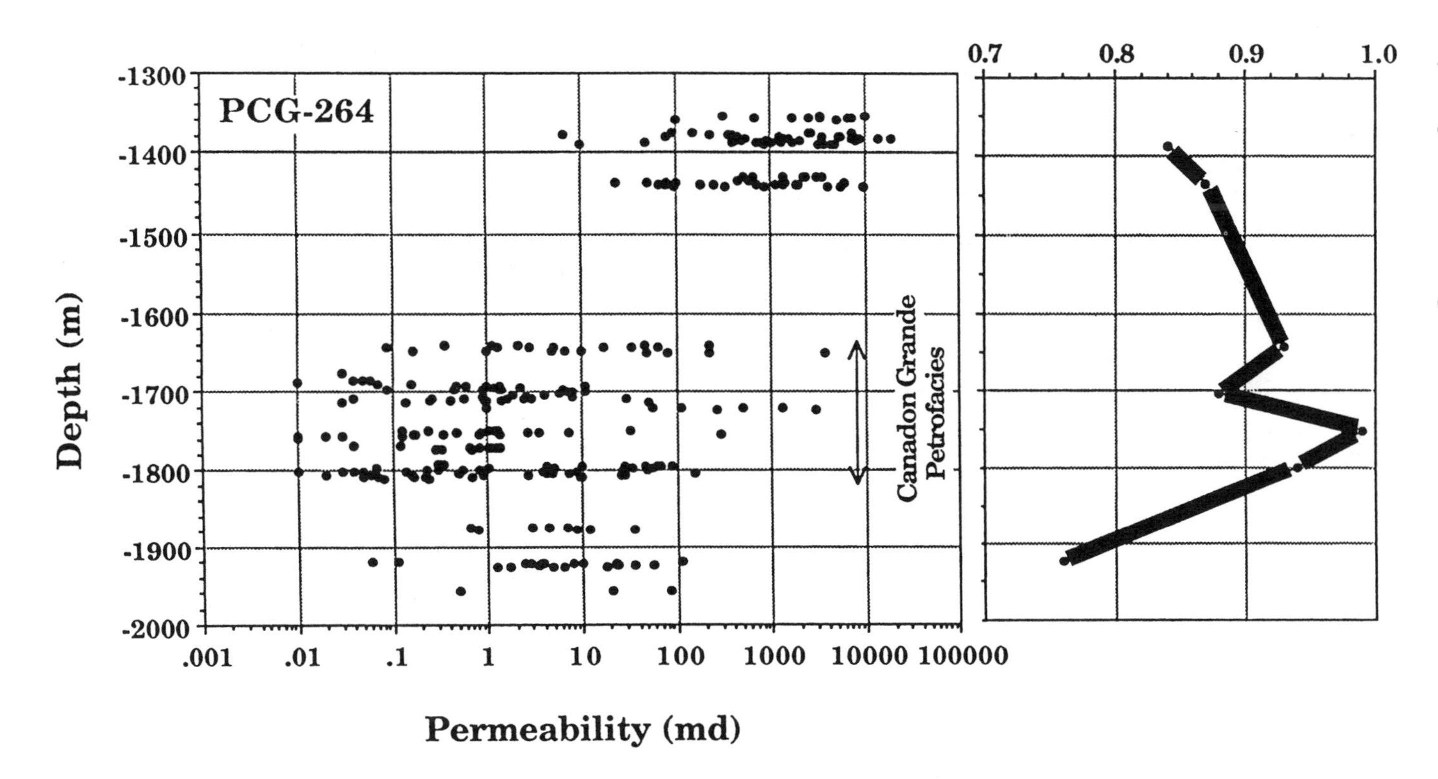

Figure 12. Permeability and Dykstra-Parsons coefficient vs. depth for the PCG-264 core. After Dunn (1992)

alteration within the leached zone. As the lithic fragments or feldspars alter, they form kaolinite-filled areas. As the alteration proceeds, these areas become both more numerous and increase in size until they interfere with one another. The permeabilty of the conventional core plug reaches a minimum when a free path through the clear, intergranular pore system ceases to exist. This variation in kaolinite pore filling is accompanied by the progressive loss of well-sorted pore apertures and an increase in porosity behind small pore apertures.

This is clearly illustrated in the ink tracings of the porosity of the Comodoro Rivadavia Formation sandstones shown in Figures 15A and 15B. Figure 15A (PCG-264, 1388.63-8.84m) shows the pore structure of the shallow-buried San Jorge petrofacies. Pores are interconnected even within the plane of the thin section. Moldic pores are visible on the left side. Below the leached zone, the San Jorge petrofacies (Figure 15B, PCG-264, 1957m) shows signs of compaction, having reduced the amount of porosity, the size of the pore apertures, and the

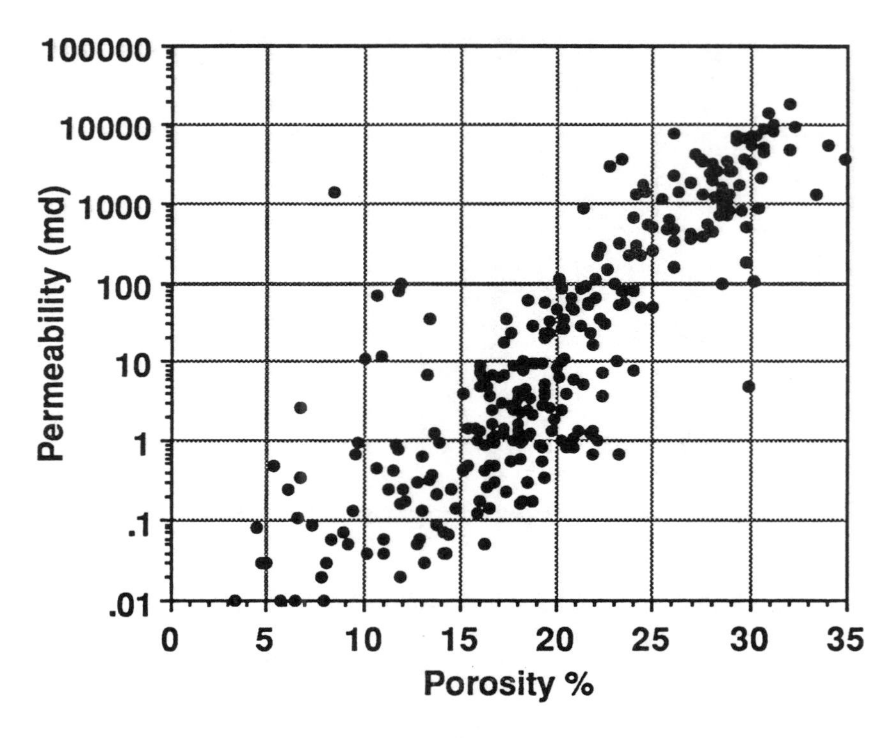

Figure 13. Porosity vs. log of permeability for the PCG-264 core.

amount of interconnected pores within the sandstone.

In contrast, ink tracings of porosity in the Cañadon Grande altered zone (Figure 16) shows how the porosity is altered to microporosity by the precipitation of kaolinite. The upper ink tracing of Figure 16 (PCG-264, 1770.33m) shows in outline the areas now occupied by kaolinite. Although the kaolinite alteration is extensive (20% of the total volume), there are still clear porosity paths through the sandstones, as indicated by the streak of clear intergranular pores (black in the figure). The lower ink tracing of Figure 16 (PCG-264, 1751.49-1.73m) shows the nearly complete filling of pores by the kaolinite. A similar sequence through the Cañadon Grande alteration is depicted in the capillary pressure curves shown in Figure 17.

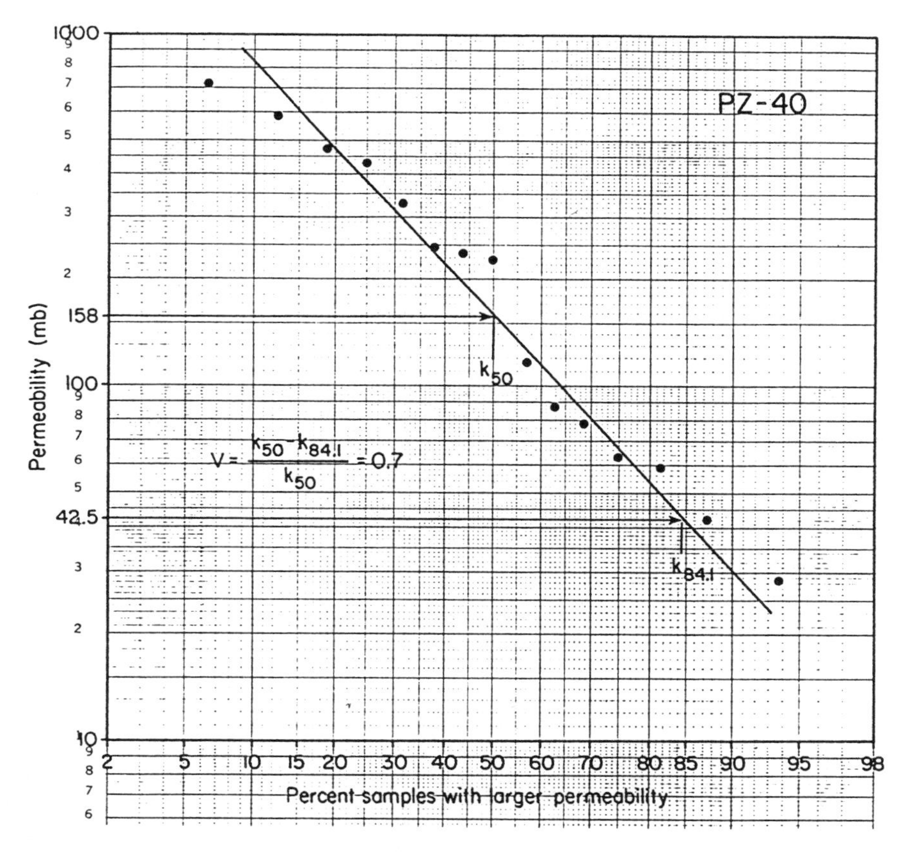

Figure 14. Dykstra-Parsons plot for the PZ-40 core, a relatively homogenous fluvial sandstone with only minor alteration and infiltration.

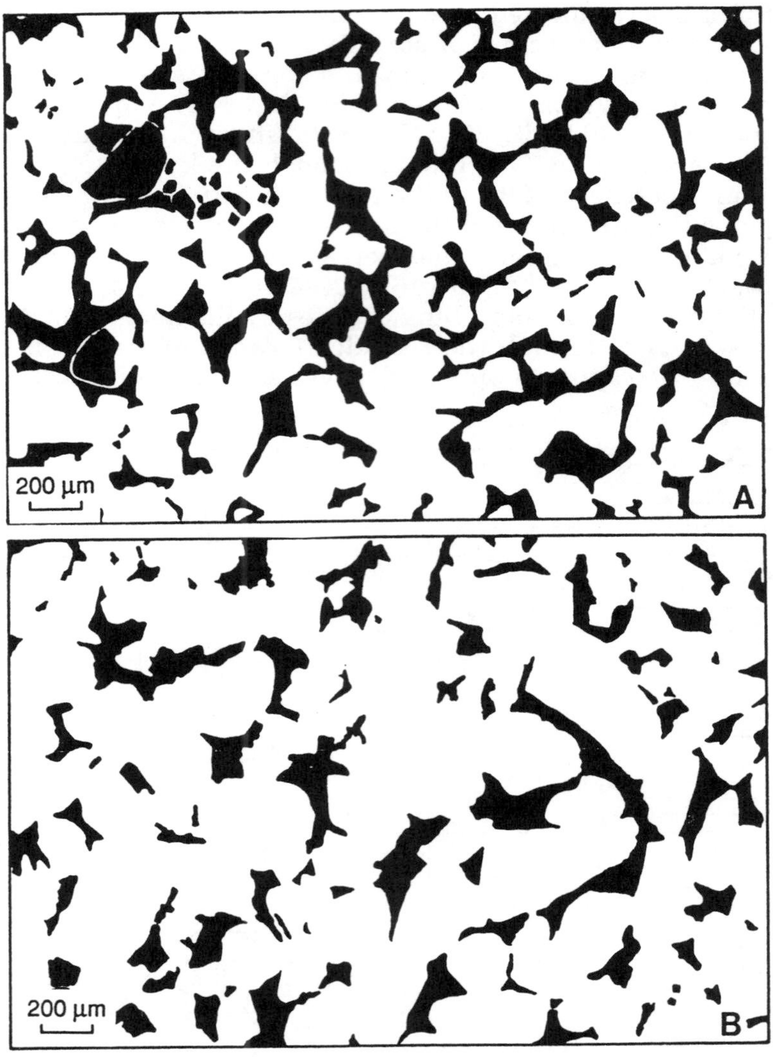

Figure 15. Drawings **A** (PCG-264, 1388.63/8.84 m: 1671 md, 29.4% porosity) and **B** (PCG-264, 1956.94/7.18m: 21 md, 19.4% porosity) are ink tracings of pore structures from photomicrographs of the San Jorge petrofacies, above (**A**) and below (**B**) the strongly altered, kaolinite-rich interval (Cañadon Grande petrofacies). Note the loss of porosity and increase in the amount of isolated pores. However, at the depth of 1957 m, the sandstones of the San Jorge petrofacies still retain interconnected pore throats. Some moldic (dissolution) porosity is visible in the left side of **A**. After Dunn (1992).

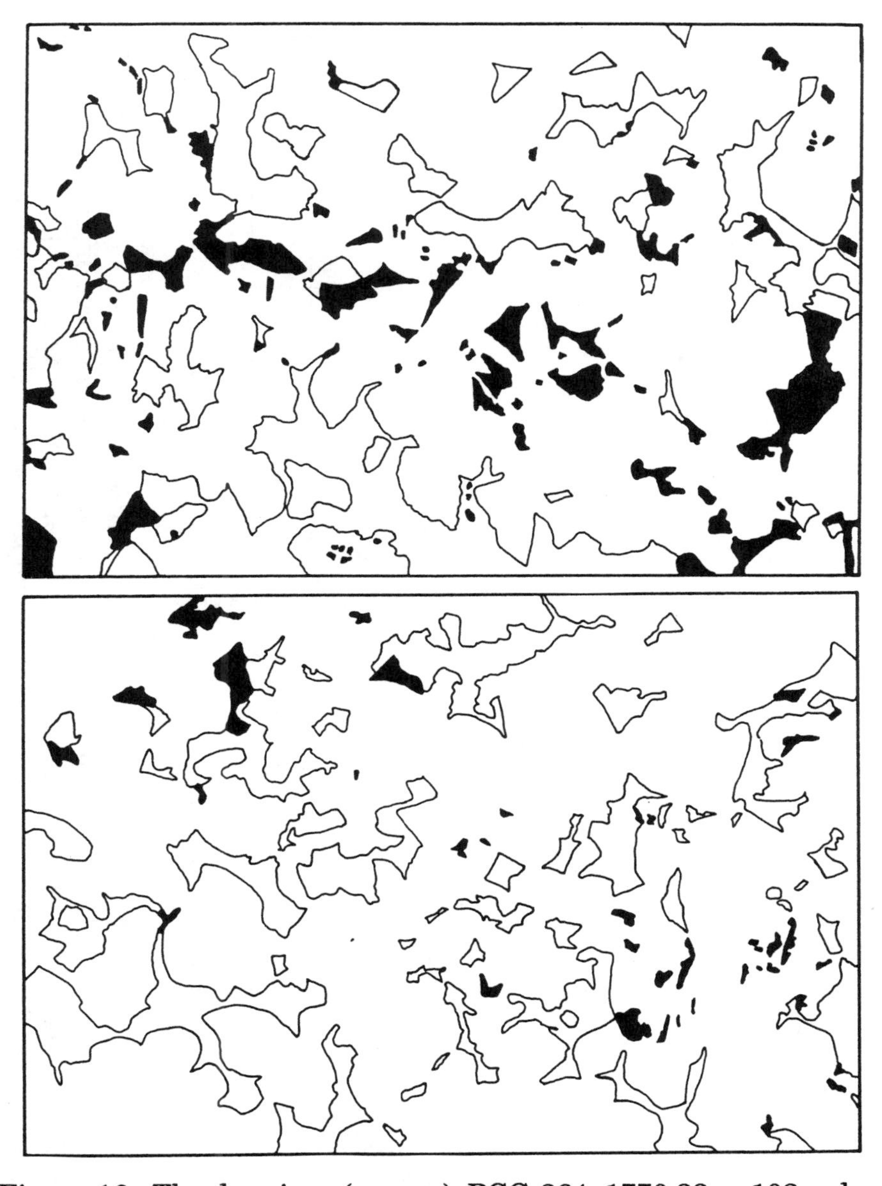

Figure 16. The drawings (**upper**) PCG-264, 1770.33m: 102 md, 22% porosity and (**lower**) PCG-264, 1751.49 m: 7.35 md, 22.4% porosity are ink tracings of pore structures and kaolinite alteration from photomicrographs of the Cañadon Grande petrofacies. The black areas are clear, intergranular porosity; the white outlined areas are pores filled with kaolinite. After Dunn (1992).

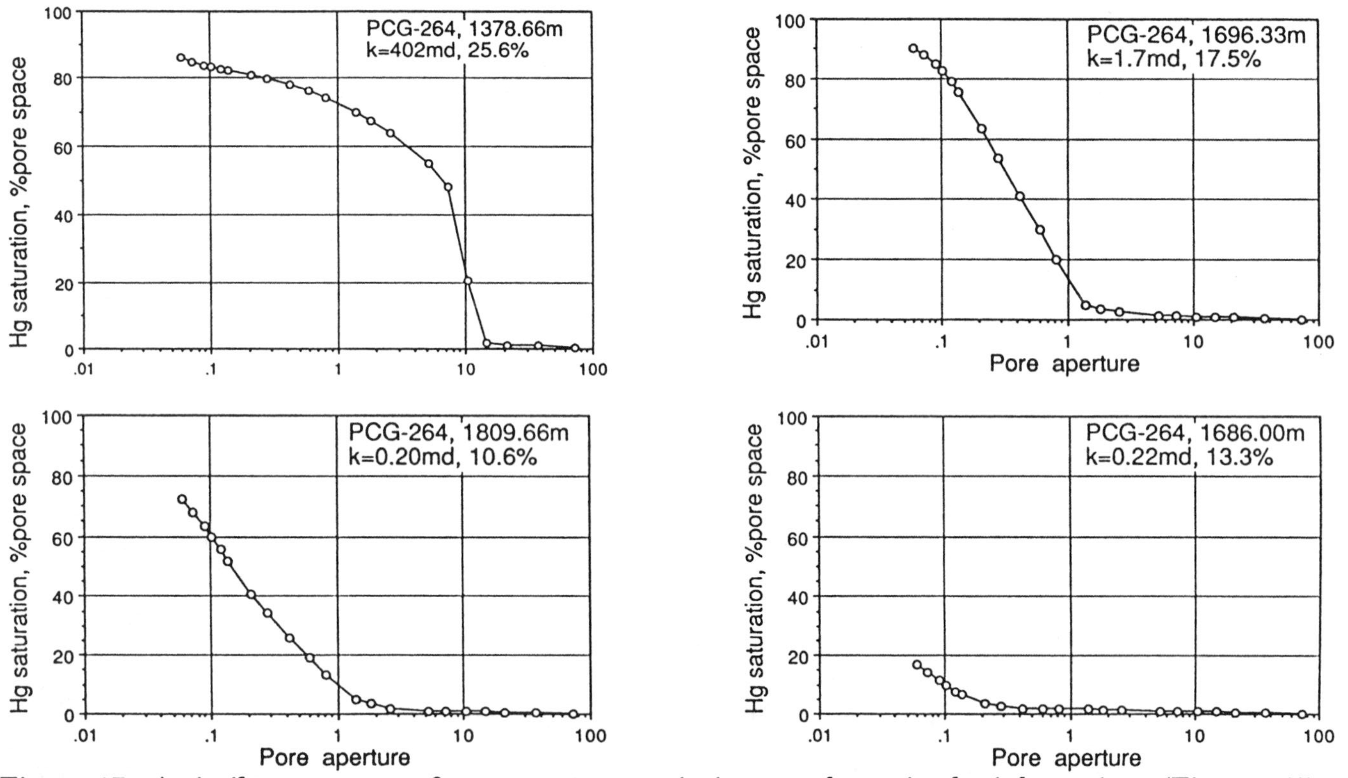

Figure 17. A similar sequence of pore aperture variations as shown in the ink tracings (Figures 15 and 16) is illustrated here in these four capillary pressure curves. Note the great differences in mercury saturation at 1800 psi for the two samples with similar porosities from the Cañadon Grande petrofacies (1686.00 m and 1696.33 m), the result of microporosity associated with kaolinite.

The grain size and sorting of the samples (generally medium to lower very coarse, moderately well sorted) do not vary systematically over this interval. Sandbed thickness and texture also do not vary systematically. Hence, the observed large variations in permeabilty heterogeneity correlates positively and solely with the extent of meteorically-induced leaching.

C. Infiltration

Horizons of mechanically-infiltrated clays within coarse clastics can be barriers to vertical permeability on a wide range of scales within reservoirs. Moraes and De Ros (1989 and 1990) indicate that for sandstone reservoirs of the Recôncavo Basin, Brazil, infiltration horizons are major features of heterogeneity. There, the zones of infiltration can reach up to tens of meters in thickness and impose a layering which is separate from the fabric of the depositional environment. These authors also discuss the difficulties encountered in formation evaluation due to low induction log response in those intervals containing abundant mechanically-infiltrated clays.

Conventional core plug permeabilities from a core are compared for infiltrated and uninfiltrated samples in Table 1. The comparison was made of calculated Dykstra-Parsons coefficient of permeability heterogeneity for the existing data and the coefficient calculated excluding samples containing abundant siliceous cutan structures. These samples containing large amounts of cutan structures and the accompanying diagenetic pore-fillling of chlorite, quartz and albite microlites have the lowest permeabilities. The V_{DP} values for the permeabilities with and without the infiltrated samples are 0.83 and 0.71, respectively (Figure 18), a change of approximately 0.1 which can be attributed to post-depositional processes.

The extent of the filling of pores by infiltrated material determines the magnitude of the anisotropy. The association of the infiltrated material with the more sinuous stream deposits within the Comodoro Rivadavia Formation suggests that the associated heterogeneities and commensurate potential difficulties in reservoir evaluation may be found within localized regions (e.g., specific fields or reservoirs).

D. Formation of Late Smectites

The Cretaceous section has been uplifted and folded into a series of north-south structures, that mark the western end of the San Jorge Basin. The evidence for uplift decreases to the east. Within the PR-816 well, the Comodoro Rivadavia sandstones contain a late-formed, filamentous smectite associated with a lack of oil stain in the core (Figure 19). The staining occurs as vertically discontinuous patches throughout a fluvial sandstone-siltstone sequence (Figure 20). The stained and unstained portions of the core are not separated by clay partings, shales or grain-size variations.

Those portions of the core that show little or no staining contain an abundance of a filamentous authigenic clay lining the pores. Where the authigenic clay is less extensive, the pore throats are clear or only partially lined, and are stained with

TABLE 1. Permeability and Porosity[a]: Infiltration-Rich vs. Infiltration-Poor Samples.

Infiltration-rich	Infiltration-poor
Permeability (md.)	
31.8	17.9
0.673	49.0
0.0531	251
5.45	124
	111
Permeability average	
(n=4)	(n=5)
geometric: 1.58	geometric: 78.7
arithmetic: 9.49	arithmetic: 110.6
Porosity average	
(n=4)	(n=5)
5.2	19.1

[a] measurements courtesy of Amoco Argentina Oil Company: air permeabilities and Boyle's Law helium porosities.

hydrocarbon. The smectites are thought to have formed prior to oil migration. They formed after the precipitation of laumontite and within sands that were well compacted. More importantly, there exist no other visible textural variation or cement which can account for the discontinuous oil staining. The filamentous clay has reduced the pore throat apertures to below the threshold size necessary for the oil column to displace water. The oil staining appears to be a valid representation of the macroscopic oil saturation heterogeneity within the cored interval. These irregular cementation

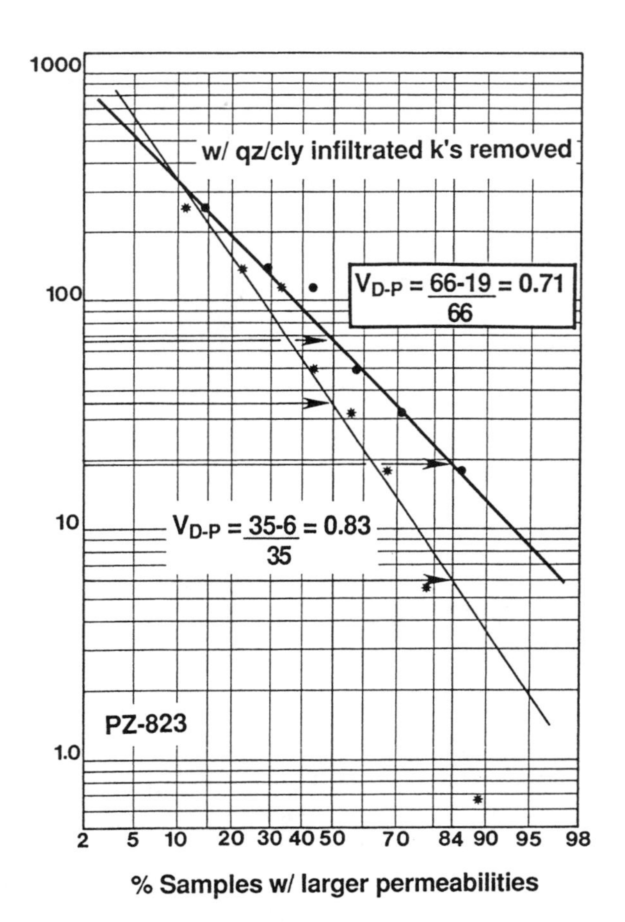

Figure 18. Dykstra-Parsons plot of permeability data from the PZ-823 core showing the change in measured permeability heterogeneity attributable to the extensive infiltration and related alteration. After Dunn (1992).

patterns reduce effective connectivity and produce irregular water saturation patterns.

A Dykstra-Parsons coefficeint for the PR-816 core was calculated to be 0.95, a high value (Figure 21). Calculation of the Dykstra-Parsons coefficient for the PR-816 core, after removing the data for samples containing late smectite, results in a value of 0.81, still high, but substantially lower. Replacing those values for the smectite-bearing samples with

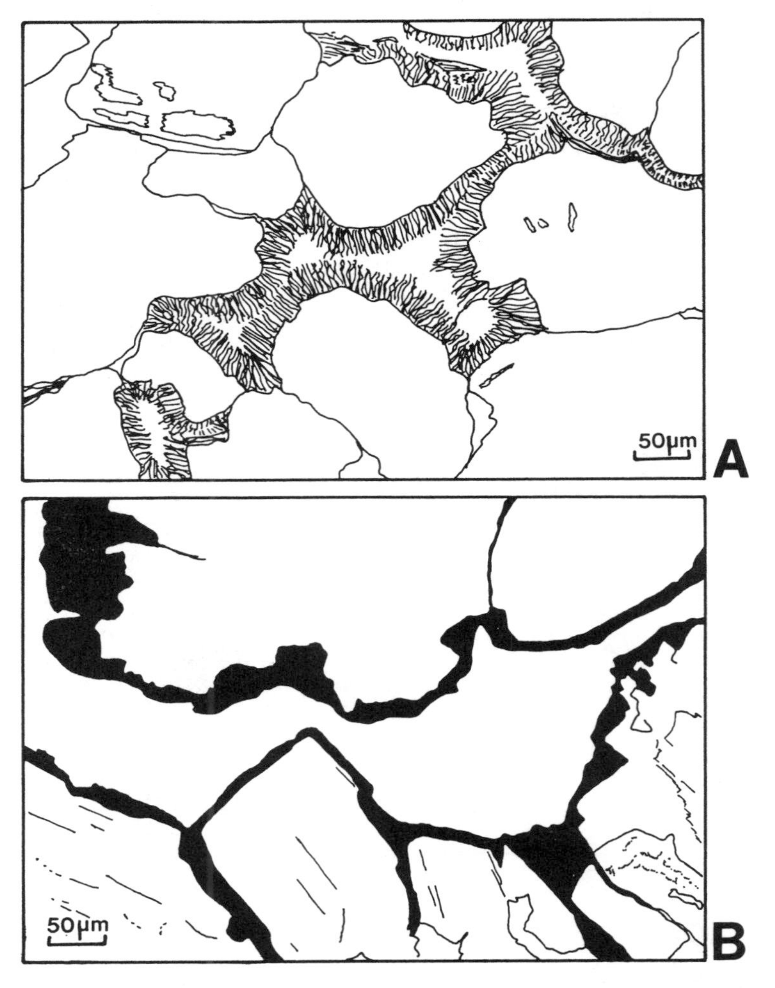

Figure 19. Ink tracings of late-forming smectite (**A**) and clear pores (**B**) where the smectite is absent and hydrocarbons have stained the pore walls.

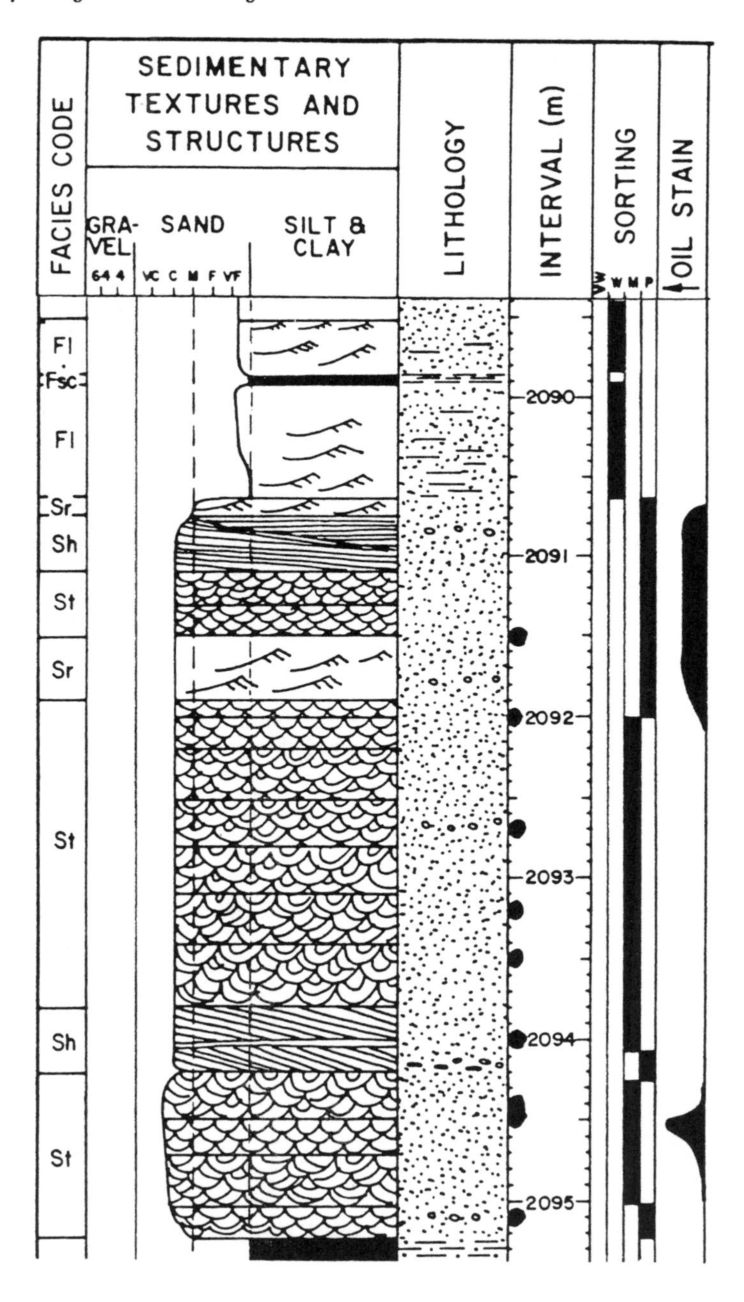

Figure 20. Graphic column of the PR-816 core, showing the areas of oil staining which was inhibited in other areas by the late forming filamentous smetite coatings.

values within the range for samples that do not contain filamentous smectite would result in a further decrease in the Dykstra-Parson's coefficient. Clearly, this authigenic phase, formed as a result of late uplift and folding of the western portion of the basin, has resulted in a substantial, measurable increase in the permeability heterogeneity of these Cretaceous clastic reservoir. The increase in hetergeneity due to late-stage smectite formation is comparable in amount to that found for the meteoric water invasion (the Cañadon Grande petrofacies) and extensive infiltration.

IV. CONCLUSIONS

Examination of post-depositional features such as infiltrated materials, cements, and dissolution textures within petrologically complex Cretaceous volcanogenic alluvial/fluvial sandstones indicates that such features effect measurable changes in vertical permeability measurements. These

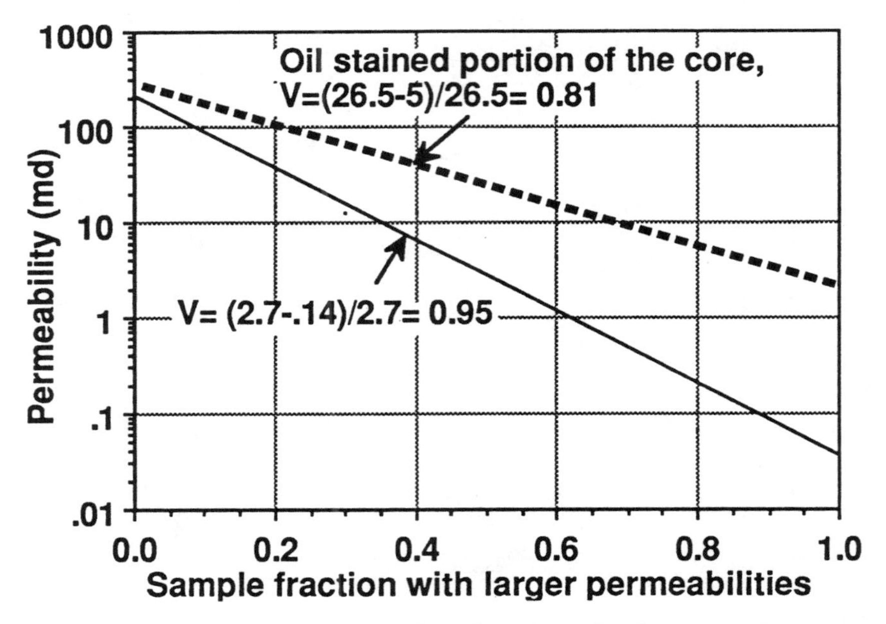

Figure 21. Dykstra-Parsons plot showing the increase in permeability heterogeneity attributable to the presence of the late-forming smectite. After Dunn (1992).

changes are systematic and effect permeability heterogeneity.
The following changes in permeability heterogeneity were observed in the San Jorge basin: (1) a decrease in heterogeneity due to simple compaction, (2) an increase in the Dykstra-Parsons coefficient of permeability heterogeneity of approximately 0.15 due each to extensive infiltration structures, wholesale dissolution and kaolinite-albite alteration, and (3) late filamentous smectite pore-lining, a feature attributed to cooling or water freshening accompanying Tertiary uplift.

V. ACKNOWLEDGEMENTS

Funds were provided by the Enhanced Oil Recovery Institute of the University of Wyoming, Amoco Production Research, Amoco Argentina Oil Company, Mobil Research and Development, and the Gas Research Institute. Access to core and samples for this study were obtained from Amoco Argentina Oil Company. The authors thank Amoco Argentina Oil Company for permission to publish certain material contained in this paper.

VI. REFERENCES

BEARD, O.C. AND WEYL, P.K., 1973, Influence of texture on porosity and permeability of unconsolidated sands, AAPG Bulletin, vol. 57, p. 349-369.

DUNN, THOMAS L., 1992, Effects of Post-depositional Processes on Reservoir Heterogeneity, University of Wyoming Ph. D. dissertation, 276p.

KHATCHIKIAN, A. and LESTA, P., 1973, Log evaluation of tuffites and tuffaceous sandstones in southern Argentina, Proceedings of the 14th Annual Logging Symposium, Society of Professional Well Log Analysts, p. 1-24.

MORAES, M. A. S. and DE ROS, L. F., 1990, Infiltrated clays in fluvial Jurassic sandstones of Recôncavo basin, northeastern Brazil, Jour. Sed. Petrol., v. 60, p. 809-819.

MORAES, M.A.S. and DEROS, L.F., 1989, Characterization and influence of mechanically infiltrated clays in fluvial reservoirs of the Recôncavo Basin, Northwestern Brazil, Bull. Geosci. Petrobras (in Portuguese), v. 2, p. 13-26.

MORGAN, J.T. and GORDON, D.T., 1970, Influence of pore geometry on water-oil relative permeability, Jour. Petr. Tech., p. 1199-1200.

MURRAY, H. H., 1988, Kaolin minerals: their genesis and occurrences, *in* S. Bailey, ed., Hydrous Phyllosilicates (exclusive of micas), Reviews in Mineralogy, Mineralogical Society of America, v. 19, p. 67-89.

PUCCI, J.C., 1987, A review of Argentina's sedimentary basins, Oil and Gas Journal, p. 52-55.

SURDAM, R.C., DUNN, T.L., MACGOWAN, D.B. and HEASLER, H.P., 1989a, Conceptual models for the prediction of porosity evolution: an example from the Frontier sandstone, Bighorn Basin, Wyoming, R.M.A.G. Annual Field Guide, p.7-28.

SURDAM, R.C., MACGOWAN, D.B. and DUNN, T.L., 1989b, Diagenetic pathways of sandstone/shale systems, Contrib. Geol., Univ. of Wyo., v. 27, no. 1, p. 21-31.

SURDAM, R.C., DUNN, T.L., HEASLER, H.P., and MACGOWAN, D.B., 1989c, Porosity evolution in sandstone/shale systems, *in* E. Hutcheon, ed., Short Course in Burial Diagenesis, M.A.C., May 1989, p. 61 - 134.

TERUGGI, M. and ROSETTO, H., 1963, Petrologia del Chubutiano del Codo Rio Senguer, Boletin de Informaciones Petroleras, p.18-35.

RESERVOIR HETEROGENEITY IN VALLEY-FILL SANDSTONE RESERVOIRS, SOUTHWEST SOCKHOLM FIELD, KANSAS

Roderick W. Tillman
Consulting Sedimentologist/Stratigrapher
Tulsa, Oklahoma

Edward D. Pittman
Consulting Petrologist
Sedona, Arizona

ABSTRACT

Valley-fill heterogeneous reservoirs in the Morrowan (Pennsylvanian) age Stateline Trend along the Kansas-Colorado border produce from the Stockholm Sandstone. Southwest Stockholm field valley-fill reservoirs are internally complex, contain geographically and vertically limited reservoirs and can only be effectively drained if the internal architecture of the reservoirs is understood. Development of detailed geologic models for fluvial, tidal and other valley-fill reservoirs in the Morrow are expected to enhance results of reservoir simulation, secondary recovery and possible tertiary production. Only by including the location and degree of impermeability of barriers to flow at the margins of and within the valley-fill reservoirs can historical simulation modeling and secondary recovery yield satisfactory economic results.

The valleys in which the reservoirs occur were cut by rivers during drops in sea level and remained as conduits (narrow open valleys trending perpendicular to the regional shoreline). At a later time, as sea level rose, the valleys were filled from the landward side by rivers or from the seaward side by shoreline processes or alternately from both the landward and seaward sides at the same time.

Detailed geologic analysis of cores from the field combined with cross sections and isopach and structure maps yielded details of reservoir heterogeneity. Reservoir production properties vary greatly between the two major producing facies, fluvial and tidal (estuarine) valley-fill sandstones in Southwest Stockholm field, Kansas. Within the approximately 150-ft thick valley-fill vertical sequence at Soutwest Stockholm field just one of the three sandstones yields significant production. Topographic highs on the valley bottom, covering an area as large as 1/2 section, reduce the depositional thickness of the reservoir sandstones. Impermeable limestones and shales commonly form lateral barriers to flow at valley margins and locally within the valleys. Erosion of portions of potentially productive sandstones within the valleys was found to be important in isolating reservoirs (flow units).

Tidally deposited (estuarine) sandstones are finer grained and more clay prone than the river sandstones and as a result have poorer reservoir properties. Fluvial deposits have average porosities of 16% (range from 11 to 20%) and average permeabilities of 700 md (range from 129 to 1890 md). Tidal-channel sandstones have average porosities of 14% (range from 11 to 17%) and permeabilities of 80 md (range 50 to 111 md). Where tidal sandstones interfinger with fluvial sandstones, vertical permeability is diminished and flow units are fragmented.

Valley-fill deposits in Southwest Stockholm field differ from deltaic deposits geologically, geometrically and in the characteristics and distribution of flow units. Overbank and fine-grained levee deposits, typical of deltas, are absent at the margins of valley-fill deposits.

Secondary (post burial) events affected the production characteristics of the reservoirs. The post-depositional events that affected the reservoirs were significantly different in some of the seven depositional lithofacies. Where millimeter thick shale lenses were deposited <u>within</u> tidally deposited sands, pressure solution of quartz in contact with the clay reduced porosity. Up to half of the porosity in Southwest Stockholm field may be leached (secondary) porosity.

Reservoir characteristics are relatively similar within sandstone units interpreted to have been deposited by the same depositional processes. All sandstones deposited by the same processes are assigned to the same lithofacies. However, reservoir characteristics are significantly different among different depositional lithofacies. Although other types of sandstones were vertically or laterally "connected" to the better reservoir lithofacies, perforated intervals were commonly limited to valley-filling <u>fluvial</u> sandstones in Southwest Stockholm field. Understanding the differences in heterogeneity characteristics between fluvial and shoreline fills of valleys, such as those studied herein, allows proper economic evaluation and more efficient reservoir management of these kinds of reservoirs.

I. INTRODUCTION

One of the most important kinds of relatively small sandstone reservoir being explored for and developed today is the valley-fill sandstone. Because of their relatively small size (0.5 to 4 miles wide), poorly understood origins and difficult to predict occurrences, valley-fill reservoirs have not always been attractive targets for drilling. However, in the last few years valley-fill reservoirs have become very important and some are currently undergoing secondary production. Because of the absence of adequate geologic descriptions, valley-fill reservoirs have, in the past, commonly not been differentiated from deltaic and river-channel sandstones. Because there are significant differences in the distribution of reservoir seals and internal heterogeneity characteristics in valley-fill reservoirs, compared to deltaic and river-channel reservoirs, different patterns of production will occur within individual fields and pools from these distinctly different geologic environments.

Southwest Stockholm field was selected for study because it appeared to be an excellent example of a valley-fill reservoir and because of the availability of cores and modern log suites. This study emphasizes geologic core interpretations, geologic mapping, detailed stratigraphic cross sections using 5"=100' logs and thin section and scanning-electron microscope analysis. Depositional processes for individual sandstone lithofacies (Seimers and Tillman, 1981) are emphasized as are secondary processes which were responsible for porosity modification.

A sequence of sometimes unrelated depositional events is involved in forming valley-fill reservoirs. The initial step in forming most, if not all, valley-fill reservoirs requires a sea-level drop (Suter and Berryhill, 1985; Hine and Snyder, 1985). When sea level drops, the shoreline moves

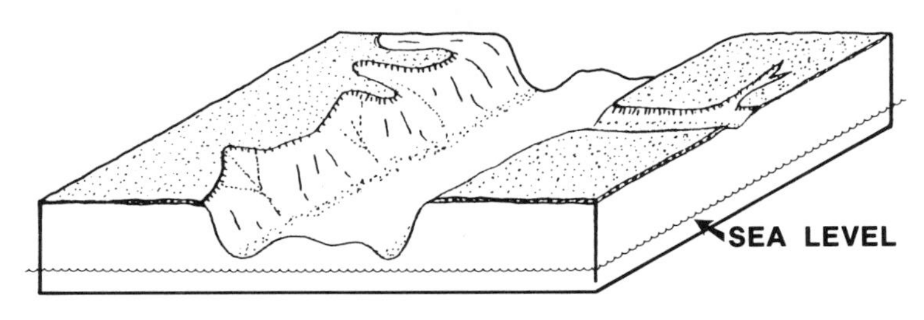

Fig. 1 - Valley cut into sediment during sea-level drop. At this time the valley is an open conduit in which no sediment is accumulated. (Farmer, 1981)

seaward and rivers cut down into underlying sediments at a rate and to a depth controlled by the drop in sea level. The rivers may erode relatively deep valleys, mostly devoid of sediment (Fig. 1). In the initial phases of valley formation the river sediment is moved seaward, bypassing the valley. As long as sea level is relatively low, little sediment will be deposited in the valley. However, with rise in sea level several processes may cause deposition of sediment in coastal portions of the previously cut valley.

Rising sea level causes the rivers flowing in the valley to aggrade (begin to deposit sediment) a short distance upstream from the ocean and the formerly open valley begins to fill with sand. Aggradation occurs when a stream is supplied with more sediment than it can transport. During periods of river overbank flooding deposits of clayey sediments may be deposited in adjacent swamps, marshes and in in the previously cut valley (Fig. 2). Because the valleys open into the ocean, marine shoreline processes may fill the seaward ends of the valleys. When the sea floods a coastal valley it is called an estuary. Because the mouths of the valleys are commonly several miles wide, normal shoreline processes may partially fill the estuary mouths. Where an embayed shoreline is present, in a direction away from the ocean, shoreline wave and current processes give way landward to tidal sandstone depositional processes in lagoons and in the most landward portions, to river deposition. A vertical sequence of intertonguing discrete fluvial, tidal and shoreline sandstones may be preserved within the valley or estuary with rise in sea level. An incomplete vertical succession of potential reservoir facies may be preserved because fluvial erosion may remove portions of the sand deposited in the valley. Another important process, which may be important in the shoreline area of the valley, is the destructive erosional process resulting from the landward

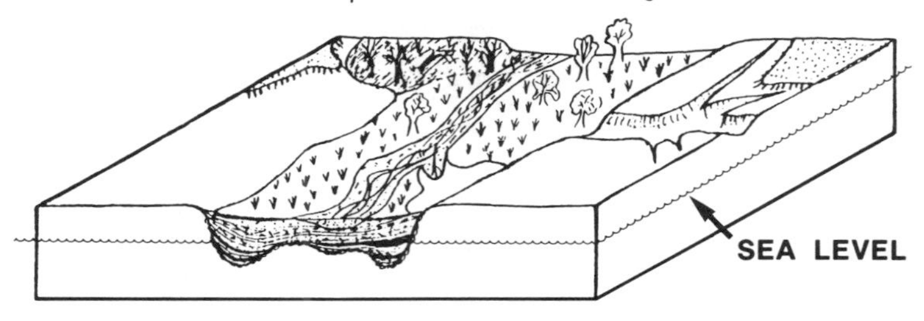

Fig. 2 - Valley-fill during period of sea-level rise. Valley is filled with fluvial (potential sandstone reservoir) and swamp and marsh deposits (non-reservoir). Fill may be a mixture of marine and river deposits deposited at or slightly below sea level. (Farmer, 1981)

movement of the sea (transgression). Transgression commonly forms a ravinement surface which removes most or all of the beach and barrier island sands. The ravinement process (Stamp, 1921; Vail et al., 1977) "scoops down into" and erodes the underlying beds. The destruction of potential reservoirs by the ravinement process is associated with the breaking of fairweather waves, especially storm waves and currents on the shore during periods of rising sea level. As the sea moves slowly over the land, much of the sand at the shoreline is carried seaward and potential reservoir sand bodies may be destroyed. This ravinement process is believed to erode only to a shallow depth (possibly less than 30') and all sediment below this depth (including significant thicknesses of valley-fill sands) will be preserved. Near the most seaward location of the shoreline, which formed during the sea-level lowstand, some valleys may be relatively straight; whereas, landward areas may be characterized by meandering valley-fill patterns, as observed in Southwest Stockholm field (Fig. 3).

Because valley-fill reservoir sandstones occur in deeply cut incisions, the preservation potential of the sandstones during periods of rising sea level is significantly greater than many other shallow marine and shoreline sandstones.

The sandstones, which form the valley-fill reservoirs in Southwest Stockholm field, are estuarine (flooded river mouths), marine shoreline deposits and river-channel deposits. The geometry and reservoir characteristics of each of these types of sandstones are different. Locally, where these three types of sandstone interfinger, complex barriers to flow may exist.

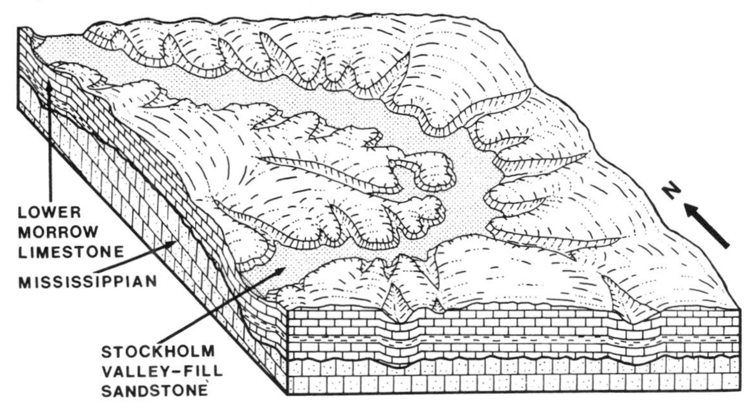

Fig. 3 - Valley-fill sandstone reservoir in Southwest Stockholm field (Shumard, 1991). Valley may be filled entirely with sandstone or partially with limestone and/or shale.

Prior to this study, the significance of the barriers to flow within and between these different types of sandstone was not understood, nor had quantitative documentation of their distribution (which permits definition of flow units) been obtained.

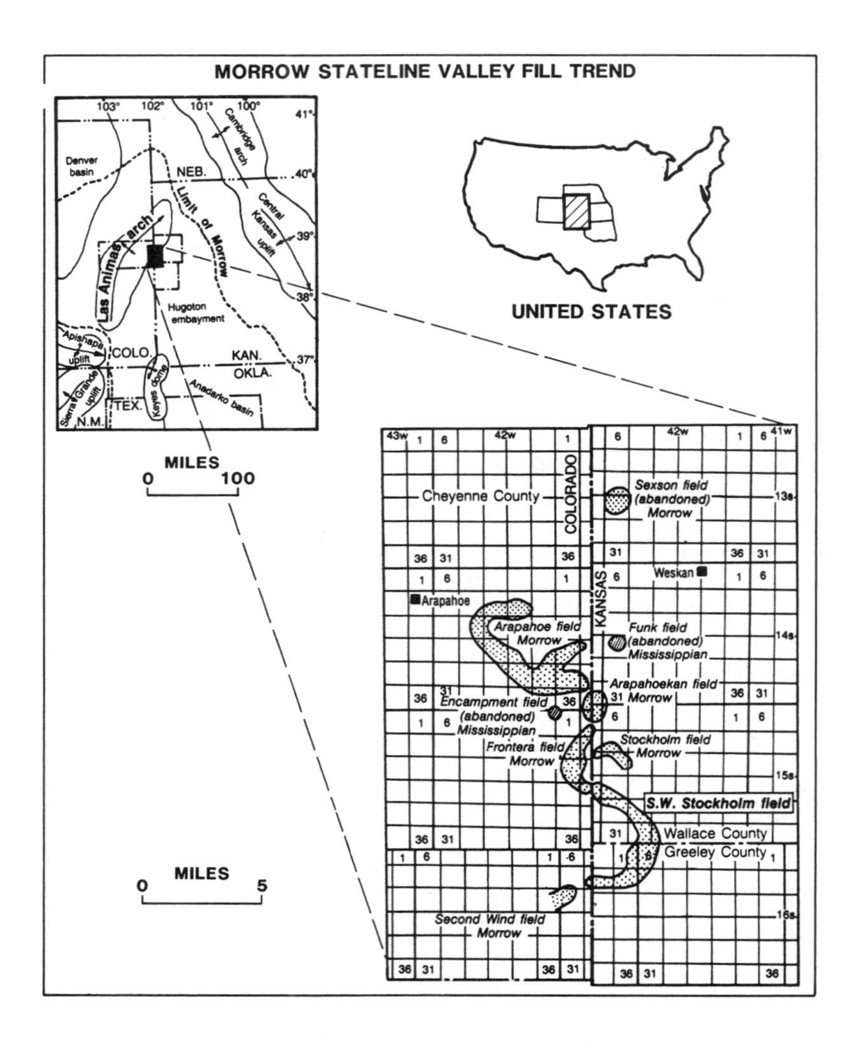

Fig. 4 - Location of 20-mile-long "Stateline Trend" Morrow Sandstone fields, which produce from valley-fill sandstone reservoirs. Southwest Stockholm field, Kansas, is located in Greeley and Wallace counties, Kansas. (Shumard and Avis, 1990)

II. SOUTHWEST STOCKHOLM FIELD

A very active play involving exploration and development of valley-fill reservoirs has been underway for the last few years in the Morrow Sandstone in "Stateline Trend" of southeastern Colorado and westernmost Kansas (Sonnenberg, 1985; Sonnenberg et al., 1990; Weimer et al., 1988; Krystinik, 1989a, b; Krystinik and Blakeney, 1990; Blakeney et al., 1990). One of the more productive valley-fill fields in the trend is Southwest Stockholm field (Fig. 4), which was discovered in 1979 and is the subject of this report. The field contains 87 wells and has produced over 5.5 million barrels of oil from the estimated 42 million barrels in place. The field has been interpreted by Shumard (1989) and Brown et al. (1990) and confirmed by the authors to consist of valley-fill fluvial and estuarine (tidal) sandstones. The composite valley-fill deposits in the field range up to 150' in thickness. The major producing reservoir in the field is the Stockholm Sandstone, which is commonly from 15 to 30 feet thick. Core porosity averages 17% (range 10-22%) whereas permeability averages 400 md (range 50-4600 md) (Shumard, 1989; Brown et al., 1990). The field covers approximately 2000 acres and ranges from 1/4 to 1/2 mile wide and is six miles long. The "Stateline Trend", which includes Southwest Stockholm field, also includes within its serpentine limits at least three other fields: Second Wind, Frontera and Arapahoe (Fig. 4). Oil in place along the "Stateline Trend" is estimated by Brown et al. (1990) to be approximately 170 million barrels. These other valley-fill fields have similar but probably slightly different production characteristics than Southwest Stockholm field.

Along the "Stateline Trend", which extends in a north-south direction on both sides of the border of Kansas and Colorado, are Morrowan Age valley-fill reservoirs, which occur in sinuous valleys that were periodically cut into marine limestones, shales and sandstones when sea level was relatively low. Each time sea level rose, reservoir sandstones and non-reservoir shales and limestones were deposited within the valleys. Some of the sandstones now are reservoirs, others have relatively poor producing characteristics and form impediments to fluid flow. A type log developed by Shumard and Avis (1990) for the field shows the vertical sequence of strata and the characteristics typical of the formations (Fig. 5).

In 1990, several geologic models were proposed for valley-fill deposition in areas such as Southwest Stockholm field. The model proposed by Wheeler et al. (1990; Fig. 6) for Morrow valley-fill deposits appears to be appropriate for at least portions of Southwest Stockholm field.

TYPE LOG

Texas Oil and Gas 4 Evans E, NW NE SE 11-16s-43w

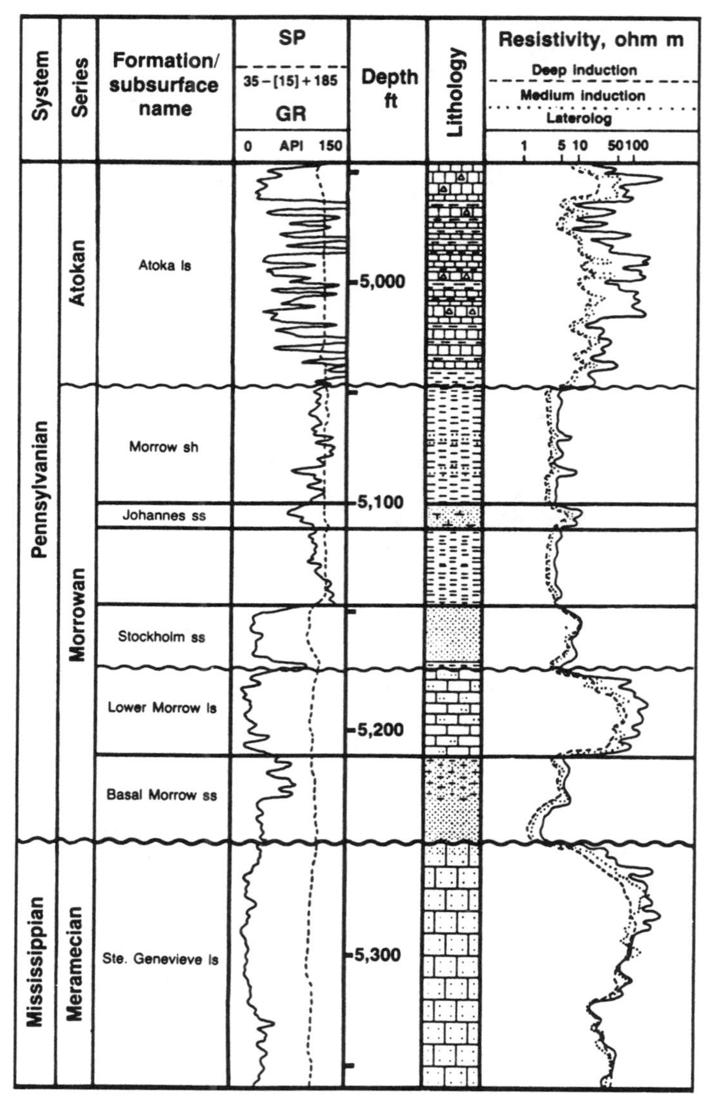

Fig. 5 - Type log for Southwest Stockholm field. Geographic location shown on Figure 9 (Well No. 124). Note that three potentially productive sandstones are present in the Morrow interval. At Southwest Stockholm field, basal Morrow Sandstone is non-productive as is the lower Morrow Limestone. The basal Morrow Sandstone is limited in distribution by limestone valley walls. The Stockholm and Johannes valley-fill sandstones are both productive in Southwest Stockholm field. The base of the Morrow is a regional unconformity, which in most areas forms the bottom of the valley that is filled with numerous lithologies including sandstone, limestone and shale. Significant relief on the unconformity at the base of the Stockholm Sandstone is limited for the most part to the width of the valley(s).

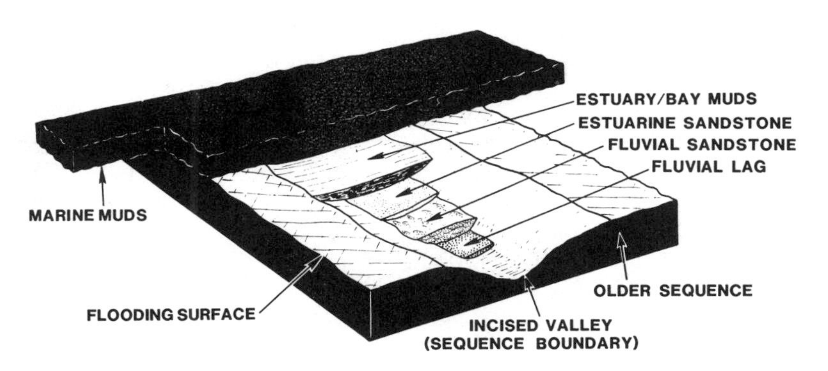

Fig. 6 - Schematic of valley-fill deposits formed in portions of Southwest Stockholm field, Kansas. Incised valley, cut into older rocks, is filled initially by coarse fluvial lag deposits. Finer grained productive fluvial sandstones fill medial portions of valley prior to relative sea-level rise into valley. As sea-level rises, estuary- and bay-deposited sandstones are locally deposited above and seaward of river sands. As sea level rises still further, the estuary becomes muddier and estuarine mud blankets and seals the reservoir sandstones. With further relative rise in sea level, marine muds blanket valley and the intervalley areas. (Modified from Wheeler et al., 1990)

A. Macroheterogeneity

In order to determine the distribution of reservoir boundaries and the distribution of large scale barriers to flow associated with the valley-fill reservoir at Southwest Stockholm field, all the available cores within the field were described and interpreted with respect to environment of deposition (Figs. 7A, 7B, 8A, 8B). The log suites for the cored wells were qualitatively calibrated to the core interpretation; log shape, gamma ray and resitivity readings were used to extend the depositional lithofacies observed in core to nearby non-cored wells. Two cross sections were constructed, which included the cored wells. Portions of these cross sections are included as figures.

The Morrowan geologic section at Southwest Stockholm field is complex, as may be observed in the type log (Fig. 5). Three Morrowan sandstone units within the valley-fill succession are potential reservoirs. The Stockholm Sandstone is the major producing unit in Southwest Stockholm field. A second sandstone is the "Johannes Sandstone", which produces locally. The basal Morrow Sandstone is non-productive within the borders of the field (Fig. 9).

The base of the valley from which Southwest Stockholm field produces is marked by a major unconformity at the top of the Mississippian limestone (Fig. 5). A significant thickness of Mississippian limestone was probably removed during the period of valley cutting prior to deposition of the basal Morrow Sandstone. The basal Morrow Sandstone is limited in distribution within the valley and may have been removed from portions of the northeast area of the field by a second period of erosion. Following erosion(?) of the basal Morrow Sandstone a significant rise in sea level occurred and a "blanket" of marine lime sediment was deposited within the valley and in areas outside the valley. A third sea-level drop occurred (top of lower Morrow Limestone, Fig. 5) and much of the limestone was eroded within the area of the valley, forming a valley within a valley.

In this valley within a valley the highly productive Stockholm Sandstone was deposited. The Stockholm Sandstone is variable vertically and horizontally. The most accurate analysis of heterogeneity within the reservoirs was derived from detailed sedimentologic description and interpretation on a unit-by-unit basis of the four cores available from the field. Geologic maps were constructed and the production characteristics of each of the units were determined and compared with other similar and dissimilar geologic units. Heterogeneity recognized in core description was compared to log parameters in the same and adjacent wells and cross

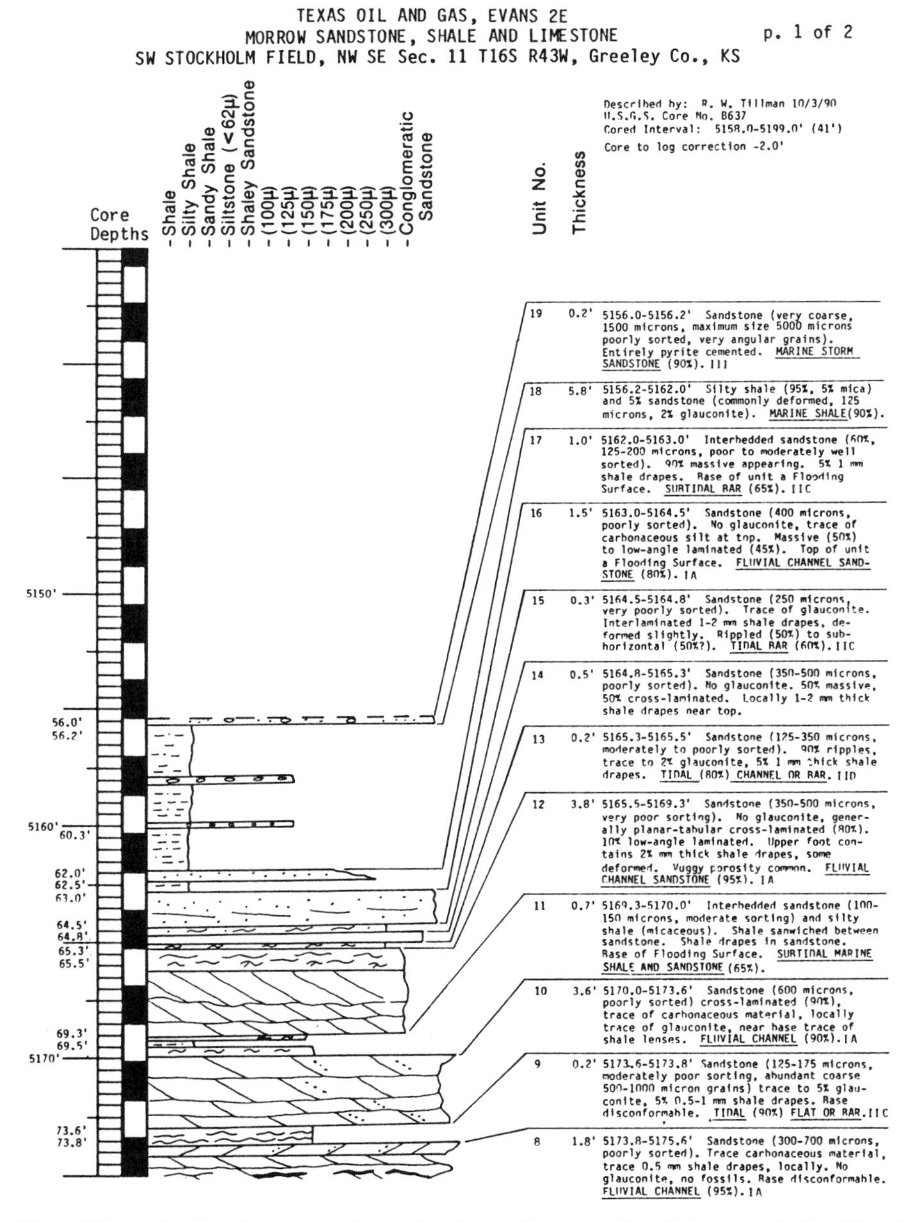

Fig. 7A - Geologic core description, Morrow Sandstone, S.W. Stockholm field portion of "Stateline Trend". For geographic location refer to Well 131 on Figure 10 base map. Reservoir is composed primarily of coarse-grained fluvial (river) deposited sandstones (Units 6, 8, 10, 12). Interbedded on 1-5' scale are thin (0.2-0.7' thick) finer grained and more clayey tidally deposited sandstones. These finer grained units impede vertical flow in the reservoir at and near this well.

TEXAS OIL AND GAS, EVANS 2E
MORROW SANDSTONE, SHALE AND LIMESTONE
SW STOCKHOLM FIELD, NW SE Sec. 11 T16S R43W, Greeley Co., KS

p .2 of 2

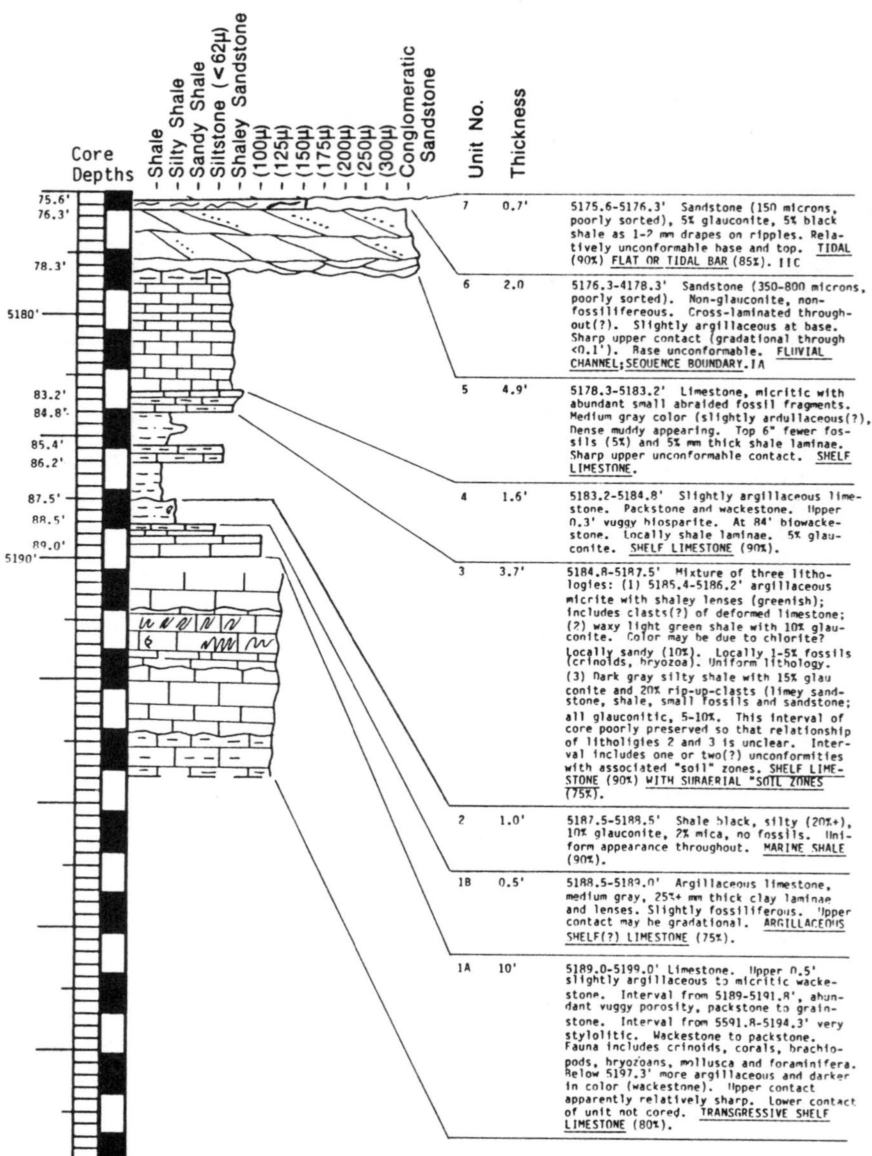

[Fig. 7B - Part 2 of 2]

sections, and geologic and production maps were completed. As may be observed in the type log (Fig. 5) and subsequently described cross sections, the major reservoir boundary seals for the Stockholm sandstones are dense non-porous cabonates below and locally lateral to the sandstone and an overlying marine shale. Within the reservoir a significant number of moderate to large scale features are recognized, which either inhibit or prevent fluid flow.

1. Discussion of TXO Evans 2E Core

The cored interval in the TXO Evans 2E well (well number 131, Fig. 9) includes, below the reservoir, approximately 21' of non-reservoir lower Morrow Limestone and, above the reservoir, 7' of non-permeable marine shales. Details of these units are described in Figure 7.

The Stockholm Sandstone in Figure 7 consists of porous and permeable medium- to coarse-grained (0.3-0.7 mm mean diameter) river-deposited sandstone with interbeds of thin, presumably laterally persistent, tidally deposited, slightly shaley and glauconitic sandstones with relatively poor reservoir characteritics.

In this core, the internal features of the fluvial sandstones are primarily steeply inclined cross laminae (Units 6, 8, 10 and 12). The bases of the fluvial deposits are sharp and the tops are either eroded (Units 6, 8 and 10) or reworked (Unit 12). The interbedded tidal- to marine-deposited units (7, 9, and 11) are much finer grained, (0.1 to 0.15 mm mean diameter) and contain lenses and beds of clay including trace amounts of the greenish mineral glauconite, which is believed to form primarily on shallow marine shelves from chemical alteration of fecal pellets. As discussed in more detail under microheterogeneity, the thin shale laminae, in addition to being partial barriers to flow within the sandstone, cause significant secondary pore-filling immediately adjacent to the laminae. Where any combination of impediments to flow occurs (tidally deposited fine-grained sands, shale layers or relatively abundant clayey laminae) in this well, the Stockholm Sandstone has poor vertical flow characteristics because the fluvial reservoir sandstones are separated every 2 to 4 feet vertically by thin (0.2 to 1.0' thick) tidal sandstones and shales. These thin barriers to flow are only barely perceptible on the log of the Evans 2E (Fig. 10) but are identifiable in the core.

2. Discussion of TXO Garrison A6 Core

The core through the Stockholm Sandstone in the TXO A6 Garrison is quite different from that encountered in the TXO Evans 2E. In the A6 Garrison, a 30' complex sequence of tidally deposited estuarine and possibly shallow marine sandstones is overlain by 8 to 10 feet of fluvial sandstone.

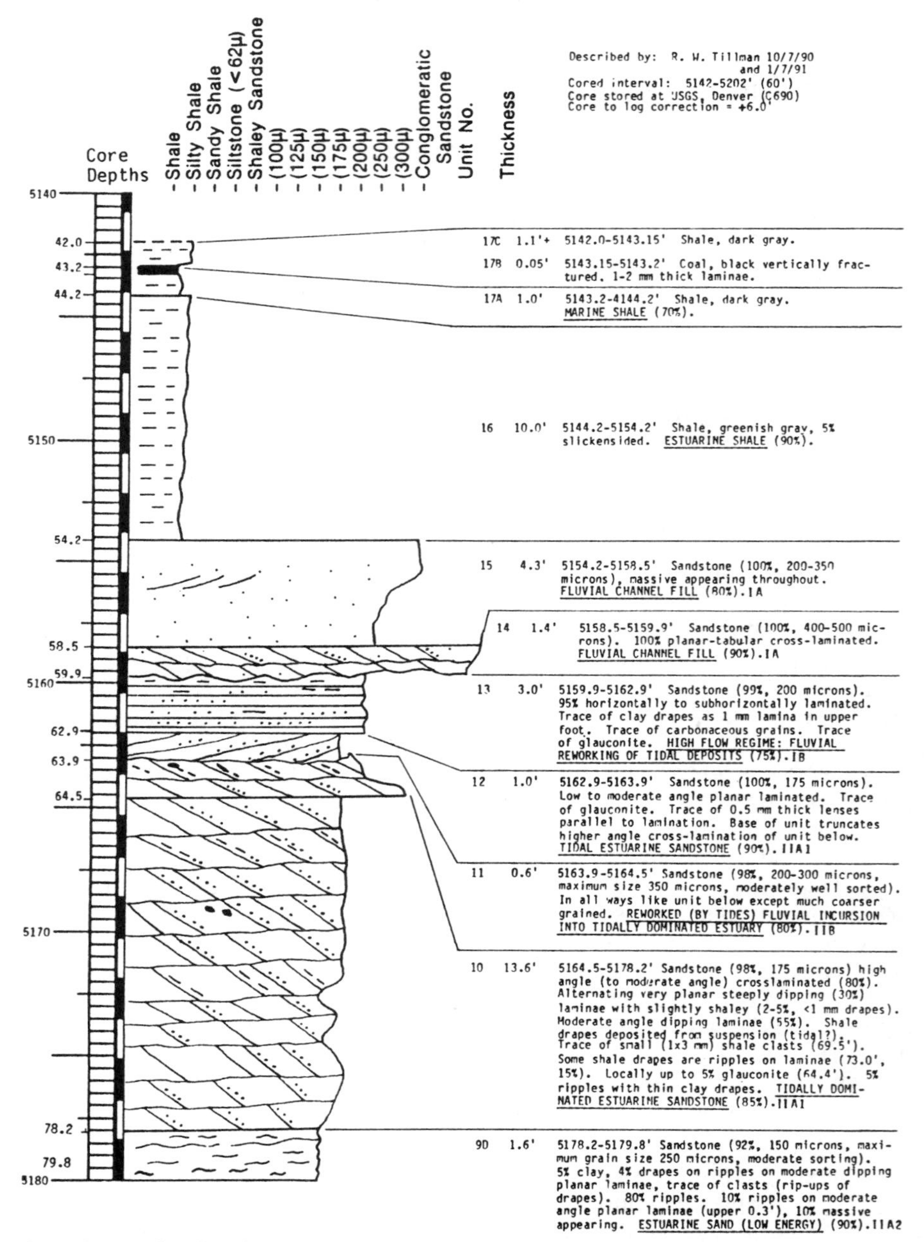

Fig. 8 - Geologic core description and interpretations, Morrow Sandstone, "Stateline Trend", Southwest Stockholm field. For geographic location refer to Well 128 on Figure 9 base map.

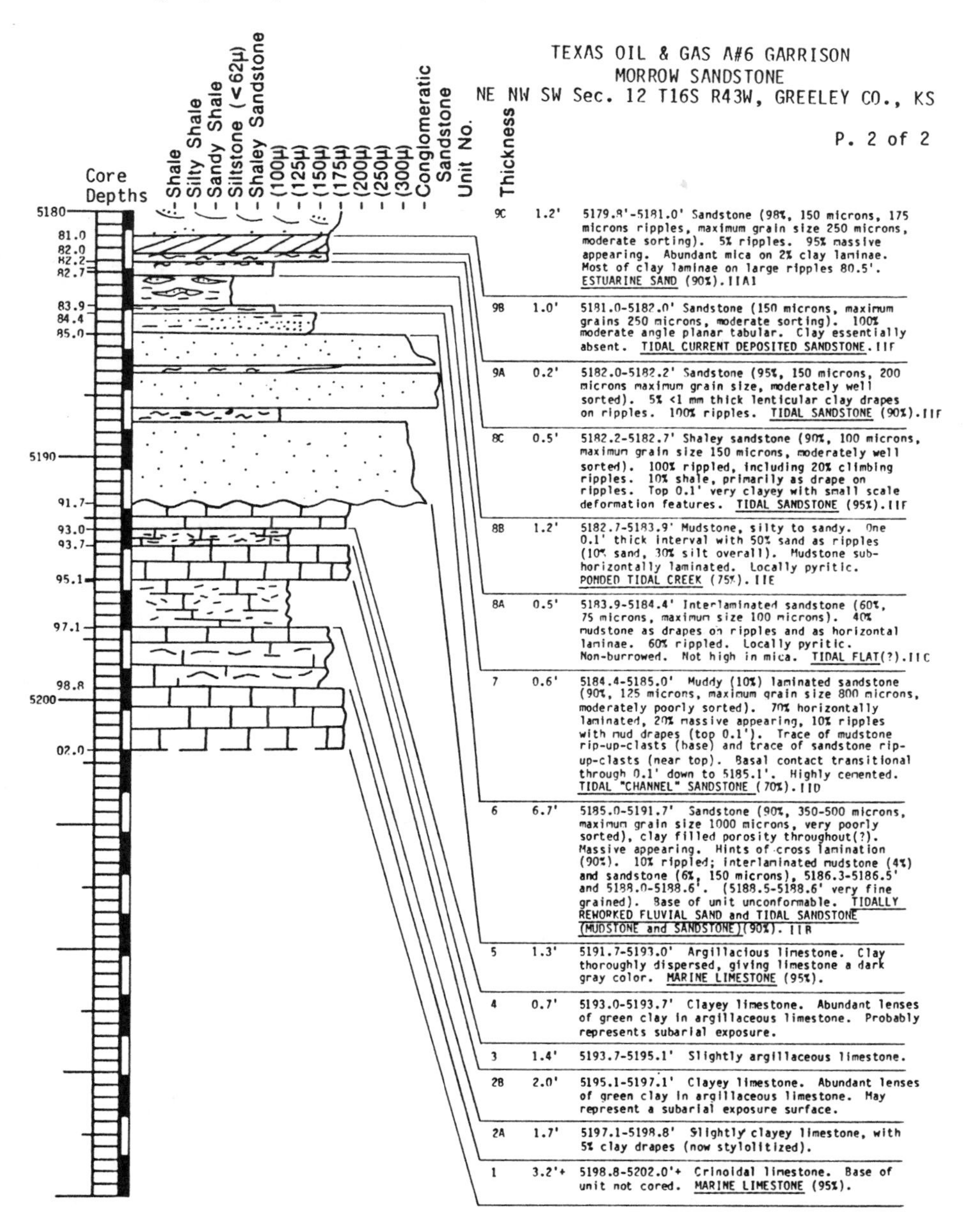

Reservoir is composed primarily of tidally deposited sandstones (Units 7-12). Perforated interval 5160-5170' (log depth), 5154-5164' (core depth). This well is included in cross section in Figure 11. River-deposited Units are 13-15.

Nonproductive Lower Morrow limestones occur below the reservoir and 10' of greenish gray estuarine shale is immediately above the reservoir (Fig. 8). These estuarine shales are overlain sharply by dark-gray marine shales. In sequence stratigraphic terms these shales are designated as a significant "flooding surface".

Only the fluvial sandstone is perforated in this well and it yielded an initial flow of 326 BOPD (compared to an inital flow of only 115 BOPD in the highly interbedded tidal and riverine TXO Evans 2E well). As shown in a portion of cross section Y-Y' (Fig. 11), in this well the fluvial sandstone is thin. As was the case in the TXO Evans 2E, the tidally deposited sandstones are generally much finer grained (0.1-0.18 mm mean size) than the fluvial sandstones. Table I lists the various sandstone facies that occur in the Stockholm Sandstone. Virtually all of the facies listed in Table I occur in the cored interval in the TXO A6 Garrison. As will be discussed in a subsequent section, sparse porosity and

TABLE I

SANDSTONE FACIES STOCKHOLM SANDSTONE SOUTHWEST STOCKHOLM FIELD

I. Fluvial Sandstones

- **A. Channel Fill**
- **B. Fluvially reworked Tidal Sandstone**

II. Tidally Deposited Sandstones

- **A. Estuary Channels**
    - **1) Sandstone Fill**
    - **2) Low-Energy Fill**
- **B. Fluvial Channel Sandstones Reworked by Tides**
- **C. Tidal Bar or Tidal Flat**
- **D. Tidal Channels**
- **E. Ponded Tidal-Creek Fill**
- **F. Tidal Sandstone**

III. Marine Shelf Sandstones

permeability data available for this well (Table II) indicate that the permeabilities and K/Ø ratios are significantly different for fluvial and tidally deposited sandstones.

TABLE II

RESERVOIR PARAMETERS FOR INDIVIDUAL SAMPLES, S.W. STOCKHOLM FIELD, STOCKHOLM SANDSTONE*

WELL	GEOLOGIC CORE UNIT	SAMPLE DEPTH (ft)	K (md)	Ø (%)	K/Ø	R_{35} (microns)
TXO Garrison A-6						
	15	5155	1400	19.5	71.8	29.3
	14	5159.5	140	12.4	11.3	11.2
Mean, River Sandstones			725	16.0	45.3	12.7
	10	5169	111	15.5	7.2	8.1
	9C	5181	80	15.3	5.2	6.7
Mean, Tidal Sandstones			95.5	15.4	6.2	7.4
TXO White AB-1						
	8	5174	1890	20.6	92.2	33.5
	8	5185	432	16.0	27.0	17.4
	8	5187	316	17.9	17.7	13.2
Mean, River Sandstones			879	18.2	45.6	25.8
TXO Bergquist 2 River						
	7B	5115.5	129	11.7	11.0	11.2
	7A	5113	50	12.3	4.1	6.2
	4	5121.5	84	14.1	6.0	7.4
	4	5122.5	83	17.8	4.7	6.0
	2	5125	77	11.5	6.7	8.4
Mean, Tidal Sandstones			73.5	13.9	5.4	7.0

* Raw data from Core Laboratory Inc. Report File 203-87028

3. Discussion of TXO Bergquist No. 2 Core

Twenty-eight feet of Well No. 60 (Fig. 9, TXO Bergquist No. 2) was cored and a geologic description and interpretation of the core was made. The core penetrated the upper two-thirds of the Stockholm Sandstone. Tidally deposited sandstones predominate in the lower 9' of the core. In the upper portion of the core, fluvial sandstones alternate with tidally deposited and/or reworked estuarine sandstones. Limited porosity and permeability data (Table III) suggest that in this well, as well as in others where porosity and permeability data were available, the fluvial sandstones have higher porosity and permeability values (average Ø = 14%, average K = 74 md). The lower values of permeability in the tidal sandstones are in part due to the presence of thin clay drapes and laminae within the sandstone. The whole Stockholm Sandstone was perforated in this well so the contribution to hydrocarbon production of tidal vs fluvial sandstone is not now determinable.

TABLE III

PETROPHYSICAL PARAMETERS FOR RESERVOIR FACIES
SOUTHWEST STOCKHOLM FIELD

Fluvial Sandstones	K (md)	Ø (%)	K/Ø	R_{35} (microns)
Mean	703	16.3	39.8	19.0
Range	129-1890	11.7-20.6	11.0-92.2	11.2-33.5
Tidal Sandstones				
Mean	80.8	14.4	5.6	7.1
Range	50-111	11.5-17.8	4.1-7.2	6.0-8.4

4. Southwest Stockholm Maps

Two isopach maps were constructed for the Southwest Stockholm field area using input supplied by the principal investigator to a Scientific Computer Application (SCA) mapping program. The maps are (1) Stockholm Sandstone, Net Sand Isopach and (2) lower Morrow Limestone, Isopach Map. These maps are similar to, but less interpretive than, those by Shumard and Avis, 1990, and Shumard (1991).

A top of Morrow structure map (Fig. 4, Shumard and Avis, 1990) indicates that a crescent-shaped structural low extends from Sec. 11, T16S, R43W to Sec. 29, T15S, R42W (Fig. 9). This crescent shaped area is the area of thickest Stockholm Sandstone reservoir sandstone (Fig. 12).

The Stockholm Sandstone is a lenticular sandstone in Southwest Stockholm field (Fig. 12). The thickness patterns are somewhat irregular and reflect the effects of both topography on the floor of the valley immediately prior to deposition of the sandstone and to the heterogeneity within the sandstone. As would be expected, the sandstone thickens away from the edges and towards the middle of the valley.

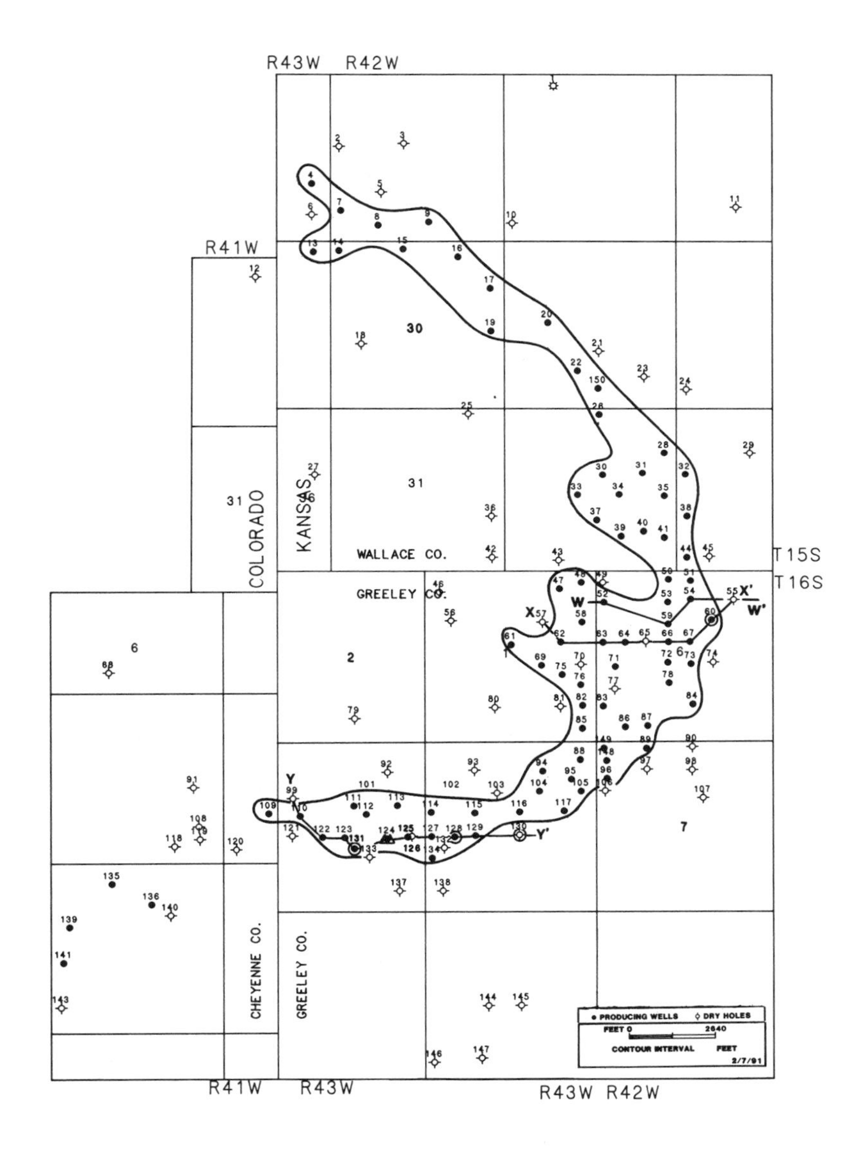

Fig. 9 - Base map, outlining borders of Southwest Stockholm field, Kansas. Location of stratigraphic cross sections X-X', Y-Y' and W-W' is indicated. Cored wells described in this study are indicated by circles. Type log well (Fig. 5) is indicated by a triangle. A list of all numbered wells on this map is available from the senior author.

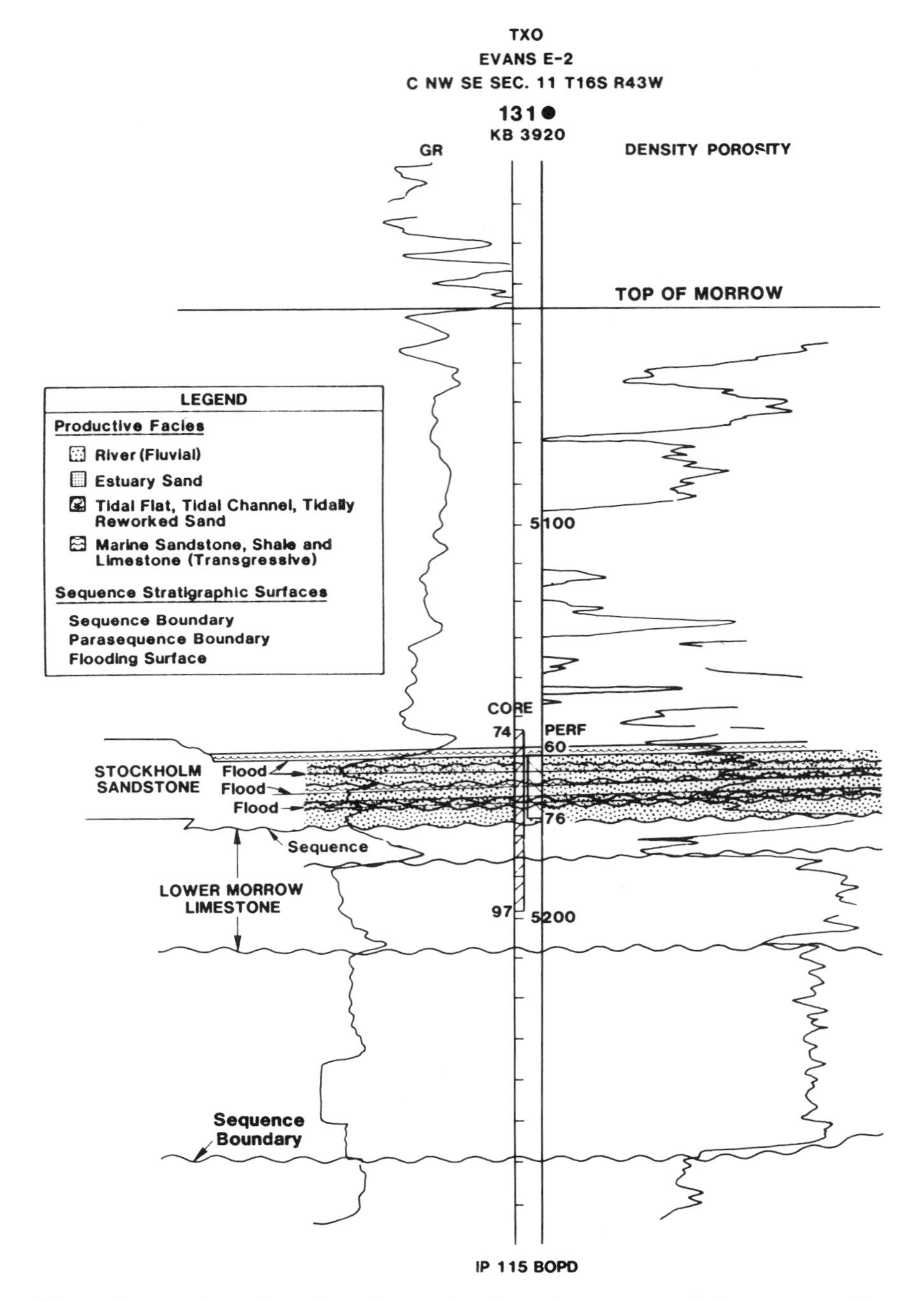

Fig. 10 - Gamma-Ray Density Log showing interpreted heterogeneity affecti fluid flow. Erosional unconformities occur at base of basal Morrow Sandsto and at base of Stockholm Sandstone. Thin, fine-grained, clayey, tidally depo ited, marine(?) sandstones are interbedded with fluvial sandstone in t Stockholm Sandstone in this well and form impediments to vertical flow this location.

The thicker net-sand intervals (30') tend to be near the center of the valley where the valley is relatively wide. It is only in the area just south of the boundary between T15S and T16S that the maximum thickness is skewed to one side of the channel (Fig. 12). At this location, an area of approximately 1/2 x 1 mile contains no net Stockholm sandstone.

The lower Morrow Limestone Isopach Map (Fig. 13) indicates that the absence of Stockholm Sandstone in the westernmost part of Sec. 11 (Fig. 12) was due to the presence of a 40-60' thick "plateau" of remnant (non-eroded) limestone. Because the valley was eroded into more or less uniformly resistant rocks during a lowstand of the sea, a roughly equal cross-sectional area for the valley was probably maintained all along the length of the field during maximum stream flow. In order to maintain a roughly equivalent cross section, the channels within the valley eroded deeper in the narrower areas. A shallow and narrow channel was also cut west of the "limestone plateau" (Figs. 13 and 14). Reactivation of basement faulting in Sec. 11, T16S, R43W prior to filling of the valley (Brown et al., 1990; Fig. 2) may have been responsible for what appears to be anomalous narrowing of the sand trend in the southern part of the field near the Colorado-Kansas border.

The "Johannes Sandstone" (Fig. 5) is a thin, discontinuous sandstone which is separated from the underlying Stockholm Sandstone by estuarine and marine shales. No cores were available from the "Johannes Sandstone" so the environment of deposition is unknown. Isopach mapping of the gross sand thickness of the "Johannes" sandstones suggests that a portion of the "limestone plateau" in the west half of Sec. 11, T16S, R43W was still a deterrent to sand deposition as late as "Johannes time" because no "Johannes Sandstone" occurs in an area of approximately 1/8 section. Rather than being a more or less continuous reservoir over portions of the field, as suggested by Shumard and Avis (1990), the "Johannes" was subdivided into at least four geographically separate pods or lenses, which were probably depositionally separate. Only a small portion of the "Johannes Sandstone" calculates as productive (Shumard and Avis, 1990).

The non-productive lower Morrow Limestone was deposited between the basal Morrow Sandstone and the Stockholm Sandstone (Fig. 5). The lower Morrow Limestone isopach map (Fig. 13) shows areas of thinning within the valley compared to a thickness of 60 to 80' outside the valley. Thinning of the limestone within the valley resulted from (1) deposition over relatively thick areas of basal Morrow Sandstone, and (2) erosion of the upper portions of the lower Morrow Limestone within the valley prior to deposition of the Stockholm Sandstone. In some area of the field (S 1/2 Sec. 6, T16S, R42W) the limestone was completely removed. The thickest valley limestones (87') occur in the area of the "limestone

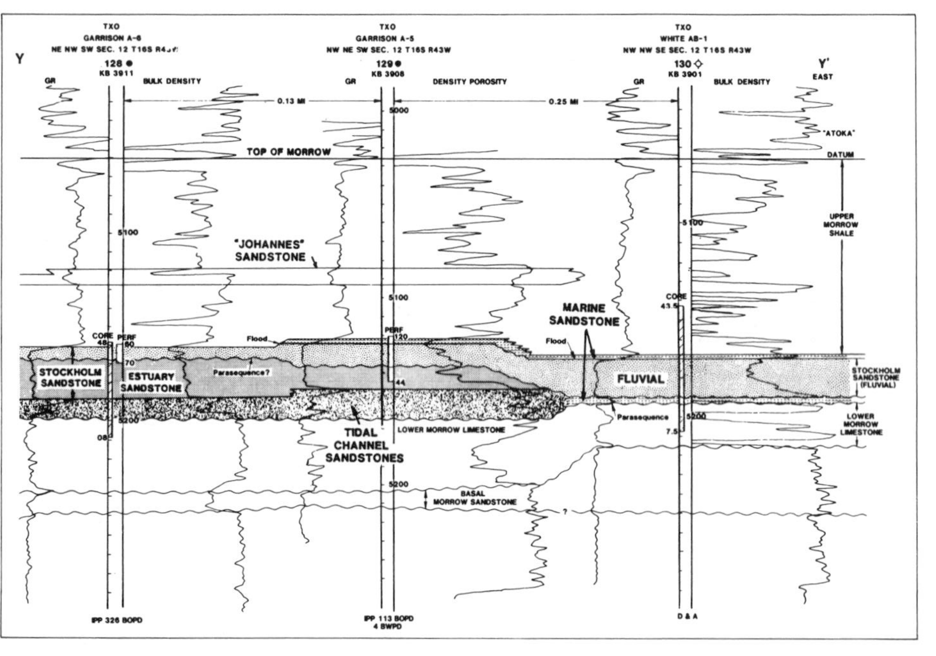

Fig. 11 - This figure and Figure 15 are portions of the stratigraphic cross section Y-Y' (Figure 9). Note lenticularity of sandstone facies identified in three cored wells (128, 130, 131). Erosional unconformities occur at base of basal Morrow Sandstone and at base of Stockholm Sandstone. Perforated intervals are primarily in fluvial sandstone intervals. Estuarine sandstones (cross bedded, clayey, fine grained) form much of the sandstone in Well 130. High clay content, etc. of tidal flat sandstones cause this facies to be marginal or subeconomic as an oil reservoir. Sharp lateral change between marine and tidal flat sandstones occurs between wells 129 and 130. Note also that a river apparently cut down through previously deposited estuarine sandstones between wells 129 and 130 and that fluvial sandstones were deposited in the river's channel. Although well No. 130 contains porous and permeable sandstone it is non-productive because it is below the oil-water contact.

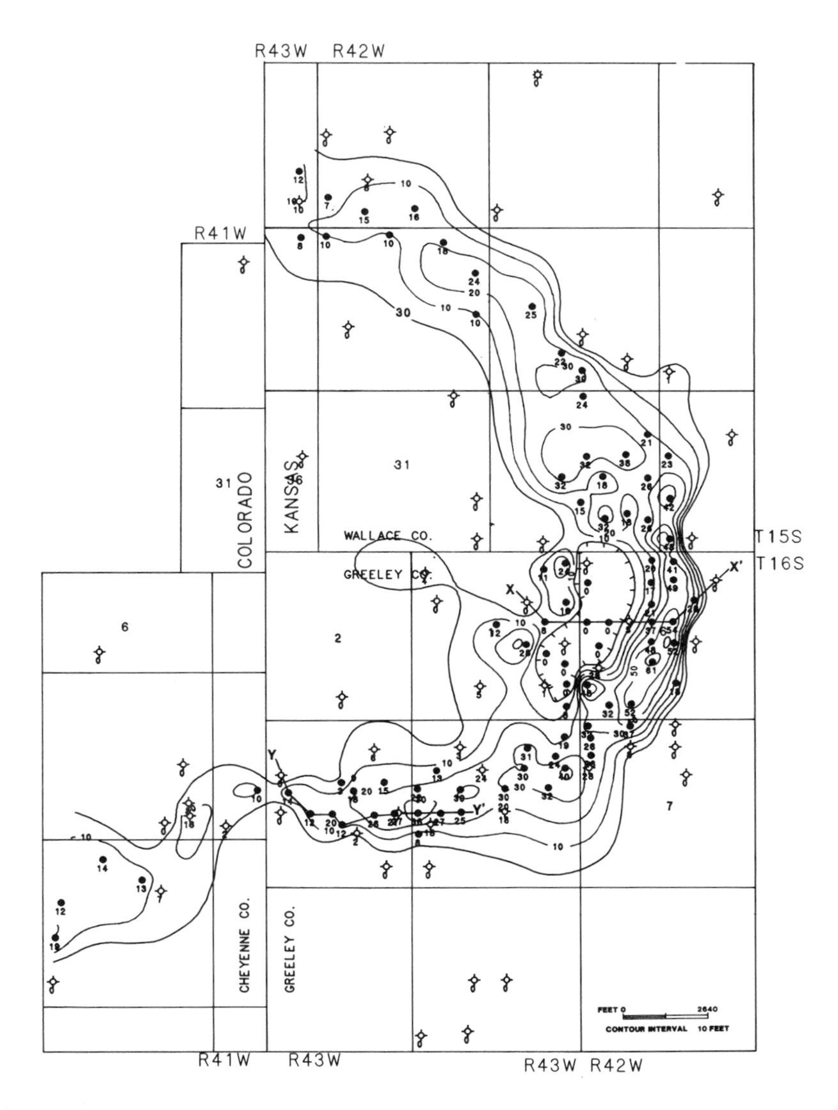

Fig. 12 - Stockholm Sandstone, Net Sand Isopach Map, Southwest Stockholm field, Kansas. Note lenticularity and sharp boundary of sand body which is constrained laterally by edges of eroded valleys. Note also that maximum sand thickness (60') occurs in a narrow portion of channel adjacent to the hachured area where no sand was deposited due to positive relief (40-60') on underlying lower Morrow Limestone (Sec. 6, T16S, R42W, Kansas). Thicknesses are posted for individual wells.

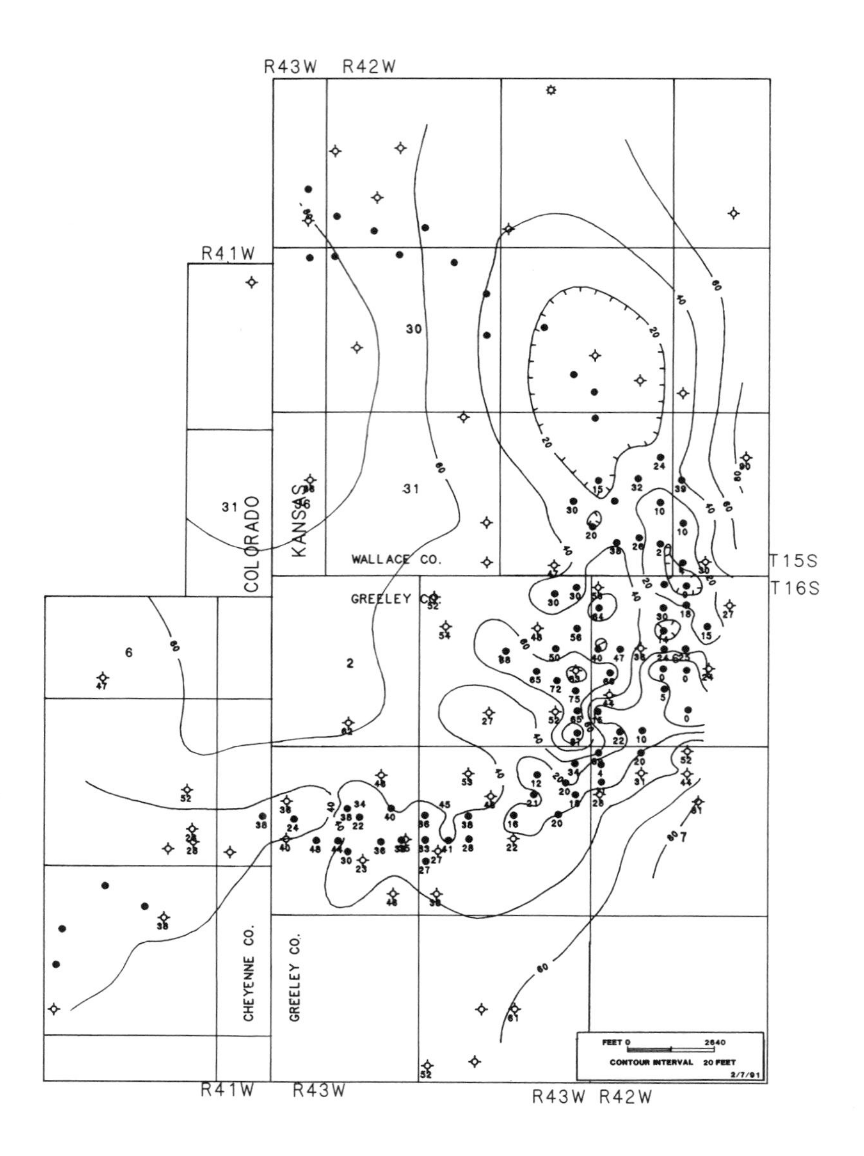

Fig. 13 - Lower Morrow Limestone Isopach Map, Southwest Stockholm field, Kansas. Refer to Fig. 11 for stratigraphic location. Thinning of limestone is limited primarily to the area of the valley. The limestone was deposited during a period of relatively high sea level and was partially eroded during a subsequent drop in sea level. Thicknesses are posted for individual wells.

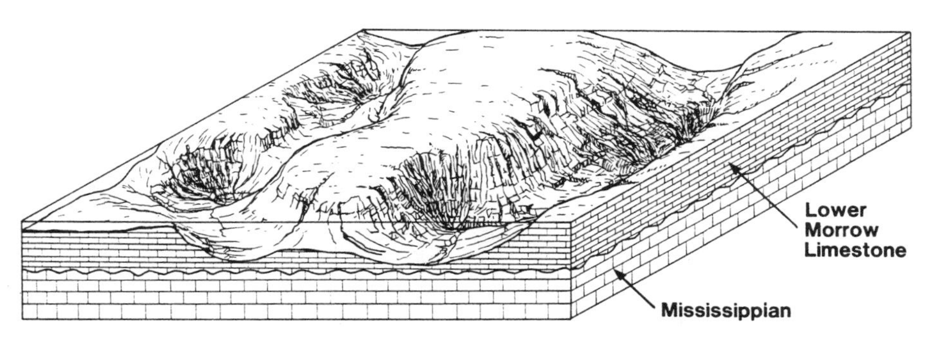

Fig. 14 - Relief in area of Southwest Stockholm field valley following erosion of lower Morrow Limestone. Upon this irregular surface the Stockholm Sandstone was deposited.

plateau", an area of little or no erosion of the limestone (Fig. 14). In the south-central portion of the field, where the underlying basal Morrow Sandstone is thickest, the lower Morrow Limestone thickness was controlled primarily by the relief on the pre-limestone depositional surfaces within the valley (i.e., between wells 129 and 130, Fig. 11). Where the valley was already filled by 50' or more of basal Morrow Sandstone, less accomodation space remained to be filled by limestone. The basal Morrow Sandstone is generally limited to the valley and thins abruptly at the edge of the valley (Fig. 15).

In order to understand the shape of the reservoir in Southwest Stockholm field, a series of cross sections were constructed. The locations of three of these cross sections (W-W', X-X', Y-Y') are shown in Figure 9. A summary version of W-W' (Fig. 16) was published by Brown et al. (1990) along with a seismic section paralleling the stratigraphic log section. This cross section shows the relief on the surface of the lower Morrow Limestone, which was filled by the Stockholm Sandstone and at least locally by shale that preceded Stockholm Sandstone deposition. The thickening of the valley is very obvious in the seismic section in Figure 16. This section also contains one of the described and interpreted cores (TXO Bergquist No. 2).

In the 80 miles along the Colorado-Kansas border, where lower Morrow Sandstones produce from valley-fill strata, at least seven unconformities occur on which valley-fill deposits have been mapped (Fig. 17). Each of these unconformities separates valley-fill deposits from underlying marine and estuarine shales and/or carbonates.

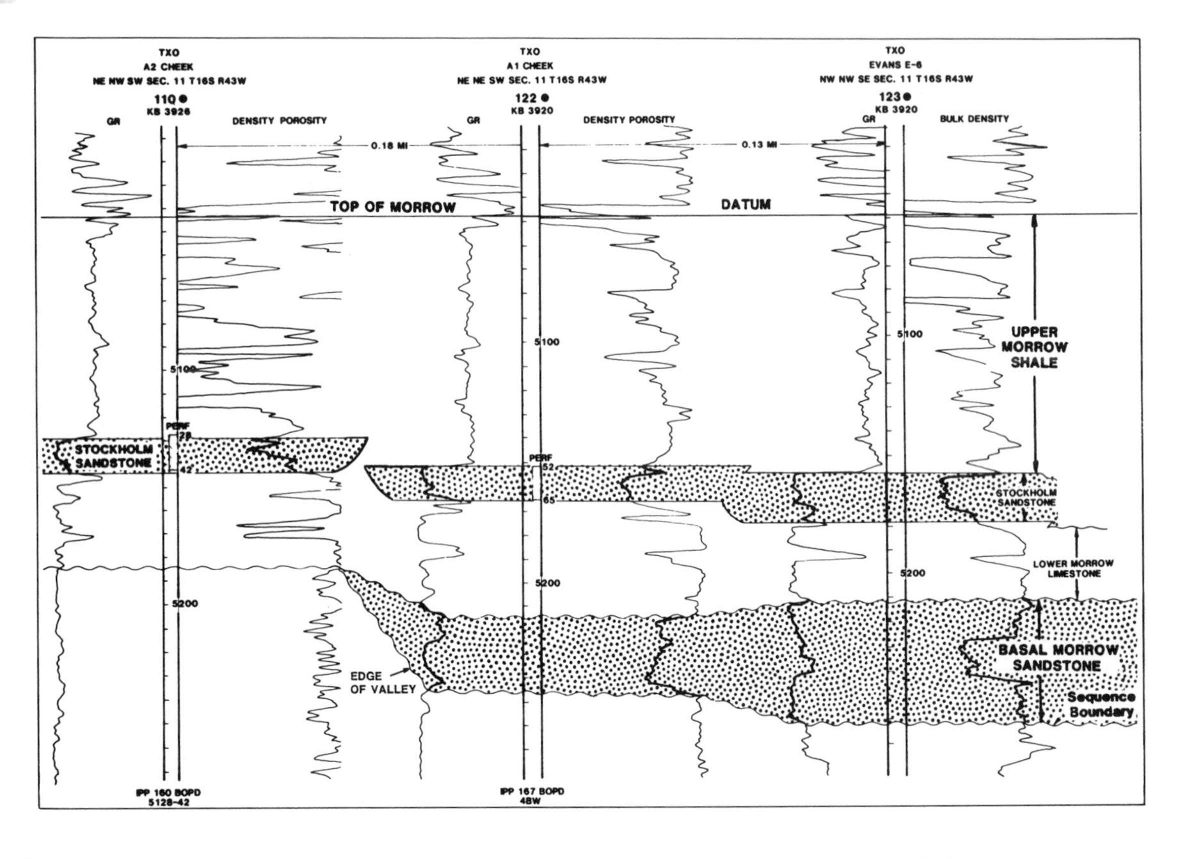

Fig. 15 - A portion of the western end of stratigraphic cross section Y-Y' (not included in paper). Erosional unconformities occur at base of basal Morrow Sandstone and at base of Stockholm Sandstone. The basal Morrow Sandstone pinches out abruptly against the northwest side of valley between wells 110 and 122. Following deposition of the basal Morrow Sandstone, minor erosion occurred at the top of the sandstone at location of well 122 prior to deposition of the lower Morrow Limestone.

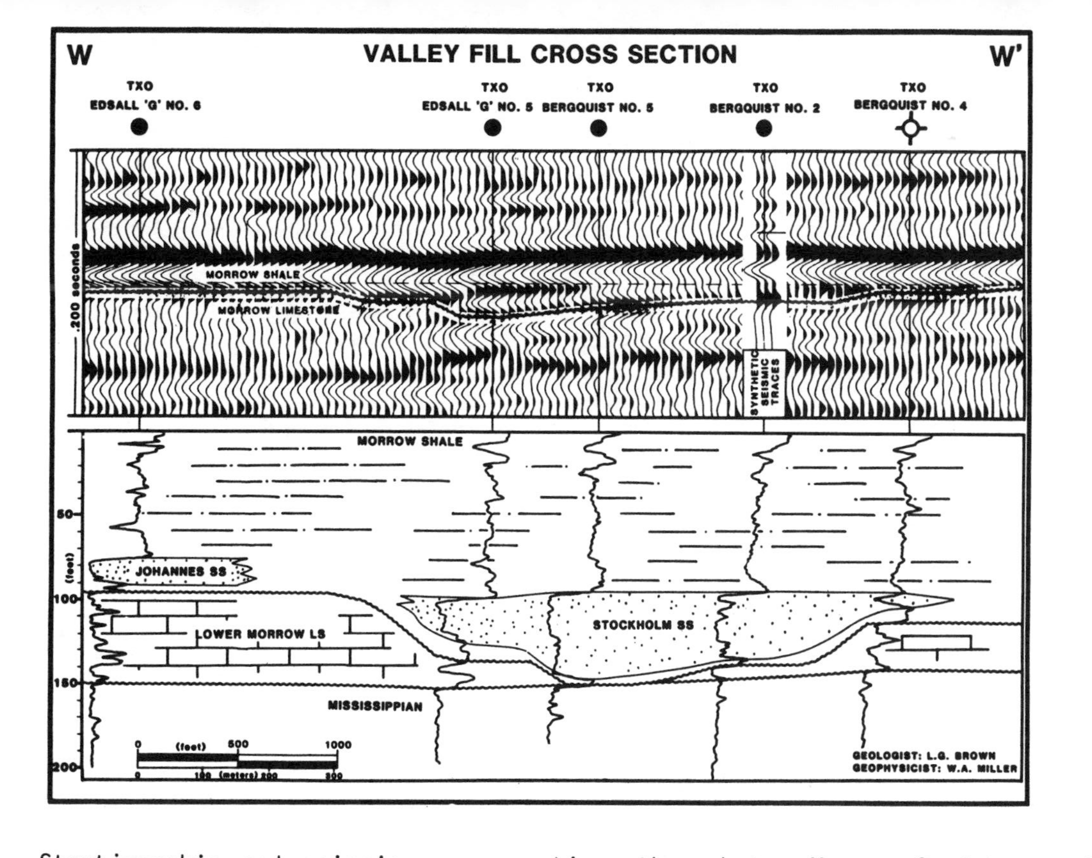

Fig. 16 - Stratigraphic and seismic cross sections through two Morrow Sandstone valley-fill reservoirs at Southwest Stockholm field, Kansas. One of cores described in this study is the TXO Bergquist No. 2. It is also included in this cross section and is the well utilized for development of a synthetic seismic trace. Note that the Stockholm Sandstone fills an erosional "channel". Flow in the "channel" was out of the diagram. The Johannes Sandstone is also limited areally by the valley walls but is more lenticular and is present in discontinuous lenses.

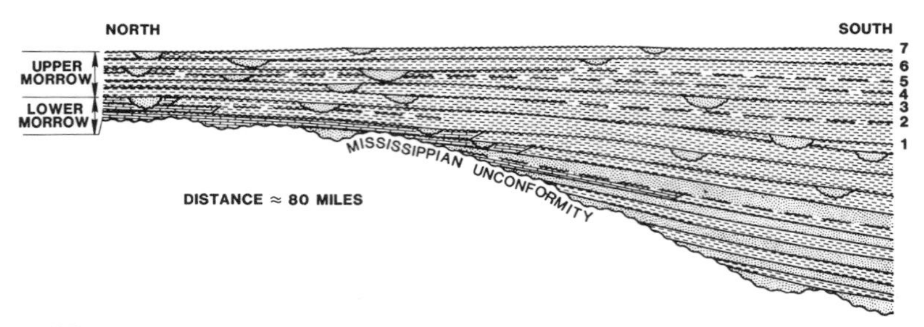

Fig. 17 - Schematic regional cross section of the Morrow Formation along the "Stateline Trend" of the Kansas-Colorado border. A subaerial unconformity is interpreted by Wheeler et al (1990) to separate each of the valley-fill deposits from the underlying marine shales and/or carbonates. The unconformities observed in the Southwest Stockholm field and to the north are numbered from the base to top, 1 through 7.

5. Discussion of Stratigraphic Cross Section X-X'

Stratigraphic cross section X-X' was constructed across the northern part of the field using gamma-ray-density logs. A portion of the cross section is shown in Figure 18. Due to confinement of the valley walls, the Stockholm Sandstone is essentially absent northwest of Well No. 64 and to the east of Well No. 60 (Fig. 12). As described above, the Stockholm Sandstone in this part of the field consists primarily of tidally deposited estuarine sandstones in the lower portion and interbedded fluvial and tidally deposited sandstones in the upper portion of the reservoir. Table 3, which is based on only a limited number of data points, suggests that significantly higher production rates should be expected from fluvial sandstones, which average 16% porosity and 700 md permeability. In contrast, tidally deposited sandstones average 14% porosity and only 80 md permeability. Cross section X-X' indicates that there is significant relief on the erosion surface at the top of lower Morrow Limestone (Figs. 11 and 18). In the center of the valley the thickness from the pre-Stockholm erosional surface to the top of the Stockholm Sandstone is 55' (Well No. 67, Fig. 9). Within 0.2 mile laterally this interval thins to 30' (Well No. 60) and within 0.4 mile from the "valley center" it thins to 15' (Well No. 55). Similar thickness relationships are found along the northwest valley margin. Where fluvial sandstones are absent the wells appear to be non-economic and were abandoned. Fluvial sandstones appear to be absent in wells No. 55 and 65, both of which are plugged. In the thicker portion of the Stockholm valley fill, shale was deposited and was preserved above the basal erosion surface and below the Stockholm Sandstone (i.e., Well No. 64, Fig. 18). These shales are

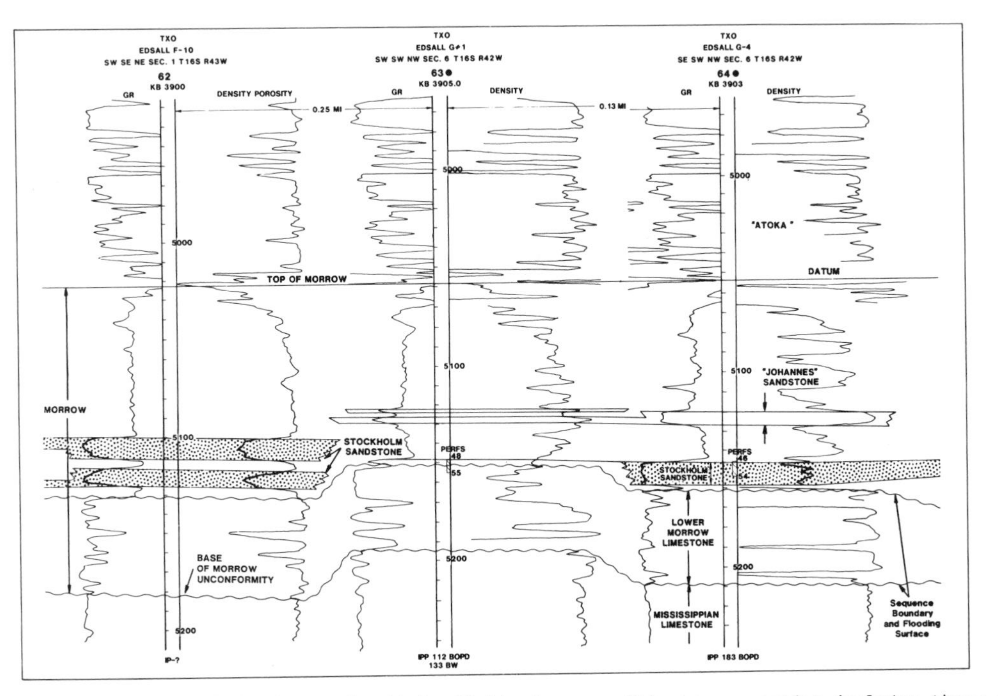

Fig. 18 - Cross section of Southwest Stockholm field, Kansas. This cross section includes three wells from near the west end of a longer (not included) east-west stratigraphic cross section X-X', which extends across the valley. Location of cross section is indicated on Figure 9. A number of significant observations concerning the cross section: (1) significant erosion has recurred on unconformities at the base and the top of the basal Morrow Sandstone; (2) the Stockholm Sandstone abruptly pinches out at the edge of the valley in the area between wells 63 and 64; (3) the thickness of the valley-fill succession deposited following deposition of the lower Morrow Limestone varies from 55' in well No. 57 (Fig. 9) to less than five feet in well No. 63. This valley in which the Stockholm Sandstone was deposited, is partially filled with reservoir-quality sandstone and partially filled with non-reservoir sandstones and shales. The "Johannes" Sandstones are lenticular and non-continuous, and adjacent lenses do not always appear to be correlative.

presumably marine, are non-productive, and form a basal and possibly locally are a lateral hydrocarbon seal.

The erosion surface separating the Mississippian limestones from the lower Morrow Limestone has, in the center of the valley, a similar profile to that which controlled the depositional thickness of the Stockholm Sandstone; however, near the edges of the valley (Wells 61, 62, 55) there is little relationship between the surfaces. Note also that the basal Morrow Sandstone (Fig. 5) is apparently absent along the transect of this cross section. The "Johannes Sandstone" is lenticular and non-productive in the wells in cross section X-X'. The "Johannes Sandstone" probably is lenticular over much of the field area and probably consists of several different sand bodies with differing origins.

6. Discussion of Stratigraphic Cross Section Y-Y'

Stratigraphic cross section Y-Y' extends across the valley in the southern portion of the field (Fig. 9). Figures 11 and 15 include portions of cross section Y-Y'. Three cored wells are included in this cross section. Cored intervals in Wells No. 128 and 131 have previously been discussed. Well No. 130, the TXO White AB-1, was cored completely through the Stockholm Sandstone (Fig. 11). At this location all but two, less than 1-foot thick sandstones at the top and base of the Stockholm Sandstone, are interpreted to be fluvial deposits. This well was plugged and abandoned even though porosity and permeability (18%, 879 md) are excellent (Table II). The Stockholm Sandstone in this well is below the oil-water contact.

The basal Morrow Sandstone occurs over the whole extent of cross section Y-Y'. This sandstone calculates on logs as non-productive and DST's of this unit are rare. The basal Morrow Sandstone pinches out against the valley wall between wells 110 and 122 (Fig. 15). The thickness of the non-productive lower Morrow Limestone is generally inversely proportional to the thickness of the underly basal Morrow Sandstone.

The Stockholm Sandstone in cross section Y-Y' is relatively well understood because three cored wells are included in the section. As discussed earlier, the three cores contain quite different percentages of fluvial, tidal (estuarine) and marine transgressive sandstones. Well No. 130 (Fig. 11) is 98% fluvial sandstone, Well No. 129 (Fig. 11) is approximately one-third fluvial sandstones and two-thirds tidal sandstones with less than 5% transgressive marine sandstones at the top of the reservoir. Well No. 131 (Fig. 10) contains highly interbedded thicker (2-4' thick) fluvial sandstones interbedded with very thin (0.5-1.0' thick) tidal (estuarine and marine) sandstones. This high degree of depositional variability is believed to be typical of the Stockholm Sandstone. Fluvial sandstones have the best reservoir characteristics in this part of the field, as they also do to the north.

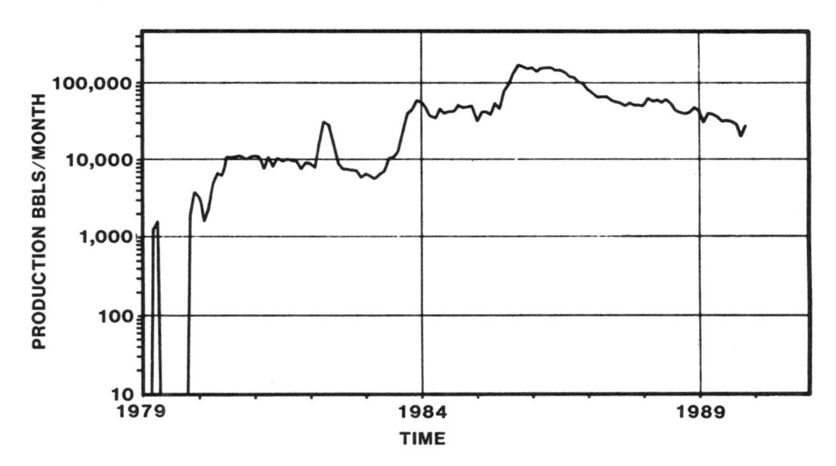

Fig. 19 - Production decline (1979-1990) for Southwest Stockholm field. The field has produced 5.5 million barrels of oil (primary production). Ultimate primary production is estimated to be approximately 7.5 million barrels, using 42 million barrels in place and a 17% recovery factor. It has been estimated by Brown et al. (1990) that 15% of the remaining oil in place can be produced using enhanced production technology from the 66 producing wells in the field.

Fluvial sandstones form the major portion of the perforated interval in Well No. 129 (Fig. 11) and all of the perforated intervals in Well No. 128 (Fig. 11). Well No. 130 (Fig. 11) contains thick fluvial sandstones but is below an oil-water contact. An apparently younger, "non-connected" portion of the Stockholm Sandstone occurs in well No. 110 (Fig. 15).

One of the most characteristic aspects of the valley-fill successions is illustrated in the section (Fig. 11) extending through wells 128, 129 and 130 where the fluvial sandstones thicken from 9' to 20' at the expense of the underlying estuarine and other tidally deposited sandstones. In this area, the base of the fluvial sandstones is interpreted to have been eroded into the tidal sandstones.

The "Johannes Sandstone" in cross section Y-Y' is a 10' thick nonproductive sandstone present in all the wells east of and including No. 124 (Fig. 9). It is absent in wells to the west.

B. Production

Southwest Stockholm field has produced 5.5 million barrels of oil from the Stockholm Sandstone during primary production. Brown et al. (1990) estimated that the field has 42 million barrels in place and using a 17% recovery factor 7.5 million

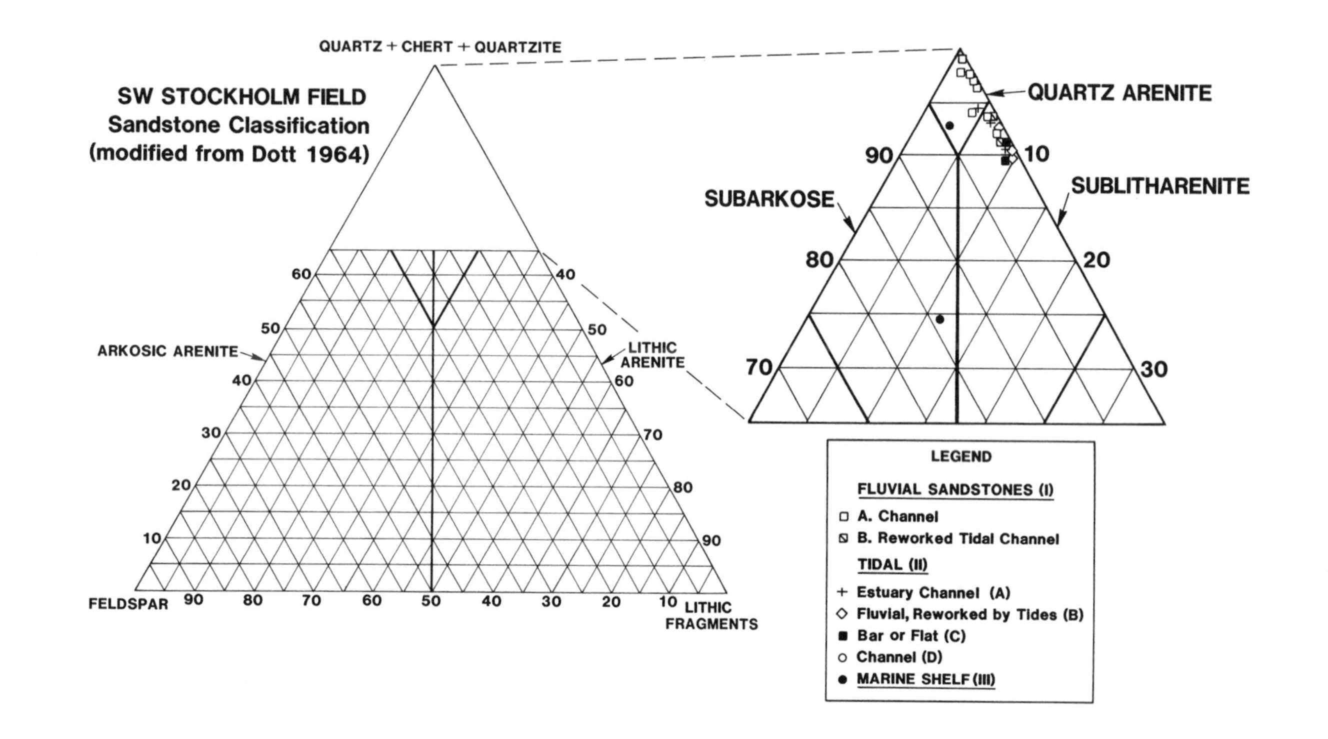

Fig. 20 - Composition of Stockholm Sandstone facies, Southwest Stockholm field, based on point-count data. Most of the samples are quartz arenites or quartz rich sub-litharenites in a modified Dott (1964) classification. The finer grained Stockholm Marine Sheld Sandstone Facies is more feldspathic than the other facies.

barrels will ultimately be produced. From mid 1980 to 1990 the field produced in excess of 10,000 barrels per month from 66 wells (Fig. 19).

C. Microheterogeneity, Petrology and Petrophysics of Stock holm Sandstone

1. Composition and Texture

The Stockholm Sandstone is quartz-rich (Figure 20). The composition was determined from microscopic thin-section analysis. The mean composition of sand grains for 24 point-counted samples was 93% quartz (Q93), 1% feldspar (F1) and 6% lithic (L6) fragments. No significant compositional difference exists among the various facies except for the Stockholm Marine Shelf Sandstone Facies III, which (based on two samples) is more feldspar-rich with a mean of $Q_{83.4}$,$F_{9.6}$, $L_{7.0}$. Tables B1 to B7 contain the composition of the various lithofacies based on modal, point-count, thin-section analysis. Detrital grains (and the range in volume percent) noted in thin section include monocrystalline quartz (25.5-68.2%), polycrystalline quartz (2.5-44.3%), plagioclase (0-10%), shale lithic fragments (0.5-7.3%), glauconite (<1%), collophane (<1%), muscovite (<1%) and trace amounts of heavy heavy minerals including tourmaline and zircon. The percentage of polycrystalline quartz increases with increasing grain grain size.

There are significant differences in grain size among the various facies. The fluvial facies is coarse-grained and typically either poorly sorted or bimodal. The Stockholm Marine Shelf Sandstone Facies III is very fine-grained with argillaceous laminae.

Sandstones in Southwest Stockholm field are unusual because there is no correlation of grain size with composition, porosity, or permeability (Figs. 21 and 22). Grain size was measured using a binocular microscope to make a visual comparison of sand grains in the core with grain size images printed on a comparator constructed on a sheet of transparent plastic (Figures 21 and 22).

2. Secondary Processes

Similar pore-filling or pore-creating processes occur in all sandstone facies in the field, however, the importance of certain processes varies among facies. The major processes of porosity destruction were compaction, quartz cementation, and kaolinite cementation. Carbonate minerals also replaced feldspar, shale lithic fragments, and the margins of quartz grains. Dissolution of carbonate cement to create secondary porosity was, at least locally, important in the reservoir at Southwest Stockholm field. Figure 23 is a general diagram showing the relative sequence of major diagenetic (secondary) events and the evidence to support the sequence for Morrowan sandstones.

Compaction in sandstones is important in reducing porosity and permeability. Intergranular volume (IGV) provides a measure of the amount of compaction that has taken place in a sandstone (Houseknecht, 1987). Figure 24 is a plot, devised by Houseknecht (1987), of intergranular volume versus intergranular cement. Samples that plot above the dashed line are dominated by cementation processes, whereas samples that plot below the dashed line are dominated by compaction processes. Intergranular volume (sometimes called minus cement porosity or precement porosity) is the sum of the intergranular cement plus intergranular porosity. It is important to exclude secondary minerals that replace grains and intragranular and moldic porosity in determining intergranular volume. In moderately well sorted sandstones the smaller the intergranular volume the greater the amount of compaction that has occurred.

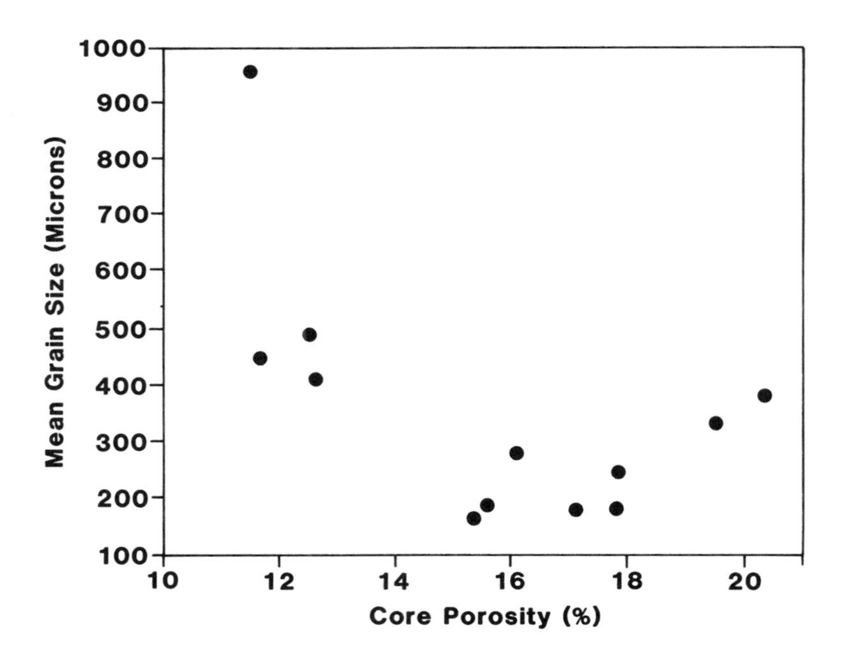

Fig. 21 - Core porosity versus mean grain size, Stockholm Sandstone. Grain size was measured using a comparator under a binocular microscope.

Note that the Stockholm Tidal Channel Facies IID and Stockholm Marine Shelf Sandstone Facies III are distinct from the other samples. The former is strongly dominated by compaction, whereas the latter is strongly dominated by cementation. The other samples overlap the dashed boundary line, but there are more samples in the compaction field than in the cementation field.

The major effects of secondary processes affecting the Stockholm Tidal Channel IID and Stockholm Marine Shelf Sandstone Facies III differ significantly from the other depositional lithofacies. The Stockholm Tidal Channel Facies IID is characterized by extensive compaction as shown by the low IGV (13.5%). If it is assumed that the initial porosity was 40%, the pore volume lost by compaction was 26.5%. The Stockholm Marine Shelf Sandstone Facies III is characterized by pervasive carbonate cement (26.3%). This cement formed relatively early as shown by the moderately high IGV (26.8%). If one assumes that the initial intergranular porosity was 40% then only 13.7% pore volume of Facies III was lost by compaction. The only other facies that has a distinctive diagenesis is Stockholm Tidal Bar Facies IIC, which is fine-grained with low porosity primarily due to quartz cement and fairly extensive compaction (IGV=17.6%). This compaction includes physical compaction as well as chemical compaction (pressure

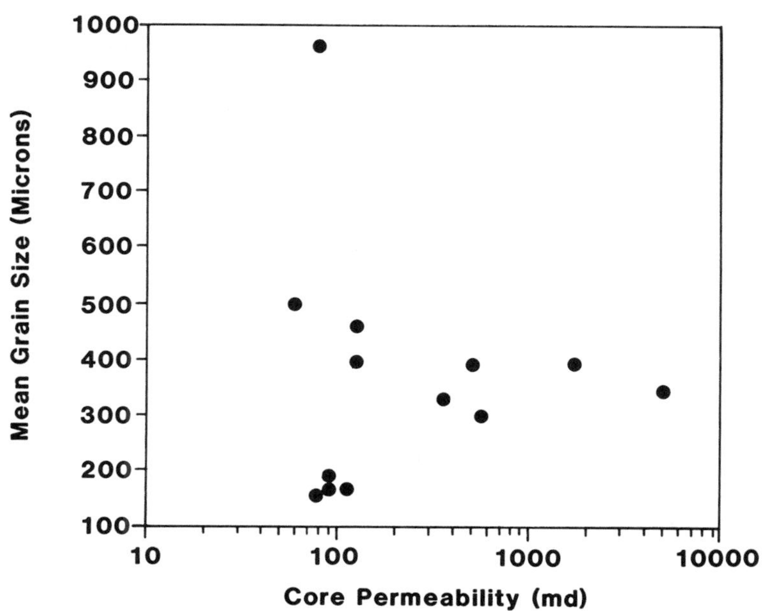

Fig. 22 - Core permeability versus mean grain size, Stockholm Sandstone. Grain size measured as described in Fig. 21.

EVENT	RELATIVE TIME SEQUENCE*	COMMENTS
Compaction	Physical — Chemical	Replacement Kaolinite deformed in coarse-grained ss; glauc. not deformed in some fine-grained ss.
Pyrite		Replacement of argillaceous material and as void-fill.
Chlorite		Patchy; underlies quartz overgrowths.
Qtz. Overgrowth		Major cement; formed relatively early; at least in fine-grained sandstone.
Calcite		Overlies quartz overgrowths; never encompasses barite.
Barite		Ferroan dolomite crystals molded around barite.
Ferroan Dolomite (Baroque)		Partially replaced feldspar shale lithic fragments and margins of quartz grains.
Siderite		Replaced ferroan dolomite cement.
Kaolinite	Replacement — Void-fill	Replaced feldspar and formed as void-fill in primary and secondary pores.
Dissolution		Some moldic pores from unknown material; some secondary intergranular pores from dissolution of carbonate cement.

*Feldspar overgrowths are too sparse to be fit into diagenetic sequence.

Fig. 23 - Diagenetic sequence of events for the Stockholm Sandstone, Southwest Stockholm field. One of the most important events in some facies after burial of sediments is dissolution of carbonate cement giving significant "secondary porosity". All events listed on figure, other than dissolution, tend to decrease porosity in Morrow valley-fill sandstones.

---

solution). Pressure solution of quartz grains has occurred along argillaceous (shaley) laminae that make up 2.6% of the sandstone. The other facies of the Stockholm Sandstone have mean intergranular volumes that range from 20.2 to 24.1% and mean total cement that ranges from 11.2 to 16.6%.

Figure 25 shows the evolution of porosity as influenced by the major diagenetic events. This is not strictly a time or depth dependent diagram because diagenetic events commonly overlap through time as the rocks subside deeper into the earth. This diagram shows that based on the thin sections studied, the present porosity is about the same for the Stockholm Tidal Channel Facies IID and the Stockholm Fluvial Facies IA, but the diagenetic history is different. The Stockholm Tidal Channel Facies IID underwent extensive porosity loss due to compaction, whereas the Stockholm Fluvial Facies IA was less affected by compaction, but developed more

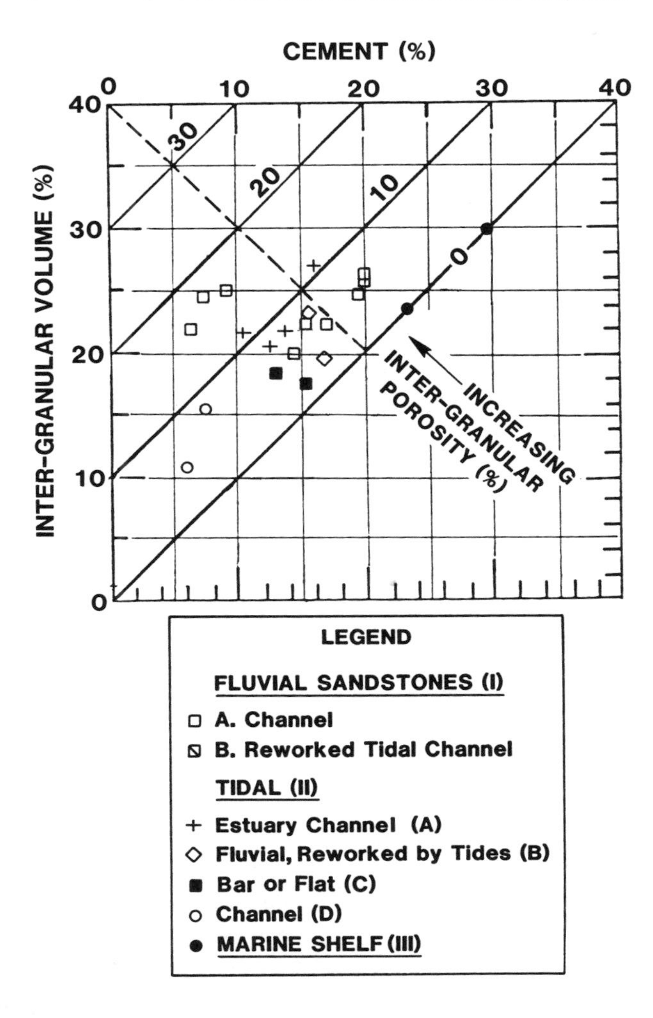

Fig. 24 - Derivation of compaction versus cementation effects, Stockholm Sandstone. This diagram is based on the assumption that the initial intergranular porosity was 40%. Intergranular porosity is destroyed by a combination of compaction and cementation. Intergranular volume is the sum of intergranular cement and intergranular porosity. A low intergranular volume indicates a strong compaction effect. Intergranular porosity can be read from the solid diagonal lines. The 45° dashed line separates fields where compaction (below the line) and cementation (above the line) are the dominant porosity-destruction processes. The Stockholm Tidal Channel Sandstone, Facies 11D, is more strongly affected by compaction than the other samples. The Stockholm Marine Shelf Sandstone Facies III was pervasively cemented by a relatively early carbonate cement and therefore has very low porosity values and is distinctly different from the other samples.

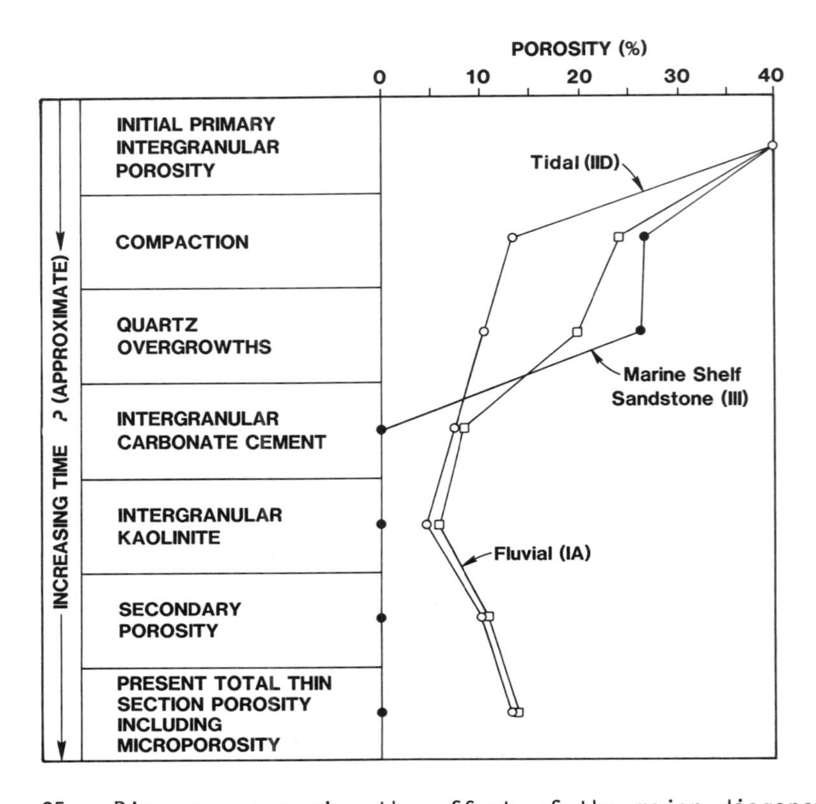

Fig. 25 - Diagram expressing the effect of the major diagenetic events on porosity for selected facies in SW Stockholm field. Interpreted sequence of porosity destroying events resulting from secondary processes in three Morrow reservoir sandstone facies: fluvial, tidal and marine shelf. Note that because rock properties differ among lithofacies, the method and amount of porosity diminution vary among the facies. Note especially the increase in porosity for fluvial and tidal sandstones due to formation of secondary porosity. Because diagenetic events commonly overlap through time with increasing depth of burial, this is not a time or depth diagram. For this diagram, however, except for compaction, depth and time generally increase downward. Comparison of the history of the Stockholm Tidal Facies IID (based on only two samples) and Stockholm Fluvial Facies IA indicates that compaction is more important in the Tidal Facies IID and that quartz overgrowths and carbonate cement are more important in the Fluvial Facies IA. However, the two facies, despite the differences in their history, have about the same total thin-section porosity. The Stockholm Marine Shelf Sandstone, Facies III, developed pervasive carbonate cement that destroyed porosity. The later dissolution event did not dissolve the carbonate, probably because the low permeability did not permit entry of the leaching solutions.

quartz and carbonate cement. In contrast, Stockholm Marine Shelf Sandstone Facies III underwent only slight compaction before pervasive carbonate cement formed to totally destroy macroporosity. This facies was not affected by the later dissolution event, perhaps because the pores were completely occluded making it difficult for fluids to enter into the sandstone to create secondary porosity.

3. Porosity and Permeability

a. General Comments. There is good evidence that the Stockholm sandstones were partially cemented early in their history by carbonate minerals, which were subsequently dissolved to create secondary intergranular porosity. Evidence for this comes from (1) preservation of significant carbonate cement with a distinctive dissolution texture in the TXO White 1-AB (Fig. 26C), (2) porous sandstone containing quartz grains with irregular outlines and embayments caused by partial replacement of the margins of quartz grains by carbonate that has been dissolved (Fig. 26D), and (3) the presence of partially replaced quartz overgrowths in porous rock, which suggests that carbonate once was present overlying the quartz (and partially replaced the quartz) and then was removed by dissolution (Fig. 27A-B). The quartz crystal facies have rough textures imparted by the replacement process (see Burley and Kantovowicz, 1986, and Pittman et al., 1992, for similar features).

These conclusions were reached knowing that the contact between carbonate cement and quartz grains in a thin section can appear to be embayed (irregular) because of an overlapping grain relationship in thin sections. The difference in hardness between quartz and carbonate minerals and the fact that carbonates have three good cleavages and quartz has no cleavage leads to partial plucking of carbonate and may give the appearance of an irregular (replacive-appearing) contact between the two minerals, which is misleading. To evaluate this effect, a portion of the thin section was photographed (Fig. 27C) and then the carbonate was dissolved with HCl acid and the same area was rephotographed (Fig. 27D). Comparison of these photographs shows that the carbonate did replace the margins of quartz grains. Therefore, it appears that a partial replacement of quartz grain margins is the most likely explanation for embayed grains in the porous sandstone that now contains only slight amounts of carbonate cement. A conservative estimate is that about half of the intergranular pores are of secondary origin.

Porosity in the Stockholm Sandstone is a mixture of intergranular macroporosity, intragranular/moldic macroporosity, and microporosity. The microporosity occurs among authigenic kaolinite crystals, within partially dissolved shale lithic fragments, and within sparse glauconite pellets. The microporosity, because of small pore throat size, does not contribute producible fluids. Intergranular porosity in Southwest Stockholm field is of primary and secondary origin. The intragranular/moldic macroporosity was formed by the partial or complete dissolution of feldspars (Fig. 26A) and lithic grains. Preserved clay infiltration rims outline framework grains of unknown mineralogy which have been dissolved to create moldic porosity (Fig. 26B).

Reservoir parameters for the general categories of fluvial and tidal sandstones are given in Table II. Mean porosities and permeabilities for fluvial lithofacies is 16.3%, 703 md; for tidal lithofacies 14.4%, 80.8 md.

b. Stockholm Sandstone Fluvial Facies IA. Subfacies of the tidal and fluvial facies may be recognized in core. The best part of the reservoir in Southwest Stockholm Field is Fluvial Facies IA of the Stockholm Sandstone. This sandstone has mean core porosity and permeability of 17.3% and 835 md, respectively. The relatively large grain size partly accounts for the high permeability. Mean point-count data for 10 samples of this facies indicates 8.6% intergranular porosity and 2.2% intragranular/moldic porosity for a total macroporosity of 10.8% (Appendix Table B1).

Fig. 26 - Photomicrographs, various facies, Stockholm Sandstone. A. Feldspar (F) was partially replaced by carbonate (C). Later, the carbonate was dissolved to create intragranular porosity. Note the remnants of feldspar (arrow) caught in the partially dissolved carbonate, TXO Garrison A-6, 5185.5 feet (core), plane polarized light. B. Infiltration clay (IC) coated sand grains of unknown origin, which were subsequently dissolved to create moldic porosity (MP), TXO Evans E-2, 5164.2 feet (core), plane polarized light. C. Carbonate cement has been partly dissolved giving the cement a "swiss-cheese" appearance, TXO Evans E-2, 5177.3 feet (core), plane polarized light. D. Quartz grains typically have embayed (arrows) grain outlines in porous sandstones. This is one line of textural evidence suggesting a secondary origin for some of the intergranular porosity, TXO Garrison A-6, 5160.7 feet (core), plane polarized light.

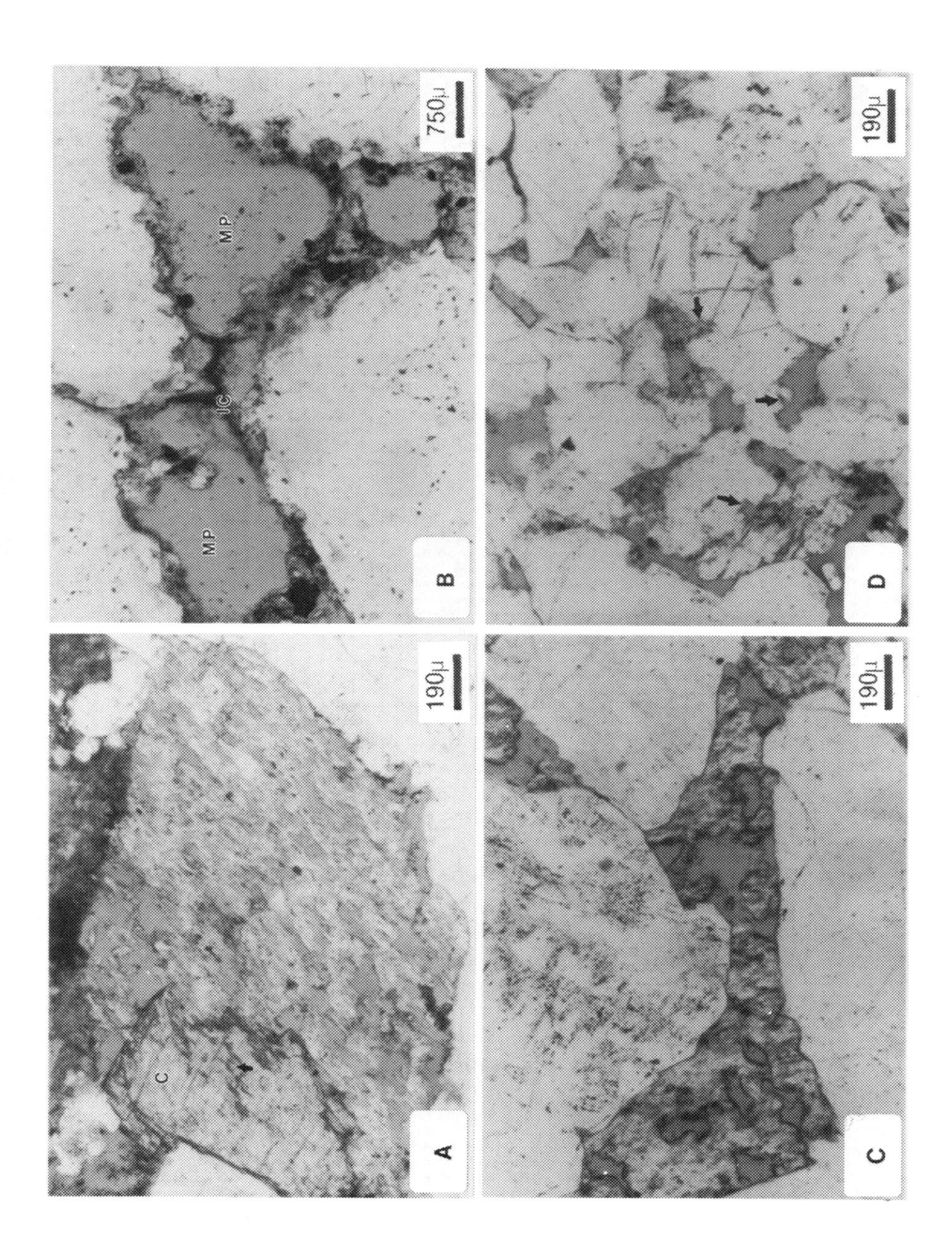
A
C
190μ
B
MP
IC
MP
750μ
C
190μ
D
190μ

c. Stockholm Sandstone, Tidal Facies (Reworked by Fluvial Processes) IB. Only two samples were point-counted from this facies and they are quite different (Table B2). This reflects the heterogeneity of the facies and generalizations are meaningless without a larger data set.

d. Stockholm Sandstone, Tidal Estuary Channel Facies IIA. This facies is petrophysically similar to the Stockholm Fluvial Facies IA, although it is finer (medium-grained). The macroporosity totals 10.9% and consists of 9.4% intergranular and 1.5% intragranular/moldic (Appendix Table B3). Permeabilities typically are about 95 md.

e. Stockholm Sandstone, Fluvial (Reworked by Tides) Facies IIB. This facies is relatively coarse (medium-grained), but the total porosity (5.5%) is relatively low based on point-counts of two samples. Most of the macroporosity is intergranular (4.8%) rather than intragranular/moldic (0.7%). This lithofacies has low porosity values, primarily because of an abundance of kaolinite (10.4%; Appendix Table B4).

Fig. 27 - Scanning-electron and thin-section photomicrographs, Stockholm Sandstone. A. Scanning-electron micrograph showing a quartz overgrowth with smooth appearing prism faces and with one end (arrow) of the quartz crystal modified by replacement contact with carbonate cement that was subsequently removed by dissolution, TXO Evans 2-E, 5177.8 feet (core). B. Scanning electron micrograph showing quartz overgrowths that were modified by contact with an overlying carbonate cement, which partially replaced the faces of the crystal. The planar faces and face contacts are recognizable despite the rough and pitted replacement texture. This evidence strongly suggests that some of the intergranular porosity is secondary in origin, TXO Evans 2-E, 5164.2 feet (core). C. and D. show the same part of a thin section before and after applying HCl acid to the thin section. Note the embayed margins (arrows) of quartz grains in C. This can be an apparent embayment related to thin sectioning of irregular grains and the difference in hardness and cleavage between quartz and carbonate. Note that after dissolving the carbonate, the embayments are still there. This indicates that the carbonate has replaced the quartz and that embayed grains in porous sandstone are evidence for secondary intergranular porosity for pores adjacent to embayed grains TXO Evans 2-E, 5190.4 feet. Photographs were taken in crossed-polarized light.

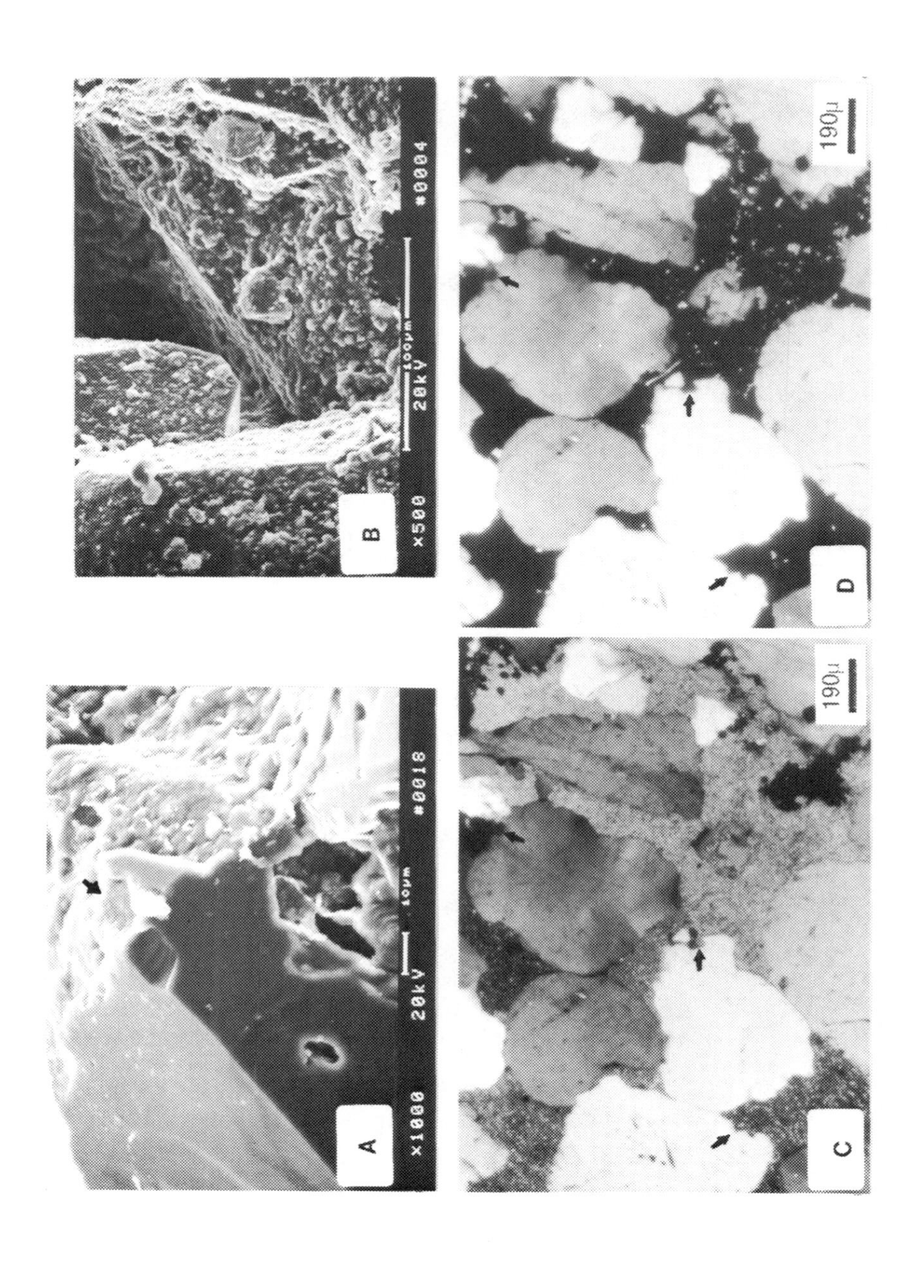
A
x1000 20kV 10µm #0018
B
x500 20kV 100µm #0004
C
190µ
D
190µ

f. Stockholm Sandstone, Tidal Bar Facies IIC. This facies is finer grained and better sorted than the fluvial facies that constitutes the better part of the reservoir. The Stockholm Tidal Bar Facies IIC has argillaceous laminae, considerable compaction, and a total porosity of only 4.6%, which consists of 3.6% intergranular porosity and 1.1% intragranular/moldic porosity (Appendix Table B5).

g. Stockholm Sandstone, Tidal Channel Facies IID. Based on limited core analyses, the Stockholm Tidal Channel Facies IID has a mean porosity and permeability of approximately 12.0% and 90 md, respectively. This facies has lost porosity largely because of compaction. Chemical compaction (pressure solution) is pronounced where illitic clay occurs along quartz-to-quartz grain contacts. A common observation of petrologists (e.g., Thomson, 1959) is that illitic clay films along grain contacts serve as a catalyst to promote pressure solution. The tight fit of some grains along their contacts has reduced the porosity and permeability compared to the Stockholm Fluvial Facies IA, which is similar compositionally and texturally except that illitic clay and pressure solution are less common.

The macroporosity in two point-counted samples totaled 10.2% and included 6.6% intergranular porosity and 3.6% intragranular/moldic porosity (Table B6). Microporosity occurs among kaolinite crystals (3.5%), within shale lithic fragments (5.6%), and within glauconite pellets (<1%).

h. Stockholm Sandstone, Marine Shelf Sandstone Facies III. Sandstones in this facies are very fine-grained with extensive carbonate cement (26.3%) and no macropores visible in the thin section (Appendix Table B7). This facies is non-reservoir.

4. Microscopic Heterogeneity

The reservoir sandstone facies in Southwest Stockholm field have considerable microheterogeneity and microporosity, but they are difficult to document because of the lack of publicly available porosity and permeability data. Ideally, thin sections for a study such as this would be cut from the plugs used for core analysis. Then, when point-counts were made of these thin sections it would be possible to determine the amount of microporosity by subtracting thin-section macroporosity from core porosity. Five thin-section samples were reasonably close to the position of the core perm plugs so these five samples were given more emphasis in our reference set, Table I.

Microheterogeneity is recognizable in thin section and under the scanning electron microscope (SEM) in the form of non-producing micropores associated with kaolinite clay and partly dissolved shale grains. Microporosity seldom yields hydrocarbons because the small pore aperatures hold bound water. Because routine commercial core analysis includes microporosity in total porosity values, an attempt was made to evaluate the distribution of micropores in the Stockholm reservoir sandstones. This was done by plotting total thin-section macroporosity versus the ratio of macroporosity to rock volume microporosity (Fig. 28). The rock volume with microporosity was derived by summing the percentages of kaolinite, shale lithic grains, and glauconite. This was done because there was no way to directly determine microporosity for the samples. What Figure 28 suggests is that samples with low porosity are dominated by microporous rock (low ratio on y axis), whereas higher porosity sandstone has a higher ratio (y axis) suggesting that micropores are less abundant relative to macropores in high porosity sandstone.

5. Reservoir Characteristics

Table II lists useful parameters for evaluating reservoir quality on a sample by sample basis. The k/Ø ratio and R_{35} have been used by Hartmann and Coalson (1990) to evaluate reservoirs. A high k/Ø ratio is indicative of a good reservoir. R_{35} is the pore aperture (throat) size on a mercury injection capillary-pressure curve where the mercury saturates 35% of the pore space. Hartmann and Coalson (1990) believe that R_{35} is useful in evaluating reservoirs. Large R_{35} values are indicative of good reservoirs. It is possible to estimate R_{35} using Winland's empirical equation, which was published by Kolodzie (1980). Winland's equation, which was developed using 322 samples of sandstones and carbonates for which porosity, permeability and mercury-injection data were available is: $\text{Log } R_{35} = 0.732 + 0.588 \text{ Log } K - 0.864 \text{ Log } Ø$. Figure 29 is a porosity-permeability plot with the solution to the Winland equation expressed graphically in the form of lines of equal pore-aperture size. Note that the largest pore apertures are at the top of the diagram. Based on the limited core-analysis data available from the Stockholm Sandstone, the fluvial sandstones generally have larger pore apertures than the tidal sandstones, although both are quality reservoirs.

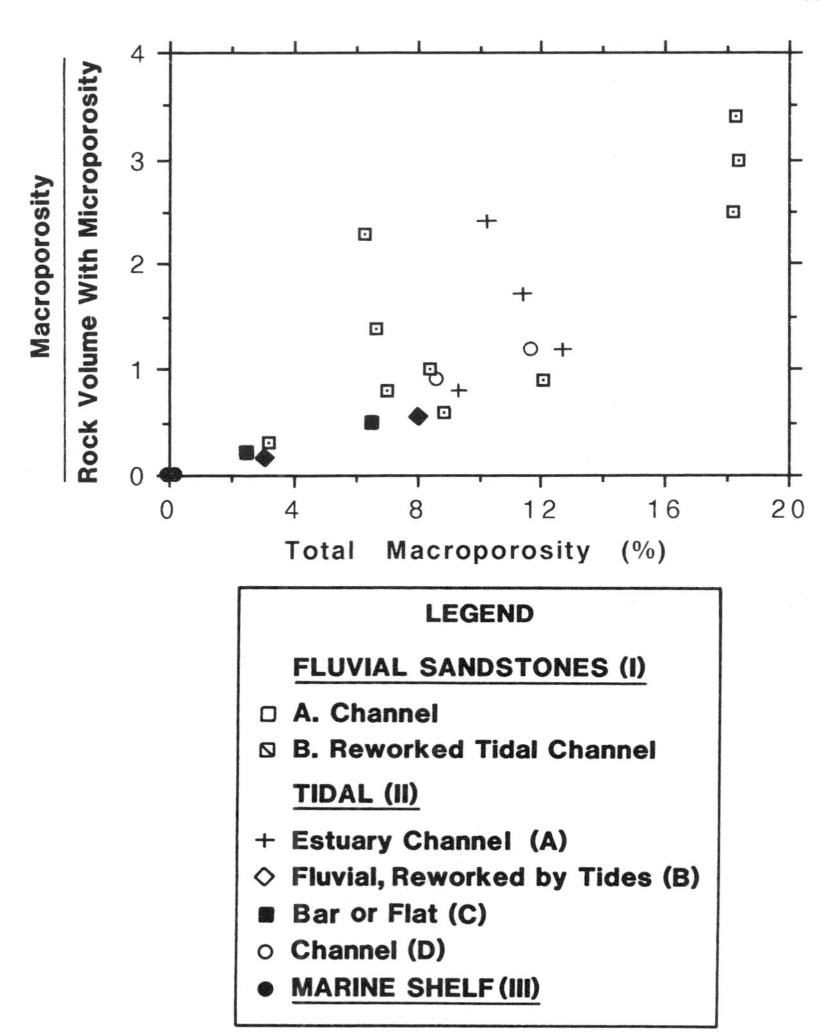

Fig. 28 - Scatter plot of total macroporosity in thin section versus the ratio of total macroporosity to the rock volume with microporosity. This approach was used because it was not possible to obtain from thin-section analysis a quantitative estimate of microporosity. This diagram suggests that sandstones with low total macroporosity have a low macro/micro ratio indicating that microporous rock (and micropores) are dominant in the lowest porosity rocks. Rocks with high macroporosity tend to have a higher macro/micro ratio, implying that microporosity may be about the same for all the samples of Stockholm Sandstone.

III. CONCLUSIONS

Southwest Stockholm field, which produces from Morrowan Stockholm and "Johannes" sandstones, is part of the Kansas-Colorado "Stateline Trend". In addition to the two producing sandstone reservoirs, the valley also was partially filled by a basal Morrow Sandstone, which is non-productive in Southwest Stockholm field, and a lower Morrow Limestone, which occurs between the basal and Stockholm sandstones. This limestone is also non-productive and forms a barrier to fluid flow between the two sandstones. Impermeable shales deposited

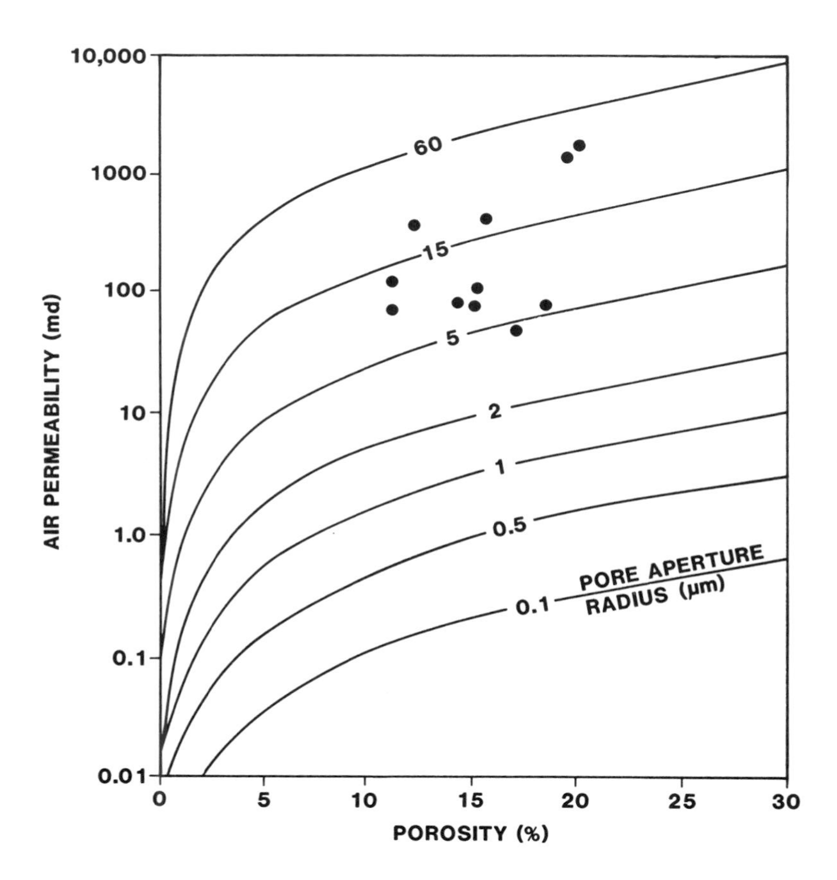

Fig. 29 - Plot of K vs Ø graphical solution to Winland's equation (from Hartmann and Coalson, 1990) showing Stockholm Sandstone samples from Southwest Stockholm Field. The Morrow Sandstones that produce from Southwest Stockholm field are primarily fluvial valley-fill sandstones.

within and outside of the valley separate the two producing sandstones (Stockholm and "Johannes"). Several of the valley-fill sandstone reservoirs abruptly terminate against the lateral margins of the valleys; some reservoirs change facies within the valley. Sediments that spilled over the edges of the valleys during late-stage flooding are generally sand-poor and non-productive.

Within the Stockholm Sandstone the reservoirs are primarily fluvial, however, tidal and estuarine sandstones deposited in the valley during minor sea-level rises also form a significant portion of the Stockholm Sandstone. These tidal and estuarine sandstones are finer grained and more clay-prone than the fluvial sandstones and as a result have poorer reservoir properties. In some wells only the fluvial portions of the sandstones are perforated. Where the finer-grained tidal sandstone interfinger with fluvial sandstones, as occurs in the TXO Evans No. 2, vertical permeability is diminished and reservoir storage capacity may be decreased.

The valleys in which the Stockholm Sandstone was deposited locally contained topographic highs, which prevented the deposition of, or diminish the thickness of, reservoir sandstones. Adjacent to these paleotopographic features, the valleys are cut much deeper and the thickest portion of the Stockholm Sandstone reservoir occurs. Commonly, these paleotopographic highs within the valley form where the underlying lower Morrow Limestone was not removed or was only slightly eroded. These paleotopographic features may have inhibited sand deposition over an area as large as 1/4 to 1/2 section.

Fluvial sandstones have average porosities of approximately 16% and average permeabilities of approximately 700 md. Tidally deposited sandstones, as a group, have decidedly inferior production characteristics; porosities average 14% and permeabilities 80 md. The coarser grained nature of the river deposits reflects their higher depositional energies and removal of most clay and very fine-grained material during deposition.

Thin-section, scanning-electron-microscope and X-ray analysis yielded information on size, origin and distribution of pores in the reservoir sandstones. A series of secondary events affected the rocks after they were deposited and although today the porosity of several of the producing facies is similar, the events that affected the rocks were different in different depositional lithofacies. The variability of these events contributed to variability in the rate of production of hydrocarbons from different facies. Tidal-Channel Facies IID sandstones underwent extensive porosity loss due to compaction and resulting interpenetration of grains, whereas the Fluvial Facies IA sandstones had initial porosity filled with quartz and calcite cement, followed by only minor losses due to compaction. In Facies

IID, as well as other tidally deposited facies where millimeter or thicker lenses of shale were deposited within the sand body by tidal currents, pressure solution of quartz grains adjacent to the shales has reduced porosity; this process is minor in the fluvial sandstones.

Many of the processes forming microheterogeneity in sandstone reservoirs decrease porosity, however, there is strong thin-section evidence that secondarily-derived (leached) porosity may form up to half of the present day intergranular (macro) porosity in the Stockholm Sandstone in Southwest Stockholm field.

Microporosity (porosity too small to yield hydrocarbons) occurs in those portions of the reservoir rocks that include secondary kaolinite and shale lenses. This porosity is included in that measured by subsurface logs but yields no producible hydrocarbons. An estimate of the percent of microporosity is useful as an aid in more accurately interpreting log porosity.

ACKNOWLEDGMENTS

This study was carried out under SBIR/DOE contract DE-FG05-90ER80976. Dorothy Swindler aided in subsurface data gathering, data analysis and data entry. Computer mapping was performed by Richard Banks of Scientific Computer Applications Inc., Tulsa, Oklahoma. Cores were provided by the U.S. Geologic Survey core repository, Denver, Colorado. Thanks to D. C. Kopaska-Merkel and W. R. Almon for their constructive editorial suggestions.

REFERENCES

Blakeney, B. A., L. F. Krystinik, A. A. Downey, 1990, Reservoir Heterogeneity in Morrow Valley Fills, Stateline Trend: Implications for Reservoir Management and Field Expansion, Morrow Sandstones of S.E. Colorado and Adjacent Areas, S. A. Sonnenberg, L. T. Shannon, K. Rader, W. F. von Drehle, G. W. Martin (eds.), Rocky Mountain Association of Geologists, p. 131-142.

Brown, L. G., W. A. Miller, E. M. Handley-Goff and S. L. Veal, 1990, Stockholm Southwest Field, S. A. Sonnenberg, L. T. Shannon, K. Rader, W. F. von Drehle, G. W. Martin (eds.), Morrow Sandstones of SE Colorado and Adjacent Areas, Rocky Mountain Association of Geologists, p. 117-130.

Burley, S. D., and J. D. Kantovowicz, 1986, Thin section and S.E.M. Textural criteria for the recognition of cement-dissolution porosity in sandstones: Sedimentology, v. 33, p. 587-604.

Dott, R. H., Jr., 1964, Wacke, graywacke and matrix - what approach to immature sandstone classification: Jour. Sedimentary Petrology, v. 34, p. 625-632.

Farmer, C. L., 1981, Tectonics and Sedimentation, Newcastle Formation (Lower Cretaceous) southwestern flank of Black Hills uplift, Wyoming and South Dakota: unpublished M. Sc. Thesis, Colorado School of Mines, 195 p.

Hartmann, D. J. and E. B. Coalson, 1990, Evaluation of the Morrow Sandstone in Sorrento Field, Cheyenne County, Colorado: in S. A. Sonnenberg and others (eds.), Morrow Sandstones of Southeast Colorado and Adjacent Areas: The Rocky Mountain Association of Geologists, Denver, Colorado, p. 91-100.

Hine, A. C. and S. W. Snyder, 1985, Coastal Lithosome Preservation: Evidence from the Shoreface and Inner Continental Shelf off Bogere Banks, North Carolina: Marine Geology, v. 63, p. 307-330.

Houseknecht, D. W., 1987, Assessing the relative importance of compaction processes and cementation to reduction of porosity in sandstones: Amer. Assn. Petrol. Geol. Bull., v. 71, p. 633-642.

Kolodzie, S., 1980, The analysis of pore throat size and the use of the Archie equation to determine OOIP in Spindle Field, Colorato: SPE 55th Annual Fall Technical Conference, Dallas, Texas, SPE Paper 9382, 4 p.

Krystinik, L. F., 1989a, Morrow Formation Facies Geometries and Rerservoir Quality in Compound Valley Fills, Central State Line area, Colorado and Kansas (abs.): Amer. Assn. Petrol. Geol. Bull., v. 73, p. 375.

Krystinik, L. F., 1989b, Sedimentary Character of Morrow Valley Fill Complexes: A Core Study and Workshop (abs.): in Exploration for Morrow Sandstone Reservoirs SE Colorado and Adjacent Areas, RMS-SEPM and RMAG Symposium, p. 5.

Krystinik, L. F. and B. A. Blakeney, 1990, Sedimentology of the Upper Morrow Formation in Eastern Colorado and Western Kansas, S. A. Sonnenberg, L. T. Shannon, K. Rader, W. F. von Drehle, G. W. Martins (eds.), Morrow Sandstones of SE Colorado and Adjacent Areas, Rocky Mountain Association of Geologists, p. 37-50.

Pittman, E. D., R. E. Larese, and M. T. Heald, Clay Coats (1992): Occurrence and reference to preservation of porosity in sandstones: in D. W. Houseknecht and E. D. Pittman (eds.), Origin, Diagenesis and Petrophysics of Clay Minerals in Sandstone, SEPM Special Publication 47, p. 241-255.

Shumard, C. B., 1989, Stockholm SW Field; a Morrow (Lower Pennsylvanian) Valley Fill Sandstone Reservoir, Hugoton Embayment, Greely and Wallace Co., Kansas, (abs.): in Tulsa Geologic Society Newsletter, Sept. 12.

Shumard, C. B., 1991, Stockholm Southwest Field, U.S.A., Anadarko Basin, Kansas, Stratigraphic Traps II, Treatise of Petroleum Geology Atlas of Oil and Gas Fields, Am. Assn. Petroleum Geologists, Tulsa, Oklahoma, p. 269-304.

Shumard, C. B. and L. E. Avis, 1990, Key Morrow predictive exploration model: Southwest Stockholm field, Oil and Gas Journal, v. 88, p. 87-94.

Shumard, C. B., 1989, Stockholm SW Field; a Morrow (Lower Pennsylvanian) Valley Fill Sandstone Reservoir, Hugoton Embayment, Greely and Wallace Co., Kansas, (abs.): in Tulsa Geologic Society Newsletter Sept. 12.

Siemers, C. T. and R. W. Tillman, 1981, Deep water clastic sediments, in C. T. Siemers, R. W. Tillman and C. R. Williamson (eds.), SEPM Core Workshop No. 2, p. 20-44.

Sonnenberg, S.A., 1985, Tectonic and Sedimentation Model for Morrow Sandstone Deposition, Sorrento field Area, Denver Basin, Colorado: The Mountain Geologist, v. 22, p. 180-191.

Sonnenberg, S. A., L. T. Shannon, K. Rader, W. F. Von Drehle, G. W. Martin (eds.), 1990, Morrow Sandstones of Southeast Colorado and Adjacent Areas, Rocky Mountain Association of Geologists, 236 p.

Stamp, L. D., 1921, On Cycles of Sedimentation in The Eocene Strata of the Anglo-Franco-Belgian Basin: Geological Magazine, v. 58, p. 108-114.

Suter, J. R. and H. L. Berryhill, Jr., 1985, Late Quaternary shelf margin deltas, northwest Gulf of Mexico: Am. Assoc. Petrol. Geol. Bull., v. 69, p. 77-91.

Thomson, A. F., 1959, Pressure solution and porosity: in H. A. Ireland (ed.), Silica in Sediments, SEPM Special Publ. 7, p. 92-110.

Vail, P. R., R. M. Mitchum, Jr., R. G. Todd, J. M. Widmier, S. Thompson, III, J. R. Sangree, J. N. Bubb and W. G. Hatfield, 1977, Seismic stratigraphy application to hydrocarbon exploration: Am. Assoc. Petrol. Geol. Memoir 26, p. 49-205.

Weimer, R. J., S. A. Sonnenberg, L. T. Shannon, 1988, Production from Valley Fill Deposits, Morrow Sandstone, South east Colorado, New Exploration Challenges and Rewards: Rocky Mountain Assoc. Geol. Outcrop, v. 37, No. 10, Oct. 1988.

Well	Depth	Mono. Qtz.	PolyQtz.	Feld.	Shale	Glauc.	Colloph.	Unk./Other	Clay lam.	Clay Infilt.	Carb. Cmt.	Carb. Replac.	Kaol.	Qtz. Cmt.
1 Garrison A-6	5156.3	47.2	23.4	0.3	1.2	0.3	0.0	0.3	0.0	0.0	2.4	0.0	4.6	2.2
2	5155.3	61.8	6.4	0.2	5.3	0.2	0.0	0.2	0.0	0.0	1.0	0.0	1.7	5.0
3 Evans 2-E	5177.8	25.5	44.3	0.0	0.5	1.0	0.7	0.5	0.0	0.5	16.9	1.0	0.7	2.2
4	5176.5	39.4	30.5	0.0	1.9	0.0	0.0	0.5	0.0	0.0	8.4	1.2	6.5	4.6
5	5175.2	39.8	29.7	0.2	4.2	0.0	0.0	0.2	0.0	1.6	0.7	0.0	10.5	4.0
6	5171.0	34.9	32.8	0.0	2.1	0.0	0.2	0.2	0.0	0.5	14.4	2.3	2.3	3.5
7	5168.2	30.2	37.0	0.0	2.2	0.0	0.0	0.2	0.0	3.4	2.9	0.7	6.8	4.4
8	5164.2	38.4	29.0	0.0	2.7	0.0	0.0	1.2	0.0	2.2	5.1	1.0	5.4	6.6
9	5163.1	29.0	36.8	0.0	2.3	0.0	0.0	0.5	0.0	-0.5	16.0	1.4	8.7	1.8
10 White AB-1	5173.2	49.5	20.5	1.2	3.4	0.0	0.0	1.0	0.0	0.0	0.7	0.0	2.0	3.4
11 MEAN		39.6	29.0	0.2	2.6	0.2	0.1	0.5	0.0	0.9	6.9	0.8	4.9	3.2

	Tot. Cmt.	Intergran. Por.	Intragran./Moldic Por.	Tot. Por.	Mean Grain Size	IGV	Macro/Micro	Core Por.	Perm.
1	9.2	15.7	2.7	18.4	508	24.9	3.0		
2	7.7	16.5	1.7	18.2	273	24.2	2.5	19.5	1400.0
3	20.8	6.3	0.0	6.3	1111	26.1	2.3		
4	20.7	5.1	1.9	7.0	546	24.6	0.8		
5	15.2	7.3	1.6	8.9	689	22.5	0.6		
6	22.5	5.8	0.9	6.7	675	26.0	1.4		
7	14.8	5.8	6.3	12.1	868	19.9	0.9		
8	18.1	5.3	3.1	8.4	687	22.4	1.0		
9	27.9	2.1	1.1	3.2	1005	28.6	0.3		
10	6.1	15.6	2.7	18.3	360	21.7	3.4		
11	16.5	8.6	2.2	10.8	672	24.1	1.6		

Table B1 - Point-count data for the Stockholm Fluvial Sandstone Facies IA. Depth is in feet. Mean grain size (µm) is the mean of the apparent long axes of 25 quartz grains measured optically in thin section. IGV is intergranular volume (%), which is the sum of integranular cement plus intergranular porosity. Macro/micro is the ratio of macroporosity to the rock volume with microporosity. Core porosity and permeability are expressed in percent and millidarcies, respectively. The other properties are in volume percent.

	Well	Depth	Mono. Qtz.	PolyQtz.	Feld.	Shale	Glauc.	Colloph.	Unk./Other	Clay lam.	Clay Infilt.	Carb. Cmt.	Carb. Replac.	Kaol.	Qtz. Cmt.
1	Garrison A-6	5160.7	60.4	7.0	0.3	4.6	0.3	0.3	0.5	0.0	0.0	1.2	0.0	2.9	5.6
2	Evans 2-E	5162.0	43.0	32.6	0.2	2.8	0.2	0.0	0.4	0.0	0.2	3.3	0.0	4.6	4.8
3	MEAN		51.7	19.8	0.3	3.7	0.3	0.2	0.5	0.0	0.1	2.3	0.0	3.8	5.2

	Tot. Cmt.	Intergran. Por.	Intragran./Moldic Por.	Tot. Por.	Mean Grain Size	IGV	Macro/Micro
1	9.7	15.5	1.7	17.2	243	25.2	2.2
2	12.7	2.4	0.4	2.8	713	15.1	0.4
3	11.2	9.0	1.1	10.0	478	20.2	1.3

Table B2 - Point-count data for Stockholm Tidal Sandstone (Reworked by Fluvial Processes) Facies IC. Depth is in feet. Mean grain size (m) is the mean of the apparent long axes of 25 quartz grains measured optically in thin section. IGV is intergranular volume (%), which is the sum of integranular cement plus intergranular porosity. Macro/micro is the ratio of macroporosity to the rock volume with microporosity. Core porosity and permeability are expressed in percent and millidarcies, respectively. The other properties are in volume percent.

	Well	Depth	Mono. Qtz.	PolyQtz.	Feld.	Shale	Glauc.	Colloph.	Unk./Other	Clay lam.	Clay Infilt.	Carb. Cmt.	Carb. Replac.	Kaol.	Qtz. Cmt.
1	Garrison A-6	5181.2	67.7	2.5	0.2	3.7	0.0	0.0	0.2	0.0	0.0	5.4	2.0	0.5	7.7
2		5174.2	62.1	7.9	0.5	6.7	0.5	0.2	0.5	0.0	0.0	0.2	0.0	4.8	7.4
3		5172.0	68.2	3.1	0.2	3.6	0.7	0.2	2.2	0.0	0.0	3.9	0.0	2.4	4.1
4		5169.0	62.9	3.2	0.5	3.7	0.3	0.3	0.3	0.0	0.0	3.0	0.3	6.2	6.7
5	MEAN		65.2	4.2	0.4	4.4	0.4	0.2	0.8	0.0	0.0	3.1	0.6	3.5	6.5

	Tot. Cmt.	Intergran. Por.	Intragran./Moldic Por.	Tot. Por.	Mean Grain Size	IGV	Macro/Micro	Core Por.	Perm.
1	15.6	7.7	2.5	10.2	248	21.3	2.4	15.3	80.0
2	12.4	7.9	1.4	9.3	325	20.3	0.8		
3	10.4	11.1	0.2	11.3	200	21.5	1.7		
4	16.1	10.9	1.7	12.6	314	26.8	1.2	15.5	111.0
5	13.6	9.4	1.5	10.9	272	22.5	1.5		

Table B3 - Point-count data, Stockholm Tidal Estuary Channel Sandstone Facies IIA. Depth is in feet. Mean grain size (μm) is the mean of the apparent long axes of 25 quartz grains measured optically in thin section. IGV is intergranular volume (%), which is the sum of integranular cement plus intergranular porosity. Macro/micr is the ratio of macroporosity to the rock volume with microporosity. Core porosity and permeability are expressed in percent and millidarcies, respectively. The other properties are in volume percent.

Well	Depth	Mono. Qtz.	PolyQtz.	Feld.	Shale	Glauc.	Colloph.	Unk./Other	Clay lam.	Clay Infilt.	Carb. Cmt.	Carb. Replac.	Kaol.	Qtz. Cmt.
1 Garrison A-6	5163.6	61.5	4.8	0.3	5.1	0.3	0.3	0.5	3.6	0.0	2.4	0.3	8.9	4.3
2 Evans E-2	5173.8	48.9	19.1	0.5	7.5	0.2	0.0	0.5	3.9	0.0	0.0	0.0	11.9	5.3
3 MEAN		55.2	12.0	0.4	6.3	0.3	0.2	0.5	3.8	0.0	1.2	0.2	10.4	4.8

	Tot. Cmt.	Intergran. Por.	Intragran./Moldic Por.	Tot. Por.	Mean Grain Size	IGV	Macro/Micro
1	15.9	7.2	0.7	7.9	299	22.8	0.6
2	17.2	2.4	0.7	3.1	471	19.6	0.2
3	16.6	4.8	0.7	5.5	385	21.2	0.4

Table B4 - Point-count data, Stockholm Fluvial Sandstone (Reworked by Tidal Processes) Facies IIB. Depth is in feet. Mean grain size (m) is the mean of the apparent long axes of 25 quartz grains measured optically in thin section. IGV is intergranular volume (%), which is the sum of integranular cement plus intergranular porosity. Macro/micro is the ratio of macroporosity to the rock volume with microporosity. Core porosity and permeability are expressed in percent and millidarcies, respectively. The other properties are in volume percent.

Well	Depth	Mono. Qtz.	PolyQtz.	Feld.	Shale	Glauc.	Colloph.	Unk./Other	Clay lam.	Clay Infilt.	Carb. Cmt.	Carb. Replac.	Kaol.	Qtz. Cmt.
1 Evans 2-E	5176.0	58.4	11.8	0.5	6.4	0.2	0.5	0.7	2.1	0.0	0.9	0.0	6.8	5.2
2	5165.5	63.5	6.6	0.2	7.6	0.2	0.0	0.9	3.1	0.0	0.5	0.0	3.6	11.1
3 MEAN		61.0	9.2	0.4	7.0	0.2	0.3	0.8	2.6	0.0	0.7	0.0	5.2	8.2

	Tot. Cmt.	Intergran. Por.	Intragran./Moldic Por.	Tot. Por.	Mean Grain Size	IGV	Macro/Micro
1	12.9	5.0	1.6	6.6	182	17.9	0.5
2	15.2	2.1	0.5	2.6	225	17.3	0.2
3	14.1	3.6	1.1	4.6	204	17.6	0.4

Table B5 - Point-count data for Stockholm Tidal Bar Sandstone Facies IIC. Depth is in feet. Mean grain size (µm) is the mean of the apparent long axes of 25 quartz grains measured optically in thin section. IGV is intergranular volume (%), which is the sum of integranular cement plus intergranular porosity. Macro/macro is the ratio of macroporosity to the rock volume with microporosity. Core porosity and permeability are expressed in percent and millidarcies, respectively. The other properties are in volume percent.

	Well	Depth	Mono. Qtz.	PolyQtz.	Feld.	Shale	Glauc.	Colloph.	Unk./Other	Clay lam.	Clay Infilt.	Carb. Cmt.	Carb. Replac.	Kaol.
1	Berquist 2	5125.6	38.4	36.2	0.4	5.4	0.2	0.0	0.2	0.0	0.0	0.5	0.0	4.0
2		5115.5	36.1	34.9	0.4	5.7	0.2	0.0	0.8	6.1	0.4	1.9	0.2	3.0
3	MEAN		37.3	35.6	0.4	5.6	0.2	0.0	0.5	3.1	0.2	2.4	0.1	3.5

	Qtz. Cmt.	Tot. Cmt.	Intergran. Por.	Intragran./Moldic Por.	Tot. Por.	Mean Grain Size	IGV	Macro/Micro	Core Por.	Perm.
1	2.7	7.4	8.3	3.4	11.7	840	15.7	1.2	11.5	77.0
2	1.5	6.6	4.9	3.8	8.7	423	11.3	0.9	11.7	129.0
3	2.1	7.0	6.6	3.6	10.2	631	13.5	1.1		

Table B6 - Point-count data for the Stockholm Tidal Channel Sandstone Facies IID. Depth is in feet. Mean grain size (µm) is the mean of the apparent long axes of 25 quartz grains measured optically in thin section. IGV is intergranular volume (%), which is the sum of integranular cement plus intergranular porosity. Macro/micro is the ratio of macroporosity to the rock volume with microporosity. Core porosity and permeability are expressed in percent and millidarcies, respectively. The other properties are in volume percent.

	Well	Depth	Mono. Qtz.	PolyQtz.	Feld.	Shale	Glauc.	Colloph.	Unk./Other	Clay lam.	Clay Infilt.	Carb. Cmt.	Carb. Replac.	Kaol.	Qtz. Cmt.
1	White AB-1	5167.0	56.2	2.6	2.8	2.1	1.6	0.2	1.2	0.5	0.0	29.7	3.2	0.0	0.0
2		5159.0	48.5	2.0	10.0	7.3	0.7	0.0	2.2	3.4	0.0	22.9	2.0	0.0	1.0
3	MEAN		52.4	2.3	6.4	4.7	1.2	0.1	1.7	2.0	0.0	26.3	2.6	0.0	0.5

	Tot. Cmt.	Intergran. Por.	Intragran./Moldic Por.	Tot. Por.	Mean Grain Size	IGV	Macro/Micro
1	32.9	0.0	0.0	0.0	106	29.7	0.0
2	25.9	0.0	0.0	0.0	78	23.9	0.0
3	29.4	0.0	0.0	0.0	92	26.8	0.0

Table B7 - Point-count data, Stockholm Marine Shelf Sandstone Facies III. Mean grain size (µm) is the mean of the apparent long axes of 25 quartz grains measured optically in thin section. IGV is intergranular volume (%), which is the sum of integranular cement plus intergranular porosity. Macro/micro is the ratio of macroporosity to the rock volume with microporosity. Core porosity and permeability are expressed in percent and millidarcies, respectively. The other properties are in volume percent.

DOMINANT CONTROL ON RESERVOIR-FLOW BEHAVIOR IN CARBONATE RESERVOIRS AS DETERMINED FROM OUTCROP STUDIES*

R. K. Senger, F. J. Lucia, C. Kerans, and M. A. Ferris

Bureau of Economic Geology
The University of Texas at Austin
Austin, Texas 78713-7508

G. E. Fogg

Department of Land, Air, and Water Resources
University of California at Davis
Davis, California 95616

I. INTRODUCTION

The investigation of carbonate-ramp deposits of the upper San Andres Formation that crop out along the Algerita Escarpment, New Mexico, is a research element of ongoing geologic and petrophysical studies conducted at the Bureau of Economic Geology's Reservoir Characterization Research Laboratory (RCRL). The primary goal of the investigation is to develop an integrated strategy involving geological, petrophysical, geostatistical, and reservoir-simulation studies that can be used to better predict flow characteristics in analogous subsurface reservoirs. Geologic investigations and detailed measurements of petrophysical parameters on continuous outcrop were used to determine not only the vertical distribution of the data but also their lateral distribution, which is typically lacking in subsurface studies.

To characterize the complex heterogeneity associated with depositional and diagenetic processes at the interwell scale, detailed permeability data were collected within the overall geologic framework from the outcrop at Lawyer Canyon, Algerita Escarpment, New Mexico (fig. 1). Geologic mapping showed a series of upward-shallowing parasequences (10 to 40 ft thick and several thousand feet long). Parasequence boundaries are typically marked by tight mudstone/wackestone beds that display variable degrees of lateral continuity ranging from several hundred feet to more than 2,500 ft and are potentially important as flow barriers (fig. 2). Within these parasequences, distinct variability of facies and petrophysical characteristics is present at scales well below those of interwell spacing typical for their subsurface counterparts (660 to 1,330 ft). Pore types and permeability-porosity relationships can also be specific to individual parasequences.

*Publication authorized by the Director, Bureau of Economic Geology, The University of Texas at Austin.

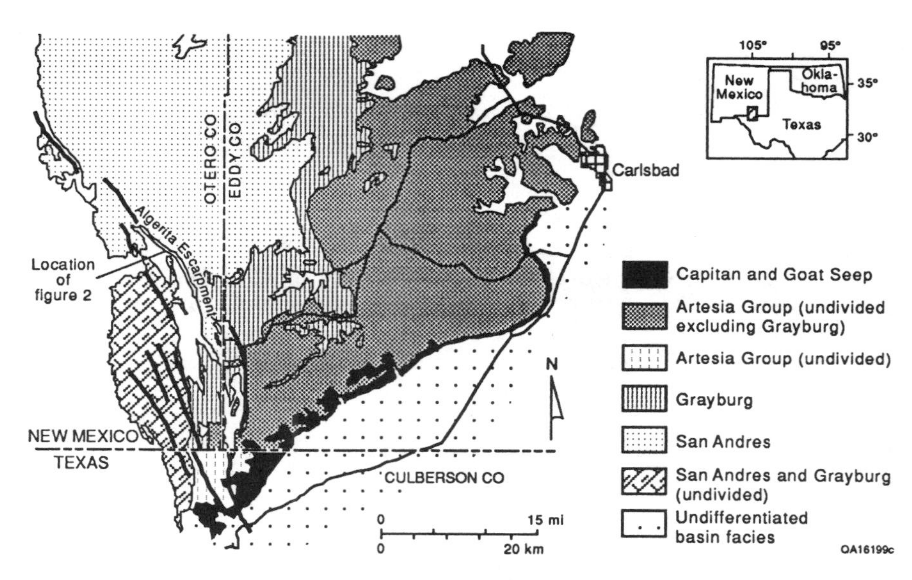

Figure 1. Geologic map of the Guadalupe Mountains compiled from Hayes (1964) and King (1948) showing the Algerita Escarpment and the Lawyer Canyon study area.

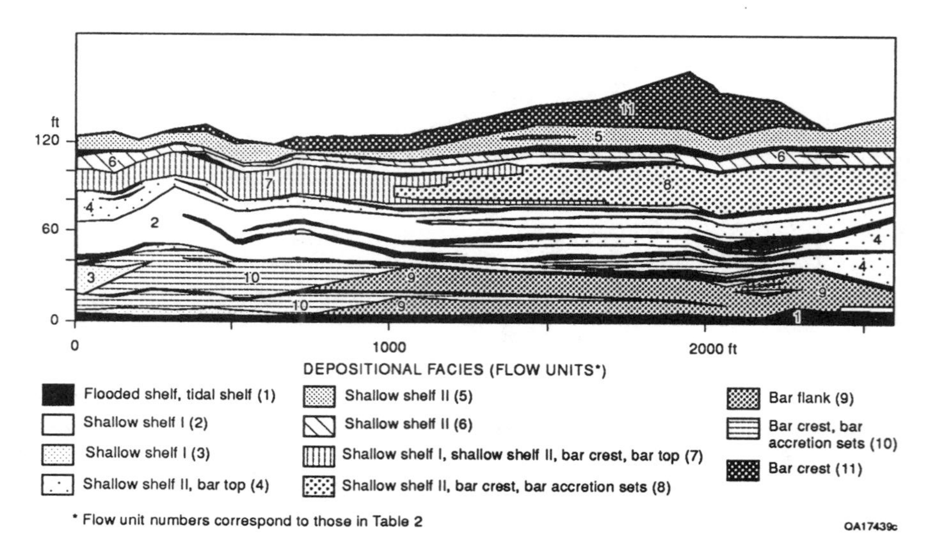

Figure 2. Distribution and geometry of depositional facies mapped in the upper San Andres parasequence window, Lawyer Canyon, Algerita Escarpment, New Mexico.

Previous studies characterizing permeability in San Andres outcrops include those of Hindrichs and others (1986) and Kittridge and others (1990). The study by Hindrichs and others was based on core plugs taken from different beds in the middle San Andres Formation. Results indicated large permeability variations within beds separated by low-permeability mudstones, with no apparent difference in permeability between horizontal and vertical cores. Kittridge and others (1990) used a mechanical field permeameter (MFP) to study spatial permeability variations at a specific site near the middle and upper San Andres Formation boundary at the Lawyer Canyon area. Geostatistical analyses of measured permeability indicated that permeability correlation lengths decreased with decreasing sample spacing and that different rock fabrics exhibited different mean permeabilities.

In this study, permeability measurements were guided by the detailed geologic mapping of continuous San Andres outcrop, which revealed a stacking pattern of parasequences and of facies and rock-fabric succession within parasequences that provide the necessary framework for petrophysical quantification of geologic models. Standard statistics were used to relate permeability to facies and rock-fabric characteristics through a petrophysical/rock-fabric approach. In addition, geostatistical analysis was applied to evaluate spatial permeability characteristics. Stochastic modeling was then used to generate a series of "realistic" permeability distributions that took into account the underlying permeability structure and the uncertainty of measurement data. Numerical waterflood simulations of selected permeability realizations were designed to characterize interwell heterogeneity and to represent heterogeneity by appropriate average properties, which can be used in reservoir-scale flow models. For the reservoir-scale flow model, a petrophysical/rock-fabric approach was used to quantify the geologic framework of the reservoir-flow model.

II. METHODOLOGY

A. Permeability Measurements

Permeability was measured with an MFP, which gauges gas-flow rates and pressure drop by pressing an injection tip against the rock surface. These data are used to calculate permeability values on the basis of a modified form of Darcy's law that incorporates effects of gas slippage at high velocity (Goggin and others, 1988). In addition, permeability and porosity were determined on the basis of conventional Hassler sleeve methods, using 1-inch-diameter core plugs taken from the outcrop. Core and MFP permeability compared reasonably well for permeabilities greater than about 1 md, which is approximately the detection limit of MFP measurements (Goggin and others, 1988).

The distribution of permeability measurements taken from the upper San Andres at Lawyer Canyon is shown in figure 3. Sampling focused on parasequence 1, composed of grainstones forming bar-crest and bar-flank facies overlying wackestones and mudstones of a flooded-shelf facies, and on parasequence 7, characterized by low-moldic and highly moldic grainstones (fig. 2). Permeability distributions were measured at scales ranging from detailed grids of 1-inch spacing to 1-ft spacing and at vertical transects that were spaced laterally between 5 and 100 ft and that contained permeability measurements at 1-ft vertical intervals.

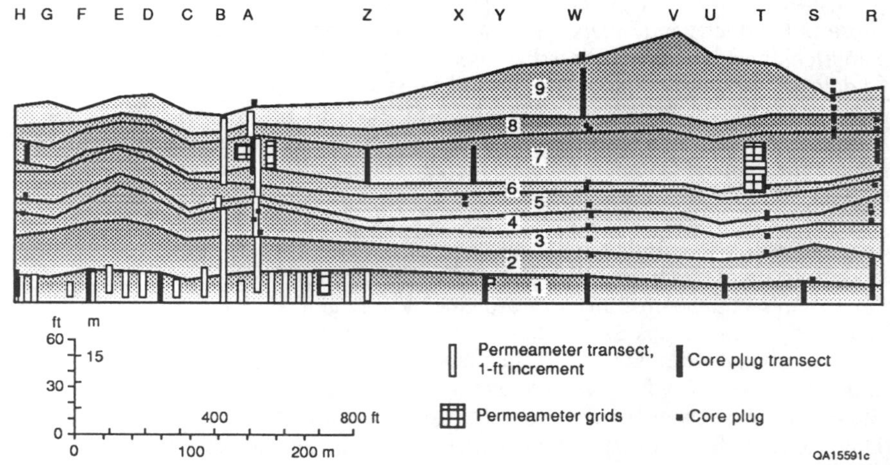

Figure 3. Location of sampling grids and transects of MFP measurements and core plugs.

The total number of MFP measurements at the Lawyer Canyon parasequence window was 1,584. Removing the outer weathering surface of the rock by chipping away an area of about 1 square inch gave the best representation of permeability (Ferris, in preparation). Preparing the sampling surface with a grinder produced permeabilities that were lower overall than those measured on chipped surfaces, owing to plugging of pore space by fines (Kittridge and others, 1990). Within each chipped area, typically several measurements were made and averaged. Depending on measurement discrepancies, as many as six different MFP readings were taken at various locations within the chipped area.

B. Geostatistics

Variography, a geostatistical technique for analyzing spatial variability of a property such as permeability, was used to help quantify the spatial permeability pattern at different scales. For further discussion of variogram analysis, refer to Journel and Huijbregts (1978) and Fogg and Lucia (1990). The variogram describes variability as a function of distance between measurements. The average variance of measurement pairs within certain distance intervals typically shows increasing variability (γ) with increasing interval range (fig. 4). Beyond a certain distance (range), γ may no longer increase. The variance that corresponds to the range is the sill, which reflects the variability where spatial correlation no longer exists; it typically corresponds to the ensemble variance of the entire data set. Small-scale heterogeneity or measurement errors can cause a variogram to originate at a high variance referred to as a nugget, representing local random variability.

Spatial permeability characteristics can often be described by the nugget, the correlation range, the sill, and the variogram model. The latter is obtained by fitting

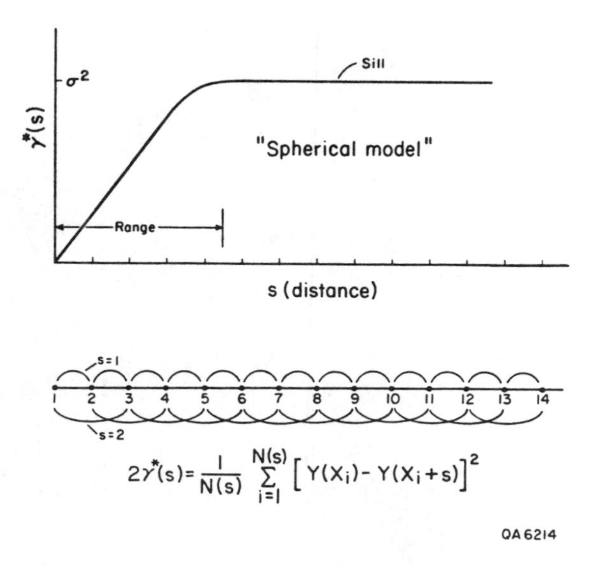

Figure 4. Schematic example of calculation of the experimental variogram.

a certain type of mathematical function to the experimental variogram. In this study the computer program GAMUK (Knudsen and Kim, 1978) was used to compute the experimental variogram. Application of the variogram to kriging or to conditional simulation usually requires an assumption of stationarity, which requires that the mean and the variogram be the same over the area of interest.

Kriging is a technique of estimating properties at points or blocks distributed over the area of interest by taking a weighted average of sample measurements surrounding a regularly spaced grid point or block. Kriging incorporates the spatial correlation structure contained in the variogram model. The kriging program is based on the program UKRIG, developed by Knudsen and Kim (1978). The permeability data were contoured with the CPS-1 contouring package (Radian Corp., 1989) and with the DI-3000 contouring package at The University of Texas Center for High Performance Computing.

Conditional simulation uses the underlying permeability structure obtained from kriging and adds the stochastic component associated with the uncertainty of the limited permeability data. Conditional simulation is performed with the program SIMPAN (Fogg, 1989). A large number of permeability realizations are screened for maximum and minimum continuity of permeable zones using the program MCSTAT (Fogg, 1989), which measures the length and thickness of domains of contiguous blocks having simulated permeability values greater than 50 md. The selection of realizations having low and high continuity of permeable zones is based on the mean horizontal continuity ($\overline{C}_h$) at the 10th and 90th percentiles, respectively. These end-member representations of "realistic" permeability distributions, conditioned on the same permeability data, are then used in waterflood simulations to evaluate

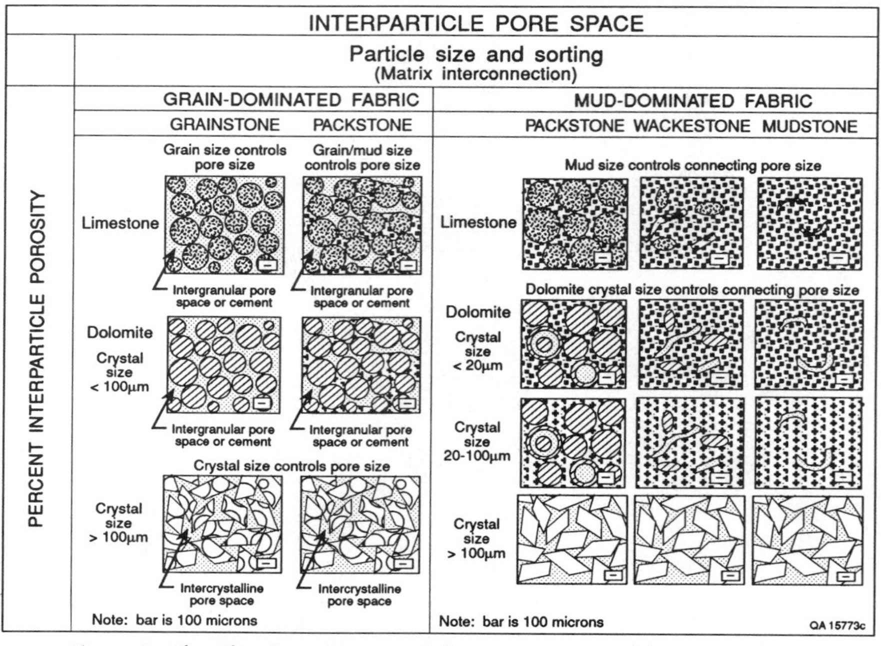

Figure 5. Classification of interparticle pore space in carbonate rocks.

reservoir-flow characteristics. For further details on stochastic reservoir simulations, refer to Fogg and others (1991). The reservoir simulator ECLIPSE (ECL Petroleum Technologies, 1990) was used in this study for two-phase waterflood simulations.

III. PERMEABILITY CHARACTERIZATION, LAWYER CANYON OUTCROP

A. Petrophysical/Rock-Fabric Approach

Petrophysical parameters such as porosity, permeability, and saturation were quantified by relating rock texture to pore-type and pore-size distribution (fig. 5). Spatial distribution of these petrophysical parameters was defined by geologic concepts of sedimentation, diagenesis, and tectonics, which provided the basis for the three-dimensional reservoir framework. Pore space in carbonate rocks can be divided into interparticle and vuggy pores on the basis of the particulate nature of carbonate rocks (Lucia, 1983). Three petrophysical/rock-fabric classes for nonvuggy pore space can be distinguished on the basis of porosity, permeability, and saturation relationships (fig. 6), and vuggy pore space can be divided into separate vugs (moldic, intrafossil, etc.) and touching vugs (cavernous, fractures, etc.) on the basis of the type of interconnection. The petrophysical classes for nonvuggy pores have distinct porosity-permeability relationships (fig. 7a), as well as relationships between porosity and water saturation (fig. 8).

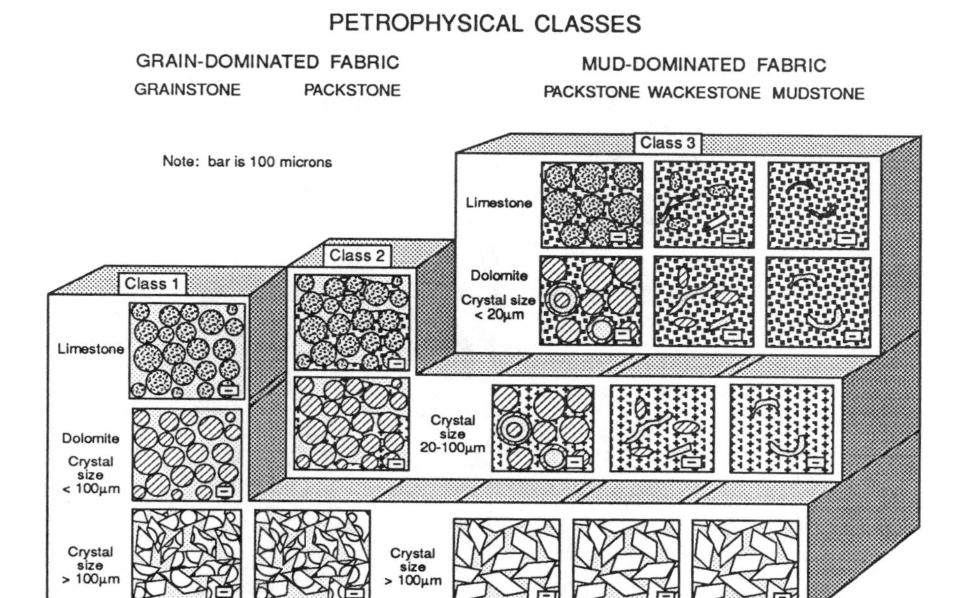

Figure 6. Petrophysical/rock-fabric classes for nonvuggy carbonates.

In addition to the nonvuggy pore types, separate vugs were important to this study. Separate vugs, such as moldic pores, are connected only through interparticle pore space and result in a lower permeability than that which would be expected if the porosity were all interparticle; that is, they do not fall into the different porosity-permeability field for nonvuggy pore types (fig. 7b). However, vuggy pore space is typically large enough that it can be assumed to be filled with hydrocarbons.

According to this rock-fabric classification, five productive rock-fabric units were recognized in the outcrop at the Lawyer Canyon parasequence window (fig. 9): grainstones, separate-vug grainstones, grain-dominated packstones, mud-dominated packstones and wackestones, and tight mudstones and fenestral caps. The information was derived from permeability and porosity measurements of core plugs and from a petrographic description of thin section taken from core plugs.

For the evaluation of permeability characteristics with respect to facies and fabrics, only those data were taken that follow the geologic measured sections for which spatial coordinates, facies, and fabric designations are available. The histogram of permeabilities using both core and MFP measurements (fig. 10) shows a roughly lognormal distribution. Core permeabilities are skewed toward lower permeabilities than are the MFP measurements, owing to the 1-md detection limit of the MFP data. Within relatively permeable facies (e.g., the bar-crest facies within parasequence 1), core and MFP measurements have statistically similar populations,

(a)

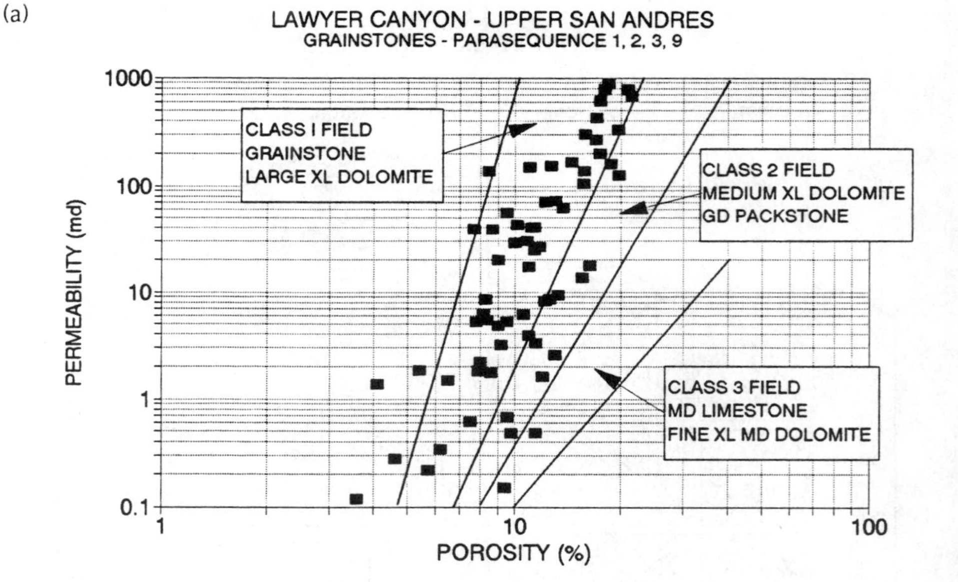

(b)

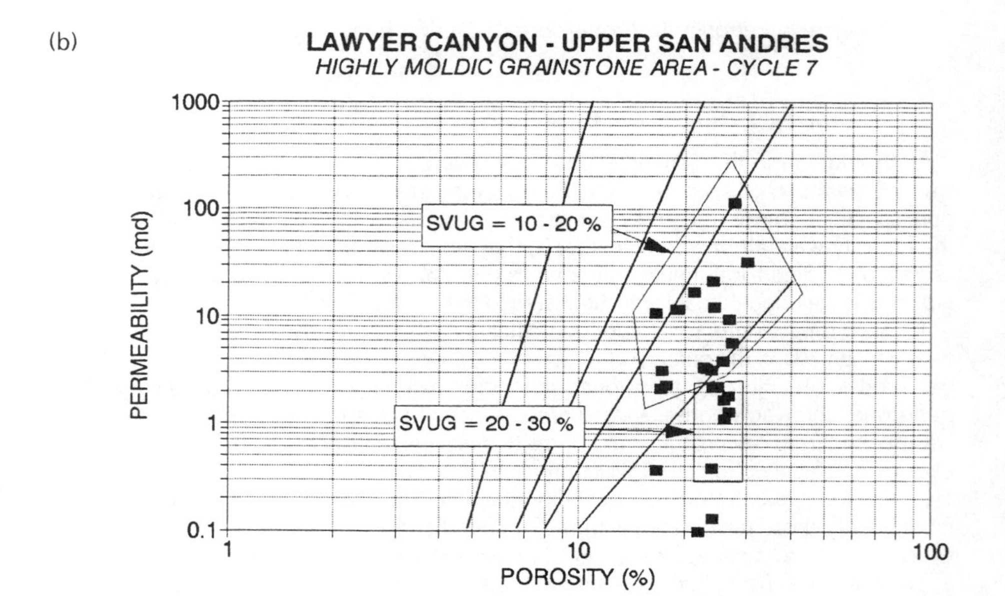

Figure 7. Porosity-permeability relationships for various rock-fabric fields in nonvuggy carbonates. Data are from core-plug measurements of (a) nonvuggy grainstones and (b) highly moldic grainstones from the Lawyer Canyon outcrop.

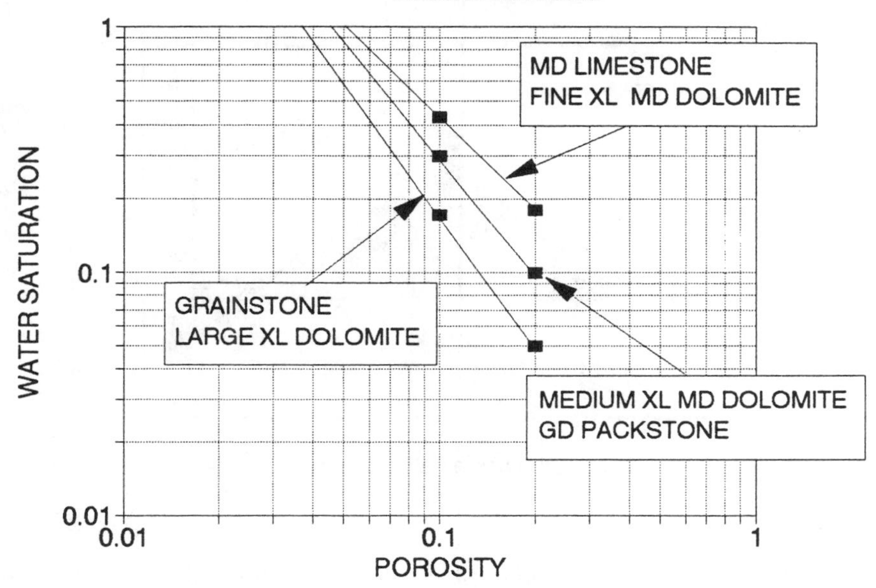

Figure 8. Rock-fabric, porosity, and water-saturation relationships from capillary-pressure curves.

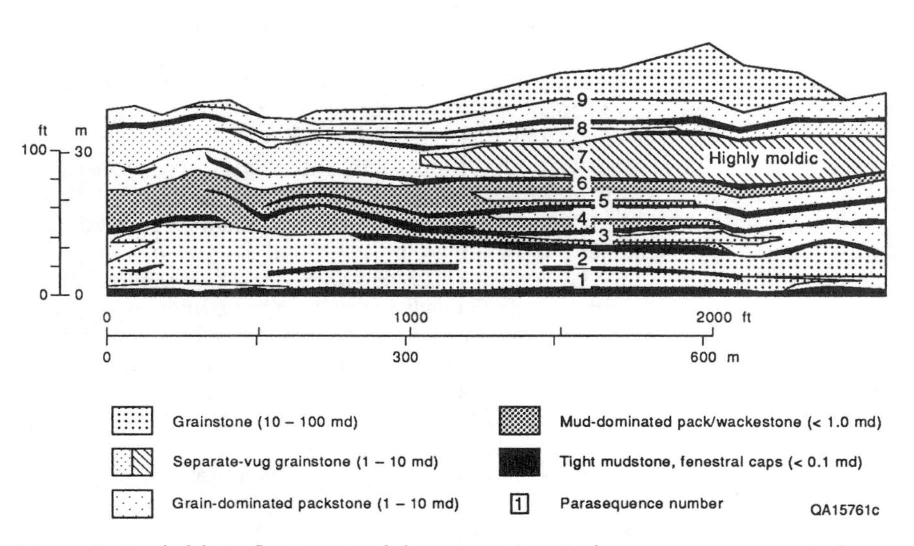

Figure 9. Rock-fabric flow units of the upper San Andres parasequence window, Lawyer Canyon, Algerita Escarpment, New Mexico.

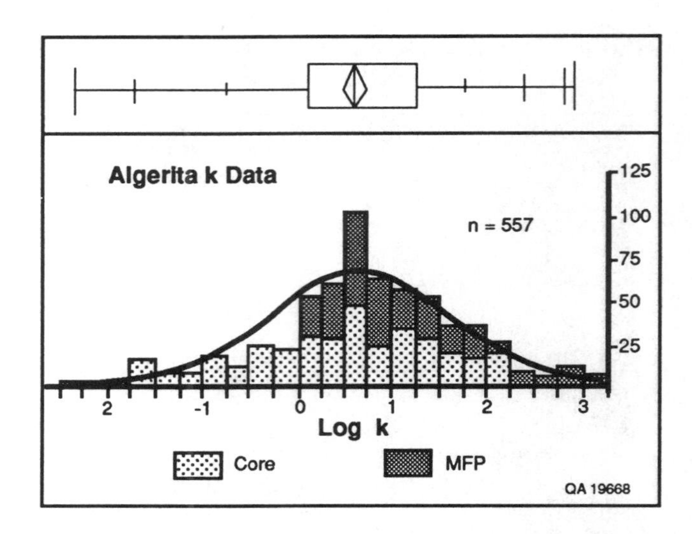

Figure 10. Histogram of permeabilities obtained along the geologically measured sections.

with similar geometric means of 1.33 and 1.34 md, respectively. Statistical comparison of the two populations using a t-test indicates that the null hypothesis cannot be rejected at the 80 percent confidence level (t-statistics = –0.1318; p-value = 0.8953).

The dominant rock fabrics exhibit significant differences in mean permeability (fig. 11), with mudstone having the lowest permeability and grainstone having the highest permeability. Most of the mapped facies are also characterized by significantly different mean permeabilities (fig. 12). Generally, shelf facies exhibit significantly lower mean permeabilities than bar facies, with the bar-crest and bar-accretion-set facies having the highest mean permeability of log k = 1.1 md. The facies characteristic (fig. 12) is consistent with the rock-fabric characteristic (fig. 11) because the bar facies consist mostly of high-permeability grainstones and the shelf facies consist mostly of low-permeability, mud-dominated fabrics. Furthermore, the mapped parasequences are characterized by different facies and fabric combinations, indicating different hydraulic properties of each parasequence. Mean permeabilities in parasequences 1, 2, 7, and 9 are significantly higher than those in parasequences 3, 4, 5, 6, and 8. The latter sequences consist mostly of packstone and wackestone, whereas parasequences 1, 2, 7, and 9 consist predominantly of grainstones (fig. 9). Grainstones in parasequence 7 are characterized by moldic-pore types with high porosity but slightly lower permeability than those in the other cycles having intergranular porosity.

Within individual facies, permeability varies by as much as five orders of magnitude (fig. 12). Characterization of spatial permeability patterns within individual facies is therefore important for predicting flow behavior in these ramp-crest grainstone bar complexes. If permeability within facies is spatially uncorrelated (i.e., random), then the two-dimensional effective permeability of that facies can be

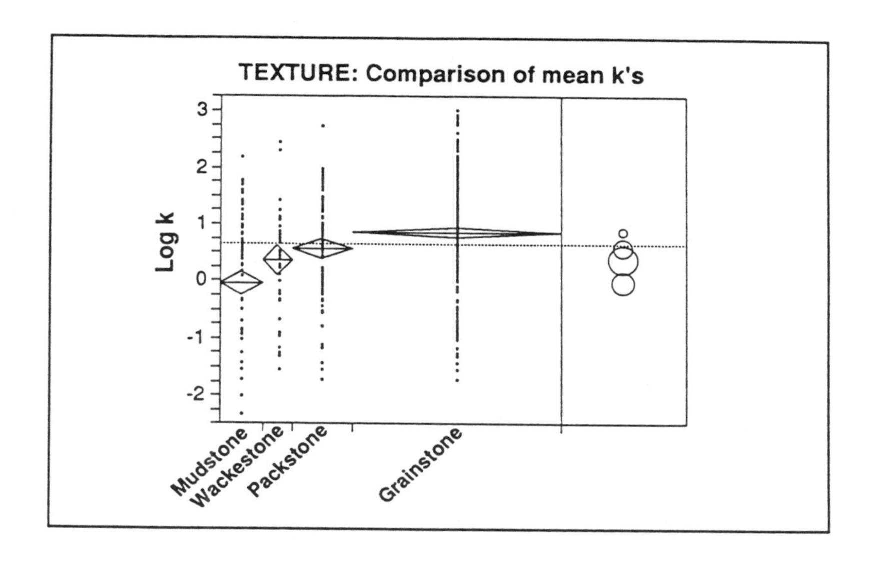

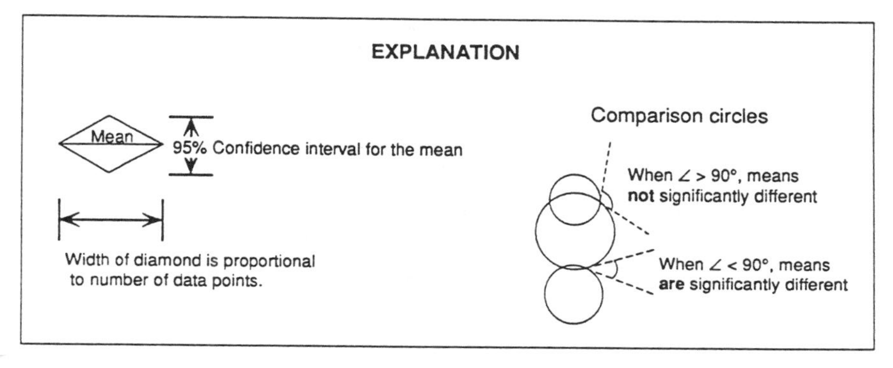

Figure 11. Statistical comparison of means of permeability for different textures. Comparison circles correspond to the 95 percent confidence intervals for the means.

estimated by taking the geometric average of the local permeabilities (Warren and Price, 1961). If permeability within facies exhibits significant spatial correlation, then effective permeability of that facies must be estimated by taking some other type of average.

B. Spatial Permeability Patterns and Variography

Spatial patterns in permeability were characterized and mapped in three steps. First, the data were contoured with an inverse-distance-squared algorithm to depict any trends or anisotropies in the data. Second, variograms were computed for

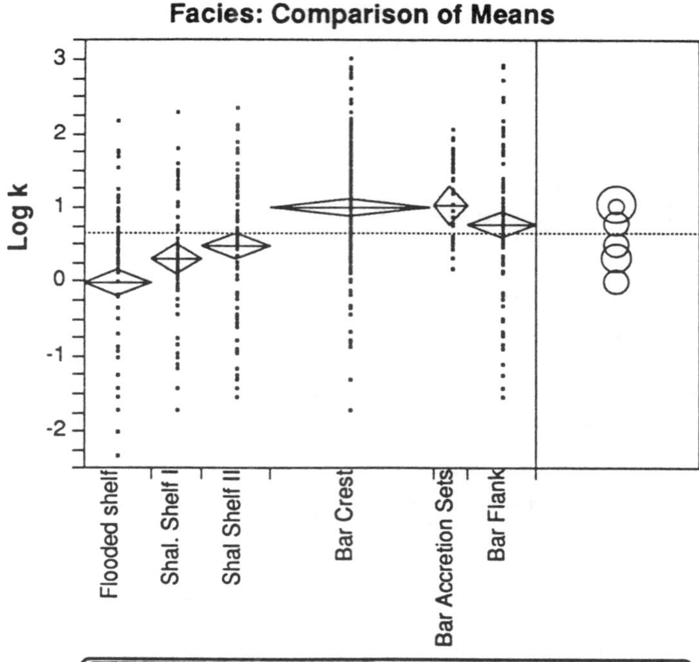

Mean Estimates

Facies	N	Mean	Std Error
Flooded shelf	82	-0.01	0.10
Shallow shelf I	48	0.25	0.13
Shallow shelf II	73	0.51	0.11
Bar crest	158	1.07	0.07
Bar top	6	0.10	0.37
Tidal flat	2	0.37	0.65
Bar accretion sets	18	1.11	0.22
Bar flank	82	0.76	0.10
Shallow shelf I (moldic)	16	0.45	0.23
Shallow shelf II (moldic)	9	0.26	0.30
Bar crest (moldic)	39	0.73	0.15
Bar accretion sets (moldic)	24	0.98	0.19

Figure 12. Statistical comparison of means of permeability for different facies. Standard error values refer to the estimated means.

different lag spacings and directions that were consistent with the data spacings and inverse-distance maps. Third, variogram models were fit to the variograms and were used to create point-kriged maps of spatial permeability patterns.

Standard contouring (inverse-distance-squared) of the detailed permeability transects spaced between 25 and 100 ft in parasequence 1 (fig. 2), using the CPS-1

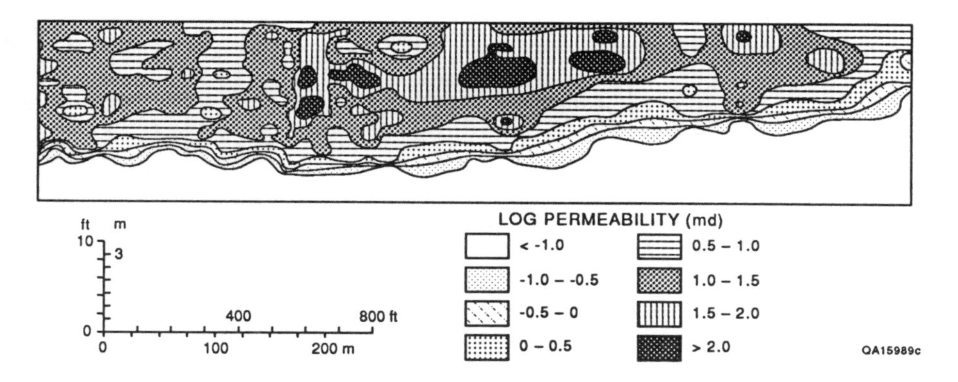

Figure 13. Permeability distribution for normalized parasequence 1 based on inverse-distance-squared contouring algorithm using CPS-1.

contouring package, shows extreme heterogeneity (fig. 13) within the bar-crest facies, bar-flank facies, and shallow-shelf facies, which are collectively referred to as the grainstone facies of parasequence 1.

To evaluate heterogeneity at different scales, permeability measurements from the different measurement grids were analyzed using variography. Within the bar-crest facies of parasequence 1, permeability transects were typically spaced 25 ft apart. Between transects A and Z (fig. 3), vertical transects were spaced 5 ft apart. Horizontal and vertical variograms of the permeability data indicate a short-range correlation range of about 3 ft in the vertical direction and a possible correlation range of about 30 ft in the horizontal direction (fig. 14a, b). In both cases, however, the spherical variogram indicates nugget constants of σ_0 = 0.15 md^2 and σ_0 = 0.35 md^2, respectively. This relatively large nugget indicates small-scale random variability of permeability. The small-scale random permeability variation is apparent in permeability patterns of the smaller scale grids, which were measured at regular 1-ft spacings (fig. 15) and at 1-inch spacings (not shown here). Variogram analysis of these small-scale permeability grids did not indicate a noticeable permeability correlation but showed a large variability in permeability. Measurement accuracy of the minipermeameter typically decreased toward the lower detection limit of 1 md and may have accentuated some of the observed noise in the permeability data. Extending the log spacing of the variogram (fig. 14c) shows the large-scale permeability pattern, characterized by nested structures. The range of these nested structures of about 400 ft is reflected in the overall permeability pattern shown in figure 13.

Using the fitted variogram models (fig. 14a, b), a kriged permeability map was constructed for the northern half of parasequence 1, consisting predominantly of bar-crest facies. Note that the kriged permeability map (fig. 16), based on the vertical permeability transects spaced 25 ft apart, shows a much smoother distribution than does the kriged permeability map based on the 1-ft grid (fig. 15).

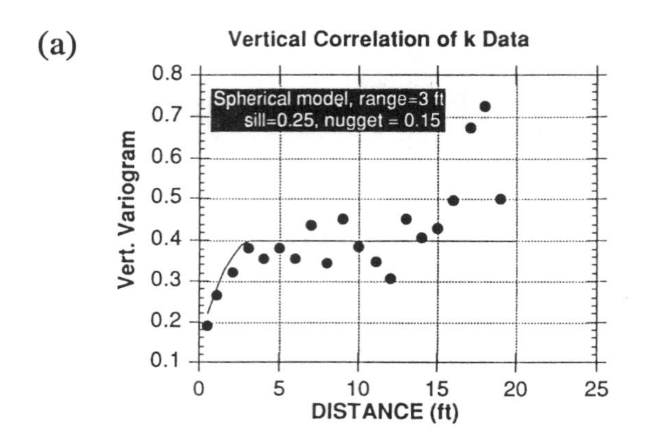

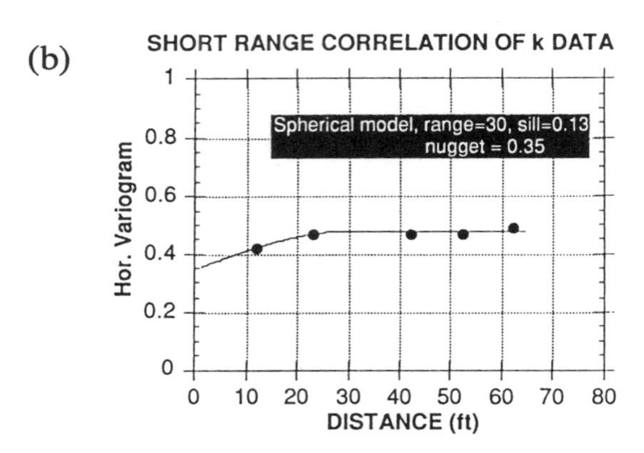

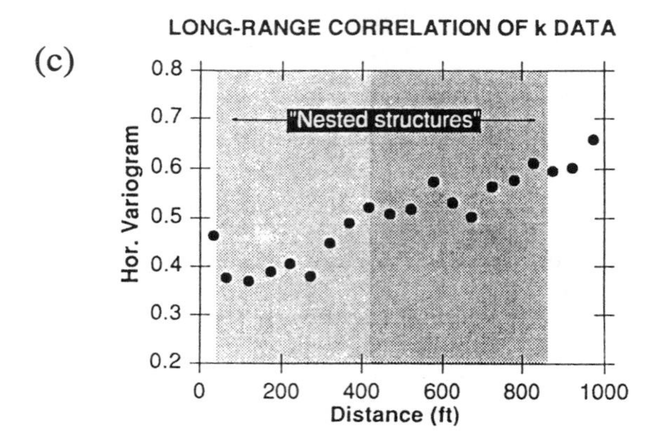

Figure 14. Sample variograms for the permeability transects from the grainstone facies in parasequence 1: (a) vertical variogram based on all transects, (b) short-range horizontal variogram, based on the 5-ft grid between sections A and Z (fig. 3), and (c) long-range horizontal variogram, based on all transects spaced between 25 and 100 ft.

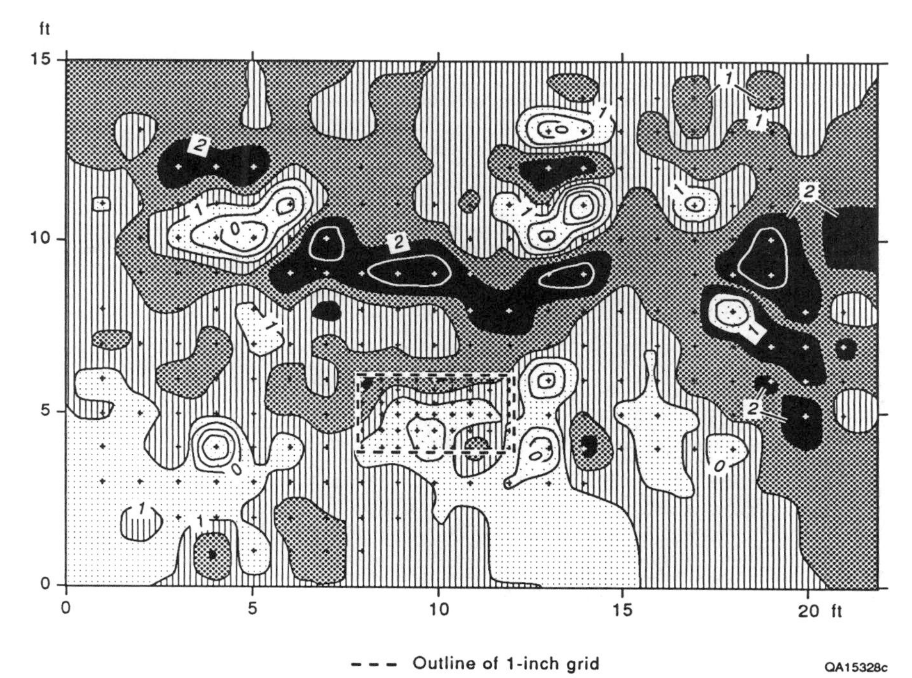

Figure 15. One-foot sampling grid in parasequence 1 with kriged permeability contours.

C. Conditional Simulation

Even though kriging can incorporate permeability correlation structures, it tends to average permeability over larger areas, ignoring small-scale heterogeneity. On the basis of the short-range correlation of permeability data (fig. 14), a series of stochastic permeability realizations were produced for the grainstone facies in parasequence 1. The model extends laterally from 0 to 1,050 ft and is 17 ft thick, with block sizes of 5 ft by 1 ft. The simulations, conditioned to permeabilities measured along vertical transects spaced approximately 25 ft apart (fig. 3), incorporate the correlation structure from the variograms (fig. 14a, b).

Two permeability distributions (realization nos. 7 and 11, table 1) out of 200 stochastic permeability realizations were selected for flow simulations, representing maximum and minimum lateral continuity (C_h) of domains having permeability values greater than 50 md (table 1). Comparison of the two permeability realizations (figs. 17a, b) does not show a noticeable difference. The ranges of 3 ft (vertical) and 30 ft (horizontal) (figs. 14a, b) are not immediately apparent in these realizations (fig. 17); the permeability patterns appear spatially uncorrelated because of the relatively large nugget, which is about the same magnitude as the sill (fig. 14). These conditional simulated realizations, however, preserve the spatial variability

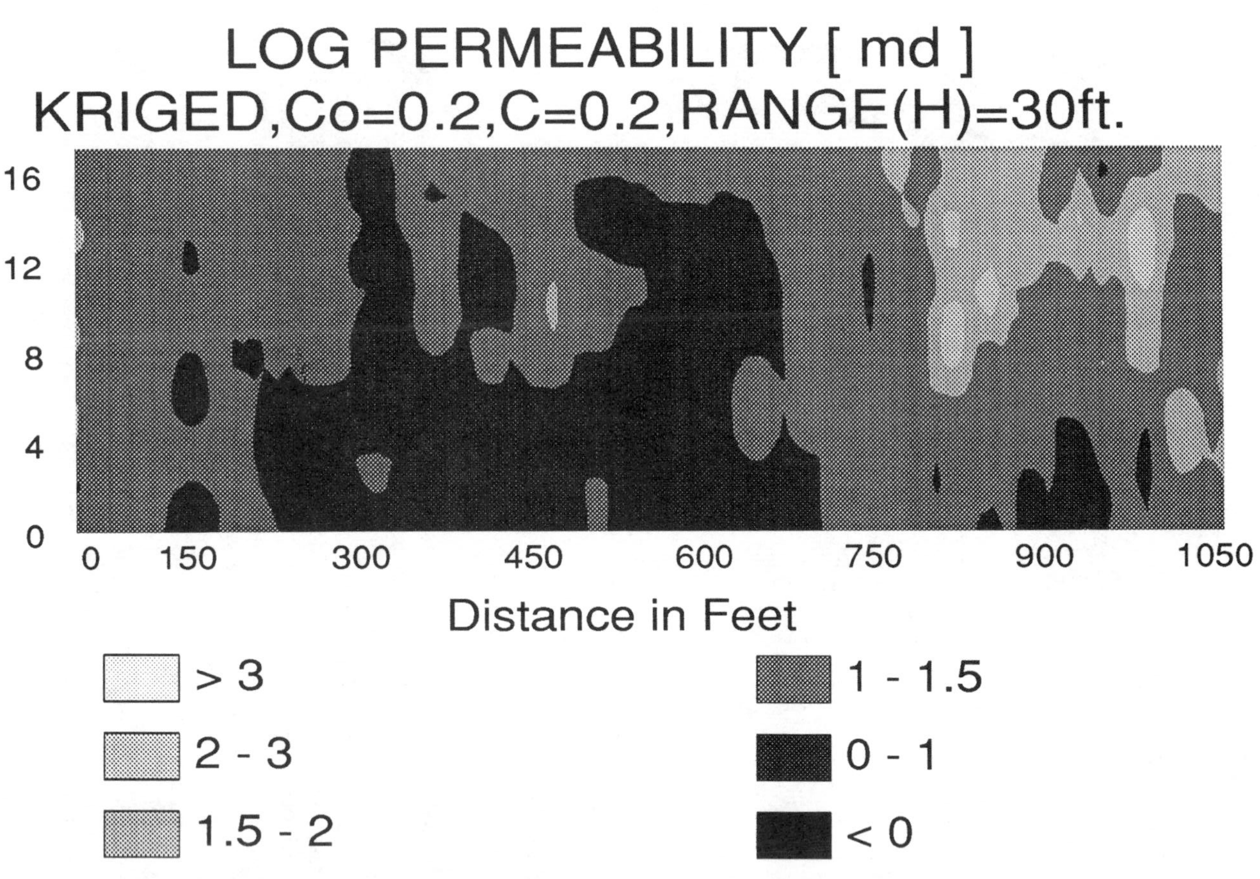

Figure 16. Kriged permeability distribution of the left part of parasequence 1, representing the bar-crest facies.

Table 1. Input parameters for waterflood simulations of grainstone facies in parasequence 1.

No.	Permeability realization	Statistics: Mean log-k (md)	Variance log-k (md^2)	Nugget (md^2)	Sill (md^2)	Mean horizontal continuity ($\overline{C}_h$) (ft)
1*	7	1.219	0.42	0.2	0.2	7.53
2	7	1.219	0.42	0.2	0.2	7.53
3	11	1.219	0.45	0.2	0.2	8.82
4	45	1.219	0.38	0.0	0.4	11.95
5	kriged	1.219	0.46	0.2	0.2	N/A
6*	facies-averaged	1.219	N/A	N/A	N/A	N/A

*Single relative-permeability and capillary-pressure curves.

Fluid properties:
Oil viscosity 1.000 cP
Oil density 55 lb/ft^3
Water density 64 lb/ft^3

exhibited in the variograms, whereas the kriged permeability map (fig. 16) averages out much of this variability.

As mentioned above, measurement uncertainty of the minipermeameter may have accentuated some of the observed noise in the permeability data, which is reflected in the relatively large nugget of the variograms (fig. 14). For comparison, unconditional permeability realizations were produced that incorporate the mean, variance, and variogram range but have a zero nugget and a sill of 0.4 md^2 and are not conditioned to the actual permeability values. All of these unconditional permeability realizations have mean continuity values that are higher than those of the conditional permeability realizations based on a 0.2-md^2 nugget. The unconditional realization 45 (fig. 18), which was selected on the basis of low continuity of relatively permeable zones, shows a smoother permeability pattern than that of a conditional permeability realization having a relatively high nugget (fig. 17). For comparison of waterflooding results, the different permeability realizations were corrected to the same mean permeability, which was equivalent to the data mean of log k = 1.219 md (table 1).

IV. WATERFLOOD SIMULATIONS OF PARASEQUENCE 1

A. Modeling Strategy

Waterflooding of the hypothetical two-dimensional reservoir was simulated by injecting water along the right boundary and producing along the left boundary. Injection and production were controlled by prescribed pressure conditions of 2,450 psi and 750 psi, respectively. A series of numerical simulations was performed to evaluate different effects associated with the observed heterogeneity on production characteristics. Flow simulations incorporating the observed heterogeneity were compared with those using a mean permeability to evaluate whether the observed

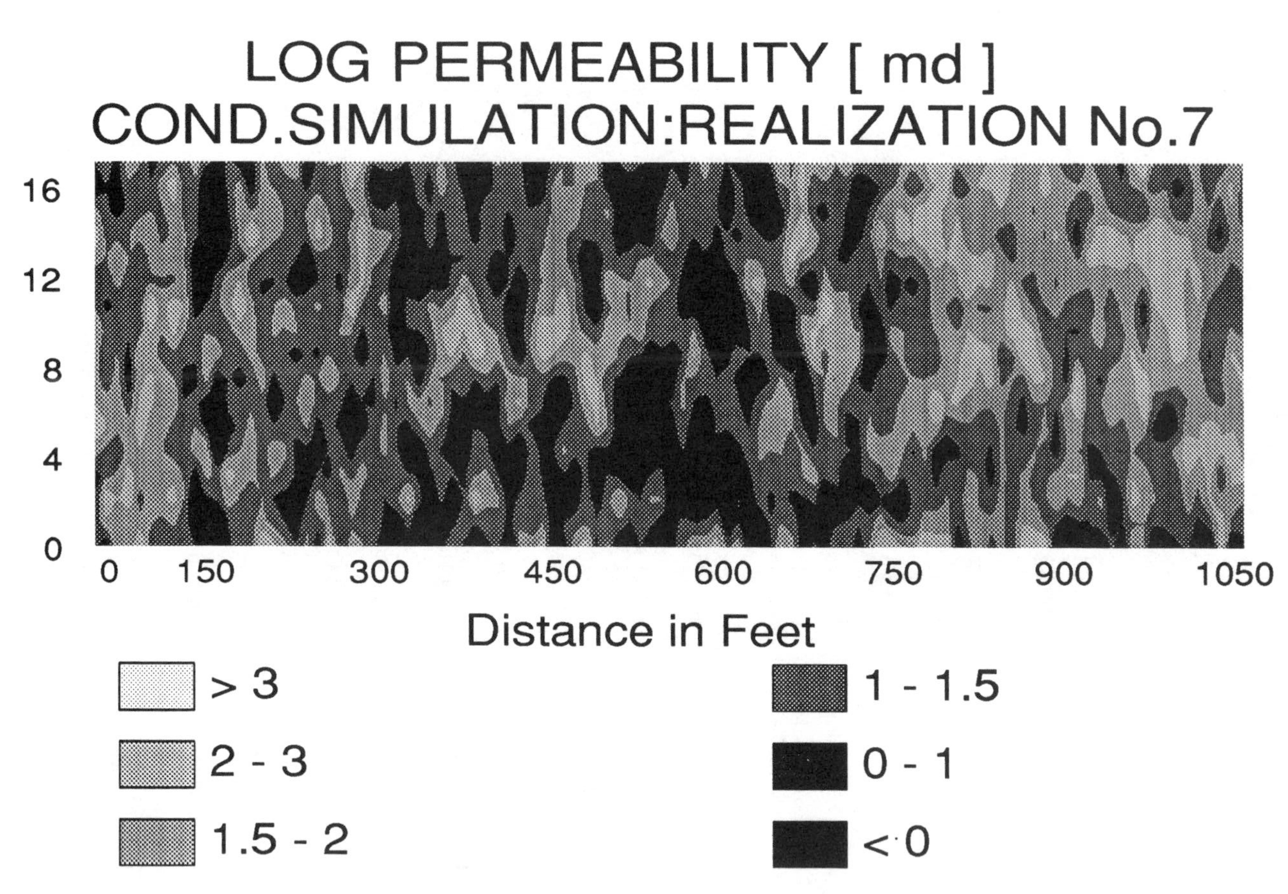

Figure 17. Conditional permeability distributions of (a) realization 7, representing lower continuity, and (b) realization 11, representing higher continuity of high-permeability zones.

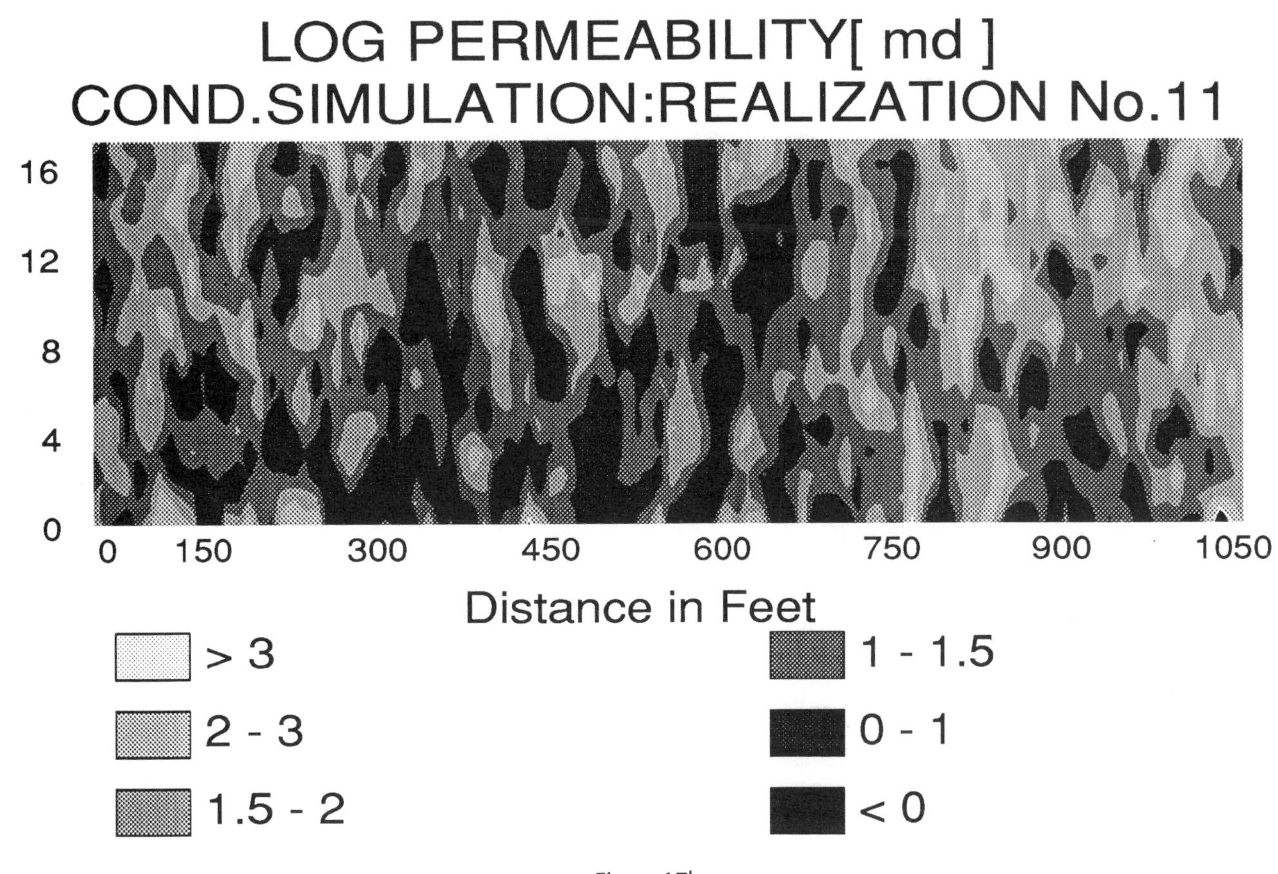

LOG PERMEABILITY[md]
COND.SIMULATION:REALIZATION No.11
16
12
8
4
0
0
150
300
450
600
750
900
1050
Distance in Feet
> 3
2 - 3
1.5 - 2
1 - 1.5
0 - 1
< 0

Figure 17b

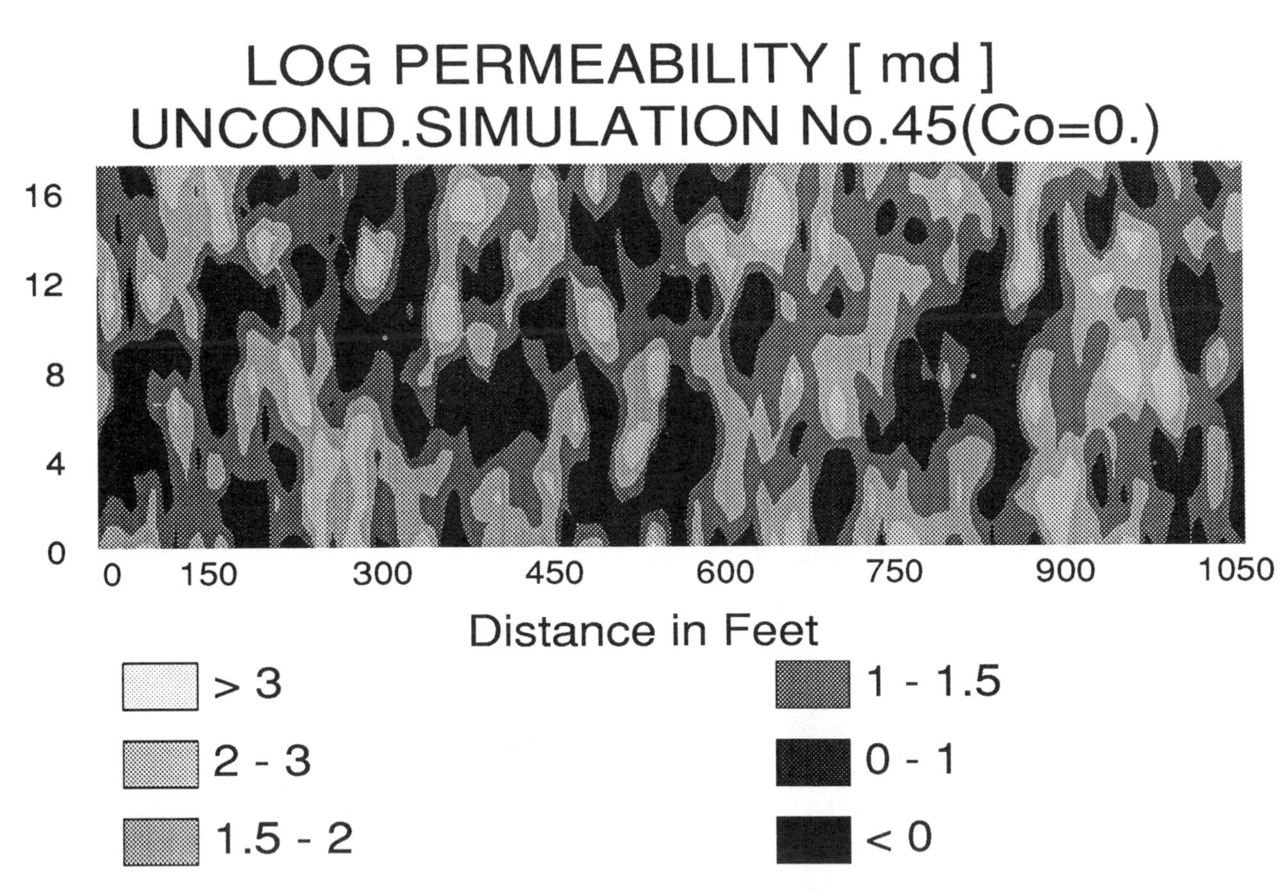

Figure 18. Unconditional permeability distribution of realization 45, representing low continuity of high-permeability zones.

permeability heterogeneity could be represented by a geometric-mean permeability and to determine the possible impact of short-range permeability correlation on production characteristics.

Input data for the simulator included the stochastic permeability distributions, porosities, and relative-permeability and capillary-pressure curves. Porosity-permeability relationships established on the basis of core-plug analyses for grainstones in parasequence 1 were used to calculate porosity distributions from the stochastic permeability realizations. The following empirical porosity-permeability relation is based on a linear transform representing intergranular pore characteristics in a grainstone (fig. 7a):

$$k = (5.01 \times 10^{-8})\, \phi^{8.33} \tag{1}$$

where k is absolute permeability (md) and ϕ is intergranular porosity (fraction). Similarly, an empirical relationship between water saturation, intergranular porosity, and capillary pressure, established for grainstones, was used to calculate capillary pressure as a function of water saturation and porosity of the grainstone facies in parasequence 1:

$$S_w = 68.581\, h^{-0.316}\, \phi^{-1.745} \tag{2}$$

where h represents capillary pressure expressed as the height of the reservoir above the free-water level. Initial water saturation as a function of porosity was computed using equation (2), assuming that the hypothetical reservoir is 500 ft above the free-water level. Residual oil saturation was assumed to be uniform at 25 percent.

The relative-permeability functions for oil and water were determined from the following equation (Honarpour and others, 1982):

$$k_{rw} = k_{rw}^{0} \left(\frac{S_w - S_{wr}}{1 - S_{or} - S_{wr}} \right)^{N_w} \tag{3a}$$

$$k_{ro} = k_{ro}^{0} \left(\frac{1 - S_w - S_{wr}}{1 - S_{or} - S_{wr}} \right)^{N_o} \tag{3b}$$

where S_{or} and S_{wr} are the residual oil saturation and the residual water saturation, respectively. The exponents N_w and N_o were derived from fitting relative permeability data obtained from grainstone fabric of two Dune Grayburg field cores. Both exponents were approximately 3 and were determined from the slope of the regression line representing the log of relative permeability versus the log of the normalized saturations in equations (3a) and (3b). Similarly, the relative-permeability endpoints k_{rw}^{0} and k_{ro}^{0} were derived from the intercepts of the log-log plots of the measured relative-permeability data versus saturation, which were 0.266 and 0.484, respectively, and correspond to residual oil saturation $S_{or} = 0.25$ and residual water saturation $S_{wr} = 0.1$, respectively.

Porosities derived from the different permeability realizations through the permeability-porosity transform in equation (1) typically range between 5 and 25 percent. Four relative-permeability and capillary-pressure curves representative of four porosity intervals were used: 5 to 10 percent, 10 to 15 percent, 15 to 20 percent, and 20 to 25 percent. For these porosity intervals, residual water saturations calculated from equation (2) were used to compute the relative-permeability curves according to equation (3). The relative-permeability and capillary-pressure curves used in the different flow simulations are shown in figures 19 and 20, respectively.

Six numerical simulations were run (table 1). Simulation nos. 1 and 6 used a single relative-permeability and a single capillary-pressure curve based on the arithmetic mean of porosity. The other simulations incorporated porosity-dependent relative-permeability and capillary-pressure curves (figs. 19, 20). The different simulations include (nos. 1 and 2) conditional permeability realization 7 (fig. 17a), representing low continuity of permeable zones; (no. 3) conditional permeability realization 11 (fig. 17b), representing high continuity of permeable zones; (no. 4) unconditional permeability realization 45 (fig. 18), representing low continuity of permeable zones assuming zero nugget; (no. 5) kriged permeability distribution (fig. 16); and (no. 6) uniform permeability distribution based on the geometric mean of measured permeability.

Computed water saturations of all simulations exhibited relatively sharp and vertical injection fronts despite the large variations in permeability, initial saturations, and capillary pressures in some simulations.

B. Simulation Results

Initial water saturation in simulation 1 (fig. 21) is uniform calculated from equation (2), using the arithmetic mean of porosity and assuming a reservoir height above the free-water level of 500 ft. Initial water saturation in simulation 2 is also calculated from equation (2), but by using variable porosity values computed from the porosity-permeability transform (fig. 7). This results in the uneven initial saturation distribution to the left of the water-injection front (fig. 22). Water saturations in the flooded zones are dependent upon capillary pressure as well as relative and absolute permeability. In simulation 1 (fig. 21) a single capillary-pressure curve and a single relative-permeability curve (figs. 19 and 20) were used, resulting in a much smoother saturation distribution than in simulation 2 (fig. 22). Despite the relatively heterogeneous permeability distribution, the water-injection front in both simulations is relatively sharp and approximately vertical.

Production characteristics of all six simulations are shown in figures 23 through 25. Each production curve for the different permeability realizations is characterized by an initial peak, followed by a gentle decline and then a rapid decline. The rapid decline represents the relatively sharp breakthrough of the water-injection front. Water breakthrough is dependent on the mean continuity of permeable zones in the stochastic permeability realizations. Water breakthrough in realization 11, representing high continuity of permeable zones, occurs earlier than that in realization 7, representing low continuity of permeable zones. Unconditional realization 45 has the highest mean horizontal continuity ($\overline{C}_h$ = 11.95 ft; table 1) and has an earlier

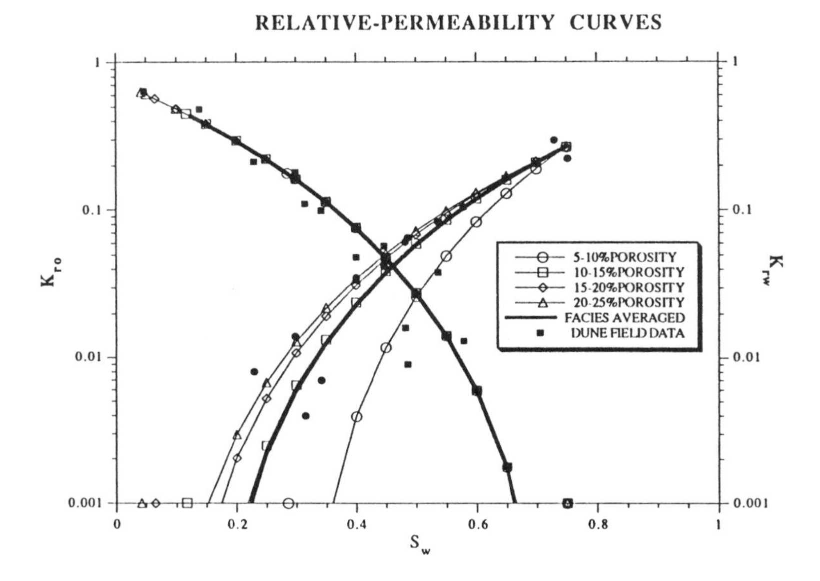

Figure 19. Relative-permeability curves used for the different waterflood simulations in parasequence 1.

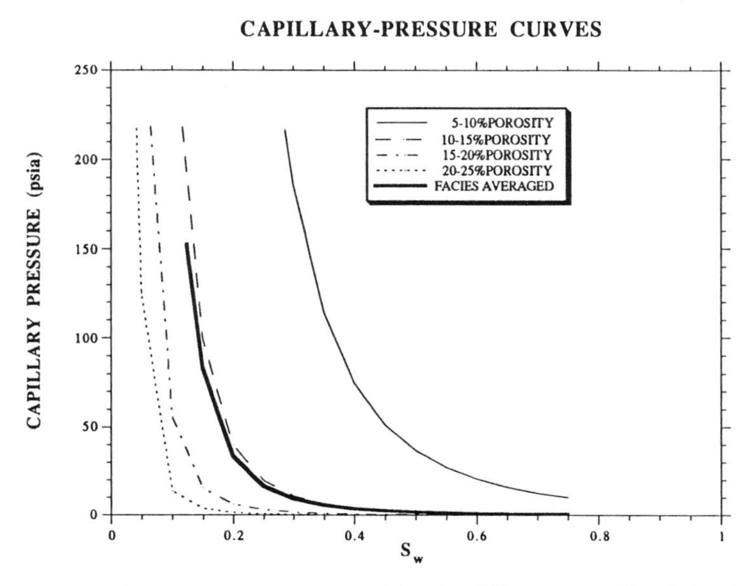

Figure 20. Capillary-pressure curves used for the different waterflood simulations in parasequence 1.

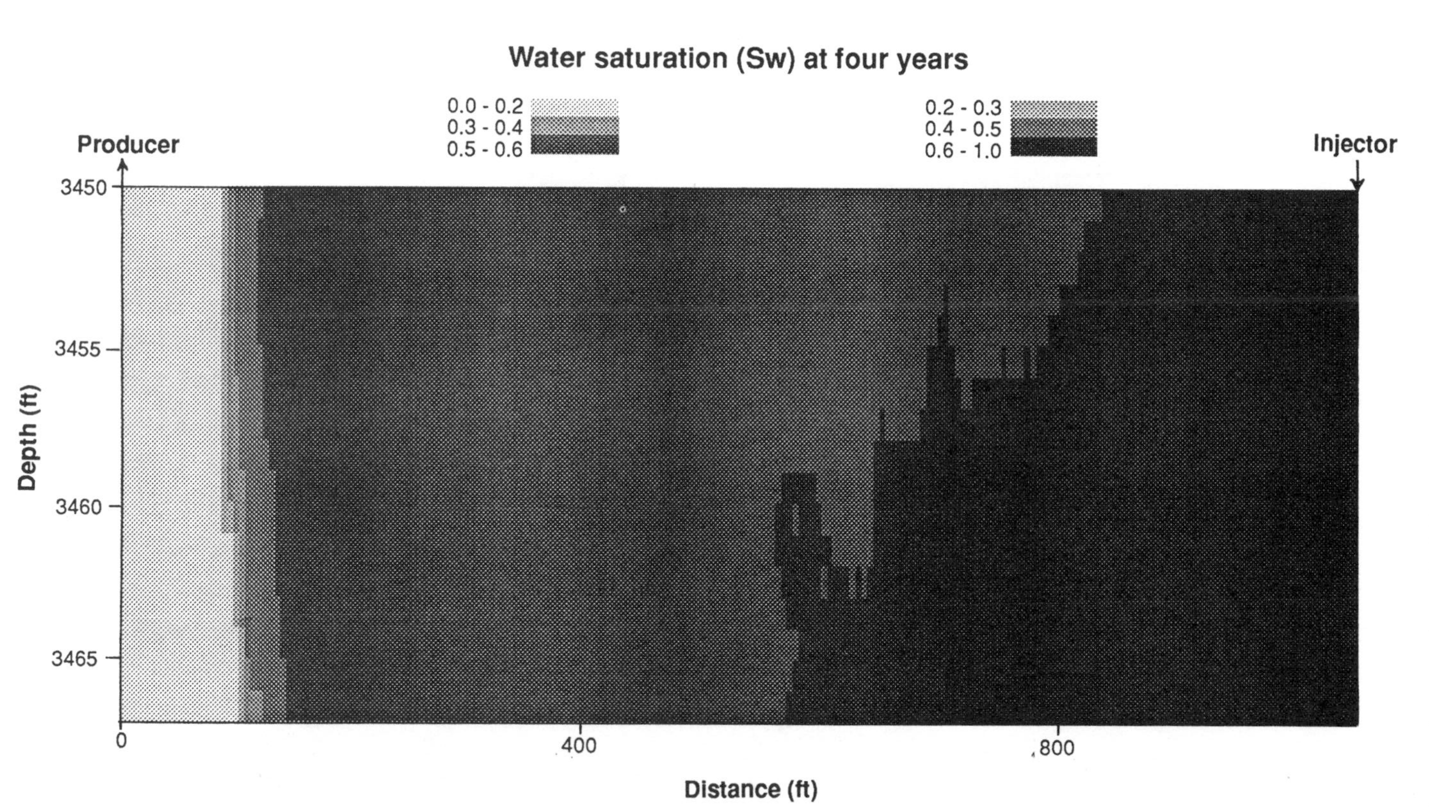

Figure 21. Computed water saturations for simulation 1 after injecting water for 4 yr, incorporating the permeability realization 7 (with uniform initial saturation and single relative-permeability and capillary-pressure curves based on arithmetic-mean porosity).

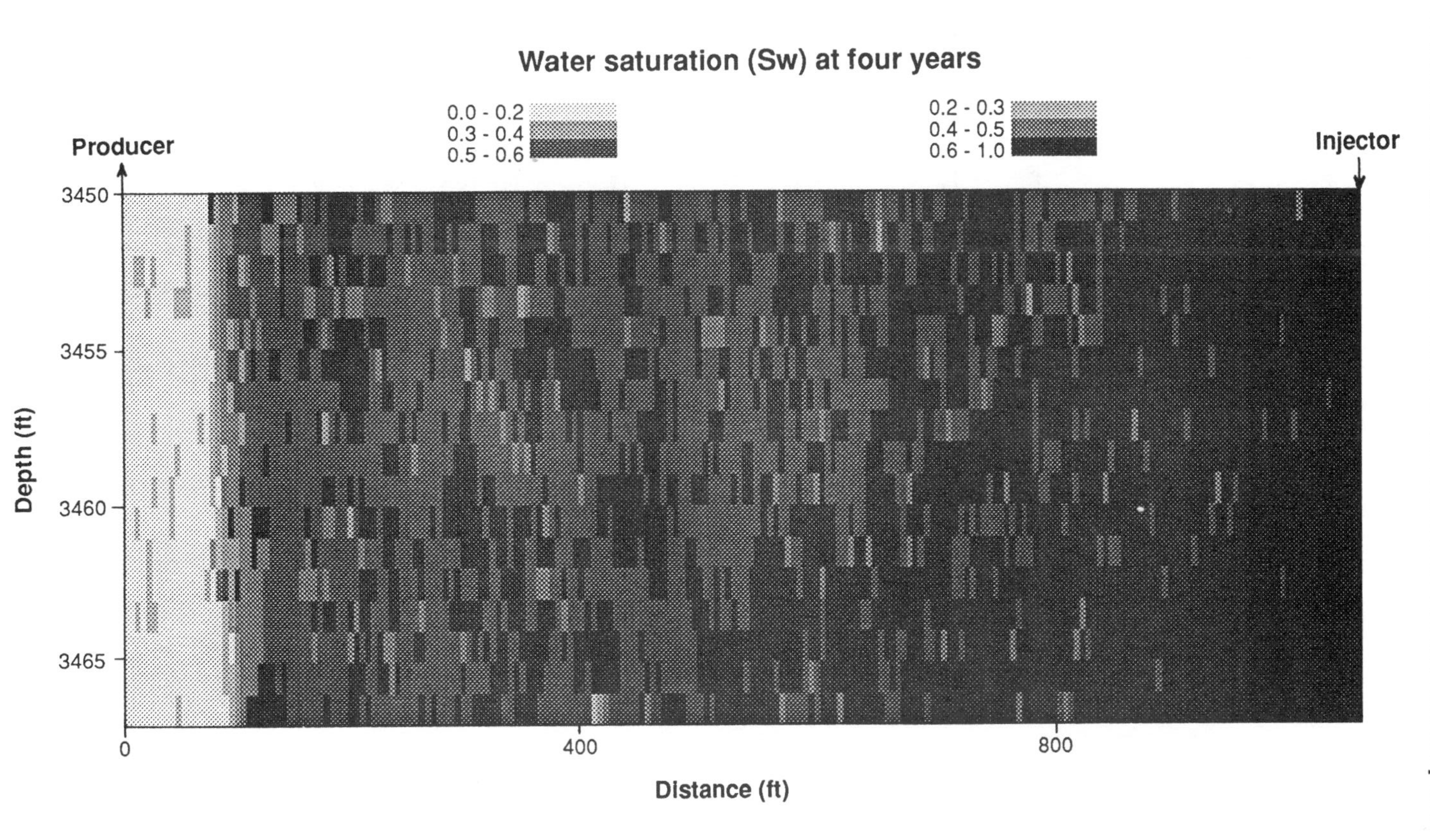

Figure 22. Computed water saturations for simulation 2 after injecting water for 4 yr, incorporating the permeability realization 7 (with relative-permeability curves and capillary-pressure curves dependent on porosity).

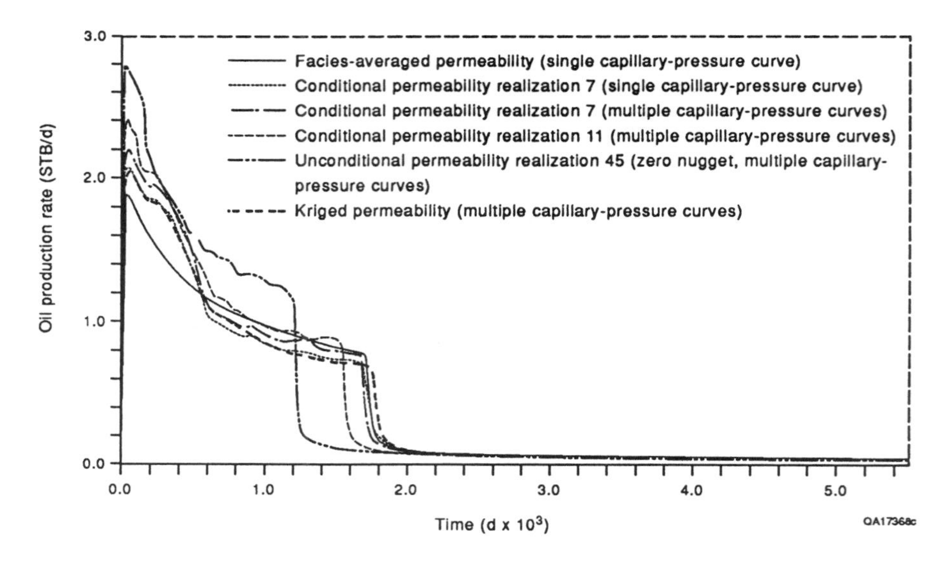

Figure 23. Oil-production rate versus time for waterflood simulations in parasequence 1.

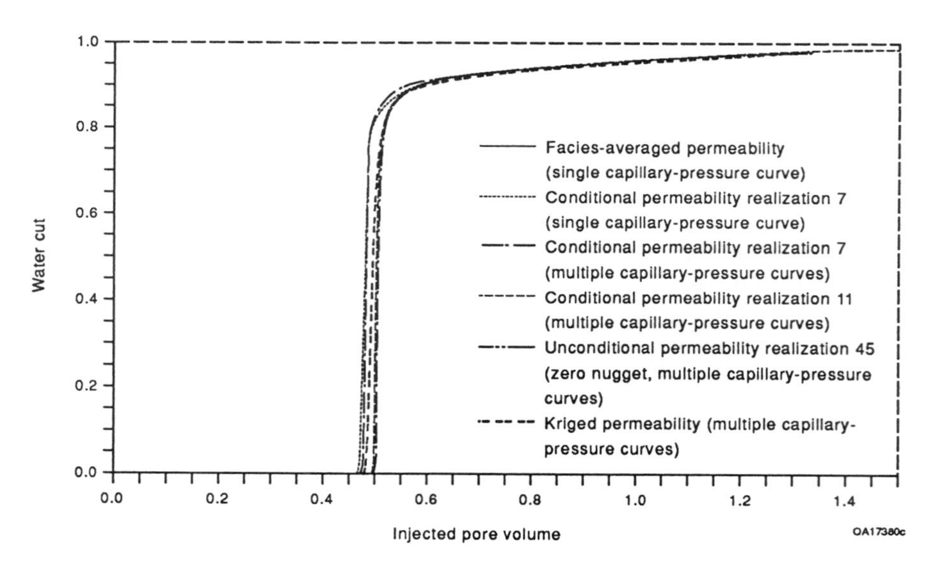

Figure 24. Water-oil ratio versus injected pore volumes for waterflood simulations in parasequence 1.

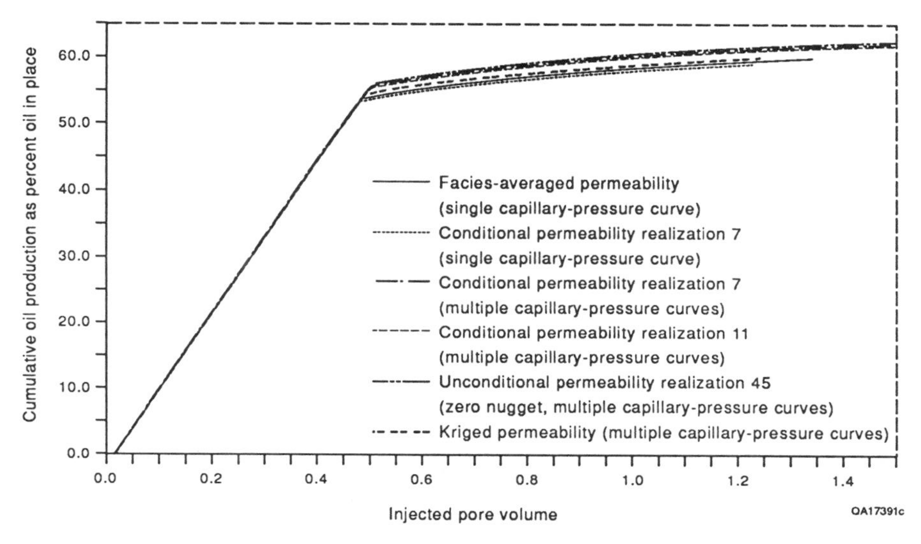

Figure 25. Cumulative oil production as percentage of original oil in place for waterflood simulations in parasequence 1.

breakthrough than any other simulation (fig. 23). Although unconditional realization 45, having a zero nugget, represents low continuity of permeable zones, its mean horizontal continuity is higher than that in realization 11, which represents high continuity of permeable zones but has a nugget of 0.2 md^2 (table 1). Higher continuity of relatively permeable zones results in higher interconnection of these zones and thus produces higher effective permeability than do permeability realizations characterized by low continuity (Fogg, 1989). As a result, unconditional permeability realization 45 shows the earliest water breakthrough because of the overall higher effective permeability. Also, the production curve of unconditional realization 45 illustrates the highest initial production rate but a subsequently steeper decline than production curves from the conditional permeability realizations (fig. 23). As with the water breakthrough, the initial production peaks of the stochastic permeability realizations are also dependent on the mean horizontal continuity.

The kriged permeability distribution shows a production curve similar to that of the conditional permeability realizations. This suggests that the larger scale permeability variation (fig. 16) controls the overall production characteristics and that the small-scale permeability variations incorporated into the stochastic permeability realizations have little importance (figs. 17 and 18). In comparison, the facies-averaged permeability distribution, which has no spatial permeability variation, indicates a smooth, approximately exponential production decline before the water breakthrough, which occurs at approximately the same time as that in the kriged and conditional permeability realizations.

Plotting water cut against injected pore volume shows nearly identical curves for the stochastic permeability realizations, incorporating porosity-dependent capillary-pressure curves (fig. 24). Simulation 1 with permeability realization 7, incorporating a single capillary-pressure curve based on an arithmetic-mean porosity, and simulation 6 with the facies-averaged permeability distribution, incorporating a single capillary-pressure curve, show earlier water-breakthrough curves than those of the stochastic permeability realizations, incorporating porosity-dependent capillary-pressure curves (fig. 24). In comparison, the kriged permeability distribution falls between the two groups (fig. 24).

The sweep efficiency is improved using porosity-dependent capillary pressures, as indicated by the cumulative oil production as percent total oil in place (fig. 25). The waterflood simulations incorporating multiple capillary-pressure and relative-permeability curves (simulations 2 to 5) indicate higher sweep efficiency than those with single capillary-pressure and relative-permeability curves (simulations 1 and 6). Note that the sweep efficiency in simulation 5, representing the kriged permeability distribution, is slightly lower than the stochastic permeability realizations (simulations 2 to 4), indicating that the small-scale heterogeneity causes an increase in sweep efficiency.

C. Discussion

The effect of small-scale heterogeneity and associated capillary-pressure phenomena on cumulative production characteristics is relatively small and accounts for less than a 2.5 percent increase in sweep efficiency (fig. 25). For practical purposes, the observed heterogeneity within the grainstone facies in parasequence 1 can be represented by a geometric-mean permeability distribution using an arithmetic-mean porosity for calculating uniform initial water saturation and concomitant capillary pressure according to equation (2). However, production-rate history cannot be represented by a facies-averaged permeability distribution. The kriged permeability distribution (fig. 16) yields the same production-rate pattern as the conditional permeability realizations (fig. 23), indicating that small-scale permeability variability has a negligible effect on the production rate and only the large-scale permeability patterns need to be incorporated into the flow model for history matching.

Most hydrocarbon reservoirs typically consist of several different facies or rock-fabric units. The detailed geologic framework of the entire upper San Andres outcrop at Lawyer Canyon (fig. 2) is more typical of the complexity of subsurface reservoirs in a ramp-crest depositional environment than is parasequence 1 alone. One might expect that the production characteristics of this reservoir, composed of nine parasequences separated by discontinuous, tight mudstone layers, are more dependent on the spatial relationship of the different facies and flow barriers than on the permeability pattern within individual facies if significant heterogeneity does not occur within the facies. The above flow simulations indicate that even though permeability heterogeneity within individual facies exhibits some short-range correlation, it can be represented by a geometric-mean permeability distribution that, overall, yields the same cumulative production characteristics as those incorporating the observed permeability variability. To study the effects of larger scale features associated with the spatial distribution of different rock-fabric units on reservoir-flow behavior, waterflood simulations of the entire Lawyer Canyon outcrop model (fig. 2) were performed.

V. RESERVOIR-FLOW CHARACTERIZATION, LAWYER CANYON OUTCROP

A. Conceptualization of the Outcrop Reservoir-Flow Model

The geologic model of the Lawyer Canyon parasequence window (fig. 2), in conjunction with the rock-fabric characterization of the depositional facies (fig. 9), is the basis of the conceptual reservoir-flow model. Initial saturations for the different flow units were calculated using the porosity-saturation transforms for the three rock-fabric classes (fig. 8) and average porosities. It was assumed that the grainstones in parasequence 7, characterized by variable amounts of separate-vug porosity, (fig. 7b) have the same porosity-saturation relationship as the nonvuggy grainstones in parasequences 1, 2, 3, and 9 (fig. 9). The effect of vuggy porosity was accounted for by assigning a higher residual oil saturation to the grainstones in parasequence 7.

The constructed reservoir flow model distinguishes 11 flow units that have different average permeability, porosity, initial water saturation, and residual oil saturation (table 2). The flow model is discretized in 4,089 irregularly shaped grid blocks, representing the spatial distribution and the petrophysical properties of the different rock fabrics and depositional facies (table 2). Reservoir block sizes are 100 ft in the horizontal direction and have variable thickness in the vertical direction, ranging from less than 0.5 ft to several feet. The constructed reservoir model incorporates the general geometry and the spatial distribution of the different facies mapped in the outcrop, as shown by the distribution of initial water saturation of the discretized flow units (fig. 26).

Relative permeabilities for the different flow units are based on the shapes of relative-permeability curves derived from fitting relative-permeability data obtained from cores in the Dune field, West Texas. Similar to the relative-permeability curves in figure 19, the fitted curves were adjusted to the computed initial water saturations and to the residual oil saturations of the different flow units (table 2). Although the shape of the relative-permeability curves was obtained by fitting relative-permeability measurements from grainstone cores from the Dune field reservoir, West Texas (fig. 19), the same curve shapes were used in this study not only for the grainstone rock fabrics but also for the grain-dominated packstone and mudstone/wackestone rock fabrics. Only the relative-permeability endpoints were adjusted according to the computed initial water saturations and assumed residual oil saturations.

Capillary-pressure curves were calculated on the basis of average porosities and rock-fabric classifications of the different flow units. In addition to the porosity-dependent saturation–capillary-pressure relationship for grainstones (eq. 2), the following relationships are used for the other two rock-fabric classes:

Grain-dominated packstone $$S_w = 106.524\, h^{-0.407}\, \phi^{-1.440} \quad (4)$$

Mudstone-wackestone $$S_w = 161.023\, h^{-0.505}\, \phi^{-1.2104} \quad (5)$$

Table 2. Properties of rock-fabric flow units for Lawyer Canyon outcrop reservoir model.

Flow units	Rock fabric	Depositional facies	Porosity (arithm. average)	Permeability (geometric average, md)	Initial water saturation	Residual oil saturation
1	Mudstone	Flooded shelf Tidal flat	0.04	0.01	0.9	0.01
2	Wackestone	Shallow shelf I	0.105	0.30	0.405	0.4
3	Grain-domin. packstone	Shallow shelf I	0.085	4.50	0.214	0.35
4	Grain-domin. packstone	Shallow shelf II Bar top	0.129	1.80	0.40	0.35
5	Grain-domin. packstone	Shallow shelf II	0.118	5.30	0.243	0.35
6	Grainstone (moldic)	Shallow shelf II	0.145	0.7	0.091	0.40
7	Grainstone (moldic)	Shallow shelf I Shallow shelf II Bar crest Bar top	0.159	2.2	0.077	0.40
8	Grainstone (highly moldic)	Shallow shelf II Bar crest Bar-accretion sets	0.23	2.5	0.041	0.40
9	Grainstone	Bar flank	0.095	9.5	0.189	0.35
10	Grainstone	Bar crest Bar-accretion sets	0.11	21.3	0.147	0.25
11	Grainstone	Bar crest	0.135	44.0	0.103	0.25

Fluid properties:

Oil viscosity	1.000 cP
Water viscosity	0.804 cP
Oil density	55 lb/ft^3
Water density	64 lb/ft^3

where h is the height of the reservoir above the free-water level (ft) and ϕ is the porosity (fraction).

In a series of waterflood simulations, various factors affecting reservoir-flow behavior were examined (table 3) using the ECLIPSE reservoir simulator. Simulation EC-A represents the base scenario, which is used to describe waterflooding in this reservoir model. Production characteristics of simulation EC-A were compared with those from other simulations to evaluate effects of capillary pressures (simulation EC-B), model conceptualization (simulations EC-N and EC-DP), and different injection practices (simulations EC-R and EC-F).

B. Simulation Results

Waterflooding of the Lawyer Canyon outcrop reservoir model was simulated by injecting water through a fully penetrating well along the right side of the model

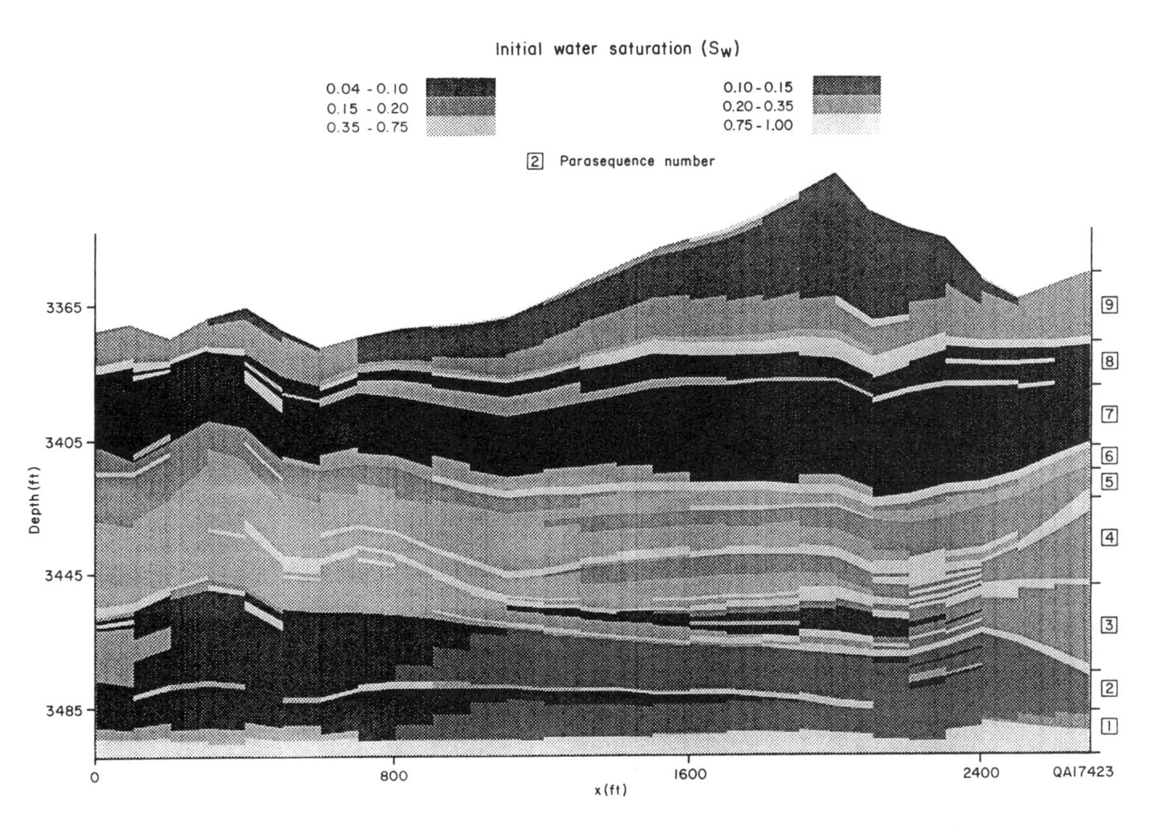

Figure 26. Initial water saturation of the flow units of the outcrop model.

Table 3. Waterflood simulations of the Lawyer Canyon outcrop model.

Sim. no.	Model scenario: Grid	Production well location	Capillary pressure	Permeability data
EC-A	Irregular	Right	Yes	Facies-averaged
EC-B	Irregular	Right	No	Facies-averaged
EC-N	Normalized	Right	Yes	Facies-averaged
EC-DP	Normalized	Right	Single	Linear interpol. between wells
EC-R	Irregular	Left	Yes	Facies-averaged
EC-F	Irregular	Middle	Yes	Facies-averaged

and by producing from a well at the left side (fig. 26) The injection and production rates were controlled by prescribed pressures of 4,350 psi and 750 psi, respectively.

The computed change in water saturation for simulation EC-A after water injection of 20 yr (fig. 27a), 40 yr (fig. 27b), and 60 yr (fig. 27c) demonstrates that the high-permeability grainstone rock-fabric units in parasequences 1, 2, and 9 are preferentially flooded. Furthermore, flooding is controlled by the relatively tight mudstone units separating most of the parasequences. The grainstone facies in parasequences 7 and 8 are characterized by lower permeabilities and higher porosities than those in parasequences 1, 2, and 9, owing to separate-vug porosity (table 2); consequently, the water-injection front does not advance as far as that in parasequence 9 (fig. 27). However, the water-injection front in parasequences 3 through 6 appears to have advanced farther than that in parasequences 7 and 8, although the permeability of the predominantly wackestone rock fabrics in parasequences 3 through 6 is lower than that of the moldic grainstone rock fabrics in parasequence 7. This indicates that the movement of the displacement front is not only affected by the permeability contrast but inversely related to porosity.

In the upper right of the model, the change in computed water saturation after 40 and 60 yr indicates cross flow of water from parasequence 9 into parasequence 7, thereby bypassing the injection front within parasequences 7 and 8. As a result, an area of unswept, mobile oil develops in the right part of the model, as shown by the computed water saturation distribution after 60 yr of waterflooding (fig. 28). This mobile oil is trapped by cross flow even though the mudstone layers, having a permeability of 0.01 md, represent the parasequence boundaries and are continuous in this area.

The production characteristics of simulation EC-A and the other simulations (table 3) are shown in figure 29 (production rate), figure 30 (water cut), and figure 31 (cumulative production as percent oil in place). Comparison of the production characteristics of the different simulations were used to evaluate various factors affecting reservoir-flow behavior.

1. Effect of Capillary Pressure

In the first test, effects of capillary pressure were studied. Simulation EC-B does not incorporate capillary pressures (table 3). Production rates in simulation

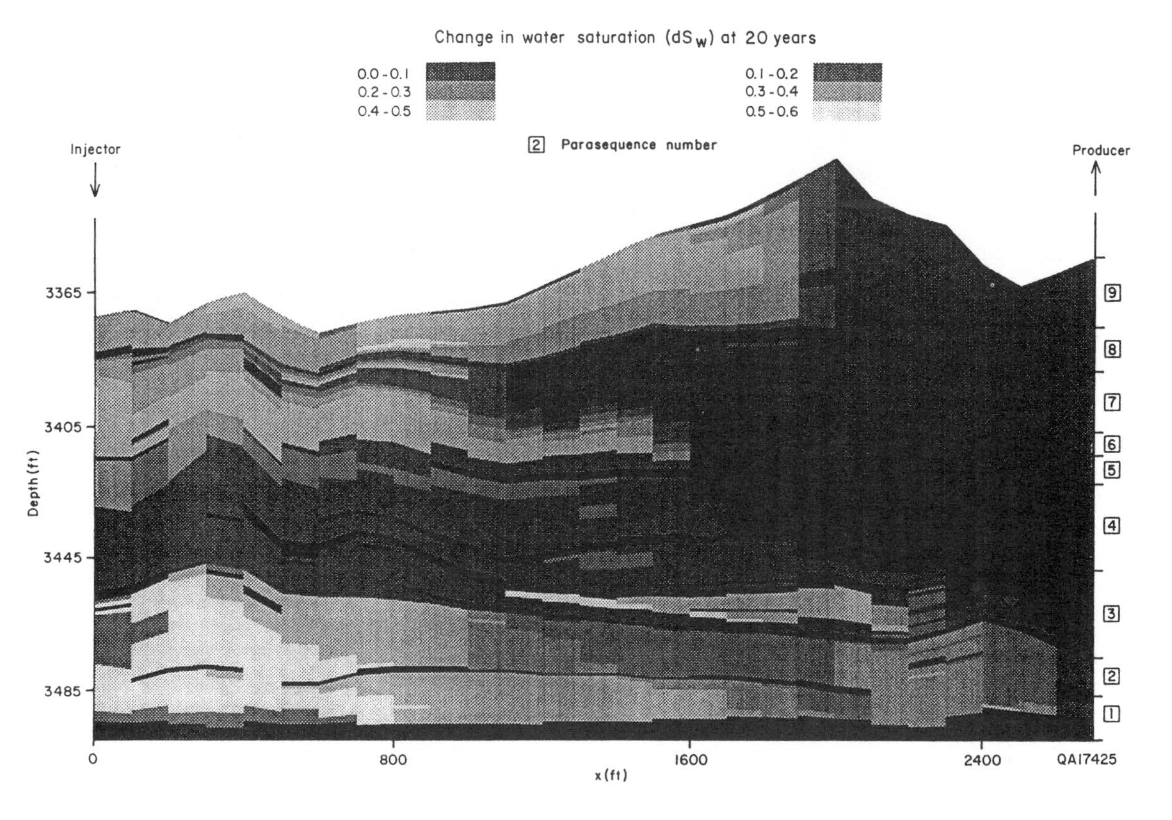

Figure 27. Computed changes in water saturation for simulation EC-A after waterflooding of (a) 20 yr, (b) 40 yr, (c) 60 yr.

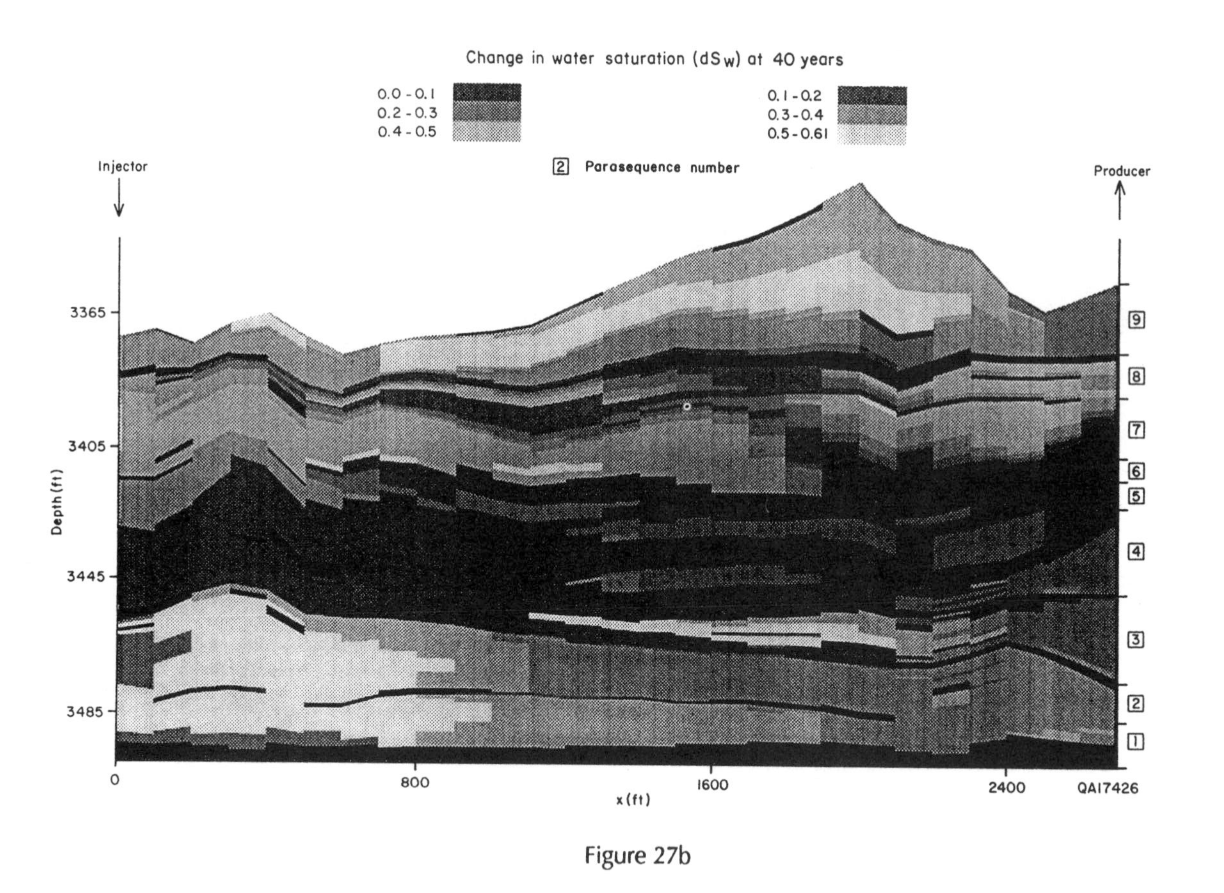

Change in water saturation (dSw) at 40 years
0.0 - 0.1
0.2 - 0.3
0.4 - 0.5
0.1 - 0.2
0.3 - 0.4
0.5 - 0.61
2 Parasequence number
Injector
Producer
Depth (ft)
3365
3405
3445
3485
9
8
7
6
5
4
3
2
1
0
800
1600
2400
QA17426
x (ft)

Figure 27b

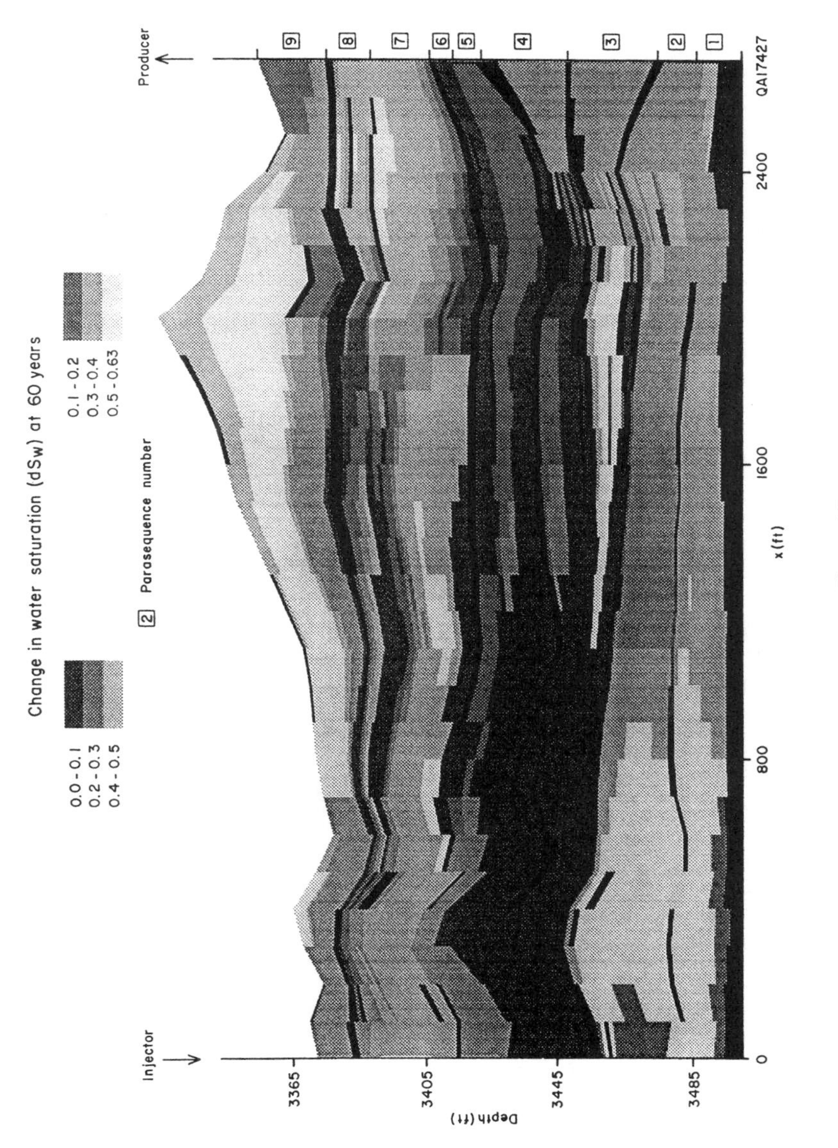

Change in water saturation (dSw) at 60 years
0.0 - 0.1
0.2 - 0.3
0.4 - 0.5
0.1 - 0.2
0.3 - 0.4
0.5 - 0.63
2 Parasequence number
Injector
Producer
1
2
3
4
5
6
7
8
9
3365
3405
3445
3485
Depth (ft)
0
800
1600
2400
x (ft)
QA17427

Figure 27c

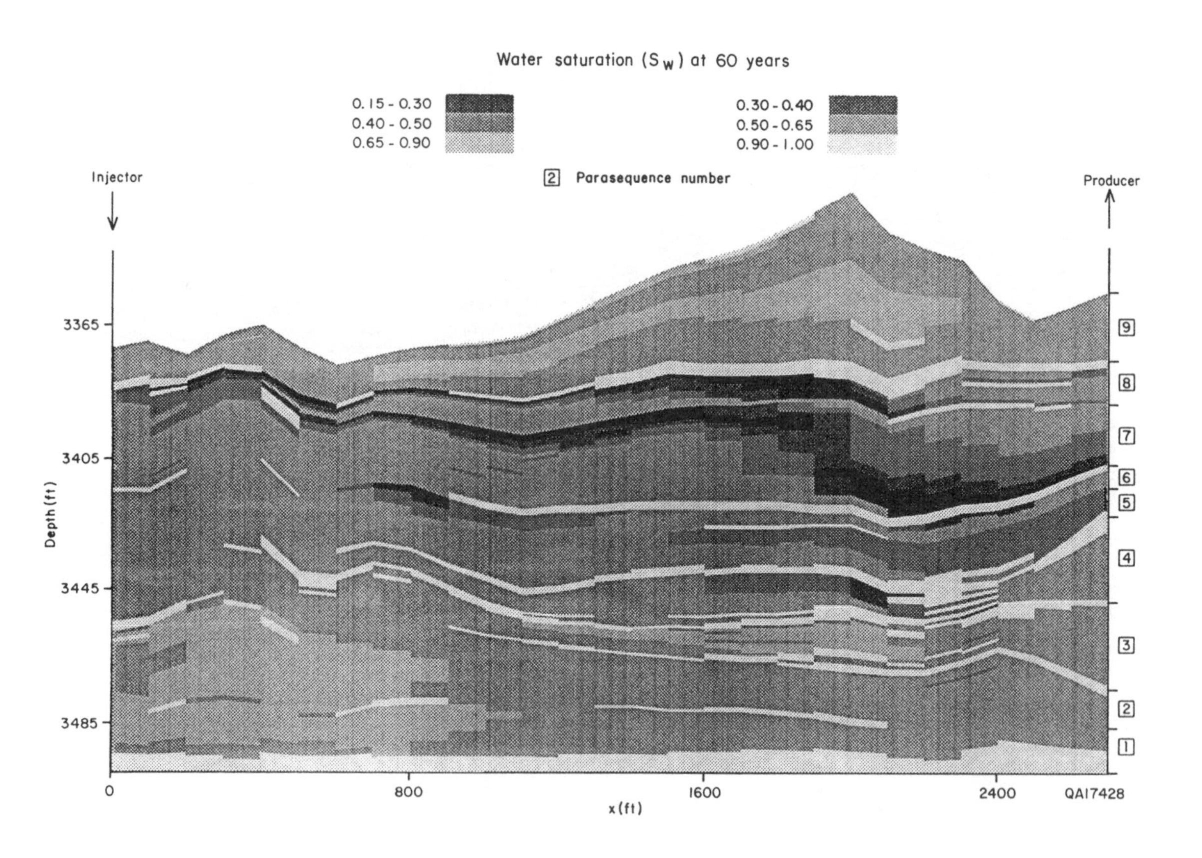

Figure 28. Computed water saturation for simulation EC-A after water injection of 60 yr.

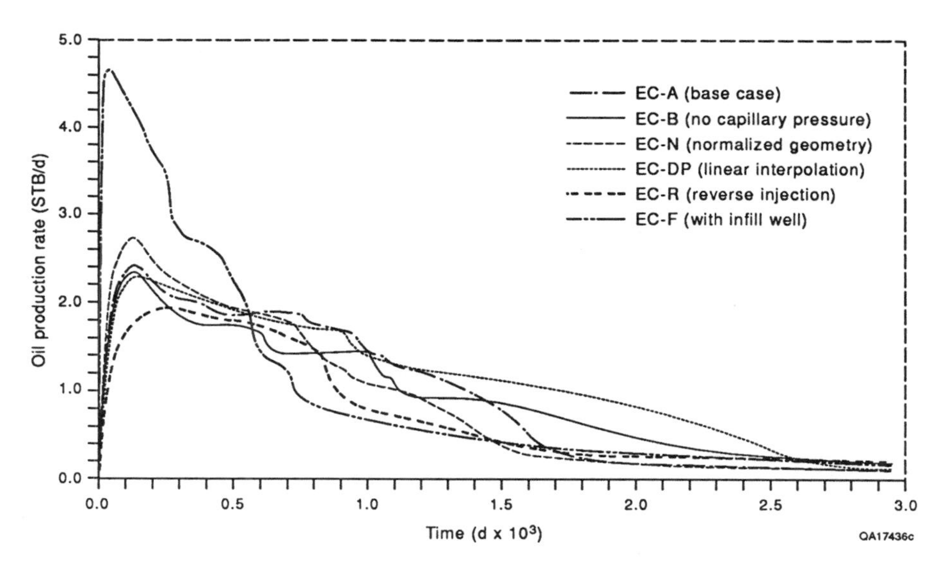

Figure 29. Oil-production rate versus time for simulations of the Lawyer Canyon outcrop model.

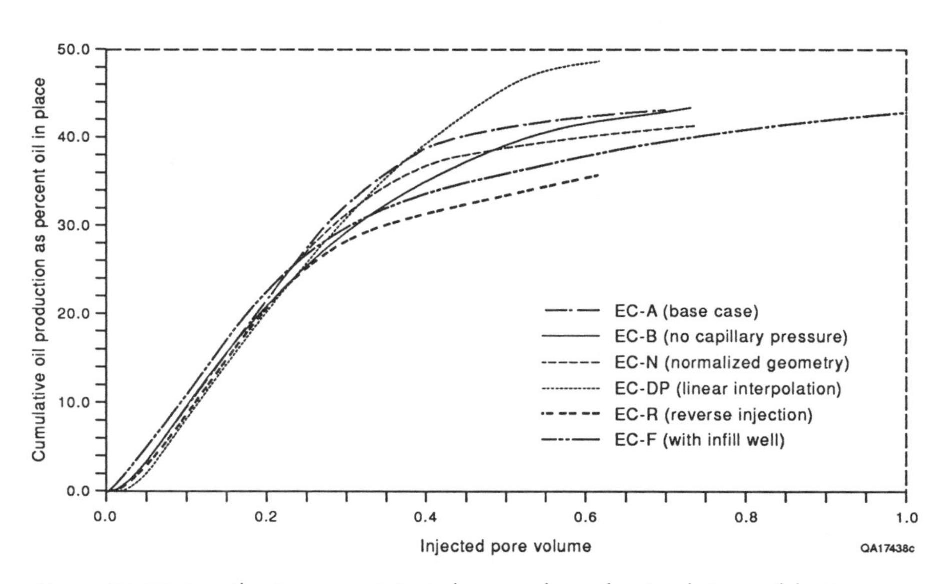

Figure 30. Water-oil ratio versus injected pore volume for simulations of the Lawyer Canyon outcrop model.

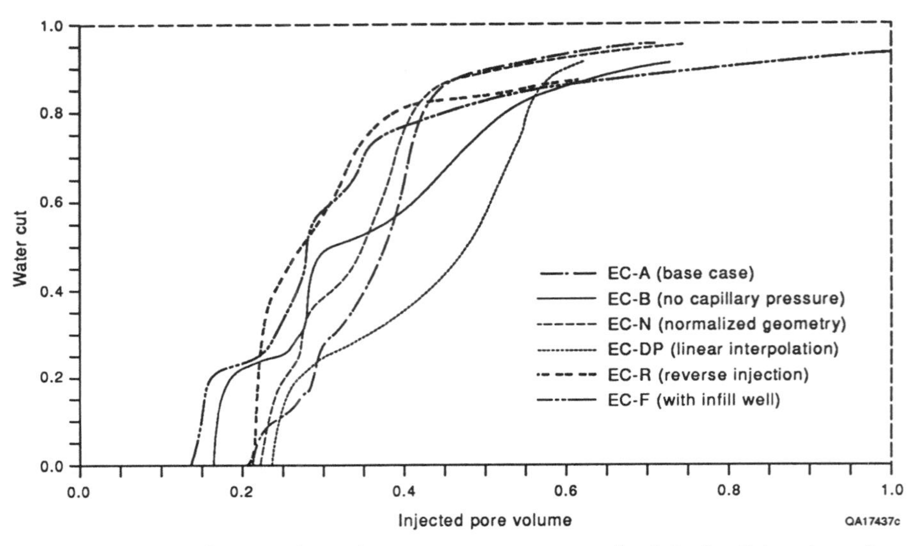

Figure 31. Cumulative oil production as percentage of original oil in place for simulations of the Lawyer Canyon outcrop model.

EC-B show a more stepwise decline with time, reflecting the flooding of the grainstone-dominated parasequences, as compared with production characteristics of the base case in simulation EC-A (fig. 29). Production rates are initially lower than those in simulation EC-A, but simulation EC-B maintains higher rates after 16,000 days. The stepwise decline in production rate is also reflected in the stepwise increase in water cut (fig. 30). Neglecting capillary pressures results in a lower seep efficiency, as shown by the cumulative production curve (fig. 31). In simulation EC-A, capillary pressure improves sweep of the less permeable zones in parasequences 3 through 6, whereas in simulation EC-B, waterflooding is restricted to the more permeable grainstone facies in parasequences 1, 2, 7, and 9. Ultimate oil recovery of the two simulations, however, is the same (fig. 31).

2. Effect of Model Conceptualization

The effect of irregular formation geometry as compared with that of normalized formation geometry was evaluated in simulations EC-N and EC-DP (table 3). In simulation EC-N, the nine parasequences were normalized to a constant thickness, approximating the spatial distribution of the mapped facies (fig. 2). Production rates are initially higher than those in simulation EC-A, but they drop off more rapidly (fig. 29). The water-breakthrough curve for the normalized reservoir model (simulation EC-N) is steeper than that in simulation EC-A (fig. 30) and shows a lower recovery efficiency (fig. 31).

In simulation EC-DP, the reservoir model was constructed using the permeabilities of the individual flow units at the injection and production wells and then linearly interpolating permeability between wells using the normalized grid from simulation EC-N. This scenario represents a typical reservoir model constructed from well data, where the facies and permeability distributions of the interwell area are unknown. When using well data from only the left and right sides of the outcrop model (fig. 2), the relatively permeable grainstone facies in parasequence 9 is not incorporated into the layered model. Furthermore, only a single relative-permeability curve and a single capillary-pressure curve are used in simulation EC-DP.

Initial production rates in simulation EC-DP are similar to those in simulation EC-A but do not show the drop off after about 16,000 days (fig. 29), which is reflected in a less steep water-breakthrough curve than that in simulation EC-A (fig. 30). More importantly, sweep efficiency is overestimated in this layered model (fig. 31).

3. Effect of Injection Practice

Two additional flow scenarios were simulated to evaluate the effects of different injection schemes (table 3). In simulation EC-R, injection and production is reversed when the reservoir is flooded from the right side. Although the reservoir model and properties are the same in simulation EC-R and in EC-A, the production characteristics are noticeably different. Initial production rates in simulation EC-R are lower than those in simulation EC-A (fig. 29), but they remain slightly higher after 16,000 days. That is, at later times simulation EC-R produces at a lower water-oil ratio, which is characterized by the water-cut curve that levels out at a lower value than that of simulation EC-A (fig. 30). Sweep efficiency is significantly lower in simulation EC-R than in simulation EC-A (fig. 31), indicating that the spatial distribution of permeable grainstone facies relative to the direction of the waterflood (fig. 2) is important for the overall reservoir-flow behavior.

Comparing the change in water saturation after 40 yr of waterflooding shows a much larger area of unswept oil in the center of the model in simulation EC-R (fig. 32) than in simulation EC-A (fig. 27b). More importantly, cross flow occurs on the left side of the model toward the production well in simulation EC-R (fig. 32), from parasequence 9 all the way to parasequences 1 and 2. Although parasequences 3 through 6 are composed predominantly of less permeable wackestones on the left side of the model, which change to grain-dominated packstones on the right side, cross flow does not occur in simulation EC-A (fig. 27). On the other hand, cross flow across parasequences 3 through 6 on the left side of the model is facilitated by the fact that the tight mudstone layers are discontinuous, whereas on the right side they are continuous (fig. 2). However, the spatial distribution of the higher-permeablity grainstone facies on the left part of the model in parasequences 1 and 2 (table 2, unit 10) and in parasequence 9 (table 2, unit 11) are crucial for cross flow through parasequences 3 through 6 on the left. As indicated in simulation EC-A (fig. 28), continuous mudstone layers do not necessarily represent flow barriers, as shown by the cross flow between parasequences 9 and 7 in the upper right of the model.

In the final simulation EC-F, the production well was located in the center of the model, and an injection well was placed at either side of the model. Prescribed pressures were adjusted in order to create the same pressure gradient between the

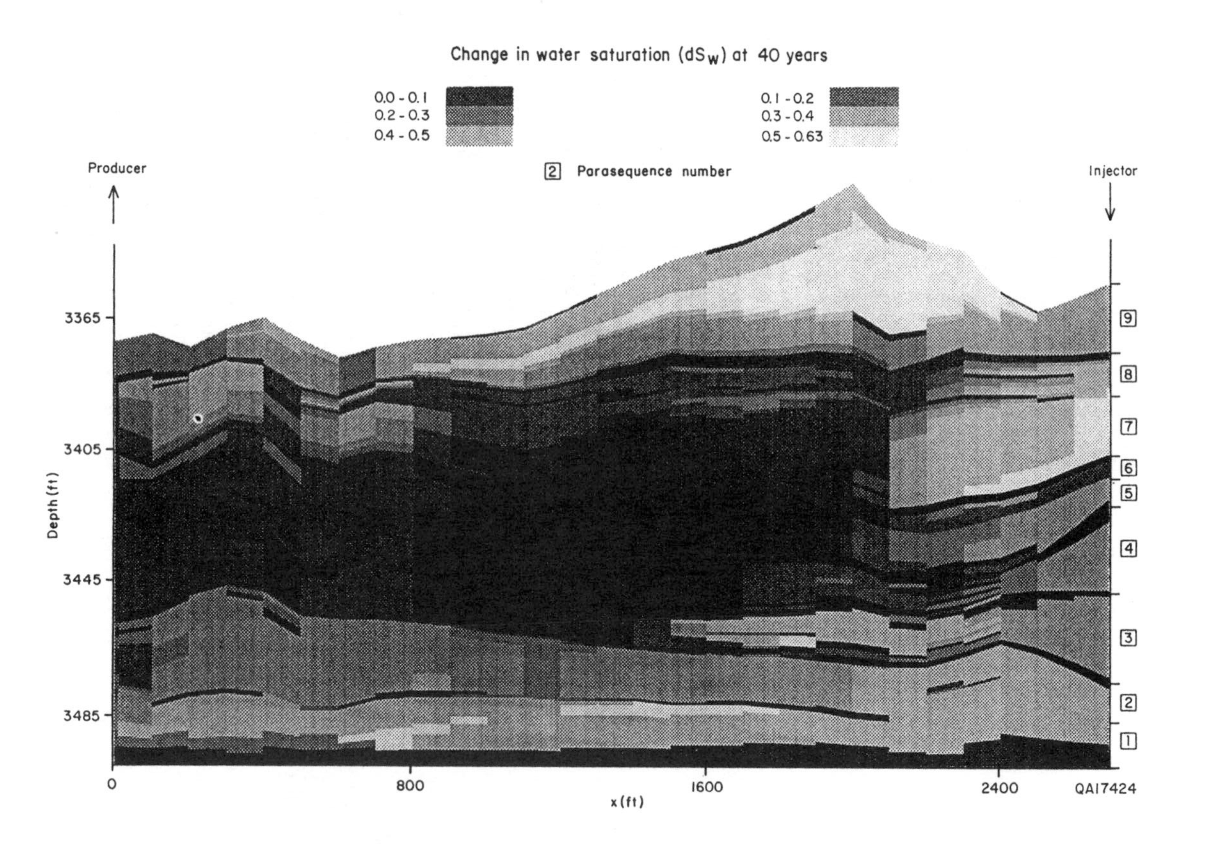

Figure 32. Computed changes in water saturation after 40 yr of waterflooding for simulation EC-R with inverse-injection pattern.

injection wells and production well as those used in the other simulations. As one would expect, initial production rates are much higher in simulation EC-F than those in the other simulations but subsequently show a much earlier and steeper decline. However, after 16,000 days, the production rate levels off at a slightly higher rate than that in simulation EC-A. Similar to the reverse-injection pattern in simulation EC-R, the water-cut curve breaks at a lower water-oil ratio than that in simulation EC-A (fig. 30). Simulation EC-I incorporates the flow pattern in the left part of the model similar to that in simulation EC-A, but on the right part of the model the flow behavior corresponds to that of simulation EC-R; consequently the sweep efficiency of simulation EC-F is lower than in simulation EC-A but slightly higher than in simulation EC-R (fig. 31).

VII. DISCUSSION

A. Implications of Outcrop Studies on Reservoir Characterization

Detailed geologic mapping of the upper San Andres Formation at the Lawyer Canyon area, Algerita Escarpment, yielded deterministic images of geologic facies architecture and corresponding porosity-permeability structure at scales ranging from a foot to tens of feet vertically and a foot to hundreds of feet laterally. The horizontal resolution of both the geologic facies mapping and the rock-property measurements allowed for detailed deterministic characterization of interwell heterogeneity. The geologic mapping within a sequence-stratigraphic framework calls for application to subsurface San Andres reservoirs in other parts of the Permian Basin, as well as to similar carbonate-ramp systems worldwide. A companion study is applying the general concepts derived from the outcrop study to the San Andres Seminole field, West Texas.

Geologic mapping revealed a series of upward-shallowing parasequences (10 to 40 ft thick and several thousand feet long). Parasequence boundaries are typically marked by tight mudstone/wackestone beds that display variable degrees of lateral continuity ranging from several hundred feet to more than 2,500 ft. Facies development within a parasequence can be highly variable. Generally, the thicker the parasequence, the more laterally variable is the resultant facies mosaic.

As documented by the flow simulations, the spatial distribution of permeable grainstone facies relative to the parasequence boundaries that show variable degrees of lateral continuity are crucial for understanding reservoir performance. Although it is difficult to extrapolate facies geometry in the interwell region in subsurface reservoirs, the outcrop study provides insight into the internal architecture of individual parasequences. Depositional processes in a ramp-crest environment produces upward-shallowing, upward-coarsening facies within a parasequence, starting with mudstone deposition associated with a rapid sea-level rise. Subsequently, carbonates are deposited as wackestone/packstone facies and finally as bar-crest and bar-flank grainstones and packstones. However, the lateral dimensions of these facies are highly variable and may not be penetrated by wells. The grainstone facies in parasequence 9 reaches a maximum thickness of 38 ft but quickly thins laterally and is completely absent at the northern and southern edges of the Lawyer Canyon parasequence window (fig. 2).

Within grainstones, permeability varies by as much as five orders of magnitude. Variogram analysis of spatial permeability in grainstone facies of parasequence 1 indicates short-range correlation but relatively high, locally random heterogeneity, reflected in a relatively high nugget compared with the sill. Similar permeability structures were obtained from the vuggy grainstone facies in parasequence 7. Flow simulations indicate that the effective permeability within the grainstone facies can be represented by the geometric-mean permeability. Note, however, that porosity-dependent capillary-pressure relationships, in combination with the small-scale heterogeneity, increase the sweep efficiency of individual facies. However, the effect is relatively small and will be outweighed by the sweep efficiency, which is largely controlled by injection practices and by the spatial distribution of individual facies relative to parasequence boundaries.

Reservoir-flow simulations of the entire outcrop model underscore the importance of knowing the facies architecture between wells. The simulation EC-DP, representing the standard approach for a subsurface reservoir model by linearly interpolating properties between wells, significantly overestimates the cumulative oil production compared with the base-case simulation EC-A, which incorporates the detailed spatial facies distribution between wells (fig. 31).

B. Determination of Remaining Oil

The changes in computed water saturation for simulation EC-A indicate that bypassing unswept oil through cross flow of injected water across low-permeability mudstones is an important mechanism. As a result, an area of unswept oil develops within parasequence 7 because of the slower advance of the water-injection front within parasequence 7. As shown in simulation EC-R, where the injection pattern is reversed, the cross-flow effect becomes more dominant and a greater area of unswept oil develops than in simulation EC-A. Comparison of production characteristics between simulation EC-A and EC-R indicates that simply changing the waterflood direction affects sweep efficiency. This implies that the spatial distribution of facies relative to the waterflood direction can significantly affect how the reservoir produces, as well as the volume and location of unswept oil.

The simulations further document that although the parasequence boundaries, represented by mudstone/wackestone units, strongly control waterflooding, these low-permeability mudstone layers do not necessarily represent flow barriers but can allow for significant cross flow. If, and when, cross flow occurs through these mudstones depends on the spatial distribution of the high-permeability grainstone facies relative to the waterflood direction.

Oil recovery in these simulations approaches 45 percent of total oil in place (fig. 31), which is high compared with oil-recovery data of San Andres reservoirs in the Permian Basin. However, the limitations of the two-dimensionality of the outcrop reservoir model overestimate sweep efficiency (Fogg and Lucia, 1990). The two-dimensionality of the cross-sectional model forces all fluid flow into the vertical plane between the wells and thus represents only the most direct flow between the injection and production wells. In three dimensions, flow away from the injection well and toward the production well is represented by radial streamlines characterized by increasing path length away from the most direct streamline between the wells. As a result, the water-breakthrough curves can be expected to be less steep in three-dimensional scenarios.

More important are the potential effects of three-dimensional heterogeneity, that is, possible circuitous flow paths perpendicular to the cross-sectional plane are not accounted for in the two-dimensional model. In addition to the effects of radial flow paths, neglecting heterogeneity in the third dimension also tends to overestimate sweep efficiency in the cross-sectional outcrop model. On the other hand, even less distinct mechanisms for trapping unswept oil indicated in the cross-sectional model may be enhanced when considering heterogeneity in the third dimension.

VII. SUMMARY

New approaches for reservoir characterization of heterogeneous carbonate-ramp deposits were presented that integrate geological, petrophysical, geostatistical, and reservoir-simulation studies of continuous outcrop of the San Andres Formation, New Mexico. Detailed geologic mapping revealed a series of upward-shallowing parasequences—10 to 40 ft thick and several thousand feet long—that form the geologic framework of the reservoir model. Parasequence boundaries, typically marked by tight mudstone/wackestone beds that display variable degrees of lateral continuity (ranging from several hundred feet to more than 2,500 ft) were shown to be important as potential flow barriers. To characterize the complex heterogeneity associated with depositional and diagenetic processes on the interwell scale, geologic and petrophysical data were collected from outcrops at the Lawyer Canyon area of the Algerita Escarpment. Detailed permeability measurements using both mini-air permeameter and core plugs were taken at different scales to characterize the spatial hierarchy of permeability patterns. Geostatistical analysis of permeability measurements indicated varying spatial correlation at different measurement scales. In all cases, however, permeability variability was characterized by substantial random heterogeneity at local scales.

Conditional simulations of permeability within individual facies showed apparent randomness owing to the large nugget effect. Waterflood simulations of conditional permeability realizations yielded results similar to those simulations using a geometric-mean permeability, indicating that the observed permeability heterogeneity within facies can be represented by a geometric mean. Mean permeabilities of most facies and rock fabrics differed significantly at the 95 percent confidence level and can be used to represent large-scale heterogeneity in reservoir-scale flow simulators.

Five basic rock-fabric units, identified in the Lawyer Canyon outcrop area, were incorporated into a geologic model describing the spatial distribution of depositional facies. Petrophysical parameters of these facies were derived on the basis of a petrophysical/rock-fabric approach using mean permeabilities, porosity, and saturation relationships characteristic of each rock fabric. Two-dimensional waterflood simulations were performed for the outcrop model to evaluate various factors affecting reservoir-flow behavior. The results indicated that the spatial distribution of high-permeability facies relative to the distribution of low-permeability parasequence boundaries, in conjunction with the waterflood direction, is crucial for predicting reservoir performance. The study also demonstrated that areas of unswept oil can result from cross flow across relatively low-permeability mudstone layers.

IX. REFERENCES

ECL Petroleum Technologies, 1990, ECLIPSE Reference Manual.

Ferris, M. A., in preparation, Permeability distribution on the upper San Andres Formation outcrop, Guadalupe Mountains, New Mexico: The University of Texas at Austin, Master's thesis.

Fogg, G. E., 1989, Stochastic analysis of aquifer interconnectedness: Wilcox Group, Trawick area, East Texas: The University of Texas at Austin, Bureau of Economic Geology Report of Investigations No. 189, 68 p.

Fogg, G. E., and Lucia, F. J., 1990, Reservoir modeling of restricted platform carbonates: Geologic/geostatistical characterization of interwell-scale reservoir heterogeneity, Dune field, Crane County, Texas: The University of Texas at Ausin, Bureau of Economic Geology Report of Investigations No. 190, 66 p.

Fogg, G. E., Lucia, F. J., and Senger, R. K., 1991, Stochastic simulation of interwell heterogeneity for improved prediction of sweep efficiency in a carbonate reservoir, *in* Lake, L. W., Carroll, H. B., Jr., and Wesson, T. C., eds., Reservoir Characterization II, San Diego, California: San Diego, California, Academic Press, p. 355–381.

Goggin, D. J., Thrasher, R. L., and Lake, L. W., 1988, A theoretical and experimental analysis of minipermeameter response including gas slippage and high velocity flow effects: In Situ, v. 12, no. 1/2, p. 79–116.

Hayes, P. T., 1964, Geology of the Guadalupe Mountains, New Mexico: U.S. Geological Survey Professional Paper 446, 69 p.

Hindrichs, S. L., Lucia, F. J., and Mathis, R. L., 1986, Permeability distribution and reservoir continuity in Permian San Andres shelf carbonates, Guadalupe Mountains, New Mexico, *in* Hydrocarbon Reservoir Studies — San Andres/ Grayburg Formations: Proceedings of the Permian Basin Research Conference, Permian Basin Section, Society of Economic Paleontologists and Mineralogists Publication No. 86-26, p. 31–36.

Honarpuour, Mehdi, Koederitz, L., and Harvey, H. A., 1982, Relative permeability of petroleum reservoirs: Boca Raton, Florida, CRC Press, 143 p.

Journel, A. G., and Huijbregts, Ch. J., 1978, Mining geostatistics: New York, Academic Press, 600 p.

King, P. B., 1948, Geology of the southern Guadalupe Mountains, Texas: U.S. Geological Survey Professional Paper 480, 183 p.

Kittridge, M. G., Lake, L. W., Lucia, F. J., and Fogg, G. E., 1990, Outcrop/subsurface comparisons of heterogeneity in the San Andres Formation: SPE Formation Evaluation, September 1990, p. 233–240.

Knudsen, H. P., and Kim, Y. C., 1978, A short course on geostatistical ore reserve estimation: University of Arizona, Department of Mining and Geological Engineering, 224 p.

Lucia, F. J., 1983, Petrophysical parameters estimated from visual descriptions of carbonate rocks: a field classification of carbonate pore space: Journal of Petroleum Technology, v. 35, no. 3, p. 629–637.

Radian Corporation, 1989, CPS-1 user's guide.

Warren, J. E., and Price, H. S., 1961, Flow in heterogeneous porous media: Society of Petroleum Engineers Journal, September 1961, p. 153–169.

APPLICATION OF OUTCROP DATA FOR CHARACTERIZING RESERVOIRS AND DERIVING GRID-BLOCK SCALE VALUES FOR NUMERICAL SIMULATION

Liviu Tomutsa, Ming-Ming Chang, and Susan Jackson
National Institute for Petroleum and Energy Research

ABSTRACT

The objective of this study was to determine the effect of the spatial arrangement and density of core-plug scale permeability data in deriving grid-block scale values for numerical simulation. Various averaging techniques were applied to closely spaced outcrop permeability data[1,2] to generate input permeability models used in simulations. The effect of permeability models with varying amounts of detail, as well as viscosity ratio and grid block size, on waterflood oil recovery was determined by numerical simulation using BEST,[3] NIPER's improved version of BOAST.

The results of this study provide a guideline for the degree of detail required for numerical simulator input permeability models for the specific type of rocks and conditions investigated. In the absence of the required detail, an indication of the magnitude and direction of errors that can be expected in oil recovery prediction is given.

Two sampled outcrop areas which differed in geological and permeability characteristics were modeled and simulated. One area, 5x10 ft, was relatively homogeneous with a permeability coefficient of variation of 0.58; while the second area, 4x21 ft, had discontinuous high permeability layers with contrasts of mean permeability approximately 1:50 between layers and a permeability coefficient of variation of 1.28.

The level of detail in the input permeability models ranged from one average permeability value to the incorporation of the maximum amount of information available. Input permeability models included: (1) profile layer models based on one vertical permeability profile; (2) kriged permeability models with 0, 1, and 2 standard deviations randomly imposed on the interpolated grid; stochastic realizations using: (3) the indicator kriging method and (4) the turning bands method; and (5) a geological model containing the maximum amount of information available from outcrop descriptions and photographs and core-plug measurements, and considered the most accurate representation of the rocks.

Results of this study indicate the following: (1) to predict waterflood oil recovery accurately, greater detail in simulator input models is required when the reservoir is heterogeneous and the viscosity ratio is adverse. However, indicator kriging and turning bands methods can adequately predict waterflood oil recovery with relatively sparse amounts of data; (2) failure to account for the distribution of discontinuous, high-permeability sand layers will result in a pessimistic prediction of waterflood production; and (3) the presence of small-scale permeability heterogeneity increases dispersivity of the water front, decreases the effective permeability, reduces fingering, and thereby improves sweep efficiency and increases waterflood oil recovery.

I. INTRODUCTION

Numerical reservoir simulation studies provide critical information for decisions on oil well commerciality, maximum efficient rate of production (MER), overall reservoir management, initiation of secondary recovery projects, and selection of optimum enhanced oil recovery (EOR) processes. Without a realistic geologic and engineering reservoir description, simulation results may be substantially in error.

Quantitative reservoir models are necessary for selecting grid block dimensions and assigning appropriate petrophysical values to the grid blocks in reservoir simulation. Depending on the degree and scale of reservoir heterogeneity, the level of detail obtainable from subsurface sampling is more often than not inadequate to assign simulation parameters accurately to predict fluid flow on the interwell scale. Quantitative descriptions for simulation at scales less than well spacing can be obtained, however, by combining available subsurface reservoir data with a detailed quantitative model of rock composition and fluid flow characteristics constructed through statistical sampling and analysis of analogous outcrops.

The need for incorporating the effect of rock heterogeneities in reservoir models had been recognized as early as 1950 when Muskat derived analytical formulas for flow in a stratified reservoir.[4] Dykstra and Parsons[5] defined their well known coefficient which characterizes the reservoir heterogeneity and estimates the waterflood sweep efficiency in a layered heterogeneous reservoir.[6] The Dykstra-Parsons coefficient, however, does not include a spatial component of variability.

The advent of computers allowed Warren and Price[7] to study the effect of random spatial distributions of permeability values on single-phase fluid flow using Monte Carlo numerical simulators. They found that the effective reservoir permeability is equal to the geometric mean of the individual permeability values.

The effect of small-scale heterogeneities on core relative permeabilities was studied by Huppler[8] who found that channel-like effects can significantly alter the measurement results. Hearn[9] calculated pseudo relative permeabilities for stratified reservoirs with a negligible vertical pressure gradient. Kortekaas[10] simulated flow in crossbedded reservoirs by using pseudo relative permeabilities and found that oil trapping takes place behind the higher permeability foreset laminae.

The geostatistical approach in various kriging interpolation schemes is often used to generate maps of simulator input parameters (permeability, porosity, saturations, etc.) and uncertainties in these values; however, these maps are typically smoother than the actual distribution in the reservoir.

To reinstate the randomness existing in a reservoir, Monte Carlo simulations have been conducted by Smith et al.[11] and Delhome[12] in geostatistical work applied to hydrology. Recently, various methods to generate random fields have been proposed and used[13] to preserve the randomness and the correlation lengths existing in an actual reservoir. A good presentation of geostatistical theory and practice is given by Journel and Huijbregts.[14] By creating reservoir models statistically compatible with existing data and by running numerical flow simulations on these models, a range of outputs is obtained, and an uncertainty in the simulation results based on the input data can be computed.

A systematic approach for handling heterogeneities from small to large scale was given by Lasseter et al.[15] by using pseudo functions. They found that while small-scale heterogeneities mainly cause oil trapping, the medium- and large-scale heterogeneities control lateral and vertical communication and ultimate oil recovery. The use of fractals in generating permeability distributions and predicting reservoir performance by use of simulations was described by Hewett et al.[16] The method described is based on the scale range (R/S) analysis of porosity in a 1,100-ft-thick reservoir and lateral extrapolation of the reservoir properties by generating synthetic logs based on the observed fractal behavior.

Another method for calculating effective grid block values was presented by Begg et al.[17], where the rock type distribution and the effect of layer geometry was studied. They found that the effective permeabilities of the various rock types were controlled by the length to width ratio of the sand layers.

A review of the application of various stochastic modeling techniques used to create reservoir representations is given by Haldorsen and Damsleth.[18] Before one selects one or another stochastic modeling technique, the structure of the reservoir should be understood, (i.e. layer cake, jig-saw or labyrinth type, Weber and Van Geuns[19]), and the statistical and spatial information regarding the reservoir rock types present should be available..

In the present work values for directional correlation lengths available from analogous outcrop measurements and depositional environment characteristics, are used to generate realizations of rock properties (permeability, k, porosity, ø) statistically and spatially compatible with the observed values by means of kriging, indicator kriging[20,21] and the turning bands methods.[22,23]

The effect of permeability models with varying amounts of detail, as well as viscosity ratio and grid block size, on waterflood oil recovery was determined by numerical simulation using BEST,[3] NIPER's improved version of BOAST.

II. GEOLOGICAL BACKGROUND

A quantitative geologic model was developed for an outcrop of the upper Cretaceous Shannon Sandstone formation, a Shelf Sand Ridge deposit that produces oil in Hartzog Draw field, Heldt Draw field, Teapot Dome field, and many other smaller fields in the Powder River Basin. The model was based on detailed sedimentological descriptions, laboratory measurements of permeability and porosity in over 1,200 outcrop core plugs and statistical analysis of the spatial distribution of permeability across the 1,000-ft-long outcrop exposure of the Shannon Sandstone exposed around the Salt Creek anticline, north of Casper, Wyoming (Figure 1). The resulting model was compared with subsurface data from Hartzog Draw field, located 40 miles away at a depth of approximately 9,300 ft. and Teapot Dome field, located 5 miles away. at a depth of 300 ft. Sedimentological features such as grain size, lithology, and sedimentary structures and petrophysical properties (permeability, k, porosity, ø) in the outcrop were used to determine the degree of depositional similarity. This work was reported elsewhere.[1,2] Major conclusions from this work were as follows:

1. Sedimentologically defined units (facies) within the Shannon Sandstone provide a good approximation of rock units with similar permeability characteristics in both the outcrop and Hartzog Draw field.
2. Stratification types and sedimentary structures can be related to permeability classes in both outcrop and subsurface and may define the spatial dimensions for scaling-up to reservoir simulation grid-block size volumes.
3. Although the magnitude of permeabilities in the outcrop and Hartzog Draw field differ by two orders of magnitude, similar decreasing downward trends in permeability exist through the same sequence of facies and stratification types. Decreasing permeabilities correspond to decreasing depositional energy and indicate that the primary depositional permeability pattern has not been totally masked by diagenesis.

A brief geological description of the two areas used in the simulation studies is presented below.

Four sedimentologically defined units (indicated in Figure 2) of differing lithology and bedding characteristics occur within the High Energy Ridge Margin (HERM) facies in the outcrop studied.

Area A is located within HERM Units 3a and 3b which consist of 1 to 2 ft thick trough cross-bedded sets, which extend laterally for approximately 20 ft. The sand contains 5 to 20% glauconite that is both disseminated and concentrated along cross-bed laminae. Clay drapes and limonite clasts also occur along bedding planes and cross-bed laminae. Units 3a and 3b are similar, except for slight differences in the amount of glauconite and clay drapes present. Because of the difficulty of core recovery from clay and shale drapes, limonite clasts and lenses, their effects on the permeability distribution were not considered. Consequently, the model developed for this area did not contain the total amount of heterogeneity present in the rocks. Analysis of the permeability data for Area A (Units 3a and 3b) indicates a relatively sharp-peaked frequency distribution (Fig. 3a) with a mean natural logarithm permeability of 7.1 (1,215 md), and a standard deviation of 0.94 (Table 1).

Area B is located within HERM Units 3c, 3d, and bioturbated shelf sand (Figure 2). The following four rock types were distinguished: (1) bioturbated 200 micron sandstone with 10% clay, 15% glauconite, and 90% burrowing (Unit 3c, bioturbated shelf-margin sand); (2) horizontal to massive bedded 200 micron sand with 5% clay (HERM Unit 3d); (3) trough and subhorizontal stratified 300 micron sand with up to 15% glauconite (HERM Unit 3d); and (4) bioturbated 125 micron sand with trace of clay, 2% glauconite, and 99% burrowing (Bioturbated Shelf Sandstone facies). The HERM unit 3d is a highly stratified unit that consists of 0.25 to 0.5 ft thick trough and subhorizontal cross-beds which are glauconite-rich alternating with glauconite-poor, finer-grained horizontal to massive bedded layers. The glauconite-rich cross-beds have a mean permeability of about 500 md while the finer-grained beds have a mean permeability of about 50 md. The lateral extent of these beds ranges between 10 and 20 ft. The permeability distribution for Unit 3d in Area B is bimodal and reflects the presence of the two stratification types and lithologies. The mean permeability of the lower permeability group is around 12 md while the mean of the higher permeability group is approximately 650 md, yielding a permeability contrast of 1:50 between the layers.

III. PERMEABILITY MODELS

A. Area A Permeability Models

Six permeability models were generated using 31 permeability values measured from 1-in. diameter core plugs drilled from the face of the outcrop. Four of these permeability models are shown in Figure 4. The arithmetic average of the samples was 1,622 md, geometric mean was 1,215 md, and the standard deviation was 946 md (Table 1). The permeability models generated ranged from a single permeability value to the incorporation of the maximum amount of available information. Two grid configurations were used in Area A: a 40X1X20 grid which contained grid block sizes of 0.25 ft, close to the size of the outcrop plug samples, and coarser grid of 10X1X5 grid with grid-block sizes of 1 ft by 1 ft, constructed by assigning the arithmetic average value of 16 grid blocks (4X1X4 blocks from the 40X1X20 grid). The permeability models developed and simulated are as follows:

1. Uniform permeability model

A constant permeability of 1,215 md, the geometric average of permeability of the 31 core plugs, was assigned to all gridblocks. The geometric average was selected based on simulations of permeability fields generated by randomly sampling the permeability distribution of the measured permeability samples in Area A. This model is the least detailed, and serves as a base case for comparison of the other more complex representations.

2. Profile layer models

Three, five-layer models were constructed based on three vertical permeability profiles (Table 2). The permeabilities assigned to each layer were from actual measured values and extended across the entire area simulated. The Dykstra-Parsons coefficients of these three permeability profiles were 0.30, 0.31, and 0.51, which indicated relatively homogeneous permeability distributions.

3. Kriged permeability model (K0 model)

The permeability distribution for this model was based on the interpolation between the 31 measured core plugs in Area A. The interpolation method used was a kriging routine which assumed isotropic correlation lengths. This is justified by the fact that strong layering is not present, and a high-permeability area (more than 2,000 md) is located in the central part of Area A, while permeability drops to less than 1,000 md in the top and bottom part of the area (Figure 4a).

4. Kriged permeability plus one standard deviation noise (K1 model)

A random perturbation was imposed on the interpolated values of permeability to compensate for the smoothing effect due to interpolation and to preserve the random component naturally found in permeability distributions. Natural logarithm values of permeability and the standard deviation (946 md) were used to preserve the geometric mean of the population. A series of random numbers ranging from -0.5 to +0.5 were generated and then multiplied by the logarithmic standard deviation value of 6.85. This product was then added to the logarithmic value of the interpolated grid block permeability value and reconverted to a non-logarithm permeability value. Figure 4b shows the permeability distribution generated using this procedure.

5. Kriged permeability plus two standard deviations noise (K2 model)

This model was constructed as described above, except that two standard deviations were added to the kriged-based permeability values (Figure 4c).

6. Coarse grid model

A coarse grid permeability model, constructed by assigning the arithmetic average value of 16 grid blocks (four by four blocks in vertical and horizontal directions) from the K0 permeability models is shown in Figure 4d.

B. Area B Permeability Models

The permeability models, constructed on 85x1x17 grids with square grid blocks 0.25ft x 0.25ft are shown in Figure 5 and are described below:

1. Three-layer model

Due to the layered nature of the area and the bimodal permeability distribution, the simplest model used in this area was a three-layer model (Figure 5a). The three layers have constant permeabilities of the geometric mean of the samples located in that layer and are from bottom to top: 7, 157, and 14 md.

2. Profile layer model

This model (Figure 5b) is based on the permeability obtained from one vertical profile selected from a grid, which was generated for the geological model and is described below. Each layer has constant permeability values across the area simulated. This model is analogous to those typically used in the field, where information from one well is extrapolated across the interwell area (Table 3).

3. Geological fine grid model

This model is the most detailed and therefore accurate permeability representation and serves as a bench mark for comparison to the other less detailed models. It incorporates geological information about the spatial distribution and lateral continuity of the layers obtained from outcrop descriptions and photographs with the 101 core-plug permeability measurements (Figure 5c).

Comparison of permeability statistical parameters from the four rock types indicated distinct permeability classes for each rock type. The top and bottom layers are included as boundary layers and are from a different facies than the HERM facies of interest. The rock type of each grid block was identified from photographs, outcrop descriptions and core-plug examination. To preserve the spatial distribution of rock types, (high permeability discontinuous layers), the dimensions and distribution of the rock types and layers measured from the outcrop were used to construct the model rather than interpolating between the data points. Permeabilities were assigned to each grid block by randomly sampling the permeability distribution for that class. The measured permeability value was assigned to those 101 grid blocks which had been sampled. This method of assigning grid-block permeability values preserved the frequency distribution characteristics within each permeability class.

4. Geological coarse grid model

A coarse-grid permeability model (21x1x4) was generated by taking the arithmetic average of 16 neighboring blocks from the geological model (Figure 5d).

5. Indicator kriging model

A dual rock indicator kriging model (Figure 5e) was generated using the software package developed at Stanford.University.[21] Based on variogram analysis, correlation lengths of 10 ft laterally and 0.25 ft vertically, were selected. Examination of outcrop photographs, were used to generate the spatial distribution of the two rock types. The correlation lengths calculated were one-half the average length of the bed sets measured from the outcrop exposure. The grid blocks for each rock type was assigned a permeability value drawn randomly from the statistical distribution for that rock type. The realizations were conditioned on the vertical profiles located at points 0 and 21 ft.

6. Turning bands model

The turning bands model used to generate unconditional realizations of permeability (Figure 5f) was the software package TUBA[23], developed at New Mexico Institute for Mining and Technology. The same lateral and vertical correlation lengths used in the indicator kriging model were applied here. The realizations that best approximated the geological model were based on the Bessel and exponential correlation functions and the log normal permeability distribution function.

IV. SIMULATION PARAMETERS

A. Viscosity Ratios

Two oil-to-water viscosity ratios: 2.78 corresponding to 35° API oil gravity and 24.5, corresponding to 20° API oil gravity were used to represent a range of viscosity ratios. The viscosity ratios used correspond to the viscosity values at 1,000 psi, which is reservoir the pressure occurring in the waterflood simulations.

B. Vertical Permeability

Vertical permeability values were measured from cubes cut from 2-in.-diameter cylindrical cores. Permeability was measured from the 1.4-in.-square cubes in the x (parallel to the outcrop face), y (normal to the outcrop face) and z (vertical to bedding plane) directions. Mean permeability values measured in the 2-in.-diameter cores were similar to those measured in the x and y direction from the cubic samples (1,080, 1,021, and 1,099 md, respectively), while those measured in the z direction were lower (640 md) (Table 4). The vertical permeability values used in the simulations were represented by the relation $k_v = 0.00038\ k_h^2$ derived from a plot of measured vertical permeability versus horizontal permeability.

C. Relative Permeability

Relative permeability measurements from Teapot Dome field were used in the waterflood simulations. Teapot Dome is located approximately 5 miles from the outcrop studied and produces oil from the Shannon Sandstone at a depth of around 350 ft. In the absence of other relative permeability data, due to the close proximity and shallow depth of the reservoir, and the similarity in porosity and permeability distributions,[19] the use of the Teapot Dome relative permeability tables is expected to be a good approximation for the outcrop simulations. In Area A, two relative permeability tables were used: (1) a table generated from the measurement of a sample with permeability-to-air of 850 md was used for rocks which have permeability-to-air above 200 md (Table 5a) and (2) a table generated from the measurement of a sample with permeability-to-air of 56.1 md was used for those rocks with air permeability less than 200 md (Table 5b). The Area B relative permeability curves were similar to those in Area A, except that the endpoints of the two relative permeability curves were 85% instead 80% for permeabilities greater than 200 md and 72% instead 50% for permeabilities less than 200 md.

D. Initial Water Saturation

Initial water saturation values were determined from the laboratory measurements of relative permeability. In Area A, initial water saturation values of 15 and 28% were assigned for rocks with air permeability greater and less than 200 md, respectively, while in Area B, initial water saturations of 20 and 50%, respectively were assigned. Slightly higher initial water saturations were assigned in Area B because of lower average permeability values.

E. Well Locations and Flow Rates

The configuration of the simulation consisted of water injection (or source) assigned to all grid blocks at one edge of the model and production (or sink) to all grid blocks at the opposite edge. A constant flowrate of 5 bbl/d was applied to both wells. This flow rate is equivalent to a velocity of 3 ft/d which is common in field waterfloods. All the pore volume (PV) values presented in the text, figures, and tables are expressed as movable pore volumes (MPV).

V. SIMULATION RESULTS

A. Waterflood Simulation of Area A

Three types of illustrations are used to analyze and compare results for 35° API (viscosity ratios V_r = 2.78) and 20° API (viscosity ratio, V_r = 29.5) cumulative oil production of movable oil in place (MOIP), (Figures 6a,b; 8a,b); water-oil ratio (Figures 7a,b; 9a,b), both of which illustrate production performance; and oil saturation distributions at certain production times (Figures 10a-f and 11a-f) which aids in understanding the waterflooding behavior.

1. Uniform Permeability Model

In the uniform permeability model, 1,215 md, the geometric average of permeability from 30 outcrop samples, was assigned to each of the 800 grid blocks. Waterfloods were simulated using two types of oil and their corresponding viscosity ratio (v_r): 35° API, v_r = 2.78; 20° API, v_r = 24.5. The simulation results of this model will be used for comparison with the other more detailed models developed.

As expected, the oil recovery curves indicates that the higher the viscosity ratio, the poorer the sweep efficiency, and the lower oil the recovery is (Figures 6a,b), due to the earlier water breakthrough (Figures 7a,b) The low viscosity ratio case exhibits a later breakthrough but faster increase in the water-oil ratio after breakthrough, than the more adverse viscosity ratio of 24.5.

Oil saturation distributions for the viscosity ratio of 2.78 after injection of 0.64 PV indicates a piston-like displacement with a slight water tongue at the bottom due to gravity effect. The oil saturation distribution for the viscosity ratio 24.5 after injection of 0.64 PV, appears also piston like. However, the width of the transition zone, defined as from 17% (residual oil) to 85% (initial oil saturation) is longer and the breakthrough is earlier than for the more favorable viscosity ratio case. The long transition zone can be considered equivalent to viscous fingering.

2. Vertical Profile (VP) Layer Models

Three, 5-layer models were constructed using three vertical profiles of permeability existing in the core-plug data (Table 2). These three cases showed similar overall waterflood performances and the lowest recovery among all models. (Figure 6a). They also showed similar water-oil-ratio values (earlier than all other models Fig. 7a) although some minor differences exist. Profile model 2, which had the lowest Dykstra-Parsons coefficient (0.30), had 3 to 5% greater oil recovery and 0.1 PV later water-breakthrough. Differences of 3% in oil production between profile models 2 and 3 which have similar Dykstra-Parsons coefficients (0.30 and 0.31, respectively) are due to the arrangement of layers rather than the degree of heterogeneity. The similarity of performances of the three layer models can be attributed to the relatively homogeneous permeability distribution (coefficient of variation 0.58) in Area A. Figures 10a-c show typical oil saturation distribution for the first of the three profile layer models after 0.64 PV of water was injected. The water front movement in each layer was proportional to the permeability of that layer. The more pessimistic oil recovery prediction for the profile models is due to the continuous channels erroneously assumed in the model.

3. Kriged Permeability Model Without Random Components (K0 Model)

After injection of 0.64 pore volumes (PV) of fluid with a 2.78 viscosity ratio, the resulting transition zone presents a relatively sharp, 7-ft wide front. (Figure 11a) while for the more adverse, 24.5 viscosity ratio case, the width was greater than 10 feet (Figure 9b). The elongated shape of the front is due to the high-permeability zone in the central part of Area A (Figure 4) which resulted in the rapid movement of the injected water.

The effect of viscosity ratio on produced water-oil-ratio is that the longest breakthrough time followed by a fast increase in the water-oil-ratio occurred for the lower viscosity ratio case, whereas an earlier increase in water-oil-ratio occurs for the more adverse viscosity ratio of 24.5.

The oil recovery from the K0 model is 6% greater than that from the three profile models (Figure 6a) and the breakthrough time occurs at 0.8 PV compared to 0.6 PV for the profile models (Figure 7a).

4. Kriged Permeability Models - Coarse Grid

Regardless of the permeability perturbations imposed on the original kriged permeability model, the 16-block averaging coarse grid showed similar oil saturation distributions. They also showed similar waterflood performance with the kriged model (KO) (Figure 6a) for the favorable viscosity ratio and more optimistic recovery for the unfavorable viscosity ratio (Figure 6b). For both viscosity ratios the coarse grid models showed a slightly earlier break through than the five good kriged model (Figures 7a,b). This is due to the loss of local permeability heterogeneity in the model through the averaging procedure.

5. Kriged Permeability Models With Random Components

The viscosity ratio has a similar effect on width of the oil saturation transition zone in the K1 and K2 models as the K0 model (Figures 11a-f). However, narrower transition zone widths occur in the K2 model indicating that better sweep efficiency occurs with greater amounts of small-scale heterogeneity. This relationship exists for all cases of mobility ratios simulated, however it is more pronounced for the more favorable viscosity ratio of 2.78 (Figures 11a,c,e).

Comparison of production performance among the three kriged models indicates that the highest oil recovery results from simulation of the K2 model for both viscosity ratios (Figures 8a,b). The greatest small-scale permeability heterogeneity exhibits the most efficient recovery. Oil recovery from the K2 model is 7% MOIP higher than from the K0 model and 3% MOIP higher than the K1 model for the viscosity ratio of 2.78 (Figure 8a). A slightly lower difference of 5% MOIP between the K0 and K2 models exists for the less favorable viscosity ratio of 24.5 (Figure 8b).

The benefits of small-scale heterogeneity are also illustrated in Figure 9a where, for a viscosity ratio of 2.78, the permeability model with the greatest small-scale variability (K2) has the latest breakthrough time. This effect is due to the presence of heterogeneity which diverts the flow from a straight line path and thereby increases the dispersivity of the water front. The increase in dispersivity reduces the fingering phenomena, delays the water breakthrough and thereby improves the sweep efficiency. In the case of the less favorable viscosity ratio of 24.5, delay in breakthrough time and the increase of water-oil ratio also occur from the addition of small-scale heterogeneity in the permeability models (Figure 9b).

An increase in small-scale permeability heterogeneity appears to reduce the effective permeability and causes a larger pressure gradient at a constant flow rate. Figures 12a-c show pressure distributions resulting from waterflood simulations of the K0, K1 and K2 models respectively, using a viscosity ratio of 2.78. A pressure drop of 12 psi was obtained across the 10 ft simulated for the K0 model (Figure 12a), while pressure drops of 15 and 18 psi were observed for the K1 (Figure 12b) and K2 (Figure 12c) models respectively, after injection of 0.17 PV.

6. Discussion

a. Viscosity ratio 2.78. In the case of the favorable viscosity ratio ($V_r = 2.78$), a maximum difference of 11% MOIP in cumulative oil recovery prediction occurs between the uniform permeability model and the vertical profile layer model (vertical profile 2), while oil recoveries from the coarse-grid and fine-grid kriged models are essentially the same (Figure 6a). The vertical profile model (profile 2), is 6% lower in oil recovery than the kriged model and gives a slightly pessimistic prediction of oil recovery.

The oil recovery curve for the uniform model lies within the same 7% range of predicted recoveries from the three kriged models. Since the range of oil recoveries from the various kriged models is assumed to encompass the accurate prediction of waterflood recovery, the similarity in cumulative oil recovery suggests that the uniform model, a single geometric average value, is an acceptable representation of permeability for this area which has a relatively homogeneous permeability distribution.

b. Viscosity ratio 24.5. In the case of the less favorable viscosity ratio ($V_r = 24.5$), a maximum difference exceeding 10% MOIP cumulative oil recovery occurs between the uniform permeability model and kriged fine-grid model (Figure 6b). Due to this relatively large difference in oil recovery, the uniform permeability model is not an adequate representation of permeability when the viscosity ratio is 24.5 as it was in the more favorable viscosity ratio case.

The range of oil recoveries among the various kriged models is 5%, which is 2% less than the 7% range found in the 2.78 viscosity ratio case. The lower range of oil recoveries indicates a decreased sensitivity to permeability variation of the water flow under an adverse mobility ratio.

Oil recoveries from the coarse grid model are more then those resulting from the K0 model (Figure 6b) but within a 10% range. This suggests that this model is an acceptable representation of permeability, as it was for the fine-grid model.

B. Waterflood Simulation of Area B

As in Area A, three types of illustrations are used to analyze and compare results for API 35° (V_r = 2.78) and API 20° (V_r = 24.5) cumulative oil production of movable oil-in-place (MOIP), (Figures 13a,b); water-oil-ratio (Figures 14a,b), and oil saturation distributions at certain production times (Figs. 15-20).

1. Three-layer model

Simulations of the three-layer model with a viscosity ratio of 2.78 indicated piston-like displacement with a sharp front due to the single, higher permeability layer. This model exhibits the latest breakthrough and the most rapid increase in water-oil-ratio after breakthrough of all models studied (Figure 14a). The three-layer coarse grid model also showed piston-like displacement; however, an earlier breakthrough and therefore a slower buildup of production water-oil-ratio occurred in the coarse grid model case (Figure 14a).

Comparison of saturation distributions of the fine-grid model with a viscosity ratio of 24.5 to that with a 2.78 viscosity ratio indicate a wider transition zone (Figures 15a-d) and less oil recovery - 67% MOIP for the 24.5-viscosity ratio case (Figure 13b) versus greater than 90% MOIP for the 2.78 viscosity ratio case (Figure 13a) after 2.5 PV water injection.

Simulation results for the 24.5 viscosity ratio, uniform coarse grid model indicate similar results as the uniform fine-grid model except for an earlier breakthrough (Figure 14b) and 4% MOIP less oil recovery after 1 PV water injection (Figure 13b).

2. Profile layer model

Three fingers were observed from the waterflood simulation for the 2.78 viscosity ratio after injection of 0.32 PV due to the three continuous permeable layers in the permeability model (Figures 16a,b). The highest permeability finger caused an early breakthrough and causes a significant quantity of oil to be left unswept in the lower permeability layers after breakthrough.

Similar channeling and early breakthrough occurred in the 24.5 viscosity ratio case (Figures 16c,d), however, differences in viscosity ratio resulted in a 10% decrease in oil recovery. After water injection of 1 PV, 52% MOIP was recovered for the $V_r = 2.78$ case, whereas 42% MOIP was recovered for the $V_r = 24.5$ case.

3. Geologic fine-grid model

The permeability distribution is shown in Figure 5. The sequence of oil saturation distribution during waterflood of the 2.78 viscosity ratio case illustrates good sweep efficiency in spite of high permeability contrast layers because the high permeability layers are not continuous across the simulation area (Figures 17a-d). One continuous, 3-in. thick flow channel and another 8-ft. long relatively high-permeability layer caused fingering which is reflected in the oil saturation distribution after 0.32 PV water injection (Figure 17a). The injected water flowed from these two "layers" to four other discontinuous, good permeability layers which ranged from 2 to 11 ft. in length. This cross-flow caused a late water breakthrough and a relatively uniform sweep that is apparent in the oil saturation distribution after 0.85 PV water injection (Figure 17b). After water injection of 1.75 PV, most oil located between the permeable sand layers were swept out. Residual oil was found on the top and the bottom of the model, areas which were bypassed by the water. In contrast to the 2.78 viscosity ratio case, waterflood results of the 24.5 viscosity ratio showed a very early breakthrough (Figure 14b) and therefore poor sweep efficiency. The adverse viscosity ratio ($V_r = 24.5$) also caused channeling (Figure 17c) and reduced the amount of cross-flow of injected water which resulted in early breakthrough. The residual oil saturation distribution after an injection of 0.85 PV is shown in Figure 17d and illustrates the lower MOIP recovery for this unfavorable viscosity ratio case when compared to the more favorable viscosity ratio production.

4. Geological coarse-grid model

Comparison of the residual oil saturation distributions resulting from the coarse-grid geologic model, 2.78 viscosity ratio case (Figures 18a-b) with that of the fine grid model (Figures 17a-b) indicates higher saturations on the bottom close to the producer. This difference is due to the averaging procedure which distorted the original permeability continuity. The permeability channel close to the bottom of the producer was eliminated by the neighboring low-permeability rock in the averaging process. Other differences include an earlier breakthrough, higher water-oil ratio for the coarse-grid model (Figure 14a), and lower ultimate oil recovery for the viscosity ratio 2.78 (Figure 13a).

For the viscosity ratio 24.5 a higher oil recovery (Figure 13b) with a corresponding lower residual oil saturation (Figures 18c-d) is observed for the coarse grid model compared to the geological fine grid model.

5. Indicator kriging model

The permeability distribution shown in Figure 5e is typical of the various realizations generated using the 10 ft lateral correlation length and 0.5 ft vertical correlation length and it shows a marked similarity to the geological model (Figure 5c). This is due both to the choice of correlation lengths and to the conditioning of the realizations on the two extreme vertical profiles. The high permeability zone present in the lower part of both the geological and indicator kriging model is the main source of similar oil saturation distributions for the two models. The indicator kriging initially overestimates the oil production for the low viscosity ratio but after 2 PV injected, both models show the same cumulative oil production (Figures 13a,b). For the high viscosity ratio case, the indicator kriging model matches the geological model for less than 0.3 PV injected but then due to less continuity under estimates the oil recovery. Again after 2 PV injected both cumulative productions coverage toward 40% MOIP.

A significant difference between the two waterfloods is in the water-oil ratio for the high viscosity case. While after breakthrough, the geological model produces a rapid rise in WOR the indicator kriging model predicts a value which increases asymptotically toward 10 (Figure 14b). The saturation distributions after 0.32 PV and 0.85 PV for the 2.78 viscosity ratio are shown in Figures 19a-b. They show a sharper front than the geological model with a more complete sweep behind the front. The satuation distributions for the unfavorable viscosity ratio $V_r = 2.45$, display a more diffuse front (Figures 19c-d) than the $V_r = 2.78$ case.

6. Turning bands model

Although the permeability distribution generated using the Bessel function correlation form (Figure 5f) is not as close to the geological model (Fig. 5c which is to be expected as this was not a conditional simulation), the cumulative oil production curves, (Figures 13a,b) are very close to the results from the geological model. The Bessel function model slightly underestimates the oil recovery for the viscosity ratio 2.78 and overstimates the initial oil recovery for the 24.5 viscosity ratio. For less than 0.5 PV injected this model behaves like the profile model but for values greater than 0.5 PV its predictions become very close to the geological model. The saturation distributions (Figures 20a-d) are also quite close to those of the geological model. The similarity of the fluid saturation distributions and recovery curves from the profile model can be traced to the presence of a higher permeability zone in both models. As in the indicator kriging model, the geometry of the higher permeability zone is directly controlled by the choice of lateral and vertical corelation lengths. Due to the higher continuity and permeability of the high permeability zone, the water breakthrough takes place earlier and the WOR has a higher value than in the case of the geological model.

7. Discussion

Comparison of the predicted oil recovery from the simplest permeability model, the uniform permeability model to the most detailed model, the geological model at two viscosity ratios indicates widely different predictions in Area B. For the more favorable viscosity ratio of 2.78, after 1 PV injection, the ratio of oil recovery from the uniform model to the geological model is 1.3 (Figure 13a); however, for the viscosity ratio of 24.5, the oil recovery ratio increases to 1.9 (Figure 13b), indicating a larger discrepancy in the oil recovery prediction. These results suggest that the more adverse the mobility ratio, the more detailed information is required for constructing the permeability model.

a. Viscosity ratio 2.78. In the case of the favorable viscosity ratio (2.78), greater oil recoveries were predicted after 0.85 PV injection for the three-layer, fine-grid uniform model (87% MOIP) and lower oil recovery for the profile layer model (49% MOIP), when compared to the oil recovery for the geologic model (63% MOIP) (Figure 13a). After injection of 1 PV, a 20% MOIP difference in oil recovery still exists between the three-layer fine-grid model and the geologic fine-grid model. When compared to the oil recovery for the fine grid geological model, the indicator kriging model overpredicts the oil recovery by 12% MOIP and the turning bands underpredicts the oil recovery by 4% MOIP. However, after injection of 2 PV, the oil recoveries for all models converge toward 90% MOIP, with the exception of the profile layer and the coarse grid geological models which converge toward significantly lower recoveries.

The similarity in production of the geologic model to that of the uniform models after 2 PV injected is due to the crossflow between the good and poor permeability sand layers in the geologic fine-grid model. The cross-flow disperses the front of injected water and results in increased sweep efficiency and eventually, after 2.4 PV of water injection, oil production from the geologic fine-grid model equals that of the uniform model. The poorer production performance of the coarse grid geological model and the profile layer model indicate that failure to account for the random distribution of high permeability sand layers will result in a pessimistic prediction of production.

Comparison of the fine grid and coarse grid geological models indicates that after injection of 0.55 PV (breakthrough time) the coarse-grid geological model provides a pessimistic prediction of oil recovery compared to the fine-grid geological model (Figure 13a). The discrepancy between the coarse-grid and fine-grid geological models increases to 22% after injection of 2 PV and is due to the creation of a low-permeability area which is left unswept in the coarse-grid geological model. These results are in contrast to Area A, where the predicted oil recoveries from the coarse-grid model are similar to those for the kriged, fine-grid models and illustrates that more detailed permeability models are required as the degree of heterogeneity increases.

The profile layer model provides the most pessimistic prediction after injection of 0.32 PV or at breakthrough time (Figures 13a,14a). After injection of 1.5 PV water, the difference in oil recovery between the profile layer model and the geologic fine-grid model increases to 25%. Unlike the fine-grid geological model, the absence of crossflow in the profile layer model results in a lower oil recovery.

Both the indicator kriging and the turning bands model generate results very close to those generated by using the geological fine grid model. This is due to the fact that both models preserve the characteristics of the statistics (mean, standard deviation) and of the spatial distribution (correlation lengths) of rock permeability. Although not shown in the present work, both stochastic models will generate different spatial distributions of permeability if other values for the means, standard deviations and correlation lengths are selected.

b. Viscosity ratio 24.5. In the case of the more unfavorable viscosity ratio (24.5), at 1 PV fluid injected simulation results from the indicator kriging, predicted the lowest oil recovery (22% MOIP), the geological model predicted 31% MOIP and the most optimistic model was the 3 layer with 59% MOIP. (Figure 13b). After injection of 2 pore volumes of water (Figure 13b) the three layer model predicted 62% MOIP recovery while the geological, indicator kriging and the turning bands models predicted recoveries in the 32-37% MOIP range. Both stochastic models and the geological model converge toward the same 40% MOIP ultimate oil recovery. While for the more favorable viscosity ratio of 2.78, after 1 PV injection, the ratio of oil recovery from the three layer model to the geological model is 1.3. For the unfavorable viscosity ratio of 24.5, the oil recovery ratio increases to 1.9, indicating a larger discrepancy in the oil recovery prediction by different models.

In both viscosity ratio cases, the fine-grid uniform model followed by coarse-grid uniform model, provides the highest oil recovery predictions. However, the order of the recovery from the fine-grid geological, coarse-grid geological and the profile layer models differs in the two viscosity ratio cases (Figure 13a,b).

The profile layer model is traditionally considered to be the most pessimistic input model, and this is illustrated in the 2.78 viscosity ratio case. However, in the 24.5 viscosity ratio case, the profile layer model as well as the other models studied,with the exception of indicator kriging model, resulted in higher oil recovery predictions than the more accurate geological model. These results illustrate the importance of generating a more detailed model when the reservoir is composed of discontinuous sand layers of contrasting permeability and the viscosity ratio is relatively unfavorable.

The difference in recoveries can be attributed to the adverse mobility ratio (24.5) that enables water to move more easily than oil. The movement of water between the discontinuous high permeability layers results in water channeling rather than the more efficient sweep created by crossflow. The water relative permeability value in the water channels increases with flooding time; therefore, the injected water follows the existing channel and leaves a large volume of formation rock unswept. This mechanism is illustrated by the increased channeling in the geological model, 24.5 viscosity ratio case which is even more pronounced than in the profile layer model.

VI. SUMMARY AND CONCLUSIONS

1. The more heterogeneous the reservoir, the more detailed information is required for constructing the permeability model. Knowledge of permeability and porosity statistics and especially correlation lengths are essential in computing correct oil recovery.

2. Comparison of the difference between the simplest permeability model and the most detailed, accurate permeability model in the two areas studied varies widely. In Area A, a relatively homogeneous area, with a permeability coefficient of variation of 0.58 and no distinct heterogeneity, the difference between the two models is 4% MOIP for the 2.78 viscosity ratio case at 1 PV and 10% MOIP for the 24.5 viscosity ratio case at 1 PV. In this type of rock, a single permeability value was an adequate representation of the area simulated. In Area B, however, a layered system with permeability contrasts of 1:50 between the layers,and a permeability coefficient of variation of 1.28, a 20% MOIP difference occurs for the 2.78 viscosity ratio case at 1 PV and 28% MOIP difference for the 24.5-viscosity ratio case, underscoring the importance of detailed information in models for oil recovery predictions.

3. The more adverse the viscosity ratio, the more detailed information is required for constructing the permeability model. Comparison of the predicted oil recovery from the simplest permeability model, the uniform permeability model, to the most detailed, the geological model at two viscosity ratios indicates widely different predictions in Area B.

4. Both indicator kriging and turning bands models can be used to accurately predict the oil recovery if the correct permeability statistics and correlation lengths are used. Geostatistical analysis of analogous outcrops can be a valuable source of such information.

5. Spatial correlations of magnitudes comparable to the interwell distances should be reflected in the models simulated to correctly predict the waterflood oil recovery.

6. The traditionally used profile layer model resulted in an highly optimistic oil recovery prediction in the case of a unfavorable viscosity ratio and a highly pesimistic prediction for a favorable viscosity ratio for a reservoir with discontinuous layers with contrasting permeabilities. This effect is mainly due

to the failure of the model to correctly account for the distribution of sand layers with high permeability contrasts.

7. Small-scale permeability uncorrelated heterogeneities increases dispersivity of the water front, decreases the effective permeability, reduces fingering, and thereby improves sweep efficiency and increases oil production and improves waterfloods.

ACKNOWLEDGMENTS

This work was sponsored by the U.S. Dept of Energy under Cooperative Agreement DE-FC2283FE60149, AGIP SpA, ARCO Oil and Gas, BP Exploration, IITRI and RIPED.

REFERENCES

1. Tomutsa, L., S. Jackson, M. Szpakiewicz, I. Palmer. Geostatistical Characterization and Comparison of Outcrop and Subsurface Facies: Shannon Shelf Sand Ridges. Presented at 56th California Regional SPE Meeting, April 2-4, 1986 SPE Paper No. 15127.

2. Jackson, S. R., L. Tomutsa, R. Tillman. Quantified Spatial Variations of Reservoir Parameters in Shelf Sandstone Ridge Deposits, First Annual Report, Feb. 1988, Contract Number B08677-1, (proprietary report).

3. Tomutsa, L. and J. Knight. Effect of Reservoirs Heterogeneities on Waterflood and EOR Chemical Performance. Depart of Energy Report No. NIPER-235, July 1987.

4. Muskat, M. The Effect of Permeability Stratification in Complete Water Drive Systems. Trans., AIME, v. 189, December 1950, pp. 349-358.

5. Dykstra, M. and R. L. Parsons. The Prediction of Oil Recovery by Waterflood. Secondary Recovery of Oil in the United States, 2nd Ed. API, 1950, pp. 160-170.

6. Willhite, G. P. Waterflooding. SPE monograph, 1986.

7. Warren, J. E. and H. S. Price. Flow in Heterogeneous Porous Media. SPEJ, v. 1, No. 3, Sept. 1961, pp. 153-169.

8. Huppler, J. D. Numerical Investigation of the Effects of Core Heterogeneities on Waterflood Relative Permeabilities. SPEJ, v. 10, No. 4, Dec. 1970, pp. 381-392.

9. Hearn, C. C. Simulation of Stratified Waterflooding by Pseudo Relative Permeability Curves. J. Pet. Tech., v. 23, July 1971, pp. 805-813.

10. Kortekaas, T. F. M. Water/Oil Displacement Characteristics in Cross Bedded Reservoir Zones. Pres. at Soc. Petrol. Eng. 58th Ann. Tech. Conf. and Exhib., San Francisco; SPEJ, v. 15, No. 6, Dec. 1985, pp. 917-926.

11. Smith, L. and F. W. Schwartz. Mass Transport 3. Role of Hydraulic Conductivity Data in Prediction. Water Resources Research, v. 17, No. 5, Oct. 1981, pp 1463-1479.

12. Delhomme, J. P. Spatial Variability and Uncertainty in Groundwater Flow Parameters A Geostatistical Approach. Water Resources Research, v. 15, No. 2, April 1979.

13. Sultan, Ahmand Junaid, John P. Heller and Allan L. Gutjahr. Generation and Testing of Random Fields and the Conditional Simulation of Reservoirs. Petroleum Society of CIM/SPE International Tech. Meeting, Calgary,1990.

14. Journel, A. G.and C. J. Huijbregts. Mining Geostatistics, Academic Press, 1978.

15. Lasseter, T. J., T. R. Waggoner and L. W. Lake. Reservoir Heterogeneities and Their Influence on Ultimate Recovery. In Reservoir Characterization. Eds., L. W. Lake and H. B. Carroll, Academic Press, 1986, pp. 545-560.

16. Hewett, T. A. Fractal Distributions of Reservoir Heterogeneity and Their Influence on Fluid Transport. Pres. at 61st Ann. Technical Conf. and Exhib of SPE in New Orleans, Oct. 1986. SPE paper 15386.

17. Begg, S. H., R. R. Carter, and P. Dranfield. Assigning Effective Values to Simulator Gridblock Parameters for Heterogeneous Reservoirs. November 1989. SPE Reservoir Engineering.

18. Haldorsen, H. H., E. Damsleth,. Stochastic Modeling, JPT, vol. 42, no. 4, April 1990 pp. 405-412.

19. Weber, K. J. and L. C. van Geuns. Framework for Constructing Clastic Reservoir Simulation Models. Pres. at 69 thSPE Annual Technical Conf and Exhibit, San Antonio, TX, 1989., SPE 19582.

20. Journel, A. G. and F. G. Alabert. New Method for Reservoir Mapping. In: JPT, vol. 42, no. 2, February 1990.

21 Gomez-Hernandez, J. J. and R. M. Srivastava. ISIM3D: An ANSI-C Three-Dimensional Multiple Indicator Conditional Simulation Program, Computers and Geosciences Vol. 16, No. 4, pp. 395-440, 199.

22. Wilson, J. L. and A. Gutjahr. Synthetic Generation of Random Permeability Fields for Heterogeneous Reservoir Simulation. Pres. at SPE Symposium on Reservoir Simulation, Houston, TX, Feb. 1989. SPE 18435.

23. Zimmerman, D. A. and J. L. Wilson. Description of and Users Manual for TUBA: A Computer Code for Generating Two-Dimensional Fields via the Turning Bands Method. Prepared by New Mexico Institute for Mining & Technology, July 1990. Distributed by GRAM, Inc, Albuquerque, New Mexico.

TABLE 1. - Permeability statistics of measured core plugs.

	Area A		Area B	
	Perm., md	Natural log perm.	Perm., md	Natural log perm.
Sample size	31	31	47	47
Average	1622	7.10	466	5.05
Geometric mean	1215	--	156	--
Standard deviation	945.75	0.94	595.10	1.94

TABLE 2. - Permeabilities (md) assigned to layers in the Vertical Profile Layer models in Area A.

Layer	Vertical profile 1	Vertical profile 2	Vertical profile 3
1	642	64	1,563
2	1,312	2,803	1,768
3	1,982	2,970	149
4	2,504	2,805	1,098
5	799	1,508	2,069

TABLE 3. - Permeabilities (md) assigned to layers in the profile layer model in Area B.

Layer	Horizontal permeability
1	35
2	47
3	46
4	36
5	23
6	50
7	392
8	358
9	74
10	11
11	548
12	559
13	14
14	450
15	14
16	6
17	13

TABLE 4. - Summary statistics of directional permeabilities.

	Two-inch cylindrical cores	Cubic samples, x direction	Cubic samples, x direction	Cubic samples, z direction (vertical)
Sample size	81	73	75	74
Average, md	1,079.51	1,020.78	1,099.12	639.52
Median, md	1,047.78	1,022.57	1,112.79	659.84
Geometric mean, md	622.78	592.53	742.52	221.22
Standard deviation, md	712.39	684.60	712.97	563.80
Minimum, md	1.92	3.44	2.73	0.78
Maximum, md	3,442.48	3,243.45	3,418.63	2,771.06

TABLE 5a. - Relative permeability for rocks with permeability greater than 200 md Area A.

Fluid saturation	Relative permeability	
	Oil	Water
0	0	0
0.15	0	0
0.24	0.03	0.01
0.30	0.06	0.02
0.40	0.18	0.04
0.50	0.37	0.07
0.60	0.58	0.11
0.70	0.83	0.17
0.76	1.00	0.31
1.82	1.00	0.31
1.00	1.00	1.00

TABLE 5b. - Relative permeability for rocks with permeability less than 200 md Area A.

Fluid saturation	Relative permeability	
	Oil	Water
0	0	0
0.28	0	0
0.32	0.03	0.0016
0.41	0.11	0.0045
0.57	0.40	0.0244
0.69	0.82	0.0481
0.72	1.00	0.0583
0.74	1.00	0.0649
1.00	1.00	0.0649

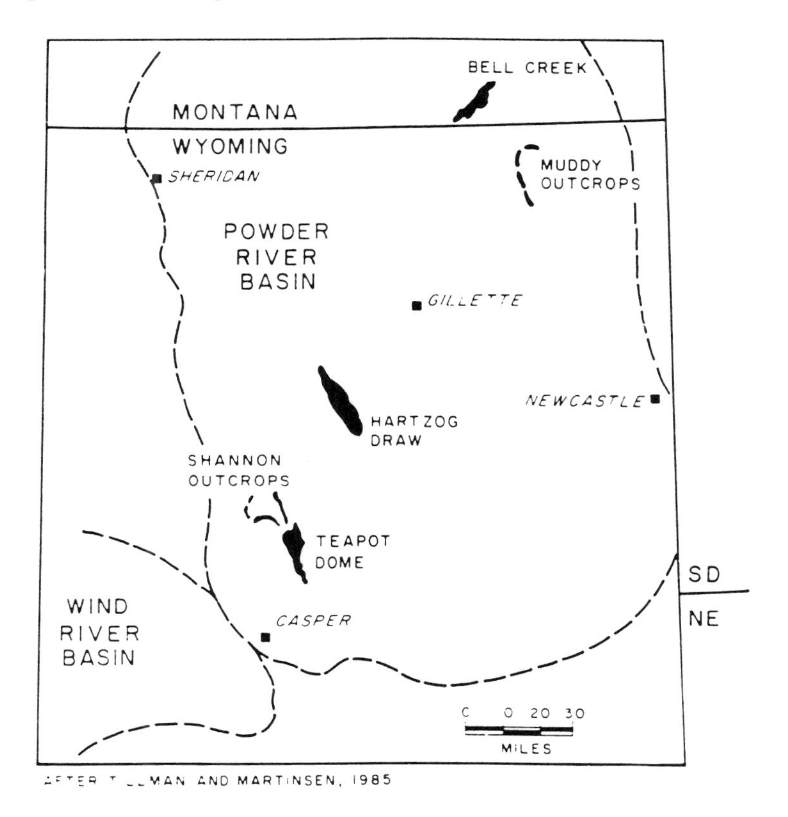

Fig. 1. Location of Shannon outcrops, Powder River Basin, Wyoming.

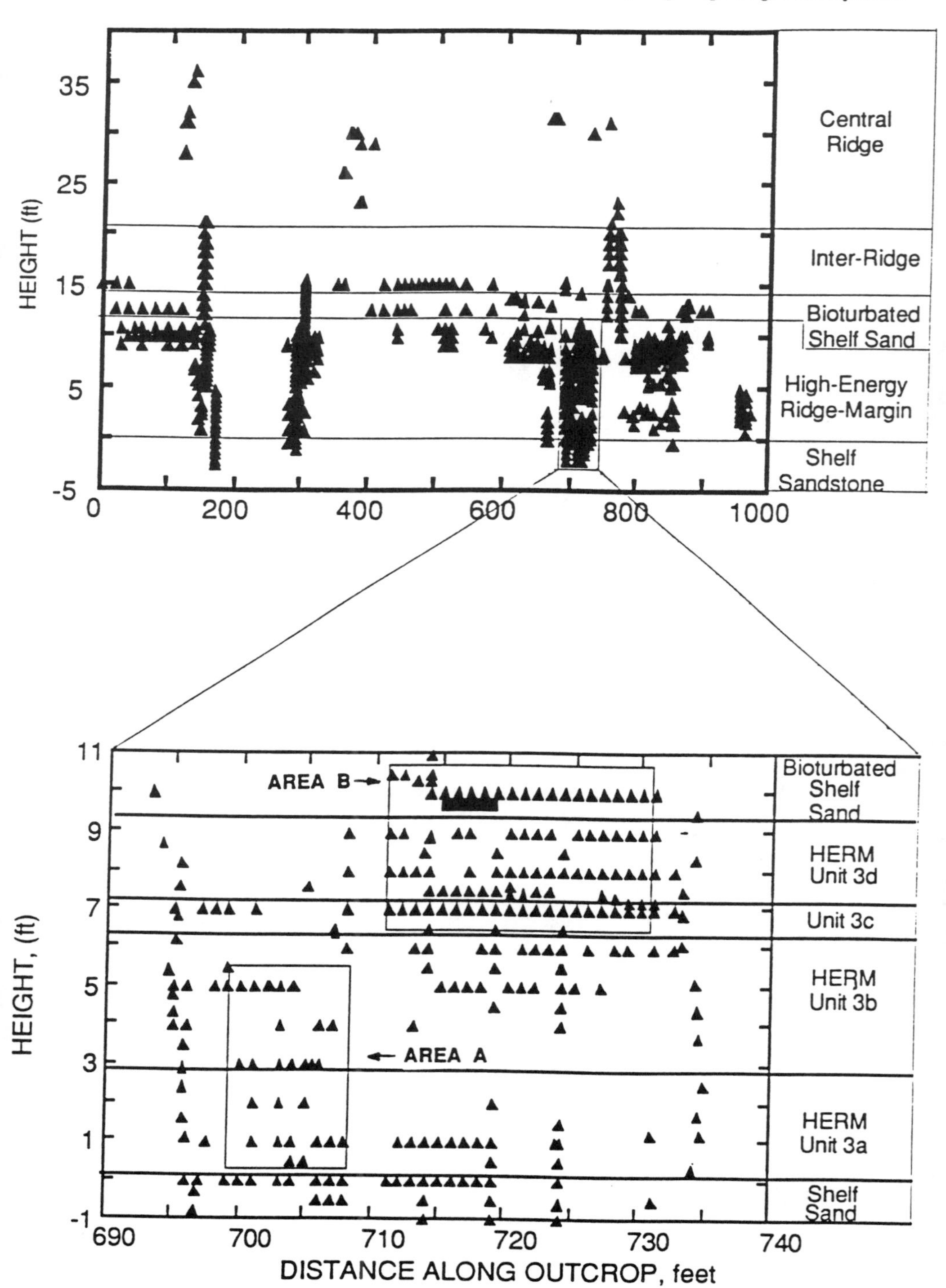

Fig. 2. Outcrop sample locations of 1-inch-diameter core plugs.

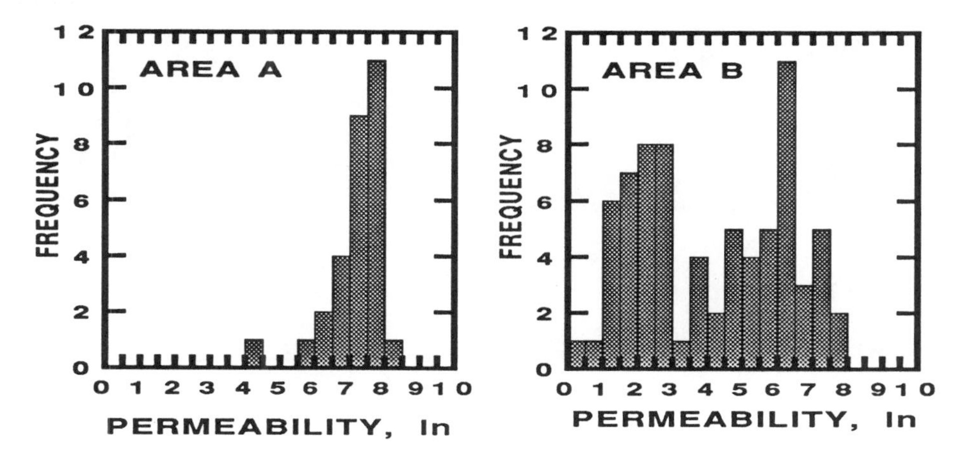

Fig. 3. Permeability histograms for Shannon outcrop data.

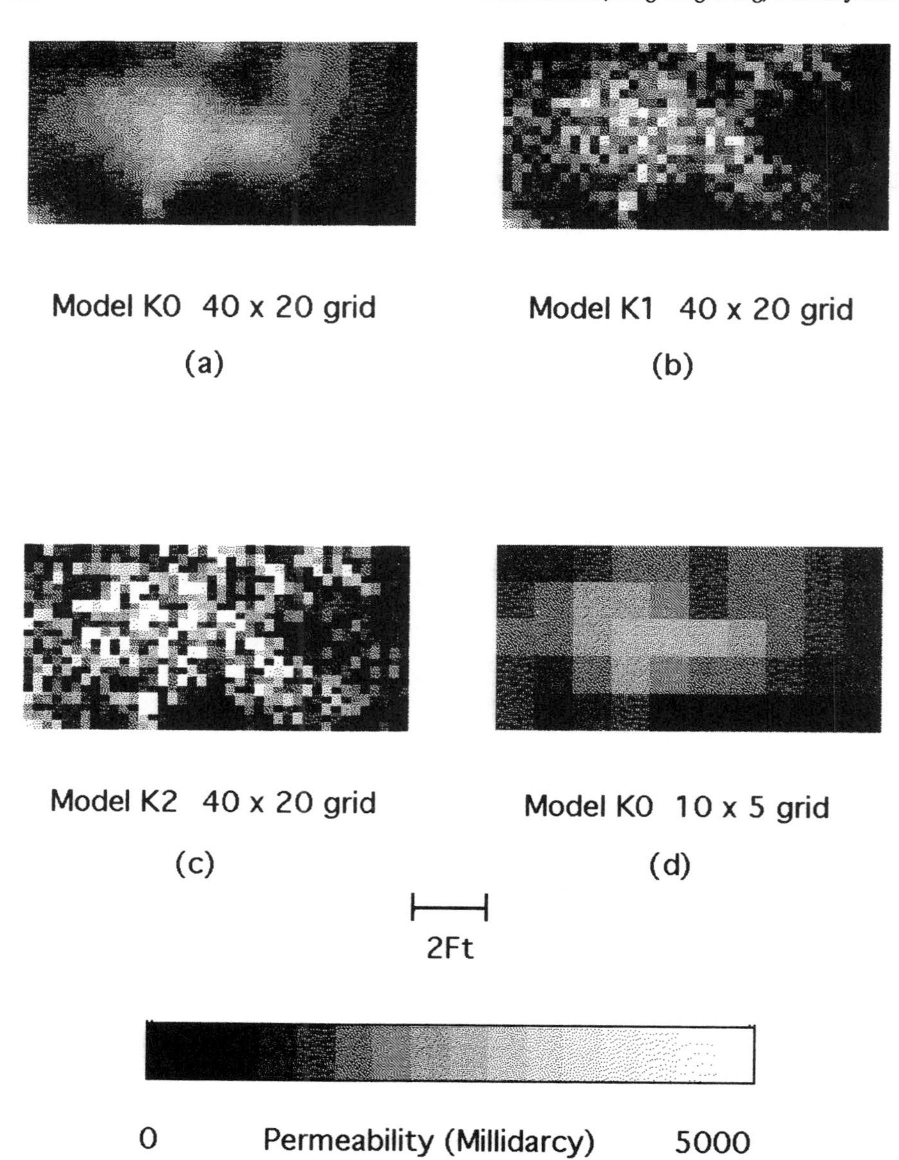

Fig. 4. Kriged Permeability Models, Area A.

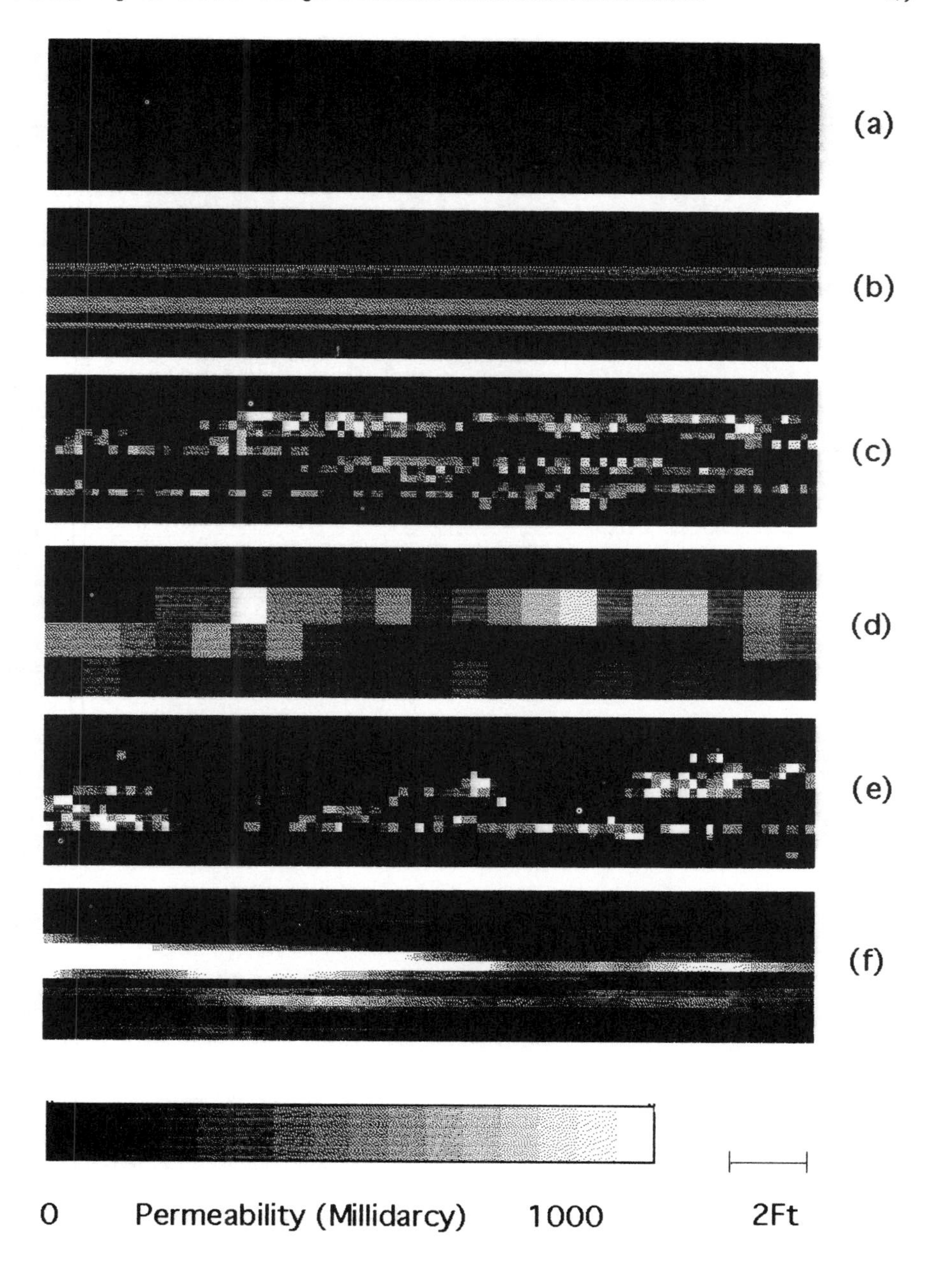

Fig. 5. Permeability Models, Area B.

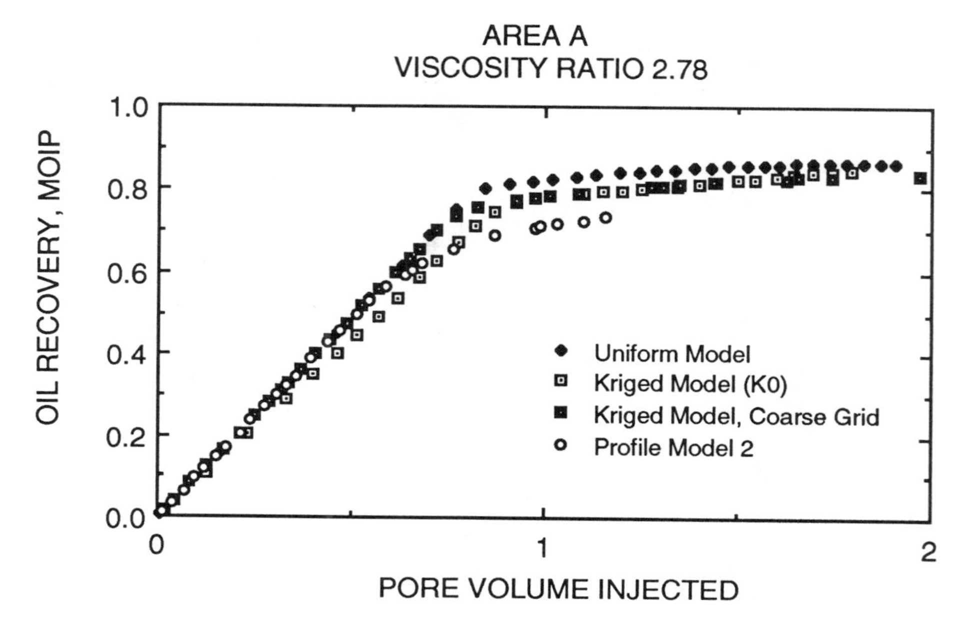

Fig. 6b. Cumulative oil production (MOIP) for various permeability models in Area A, API 20°, viscosity ratio 24.5.

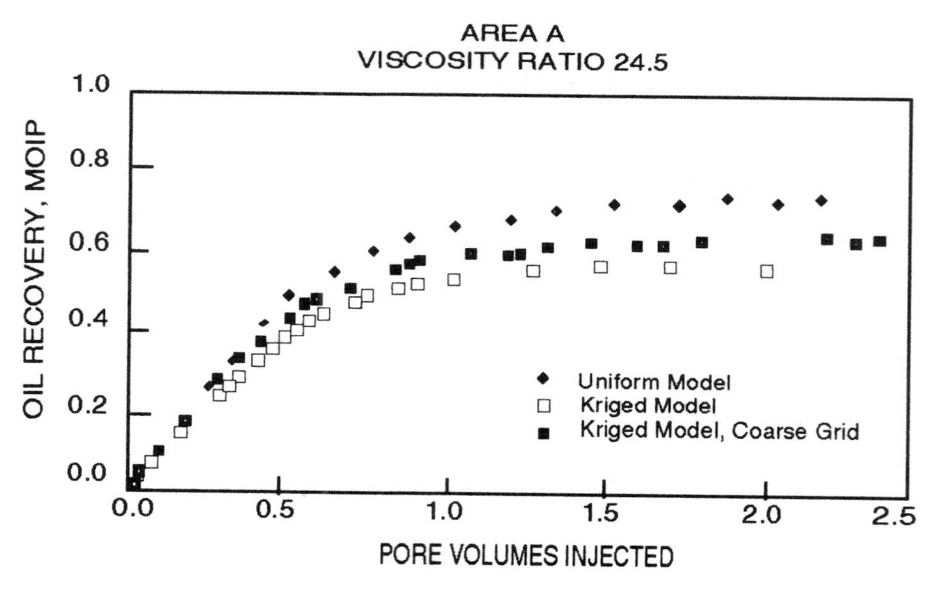

Fig. 6a. Cumulative oil production (MOIP) for various permeability models in API 35°, viscosity ratio 2.78, Area A.

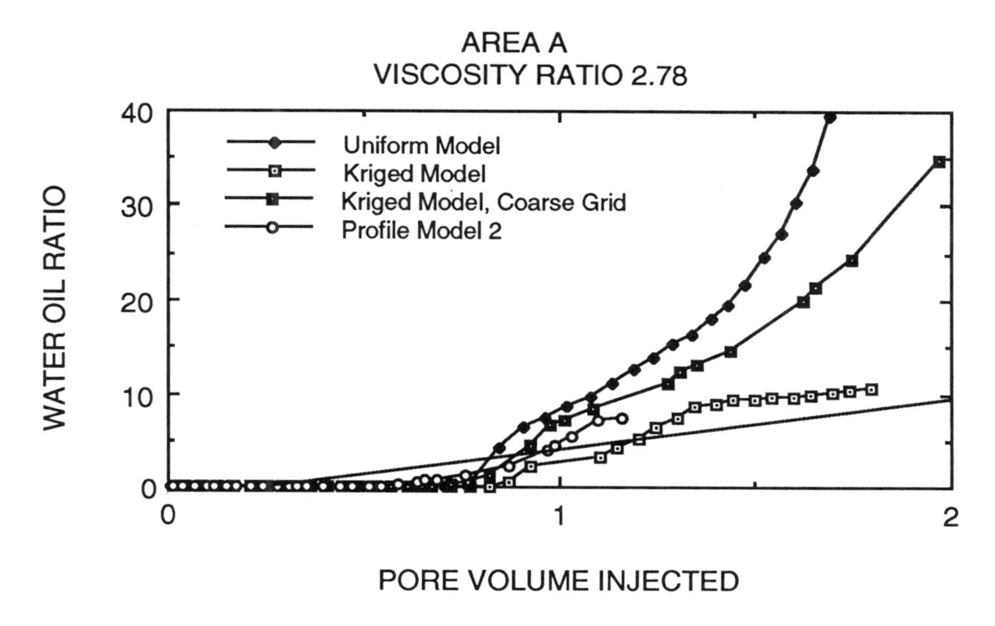

Fig. 7a. Water-oil ratio for permeability models, API 35°, viscosity ratio 2.78, Area A.

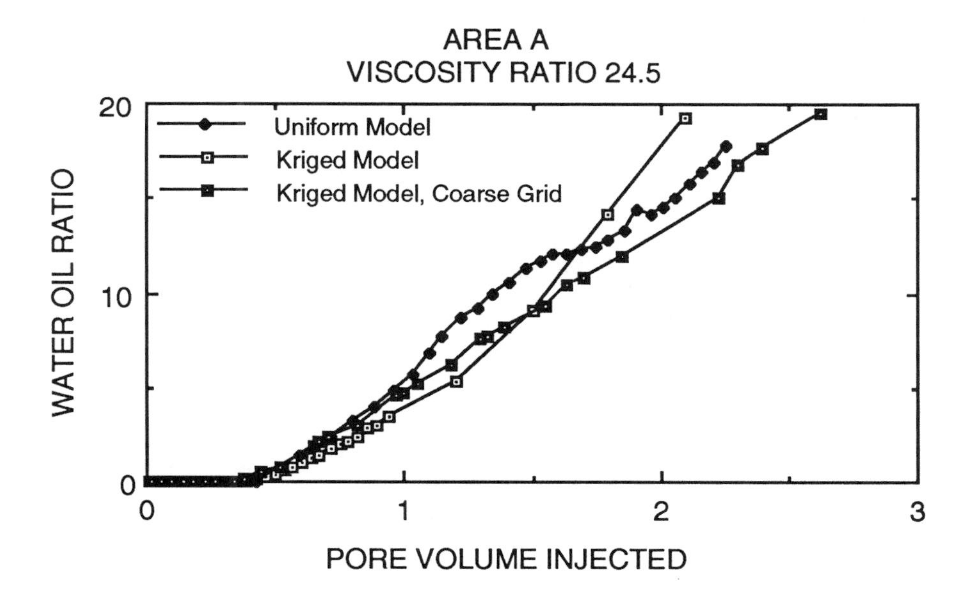

Fig. 7b. Water-oil ratio for permeability models API 20°, viscosity ratio 24, Area A.

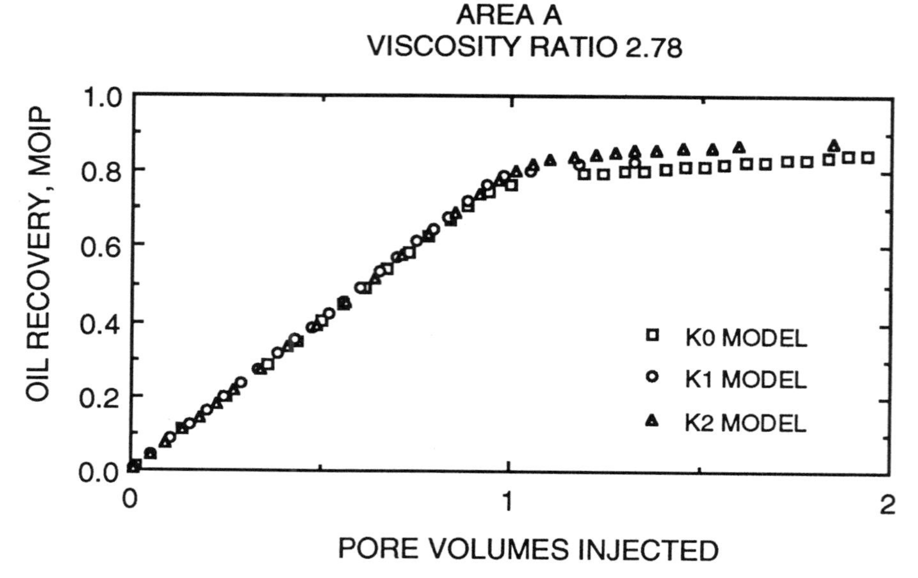

Fig. 8a. Effect of random components on cumulative oil production (MOIP), Kriged Permeability Models, viscosity ratio 2.78, Area A.

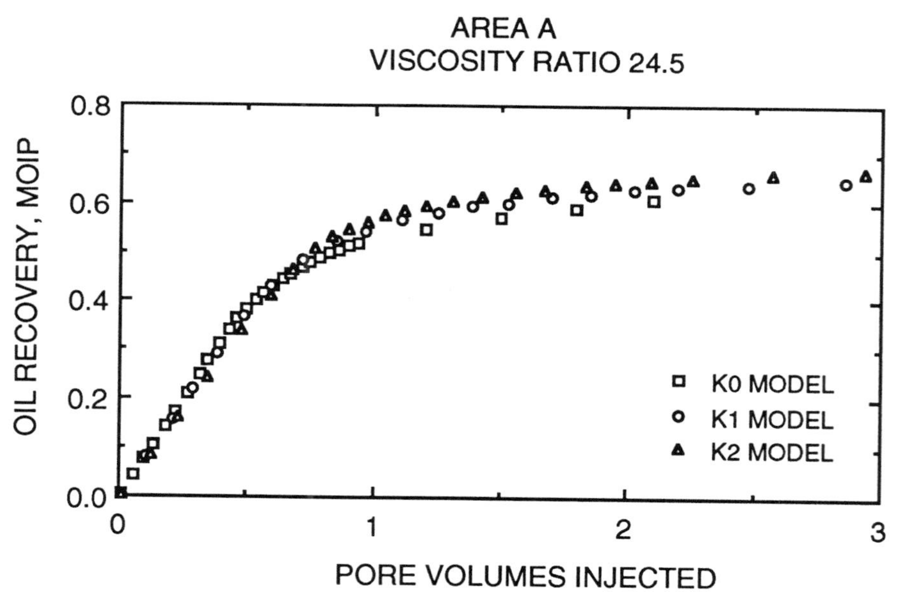

Fig. 8b. Effect of random components on cumulative oil production (MOIP), Kriged Permeability Models, viscosity ratio 24.5, Area A.

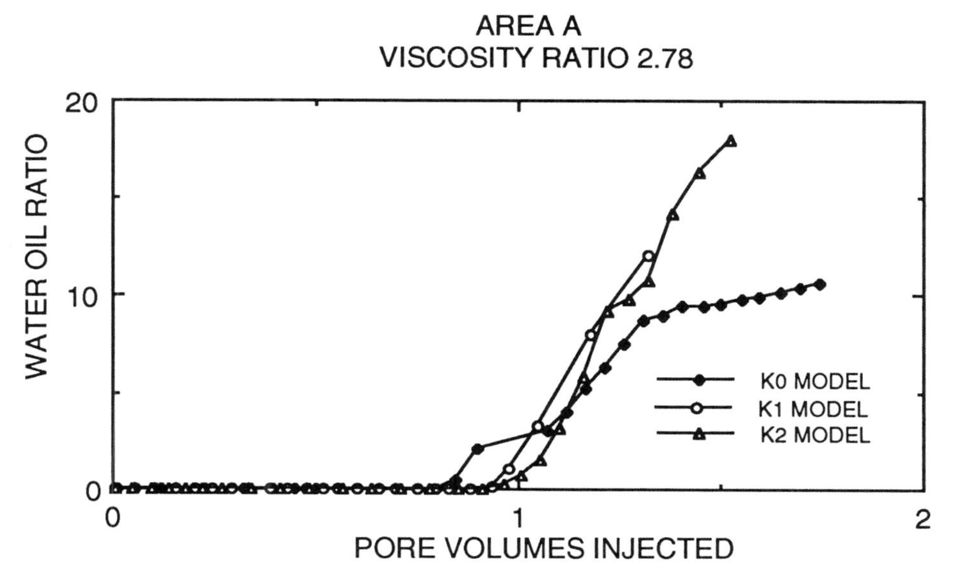

Fig. 9a. Effect of random components on water-oil-ratio, Kriged Permeability Models, viscosity ratio 2.78, Area A.

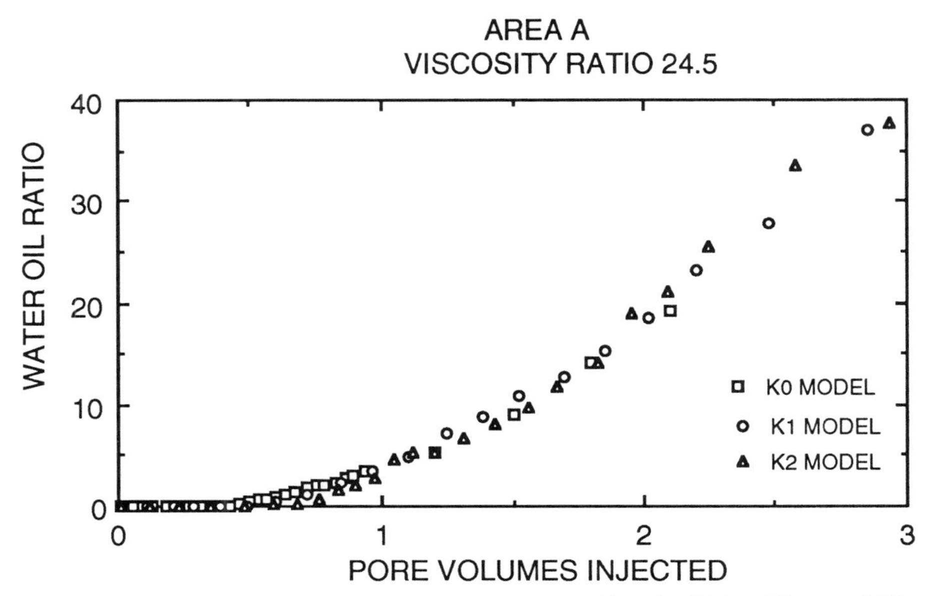

Fig. 9b. Effect of random components on water-oil-ratio, Kriged Permeability Models, viscosity ratio 24.5, Area A.

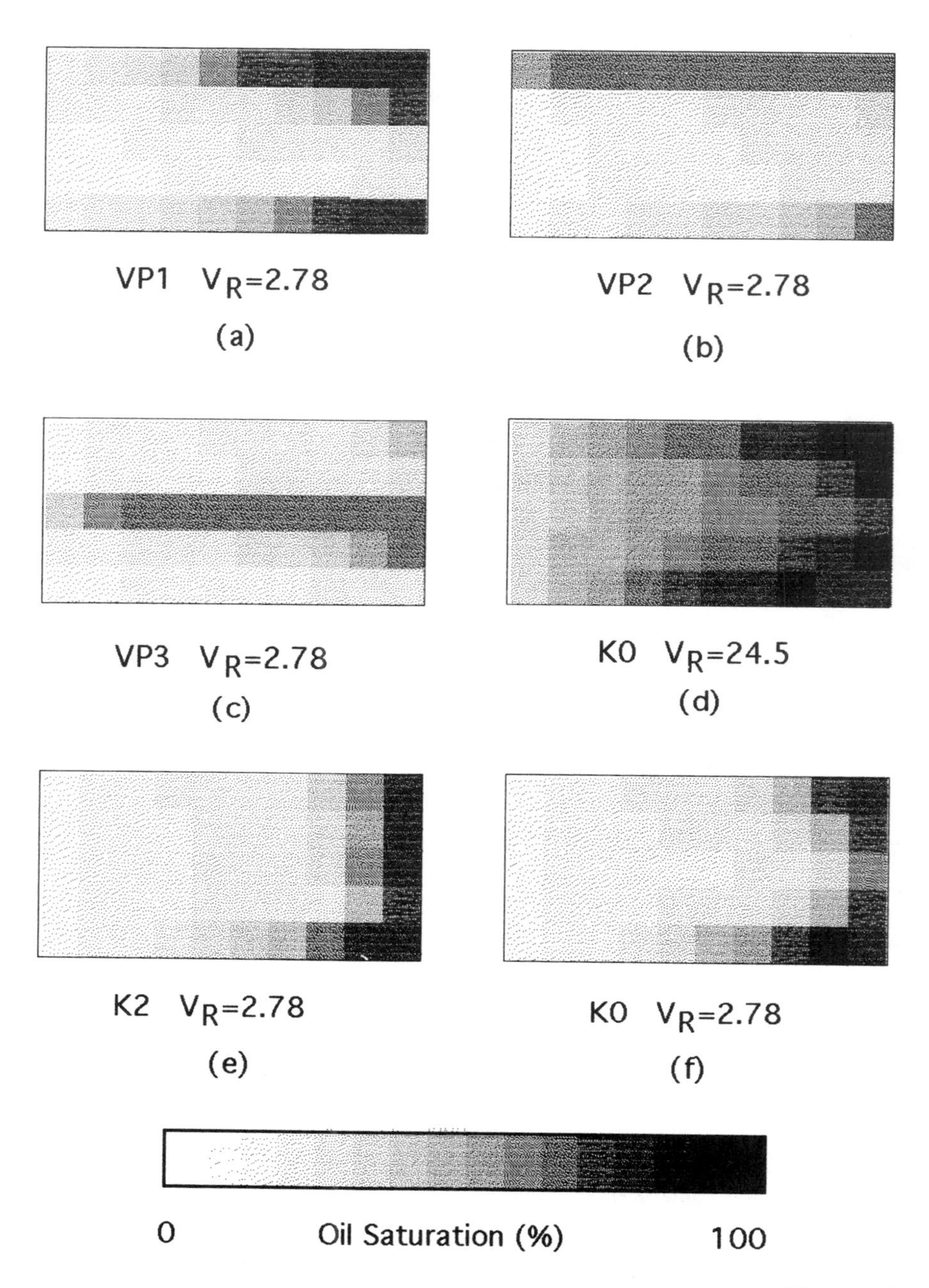

Fig. 10. Oil saturation distribution after 0.64 PV injection,Coarse grid, Area A, Profile Layer and Kriged Models (a-c).

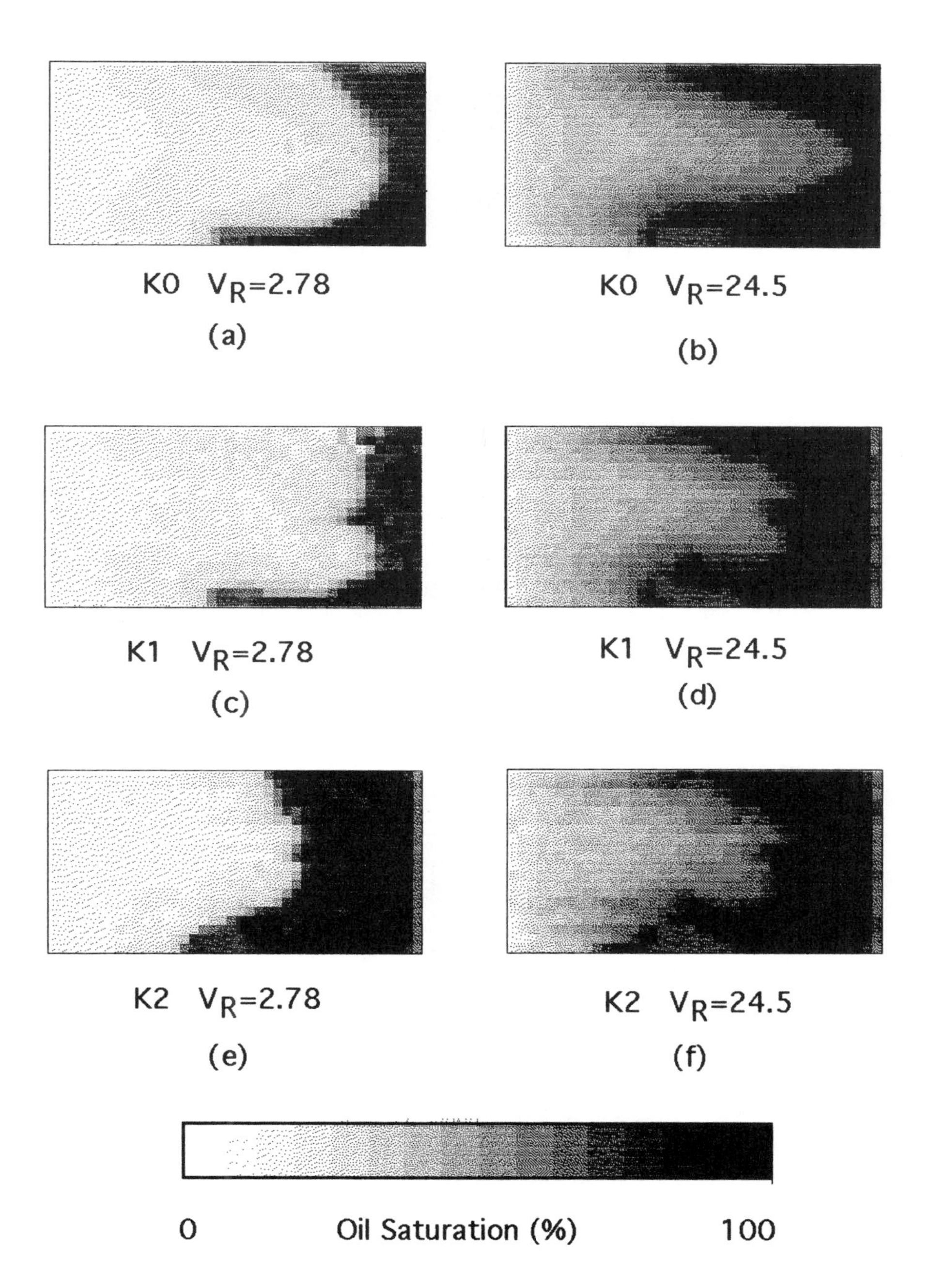

Fig. 11. Oil saturation distribution after 0.64 PV injection, Kriged Permeability Models, Area A.

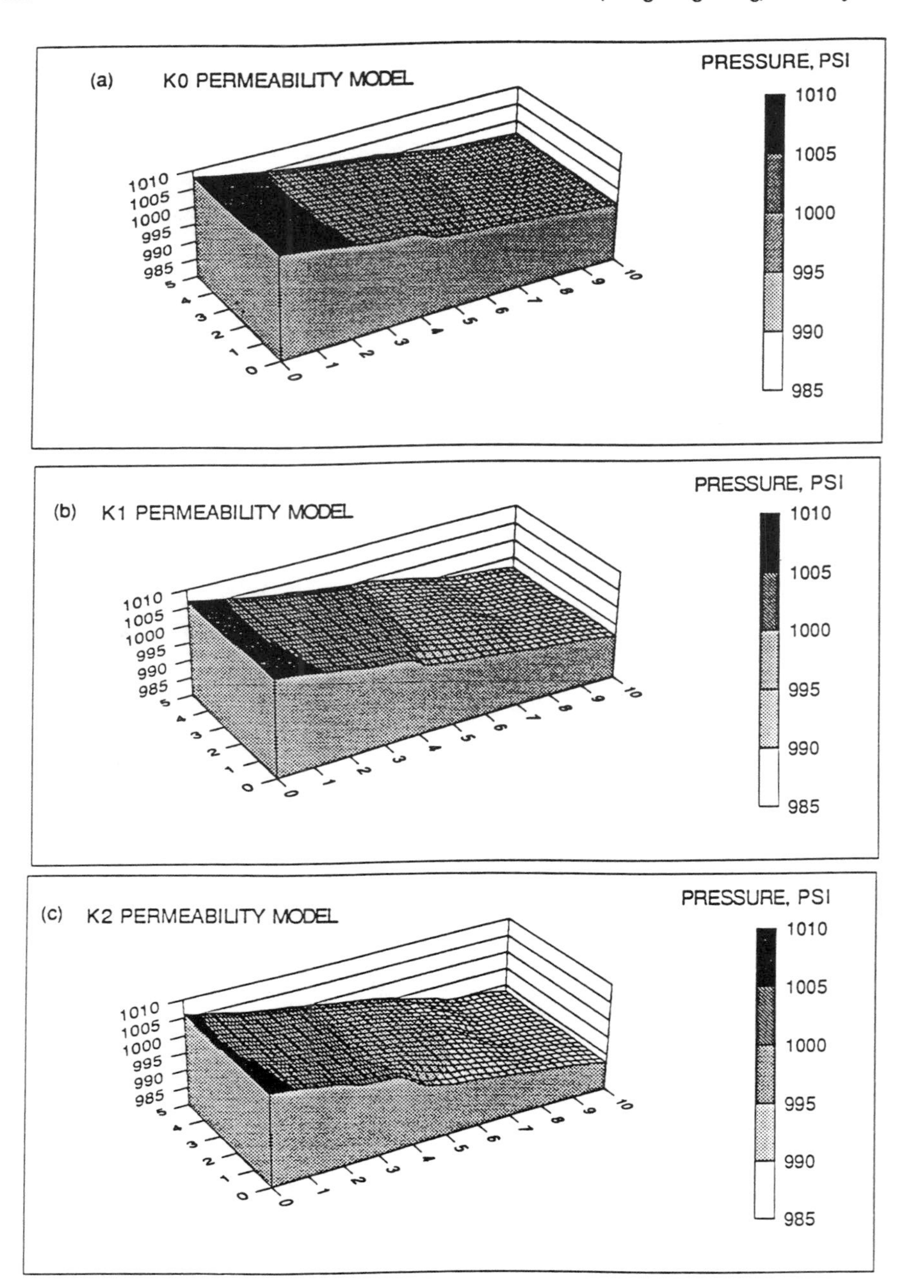

Fig. 12. Pressure distributions resulting from waterflood simulations of (a) K0 permeability model, (b) K1 Permeability Model, and (c) K2 Permeability Model, Area A.

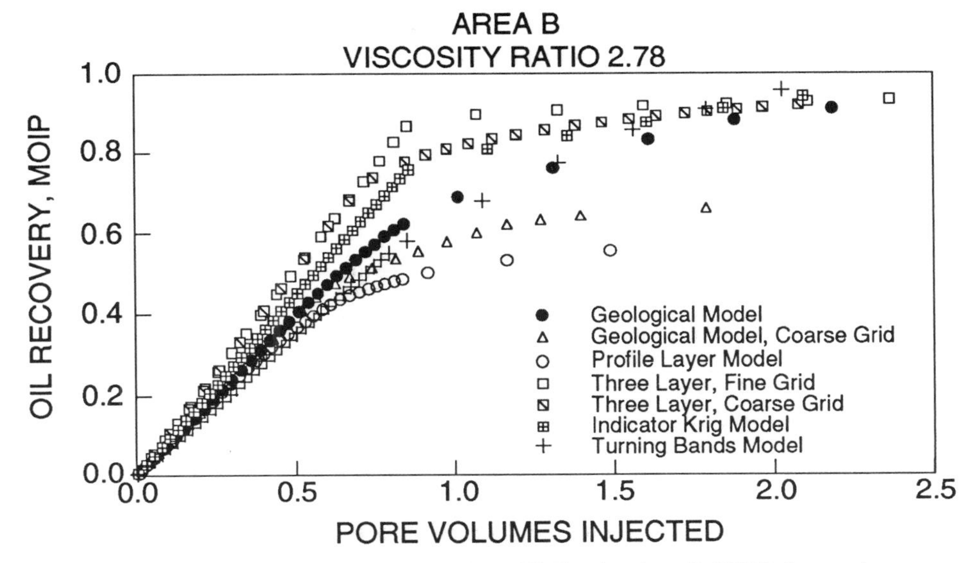

Fig. 13a. Comparison of cumulative Oil Production (MOIP) for various permeability models, viscosity ratio 2.78, Area B.

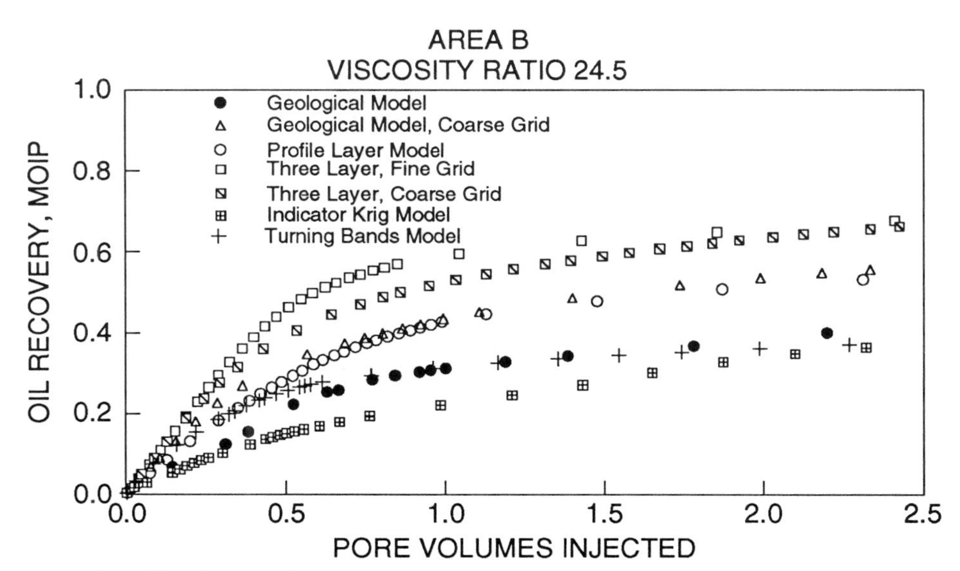

Fig. 13b. Comparison of cumulative Oil Production (MOIP) for various permeability models, viscosity ratio 24.5, Area B.

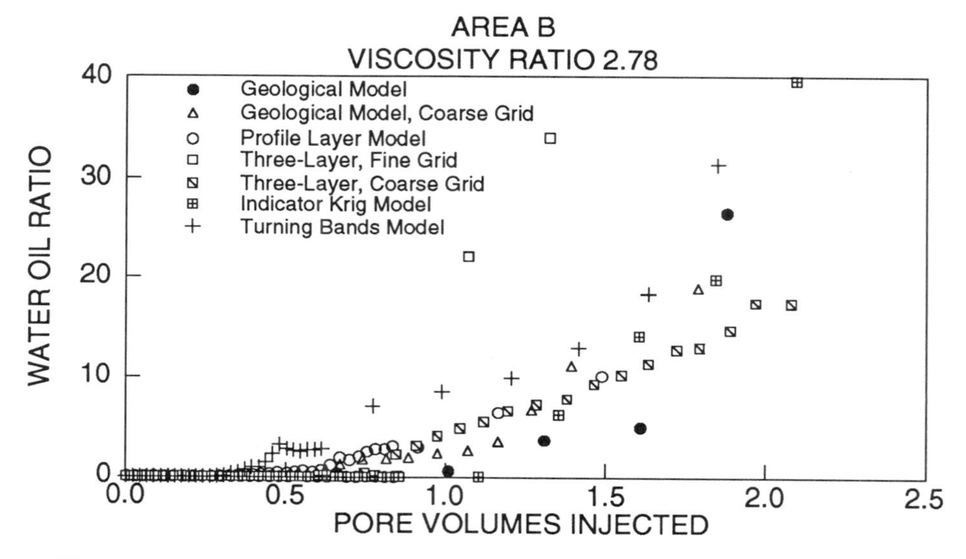

Fig. 14a. Comparison of the water-oil-ratio for various permeability models viscosity ratio 2.78, Area B.

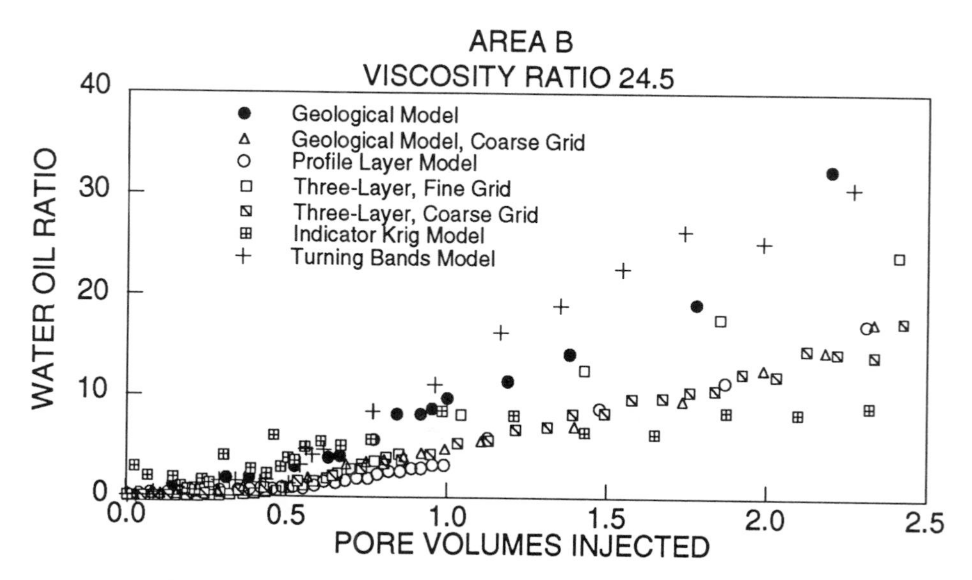

Fig. 14b. Comparison of the water-oil-ratio for the Area B Models, viscosity ratio 24.5, Area B.

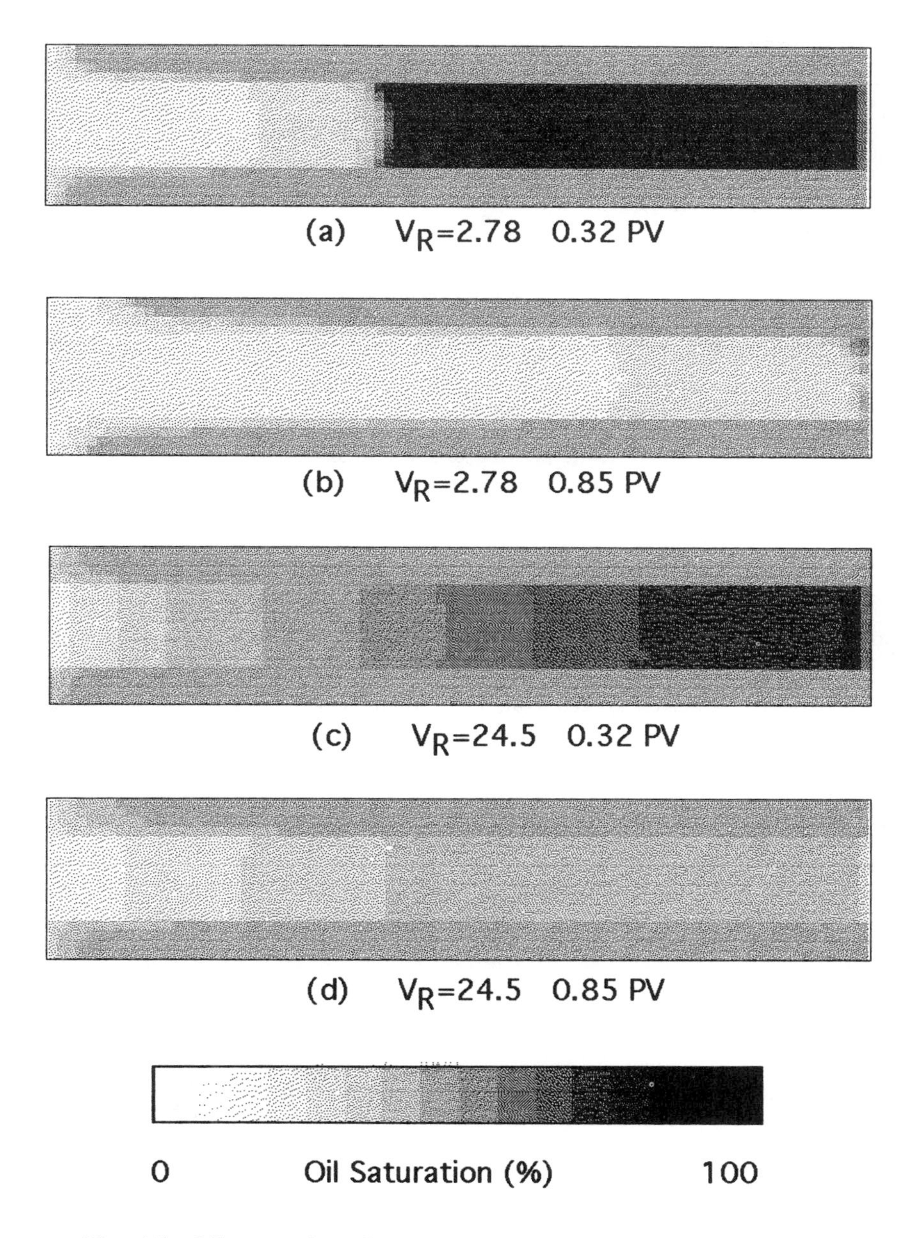

Fig. 15. Oil saturation distributions, Three-layer Model, viscosity ratios 2.78 (a,b) and 24.5 (c,d), Area B.

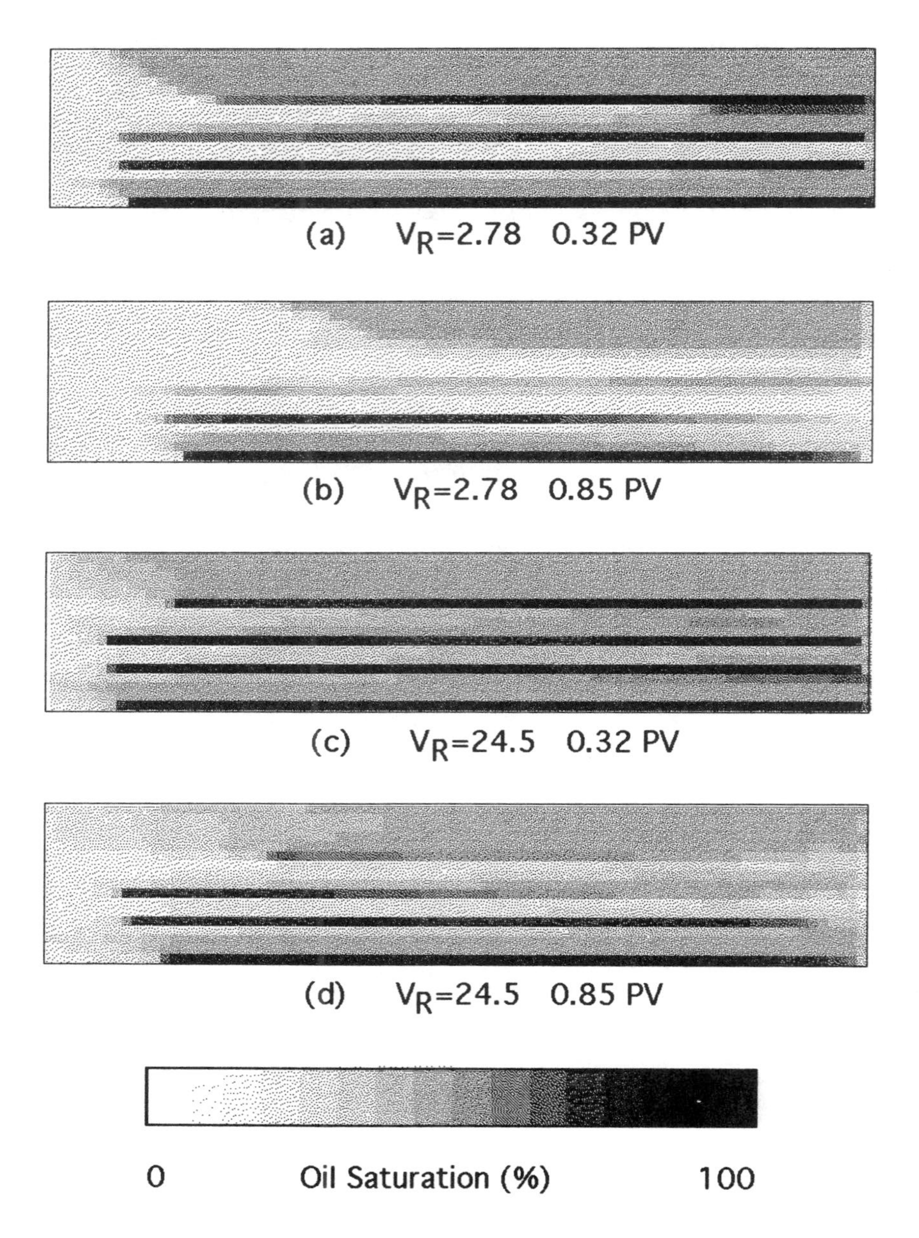

Fig. 16. Oil saturation distributions, Profile Layer Model, viscosity ratio 2.78 (a,b) and 24.5 (c,d), Area B.

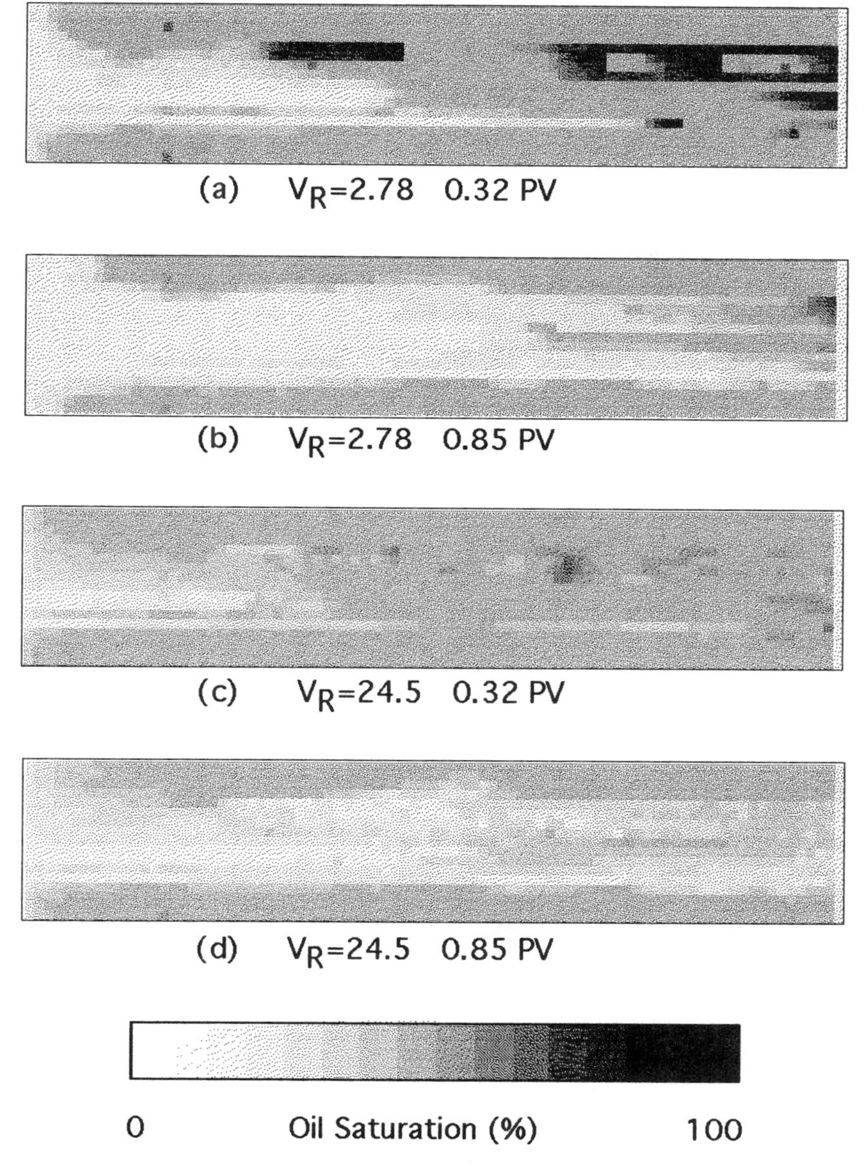

Fig. 17. Oil saturation distributions, Fine-Grid Geological Model, viscosity ratios 2.78 (a,b) and 24.5 (c,d), Area B.

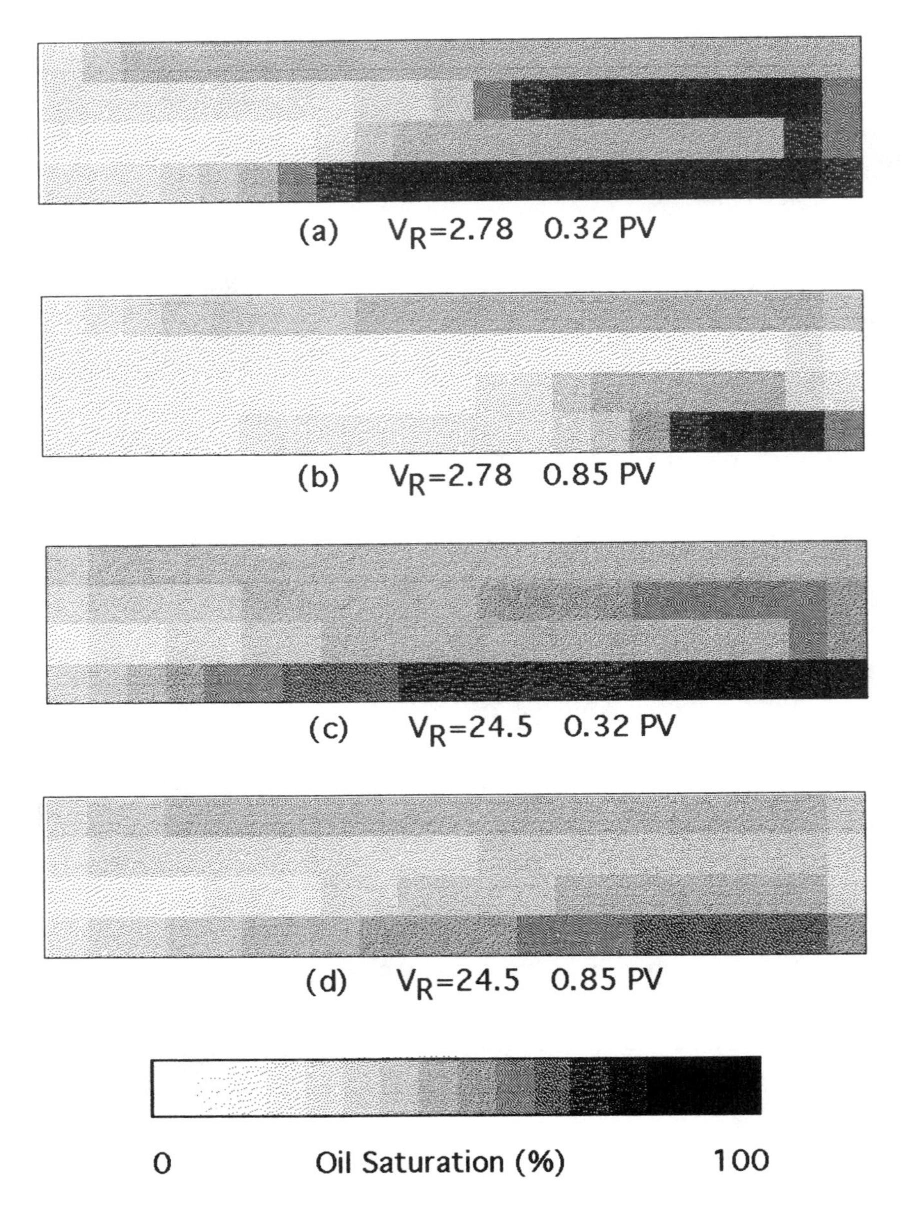

Fig. 18. Oil saturation distributions, Coarse-Grid Geological Model, viscosity ratios 2.78 (a,b) and 24.5 (c,d), Area B.

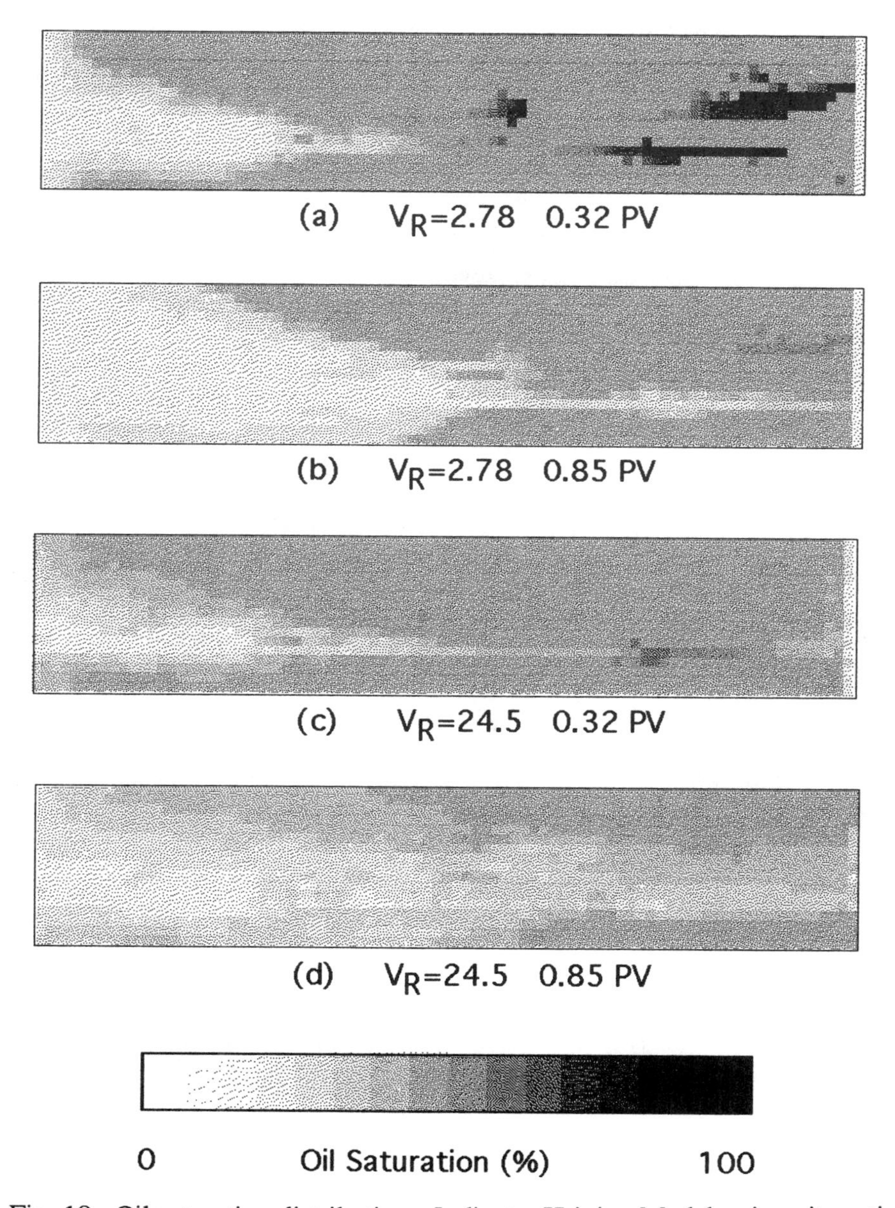

Fig. 19. Oil saturation distributions, Indicator Kriging Model, viscosity ratio 2.78 (a,b) and 24.5 (c,d), Area B.

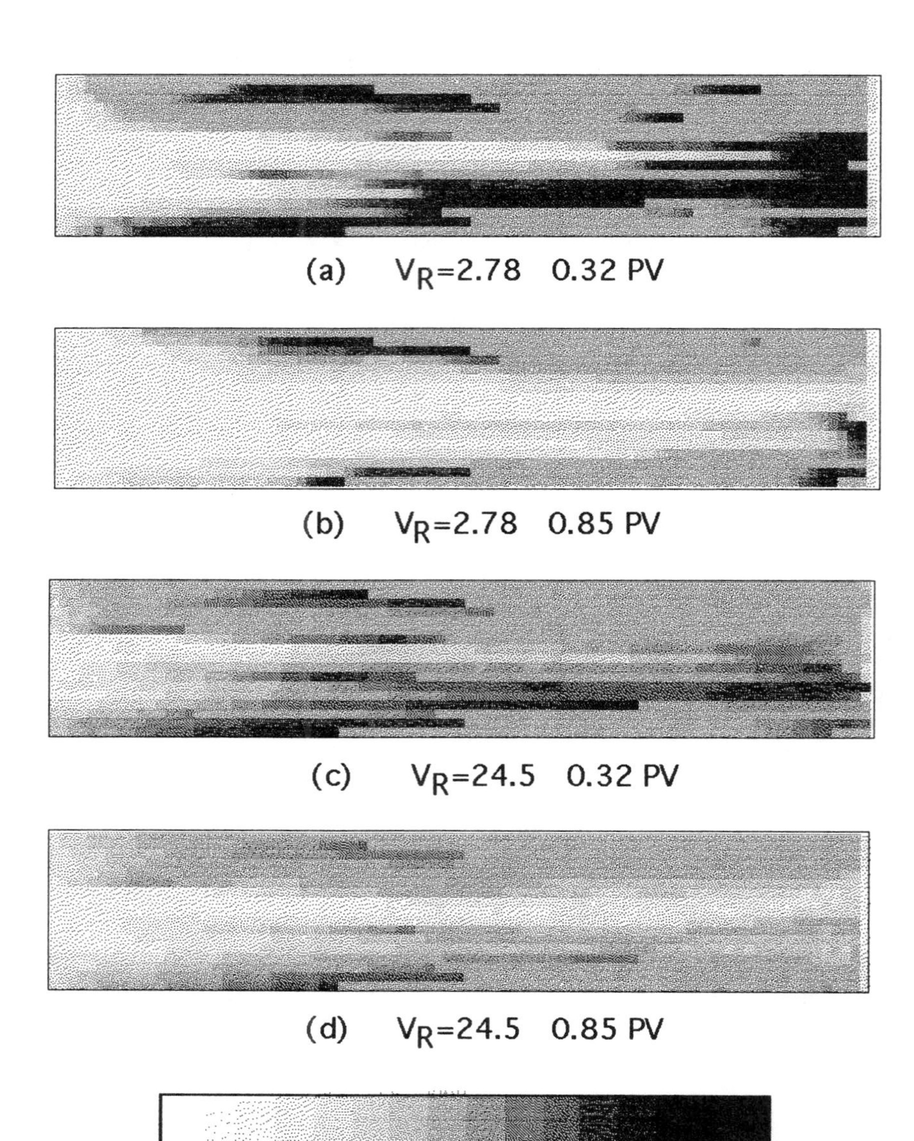

Fig. 20. Oil saturation distributions, Turning Bands Model, viscosity ratios 2.78 (a,b) and 24.5 (c,d), Area B.

OPTIMAL SCALES FOR REPRESENTING RESERVOIR HETEROGENEITY*

R. C. Wattenbarger
Khalid Aziz
F.M. Orr, Jr.

Department of Petroleum Engineering
Stanford University
Stanford, California

Abstract

The subject of this paper is the improvement of the representation of permeability heterogeneity in reservoir models. Two numerical studies are presented. The first explores the effects of changes in the scale at which permeability variations are explicitly represented—the geologic representation scale. It shows that reservoir response may continue to change as the representation scale is decreased, and suggests that mixing due to sub-representation scale heterogeneity must be represented implicitly through effective mixing parameters, such as dispersivity. The second numerical study investigates whether this is possible. It demonstrates that the effects of small and large scale heterogeneities can be represented by a combination of explicit representation through varying gridblock permeabilities and implicit representation through effective mixing parameters.

However, at a given scale the effective mixing parameters may not be able to accurately represent the effects of mixing caused by sub-representation scale heterogeneity. If this is the case, we suggest that the robustness of the effective mixing parameters can best be improved by reducing the geologic representation scale while holding constant the averaging scale of the effective mixing parameters. The geologic representation scale is chosen as the maximum scale at which the effective mixing parameters are robust—that is, at which the effective mixing parameters are accurate over a wide range of flow conditions.

*Supported by the Stanford Center for Reservoir Forecasting (SCRF) and the Stanford Reservoir Simulation Industrial Affiliates Program (SUPRI-B).

I. INTRODUCTION

The goal of reservoir simulation is to provide accurate models of reservoir behavior. Often a reservoir simulator is complex enough to match the past known response of a real system, but is not robust enough to predict the future unknown response. Since the future is where our interest lies, what is desired is a robust reservoir simulator that can be used with confidence in a predictive mode. Here, robustness refers to the ability of a reservoir simulator to model a wide range of flow conditions with the same degree of accuracy. Thus, a robust simulator can accurately model not just a single, history-matched scenario, but a variety of scenarios with different boundary conditions and recovery processes.

Often models are not robust because of problems associated with the modeling of permeability induced mixing.[1,2] Such mixing is traditionally modeled by the explicit representation of permeability heterogeneity through varying gridblock permeabilities and/or through the use of effective mixing parameters such as dispersivity or relative permeability. Considerable effort has been applied toward improving explicit representation of heterogeneities[3,4,5,6] or toward improving effective mixing parameters.[7,8,9,10] Although the problem of representing permeability heterogeneities is often broken into separate problems of explicit and implicit representation, this paper treats the problem as a whole. This is done by determining an adequate scale for the explicit representation of permeability.

Intuitively, the robustness of a reservoir model should increase as the geologic representation scale (GRS) decreases; unfortunately, so to should the computational costs. Furthermore, as the GRS decreases the amount of explicit heterogeneity representation increases, thereby decreasing the need for the implicit representation of permeability induced mixing. Thus, the GRS represents a tradeoff between explicit and implicit heterogeneity representation, between robustness and computational cost. To reduce computational expense, it is desirable to use as coarse a GRS as possible while retaining model accuracy and robustness. This paper investigates issues associated with choosing this maximum, or optimal, GRS.

II. PREVIOUS WORK

In a field study were the number of gridblocks is clearly constrained by computer limitations, gridblocks are often selected to delineate geologic zones and oil, gas and water zones.[11] If not, Coats[12] recommends determining a suitable block size by "repeated runs using fewer blocks until resolution is lost concerning the facets of field production being estimated". Unfortunately, the block size at which response resolution is lost is somewhat subjective. Several researchers[2,13,14,15] follow this procedure for both homogeneous and heterogeneous reservoirs with a variety of results. Warren and Price,[15] and Stalkup[14] investigate the sensitivity to block size by plotting the response versus block length raised to a power. The power being an estimate of the global order of numerical approximation. Often this plot results in a linear relation which may be extrapolated to determine the response at zero block size. In all cases, the response continues to change as block size is reduced, giving no clear indication of a best block size. At best, one can choose a block size according to some acceptable error. Warren and Price, and Stalkup's studies were concerned with analyzing numerical errors. In their studies, all reservoir properties, including the GRS, were held constant as the computation scale was varied. Normally, no distinction is made between the computation and representation scales.

A different approach was used by Hewett and Behrens.[2] Starting with a very fine cross sectional representation of a fractal permeability field, they studied changes in reservoir response due to a varying number of model layers. This was done by performing waterflood simulations with fewer and fewer model layers, using appropriately scaled block permeabilities at each level while holding all other reservoir properties constant. The computational scale and the representation scale were the same. The results were summarized by plotting the response, oil recovery at one pore volume injected and breakthrough time, versus the logarithm of the number of layers. As the number of layers increased from one, the response steadily changed up to about 30 layers. Increasing the number of layers past 30 had little effect—the response remained approximately constant. The number of layers at which the response stabilized appears to occur rather abruptly. Hewett and Behrens deem this the critical resolution, beyond which macroscopic media flow properties can no longer be used.

Another study, with different results was performed by Davies and Haldorsen.[16] Their fine scale geologic field consisted of a cross section containing 150 stochastically-generated shale barriers of varying lengths and zero thickness. A grid size study was performed by holding the number of layers constant, while varying the number of columns. At each discretization level pseudo functions were derived from the response of a reference, fine scale, water injection run. They found that the water breakthrough time fluctuated and was lower than that of the fine scale run for small block widths, eventually stabilizing and becoming more accurate as the block size increased. They concluded that the horizontal block dimension of a coarse model should be greater than the largest shale expected within a block.

A number of studies are based on a layered permeability model. An early study by Testerman,[17] used a downsizing procedure which grouped individual layers into zones. The zones were defined to minimize the permeability variation within zones and maximize the variation between zones. The downsizing procedure was stopped, thus setting the number of zones, when further division produced no significant improvement in an objective function. Craig[18] performed a layer sensitivity study based on the effluent response to steady water injection in a five spot-pattern. At each discretization level, layer permeabilities were defined by drawing from a log-normal distribution at equal percentile increments. Craig found that the accuracy of the cumulative production prediction increased with the number of layers and time and decreased with mobility ratio and permeability variance. Based on a given level of accuracy, Craig recommended the number of layers to be used in a study. Lake and Hirasaki[19] developed a procedure for grouping layers based on the degree of crossflow between layers due to transverse dispersion. They found that for sufficiently high transverse dispersion numbers, the longitudinal spreading averaged over two or more layers could be represented by a single layer with a greater longitudinal dispersion coefficient. Similar arguments based on capillary crossflow and/or gravity equilibrium have been used in multiphase studies to select the number of layers, or to justify 2D areal models.[7,20]

III. EFFECTS OF GRS

In this section the effects of the GRS on block permeabilities and reservoir response are explored through an example permeability field and transport problem. Throughout this section, it is assumed that only permeability changes with the GRS; that is, all other model parameters are held constant at their macroscopic levels. In addition to illustrating the effects of the GRS, this example serves to determine whether or not the optimal resolution observed by Hewett and Behrens exists when the computation and representation scales are decoupled.

A fine scale multi-normal log-permeability field was generated on a 128x128 grid using gaussian sequential simulation.[21] The *a priori* log-permeability covariance model used in the simulation was exponential with an isotropic integral correlation range of 0.5 and a variance of 1. The resulting realization has a log-permeability variance of 0.729. The experimental variogram was anisotropic. At a lag of 0.5, the semi-variogram reached a value of 1.12 in the x direction and 0.36 in the y direction.

A. *Effects of GRS on Permeability*

Several permeability fields with coarse representation scales were defined from the "point" (128x128) permeabilities. For simplicity, only representation grids of 64x64, 32x32, 16x16, 8x8, 4x4, 2x2 were investigated. Thus, the number of representation blocks is the same in each direction, $N_{repx} = N_{repy}$, and each coarse representation block contains the same integral number of point blocks. The permeability of each coarse block was taken to be the geometric average of the point permeabilities within the coarse block. Geometric averaging is chosen because it is reasonably accurate and simple.[15,22]

Grayscale images of the permeability at three GRS's are shown in Figure 1. These images qualitatively summarize the combined effects of averaging and discretization. As the GRS increases, small correlated regions of the point permeability lose distinction within the large homogeneous regions of the representation blocks.

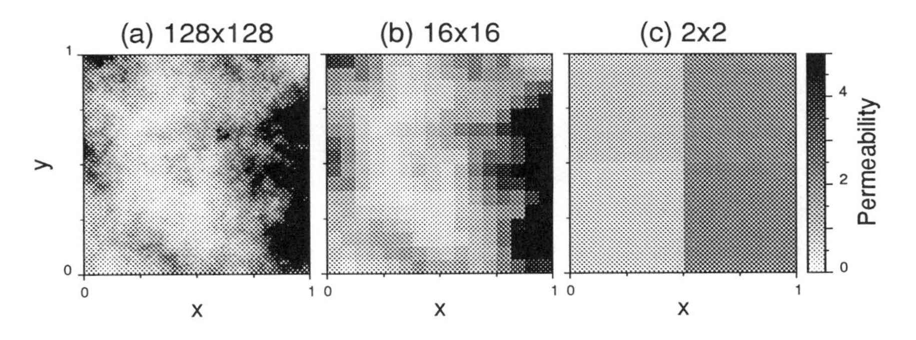

Figure 1: Permeability maps for GRS of (a) 128x128 (b) 16x16 (c) 2x2.

Statistics of the log-permeability for each GRS are shown in Figure 2. Plots of the cumulative distribution function, Figure 2a, show that the spread of the distribution about the mean decreases as the GRS increases. Furthermore, as the GRS increases the mean remains constant at 0.258, the minimum monotonically increases from a point value of -1.85, and the maximum monotonically decreases from a point value of 3.11. As shown in Figure 2b, the variance of the log-permeability decreases approximately linearly with an increase in the representation length, $L_{rep} = 1/N_{repx}$.

The experimental semi-variogram and covariance of the log-permeability for separation vectors parallel to the y-axis are shown in Figures 2b and c. The GRS has a strong impact on the experimental semi-variogram. The value of the semi-variogram, and hence permeability variability, is reduced at all sep-

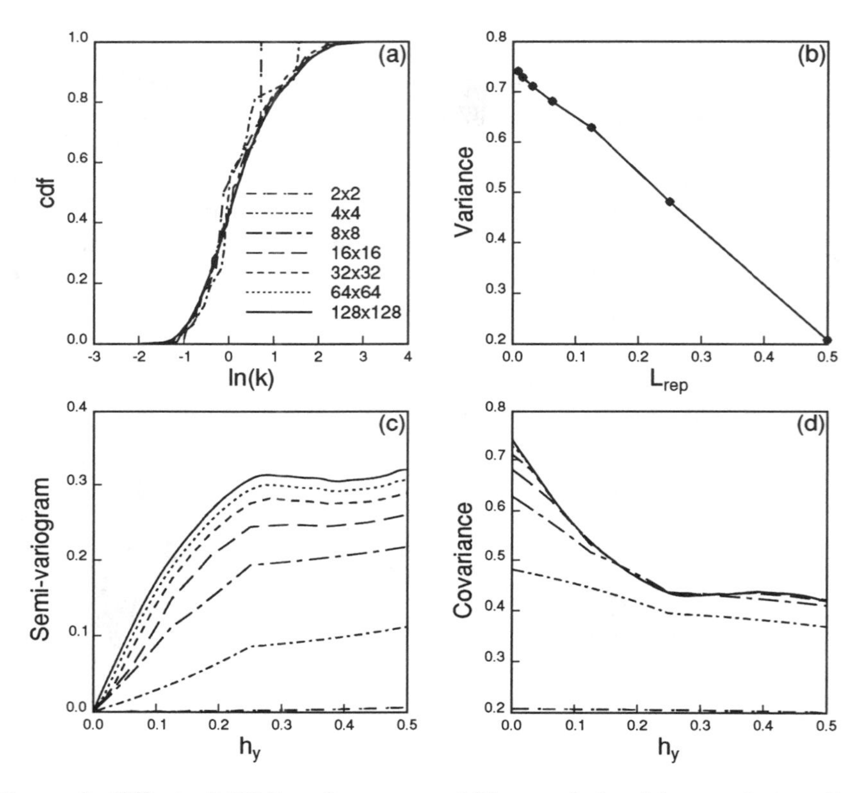

Figure 2: Effect of GRS on log-permeability statistics (a) cumulative distribution function (b) variance versus representation length (c) semi-variogram versus separation distance, with separation vector parallel to y axis (d) covariance versus separation distance, with separation vector parallel to y axis.

aration distances as the GRS is increased. When L_{rep} is less than the range of the point semi-variogram ($L_{rep} < L_{\kappa}$, where $L_{\kappa} \approx 1/4$), the range of the coarse semi-variogram is approximately that of the point semi-variogram. Otherwise, the range of the coarse semi-variogram is approximately equal to the L_{rep}.

When L_{rep} is less than the range of the point covariance ($L_{rep} < L_{\kappa}$), the coarse GRS covariance is primarily affected at small separation distances. In this case, the coarse GRS covariance is approximately equal to the point covariance for separation distances greater than the representation scale ($h > L_{rep}$). For separation distances less than the representation scale ($h < L_{rep}$), the coarse GRS covariance interpolates, approximately linearly, the point covariance from a separation distance of zero to L_{rep}. Indeed, the coarse GRS covariance and semi-variograms are approximately piecewise linear for all separation distances. This linearity is a result of discretization and is expected whenever the separation vector is aligned with one of the axes of the representation grid.[23] When L_{rep} is greater than the range of the point covariance ($L_{rep} > L_{\kappa}$), the entire coarse GRS covariance falls below the point covariance.

The effects of the GRS on the statistical properties of spatially averaged random fields is dealt with extensively in the geostatistics literature.[24,25,26] In particular, the effects of sample size (representation scale) and sampling region (reservoir size) on univariate statistics are referred to as the *support effect*, while the effect of spatial averaging on semi-variogram and spatial covariance is referred to as *regularization*. Rubin and Gómez-Hernández[27,6] investigated the effects of block size on the covariance of effective log-transmissivities both through analytic stochastic analysis and numerical simulation. The qualitative features of their results are similar to those given here.

The primary effects of spatial averaging and discretization are to reduce univariate and spatial variability. While interesting, the effects of GRS on permeability are important only inasmuch as they impact reservoir response. The effect of GRS on reservoir response and its relation to changes in permeability are discussed in the following sections.

B. *Effect of GRS on Reservoir Response*

To investigate the effects of the GRS on species transport, a model transport problem was solved for each GRS. The process is that of a non-reactive tracer flowing in a single, constant density, constant viscosity phase and incompressible system. Under these assumptions, the governing equations are:[28]

$$\phi \frac{\partial C}{\partial t} + \vec{\nabla} \cdot (C\vec{u}) - \vec{\nabla} \cdot (\phi \vec{\vec{K}} \cdot \vec{\nabla} C) = 0 \tag{1}$$

$$\vec{\nabla} \cdot (k \vec{\nabla} \Phi) = 0 \tag{2}$$

$$\vec{u} = -k\vec{\nabla}\Phi \tag{3}$$

$$\vec{\vec{K}} = \left(D + \frac{\alpha_T}{\phi}|\vec{u}|\right)\vec{\vec{\delta}} + \left(\frac{\alpha_L - \alpha_T}{\phi|\vec{u}|}\right)|\vec{u}\vec{u}| \tag{4}$$

These equations are the component mass balance equation, the potential equation, Darcy's law, and an expression for the dispersion tensor. For the runs in this section the dispersion tensor was taken to be zero. In addition to the governing equations, the following initial and boundary conditions were applied:

$$\begin{aligned}
C(x,y,0) &= 0\\
C(0,y,t) &= 1\\
\frac{\partial C}{\partial x}(1,y,t) &= 0\\
\\
\Phi(1,y,t) &= 0\\
\Phi(0,y,t) &= \Phi'\\
\int_0^1 u_x(0,y,t)dy &= 1\\
u_y(x,0,t) &= u_y(x,1,t) = 0
\end{aligned} \tag{5}$$

Thus, a fluid of unit tracer concentration is injected into a constant potential inlet at $x = 0$ and displaces a similar fluid of zero tracer concentration toward a constant potential outlet at $x = 1$. The inlet boundary conditions insure that potential is constant along the inlet face and that the volumetric injection rate is unity. Thus, time has units of pore volumes injected (PVI).

This problem was solved numerically using finite differences. The steady-state potential equation was discretized using central differencing (a five-point stencil in 2D) and the resulting linear system of equations was solved with an iterative solution technique.[29] Convergence of the potential solution was based on the accumulated concentration error of each block.[30] The convection-dispersion equation was solved using a high-order, total variations diminishing, finite difference scheme[31] along with a high-throughput timestepping scheme.[23]

This work is concerned with improving the mathematical model through better reservoir characterization, and not with developing corrections for inadequate numerical solution techniques. To this end, the distinction between the mathematical model and solution technique is made as clear as possible. This is accomplished by *decoupling the geologic representation from the computational scale* — two scales which are normally paired in a common gridblock size. Unless otherwise specified, a computational scale of 256x128 is used for each simulation run, even though the GRS changes from 128x128 to 2x2. This is done to hold the numerical errors approximately constant while the GRS is varied, thus isolating the impact of the changing GRS.

Grayscale images of the concentration at 0.5 pore volumes injected are shown in the top row of Figure 3 for three representation scales. These images show that the large features of the point GRS concentration front are preserved at a GRS of 16x16 but are completely lost at a GRS of 2x2. However, the 16x16 GRS front appears somewhat less smooth than the point front and has different small scale features than the point front. The bottom row of Figure 3 displays average concentration profiles at time intervals of 0.1 PVI for the same three GRS's. The average concentration, $\overline{C}(x,t) = \frac{1}{L_y}\int_0^{L_y} C(x,y,t)dy$, is the concentration of an equivalent 1D system. The average concentration profiles further indicate the similarity between the 128x128 and 16x16 GRS runs, and their dissimilarity to the 2x2 GRS run. Moreover, the irregular character of these profiles indicates the inappropriateness of an equivalent 1D convection-dispersion model.

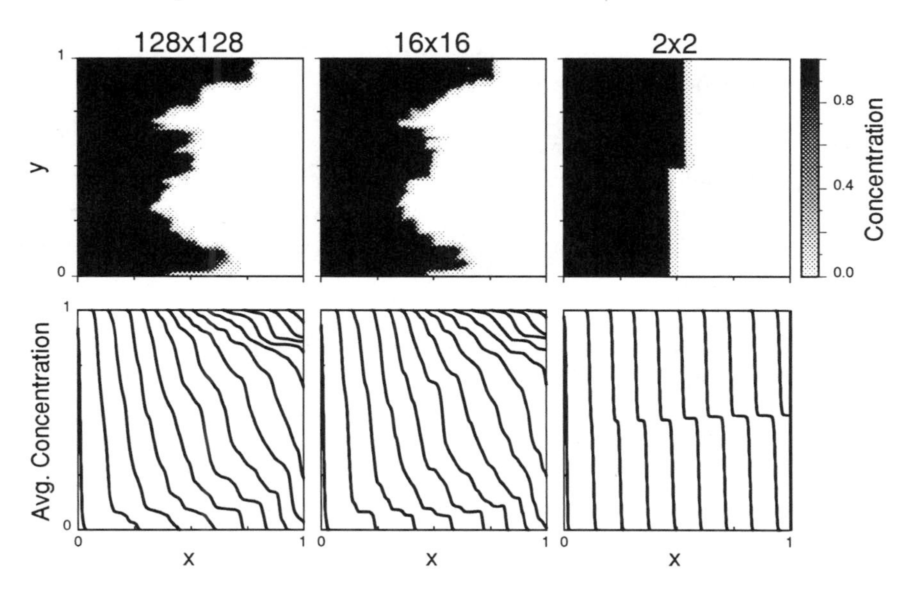

Figure 3: Effect of GRS on reservoir response using a 256x128 computational grid (top row) tracer concentration maps at 0.5 pore volumes injected (bottom row) profiles of concentration averaged in the y-direction at 0.1 PVI time intervals.

The time evolution of the concentration profiles is summarized by a plot of the mixing zone length versus time, shown in Figure 4a. Here the mixing zone is defined as the distance between average concentrations of 0.1 and 0.9.

For early times, the mixing zone grows approximately linearly with time, indicating flux-induced mixing (anomalous diffusion).[1,32] The length of the mixing zone, and hence the amount of convective mixing, is smallest for the 2x2 GRS and exhibits a maximum for the point GRS. At early times the 8x8 GRS has the greatest mixing zone length; however, the final mixing zone length increases monotonically from the 2x2 GRS case to the point GRS case.

Figure 4c shows the effluent concentration profiles for each GRS. The effluent profile of a purely homogeneous reservoir with no dispersion would jump from zero to one at one pore volume injected, with more heterogeneous reservoirs displaying greater spread and earlier breakthrough times. The effluent profiles for the coarser GRS's have less spread than the point GRS, indicating a less heterogeneous system. However, as with the mixing zone, the results are somewhat ambiguous. For example, the 2x2 GRS reaches a concentration of 0.82 before the point GRS, while the 8x8 GRS reaches that concentration after the point GRS. The cumulative tracer production plots shown in Figure 4b are less ambiguous. The cumulative tracer production at any time increases monotonically with the GRS, with the coarser GRS curves having a more homogeneous character.

Figures 4d-f best illustrate the effects of GRS on the reservoir response. In these plots, a single value which characterizes the response is plotted versus the representation length. Figures 4e and f plot the same responses as Hewett and Behrens: cumulative production at one pore volume injected and breakthrough time. Hewett and Behrens observed a critical resolution where the response did not change as the number of representation/computation layers increased (about 30). The cumulative production at one PVI does not show such a critical resolution. Indeed the cumulative production at one PVI varies approximately linearly with L_{rep}, continuing to vary at all representation lengths. For the 256x128 computational grid the breakthrough time also varies monotonically with L_{rep}. However, the rate at which the breakthrough time changes with the GRS decreases significantly below the 32x32 GRS. Furthermore, when a 128x128 computational grid is used the breakthrough time is greater for the point GRS than for the 64x64 GRS. Finally, Figure 4d shows the error in the effluent curves of the coarse GRS runs. The effluent error was calculated as the overall area between the coarse GRS effluent curve and the point effluent curve: $\int_0^{1.5} \left|\overline{C}_{128\times128}(1,t) - \overline{C}(1,t)\right| dt$. Like the cumulative production at one PVI, the effluent error also varies approximately linearly with L_{rep}.

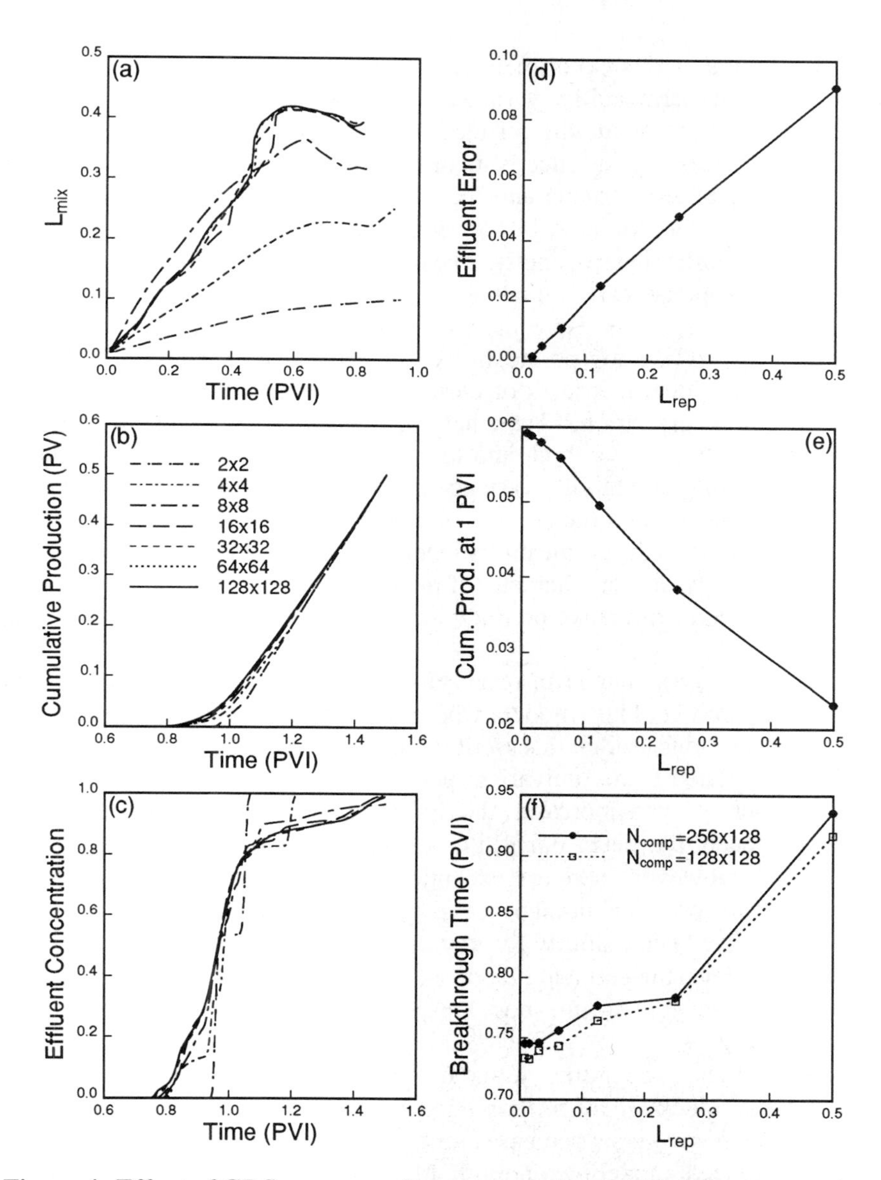

Figure 4: Effect of GRS on reservoir response using a 256x128 computational grid (a) mixing zone length (b) cumulative tracer production (c) effluent tracer concentration (d) effluent concentration error (e) cumulative tracer production at one pore volume injected (f) tracer breakthrough time using a 256x128 computational grid and a 128x128 computational grid.

C. Discussion

The results of this experiment are not too surprising. As the GRS decreases, the log-permeability variance and measures of convective mixing continue to increase, while the correlation range remains approximately constant. The decrease in variance is a consequence of the spatial averaging of a spatially correlated variable and the discretization of a spatial variable in a finite region. The increase in convective mixing can be attributed to increased permeability heterogeneity. The impact of permeability heterogeneity on reservoir response can be quantitatively summarized by the heterogeneity index, HI $= \sigma_\kappa^2 L_\kappa$, where σ_κ^2 is the variance of the log-permeability and L_κ is the range of the log-permeability in the mean flow direction.[33,34] Unlike, some traditional measures of permeability heterogeneity which do not consider spatial correlation,[35,36] the heterogeneity index reflects both the size of and variability between high and low permeability regions. As the GRS decreases the log-permeability variance increases, thus increasing the permeability heterogeneity and the amount of convective mixing. This view of the relationship between the permeability heterogeneity and the reservoir response is supported by the similar character of the plots of log-permeability variance, effluent error, and cumulative production at one PVI versus the representation length.

The continuing change in reservoir response with GRS is particularly true for a layered geologic model. Under the same governing equations and boundary conditions used in this section, the breakthrough times of each layer are directly related to the univariate permeability distribution.[36] In general, as the number of layers increases, the discrete/layer permeability distribution should approach, but never equal, that of the continuous/point permeability distribution. However, there are exceptions. If the underlying permeability is a spatially uncorrelated point process, spatial averaging yields the expected value of the point permeability for any finite number of layers. Thus, at any discretization level the permeability of each layer is equal to the expected value of the point permeability—the model response does not change as the number of layers changes.

The character of Figures 4d and e is similar to plots of response versus computational block length for homogeneous systems, except that mixing in the homogeneous systems decreases rather than increases as the block length decreases. Indeed, modeling a point field using a coarse GRS can be viewed as neglecting a term in the governing equations. Just as the manner in which the numerical error changes with block size depends on the numerical method, the manner in which the representation error changes with the GRS depends on the method of calculating block permeabilities.

The critical resolution observed by Hewett and Behrens was not observed in this study. Possible explanations for their observation of a critical resolution is their coupling of the numerical and computational scales, the significant anisotropy of their permeability field, and/or their having varied only the number of layers while the number of columns (horizontal blocks) was held constant. Indeed, the number of layers at the critical resolution (30) was on the order of the number columns (20). In this study, the representation length at which the rate of decrease in the breakthrough time with GRS changes significantly (0.4) is approximately the equal to the separation distance at which the x and y directional semi-variograms no longer coincide (0.5).

IV. MODELING SUB-REPRESENTATION SCALE HETEROGENEITY

The possibility that a finite GRS model may be in error, regardless of the representation scale, suggests the use of effective properties to improve model accuracy. As demonstrated in the preceding section, the error existing at a finite GRS is due to inadequate representation of permeability heterogeneity and consequently, inadequate representation of convective mixing. For miscible transport, small scale convective mixing is traditionally modeled through dispersivity. In this section, a numerical experiment is performed to determine if mixing due to small *and* large scale heterogeneities can be modeled by a combination of explicit representation using a discrete geologic model and implicit representation using dispersivities. This experiment also introduces the use of the non-represented permeability to evaluate mixing parameter robustness.

A. *Uncorrelated, Layered Permeability Experiment*

To investigate the separation of permeability representation into explicit and implicit components, an idealized geologic model was constructed. The point log-permeability of this model is the sum of a spatially uncorrelated field and a layered field:

$$\kappa = \kappa_l + \kappa_w \tag{6}$$

where κ is the point log-permeability, κ_l is the layered log-permeability, and κ_w is the spatially uncorrelated (white-noise) log-permeability. The layered field was constructed by randomly drawing eight log-permeabilities from a normal distribution with a variance of one and a mean of zero. The white-noise field was constructed by randomly drawing log-permeabilities from a normal distribution with a variance of 0.5 and mean of zero on a 128x128 representation grid. The realization statistics are summarized in Table 1.

Table 1: Summary of point, layer and white-noise log-permeability statistics.

	Point	Layer	White-Noise
Mean	-0.03	0.00	-0.02
Variance	1.09	0.50	0.57
Minimum	-3.59	-2.39	-1.21
Median	-0.06	0.00	0.05
Maximum	3.77	2.59	1.44

Grayscale images of the point permeability field and its component fields are shown in the top row of Figure 5. Because the model permeabilities are constant within representation blocks, the permeabilities are spatially correlated for separation distances less than the representation length. Indeed, the white-noise permeability is actually spatially correlated for separation distances up to $1/128$ in the x and y directions. Similarly, the layer permeability is perfectly correlated in the x-direction and correlated for separation distances up to $1/8$ in the y-direction.

Simulations of flow through the point permeability field and each of its components were performed using the same governing equations, boundary conditions, and numerical techniques as in Section III. A 128x128 computational grid was used for all simulations in this section. Grayscale images of the concentration at a time of 0.5 PVI are shown in the middle row of Figure 5. These images illustrate the character of the mixing induced by each permeability field. The concentration front of the layered permeability is dominated by eight large fingers of various sizes corresponding to each permeability layer. The concentration front of the white-noise permeability, however, is composed of a large number of fingers of similar transverse and longitudinal dimensions. The front of the point permeability is dominated by the same large fingers as the layered model, but is also affected by the white-noise permeabilities—particularly by the crossflow they induce. The relative impact of the component permeability fields on the response of the point permeability field is further demonstrated by effluent concentration and cumulative tracer production plots, top row of Figure 6.

The average concentration profiles indicate the general character of the mixing induced by each of the permeability fields. The bottom row of Figure 5 shows the average concentration profiles at 0.1 PVI time intervals for each permeability field. The point and layered permeability field have complex

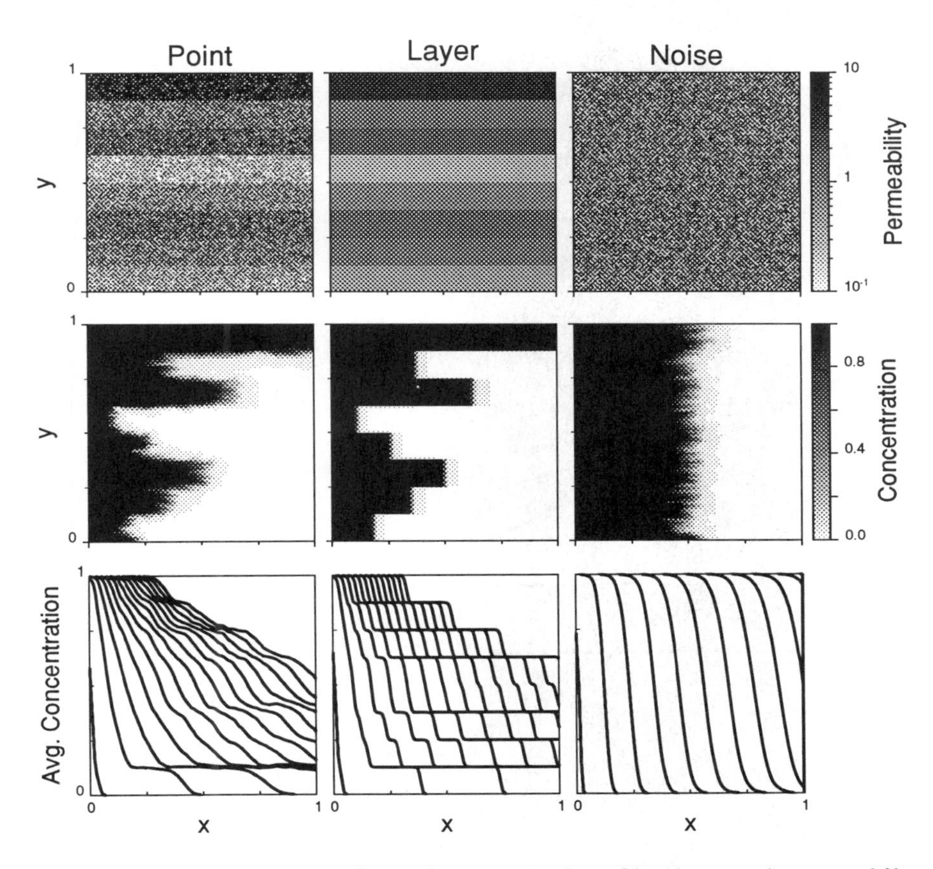

Figure 5: Point, layer, and white-noise maps and profiles (top row) permeability maps (middle row) tracer concentration maps at 0.5 pore volumes injected (bottom row) profiles of concentration averaged in the y-direction at 0.1 PVI time intervals.

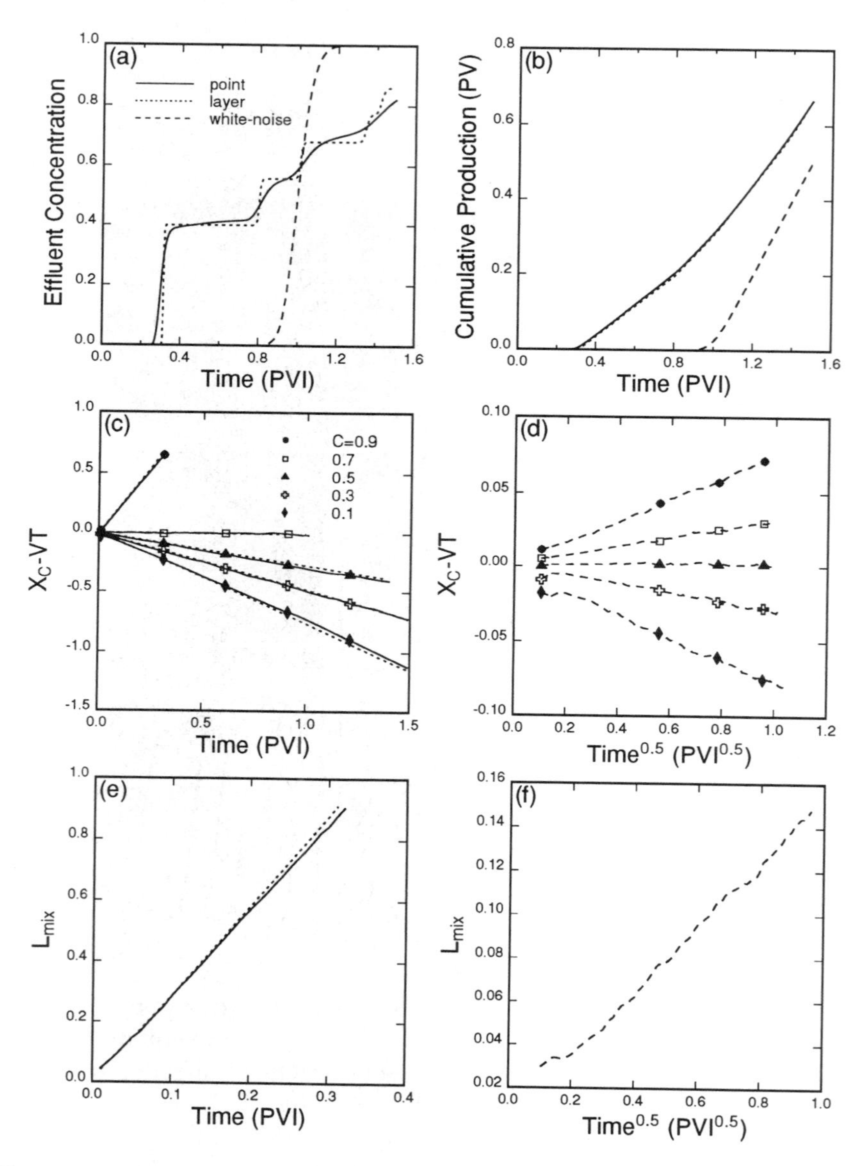

Figure 6: Point, layer, and white-noise response plots (a) effluent concentration (b) cumulative production (c) $(x_c - vt)$ for point and layer runs (d) $(x_c - vt)$ for white-noise run (e) mixing zone length for point and layer runs (f) mixing zone length for white-noise run.

profiles which cannot be represented by a 1D convection-dispersion model.† However, the profiles of the white-noise permeability field have the distinct complementary error function shape which can be represented by a 1D model convection-disperion model.

The $(x_c - vt)$ versus t plots in the middle row of Figure 6 describe the movement of a particular average concentration with time. The form of these plots derives from the analytic solution of the 1D convection-dispersion equation and the analytic solution of the purely convective layered reservoir problem. An approximate solution of the convection-dispersion equation under these boundary conditions is:[28]

$$C = \frac{1}{2}\text{erfc}\left(\frac{x - vt}{2\sqrt{Kt}}\right) \quad (7)$$

from which,

$$x_c - vt = \left[2\text{erfc}^{-1}(2C)\sqrt{K}\right]\sqrt{t} \quad (8)$$

where x_c is the location of a concentration C at a given time t and v is the fluid velocity (u/ϕ). Thus, $(x_c - vt)$ varies linearly with the square root of time. For a layered system under boundary conditions described by Equations 5, an average concentration moves at a constant velocity, thus:

$$x_c = v_c t \quad (9)$$

from which

$$x_c - vt = (v_c - v)t \quad (10)$$

where v_c is the concentration velocity. In this case, $(x_c - vt)$ varies linearly with time. As expected, Figure 6c shows $(x_c - vt)$ varying linearly with time for the layered system and approximately so for the point permeability field. Such is not the case for the white-noise permeability field, where $(x_c - vt)$ varies linearly with the square root of time, Figure 6d. The mixing zone length at a given time can be found from Figures 6d and e as the distance between the $(x_c - vt)$ curves for 0.9 and 0.1 concentrations; that is, $L_{mix} \equiv (x_{0.9} - x_{0.1}) = (x_{0.9} - vt) - (x_{0.1} - vt)$. Figure 6e shows the mixing zone length of the point permeability case falling slightly below that of the layered case. Thus, the addition of small scale heterogeneity to the layered system causes transverse mixing between layers which decreases the impact of longitudinal mixing.

---

†It may be possible to represent the layered system behavior with a homogeneous model and a relative-permeability like function, but only for one set of boundary conditions.[37]

1. Matching White-Noise Response With a Homogeneous/Dispersivity Model

As indicated by the plots for the white-noise run, the behavior of the average concentration can be modeled by an equivalent homogeneous system and dispersion tensor. The longitudinal dispersivity for this system was estimated to be 0.0020 through an analysis of the effluent concentration of the white-noise run.[38] Using this dispersivity, both the concentration profiles and the effluent concentration profiles are closely reproduced by a 1D convection-dispersion model. Similarly, the transverse dispersivity for the homogeneous system was estimated to be 0.00015. The transverse dispersivity was found by simulating a flow problem through the white-noise permeability field which differed from the first case only in the inlet concentration boundary condition. In this case, the inlet concentration was specified to be 1 for $y \in (0, 0.5)$ and 0 for $y \in (0.5, 1)$. After injecting approximately one pore volume, the concentration in the system remains constant. The transverse dispersivity was found by analyzing the concentration profile at the outlet in a manner similar to that used for effluent concentration.

2. Matching Point Response With a Layer/Dispersivity Model

Because the white-noise response can be modeled using lumped parameters, it seems likely that the point response can be modeled using lumped parameters in conjunction with a layered permeability field: a layer/dispersivity model. To determine if this is true, an attempt was made to build such a model. Initial attempts at matching the point response using a model with the original layer permeabilities and various dispersivities indicated that some adjustment of the layer permeabilities was necessary. Hence, new layer permeabilities were calculated using the velocity field of the point model. Each new layer permeability was found such that the effluent volumetric flow rate of each layer matched that of the point model. The maximum percentage change in a layer permeability was 6 percent. Using the adjusted layer permeabilities, optimal longitudinal and transverse dispersivities were found using non-linear regression. The regression minimized the absolute difference (as defined in Section III) between the effluent concentration curve of the point model and that of the layer/dispersivity model. Regression gave a value of 0.002168 for the longitudinal dispersivity and 0.000152 for the transverse dispersivity.

Figure 7 shows a comparison of the point model response and the layer/dispersivity model response. The match is close, but not perfect. An improved match could probably be obtained by determining the layer permeabilities through regression.

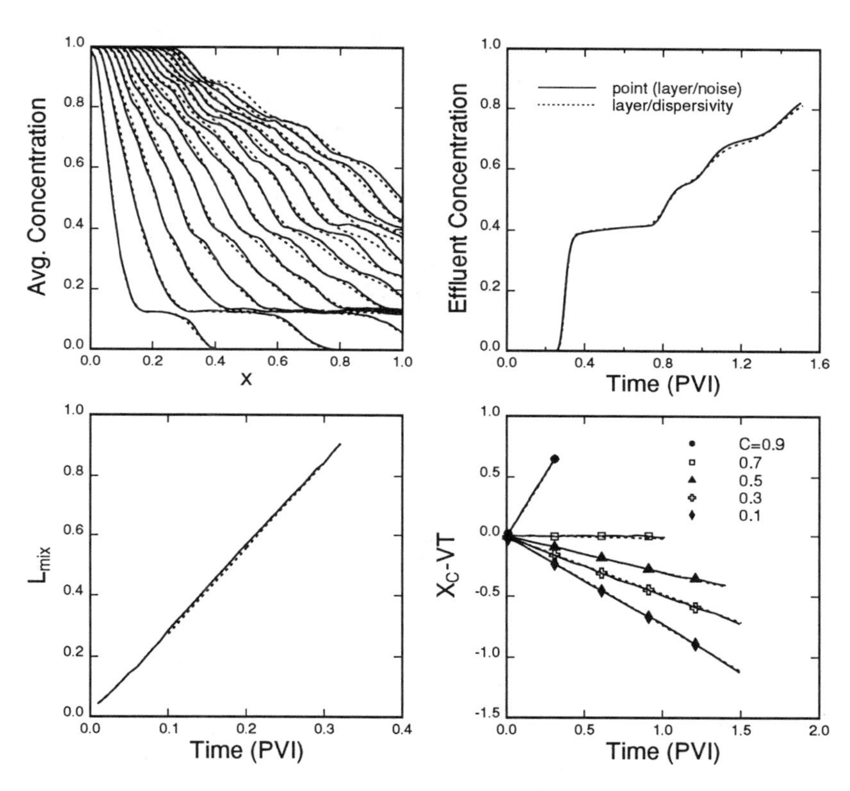

Figure 7: Comparison of point (layer/white-noise) model response to the layer/dispersivity model response (a) average concentration profiles (b) effluent concentration (c) mixing zone length (d) $(x_c - vt)$.

B. Discussion

The results of this experiment indicate a successful, though not perfect, attempt at representing large and small scale heterogeneity using a combination of explicit (layer permeabilities) and implicit (dispersivities) representation. Moreover, the longitudinal and transverse dispersivities calculated from simulations through the white-noise permeability field are approximately equal to the optimal dispersivities of the layer/dispersivity model. The longitudinal dispersivity needed to model flow through the white-noise field can also be estimated from the covariance of the white-noise log-permeability.[22,39,40] Under various assumptions, the asymptotic longitudinal dispersivity for steady flow through a stationary permeability field is given by:

$$\alpha_L = \int_0^\infty C_\kappa(\vec{v}_m h)\, dh \tag{11}$$

where $\vec{v}_m$ is a unit vector in the direction of mean flow. The covariance of the white-noise log-permeability in the x-direction is σ_κ^2 at $h = 0$ and decreases linearly to 0 at $h = L_{rep}$, remaining 0 for $h > L_{rep}$. In this case, the integral in Equation 11 is the area of a triangle and $\alpha_L = \frac{1}{2}\sigma_\kappa^2 L_{rep} = 0.0019$. This value is close to that determined from the simulations. The values of the various dispersivities are summarized in Table 2.

Table 2: Summary of Calculated Dispersivities

	α_L	α_T
Layer/Dispersivity Optimal	0.002168	0.000152
White-Noise Response	0.0020	0.00015
White-Noise Covariance	0.0019	

1. Non-Represented Permeability

The relationship between the dispersivities derived from the white-noise permeability field and those of the layer/dispersivity model indicate the usefulness of separating the point permeability into represented (layer) and non-represented (white-noise) components. A model for separating a general permeability field into represented and non-represented components is:

$$\begin{aligned} k(\vec{x}) &= k_r(\vec{x})k_{nr}(\vec{x}) \\ \kappa(\vec{x}) &= \kappa_r(\vec{x}) + \kappa_{nr}(\vec{x}) \end{aligned} \tag{12}$$

where $k(\vec{x})$, $k_r(\vec{x})$, and $k_{nr}(\vec{x})$ are the point, represented and non-represented permeabilities at location $\vec{x} = (x, y, z)$; $\kappa(\vec{x})$, $\kappa_r(\vec{x})$, and $\kappa_{nr}(\vec{x})$ are the point, represented and non-represented log-permeabilities at $\vec{x}$. This separation model is similar to ones used in geostatistics and stochastic analysis of effective properties which separate a random field into its estimated or expected value and a fluctuation.[24,22] It is also related to the series representation of a random variable as the sum of component random variables.[41,24,32] If the non-represented and represented permeabilities are uncorrelated, the covariance of the non-represented can be determined from the point and represented log-permeability covariances.

$$\begin{aligned} C_\kappa(\vec{h}) &= Cov[\kappa_r(\vec{x}) + \kappa_{nr}(\vec{x}), \kappa_r(\vec{x}+\vec{h}) + \kappa_{nr}(\vec{x}+\vec{h})] \\ &= C_{\kappa_r}(\vec{h}) + C_{\kappa_{nr}}(\vec{h}) \;, \text{ if } Cov[\kappa_r(\vec{x}), \kappa_{nr}(\vec{x}+\vec{h})] = 0 \qquad (13) \end{aligned}$$

When appropriate, this relation may be used with Equation 11, to estimate the asymptotic longitudinal dispersivity associated with non-represented permeability:

$$\alpha_{L_{nr}} = \int_0^\infty [C_\kappa(\vec{v}_m h) - C_{\kappa_r}(\vec{v}_m h)] \; dh \qquad (14)$$

The explicit separation of the point permeability into represented and non-represented components is appealing because it gives a concrete meaning to what is being represented by the effective mixing parameters. The experiment of this section suggests that the non-represented permeability can be used to analyze the robustness of effective mixing parameters and, perhaps, to estimate parameter values. Indeed, it is unlikely that a coarse model, consisting of effective mixing parameters and represented permeability, could robustly represent the point permeabilities if flows through the non-represented permeabilities could not be represented by effective mixing parameters. Thus, the difficult problem of determining whether a point permeability field can be represented by a given coarse permeability field and effective mixing parameters is transformed into a simpler, more restrictive problem.

2. Robustness of Effective Mixing Parameters

The simpler problem is that of determining whether flow through a point permeability field, the non-represented permeability field, can be modeled with a homogeneous permeability field and effective mixing parameters. This problem is a common one and is the subject of continuing research. In particular, much work has been conducted on determining whether the spreading characteristics of tracer flows through heterogeneous permeability fields can be represented by dispersivities.[22,42,39,1,2,32,40] This research and field studies indicate that the apparent longitudinal dispersivity necessary to characterize the spread of a front or a plume, may continue to change with distance traveled (time). However, if the spatial correlation range of the permeability field is finite, the apparent longitudinal dispersivity should reach a constant value after traveling several correlation lengths, perhaps asymptotically. In this case, the spread of a front can be correctly modeled by a finite dispersivity for travel distances greater than some multiple of the correlation length. For travel distances less than this multiple of the correlation length, the front spread is incorrectly modeled by a constant dispersivity.

The dependency of apparent dispersivity on travel distance can cause a model to be non-robust. Consider, for example, a finite system where the permeability correlation length is approximately one fifth of the system length. To begin, the system is subjected to a tracer displacement in the x-direction. As the tracer front moves through the field, the apparent longitudinal dispersivity continues to change, never reaching an asymptotic value. Despite this, it is possible to approximately match the observed effluent concentration using a homogeneous model with a constant dispersivity. This model, however, is only valid for the particular boundary conditions and system length for which it was derived. At early times, the model over predicts the amount of front spreading, with the error decreasing as the front approaches the outlet.

Next, consider a displacement through the same system with mean flow initially in the x-direction. However, the boundary conditions are changed at an early time to induce mean flow in the y-direction. At the time when the flow direction is changed, the predicted front spread in the x-direction is in error. Unlike the first case, this error will not decrease with time but will persist. Thus, the accuracy of the model is sensitive to changes in boundary conditions and is not robust. In summary, the sooner the apparent dispersivity nears its asymptotic value, the more accurately a model predicts early time front spreading. This decreases the sensitivity of the model to changes in boundary conditions and system size and leads to a more robust model.

The robustness of a dispersivity based model can be determined either indirectly through knowledge of the statistical character of the permeability field, or directly by performing simulations of flow through the permeability field. For permeability fields whose statistical character is known, it may be possible to determine the robustness of a dispersivity model *a priori* by comparing the correlation length to the system length. If statistical analysis alone is not adequate, the character of concentration front spreading induced by the permeability field can be determined by flow simulations. As discussed above and elsewhere, examining the effluent concentration alone is not a good indication of model robustness.[42,2,1] Instead, it is important to examine the rate at which a front or plume changes with time.

V. OPTIMAL REPRESENTATION SCALE

The first experiment showed that an optimal GRS cannot generally be defined by a straightforward analysis of model response versus GRS. The second experiment supports the common belief that a fine scale permeability field can be represented by a combination of explicit representation and representation through effective mixing parameters. However, if the correlation range of the non-represented permeability is large with respect to the system size, this scheme probably will not work. Fortunately, as demonstrated in the first experiment, the correlation length of the non-represented permeability may be reduced by decreasing the GRS. Ideally then, the GRS and the correlation length of the non-represented permeabilities can be reduced to a size which is small enough to allow effective mixing parameters to robustly model the mixing induced by the sub-representation scale heterogeneity. Hence, the optimal GRS is the largest GRS for which the effective mixing parameters satisfy a desired level of robustness.

This method of choosing a GRS differs from other approaches in that it focuses on the ability of the resulting model to accurately model a variety of conditions, not just the accuracy of the model for a particular scenario. The success of this method depends on a number of things. Primarily, this method requires a means of accessing model robustness (such as the use of the non-represented permeability). Additionally, model robustness must increase as the GRS decreases. This issue and possible complications are discussed below.

A. *Averaging Volume for Effective Mixing Parameters*

In Section IV, a single longitudinal dispersivity was used to model the mixing behavior of a tracer flowing through the non-represented permeability field. The robustness of the dispersivity was determined by analyzing a concentration averaged across the entire system, $\overline{C}(x)$. Consider a concentration which is averaged over a smaller distance, $\overline{C}_{L_{avg}}(x, y, t) = \frac{1}{L_{avg}} \int_{y-L_{avg}/2}^{y+L_{avg}/2} C(x, \alpha, t)\, d\alpha$. As L_{avg} is reduced from L_y to L_{rep} (1 to 1/128) the size of the fingers in the white-noise field, Figure 5b, become large with respect to L_{avg}. As a consequence, the behavior of $\overline{C}_{L_{avg}}(x, y, t)$ can no longer be modeled with a dispersivity. This is because the permeability correlation scale has become large compared to the averaging scale. As such, an individual permeability flow path or barrier is more likely to significantly affect the average concentration. Consequently, the average concentration is less predictable, more variable, and more difficult to model with an effective mixing parameter. Similar discussions on the relation to averaging volumes and variability can be found elsewhere.[42,43]

Consider, then, the commonly used approach of defining a different effective parameter for each representation block.[9,44,45] Such block effective parameters are defined to match the response of fine scale simulations. The idea is to match as closely as possible the mixing occurring within each block. Unfortunately, if the permeability correlation scale is on the order of the representation scale or larger, the mixing behavior occurring within an individual block cannot be accurately modeled by a dispersivity. Hence, the coarse model has a different dispersivity for each block, none of which is accurate (able to match a single response) or robust (able to model under various flow conditions with the same accuracy). The only hope is that the errors occurring within each block are not systematic; thus, canceling one another through the course of a simulation. Because heterogeneity is likely to exists at all scales, the situation does not improve as the GRS decreases. The presence of correlated permeability regions as large or larger than the representation block continue to exist. Thus, the number of dispersivities increases without necessarily increasing the robustness of the dispersivities or the robustness of the model as a whole. This partially explains the behavior observed by Davies and Haldorsen,[16] whose coarse model only became accurate when the representation scale (hence, the scale over which pseudo functions were defined) increased beyond that of the largest shale.

Ideally then, the scale over which an effective parameter is constant, the parameter averaging scale, is decoupled from the GRS. Decoupling the parameter averaging scale from the GRS has several advantages. First, it results in a simpler model with fewer parameters. More importantly, if the parameter averaging scale is held constant while the GRS is decreased, the correlation range of the non-represented permeabilities should decrease with respect to the parameter averaging scale. Thus, the accuracy and robustness of the effective parameter should increase as the GRS decreases. This, is of coarse, crucial to applying the definition of the optimal GRS. Finally, the robustness of the dispersivity is an indication of the robustness of the model as a whole. Ideally, only a single mixing parameter is needed and the parameter is averaged over the entire system, as in the second experiment of this work. For less ideal permeability fields, the non-represented permeability may not be statistically homogeneous. In this case, spatially varying effective parameters may be required.

B. Multi-Phase Flow, Diffusion

The term "effective mixing parameter" is used in this paper to emphasize that much of the discussion about the representation of permeability heterogeneity may be applicable to processes other than tracer flow. The primary feature of multi-phase flow is that functions, rather than scalar parameters may be employed in matching fine scale results. Thus, a single fine scale simulation can often be matched. The danger is to confuse the ability to match a single scenario with model robustness. The robustness of a relative permeability could possibly be determined by examining x, t diagrams of flow through the non-represented permeability.[2] The validity of such a test of robustness is not known. It is possible that the non-linearity of the governing equations for multiphase flow lessen the usefulness of the non-represented permeability concept.

Diffusion was assumed to be zero in both of the studies performed herein. Large amounts of diffusion lessen the growth of heterogeneity induced fingers by causing mass transfer between the fingers and adjacent zones.[1,19] For a finite diffusion coefficient, the importance of the diffusion term depends on the flow rate. A higher the flow rate decreases the importance of the diffusion term and increases the size of heterogeneity induced fingers. Thus, a large amount diffusion should ease the representation of permeability heterogeneity. However, because this effect depends on the flow rate, the highest flow rate expected in the reservoir should be used when determining the GRS.

VI. CONCLUSIONS

- When holding all reservoir parameters but permeability at their point values, reservoir response may continue to vary as the GRS decreases.

- The optimal GRS is defined as the maximum scale at which effective mixing parameters model the effects of non-represented heterogeneities to a desired level of robustness.

- The non-represented permeability is useful in determining the robustness of effective mixing parameters and perhaps in determining their value.

- To insure that the robustness of an effective mixing parameter improves as the geologic representation scale is reduced, the parameter averaging scale should be decoupled from the GRS and held constant.

NOMENCLATURE

C,$\overline{C}$	=	concentration, spatial average concentration
$C(\vec{h})$	=	spatial autocovariance for separation vector $\vec{h}$
$Cov(X, Y)$	=	covariance of X and Y, $E[(X - E(X))(Y - E(Y))]$
D	=	porous media diffusion coefficient
$E(X)$	=	expected value of X
erfc(x)	=	complementary error function
GRS	=	geologic representation scale
h	=	separation distance (lag)
k	=	permeability
$\vec{\vec{K}}$	=	dispersion tensor
$L_{avg}, L_{mix}, L_{rep}$	=	averaging, mixing zone, and representation length
L_x, L_y, L_z	=	system length in x,y and z directions
L_κ	=	correlation length of log-permeability
N_{comp}	=	number of computation blocks
N_{rep}	=	number of representation blocks
PV, PVI	=	pore volumes, pore volumes injected
t	=	time
$\vec{u}$	=	Darcy velocity, volumetric flux
$\vec{v}$	=	interstitial velocity, $\vec{u}/\phi$
$\vec{v}_c$	=	velocity of a concentration
x, y, z	=	Cartesian directions
x_c	=	x location of a concentration
α_L,α_T	=	longitudinal and transverse dispersivity
$\vec{\vec{\delta}}$	=	identity tensor
κ	=	log-permeability
$\gamma(h)$	=	semi-variogram for separation distance h
ϕ	=	porosity
$\mathbf{\Phi}$	=	flow potential
σ^2	=	variance

Subscripts

l	=	layer property
m	=	mean
nr	=	non-represented
r, rep	=	represented, representation
w	=	white-noise property

ACKNOWLEDGEMENTS

The authors wish to thank André Journel and Tom Hewett for their criticisms and suggestions. The authors also thank the Stanford Center for Reservoir Forecasting (SCRF) and the Stanford Reservoir Simulation Industrial Affiliates Program (SUPRI-B) for their financial support.

REFERENCES

1. Arya, A., Hewett, T. A., Larson, R. G., and Lake, L. W.: "Dispersion and Reservoir Heterogeneity," *SPERE* (February 1988) 139–48.

2. Hewett, T. A. and Behrens, R. A.: "Conditional Simulation of Reservoir Heterogeneity with Fractals," paper SPE 18326 presented at the 1988 SPE Annual Technical Conference and Exhibition, Houston.

3. Desbarats, A. J.: *Stochastic Modeling of Flow in Sand-Shale Sequences*, PhD dissertation, Stanford University, Stanford, CA (February 1987).

4. White, C. D.: *Representation of Heterogeneity for Numerical Reservoir simulation*, PhD dissertation, Stanford University, Stanford, CA (June 1987).

5. Haldorsen, H. H. and Lake, L. W.: "A New Approach to Shale Management in Field-Scale Models," *SPEJ* (August 1984) 447–457.

6. Rubin, Y., Gómez-Hernández, J. J., and Journel, A. G.: "Analysis of Upscaling and Effective Properties in Disordered Media," *Reservoir Characterization, II*, L. W. Lake and H. B. Carroll Jr. (eds.), Academic Press, Inc., London (1991).

7. Coats, K. H., Nielson, R. L., Terhune, M. H., and Weber, A. G.: "Simulation of Three Dimensional, Two-Phase Flow in Oil and Gas Reservoirs," *SPEJ* (1967) 377–388.

8. Hearn, C. L.: "Simulation of Stratified Waterflooding by Pseudo Relative Permeability Curves," *JPT* (July 1971) 805–813.

9. Kyte, J. R. and Berry, D. W.: "New Pseudo Functions to Control Numerical Dispersion," *SPEJ* (August 1975) 269–276.

10. Jacks, H. H., Smith, O. J. E., and Mattax, C. C.: "The Modeling of a Three-Dimensional Reservoir with a Two-Dimensional Reservoir Simulator-The Use of Dynamic Pseudo Functions," *SPEJ* (June 1974) 175–185.

11. Tollas, J. M. and McKinney, A.: "Brent Field 3-D Reservoir Simulation," paper SPE 18306 presented at the 1988 SPE Annual Technical Conference and Exhibition, Houston.

12. Coats, K. H.: "Use and Misuse of Reservoir Simulation Models," *JPT* (November 1969) 1391–1398.

13. Smith, P. J. and Brown, C. E.: "The Interpretation of Tracer Test Response from Heterogeneous Reservoirs," paper SPE 13262 presented at the 1984 SPE Annual Technical Conference and Exhibition, Houston.

14. Stalkup, F. I., Lo, L. L., and Dean, R. H.: "Sensitivity to Gridding of Miscible Flood Predictions Made With Upstream Differenced Simulators," paper SPE 20178 presented at the 1990 SPE/DOE Symposium on Enhanced Oil Recovery, Tulsa.

15. Warren, J. E. and Price, H. S.: "Flow in Heterogeneous Porous Media," *SPEJ* (September 1961) 153–56.

16. Davies, B. J. and Haldorsen, H. H.: "Pseudofunctions in Formations Containing Discontinuous Shales: A Numerical Study," paper SPE 16012 presented at the 1987 SPE Reservoir Simulation Symposium, San Antonio.

17. Testerman, J. D.: "A Statistical Reservoir-Zonation Technique," *JPT* (August 1962) 889–893.

18. Craig, Jr., F. F.: "Effect of Reservoir Description on Performance Predictions," *JPT* (October 1970) 1239–1245.

19. Lake, L. W. and Hirasaki, G. J.: "Taylor's Dispersion in Stratified Porous Media," *SPEJ* (August 1981) 459–468.

20. Silva, R. J., Niko, H., van den Bergh, J. N., and Sancevic, Z. A.: "Numerical Modeling of Gas and Water Injection in a Geologically Complex Volative Oil Reservoir in the Lake Maracaibo Area, Venezuela," paper SPE 18322 presented at the 1988 SPE Annual Technical Conference and Exhibition, Houston.

21. Gómez-Hernández, J. J.: *A Stochastic Approach to the Simulation of Block Conductivity Fields Conditioned Upon Data Measured at a Smaller Scale*, PhD dissertation, Stanford University, Stanford, CA (March 1991).

22. Gelhar, L. W. and Axness, C. L.: "Three-Dimensional Stochastic Analysis of Macrodispersion in Aquifers," *Water Resour. Res.* (February 1983) **19**, No. 1, 161–180.

23. Wattenbarger, R. C.: *Robust Representation of Permeability Heterogeneity*, PhD dissertation, Stanford University, Stanford, CA (*to be completed* 1992).

24. Journel, A. G. and Huijbregts, C. J.: *Mining Geostatistics*, Academic Press Inc., London (1978).

25. Clark, I.: *Practical Geostatistics*, Applied Science Publishers Ltd., Essex (1979).

26. Isaaks, E. H. and Srivastava, R. M.: *Applied Geostatistics*, Oxford University Press, New York (1989).

27. Rubin, Y. and Gómez-Hernández, J. J.: "A Stochastic Approach to the Problem of Upscaling of Transmissivity in Disordered Media. 1. Theory and Unconditional Simulations," *Water Resour. Res.* (April 1990) **26**, No. 4, 691–701.

28. Lake, L. W.: *Enhanced Oil Recovery*, Prentice Hall Inc., New Jersey (1989).

29. Saad, Y. and Schultz, M. H.: "GMRES: A Generalized Minimal Residual Algorithm for Solving Non-Symmetric Linear Systems," *SIAM J. Sci. Comput.* (1986) No. 7, 856–869.

30. Wattenbarger, R. A.: "Convergence of the Implicit Pressure-Explicit Saturation Method," *JPT* (November 1968) 1220.

31. Roe, P. L.: "Some Contributions to the Modelling of Discontinuous Flows," Lectures in Applied Mathematics (1985) **22**, 163–193.

32. Furtado, F., Glimm, J., Lindquist, B., and Pereira, F.: "Multi-Length Scale Computations of the Mixing Length Growth in Tracer Flows," Emerging Technologies Conference, F. Kovarik (ed.), Houston, TX (June 1990).

33. Mishra, S.: *On the Use of Pressure and Tracer Test Data for Reservoir Description,* PhD dissertation, Stanford University, Stanford, CA (September 1987).

34. Araktingi, U. G.: *Viscous Fingering in Heterogeneous Porous Media,* PhD dissertation, Stanford University, Stanford, CA (June 1988).

35. Jensen, J. L. and Lake, L. W.: "The Influence of Sample Size and Permeability Distribution on Heterogeneity Measures," *SPERE* (May 1988) 629–37.

36. Dykstra, H. and Parsons, R. L.: "The Prediction of Oil Recovery by Waterflooding," *Secondary Recovery of Oil in the United States,* API (1950).

37. Pande, K. K., Ramey, Jr., H. J., Brigham, W. E., and Orr, Jr., F. M.: "Frontal Advance Theory for Flow in Heterogeneous Porous Media," paper SPE 16344 presented at the 1987 SPE California Regional Meeting, Ventura.

38. Correa, A. C., Pande, K. K., Ramey, Jr., H. J., and Brigham, W. E.: "Prediction and Interpretation of Miscible Displacement Performance Using a Transverse Matrix Dispersion Model," paper SPE 16704 presented at the 1987 SPE Annual Technical Conference and Exhibition, Dallas.

39. Dagan, G.: "Time-Dependent Macrodispersion for Solute Transport in Anisotropic Heterogeneous Aquifers," *Water Resour. Res.* (September 1988) **24**, No. 9, 1491–1500.

40. Neuman, S. P.: "Universal Scaling of Hydraulic Conductivities and Dispersivities in Geologic Media," *Water Resour. Res.* (August 1990) **26**, No. 8, 1749–1758.

41. Serra, J.: "Les Structures Gigognes: Morphologie Mathematique et Interpretation Metallogenique," *Mineralium Deposita* (1968) **3**, 135–154.

42. Dagan, G.: "Solute Transport in Heterogeneous Porous Formations," *J. Fluid Mech.* (1984) **145**, 151–177.

43. Haldorsen, H. H.: "Simulator Parameter Assignment and the Problem of Scale in Reservoir Engineering," *Reservoir Characterization,* L. W. Lake and H. B. Carroll Jr. (eds.), Academic Press, Inc., London (1986) 293–340.

44. Lasseter, T. J., Waggoner, J. R., and Lake, L. W.: "Reservoir Heterogeneities and their their influence on ultimate recovery," *Reservoir Characterization*, L. W. Lake and H. B. Carroll, Jr. (eds.), Academic Press, Inc. (1986) 545–559.

45. Kossack, C. A.: "Scaling-Up Laboratory Relative Permeabilities and Rock Heterogeneities With Pseudo Functions for Field Simulations," paper SPE 18436 presented at the 1989 SPE Annual Technical Conference and Exhibition, Houston.

DETERMINING ORIENTATION AND CONDUCTIVITY OF HIGH PERMEABILITY CHANNELS IN NATURALLY FRACTURED RESERVOIRS

H. Kazemi
Marathon Oil Company
Littleton, Colorado

A. A. Shinta
Colorado School of Mines
Golden, Colorado

SUMMARY

All reservoirs are likely to contain natural fractures. These fractures, in general, do not form an interconnected network, but they do affect primary production and the displacement of oil by water, gas, or other displacing agents. To improve the displacement efficiency of the secondary and tertiary oil recovery processes, we need to know fracture orientation and conductivity in a reservoir, both locally and globally. Once this information becomes available, remedial measures to improve oil recovery can be accomplished more prudently.

In this paper, two techniques to determine the effective orientation and conductivity of fractures in the drainage area of a cluster of wells will be presented. One technique is the classical pressure interference testing and the second technique is the use of chemical tracers. Tracer techniques should provide more resolution than the interference testing. Mathematical models will be presented for both applications.

The mathematical model for the tracer response analysis is new and uses a simple and flexible matrix-fracture transfer function. The computer code for the model executes one to two orders of magnitude faster than the conventional dual-porosity/dual-permeability formulations. Furthermore, two different nine-point finite difference schemes are used to appropriately account for the flow channeling and the permeability tensor.

I. INTRODUCTION

Folding, faulting, and subsidence of sediments over geologic time cause fracturing. The more brittle the rock, the more intensely it fractures. Carbonate reservoirs are typically more fractured than sandstone reservoirs for this reason. Solution enlargement of the pores near fractures is possible, as is also partial or full sealing of fractures by mineral precipitation.

Natural fractures affect all phases of petroleum reservoir life from the accumulation of oil to the techniques used to manage oil production. Since the 1860's fractures were known to exist in many oil reservoirs. However, only in the last thirty or forty years has a significant interest in the effect of fractures on oil production emerged.

Naturally fractured reservoirs are generally classified as "dual-porosity" where one porosity is associated with the matrix (reservoir host rock) and another porosity represents the fractures and vugs. In dual-porosity oil reservoirs, fractures provide the main path for fluid flow from the reservoir, then oil from the matrix blocks flows into the fracture space and the fractures carry the oil to the wellbore. For water-wet reservoirs, when water comes in contact with the oil zone, water imbibes into the matrix blocks to displace oil. Combinations of large flow rates, low matrix permeability, and weak imbibition result in water fingering through the fractures into the wellbore. Once fingering of water occurs, the water-oil ratio increases to large values. In oil reservoirs containing free gas, gas fingering and high gas-oil ratio can also be expected.

Fractured reservoirs come in a wide variety of rock mineralogy (carbonate, diatomite, granite, schist, sandstone, shale and coal), porosity, and permeability. Carbonates include limestone, dolomite, and chalk. Fractured limestones are prevalent in the giant and prolific fields of the Middle East. Fractured dolomites are exemplified by the San Andres formation in many West Texas fields and fractured chalks are found in Texas (Austin Chalk), North Sea (Ekofisk), and other parts of the world. Fractured sandstone/siltstone is typified by Spraberry Field in Texas and diatomite is the rock mineralogy of a producing formation in the Belridge Field in California. Fractured Monterey formation in coastal California is composed of chert, dolostone, porcellanite and organic-rich marl.

The presence of open fractures, the degree of fracture interconnectedness, and permeability anisotropy significantly affect the mechanisms responsible for oil recovery from the matrix. Viscous displacement of oil by water or gas in the matrix is often not very effective because only very small pressure gradients are generated across the individual matrix blocks that are surrounded by fractures. Solution gas drive, gravity drainage, and water imbibition are the most prevalent recovery mechanisms, but depending on the effective size of matrix blocks much of the oil can remain in the matrix. And, this remaining oil is the target for improved oil recovery.

Many old and new ideas are being applied or considered for producing the unrecovered mobile oil from fractured reservoirs. These include the use of horizontal wells, infill wells, pattern realignment, conformance improving gels and foams, surface active agents for wettability alteration, dilute surfactant, steam, and CO_2. Infill drilling can provide a means of accessing the poorly drained segments (or compartments) of a reservoir.

To improve oil recovery from naturally fractured reservoirs one needs to quantify the channelling of injected fluids and develop a map of the flow channel distribution for remedial measures. This paper is intended to provide a viable technique to accomplish this when tracers are used.

II. DIRECTIONAL FLOW

A. Pressure Interference Testing

Directional flow results from either permeability anisotropy in an otherwise homogeneous reservoir or the presence of high permeability channels in a heterogeneous reservoir. The latter, under favorable conditions however, could effectively produce flow characteristics of an isotropic homogeneous reservoir. In fact, pressure interference testing conducted in many domestic reservoirs has produced results that support this claim. One example of such a test was reported by Gogarty as shown by Fig. 1.[1] It can be seen that the permeability magnitude varies from one part of the field to another, but the permeability major axis is southwest-northeast direction.

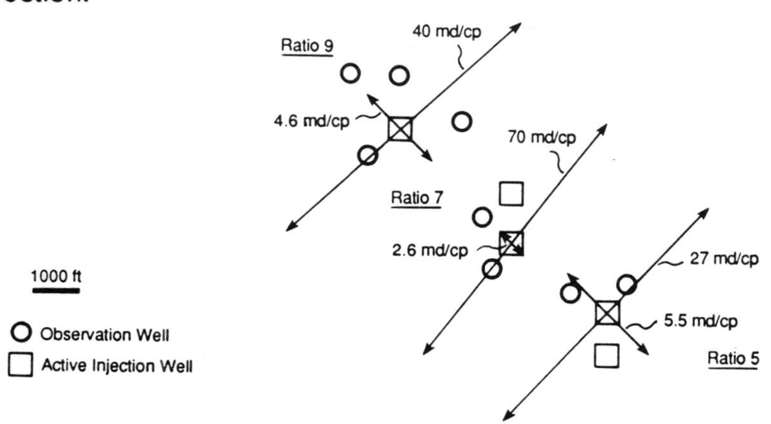

FIGURE 1. INTERFERENCE TESTING

(From Gogarty, W.B., JPT, 1983, Courtesey SPE)

Earlougher presents a technique to determine the direction and magnitude of the permeability tensor from pressure interference testing data.[2] The technique requires measurement of pressure response in at least three observation wells. The analysis of these pressures by type-curve matching (or least-squares regression technique) yields k_{xx}, k_{yy} and k_{xy} which are used to calculate k_{max}, k_{min} and θ. Equations are summarized in Appendix A.

B. Tracer Response Analysis

In this section we develop the equations for the water soluble tracers (e.g., B_r^-, I^-, and SCN^-) in waterflooding operations. Similar equations can be developed for gas tracers in gas injection operations.

In an earlier paper Kazemi, et al.,[3] presented a numerical scheme for solving an analog of the Buckley-Leverett waterflooding problem for fractured reservoirs. The waterflood equations plus tracer transport equations are presented below.

The waterflood equation is:

$$-u\frac{\partial f_{wf}}{\partial x} = \frac{\partial S_{wf}}{\partial t} + \frac{\lambda R_\infty}{\phi_f}\int_o^t e^{-\lambda(t-\tau)}\frac{\partial S_{wf}}{\partial \tau}d\tau \tag{1}$$

Equation (1) was solved numerically as shown:

$$-u\left(f^n_{wf,i} - f^n_{wf,i-1}\right)/\Delta x_i = \left(S^{n+1}_{wf,i} - S^n_{wf,i}\right)/\Delta t + \frac{R_\infty\lambda}{\phi_f}\sum_{j=0}^{n}\left[\left(S^{j+1}_{wf,i} - S^{i}_{wf,i}\right)\prod_{k=j}^{n} e^{-\lambda\Delta t_k}\right];$$

$$i = 1,2,\ldots,IMAX \tag{2}$$

$$S^{n+1}_{wf,i} - S^n_{wf,i} = \left(\frac{1}{\Delta t_n} + \frac{R_\infty\lambda}{\phi_f}e^{-\lambda\Delta t_n}\right)^{-1} \times \left\{-u\frac{f^n_{wf,i} - f^n_{wf,i-1}}{\Delta x_i} - \frac{R_\infty\lambda}{\phi_f}SUM^{n-1}e^{-\lambda\Delta t_n}\right\} \tag{3}$$

$$SUM^{n-1} = \left[SUM^{n-2} + \left(S^{n}_{wf,i} - S^{n-1}_{wf,i-1}\right)\right]e^{-\lambda\Delta t_{n-1}} \; ; \; n \geq 1 \tag{4}$$

$$SUM^{-1} = 0$$

Equation (1) uses a matrix-fracture transfer function of the form

$$R = R_{\infty}\left(1 - e^{-\lambda t}\right) \tag{5}$$

For more flexibility in matching reservoir performance, one can easily use the following form

$$R = R_{\infty} - R_1 e^{-\lambda_1 t} - R_2 e^{-\lambda_1 t} - \ldots \tag{6}$$

where

$$R_1 + R_2 + \ldots = R_{\infty}. \tag{7}$$

To use Eq. (6) in Eq. (1), one should replace the transfer function

$$\tau_w = \frac{\lambda R_{\infty}}{\phi_f} \int_o^t e^{-\lambda(t-\tau)} \frac{\partial S_{wf}}{\partial \tau} d\tau$$

with

$$\tau_w = \frac{\lambda_1 R_1}{\phi_f} \int_o^t e^{-\lambda_1(t-\tau)} \frac{\partial S_{wf}}{\partial \tau} d\tau + \frac{\lambda_2 R_2}{\phi_f} \int_o^t e^{-\lambda_2(t-\tau)} \frac{\partial S_{wf}}{\partial \tau} d\tau + \ldots \tag{8}$$

In the one-dimensional setting, the tracer transport equation for a given non-adsorbing tracer will be

$$-u \frac{\partial S_{wf} C}{\partial x} = \frac{\partial S_{wf} C}{\partial t} + \left[\frac{\lambda R_{\infty}}{\phi_f} \int_o^t e^{-\lambda(t-\tau)} \frac{\partial S_{wf}}{\partial \tau} d\tau\right] C \tag{9}$$

To solve Eq. (9) numerically, we follow the format of Eq. (2) as follows:

$$-u\left(f_{wf,i}^{n}\,C_i^n - f_{wf,i-1}^{n}\,C_{i-1}^n\right)/\Delta x_i = \left(S_{wf,i}^{n+1}\,C_i^{n+1} - S_{wf,i}^{n}\,C_i^n\right)/\,\Delta t$$

$$+\left\{\frac{R_\infty \lambda}{\phi_f}\sum_{j=0}^{n}\left[\left(S_{wf,i}^{j+1} - S_{wf,i}^{j}\right)\prod_{k=i}^{n} e^{-\lambda \Delta t_k}\right]\right\} C_i^{n+1}\,;$$

$$i = 1, 2, \ldots, \text{IMAX} \qquad (10)$$

Equation (2) is first solved for $S_{wf,i}^{n+1}$; then it is substituted in Eq. (10) to solve for C_i^{n+1} .

For three-dimensional field simulations, the following equations are solved simultaneously by standard fully implicit reservoir simulation techniques:

Water:

$$\Delta\left[T_{wf}\left(\Delta p_{wf} - \gamma_{wf}\,\Delta D_f\right)\right] - \tau_w + q_{wf} = \frac{V_R}{5.6146\Delta t}\,\Delta_t\left(\frac{\phi_f\,S_{wf}}{B_{wf}}\right) \qquad (11)$$

Oil:

$$\Delta\left[T_{of}\left(\Delta p_{of} - \gamma_{of}\,\Delta D_f\right)\right] - \tau_o + q_{of} = \frac{V_R}{5.6146\Delta t}\,\Delta_t\left(\frac{\phi_f\,S_{of}}{B_{of}}\right) \qquad (12)$$

where the x-components of the transmissibilities T_{wf} and T_{of} are shown below:

$$T_{wfx} = 0.001127\,\frac{\Delta y \Delta z}{\Delta x}\,k_{fxx}\left(\frac{k_r}{\mu B}\right)_{wf} \qquad (13)$$

$$T_{ofx} = 0.001127\,\frac{\Delta y \Delta z}{\Delta x}\,k_{fxx}\left(\frac{k_r}{\mu B}\right)_{of} \qquad (14)$$

The water and oil transfer functions are:

$$\tau_w = \frac{R_\infty \lambda\, V_R}{5.6146} \sum_{j=0}^{n} \left\{ \Delta_t \left(\frac{S_{wf}}{B_{wf}} \right) \prod_{k=j}^{n} e^{-\lambda \Delta t_k} \right\} \tag{15}$$

$$\tau_o = \frac{R_\infty \lambda\, V_R}{5.6146} \sum_{j=0}^{n} \left\{ \Delta_t \left(\frac{S_{of}}{B_{of}} \right) \prod_{k=j}^{n} e^{-\lambda \Delta t_k} \right\} \tag{16}$$

Tracer in water:

$$\Delta\left[T_{wf}\left(\Delta p_{wf} - \gamma_{wf}\Delta D_f\right) C + \widehat{K}_{wf}\Delta C\right] - \tau_w C + q_{wf} C$$

$$= \frac{V_R}{5.6146\, \Delta t} \Delta_t \left[\left(\frac{\phi_f S_{wf} C}{B_{wf}} \right) + \left(1 - \phi_f\right)(SG)_s\, a \right] \tag{17}$$

where $\widehat{K}_{wf}$ is the dispersivity coefficient multiplier, and its x-component is given by Eq. (18):

$$\widehat{K}_{wfxx} = \frac{\Delta y\, \Delta z}{\Delta x} K_{wfxx} \tag{18}$$

K_{xx}, K_{xy}, and K_{yy} are the components of the dispersivity coefficient tensor as shown below:

$$K_{wfxx} = \frac{D_{wf}}{\tau_f} + \frac{1}{\phi_f S_{wf} |\vec{v}_{wf}|} \left[\alpha_{Lwf} v_{wfx}^2 + \alpha_{Twf} v_{wfy}^2 \right] \tag{19}$$

$$K_{wfxy} = K_{wfyx} = \frac{1}{\phi_f S_{wf} |\vec{v}_{wf}|} \left[(\alpha_{Lwf} - \alpha_{Twf})\, v_{wfx} v_{wfy} \right] \tag{20}$$

$$K_{wfyy} = \frac{D_{wf}}{\tau_f} + \frac{1}{\phi_f S_{wf} |\vec{v}_{wf}|} \left[\alpha_{Lwf} v_{wfy}^2 + \alpha_{Twf} v_{wfx}^2 \right] \tag{21}$$

$$\vec{v}_{wf} = -\,\overline{\overline{k}}_f \frac{k_{rwf}}{\mu_w} \left(\nabla p_{wf} - \gamma_w \nabla D \right) \tag{22}$$

$$|\vec{v}_{wf}| = \left(v_{wfx}^2 + v_{wfy}^2\right)^{1/2} \qquad (23)$$

The mathematical details and program code will appear in Ref. 4.

C. Nine-Point Difference

To properly account for either reservoir channelling or permeability anisotropy, and to reduce grid orientation we used two different nine-point schemes in the xy-plane that will be described below.

If flow channelling in a fractured reservoir is the result of reservoir heterogeneity (and not a result of permeability anisotropy) then a five-point connection grid as shown by Fig. 2 is inadequate. This problem can be greatly alleviated by using a nine-point connection grid as shown by Fig. 3. Then we superimpose a nine-point finite-difference on this grid with a weight of 1/2 for each of the eight connections to a given node; the center node will have, therefore, a weight of 4. This contrasts the 2/3 weight for the x and y directions and 1/6 weight for the diagonal direction in a standard 9-point formulation.[7] Our 1/2 weights are based on the assumption that each of the eight channels emanating from a node acts independent of each other and each channel's permeability, viscosity, and pressure gradient control the amount of flow in that channel according to Darcy's equation. It can be shown by a physical argument that the 1/2 weight is actually $\sqrt{2}/(2+\sqrt{2})$ or 0.4142. But, the use of 1/2 does not make much difference in the outcome of the calculations.

If flow channelling in a fractured reservoir is the result of permeability anisotropy, then the nine-point connection of Fig. 3 plus the use of Shiralker's nine-point coefficients[5] should produce accurate results. Shiralker's coefficients also greatly reduce grid orientation sensitivity for high mobility displacements. More information on this subject is reported by Wolcott.[6] As in Wolcott's work, we use a modified set of Shiralker coefficients (please see Appendix B) that we developed for harmonic-averaged permeability in a block-centered grid in contrast to arithmetic-averaged permeability used by Shiralkar in a point-distributed grid.

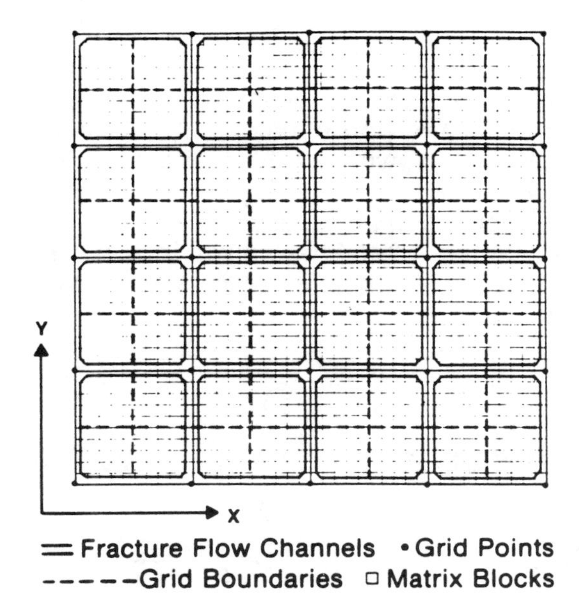

FIGURE 2. FIVE-POINT CONNECTION GRID.
(from Gilman and Kazemi, 1983, courtesy SPE)

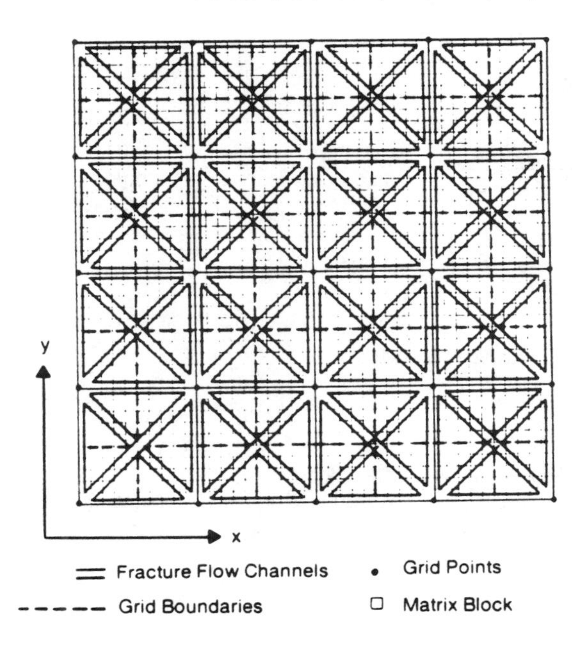

FIGURE 3. NINE-POINT CONNECTION GRID
(from Gilman and Kazemi, 1983, courtesy SPE)

III. RESULTS

We first present numerical results for a tracer injection in an inverted five-spot using two grid orientations—the diagonal and the parallel—as described elsewhere.[7] In these experiments, we use both the Shiralkar and the 1/2-weight coefficients both for a homogeneous isotropic example and an anisotropic example. The data used to evaluate the simulatorare given in Table 1. Figures 4-7 are for the homogeneous isotropic example and Figures 8 and 9 are for the anisotropic example.

TABLE I

DATA USED TO EVALUATE SIMULATOR

B_o = 1 RB/STB
B_w = 1 RB/STB
p_i = 100 psia
S_{wi} = 0.275
μ_o = 1.67 cp
μ_w = 0.5 cp
$c_w = c_o$ = 10^{-5} psi^{-1}
p_{cf} = 0
$\Delta x = \Delta y$ = 187 ft
Δz = 20 ft

$$k_{rwf} = k^*_{rwf}\left(\frac{S_{wf} - S_{wrf}}{1 - S_{wf} - S_{orf}}\right)^{nw}$$

$$k_{rof} = k^*_{rof}\left(\frac{S_{wf} - S_{orf}}{1 - S_{wf} - S_{orf}}\right)^{no}$$

k^*_{rwf} = 0.3

k^*_{rof} = 1.0
S_{wrf} = S_{wrm} = 0.275
S_{orf} = S_{orm} = 0.375
nw = 3.4
no = 2.0
M = 1 (using end-point relative permeabilities)
R_∞ = 0.07
λ_1 = 0.005 day^{-1} (0-2 MOPV)
λ_2 = 0.25 day^{-1} (2- ∞ MOPV)
ϕ_f = 0.01
ϕ_m = 0.20
t_{inj} = 10 days

Onset of tracer injection at WOR = 10-15
Grid Network:
  Diagonal 6x6
  Parallel 8x8
Permeability:
  Homogeneous: $k_{xxf} = k_{yyf} = k_{45°}$ = 39,528 md
  Anisotropic: Parallel, k_{xxf} = 125,000 md, k_{yyf} = 12,500 md, k_{xyf} = 0
  Diagonal, k_{xxf} = 68,750 md, k_{yyf} = 68,750 md, k_{xyf} = 56,250 md
Injection Rate:
  Diagonal 600 B/D
  Parallel 553 B/D
Injected Tracer Concentration 20,000 ppm
Dispersion Coefficient = 0

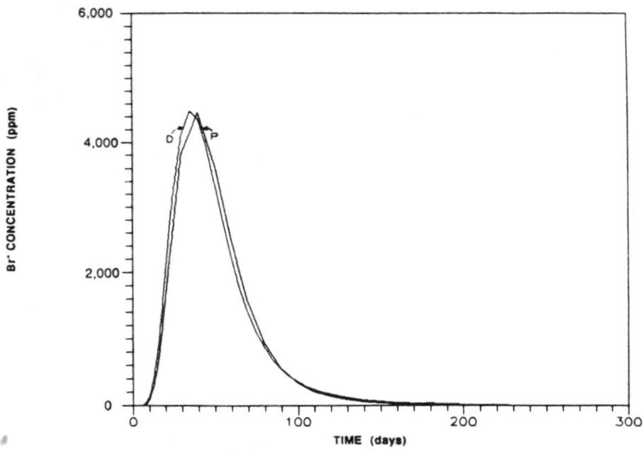

FIGURE 4. TRACER RESPONSE FOR HOMOGENOUS RESERVOIR USING SHIRALKAR NINE-POINT COEFFICIENTS.
(R_{∞}=0.07, M=1, P=parallel 8x8 grid, D=diagonal 6x6 grid)

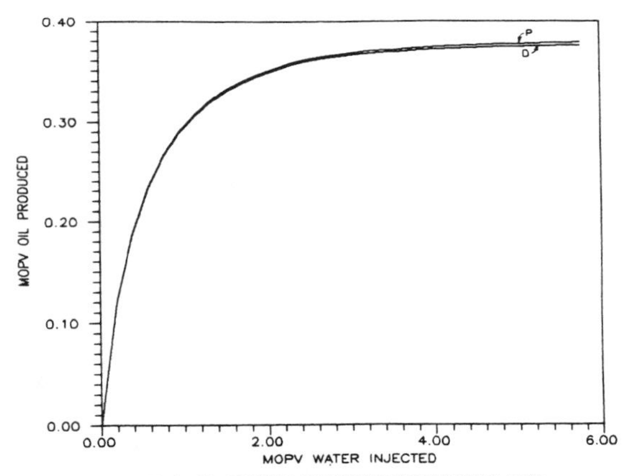

FIGURE 5. OIL RECOVERY FOR HOMOGENOUS RESERVOIR USING SHIRALKAR NINE-POINT COEFFICIENTS.
(R_{∞}=0.07, M=1, P=parallel 8x8 grid, D=diagonal 6x6 grid)

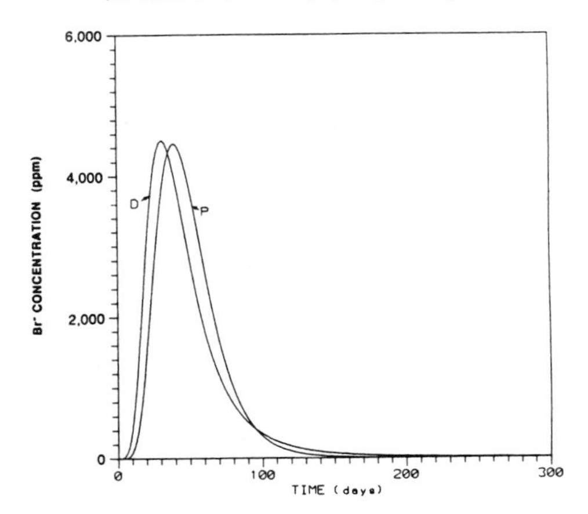

FIGURE 6. TRACER RESPONSE FOR HOMOGENOUS RESERVOIR USING 1/2 WEIGHTING NINE-POINT COEFFICIENTS.
(R_{∞}=0.07, M=1, P=parallel 8x8 grid, D=diagonal 6x6 grid)

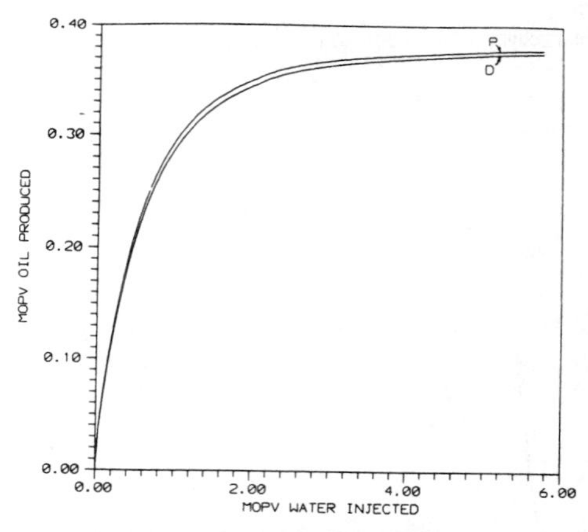

FIGURE 7. OIL RECOVERY FOR HOMOGENOUS RESERVOIR USING 1/2 WEIGHTING NINE-POINT COEFFICIENTS.
(R_{∞}=0.07, M=1, P=parallel 8x8 grid, D=diagonal 6x6 grid)

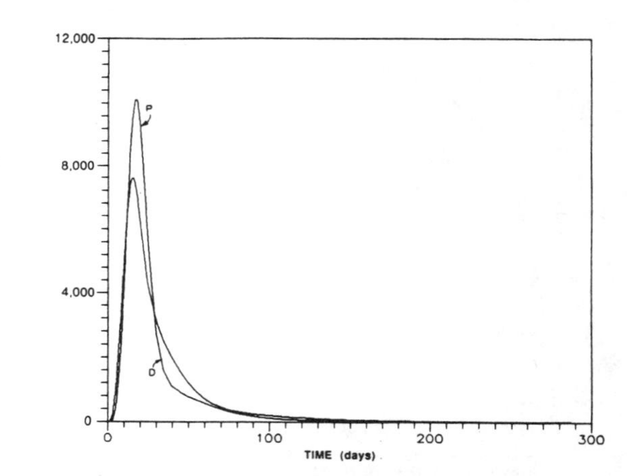

FIGURE 8. TRACER RESPONSE FOR ANISOTROPIC RESERVOIR USING SHIRALKAR NINE-POINT COEFFICIENTS.
R_{∞}=0.07, M=1, P=parallel 8x8 grid, D=diagonal 6x6 grid)

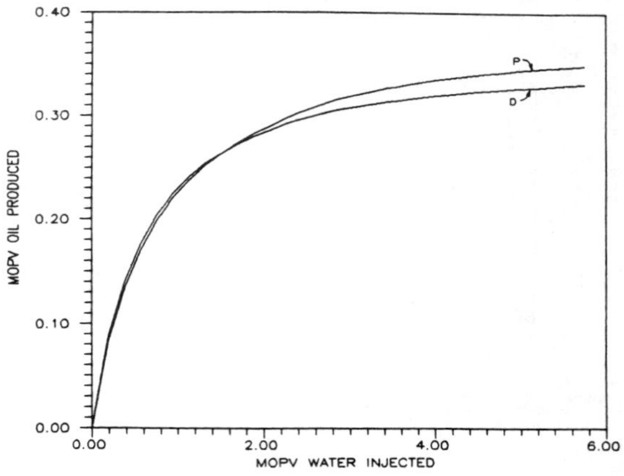

FIGURE 9. OIL RECOVER FOR ANISOTROPIC RESERVOIR USING SHIRALKAR NINE-POINT COEFFICIENTS.
(R_{∞}=0.07, M=1, P=parallel 8x8 grid, D=diagonal 6x6 grid)

Next we present a field example for an inverted five-spot waterflood pattern with data shown in Table 2. The data were modified for the purpose of this paper. The original field waterflood pattern resembles Fig. 10. Tracer was injected in the center well for 30 days. The tracer response measured in Wells A, B, C and D are the results of tracer coming from the center injector diluted by water containing no tracer from other water injectors outside the center pattern and water entering the wells from the underlying aquifer. The tracer response shown for Wells A, B, C, and D were, therefore, adjusted by material balance for the dilutions (please see Appendix C); and the simulation was performed on the shaded area as a confined five-spot pattern.

TABLE II

FIELD DATA

B_o	=	1.044 RB/STB	k_{rwf}	=	$S_{wf}^{\ 1.2}$
B_w	=	0.998 RB/STB	k_{rof}	=	$(1 - S_{wf})^{1.2}$
p_i	=	1000 psia	$\Delta x = \Delta y$	=	54.4 ft
S_{wfi}	=	0.0	Δz	=	40 ft
μ_o	=	8 cp	ϕ_f	=	0.01 (initial estimate)
μ_w	=	0.9 cp	ϕ_m	=	0.18
p_{cf}	=	0	S_{wrm}	=	0.2
$c_w = c_\phi$	=	3.5×10^{-6} psi^{-1}	S_{orm}	=	0.3
c_o	=	1×10^{-5} psi^{-1}	k_m	=	30 md

Grid Network:
15x15x1

Injection Data:
q_{inj} = 1600 B/D
c_{inj} = 900 ppm for 30 days; tracer injection started after 10 years of production

Production Data:
Well A: q_t = 64 B/D
Well B: q_t = 700 B/D
Well C: q_t = 214 B/D
Well D: q_t = 622 B/D

Dispersion Coefficient = 0

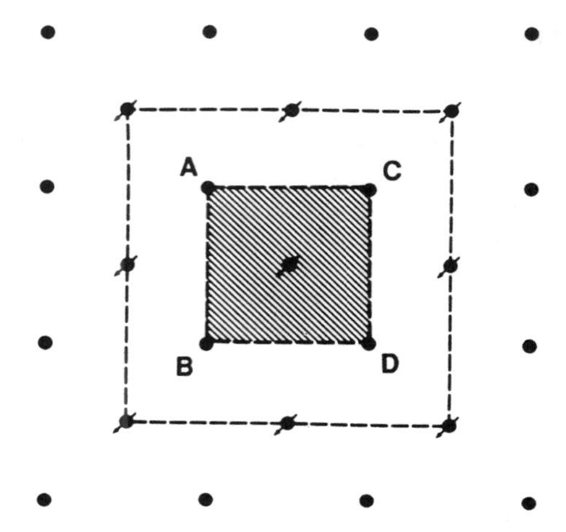

FIGURE 10. A CONCEPTUAL CONFINED FIVE SPOT PATTERN.

Input data for history matching are shown in Table 3 and the history match results are presented in Table 4. Figure 11 illustrates the concept underlying the history matching—which is based on the fact that (1) water cut is large at the time of traced injection, and, (2) saturation change in the reservoir during the tracer test is negligible.

The history match of the tracer response is shown on Fig. 12. The match is excellent and it took 10 runs to obtain the match. Table 4 shows that permeability and porosity, obtained as a consequence of the history match, vary significantly in the vicinity of each production well (i.e., in each quadrant of the pattern).

TABLE III

INPUT DATA TO START HISTORY MATCH

R_∞ = 0.09
λ = 0.00576 day^{-1}
$k_{xxf} = k_{yyf}$ = 30,000 md
ϕ_f = 0.01
q_t = 400 B/D for Wells A, B, C and D

TABLE IV

HISTORY MATCH RESULTS

	Well A	Well B	Well C	Well D
q_t, B/D	64	700	214	622
WOR at injection time	10	56	11	39
ϕ_f	0.012	0.023	0.008	0.024
k_{xxf}, md	500	34,000	3,500	17,000
k_{yyf}, md	2500	28,000	500	10,000

R_∞ = 0.12
λ = 0.0005 day^{-1}

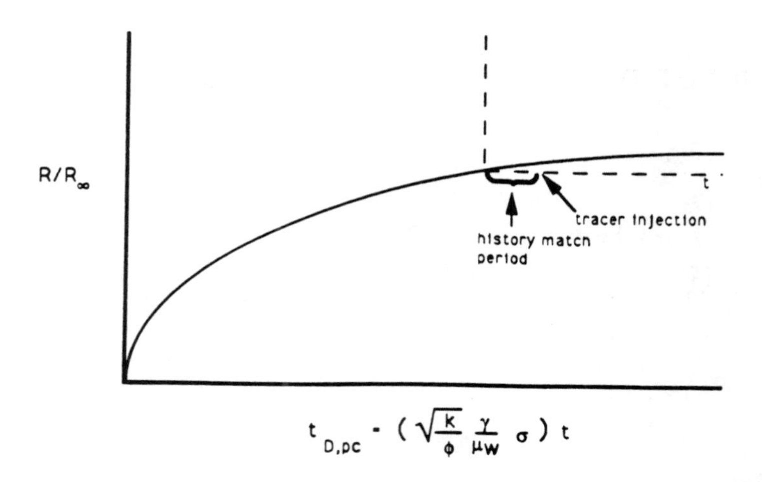

FIGURE 11. CONCEPTUAL OIL RECOVERY CURVE FROM MATRIX

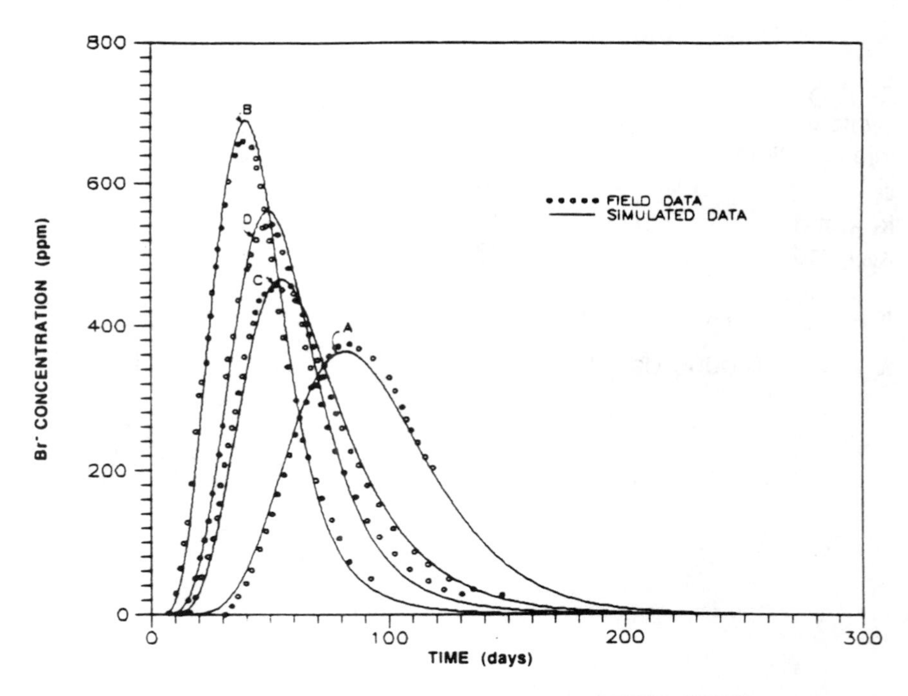

FIGURE 12. INVERTED FIVE SPOT CONFINED PATTERN TRACER PERFORMANCE HISTORY MATCH FOR WELLS A, B, C, AND D.

IV. CONCLUSIONS

From the results of this study, we conclude that:

1. The tracer transport model presented in this paper provides a practical and viable tool for determining orientation and conductivity of high permeability channels in naturally structured reservoirs. The information should be quite useful in planning remedial measures for conformance improvement, and ascertaining reservoir data useful for use in dual-porosity simulators in conjunction with improved or enhanced oil recovery modeling. The tracer transport equations for gas injection operations can be formulated similar to the equations used for waterflooding operations presented in this paper.
2. The numerical algorithm is considerably less complex and executes one to two orders of magnitude faster than the code for standard dual-porosity simulators. This permits more nodes and smaller grid spacing that should lead to more accurate numerical results. Furthermore, single-porosity simulators can be readily modified to include the matrix-fracture transfer function and the tracer transport equations.
3. History matching with this model is much simpler than the standard dual-porosity simulation since it requires considerably less complex input data.
4. While the technique of this paper can be used with non-uniform grid, a grid as close to uniform square grid as possible should be used with the nine-point finite-difference formulations presented here. For non-uniform grid, we recommend the use of controlled-volume finite-element technique. This technique is reported in Ref. 8.

V. NOMENCLATURE

B	=	formation volume factor, RB/STB
c	=	compressibility, psi^{-1}
C	=	chemical concentration, ppm
D	=	depth, positive down, ft
D	=	molecular diffusion coefficient, ft^2/D
f	=	fractional flow
h	=	height, ft
IMAX	=	maximum number of nodes
k	=	permeability, md

k_{max} = principal maximum permeability, md
k_{min} = principal minimum permeability, md
k_r = relative permeability, dimensionless
k_{xx} = xx-component of permeability tensor, md
k_{xy} = xy-component of permeability tensor, md
k_{yy} = yy-component of permeability tensor, md
$\bar{k}$ = geometric average permeability, md
$\bar{\bar{k}}$ = permeability tensor, md
K = dispersivity coefficient, ft^2/D
K_{xx} = xx-component of dispersivity coefficient tensor, ft^2/D
K_{xy} = yy-component of dispersivity coefficient tensor, ft^2/D
K_{yy} = yy-component of dispersivity coefficient tensort, ft^2/D
K = dispersivity coefficient multiplier, ft^3/D
L = matrix block size, ft
M = mobility ratio
MOPV = moveable oil pore volume
p = pressure, psi
p_D = dimensionless pressure
q = flow rate, STB/D
R = recovery, fraction
R_∞ = ultimate recovery, fraction
S = saturation, fraction
SG = specific gravity
t = time, days
t_D = dimensionless time for pressure transient analysis calculations
$t_{D,pc}$ = dimensionless time for waterflood imbibition calculations
T = transmissibility, STB/psi-D
u = interstitial velocity, ft/D
v = Darcy velocity, ft/D
V_R = node volume ($\Delta x \Delta y \Delta z$), ft^3
w = fractional amount

α = dispersivity, ft
α_L = longitudinal dispersivity, ft
α_T = transverse dispersivity, ft
γ = static pressure gradient, psi/ft
γ = interfacial tension, dyne/cm
Δp = pressure change, psi
Δt = time-step size, days
Δt^* = intersection time in pressure buildup, hr
Δx = gridblock length, ft
Δy = gridblock width, ft
Δz = gridblock thickness, ft
Δ = finite difference operator
Δ_t = value at time n + 1 minus value at time n
λ = exponential recovery constant, day^{-1}
μ = viscosity, cp
ρ = density, lbm/ft^3
σ = shape factor, ft^{-2}
τ = fracture/matrix transfer term, STB/D
τ = integration variable in the convolution integral, day
θ = contact angle
ϕ = porosity, fraction

Subscripts

c = capillary
c = connate
D = dimensionless
f = fracture
i = initial
m = matrix
o = oil
r = relative
r = residual
s = solid

w	=	water
x	=	x direction
y	=	y direction
z	=	z direction
β	=	phase (water, oil, gas)
ϕ	=	rock
∞	=	infinite time

Superscripts

n	=	time level
0	=	zero time

VI. APPENDIX A

Equations for Pressure Interference Testing in Anisotropic Reservoirs

From Reference 2 and a knowledge of tensor transformation rules the following working equations were compiled for use in pressure interference test analysis:

$$\Delta p = p_i - p(t,x,y) = \frac{141.2\ qB\mu}{\sqrt{k_{max}k_{min}}\,h} P_D\left(\left[t_D/r_D^2\right]_{dir}\right) \tag{A-1}$$

$$\left(\frac{t_D}{r_D^2}\right)_{dir} = \frac{0.0002637t}{\phi\mu c_t}\left[\frac{k_{max}k_{min}}{\left(y^2k_{xx} + x^2k_{yy} - 2xyk_{xy}\right)}\right] \tag{A-2}$$

$$\overline{k} = \sqrt{k_{max}k_{min}} = \frac{141.2\ qB\mu(P_D)_M}{h\Delta P_M} \tag{A-3}$$

$$y^2k_{xx} + x^2k_{yy} - 2xy\ k_{xy} = \frac{(0.0002637)\ k_{max}k_{min}}{\phi\mu c_t}\ \frac{t_M}{\left(t_D/r_D^2\right)_M} \tag{A-4}$$

$$k_{xx}k_{yy} - k_{xy}^2 = k_{max}k_{min} = \overline{k}^2 \tag{A-5}$$

$$k_{xx} = k_{min} + (k_{max} - k_{min}) \cos^2\theta \tag{A-6}$$

$$k_{yy} = k_{min} + (k_{max} - k_{min}) \sin^2\theta \tag{A-7}$$

$$k_{xy} = k_{yx} = (k_{max} - k_{min}) \sin\theta \cos\theta \tag{A-8}$$

$$k_{max} = 0.5 \left\{(k_{xx} + k_{yy}) + \left[(k_{xx} - k_{yy})^2 + 4k_{xy}{}^2\right]^{1/2}\right\} \tag{A-9}$$

$$k_{min} = 0.5 \left\{(k_{xx} + k_{yy}) - \left[(k_{xx} - k_{yy})^2 + 4k_{xy}{}^2\right]^{1/2}\right\} \tag{A-10}$$

$$\theta = \arctan\left(\frac{k_{max} - k_{xx}}{k_{xy}}\right) \tag{A-11}$$

$$\frac{k_\theta \cos^2\theta}{k_{max}} + \frac{k_\theta \sin^2\theta}{k_{mim}} = 1 \tag{A-12}$$

$$v_\theta = -0.001127 \frac{1}{\mu}\left[(k_\theta \cos_\theta)\frac{\partial p}{\partial x} + (k_\theta \sin\theta)\frac{\partial p}{\partial y}\right] \tag{A-13}$$

$|\Delta p(t,x,y)|$ is plotted vs. t on log-log scale for all the observation wells. The resulting plots are matched against a P_D vs. $(t_D/r_D^2)dir$ log-log type curve. Once a satisfactory match is obtained, $\bar{k}$ is calculated using Eq. (A-3) at any match point M.

Next, for a fixed pressure match, Δp_M, the corresponding three time matches t_M and $(t_D/r_D^2)_M$ are read and substituted in Eq. (A-4) to form three equations in these unknowns k_{xx}, k_{xy} and k_{yy}. These are calculated and the values are substituted in Eqs. (A-9) through (A-11) to calculate k_{max}, k_{min} and θ.

VII. APPENDIX B

Modification of Shiralkar's Nine-Point Coefficients for use with Block-Centered Finite-Difference Formulation.

Shiralkar developed a robust nine-point scheme for anisotropic heterogeneous reservoirs where the transmissibilities yield positive values and reduce grid orientation effects for adverse mobility ratios. Shiralkar formulation was for point-distributed grid. We have extended his formulation for use with block-centered grid; the results yield non-negative transmissibilities.

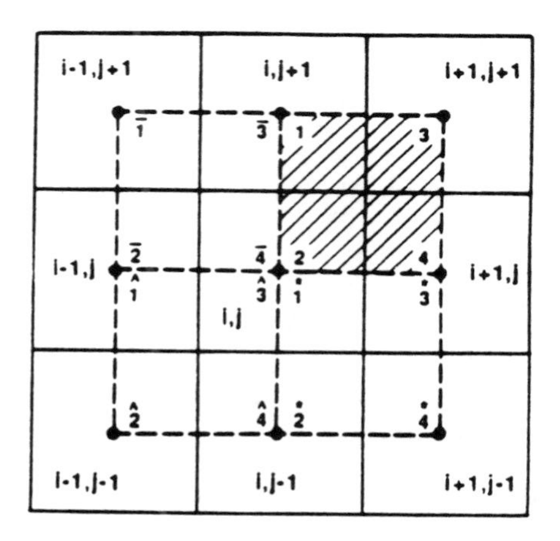

FIGURE B-1. NINE-POINT BLOCK-CENTERED GRID AND ITS RELATION TO SHIRALKAR'S POINT-DISTRIBUTED NUMBERING SCHEME.

The four-node computational element used by Shiralkar is shown by the shaded square with corner points labeled 1, 2, 3, and 4 on Fig. B-1. The conventional block-centered transmissibilities $\left(T_{xi+1/2,j}, T_{yi,\,j+1/2}, \text{etc.}\right)$ and Shiralkar point-distributed transmissibilities (T_{24}, T_{12}, T_{23}, and T_{14}) were used to generate the modified transmissibilities $\left(\widehat{T}_{i+1/2\,j}, \widehat{T}_{i,\,j+1/2}, \text{etc.}\right)$. These modified transmissibilities were used for our work as shown below:

$$\widehat{T}_{i+1/2,\,j} = T_{24} + T_{1^*3^*} \tag{B-1}$$

$$\widehat{T}_{i,\,j+1/2} = T_{12} + T_{\bar{3}\bar{4}} \tag{B-2}$$

$$\widehat{T}^{23}_{i+1/2,\,j+1/2} = T_{23} \tag{B-3}$$

$$\widehat{T}^{14}_{i+1/2,\,j+1/2} = T_{14} \tag{B-4}$$

where

$$T_{24} = \left\{\frac{T^{24}_{xi+1/2,\,j}}{T^{24}_{xi+1/2,\,j} + T^{13}_{xi+1/2,\,j+1}}\left(q_x - T_{14} - T_{23}\right)\right\} \tag{B-5}$$

$$T_{1^*3^*} = \left\{\frac{T^{1^*3^*}_{xi+1/2,\,j}}{T^{1^*3^*}_{xi+1/2,\,j} + T^{2^*4^*}_{xi+1/2,\,j-1}}\left(q^*_x - T_{1^*4^*} - T_{2^*3^*}\right)\right\} \tag{B-6}$$

$$T_{12} = \left\{\frac{T^{12}_{yi,\,j+1/2}}{T^{12}_{yi,\,j+1/2} + T^{34}_{yi+1,\,j+1/2}}\left(q_y - T_{14} - T_{23}\right)\right\} \tag{B-7}$$

$$T_{\bar{3}\bar{4}} = \left\{\frac{T^{\bar{3}\bar{4}}_{yi,\,j+1/2}}{T^{\bar{3}\bar{4}}_{yi,\,j+1/2} + T^{\bar{1}\bar{2}}_{yi-1,\,j+1/2}}\left(\bar{q}_y - T_{\bar{1}\bar{4}} - T_{\bar{2}\bar{3}}\right)\right\} \tag{B-8}$$

$$T_{23} = \left\{\frac{1}{3}\frac{(q_x - q_{xy})(q_y - q_{xy})}{(q_x + q_y - 2q_{xy})} + q_{xy}\right\} \text{ for } q_{xy} \geq 0 \tag{B-9}$$

$$T_{23} = \left\{\frac{1}{3}\frac{(q_x - q_{xy})(q_y - q_{xy})}{\left(q_x + q_y - 2q\delta_{xy}\right)}\right\} \text{ for } q_{xy} < 0 \tag{B-10}$$

$$T_{14} = \left\{ \frac{1}{3} \frac{(q_x - q_{xy})(q_y - q_{xy})}{(q_x + q_y - 2q_{xy})} \right\} \text{ for } q_{xy} \geq 0 \quad \text{(B-11)}$$

$$T_{14} = \{T_{23} - q_{xy}\} \text{ for } q_{xy} < 0 \quad \text{(B-12)}$$

$$q_x = \frac{1}{2}\left(T^{24}_{xi+1/2,\,j} + T^{13}_{xi+1/2,\,j+1}\right) \quad \text{(B-13)}$$

$$q_y = \frac{1}{2}\left(T^{12}_{yi,\,j+1/2} + T^{34}_{yi+1,\,j+1/2}\right) \quad \text{(B-14)}$$

$$q_{xy} = \frac{1}{2}\left(T^{23}_{xyi+1/2,\,j+1/2} + T^{14}_{xyi+2,\,j+1/2}\right) \quad \text{(B-15)}$$

$T^{24}_{xi+1/2,\,j}$, $T^{13}_{xi+1/2,\,j+1}$, $T^{12}_{yi,\,j+1/2}$, $T^{34}_{yi+1,\,j+1/2}$, $T^{1^*3^*}_{xi+1/2,\,j}$, $T^{2^*4^*}_{xi+1/2,\,j-1}$, $T^{\bar{3}\bar{4}}_{yi,\,j+1/2}$, $T^{\bar{1}\bar{2}}_{yi-1,\,j+1/2}$, $T^{23}_{xyi+1/2,\,j+1/2}$, $T^{14}_{xyi+2,\,j+1/2}$ are harmonic-averaged five-point functions. To calculate T^{24}_x, T^{13}_x, etc. use grid size Δx and Δy as in standard five-point formulations. To calculate T^{23}_{xy} and T^{14}_{xy} use appropriate diagonal distances between points 2 and 3 and 1 and 4, respectively.

VIII. APPENDIX C

Equations to Convert Tracer Response Concentrations in a Non-Confined Five-Spot Pattern to Concentrations For Use in a Five-Spot Window Model

As can be seen from Fig. 10, when tracer is injected in the center well of the gray-shaded area, the tracer response in wells A, B, C, and D will be diluted by water from the water injectors outside the center pattern. If we wish to isolate the gray-shaded area as a confined window for modeling purposes, the tracer dilutions have to be subtracted out. The following equations provide a procedure to accomplish this. This method should be used with caution because reservoir heterogeneities can cause fluid channeling to the outside of the "perceived" confined window area.

The cumulative amount of tracer produced in Wells A, B, . . ., are given below:

$$AMT_A = \sum_{n=1}^{N} q_{wA}^{n} C_A^{n} \Delta t_n \tag{C-1}$$

$$AMT_B = \sum_{n=1}^{N} q_{wB}^{n} C_B^{n} \Delta t_n \tag{C-2}$$

The fractions of tracer produced by Wells A, B, . . ., are:

$$w_A = AMT_A /(AMT_A + AMT_B + \ldots) \tag{C-3}$$

$$w_B = AMT_B /(AMT_A + AMT_B + \ldots) \tag{C-4}$$

The flow rates of water and oil produced by Wells A, B, . . . , in the confined window model are $\hat{q}_{wA}$, $\hat{q}_{oA}$, etc., and are given below:

$$\hat{q}_{wA} = w_A \, q_{inj} \, WOR_A \tag{C-5}$$

$$\hat{q}_{oA} = \hat{q}_{wA} / WOR_A \tag{C-6}$$

The concentration of tracer produced by Wells A, B, . . ., in the confined window model $\hat{C}_A$, $\hat{C}_B$, . . ., are as shown:

$$\hat{C}_A = C_A \left(q_{WA}/\hat{q}_{WA}\right) \tag{C-7}$$

$$\hat{C}_B = C_B \left(q_{WB}/\hat{q}_{WB}\right) \tag{C-8}$$

IX. ACKNOWLEDGEMENT

The authors thank Marathon Oil Company for permission to participate in preparation and publication of this paper. In addition, the authors are grateful to Mrs. L. L. Fitzpatrick for her creativity in formatting the equations and for typing the manuscript.

X. REFERENCES

1. Gogarty, W. B.: "Enhanced Oil Recovery Through the Use of Chemicals - Part 2," ***JPT,*** Oct. 1983, 1767-1775.

2. Earlougher, R. C., Jr.: **Advances in Well Test Analysis**, Monograph Series, SPE, Dallas, 1967, Vol. 5.

3. Kazemi, H., Gilman, J. R. and El-Sharkawy, A. M.: SPE 19849, "Analytical and Numerical Solution of Oil Recovery from Fractured Reservoirs Using Empirical Transfer Functions," 64th SPE Annual Technical Conference and Exhibition, San Antonio, TX, Oct. 8-11, 1989.

4. Shinta, A. A.: PhD Dissertation, Petroleum Engineering Department, Colorado School of Mines, February 11, 1992.

5. Shiralkar, G. S.: SPE 18442, "Reservoir Simulation of Generally Anisotropic Systems" 10th SPE Symposium on Reservoir Simulation, Houston, TX, Feb. 6-8, 1989.

6. Wolcott, K. D.: "Mixed Five-Point/Nine-Point Formulation of Multiphase Flow in Petroleum Reservoirs Using the Fully Implicit Solution Technique," M.S. Thesis, Petroleum Engineering Department, Colorado School of Mines (April 1991).

7. Ostebo, B. and Kazemi, H.: SPE 21227, "Mixed Five-Point Nine-Point Finite-Difference Formulation of Multiphase Flow in Petroleum Reservoirs," 11th SPE Symposium on Reservoir Simulation, Anaheim, CA, Feb. 17-20, 1991.

8. Fung, L. S., Hiebert, A. D. and Nghiem, L. X.: SPE 21224, "Reservoir Simulation with a Controlled-Volume, Finite-Element Method," 11th SPE Symposium on Reservoir Simulation, Anaheim, CA, Feb. 17-20, 1991.

XI. METRIC CONVERSION FACTORS

bbl x 1.589 873	E–01	=	m^3
cp x 1.0	E–03	=	Pa•s
ft x 3.048	E–01	=	m
md x 9.869 233	E–04	=	μm^2
psi x 6.894 757	E+00	=	kPa
psi^{-1} x 1.450 377	E–01	=	kPa^{-1}

SESSION 2

Field Studies and Data Needs

Co-Chairmen

Neil Humphreys, Mobil E&P Services

Charles Kerans, Bureau of Economic Geology,
The University of Texas at Austin

QUALITY CONTROL OF SPECIAL CORE ANALYSIS MEASUREMENTS

Eve S. Sprunt

Mobil Research and Development Corporation
Dallas Research Laboratory
Dallas, Texas

I. ABSTRACT

One of the problems in field studies utilizing laboratory data is determining whether or not the laboratory core measurements are accurate. This is particularly true of expensive special core analysis measurements. The high cost of such measurements means that usually only a few measurements are available for a given field and few or no repeat measurements are available. Our experience has shown that some of the simple and inexpensive components of special core analysis measurements may be responsible for significant errors. Separate quality control on these inexpensive measurements can improve the reliability of the costly, special core analysis measurements.

In basic core analysis measurements, porosity measurements are usually considered acceptable if they are accurate to half a porosity unit. However, in special core analysis tests in which the porosity is used to compute saturation, an error of half a porosity unit can translate into a large error in water saturation. Such errors are particularly troublesome for low-porosity rocks at low water saturations. Laboratory measurements whose interpretation depends on accurate determination of the water saturation include electrical resistivity, capillary pressure, and relative permeability. Inaccurate determination of the saturation can introduce curvature of the resistivity index versus water saturation crossplot and could lead to erroneous identification of a sample as having a conductive rock matrix or bound water. This is true regardless of how accurately the resistivity itself is measured. While it is prohibitively expensive to run repeat full resistivity measurements for a sample showing curvature in the resistivity index plot, the porosity should be run by several different methods to determine whether an error in porosity could be causing the curvature. Careful evaluation of the porosity measurements can bracket the correct porosity and determine whether a significant error in porosity could be affecting the special core analysis results.

II. INTRODUCTION

In school most of us are taught to assign error estimates to laboratory measurements. However, in practice in the oil industry most laboratory measurements are reported without any confidence limits. Commercial laboratories are loathe to report error bars because their competitors will assert that they can do better. Also, few published papers contain error analysis of the reported experimental test results. Thus, many in the oil industry do not have a sense of the magnitude of errors in laboratory data.

Although error analysis is seldom presented in experimental papers, there are many papers on errors analysis of core analysis measurements. Amaefule and Keelan (1989) present uncertainty equations for many basic and special core analysis measurements. However, my experience (e.g. Sprunt et al., 1990) indicates that the magnitude of the errors they suggest is optimistic.

Quite a few papers focus on error analysis of electrical measurements (Chen and Fang, 1986; Freedman and Ausburn, 1985; Hook, 1983; Hoyer and Spann, 1975; Worthington, 1975). Of these papers, Worthington (1975) and Hoyer and Spann (1975) approach the problem from an experimentalist's viewpoint examining some of the individual components of the laboratory procedures and why they cause problems.

This paper approaches the error analysis from an experimentalist's viewpoint. In particular, it will be shown how the ability to accurately determine pore volume affects special core analysis measurements that depend on porosity and saturation measurement. Particular attention is given to electrical resistivity measurements. Measurements whose interpretation depends on saturation determination are much more sensitive to errors in the pore volume than measurements that rely on the porosity. Examples of such laboratory measurements include the electrical resistivity measurements needed to obtain the saturation exponent, relative permeability, and capillary pressure.

III. PORE VOLUME

In understanding how the accuracy of pore-volume measurements affects special core analysis measurements, it is important to understand how porosity and pore volume are determined. Measurement or calculation of pore volume is necessary for measurement of porosity. Whereas most people do not know how accurate pore-volume measurements are, they assume porosity measurements are good to within 0.5 porosity units (pu).

Porosity ϕ is determined by measuring two of the following three quantities: pore volume (V_p), grain volume (V_g), and bulk volume (V_b).

$$\phi = \frac{V_p}{V_b} = \frac{(V_b - V_g)}{V_b} = \frac{V_p}{(V_p + V_g)} \quad (1)$$

In an inter-laboratory comparison, Dotson et al. (1951) showed that the average deviation in porosity was 0.5 pu, but the range was from 0.3 to 1.0 pu. Thomas and Pugh (1989) use ± 0.5 pu as their "experience-based" maximum acceptable deviation in standard core-plug analysis and found 65% of labs in 1987 met their quality assurance criteria, which included that standard.

If porosity measurements are used to characterize a reservoir, for example to determine average porosity of a zone or to calibrate well logs, ± 0.5 pu accuracy is usually adequate. Additional uncertainty is introduced in scaling and averaging the laboratory data. However, if the porosity measurement is used in characterization of a core plug for special core analysis measurements, a 0.5 pu error in porosity can introduce significant errors.

There are many methods of obtaining the quantities in Equation (1) (API, 1960). If the pore volume rather than the porosity is the true quantity of interest, it is best to measure the pore volume directly rather than to calculate it from the grain volume and bulk volume. Ideally, when pore volume is required for a special core analysis measurement, the pore volume should be measured directly under the same conditions as the special core analysis test. This is important because pore volume changes with pressure and because sleeve material may progressively intrude into surface porosity with increasing pressure. In practice, however, this is not always done; and ambient pore volume measurements or calculated ambient pore volumes may be corrected to overburden pressure conditions by a variety of techniques including correlations based on measurements on other samples.

Porosity errors are important when compressibility is a significant factor or when pore volume is difficult to accurately quantify. In low-porosity rocks, it is especially important to properly account for pore compressibility. Cases in which accurate determination of the pore volume is difficult include vuggy carbonates (Sprunt, 1989), low-permeability samples with high porosity, low-porosity samples, and samples containing gypsum.

IV. SATURATION

The most common methods of determining water saturation during special core analysis are based on volumetric and gravimetric measurements (Maerefat et al., 1990; Sprunt, 1990), and both these methods depend directly on accurate measurement of the pore volume. Imaging techniques such as x-ray absorption, microwave absorption, and nuclear magnetic resonance do not require measurement of the pore volume.

The equation for determining water saturation (S_w) volumetrically is:

$$S_w = \frac{(V_p - V_o)}{V_p} \quad (2)$$

where V_o is the volume of water expelled from the sample.

If the sample is small and has low porosity, it is very difficult to accurately measure changes in saturation because very small amounts of fluid are expelled. For example, a 1-inch diameter by 1-inch long sample of a 10-percent porosity rock expels only 0.13 cc for a 10-percent change in water saturation. If a 1.5-inch diameter, 1-inch long sample is used, 0.29 cc corresponds to the same 10-percent change in water saturation. The water saturation of the larger sample can be measured more accurately.

When samples are selected for special core analysis measurements, it is a good idea to check the pore volume of the sample. If preserved samples are to be used for the test, the pore volume can be estimated using the porosity of an adjacent sample. Using the accuracy to which fluid changes can be determined, you can then determine whether or not the saturation changes during the special core analysis test can be determined sufficiently accurately. In most cases a pore volume of 5 cc or more is desirable.

Figures 1 and 2 illustrate how the error in water saturation increases with decreasing porosity and decreasing water saturation. These figures were generated, assuming that the entire porosity error is attributable to an error in the pore volume. Frequently, bulk volume determination is a source of errors (Thomas and Pugh, 1989), but if pore volume is not directly determined, the error in bulk volume translates into an error in pore volume. The main warning of these figures is that accurate measurements are required on low-porosity samples, especially at low-water saturations, if special core analysis measurements that depend on water saturation are to be meaningful.

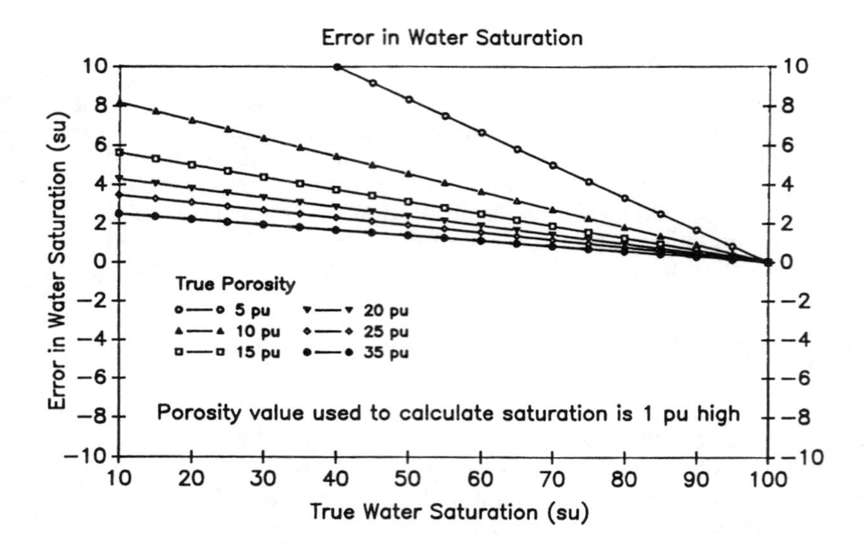

Figure 1: The error in water saturation was calculated as a function of water saturation assuming that all the error in the porosity measurement is in the pore volume.

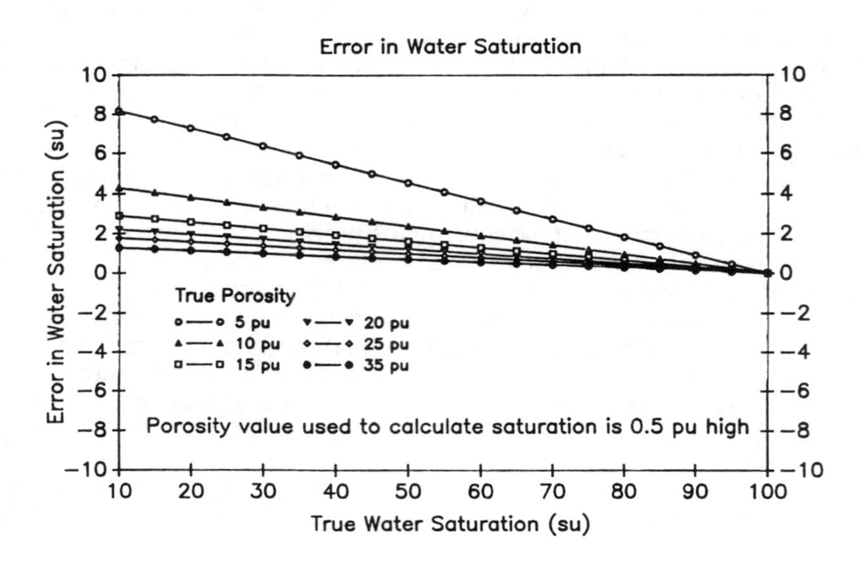

Figure 2: The error in water saturation was calculated as a function of water saturation assuming that all the error in the porosity measurement is in the pore volume. Porosity is commonly assumed to be accurate within 0.5 pu.

V. CEMENTATION EXPONENT

Cementation exponent, *m*, was defined by Archie (1942) in the equation $F = \phi^{-m} = R_o/R_w$, where F is the formation factor, R_o is the resistivity of the fully brine saturated rock, R_w is the resistivity of the brine, and ϕ is porosity. Assuming the correct cementation exponent was 2.0, the formation factors were calculated for a range of porosities from 5 to 35 percent. These values of the formation factor were then used to calculate cementation exponents using porosities that were in error by ± 0.5 and ± 1.0 pu (Figure 3).

The error in cementation exponent from a 1 pu error in porosity is less than 5 percent if the porosity is 10 pu or greater. For a 0.5 pu error, the error in cementation exponent is less than 2.5 percent for samples with porosity of 10 pu or more. Except for low porosity rocks, normally acceptable errors in porosity do not introduce serious errors into the determination of the cementation exponent.

VI. SATURATION EXPONENT

The saturation exponent is defined by Archie(1942) in the equation $R = R_o/S_w^{-n}$ where R is the resistivity of the rock at some partial saturation and S_w is the water saturation. If the correct value of the saturation exponent is assumed to be 2.0, the value of the resistivity index, R/R_o, can be calculated for different values of water saturation. Assuming a perfect Archie rock, the resistivity index was calculated for values of saturation between 10 and 100 percent.

In an Archie (1942) rock, the saturation exponent is the slope of the straight line fitted to the bi-logarithmic crossplot of resistivity index versus water saturation. On the bi-logarithmic crossplot of true resistivity index versus apparent water saturation (Figure 4), only the saturations corresponding to the correct porosity are perfectly fitted by a straight line. If the measured porosity is too low, connecting the data points gives a line which curves downward, because the apparent water saturation for any resistivity reading is too low. If the measured porosity is too high, the line connecting the data points curves upward, because the apparent water saturation for any resistivity reading is too high.

Volumetric determination of the saturation is commonly used to monitor saturation during resistivity or other special core analysis measurements which depend on saturation; a crosscheck of the final volumetric saturation with a Dean-Stark extraction saturation is recommended. If the extraction saturation is greater than the final volumetric saturation, the pore volume may be erroneously low.

Determination of the saturation exponent is complicated because the errors in porosity cause curvature in the bi-logarithmic crossplot of

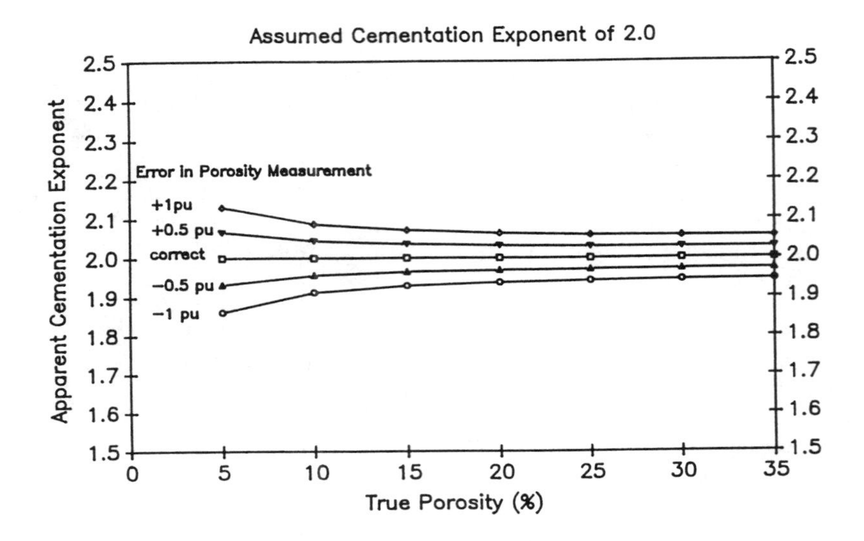

Figure 3: The cementation exponent was calculated for different errors in the porosity assuming that the true cementation exponent was 2.0.

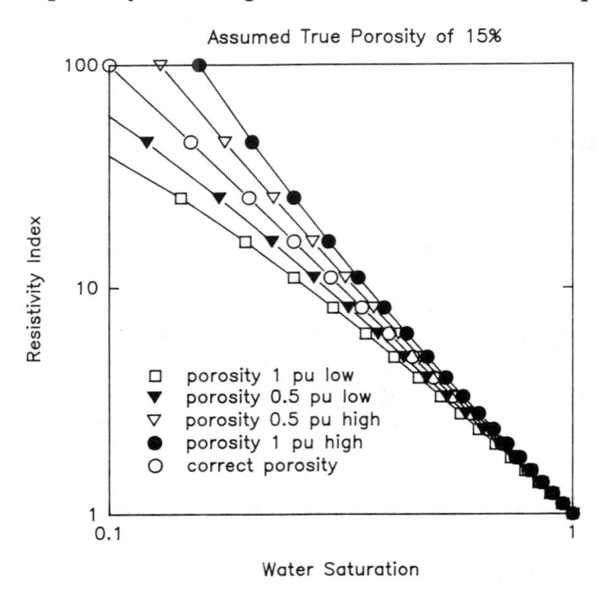

Figure 4: In this crossplot, only the assumed true porosity of 15% gives a straight-line relationship. If the porosity used to calculate water saturation is too low, the line curves downward. If the porosity used to calculate the water saturation is too high, the line curves upward.

resistivity index versus water saturation. If the saturation exponent is determined by fitting a least-squares line to the data and forcing the line through the 100 percent water saturation and resistivity index equal to 1.0 point, the slope of that least-squares line and, hence, the saturation exponent depend on the minimum water saturation used. The error in saturation exponent decreases as the data analysis is limited to higher water saturations (Figures 5, 6, and 7). When only the data above an apparent water saturation of 50 percent are used, an error in porosity of 1 pu causes an error of almost 10 percent in the saturation exponent of a sample with 15 pu. The error in saturation exponent increases rapidly with decreasing porosity for porosities less than 15 pu.

The objective of measuring the saturation exponent in the laboratory is to use that exponent in the Archie equation to calculate water saturation from log data. However, the error in the determination of water saturation of the plug sample due to the erroneous pore volume may not be the same as the error in the water saturation calculated using the laboratory-determined saturation exponent. The error is altered because the saturation exponent is determined from a linear fit to nonlinear data (Figure 4). Depending on how the straight line is fitted to the nonlinear data (Figures 5, 6, and 7), the error in water saturation calculated using the apparent saturation exponent may be larger or smaller than the error in the laboratory measurement of saturation.

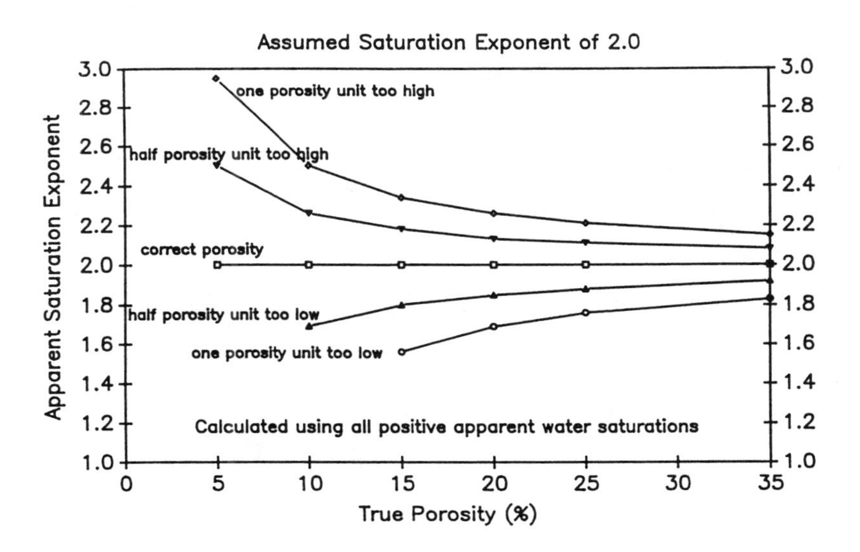

Figure 5: The saturation exponent was obtained from a least squares fit to the resistivity index/water saturation plot with the line forced through the point of 100 percent water saturation and resistivity index of 1.0.

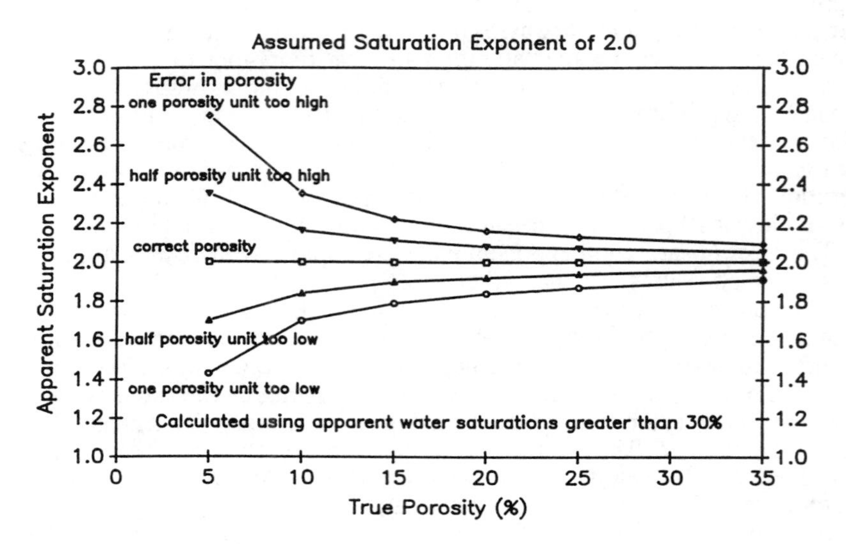

Figure 6: The same plot as in Figure 5, but this time only apparent water saturations greater than 30 percent were used to determine the saturation exponent.

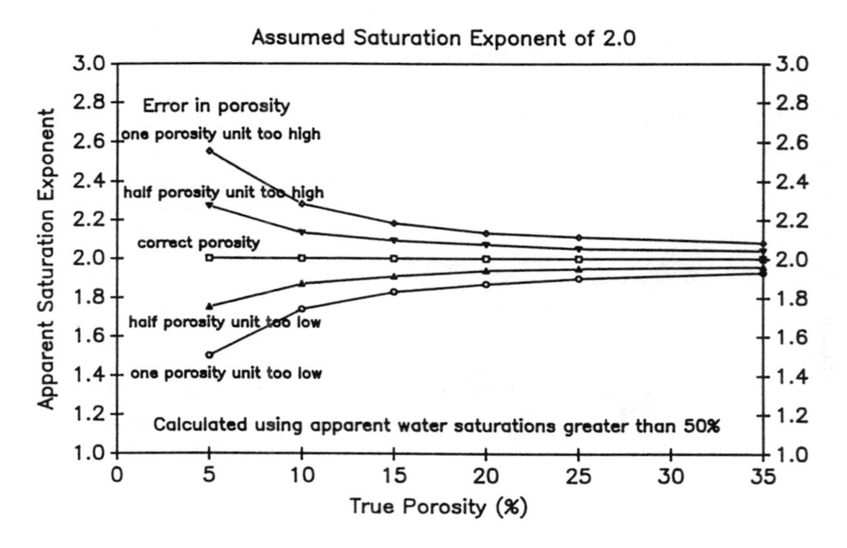

Figure 7: The same plot as in Figure 5, but this time only apparent water saturations greater than 50 percent were used to determine the saturation exponent.

VII. SAMPLE DATA

Determination of the correct pore volume to use in tests on vuggy carbonates is difficult (Sprunt, 1989). For each of eight samples with varying vug sizes and amounts of vuggy porosity, a service company reported two different values of porosity (Table 1). Both porosity values were calculated using the same Boyle's Law grain volume measurement. The "caliper" porosity used a bulk volume determined by measuring the length and diameter of the plug samples with calipers. The caliper porosity is frequently erroneously high because the plugs are not perfect cylinders and may have chipped edges. The "summation" porosity was calculated using a pore volume determined from the volume of fluid required to saturate the samples. The summation porosity is often too low because surface vugs will not retain water. This "summation" porosity is not the same as the "summation of fluids" porosity in which the gas, oil, and water contents are determined independently and summed to determine the pore volume. The true porosity of these samples probably falls between the caliper and summation porosity values.

Table I: Vuggy Carbonate Samples

	Porosity (%)		Cementation Exponent		Saturation Exponent	
Sample	Caliper	Summation	Caliper	Summation	Caliper	Summation
3	17.1	15.7	1.93	2.03	1.22	1.74
5	12.6	11.5	1.80	1.88	1.24	1.70
8	11.6	11.3	1.77	1.80	1.14	1.26
10	12.4	11.4	1.83	1.90	1.14	1.78
12	8.4	7.7	1.73	1.79	1.21	1.54
19	14.2	13.7	1.83	1.86	1.30	1.59
22	19.7	19.1	1.97	2.01	1.53	1.71
29	24.4	24.1	2.06	2.08	1.54	1.63
Average	15.1	14.3	1.86	1.92	1.29	1.62

For these samples we do not know the correct porosity, the correct cementation exponent, or the correct saturation exponent. The electrical resistivity data were interpreted using both sets of porosity measurements. As predicted by the hypothetical calculations, the differences in porosity make little difference in the cementation exponent (Table 1), even though the porosity differences are as high as 1.4 pu. The porosity differences corresponded to large differences in the saturation exponents (Table 1).

Some pore geometries, such as those associated with clay coatings or microporosity, can cause curvature in the resistivity index/water saturation relationship (de Waal et al., 1989; Dicker and Bemelmans, 1984; Diederix, 1982; Givens, 1987; and Swanson, 1985). The analysis here has shown that curvature may also be due to errors in the porosity and, hence, in the water-saturation determination. In some cases it may be difficult to differentiate between curvature due to errors in porosity measurement and curvature due to rock properties or other effects. However, for these clean, carbonate samples, curvature in the resistivity crossplot is probably due to errors in porosity measurement.

In the data for these clean carbonates, the curvature appears to be due to porosity-measurement errors because the direction of curvature depends on the type of porosity measurement used to calculate saturation. The direction of curvature due to porosity errors indicates whether the porosity is erroneously high or low. The data generated using the caliper porosity form either a linear or an upward curving trend on the bi-logarithmic resistivity index/water saturation plot because that porosity, when in error, is too large (Figures 8 and 9). The corresponding data based on summation porosity form either a linear or a downward curving trend because that porosity, when in error, is too low (Figures 8 and 9). Since this formation is not clay bearing and we know that porosity measurement is a problem for this formation, analysis of these crossplots on a sample-by-sample basis can aid in determining which value of porosity is probably more reliable for each sample.

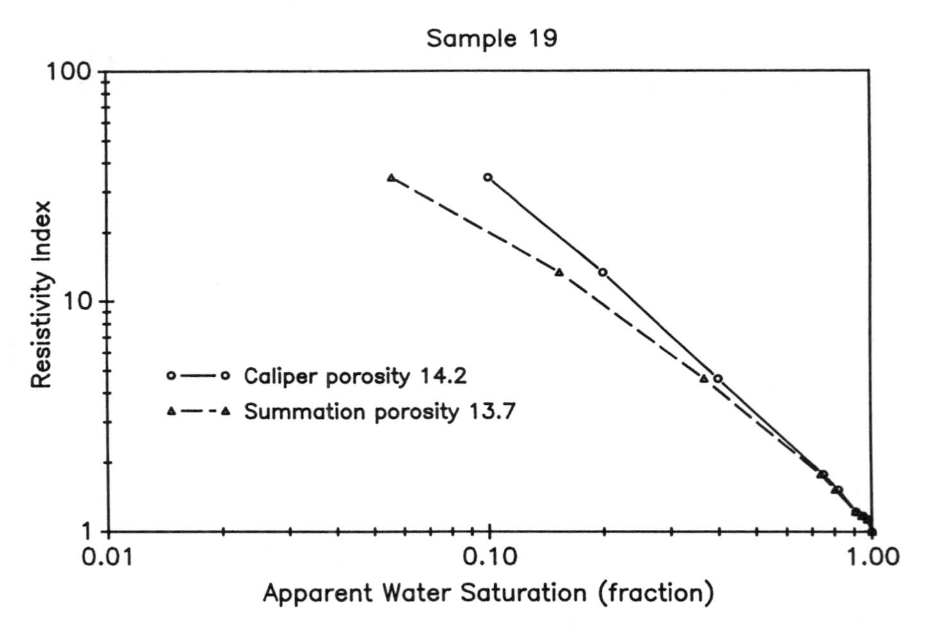

Figure 8: Each laboratory measurement of the fluid content of sample 19 was used to calculate apparent water saturation using the pore volumes obtained from the caliper porosity and the summation porosity. These apparent water saturations were then used in the resistivity index versus apparent water saturation crossplots. Table 1 gives the saturation exponents.

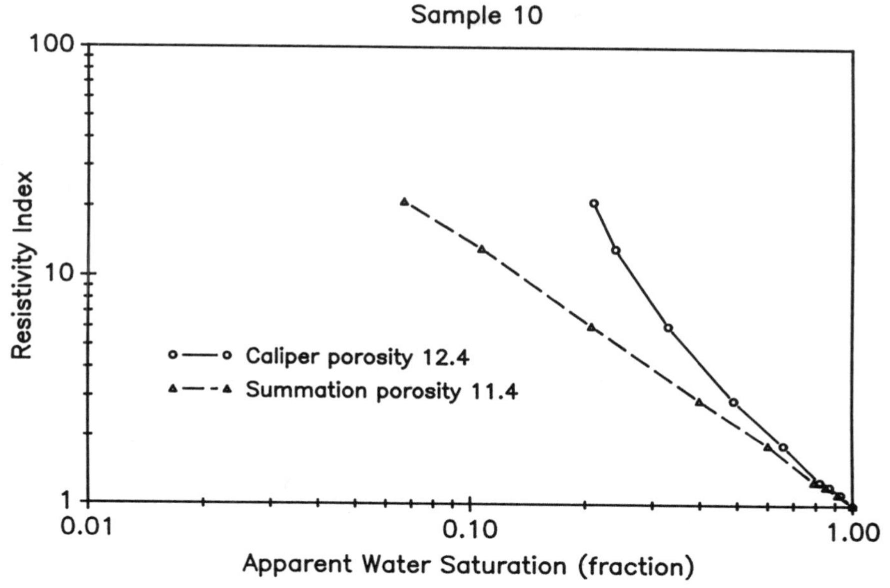

Figure 9: Each laboratory measurement of the fluid content of sample 10 was used to calculate two values of apparent water saturation using the pore volumes obtained from the caliper porosity and the summation porosity. These apparent water saturations were then used in the resistivity index versus apparent water saturation crossplots. Table 1 lists the saturation exponents, which were determined by fitting a least-squares line through the data points and forcing the line through the point 1,1.

VIII. DISCUSSION

Quality control programs are generally restricted to routine core analysis. Even in routine measurements, which include porosity and single-phase permeability, the importance of accurate measurements is sometimes not recognized because a 0.5 pu error in routine porosity may not be significant when the core values are averaged to evaluate reserves. This paper shows that errors in porosity that are acceptable for routine core analysis can introduce significant errors in special core analysis measurements.

The cementation exponent is less sensitive to errors in porosity measurement than the saturation exponent (Figure 10). The differences between the values of the cementation exponents increase as the difference in porosities increase (Figure 11), but porosity measurement will probably not be the major source of error in measurement of the cementation exponent. Errors, such as mis-measurement of the brine salinity or

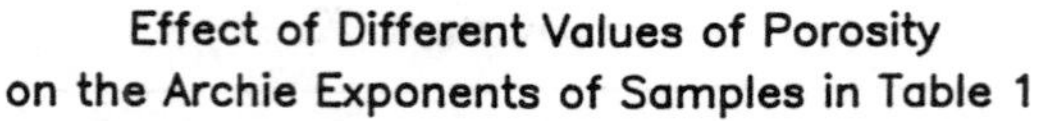

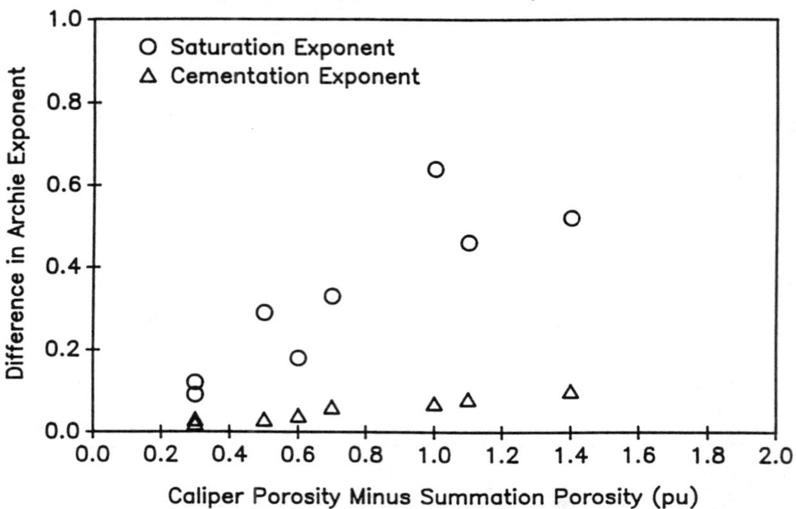

Figure 10: The effect of differences in porosity is much smaller on the cementation exponent than on the saturation exponent.

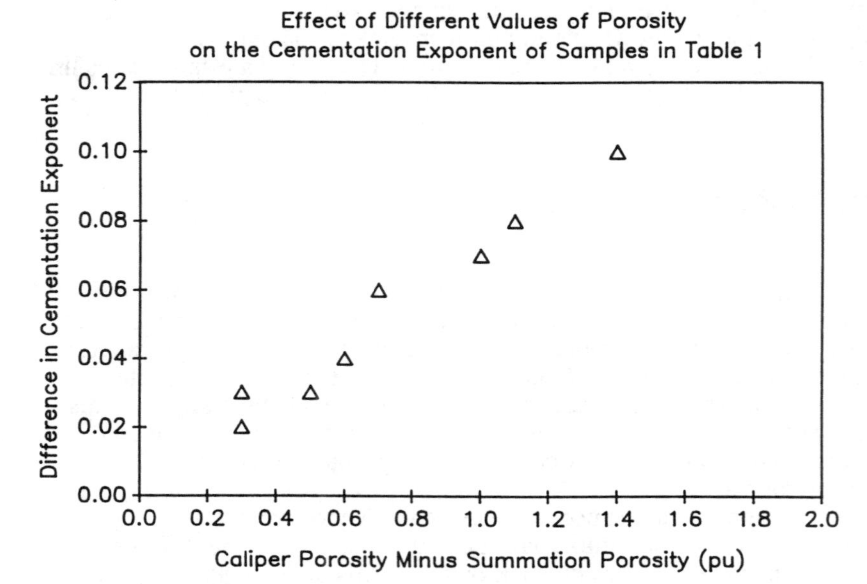

Figure 11: This figure shows on an expanded vertical scale that the difference in cementation exponent increases with the difference in porosity.

mis-measurement of the electrode spacing as discussed in Sprunt et al. (1990), will probably be more significant than porosity errors.

IX. SUMMARY AND CONCLUSIONS

Errors in porosity can introduce large errors into the determination of the saturation exponent. Inaccurate pore-volume measurements can cause curvature of data on the bi-logarithmic resistivity index water saturation crossplot. However, this curvature is not necessarily due to errors in porosity measurement. Curvature has been attributed to various pore geometries (Dicker and Bemelmans, 1984; Diederix, 1982; Givens, 1987; and Swanson, 1985). Other nonlinearities may be introduced by nonuniform fluid distributions (Sprunt et al., 1991). It is important to try to differentiate between laboratory errors and physical phenomena.

If curvature is observed in laboratory measurements, it would be prudent, if possible, to carefully measure the porosity/pore volume by more than one technique. Measuring porosity is far less expensive than measuring electrical resistivity as a function of water saturation. A single set of electrical measurements can be evaluated using porosity/pore volume measurements obtained by different techniques. In evaluating the porosity/pore volume measurement techniques, one should determine whether the porosities obtained will tend to be above or below the true value. If different techniques are used and the resistivity measurements curve upward or downward depending on the technique as in the sample data presented in this paper, the curvature is probably due to errors in the porosity measurement and not to the underlying rock properties. If the data always curve in the same direction with only the magnitude of the curvature changing, the curvature is probably due to the rock properties.

In some reservoirs, such as tight gas sands or fractured carbonates, the average formation porosity may be around 10 pu or less. In planning special core analysis tests on such formations, it is important to consider whether the results will be sufficiently accurate (Figures 1 and 2). It may be necessary to impose stringent quality control and/or revise the test procedures. In some cases, sufficiently accurate data may not be obtainable with existing techniques. The accuracy of the measurements for low-porosity samples should be kept in mind in planning laboratory testing. Errors in pore-volume measurement are most serious for low-porosity samples and when the pore volume is used to calculate saturation. For such samples commonly accepted accuracy tolerances for routine core analysis measurements may introduce unacceptably large errors in the parameters obtained from special core analysis.

REFERENCES

Amaefule, J. O., and D. K. Keelan, 1989, Stochastic Approach to Computation of Uncertainties in Petrophysical Properties, Transactions of the Society of Core Analysts, v. 1, paper SCA-8907.

American Petroleum Institute, 1960, API Recommended Practice 40 (RP40), Recommended Practice for Core Analysis Procedure, 55 pp.

Archie, G. E., 1942, The Electrical Resistivity Log as an Aid in Determining Some Reservoir Characteristics, Trans. AIME, v. 198, pp. 54-62.

Chen, H. C., and J. H. Fang, 1986, Sensitivity Analysis of the Parameters in Archie's Water Saturation Equation, *The Log Analyst,* v. 27, n. 5, pp. 39-44.

de Waal, J. A., R. M. M. Smits, J. D. de Graaf and B. A. Schipper, 1989, Measurement and Evaluation of Resistivity Index Curves, SPWLA 30th Annual Symposium, Paper II.

Dicker, A. I. M., and Bemelmans, W. A., 1984, Models for simulating electrical resistance of porous media, SPWLA 25th Annual Logging Symposium, June 10-13, 1984, Paper HH.

Diederix, K. M., 1982, Anomalous relationships between resistivity index and water saturations in the Rotliegend sand (The Netherlands), SPWLA 23rd Annual Logging Symposium, July 6-9, 1982, Paper X.

Dotson, B. J., R. L. Slobod, P. N. McCreery, and J. W. Spurlock, 1951, Porosity Measurement Comparisons by Five Laboratories, Petroleum Transactions, AIME, v. 192, pp. 341-346.

Freedman, R., and B. E. Ausburn, 1985, The Waxman-Smits Equation for Shaley Sands: I. Simple Methods of Solution; II. Error Analysis, *The Log Analyst*, v. 26, n. 2, pp. 11-24.

Givens, W. W., 1987, A Conductive Rock Matrix Model (CRMM) for the Analysis of Low-Contrast Resistivity Formations, *The Log Analyst*, v. 28, n. 2, pp. 138-151.

Hook, J. R., 1983, The Precision of Core Analysis Data and Some Implications for Reservoir Evaluation, SPWLA 24th Annual Logging Symposium, Paper Y.

Hoyer, W. A., and M. M. Spann, 1975, Comments on Obtaining Accurate Electrical Properties of Cores, SPWLA 16th Annual Logging Symposium, Paper B.

Maerefat, N. L., B. A. Baldwin, A. A. Chaves, G. A. LaTorraca, B. F. Swanson, 1990, SCA Guidelines for Sample Preparation and Porosity Measurement of Electrical Resistivity Samples, Part IV - Guidelines for Saturation and Desaturating Core Plugs during Electrical Resistivity Measurements, *The Log Analyst*, v. 31, n. 2, pp. 68-75.

Sprunt, E. S., 1989, Arun Core Analysis: Special Procedures for Vuggy Carbonates, *The Log Analyst,* v. 30, n. 5, pp. 353-362.

Sprunt, E. S., 1990, Proposed Electrical Resistivity Form, *The Log Analyst*, v. 31, n. 2, pp. 89-93.

Sprunt, E. S., K. P. Desai, M. E. Coles, R. M. Davis, and E. L. Muegge, 1991, CT Scan-Monitored Electrical Resistivity Measurements Show Problems Achieving Homogeneous Saturation, *SPE Formation Evaluation*, v. 6, n. 2, pp. 134-140.

Sprunt, E. S., R. E. Maute, and C. L. Rackers, 1990, An Interpretation of the SCA Electrical Resistivity Study, *The Log Analyst*, v. 31, n. 2.

Swanson, B. F., 1985, Microporosity in Reservoir Rocks: Its Measurement and Influence on Electrical Resistivity, *The Log Analyst*, v. 26, n. 6, pp. 42-52.

Thomas, D. C., and V. J. Pugh, 1989, A Statistical Analysis of the Accuracy and Reproducibility of Standard Core Analysis, *The Log Analyst*, v. 30, n. 2, pp. 71-77.

Worthington, A. E., 1975, Errors in the Laboratory Measurement of Formation Resistivity Factor, SPWLA 16th Annual Logging Symposium, Paper D.

COMPARISON OF DIFFERENT HORIZONTAL FRACTAL DISTRIBUTIONS IN RESERVOIR SIMULATIONS

R. A. Beier
H. H. Hardy

Conoco Inc.
Ponca City, Oklahoma

ABSTRACT

Past studies have demonstrated that porosity logs in vertical wells have a fractal character called fractional Gaussian noise (fGn). This observation triggered the application of fractal geostatistics to describe reservoir heterogeneity. These applications require knowledge of not only the vertical statistical character but also the horizontal character. In early studies, workers assumed a different fractal character in the horizontal direction called fractional Brownian motion (fBm). More recent analyses of outcrop and core photos and porosity logs in horizontal wells have found fGn in the horizontal direction. The purpose of the present study is to compare reservoir simulations using fGn and fBm horizontal distributions. A mature waterflood in a carbonate reservoir serves as a test case. Simulated water cuts in both types of distributions match field data. Still, the fGn distributions should be more representative of reservoirs, because they are consistent with outcrop and core photographs, and horizontal well data.

I. INTRODUCTION

Fractal geostatistics is a numerical technique to generate distributions of reservoir properties, such as porosity and permeability, for reservoir simulation. Conventional reservoir simulations often embed reservoir heterogeneity in relative permeability curves that must be determined by trial and error methods. Workers[1-6] claim fractal distributions greatly reduce the man-hours required for history matches and still accurately predict reservoir performance. Fractal geostatistics have been applied to both waterflood and CO_2 field projects.

The application of fractal geostatistics to reservoirs is based on the observation that porosity logs in vertical wells have a fractal character. Hewett[1] concluded the fractal structure found in vertical well logs is fractional Gaussian noise (fGn). Well spacing is usually much too large to determine the statistical structure in the horizontal direction. To carry his work forward, Hewett postulated that the horizontal structure is a different distribution called fractional Brownian motion (fBm).

Recently, other workers have analyzed the horizontal structure of sedimentary layers. Hardy[7] analyzed distributions of dark and light regions in photographs of cores and outcrops. Values within any vertical trace were found to have the same distribution as found in vertical well logs - fGn. These results suggest the photos capture the same statistics of reservoir properties as that seen in well logs. Horizontal traces were also found to be fGn, not fBm. In the core photos, the horizontal layering is due to different standard deviations of values in the horizontal and vertical directions rather than a difference in the fractal character.

Crane and Tubman[8] and Hardy[9] recently analyzed porosity logs from horizontal wells in both sandstone and carbonate formations. All cases show fGn structure, the same as found in the photograph study.

The purpose of the present study is to compare vertical cross sections based on either fGn or fBm in the horizontal direction. The cross sections are tested by performing flow studies and comparing the simulated water cuts with field production data. A mature waterflood in a carbonate reservoir serves as a case study. A hybrid finite difference/streamtube technique[6] provides an areal simulation of the waterflood.

II. HISTORY OF MATURE WATERFLOOD

Our case study is a carbonate reservoir in southeastern New Mexico. The majority of development drilling occurred in

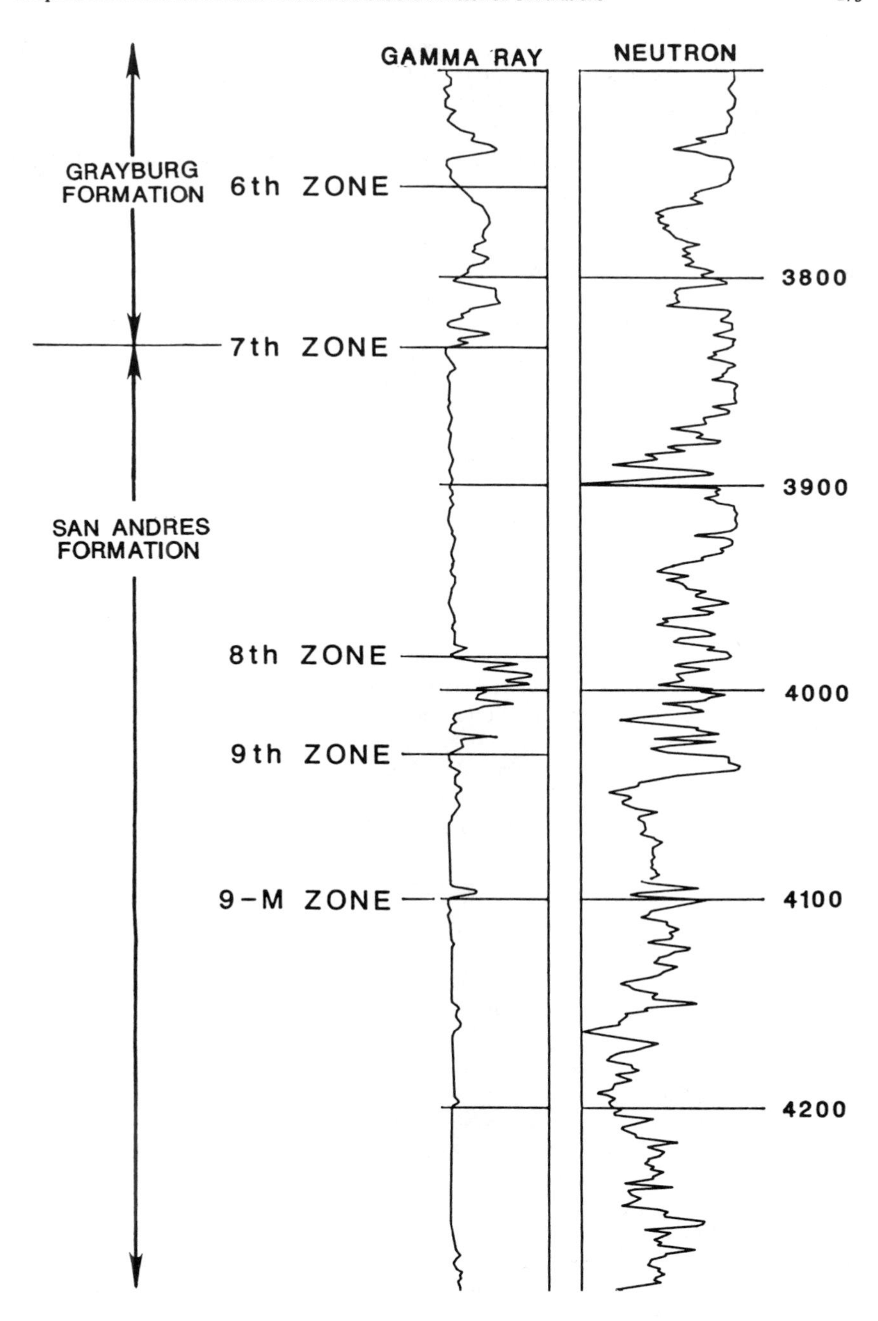

Fig. 1. Typical porosity log for Grayburg and San Andres formations.

the early 1940's. Wells produce from Grayburg dolomitic sands(mainly the Sixth zone) and three San Andres dolomite pay zones (Seventh, Upper Ninth, and Ninth Massive) at depths ranging from 3,600 to 4,300 feet as shown in Figure 1. The primary production mechanism of the reservoir was solution gas drive.

Waterflood operations on 5-spot patterns were phased in during the mid to late 1960's. The original 40-acre well spacing was infilled to 20 acres to form inverted 9-spot patterns in the early 1970's. Operations of the inverted 9-spot patterns generally remained unchanged for nearly 20 years. This long period of nearly constant operating conditions supports our use of a typical well pattern to represent an entire section (1 square mile). We focus on the waterflood performance in two adjacent sections that we call Area A and Area B.

III. FRACTAL DISTRIBUTIONS

The fractal character found in well logs describes a scaling relationship in sedimentary structures. The scaling relationship says that the small variations of reservoir properties seen at the core size are a scaled down version of the large scale variations seen at the reservoir scale. This scaling relationship can occur in several ways. Two common fractal relationships are fGn and fBm. References 1 and 9 give examples of both.

A. fGn Cross Sections

The method to generate fGn cross sections is based on previous studies[7,9] of core and outcrop photographs. The method generates good forgeries of core photos. Here we use the method to generate property distributions between wells. The resulting distributions are consistent with the core photo study and horizontal well data. The mathematical details of this process are found in the Appendix. A brief verbal description follows.

The first step is to identify reservoir units. In our field case, the production interval is divided into the Grayburg and San Andres formations. We treat each formation separately and later stack the two generated cross sections to form the final section. A representative well pair is chosen from each study area.

Digitized Sidewall Neutron Porosity (SNP) logs provide porosity measurements. The log readings are calibrated to core measured porosities with linear scaling. The corrected porosity log values are depth adjusted by stretching and compressing the depth values to align the tops and bottoms of each formation. Linear interpolation is used to fill in the array, giving a layered image. Fractal noise is then added to the image.

The amount of fractal noise added depends on how closely the depth adjusted well logs correlate. At each depth, the difference between the two log porosities is computed. Then, the total variance of these differences is calculated for the entire log section under study. If the two logs are identical, the total variance is zero and no fractal noise needs to be added. In practice, the variance is nonzero and the amount of added fractal noise increases as the total variance increases. Generation of fractal noise requires a value of the intermittency exponent, H. To evaluate H, we apply R/S analysis, spectral density analysis, and variogram analysis to the SNP logs. (See References 1 and 10.) A value of 0.87 applies to both formations. The SNP logs provide readings at one-half-foot intervals. However, the vertical spatial resolution of the SNP tool is about two feet. For this reason, only readings at two-foot intervals are used to generate a porosity cross section with square grids (2 ft by 2 ft).

It is important that the square grids in the generated cross section have sides corresponding to the vertical spatial

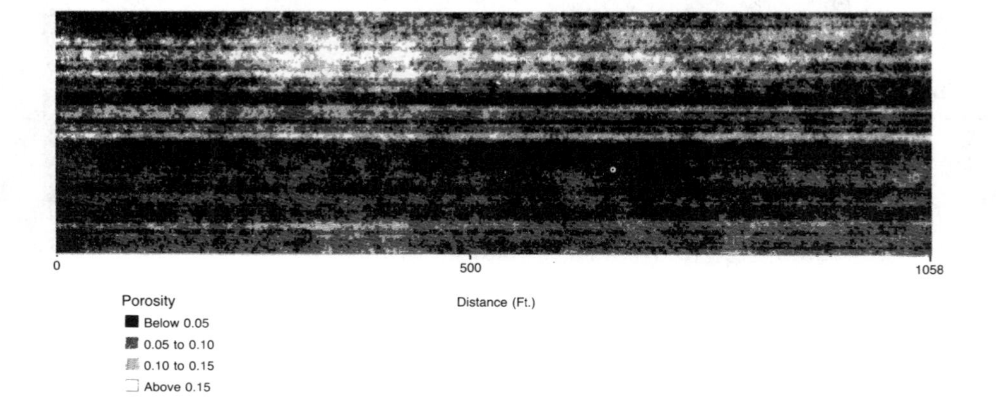

Fig. 2. Fractal distribution of porosity for Area A based on fGn in the horizontal direction (fine grid).

resolution of the logging tool. This ensures the proper amount of fractal noise is added to the layered image. If larger grid blocks are used in the generated cross section, the amount of noise measured at the porosity log scale (2 ft by 2 ft) is incorrectly added at the larger grid block scale. In geostatistical terminology, the support size of the measured values should correspond to the support size of the generated cross section.

A fine scale cross section (2 ft by 2 ft grids) for Area A is shown in Figure 2. The number of grids in this 512 by 150 array is too large for flow simulations. For this reason, a coarser grid must be obtained. This is done by reducing the number of grids in the horizontal direction. The fine scale grid values for porosity are scaled up using arithmetic averaging. The number of vertical grids remains unchanged at 150 (Figure 3).

Horizontal permeabilities are treated in the following way. Horizontal permeabilities are first assigned to the fine grid as a function of porosity by the formula

$$k = a\, 10^{b\phi} \tag{1}$$

Values of a and b are set to 0.001 and 33.3, respectively. With these values, the permeability covers 5 log cycles as the porosity ranges from 2 to 17 porosity units. A coarser grid of horizontal permeability is obtained by applying harmonic

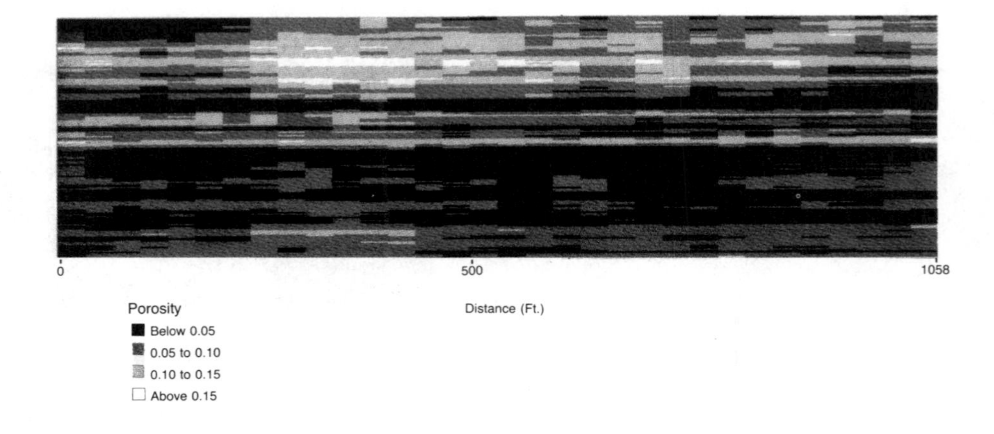

Fig. 3. Fractal distribution of porosity for Area A based on fGn in the horizontal direction (coarse grid).

averaging only in the horizontal direction. Figure 4 is the final coarse cross section for permeability. The permeability ranges in Figure 4 correspond to the porosity ranges in Figure 3 through Equation 1.

B. fBm Cross Sections

The method outlined by Hewett and Behrens[3] generates the fBm cross sections. Again, the corrected and depth adjusted log porosities are placed on the vertical edges of the cross section, followed by linear interpolation to produce a layered image. But now the added fractal noise is a different type and is added in a different manner. The result is fBm in the horizontal direction and fGn in the vertical direction. Details are provided in the Appendix. The resulting porosity

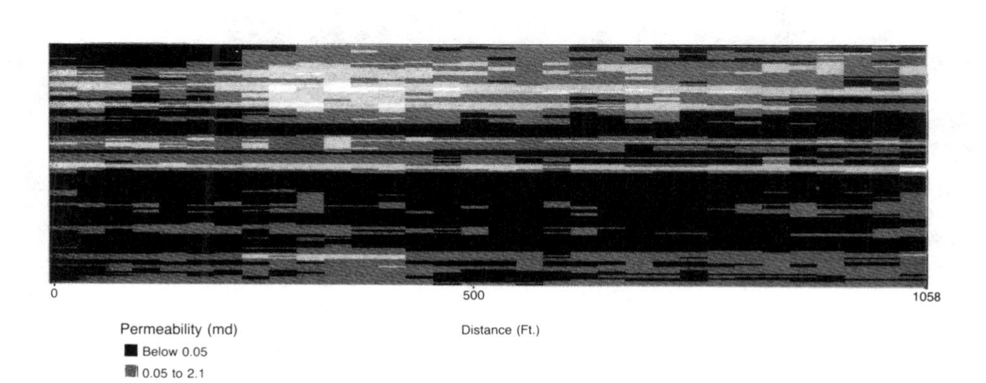

Fig. 4. Fractal distribution of permeability for Area A based on fGn in the horizontal direction (coarse grid).

cross section (Figure 5) is more layered than the fGn cross section (Figure 3). With the fBm method, a coarse porosity grid is generated directly. Equation 1 is applied directly to the coarse porosity grid to get the corresponding horizontal permeability map.

C. Other Cross Sections

For comparison in reservoir simulations, two simple porosity distributions are also considered--linear interpolation and layer cake. Figure 6 shows a cross section based

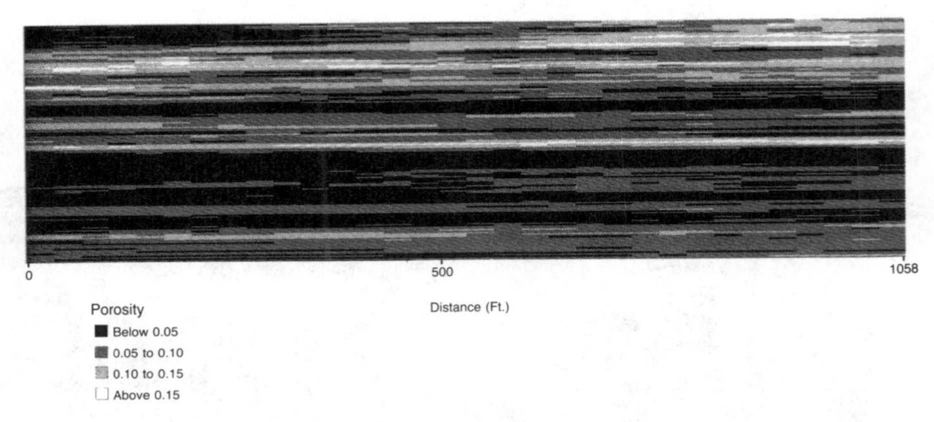

Fig. 5. Fractal distribution of porosity for Area A based on fBm in the horizontal direction (coarse grid).

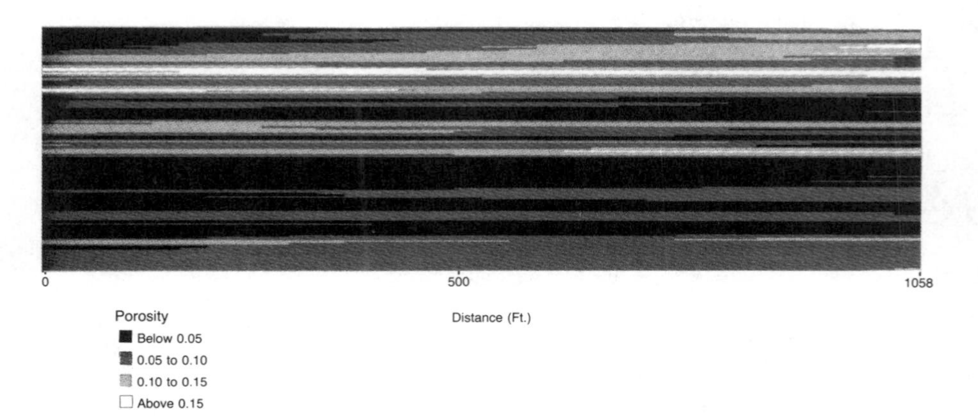

Fig. 6. Linear interpolation of porosity for Area A (coarse grid).

on linear interpolation between wells. A figure is not included for the layer cake case, but it is created by taking the porosity values from the well at the right-hand edge and extending them across the cross section. In all cross sections, the vertical permeability is set to 0.1 times the horizontal permeability at the coarse grid scale.

IV. COMPARISON WITH FIELD PRODUCTION DATA

The above cross sections are tested by performing flow studies using a hybrid finite difference/streamtube method.[6] The generated fractal distributions are input to a conventional reservoir simulator to make fractional flow curves for the cross section. Fluid displacements in the field project are calculated by incorporating the fractional flow curves in a streamtube model. The hybrid finite difference/streamtube method predicts water cuts as a function of pore volumes of water injected but not rates of injection or production. The time coordinates in the following graphs are obtained by simply honoring historical production rates.

Fluid displacement calculations are performed without adjusting any available parameters except initial water saturations. For the waterflood field case, water cuts are as large as 35 percent at the start of the waterflood. Evidently, these high initial water cuts are due to water transition zones with some possible water encroachment from edges and underlying aquifers. These high initial water cuts made it necessary to adjust the initial water saturations in the simulations to produce correspondingly high initial water cuts.

As shown below, flow simulations based on either fGn or fBm cross sections match field water cut data. Although the initial water saturations assumed for each type of cross section are not always identical, both sets are reasonable. If field water cuts are the only criteria, fGn and fBm type cross sections are equally good for this case with close well spacing (20 acres per well). Close well spacing may diminish the differences between the two types of cross sections. As the well spacing decreases, simple linear interpolation becomes more valid and the importance of fractal noise diminishes. Still, the fGn distributions should be more representative of reservoirs, because they are consistent with outcrop and core photographs, and horizontal well data.

A. Waterflood in Area A

Simulation results based on a single fBm cross section show good agreement with field water cuts (Figure 7). This result is consistent with earlier studies.[1-6] The initial water saturations are chosen to give a good match of the fBm curve to the field data. The chosen water saturations give an initial water fractional flow of 7 percent in both the Grayburg and San Andres formations. Water cuts based on the linear interpolation cross section are near the fBm cross section curve. On the other hand, a single fGn cross section gives lower water cuts throughout the waterflood.

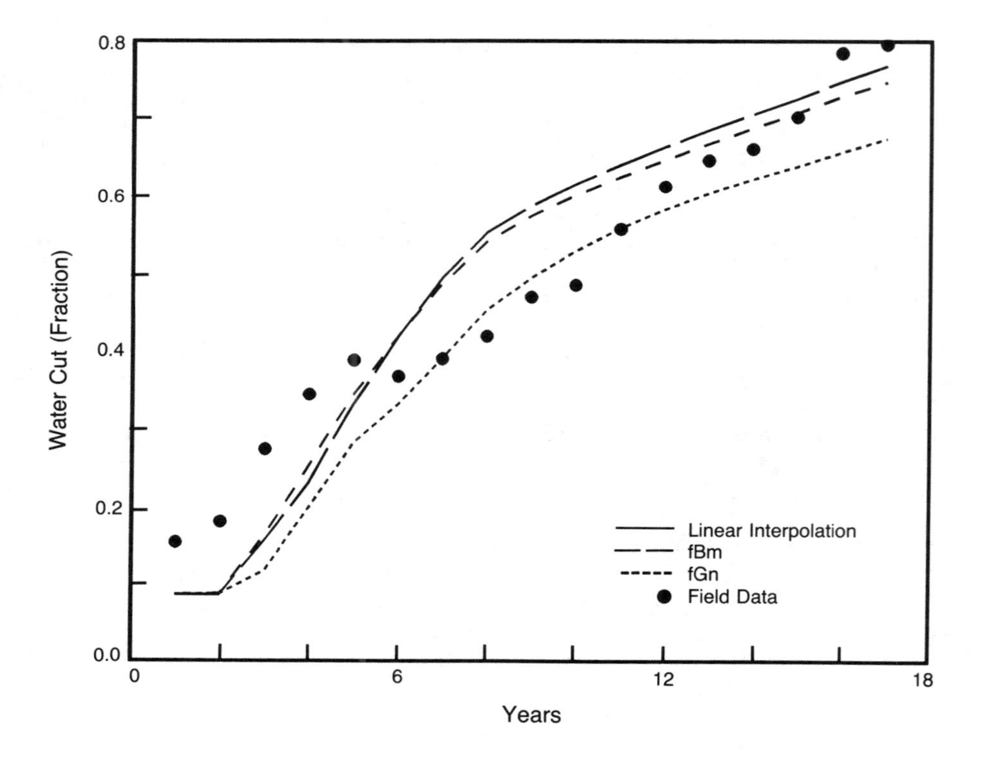

Fig. 7. Simulated water cuts in Area A compared to field data.

Increasing the initial water saturations improves the fGn curve match (Figure 8). Here the initial water saturations are set to give an initial water fractional flow of 12 percent. This value is closer to the initial water cut seen in the field than the 7 percent used in the fBm case.

The statistical model can generate an unlimited number of realizations. For Area A, flow simulations on 20 fGn realizations give a range of water cut curves represented by the shaded region in Figure 8. For reference, water cut curves are also shown for linear interpolation and layer cake cross

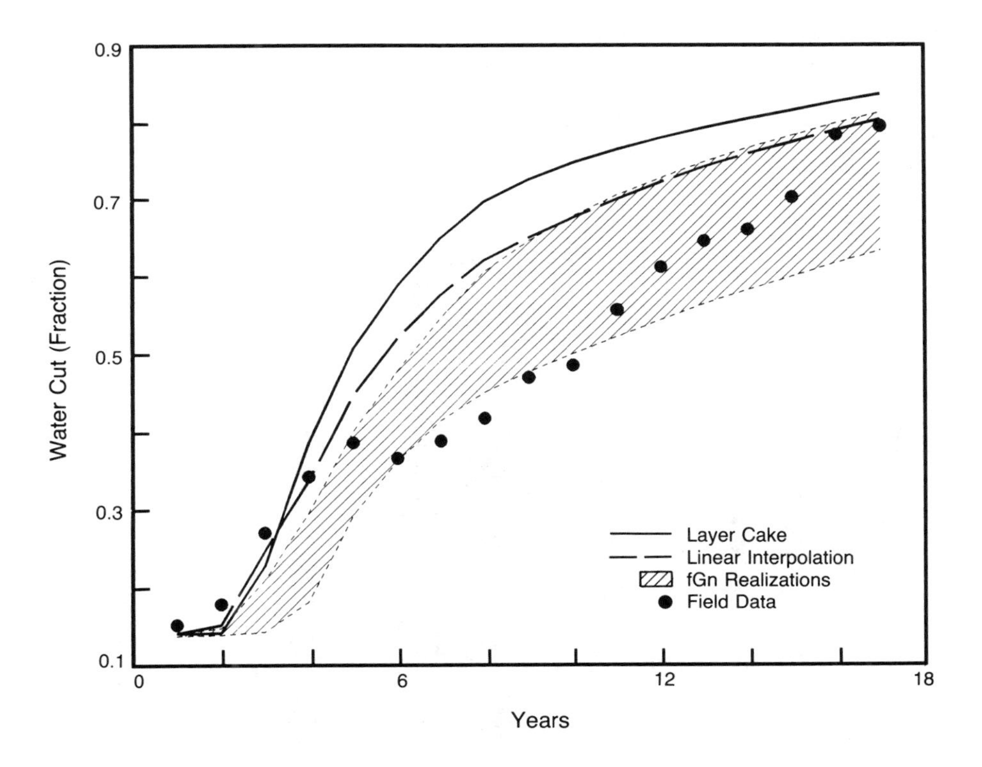

Fig. 8. Simulated water cuts for 20 fGn realizations for Area A (higher S_{wi}).

sections. The frequency distribution of cumulative oil production is displayed in Figure 9. For a 10-acre pattern element, the cumulative oil production ranges from 109 to 135 MSTB with a mean of 122 MSTB. For the layer cake case, the cumulative oil production of 81 MSTB is substantially less than all fGn cases. Correspondingly, the water cut curve is above all the fGn curves (Figure 8). The linear interpolation case has a cumulative oil production of 104 MSTB, which is slightly less than the range of all the fGn cases. Flow simulations on 20 fBm type cross sections give a broader histogram (Figure 9) where the cumulative oil production ranges between 105 to 143 MSTB with a mean of 125 MSTB. In the water cut curves shown in Figure 10, the linear interpolation case lies within the band corresponding to fBm cross

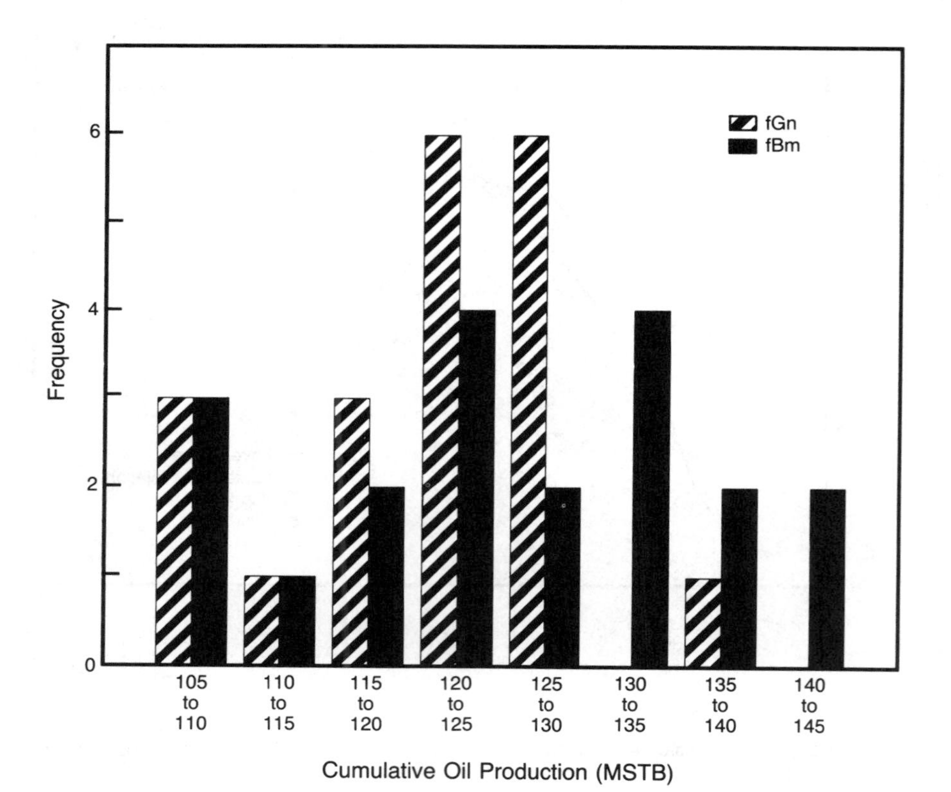

Fig. 9. Histogram of cumulative oil production for 20 fGn and 20 fBm realizations (10 acre pattern element in Area A).

sections. Recall the fBm cross sections have higher initial water saturations in order to match the field water cuts. With these higher saturations, the linear interpolation case gives a cumulative oil production of 120 MSTB. This value is slightly below the mean of the fBm type cross sections. Again, the layer cake cross section gives the lowest cumulative oil production of 95 MSTB.

As mentioned above, the initial water saturations used in the simulations are different for the fGn and fBm cross sections. Within the uncertainty of the field data, both sets are reasonable. Therefore, flow simulations based on either fGn or fBm cross sections match field water cuts.

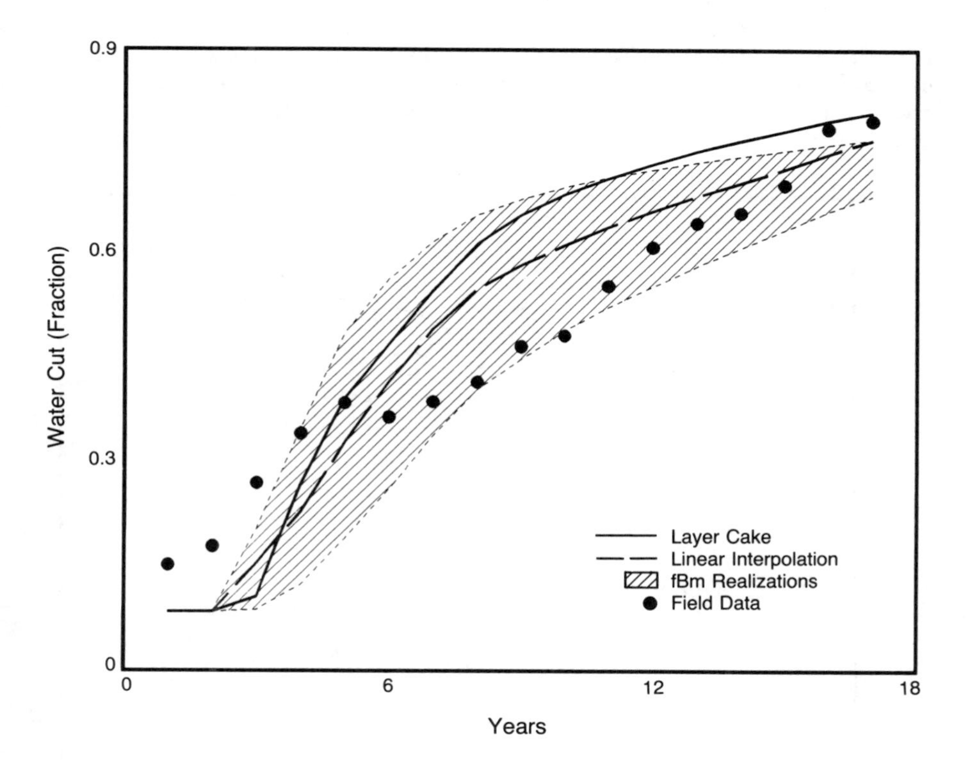

Fig. 10. Simulated water cuts for 20 fBm realizations for Area A (lower S_{wi}).

B. Waterflood in Area B

Flow simulations for Area B give similar agreement with field water cut data for both fGn and fBm cross sections (Figure 11). For both cross sections, the initial water saturations are set to give an initial water fractional flow of 17 percent in all zones at reservoir conditions.

V. SINGLE PHASE LINEAR FLOW

All the above comparisons deal with water cut predictions but do not consider flow rates. The hybrid finite difference/ streamtube method does not predict injection/production rates or effective permeabilities. A simple evaluation of the sensitivity of effective permeability is made below by comparing pressure drops across different vertical cross sections for the same flow rates with a single phase. The width of the cross section is constant.

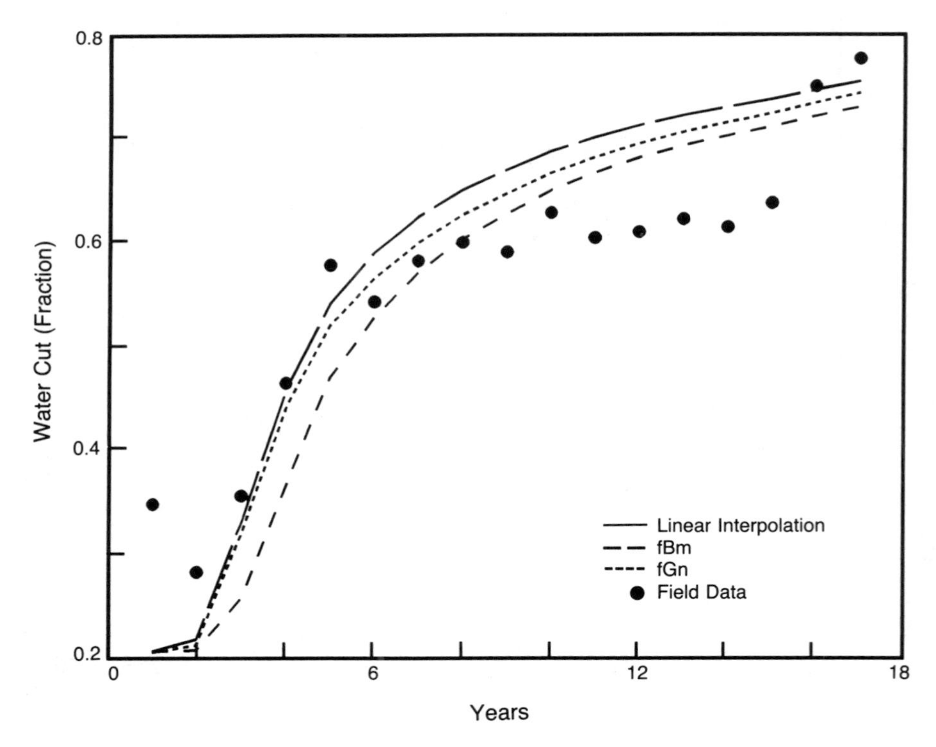

Fig. 11. Simulated water cuts in Area B compared to field data.

The permeabilities for linear flow cannot be used directly to predict injection/production rates in the field. The flow patterns in the field case are more complicated. In the present study, estimates are limited to 2D reservoir descriptions. The connectivity may be different in a full 3D distribution of reservoir properties. Still, variations in 2D permeabilities should give some indication of the corresponding 3D variations.

As shown below, the permeability can vary by a factor of 10 among the different types of vertical cross sections. At least for this waterflood case study, the permeability is much more sensitive than water cut to the type of reservoir property distribution.

A. fGn Cross Sections

The permeabilities in Figure 12 have been normalized by dividing the effective permeability of each cross section by

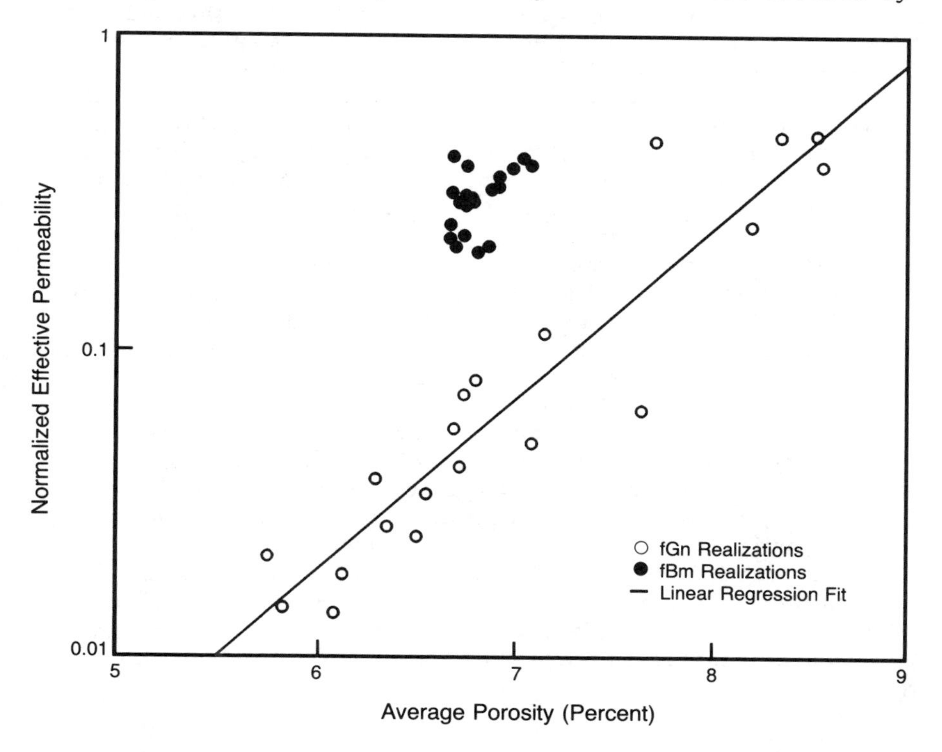

Fig. 12. Effective permeabilities for fGn and fBm realizations for Area A.

the effective permeability of the layer cake cross section. The normalized permeabilities are shown for 20 fGn realizations. The average porosity varies among the different realizations. In generating the cross sections, the average porosity is allowed to fluctuate about the average of the input porosities from the well data (see Appendix). As would be expected from the porosity/permeability relationship (Equation 1), the permeability increases with increasing average porosity. For the average well porosity (6.85 percent), the normalized permeability is 0.055. This permeability is nearly a factor of 20 less than the layer cake permeability. The normalized permeability for the linear interpolation cross section is 0.48, or about a factor of 2 less than the layer cake permeability. Effective permeabilities vary greatly among the different types of cross sections.

B. fBm Cross Sections

Although the field water cut data can be matched with either fGn or fBm cross sections, the corresponding permeabilities are different (Figure 12). The permeability for the fGn cross sections is typically a factor of 5 smaller. This difference can be partially explained by the application of the permeability/porosity transform (Equation 1) at different grid scales in the two methods. For the fGn cross section, Equation 1 is applied at the fine grid scale in Figure 2. The subsequent harmonic averaging to coarsen the grid puts the most weight on the lowest permeability. Since this harmonic averaging step is not included in the fBm procedure, the fGn procedure may give lower permeabilities. Even so, the harmonic averaging does not account for the entire permeability difference. If we apply Equation 1 at the coarser scale for both types of distributions, the fGn cross sections have permeabilities that are still about a factor of 2 to 3 lower.

The fBm methodology for producing cross sections gives much less variation in the average porosity among realizations (Figure 12). The fBm procedure constrains the mean porosity of the <u>entire</u> cross section to the mean value of the well data. On the other hand, the fGn methodology matches the mean porosity of the <u>edges</u> with the mean value of the well data. (See the Appendix for details.) For the fGn cross sections, the greater scatter in porosity carries over to effective permeabilities.

VI. CONCLUSIONS

Flow simulations based on either fGn or fBm cross sections match field water cut data from a mature carbonate reservoir. The initial water saturation is the only parameter adjusted to obtain a match. This adjustment is necessary due to initial water production from transition zones. Although the initial water saturations assumed for each type of distribution are not always identical, both are reasonable. Still, the fGn distributions should be more representative of reservoirs, because they are consistent with outcrop and core photos and horizontal well data from other reservoirs.

Flow simulations on 20 fGn realizations give cumulative oil versus cumulative injection curves that fall within a band of 11 percent about their mean curve. Similarly, the fBm cross sections give a band of 16 percent about their mean curve. Thus, the band is slightly narrower for the fGn cross sections. In addition, the fGn distributions have lower effective permeabilities (linear flow) than the fBm distributions.

The field case studied has close well spacing (20 acres per well). We suspect larger well spacing will increase the differences between the two types of cross sections. As the well spacing increases, simple linear interpolation and fBm are expected to become poorer representations for reservoir heterogeneity.

APPENDIX

A. fGn Cross Section Generation

To build fGn cross sections, the left and right edges of a cross section, $a_{j,k}$, are filled with well log data (L_k for the left edge well log data and R_k for the right edge well log data):

$$a_{1,k} = L_k \ \& \ a_{NX,k} = R_k \ , \ for \ k = 1, NY \qquad \text{(A-1)}$$

where

NX = number of grids in the x direction (horizontal)

NY = number of grids in the y direction (vertical)

A linear interpolation is made between the edges of the array:

$$A_{j,k} = a_{1,k} + \frac{(a_{NX,k} - a_{1,k})}{(NX - 1)} \times (j - 1) \qquad \text{(A-2)}$$

The amount of fGn noise to be added is calculated. It is assumed that the well logs have been put on depth, and therefore, differences between the left and right log values (other than an overall trend) are the result of noise. To calculate the amount of noise, first any overall trends are removed by subtracting the average of each log from each log value. This gives

$$\phi_k(L) = L_k - \overline{L} \qquad \text{(A-3)}$$

where

$$\overline{L} = \frac{1}{NY}\sum_{k=1}^{NY} L_k \qquad \text{(A-4)}$$

and

$$\phi_k(R) = R_k - \overline{R} \qquad \text{(A-5)}$$

where

$$\overline{R} = \frac{1}{NY}\sum_{k=1}^{NY} R_k \qquad \text{(A-6)}$$

The variance, V, of the difference in the two logs without trends is calculated,

$$V = \frac{1}{2\,NY}\sum_{k=1}^{NY} (\phi_k(L) - \phi_k(R))^2 \qquad \text{(A-7)}$$

V corresponds to the variogram at lag NX of the noise that needs to be added.

A noise array is now constructed from the appropriate spectral density, S, of fGn noise. The spectral density of the noise to be added is

$$S(\omega_x, \omega_y) = \frac{A}{[(\omega_x)^2 + (\omega_y)^2)]^{\beta}} \qquad \text{(A-8)}$$

with

$$\omega_x = 2\pi f_x \tag{A-9}$$

$$\omega_y = 2\pi f_y \tag{A-10}$$

$$\beta = H \quad where \quad 0.5 < H < 1 \tag{A-11}$$

Here f_x and f_y are the x and y frequencies in Fourier space.

The noise array, $\eta_{j,k}$, is generated with H determined from a fractal analysis of the well logs. As a discrete inverse Fourier transform, the noise array, $\eta_{j,k}$, is generated from Equation A-8 as follows:

$$\eta_{i,j} = \sum_{m=1}^{NX} \sum_{n=1}^{NY} a_{m,n} e^{2\pi i[(j-1)(m-1)/NX+(k-1)(n-1)/NY]} \tag{A-12}$$

Here,

$$a_{m,n} = R_{m,n} e^{i\Theta_{m,n}} \tag{A-13}$$

where

$$R_{m,n}^2 = \frac{1}{[(m-1)^2 + (LX/LY)^2(n-1)^2]^\beta}$$

$$for \ \frac{NX}{2} \geq m-1 \geq 0 \ \ \& \ \ \frac{NY}{2} \geq n-1 \geq 0 \tag{A-14}$$

LX = separation of the wells (in feet)
LY = thickness of the cross section (in feet)

except

$$R_{1,1} = 0.0 \tag{A-15}$$

Because of symmetry requirements,

$$R_{m,NY+2-n} = R_{m,n} \ \ for \ \ \frac{NX}{2} \geq m > 1 \ \ \& \ \ \frac{NY}{2} \geq n > 1 \tag{A-16}$$

The phase angle, $\Theta_{m,n}$, is defined as a random number between $-\pi$ and $+\pi$ for the range of m and n values satisfying

$$\frac{NX}{2} \geq m-1 \geq 0 \ \ \& \ \ \frac{NY}{2} \geq n-1 \geq 0 \tag{A-17}$$

and

$$\frac{NX}{2} > m - 1 > 0 \quad \& \quad NY - 1 \geq n - 1 > \frac{NY}{2} \tag{A-18}$$

Finally, the 2-D Fourier transform requires the following:

$$a_{NX+2-m,1} = a^*_{m,1} \quad for \quad \frac{NX}{2} \geq m > 1 \tag{A-19}$$

$$a_{1,NY+2-n} = a^*_{1,n} \quad for \quad \frac{NY}{2} \geq n > 1 \tag{A-20}$$

$$a_{NX+2-m,NY+2-n} = a^*_{m,n} \quad for \quad \frac{NX}{2} \geq m > 1 \;\&\; NY \geq n > 1 \tag{A-21}$$

$$a_{\frac{NX}{2}+1,NY+2-n} = a^*_{\frac{NX}{2}+1,n} \quad for \quad \frac{NY}{2} \geq n > 1 \tag{A-22}$$

Here a^* is the complex conjugate of a. For a most accurate distribution, only one-half to one-fourth of the original distribution should be retained (Reference 11), but for this work, the entire array is used.

Note $\eta_{j,k}$ only needs to be generated to agree with Equation A-8 to within a multiplicative constant, because $\eta_{j,k}$ will be rescaled to the well data before it is used. Equation A-15 makes the average of the noise array to be equal to zero.

The desire now is to adjust the variogram of lag NX of the generated noise array to be equal to V. An indirect approach to this produces a more stable solution than a direct one. A direct approach would be a linear scaling that would set the variogram of lag NX equal to V. But if it is assumed that the variogram of the generated noise array has reached its sill for a lag equal to the well separation, then the variance of the noise array, v_n, is equal to the variogram of lag NX. Since there are many more values in the variance than in the variogram of lag NX, the variance produces a more stable estimate of the variogram of lag NX. The variance of the array is therefore used for scaling the noise. The variance of the noise array, v_n, is calculated as follows:

$$v_n = \frac{1}{NX\ NY} \sum_{j=1}^{NX} \sum_{k=1}^{NY} (\eta_{j,k} - \bar{\eta})^2 \tag{A-23}$$

with

$$\overline{\eta} = \frac{1}{NX\ NY}\sum_{j=1}^{NX}\sum_{k=1}^{NY}\eta_{j,k} \tag{A-24}$$

Since the average value in the generated noise array is zero, the following linear scaling gives an array, $H_{j,k}$, with the desired amount of noise and zero mean (see the "linear scaling" section at the end of this Appendix):

$$\mathbf{H}_{j,k} = \eta_{j,k}\sqrt{V/V_n} \tag{A-25}$$

The scaled noise array, $H_{j,k}$, is then added to the linear interpolation array, $A_{j,k}$:

$$B_{j,k} = A_{j,k} + \mathbf{H}_{j,k} \tag{A-26}$$

The inclusion of the linear interpolation array introduces trends in any horizontal trace. These trends alter the statistical structure of the fGn in any horizontal trace. An alternative method is to replace the linear interpolation array in Equation A-26 with a layer cake array. Then the porosity value at a given depth would be the average of the two edge values. The sum of the layer cake array and the fGn noise array would still have fGn statistical structure. However, the resulting cross section would not honor the input well log values at the edges as well as the procedure with linear interpolation. We use linear interpolation in the present study.

With linear interpolation, $B_{j,k}$ in Equation A-26 has a total variance greater than the variance of the original log data. The average of the resulting array, $B_{j,k}$, is unchanged from the linear interpolation array, since the average of the noise array, $H_{j,k}$, is zero.

One final linear scaling is made. Its purpose is to adjust the variance and average of the edges of the array to agree with the variance and average of the log data. This is done so that the edges of the final cross section will closely match the input well log data.

The linear scaling which gives the desired variance and average is the following:

$$C_{j,k} = \left[\frac{V_L}{V_H}\right]^{1/2}(B_{j,k} - \overline{B_H}) + \overline{B_W} \tag{A-27}$$

where

$$\overline{B_W} = (\overline{L} + \overline{R})/2 \tag{A-28}$$

$$\overline{B_H} = [\, \overline{B_1} + \overline{B_{NX}}\,]/2 \qquad \text{(A-29)}$$

$$v_L = \left[\frac{1}{NY} \sum_{k=1}^{NY} (L_k - \overline{L})^2 + \frac{1}{NY} \sum_{k=1}^{NY} (R_k - \overline{R})^2 \right] / 2 \qquad \text{(A-30)}$$

$$v_H = \left[\frac{1}{NY} \sum_{k=1}^{NY} (B_{1,k} - \overline{B_1})^2 + \frac{1}{NY} \sum_{k=1}^{NY} (B_{NX,k} - \overline{B_{NX}})^2 \right] / 2 \qquad \text{(A-31)}$$

$$\overline{B_1} = \frac{1}{NY} \sum_{k=1}^{NY} B_{1,k} \qquad \text{(A-32)}$$

$$\overline{B_{NX}} = \frac{1}{NY} \sum_{k=1}^{NY} B_{NX,k} \qquad \text{(A-33)}$$

The variance and average value of the edges of the final array, $C_{j,k}$, are now the same as the variance, v_L, and average, B_W, of the input well log data from both wells. As a result of scaling the edges rather than the overall array average to the well log data, the overall average and variance of the final array may be greater than or less than the average and variance of the input well log data for a particular realization.

When fGn cross sections are generated for the field study, they are generated using log porosities and square grids, 2′ x 2′. Horizontal permeabilities are assigned using Equation 1. The grids are then made more coarse (2′ x 16′). This is done with three calculations. First, the horizontal permeability of the larger grid is set equal to a harmonic average of the small grids that make it up. Second, the vertical permeability of the larger grid is set equal to 0.1 times the large grid horizontal permeability. Third, the porosity of each large grid is set equal to the arithmetic average of the porosities of the small grids.

B. fBm Cross Section Generation

Hewett and Behrens[3] describe the method to build fBm cross sections. Following their method, the left and right edges of a cross section, $a_{j,k}$, are filled with well log data as shown in Equation A-1. A linear interpolation is made between the edges of the array as shown in Equation A-2.

fBm noise is now generated. This noise is actually fBm in the horizontal direction and fGn in the vertical direction. The noise is generated from an equation for the spectral density similar to the way fGn noise was generated. As a discrete inverse Fourier transform, the noise array, $\eta_{j,k}$, is generated in the same manner as fGn (Equations A-12 through A-22) except

$$R^2_{m,n} = \frac{1}{[(m-1)^2 + c^2(n-1)^2]^{\beta}\,(m-1)^2} \tag{A-34}$$

with

$$c = NX/NY \tag{A-35}$$

and

$$\beta = H + 1 \quad where \quad 0.5 < H < 1 \tag{A-36}$$

Note that this aspect ratio, c, is based on the number of grids, not the physical dimensions of the cross section. As with fGn, $\eta_{j,k}$ only needs to be generated to within a multiplicative constant, because $\eta_{j,k}$ will be rescaled to the well data before it is used.

The noise array, $\eta_{j,k}$, is now scaled so that the total variance of the noise array, v_T, matches the variance of the well logs, v_L. This is done by the following linear scaling of the noise array:

$$\psi_{j,k} = \left[\frac{v_L}{v_T}\right]^{1/2} \eta_{j,k} \tag{A-37}$$

where

$$v_T = \frac{1}{NX\ NY}\sum_{j=1}^{NX}\sum_{k=1}^{NY}(\eta_{j,k} - \overline{\eta})^2 \tag{A-38}$$

$$\overline{\eta} = \frac{1}{NX\ NY}\sum_{j=1}^{NX}\sum_{k=1}^{NY}\eta_{j,k} \tag{A-39}$$

and v_L is defined by Equation A-30.

The next step is to add the noise array, $\eta_{j,k}$, to the linear interpolation array, $A_{j,k}$, in such a way that the edges of the final array are similar to the original log data. This is done by a procedure outlined by Journel.[12] First, a linear interpolation between the edges of the noise array is made,

$$d_{j,k} = \eta_{1,k} + \frac{(\eta_{NX,k} - \eta_{1,k})}{(NX - 1)} \times (j - 1) \tag{A-40}$$

This array, $d_{j,k}$, is subtracted from the original noise array so that the edges of the new noise array are zero:

$$e_{j,k} = \eta_{j,k} - d_{j,k} \tag{A-41}$$

The resulting array, $e_{j,k}$, is now added to the linear interpolation array, $A_{j,k}$,

$$f_{j,k} = A_{j,k} + e_{j,k} \tag{A-42}$$

As with fGn, one final linear scaling is made. The purpose is to adjust the variance of the final array to the variance of the well logs while preserving the average of the edges of the array to the average of the well log data. The linear scaling which gives the desired result is the following:

$$g_{j,k} = \left[\frac{v_L}{v_F}\right]^{1/2} (f_{j,k} - \overline{f_e}) + \overline{f_L} \tag{A-43}$$

where

$$v_F = \frac{1}{NX\ NY}\sum_{j=1}^{NX}\sum_{k=1}^{NY} (f_{j,k} - \overline{f})^2 \tag{A-44}$$

$$\overline{f} = \frac{1}{NX\ NY}\sum_{j=1}^{NX}\sum_{k=1}^{NY} f_{j,k} \tag{A-45}$$

$$\overline{f_e} = \left[\frac{1}{NY}\sum_{k=1}^{NY} f_{1,k} + \frac{1}{NY}\sum_{k=1}^{NY} f_{NX,k}\right]/2 \tag{A-46}$$

$$\overline{f_L} = [\,\overline{L} + \overline{R}\,]/2 \tag{A-47}$$

and v_L is defined by Equation A-30. $\overline{L}$ and $\overline{R}$ are defined by Equations A-4 and A-6, respectively.

When fBm cross sections are generated for the field case, they are generated from porosity log data at the scale they are to be used (here 2′ x 16′). No coarsening up of the grid is performed. As with fGn, horizontal permeabilities are assigned to the grids using Equation 1. Vertical permeabilities are set to 1/10 the horizontal permeabilities.

C. Linear Scaling

Linear scaling is used several times in generating cross sections. The basis of this is described here. One Gaussian distribution can be changed into another Gaussian distribution while preserving the relationship of all of the point values if a linear transformation is made. To go from a Gaussian distribution with mean, μ_1, and standard deviation, σ_1, to another Gaussian distribution with mean, μ_2, and standard deviation, σ_2, each value in the first array, x_j, is changed into each value in the second array, y_j, as follows:

$$y_j = \left[\frac{\sigma_2}{\sigma_1}\right](x_j - \mu_1) + \mu_2 \qquad \text{(A-48)}$$

The linear scaling equations used for fGn and fBm cross section generation can be derived from Equation A-48 by noting that the variance of a distribution is equal to its standard deviation squared.

ACKNOWLEDGEMENTS

A computer program from B. H. Caudle at the University of Texas at Austin was used in part of the streamtube calculations. The authors also thank Conoco Inc. for permission to publish this paper.

REFERENCES

1. Hewett, T. A.: "Fractal Distributions of Reservoir Heterogeneity and Their Influence on Fluid Transport," paper SPE 15386 presented at the 1986 SPE Annual Technical Conference and Exhibition, New Orleans, Oct. 5-8.

2. Emanuel, A. S., Alameda, G. K., Behrens, R. A., and Hewett, T. A.: "Reservoir Performance Prediction Methods Based on Fractal Geostatistics," SPE Reservoir Engineering (August 1989) 311-318.

3. Hewett, T. A. and Behrens, R. A.: "Conditional Simulation of Reservoir Heterogeneity With Fractals," SPE Formation Evaluation (September 1990) 217-225.

4. Mathews, J. L., Emanuel, A. S., and Edwards, K. A.: "A Modeling Study of the Mitsue Stage 1 Miscible Flood Using Fractal Geostatistics," paper SPE 18327 presented at the 1988 SPE Annual Technical Conference and Exhibition, Houston, Oct. 2-5.

5. Tang, R. W., Behrens, R. A., and Emanuel, A. S.: "Reservoir Studies With Geostatistics To Forecast Performance," SPE Reservoir Engineering (May 1991) 253-258.

6. Hewett, T. A. and Behrens, R. A.: "Scaling Laws in Reservoir Simulation and Their Use in a Hybrid Finite Difference/Streamtube Approach to Simulating the Effects of Permeability Heterogeneity," Reservoir Characterization II, L. W. Lake, H. B. Carroll, Jr., and T. C. Wesson (eds.), Academic Press, San Diego, California (1991) 402-441.

7. Hardy, H. H.: "The Fractal Character of Photos of Slabbed Cores," Mathematical Geology, Vol. 24, No. 1, (January 1992), 73-97.

8. Crane, S. D. and Tubman, K. M.: "Reservoir Variability and Modeling With Fractals," paper SPE 20606 presented at the 1990 SPE Annual Technical Conference and Exhibition, New Orleans, LA, September 23-26.

9. Hardy, H. H.: "Fractal Analysis of Core Photographs with Application to Reservoir Characterization," presented at the NIPER/DOE Third International Reservoir Characterization Technical Conference, Tulsa, Nov. 3-5, 1991.

10. Feder, J.: Fractals, Plenum Press, New York, 1988.

11. Saupe, D.: "Algorithms for Random Fractals," The Science of Fractal Images, Heinz-Otto Peitgen and D. Saupe (eds.), Springer-Verlag, New York (1988) 71-136.

12. Journel, A. G. and Huijbregts, Ch. J.: Mining Geostatistics, Academic Press, New York, 1978.

Investigating Infill Drilling Performance and Reservoir Continuity Using Geostatistics

D. S. Wolcott, ARCO Alaska, Inc.
A. K. Chopra, ARCO EPT

I. INTRODUCTION

One objective of a field development program is to determine the maximum well spacing that will effectively drain the reservoir. While widely spaced wells have proven to be effective, field experience is showing that infill drilling produces substantial additional recovery.[1-3] In heterogeneous reservoirs, infill drilling increases reservoir continuity and enhances areal and vertical sweep efficiencies. For reservoir management and economic reasons it is important to predict the performance and additional recovery from infill drilling. Traditionally, reservoir descriptions have been generated, at interwell locations, using algorithms which result in smoothed predictions. The use of homogeneous or smoothed descriptions is not adequate because important mechanisms are not considered. Incorporation of heterogeneities in reservoir simulations is important but cannot be conducted with conventional mapping techniques. The use of geostatistics provides a framework for incorporating reservoir heterogeneities at interwell locations in the description.

For many years, the mining industry has used geostatistics to evaluate the quality, quantity and orientation of ore bodies.[4-9] The term geostatistics refers to statistical techniques that account for spatial continuity and variability. Continuity is an essential feature of natural phenomena, but not easily quantified using traditional deterministic techniques. The importance of incorporating spatial variability in an optimal weighting system was first recognized by Drozdov, a Russian hydrometeorologist, in the early 1940's. The first elementary kriging method of regression was done by Krige,[9] who applied his techniques to gold mines in South Africa. Matheron[4] provided a foundation for geostatistics and gave the name of "regionalized variables" to the variables typical of a phenomenon having a spatial correlation structure. More recently, Lorenz,[10] Tsonis,[11] and Tennekes[12] have applied stochastic methods to weather forecasting. Geostatistics provides methods for predicting uncertainty by creating multiple realizations of the reservoir description. The probabilistic framework is useful because it leads to a system of equations for determining the estimate and uncertainty.

In recent years, the petroleum industry has placed greater emphasis on these techniques for reservoir characterization because of their promise to integrate data and provide a framework for describing interwell heterogeneities.[13-19] The key issues in reservoir characterization are the estimation and representation of key reservoir properties at interwell locations, prediction of reservoir performance, and the understanding of reservoir heterogeneity. The precise level of detail required and suitable techniques are not well-defined in the industry. The detail that must be described in a reservoir description depends on its intended use and varies for different processes and reservoirs.[20-23]

Several estimation approaches have relied on deterministic interpolation of key reservoir properties based on a geological model of the reservoir. The geological model is based on thin sections, core descriptions, sedimentology of the region, and the associated interpretation.[23-30] In some of these approaches, boolean methods are applied successfully to represent geology. However, in other instances, contour maps are drawn manually or by computer to honor the well data. These contour maps are then translated to a grid suitable for reservoir simulation. Currently, many traditional mapping techniques minimize the inherent geological heterogeneities or do not consider them. These descriptions may generate inaccurate results when used with flow simulators. The industry-wide application of geostatistics is still limited in reservoir engineering because of lack of understanding of how reservoir heterogeneity, generated using geostatistical techniques, affects fluid flow and reservoir performance.[31] Most previous work on flow simulation using geostatistical realizations has focused on using two-dimensional cross sections with streamtube models.[17,18]

This paper presents an investigation of infill drilling performance and reservoir continuity using geostatistics and a reservoir simulator. The study area is the A4 Sandstone Formation, Drillsite 1E (DS-1E), in the Kuparuk River Field (Fig. 1). The A4 Sandstone, particularly DS-1E, was chosen because of the higher density of wells in this area. DS-1E has variable (40 to 160-acre) spacing compared to the remainder of the field developed on 160-acre spacing. Also, previous geologic studies have correlated the lithofacies of subunits of the A4 interval resulting in a high resolution geologic description of this drillsite.[32]

Maps generated from kriging and conditional simulations are used in a three-dimensional black oil reservoir simulator. We show that inclusion of heterogeneities using conditional simulation predicts recovery efficiencies different from those predicted by kriging, a smoother mapping technique. Also presented are reservoir performance predictions showing lower recovery efficiencies for thinner beds and additional recovery from infill drilling.

II. RESERVOIR GEOLOGY AND SUBUNIT DATA

The Kuparuk River Field, discovered in 1969, is on the North Slope of Alaska, near the western edge of the Prudhoe Bay Unit. The field produces from two distinct intervals. The lower of these is the A interval which is composed of five sand bodies (A1 through A5, Fig. 1).[33] The interval is characterized by lenticular, shingled, sheet-like sandstone bodies. The A4 sand body is the major producing sand of the A interval in DS-1E. These sands were deposited in a regressive shelf setting during the Lower Cretaceous age. The best reservoir-quality rocks are amalgamated deposits of multiple episodic storms. The thinner, lower quality reservoir sands represent isolated storm events. Sandstone beds in these facies range from 0.5 to 8 feet thick.

The A4 sand is divided into eight subunits named according to geologic nomenclature: A, B1, B2, B3, B4, C1, C2, and C3 (Fig. 2). The A subunit is the lower sand and C3 subunit the upper. The eight subunits are characterized as hummocky, cross stratified and flaser bedded sandstone.[34] They are discrete flow units within the A4 sandstone body and are separated by mappable shale sequences. These shale sequences restrict vertical flow during waterflood.

Geologic experience in this area indicates higher permeability typically occurs in the thicker zones.[32,35] A correlation of increasing permeability with thickness was obtained from core data. Shown in Fig. 3 is the geometric mean of permeability versus total thickness for each subunit . An example linear model is shown demonstrating the trend. The higher permeabilities are generally associated with the thicker amalgamated beds and lower permeabilities with the thin single event beds. A similar relationship has been observed by other investigators for another sandstone deposit.[36]

Ankerite, a carbonate authigenic mineral, is a common diagenetic cement in the A4 Sandstone and is present in five of the subunits. This cement fills the porosity and creates permeability barriers. The effect of ankerite zones was included in the reservoir descriptions.

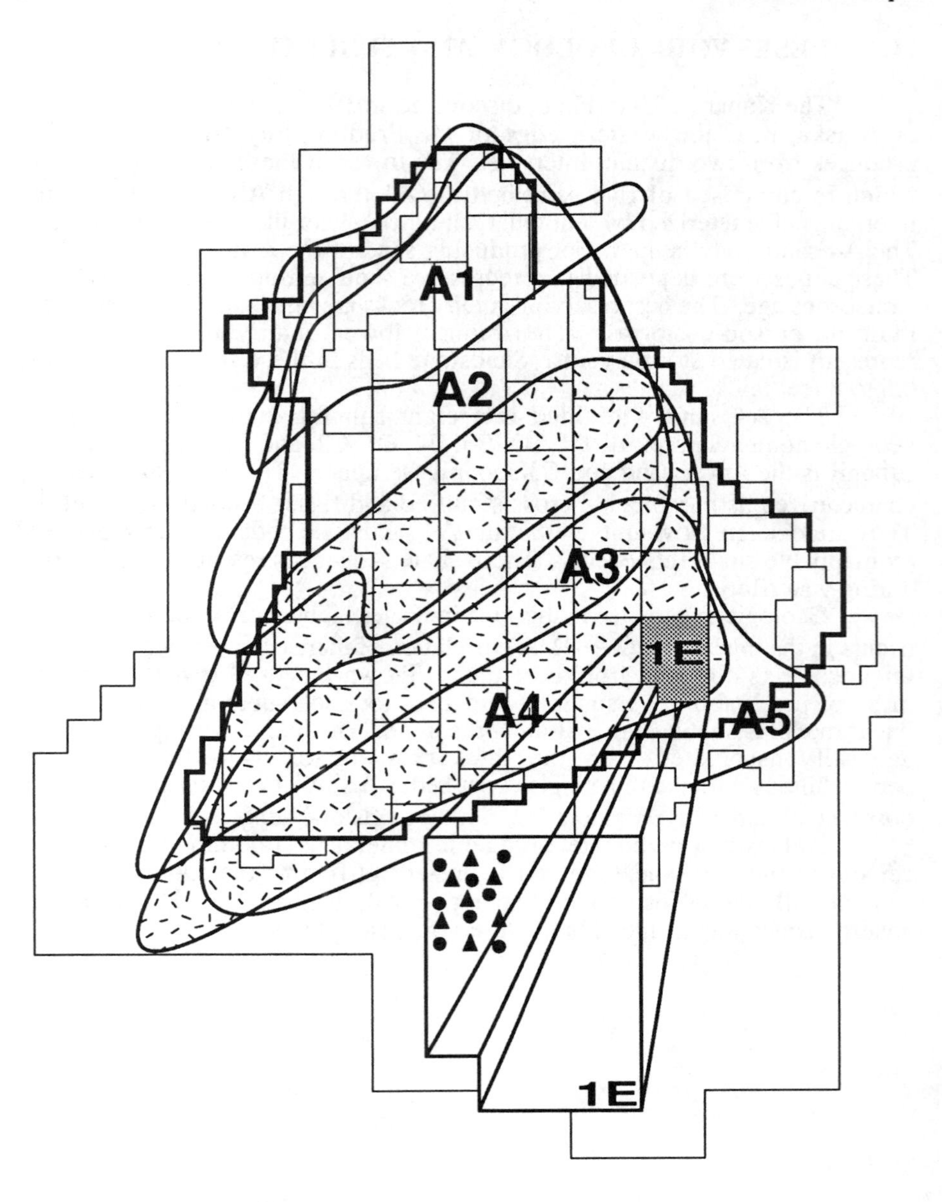

Fig 1. Locality map of the A Sands and Drillsite 1E, Kuparuk River Field.

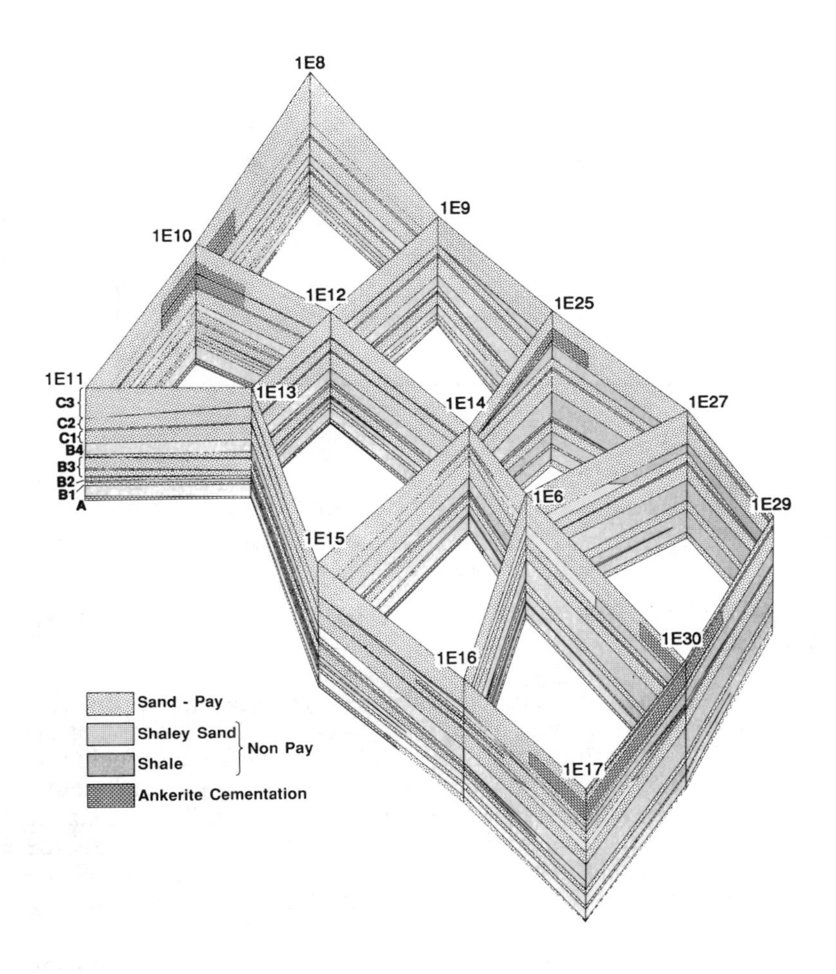

Fig. 2. Fence diagram of the Kuparuk A4 sandstone body in Drillsite 1E, Kuparuk River field (Copyright 1991 SPE, SPE 22164, International Arctic Technology Conference, May 29-31, 1991 in Anchorage AK).

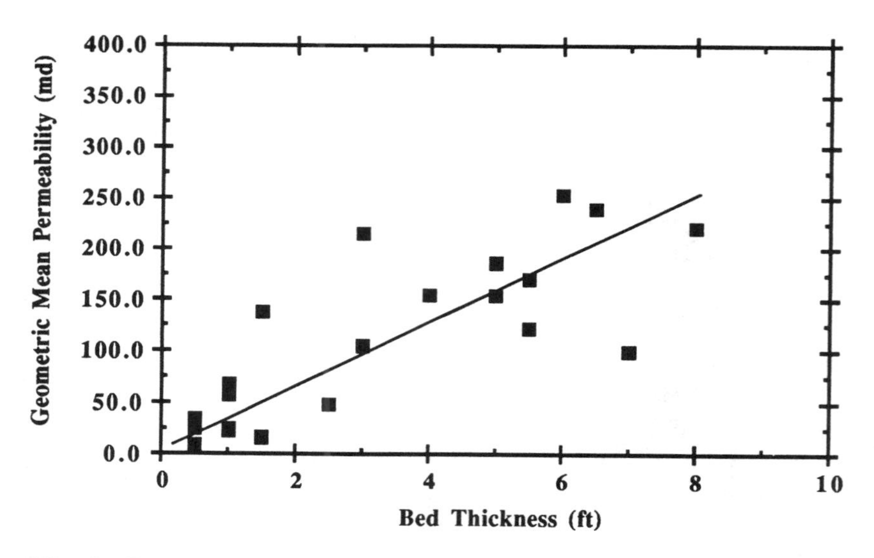

Fig. 3. Geometric mean of permeability for each subunit by thickness with an example linear model.

III. VARIOGRAM EVALUATION AND INFERENCE USING ANALOGS

The first step in evaluating reservoir heterogeneity was to construct semi-variograms for each subunit, using gross thickness data for each of the eight subunits. The gross thickness is the total subunit thickness including ankerite zones. The important values derived from an experimental variogram are its shape, nugget, sill and range. The range is the distance after which the data are not correlated, nugget generally represents the measurement error or lack of small scale correlation of the data, and sill is a measure of sample variance. The Appendix shows the method to calculate a semi-variogram and Clark[37] gives more detail.

For each of the eight subunits in DS-1E, thickness data are available from 15 wells. To confidently infer a semi-variogram with information from only 15 data points is difficult. Therefore, information from an analog formation was included to interpret the semi-variograms. This improved the understanding of the continuities and thickness distributions occurring in storm-generated shelf sandstones.

The analog studied was the Miri Formation[38] from the Seria Field in Brunei, Borneo. This field was chosen as an analog because it is a storm-generated shelf sandstone and considered to be depositionally similar to the Kuparuk River Field.[32] The Seria Field has wells drilled on two to seven acre spacing and contains six sand layers. An example

Miri thickness map for the B sand is in Fig. 4. Since this field was sampled on such a close spacing and the number of wells is large, the confidence in the resulting semi-variogram is greater. Therefore, the character of the semi-variograms at interwell scale is better-understood.

The most notable contributions from the Miri semi-variograms were the approximate values for nugget, range and sill. In Fig. 5, the normalized variograms of four Miri sands and a spherical model are shown. The normalized semi-variograms were obtained by dividing each semi-variogram by its variance. Two significant observations are: a) the nugget values are approximately zero, and b) correlations exists for the first 2000 ft. Although the semi-variograms do show a nested structure at 2500 ft, this structure was not observed from the limited data in Kuparuk DS-1E. This lack of similarity is not significant because the scale of interval heterogeneities for DS-1E wells is less than the distance at which the nested structure is observed. Other possible methods for deducing the semi-variogram are the use of outcrop analogs[39-40] and soft seismic derived correlations.[41-44] The data can be further conditioned by pressure transient tests.[45] These methods were not used in the present study.

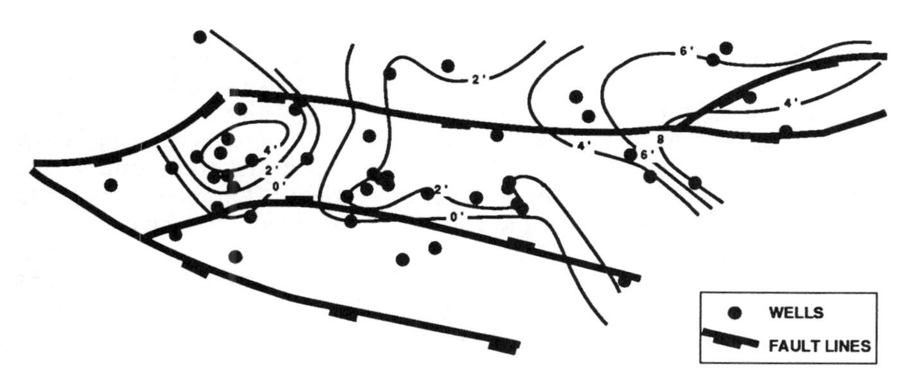

Fig. 4. Thickness map, B-Sand, Miri Formation, Seria Field, Borneo (Atkinson et al., Canadian Society of Petroleum Geologists, 1986).

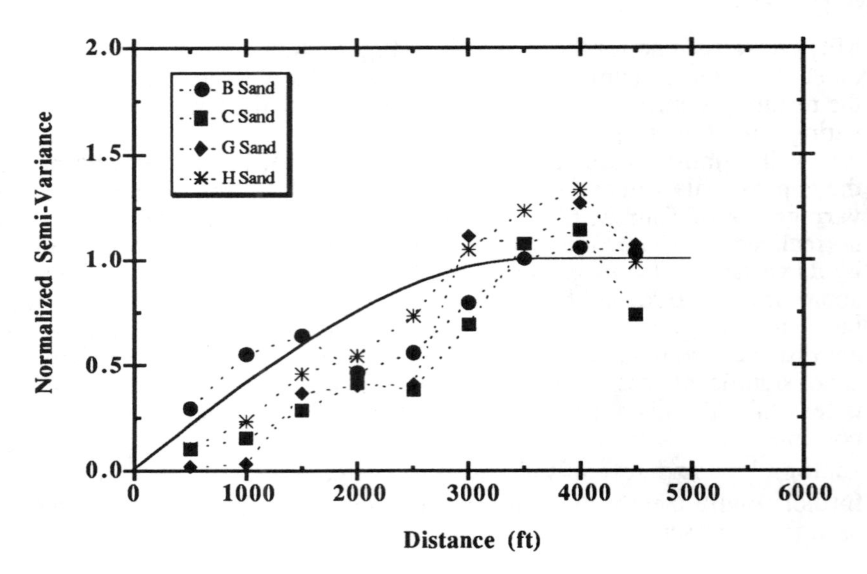

Fig. 5. Example spherical model with the normalized omni-directional semi-variograms of thickness, Miri Formation, Seria Field, Borneo.

The information from the Miri analog was used when fitting the spherical models to the Kuparuk semi-variograms. Because of the large distance between wells and less number of wells in the Kuparuk DS-1E, sufficient statistics were not always available for semi-variance values below the range of the semi-variogram. In such cases, as a first approximation to the Miri analog, the sill was set equal to the variance, the range was set to 3000 feet and the nugget was set to zero. Then, the model was fine tuned by raising or lowering the sill until the best fits through the Kuparuk experimental semi-variograms were obtained. Different spherical models were fit to the experimental semi-variograms of thickness for all the subunits. Fig. 6 is a histogram of the C3 subunit. All the normalized semi-variograms and an example spherical model for subunit C3 are shown in Fig. 7.

For subunits A, B1, B3, C2 and C3, a semi-variogram of ankerite thickness was also generated. Because limited ankerite data were available for each layer, only one composite ankerite semi-variogram was generated using the data from all five layers. A spherical model was fit to the composite variogram.

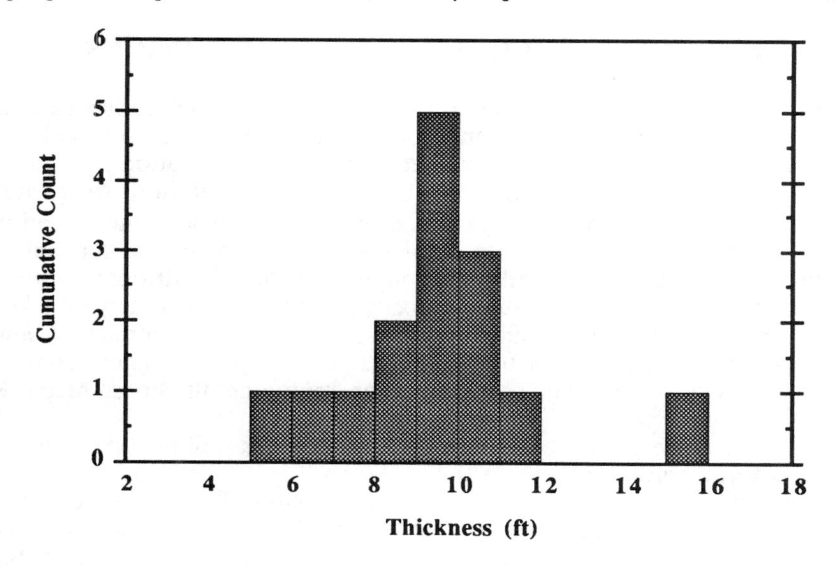

Fig. 6 - Histogram of thickness for Subunit C3.

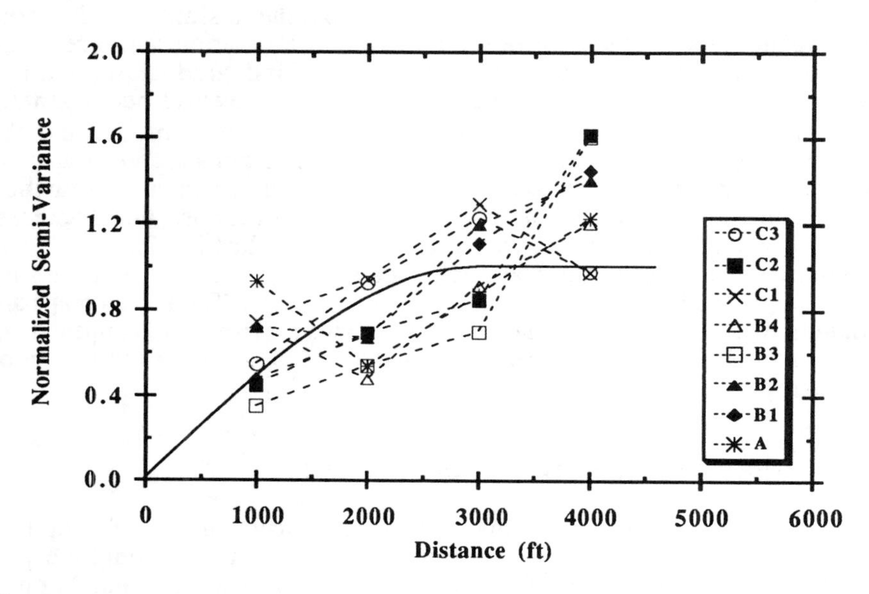

Fig. 7 - Normalized omni-directional semi-variograms for different subunits and the spherical model for subunit C3.

IV. KRIGING AND CONDITIONAL SIMULATIONS

The variogram model and thickness data, from each subunit and well respectively, were input in the conditional simulator to predict the thicknesses at interwell reservoir gridblocks. The conditional simulation package is a commercial program using the turning bands method to generate realizations.[46] Ten conditional simulation realizations and one kriged map were generated. The ankerite semi-variogram model and the ankerite thickness data from the wells were input into the conditional simulator. Again, ten conditional simulation realizations and one kriged map of ankerite were generated for the five subunits containing ankerite. For these subunits, the ankerite thicknesses for maps 1 through 10 were subtracted from their gross thickness maps. The resulting maps are the net thickness maps (Fig. 8).

The permeability values for each of the descriptions were computed using a permeability-net thickness correlation (Fig. 9). Porosity values were determined using a porosity-permeability correlation. This method may result in slightly different total pore volume for each realization. The kriged pore volume was the average of all-different realizations. To account for these differences, the pore volume injected for each realization was normalized by its total pore volume.

For each of the eight subunits, ten conditional realizations and the kriged map are possible reservoir descriptions honoring the known data and spatial correlation structure. The kriging technique is similar to the more traditional mapping techniques minimizing the estimation variance. The resulting kriged maps have smooth contours and gradual transitions. However, the conditional simulations incorporate a level of heterogeneity known to exist.[34] These maps do not have smooth contours and the transitions may be more abrupt. Conditional realizations are generated by using the kriged map as the base map and using semi-variogram information to infer the possible ranges of values at interwell locations. For thickness, the kriged map is compared with a conditional realization in Fig. 8. A similar comparison for permeability is shown in Fig. 9. The permeability maps are derived from their respective thickness maps. The kriged maps are smooth, whereas the conditional realizations have a more ragged appearance. Ten conditional realizations were generated to approximate the total range of geologic possibilities.

V. RESERVOIR FLOW SIMULATION

The recovery variations among different realizations and effect of infill drilling were studied using the flow simulator. For infill drilling, the well configurations and number of patterns need to be considered in addition to reservoir heterogeneities. Four producers at the corners of a pattern flood are equivalent to one producing well in a homogeneous reservoir description.

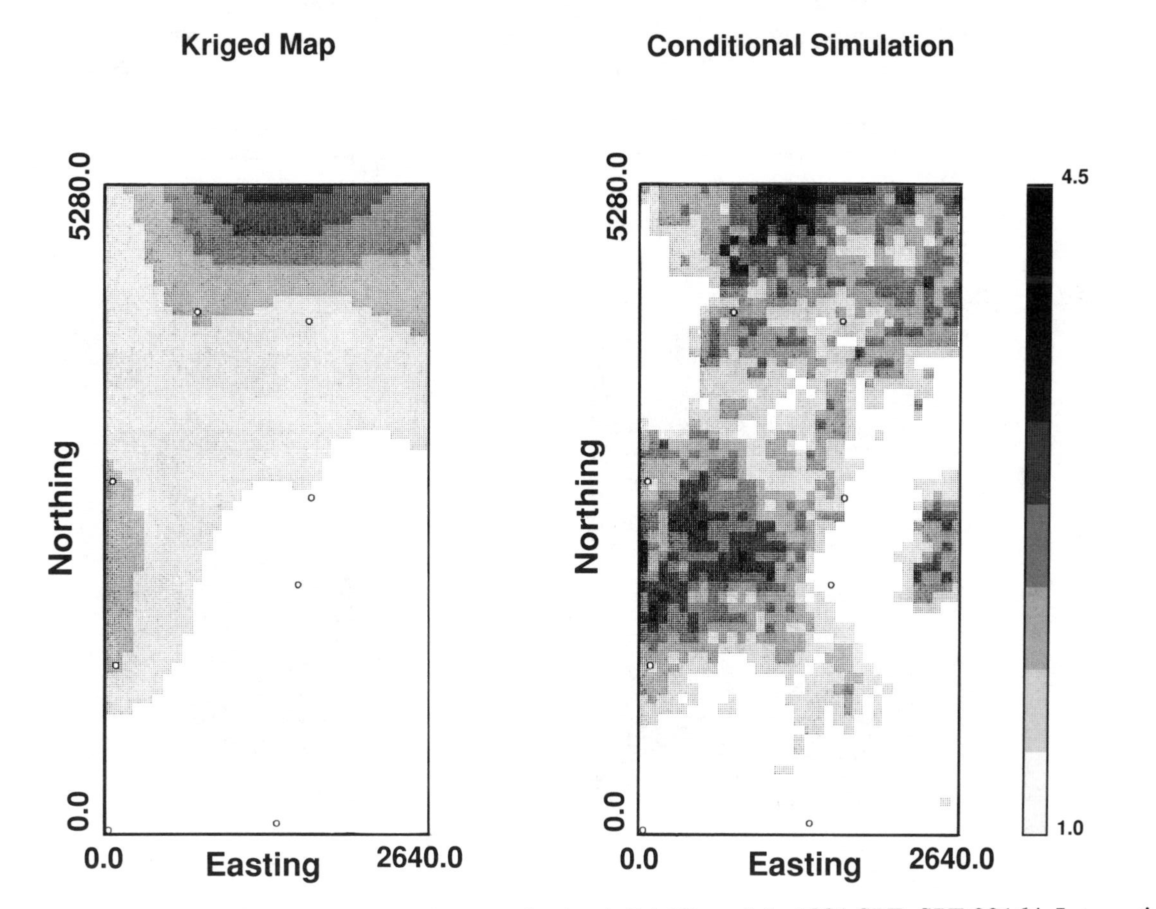

Fig. 8. Example kriged and conditionally simulated maps of subunit B4 (Copyright 1991 SPE, SPE 22164, International Arctic Technology Conference, May 29-31, 1991 in Anchorage AK).

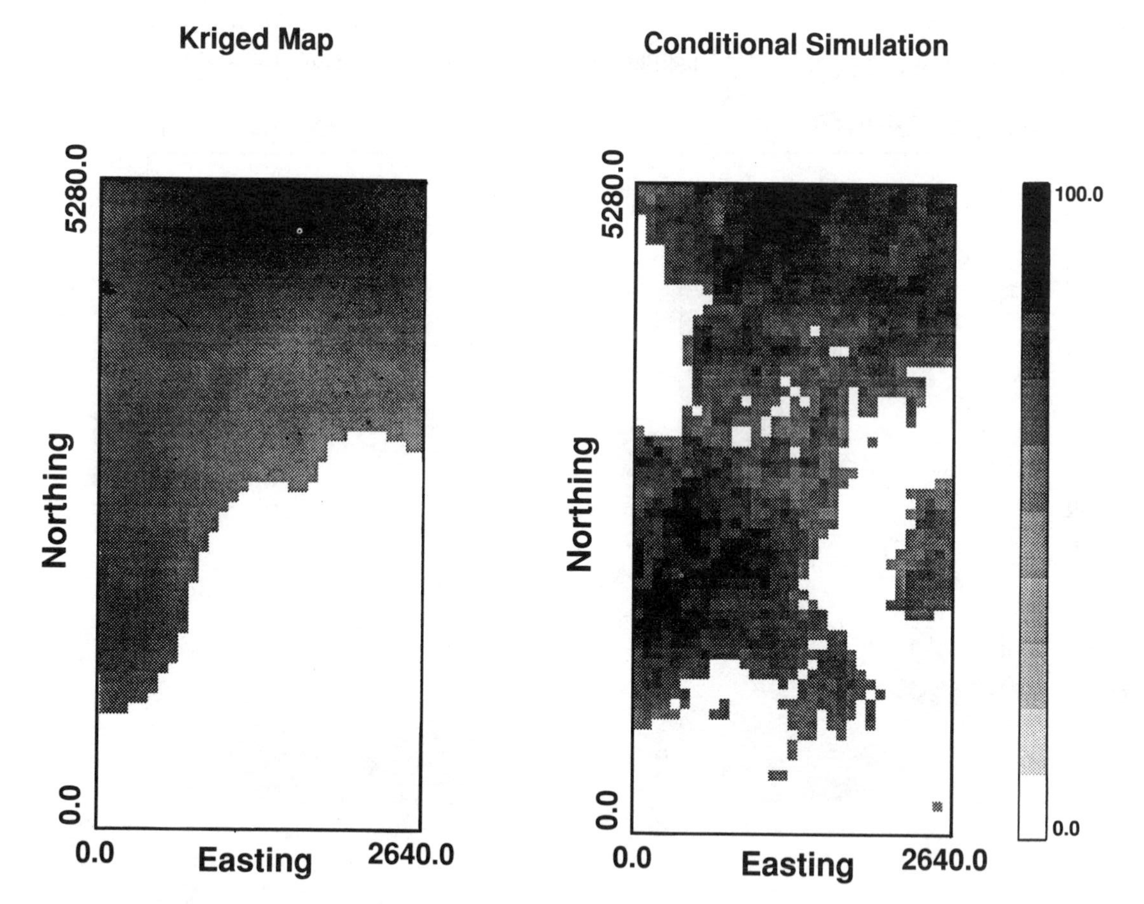

Fig. 9. Example permeability maps for the kriged and conditionally simulated maps for subunit B4 (Copyright 1991 SPE, SPE 22164, International Arctic Technology Conference, May 29-31, 1991 in Anchorage AK).

Studying one specific configuration would not provide a realistic measure of the benefit of infill drilling for heterogeneous geology. Scaling up the performance of one or a few patterns with pre-defined configurations may not represent full field performance. For example, conditional simulations generated using well data from only one pattern would retatin the same gross features as a kriged map. Thick areas from map to map would be in similar locations as would pinchouts. Well placement would have a significant effect on recovery and thereby complicate scaleup. One good alternative may be to use descriptions from unconditional simulations to evaluate the effect of infill drilling. When the reservoir description is conditioned to well data, we can study different well configurations and choose the best one for additional recovery from infill drilling on the specific drillsite.

The configurations shown in Fig. 10, for 160, 80 and 40-acre spacings, were found to be the best for additional recovery from infill drilling of DS-1E. For the 160-acre spacing, there are three water injectors and three oil producers. For the 80-acre, all 160-acre injectors are converted to producers and two injectors added. The 160 and 80 acre wells are converted to injectors for the 40-acre case, and seven producers added. The ten conditional realizations and one kriged map were incorporated in a black oil reservoir simulator for each of the configurations.

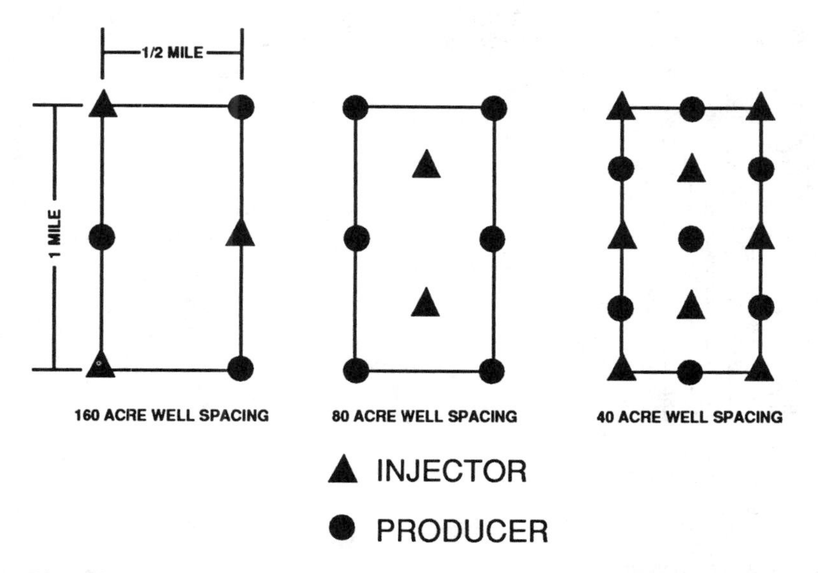

Fig. 10. Location of the simulation wells for 160, 80 and 40 acre spacing (Copyright 1991 SPE, SPE 22164, International Arctic Technology Conference, May 29-31, 1991 in Anchorage AK).

Each layer in the reservoir flow simulation represents a geological subunit. The 1.0 by 0.5 square mile area of each layer is divided into 2145 gridblocks of 82.5 by 82.5 square feet (Fig. 11). There are 33 gridblocks in the easting direction and 65 gridblocks in the northing direction. The total number of gridblocks for eight layers is 17,160. Because of the continuous shales between the subunits, the vertical transmissibilities are set to zero. Hence, comparisons of the areal sweep efficiencies of individual subunits are easily conducted.

The fluid property data are based on fluid samples from the A sand in DS-1E. The gravity of the fluid is 21.8^{o} API. The initial reservoir pressure and the bubble point pressure are 3150 and 2535 psia, respectively. The relative permeability data are also for the A-sand. The average connate water saturation is 18% and the average residual oil saturation to water is 16.9%. The average porosity is 22% and the absolute permeability to air varies between 20 md and 300 md.

The shut-in logic in the simulator was such that a well was shutin when its producing WOR reached 19. Once all the producing wells were shut-in, the simulation was complete. A prudent field practice is to cement squeeze the high WOR layers. However, this was considered impractical for closely spaced thin subunits within the A4 sand body. Therefore, subunits were not shut-in selectively in the simulator.

Production and injection for the 160, 80 and 40-acre cases for 11 descriptions were examined by layer and well. The individual layer production was examined because thicker continuous layers masked the recovery efficiency from the thinner less continuous layers.

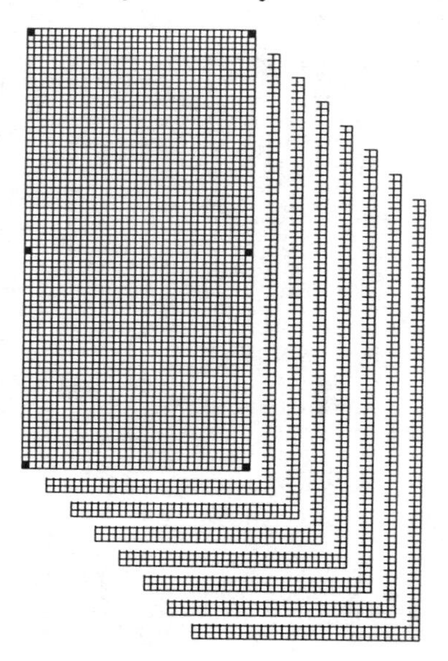

Fig. 11. High resolution description of the eight subunits in DS-1E. 33 by 65 gridblocks per layer and 8 layers, 17,160 gridblocks total.

VI. RESULTS

A. Recovery Variation

Normalized recovery efficiency is plotted versus pore volume injected for the 11 realizations for each well spacing. The normalized recovery efficiency is the recovery from a particular realization divided by the recovery obtained from the kriged reservoir description for the 160 acre spacing. For the Kuparuk DS-1E, there are significant recovery differences between kriged and conditionally simulated descriptions (Figs. 12 through 14). As seen, the kriged recoveries are always higher than the conditional recoveries through all ranges of pore volumes injected. This is true for 160, 80 and 40-acre well spacings. The conditional recoveries are lower because conditional simulations generate a more heterogeneous picture of the reservoir. Also, conditional simulations provide a technique to assess uncertainty in the performance predictions based on limited available data.

By definition the average of the conditional thickness maps should equal the kriged thickness maps. However, this is not true for flow simulated recovery results. The equations describing fluid flow in porous media are nonlinear and therefore may or may not produce the same result after flow simulation. This agrees with intuition and experience which has shown reservoir heterogeneities effect recovery.

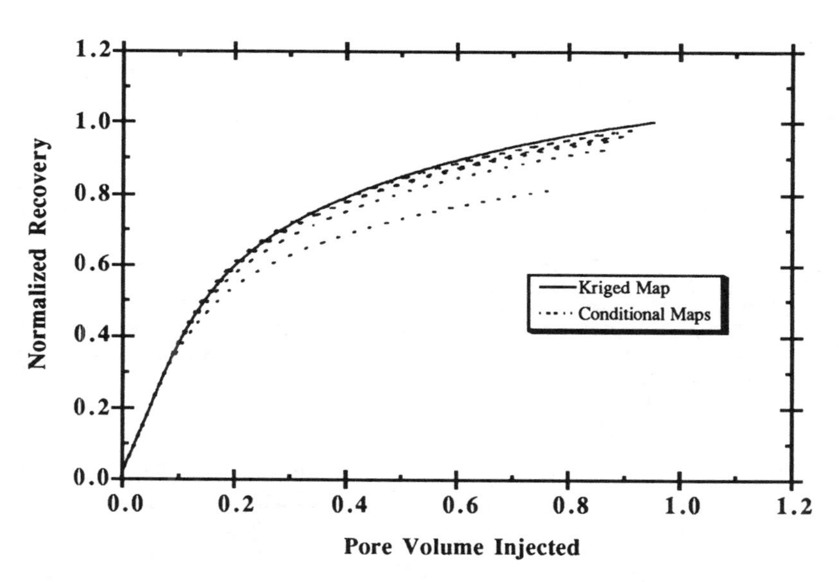

Figure 12 - Normalized recovery response for eleven realizations on 160 acre spacing.

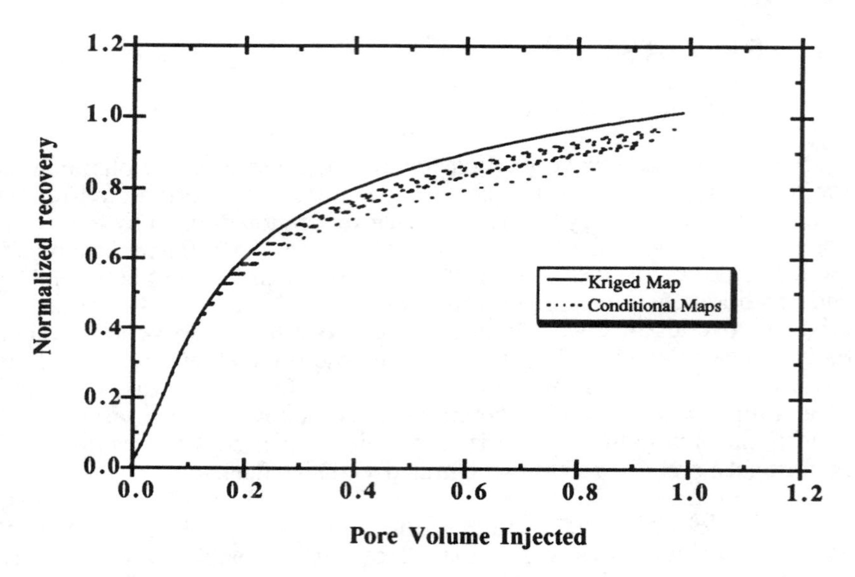

Figure 13 - Normalized recovery response for eleven realizations on 80 acre spacing.

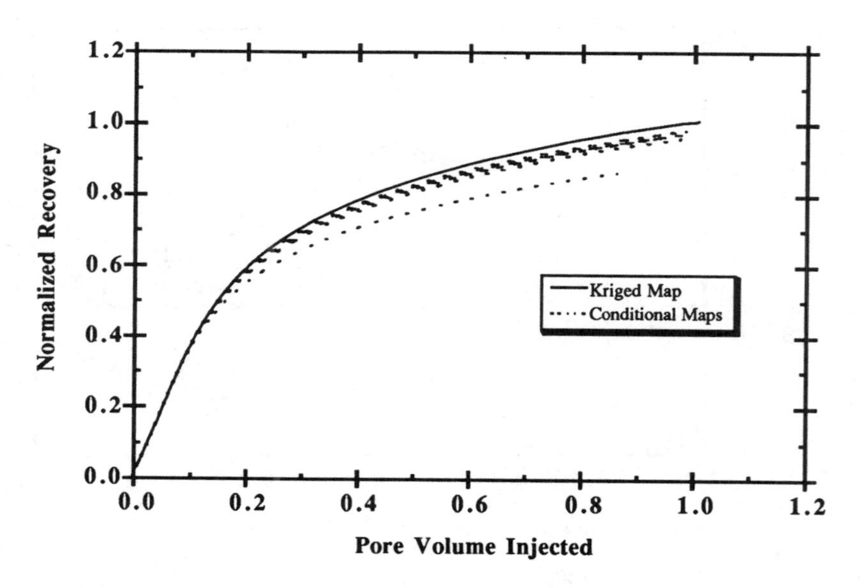

Figure 14 - Normalized recovery response for eleven realizations on 40 acre spacing.

The previous discussion is on recovery predictions for all eight layers. The normalized recovery efficiency for each layer was examined to understand the effect of conditional simulation on characterization and response of individual layers. Examples of the thickness maps for both a thin (A) and a thick (C1) subunit are in Figs. 15 and 16, respectively. The thin subunit has more discontinuities and the thick subunit is more homogeneous.

The results show the conditional simulation has a greater effect on recovery of thinner subunits. For example, in Fig. 17, subunit A has a recovery efficiency for the kriged map which is much larger than the conditional realization recoveries. This is also true for the thin subunit B4 as shown in Fig. 18 (example thickness maps Fig. 8). Similarly, the other thin subunits (B1, B2, and B3) have recovery efficiencies more affected by the discontinuities incorporated in the mapping. The inclusion of the heterogeneities in the thin sands results in a lower predicted recovery efficiency. The conditional realizations incorporate believed heterogeneities which are smoothed by a kriged map. The normalized recovery efficiency, hence the sweep efficiency, depends on heterogeneity and mobility ratio. The recovery efficiency in the thinner subunits would be affected even more for adverse mobility ratio displacements.

The recovery efficiency predictions for the thicker subunits C1, C2, and C3, are not significantly affected by the description method. For the thick C1 subunit, the kriged recovery is approximately the average of the conditional realization recovery predictions (Fig. 19). This implies that conditional simulation for the thicker subunits does not introduce discontinuities adversely impacting recovery. The fluctuations around the kriged recovery curve reflect the uncertainty in the reservoir description.

Based on these results, a correlation is suggested between recovery efficiency and subunit thickness. The average of recovery of all realizations for each subunit was plotted against the corresponding average thickness for the 160 acre well spacing (Fig. 20). The figure shows a strong correlation between the two variables. Moreover, the recovery efficiency begins to fall significantly for layers with average thickness less than four feet. The thickest subunit, C3, consistently has the highest recovery efficiency. As thinning progresses, so does the reduction in recovery efficiency. For subunit thickness less than 4 feet, the trend is very abrupt. The correlation between recovery efficiency and thickness (not shown in this paper) also occurs in the 80 and 40-acre well spacings. The results show the reduction in overall recoveries, when conditionally simulated maps are used, is caused by low recovery efficiency in the thinner subunits.

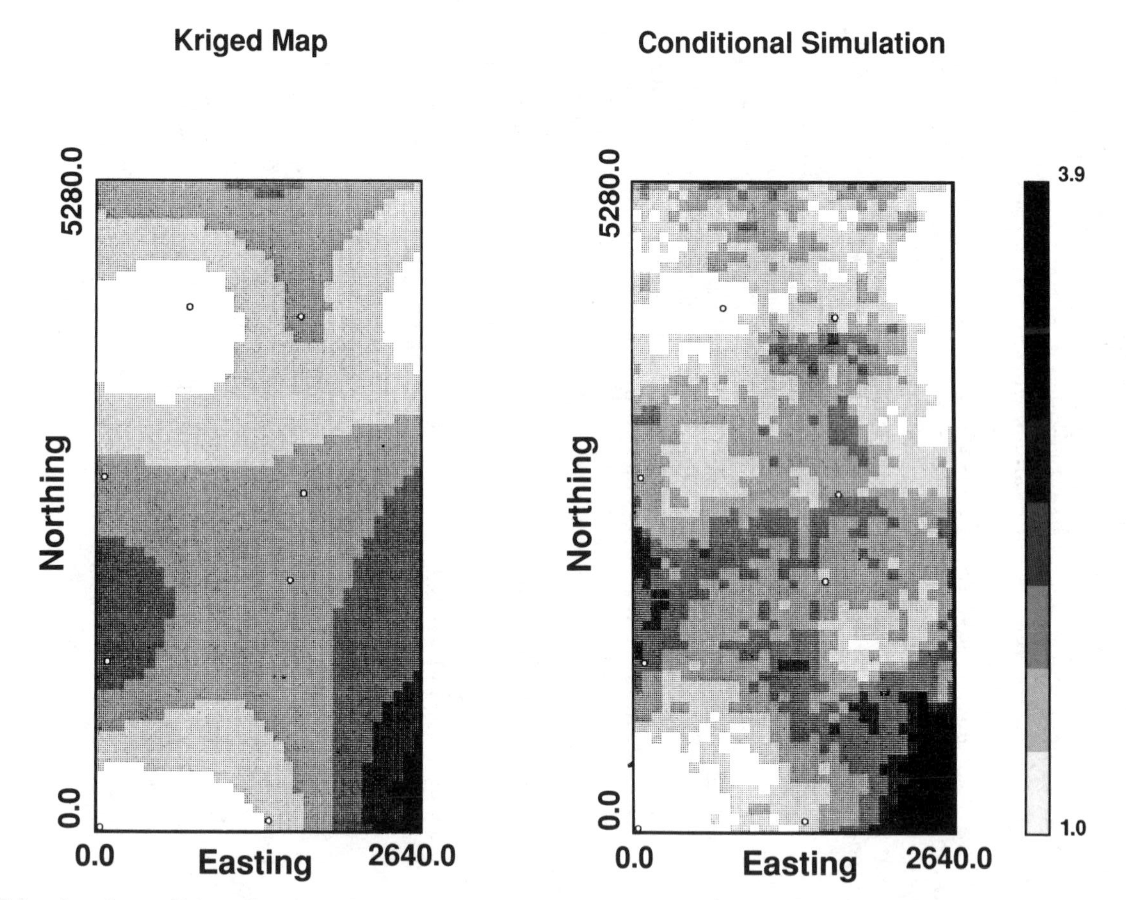

Fig. 15. Kriged and conditionally simulated map of subunit A (Copyright 1991 SPE, SPE 22164, International Arctic Technology Conference, May 29-31, 1991 in Anchorage AK).

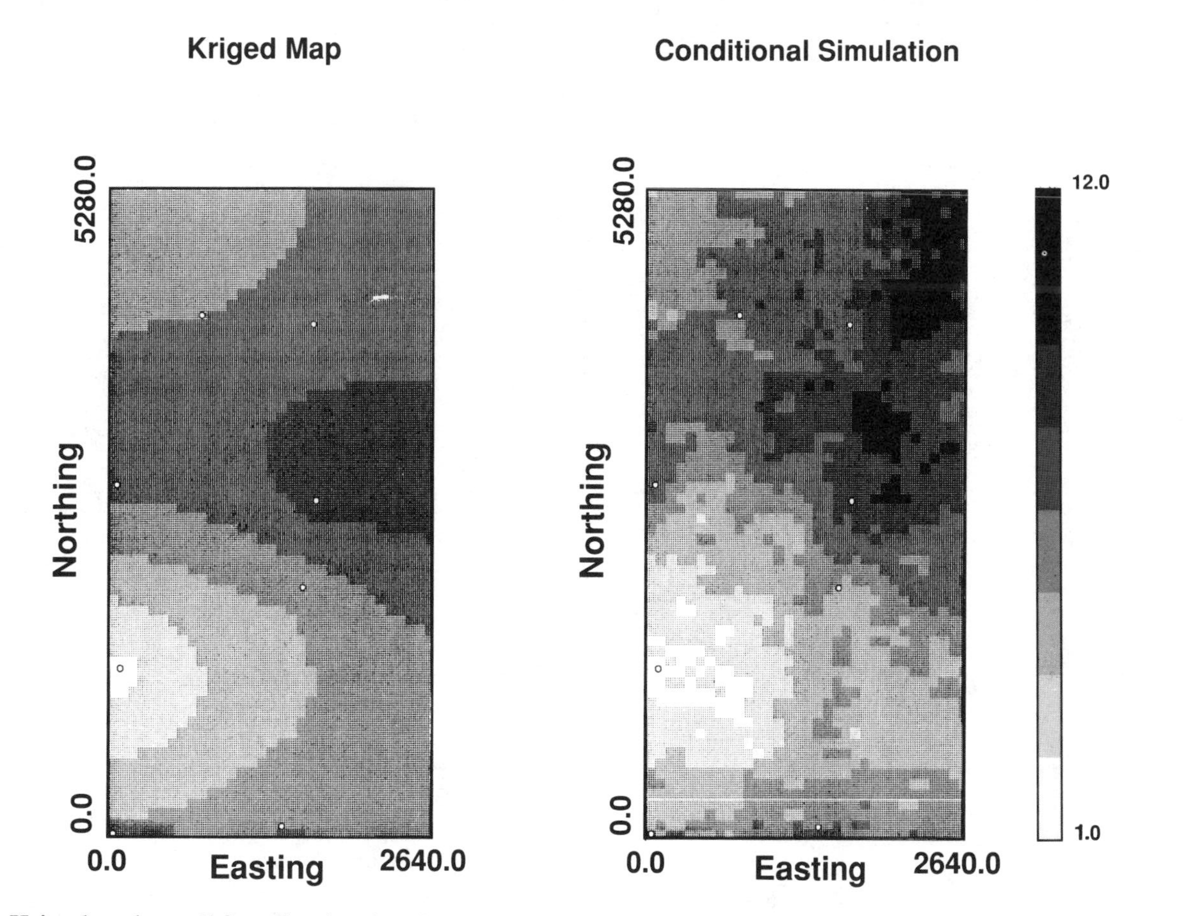

Fig. 16. Kriged and conditionally simulated map of subunit C1 (Copyright 1991 SPE, SPE 22164, International Arctic Technology Conference, May 29-31, 1991 in Anchorage AK).

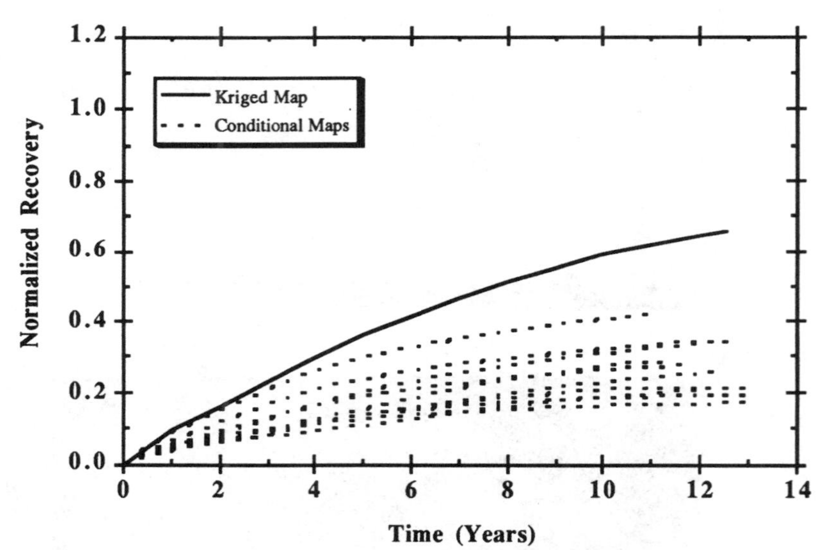

Figure 17 - Normalized recovery response of eleven realizations on 160 acre spacing for subunit A.

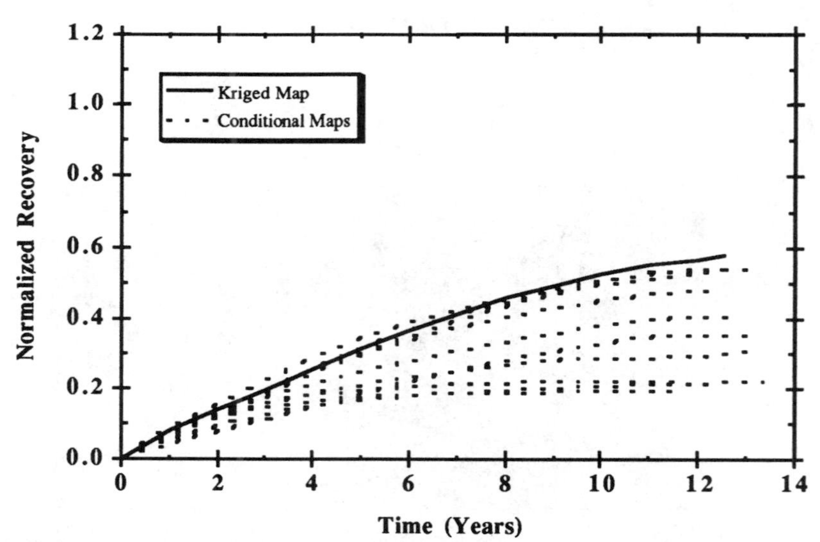

Figure 18 - Normalized recovery response of eleven realizations on 160 acre spacing for subunit B4.

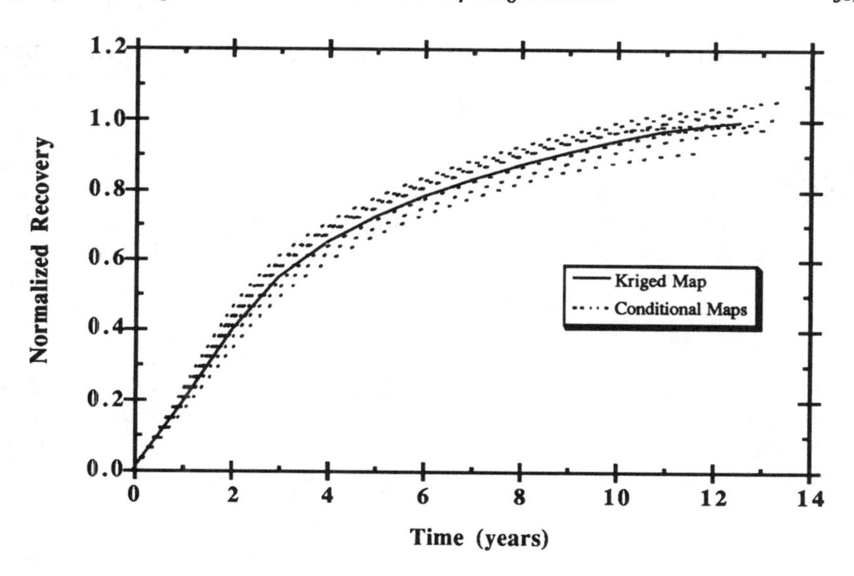

Figure 19 - Normalized recovery response of eleven realizations on 160 acre spacing, subunit C1.

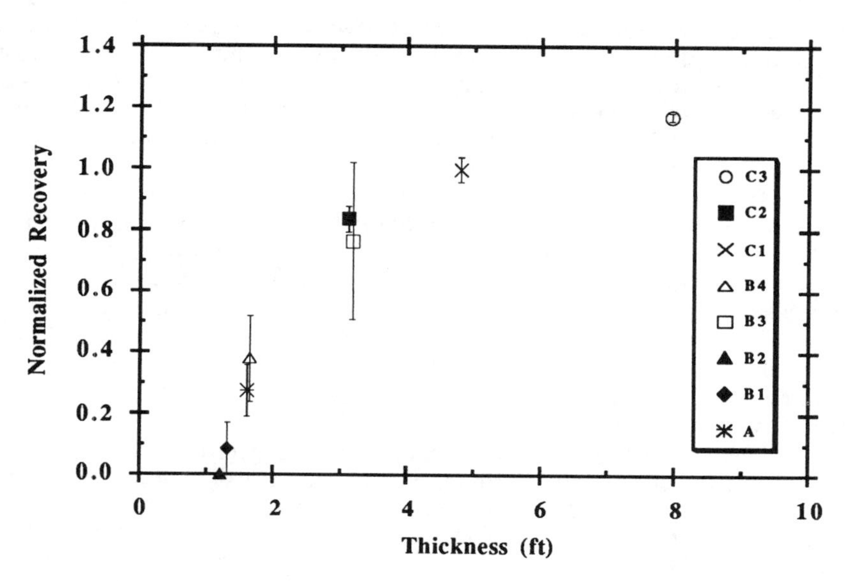

Figure 20 - Average normalized recovery with error bars from 160-acre flow simulated realizations.

Average layer thickness alone, however, is not a complete estimator of recovery efficiency. Other factors, such as variance and correlation structure, influence recovery. For instance, in Fig. 20, the C2 and B3 subunits have nearly the same average thickness, but the recovery efficiencies for the B3 subunit are lower and distributed more widely than the C2 subunit. This is because the B3 subunit, where it exists, can be very thick but then thins out quickly. The C2 subunit is areally continuous, and does not vary greatly in thickness. An example of each is shown in Fig. 21.

B. Additional Recovery from Infill Drilling

The recoveries from flow simulations of 40 acre well spacing were subtracted from those for the 160-acre well spacing for each of the subunits to evaluate the additional recovery from infill drilling. Fig. 22 shows the additional recovery versus average subunit thickness and an approximate curve drawn through the results. The additional recovery from infill drilling increases as the average thickness decreases. The error bars show the uncertainty in the predicted recovery prediction within plus or minus one standard deviation. The increase in additional recovery increases rapidly for average thicknesses less than four feet. The lower extent of the error bars in some cases, i.e. A, B3 and C3, predicts negative additional recoveries. Negative recovery predictions from infill drilling can occur because the description favors a particular well pattern. This emphasizes why flow simulating multiple probable descriptions is important. When simulating more homogenous sands (e.g., C3) negative recoveries can occur because the flood progresses much faster leaving behind higher oil saturations. Physical diffusion, which is a function of the square root of time, is less and because of fewer time steps so is numerical dispersion. Less dispersion results in less fluid mixing which for the homogenous examples shown in this work reduces predicted recovery.

The additional recovery from infill drilling was also calculated for the 160-acre to 80-acre case. Though not shown in this paper, the same basic trends are observed. However, the magnitude of additional recovery for subunits less than three feet is not as large as observed for the 160-acre to 40-acre case. This is because the discontinuities in the thin subunits limit recovery at larger well spacing. Such trends were less pronounced and not easily discernible using kriged maps. As shown in Fig. 23, the kriged map estimates are more widely scattered and attempting to fit a correlation through the data without conditional map results would be difficult.

The average of the normalized recoveries from the 10 conditional realizations for each of the infill cases is in Fig. 24. There is a correlation between higher recovery efficiencies and closer well spacing. The average 80-acre well spacing recovery efficiency is higher than the 160-acre recovery efficiency and the 40-acre recovery efficiency is higher than the 80-acre recovery efficiency.

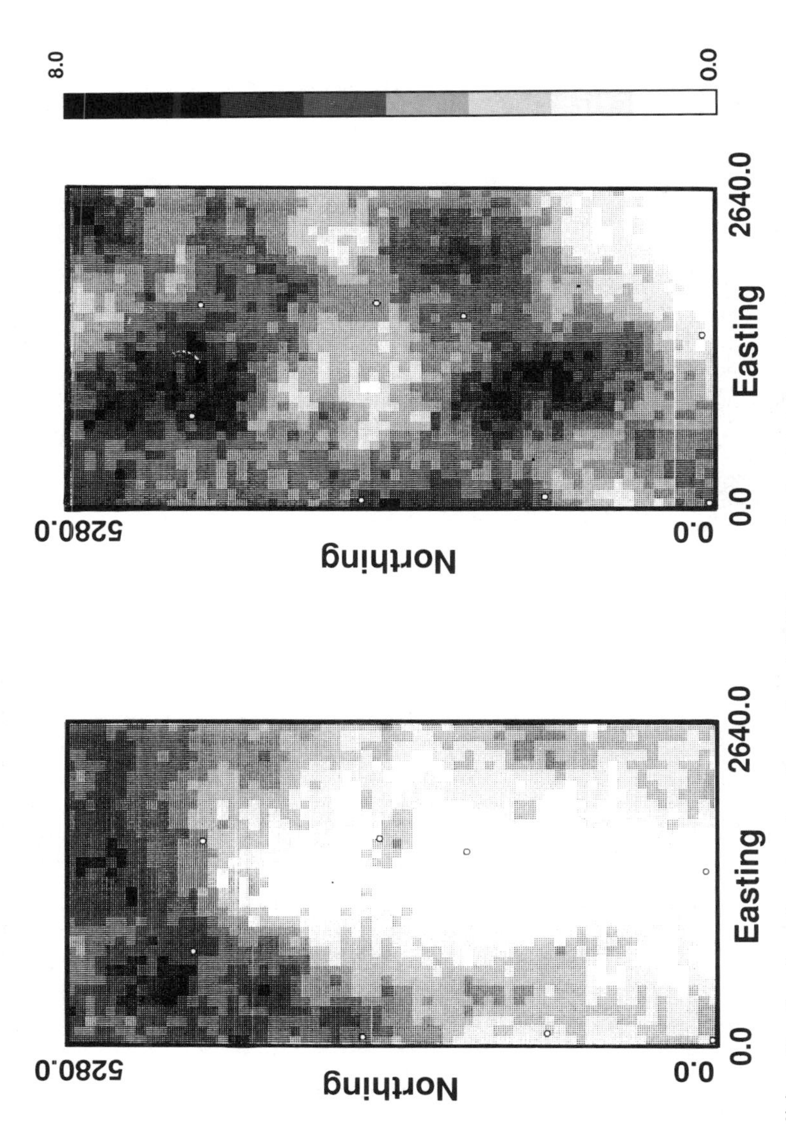

Fig. 21 Conditionally simulated maps showing subunit B3 is relatively thick but also has thin or no sand areas whereas subunit C2 is relatively thick with less thinning. The simulated recoveries for subunit B3 vary more than subunit C2 and the overall average is less.

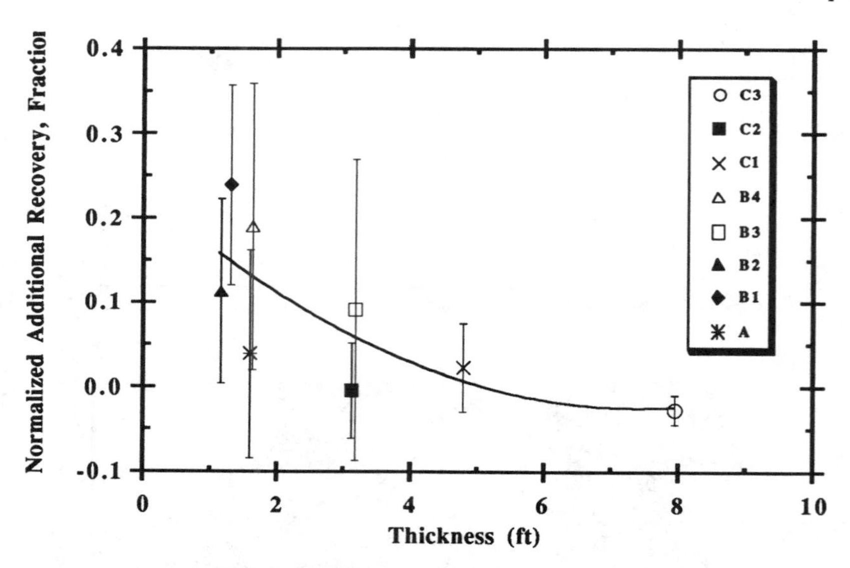

Figure 22 - Normalized conditional average additional recovery (fraction of 160 acre kriged map recovery) with error bars of 160 to 40 acre spacing.

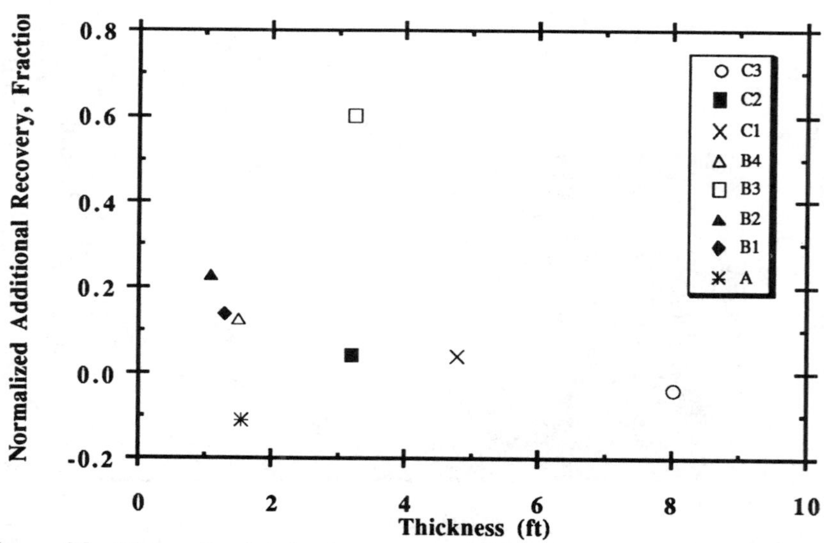

Figure 23 - Normalized kriged average additional recovery (fraction of 160 acre kriged map recovery) of 160 to 40 acre spacing.

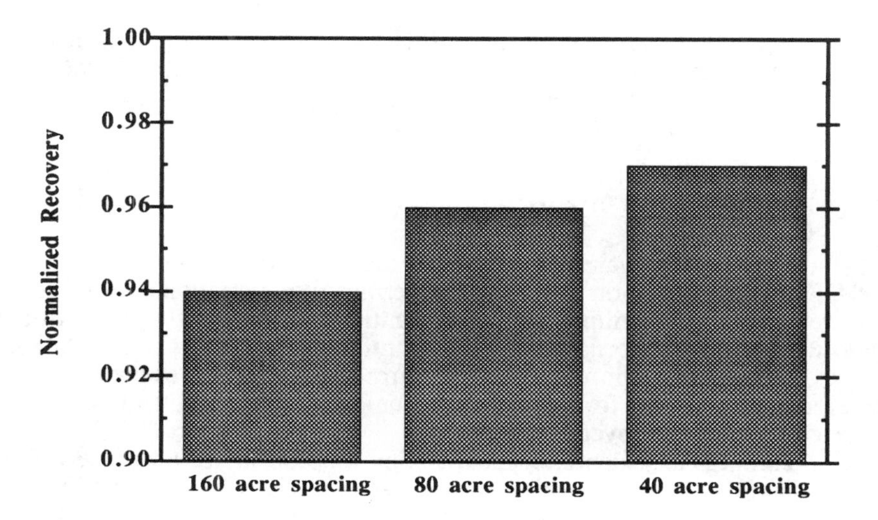

Figure 24 - Average recovery from conditional realizations.

VII. CONCLUSIONS

1) Recovery efficiencies from conditional simulation realizations are lower than smooth kriged map recovery efficiencies.

2) A correlation between recovery efficiency and thickness is observed. Thin beds (less than four feet) have lower recovery efficiency, whereas thick beds (more than seven feet) have higher recovery efficiency.

3) Reservoir performance predictions of the conditional simulation realizations for different well spacings show additional recovery benefits from infill drilling. Smaller well spacing gives higher recovery efficiency, agreeing with field experience. Other sand bodies and other areas of the field having thinner subunits will have even higher additional recovery from infill drilling.

4) The additional recovery predictions from infill drilling from 160-acre to 40-acre spacing is significantly higher for subunits less than five feet thick. The same trend is observed in the 160-acre to 80-acre case, except the additional recovery for subunits less than three feet is not as large. This is caused by the scale of heterogeneity generated by conditional simulations.

5) Geostatistics provides an effective framework to quantify the uncertainty in reservoir performance and allows us to develop an understanding of the additional recovery from infill drilling from different subunits.

VIII. FUTURE RESEARCH

The recovery variations and additional recoveries, predicted for the DS-1E area, are based on thickness and permeability variations. Additional recovery from infill drilling can be substantially higher in other parts of the field with higher proportions of thin discontinuous beds. Other correlations (e.g., permeability-porosity) may introduce a different nature of heterogeneity. Other forms of heterogeneities, such as faulting, can also increase additional recovery estimates.

Turning bands conditional simulation was considered adequate within the scope of the present investigation because it provides insight regarding the effect of heterogeneity vis-a-vis smooth maps. From geologic experience, these predictions show a level of heterogeneity believed to be more realistic than the kriged maps. Future work will also address the modern techniques of indicator simulation,[13] simulated annealing.[19,20] and scaleup techniques.[51]

IX. NOMENCLATURE

a	= range (ft)
C	= sill or variance
g(h)	= variance as a function of h
h	= distance between points (lag)
N(h)	= pairs of Z values separated by h
$Z(x_i)$	= Z value at point x_i
$Z(x_i+h)$	= Z value at distance h from point x_i

X. ACKNOWLEDGEMENTS

We thank ARCO and the Kuparuk Unit Co-owners for permission to publish this paper. Technical reviews from John Bolling, Jim Lorsong and Wesley Monroe were helpful and appreciated. We especially thank Mark Scheihing for providing his geologic interpretations. Thanks to Dave Bell, Windsong Fong and Gary Woodling for their help. Thanks are also due to SPE for granting use of the text and figures contained in SPE paper 22164, presented at the International Arctic Technology Conference, May 29-31, 1991 in Anchorage, AK.[35]

XI. REFERENCES

1. Barber, A.H., George, C.J., Stiles, L.H., and Thompson, B.B.:"Infill Drilling To Increase Reserves-Actual Experience in Nine Fields in Texas, Oklahoma, and Illinois," JPT (Aug. 1983) 1530-38.
2. Newn, K.T.C., Slik, P., and Tan, B.C.: "Infill Development in an Old Field," OSEA 88188, February 1988.
3. Gould, T.L., and Sarem, A.M.S.: "Infill Drilling for Incremental Recovery,"JPT (March 1989) 229-37.
4. Matheron, G.:"The Theory of Regionalized Variables and its Applications", Paris School of Mines, Cah. Cent. Morphologie Math., 5 Fontainebleau.
5. Ripley, B. D.: *Spatial Statistics,* John Wiley & Sons, New York City (1981).
6. Matheron, G.: "The Intrinsic Random Functions and Their Applications," *Advances in Applied Probability* (1973) **5**, 439-68.
7. Journel, A. G. and Huijbregts, C. J.: *Mining Geostatistics,* Academic Press (1978).
8. Isaaks, E. H. and Srivastava, R. M.: *Applied Geostatistics,* Oxford University Press, New York (1989).
9. Krige, D.G.: "A Statistical Analysis of Some of the Borehole Values in the Orange Free State Goldfield," J. Chem. Metall Min. Soc., South Africa, J3, 47-70.
10. Lorenz, E.N.: "Nonlinear Statistical Weather Predictions," paper presented at the 1980 World Meteorological Organization Symposium on Probabilistic and Statistical Methods in Weather Forecasting, Nice, France, Sept. 8 - 12.
11. Tsonis, A.A.: "Chaos and Unpredictability of Weather," Weather (June, 1989) 449 No. 6, 258-63.
12. Tennekes, H.: "Outlook: Scattered Showers," Bull., American Meteorological Soc. (April, 1988) 6g, No. 4, 368-72.
13. Journel, A. G. and Alabert, F. G.: "New Method for Reservoir Mapping," *JPT* (Feb. 1990) 212-18.
14. Journel, A. G. and Gomez-Hernandez, J. J.: "Stochastic Imaging of the Wilmington Clastic Sequence," paper SPE 19857 presented at the 1989 SPE Annual Technical Conference and Exhibition, San Antonio, Oct. 8-11.
15. Dubrule, O. and Haldorsen, H. H.: "Geostatistics for Permeability Estimation," *Reservoir Characterization*, L. W. Lake and H. B. Carrol Jr. (eds.), Academic Press, Orlando, FL (1986) 223-48.
16. Delfiner, P., Delhomme, J. P., and Pelissier-Combescure, J.: "Application of Geostatistical Analysis to the Evaluation of Petroleum Reservoirs with Well Logs," paper presented at the 1983 SPWLA Annual Logging Symposium, Calgary, July 27-30.
17. Hewett, T. A.: "Fractal Distributions of Reservoir Heterogeneity and Their Influence on Fluid Transport," paper SPE 15386 presented at the 1986 SPE Annual Technical Conference and Exhibition, New Orleans, Oct. 5-8.
18. Matthews, J. L., Emanuel, A. S., and Edwards, K. A.: "Fractal Methods Improve Mitsue Miscible Predictions," *JPT* (Nov. 1989) 1126-42.

19. Goggin, D. J. *et al.*: "Permeability Transects in Eolian Sands and Their Use in Generating Random Permeability Fields," paper SPE 19586 presented at the 1989 SPE Annual Technical Conference and Exhibition, San Antonio, Oct. 8-11.
20. Chopra, A. K., Stein, M. H., and Ader, J. C.: "Development of Reservoir Descriptions to Aid in Design of EOR Projects," SPERE (May 1989) 143-150.
21. Chopra, A. K., Stein, M. H., and Dismuke, C. T.: "Prediction of Performance of Miscible Gas Pilots," paper SPE 18078 presented at the 1988 SPE Annual Technical Conference and Exhibition, Houston, Oct. 2-5.
22. Chopra, A. K.: "Reservoir Descriptions Via Pulse Testing: A Technology Evaluation," paper SPE 17568 presented at the 1988 SPE International Meeting on Petroleum Engineering, Tianjin, China, Nov. 1-4.
23. Weber, K. J. and van Geuns, L. C.: "Framework for Constructing Clastic Reservoir Simulation Models," Paper SPE 19582 presented at the 1989 SPE Annual Technical Conference and Exhibition, San Antonio, Oct. 8-11.
24. Allen, J. R. L.: "Studies in Fluviatile Sedimentation: An Exploratory Quantitative Model for the Architecture of Avulsion-Controlled Alluvial Suites," *Sedimentary Geology* (1978) **21**, 129-47.
25. Knutson, C. F.: "Modeling of Noncontinuous Fort Union Mesaverde Sandstone Reservoirs, Piceance Basin, Northwestern Colorado," *SPEJ* (Aug. 1976) 175-88.
26. Bridge, J. S. and Leeder, M. R.: "A Simulation Model of Alluvial Stratigraphy," *Sedimentology* (1979) **26**, 617-44.
27. Augedal, H. O., Stanley, K. O., and Omre, H.: "SISABOSA, a Program for Stochastic Modelling and Evaluation of Reservoir Geology," Report SAND 18/86 presented at the 1986 Conference on Reservoir Description and Simulation with Emphasis on EOR, Oslo, Sept. 5-7.
28. Gundesf, R. and Egeland, O.: "SESIMIRA–A New Geological Tool for 3-D Modeling of Heterogeneous Reservoirs," paper presented at the 1989 Intl. Conference on North Sea Oil and Gas Reservoirs, Trondheim, May 8-11.
29. Wadsley, A. W., Erlandsen, S., and Geomans, H. W.: "HEX, A Tool for Integrated Fluvial Architecture Modeling and Numerical Simulation of Recovery Processes," paper presented at the 1989 Intl. Conference on North Sea Oil and Gas Reservoirs, Trondheim, May 8-11.
30. King, P. R.: "The Connectivity and Conductivity of Overlapping Sandbodies," paper presented at the 1989 Intl. Conference on North Sea Oil and Gas Reservoirs, Trondheim, May 8-10.
31. Chopra, A. K., Severson, C. D. and Carhart, S. R., "Evaluation of Geostatistical Techniques for Reservoir Characterization," SPE paper 20734 presented at the 65th Annual Technical Conference held in New Orleans, LA., Sept. 23-26, 1990.
32. Scheihing, M.H.: "Reservoir Anatomy of the Kuparuk A4-Sandstone Body, Kuparuk Field, North Slope of Alaska," Internal ARCO Report, AOGC, 1988.

33. Masterson, W. D. and Paris, C. E.: "Depositional History and Reservoir Description of the Kuparuk River Formation, North Slope, Alaska," Alaskan North Slope Geology, Pacific Section SEPM, Guidebook 50, Tailleur, I. and Weimer, P. (Eds.), 1987.
34. Gaynor, Gerard C., and Scheihing, Mark H.:"Shelf Depositional Environments and Reservoir Characteristics of the Kuparuk River Formation (Lower Cretaceous), Kuparuk Field, North Slope, Alaska," Giant Oil and Gas Fields: A Core Work Shop, Lamando, A. J. and Harris, P. M., eds., SEPM Core Work Shop #12 V. 1, March, 1988, 333-389.
35. Wolcott, D.S. and Chopra, A. K.; "Incorporating Reservoir Heterogeneity Using Geostatistics to Investigate Waterflood Recoveries for Drillsite 1E, A4 Sandstone Body, Kuparuk River Field, Alaska," paper SPE 22164 presented at the International Arctic Technology Conference held in Anchorage, AK., May 29-31, 1991.
36. Jacobsen, T. and Rendall H.: "Permeability Patterns In Some Fluvial Sandstones. An Outcrop Study From Yorkshire, North East England." Continental Shelf and Petroleum Technology research Institute A/S, IKU (1989), No. 89.055.
37. Clark, Isobel: *Practical Geostatistics*, Elsevier Publishers, Essex,(1984).
38. Atkinson, C.D., Goesten B.G., Speksnijder A., and Van der Vlugt W: "Storm-generated sandstone in the Miocene Miri Formation, Seria Field, Brunei (N.W. Borneo)," R. J. Knight and J. R. McLean, (eds.), Shelf Sands and Sandstones: Can. Soc. Petrol. Geol. Memoir 11 (1986), 213-240.
39. Kittridge, M. G. *et al*: "Outcrop/Subsurface Comparisons of Heterogeneity in the San Andres Formation," paper SPE 19596 presented at the 1989 SPE Annual Technical Conference and Exhibition, San Antonio, Oct. 8-11.
40. Guerrillot, D. *et al.*: "An Integrated Model for Computer Aided Reservoir Description: From Outcrop Study to Fluid Flow Simulations," paper presented at the 1989 European Symposium on Improved Oil Recovery, Budapest, April 25-27.
41. Robertson, J. D.: "Reservoir Management Using 3-D Seismic Data," *The Leading Edge* (Feb. 1989) **8**, No. 2, 25-32.
42. Nolen-Hoeksema, R. C.: "Future Role of Geophysics in Reservoir Engineering," paper presented at the 1988 Joint SEG/EAEG Research Workshop on Reservoir Geophysics, Dallas, July 31-Aug. 3.
43. Doyen, P. M., Guidish, T. M., and de Buyl, M. H.: "Monte Carlo Simulation of Lithology From Seismic Data in a Channel-Sand Reservoir," paper SPE 19588 presented at the 1989 SPE Annual Technical Conference and Exhibition, San Antonio, Oct. 8-11.
44. Martinez, R. D., Cornish, B. E., and Seymour, R. H.: "An Integrated Approach for Reservoir Description Using Seismic, Borehole, and Geologic Data," paper SPE 19581 presented at the 1989 SPE Annual Technical Conference and Exhibition, San Antonio, Oct. 8-11.
45. Alabert, F. G.: "Constraining Description of Randomly Heterogeneous Reservoirs to Pressure Test Data: A Monte Carlo Study," paper SPE 19600 presented at the 1989 SPE Annual Technical Conference and Exhibition, San Antonio, Oct. 8-11.

46. Delhomme, J. P.and Chiles, J. P., "Conditional Simulation Package for BLUEPACK Geostatistics System (SIMPAC)," Centre De Geostatistique et de Morphologie Mathematique, Fontainbleu, France, 1976.

47. Gomez-Hernandez and Journel, A. G.: “Stochastic Characterization of Grid-Block Permeabilities: From Point Values to Block Tensors,” paper presented at the 1990 2nd European Conference on the mathematics of Oil Recovery, Arles, France, Sept. 11-14.

APPENDIX

The semi-variograms were calculated by:

$$\gamma(h) = \frac{\Sigma[Z(x_i) - Z(x_i + h)]^2}{2 * N(h)} \tag{1}$$

The spherical models were calculated by:

$$\begin{aligned} \gamma(h) &= C * \left(\frac{3h}{2a} - \frac{h^3}{2a^3} \right), & h < a \\ &= C, & h \geq a \end{aligned} \tag{2}$$

A COMPARISON OF OUTCROP AND SUBSURFACE GEOLOGIC CHARACTERISTICS AND FLUID-FLOW PROPERTIES IN THE LOWER CRETACEOUS MUDDY J SANDSTONE

David F. Mayer[1]

Department of Petroleum Engineering
Colorado School of Mines
Golden, Colorado

Mark A. Chapin[2]

Department of Geology and Geological Engineering
Colorado School of Mines
Golden, Colorado

I. INTRODUCTION

Outcrop studies have been made to help define the external geometry of reservoirs, their internal architecture, and the distribution of fluid-flow properties between wells. In these types of studies analogous outcrops in the same formation or in similar depositional environments as producing fields are used .

This paper compares characteristics of the Muddy (J) Sandstone observed in outcrops and in cores from the Peoria Field in Colorado. These comparisons were made to determine if the outcrops were analogous to equivalent strata in the subsurface and if outcrop observations and measurements could assist in the construction of fluid-flow models.

---

[1]Present address: Evans, Carey, and Crozier, Bakersfield, California.

[2]Present address: Shell Offshore, Inc., New Orleans, Louisiana.

II. STRATIGRAPHY OF THE J SANDSTONE IN THE DENVER BASIN

This study compares reservoir properties and geologic characteristics of outcrops and subsurface reservoir rocks from the J Sandstone which is the upper most unit of the Dakota Group in the Denver Basin in Colorado. The Lower Cretaceous Dakota Group was deposited over the Jurassic Morrison Formation and consists of both marine and nonmarine rocks (Figure 1).

Geologic Age		Stages	Stratigraphic Units: Surface Outcrops (Front Range)			Stratigraphic Units: Subsurface (Denver Basin)	
Cretaceous	Upper Cretaceous		Niobrara Formation			Niobrara Formation	
Cretaceous	Upper Cretaceous	Turonian	Benton Group	Carlile Shale		Benton Group	Carlile Shale
Cretaceous	Upper Cretaceous	Cenomanian	Benton Group	Greenhorn Limestone		Benton Group	Greenhorn Limestone
Cretaceous	Upper Cretaceous	Cenomanian	Benton Group	Graneros Shale		Benton Group	Graneros Shale
Cretaceous	Upper Cretaceous					Benton Group	"D" Sandstone
Cretaceous	Upper Cretaceous	Albian	Benton Group	Mowry Shale		Benton Group	Huntsman Shale
Cretaceous	Lower Cretaceous	Albian	Dakota Group	South Platte Formation	Muddy Sandstone	Dakota Group	"J" Sandstone
Cretaceous	Lower Cretaceous	Albian	Dakota Group	South Platte Formation	Skull Creek Shale	Dakota Group	Skull Creek Shale
Cretaceous	Lower Cretaceous	Albian	Dakota Group	South Platte Formation	Plainview Sandstone	Dakota Group	Fall River Formation
Cretaceous	Lower Cretaceous	Aptian	Dakota Group	Lytle Formation		Dakota Group	Lakota Formation
Jurassic			Morrison Formation			Morrison Formation	

Fig. 1. Surface and subsurface stratigraphic units of the Denver Basin and Front Range outcrops.

A. Fort Collins Member

The Skull Creek Shale of the Dakota Group was deposited in the Cretaceous Western Interior Seaway in the Peoria Field area and on its margins as seen in outcrops near Morrison, Colorado (Figure 2) (1). The Skull Creek Shale contact with the overlying Muddy (J) Sandstone is transitional and indicates a rapid regression of the sea (2). This lower portion of the Muddy (J) Sandstone has been called the Fort Collins Member or the J-3 unit in the Denver Basin (3). The J-3 unit includes the Fort Collins Member and also floodplain and crevasse splay deposits above the regional unconformity but below channel sands (J-2 interval) in the Horsetooth Member discussed below.

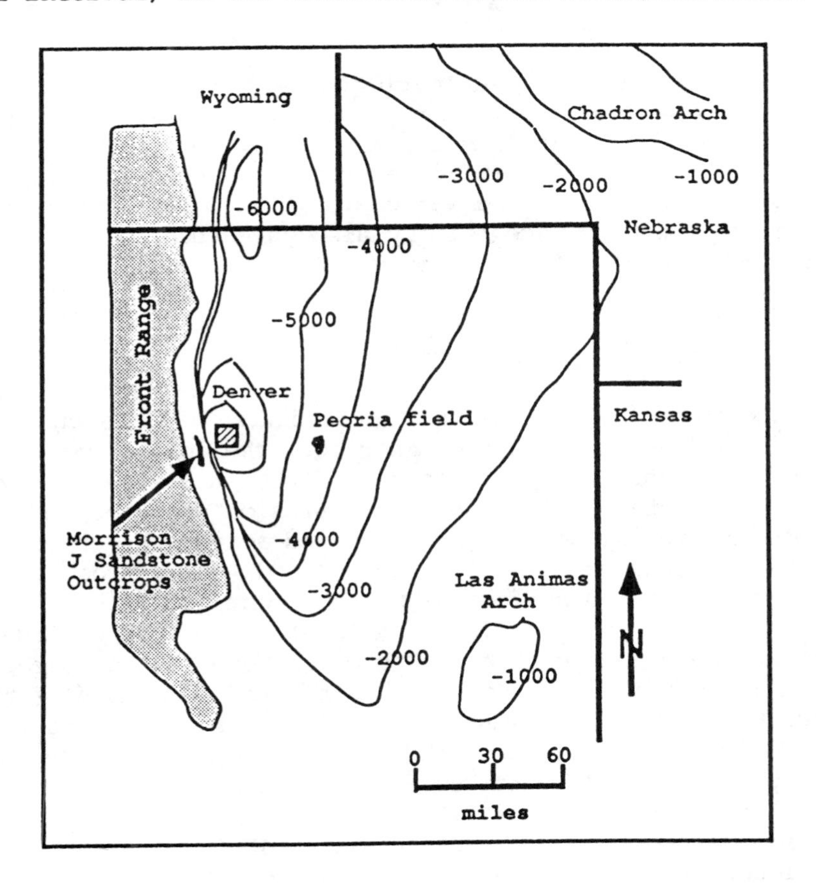

Fig. 2. Structure contours on top of the Precambrian (modified from Matuszczak (4)).

The Fort Collins Member is composed of an upward coarsening, bioturbated sandstone which has been interpreted to be a delta front facies deposited as a result of the regression of the Western Interior Sea (5). These marine sandstone deposits are truncated by a regional unconformity caused by the continued drop in sea level which allowed erosion of marine sediments (3). Basement block movement formed topographic lows that created drainage paths to the regressing sea which incised valleys into the underlying strata (3). The Fort Collins Member was partially removed from higher elevations and in some cases was totally removed from the valleys.

B. Horsetooth Member

The rise in sea level which followed caused backfilling of the incised valleys with floodplain, crevasse splay, distributary channel, fluvial, estuarine, and overbank deposits. These valley-fill deposits have been referred to as the J-2 unit in the Denver Basin (6). Strata deposited above the lowstand unconformity also have been called the Horsetooth Member of the Muddy (J) Sandstone (3).

C. Transgressive Marine Deposits

The continued rise in sea level filled valleys and deposited coastal plain sediments over the entire region. This transgression of the sea caused the shoreline to move over the coastal plain. Shoreface erosion occurred and a transgressive surface of erosion moved across the coastal plain sediments (3). A bioturbated sandstone was deposited above this erosion surface as the sea continued its transgression. This sand, which is at the top of the Muddy (J) Sandstone and the Dakota Group, is referred to as the J-1 unit in the Denver Basin. Figure 3 is a summary of the sequence of sediments which resulted from these events (7).

As the sea level continued to rise marine silt and mud (Mowry Shale and Huntsman Shale of the Benton Group) were deposited over the Muddy (J) Sandstone. Deposition and subsidence caused burial of the Muddy (J) Sandstone to continue through the late Cretaceous and into the Cenozoic when the Laramide Orogeny ended burial and deformed the Denver Basin to its current condition. The uplift allowed erosion of overlying strata exposing sediments of the Lower Cretaceous along the western edge of the Denver Basin (3).

The Peoria Field produces from fluvial channel and splay sands in the J-2 unit of the Muddy (J) Sandstone in the Denver Basin (7). The J-1, J-2, and J-3 units have not been defined in outcrops. However, the rocks studied in outcrop are channel sands deposited on an erosion surface cut into the Skull Creek Shale (1). The characteristics of these channel sand bodies are compared below.

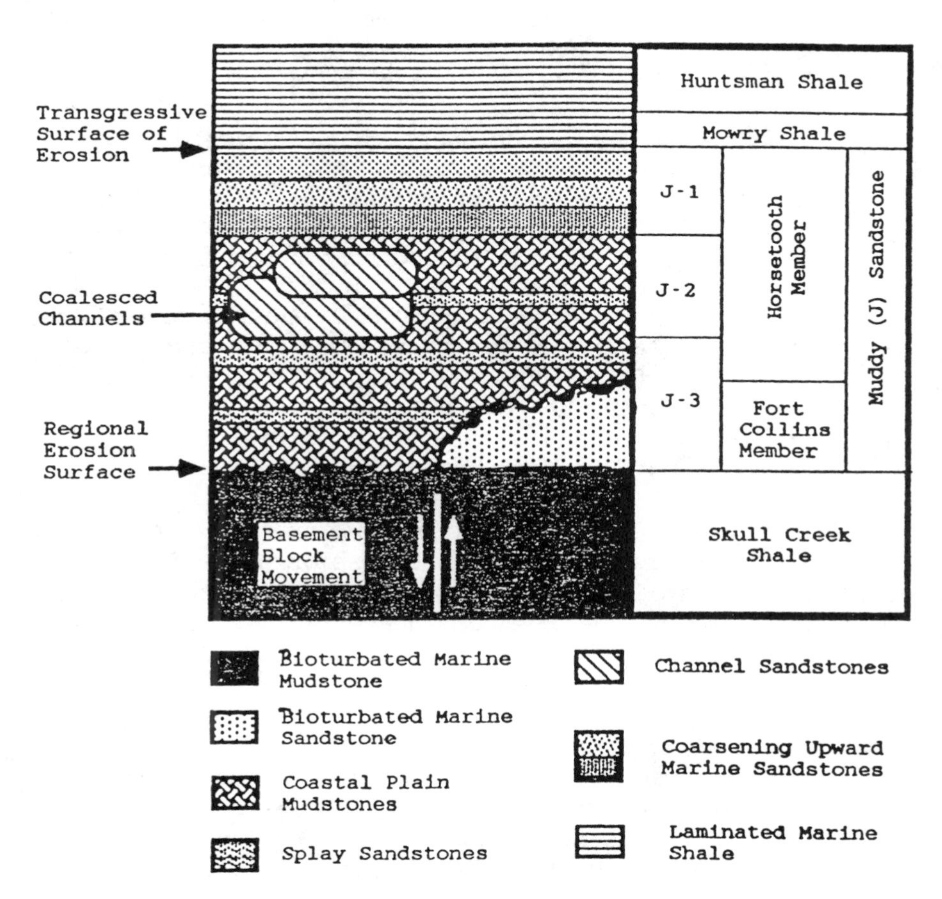

Fig. 3. Muddy (J) Sandstone stratigraphy in the Denver Basin (modified from Chapin (7)).

III. A COMPARISON OF LITHOFACIES IN THE PEORIA FIELD AND IN OUTCROPS

Cores and outcrop sections were described and small-scale lithologic units with associated grain sizes, clay content, and biological and physical sedimentary structures were grouped together into facies. Five facies were distinguished from these descriptions of which two were reservoir facies (7). Similar facies were found in outcrop sections and near-surface cores and in cores from the Peoria Field. These five facies are described below.

1. Crevasse splay and splay channels composed of coarsening-up units of planar and rippled laminae, with thin, trough cross-stratified beds at the top. This unit is productive in several wells in the Peoria Field but has not contributed a significant amount of production.
2. Fine-grained vertical accretion deposits made up of parallel-laminated and convolute-laminated overbank mudstones from marshes, lakes, and brackish bays were found in the Peoria Field area. Outcrop overbank facies are composed of tidal-flat sandstones and mudstones.

Three facies were used to describe point bar deposits in the Peoria Field. This same succession of facies was observed in channels found in outcrops. These point bar and channel facies have been designated Facies I, II, and III with the numbers indicating their succession from the base to the top of a channel.

3. Facies I consists of trough cross-stratified and massive sandstone and a minor amount of planar-laminated sandstone. Trough cross-stratified structures usually range from .5 to 1.5 feet in thickness in Peoria Field cores. Facies I represents the basal channel fill of a point bar. Deposits above basal scours commonly consist of mudclast-rich sandstones made up of clay rip-up clasts or mudstone break-up structures. Lateral accretion surfaces with mudstone drapes separating trough cross-stratified sandstone units have been observed in several wells. Facies I in outcrop is

comprised mainly of cross-stratified structures, both trough and planar-tabular sets. Large sand bars also can be found in outcrops and are included with cross-stratified structures in the comparisons which follow. This facies contains the best reservoir rock and the largest grain sizes.

4. Ripple-laminated sandstone is the predominant stratification type found in Facies II in the Peoria Field. Mudstone draped ripple laminations as well as thin, trough cross-stratified and massive beds also can be found in this facies. Facies II typically consists of lower energy sediments deposited over Facies I. These are deposits of the middle to upper point bar, but Facies II also can occur as the lowest facies in a channel above the scour surface in partial abandonment fills. In outcrop Facies II contains more trough cross-stratified sandstone than in the Peoria field. Facies II also is considered to contain reservoir sands but these are of poorer quality than those of Facies I.
5. Facies III consists mainly of ripple-laminated sandstone which is finer grained than that found in Facies II and contains a much greater quantity of mud drapes. Wavy or flaser bedding and convolute bedding also can be found. This facies can be present either as a thin interval at the top of point bar deposits or as a thick, abandoned channel fill (7).

Over 200 sieve analyses were performed on outcrop and Peoria Field samples. Most of the samples ranged from fine-grained to very-fine-grained sand. Several samples from Facies I were lower-medium grained and vertical accretion deposits were made up of silt-sized grains.

The average of the mean grain diameters of Facies I samples from the Peoria Field was found to be 2.63 phi units and the average Trask sorting coefficient was 1.14 (lower-fine grained, very-well-sorted sand). The average Facies II sample was well-sorted, upper-very-fine-grained sand with an average Trask sorting coefficient of 1.24 and a mean grain diameter of 3.10 phi units. Facies III samples were well sorted, lower-very-fine-grained sand.

Peoria Field samples from Facies I were finer grained and better sorted than those from outcrop core plugs. The average Facies I outcrop sample was composed of well-sorted, upper-fine-grained sand with a Trask sorting coefficient of 1.21 and an average grain diameter of 2.40 phi units. Outcrop samples from Facies II were found to have a mean grain diamater of 3.14 phi units and a Trask sorting coefficient of 1.26 making the average sample well-sorted, upper-very-fine-grained sand.

The differences in the mean grain diameters between the facies exhibits an upward-fining pattern. This upward-fining pattern within a channel was found to occur in steps between facies but within facies grain size changes were erratic with intervals of larger or smaller grain sizes dispersed through a generally upward-fining trend. Within Facies I it was common to find smaller grain sizes immediately above the basal scour surface.

Chapin determined the percentage of each facies within channelbelts in the Peoria Field and in outcrops (7). This comparison is shown in Figure 4.

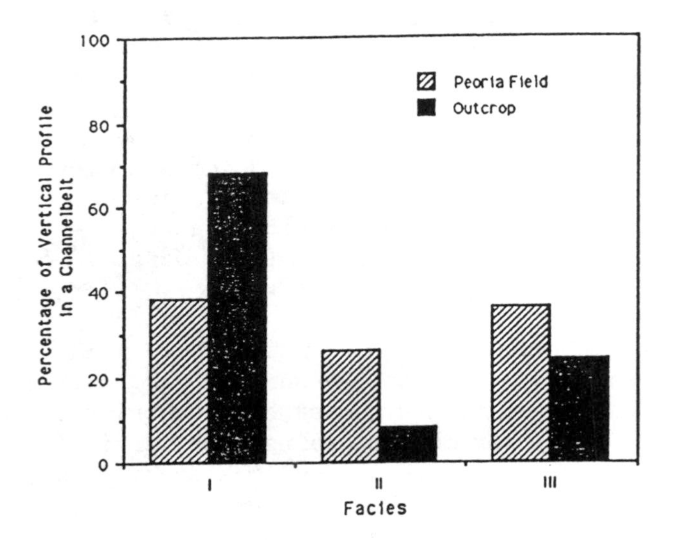

Fig. 4. Distribution of facies within channelbelts (from Chapin (7)).

IV. COMPARISON OF EXTERNAL SANDBODY GEOMETRIES

Genetic units have distinct internal arrangements of facies and characteristic dimensions and geometries. This enables the distribution of facies identified in cores or outcrops to be used to infer the geometry of a sedimentary body. Comparisons of the distribution of stratification types within similar facies, of the proportion of facies within channelbelts, and the nature of overbank deposits between outcrops and the Peoria Field indicate differences in stream geometries (7).

Sandstone bodies observed in outcrop exhibit both downstream accretion in the form of large cross-bed sets with steep foresets deposited as bars within the channels and lateral accretion identified by the existence of epsilon cross-bedding with a upward-fining sequence occurring in steps. Deposits outside the channels contain tidal-flat sandstones and mudstones which indicate the channels were located in a nearshore environment. These facies relationships indicate Facies I, II, and III in outcrop were probably deposited by low-sinuousity distributary channels (7). Others have found that low-sinuousity channels can contain downstream-accreting, longitudinal and transverse bars (5) and also point bars (8).

Cores from the Peoria Field were found to have a fining-up sequence of sediments occurring in steps, a decrease in the scale of the sedimentary structures moving up through a facies, and a thicker and more developed Facies II sandstone than the outcrop sections. These stepwise changes were abrupt and could be identified by distinct changes in grain size, permeability, and stratification types. The facies forming the steps could be correlated between wells when the wells penetrated the same channel.

Outcrops contain less Facies II than the Peoria Field and include more cross-stratified sandstone in this unit. The overbank deposits in the Peoria Field do not contain evidence of a nearshore environment and mainly consist of coarsening-up vertical accretion to crevasse splay sequences. Large cross-bed sets and planar-tabular cross-stratification were not observed in the Peoria Field cores. The sandstone from Facies I, II, and III in the Peoria

Field were probably deposited by moderate- to high-sinuousity meandering streams of the coastal plain (7).

Within the Peoria Field two levels of meanderbelt sandbodies were found. These two levels of coalesced meanderbelts exhibited varying degrees of interconnectedness (7). Channels and the distribution of facies within the channels were mappable sandbodies in the Peoria Field (7).

V. DIAGENETIC CHANGES AND THE EFFECT OF CONFINING STRESS

Quartz overgrowths, authigenic kaolinite, and feldspar dissolution creating secondary porosity are the most prominent diagenetic features found in samples from the Peoria Field and from outcrops. Samples from both locations showed similar diagenetic features when observed using a scanning electron microscope.

Both surface and subsurface rocks from Facies I and II are texturally mature, quartz-rich sandstones. Quartz overgrowths are the most common cement and cause a significant reduction of the original porosity. Kaolinite which occurs as a pore-filling clay was found to be the most abundant authigenic clay making up approximately 3 to 4 percent of the rock (9). Other clays present in small amounts in Peoria Field samples include illite, smectite, and mixed-layer clays (9). Outcrop samples contain small amounts of illite and pyrite (10).

The dissolution of feldspars has caused the formation of secondary porosity. Thin sections from outcrop samples generally exhibited greater amounts of kaolinite and secondary porosity than Peoria Field samples (7). This is probably due to fresh water moving through the outcrops.

The initial burial history of both outcrop and Peoria Field rocks were similar but the Laramide Orogeny and subsequent erosion of the overburden exposed the Muddy (J) Sandstone along the western edge of the Denver Basin. Because of their similar initial burial histories, diagenetic modifications, and cementing, outcrop and subsurface reservoir rocks were believed to have similar compressibility characteristics. This was confirmed by measurements of pore volume and permeability on outcrop and

Peoria Field samples at different values of confining stress. Figure 5a and b show that similar changes occur to surface and subsurface samples with the application of confining stress.

(a)

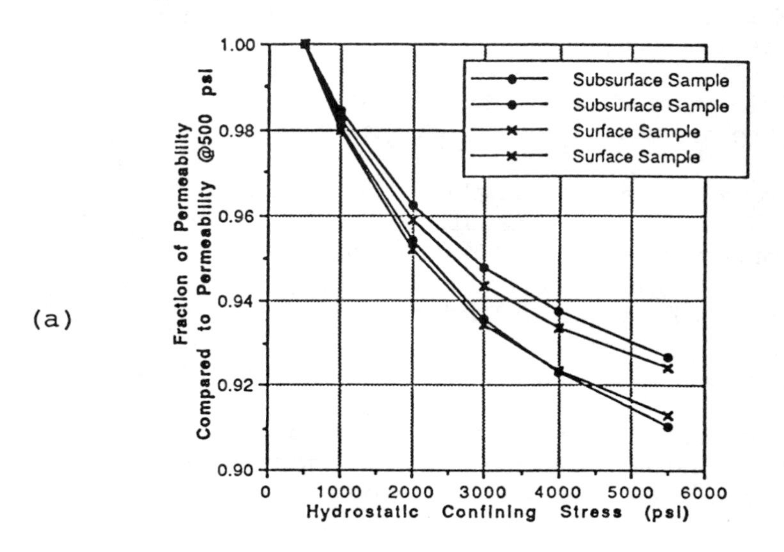

(b)

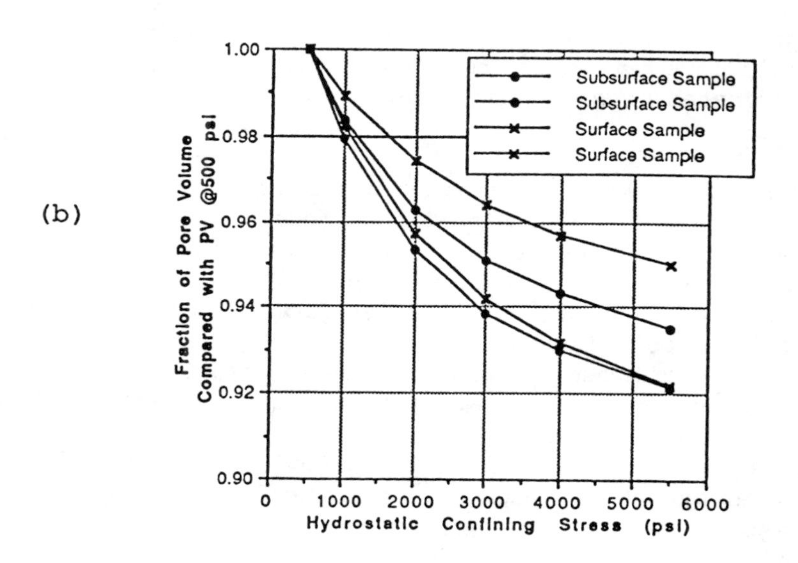

Fig. 5. (a) Permeability versus net stress. (b) Pore Volume versus net stress.

VI. PORE TYPES IN OUTCROP AND SUBSURFACE ROCKS

Based on sieve analysis, mercury injection curves, and air permeabilities it was found that the two reservoir facies contained six pore types, three within Facies I and three within Facies II. These pore types are associated with specific stratification types. The following list describes the six pore types which were found:

1) Lower-medium-grained through upper-fine-grained trough and planar-tabular cross-bedded strata and horizontal-laminated beds within Facies I with grain sizes ranging from 1.8 to 2.4 phi units.
2) Massive beds within Facies I.
3) Lower-fine-grained, cross-bedded strata within Facies I with grain sizes between 2.4 and 2.9 phi units.
4) Cross-bedded and massive beds within Facies II with grain sizes between 2.9 and 3.1 phi units.
5) Ripple-laminated beds which do not have mud drapes. This pore type from Facies II was found to have mean grain sizes between 3.0 and 3.4 phi units.
6) Ripple-laminated beds with mud drapes and mean grain sizes between 3.4 through 3.8 phi units. This pore type also includes wavy and convolute laminated beds and mud break-up beds within Facies II.

Figures 6a and b show mercury injection curves from outcrop and Peoria field samples for different stratification types. Mercury injection curves measured on outcrop samples were found to have wider pore throat distributions than Peoria Field samples from the same stratification type. In most cases the outcrop samples also had lower displacement pressures than subsurface samples. Because grain sizes and sorting factors were similar for surface and subsurface samples the wider distribution of pore throat sizes is believed to be caused by the greater amount of secondary porosity found in outcrop samples.

Channels and facies within the channels were mappable but the vertical and areal distribution of pore types within each facies could not be mapped with any degree of certainty. For this case geostatistical techniques are required to map the distribution of pore types within each facies.

(a)

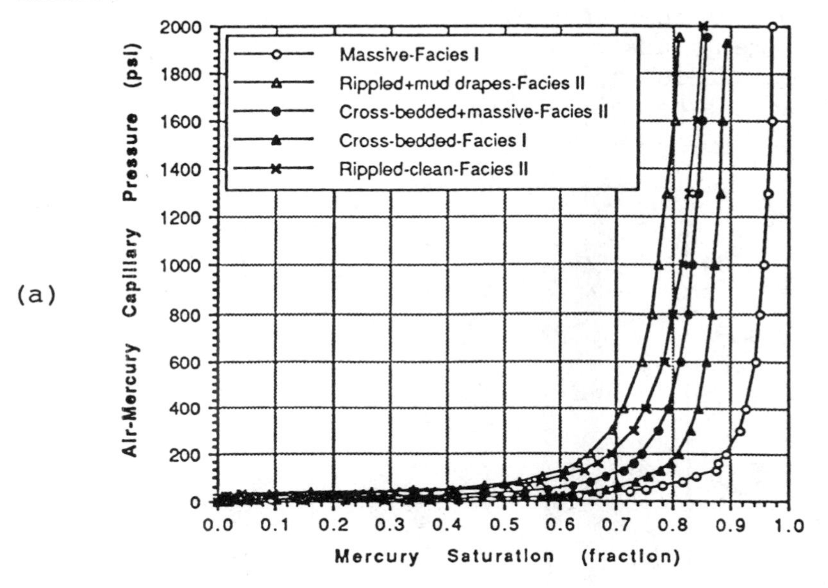

(b)

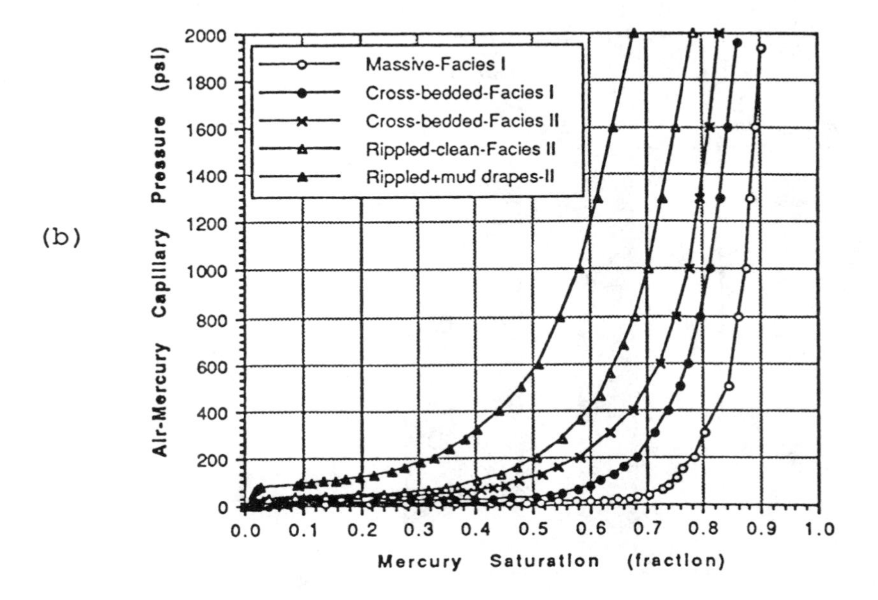

Fig. 6. (a) Mercury injection curves measured on subsurface samples. (b) Mercury injection curves measured on outcrop samples.

Figures 7a and b show the distribution of the pore types found in outcrops and in Peoria Field cores. Outcrops contained greater amounts of coarser-grained, cross-bedded sandstone in Facies I and more cross-bedded sandstone within Facies II than the Peoria Field. Peoria Field samples contained more ripple laminated sandstone in Facies II than outcrops.

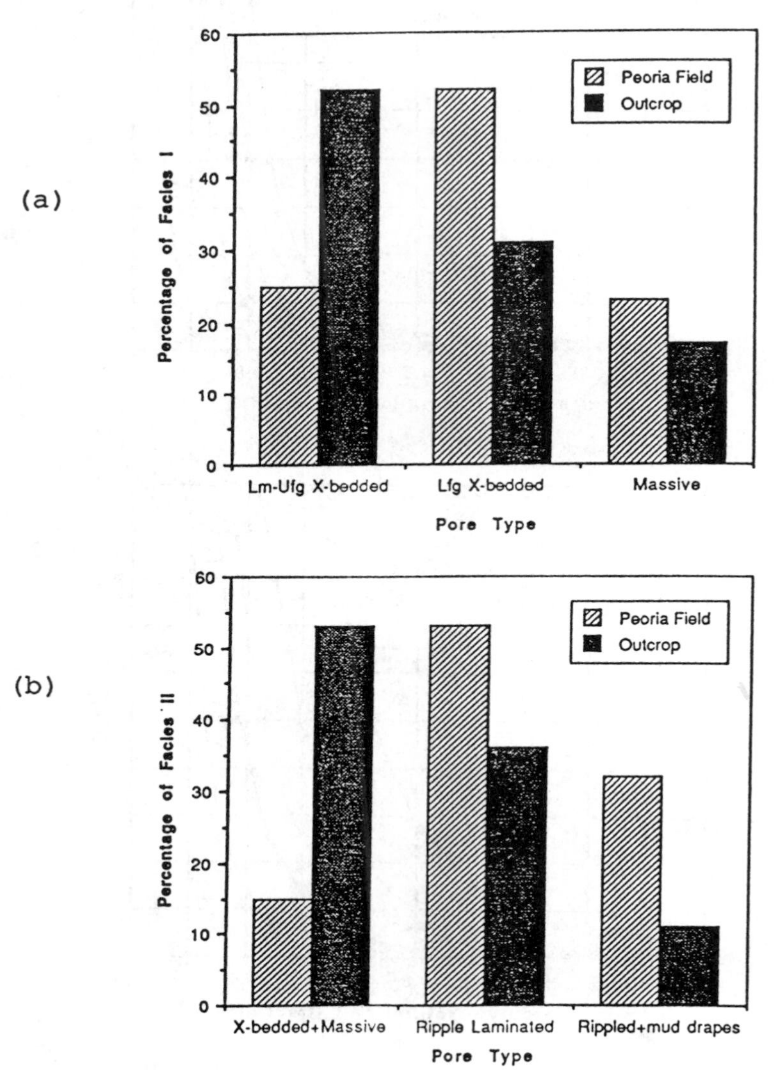

Fig. 7. (a) Distribution of pore types within Facies I. (b) Distribution of pore types Facies II.

VII. PERMEABILITY CHARACTERISTICS

Porosity and permeability values measured on outcrop samples from Facies I and II were greater than those from the Peoria Field. Permeability-porosity crossplots using core plug data from outcrop and Peoria Field samples are shown in Figure 8a and b. Data from each facies are identified on these graphs. These data indicate that each of the reservoir facies is associated with a different permeability population. Histograms of these populations show permeability values for each facies to be lognormally distributed and porosity to be normally distributed.

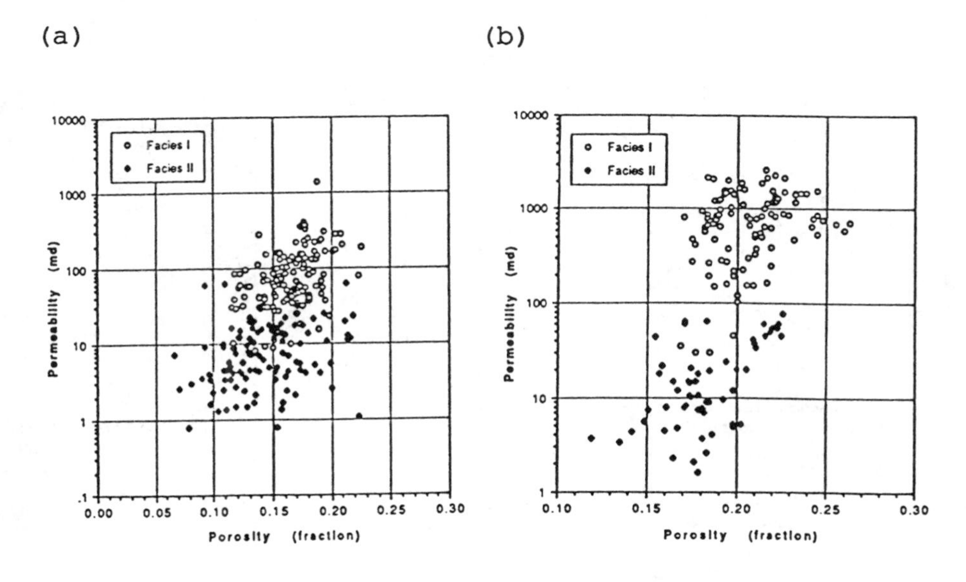

Fig. 8. (a) Permeability-porosity crossplot of Peoria Field data. (b) Crossplot of outcrop data.

Minipermeameter measurements were made on cores from nine wells and core plug data were measured on cores from twenty wells in the Peoria Field. Figure 9 shows the distribution of facies and pore types and their corresponding permeabilities in Peoria Unit #67. A Dykstra-Parsons plot of minipermeameter data from this well is shown in Figure 10. A Dykstra-Parsons coefficient computed from the best-fit straight line representing all the permeability data for this well was equal to .86 indicating a heterogeneous reservoir. However, the data shown in Figure 10 do not fall on a single line; instead, several straight-line segments can be identified.

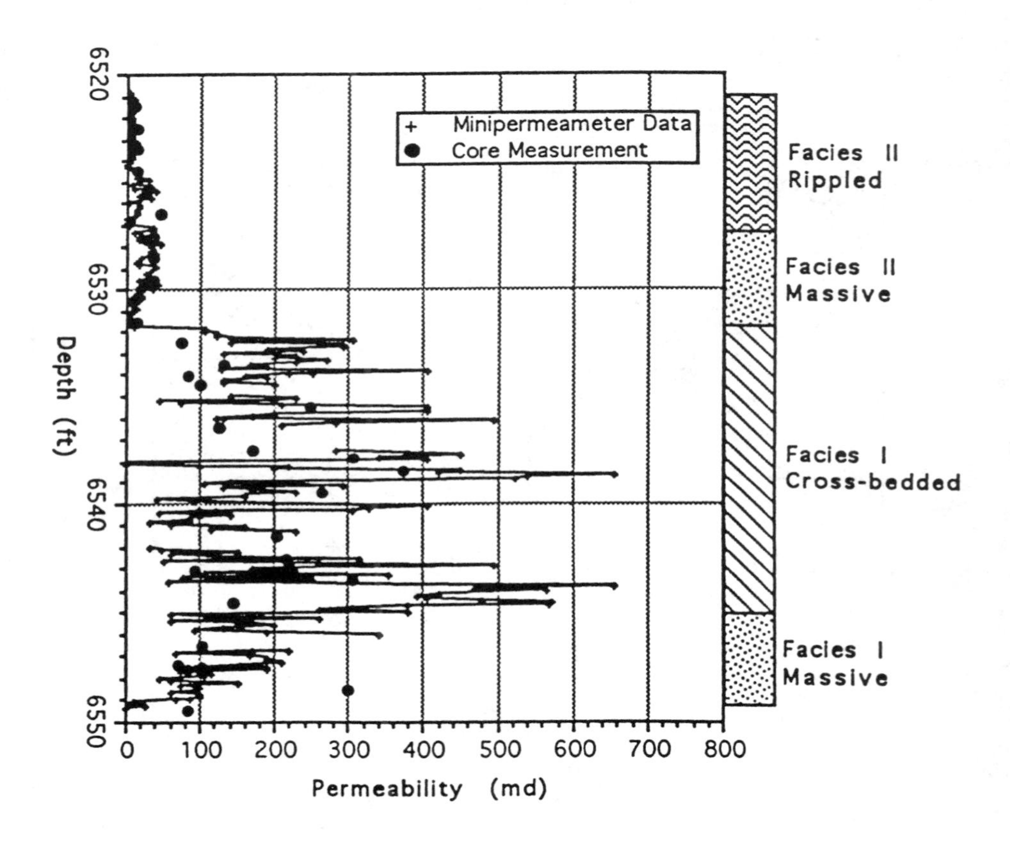

Fig. 9. Distribution of permeability, pore types, and facies in Peoria Unit #67.

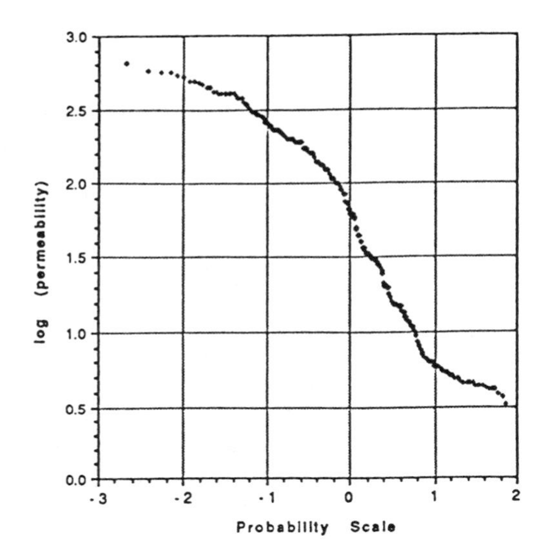

Fig. 10. Dykstra-Parsons plot of minipermeameter data from Peoria Unit #67.

In order to determine if permeability was related to smaller reservoir units the permeability data first were classified by facies and then by pore type. Each of the straight-line segments observed in Figure 10 was found to be associated with a particular facies or pore type. All of the sampled wells exhibited this behavior. Population statistics were computed and Dykstra-Parsons plots constructed for each facies and pore type present in each sampled well. Histograms of permeability data from each pore type were lognormal distributions. Core plug permeability values from Facies I and II were used to construct the Dykstra-Parsons plot shown in Figure 11. Only 40 permeability measurements were available for this graph. From this plot it is difficult to detect the existence of different permeability populations. The greater number of permeability measurements obtained using a minipermeameter enabled more detailed features to be identified on the Dykstra-Parsons plots.

The Dykstra-Parsons coefficients and geometric mean permeabilities computed for each facies and pore type in this well are shown below:

	Dykstra-Parsons Coefficient	Geometric Mean Permeability (md)
1. All core plug permeability data =	.74	51.
2. All miniperm data =	.86	53.
a. Facies I =	.58	163.
i. Lower-fine-grained cross-bedded =	.47	192.
ii. Massive =	.43	87.
b. Facies II =	.62	11.4
i. Massive =	.37	23.7
ii. Ripple laminated =	.51	8.0

From these data it can be seen that Peoria Unit #67 is composed of two permeability populations each related to a reservoir facies. Each facies is more homogeneous than the total reservoir thickness. The large difference in permeabilities between Facies I and II is due to the difference in grain size between these two facies.

The permeability population of each facies contains several smaller permeability populations which are related to different pore types. Pore types are more homogeneous than the facies in which they are found. Differences in permeability between different pore types are believed to be caused by differences in packing and grain sizes in different stratification types. These observations indicate that a single Dykstra-Parsons coefficient can be a misleading indication of heterogeneity when a reservoir is composed of distinct zones with different permeability populations.

A Dykstra-Parsons plot of 165 core plug permeabilities taken from outcrop samples is shown in Figure 12. Facies and pore types in outcrops exhibited similar characteristics to those observed in the Peoria Field. The average permeability of each pore type was greater in outcrops than in the Peoria Field.

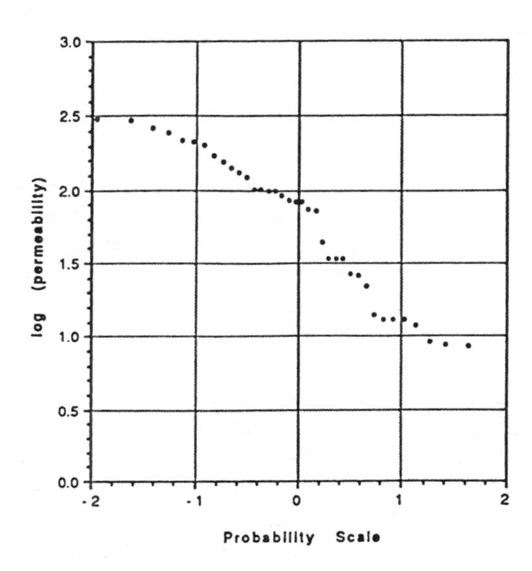

Fig. 11. Dykstra-Parsons plot of core plug permeabilities from Peoria Unit #67.

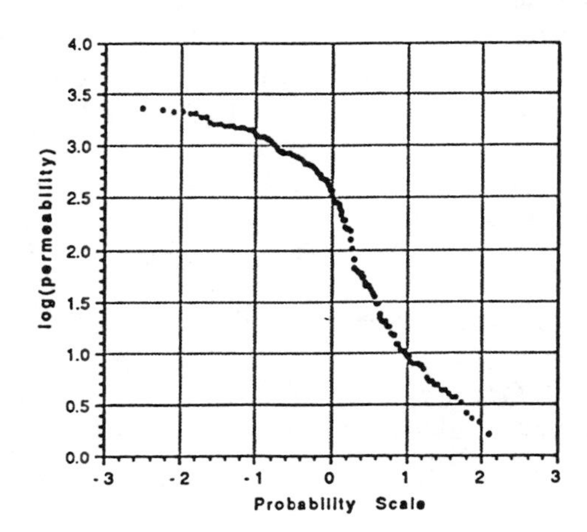

Fig. 12. Dykstra-Parsons plot of outcrop permeability data from core plug measurements.

VIII. CAPILLARY PRESSURE AND RELATIVE PERMEABILITY CHARACTERISTICS

Restored-state, oil-water relative permeability and capillary pressure measurements were made on outcrop and Peoria Field samples from each pore type. Oil-water capillary pressure curves measured during primary drainage are displayed in Figure 13a and b. These measurements were performed at reservoir temperature using a synthetic brine and Peoria Field crude oil in a centrifuge. The oil-water capillary pressure curves do not follow the same patterns observed in mercury injection measurements.

The initial water saturations from Peoria Field samples ranged from 17 to 30 percent. Capillary pressure data were not measured on a ripple-laminated sample with mud drapes from the Peoria Field. However, the other Peoria Field curves indicate this pore type probably would contain oil but at a high water saturation. Outcrop samples had a much broader range of initial water saturation values. Both of the ripple-laminated pore types from outcrops exhibited very large initial water saturations and probably would not be productive in the subsurface.

Forced imbibition and secondary drainage curves provided data to compute the USBM wettability index of each core plug (11). This index varies from +1.0 for strongly water-wet rock to -1.0 for strongly oil-wet samples. The wettability index measured for each pore type is listed below:

1. Peoria Field samples
   a. Facies I
      - Massive = .1932
      - Cross-bedded = -.0043
   b. Facies II
      - Massive = -.2151
      - Ripple laminated = -.4424
2. Outcrop samples
   a. Facies I
      - Massive 1.153
      - Cross-bedded = -.0259
   b. Facies II
      - Cross-bedded = .180
      - Ripple laminated (no mud drapes) = .127

(a)

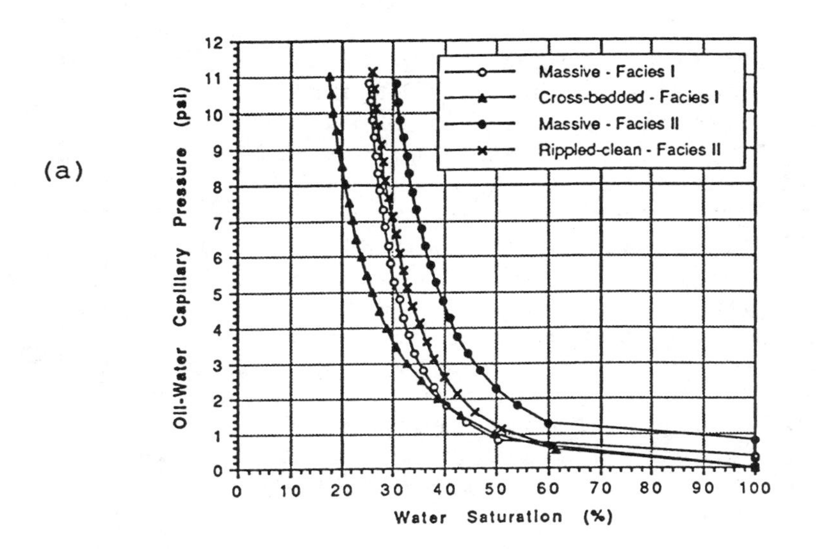

(b)

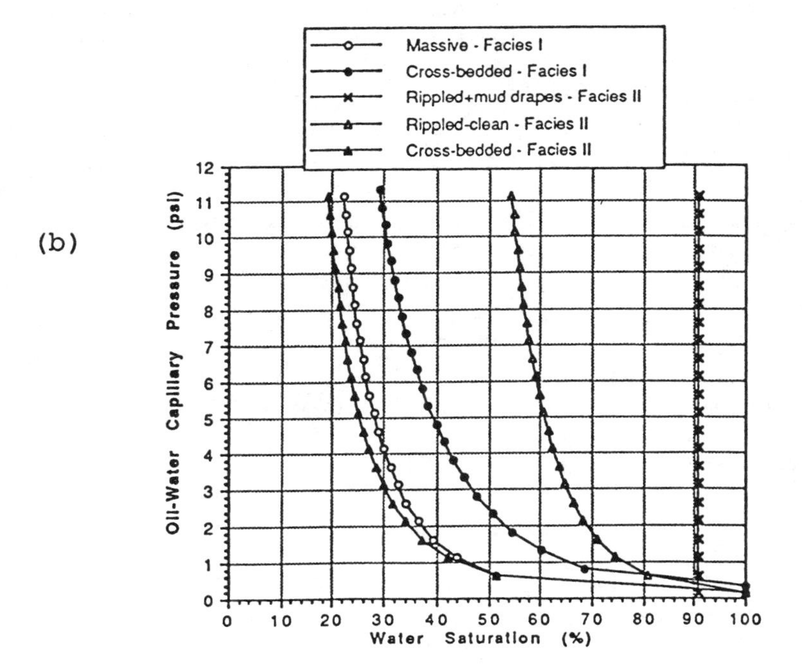

Fig. 13. (a) Oil-water capillary pressure curves from Peoria Field samples. (b) Oil-water capillary pressure curves from outcrop samples.

Oil-water relative permeability measurements were performed at reservoir conditions on outcrop and Peoria Field samples. Peoria Field data from Facies I and II are shown in Figures 14a and b and outcrop measurements are displayed in Figures 15a and b. These curves were measured during an imbibition process.

Relative permeability curves from pore types in Facies II measured on both surface and subsurface samples show a very similar character. The Peoria Field samples from Facies II were moderately oil wet and the outcrop samples were slightly water wet. These curves show oil to be mobile over a wide range of water saturations. However, small oil relative permeability values combined with low absolute permeabilities measured in Facies II indicate this facies would not provide a significant contribution to the oil production rate.

Each pore type within Facies I exhibited unique relative permeability characteristics. Peoria Field samples from Facies I had neutral to slightly water-wet wettability characteristics. Residual oil saturations of samples from Facies I in the Peoria Field were greater than those from Facies II.

Large differences in the wettability of different pore types in outcrop samples from Facies I are believed to have caused the differences in their relative permeability characteristics. Massive samples from outcrops were strongly water wet, had large values of oil relative permeability at initial water saturation, and had the largest residual oil saturation of any sample. Cross-bedded outcrop samples were found to have neutral wetting and extremely low residual oil saturations. The relative permeability characteristics of this pore type were similar to those found in Facies II samples.

The differences in relative permeability characteristics of pore types within Facies I should be accounted for in fluid flow models.

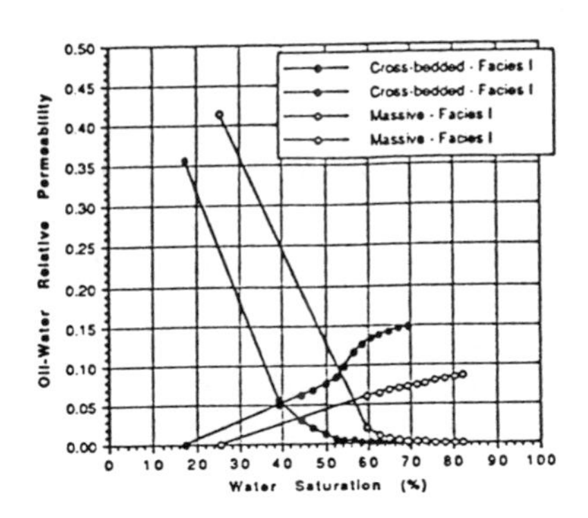

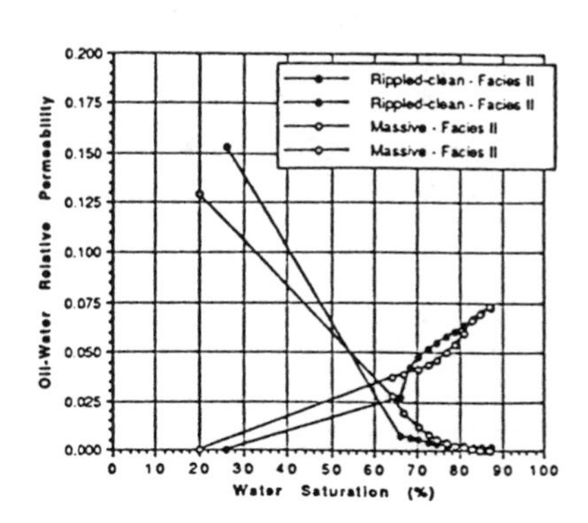

Fig. 14. (a) Peoria Field oil-water relative permeability curves from Facies I. (b) Peoria Field oil-water relative permeability curves from Facies II.

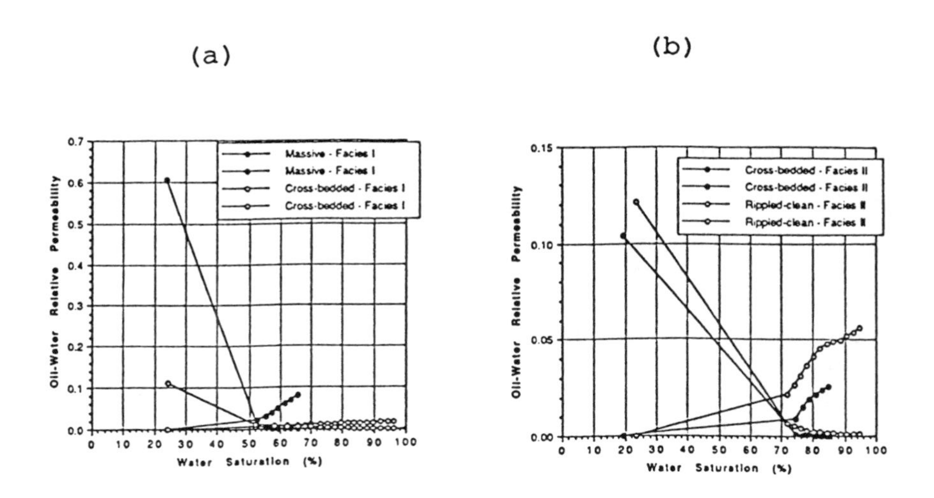

Fig. 15. (a) Outcrop oil-water relative permeability curves from Facies I. (b) Outcrop oil-water relative permeability curves from Facies II.

IX. SUMMARY AND CONCLUSIONS

Comparisons of outcrop and subsurface rocks from equivalent strata in the Muddy (J) Sandstone were made to help define the interwell character of channel sands in the Peoria Field for fluid-flow models. Outcrops were found to have grain sizes, sedimentary structures, and facies successions within channel deposits similar to those in the Peoria Field. From these observations two reservoir facies were identified in both surface and subsurface rocks. The proportion of each facies within channels and the distribution of stratification types comprising each facies were different in outcrops and in the Peoria Field.

Channels observed in outcrops were interpreted to be low-sinuousity distributary channels and Peoria Field channel facies were thought to be moderate- to high-sinuousity, fluvial, point bar sands deposited on the coastal plain.

Mercury injection characteristics, grain size analyses, and permeability populations showed that the two facies were each composed of three different pore types. Each pore type was associated with one or several stratification types. Similar pore types were found in outcrops and Peoria Field cores. However, the properties of each pore type differed between the surface and subsurface. Also, the amount of each pore type within the two reservoir facies was different in outcrops and Peoria Field cores. Oil-water relative permeability and capillary pressure curves were different for each pore type and between surface and subsurface samples from similar stratification types.

For this case an outcrop study was useful for determining how fluid-flow properties are related to rock types and for establishing the number of units (pore types) needed for reservoir description. These results indicate the need to consider the distribution of stratification types or pore types within a facies when scaling up reservoir properties for fluid-flow models in this type of reservoir.

ACKNOWLEDGMENTS

The authors would like to thank Marathon Oil Company for performing the relative permeability and capillary pressure measurements and the porosity and permeability measurements at different values of confining stress.

REFERENCES

1. Weimer, R.J. and Land, C.B., Jr.: "Field Guide to Dakota Group (Cretaceous) Stratigraphy Golden-Morrison Area, Colorado," The Mountain Geologist, v. 9, no. 2-3, p.241-267.

2. MacKenzie, D.B.: "Post-Lytle Dakota Group on West Flank of Denver Basin, Colorado," The Mountain Geologist, v. 8, no. 3, 1971, p.91-131.

3. Weimer, R.J. and Sonnenberg, S.A.: "Sequence Stratigraphic Analysis of Muddy (J) Sandstone Reservoir, Wattenberg Field, Denver Basin, Colorado," in Petrogenesis and Petrophysics of Selected Sandstone Reservoirs of the Rocky Mountain Region, E.B. Coalson, ed. RMAG, 1989, Denver, CO, p. 197-220.

4. Matuszczak, R.A.: "Wattenberg Field, Denver Basin, Colorado," The Mountain Geologist, v. 10, 1973, p. 99-105.

5. Land, C.B. and Weimer, R.J.: "Peoria Field, Denver Basin, Colorado - J Sandstone Distributary Channel Reservoir," in Depositional Modeling of Detrital Rocks, SEPM Core Workshop No. 8, August 1985, Golden, CO, p. 59-82.

6. Harms, J.C.: "Stratigraphic Traps in a Valley Fill, Western Nebraska," AAPG Bull., v. 50, no. 10, Oct. 1966, p.2119-2149.

COMPARISON OF GEOSTATISTICAL CHARACTERIZATION METHODS FOR THE KERN RIVER FIELD©

Joseph P. Vogt

Exploration & Production Technology Department
Texaco, Inc.
Houston, Texas

ABSTRACT

Three dimensional conditional simulation (3DCS) is a method to interpolate reservoir characteristics such as porosity, permeability, and facies between wells. 3DCS is usually thought to be the best characterization technique for flow studies. However, this method is also quite CPU intensive. Other geostatistical techniques, such as pseudo 3D conditional simulation (P3DCS), 3D kriging (3DK), and pseudo 3D kriging (P3DK) are less CPU intensive. Therefore it is important to know when P3DCS, 3DK, and P3DK can be substituted for 3DCS. In this paper, we will answer this question for the Kern River Field in California.

© A substantial portion of this paper is from SPE 22338, which will be published in the proceedings of the SPE International Meeting on Petroleum Engineering to be held in Beijing, China, March 24-27, 1992.

I. INTRODUCTION

The Kern River Field is located northeast of Bakersfield, California on the southeastern edge of the San Joaquin Valley (Figure 1). The field was discovered in 1899 and ranks as the fifth largest in the United States in original oil in place.[1,2]

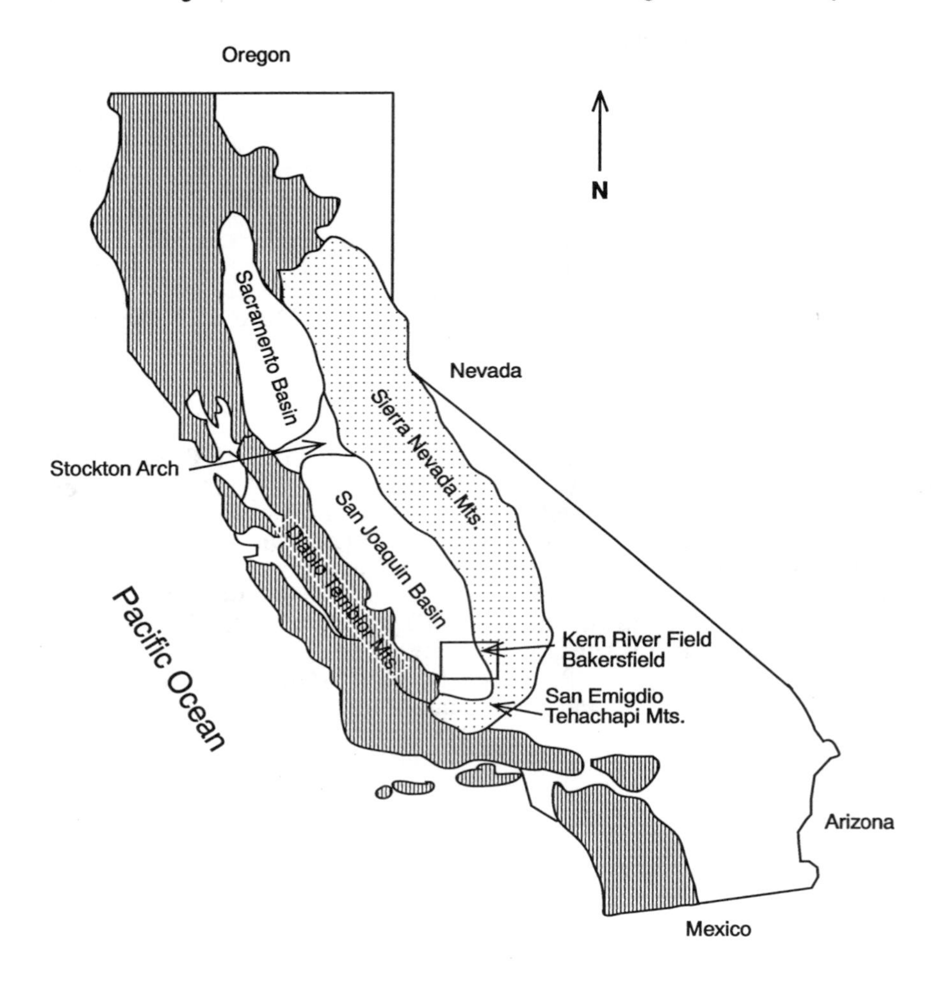

Figure 1. Kern River location.

For the first fifty years of its life, Kern production was slowed by the high viscosity of the crude (12 Pa.s (12,000 cp)) at the reservoir temperature of 20-30°C (68-86°F). Beginning in the mid-1950's, bottomhole heaters, steam stimulation, and finally steam displacement rejuvenated the field in a development program that continues to this day. Overall field production has plateaued and now appears to be slowly decreasing. Intensive geological and engineering efforts are being made to produce as much of the remaining oil as practical. These efforts include the use of geostatistics to help develop geological models at both the well-to-well and field-wide scales.

Even with today's computers, geostatistical software can be expensive to run. Therefore it's important to use the least CPU intensive modeling technique that will still accurately characterize the reservoir. Previous work[3] has shown that conditional simulation is better than kriging for characterizing the permeability of a 2 ft by 2 ft Berea sandstone slab. By comparing outcrops with geostatistical results, we've found that conditional simulation is also better than kriging for predicting the shale continuity at Kern River. In this paper, therefore, we'll regard 3DCS as reality. The other three methods will be compared to 3DCS on the basis of (1) shale continuity predictions and (2) CPU time.

In addition to comparing geostatistical techniques, this paper points out two important limitations of most commercial gridding software that claims to be 3D. First, most software works in only one plane at a time. The resulting 3D grid, then, is really made up of a series of independent 2D grids (i.e., pseudo 3D). For reservoirs that have small variability (high correlation) perpendicular to these 2D grids, the resulting 3D grid is not realistic. Second, most (all?) commercial software uses algorithms that, like kriging, do not reproduce the statistics and texture of the original data. Rather, these algorithms underpredict the extreme values of the data and smooth it out spatially. The physical results of this second shortcoming will be seen in the course of this paper.

II. SOFTWARE

The geostatistics software we used uses conventional kriging[4] and two different types of simulation: the turning bands method (tbm)[4] and the matrix decomposition method (mdm).[5] Some general comments are offered below.

For conventional kriging, Gaussian elimination is used to solve the kriging equations.

For single conditional simulation realizations, the tbm is faster than the mdm. The tbm is not used for three dimensional (3D) problems, however, because in 3D, mathematical difficulties can be encountered when using fractal variograms. These difficulties do not occur with the tbm in 2D.

For single conditional simulation realizations, the mdm is slow compared to the tbm because the decomposition of the covariance matrix in the mdm takes a long time. However, as the number of realizations increases, the mdm becomes relatively more efficient because the same matrix decomposition can be used for all the realizations.

In summary, it's best to use the tbm for a small number of 2D realizations. For a large number of realizations or for a 3D problem, mdm is best.

III. STATISTICS

The short normal resistivity data used in this paper came from two "typical" wells (5510053 and 5510076, also called Reed Crude A53 and Reed Crude A76, respectively) in the central portion of the Kern River Field. The sample probability density function (pdf) from the wells is shown in Figure 2.

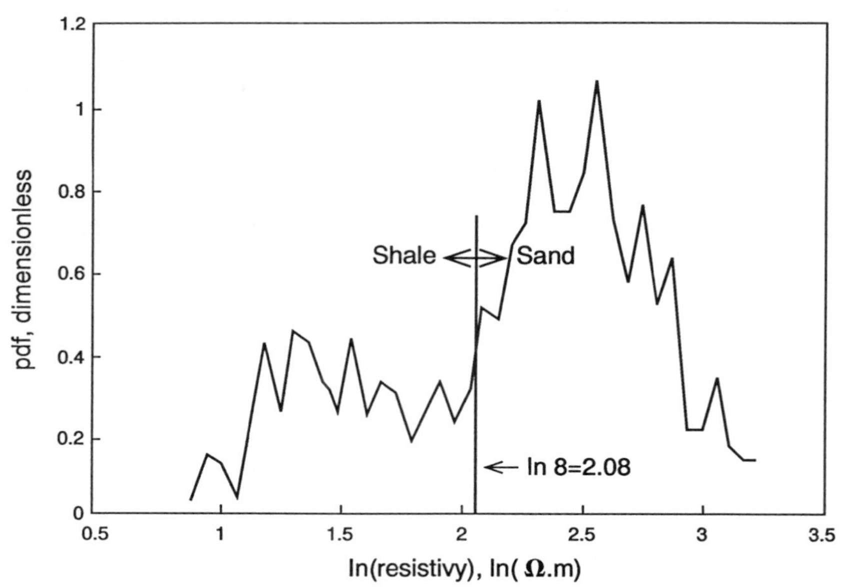

Figure 2. Sample probability density function (PDF)

Notice that the sample pdf is bimodal, that is, it has two humps. The first hump (low resistivity values) comes from shale, and the second hump (high resistivity values) comes from sand.

Normally, data from more than two wells would be used to generate resistivity statistics and variograms. The purpose of this paper, however, is to demonstrate techniques rather than create a highly accurate Kern River model. For demonstration purposes, then, data from only two wells is sufficient.

IV. VARIOGRAPHY

Kriging as well as the tbm and the mdm require a variogram as input.

The first step in making a variogram is to stratigraphically correlate the wells that are going to be used.[6] This was done for 5510053 and 5510076.

After plotting the vertical variogram, it was decided to fit it with a fractional Gaussian noise fractal model, whose equation is[7]

$$2\gamma(h,\delta) = V\delta^{2H-2}\left(2 - \left(\frac{h}{\delta}+1\right)^{2H} + 2\left(\frac{h}{\delta}\right)^{2H} - \left|\frac{h}{\delta}-1\right|^{2H}\right) \quad (1)$$

In this equation, V is a fitting constant, δ is a length which is often taken as the data interval, and H is the fractal exponent. In this study, instead of taking δ as the data interval, it is taken as 0.01 times a correlation length, which is discussed in the next paragraph.

By definition, fractal variograms do not have ranges or correlation lengths.[7] Random field generation methods such as the tbm or the mdm, however, require a correlation length as input.[4,5] So in order to use such methods to generate random fractal fields, we defined artificial correlation lengths that are equal to or greater than the ranges of the data. This will not affect the problem physically because the variogram is strictly fractal within the range of the problem. However, it does satisfy the mathematical need to have a correlation length.

For the vertical direction, only 305 m (1000 ft) of data was available, so the vertical correlation length was taken as 305 m (1000 ft). An R/S analysis[8] gave the value of the fractal exponent H as 0.85. The constant V was obtained from the condition that the variogram at 305 m (1000 ft) equal the variance of the resistivity, which is 100. This gave V equal to 117.29 $(\Omega.m)^2$.

A picture of the vertical variogram is shown in Figure 3.

Next a plot was made of the horizontal variogram. It was also fitted with the fractional Gaussian noise fractal model of Equation (1). It was assumed that both the variance and fractal exponent H of the resistivity in the horizontal direction are equal to their values in the vertical direction. Then the three constants V, δ, and correlation length for the horizontal direction were determined from the three conditions that (1) δ for the horizontal direction is equal to 0.01 times the horizontal correlation length, that (2) the horizontal variogram curve at the well spacing (which is 122 m (400 ft)) passes through the calculated variogram data point (which is 100 $(\Omega.m)^2$), and that (3) the value of the horizontal variogram at its correlation length is equal to the variance of the resistivity in the horizontal direction. This gives values of V of 354.73 $(\Omega.m)^2$, δ of 122 m (400 ft), and correlation length of 12,192 m (40,000 ft) for the horizontal direction.

A picture of the horizontal variogram is shown in Figure 3.

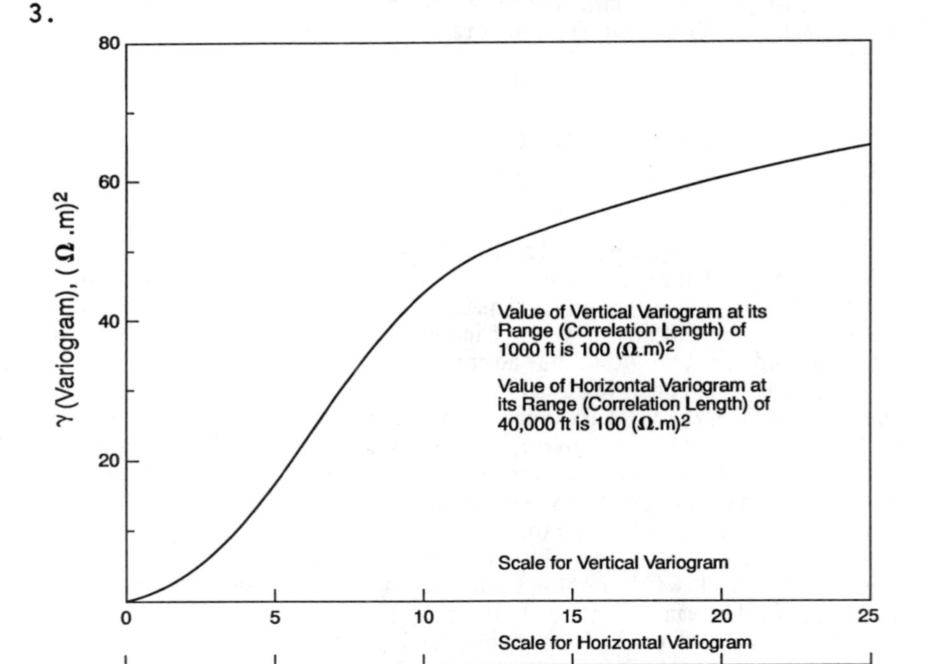

Figure 3. Vertical and horizontal resistivity variograms.

V. WELLS

Studies subsequent to this one are being conducted to arrive at geological and engineering conclusions for specific portions of the Kern River Field. Our purpose here, however, is to compare how four different geostatistical methods interpolate properties between wells.

The simplest and most conclusive way to compare geostatistical methods is to first generate artificial wells and then to use the different methods to interpolate between them. Using generated wells insures that the random variable being studied is stationary, assures that the data set contains no outliers, avoids the step of scale-up averaging of resistivity values, allows us to put the wells exactly where we desire, and generally makes the study easier to conduct and interpret.

In this study, we generated nine wells with the 61 m (200 ft) spacing shown in Figure 4. Each well has twenty equally spaced resistivity values. These values were generated by using the 3DCS method. The fractional Gaussian noise variograms in Figure 3 were used as input. Even though the wells were generated artificially, they cannot be qualitatively distinguished from actual Kern River wells.

VI. ESTIMATION AND SIMULATION

A. Introduction

Resistivity values between wells can be obtained by estimation (e.g., kriging) or simulation (e.g., the tbm or the mdm).

Kriging estimates a value at a point by taking a weighted linear combination of the values at neighboring points. The weights are chosen to make the estimate unbiased (the average error of a series of estimates is zero) and to minimize the variance of the errors. Kriging always honors the data.

Because of the assumptions on which it is based (zero average error and minimum variance of the error), kriging is a relatively accurate interpolation method. However, in general, the statistics of kriged values are different than the data used as input to the kriging procedure. In particular, the kriged results have fewer extreme values (i.e., have a smaller standard deviation) than the input data. Further, the variogram is not preserved. Kriged maps are usually smoother than reality. Thus kriging is most appropriate for such tasks as calculating an average reservoir property. An example is average porosity values to be used in reserves calculations.

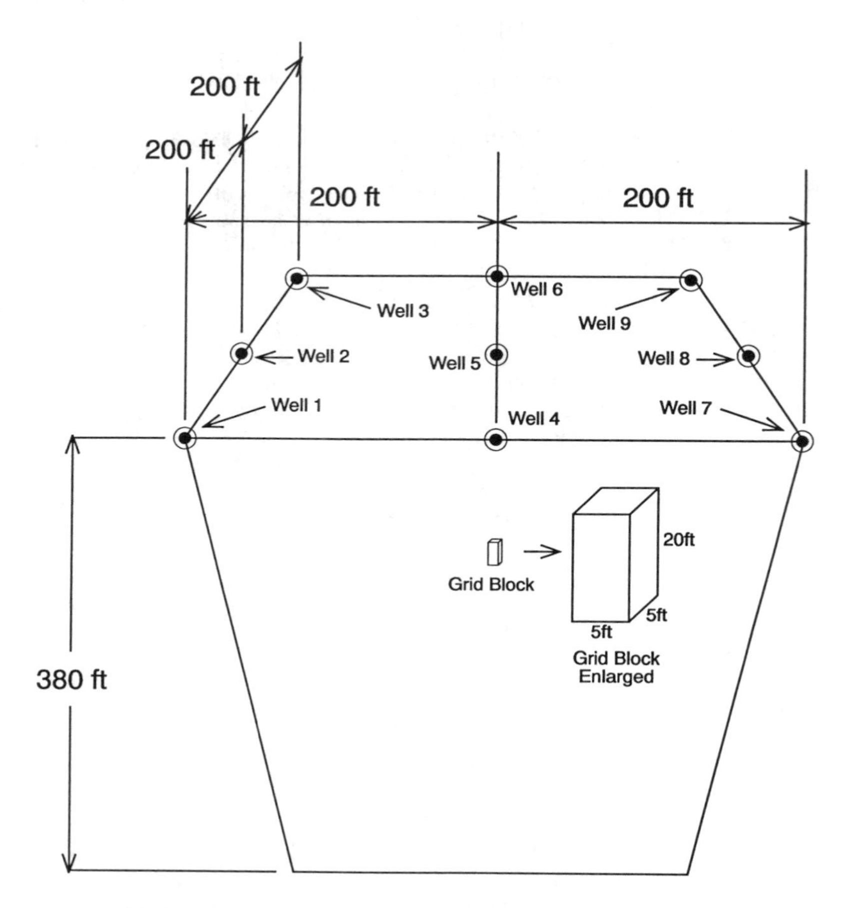

Note:

(1) The 200 and 300 ft Dimensions are Measured Between th Middle of Grid Blocks

(2) The Grid Contains 81x81x20=131,220 Blocks

Figure 4. Grid dimensions and well locations.

Simulation generates points in space that have the same histogram and variogram as the input data. In gridding applications, it duplicates the texture of the exhaustive data set. Thus simulation is most appropriate for calculating a property, such as permeability, that will be used in flow simulations. This is because flow is controlled by the extreme values of permeability rather than its average value. The disadvantage of simulation is that its average error is larger than kriging.

In this paper, the pseudo 3D techniques are mathematically the same as the true 3D methods except that the pseudo techniques are applied to one layer at a time while the true 3D methods are applied to all the layers simultaneously.

Using the resistivity values from the nine wells shown in Figure 4 as input, the P3DK, 3DK, P3DCS, and 3DCS methods were all applied. In all four cases, the value of the variogram (60.3 $(\Omega.m)^2$) at the layer thickness (20 ft) is 60.3% of the sample variance (100 $(\Omega.m)^2$). Qualitatively, this means that there is a large variability (small correlation) between layers. In such cases, pseudo 3D techniques will work almost as well as the corresponding true 3D method.

B. p3dk

The 3DK method (using a search ellipse with 1000 ft horizontal semi-axes) was applied at each of the grid nodes indicated in Figure 4. This gave a total of 81x81x20 = 131,220 resistivity values. Then, based on experience at Kern River, all nodes with resistivity values greater than 8 Ω.m were taken as sand, and all nodes with resistivity values less than 8 Ω.m were taken as shale.

C. 3dk

This method is the same as P3DK except (1) a search ellipsoid with horizontal and vertical semi-axes of 1000 ft and 40 ft, respectively, replaces the search ellipse of P3DK and (2) data in all layers within the search ellipse is used simultaneously for kriging.

D. p3dcs

Same as P3DK except a conditional simulation technique[4] is used instead of kriging.

E. 3dcs

Same as 3DK except a conditional simulation technique[4] is used instead of kriging.

VII. COMPARISON OF RESULTS

A. Physics

Current wisdom at Kern River is that permeability differences within the sands do not significantly affect fluid flow. Rather, shale continuity is what governs fluid movement.

For some layers (e.g., layers 1, 2, 3, 5, and 8), all four geostatistics results predict about the same shale continuity. To save space, these layers are not pictured. For other layers (e.g., layers 4, 6, 7, and 9), the four methods predict different results. The differences (see Figures 5 and 6) are:

(1) fourth layer: 3DCS predicts a shale in the lower right hand corner that the other methods do not predict

(2) sixth layer: P3DCS and 3DCS show more sand continuity than P3DK and 3DK

(3) seventh layer: P3DCS shows a break in the shale in the lower left hand corner, but the other three methods do not

(4) ninth layer: P3DK and 3DK predict very small holes in the shale only at the middle of the upper edge, whereas P3DCS and 3DCS predict larger holes at these locations as well as holes at the middle of the right hand edge; in addition, P3DCS predicts holes at the middle of the front edge.

The most important difference between the four geostatistical methods is illustrated by layer nine. Both kriging methods predict that this layer is a shale barrier. Both conditional simulation methods, however, predict that this shale has holes in it. This shows that the kriging methods sometimes obliterate critical features because kriging tends to give overly smooth results. For this reason, kriging methods (or any of the other commonly used gridding techniques that give overly smooth results) are not recommended for simulating the shales at Kern River.

B. Computer Time

Table 1 gives CPU time to generate an 81x81x20 grid block model using each of the four geostatistics methods. The 3DCS method, which is the best technique for predicting shale continuity, also takes the most CPU time. It is followed, in descending order of time, by 3DK, P3DCS, and P3DK.

Table 1 also gives some indication of how the number of layers affects CPU time for the various methods. As expected, for only one layer, there is little or no CPU time advantage in using the pseudo 3D methods. For twenty layers, however, there is a substantial advantage.

Table 1 shows that kriging is much faster than conditional simulation for one layer, but is not so much faster for twenty layers.

TABLE I. CPU time for four geostatistical methods.*

Interpolation Method	User CPU Time† (sec)	System CPU Time† (sec)	Total CPU Time† (sec)
P3DK	72.32/ 3.70	3.86/0.30	76.18/ 4.00
3DK	290.14/ 3.03	12.61/0.23	302.75/ 3.26
P3DCS	100.34/29.16	3.15/0.93	103.49/30.09
3DCS	330.94/35.75	19.13/8.79	350.07/44.54

* Run on a Cray XMP 216 with a UNICOS operating system

† (Time to Run a 20 Layer Model)/(Time to Run a 1 Layer Model)

Finally, notice that the time it takes to run twenty layers using the P3DCS method is much less than twenty times the time for one layer. This is because the spectral density function in the tbm only has to be calculated one time, regardless of the number of layers in the model.

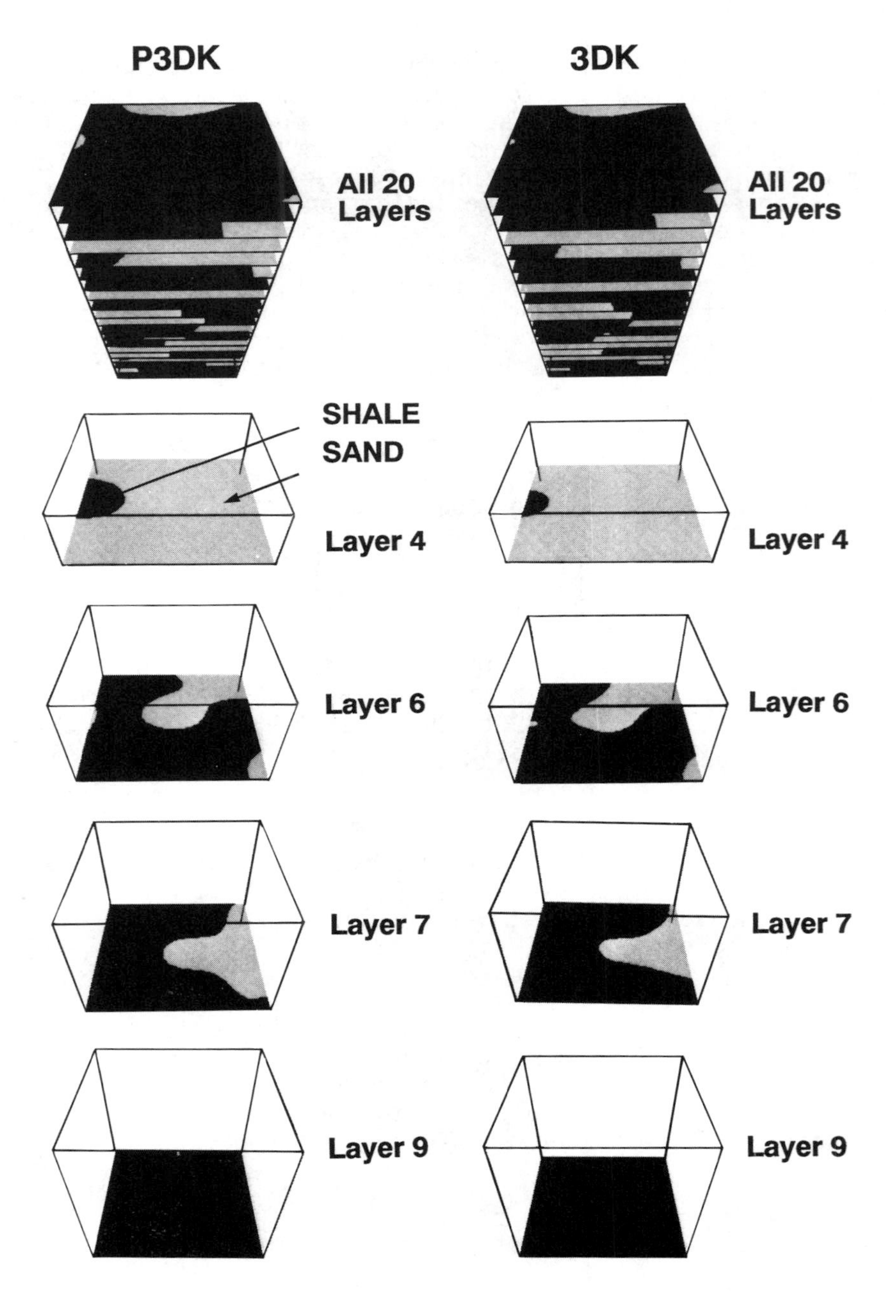

Figure 5. P3DK and 3DK results.

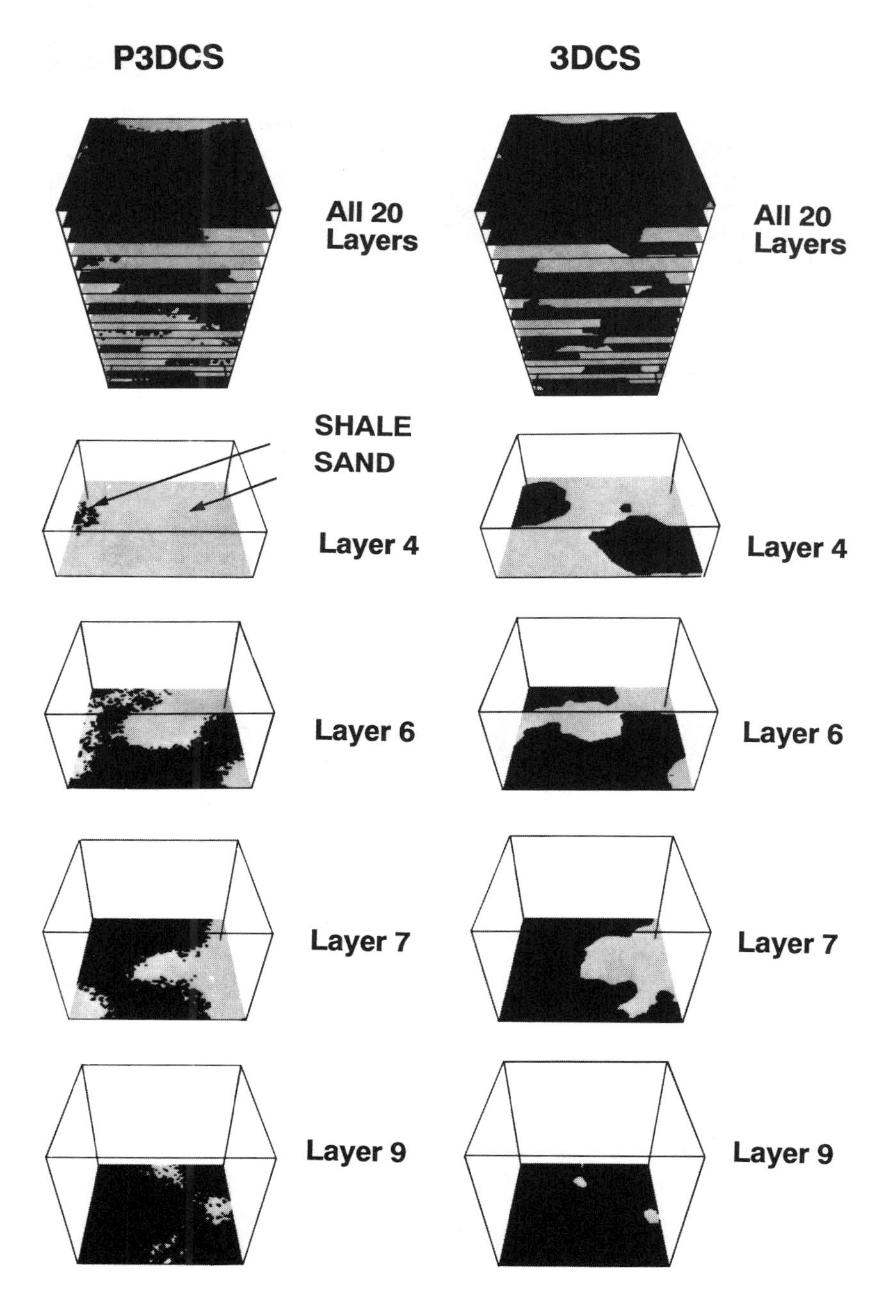

Figure 6. P3DCS and 3DCS results.

C. Cost Effectiveness

To determine the most cost effective geostatistics method to characterize the Kern River Field for flow simulation, we can eliminate the P3DK and 3DK methods because, as pointed out in the physics section, these techniques can give unrealistic results. The P3DCS results are qualitatively similar enough to 3DCS that they can usually be used for screening purposes at a considerable cost reduction (70% for twenty layers and 32% for one layer).

VIII. CONCLUSIONS

1. P3DK and 3DK (and all other algorithms used in commercial gridding packages) overpredict shale continuity at Kern River.

2. 3DCS is the best method for predicting shale continuity in the Kern River Field. P3DCS is a good screening method because it gives results similar to 3DCS but uses less CPU time (32% less for a one layer model and 70% less for twenty layers).

3. In general, P3DCS will give results similar to 3DCS whenever the vertical variability between layers is large (vertical correlation between layers is small).

IX. GLOSSARY

conditional simulation: A simulation that honors the input data.

correlation length: The maximum distance over which the value of a variable influences neighboring values.

estimation: A process that uses sample information to predict values in areas that have not been sampled.

fractal: A property is called fractal if it looks similar at all scales.

fractional Gaussian noise: A generalization of Gaussian noise (H = 1) in which the exponent H can take on fractional values from 1/2 to 1.

histogram: A curve that indicates the probability that a variable has a given value.

kriging: An estimation method that minimizes the expected value of the squared error of an estimate.

mdm (matrix decomposition method): An averaging method that generates a random field of numbers in space with a given histogram and variogram.

pseudo 3D: When a gridding process for a series of parallel planes is done independently in each plane, the collective grid is called pseudo 3D.

realization: One possible set of outcomes of a series of values in space.

probability density function: A type of histogram.

R/S analysis: An analysis which uses the range R in the values of a random variable and the standard deviation S of the variable to determine the similarity exponent H.

search ellipse or ellipsoid: An ellipse (2D) or ellipsoid (3D) within which one searches for points that are to be used for some purpose, such as kriging, etc.

simulation: A process that generates a random field of numbers in space with a given histogram and variogram.

tbm (turning bands method): A spectral method that generates a random field of numbers in space with a given histogram and variogram.

variability: High variability (low correlation) means that the value at a point is not related to the value of neighboring points.

variance: The average of the squared difference between a variable and its mean.

variogram: A curve that describes how a property can vary in space as a function of the distance between two points.

ACKNOWLEDGEMENTS

I thank Dr. A. P. Yang for using his software to generate the P3DK, 3DK, P3DCS, and 3DCS grids discussed in this paper.

REFERENCES

1. Brelih, D. A. and Kodl, E. J.: "Detailed Mapping of Fluvial Sand Bodies Improves Perforating Strategy at Kern River Field", paper SPE 20080, presented at the 1990 California Regional Meeting, Ventura, April 4-6.

2. Kodl, E. J., Eacmen, J. C., and Coburn, M. G.: "A Geologic Update of the Emplacement Mechanism within the Kern River Formation at the Kern River Field", from Structure, Stratigraphy, and Hydrocarbon Occurences of the San Joaquin Basin, California, J. G. Kuespert and S. A. Reid (eds.), Publication of SEPN Pacific Section and AAPG, June 1, 1990, pp. 59-71.

3. Journel, A. G. and Alabert, F. G.: "Focusing on Spatial Connectivity of Extreme-Valued Attributes: Stochastic Indicator Models of Reservoir Heterogeneities", paper SPE 18324, presented at the 1988 Annual Technical Conference and Exhibition, Houston, Oct. 2-5.

4. Journel, A. G. and Huibregts, Ch. J.: Mining Geostatistics, Academic Press, New York, 1978.

5. Fogg, G. E., Lucia, F. J., and Senger, R. K.: "Stochastic Simulation of Interwell-Scale Heterogeneity for Improved Prediction of Sweep Efficiency in a Carbonate Reservoir", Reservoir Characterization II, L. W. Lake, H. B. Carroll, Jr. and T. C. Wesson (eds.), Academic Press, New York, 1991: pp. 355-381.

6. Journel, A. G. and Gomez-Hernandez, J. J.: "Stochastic Imaging of the Wilmington Clastic Sequence", paper SPE 19857, presented at the 1989 Annual Technical Conference and Exhibition, San Antonio, Oct. 8-11.

7. Hewett, T. A.: "Fractal Distributions of Reservoir Heterogeneity and Their Influence on Fluid Transport", paper SPE 15386, presented at the 1986 Annual Technical Conference and Exhibition", New Orleans, Oct. 5-8.

8. Mandlebrot, B. B. and Wallis, J. R.: "Robustness of the Rescaled Range R/S in the Measurement of Noncyclic Long Run Statistical Dependence", Water Res. Res. (Oct. 1969), 5, 5, 967-988.

SESSION 3

Modeling/Description of Interwell Region

Co-Chairmen

Steve Begg, BP Exploration Alaska

Mohan Kelkar, University of Tulsa

AN APPLICATION OF THE ROCKIT GEOLOGICAL SIMULATOR TO THE FRONTIER FORMATION, WYOMING

Paul Armitage

PEDSU
Winfrith Technology Centre
Dorchester, UK

Richard J. Norris

Department of Mineral Resources Engineering
Imperial College
London, UK

I. INTRODUCTION

A major obstacle to producing good reservoir simulation models is the areally limited data that are obtained from well core and well log analysis. These limitations are particularly severe in the North Sea due to the very large well spacing resulting from the high costs of offshore development. The well data are of a high quality, but only cover a small part of the reservoir. Large scale seismic data are also collected, but this is of poorer resolution.

The rock properties observed at each well only give a very limited view of the reservoir as a whole. The real problem is trying to ascertain which structures are present between wells. It is common practice to interpolate linearly the reservoir structure from the rock sequences seen at wells. This simple assumption is not guaranteed to yield acceptable results.

Methods have been developed, therefore, which use statistical representations of the likely geological structure to construct reservoir models. In this way less continuous geological structures can be included in the resulting model, such as sand body stacking sequences and isolated geological formations. In reference (1) Farmer presented the detailed mathematical background to the theory of a new technique for generating statistical representations of reservoirs. The ROCKIT program applies this theory giving a practical tool useful for actual reservoirs.

The work presented here looks at the effectiveness with which ROCKIT's statistical representation of reservoir structure can synthesize a geology which has the same physical and flow properties as the original data. To achieve this, outcrop data from the Frontier Formation in Wyoming were used. The sand/shale (heterolithic) facies of this shallow marine sandstone sequence show a variety of structures, including isolated sand units and correlation over distances large compared with the sample spacing. These, and similar, features are thought to provide a good test of ROCKIT's statistical method.

In addition, the study seeks to verify that the theoretical advantages stated in reference (1) do represent real and practical advantages in an operational sense. The advantages stated by Farmer are that the method (1) can be proved to converge, (2) involves no inverse problems, and (3) has control parameters with an intuitive interpretation.

As with alternative methods, ROCKIT's input is a statistical description of the reservoir rock. However, this input is different from that for other methods, and somewhat simpler. Firstly, the global or single-point histogram is required; this may be just the proportion of the total number of grid cells in each permeability range or lithology, for example. Secondly, the two-point histogram is needed; this is the number of transitions between, say, lithologies in selected directions (e.g., horizontal and vertical nearest neighbours). The input consists, therefore, of a displacement vector , a lithology pairing and the number of such pairings required in the array for the given displacement. There is no theoretical limit on the number of two-point histograms that can be defined. The code can compute the single-and two-point histograms if a 'training pattern' such as a detailed description of an outcrop is available. These ideas are explained more fully in section II.

A synthetic geology is generated from the input statistics using the global minimisation technique of simulated annealing (2). This method ensures that a solution is converged that is the best possible match to the two-point histograms.

II. THEORY

The detailed mathematics of the two-point histogram technique as a statistical representation of geological structure is developed in reference (1). It is not our intention here to present a rigorous mathematical description of the method, but rather to set down the background to the technique, and show its practical application by use of a simple example.

The two-point histogram method was developed in an attempt to overcome some of the drawbacks of existing statistical methods for representing geological structure. Such considerations were the adequacy of correlation functions as a way of representing local structure, and the fact that standard correlation function methods could only be made to converge by increasing the degree of conditioning of the input data. Additionally, techniques which are an improvement on the correlation function method, such as Markov and Boolean methods, require the solution of difficult inverse problems to calculate their control parameters. An example of the latter is the calculation of Markov correlation coefficients from a 'training pattern' in order to generate synthetic geologies. In reference (1), Farmer presents the two-point histogram technique as a method which (1) can be proved to converge, (2) does not involve the solution of inverse problems, and (3) has intuitive control parameters.

The basis of the method is best illustrated through use of a simple example.

A. Training Pattern

Figure 1 illustrates a pattern of zeros and ones which could represent a geological property such as permeability, porosity or lithology. In this example let zero represent the presence of a shale and one the presence of sand.

Figure 1 is a training pattern for ROCKIT and could represent any, or all, of the following:

1. measured data on an outcrop,

2. mapped/photographic data,

3. a geologist's drawing.

B. The Global and Two-Point Histograms

The global histogram is simply the proportion of each lithology present in the pattern. Figure 1 has twelve zeros and thirteen ones, giving the global histogram shown in figure 2a. Here, the histogram has been normalised to give the total area under the plot to be one. The two-point histogram is the number of transitions between lithologies in specified directions and at specified spacings. These specifications are normally referred to as control directions. Two examples of two-point histograms are given for the data in figure 1:

1. when directions are horizontal and vertical and the spacing is nearest-neighbour grids,

0	1	1	0	0
1	0	0	1	1
1	1	1	1	1
0	0	0	0	0
0	1	0	1	1

Figure 1. An example of a training pattern

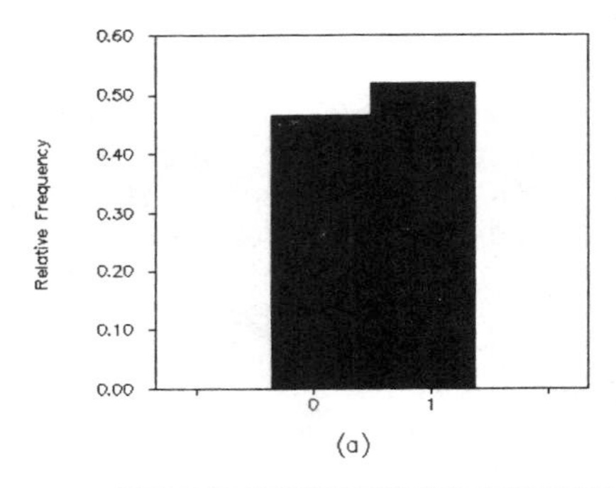

(a)

	Number of Pairs	
Pairs	Horizontal Nearest Neighbour	Vertical Nearest Neighbour
0–0	6	2
0–1	4	8
1–0	3	7
1–1	7	3

(b)

	Number of Pairs		
Pairs	Horizontal Nearest Neighbour	Vertical Nearest Neighbour	Diagonal Nearest Neighbour
0–0	6	2	2
0–1	4	8	6
1–0	3	7	5
1–1	7	3	3

(c)

Figure 2. Example histograms (a) global (single-point), (b) two-point for control directions (1,0) and (0,1), (c) two-point for control directions (1,0), (0,1) and (1,1).

2. when directions are horizontal, vertical and diagonal and the spacing is still nearest-neighbour.

The two-point histograms for (1) and (2) are shown in figures 2b and c.

The two-point histogram is essentially a statistical description of the correlation structure in the data.

There is a very large number of possible control directions, and hence a large number of two-point histograms, even for a small grid in the example used here. There is a distinct computational advantage in keeping the number of control directions to a minimum. In practice, the aim is to calculate a synthetic geology, using only a small number of control directions, which has the same flooding or displacement characteristics as the training pattern.

ROCKIT will generate a synthetic geology which has the same two-point histogram as the training pattern.

C. Solution Method

Once a global histogram and two-point histograms have been specified, ROCKIT proceeds by generating a random distribution of the property, which, initially, satisfies only the global histogram. The program then chooses pairs of cells at random and swaps their properties. If the new pattern conforms better to the specified two-point histograms than the pattern before the swap, then the swap is accepted. If the new pattern does not improve the match, the swap may still be accepted with a probability, p, where,

$$p = e^{-\Delta E/kT}$$

where

ΔE is a measure of the change in the two-point histograms produced by the swap

k is a constant

T is user-supplied and can be thought of as a convergence control parameter.

Note that,

$$\lim_{T \to \infty} p = 1$$

and $$\lim_{T \to 0} p = 0$$

This part of the algorithm, known as simulated annealing, is designed to ensure that the final pattern achieved is the global optimal solution rather than only a local optimum. The technique is discussed in reference 2. The use of such an algorithm overcomes the problem that if swaps are accepted or rejected simply on the basis of whether or not they improve the match to the two-point histogram a pattern may be obtained in which it is not possible to make any single swap which leads to an improvement, but which is, nevertheless, not the best possible match to the two-point histogram. In other words, a local optimum has been found rather than a global one. By accepting a certain proportion of swaps which do not improve the match, the convergence proceeds more slowly and ensures that the global optimal solution is found.

The steps in generating the synthetic geology from the global and two-point histograms may be summarised as follows:

1. Generate a random distribution satisfying the global histogram.

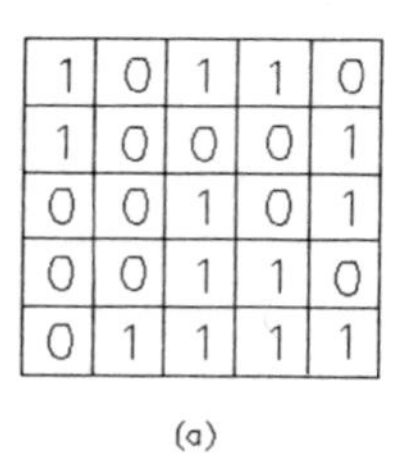

(a)

0	0	0	1	1
1	1	0	0	0
0	0	1	1	1
1	1	1	0	0
0	1	1	0	1

(b)

0	1	1	0	0
0	0	0	1	1
1	1	1	1	0
0	0	0	1	1
1	0	1	1	0

(c)

Figure 3. Solutions for the example of figure 1, (a) random, (b) ROCKIT solution with control directions (1,0), (0,1), (c) ROCKIT solution with control directions (1,0), (0,1), (1,1)

2. Select large T to give a high probability of accepting swaps which do not improve the match to the two-point histograms.

3. Swap pairs of cells at random. If the number of swaps accepted is small then the final solution has been found. If not, repeat with a smaller value of T. Figure 3 shows an initial random distribution and final optimal solutions for both two-point histograms presented in figure 2.

III. INPUT GEOLOGICAL DATA

The data used for the study presented in this paper were collected for a research project aimed at quantifying permeability variation in heterolithic facies (3). One outcrop selected for study was the Wall Creek Member of the Frontier Formation, Wyoming. The Frontier Formation was deposited during the late Cretaceous, in the Western Interior Seaway of the USA. The regional stratigraphy is characterized by a series of coarsening-up sequences, representing major prograding events. The Wall Creek member of the Frontier Formation outcrops on the eastern flank of the Tisdale Anticline, north of Casper. This locality provides excellent exposures of heterolithic facies. The section used for this study is a very high shale-content sand-streaked mudstone. The section is orientated approximately parallel to ripple-crest.

The data set used was developed as a binary indicator map. That is, the continuously varying sand and shale geometries in heterolithic lithofacies were converted into binary data, with 1's representing sands and 0's representing shales. Fine grids (5mm spacing) were imposed on detailed photomosaics of the lithofacies. These grids were sampled at every node, producing contiguous data that accurately model the spatial variation of the sands and shales. The spacing of the grid nodes was selected as a result of sensitivity testing regarding the minimum size of shale-breaks and sand-connections. Five millimetre spacing provides sufficient detail, whilst keeping the volume of data produced to a reasonable size.

A 70 x 38 portion of the full data set was used as the training pattern for the studies described here, figure 4 where shale is shown as black and sand as white. This size of array was chosen for computational reasons, to give reasonably short run times.

IV. MODELLING STRATEGY

In order to compare the physical and flow characteristics of the ROCKIT generated synthetic geologies with those of the training pattern, three means of analysis are employed. They are used in conjunction with each other:

1. The first method consists merely of the production of graphics in the form of simple block fill grids; see, for example, figure 4. A visual inspection of the training pattern and ROCKIT geologies can then be carried out. Similarities looked for include sand unit shape and connectivity.

2. Calculation of the effective permeability of the training pattern and synthetic geology is performed. The method employed is to equate Darcy's Law for flux through the heterogeneous medium and an equivalent homogenous medium and hence derive a single effective permeability which gives the same flux for a given pressure drop.

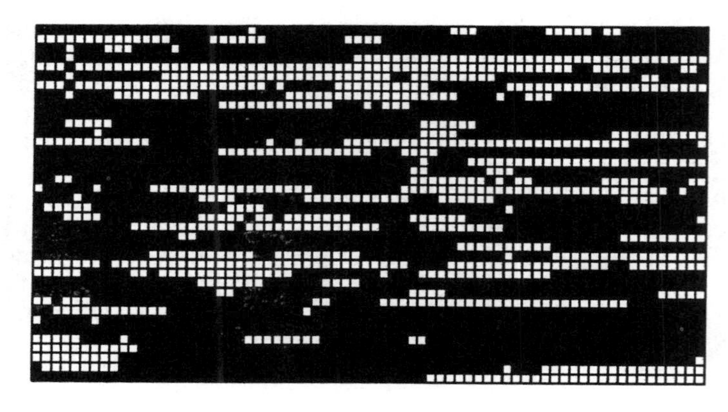

Figure 4. Section of the Frontier Formation outcrop data used as the training pattern for ROCKIT.

3. Unit mobility ratio waterfloods through the formations are simulated. This is done by setting up models of the geologies and injecting along one face and producing at the other, figure 5. Neither gravity nor capillary pressure effects are included in the models. The relative permeabilities used are linear between 0 and 1. The simulations were configured in this manner in order to assess the geological similarities between different cross-sections without also having to account for effects of the fluid flow process. (In general, the heterogeneity cannot be considered separately to the particular fluid flow process involved). The production well was constrained by a minimum bottomhole pressure (BHP) of 1500 psia and injection rates were determined by voidage replacement.

For each of 2 and 3 above, high permeability was assigned to the shale and low to the sand to give connected high permeability routes across the arrays. For the purposes

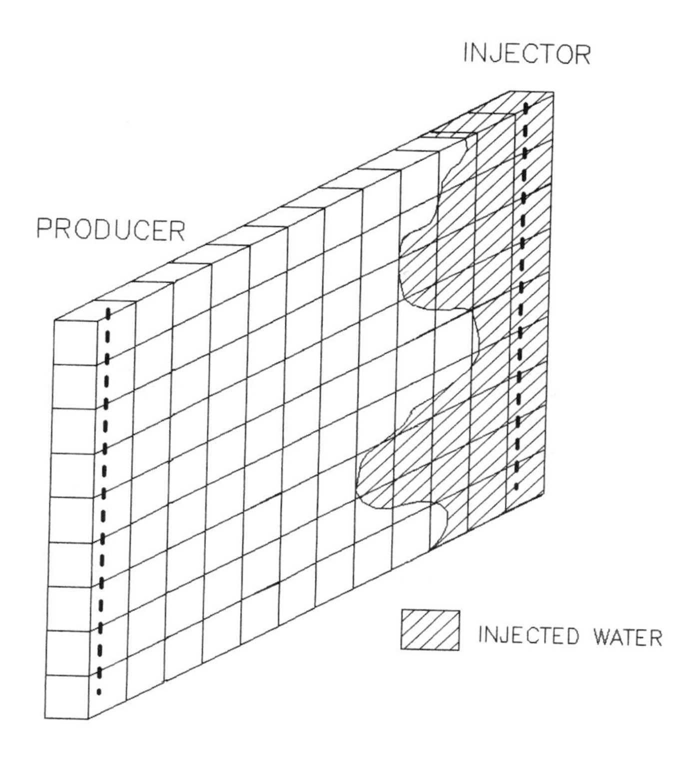

Figure 5. Schematic showing model configuration for waterflood simulation.

of comparison between synthetic geologies and the training pattern this does not present any difficulty since we are only interested in the variation of properties, rather than absolute values. The reason for doing this is that only 34% of the training pattern is sand, which is insufficient to give connected high permeability pathways across the section. High permeability is represented in the calculations by a value of 1 permeability unit and low by a value of 10^{-6} permeability units (a value of zero is not permitted by the numerical routines used.)

ROCKIT synthetic geologies were generated for several sets of control directions; three realisations being produced in each instance. Each realisation was conditioned to those values seen in the training pattern in the right and left-hand columns to represent wells at which data have been collected. As well as the ROCKIT results, analyses were carried out on:

1. The training pattern.

2. Three completely random permeability distributions generated from the global histogram of the training pattern.

3. Three random distributions generated from the global histogram of the training pattern, with no diagonal high permeability "connections". In the course of generating a random distribution a situation such as is shown in figure 6a, which shows a 3 x 3 array, could arise. If a

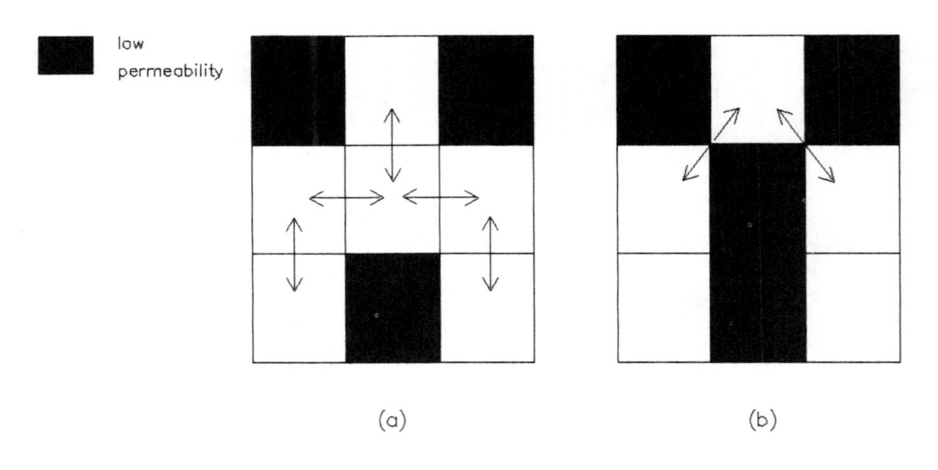

Figure 6. Arrangements of high and low permeability grid cells (a) allowed, (b) disallowed.

low permeability value is then placed in the central cell, as in figure 6b, only diagonal "connections" between the high permeability cell in the top row and other high permeability cells occur. This makes the upper high permeability cell inaccessible. Thus, in the generation of this second set of random distributions, the emplacement of low permeability in the middle cell of the grid shown in figure 6b would be disallowed. These distributions were generated in order to reduce the large inaccessible pore volumes seen in the realisations generated in 2 above.

Additionally, comparisons are made with the analytical solutions for a homogeneous permeability equivalent to the effective value calculated for the training pattern, and with a layered distribution obtained by interpolating the high and low permeability between wells.

The main result from the waterfloods used for comparative purposes is the watercut development as a function of time and as a function of pore volumes injected.

V. RESULTS

The section of the Wyoming data set used in this study as the training pattern for ROCKIT is shown in figure 4. Several features are worth noting. Firstly, an almost continuous zone of sand extends from one side of the section to the other towards the top of the figure. Since, for the purposes of effective permeability calculations and waterflood simulations, sand is assigned a low value of permeability, this acts as a barrier to flow in the vertical direction. Ideally, correlation should be on a scale relatively small compared with the size of the section for optimum use of statistical methods such as those incorporated in ROCKIT. However, in the case of many real data sets this is not always the case. Features such as this could be incorporated in the geological model deterministically, but this has not been done here in order to provide ROCKIT with a particularly difficult problem to solve. The correlation in the low permeability material is obviously greater in the horizontal direction (parallel to ripple crests) than in the vertical. Some areas of high permeability material exist that are not connected across the full width of the section;

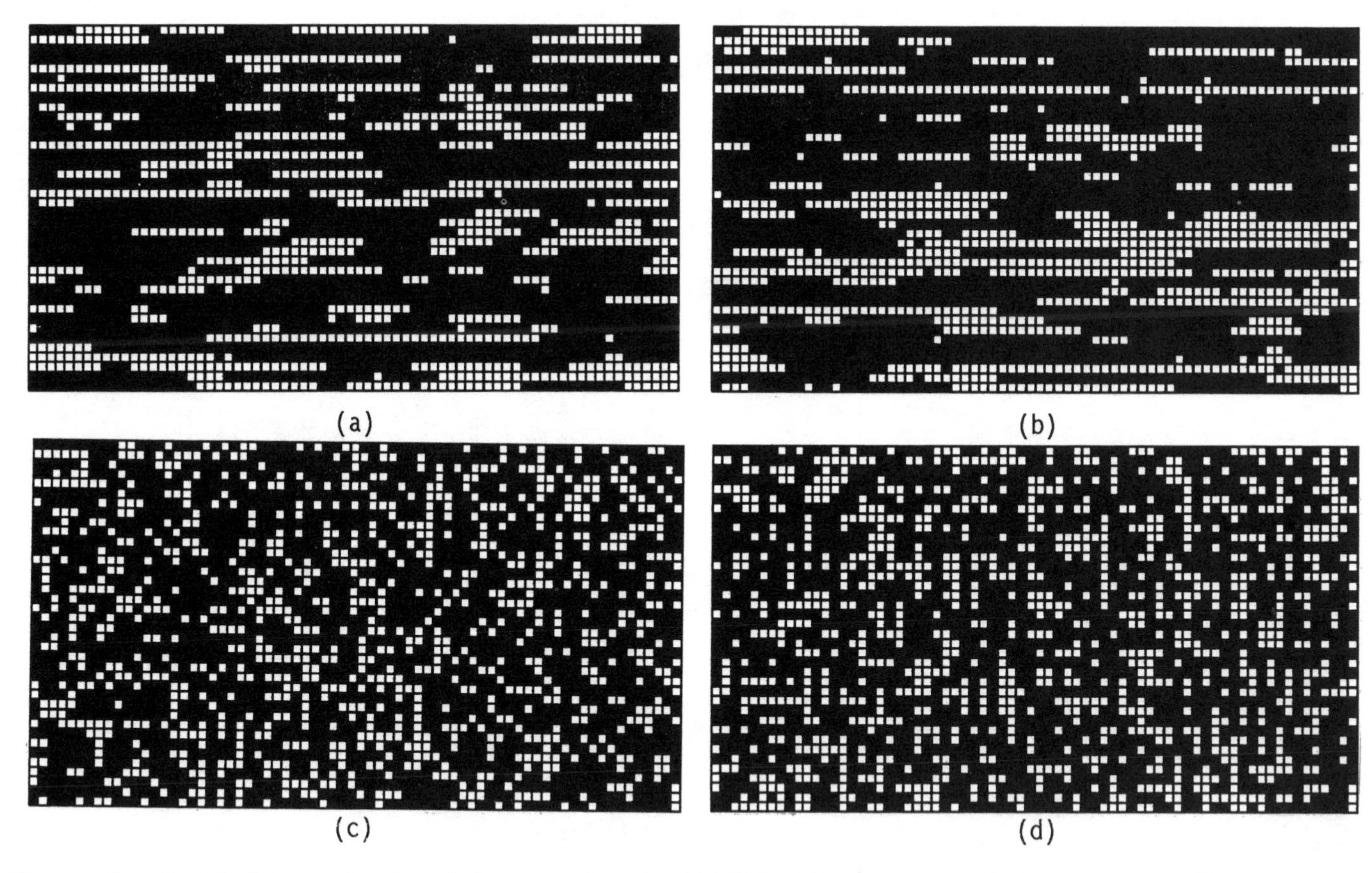

Figure 7. Synthetic geologies (a) generated by ROCKIT with one control direction , (b) generated by ROCKIT with nine control directions, (c) random, (d) random with no "diagonal connections"

TABLE I. Control directions used by ROCKIT in the generation of the synthetic geologies shown in figures 7(a) and (b).

No. of Control Directions	Control Directions Used
1	(1,0)
9	(1,0), (0,1), (2,0), (0,2), (4,0), (8,0) (10,0), (15,0), (20,0)

these isolated areas have an impact on the total recovery possible and on well completion locations in the waterflood simulations.

A. Visual Inspection

Figures 7a and b show a realisation generated by ROCKIT for each of one and nine control directions respectively, the particular directions used are given in Table I. These are representative of several sets of control directions studied, but neither set should be considered the optimum for characterising the training pattern. Also shown in figure 7c and d are realisations for each of the random cases discussed in section IV. These should be compared with figure 4 which shows the training pattern. Neither random case contains any features similar to the training pattern. However, the two ROCKIT realisations show many similar features, with even the single search direction realisation containing significant correlation in the horizontal direction. There is little correlation in the vertical direction. The nine search direction realisation shows sand streaks continuous across the full width of the section and ripple features similar to those in the training pattern. The former resemble those features in the training pattern that might be considered for incorporation deterministically.

TABLE II. Effective permeabilities calculated for synthetic geologies.

Realisation	k_{xeff} (perm units)	k_{zeff} (perm units)
Frontier Formation Data	0.437	1.7×10^{-5}
1 control direction #1	0.487	3.3×10^{-2}
" " " #2	0.324	2.8×10^{-5}
" " " #3	0.438	3.9×10^{-2}
9 control directions #1	0.474	2.2×10^{-5}
" " " #2	0.436	2.9×10^{-2}
" " " #3	0.477	1.4×10^{-5}
Random (1st Set) #1	0.124	1.39×10^{-1}
" #2	0.154	1.42×10^{-1}
" #3	0.087	1.31×10^{-1}
Random (2nd Set) #1	0.217	2.62×10^{-1}
" #2	0.209	2.49×10^{-1}
" #3	0.240	2.26×10^{-1}
Layered	0.658	9.5×10^{-6}

B. Effective Permeability

For the purposes of effective permeability calculations high permeability was taken as 1 permeability unit and low as 10^{-6} permeability units. The effective values calculated for the training pattern are 0.437 permeability units in the x-direction (horizontal) and 1.7×10^{-5} permeability units in the z-direction (vertical). This latter value is very low (practically zero) due to the almost continuous low permeability barrier towards the top of the section. Table II shows the effective permeability values calculated for three realisations for each of the sets of control directions shown in Table I, and the random cases discussed earlier. The randomly generated distributions all have an effective permeability in the x-direction which is much smaller than the training pattern value. The values in the z-direction are similar to the horizontal values. The random distributions which have no diagonal "connections" have larger effective permeabilities due to the increased proportion of connected high permeability facies. For both sets of control directions ROCKIT gives a good approximation to the x-direction effective permeability, but only in some

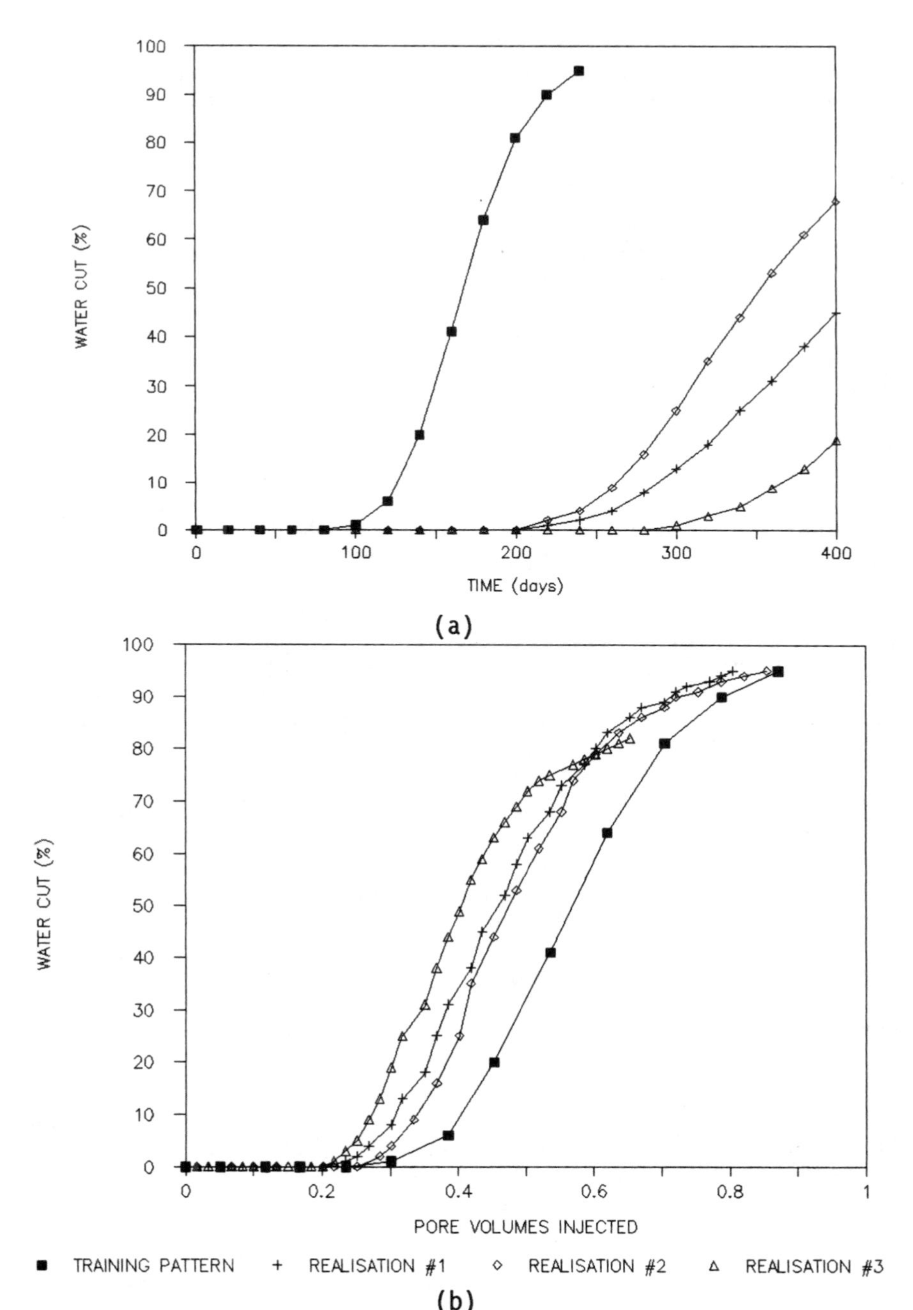

Figure 8. Water cut development for random distributions (a) against time and (b) against pore volumes injected.

realisations is the z-direction k_{eff} very low. This is because only some of the realisations have low permeability barriers extending completely across the section.

C. Unit Mobility Waterflood Simulations

Each of the ROCKIT and random synthetic geologies were incorporated into a two-dimensional cross-sectional numerical model as described in section IV. The important features of the numerical model to note are that capillary pressure and gravity effects are excluded and relative permeabilities are simplified and are linear. This was done so as to give the best indication of the impact of the heterogeneity on the flood front behaviour, without having to take into account factors introduced by the particular fluid flow process. Well controls were such that the pressure drop across the section was approximately the same from simulation to simulation. As the production and the injection rates varied, the results, particularly the water cut development, were analysed as a function of pore volumes injected as well as a function of time.

Figures 8, 9, 10 and 11 show water cut profiles against time and pore volumes injected for the random and ROCKIT realisations described in B above. Each figure also shows the profile for the training pattern for comparison.

The purely random realisations, figure 8, give a poor match to the water cut profile, against both time (figure 8a) and pore volumes injected (figure 8b). The very low effective permeability in the horizontal direction compared to the training pattern results in lower rates for the same pressure difference giving later breakthrough times and a slower increase in water cut, figure 8a. The breakthrough time when expressed in pore volumes injected is "earlier" than for the training pattern, figure 8b, since for these purely random distributions much of the pore volume is inaccessible, and less volume is being swept.

For the second set of random realisations, with no diagonal "connections", the match to water breakthrough time is slightly better (figure 9a), but still too late; the horizontal effective permeability is still low compared with the training pattern. However, the profile as a function of pore volumes injected is an excellent match (figure 9b). This is because each of these realisations contains about the same inaccessible pore volume as the training pattern (the volume is small). It should be noted that despite the fact

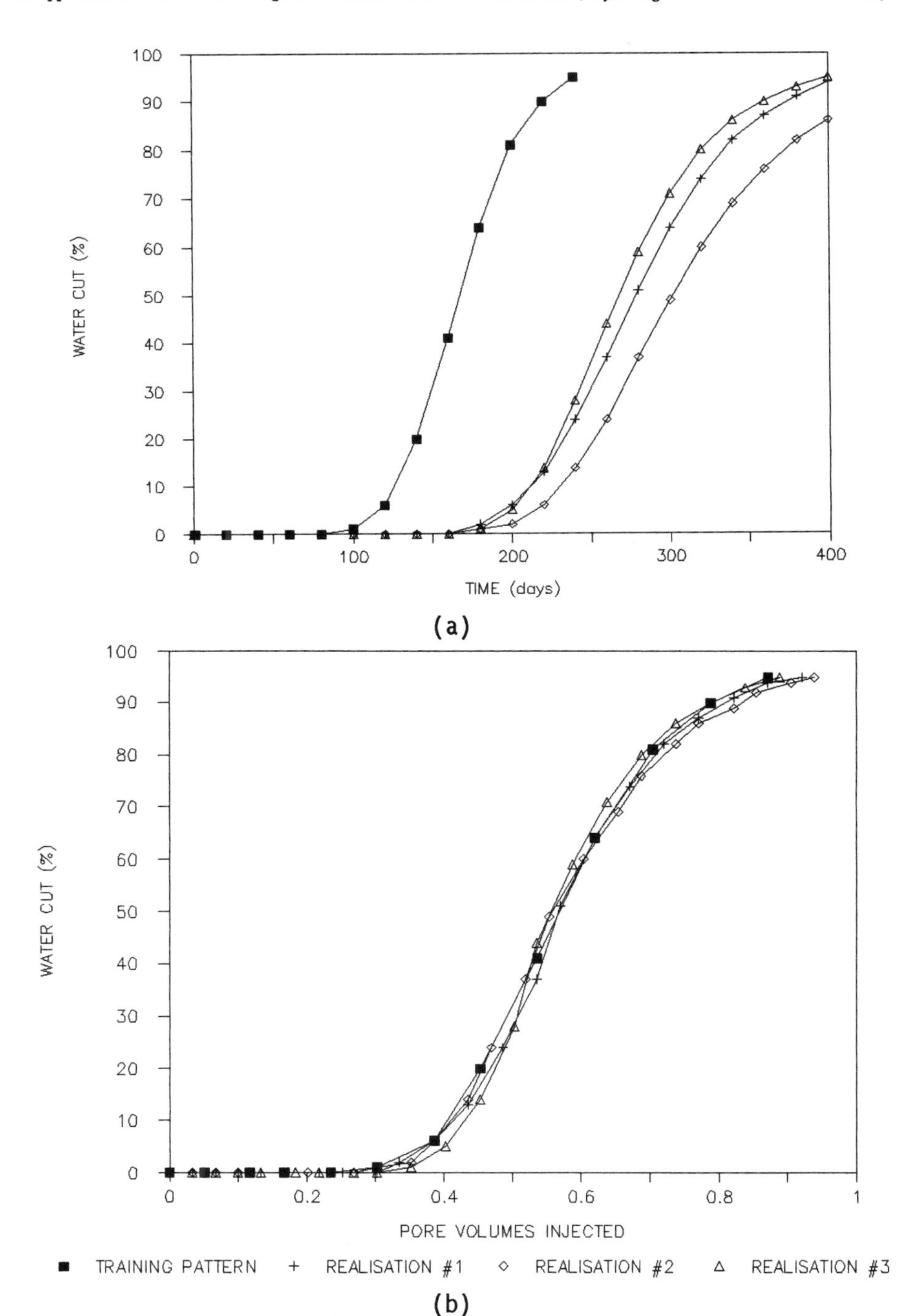

Figure 9. Water cut development for random distributions with no "diagonal connections" (a) against time and (b) against pore volumes injected.

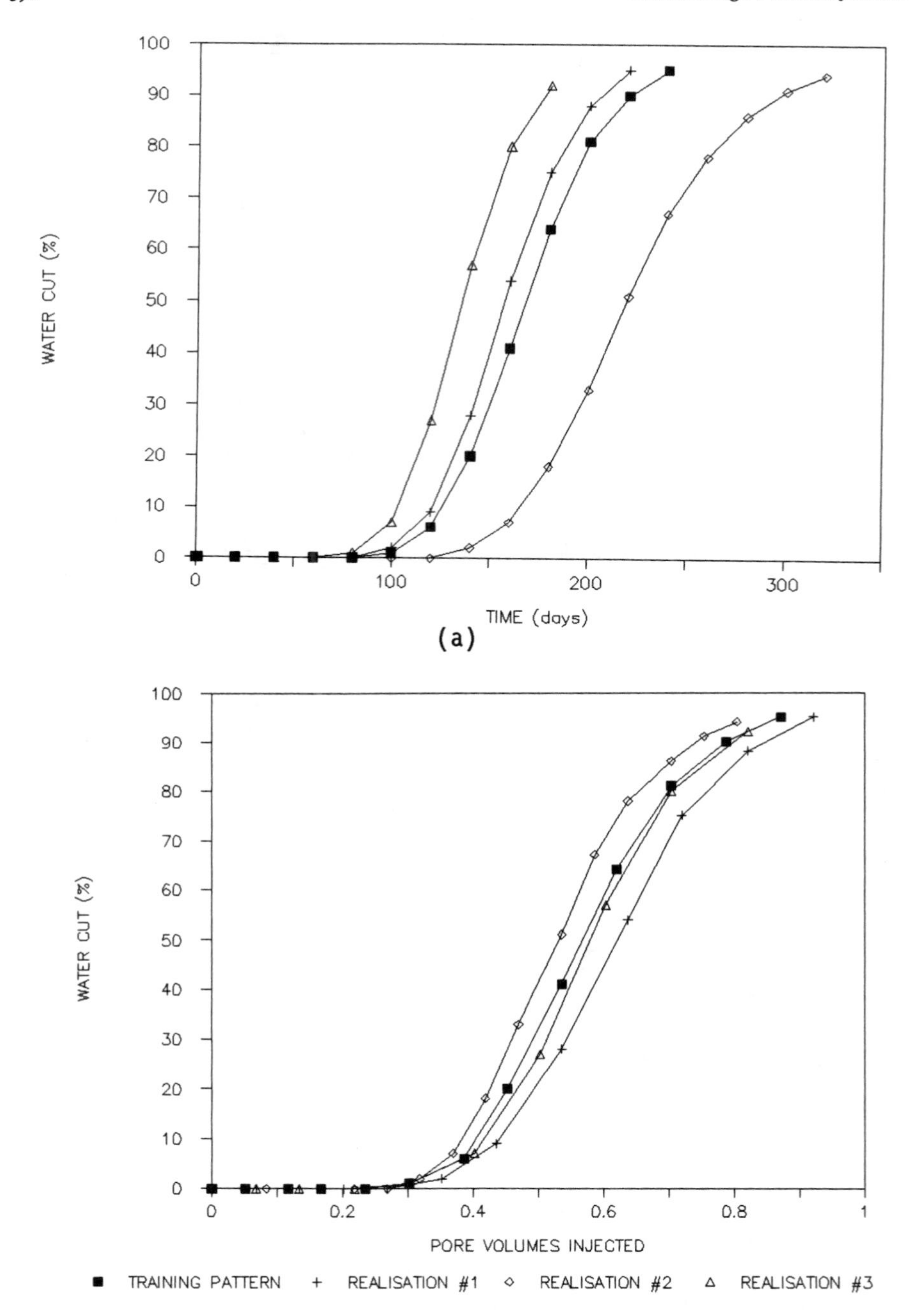

Figure 10. Water cut development for ROCKIT realisations with one control direction (a) against time and (b) against pore volumes injected.

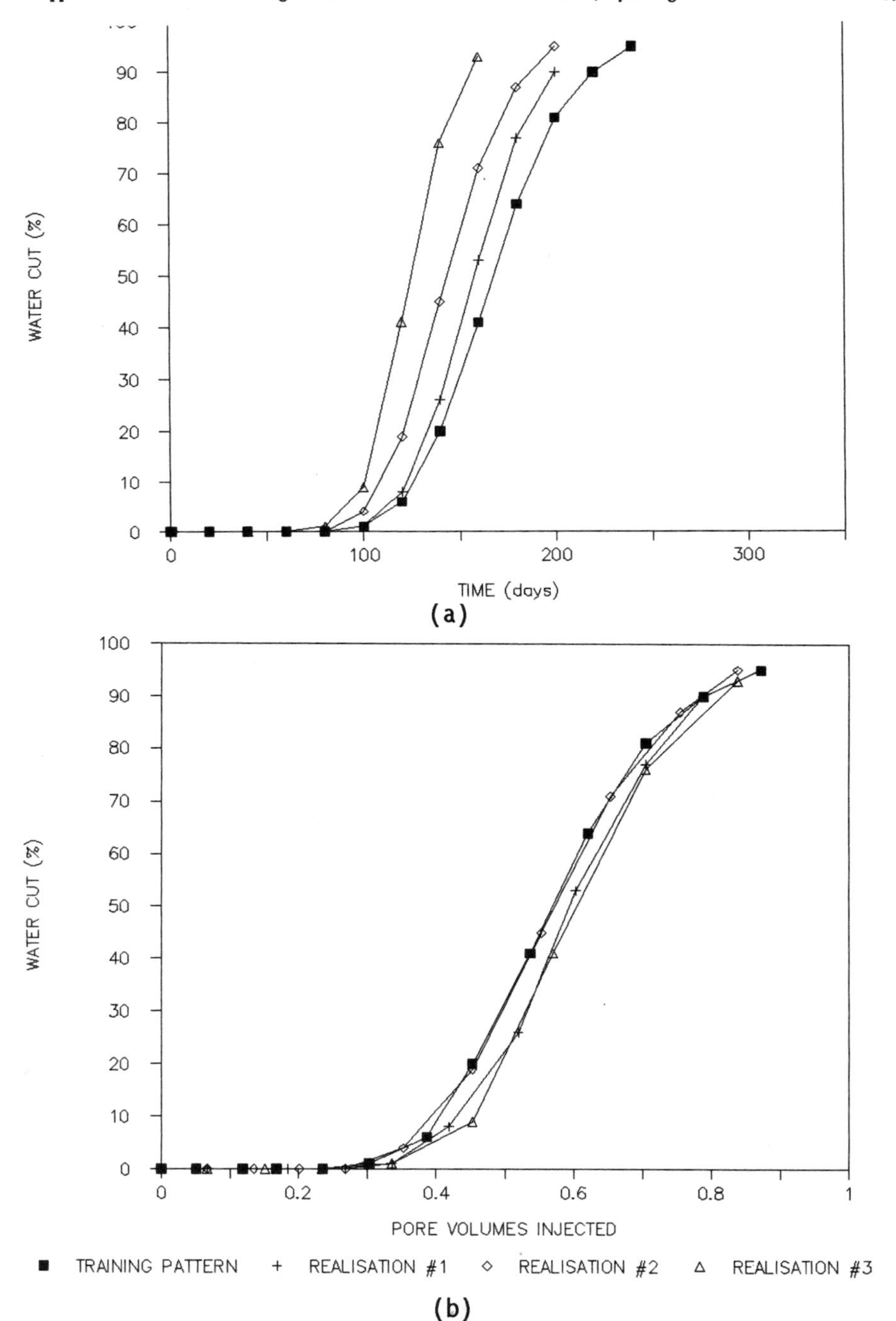

Figure 11. Water cut development for ROCKIT realisations with nine control directions (a) against time and (b) against pore volumes injected.

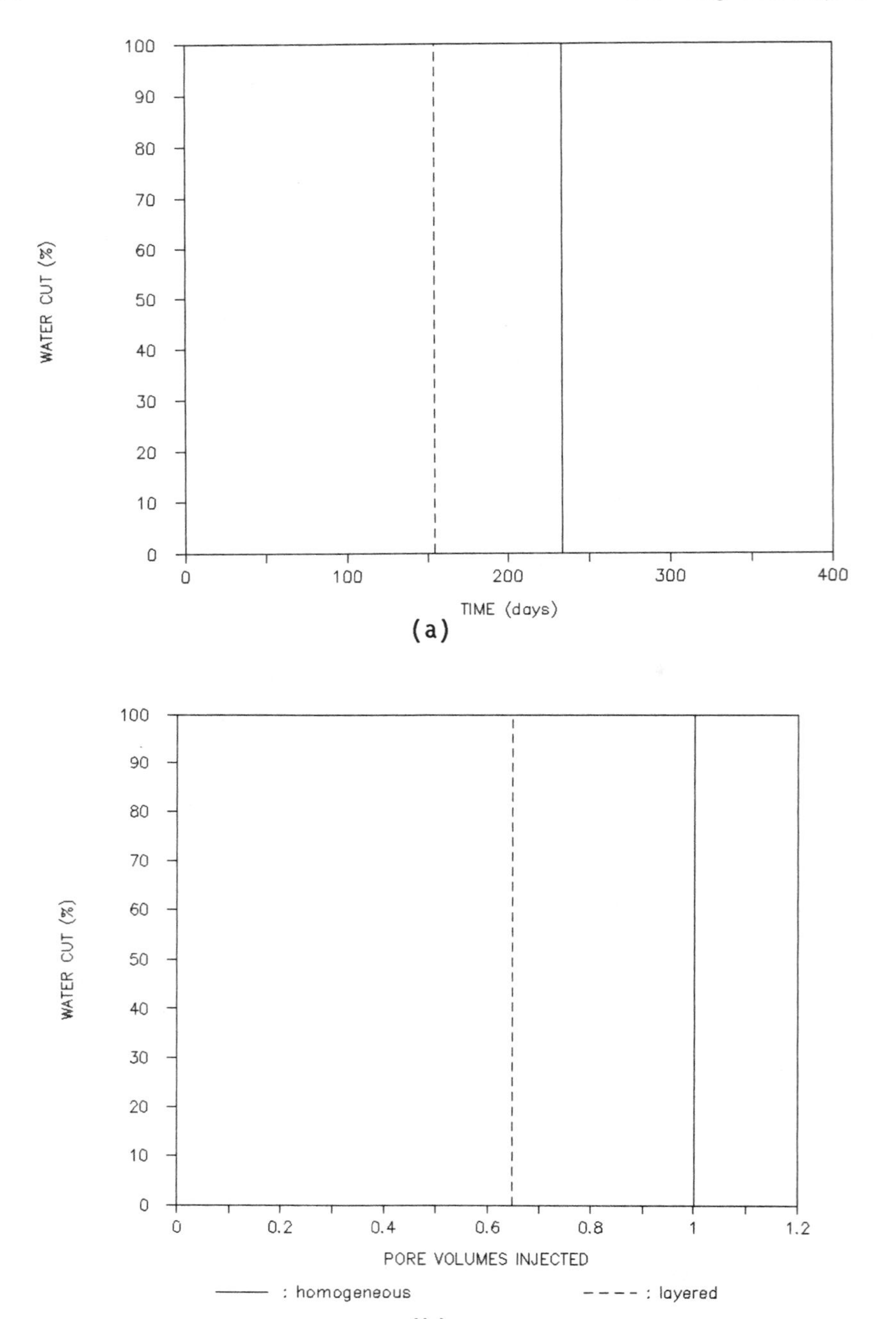

Figure 12. Water cut development for homogeneous and layered systems (a) against time and (b) against pore volumes injected.

that the match for these runs is the best of all the realisations considered here, the other results for these realisations are not as good, as we have seen. This emphasises the importance of considering several criteria for the determination of a "good match".

The ROCKIT realisations, figures 10 and 11, give a better match to water breakthrough time than the random distributions. However, in the case of one control direction, figure 10a, there is a wide variation between realisations in time to water breakthrough. The variation for the nine control direction case is much less, figure 11a, although all breakthrough times are a little early and the subsequent water cut development a little too steep. In terms of pore volumes injected, for one control direction, figure 10b, the variation in breakthrough time is reduced and each realisation gives a fair match to the training pattern's behaviour. For nine search directions, figure 11b, the variation in breakthrough times and watercut development are smaller again and the match to the training pattern's profile is very good.

Figure 12 shows the water cut profile obtained analytically for a homogeneous system of permeability equivalent to the effective values calculated for the training pattern. This figure also shows the profile for a layered system which honours the training pattern's global histogram. The effective permeabilities for this latter system are 0.657 permeability units in the horizontal direction and 10^{-5} in the vertical. These are equivalent to the arithmetic and harmonic means of the cell permeability values, respectively. The watercut curves are vertical lines since the straight line relative permeabilities give a non-dispersed flood front. Thus, the flood front saturation is 100% and displaces the oil in a piston-like manner to give an instantaneous increase from 0% to 100% in water cut at breakthrough. The breakthrough times are calculated assuming a pressure drop across the system typical of those observed in the simulations using random and ROCKIT geologies. Note that the breakthrough time for the homogeneous system expressed in pore volumes injected would reduce to that for the layered system if the system were considered to have a net-to-gross ratio as calculated from the training pattern (66%). The water cut profile would then be equivalent to that of the layered system. It is significant that for the layered system breakthrough occurs earlier than the 50% water cut in the training pattern. The x-direction effective permeability is higher in the layered system than in the training pattern.

Clearly the assumption of a layered or homogeneous system is not going to give the detail on the watercut profile of a truly heterogeneous system. Water breakthrough times cannot be estimated easily.

VI. DISCUSSION

The results show that the statistical method incorporated in the ROCKIT program is able to reproduce many of the visual, effective permeability and waterflood characteristics of the Frontier Formation data set studied here. In this sense the realisations generated by ROCKIT can be considered superior to the random, homogeneous or layered permeability distributions more usually employed to interpolate between observed data at wells. However, control parameters must be chosen in an intelligent manner in order to ensure good results. Control directions should be chosen to take into account any obvious geological features. At the present time it is not clear how to define an "optimum" set of control directions to characterise a training pattern other than by trial and error. However, it appears to be the case that very good realisations can be generated by ROCKIT using only a small number of control directions.

The geological features of the training pattern that could be incorporated deterministically are reproduced by ROCKIT in some realisations. Correlation on a scale larger than that of the section is difficult to reproduce with geostatistical methods, but ROCKIT performs well in this respect, with two out of the three realisations generated from nine control directions containing these features.

It would appear that the variation in characteristics between realisations decreases as the number of control directions increases. This seems to follow intuitively from the theoretical result that the training pattern can be completely defined by specifying all possible two-point histograms. This reduced variation increases confidence that the results from any one realisation are typical of the full set of possible realisations.

In assessing the technique each of the criteria used (visual inspection, k_{eff} calculation and waterflood performance) cannot be used in isolation to judge the adequacy of a particular realisation; they must be used together. For example, a good match to effective permeability does not guarantee a good match to waterflood behaviour (although a very bad match to k_{eff} will also be a bad match to waterflood behaviour). In the light of this, the criteria used for assessing a realisation should be somehow indicative of the ultimate use to which the synthetic geology will be put. For example, the "best" realisations for the unit mobility waterflood considered here are not necessarily the "best" for any other flow process, such as a gas flood.

In section II three theoretical advantages of the two-point histogram method were given. These had been presented in reference (1). In the light of the studies performed with ROCKIT, which are presented in this paper, it can be said that these theoretical advantages represent practical advantages when the method is used in an operational sense. Firstly, the fact that the method can be proved to converge simply by increasing the number of specified two-point histograms, although indirect in application, suggests that less variation between realisations will be observed for larger numbers of two-point histograms. This is seen to be the case in the results presented here. However, there are definite advantages to keeping the number of two-point histograms to a minimum, especially for computational reasons. This makes it crucial to define an optimum set. As seen here, ROCKIT gives very good results with only a small number of control directions, anyway.

Secondly, there are no difficult inverse problems to solve. Computationally, this results in converged solutions being generated in very short times. Each realisation was generated in the order of a minute on a CRAY-2; on a work-station realisations can be generated in under five minutes. Clearly, these times are problem dependant and will vary with array size.

Finally, intuitive control parameters are used by the method. Two-point histograms are easy to generate from training patterns, and can be easily calculated from the format in which data bases are collected by field geologists. The method has much to recommend it, but it must be used in a sensible fashion as the choice of control directions for the calculation of the two-point histograms is crucial.

VII. CONCLUSIONS

The main conclusions are:

1. The two-point histogram method has proved to be exceptionally effective at generating synthetic geologies that match the visual characteristics, effective permeabilities and unit mobility waterflood behaviour of a given training pattern.

2. Choice of control directions should take into account any obvious geological features, such as the geometry of correlated permeability structures.

3. A good match to any one criterion, such as effective permeability, does not necessarily guarantee a good match to any other criterion, such as waterflood performance. Obviously, the overall match would be unacceptable if the match to any one criterion was poor.

4. Different realisations for the same set of control directions give similar results. Variation between realisations decreases with increasing number of control directions.

5. The theoretical advantages of the method identified by Farmer in reference (1) have been shown to have practical significance when the method is used operationally. The convergence of the method leads to the definition of an optimum set of control directions. The absence of difficult inverse problems makes the program computationally fast allowing realisations to be generated in a short time. The control parameters (the two-point histograms) are intuitive, allowing easy definition of sensible control directions, and are easily calculated from the type of data collected by field geologists.

6. Extension of the assessment of the method should include generation of synthetic geologies in three dimensions. Other fluid flow processes with differing relative permeability, capillary pressure, gravity and viscosity properties could be considered in analysing the quality of the synthetic geologies.

ACKNOWLEDGEMENTS

The authors would like to thank Dr R Bibby, Messrs J M D Thomas, S A Randall and K I Rollett of the Winfrith Technology Centre for their help in the study of the two-point histogram method; Dr J J M Lewis of the Royal School of Mines, Imperial College (now at Heriot-Watt University) for his help in data gathering and assessment. The data were collected under the terms of a project sponsored by British Petroleum Development Limited at Imperial College. The research at Winfrith was funded by the UK Department of Energy.

REFERENCES

1. Farmer C L. The Mathematical Generation of Reservoir Geology. Presented at the Joint IMA/SPE European Conference on the Mathematics of Oil Recovery, Robinson College, Cambridge University, 25th - 27th July 1989.

2. Radcliffe N and Wilson G. Natural Solutions Give Their Best. New Scientist, No. 1712. April 1990.

3. Norris R J and Lewis J J M. The Geological Modelling of Effective Permeability in Complex Heterolithic Facies. SPE Paper 22692, to be presented at the 66th Annual Technical Conference, Dallas, Tx. October 6th - 9th, 1991.

ASSESSING DISTRIBUTIONS OF RESERVOIR PROPERTIES USING HORIZONTAL WELL DATA

Godofredo Perez[1]
Mohan Kelkar

The University of Tulsa
Tulsa, Oklahoma 74104

This paper presents a comprehensive geostatistical evaluation of the distribution of porosity in a carbonate reservoir using logs of a horizontal well and several closely spaced vertical wells. In many field cases, a significant part of the uncertainty in descriptions of reservoir properties can be due to limited information about the nature of the spatial distribution of properties in inter-well regions. The major contribution of this paper is to provide guidelines to assess information about, rarely available, small scale variability of reservoir properties in inter-well regions, from the widely available measurements at vertical wells.

I. INTRODUCTION

---

[1]Currently at ARCO Oil and Gas Company, Plano, Texas 75075.

Proper quantification of the spatial distribution of properties and associated uncertainties is important for most reservoir engineering applications. For example, descriptions of reservoir properties are needed for oil-in-place evaluation, displacement mechanisms models and infill well selection. As a result, a major goal of recent reservoir characterization practices has been to develop methods to describe spatial distributions of properties which incorporate observed heterogeneity and variability due to the geologic complexities. Geostatistics is a methodology to quantify spatial variability and to generate descriptions at the unsampled inter-well regions which conform with observed statistical attributes of a property.

In several field cases, porosity logs (neutron, density and acoustic) are the only source available to estimate distributions of reservoir properties. A conventional approach to develop descriptions of properties is to divide the reservoir into layers of constant properties and/or to include lateral variability by interpolation of *average* values from well logs. The assumptions behind this conventional approach can result in significant errors, because it ignores the geologic complexities, inherent heterogeneity of reservoir properties, and the uncertainties associated with the relatively small amount and irregular density of the sample data.

Distribution of properties, such as porosity and permeability, are affected by depositional and post-depositional geologic processes and must possess certain spatial continuity in order for the hydrocarbon accumulations to be productive reservoirs. This nature of reservoirs controls the spatial continuity of properties along different directions in the reservoir.

Information about the magnitude and variation of properties at small scales in the inter-well regions of reservoirs is limited, because sources of direct measurements, such as horizontal wells and seismic data, may not be available. Only a few studies, based on data from outcrops and horizontal wells, provide analyses and comparisons of the statistics and the spatial correlation of properties along different directions in reservoirs. Measurements of rock properties in outcrops may provide a viable alternative to overcome the difficulties of sampling reservoirs along different directions.

The work of Smith[1] includes a statistical analysis of core measurements of porosity, permeability, rock compressibility and grain size in a stratified, unconsolidated sandstone outcrop. The sampling configurations in this study include transects parallel and perpendicular to the direction of stratification. The observations of Smith[1] indicate that the means of porosity and permeability is the transects parallel and perpendicular to the direction of stratification are close. However, the standard deviations of porosity and permeability for the perpendicular transect are about two times greater than for the parallel transect. Smith's[1] investigation indicates that the autocorrelation functions of porosity and permeability are similar for the parallel transect, but these have different character for the perpendicular transect. Even though the character of the autocorrelation functions are different for the parallel and the perpendicular transects, the extent of the correlation is of the same order of magnitude for these two transects.

Goggin et al.[2] presented a geologic description and a statistical analysis of permeability for an eolian sandstone outcrop. The permeability data in this study consisted of core calibrated mini-permeameter measurements in five concentric outcrop grids. The dimensions of the grids ranged from a few feet to hundreds of feet. The observations of Goggin et al.[2] indicate that permeability distribution varies for different stratification types in this eolian outcrop and the coefficient of variation of permeability follows an irregular trend as a function of the area of the concentric grids. The correlation ranges of permeability for the horizontal and vertical directions observed in the spatial analysis are significantly different for the larger and the smaller grids. The ratios of horizontal to vertical correlation ranges for the largest and the intermediate grids are 1.5 and 3.0, respectively, while, for the smallest grid this ratio is 1/6. Goggin et al.[2] explain that these differences are a result of the distinct character of the heterogeneities for the different types of stratification.

Kittridge et al.[3] conducted a study to evaluate the statistical properties of permeability in an outcrop of a carbonate formation. In addition, observations from the outcrop were compared to statistical analyses of core data of a field producing from the same formation. The permeability sample data for the outcrop consisted of mini-permeameter measurements in a large grid (100 by 80 feet), six small grids within the large grid and a vertical transect. The univariate statistical analysis of permeability indicates that the mean from the outcrop data differs by several orders of magnitude from the field measurements, but the coefficients of variation are similar for these two sources. On the spatial analysis of permeability, Kittridge et al.[3] found that the correlation range of the vertical transect is in close agreement with the average correlation range of the core data from several wells in the field. For the field data, the correlation range of permeability and porosity are similar.

More recently, Crane and Tubman[4] evaluated the spatial correlation of density, neutron and sonic logs of horizontal and vertical wells with fractal models in a carbonate and a sandstone reservoir. In both reservoirs, Crane and Tubman[4] used models of fractional Gaussian noise processes to study the spatial correlation of the logs. For the carbonate reservoir, the intermittency exponent for three horizontal logs range between 0.85 and 0.93, and for four vertical logs range between 0.88 and 0.89. For the sandstone reservoir, the average intermittency exponents of the logs for five horizontal wells is 0.90 and for seven vertical wells is 0.83. A conclusion of this investigation[4] is that the depositional and post-depositional environments appear to influence the differences or similarities between the intermittency exponents of the horizontal and vertical logs. According to Crane and Tubman,[4] for the uniform marine depositional environment of the carbonate reservoir, values of the horizontal and vertical intermittency exponents are similar, while for the more complex braided-stream environment of the sandstone reservoir, the horizontal and vertical intermittency exponents are different.

The first section of this paper is a background about fractal models used in this investigation to analyze the spatial correlation of reservoir properties. The Study Region section provides a brief description of the geology and the sources of well data for the carbonate field discussed in this paper. The following sections are univariate and spatial statistical analyses of porosity and highlight comparisons among the statistical attributes of porosity for the horizontal and the vertical logs. The analysis is extended to investigate the effect of sample volume size on the spatial statistics. Finally, three-dimensional porosity distributions are generated with a conditional simulation method. Distributions are generated using different models derived from the vertical logs and compared to the distribution observed in the horizontal log.

II. FRACTAL MODELS

Statistical fractal models are useful to represent spatial records and time series of natural phenomena. Hewett[5] proposed an approach to model the spatial variations of reservoir properties based on the concept of statistical fractals. The two statistical fractal processes used in this investigation are the fractional Brownian motion (fBm) and the fractional Gaussian noise (fGn). A Comprehensive description and theoretical developments of statistical fractal models are given by Mandelbrot and Van Ness.[6] More general reviews on the application of fractals to model porous media are given by Williams and Dawe[7] and Sahimi and Yortos.[8]

The spatial correlation model of a fBm process is a power-law relation given by[6]

$$\gamma(h) = V_H h^{2H} \tag{1}$$

fGn is approximately the derivative of fBm. An expression for the spatial correlation model of fGn processes has been derived by Mandelbrot and Van Ness[6] and Mandelbrot,[9] and it is given by

$$\gamma(h) = \frac{1}{2} V_H \delta^{2H-2}\left[2 - \left(\left|\frac{h}{\delta}\right| + 1 \right)^{2H} + 2 \left|\frac{h}{\delta}\right|^{2H} - \left(\left|\frac{h}{\delta}\right| - 1 \right)^{2H} \right] \tag{2}$$

The intermittency exponent, H, in Equations 1 and 2 is related to the fractal dimension and it varies between zero and one. Different ranges of H represent different types of spatial correlation. For $0 < H < 0.50$, the correlation is anti-persistent (high values tend to be followed by low values and vice versa) and for $0.50 < H < 1$, the correlation is persistent (successive values tend to be of similar magnitudes). For the case where H = 0.50, Equation 1 represents the semi-variogram of a Brownian motion or a random walk and Equation 2 is a constant corresponding to a Gaussian or white noise. In Equation 2, the factor $V_H\delta^{2H-2}$ is the variance of a fGn process. The parameter δ in Equation 2 is a smoothing factor which is required to differentiate fBm in order to arrive at the spatial correlation model for fGn.[6]

In theory, the correlation ranges of fBm and fGn processes are infinite (except for a fGn process with H = 0.50). The semi-variogram models of fBm and fGn in Equations 1 and 2 are shown in Figure 1 for different intermittency exponents. The semi-variograms of fBm for intermittency exponents greater than 0.50 increase without bounds as the lag increases, while for fGn the semi-variograms tend to a constant value. The semi-variograms of fBm for small intermittency exponents (close to zero) are similar to the semi-variograms of fGn with large intermittency exponents (close to one). These similarities are illustrated in Figure 2 for the semi-variograms of fBm for $0.05 \leq H \leq 0.50$ and of fGn for $0.50 \leq H \leq 0.95$.

The smoothing factor, δ, in Equation 2 accounts for the influence of the sample size on the spatial correlation structure.[5,6] An example of this sample size effect on the semi-variogram of a fGn process is illustrated in Figure 3. In this example, the semi-variograms of fGn are compared for two cases with sample spacings equals to 1 and 5 units and smoothing factors equal to corresponding sample spacing. For both cases, the intermittency exponent and the factor V_H have the same value. Figure 3 shows that for the same lag, the magnitude of the semi-variogram is smaller for the larger sample spacing. This observation indicates that the semi-variogram of fGn accounts for the expected reduction on the variability of the processes as the sample size increases.

Several methods are available to estimate the intermittency exponent of fBm and fGn processes. For a fBm process, the power-law relation given by Equation 1 can be used to estimate the intermittency exponent directly from the

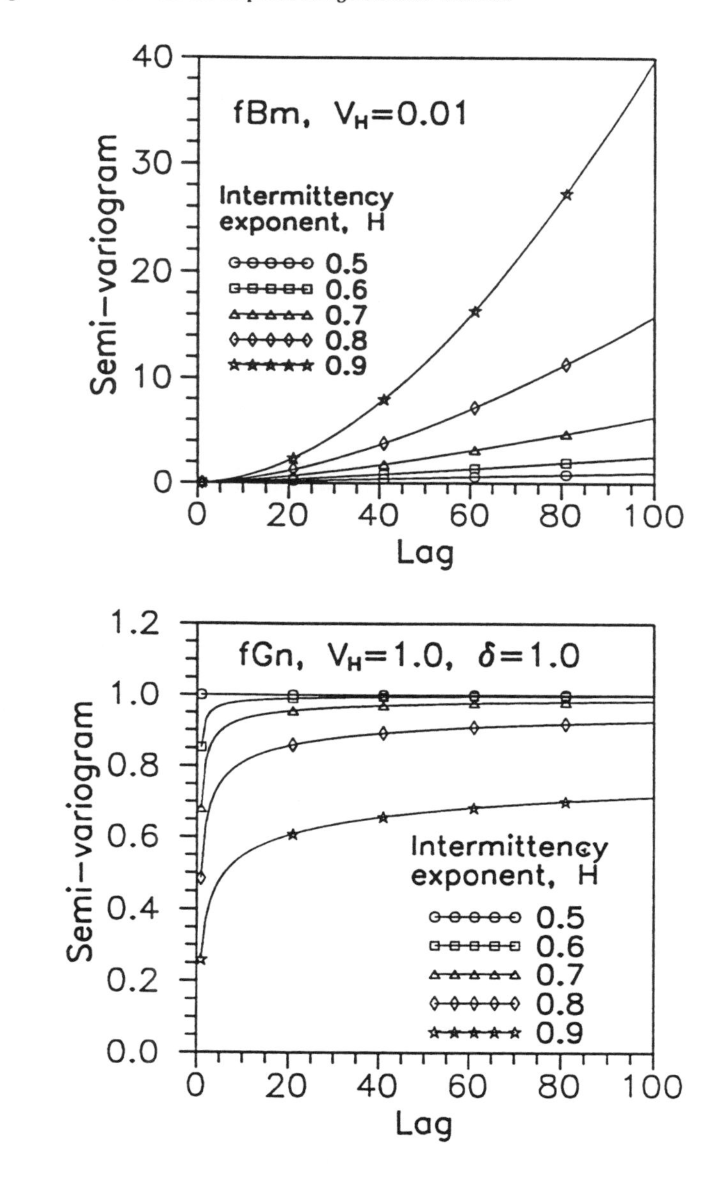

Figure 1: Semi-Variogram Models of fBm for $V_H = 0.01$ and of fGn for $V_H = 1.0$ and $\delta = 1.0$

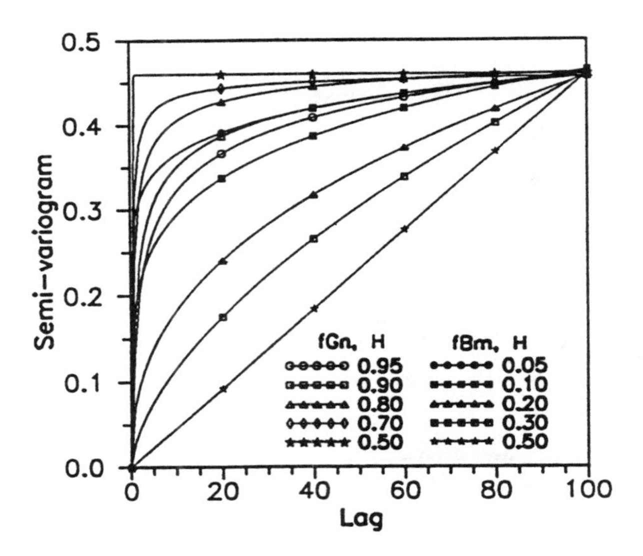

Figure 2: Comparison of Semi-Variogram Models of fBm for $0.05 \leq H \leq 0.50$ and fGn for $0.50 \leq H \leq 0.95$ and $\delta = 1.0$

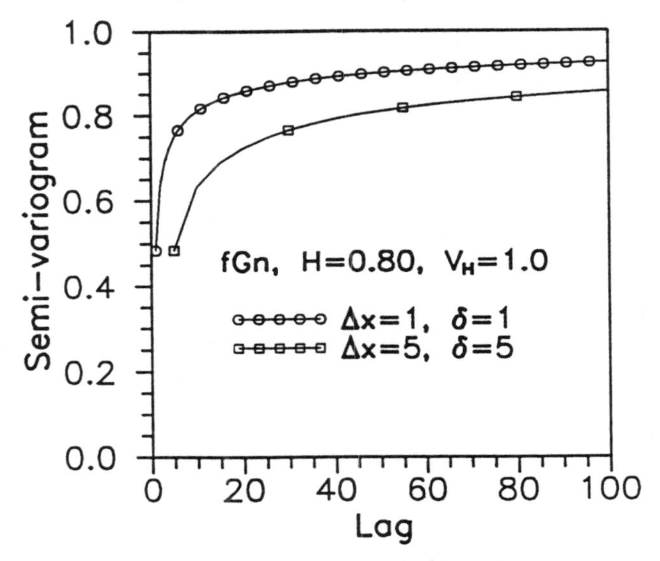

Figure 3: Comparison of Semi-Variogram Models of fGn for Different Sample Spacings and Size

slope of a plot of sample semi-variogram versus lags on logarithmic coordinates. Hewett[5] describes another method to evaluate the intermittency exponent of fBm using an expression for the spectral density representation of the semi-variogram given by Equation 1. The methods to evaluate the intermittency exponent of fGn include the Re-scaled range (R/S) analysis, spectral density and the grading methods. An extensive description of these methods is given by Hewett.[5] The R/S analysis is based on a scaling expression developed by Hurst[10] which relates the sequential range of the cumulative departures from the mean, lag, and intermittency exponent. In the spectral density method the intermittency exponent is calculated from an approximation of the spectral density of the semi-variogram of fGn. The grading methods consist of applying the methods developed to analyze fBm, such as the semi-variogram and the spectral density methods, to a graded fGn process. The graded sequence is the cumulative value or the discrete integral of a fGn process. Another method to estimate the intermittency exponent of a graded fGn sequence is the box-counting method described by Feder.[11] The box-counting method is based on the scaling expression for deterministic fractals which relates the number of boxes required to cover the trace of the process, the size of the boxes and the intermittency exponent. According to Hewett,[5] an advantage of the grading methods is that a graded sequence is more regular and may be easier to evaluate than a more erratic fGn process.

III. STUDY REGION

This section provides a brief geologic description of the carbonate reservoir studied in this paper. Then, the locations of the wells within the study region and the available well data are described.

A. Geology

The data set used in this investigation consists of porosity logs from eight wells in a tight carbonate reservoir. The predominant lithology of the reservoir is dolomite. The oil producing formation is very complex and heterogeneous due to the post-depositional processes, such as cementation and several re-dolomitization cycles. Other heterogeneities observed in the cores include cement filled Burrows and Vugs. As a result, the formation is a very tight rock (low porosity and permeability) and includes non-porous intervals.

The producing formation has an average thickness of fifty feet and it is overlayed by a dense shale (the oil source rock). At the bottom of the formation, a tight shale streak is present in many wells which probably prevented the migration of oil to a high porosity region located immediately below.[12]

B. Well Data Set

The horizontal well and the surrounding vertical wells used in this study are within a surface area of approximately one square mile. Figure 4 shows a map of the location of all the wells. The relative position and length of the horizontal log with respect to the vertical logs can be seen in the vertical profile shown in Figure 5. The horizontal section of the horizontal well (dotted line in Figure 5) is 1481 feet long and it has been selected by ensuring that the angle of inclination at each point does not deviate by more than five degrees from a horizontal reference. The vertical logs vary in length from 365 feet (Well S5) to 721 feet (Well S2) and these logs include reservoir regions above and below the oil production formation.

The porosity logs have been derived from neutron and density logs at a one foot spacing. These porosities include conventional log corrections, such as lithology and environmental corrections.

IV. ANALYSIS OF UNIVARIATE STATISTICS

The univariate parameters used in the statistical analysis of the log sample data are the mean, the variance, the coefficient of variation (standard deviation divided by mean), the range, the median, and the quartiles. Table 1 summarizes these parameters for the porosity logs of the horizontal well, all combined

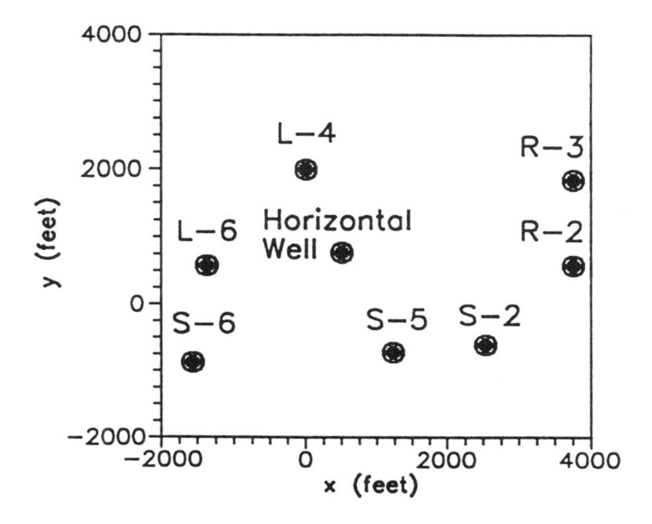

Figure 4: Map of Well Locations in the Study Region

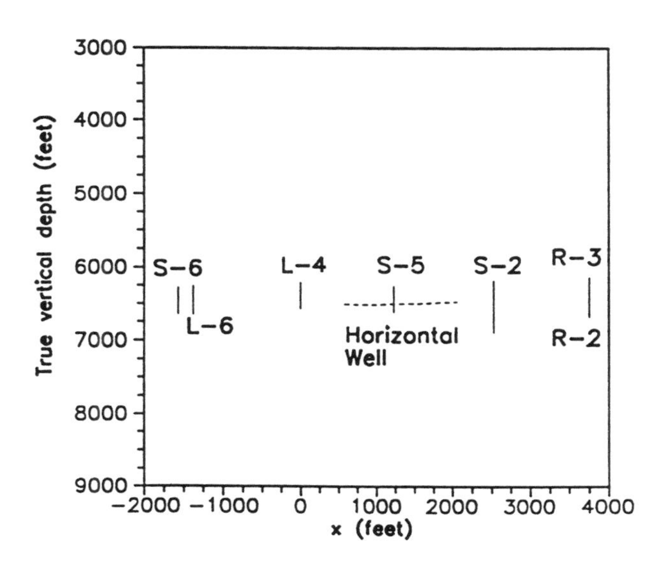

Figure 5: Vertical Profile of Wells in the Study Region

vertical wells and the individual vertical wells. Figure 6 shows the porosity log of the horizontal well and Figure 7 shows the logs of two vertical wells. In order to facilitate the comparison of the relative variability among the logs porosities in Figures 6 and 7 are standardized to zero mean and unit variance.

Overall mean porosity of the logs is low because of the tight nature of the reservoir. The mean porosity of all the vertical logs is greater than the mean porosity of the horizontal log. The mean porosity of only two vertical logs (Wells L6 and S5) is lower than the mean porosity of the horizontal log. The mean porosity of the horizontal logs is low because it intercepts a large fraction of non-porous regions (zero porosity measurements) as indicated in Figure 6 and as reflected in the low values of the first quartile and the median. The first quartile of porosity for the vertical logs is higher than for the horizontal log.

Table 1
Univariate Statistics of Porosity Logs

Well Name	Number	Mean (%)	Variance (%^{2})	Coefficient of Variation	Range* (%)	Median (%)	1st Quartile (%)	3rd Quartile (%)
Horizontal	1461	1.389	3.518	1.350	9.780	0.380	0.000	2.470
All Vertical	3183	1.752	5.392	1.326	17.74	1.060	0.360	2.180
L4	371	1.470	1.771	0.905	8.000	1.240	0.540	2.010
L6	401	1.243	2.008	1.140	9.930	0.860	0.360	1.510
R2	521	1.905	3.791	1.022	11.19	1.390	0.620	2.540
R3	434	2.057	4.962	1.083	10.68	1.555	0.470	2.690
S2	721	2.222	13.02	1.624	17.74	0.860	0.190	2.230
S5	365	1.350	2.155	1.087	7.170	0.880	0.350	1.810
S6	370	1.491	2.694	1.101	10.02	1.040	0.310	2.200

*The range is also equal to the maximum porosity since the minimum is equal to zero for all the wells.

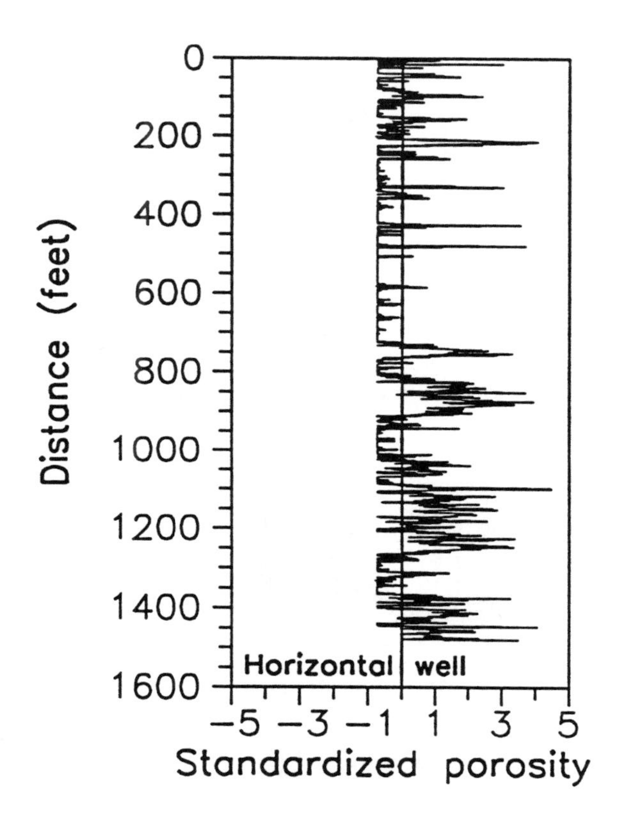

Figure 6: Standardized Porosity Log of Horizontal Well
(Zero Mean and Unit Variance)

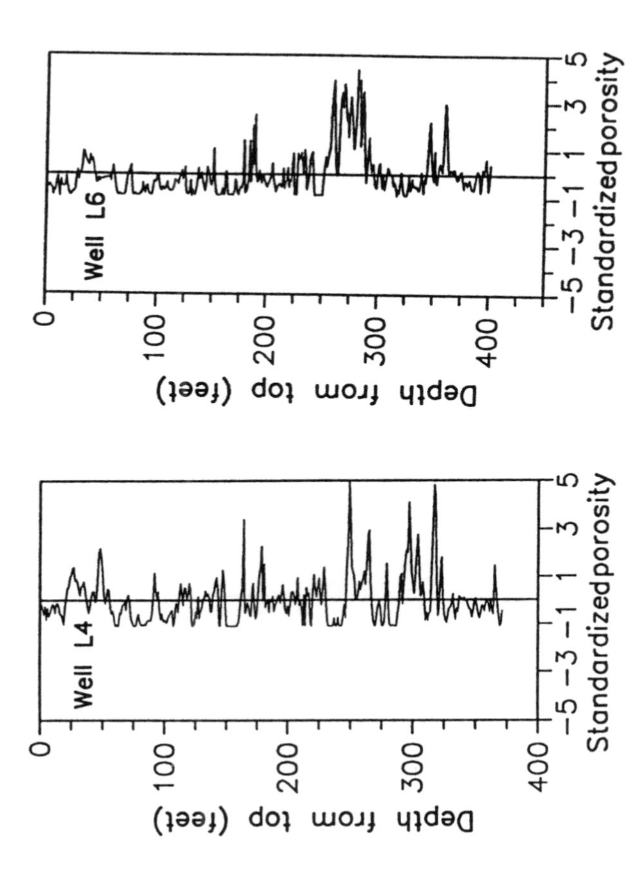

Figure 7: Standardized Porosity Logs of Wells L4 and L6 (Zero Mean and Unit Variance)

The porosity variance of the horizontal log is smaller than the variance of all combined vertical logs. However, the variance of the horizontal log is within the range of variances of individual vertical logs. Referring to Figure 4, it can be noted that the three wells (Wells L4, L6, and S6) with a lower variance are on the west side of the horizontal well, while the three wells (Wells R2, R3 and S2) with a higher variance are in the east side of the horizontal well. The remaining well, Well S5, has a porosity variance smaller than the variance of the horizontal well and it is located between the groups of wells with low and high porosity variance.

The coefficients of variation of the porosity logs from the horizontal well and all the vertical wells logs are close in magnitude. For the horizontal well, the coefficient of variation is slightly higher (greater than one) due to the low magnitude of the mean porosity. Except for Well S2, the coefficients of variation of individual vertical logs are close to one. The coefficient of variation is high for Well S2 because of a small region in the lower part of the formation with abnormally high porosities.

The histogram of porosity of the horizontal log is shown in Figure 8 and of all the vertical logs is shown in Figure 9. It can be noted that the distributions shown in Figures 8 and 9 are skewed to the right due to the large fraction of low porosity measurements. For the horizontal well, the high frequency (about 43 %) of the first class is due to the large number of zero porosity values. The small frequency of the last porosity class for the vertical logs is due to the high porosity region of Well S2 (with a maximum porosity of 17.74 %), which is also responsible for the irregularities in the coefficient of variation, as explained above.

V. ANALYSIS OF SPATIAL STATISTICS

Spatial correlation and variability of the porosity logs from the horizontal and the vertical wells are evaluated with fractal models and the semi-variogram analysis. The intermittency exponents of the fractal models are evaluated with

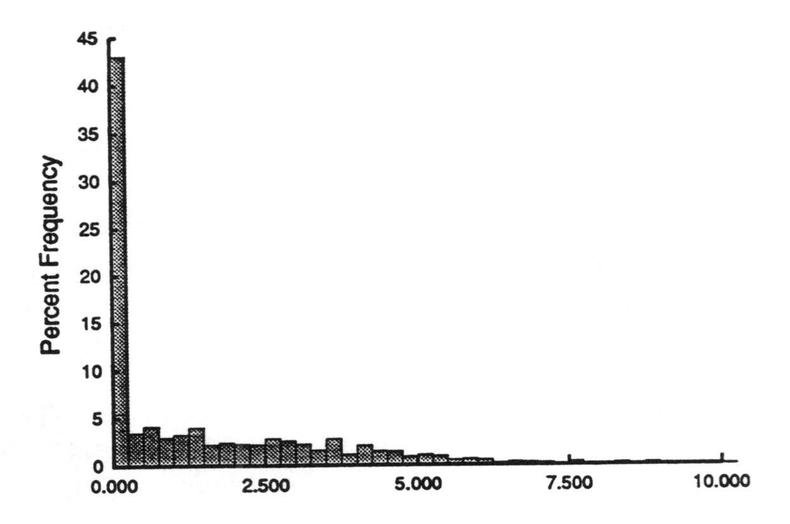

Figure 8: Histogram for Porosity (%) Log of Horizontal Well

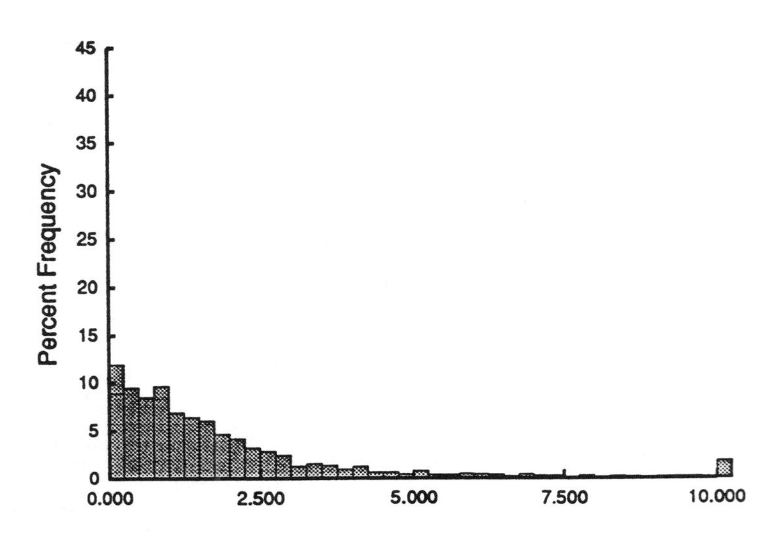

Figure 9: Histogram for Porosity (%) Logs of Vertical Wells

the re-scaled range analysis described in the Fractal Models section. The semi-variogram analysis is mainly used in this section to evaluate the correlation range or the distance over which the property is correlated. The semi-variogram analysis is extended to investigate the effect of the sample volume on the spatial statistics of porosity.

A. Fractal Analysis

The results of the R/S analysis of the porosity logs for all the wells are summarized in Table 2. These results are the slopes of the line fitted to the plots of R/S versus lag in log-log coordinates, as shown in Figure 10 for the horizontal log and for the log of Well L4. The horizontal spatial correlation can be represented with a power-law semi-variogram model of a fBm process (Equation 1). In the case of a fBm process, the slope of the R/S plot is approximately equal to one plus the intermittency exponent; thus, the intermittency exponent for the horizontal log is equal to 0.073. Indeed, a plot in logarithmic coordinates of the sample semi-variogram versus lag for the horizontal log follows a well defined linear trend, as illustrated in the Semi-Variogram Analysis section. The intermittency exponent calculated from the semi-variogram plot of the horizontal log is equal to 0.13 and it is in a close agreement with the intermittency exponent calculated in the R/S analysis. For all the vertical wells, Table 2 shows that the R/S slope is smaller than one. This implies that the spatial correlation of the vertical logs can be represented with models of fGn processes.

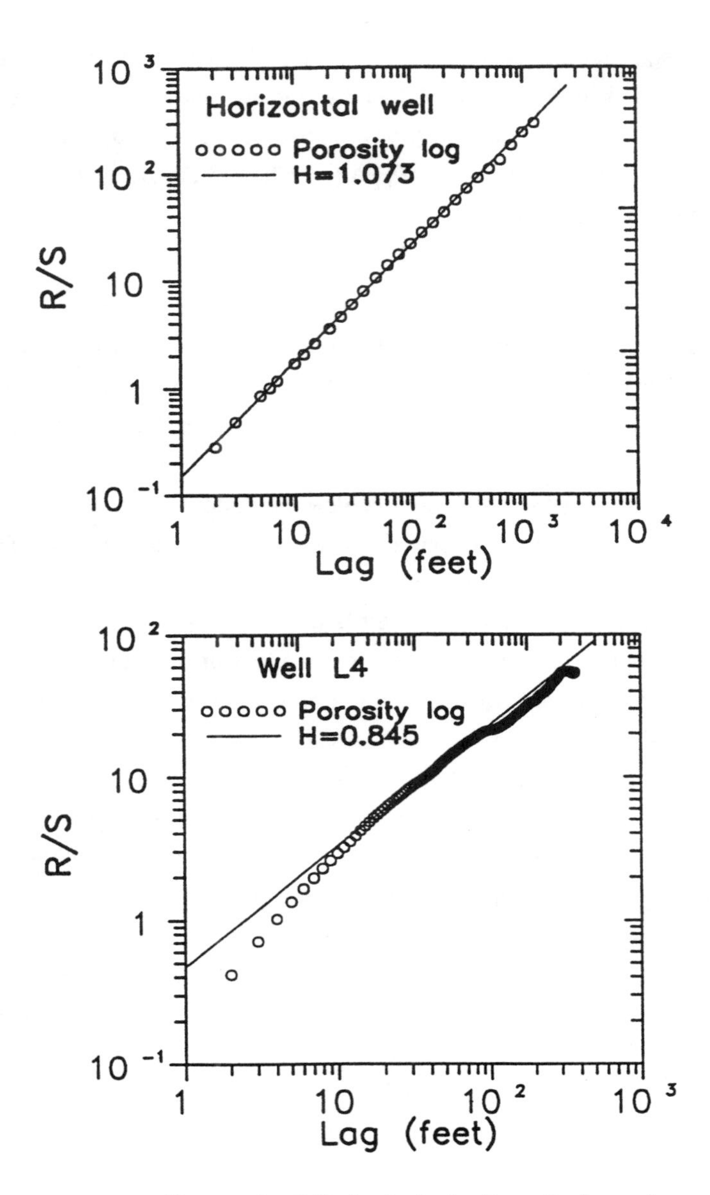

Figure 10: R/S Analysis for Porosity Log of Horizontal Well and Well L4

Table 2

Spatial Statistics of Porosity Logs

Well Name	R/S Slope*	Apparent Range (Feet)	Apparent Range/ Horizontal Range
Horizontal	1.073	143	1.000
All Vertical	-	56	0.392
L4	0.845	9	0.0621
L6	0.965	29	0.200
R2	0.778	40	0.276
R3	0.928	32	0.221
S2	0.970	59	0.407
S5	0.928	26	0.179
S6	0.868	17	0.117

*If the R/S slope is less than one, then this slope is equal to the intermittency exponent of fGn model. Otherwise, if the R/S slope is greater than one, then the slope minus one is equal to the intermittency exponent of a fBm model.

In the case of the horizontal well, a fBm intermittency exponent smaller than 0.5 indicates an anti-persistent behavior or that deviations from the mean are likely to be followed by deviations of the opposite sign. This behavior occurs because the horizontal well intercepts many times non-porous regions (zero porosity) resulting in many fluctuations of opposite sign around the mean line, as indicated in Figure 6. This characteristic of the formation may be a reason why a fBm model is obtained to describe horizontal well data instead of the fGn models observed by Crane and Tubman[4] for three horizontal wells in a carbonate reservoir. The intermittency exponents of the vertical well logs are in the range of values observed by Hewett[5] and Crane and Tubman[4] in carbonate and sandstone reservoirs.

B. Semi-Variogram Analysis

The porosity semi-variograms of the horizontal and two of the vertical porosity logs are compared in Figure 11. The apparent correlation ranges calculated from these semi-variograms are given in Table 2. Here, the apparent

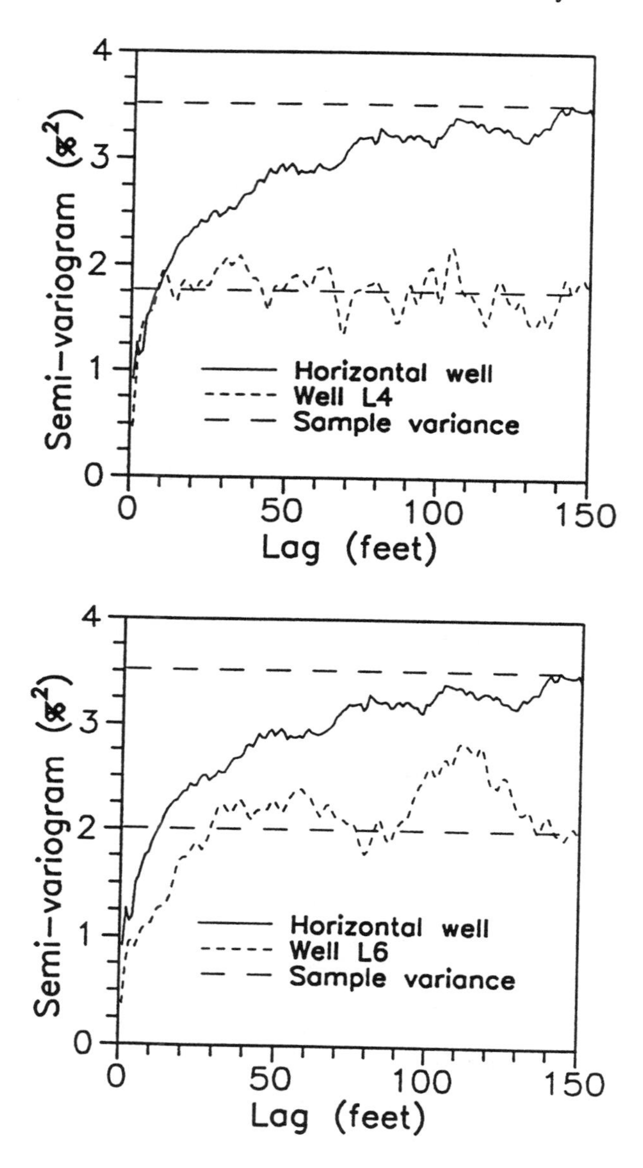

Figure 11: Comparison of Semi-Variograms of Porosity Log of Horizontal Well and of Wells L4 and L6

correlation range is defined as the smallest lag at which the semi-variogram becomes greater or equal to the variance of the samples. This definition of the apparent correlation range, depending on the behavior of the semi-variogram, may be slightly different than a correlation range calculated by fitting a model. However, the apparent correlation range has the advantage that it is independent of a model and it is not subject to the bias involved in fitting a model.

The apparent correlation range of the horizontal log is more than two times longer than any of the apparent ranges of the vertical logs. The vertical wells with porosity variance greater than the horizontal well have an apparent correlation range longer than the vertical wells of lower porosity variance. The anisotropy ratio (vertical divided by horizontal apparent correlation range) of the vertical wells is between 0.0621 (Well L4) and 0.407 (Well S2).

The shape of the semi-variogram is a measure of the degree of relative variability of the spatial distribution of a property. The extreme shapes correspond to a completely uncorrelated distribution (constant semi-variogram with magnitude approximately equal to the sample variance) and a distribution consisting of a constant value (semi-variogram equal to zero). In order to compare the shape of the semi-variograms, the sample semi-variograms have been normalized by dividing the lag by the apparent correlation range and the semi-variogram by the sample variance of each porosity log. These normalized semi-variograms for the horizontal and vertical logs are shown in Figure 12. The vertical wells with a low porosity variance have similar semi-variogram shapes and are slightly more continuous than the shape of the horizontal well semi-variogram. For the vertical wells of high porosity variance, only Well R2 has a semi-variogram shape which indicates more relative variability than the horizontal well semi-variogram; the other two wells (Wells R3 and S2) have semi-variogram shapes similar to the low porosity variance wells.

The intermittency exponents calculated in the Fractal Analysis section are another measure of the degree of relative spatial variability of a variable around the mean. The results of the Fractal Analysis section indicate that the horizontal

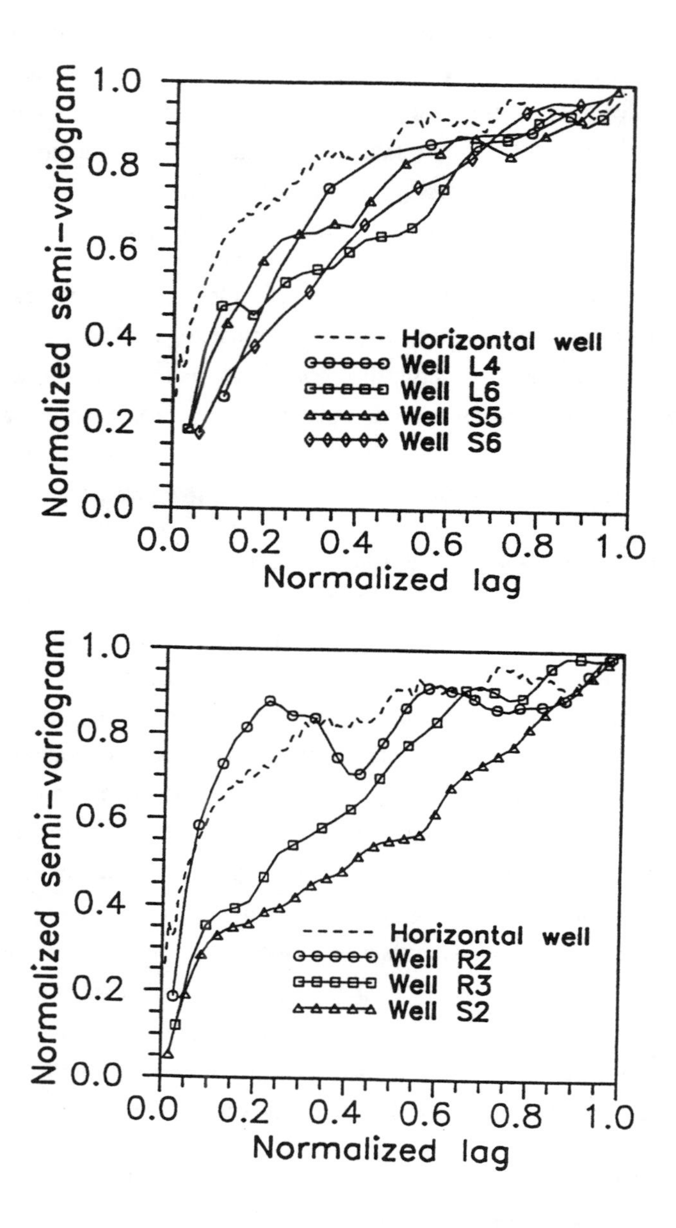

Figure 12: Comparison of Normalized Semi-Variograms of Porosity Logs of Horizontal and of Vertical Wells

and the vertical logs resemble different fractal processes. However, it is not surprising that the shape of the normalized semi-variograms for the horizontal and vertical wells in Figure 13 are similar because the shape of the semi-variograms of fBm and fGn can be similar, as explained in the Fractal Models section, for the range of intermittency exponents considered (Table 2).

The porosity semi-variogram of the horizontal log is compared in Figure 13 to the semi-variograms along the vertical and horizontal directions derived from the sample data of all the vertical logs. A comparison of the semi-variograms of the horizontal and all the vertical logs indicates that there is a greater spatial variability (except for the first lag) in the vertical than in the horizontal directions. The semi-variogram in the horizontal direction can be computed from the vertical logs only for a few large lags corresponding to the separation distance among the vertical wells. As shown in Figure 13, for the configuration of the vertical wells in this study region (Figure 4), the shortest horizontal lag is 1200 feet and the longest horizontal lag is 5900 feet. The magnitude of these semi-variogram values at the smaller lags is close to the porosity variance of the horizontal log while for the larger lags it is close to the variance of the sample data of all the vertical logs. The values of this semi-variogram along the horizontal direction fluctuate significnatly and there are too few points to be able to infer a correlation structure.

The observations from Figure 13 can be used to derive practical guidelines to assess information about the actual correlation range for the horizontal direction from the vertical well data. Based on the semi-variogram of the vertical data in the horizontal direction, it can only be inferred that the actual correlation range for the horizontal direction is smaller than the minimum distance between the vertical wells. For the reservoir studied in this investigation, it is known that the actual correlation range in the horizontal direction is longer than the correlation range along the vertical direction. Generally, it can be expected that the correlation range in the horizontal direction is longer than the range in the vertical direction. Therefore, the correlation ranges calculated from the vertical well data for the horizontal and vertical directions provide upper and lower limits for the actual horizontal correlation range, respectively. For the carbonate reservoir considered in this investigation, these upper and lower limits for the horizontal correlation range are 1200 and 56 feet, respectively. Since, the range of these limits can be large, in practice the sensitivity of the conditional simulations or estimations to the horizontal correlation range within these limits can be included as an additional source of uncertainty.

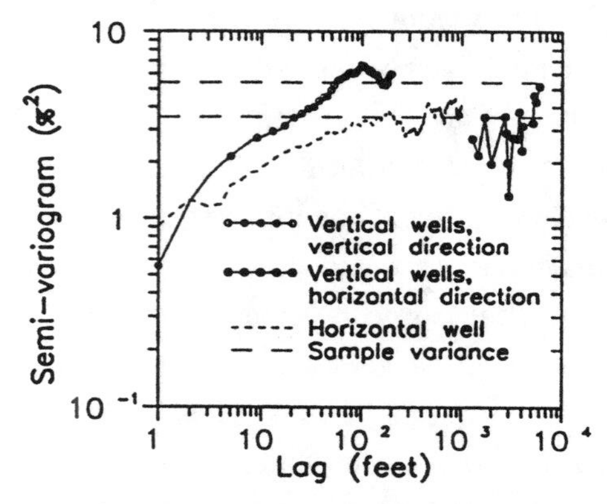

Figure 13: Comparison of Semi-Variograms of Porosity Logs of Horizontal Well and of Vertical Wells for the Vertical and Horizontal Directions

Even though, the nature of geologic setting is not part of the analyses in this paper, additional research might show that it is possible to narrow even further the limits of the spatial statistical parameters by incorporating information about the depositional environment. For example, in a sandstone field,[13] a horizontal correlation range equal to the separation distance between the wells (3030 feet) which is significantly longer than the vertical correlation range (17 feet) was adequate to represent the lateral spatial correlation of porosity, while in the carbonate reservoir discussed in this paper, the magnitude of the horizontal correlation range is closer to the vertical correlation range of the vertical logs than to the separation distance among the vertical wells.

C. Sample Volume Effect

The effect of the size of sample volumes on the statistics of porosity is investigated using the horizontal log. In this section, support volume refers to the volume of rock around the well sampled by the measurements of the logging tool. For example, neutron logs in vertical wells have a depth of investigation into the formation of about 1 to 2 feet[14] and a vertical resolution of about one foot. The practical implication of the effect of the support volume in the statistics of a property is that the volumes considered in most engineering calculations, such as simulation grid blocks, are significantly greater than the volumes sampled by the logging tools.

The support volume of porosity from the horizontal well log has been increased by averaging consecutive measurements and assigning this average porosity to the middle location of these measurements. In this case, the support volume per unit area of the plane normal to the direction of the well is equal to the log length which includes the measurements that were used in computing the average porosity. Figure 14 shows the porosity logs of the horizontal well for five support volumes per unit area ranging from 1 to 50 feet. Most of the high porosity regions observed in the original 1 foot log are preserved in the 4 feet volume per unit area log but almost disappeared (or averaged out) in the 50 feet volume per unit area log.

The semi-variograms of the horizontal logs for different support volumes are compared in Figure 15 and the statistical parameters of these logs are summarized in Table 3. These semi-variograms indicate that as the support volume increases the porosity variance becomes smaller and the degree of continuity increases. (The porosity means in Table 4 vary slightly because some samples at the ends of the logs were not included in the calculations.) The apparent correlation ranges do not change significantly for the semi-variograms of the logs for different support volumes considered, except for the 50 feet volume per unit area log. For the large support volumes, the observed trend of increasing correlation range may be due to the bias in the estimates of the semi-variograms as a result of the reduction of the number of samples in the logs. The logs with a support volume per unit area equal to 25 and 50 feet have only 59 and 29 sample data points, respectively.

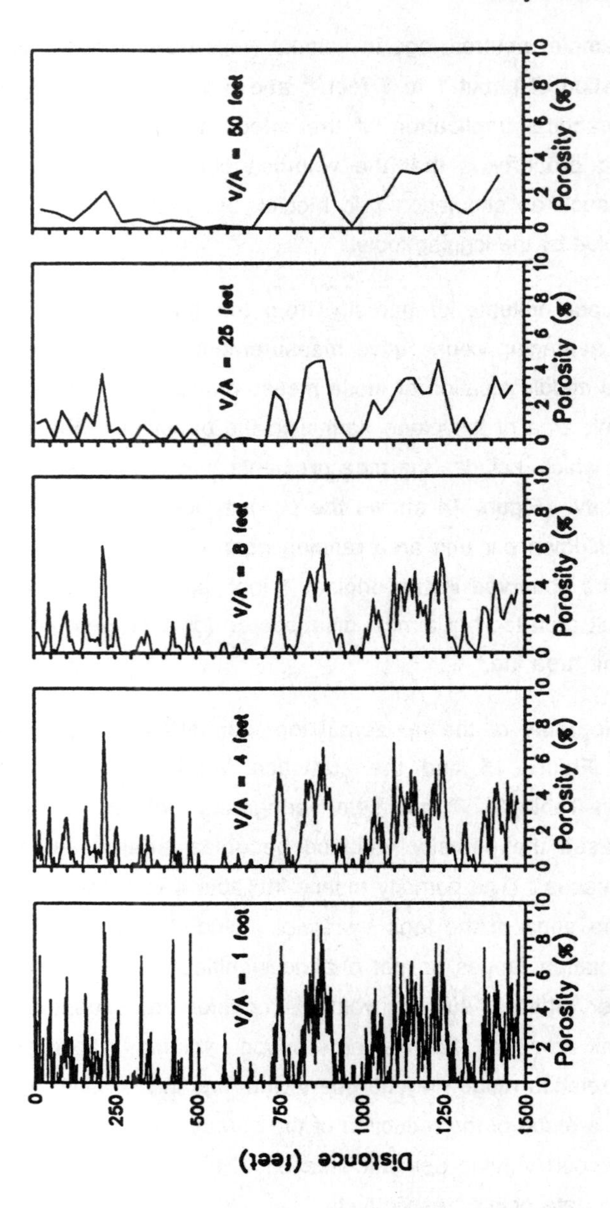

Figure 14: Porosity Log of Horizontal Well
for Different Support Volume

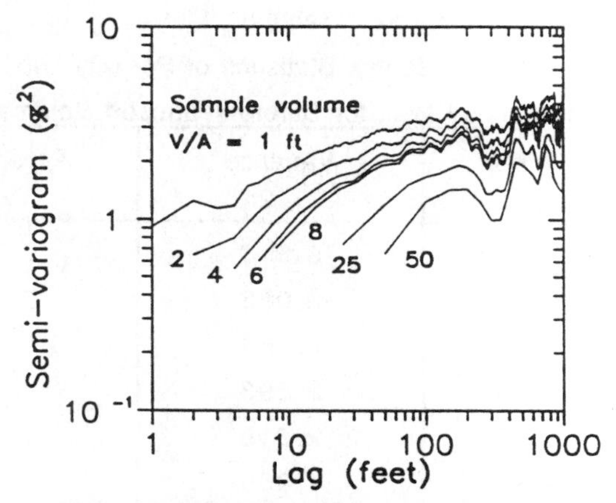

Figure 15: Semi-Variograms of Porosity Log of Horizontal Wells for Different Support Volumes

VI. CONDITIONAL SIMULATIONS

In this section, three-dimensional distributions of porosity are simulated in the study region of the carbonate field using the simulated annealing conditional simulation method described in Reference 13. The objective of these simulations is to evaluate the simulated horizontal porosity distributions for different spatial correlation models in the horizontal direction which might be used when the available information consists of porosity logs along the vertical direction.

A. Simulation Specifications

The conditional simulation region includes the horizontal well and portions of all the vertical wells shown in Figure 4 for the study region considered in this field. The specifications for the simulations are given in Table 4. The conditioning data consist of 50 porosity values at the location of each of the seven vertical wells. The cumulative distribution function is calculated from the conditioning data. The porosity conditioning data have a mean equal to 2.040% and a standard deviation equal to 2.037%. As noted in the Analysis of Univariate Statistics section, the distribution of porosity from the vertical well logs underestimates the large proportion of low porosity values observed in the distribution of the horizontal log. The semi-variogram models specified for different conditional simulation cases are described in the next section.

Table 3

Univariate and Spatial Statistics of Porosity Log of the Horizontal Well for Different Support Volumes

Volume/Area (feet)	Mean (%)	Variance ($\%^2$)	Apparent Range (feet)
1	1.389	3.518	143
2	1.388	3.052	144
4	1.388	2.758	164
6	1.382	2.593	144
8	1.389	2.518	160
25	1.386	1.869	175
50	1.360	1.474	450

B. Spatial Correlation Models

In this section, semi-variogram models for four conditional simulation cases are developed using the porosity log data of the vertical wells. The effects of the different sets of semi-variogram models on the simulated porosity distributions and the reasons for selecting these models are discussed in the following section.

The semi-variogram models developed for the horizontal and vertical directions are fGn and exponential models. The fGn and exponential models fitted to the sample semi-variogram of the vertical log data are shown by the solid lines in Figure 16. The parameters of the vertical fGn model (Equation 2) are H= 0.897, V_H = 12.0 and δ = 4.0 feet. The intermittency exponent, H, is the mean value of the intermittency exponents of the porosity logs of the vertical wells (Table 2). The vertical exponential model has a sill equal to 6.5 and a practical correlation range equal to 56 feet. For these models, the value of the sill is slightly greater than the porosity variance of the vertical wells due to the influence of the large variance of Well S2 (Table 1). Both, the fGn and the exponential models in Figure 16 provide a close fit to the vertical sample data. However, the exponential model is primarily used because it allows to explicitly specify a correlation range.

Table 4

Specification for Three-Dimensional Conditional Simulations of Porosity in the Study Region of the Carbonate Reservoir

Grid Geometry			
Direction	x	y	z
Spacing	50.0	50.0	1.0
Grid Points	108	59	50
Total Points	318600		
Conditioning Data			
Source		Porosity Logs of Wells L4, L6, R2, R3, S2, S5 and S6	
Number		350	
Distribution Function			
Source		Conditioning Data	
Number of Classes		15	
Subclass Distribution		Uniform	

Four cases with different semi-variogram models for the vertical and horizontal directions are considered in the conditional simulations. Case A is an uncorrelated simulation, where the semi-variograms are a constant value and equal to the variance of the distribution. For Case B, a semi-variogram model is specified only for the vertical direction and it is equal to the fGn model shown by the solid line in Figure 16. In Case C, the fGn semi-variogram models shown in Figure 16 are specified for the vertical and horizontal directions. The fGn model for the horizontal (x and y) directions, shown by the dashed line in Figure 16, has similar parameters as the model for the vertical direction, except that the smoothing factor is adjusted to $\delta = 4 \times 50.0$ feet, in order to account for the different grid spacing in the lateral directions. For Case D, the exponential models shown in Figure 16 are specified for the vertical and horizontal directions. The practical correlation range of the exponential model for the horizontal direction, shown by dashed the line in Figure 16, is set equal to two times the range of the model for the vertical direction and it is equal to 112 feet.

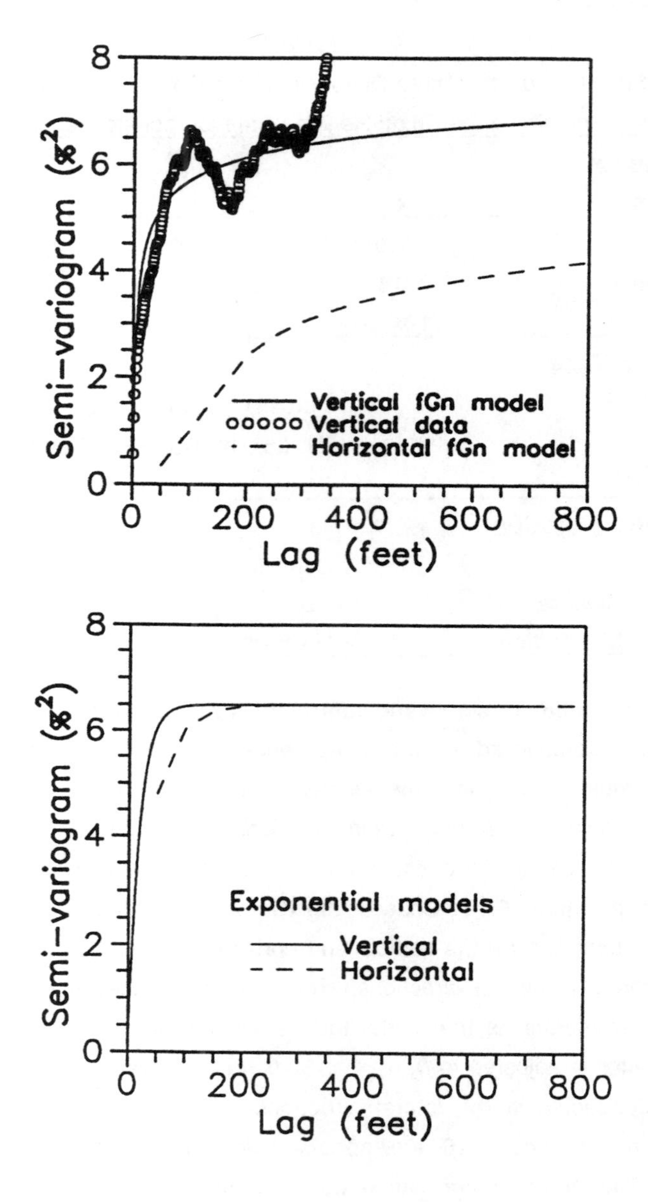

Figure 16: Semi-Variogram of Vertical Wells and Models for Conditional Simulations

C. Simulated Distributions

The results of the three-dimensional conditional simulations analyzed in this section are horizontal porosity sections for each of the four simulation cases described in the previous section. These horizontal sections are simulated porosity values along the x direction of the simulation grid at a location near to the horizontal well. The length and spacing of the simulated horizontal sections are different from the porosity log of the horizontal well. The total length of the horizontal sections is 5350 feet and it is approximately equal to the size of x direction of the study region shown in Figure 4. The spacing of simulated values in the horizontal sections is 50 feet (Table 4).

The simulated horizontal porosity sections for Cases A and B are shown in Figure 17 and it can be noted that for both cases the sections appear to be uncorrelated. The sample semi-variograms shown in Figure 18 correspond to the sections in Figure 17 and confirm the uncorrelated character of these sections. These results show that a poorly constrained simulation such as Case B which does not include explicit information about the horizontal spatial correlation yields horizontal distributions similar to an uncorrelated distribution such as Case A. Even though, the porosity section for Case A is expected to be uncorrelated, for Case B, a horizontal correlation does not develop due to any implicit influence of the conditioning data at vertical locations.

The result of the conditional simulation for Case C and the horizontal porosity log for a support volume equal to 50 feet per unit area are shown in Figure 19. The porosity log of the horizontal well shown in Figure 19 is the same log shown in Figure 14 but it is plotted with the same scale used for the simulated section. Figure 19 indicates that the simulated horizontal porosity section for Case C resembles the horizontal well porosity log of a large support volume. The sample semi-variograms indicate that there is a reasonably close agreement between the spatial correlation of the simulated section and the horizontal well porosity log. Therefore, the horizontal fGn model with a smoothing parameter which accounts for the grid spacing provides an effective representation of the lateral spatial correlation for large support volumes.

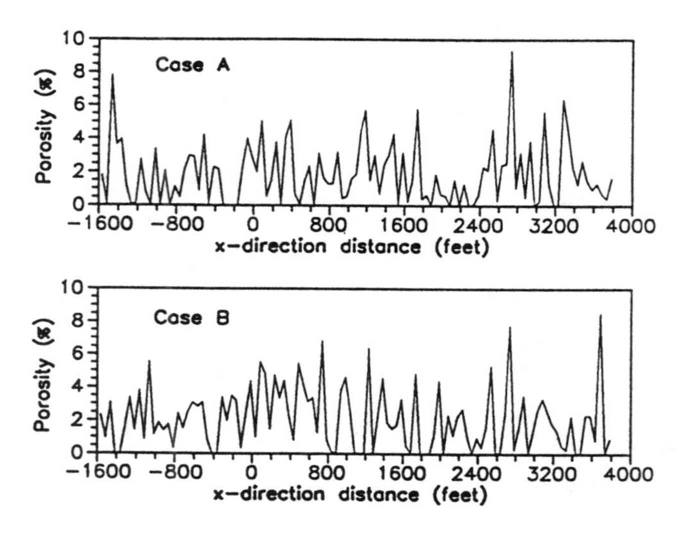

Figure 17: Horizontal Porosity Sections for Conditional Simulation Cases A and B

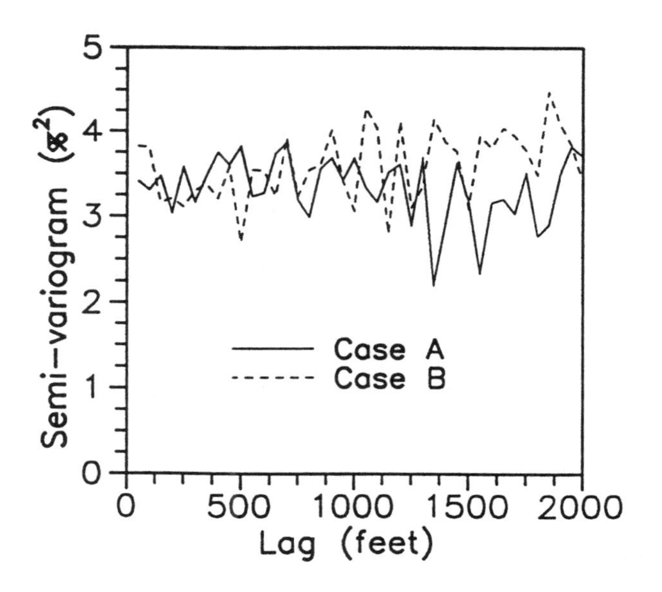

Figure 18: Semi-Variograms of Horizontal Porosity Section for Conditional Simulation Cases A and B

The simulated horizontal porosity section for Case D and a sample of the horizontal well porosity log with values spaced by 50 feet are shown in Figure 20. The appearance of the simulated section for Case D is similar to the horizontal well sampled at a spacing equal to the grid spacing specified for the lateral directions in the conditional simulations. These porosity sections are comparable, because the exponential semi-variogram model for the horizontal direction used in Case D is similar to the sample semi-variogram of the horizontal log at one foot spacing. The sample semi-variograms of the porosity sections in Figure 20 are shown in Figure 21. These results indicate that the horizontal spatial correlation of the simulated section and the horizontal well log are of the same magnitude and the large fluctuations in the sample semi-variograms are due to the fact that the horizontal well porosity log at 50 feet intervals contains only 29 sample points.

Even though, the objective of this section is not to attempt a close match between the simulations and the actual data, the results emphasize the importance of using spatial correlation models which properly account for the correlation range and the scale.

VII. CONCLUSIONS

A practical problem in most petroleum reservoirs is the availability of the sample data necessary to evaluate the spatial correlation of a property because the data are scarce in the areal directions and abundant in the vertical direction. Therefore, in this paper, log data from a horizontal and several vertical wells in a carbonate reservoir are used to assess the nature of the small-scale inter-well spatial distribution of porosity. Based on extensive analyses and comparisons of the univariate and spatial statistical attributes of the porosity logs, guidelines are developed to infer the horizontal spatial correlation from the vertical well log data in this particular carbonate reservoir. Three-dimensional conditional simulations of porosity distributions were evaluated using different models for the horizontal spatial correlation. These results indicate that the size of the support volume should be included in the conditional simulations procedures.

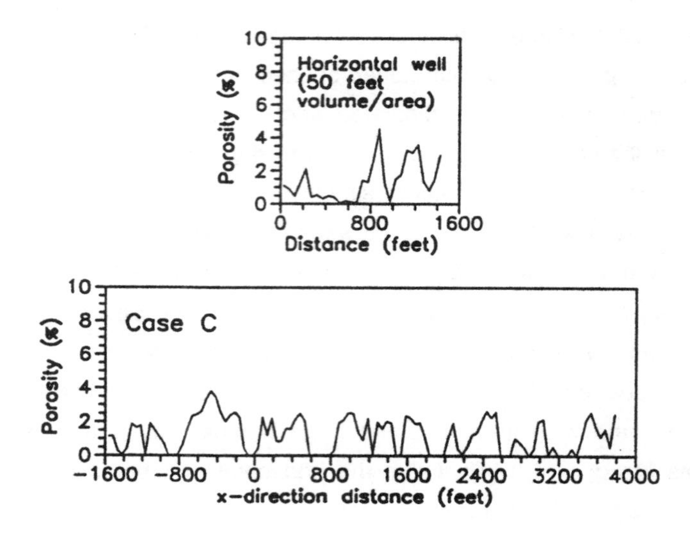

Figure 19: Horizontal Porosity Section for Conditional Simulation Case C and Horizontal Well Log with a 50 Feet Support Volume Per Unit Area

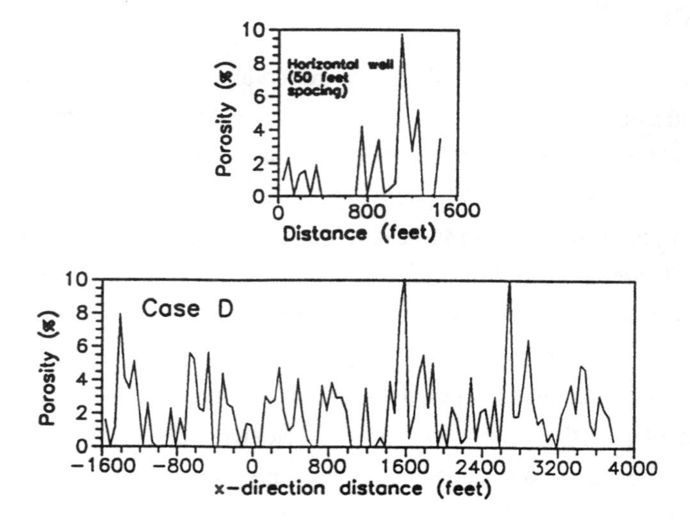

Figure 20: Horizontal Porosity Section for Conditional Simulation Case D and Horizontal Well Log Sampled at a 50 Feet Spacing

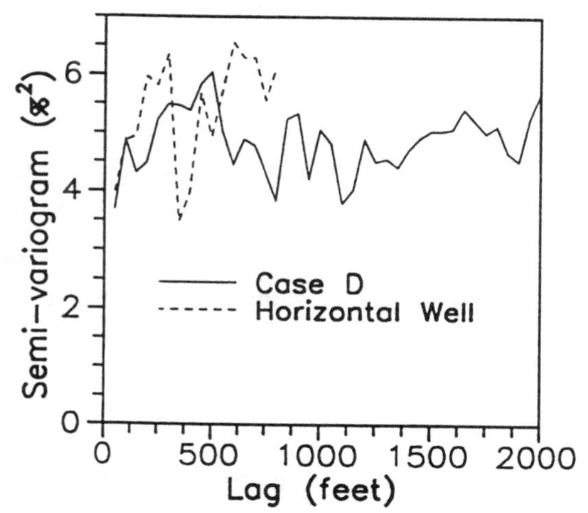

Figure 21: Semi-Variograms of Horizontal Porosity Section for Conditional Simulation Case D and of Horizontal Well Log Samples at a 50 Feet Spacing

The conclusions of this investigation are:

•The univariate analysis of porosity logs for a horizontal well and several surrounding vertical wells shows that the porosity variance of the horizontal log is slightly smaller than the variance of all the vertical logs but it is within the variance of individual vertical logs. The spatial analysis indicates that the porosity correlation range of the horizontal well log is 2.5 times greater than the vertical correlation range of all the vertical logs.

•The fractal analysis of the porosity logs indicates that the vertical logs follow fGn processes with intermittency exponents ranging between 0.78 and 0.97, while the horizontal well log follows a fBm process with an intermittency exponent close to 0.10.

•The observations from the spatial analysis of porosity indicate that the sample data from vertical wells can provide upper and lower limits for the actual horizontal correlation range.

•The three-dimensional conditional simulations of porosity for the carbonate field show that simulations with information only about the vertical spatial correlation are poorly constrained and yield uncorrelated horizontal porosity sections.

•The evaluations and comparisons in the carbonate field between the horizontal porosity log and horizontal sections from three-dimensional conditional simulations indicate that different models for the horizontal spatial correlation can account for the sample volume size. Specifically, by adjusting the smoothing factor, δ, in a fGn model, same models can be used to describe the vertical and horizontal correlations.

VIII. ACKNOWLEDGEMENTS

The authors gratefully acknowledge the data provided by Oryx Energy Company for this research. Additionally, support for this research is provided by member companies of Tulsa University Petroleum Reservoir Exploitation Projects (TUPREP) and The University of Tulsa.

IX. NOMENCLATURE

h = lag distance (feet)
H = intermittency exponent
V_H = Scaling factor for semi-variograms of fGn and fBm
∂ = Smoothing factor for semi-variogram of fGn
$¥(h)$ = Semi-variogram for lag h

X. REFERENCES

1. Smith, L.: "Spatial Variability of Flow Parameters in a Stratified Sand," *Mathematical Geology* (1981) V. 13, No. 1.

2. Goggin, D.J., Chandler, M.A., Kocurek, G. and Lake, L.W.: "Patterns of Permeability in Eolian Deposits: Page Sandstone (Jurassic), Northeastern Arizona," *SPE Formation Evaluation Journal* (June 1988) 297-306.

3. Kittridge, M.G., Lake, L.W., Lucia, F.J. and Fogg G.E.: "Outcrop/Subsurface Comparisons of Heterogeneity in the San Andres Formation," *SPE Formation Evaluation Journal* (Sept. 1990) 233-40.

4. Crane, S.D. and Tubman, K.M.: "Reservoir Variability and Modeling With Fractals," paper SPE 20606 presented at the 1990 Annual Technical Conference and Exhibition, New Orleans, Sept. 23-26.

5. Hewett, T.A.: "Fractal Distribution of Reservoir Heterogeneity and Their Influence on Fluid Transport," paper SPE 15386 presented at the 1986 Annual Technical Conference and Exhibition, New Orleans, Oct. 5-8.

6. Mandelbrot, B.B. and Van Ness, J.W.: "Fractional Brownian Motions, Fractional Noises and Applications," *SIAM Review* (Oct. 1968) 422-37.

7. Williams, J.K. and Dawe, R.A.: "Fractals - An Overview of Potential Applications to Transport in Porous Media," *Transport in Porous Media*, Reidel Publishing Company, vol. 1 (1986) 201-9.

8. Sahimi, M. and Yortsos, Y.C.: "Applications of Fractal Geometry to Porous Media: A Review," paper SPE 20476 presented at the 1990 Annual Technical Conference and Exhibition, New Orleans, Sept. 23-26.

9. Mandelbrot, B.B.: "A Fast Fractional Gaussian Noise Generator," *Water Resources Research* (June 1971) 543-53.

10. Hurst, H.E.: "Long-Term Storage Capacity of Reservoirs," *Trans. Amer. Soc. Civil Eng.,* vol. 116 (1951) 770.

11. Feder, J.: *Fractals,* Plenum Press, New York (1988).

12. Deines, T.: *Personal Communication* (Feb. 1991).

13. Perez, G.: "Stochastic Conditional Simulation for Description of Reservoir Properties," Ph. D. Dissertation, The University of Tulsa (1991).

14. Schlumberger: *Log Interpretation, Volume I-Principles,* Schlumberger Limited, New York (1972).

VARIOGRAMS AND RESERVOIR CONTINUITY[1]

Li-Ping Yuan[2] and Rudy Strobl[2]

1. AOSTRA/ARC Joint Geological Research Program.
2. Alberta Research Council, Edmonton, Alberta, Canada

I. ABSTRACT

Variograms have been used to characterize the spatial distribution of variables such as permeability and lithology, for interpolation and stochastic simulation of these properties in a reservoir. A variogram measures the similarities in the values of a variable at various distances and is a type of size measure. It can also measure sizes in different directions to identify reservoir size anisotropy, however, a variogram does not examine whether a particular phase (lithofacies or region with distinct permeability) is connected. As a result, stochastic simulations designed to reproduce a particular variogram usually produce phases with high proportions that are connected and low proportion phases disconnected. In reality, a small portion of shale can be connected and vice versa. A practical tool to measure connectivity is essential for reservoir characterization.

Two methods from mathematical morphology and image analysis are proposed to quantify connectivity. They are tested on synthetic images and an outcrop example. The results of these two methods are compared. The first measure, connectivity number, may be used on both categorical and continuous variables. The second, connectivity indicator, is limited to categorical variables but is less affected by variation associated with small objects and noise. These two methods rely on sedimentological studies to link appropriate outcrop analogues and subsurface reservoirs.

II. INTRODUCTION

Reservoir continuity is one of the major reservoir properties which significantly affects production strategy and ultimate oil recovery. Reservoir continuity commonly refers to the spatial continuity of the more permeable rocks, such as sandstones, within a reservoir. The spatial continuity of less permeable rocks, such as shales, is usually different from the continuity of more permeable rocks. This is illustrated in Figure 1a, which depicts channel sands embedded in a shaly formation. In this example, the reservoir sands are discontinuous and the shale is highly continuous. Figure 1b illustrates the reverse case where discontinuous shales are contained within continuous sand.

Variograms are the most familiar quantitative tool to characterize spatial correlation which is closely related to spatial continuity. The variogram can be applied to continuous (versus categorical or discretized) variables, such as the ore grade in mineral deposits (Journel and Huijbregts, 1978) and permeability in oil reservoirs (e.g. Dimitrakopoulos and Desbarats, 1990; Fogg et al, 1991); and the indicator variogram can also be used to characterize categorical variables, such as lithofacies (e.g. Desbarats, 1987) and permeability intervals (e.g. Journel and Alabert, 1988). Variograms represent spatial continuity in the sense of similarity in values at nearby locations; however, another aspect of the spatial continuity, which may be described

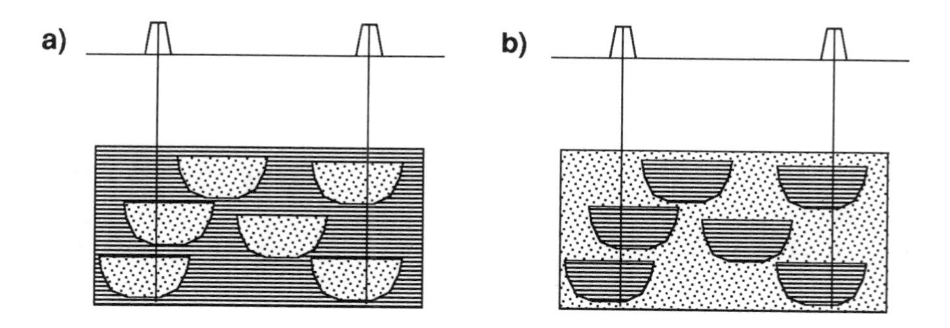

Figure 1. Two hypothetical reservoirs have different sand continuities but identical indicator variograms.

as 'connectivity', is not so well captured in the variogram. Using Figure 1 as an example, the semi-variogram (γ) is calculated as:

$$\gamma(h) = \frac{1}{2\,N(h)} \sum \left(X_i - X_{i+h}\right)^2 \qquad (1)$$

where the summation is over all pairs of points separated by h, and N(h) is the number of such pairs. In the calculation of indicator variograms, the variable, X, is either one or zero (Journal, 1983). Because Figures 1a and 1b are reverse cases, if a point in Figure 1a has a value one, the same position in Figure 1b must be zero; and vice versa. For a pair of points with a distance of h, if the difference of the two points in Figure 1a is one, then the difference for the same points in Figure 1b will be minus one; however, the squares of these differences are always the same. Therefore, Figures 1a and 1b have the same indicator variograms yet the connectivities of the sands are clearly different. Evidently, some additional tools are needed to quantify the spatial connectivity shown in Figure 1.

In this paper, examples will be used to explain how the information on both size and (size) anisotropy are related to spatial correlation and are contained in a variogram. Two connectivity measures are proposed and tested on synthetic reservoir images and on one outcrop analogue. These connectivity measures are used as robust statistics rather than being explained with detailed theoretical derivations.

III. INFORMATION IN VARIOGRAMS

A variogram measures the similarities of a parameter at various distances. Many rigorous mathematical descriptions and geological examples have been published (e.g. Journel and Huijbregts, 1978; Isaaks and Srivastava, 1990). In this section, several synthetic images will be used to show how variograms can characterize certain spatial properties. For simplicity, most of this paper discusses only binary variables; that is, categorical variables with two possible outcomes. The spatial distribution of such binary variables can be represented visually as black and white images. In the following synthetic images, we consider the white areas to be permeable sands and black areas relatively impermeable shales.

Figures 2a, 2b, and 2c are 100 x 50 grid images produced by assigning random numbers to each grid cell and thresholding (dividing cells based on whether their values are greater or less than the median) into 50% sand and 50% shale. Equal amounts of sand and shale are assigned in order to remove the effects due to varying sand/shale fractions. The differences between images are produced by applying different numbers of two-dimensional (2-D) moving averages before thresholding, in order to create different degrees of spatial correlation. Figure 2a has no moving average applied and therefore, has no spatial correlation between cells. Each cell is independent to its neighbors. The semi-variogram (solid line, Figure 2d) clearly illustrates no spatial correlation with a range equal to one. Figure 2b has applied three moving averages with a 3 x 3 template (nine cells being equally weighted) and Figure 2c has 16 3 x 3 moving averages. Each moving average increases the spatial correlation and the similarity between neighboring cells. Semi-variograms for Figures 2b and 2c, dotted and dashed lines respectively in Figure 2d, show the increasing spatial correlation as the ranges increase. Visually, the texture is becoming coarser and the sizes of sand/shale bodies are becoming larger as the spatial correlation increases.

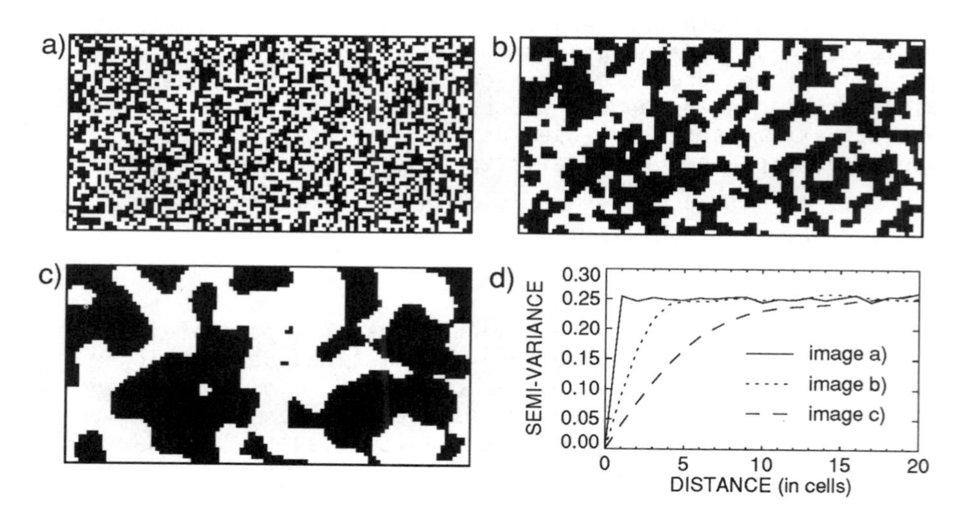

Figure 2. Three synthetic 100 x 50 grid images with different degrees of spatial correlations and their variograms characterizing isotropic conditions.

The semi-variograms (Figure 2d) all have the same theoretical sill, 0.25, equal to the standard deviation of the variable and, for binary variables, also equal to the sand fraction multiplied by the shale fraction. However, they have different ranges varying with the 'sizes' of features. In this paper, the term 'size' is used in a generalized sense to describe the sizes of any features (may not be distinct objects) i.e. the coarseness of the texture. The greater the range of the variogram, the larger the sizes of features. Variograms reflect the changes in size and therefore, variograms can be a measure of size.

Images in Figure 2 are created similarly in both the horizontal and the vertical directions to approximate isotropic conditions. In contrast, Figure 3 shows two anisotropic images which are generated by varying one-dimensional moving averages (average of three cells on a row or a column) in the horizontal and vertical directions. Figure 3a has two moving averages in the vertical direction and eight in the horizontal direction before thresholding. Figure 3b has no vertical moving average and 16 horizontal moving averages. The results show flattened, anisotropic features which have different degrees of spatial correlation

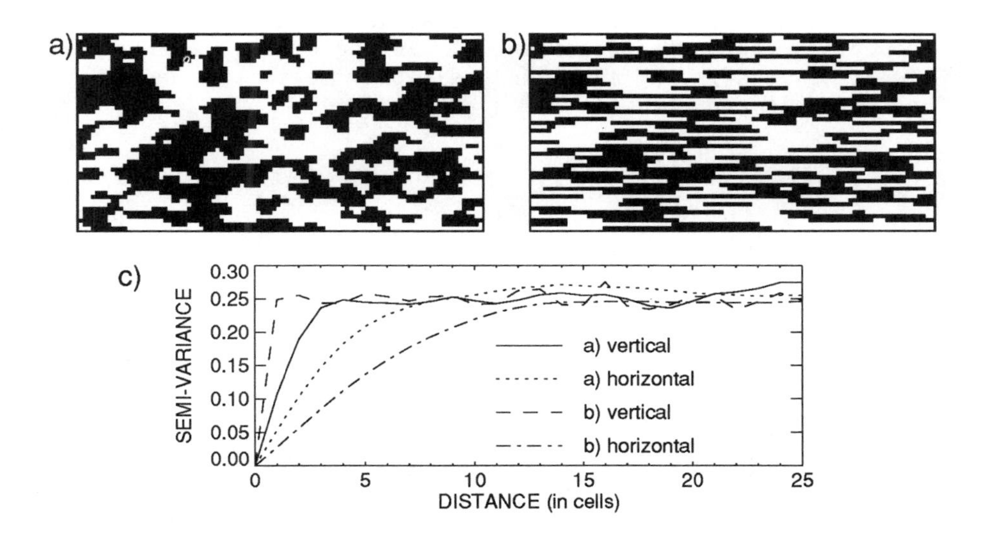

Figure 3. Two images with different spatial correlations between vertical and horizontal directions and the variograms for each direction and image characterizing anisotropic conditions.

and linear sizes between the vertical and horizontal directions. Their variograms (Figure 3c) also measure such differences that the greater the linear size, the longer the range. The differences between vertical and horizontal variograms, such as the ratio of the ranges, can be used to characterize the anisotropy of a reservoir (e.g. Fogg et al, 1991). In this paper, the term 'anisotropy' is used only for the 'size anisotropy' which may be different from other anisotropic properties such as permeability anisotropy.

The ability to characterize anisotropy is very important because most sedimentary rocks are commonly layered and, therefore, highly anisotropic. This can be shown by comparing Figure 1a and Figure 4 based on tha same horizontal and vertical scales. Figure 4 is a horizontally stretched version of Figure 1a. The linear sizes in the horizontal direction become much greater and so does the horizontal continuity. Interwell continuity has changed significantly and variograms can characterize such changes with different ranges. Many linear directional continuity features can thus be characterized as anisotropy by variograms. However, the connectivity differences between Figures 1a and 1b are not recognized due to the identical variograms.

It is possible to divide the reservoir continuity into at least two components, size (including anisotropy) and connectivity. From the previous examples, these two properties can vary independently. Variograms may adequately characterize size; however, an additional measure is needed to characterize connectivity.

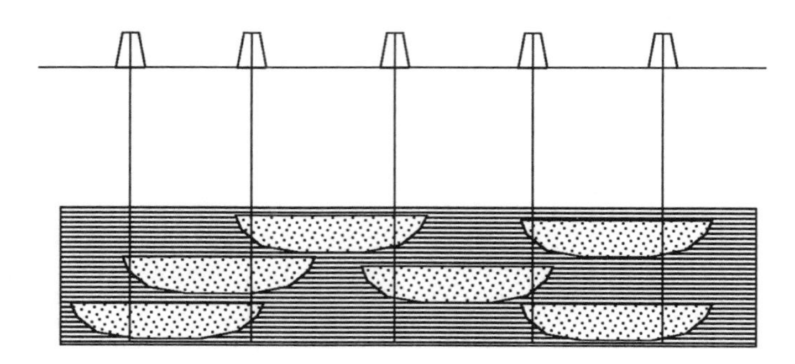

Figure 4. Horizontally stretched image of Figure 1a to simulate a highly anisotropic reservoir with good interwell continuity.

IV. CONNECTIVITY MEASURES

The two connectivity measures proposed in this section were originally developed and applied in the fields of mathematical morphology and petrographic image analysis. In order to compare and contrast these two measures, synthetic images (Figure 5) are generated for testing these methods. Four images in Figure 5 have similar amounts of sand (white) and shale (black) to minimize the dependence of sand/shale fractions. Circles having a lognormal distribution of diameters are used to reduce the effect of anisotropy and are randomly placed on the images with different degrees of overlap. Increasing overlap of the circles has the effect of increasing connectivity. The circles in Figure 5a are placed without overlapping but allowing touching to simulate poorly connected circular sand bodies. The circles in Figure 5b can have up to 40% of the area of the larger circle overlapped by a small one to simulated better connected sands. Figure 5c allows any degree of overlap between circular sands. Figure 5d is the reverse image of 5a to simulate highly connected sands.

The first measure is called 'connectivity number' (Serra, 1982) or 'Euler number' (Russ, 1990) and is equal to the number of disconnected objects minus the number of holes in those objects. For example, the connectivity number for sands in Figure 1a is six (six objects and no holes) and it is minus five (one object minus six holes) for the sand in Figure 1b. The lower the connectivity number, the greater the connectivity (compare Figure 5a to 5d). This method is simple and has been developed since 1751 (Serra, 1982). However, it treats all objects with the same importance regardless of their sizes, therefore, it is highly sensitive to small objects and noise.

The second measure is the 'connectivity indicator' (Yuan, 1990). Figure 6 shows that each sand element (a connected sand area) in Figure 5b is separated into two parts, smooth and rough areas. A smooth area (Figure 6a) approximates the largest inscribed circle and the rest is rough areas (Figure 6b) for each sand element. The connectivity indicator is equal to the total amount of rough area divided by the total sand area. It varies between zero and one where zero indicates all disconnected circular objects and as the indicator approaches unity, the object areas are observed highly connected. Figure 5 shows that as the sands become more connected, the number of sand elements (disconnected sand areas) as well as the number of largest inscribed circles decrease. In addition, as the number of holes increases, the

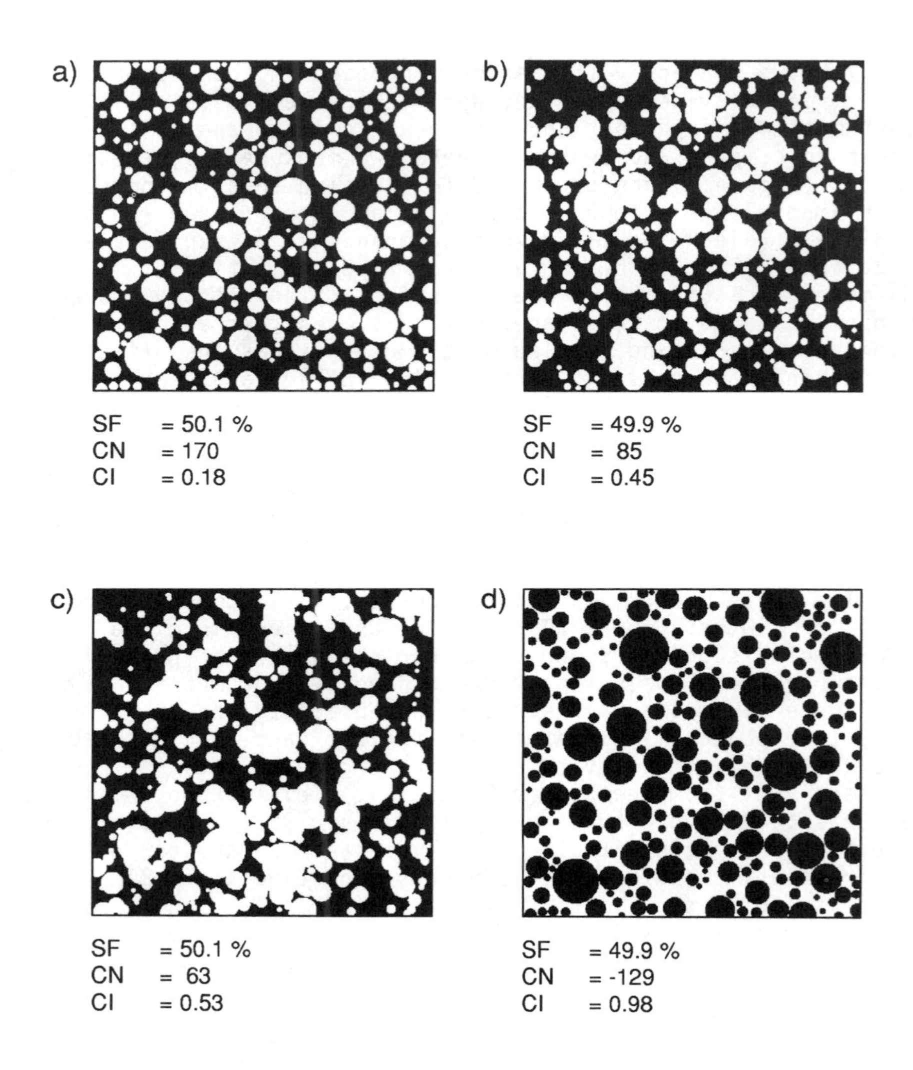

Figure 5. Four near isotropic images with similar sand (white) fraction but different connectivities on sands. Some measures for the images: SF, sand fraction; CN, connectivity number; CI, connectivity indicator.

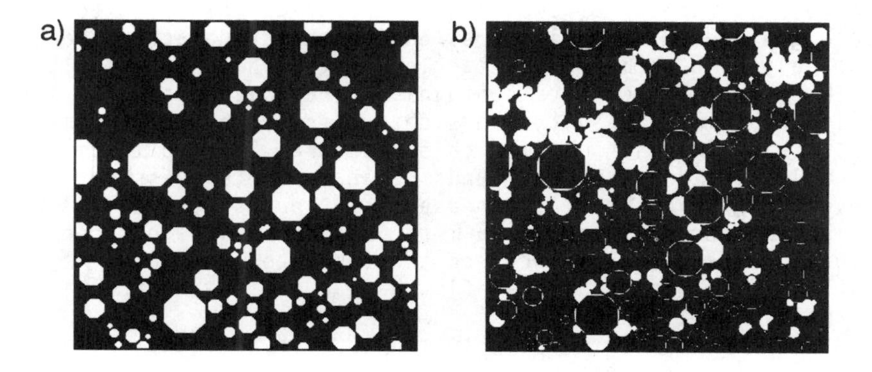

Figure 6. a) Smooth areas approximate the largest inscribed circle in each sand body and b) rough areas are the remainder sands. The combined sand image is in Figure 5b.

connectivity increases but the size of the largest inscribed circle may decrease. Therefore, the better the connectivity, the less the smooth area and the more the rough area. This connectivity indicator method uses area proportions and is less affected by small sand bodies.

From Figure 5, these two measures show the same trend of sand connectivity increasing from 5a to 5d, which agrees very well with visual perceptions.

V. OUTCROP EXAMPLE

An example from an outcrop located in northeastern Alberta's Lower Cretaceous McMurray Formation was selected to test the proposed methods to measure connectivity. The entire outcrop is approximately 80 x 5 m, with virtually complete exposure over its length. The laterally continuous exposures allowed accurate and detailed measures of all major lithofacies and was sampled to characterize permeability variation.

The lithofacies contained in this outcrop represent part of an extensive sand wave channel complex. Regional studies of both outcrop and subsurface data suggest that this succession of highly permeable cross bedded sands extended over an area of over 15 km^2. Detailed characterization and permeability sampling of the outcrop suggests that the reservoir is made up of two major lithofacies: (1) fine to coarse grained, high angle cross bedded sands characterized by relatively high permeability ranging in thickness between 10 and 80 cm and (2) toe sets containing fine grained to argillaceous massive silty sands characterized by relatively low permeability with maximum thickness from 2 cm to 10 cm. Generally, the cross bedded sands dominate the succession. However, the thin toe sets represent continuous vertical flow barriers extending laterally 5 m to 60 m across the outcrop. These toe sets are also inclined and connected in many locations as minor horizontal flow barriers with an effective thickness of 0.1 to 0.5 m.

A portion of this outcrop is mapped in detail and shown in Figure 7. This example points out that, although the toe sets make up only a small portion of the reservoir, they represent significant flow baffles. Accurate estimates of the continuities of cross beds and toe sets are needed for generating realistic models of the equivalent reservoir in the subsurface.

Consistent with the synthetic images shown earlier, the outcrop map was digitized into a 1116 x 317 grid binary image consisting of low permeability toe sets in black and high permeability cross bed sets in white (Figure 8a). The binary image is characterized by the traditional variogram and with methods discussed in this paper, connectivity number and connectivity indicator, on both cross bedded sets and toe sets. For variograms, the results for the two lithofacies are the same. The area fractions of the cross bedded and toe sets are 0.84 and 0.16 respectively, thus the theoretical sill of the semi-variograms is 0.13 (Figure 8b). The shapes of the two semi-variograms show a significant anisotropy. The range of the vertical semi-variogram is only about six cells (approximately 6 cm) long, while the horizontal semi-variogram does not reach the theoretical sill at even half of the image width (approximately 5 m). From Figure 8a, the sizes of the two lithofacies in the vertical direction are relatively small. The two lithofacies are alternating many times within the height of the image. On the other hand, the horizontal size is very large. Most of the cross bedded sets and toe sets run across the image without terminating. A longer (wider) area needs to be used to calculate a complete horizontal variogram and to better estimate anisotropy.

The connectivity number is 14 for cross bedded sets and is 7 for the toe sets (Figure 8) which suggests that the toe sets are more connected than the cross bedded sets. From Figure 8a, one can find more isolated cross bedded set areas than

Figure 7. An outcrop picture (approximately 5 by 13 m) shows two major distinct lithologies, cross-bedded sandstone and fine grained sandy to argillaceous toe sets (overlaid in black).

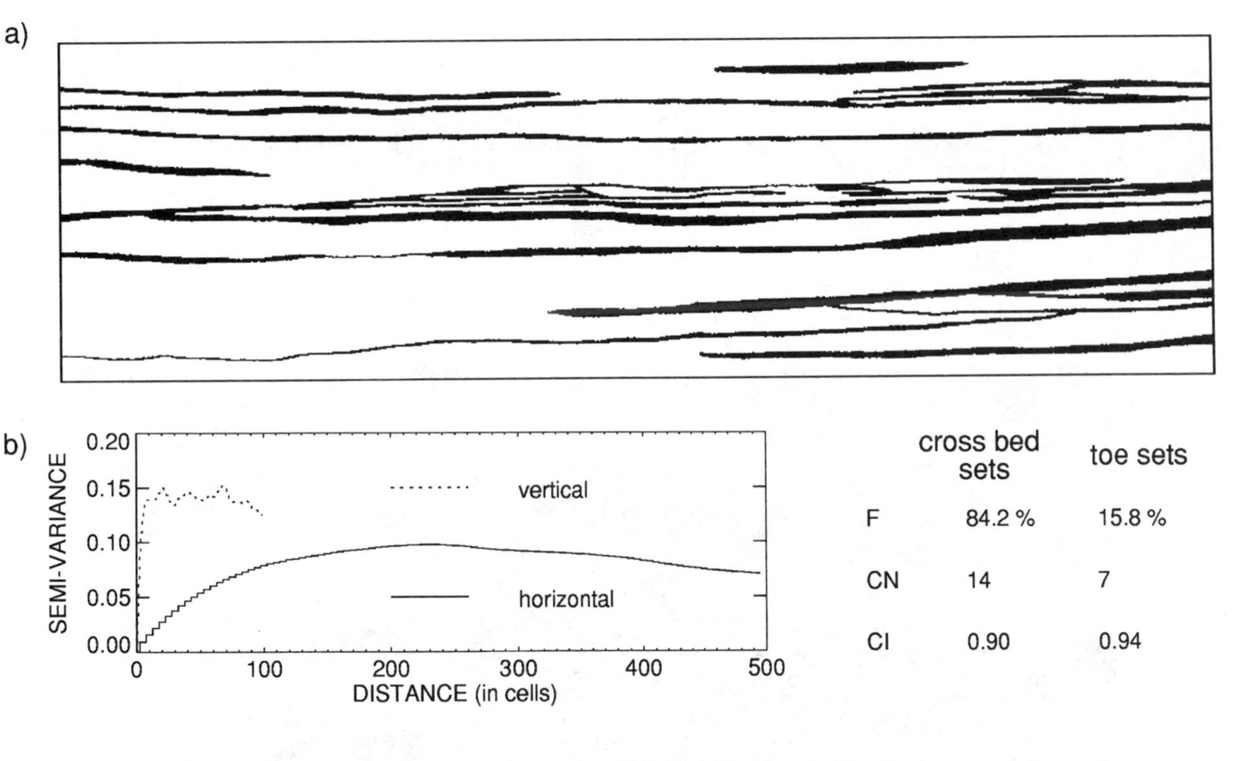

	cross bed sets	toe sets
F	84.2 %	15.8 %
CN	14	7
CI	0.90	0.94

Figure 8. a) A portion (approximately 3.2 by 12 m) of the outcrop, shown in Figure 7, is digitized into a 317 x 1116 binary image.
b) Semi-variograms, fractions (F), connectivity numbers (CN), and connectivity indicators (CI) of the digitized outcrop.

disconnected toe set areas. Connectivity indicators also indicate that the same toe sets are slightly better connected with a value of 0.94 while the cross bedded sets have a value of 0.90. This example illustrates that minor lithofacies can be equally or better connected than major lithofacies. Such reservoir continuity information is important and needs to be quantified. Assuming that lithofacies which make up a relatively small percentage of the reservoir are disconnected may lead to unrealistic estimates.

VI. DISCUSSION

Implementing the two methods proposed in this paper can be complex. The number of neighbors of a cell can be four or eight on a rectangular grid or six on a hexagonal grid. This will affect the computation of connectivity number (Serra, 1982) and the distance measure of circles (Yuan, 1991) for the connectivity indicator. In this paper, the four neighbor rule is used for connectivity number. A four-eight alternating rule is used to compute inscribed circles for the connectivity indicator. Figure 6a shows that the inscribed circles are approximated by octagons which is a result of using this four-eight alternating rule to compromise the speed and accuracy of the computation. Edge effects also exist in both methods and need to be minimized in the computer program. More details of the computer algorithms can be found in Rosenfeld and Kak (1976), Serra (1982), and Russ (1990).

For categorical variables, both methods are applicable. However, only connectivity number has been extended to continuous variables. This is important in characterizing the connectivity of permeability -- the most important variable for reservoir continuity. Complete permeability data covering a 2-D outcrop are very rare; however, land elevation is a similar continuous variable and may be used as an example to explain the connectivity of a continuous variable. For land elevations, the connectivity number is defined as the number of summits plus the number of sinks minus the number of saddles (Serra, 1982). For permeability, then, the connectivity number is equal to the number of isolated high permeability areas plus the number of isolated low permeability areas minus the number of saddle points.

However, this extension may not work well for the purpose of characterizing the connectivity of permeability. This can be explained by the land elevation example again. In a fluvial process dominated landform, high valued points are usually isolated as mountain peaks and low valued points are usually continuous and connected as river beds. Assuming an inverse landform existed, where high values were continuous as ridges and low values were disconnected as lake bottoms, the connectivity number would be the same as for the original fluvial landform. This extension of the connectivity number cannot distinguish continuous high values from isolated high values in such cases; this is analogous to the problems with the variogram in the cases in Figure 1 . Therefore, the definition of connectivity number needs to be modified for permeability problems. One suggestion may be to count only the number of summits but not the sinks. Another alternative is to divide permeability values into intervals as categorical variables and then to measure connectivity.

In certain situations, a connectivity measure is desired to be independent from size and anisotropy. The connectivity number inherently satisfies such a requirement. The connectivity indicator, however, can be affected by anisotropy. For example, sands in Figures 1a and 4 have the same connectivity number but different connectivity indicator. The stronger the anisotropy, the smaller the inscribed circles and the greater the connectivity indicator. Therefore, if it is important that connectivity measure and anisotropy be independent, a normalizing procedure should be performed for connectivity indicator. An anisotropic image can be rescaled (normalized) to be isotropic and then the connectivity indicator may be calculated.

Connectivity, as well as continuity, does not have a clear, well accepted definition. The two measures used in the previous section are, in fact, two different definitions for connectivity. The question of 'Which one is better?' is not addressed in this paper and probably will depend on the purpose and available data of a particular study.

A major limitation of these two methods is the requirement of complete 2-D data which are usually available only from well exposed outcrops. Drill hole data are sparse and cannot be used directly. Seismic data generally do not provide the resolution nor the properties (permeability and lithology) directly for detailed reservoir continuity studies. Therefore, appropriate outcrop analogues have to be found with a similar reservoir continuity to the subsurface reservoir. Sedimentological studies become crucial in finding outcrops with similar deposition environment and diagenetic history to the target reservoir. After finding adequate outcrops, connectivity can be quantified. A later goal, not addressed in this study, will be to perform stochastic simulations which honor the connectivity measure as well as other statistics (e.g. frequency distribution) and geostatistics (i.e. variogram) measures, and conditioned to the drill hole data and deterministic knowledge.

VII. CONCLUSIONS

Reservoir continuity may be separated into at least two components, size (as a generalized term) and connectivity. Variograms can characterize the sizes of various features or the coarseness of texture. In addition, variograms can also characterize size in different directions to quantify the size anisotropy in a reservoir.

Connectivity can be measured by connectivity number or connectivity indicator. The connectivity number may be applicable for both continuous and categorical variables in two or three dimensional space. The connectivity indicator is limited only to categorical variables, but is less affected by small features or noise. The two measures require complete 2-D data, which are only available from outcrops. Geological studies are then needed to find outcrops analogous to the subsurface reservoirs.

ACKNOWLEDGMENTS

The authors thank the Alberta Research Council and the Alberta Oil Sands Technology and Research Authority for funding this research program and for allowing the publication of these results. We also wish to acknowledge David Cuthiell for providing software, strong encouragement, and very helpful discussions to the research project and this manuscript. Syncrude Canada is also acknowledged for their strong support by providing valuable data and discussions to this project.

REFERENCES

Desbarats, A.J., 1987, Numerical Estimation of Effective Permeability in Sand-Shale Formations, Water Resources Research, v. 23, no. 2, p. 273-286.

Dimitrakopoulos, R. and Desbarats, A.J., 1990, Geostatistical Modelling of Grid Block Permeabilities for Three-Dimensional Reservoir Simulators, SPE #21520.

Fogg, G.E.; Lucia, F.J.; and Senger, R.K., 1991, Stochastic Simulation of Interwell-Scale Heterogeneity for Improved Prediction of Sweep Efficiency in a Carbonate Reservoir, in Reservoir Characterization II, L.W. Lake, H.B. Carroll, Jr., and T.C. Wesson, eds., Academic Press, San Diego, p. 355-381.

Isaaks, E.H., and Srivastava, R.M., 1989, An Introduction to Applied Geostatistics, Oxford University Press, 561p.

Journel, A.G., 1983, Non parametric estimation of spatial distributions, Math. Geol., v. 15, no. 3, p. 445-468.

Journel, A.G., and Alabert F.G., 1988, Focusing on Spatial Connectivity of Extreme-Valued Attributes: Stochastic Indicator Models of Reservoir Heterogeneities, SPE paper #18324.

Journel, A.G., and Huijbregts, C., 1978, Mining Geostatistics, Academic Press, Orlando, Fla., 600 p.

Rosenfeld, A., and Kak, A.C., 1982, Digital Picture Processing, V. II, Academic Press, Orlando, Fla., 349 p.

Russ, J.C., 1990, Computer-Assisted Microscopy: The Measurement and Analysis of Images, Plenum Press, New York, 453 p.

Serra, J., 1982, Image Analysis and Mathematical Morphology, Academic Press, 610 p.

Yuan, L-P., 1990, Pore Image Characterization and Its Relationship to Permeability, Society of Core Analysts Conference Preprints, Paper Number 90002, 18p.

Yuan, L-P., 1991, A Fast Algorithm for Size Analysis of Irregular Pore Areas, in Nonlinear Image Processing II, E.R. Dougherty, G.R. Arce, and C.G. Boncelet, Jr., Eds., Proceedings SPIE 1451, p. 125-136.

ARTIFICIAL INTELLIGENCE DEVELOPMENTS IN GEOSTATISTICAL RESERVOIR CHARACTERIZATION

Roussos Dimitrakopoulos

Dept. of Mining and Metallurgical Eng., McGill University
and GEOSTAT Systems International Inc.
Montreal, Quebec, Canada

ABSTRACT: This paper presents recent developments in integrating techniques and ideas from Artificial Intelligence in geostatistical reservoir characterization. First, the integration of geostatistical methods with symbolic, non-algorithmic techniques is demonstrated in the context of intelligent computer systems. Next, the technique of qualitative spatial simulation (QSS) is presented as an alternative approach in merging purely qualitative reservoir information/descriptions with stochastic simulations.

INTRODUCTION

Geostatistical or spatial stochastic techniques (David, 1977; Journel and Huijbregts, 1978) are used and developed to accommodate the needs for detailed petroleum reservoir characterization, forecasting, and management. However, difficulties may arise from a number of factors, such as:(i) the type of the deposit and problem to be tackled; (ii) the stage of reservoir development or production; (iii) the variables of interest; (iv) the data and information available; (v) the choice of proper techniques; and (vi) the mathematical complexities as well as practical intricacies of the geostatistical techniques. To provide wider access of geostatistical expertise for reservoir characterization, simplify and enhance the usability as well as effectiveness of stochastic methods, technological advances in the field of Artificial Intelligence (AI) may be considered.

The use of symbolic non-algorithmic methods (Newell and Simon, 1976; Nilsson, 1980) in encoding and applying human knowledge and expertise in petroleum related areas is not new. Intelligent computer programs or expert systems have been developed and are used in areas such as dipmeter log analysis (Smith and Baker, 1983), geophysical log environments (Shultz et al., 1987), thermodynamic modelling of oil and gas properties (Barreau et al., 1991), and others. The integration of geostatistics and AI techniques, and specifically the explicit coding of geostatistical knowledge-expertise-information in computer systems with numerical capabilities has already been suggested (Dimitrakopoulos, 1989; Dimitrakopoulos, 1992). Relevant work and examples of intelligent geostatistical computer systems in the context of mineral reserve assessment have also been presented (Dimitrakopoulos and David, 1991).

The transfer of geostatistical expertise is only the starting point and one form of a potentially polymorphic integration of geostatistics and symbolic, non-algorithmic techniques. An additional form of this integration may be seen in the context of qualitative type simulations and related computer systems, which can check, reason and reconstruct the geological - and, in general, physical - consistency of stochastic simulation results.

Qualitative simulation (Kuipers, 1986) was originally developed to simulate the behaviour of a continuous physical system by using qualitative descriptions and constraints derived from the differential equations governing the behaviour of the system. In a similar fusion, qualitative descriptions can assist discrete-event simulation (O'Keefe, 1986), and, as suggested in the present study, can be used to enrich and enhance the Monte-Carlo type simulations such as the spatial stochastic conditional simulations used in reservoir characterization. In geosciences, the only related work includes the system GORDIUS (Simmons, 1988) which can be used to generate interpretations of geological events, based on qualitative type simulations.

The present study consists of two main parts. The first includes basic concepts, requirements and characteristics of geostatistical intelligent systems, as well as two such experimental systems. Both systems are presented together with examples from applications in reservoir characterization. The second part focuses on the idea of qualitative simulation and the need for its integration with spatial stochastic simulations. This integration is specifically discussed in the context of simulating reservoir lithofacies and the constraints that geological principles and models can impose on stochasticaly generated images of reservoir lithofacies. A first simple system imposing geological constraints on a section of simulated reservoir lithologies is presented as the means to demonstrate the idea. Finally, conclusions from the work presented are outlined.

INTELLIGENT SYSTEMS FOR GEOSTATISTICAL OPERATIONS

Intelligent computer systems may be considered as generalizations of conventional programs containing, in addition to numerical capabilities, the knowledge and expertise required to guide, undertake, evaluate, and reason about geostatistical operations.

Concepts and Requirements

Geostatistical modelling of reservoir attributes may be seen as the application of relevant aspects of domain knowledge. The latter may include a number of related components: (a) the geostatistical theory, i.e. definitions, properties, relations, abilities and limitations; (b) practical and numerical intricacies of different techniques; (c) practical intricacies related to different reservoir types and variates, operational objectives, etc.; (d) reservoir engineering requirements; and, finally, (d) the knowledge of how all the previous components may be combined to effectively serve the needs of reservoir characterization. It should be emphasized that having both the relevant expertise and the ability to combine relevant pieces of knowledge are aspects of what one may consider as intelligence, and should be exhibited by intelligent computer systems in the domain.

An additional practical aspect of stochastic modelling is the use of computer subroutines which carry out the numerical calculations involved in characterizing attributes of a reservoir. Obviously, which subroutines will be used in each case, how the input to a subroutine will be specified, and how the output will be evaluated, interpreted and further related to other operations, is controlled by domain knowledge as already discussed.

Geostatistical problem-solving has two distinct components. The first refers to all aspects of geostatistical knowledge and expertise in reservoir characterization. The second corresponds to numerical data processing for geostatistical operations. The building of intelligent geostatistical computer systems requires an additional ingredient: the symbolic, non-algorithmic techniques developed in AI, which can facilitate geostatistical knowledge modelling, representation, inference and reasoning.

The use of AI techniques in building intelligent systems may be demonstrated at the most fundamental level: the conceptual modelling of the geostatistical entities involved in problem-solving. This type of modelling is based on the distinction of conceptual units in the domain and their semantics using abstraction mechanisms (Levesque and Mylopoulos, 1979; Borgida et al., 1985). Accordingly, five distinct, general, and associative conceptual geostatistical units may be

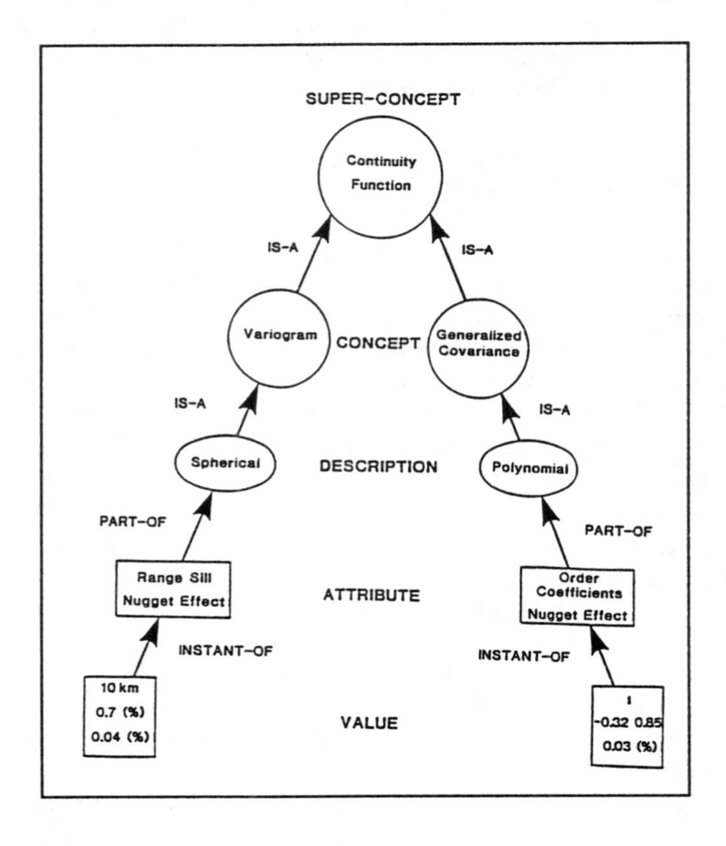

Fig. 1. Geostatistical conceptual units and their relations.

distinguished. These are super-concept, concept, description, attribute, and value. Well established semantics link the five conceptual units generating a hierarchy-taxonomy as shown in Figure 1.

Super-concepts are broad conceptual geostatistical units, for instance, continuity function or interpolation. Concepts are more specific conceptual units, such as variogram, covariance, generalized covariance, and are linked to super-concepts with an IS-A relation. Descriptions are conceptual units characterizing concepts and linked to them with an IS-A relation as well. For example, spherical is a specific description of a variogram. Attributes are characteristics of a description and linked with it through a PART-OF relation. For example, a spherical variogram has as attributes its range, sill and nugget effect. Conversely, these attributes are parts of the description of a spherical variogram. Values correspond to every attribute and are specific instances of each one of them (INSTANT-OF relation). The specific numerical values of the conceptual unit value are generated using attached procedures, which may be standard geostatistical subroutines, sets of heuristics, or values used in similar situations.

Value is the conceptual unit which links theoretical concepts with practice through procedural attachments. Consider, for instance, the concept variogram and a set of permeability data. The variogram model of the data set will be obtained through a procedural attachment. The latter may be the combination of a standard numerical subroutine calculating the experimental permeability variogram and, say, a sequence of heuristics fitting the appropriate model. Thus, while the hierarchy of geostatistical concepts provides the means to understand what a variogram is and what its characteristics are, specific instances of a variogram such as the one of the permeability data set are generated through the procedural attachments linked with the conceptual unit value. Note that a procedural attachment may simply be a mechanism recalling stored numerical values, such as the variogram characteristics of well studied reservoirs, outcrops, sedimentary environments, etc.

The conceptualization presented above captures both the basic notions of geostatistical techniques and the practical intricacies involved, thus providing the basis for the development of intelligent systems in the domain.

System Architecture, Knowledge Representation and Inference

Ongoing research has produced two experimental intelligent systems, based on the conceptualization already presented and implemented in

LISP (Winston and Horn, 1986). The first system, GEOSTAT-1, undertakes variogram calculations. The second one, GEOSTAT-2, performs geostatistical estimation of reservoir grid block properties in two dimensions.

Both systems consist of three major parts, following a generalized architecture for geostatistical intelligent systems presented in Figure 2. The three major parts of the systems are the (i) knowledge base and inference engine, (ii) interface to the user, and (iii) interface with a standard geostatistical subroutines for numerical calculations.

In the knowledge base, multiple representations (Mylopoulos and Levesque, 1984) are used. Knowledge is represented by frames and rules. Frames may be seen as data structures consisting of a name and a number of nested slots or lists containing descriptions or relations or procedures. Frames are used to control the flow of the programs, provide hierarchial classification of geostatistical concepts and procedures, fire attached procedures, group contextually or operationally related rules, and, finally, store values of parameters. Rules consist of antecedent - consequent pairs. They are used to capture empirical relations among relevant geostatistical entities related to the initialization of numerical values of the parameters needed to execute external subroutines. In addition, they evaluate the results the subroutines return, where appropriate. An example of a frame from GEOSTAT-1 and a rule from GEOSTAT-2 are shown in Figure 3.

The knowledge inference is typically performed by the inference engine, consisting of a sequence of interrelated functions. These

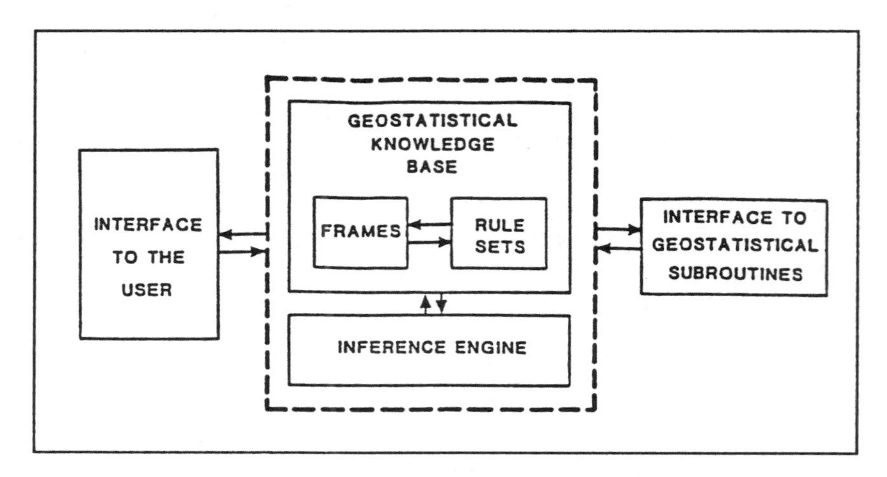

Fig. 2. General architecture of an intelligent geostatistical system.

```
(VARIOGRAM
        (SPHERICAL)
         ((NUGGET-EFFECT) (VALUE-1))
              ((SILL) (VALUE-2))
                    ((RANGE) (VALUE-3))
                          ((ANISOTROPY) (VALUE-4)))

(RULE 3-Block-size
     (IF      (PREFERENCE is NO)
              (DATA-EVENLY-SPREAD)
              (AVERAGE-SPACING is less than SHORT-RANGE))
     (THEN (BLOCK-SIZE is 0.66 TIMES AVERAGE-SPACING)))
```

Fig. 3. Examples of a frame (top) and a rule (bottom).

perform the triggering and updating of frames and the matching of rule antecedents against facts. Conflicting facts, if deduced during inference, are treated by always retaining the last deduction. The inference process proceeds in a forward-chaining fashion, because this is natural in geostatistical problem-solving.

The interface to the user employs simple networks, i.e. "tree-like" structures, which are used to ask the appropriate questions to the user and, subsequently, deduce facts from the user's responses. These facts are then processed by the system.

Example Sessions with Intelligent Systems

Practical characteristics of intelligent geostatistical computer systems can be demonstrated using as examples GEOSTAT-1 and GEOSTAT-2. Although these systems are experimental and have limited knowledge bases, i.e. can presently solve a limited type of problems, they demonstrate all required features of intelligent systems.

A session with either system starts with a series of simple questions, such as the ones shown in Figure 4. Then the collected information is processed and decisions are taken regarding the initialization of parameters needed to run external subroutines. The user may, at this point, request explanations regarding the decisions taken as shown, for example, in Figure 5. In the case of variography and GEOSTAT-1 the results of the external subroutine are further evaluated and, if

Are we dealing with more than one variable?	(Y or N)	No
Are we dealing with two-dimensional data?	(Y or N)	Yes
How many samples are involved?		26
Give me the minimum X coordinate:		2250
Give me the minimum Y coordinate:		1200

Fig. 4. Example of questions asked by GEOSTAT-1.

Do you want to know why? (Y or N) Yes

RULE-19: 2-SIZE-OF-KRIGING-SEARCH-RADIUS

concludes: (SEARCH-RADIUS is 1.5 TIMES AVERAGE-SPACING)

because: (AVERAGE-SPACING is less than SHORT-RANGE)

and

(SHORT-RANGE is grater than 1.5 TIMES AVERAGE-SPACING)

Fig. 5. Example of explanations provided by GEOSTAT-2.

I have concluded PART-2 and now I will give you my conclusions.

The variogram model is: 3D-ISOTROPIC-2DIM

WHERE

the value of the NUGGET-EFFECT is:	0.15
the value of the SILL is:	3.30
the value of the RANGE is:	2500.0

Fig. 6. Example of conclusions reached by GEOSTAT-1.

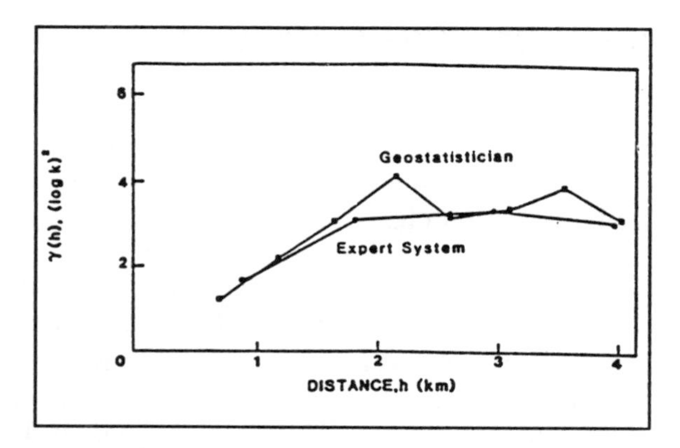

Fig. 7. Example of a variogram calculated by GEOSTAT-1.

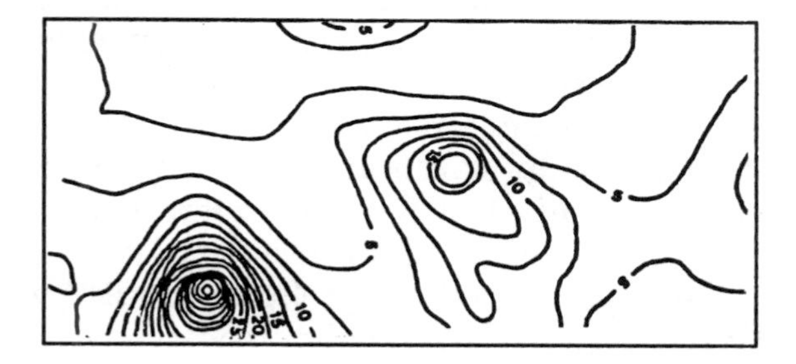

Fig. 8. Example of results derived from GEOSTAT-2.

satisfactory, conclusions are reached (Figure 6). Otherwise, explanations are provided and the inference process is repeated.

Figure 7 shows the variogram of a 2D permeability data set from a clastic reservoir as calculated using GEOSTAT-1. For comparison, the variogram of the same data set as calculated by an independent geostatistician is also shown.

Figure 8 shows a map of permeability modelled using GEOSTAT-2 and the variogram derived from GEOSTAT-1.

Future work on intelligent geostatistical systems for reservoir characterization should focus on both classification and representation of reservoir geology as well as system engineering.

ASPECTS OF QUALITATIVE SPATIAL SIMULATION

The interaction of AI and geostatistical methods may also be expressed in the context of enhancing the results of the spatial stochastic simulations used in reservoir characterization. Geostatistical simulations are a known tool for generating images of reservoir attributes. The exclusive consistency criteria for these images are the reproduction of statistical characteristics and available data. However, these criteria may not be sufficient for a simulated reservoir image to fully reproduce the geological consistency expected from the geological analysis and interpretation of the reservoir. This may be apparent when the simulation task is dealing with the modelling of reservoir sedimentary facies. In this case, the simulated lithology of a reservoir block may be found inconsistent with the lithologies generated in the surrounding blocks. It should be clear that these kinds of geological inconsistencies, if present, are not necessarily inherent to the specific spatial stochastic technique that may be used to generate lithofacies. Rather, they may be attributed to the inference of statistical characteristics from a limited set of data and information, numerical approximations used in computations, possible errors in data coding, etc.

The means to impose additional geological consistency criteria on simulated reservoir images is addressed in this section. It is seen as a complementary step to spatial stochastic simulation operations, and is based on qualitative reasoning principles as developed in AI. The name "Qualitative Spatial Simulation" or QSS seems, therefore, adequate. It should be noted that QSS is at an early stage of development.

Basic Ideas, Principles and Criteria

In AI, qualitative simulations are a key inference process in reasoning about physical systems (e.g., Bobrow, 1986). In a simulation context, they are seen as the use of non-numerical methods to provide alternatives and enhance the simulation results (e.g. Reddy et al., 1986). Independently of context, however, it may be suggested that the fundamental idea of qualitative simulation is that physical processes are governed by general laws and initial conditions from which subsequent conditions may be generated. This fundamental idea is the basis for developing qualitative spatial simulation and is further explored.

The formation of geological phenomena is governed by universal geological laws. In the case of a petroleum reservoir, these are broadly expressed in a geological interpretation which represents a general and definitive qualitative model (Figure 9). A major characteristic of a qualitative model is that it incorporates causal relationships among its attributes which can be explicitly stated. For instance, one may consider the very simple example of a reservoir whose qualitative model states that it is an upward fining sandstone sequence with four distinct sandstone facies, say A, B, C and D, representing a single uninterrupted depositional cycle. From this qualitative model and the specific descriptions of the four facies (e.g., grain size), a facies association is defined suggesting which facies are followed by which, say A → B → C → D. This facies association, based on a given model and its underlying geological laws, is a simple example of what will be called a causal relationship. In general, causal relationships are derived from a qualitative model and provide guidelines that the reservoir description should follow. For example, the reservoir described above as an upward fining sequence has specific causal relationships describing how facies A, B, C and D relate, and these relationships should be reproduced in any simulated image of the reservoir. If this is not the case, causal relationships can be used to generate the regulations about what should be done to correct the situation. These regulations will be called consistency rules, since their role is to impose geological consistency on a reservoir image. In the example reservoir used previously, a consistency rule may be:

If facies D succeeds facies B
and
no unconformity present

then correct by replacing facies D with facies C

It should be noted that consistency rules are not heuristic rules. Rather, they are consequences of geological laws and can be generated from any qualitative model and related causal relationships characterizing a reservoir. The implication in terms of intelligent computer systems is that such systems should the be able to generate these rules when a qualitative model is provided.

The qualitative model, causal relationships and consistency rules include, express and can apply geological principles on an image of reservoir geology, which has been generated using a stochastic simulation technique. This reservoir image is a typical grid-block model with a lithology assigned to each grid block. It represents an initial state upon which qualitative constraints will be tested and used to generate a final

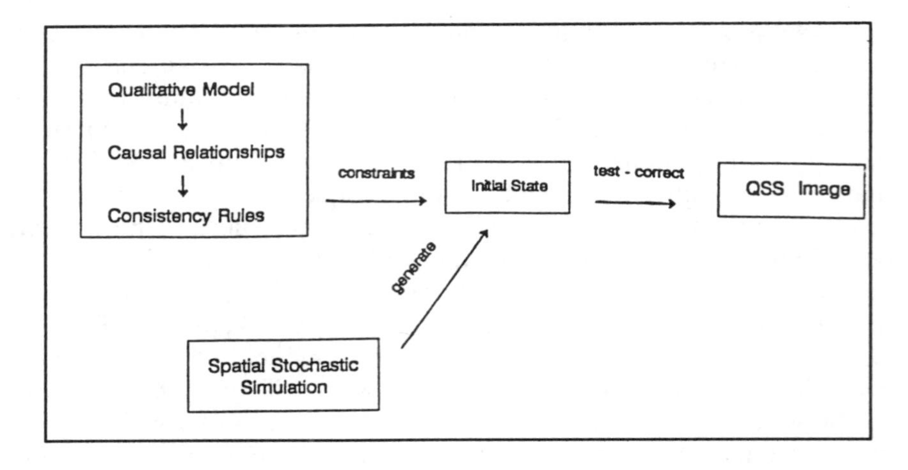

Fig. 9. Schematic representation of the proposed qualitative spatial simulation (QSS).

reservoir image (QSS image, Figure 9). A QSS image is expected to be both geologically and statistically consistent.

The QSS components including their relation to stochastic spatial simulations and the suggested mechanism for geologically constraining reservoir images is shown in Figure 9. The development of QSS is discussed next.

Requirements and Development Strategies

The implementation of a QSS mechanism in terms of a computer system requires the consideration of aspects of conceptual modelling and representation, similarly to the building of intelligent systems presented in a previous section. In a first attempt to tackle implementation problems, the abstraction scheme previously developed and shown in Figure 1 can be employed. The five conceptual units defined previously for geostatistical concepts generate the means to interrelate the entities involved in reservoir geology and QSS. For instance, (a) the conceptual unit super-concept can now be the entity "petroleum reservoir;" (b) concept may be a "clastic reservoir;" (c) description corresponds to the reservoir interpretation and includes the principal elements of the qualitative model, e.g. "fining upwards sequence;" (d) attributes are the key elements of the description such as "facies," "discontinuities," "faults" etc.; and (e) values are the lithofacies corresponding to each location of the reservoir. A reservoir geological image, generated using a geostatistical simulation technique, is used by the conceptual unit value as an initial state. Then, the consistency rules generated from the causal relationships of the reservoir attributes and as dictated by the qualitative model of the reservoir are applied, to generate the final QSS image. This conceptualization attempts to capture the understanding of the elements of the geological world that is being modelled.

A fundamental characteristic of a QSS approach, as well as the validation of a QSS image, is causal reasoning based on geological principles. Computer systems knowing and operating on model principles and causal relationships are generally referred to as a 'deep' systems. The implementation of QSS clearly calls for building 'deep' systems. It remains, however, to consider the extent of such systems. Two overlapping approaches seem to exist. The first could be the building of systems that contain geological laws and principles, and use as input the qualitative description of the reservoir to generate the final consistency rules. Subsequently, the consistency rules could be used on geostatisticaly simulated reservoir images. The second approach could consider implementing QSS systems that include series of detailed models of different sedimentary environments that could then be used to generate consistency rules for a given reservoir. The delineation of requirements for the implementation of a QSS as well as the use of different techniques, for instance diagrams of reservoir geology and case histories are currently under investigation.

Present experiments on the implementation of QSS systems are focused on limited systems which use consistency rules as input. This type of a system is referred to as a 'shallow' system, because it lacks the understanding of the domain of application and is unable to examine the validity of the consistency rules it uses. However, this type of approach is a useful research tool in understanding implementation related problems of QSS.

A QSS Example

A first and simple example of the QSS idea may be demonstrated in a small computer system implemented in LISP. The system uses as input a basic description of the reservoir, a set of consistency rules, and a section of a grid block model with lithologies assigned to each reservoir block. The system uses frames to represent descriptions, and the standard antecedent-consequent pairs to represent consistency

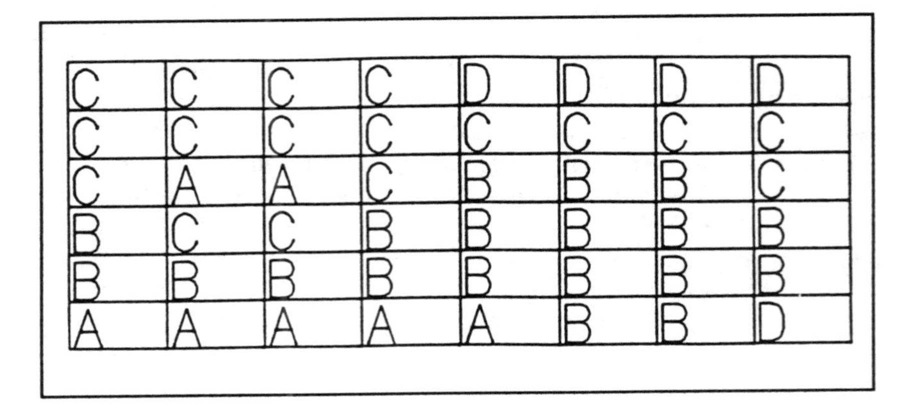

Fig. 10. Lithofacies image of a reservoir grid block model.

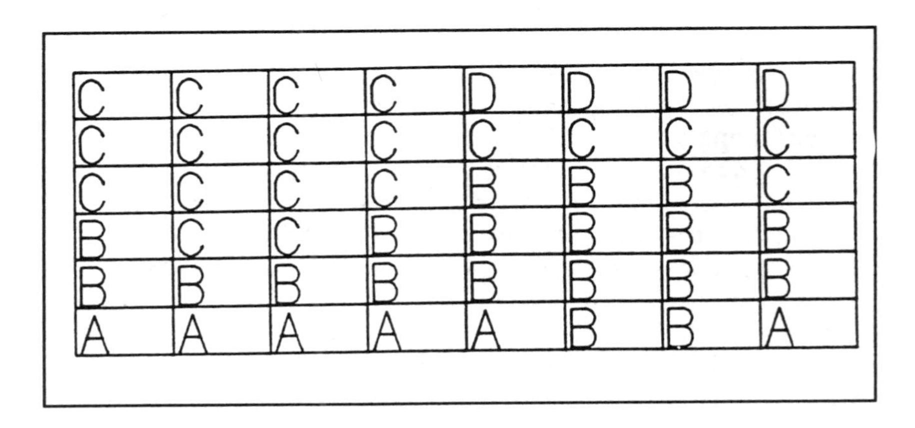

Fig. 11. The QSS lithofacies image corresponding to the reservoir shown in figure 10.

rules. An inference engine processes the input to generate a final image of the input reservoir section. If inconsistencies are found, then they are corrected and explanations are provided using only the consistency rules.

The function of the system can be seen in the example previously used. The reservoir is described as an upward fining sandstone sequence with four distinct sandstone facies, say A, B, C and D, representing a single uninterrupted depositional cycle. The reservoir is represented by a 8 x 6 grid and has been generated to show two geological inconsistencies that need to be corrected. Figure 10 depicts the reservoir lithofacies

```
RESERVOIR-BLOCK: 4,2      replace facies A with C
.........
        Do you want to know why?  (Y or N)  Yes
        RULE-4: 2-UPWARDS-FACIES-CONSISTENCY
          concludes:  (REPLACE-FACIES not C with C)
          because:    (INCONSISTENCY in GRID-BLOCK x)
              and
                      (FACIES C not-followed-by FACIES C)
```

Fig. 12. Reasoning provided for the corrections on the image in Figure 10.

grid block model. The geological inconsistencies are in (i) the left middle part of the figure, where facies A appears within facies C; and (ii) the lower right part, where facies D is below facies B. Both are obvious inconsistencies considering the facies association A → B → C → D and the absence of unconformities.

Figure 11 presents the reservoir image in Figure 10, as constrained by the QSS based system. Both inconsistencies have been corrected and the system may provide justifications if requested, as shown in Figure 12. Note that a QSS lithofacies image satisfies geological consistency criteria. However, statistical consistency criteria such as lithofacies correlations, relative percentages, etc., should be checked and should not unreasonably deviate from the statistical characteristics inferred from the data and available information.

Enhancing and Enriching Simulation of Reservoir Geology with QSS

Although the very idea of QSS is new, needs to be further explored, and issues related to implementation need further examination, two points can be made. Firstly, QSS can enhance the results of geostatistical simulation. Enhancement is seen in the context of imposing purely qualitative-descriptive geological criteria that otherwise are not necessarily followed, to improve the quality of spatial stochastic simulation results. Secondly, QSS can enrich reservoir characterization and management. This last point needs some further consideration.

Although not explicitly discussed, QSS can assist the generation of different reservoir geological images from the same geostatistically simulated reservoir. This is possible when more than one interpretation or description is available - not an impossible situation given the subjectivity of geological interpretations. In addition, it is common that due to limited or incomplete information on a reservoir, more than one geological scenario is possible. In these cases, QSS increases the number of alternatives, enriches simulation approaches and opens the way to account for the uncertainties arising from incomplete information and/or subjective interpretations in reservoir forecasting and management.

CONCLUSIONS

From the work outlined in this study, some general conclusions may be drown regarding the integration of the spatial stochastic techniques used in reservoir characterization and the symbolic non-algorithmic methods developed in the field of Artificial Intelligence, as follows:

a). The technological transfer of knowledge and expertise is attainable and the building of intelligent systems for geostatistical reservoir characterization is feasible. Future work should particularly focus on developing knowledge bases relevant to geological characteristics of known reservoirs, grouping of reservoirs regionally, genetically, etc., to assist geostatistical inference and modelling.

b). Qualitative spatial simulation based on qualitative models and causal relationships seems a promising concept in constraining and enhancing the results of geostatistical simulations. In addition, QSS enriches the possibilities and effectiveness of reservoir geology simulations in reservoir forecasting and management decision making.

c). AI methods provide the means to model and use qualitative or descriptive knowledge and information that statistical techniques are not designed to handle and can not account for.

d). Working with AI techniques leads to the rationalization of operations and the understanding of the role of subjectivity and incomplete information in geostatistical decision making involved in reservoir characterization.

As a final note, it should be stated that stochastic techniques have already contributed in different aspects of AI. AI techniques, it seems, can substantially contribute to and enrich geostatistics. Perhaps, the first AI contribution is the introduction of different types of models and associated thinking that geostatisticians are not familiar with.

ACKNOWLEDGEMENT

The comments and criticism of the two anonymous reviewers of this manuscript should be acknowledged. Support was provided by NSERC Operating Grant No. OGP0105803.

REFERENCES

Barreau, A., Braunschweig B., Emami, E. and Behar, E.: "A Knowledge-Based System for the Automation of Thermodynamic Models Adjustment," *1991 Conference on Artificial Intelligence in Petroleum Exploration & Production*, May 15-17, Texas A&M University, College Station, TX (1991), 55-62.

Bobrow, D.G.: "Qualitative Reasoning about Physical Systems: An Introduction," *Qualitative Reasoning about Physical Systems*, D.G. Bobrow (Ed.), MIT Press, Cambridge, MA (1986) 1-5.

Borgida, A., Mylopoulos, J. and Wong, K.T.: "Generalization / Specification as a Basis for Software Specification," *On Conceptual Modelling*, M.L. Brodie et al. (eds.), Springer-Verlag, New York, NY (1985) 87-118.

David, M.: *Geostatistical Ore Reserve Estimation*, Elsevier, Amsterdam (1977).

Dimitrakopoulos, R.: "Artificially Intelligent Geostatistics: A Framework Accommodating Qualitative Knowledge-Information," *Mathematical Geology* (submitted).

Dimitrakopoulos, R.: "Conditional simulation of IRF-k in the Petroleum

Industry and the Expert System perspective," PhD thesis, Ecole Polytechnique, Montreal, PQ (1989).

Dimitrakopoulos, R. and David, M.: "Artificial Intelligence in Geostatistical Ore Reserve Assessment," *Geoinformatics* (1991), **2**, 211-218.

Journel, A.G. and Huijbregts Ch.J.: *Mining Geostatistics*, Academic Press, New York (1978).

Kuipers, B.: "Qualitative Simulation," *Artificial Intelligence* (1986) **29**, 289-388.

Levesque, H.J. and Mylopoulos, J.: (1979) "Procedural Semantics for Semantic Networks," *Associative Networks*, N.V. Fidler (ed.), Academic Press, New York (1979), 93-120.

Mylopoulos, J. and Levesque, H.J.: "An Overview of Knowledge Representation," *On Conceptual Modelling*, M.L. Brodie, J. Mylopoulos and J.W. Schmidt (eds.), Springer-Verlag, New York (1984), 3-17.

Newell, A. and Simon, H.A.: "Computer Science as Empirical Enquiry: Symbols and Search," *Communications of the ACM* (1976) **19**, 113-126.

Nilsson, N.J.: *Principles of Artificial Intelligence*. Troga Publishing Co., Palo Alto, CA (1980).

O'Keefe, R.: "Simulation and Expert Systems - A Taxonomy and some Examples," *Simulation* (1986) **46**, 10-16.

Olea, R.A.: *CORRELATOR An interactive Computer System for Litho-stratigraphic Correlation of Wire Logs*, Petrophysical Series **4**, Kansas Geological Survey, Lawrence KN (1988).

Reddy, Y.V., Fox, M.S., Husain N. and McRoberts: "The Knowledge-Based Simulation System." *IEEE Software* March 1986, 26-37.

Shultz, A.W., Fang, J.H., Burston, M.R., Chen, H.C. and Reynolds, S.: "XEOP: An Expert System for Determining Clastic Depositional Environments," *Geobyte* (1987) **3**, 22-26.

Simmons, R.G.: "Combining Associational and Causal Reasoning to solve Interpretation and Planning Problems," PhD thesis, MIT, Cambridge, MA (1988).

Smith, R.G. and Baker, J.D.: "The Dipmeter Advisor System," *Eighth Intl. Joint Conf. on Artificial Intelligence* (1983), 122-129.

Winston, R.H. and Horn, B.K.P.: *LISP*, Addison Wesley, Menlo Park, CA (1986).

Modeling Heterogeneous and Fractured Reservoirs with Inverse Methods Based on Iterated Function Systems

Jane C. S. Long, Christine Doughty,
Kevin Hestir and Stephen Martel

Earth Sciences Division
Lawrence Berkeley Laboratory
Berkeley, CA 94720

ABSTRACT

Fractured and heterogeneous reservoirs are complex and difficult to characterize. In many cases, the modeling approaches used for making predictions of behavior in such reservoirs have been unsatisfactory. In this paper we describe a new modeling approach which results in a model that has fractal-like qualities. This is an inverse approach which uses observations of reservoir behavior to create a model that can reproduce observed behavior. The model is described by an iterated function system (IFS) that creates a fractal-like object that can be mapped into a conductivity distribution. It may be possible to identify subclasses of Iterated Function Systems which describe geological facies. By limiting the behavior-based search for an IFS to the geologic subclasses, we can condition the reservoir model on geologic information. This technique is under development, but several examples provide encouragement for eventual application to reservoir prediction.

1.0 INTRODUCTION

Most of the established techniques for modeling heterogeneous and fractured reservoirs are based on the assumption that the reservoir acts as an equivalent continuum on some scale, often called the representative elementary volume (REV) (Toth, 1967). Further, a common assumption is that the reservoir can be modeled by tesselating the entire region of interest with blocks of equivalent continua that are at least as large as the REV. However, it has become increasingly apparent that reservoir heterogeneities occur on every scale (Freeze, 1975), and that the concept of the REV may not always be appropriate. For example, in a fractured rock, we find fractures on every scale from the micro-fracture to major fault. Dominant flow paths develop where open, conductive fractures intersect. Flow may completely bypass parts of the reservoir and connected regions may be complex and hard to define. A similar case may be made for sand-body reservoirs. In these cases, it is difficult to define a heterogeneous model for reservoir behavior.

Two modeling approaches are commonly used to include the effect of heterogeneities. (For this discussion we can consider fractures as just another type of heterogeneity.) The first approach we will call the ''forward'' approach. In this approach one tries to infer the distribution of heterogeneities and the spatial relationships from conductivity measurements. Tools such as geostatistics can be used to create realizations which match both the local measurements and the inferred spatial correlation. The primary difficulty with this approach is that such models rely on an estimate of the geometry of the heterogeneities to predict the behavior. A model which reproduces the geometry may not match observed behavior, much less correctly predict new behavior (Long et al., 1991). Secondly, most of these techniques are restricted to producing smooth models of heterogeneities. Physical systems that are highly convoluted or poorly connected such as meander belts or fracture networks, may be extremely difficult to simulate with geostatistics.

The second approach is the inverse method. In this approach we search for a pattern of heterogeneity which matches the observed behavior of the reservoir, usually observed heads under assumed steady flow conditions. Such models have been developed by Carrera and Neuman (1986a,b,c) and Kitanidis and Vomvoris (1983), for example. The latter is particularly interesting because the inversion method is used to determine a relatively small set of geostatistical parameters, which are then used to generate heterogeneous hydrologic property distributions via kriging.

In the inverse techniques we have developed at Lawrence Berkeley Laboratory, we search for equivalent models which are based on a geologic understanding of flow. For example, Simulated Annealing (Davey et al., 1989) is an inversion technique that has been applied to fractured rock to find an Equivalent Discontinuum model (Long et al., 1991). Simulated Annealing is applied to a partially filled lattice of one-dimensional conductors, called a template, which is in effect a geologically based conceptual model for the fracture system. The algorithm searches for a configuration of lattice elements which can reproduce observed hydrologic data. At each iteration, one calculates the ''energy,'' E, of the configuration, which is a function of the difference between model predictions and observed behavior. Then a random change is made in the lattice and the new energy is computed and compared to the old energy. If the energy is decreased, the change in the configuration is kept. If the energy is increased by the change, the choice of whether or not to keep the new configuration is made randomly based on a probability which decreases with the amount of energy increase, allowing the algorithm to ''wiggle'' out of local minima. Use of the annealing algorithm is more completely documented by Davey et al. (1989) and Long et al. (1991).

The Iterated Function System (IFS) inversion method described in this paper is similar to the application of Simulated Annealing to Equivalent

Discontinuum models. An IFS is used to create a fractal-like object (an attractor) which describes reservoir heterogeneities or fractures. The inverse analysis optimizes the parameters of the IFS until the attractor-based hydrologic model matches the observed behavior of the hydraulic data.

Below we briefly describe the IFS concept and explain how these functions are used in an inversion. Then we discuss the well test data that can be used in an inversion. We provide two examples of inversions based on synthetic data generated from numerical models and two preliminary field based examples, one for heterogeneous porous materials and one for fractured rock. A third preliminary example shows how we might find classes of IFS which produce fracture geometries similar to those observed in nature. Finally we discuss future directions.

2.0 ITERATED FUNCTION SYSTEMS

An iterated function system (IFS) is a standard way to model self similar geometrical structures (Barnsley, 1988) which was developed for use in computer graphics in order to find efficient means for storing the information describing each pixel of a complex picture. In this application, one identifies an iterative process that will create the picture rather than storing the information for each pixel. The iterative process is defined by an IFS which has a relatively small number of parameters. The use of an IFS essentially exchanges the use of computer storage for the use of computer time. In our application, we want to create a model of a complex heterogeneous geologic system. Instead of trying to describe this system "pixel by pixel," we look for an IFS that can describe the geometry of the system with a small number of parameters.

An IFS creates a picture starting with an initial set of points and a set of iterative functions. At each iteration, each function in the system operates on the set of points and according to the parameters in the function translates, reflects, rotates, contracts or distorts the set of points. Over many iterations, the points in the picture coalesce towards an "attractor" which is a fractal-like object. The shape of this attractor changes gradually when the parameters of the IFS gradually change.

To create an IFS one first specifies a function f, which maps sets to sets:

$$f(A_0) = A_1 \tag{2.1}$$

where A_0 and A_1 are (compact) subsets of two (or three) dimensional space. A set A_∞ can then be defined by

$$A_{n+1} = f(A_n) \quad n = 0, 1, \ldots \tag{2.2}$$

$$A_\infty = \lim_{n \to \infty} A_n \ .$$

Given certain restrictions on the set function f, one can show (Barnsley 1988) that A_∞ exists, is independent of the starting set A_0, and generally has a fractional Hausdorff dimension. Hence f determines a fractal, A_∞. If we have a function f that is easily parameterized, then the fractal A_∞ is parameterized as well. This leads to a nice setup for modeling real-world problems, because a small number of parameters can be used to characterize a complex geometry.

A wide variety of Iterated Function Systems can be defined, but they fall into two main categories: deterministic and probabilistic. A deterministic IFS has uniquely determined parameters, and thus creates a unique attractor A_∞. A random IFS chooses some or all of its parameters randomly from probability distributions, so multiple realizations of A_∞ differ. The Iterated Function Systems used in the hydrologic inversions given in this paper are deterministic and of the form of Equations (2.3) and (2.4); those used in the fracture growth scheme (Section 8) are random.

One important example of a deterministic f used extensively by Barnsley (1988) is:

$$f(A) = g_1(A) \cup g_2(A) \cup \ldots g_k(A) \quad . \tag{2.3}$$

Here the g_i's are so called affine transforms:

$$g_i(A) = \bigcup_{\vec{x} \in A} g_i(\vec{x}) \tag{2.4}$$

$$g_i(\vec{x}) = B_i\vec{x} + \vec{b}_i$$

where B_i is a matrix and $\vec{b}_i$ a vector. The parameters characterizing f are the entries in the B_i's and $\vec{b}_i$'s. The matrix, B_i, serves to rotate, reflect, distort, and contract and the vector, $\vec{b}_i$, translates. An example IFS using k = 3 affine transformations which contract and translate, resulting in a fractal called a Sierpinski's gasket, is shown in Figure 2.1. The IFS is specified by

$$B_1 = B_2 = B_3 = \begin{bmatrix} 0.5 & 0.0 \\ 0.0 & 0.5 \end{bmatrix} , \tag{2.5}$$

$$\vec{b}_1 = (0.0, 0.0) , \quad \vec{b}_2 = (0.5, 0.0) , \quad \vec{b}_3 = (0.0, 0.5) \quad .$$

Figure 2.2 shows the attractors generated by a sequence of functions $f_1, f_2, \cdots, f_6$, where f_1 is the Sierpinski's gasket, and for j = 2, 6 every parameter of f_j differs from the corresponding parameter of f_{j-1} by a small increment. The continuous change in parameters is manifested as a continuous change in the attractors, which is a useful but not necessary condition for an IFS-based inversion procedure to work.

The most general affine transforms that operate in two-dimensions have four arbitrary entries in each B_i matrix, and two arbitrary values in each $\vec{b}_i$ vector, which gives a total of 6 parameters for each affine transformation. By understanding how the different parameters affect the shape of the attractor, we can constrain parameters to produce attractors that have desired properties, for example mimicking certain geological facies. As well as making the inversion procedure more efficient by reducing the dimensionality of the parameter space,

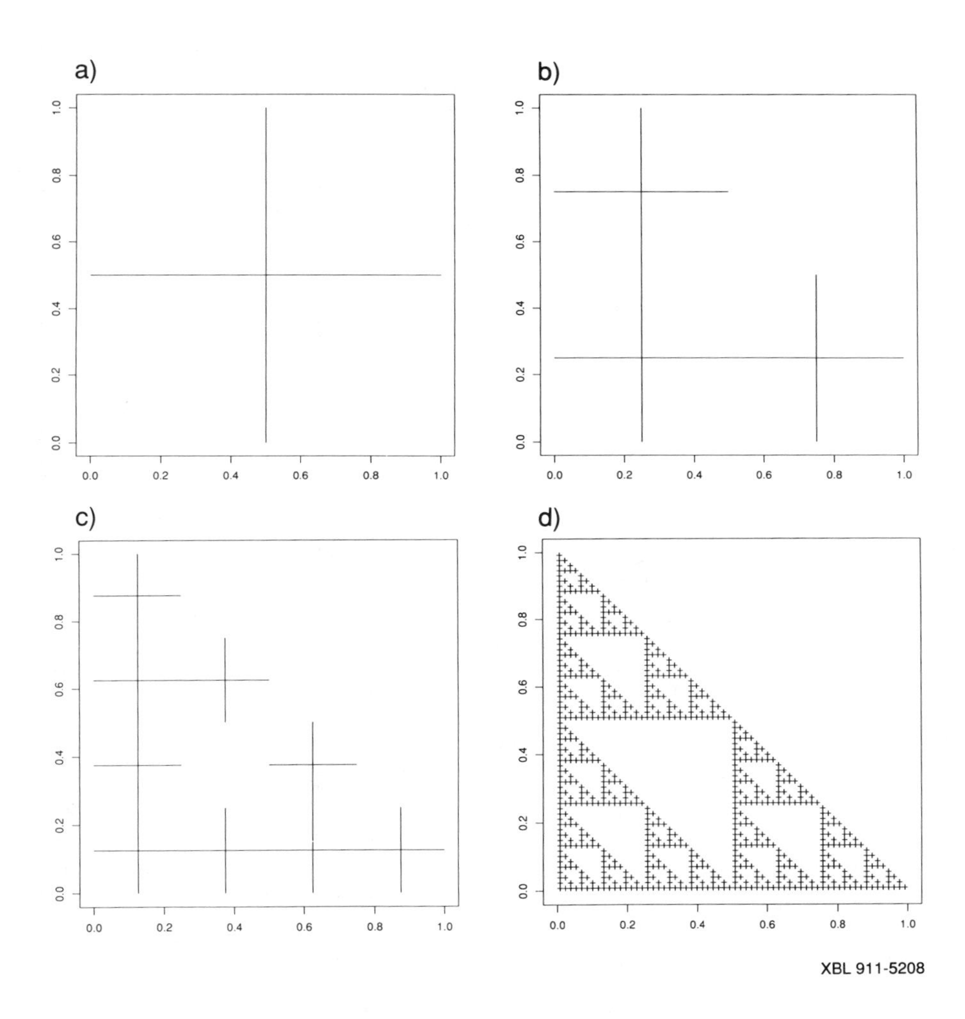

Figure 2.1. Generation of a Sierpinski's gasket using three affine transformations.

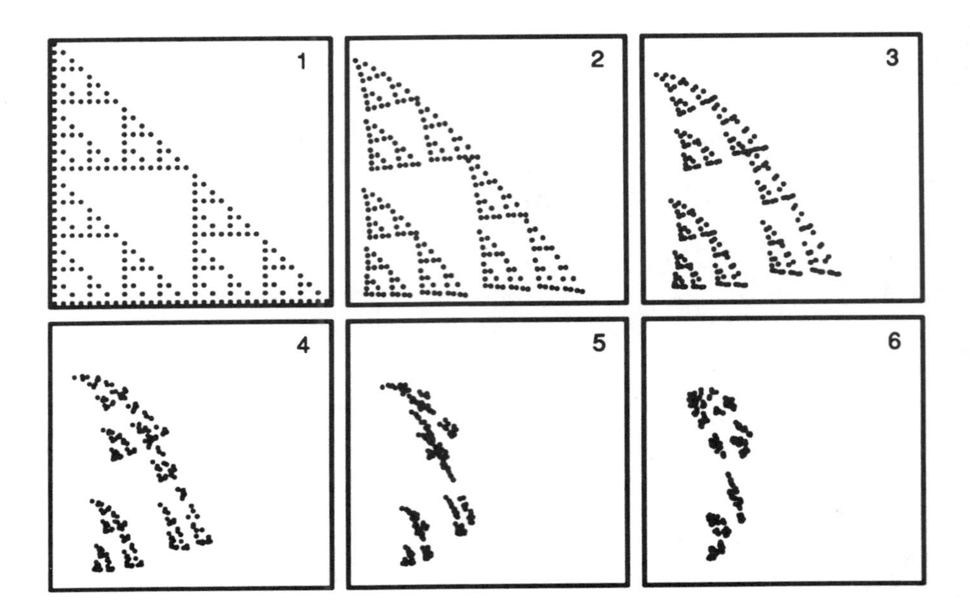

Figure 2.2. A series of attractors generated by functions whose parameters differ by small increments.

these constraints make the inversion more robust by conditioning it on known geological conditions. One simple example is to construct each B_i as a rotation matrix

$$B = S\begin{bmatrix}\cos\theta & -\sin\theta \\ \sin\theta & \cos\theta\end{bmatrix} \qquad (2.6)$$

where S is a contractivity factor $(0 < S < 1)$ and θ is a rotation angle. This formulation reduces the number of parameters of the IFS from 6 to 4 per affine transformation. By restricting θ and $\vec{b}$ to a limited range, directional trends observed in geologic media can be reproduced in the attractors.

3.0 INVERSION BASED ON ITERATED FUNCTION SYSTEMS

To use the IFS as a basis for hydrologic inversions we map the points of the attractor into a hydrologic property (conductivity, storativity etc.) distribution and use the finite element code TRINET (Karasaki, 1987) to simulate a well test. In the examples given here, the finite element mesh consists of a lattice of one dimensional conductors. We superimpose the attractor on the lattice and increment the conductance and storativity of the lattice elements that are close to each point on the attractor, as shown in Figure 3.1. The hydrologic properties of a lattice element can be incremented as many times as there are points on the attractor near by. In this way the small number of parameters of an IFS define the conductance and storativity distribution for thousands of elements.

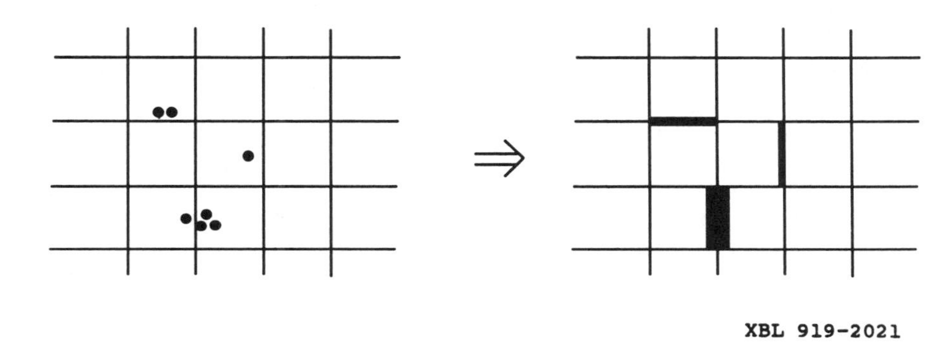

Figure 3.1. "Step" mapping between points on the attractor and increments in hydrologic properties of the lattice.

The inversion algorithm searches for IFS parameters which define a heterogeneous system that behaves like the observed well tests. We first construct a model of the flow system using a lattice of elements modified by an arbitrary IFS. We then optimize the parameters of the IFS until the model produces a good match to the well test data. The match is quantified by the energy, E, which represents, in a single number, the total amount of mismatch between the observed and modeled drawdowns, and is a convenient way of quantifying the "goodness of fit" of the model to the data during the course of an inversion. We define E as

$$E = \sum \left[\ln(h_o) - \ln(h_c) \right]^2 \tag{3.1}$$

where h_o is the observed head (or drawdown) and h_c is the head (or drawdown) calculated using the hydrologic properties mapped from the attractor. E can also be a function of flow differences or any other pertinent measure of behavior. The sum is taken over a discrete set of observation times and all observation wells. E is not normalized by the number of data points or by the magnitude of the observed drawdowns, so one cannot say *a priori* that a certain value (e.g., E = 10) is "good" or "bad" for all problems. At this point we use judgment to decide whether the mismatch between observed and calculated drawdowns is sufficiently small to be insignificant.

The optimization can be done in a variety of ways. We have used several routines available in standard numerical libraries, including downhill simplex and direction set methods (Press et al., 1986). One optimization technique that seems to work well is Simulated Annealing. In this case, we randomly choose new values of the IFS parameters and accept or reject these new choices according to the annealing algorithm as described above.

Some parts of the inversion algorithm are arbitrary. For example, we choose the number of affine transforms, k, that make up the IFS. We also choose the number of points, M, for the IFS to use in creating the attractor. The larger M is, the greater the contrast in permeability can be. One could use a high value of M to model highly conductive features in a relatively impermeable matrix or a lower value of M to model conductive features in a slightly impermeable matrix. Further, we also arbitrarily choose how to relate the increment in conductance and storativity represented by each point of the attractor. Another arbitrary choice is exactly how to map the attractor into the hydrologic parameters. One possibility is to modify the properties of just the single element closest to each attractor point (a "step" map). Alternatively, properties for all elements near an attractor could be affected, with the magnitude of the change decreasing as a function of distance from the attractor point (a distributed map). We have just begun to study the effects of such choices.

One of the attractive features of this approach is that it may be possible to choose sub-classes of Iterated Function Systems which tend to produce features observed in a geologic investigation. For example, we may be able to find Iterated Function Systems that always produce a specific type of brittle shear zone or meander belt structure. In these cases we could confine the search for hydrologic behavior to the sub-class of IFS that represents the geology. Along the same lines, once we have identified the form of the IFS that best explains all the data, the model will have fractal-like properties that may help to extrapolate behavior to scales that can not be tested in reasonable time frames.

4.0 HYDROLOGIC DATA FOR INVERSION

One of the significant problems associated with applying these techniques is the choice of data set to invert. In principle, any physical phenomena of interest which can be numerically modeled and also monitored in the field can be used in the inverse method. In practice, it can be quite difficult to pick a good data set for analysis. Some of the difficulties arise from the usual problems with field data: poorly known boundary conditions, incomplete or insufficient data, etc. Another problem faced in the inversion process is comparing model results to data. In the model we choose base values of conductance and storativity. Then, these parameters are incremented using the IFS map and the model is used to simulate well tests. The model results and the data may differ for two reasons: (1) the base values of the parameters are wrong or (2) the distribution of heterogeneities is wrong. The inversion process defined above is only designed to address the second reason. In some cases an incorrect choice of base values can be treated by shifting the model results on a log drawdown – log time plot. The y-shift corresponds to scaling the conductances of the elements uniformly up or down and the x-shift scales the diffusivity (conductivity divided by storativity). Thus at each iteration, the model results are first shifted to obtain the best fit to the data and then the energy is calculated. However, as discussed below, it is not always appropriate to shift the model results. Further, if the base parameters chosen are very bad estimates, so that large shifts are required, numerical problems can occur. For example, a large over estimate of base conductance can cause drawdowns to be so small that they are swamped by round off errors. A large log-time shift may either cause the temporal resolution of the model to be inadequate at early times, or the period of time modeled to be too short. Although shifting is an elegant way to avoid repetitious modeling, for practical purposes it may be preferable to incorporate the choice of base parameters into the optimization process. Some of the considerations necessary for using different types and amounts of well test data are discussed below.

4.1 Steady-State Tests

The simplest approach has been to use the steady-state head distribution resulting from a pumping test with a constant flow boundary condition applied at the pumping well. The energy function is constructed as a function of the differences between modeled and measured heads or drawdowns. Drawdowns induced by such a test are relatively simple to measure, and steady flow is easy and quick to model, allowing many iterations of the model to be practical. However, for steady-state flow, the pattern of drawdowns does not change when conductances of the medium are uniformly scaled up or down. So, using a single steady-state test will only give a pattern of conductance contrasts which matches the head distribution. The value of these conductances can then be scaled up or down until the applied flow boundary condition is matched. This means, not surprisingly, that models obtained largely by matching drawdowns should be more sensitive predictors of drawdown than they are of flow.

Greater sensitivity to flow can be gained by combining a series of steady-state tests. If constant flow is applied at the pumping well, the energy function can include the head at the pumping well treated as any other observed head. In this case, each of the separate tests is modeled at each iteration and the factor incrementing all the element conductances is chosen to best fit all the flow boundary conditions. If constant head boundary conditions are applied at the pumping well, the energy function can be constructed as an appropriately scaled combination of squared head differences at observation wells and squared flow differences at the pumping well. In this case, no overall scaling of conductances is needed.

Generally, multiple steady tests may provide the best data for inversion because there is no dependence on storage coefficient and the time required for steady flow calculations is very small. However, in the field each steady test is very time consuming and consequently few are usually available.

4.2 Transient Tests

Alternatively, one can use the transient interference data resulting from a constant flow boundary condition. The flow rate used in the field is specified in the model in order to predict the transient drawdown response. At each iteration, the model predicts curves of drawdown versus time that can be shifted in both the x- and y-directions in log-log space until a best match is obtained to the real curves. This process is similar to matching data to a Theis curve but in this case the shift corresponds to scaling the conductances and storage coefficients for the elements in the model. With multiple observation points, it becomes necessary to find the best shift on average. Although this process is conceptually simple, the vagaries of numerical calculation combined with the vagaries of real data can make curve matching extremely difficult to do automatically for thousands of iterations. The energy at each iteration is the sum of the squared differences in log of head for each observation point at selected times. The advantage of using this type of data in inversion is that the transients reflect the distribution of heterogeneities in space, where as a steady test is more likely to reflect the biggest bottle-neck, irrespective of where it is. The disadvantage of transient data is that we are forced to make an assumption about the relationship between storage and conductance; in other words we have more information, but another parameter to specify.

As in the steady-state case, a slightly different procedure must be used if the transient test has a constant head boundary condition. In this case we should use both the transient drawdown data at the observation wells and the transient flow rate at the pumping well in the energy function. Under these conditions the y-shift is not needed because head is pegged by the constant head boundary condition. Consequently, constant head tests are somewhat more sensitive to the initial estimate of the element conductances and in practice are more difficult to invert than constant flow test data.

4.3 Combining Different Types of Tests

If several different tests are available, these can be combined. In principle any combination of steady, transient, constant-flow or constant-head data can be combined. The main drawback for combining a large number of transient tests is the possibility of using an enormous amount of cpu time. The calculation time scales with the number of tests times the number of time steps times the number of iterations. It is not difficult to conceive of a problem that could take on the order of a month to invert.

For multiple constant-flow transients, the procedure is straight forward. At each iteration, each test is modeled and the best-fit x- and y-shifts for all the curves are identified. Theoretically, a steady-state test is a subset of a transient test and the steady drawdowns predicted by the model can be matched to the data by a shift in the y-direction, with the x-shift irrelevant for steady-state conditions. On the other hand, if we include a constant head test (transient or steady-state), we cannot use a y-shift to match the drawdown data from the constant head test.

One approach is not to use the y-shift on any of the drawdown curves. In this case, we need to have a good *a priori* estimate of element conductance. Again, flow data should be included in the energy function for the constant head case.

In general, the inversion of well test data yields a non-unique solution. Fundamentally, one rarely has enough data to specify a unique solution. The advantage of using multiple well tests is that each additional test provides more information about the system. We can use a single well test to predict a second well test, and the first two well tests to predict the third. In this way we can see if our ability to make predictions improves which implies an improvement in the uniqueness of the solution.

5.0 SYNTHETIC EXAMPLES

One way to see how well the inversion algorithm works is to generate synthetic data from a prescribed model and see if the model used to generate the data can be recovered by the inversion. At this point, we have completed a few simple cases which serve to provide encouragement for the concepts as well as point out the limitations of the method. Many issues have not yet been addressed and will be the object of further study.

5.1 A Linear High-Conductivity Feature

The first synthetic case is a simplified model which might represent the hydraulic conditions imposed by a buried stream channel or the trace of a conductive fault. We construct a two-dimensional model with a highly conductive linear feature and use IFS inversion to see if we can find the location of this feature.

An IFS composed of two affine transforms of the form

$$f(A) = g_1(A) \cup g_2(A) \tag{5.1}$$

where

$$g_i(\vec{x}) = \begin{bmatrix} 0.5 & 0.0 \\ 0.0 & 0.5 \end{bmatrix} \vec{x} + \vec{b}_i \tag{5.2}$$

has only four parameters, the two components of each of $\vec{b}_1$ and $\vec{b}_2$. This f always produces a linear attractor, with the length and orientation of the line segment depending on the $\vec{b}_i$'s. A linear high conductivity feature provides a simple demonstration of the IFS inversion procedure for several reasons: The inversion is fast because the dimension of the parameter space is small (4 instead of the usual 6 parameters per affine transformation); the evolution of the attractor as the inversion progresses is easy to visualize; and the linear high conductivity feature has a clear "signature" on the pressure transients, making the inverse problem better posed.

A synthetic data set was generated for a constant flow pump test conducted in a medium with a linear feature that has a conductivity 500 times higher than the background (see Figure 5.1). The central well pumps at a constant rate and transient heads are calculated for four surrounding observation wells. A two-dimensional finite element mesh composed of a regular 20 by 20 grid of linear elements is used; head is held constant at the outer boundary. Figure 5.2 shows the transient heads calculated for this conductivity distribution (the synthetic data). The effect of the high conductivity feature is clearly seen in the earlier, larger response of the upper well in Figure 5.2.

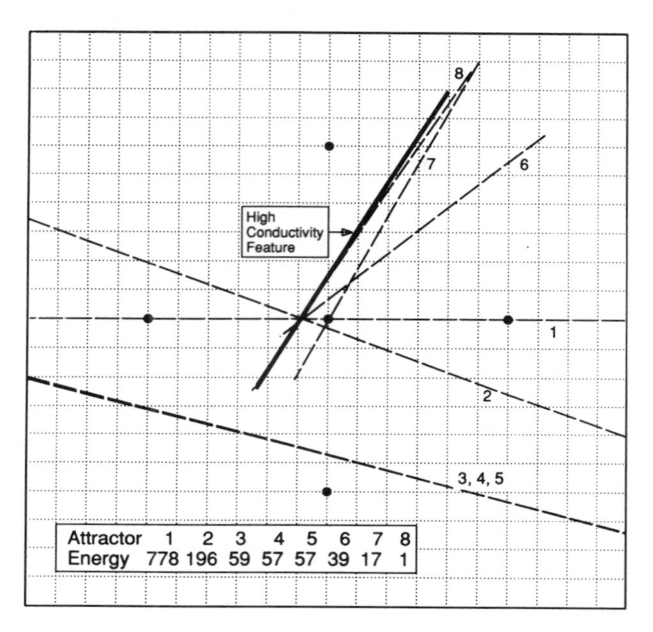

Figure 5.1. Model of synthetic case 1: the heavy black line shows the region of enhanced conductivity, the wells are marked as large black dots, and the mesh is shown as dotted lines. The initial attractor for inversion is the dashed line labeled 1; the dashed lines labeled 2-8 are a sequence of attractors found during the inversion.

The attractors found at various points during the inversion are shown in Figure 5.1, along with the corresponding energies. Note that this figure shows the attractor, each point of which is used to increment the nearest element conductance, not the conductance distribution itself. Figure 5.2 shows the pressure transients for a uniform medium (no attractor, $E = 90$) and for the final attractor determined by the inversion ($E = 1.2$). The small energy of the final attractor is due to the excellent match of all the pressure transients and is not surprising in light of the similarity of the final attractor (labeled 8 in Figure 5.1) to the original high-conductivity feature. This synthetic case illustrates the IFS inversion working very well, but it should be emphasized that real-world problems are likely to be much more complicated.

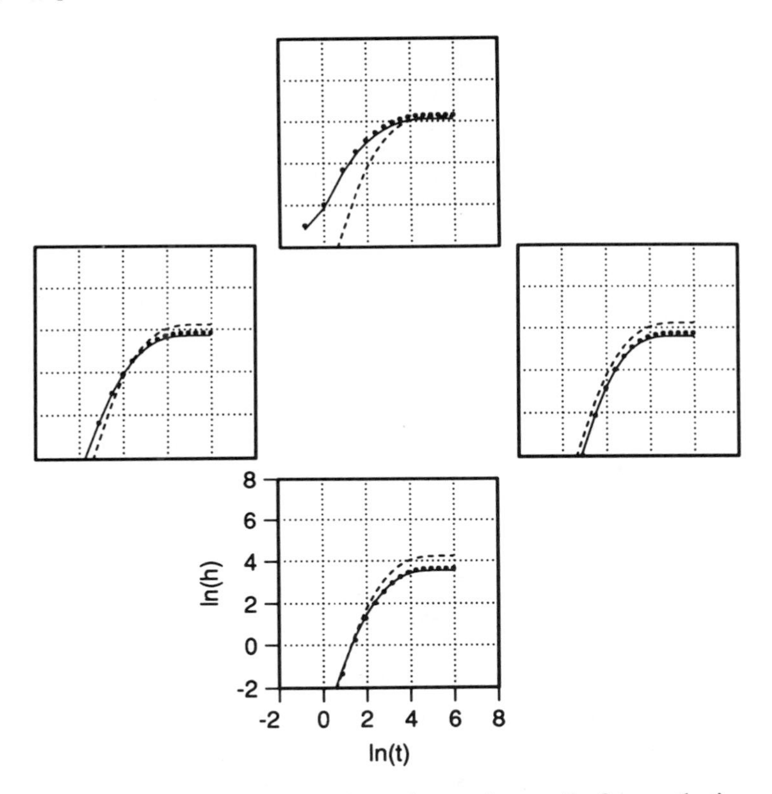

Figure 5.2. Transient heads at the four observation wells for synthetic case 1: data (black dots), a uniform medium with no attractor (dashed lines, E = 90), and the attractor labeled 8 in Figure 5.1 (solid lines, E = 1.2). The arrangement of the plots on the page follows the locations of the observation wells in the well field.

5.2 A Square Zone of Contrasting Conductivity

A second synthetic case consists of a central square region with hydrological properties significantly different than the surrounding region (Figure 5.3). First we allowed the center region to have a conductivity and storativity 100 times higher than the outer region. Then, we reversed these ratios. In each case, interference data was generated by pumping from well 0 and monitoring the response at the other five side and corner wells. The five drawdown vs. time curves are inverted using Simulated Annealing to find an optimal IFS composed of three affine transforms.

The IFS chosen to provide a starting point for the inversion is the Sierpinski's gasket shown in the first frame of Figure 2.2.

Figure 5.4 shows two different solutions for the inversion with the high conductivity and storativity in the center. Figure 5.5 shows the match between the synthetic well test data and the model results for the first solution. Figures 5.6

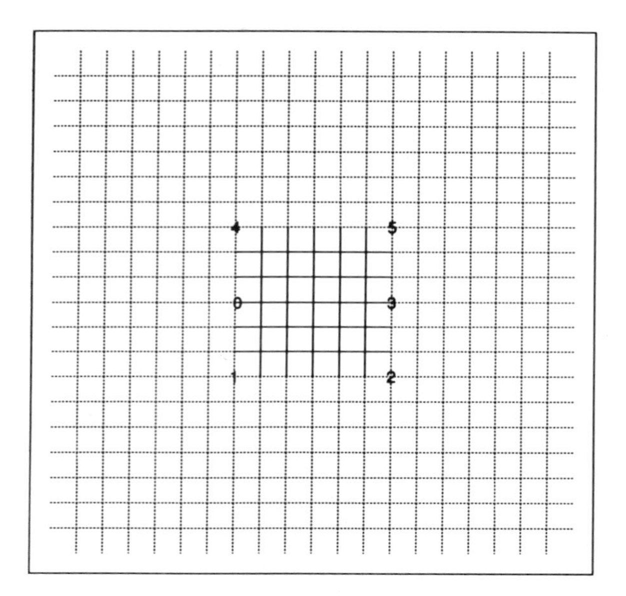

Figure 5.3. Model for the second synthetic case. The central region is at first 100 times more conductive than the outer region, then 100 times less conductive. The numbers show the locations of six wells.

and 5.7 show the corresponding information for the low conductivity and storativity in the center. The energy associated with a homogeneous lattice is about E = 200 and the energy of all the IFS solutions is about E = 10.

The algorithm is clearly able to find a central high conductivity zone and this is very encouraging. This case should be extended to see how large a contrast and how small an inhomogeneity can be detected. Also, we should investigate how far away the wells can be from the anomaly and still detect it.

Interestingly, the reverse case does not recover the geometry of the original model as well. When the high conductivity is on the outside, the algorithm puts a small region of high conductivity on the outside, but does not spread it around the anomaly. We suspect that if we based the inversion on a combination of well tests from different wells, we would have a better chance to resolve the anomaly. Also, we could use the attractor to decrement the conductance instead of increment it. This may give a solution similar to the case above.

5.3 Conclusions Drawn from Synthetic Cases

The synthetic cases we have conducted so far have provided some general confidence in the approach we are taking and show that an extensive study of synthetic cases is warranted to help refine the algorithm. Furthermore, it will be useful to corrupt the synthetic data by adding noise or by varying the boundary conditions to see how best to develop robust inversion techniques.

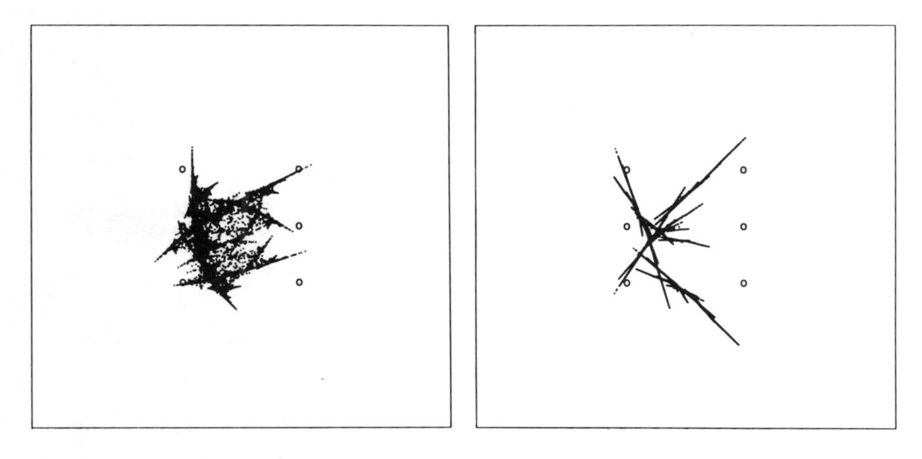

Figure 5.4. Two different attractors found by inversion of the second synthetic case for the high conductivity in the center.

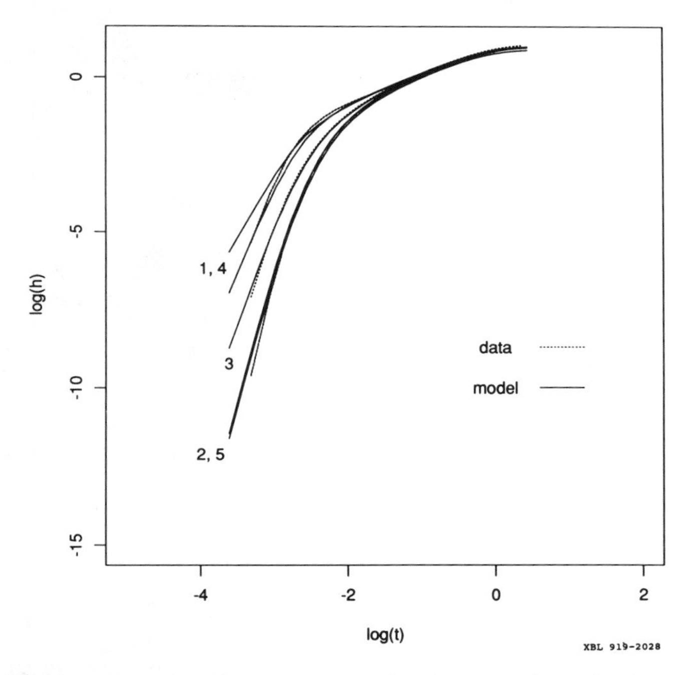

Figure 5.5. Transient drawdown response for the second synthetic case for the high conductivity in the center.

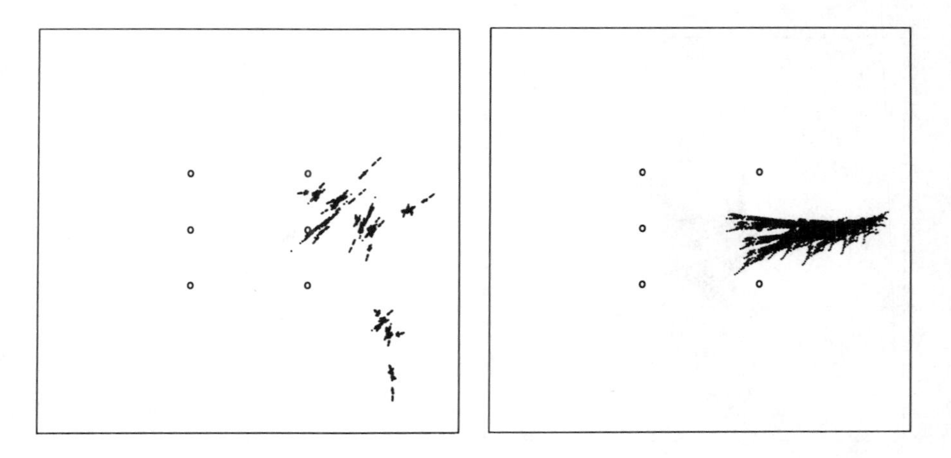

Figure 5.6. Two different attractors found by inversion of the second synthetic case for the low conductivity in the center.

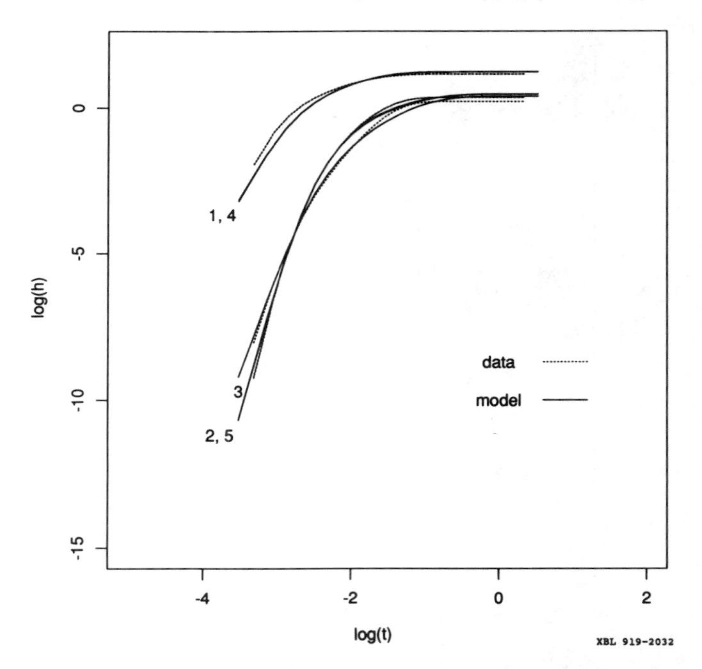

Figure 5.7. Transient drawdown responses for the second synthetic case for the low conductivity in the center.

6.0 INVERSION OF DATA FROM HETEROGENEOUS POROUS MATERIALS

A variety of well tests have been conducted on a shallow aquifer system composed of interbedded sands, silts, and clays at Kesterson Reservoir, located in the San Joaquin Valley in central California (Yates, 1988). The hydrological properties of the aquifer/aquitard system are needed in order to study the transport of various forms of selenium and other salts between surface waters and underlying aquifers. The aquifer studied in the present example is about 18 m thick, and is underlain by an impermeable clay layer and overlain by a leaky aquitard. A multi-well transient pump test is analyzed to infer the spatial distribution of permeability in the aquifer.

In the test under consideration, a central well was pumped at a constant rate and, transient drawdowns were measured at eight observation wells located 15 to 107 m away from the pumping well (see Figure 6.1a). A two-dimensional areal finite element model is used to represent the aquifer.

Figure 6.2 shows the observed drawdown vs. time curves and those calculated assuming a medium with uniform conductivity and storativity (no attractor). The energy of the uniform-medium solution is E = 38.

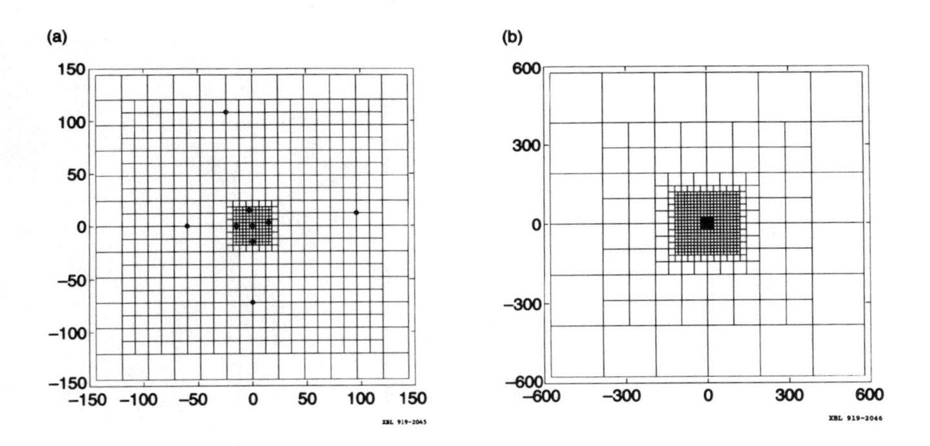

Figure 6.1. The two-dimensional finite element mesh used for the Kesterson calculation. Frame (a) shows the central part of the mesh with the well field superposed, frame (b) shows the entire mesh.

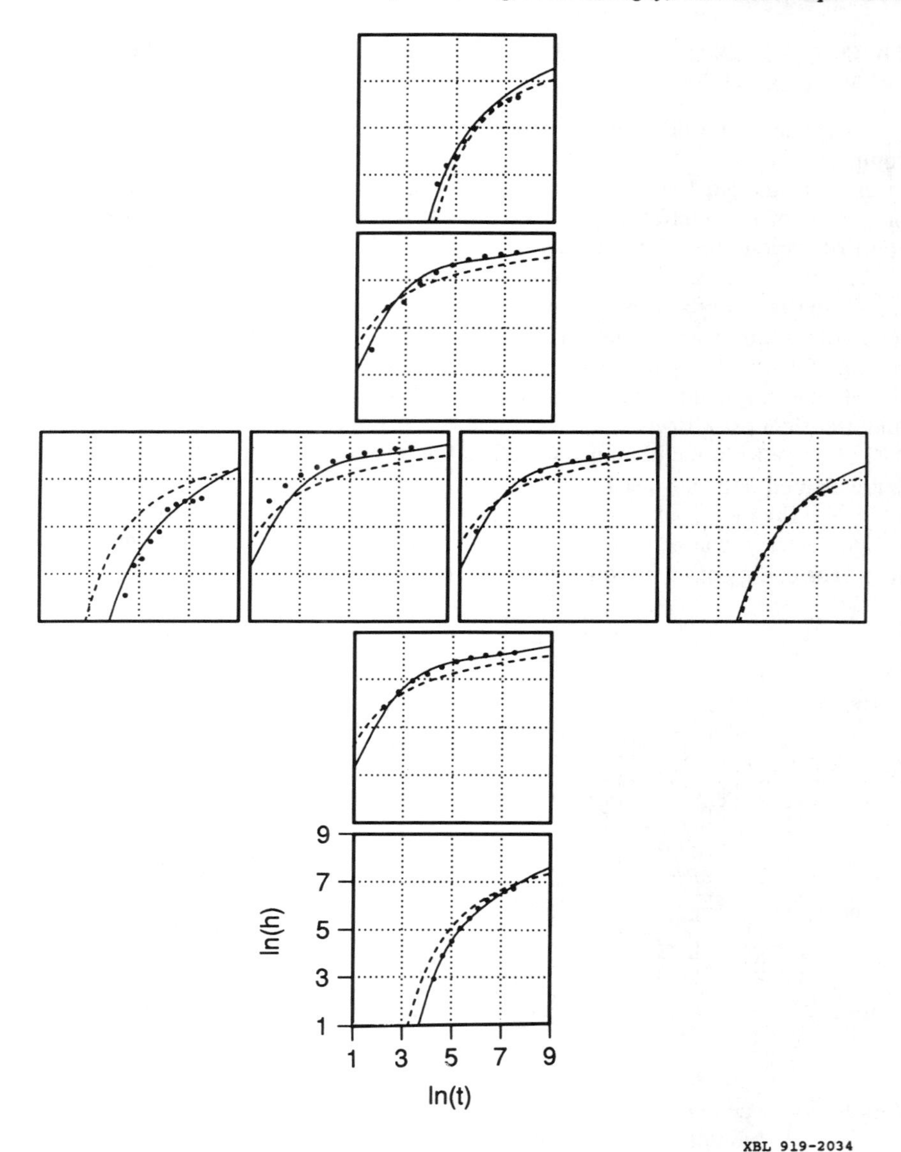

Figure 6.2. The transient drawdown response in the Kesterson observation wells: observed data (black dots), calculated response assuming a uniform medium (dashed lines, E = 38), and the calculated response that produces the minimum energy (solid lines, E = 6). The arrangement of the plots on the page follows the locations of the observations wells in the well field (Figure 6.1a).

During the inversion the attractor is constrained to remain within the region $|x| < 150$, $|y| < 150$ m shown in Figure 6.1a. If this constraint is not included, the inversion tends to waste effort changing the attractor far from the well field, where changes have little impact on the observed drawdowns. Each point of the attractor increments the conductance of the nearest mesh element. The final attractor resulting from the inversion is shown in Figure 6.3. The drawdown vs. time curves for this attractor are shown in Figure 6.2 and correspond to an energy of E = 6.

Figure 6.3. The attractor that yields the minimum energy (E = 6) for the Kesterson data.

In conclusion, the inverse method has worked well to match the observed head data. However, we have used a two-dimensional model for a three-dimensional problem, and in so doing may have obscured the effects of some geologic heterogeneities and not represented leakage from over and underlying strata.

To correctly analyze this well test, we should use a three-dimensional conceptual model. The IFS inversion method can be easily applied to these dimensions, but the computational effort will be greatly increased. Not only will the flow problem require far more computational time due to larger meshes, but a general three-dimensional attractor has 12 parameters for each affine transformation, compared to 6 for a two-dimensional attractor, doubling the dimension of the parameter space that must be searched by the inversion. Never the less, large three-dimensional inversions have been successfully computed using Simulated Annealing and it is well within the realm of possibility to perform three-dimensional IFS inversions.

7.0 INVERSION OF DATA FROM FRACTURED ROCK

At the Stripa mine in Sweden, we have been investigating the hydrology of a subvertical fracture zone called the H-zone within a 150 m × 100 m × 50 m block of rock. A series of seven wells (C1, C2, C3, C4, C5, W1, W2) penetrate this zone. An interference test, called the C1-2 test, was conducted in these holes. In this test, the C1 hole was pumped at a constant rate from a packed-off interval (interval 2) in the H-zone. Responses were measured in the other holes in intervals packed-off around the H-zone. A second experiment, called the Simulated Drift Experiment (SDE), measured the steady-state flow rate from the H-zone into an additional six parallel holes drilled within a 1 m radius (the D-holes). The entire data set is described in Olsson et al. (1989) and Black et al. (1991). An inversion of this data using Simulated Annealing is given in Long et al. (1991).

Here we present an IFS inversion based on the C1-2 cross-hole test. We then use the model produced by the inversion to predict the flow rate into the D-holes in the SDE. We treat the H-zone as a two-dimensional feature. The C-, D-, and W-boreholes penetrate the plane of the H-zone. The IFS on the plane of the H-zone describes the high conductivity regions within the plane of the fracture zone.

We first find an IFS which creates a hydrologic property distribution that reproduces the C1-2 interference data, in which a constant-flow boundary condition is applied at the C1 hole. Then we use this model to calculate the steady-state inflow to the SDE, in which a constant drawdown is imposed at the D-holes.

A two-dimensional variable-density mesh was used to model the H-zone, in order to maximize detail in the vicinity of the D-holes, provide a large enough mesh to prevent the transients from reaching the boundary too soon, and minimize the number of elements and bandwidth. The outer boundary conditions in the model were chosen to represent the estimated equilibrium head values.

7.1 IFS Inversion Based on C1-2

The inversion was done using three affine transformations in which each point on the attractor incremented both the conductance and the storativity of the nearest element. Figure 7.1 shows the well test data and the model results for the 799th iteration where the energy had dropped to about E = 13 from an initial value of about E = 45. Figure 7.2 shows the attractor at iteration 799.

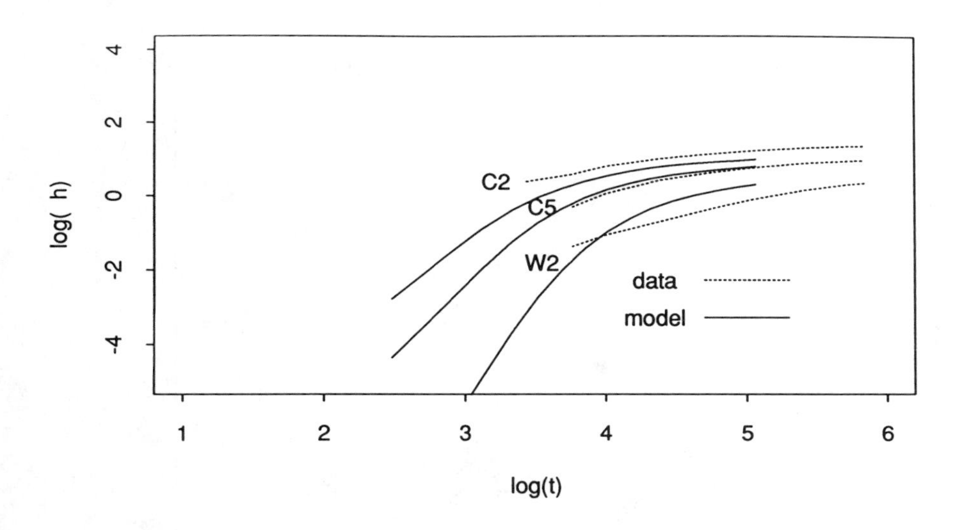

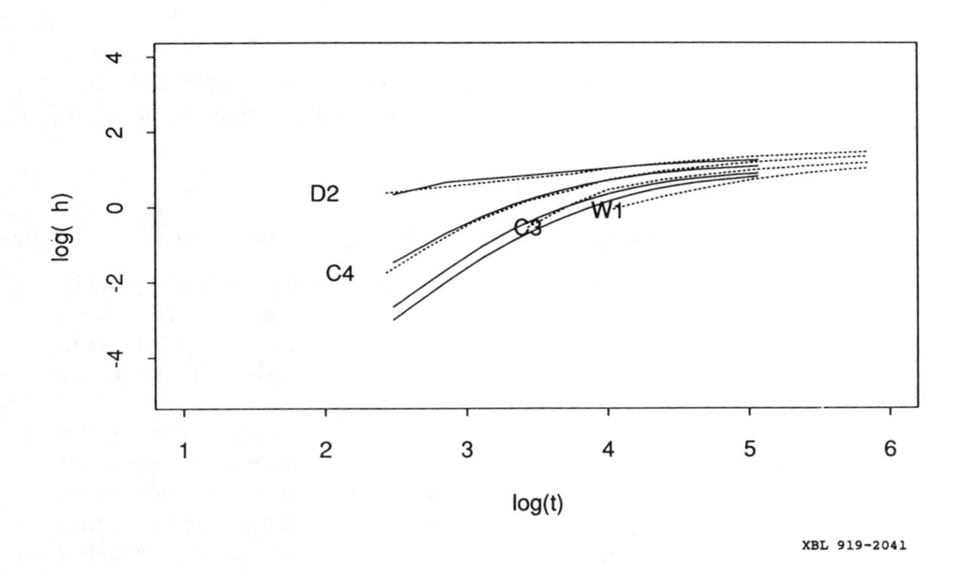

Figure 7.1. The well test data from the Stripa C1-2 test compared to the model results for iteration 799.

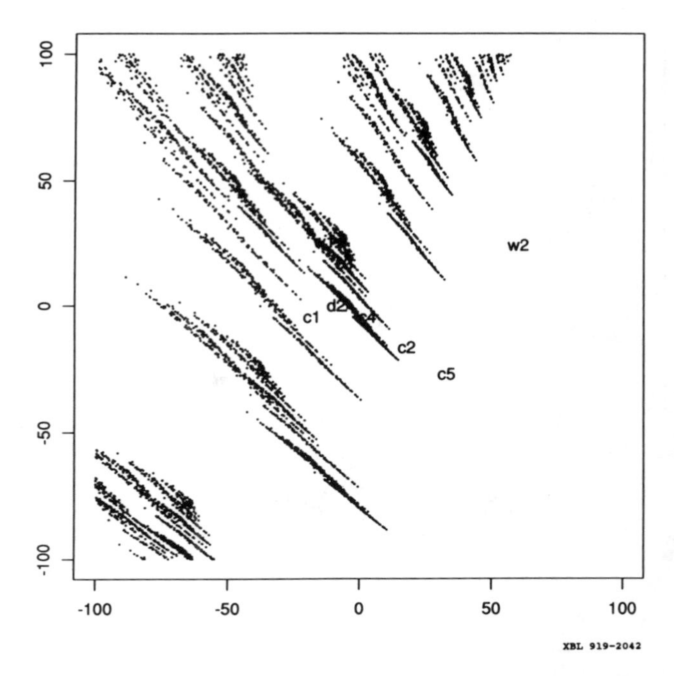

Figure 7.2. The attractor found at iteration 799 for the central 200 × 200 m section of the mesh. The six D-holes are in the immediate vicinity of D_2.

7.2. Prediction of Flow for the Simulated Drift Experiment (SDE)

We then predicted the flow rate into the D-holes during the SDE by applying a constant head at the D-holes such that we impose the same drawdown at the D-holes (220 m) as was imposed during the SDE. The actual flow rate to the D-holes from the H-zone during the SDE was estimated to be about 0.7 l/min. Our calculation gives 0.4 l/min which is low, but reasonably close.

Some interesting attributes of this inversion are that the attractor first resided entirely in the upper left-hand corner of the mesh and consistently migrated to the lower right with each iteration. Thus, the conductance in the vicinity of the C- and D-holes consistently increased, which means that further iterations may continue to move the attractor down and improve the solution. However, another possibility is that we are using too many points in the attractor which results in too high a conductance contrast, effectively forcing the attractor to stay away from the center of the well field. These possibilities are under investigation.

8.0 ITERATED FUNCTIONS TO DESCRIBE FRACTURE PATTERNS (JOINTS)

One exciting possibility for IFS inversion is that we may be able to condition the inversion on geologic information. If we can find Iterated Function Systems that reproduce the geometry of a geologic system, then inversion searches could be restricted to this class of functions. Probably the best way to find such classes of iterated functions is to base the functions on an understanding of how the system in question develops. For example, a meander belt might be described based on a physical understanding of its depositional history. For a set of joints, the functions could reflect the growth mechanics of the joints. A preliminary example of such a description of joint growth is given below.

The IFS scheme considered here is stochastic but is based on fracture mechanics concepts. The cases we consider pertain to two-dimensional fracture growth in homogeneous, isotropic elastic materials under plane strain conditions.

A commonly used criterion for fracture propagation is that fracture growth will be in the direction that minimizes the energy release rate G. For the two-dimensional case considered here

$$G = (K_I^2+K_{II}^2)\frac{1-\nu^2}{E} \tag{8.1}$$

where K_I and K_{II} are the mode I and mode II stress intensity factors, respectively (Lawn and Wilshaw, 1975). The terms ν and E are elastic constants: ν is Poisson's ratio, and E is Young's modulus. For an isolated mode I crack (dilatant fracture or joint), K_{II} equals zero and

$$K_I = \sigma_d\sqrt{\pi L/2} \tag{8.2}$$

where σ_d is the driving pressure and L is the length of the crack. Expressions (8.1) and (8.2) together show that

$$G \propto L. \tag{8.3}$$

According to classical fracture mechanics, a fracture subject to a constant driving pressure will grow rapidly if G exceeds a critical level, G_c, which is a material property. By this criterion, once a fracture reaches a critical length it should continue propagating with no increase in the driving pressure. Subcritical (slow) fracture growth can occur if G is below G_c. The subcritical fracture growth rate v is commonly described by a power law, $v \propto G^n$, (Atkinson and Meredith, 1987a). Although experimental values of the exponent n usually exceed 10 (Atkinson and Meredith, 1987b), Olson (1990) argues that field evidence suggests that n commonly is near 1 under natural conditions. If so, this would mean that $v \propto L$ for subcritical growth.

We assume that the relative probability of fracture growth (either by propagation of an existing fracture or by growth of a new "daughter fracture" near the tip of a pre-existing parent) is proportional to G. Based on Equation (8.3) we scale the relative growth probabilities to the fracture length L:

$$P_1(\text{fracture growth}) \begin{cases} =(1/L_c)\,L & L < L_c \\ =1 & L \geq L_c \end{cases} \tag{8.4}$$

The constant of proportionality ($1/L_c$) acts as a critical length in Equation (8.4). During a given iteration through the fracture-generating program, growth *will* occur for fractures longer than L_c. Growth *may* occur for fractures shorter than L_c; this condition corresponds to subcritical crack growth.

The IFS algorithm proceeds in four steps, with each fracture checked in a given iteration. First, a decision is made regarding fracture growth. The probability of fracture growth P_1 is calculated using Equation (8.4), and a random number Q_1 between 0 and 1. If P_1 is greater than Q_1, there will be growth; if not, another fracture is checked for growth. Second, a decision is made whether the pre-existing "parent" will grow or a new "daughter" crack will form. The parameter that defines the relative probability of in-plane propagation of a parent is P_2; the relative probability of a daughter nucleating is therefore ($1-P_2$). Another random number Q_2 between 0 and 1 is selected. If P_2 is greater than Q_2, the parent will grow; otherwise, a daughter will form. Third, the increment of growth ΔL is calculated according to the expression

$$\Delta L = LBQ_3 \tag{8.5}$$

where L is the parent length, B is a maximum growth increment parameter set by the user ($0 < B < 1$), and Q_3 is a random number between 0 and 1.

Fourth, if a daughter crack forms, its location must be determined. The coordinates (r,θ) of the center of a daughter crack are set relative to the tip of the parent crack (Figure 8.1). They are determined stochastically using two random numbers (Q_4 and Q_5) and a probability density distribution based on the stress state near a crack tip. The equation for the crack-perpendicular stress (σ_{yy}) near the tip of a crack is (Lawn and Wilshaw, 1975):

$$\sigma_{yy}(r,\theta) = (K_I/\sqrt{2\pi r})\cos(\theta/2)[1+\sin(\theta/2)\cos(3\theta/2)] \tag{8.6}$$

$$-\pi < \theta < \pi, \quad 0 < r < B$$

The contributions that contain r and θ in Equation (8.6) can be isolated and normalized to yield probability density distributions for daughter crack locations as a function of r and θ. These distributions show that the probability density tails off with distance from the crack tip and has maxima near $\theta = \pm 60°$ instead of directly ahead of the crack tip. This causes a daughter crack to preferentially grow near the tip of a parent crack but off to the side.

The number of iterations through the algorithm is set by the user. A larger number of iterations allows longer and more numerous fractures to be grown.

There are four key aspects of this approach that should make it useful. First, it can generate fracture patterns in much less time than approaches that explicitly account for the mechanical interaction between fractures. Second, fracture

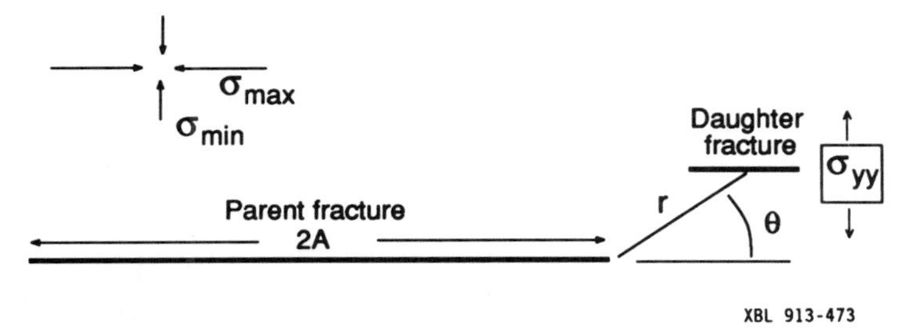

Figure 8.1. Diagram showing the relative positions of a parent fracture and a daughter fracture and the orientations of the most-compressive and least-compressive far-field stresses.

growth occurs only near crack tips (where stresses are particularly favorable), so in this regard it is consistent with fracture mechanics principles. Third, new cracks can develop; techniques that explicitly account for fracture interaction usually only allow pre-existing fractures to grow. Fourth, there are only a few parameters to manipulate (the starter crack distribution, P_2, B, L_c, and the number of iterations).

An important assumption incorporated in the rules outlined above is that fracture interaction is weak and that every crack grows as though it were isolated. As a result the algorithm only permits a parallel set of fractures to grow. This approach most appropriately applies to cases where the far-field principal stresses (rather than crack interactions) dictate fracture shapes and the stress perturbations due to fracture interaction are weak (i.e. the difference in magnitude between the remote principal stresses is large relative to the driving pressure in the fractures; driving pressure equals internal fluid pressure minus remote least compressive principal stress). Although this is a significant restriction, it should not invalidate the approach. The fracture traces in many natural sets are fairly straight, indicating that fracture interaction commonly is not strong.

From the simulations conducted to date, three main points emerge. First, this approach can generate realistic-looking fracture growth sequences (Figure 8.2) that compare favorably with detailed outcrop maps (e.g. Figure 8.3). Second, many starter cracks are needed to produce realistic-looking patterns. Third, most realistic-looking patterns are produced if the probability of daughter fracture generation is very low. If $P_2 = 1$ (i.e. only pre-existing cracks can grow) and the length of the starter cracks is is greater than L_c, then the resulting fracture length distribution approaches a log-normal distribution as the number of iterations becomes large (Figure 8.4). Even a very small probability of daughter growth can cause a tremendous change in the fracture length distribution. Figure 8.4 also shows a distribution produced when the probability of daughter growth $1\text{-}P_2 = 0.01$. This distribution would be better described by a power-law function. For cases such as this, the shortest cracks are concentrated in belts along the largest fractures. This type of pattern resembles joint zones.

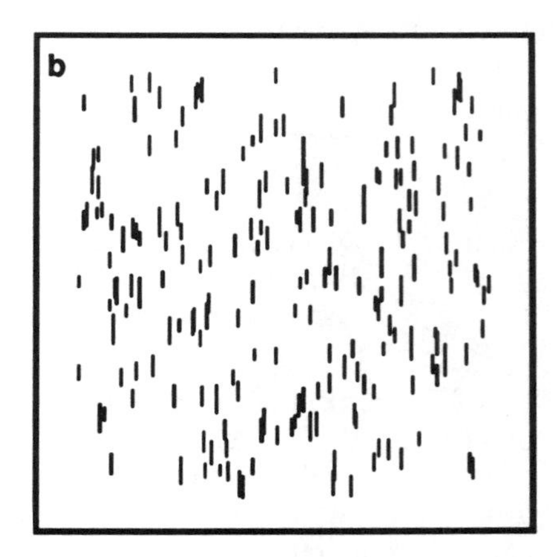

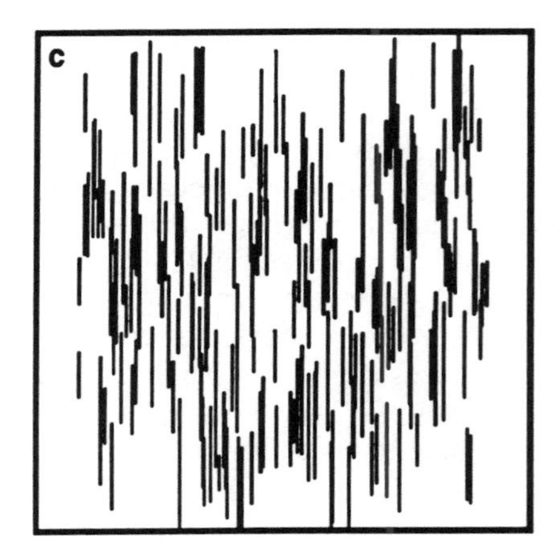

XBL 919-2043

Figure 8.2. Development of a fracture pattern from 200 randomly located starter cracks after (a) 0, (b) 110, and (c) 140 iterations. Each starter crack in (a) is 1 cm long, a distance corresponding to the likely initial crack length of Figure 8.3.

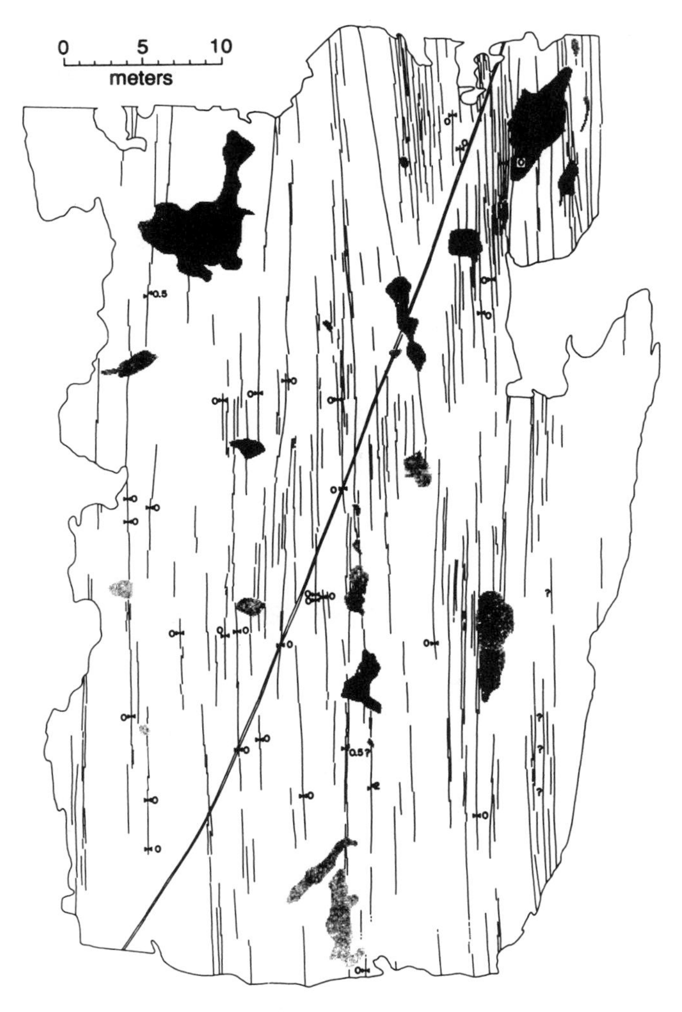

Figure 8.3. Map of fracture traces in a granite outcrop (modified from Segall and Pollard, 1983). The fractures dip steeply. Numbers indicate amount of lateral separation (in centimeters) across fractures. The feature marked by a double line is vein. Areas where the outcrop is covered are shown in gray. The pattern here is visually similar to that in Figure 8.2c.

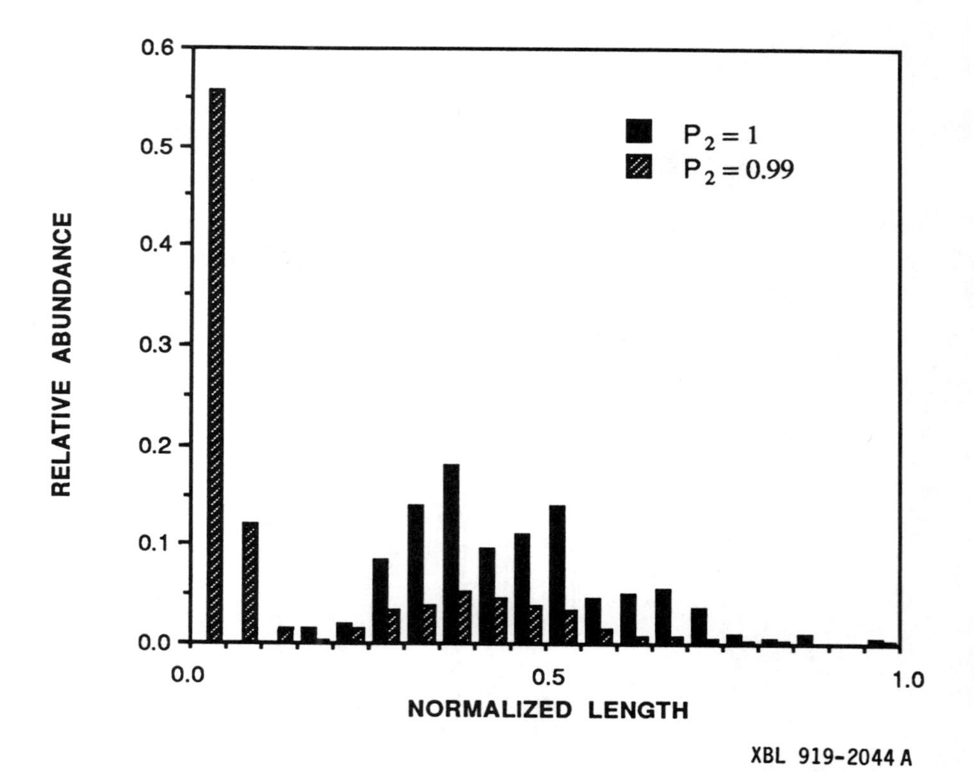

Figure 8.4. Comparison of two fracture length distributions developed from 200 starter cracks and 150 iterations. For P_2 = 1.0 the distribution is approximately lognormal; for P_2 = 0.99 the distribution is better described by a power law. Fracture lengths are normalized by the longest fracture length.

9.0 CONCLUSIONS AND RECOMMENDATIONS

This paper gives some preliminary applications of a new inverse approach to modeling heterogeneous and fractured reservoirs. We believe this approach has great possibilities as a practical tool. However, to reach this goal, much remains to be done. We need to learn more about how the inversion process works. The techniques we have started with could be extended to apply to more complex cases and data sets. To do this we will have to improve the ''intelligence'' of the search for good solutions. In the real world, where we never know enough about the subsurface environment, this method can provide a series of comparable solutions that reproduce the behavior we know about and extend our abilities to predict behavior in the future. Below, we discuss these points.

9.1 The Inversion Algorithm

We are just beginning to see how the inversion algorithm works. The technique has a strong scientific basis, but at this point the application requires arbitrary decisions. We need to begin to dissect these decisions to see how they affect the solutions. For example, we assign changes in conductance to the elements of the background lattice in an arbitrary way. We have chosen to increment conductance but if, for example, we are looking at clay lenses in sand, we might want to decrement conductance. In some cases we only incremented the element nearest the attractor point, in other cases we proportioned the increment according to proximity to the point in order to produce a smoother distribution. Different geologic systems may provide a reason for doing one or the other. The way in which we assign conductance increments will effect the resolution of anomalous features.

We also connect the increment in conductance to an increment in storativity. These two parameters are not necessarily uniquely related and we might need to determine when a more complex relationship is required. For example, clays may have high storativity and low conductance, whereas the opposite can hold for sands. We might have to define several conductivity and storativity relationships and a set of rules for using one or the other. Similarly, we arbitrarily choose M, the number of points in the attractor. We could begin to include this parameter in the inversion, possibly by stopping the process after a certain number of iterations, optimizing M and then continuing, etc.

9.2 Extensions of the Method

At the present time we are able to look at two-dimensional systems and invert based on either one steady or transient well test. Obvious extensions include the ability to include more than one well test simultaneously. Preliminary work with Simulated Annealing indicated that predictions based on two well tests significantly improved over those using one. Adding more well tests is analogous to increasing the ray coverage in producing a geophysical tomogram.

Although we have never run a fully three-dimensional IFS inversion, there is in principle nothing preventing us from doing so even though it may be time consuming. For example, we might model the Kesterson case as several layers and in this way be able to include some of the partial penetration and leakage effects that we have had to neglect in our two-dimensional model.

Use of diffusive phenomena such as pressure transients to resolve permeability anomalies has some inherent difficulties. When we receive a pressure transient, we commonly have little idea of the geometry of the flow path between the source and the observation. The fact that this path may be significantly different than a straight line means that the pressure transient data is inherently hard to interpret. One might say that the information in the signal is "diffused." This fact has always pointed to the use of tracer tests as an alternative data source for inversion. This is under consideration, but will probably introduce as many complications as it removes. It is probably true that one can do better predicting head by inverting head measurements, predicting flow by inverting flow measurements, and predicting tracer arrival times by inverting arrival time measurements. The task of building one model that can predict all of these simultaneously is a research program in itself.

9.3 Efficiency

Clearly if we want to make IFS inversion a practical tool we must find efficient algorithms. The inversions done for this paper were completed with very crude, simple codes which in no way optimized the calculations. They were run on a Solbourne 500 series workstation, with CPU times ranging from 25 minutes for the first synthetic case (which was specifically designed to run quickly) to about two days for the Kesterson inversion.

The computer science aspects of this problem are important. These can include simply better programming and optimization algorithms, but might also include the use of chip design such that the relevant equations are hard-wired into the computer. Solvers based on computer architecture are very attractive for these problems where we expect to make many thousands of iterations. Another interesting possibility is to learn to solve the diffusion equation analytically directly on the attractor, thus obviating the need for extensive numerical analysis of each iteration.

A more down to earth way to improve efficiency is to be smarter about the way that we search for solutions. We can incorporate *a priori* information such as geophysical data to force the search to look for permeability anomalies where there are geophysical anomalies. This is conceptually very simple and could be incorporated very easily simply by lowering the energy when an attractor point falls inside the geophysical anomaly. Co-inversion of both geophysical data and hydrologic data might be useful, but our experience is that it may be better to use the interpretation of the geophysical results as *a priori* information in the hydrologic inversion. This is because a significant amount of expert judgment is called on to interpret geophysical measurements and this judgment would be overlooked in a co-inversion.

9.4 Geologic Approach

The work on fracture growth schemes has tremendous promise for being able to reproduce fracture patterns. Clearly, similar work could be done to describe other hydrologically important geologic features. Sites which have been exhaustively explored will be critical for learning to build functions that describe heterogeneities for specific geologic conditions. Several such sites are being developed for the purpose of understanding heterogeneity and may be very useful for this work.

The work on graphics using Iterated Function Systems has included development of techniques for finding the IFS that describes a given pixel plot. This work could be extended to three-dimensions in order to find the IFS that describes the geology of a given quantified site. If we can then begin to examine the nature of these functions, we may characterize classes of IFS that represent geologic situations.

If hydrologic inversions can be limited to geologically determined classes of Iterated Function Systems, this would produce results that à priori resolve realistic features. In this way it may be possible to improve the efficiency, resolution and extrapolation of hydrologic inversions.

9.5 Uniqueness and Prediction

The problem of specifying the uniqueness of solutions always arises in the inverse problem, especially in the earth sciences. The fact is that we rarely if ever have enough data to completely specify an underground system and we have to accept uncertainty. What is especially attractive about the IFS inverse approaches we are developing is that they produce a range of solutions and thus can produce a range of predictions.

We think it is important to design approaches to the reservoir characterization problem that recognize from the beginning that the solution to the inverse problem is non-unique and that predictions made with these models have errors which should be quantified in some way. A good research program in reservoir characterization should include a sequence of predictions and measurements in order to determine if the model is converging to a useful predictive tool. A simple example of this would be to use an inversion based on one well test to predict the results of a second; then the two tests to predict the results of a third, etc. In this way we can see how much data is needed to make predictions sufficient for the purpose at hand.

9.6 Evaluation

The IFS inversion scheme seems to be a promising line of research. The approach is inherently interdisciplinary in nature and should be able to produce models that incorporate the many types of information that are available for a reservoir. The models use behavior to predict behavior and are consequently inherently consistent. Some encouraging initial results have been obtained, but there is much left to do.

10.0 REFERENCES

Atkinson, B. K., and Meredith, P. G., 1987a. The Theory of Subcritical Crack Growth with Applications to Minerals and Rocks, *Fracture Mechanics of Rock,* Atkinson, B. K., ed., Academic Press, London, p. 111-166.

Atkinson, B. K., and Meredith, P. G., 1987b. Experimental Fracture Mechanics Data for Rocks and Minerals, *Fracture Mechanics of Rock,* Atkinson, B. K., ed., Academic Press, London, p. 477-525.

Barnsley, M., 1988. *Fractals Everywhere,* Academic Press, Inc., Boston, Ch. 3.

Black, J. H., Olsson, O., Gale, J. E., Holmes, D. C., 1991. Site Characterization and Validation, Stage 4, Preliminary Assessment and Detail Predictions, Report in preparation, Swedish Nuclear Fuel and Waste Management Co., Stockholm, Sweden.

Carrera, J. and Neuman, S. P., 1986a. Estimation of Aquifer Parameters under Transient and Steady State Conditions: 1. Maximum Likelihood Method Incorporating Prior Information, *Water Resources Research, 22,* (2), 199-210.

Carrera, J. and Neuman, S. P., 1986b. Estimation of Aquifer Parameters under Transient and Steady State Conditions: 2. Uniqueness, Stability, and Solution Algorithms, *Water Resources Research, 22,* (2), 211-227.

Carrera, J. and Neuman, S. P., 1986c. Estimation of Aquifer Parameters under Transient and Steady State Conditions: 3. Application to Synthetic and Field Data, *Water Resources Research, 22,* (2), 228-242.

Davey, A., Karasaki, K., Long, J. C. S., Landsfeld, M., Mensch, A. and Martel, S. 1989. Analysis of the Hydraulic Data of the MI Experiment, Report No. LBL-27864, Lawrence Berkeley Laboratory, Berkeley, California.

Freeze, R. A., 1975. A Stochastic-Conceptual Analysis of One-Dimensional Groundwater Flow in Nonuniform Homogeneous Media, *Water Resources Research, 11,* (5), 725-741.

Karasaki, K, 1987. A New Advection-Dispersion Code for Calculating Transport in Fracture Networks, Earth Sciences Division 1986 Annual Report, LBL-22090, Lawrence Berkeley Laboratory, Berkeley, California, pp. 55-58.

Kitanidis, P. K. and Vomvoris, E. G., 1983. A Geostatistical Approach to the Inverse Problem in Groundwater Modeling (Steady State) and One-Dimensional Simulations, *Water Resources Research, 19,* (3), 677-690.

Lawn, B. R. and Wilshaw, T. R., 1975. *Fracture of Brittle Solids,* Cambridge University Press, 204 pp.

Long, J. C. S., Mauldon, A. D., Nelson, K., Martel, S., Fuller, P. and Karasaki, K, 1991. Prediction of Flow and Drawdown for the Site Characterization and Validation Site in the Stripa Mine, Report No. LBL-31761, Lawrence Berkeley Laboratory, Berkeley, California.

Olson, J. E., 1990. Fracture Mechanics Analysis of Joints and Veins, Stanford University, Stanford, California [Ph.D. dissertation], 187 pp.

Olsson, O., Black, J. H., Gale, J. E. and Holmes, D. C., 1989. Site Characterization and Validation, Stage 2, Preliminary Predictions, Report TR 89-03, Swedish Nuclear Fuel and Waste Management Co., Stockholm, Sweden.

Press, W. H., Flannery, B. P, Teukolsky, S. A., Vetterling, W. T., 1986. *Numerical Recipes: The Art of Scientific Computing,* Cambridge University Press, New York, Ch. 10.

Segall, P., and Pollard, D. D., 1983. Joint Formation in Granitic Rock of the Sierra Nevada. *Geological Society of America Bulletin, 94,* 454-462.

Toth, J., 1967. Groundwater in Sedimentary (Clastic Rocks). Proc. National Symposium on Groundwater Hydrology, San Francisco, CA, Nov. 6-8, p. 91-100.

Yates, C. C., 1988. Analysis of Pumping Test Data of a Leaky and Layered Aquifer with Partially Penetrating Wells, University of California, Berkeley [M.Sc. dissertation], 198 pp.

ACKNOWLEDGMENT

The fracture related work was supported by the Director, Office of Civilian Radioactive Waste Management, Office of External Relations and and the porous media work was supported by the Office of Health and Environmental Research, Ecological Research Division, Subsurface Science Program at the Lawrence Berkeley Laboratory which is operated by the University of California under U. S. Department of Energy Contract No. DE-AC03-76SF00098. The authors are grateful to Robert Zimmerman and Ernest Majer for their thoughtful reviews.

HIERARCHICAL SCALING OF CONSTITUTIVE RELATIONSHIPS CONTROLLING MULTI-PHASE FLOW IN FRACTURED GEOLOGIC MEDIA

Carl A. Mendoza
Edward A. Sudicky

Waterloo Centre for Groundwater Research
University of Waterloo
Waterloo, Ontario, Canada

1. INTRODUCTION

The behaviour of two or more immiscible liquids in fractured geologic materials is of great interest for several reasons. In petroleum and coal-gas reservoirs fractures often control the production of oil and/or gas, while near the ground surface fractures may permit toxic non-aqueous phase liquids (NAPLs) to penetrate to great depths where they may contaminate drinking water supplies. In addition, the migration of water through partially water saturated fractures may affect the isolation of nuclear waste and multi-phase flow through fractures commonly plays a dominant role in geothermal energy production. Despite the important influence of fractures on multi-phase fluid flow, there is limited quantitative knowledge about the flow of two immiscible liquids through fractured geologic materials. In particular, the capillary-pressure/saturation and relative-permeability/saturation constitutive relationships, which are necessary for detailed numerical simulation of multi-phase flow processes, are poorly quantified.

The ultimate objective of this study is to use theoretically based numerical models to determine the form of, and the relationship between, the multi-phase flow constitutive relationships for fractures in geologic media. Toward this end, we consider the case of two incompressible, immiscible fluids flowing through a single rough-walled fracture plane, and we employ stochastic principles to derive constitutive relationships for higher orders of description from the micro-scale to the macro-scale. In this way, the essential physics of the micro-scale flow mechanisms are retained at higher levels of description. In this paper we outline

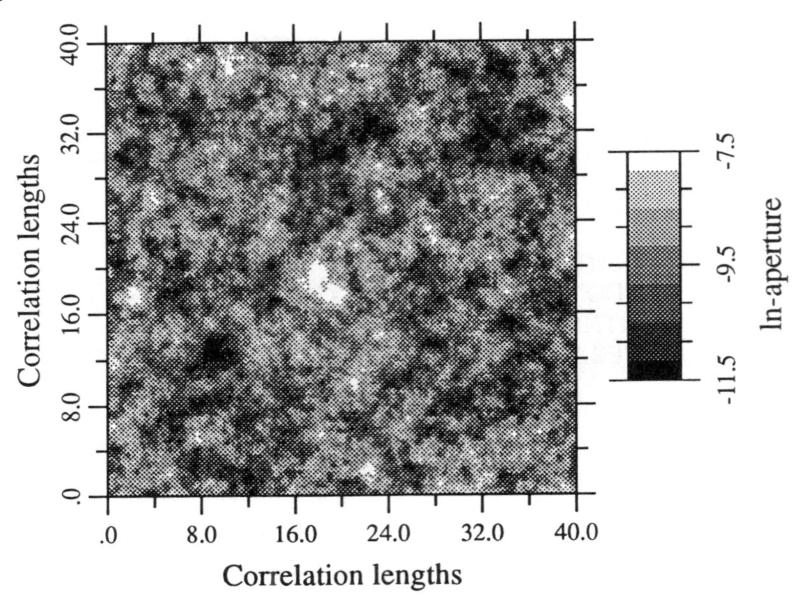

Figure 1. An example of a statistically-isotropic, spatially-correlated, random aperture field. The log-normal mean is -9.5 with a variance of 0.5. There are 40 correlation lengths across the field and 5 cells per correlation length. Darker shades indicate smaller apertures.

the theory employed and present some preliminary results that demonstrate the types of analyses possible. In-depth analysis and a wider-ranging suite of simulations are left for further study.

2. METHODOLOGY

Recent studies of two-phase flow in single rough-walled fracture planes, notably those of Pruess and Tsang [1990] and Pyrak-Nolte et. al. [1990], have employed traditional percolation theory to determine the distribution of fluids within a fracture plane. This study differs significantly from these previous works by using invasion percolation concepts [Wilkinson and Willemsen, 1983] to account for invading fluid accessibility and resident fluid trapping. The inclusion of these processes results in solutions that exhibit hysteresis and residual saturations in the constitutive relationships for two-phase flow.

The numerical method is designed to mimic, as closely as possible, procedures that might be used in the laboratory. Gash [1991] described laboratory techniques for measuring the two-phase flow properties of fractured coal samples. For the numerical model, we generate synthetic spatially-correlated random field to represent the distribution of apertures within a fracture plane. Here we assume that the aperture distribution is adequately represented by a log-normal distribution with an exponential correlation function describing the spatial persistence of the aperture distribution. Such a description has been found to be representative of several fracture aperture distributions measured in the field and laboratory [Pruess and Tsang, 1990]. Of course, other statistical distributions or actual aperture measurements may also be used within the theoretical framework. The random aperture fields used here are generated using spectral techniques incorporated in an algorithm developed by Robin [1991]. Figure 1 illustrates one of the fields used in later simulations.

The fracture plane is discretized at the micro-scale so that locally each cell may be considered to have a constant aperture. With this assumption the volumetric flow rate (Q) through each cell is described by the solution of the Navier-Stokes equation for flow between two parallel plates. For laminar flow the result is the cubic law:

$$Q = W \frac{b^3}{12\mu} \frac{\partial P}{\partial x} \tag{1}$$

where W is the fracture cell width, b is the aperture, μ is the fluid viscosity and the derivative represents the pressure gradient. By analogy to Darcy's Law, the cell permeability is described by:

$$k = \frac{b^2}{12} \tag{2}$$

The permeability depends only on the square of the fracture aperture because the cross-sectional area available for flow also depends upon the aperture.

To determine the effective, macro-scale, single-phase permeability of the fracture plane, a specified pressure gradient is applied across the discretized fracture plane and the sides are considered to be impermeable. With these boundary conditions, the single-phase, steady-state flow equation is solved using a five-point finite volume scheme with harmonic weighting of the cell permeabilities and a preconditioned conjugate gradient matrix solver. The fluid flux across the fracture plane is calculated by back-substitution of the pressure solution into the matrix equation and the effective, macro-scale, single-phase permeability is calculated from Equations 1 and 2.

Drainage capillary-pressure/saturation measurements are obtained by allowing an invading non-wetting fluid to displace resident wetting fluid at increasingly higher capillary pressure steps. For imbibition, the invading and resident fluids are exchanged and the capillary pressure is decreased. Because we have discretized the fracture at the micro-scale and assumed that each cell behaves as if the walls were parallel plates, it is reasonable that a cell will be completely occupied by either one fluid or the other. If the boundary between cells can be considered to have an elongated rectangular cross-section, the possible fluid occupancy will be determined by the local capillary pressure:

$$P_c = P_n - P_w = \frac{2\sigma}{b} \cos\alpha \tag{3}$$

where P_n and P_w are the non-wetting fluid and wetting fluid pressures, σ is the interfacial tension between the two fluids and α is the interface contact angle with the solid phase.

From this relationship, for a given aperture (b^*) there must be a corresponding "critical" capillary pressure (P_c^*) which controls which of the two fluids may occupy the aperture segment: if the local capillary pressure (P_c) is greater than P_c^* the non-wetting fluid may occupy the aperture segment; if, on the other hand, P_c is less than P_c^* wetting fluid may occupy the aperture segment. This criterion was used by Pruess and Tsang [1990]; however, their analysis used traditional percolation theory where *all* apertures above the critical size were occupied by the non-wetting fluid. Their approach ignored the facts that an aperture segment may only be occupied by the invading fluid if the segment is accessible from the inlet end, and that resident fluid may become isolated and trapped within the fracture plane. These latter two processes, which allow for hysteretic constitutive relationships and the formation of residual saturations, are incorporated in this study by the use of invasion percolation theory [Wilkinson and Willemson, 1983]. In order to accommodate these additional processes, we proceed as follows. For a given applied capillary pressure, each cell is characterised as possibly being occupied or not occupied by the invading fluid; however, only cells that have a continuous pathway, within the invading fluid, from the source of fluid (inlet reservoir) to the cell of interest may actually be occupied. In addition, if any region of resident fluid becomes surrounded by invading fluid, and thus isolated from both the inlet and outlet ends of the fracture, it is designated as being trapped and that region cannot subsequently be occupied by the invading fluid. To generate a capillary-pressure/saturation curve this procedure is repeated many times with increasing or decreasing capillary pressure for the drainage and imbibition processes respectively.

Each capillary pressure step results in a defined fluid distribution within the fracture plane and, if a fluid is continuous from one end of the domain to the other, the application of a pressure gradient will result in flow through the fracture. By requiring that the applied pressure gradient be small, the capillary pressure, and thus the fluid distribution, will not be significantly affected by flow and we may calculate steady-state, single-phase flow solutions for each continuous fluid. The procedure for this is essentially the same as for a single-phase flow solution; the only difference is that flow does not occur throughout the entire domain — only cells occupied by the fluid of interest are considered to participate in the flow process and all cells occupied by the other fluid are treated as having zero permeability. Once the flow solution is obtained and fluid fluxes back-calculated, the effective permeability, and subsequently the relative permeability, of the fracture may be determined. This approach to determining relative permeabilities was also used by Pruess and Tsang [1990].

In order to obtain the mean, effective behaviour of the constitutive relationships and their uncertainty, the above procedures are repeated for many different, but statistically equivalent, realizations of the fracture aperture fields. This Monte Carlo approach has the effect of averaging the variations in results that arise from having different unique realizations of the aperture field.

3. *SIMULATIONS*

The results of several different simulations are presented here to demonstrate the utility of the method and the effects of trapping. All of the simulations have some common properties: the mean log-aperture is -9.5 (aperture in metres) with a variance of 0.5; the isotropic correlation length of the log-aperture field is 0.25 length units with spatial discretizations of 0.05 length units in each direction; and, 20 Monte Carlo realizations are performed in each case. For presentation purposes, capillary pressures are represented by the inverse of the aperture expressed in microns. Equation 3 may be used to scale the results for any given pair of immiscible fluids. To highlight trends in the capillary-pressure vs. saturation and relative-permeability vs. saturation data, average curves derived from a moving average calculated on sorted values have been drawn through the composite results; these curves should not necessarily be taken as the ensemble mean average.

Figure 2 shows a series of fluid distributions for the drainage process where non-wetting fluid is invading a fracture that was initially filled with wetting fluid. The non-wetting fluid starts to invade the top of the domain and progresses downward as the capillary pressure is increased. By comparing the fluid distributions to the aperture distribution shown in Figure 1, we see that as the capillary pressure is increased smaller and smaller aperture segments are invaded and wetting fluid becomes isolated and trapped. At low capillary pressures the non-wetting fluid does not span the length of the fracture, and hence only the wetting fluid may flow, whereas at high capillary pressures only the non-wetting fluid may flow. For this particular realization, both fluids are continuous and may flow simultaneously for a small range of intermediate capillary pressures. In other realizations, simultaneous flow may not occur due to interference between the two fluids.

Several sets of simulations were performed using 200 by 200 grids. The first set assumed that the fracture was initially filled with wetting fluid and the non-wetting fluid was then allowed to invade. After residual saturation was achieved, the process was reversed and wetting fluid was allowed to invade the fracture. In the second set of simulations, the role of the fluids was reversed such that the fracture was initially occupied by non-wetting fluid and wetting fluid was allowed to imbibe and then drain. The results of the 20 Monte Carlo simulations for these four different limbs of the constitutive relationships are presented in Figures 3 and 4. The capillary pressure curves exhibit significant hysteresis and the average residual saturations are seen to be about 25% for the wetting fluid and about 30% for the non-wetting fluid. In general, these residual saturation values depend on the variance of the aperture distribution. The relative permeability curves are highly

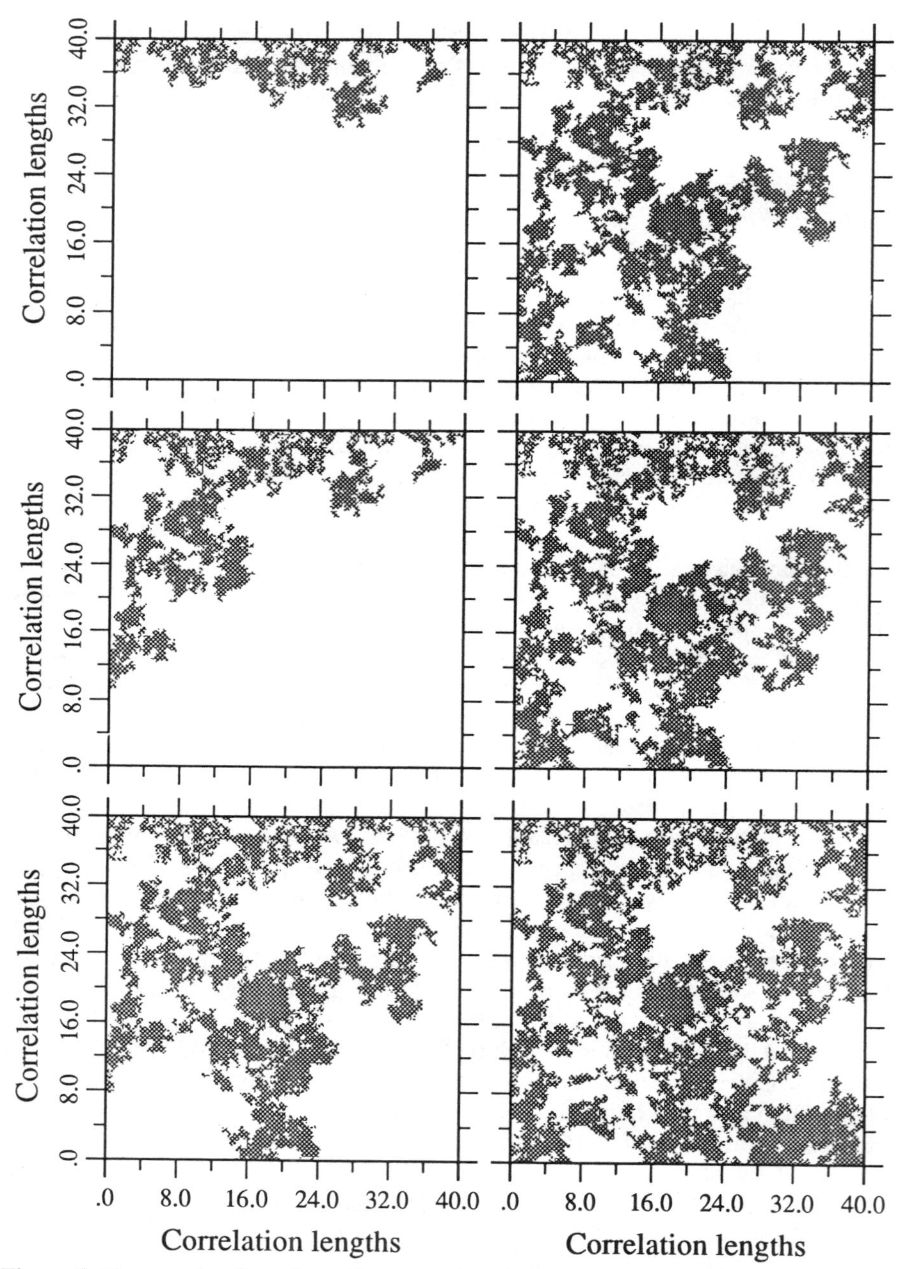

Figure 2. Progressive invasion of non-wetting fluid at increasingly higher capillary pressures into a fracture initially saturated with wetting fluid. Shaded areas indicate regions occupied by non-wetting fluid. Results are for the aperture distribution shown in Figure 1.

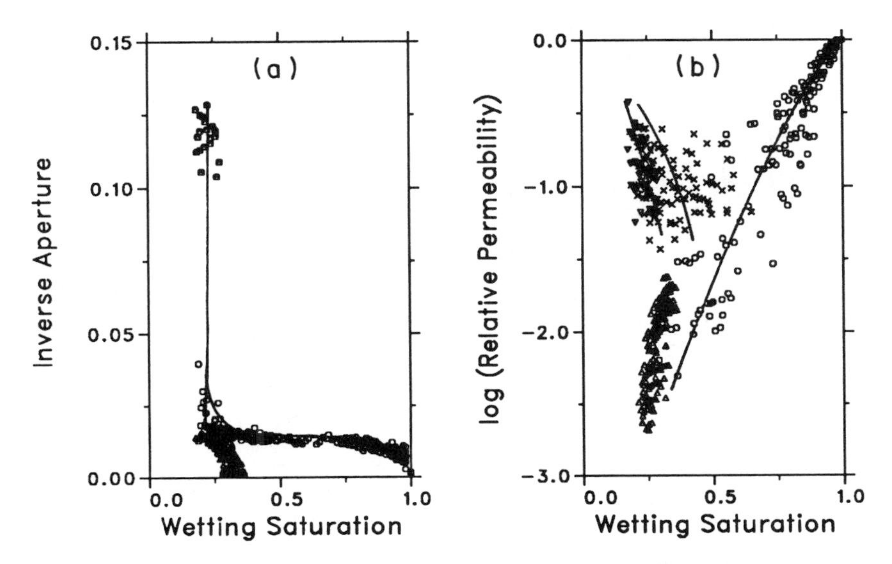

Figure 3. Composite data and mean curves for a) capillary-pressure vs. saturation and b) relative-permeability vs. saturation for the case of drainage followed by imbibition.

uncertain, but indicate that there is a great deal of interference between the two fluids and that the non-wetting fluid may only flow once it occupies a significant fraction of the fracture plane.

In order to illustrate the effects of the trapping mechanism on the constitutive relationships, drainage and imbibition simulations were performed, using the same aperture realizations as above, but with the trapping mechanism excluded from the model. Fluid accessibility from the inlet end was, however, still considered in these simulations. The composite results for the 20 realizations are illustrated in Figure 5. By comparing Figure 5 to Figures 3 and 4, it can be seen that the results obtained with trapping excluded are very much different. The main differences are that no residual is formed, there is less hysteresis in the relationships and the non-wetting fluid may flow over a greater range of saturations. If accessibility had not been considered in these simulations, there would be no hysteretic effect and the drainage and imbibition curves would lie on top of one another.

4. *DISCUSSION*

The simulations presented here illustrate that the methodology is capable of producing physically realistic results in that the constitutive relationships are

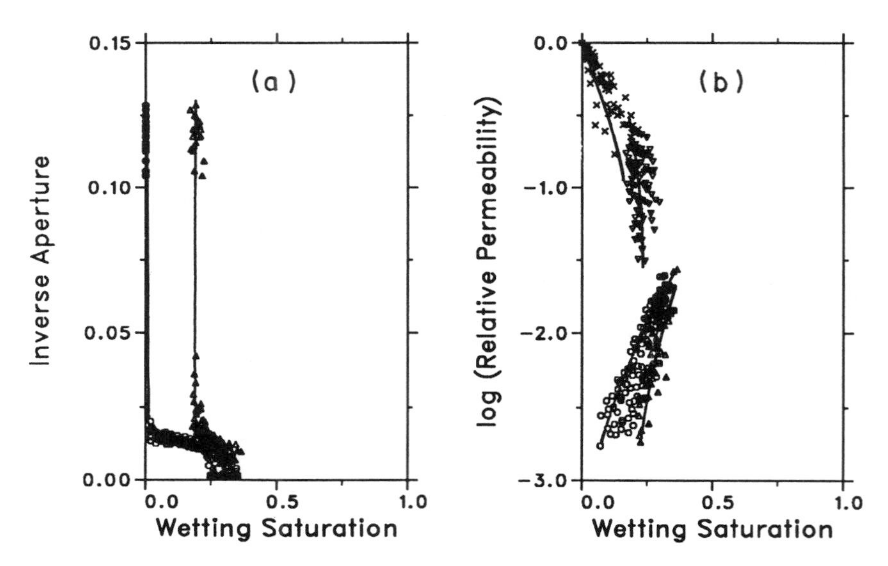

Figure 4. Composite data and mean curves for a) capillary-pressure vs. saturation and b) relative-permeability vs. saturation for the case of imbibition followed by drainage.

hysteretic when accessibility is considered and they possess a residual saturation when trapping is considered. It is also demonstrated that the results are quite different when these two processes are excluded from the analysis. It remains to be confirmed experimentally which of the conceptual models is correct.

Some of the mean effective curves generated in this study, particularly those for relative permeabilities, have a large amount of uncertainty associated with them. This is due to several factors, including the fact that the results for only 20 Monte Carlo simulations have been presented. Performing additional Monte Carlo simulations might help to reduce the uncertainty to some extent. On the other hand, this large uncertainty in the structure of the constitutive relationships emphasizes the fact that predicting multi-phase flow through rough-walled fractures will in general be problematic because the point-to-point variations in the aperture values of a real fracture will usually be immeasurable in the field.

There are reports in the literature [eg., Pyrak-Nolte et al., 1990] that the results of computations such as those presented here are dependent on the size of the computational domain. To investigate this possibility, simulations were performed using grids of 100 by 100 cells and 300 by 300 cells in addition to the 200 by 200 cell problem already discussed. At this scale we observed very little, if any, dependence of the results on domain size; however, subsequent simulations on the order of millions of cells have indicated that there is indeed a dependence of the residual saturation on domain size. This aspect of the problem is the subject of an ongoing investigation.

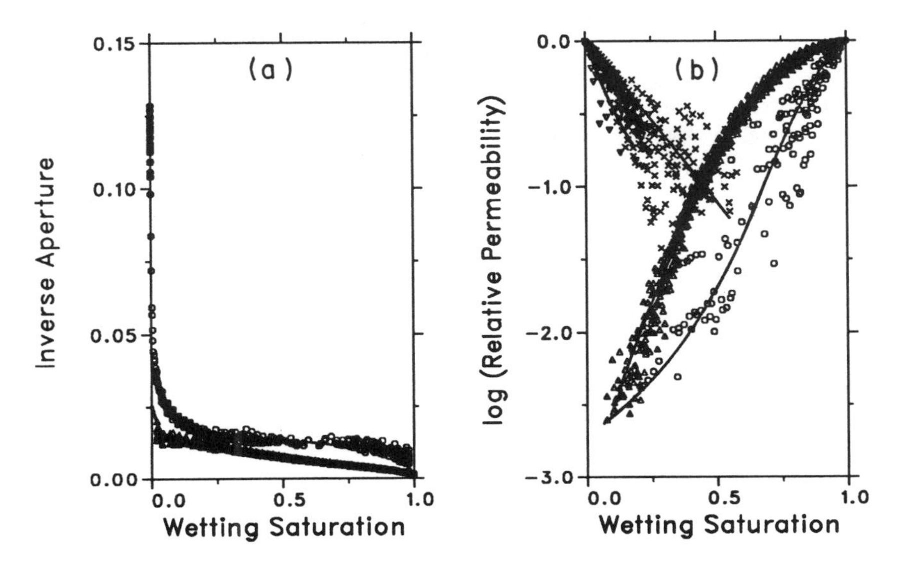

Figure 5. Composite data and mean curves for a) capillary-pressure vs. saturation and b) relative-permeability vs. saturation with no trapping mechanism.

ACKNOWLEDGEMENTS

This research was supported by a Natural Sciences and Engineering Research Council operating grant to E.A. Sudicky, the University of Waterloo Porous Media Research Institute and the University Consortium for Solvents-in-Groundwater Research, which is sponsored by: the Ciba-Geigy, Eastman Kodak and General Electric companies; the Ontario University Research Incentive Fund; and, the Natural Sciences and Engineering Research Council.

REFERENCES

Gash, B.W., Measurement of rock properties in coal for coalbed methane production, SPE Paper 22909 presented at *SPE Annual Technical Conference and Exhibition*, Dallas, October 6-9, 1991.

Pruess, K. and Y.W. Tsang, On two-phase relative permeability and capillary pressure of rough-walled fractures, *Water Resources Research*, 26(9), 1915-1926, 1990.

Pyrak-Nolte, L.J., D.D. Nolte, L.R. Myer and N.G.W. Cook, Fluid flow through single fractures, in *Rock Joints*, Barton and Stephansson (eds.), Balkema, Rotterdam, 405-412, 1990.

Robin, M.J.L., Migration of Reactive Solutes in Three-Dimensional Heterogeneous Porous Media, Ph.D. thesis, Dept. of Earth Sciences, University of Waterloo, 1991.

Wilkinson, D. and J.F. Willemson, Invasion percolation: a new form of percolation theory, *Journal of Physics, A*, 16, 3365-3376, 1983.

INTEGRATION OF SEISMIC AND WELL LOG DATA IN RESERVOIR MODELING

U. G. Araktingi
W. M. Bashore
T. T. B. Tran

Chevron Oil Field Research Co.
La Habra, California

T. A. Hewett

Department of Petroleum Engineering
Stanford University
Stanford, California

I. ABSTRACT

Procedures for integrating seismic data with well log data in the construction of reservoir models are reviewed. The processing requirements for the seismic data for this application and potential pitfalls which may be encountered are discussed. A case study dataset based on the inversion of a series of two dimensional seismic lines and a set of synthetic well logs produced by a spectral extrapolation of those inversions provides the basis for a comparison of the different methods. Displays of a single stratigraphic horizon show the strong influence of the data sampling configuration on maps constructed using either the seismic or well log data alone. Two methods for integrating seismic and well log data, kriging with an external drift and indicator cokriging under the Markov-Bayes hypothesis, are compared. Both show modifications of the distributions produced using well log data alone. These modifications incorporate features seen in the seismic data which were not evident from the well log data alone. The calibration step performed in the Markov-Bayes procedure allows for a more quantitative evaluation of the influence of the seismic data on the local probability distributions of values than the uncalibrated external drift method.

II. INTRODUCTION

The use of detailed, geologically realistic, quantitative reservoir models for simulating the effects of reservoir heterogeneity on oil recovery processes has come into increasing practice in recent years. In using these models, petrophysical and flow properties (permeability, porosity, dispersivity, etc.) which are defined at the scale of laboratory measurements must be specified at every location represented in a numerical fluid flow simulation. Although measurements made on cores or derived from well logs are available only at discrete sample locations, namely wells, there is a need to assign flow properties to other, unsampled, locations. Smooth interpolation of properties to unsampled regions is known to produce biased results in flow simulations (Hewett and Behrens, 1990; Omre, 1991).

In addition, the scale of representation in most field-scale flow simulations is much larger than the scale of laboratory measurements, and single property values are assigned to spatial volumes which are much larger than the scale of the measurements which define the flow properties. The volumes associated with the assignment of these properties may include considerable variations in all of the properties, and interactions among variations in properties like absolute and relative permeability make the assignment of single property values difficult. Rules for averaging flow properties are simply not known. Detailed, geologically realistic flow models are needed to determine the flow response of reservoir volume elements the size of conventional simulator grid blocks.

It is generally recognized that realistic numerical geologic models must accurately reproduce the large scale structure and continuity of bedding in a reservoir, as this constitutes the basic "plumbing" of the reservoir (Haldorsen and Damsleth, 1990). The large scale geometry of the reservoir frequently controls the degree of mobile fluid continuity between wells and can control the access of drainage and displacement processes to significant fractions of the reservoir volume. The overall architecture of a reservoir may be introduced into a simulation through the use of a stratigraphic coordinate system based on interpreted geologic horizons (Rendu and Ready, 1982; Dagbert et al., 1984), or by the definition of contiguous flow facies introduced by Boolean or mosaic stochastic simulations (Omre, 1991). Alternatively, the large scale structure can be derived from seismic data. In this paper we will focus on several alternative ways of integrating seismic data into a reservoir model.

In addition to the large scale features, the role of interwell-scale variations of properties is known to be important in the simulation of displacement processes (Hewett and Behrens, 1990). The dominance of reservoir heterogeneity over fluid mechanical instabilities is well established in understanding the efficiency of miscible displacement processes (Araktingi and Orr, 1988; Waggoner et al., 1991). In immiscible displacements, permeability heterogeneity causes the shape of displacement fronts to change in a way which modifies the effective, or "pseudo", relative permeabilities required to represent the displacement on coarse simulation grids which average away subgrid scale heterogeneity (Kyte and Berry, 1975; Hewett and Behrens, 1990). Multicomponent, multiphase displacement processes also require new component fractional flow functions and effective compositions to be represented properly in coarse grid simulations (Barker and Fayers, 1991). The actual effects of permeability heterogeneity on these flow processes must be determined before deriving the parameters required in empirical mixing models or deriving effective flow properties. This behavior can be determined with high resolution flow simulations on simulation grids which realistically mimic the character of flow property variations over a range of scales (Fayers, Blunt, and Christie, 1990). Ideally, these flow simulations should have the resolution of property variations observed in well logs and core measurements and should honor those data where they are available.

The task, then, is to construct high resolution quantitative reservoir models which honor data available at wells, maintain the character of property variations observed in well log and core measurements in the interwell region, and preserve the large scale structure and continuity of the reservoir observed in seismic data.

III. METHODS OF INCORPORATING SEISMIC DATA

The integration of seismic data with well log and core data presents several interesting challenges. These arise from the vastly different scales of spatial resolution and spatial sampling of the two kinds of data. Seismic data have a vertical resolution typically given in the tens of meters, while well log measurements typically have a vertical resolution on the order of a meter or less. Seismic data densely sample the entire interwell region, while well log data are available only at wells, and the measurements only probe centimeters into the near wellbore region. Several different approaches for combining the two types of data have been proposed.

A. Kriging and Regression Analysis

Perhaps the most straightforward method of using seismic data is to establish a correlation between a well log measurement and a seismic attribute, and use that correlation to estimate well log values at the locations of the seismic data based on the correlation (Doyen, 1988). The correlation is typically established by regression analysis and frequently shows a large amount of scatter. When well log properties are assigned on the basis of a curve representing the observed correlation, there is no guarantee that the data available at well locations will match those established by the correlation curve, so well data are usually not honored. In addition, this approach uses only the mean of the correlation without any information about the spread, incorporates no measures of spatial continuity, and provides no measures of uncertainty beyond a correlation coefficient which applies uniformly to all estimated values independent of their proximity to actual measurements. The values of the well data are only used in establishing the correlation. The resulting field of interpreted well log values is determined solely by the spatial distribution of the seismic data, independent of the spatial sampling of the original well log data. The scale of variations in properties is limited to the scale of resolution of the seismic data.

When regression analysis is used to construct models on grids which do not coincide with the configuration of the seismic data, the seismic data are first kriged to the computation grid and the well log values are then assigned on the basis of the regression curve. With the proper choice of contour levels, maps of the well log values derived by this method will look identical to kriged maps of the seismic data.

B. Kriging With an External Drift

In this approach, the well log data is regarded as the primary data, and is analyzed to determine the statistical structure of property variations using variograms or covariances. As with ordinary kriging, optimal weights for combining the well data to make an estimate at an unsampled location are determined subject to the constraint that they be unbiased (i.e., they sum to unity). The seismic data are introduced by adding an additional constraint in the calculation of the optimal weights. This constraint requires that the weights reproduce the seismic data measured at the location being estimated when applied to the seismic data available at the well locations (Marechal, 1984; Galli and Meunier, 1987; Deutsch, 1991). This has the effect of reproducing the shape, or large-scale structure, of the seismic data in the estimated field of well log values. To the extent that the seismic attribute chosen correlates with the well log value being estimated, the seismic data can introduce local trends into the kriging system and account for interwell-scale variations not observed in the well data. The use of this approach requires the availability of seismic data at all of the well locations and at all of the estimation points. When the locations of the seismic data do not correspond to the data and computation grid locations, the seismic data values can be estimated at those locations by kriging. Given the high spatial sampling of seismic data and its low spatial resolution, the smoothing effects of this interpolation will have little effect on the results, as the seismic data is only being used to constrain the smoothly varying component of the total variability anyway.

The smoothing effect in the interpolation of the well data, however, can bias the results of flow simulations to determine the effects of heterogeneity on process performance. To reproduce the local variability of flow properties in the interwell regions, conditional simulations must be used. The most appropriate simulation technique in this instance is Gaussian Sequential Simulation (Deutsch and Journel, 1991). In this approach, the grid values are simulated sequentially along a random path through the grid. At each grid location, the kriged estimate and kriging variance are calculated as described above and a random number is drawn from a normal distribution with a mean and variance equal to the kriged estimate and kriging variance, respectively. This simulated value is then treated as real data in the simulation of subsequent grid values. If the well data are not normally distributed, the simulation can be constructed using the normal transforms of the well data, and then be back transformed to reproduce the histogram of the well data. Examples of Gaussian Sequential Simulations based on kriging the well data, with seismic data treated as an external drift, will be shown.

C. Cokriging

The integration of seismic and well log data with cokriging involves the treatment of the seismic data as a covariate, with its own internal spatial correlations as well as cross-correlations with the primary well data (Doyen, 1988). It uses more information about the seismic data, in that the data values themselves contribute to the estimates, and the degree of correlation between the seismic data and the well values is quantified through the use of a cross-variogram of the seismic data with the well data. It also does not require the availability of seismic data at every well and grid location.

The attractive features of cokriging described above are offset in practice by the need to model two variograms and a cross-variogram. Not only is the amount of data analysis increased, but the allowed forms of the variograms and cross-variogram that can be fit to the data must satisfy the requirements of a linear model of coregionalization (Luster, 1985; Deutsch, 1991). These rather restrictive requirements can make the simultaneous fitting of the three correlation measures difficult. As a consequence, cokriging has not been extensively used for combining well log and seismic data, except with synthetic data sets having known correlations and cross-correlations (Doyen, 1988). Recently, a new approach to integrating seismic and well data has been described that retains the desirable features of cokriging while eliminating the need to simultaneously model the three variograms.

D. Cokriging with the Markov-Bayes Indicator Formalism

In this approach to integrating seismic and well log data, the calibration of the seismic data against the well log data is treated in a Bayesian framework as "prior local information" about the well data (Journel and Zhu, 1990). Wherever seismic data are available, the calibration scattergram of collocated well and seismic data is consulted to determine the distribution of corresponding well data in the calibration, which is then treated as prior local information at the location of the seismic data where well data is not available. This prior distribution is coded as a series of indicators corresponding to different classes of well data values. The values of the indicators correspond to the value of the cumulative distribution function (cdf) of the well data for that value of seismic data, i.e., they give the probability that the well data is less than the well data values represented by the indicator. These prior distributions may be quite different for each range of seismic data, having not only different means, but also different skewness and a different spread around the mean.

This prior distribution is then updated to account for the additional information provided by the presence of well data and other seismic data in its neighborhood and knowledge of the spatial correlations in all of the data. This is done by first coding the well data into indicators using a step function cdf for the well data values. All of the indicator data is then cokriged to update the prior distribution into a conditional posterior distribution representing all of the available information and its reliability and spatial relations. At locations where no seismic data are available, but well and seismic data are available within its neighborhood, the conditional posterior distribution is obtained by cokriging all of the indicator data in its neighborhood.

The problem of inferring and joint modeling of the variograms and cross-variograms required for cokriging is solved by assuming the validity of a Markov approximation. This entails the hypothesis that "hard (well) information always prevails over any soft (seismic) collocated information" (Journel and Zhu, 1990). Under this hypothesis it can be shown that the seismic data variograms and the well and seismic cross-variograms can be derived from a knowledge of the well data variograms and the calibration of the seismic against the well data. The cross-variograms are proportional to the well data variograms with the constant of proportionality derived from the scatter of the calibration data. The shape of the seismic data variogram is also proportional to the well data variogram for positive lags, but it has, in addition, a nugget effect at the origin that is also determined from the calibration data. In this way, all of the information available from the calibration data and the spatial relations of the hard and soft data are used in determining the probability distribution of well values at undrilled locations.

With a method for determining the probability distribution of well values at unsampled locations, conditional simulations of the well values can be generated by the method of Sequential Indicator Simulation (Journel and Alabert, 1988). In this method, well values are simulated sequentially along a random path through the computation grid. At each node the probability distribution of well values at that node is determined as described above. A uniform random number is then drawn, and the well log value corresponding to that value of the cdf is used as the simulated value for the grid node. In the simulation of subsequent grid values, the previously simulated values are treated as actual well values.

IV. CASE STUDY DATASET

A case study comparing several of these methods was made to demonstrate their use in practice and to elucidate some of the practical problems in dealing with datasets from the field. An orthogonal grid of fourteen seismic lines was selected for this study, and the basemap is shown in Figure 1. The seven east-west oriented lines are coincident with the primary depositional dip direction, and the seven north-south lines follow depositional strike. On average they are spaced about 4 kilometers.

The original seismic data were acquired at a surface interval of 25 meters, but the tremendous volume of data necessitated decimating down to every eighth trace, or 200 meters. This still resulted in 132 traces and 141 traces on the North-South and East-West lines, respectively, for a total of 1911 traces. As will be demonstrated subsequently, the horizontal correlation lengths in the seismic data are on the order of 1000-1200 feet, and so, the decimation should not cause aliased or lost spatial information. Similarly, the data were also pared to a 2000 millisecond window, which at 4 millisecond sampling yields 501 samples per trace, or nearly 1 million total data points.

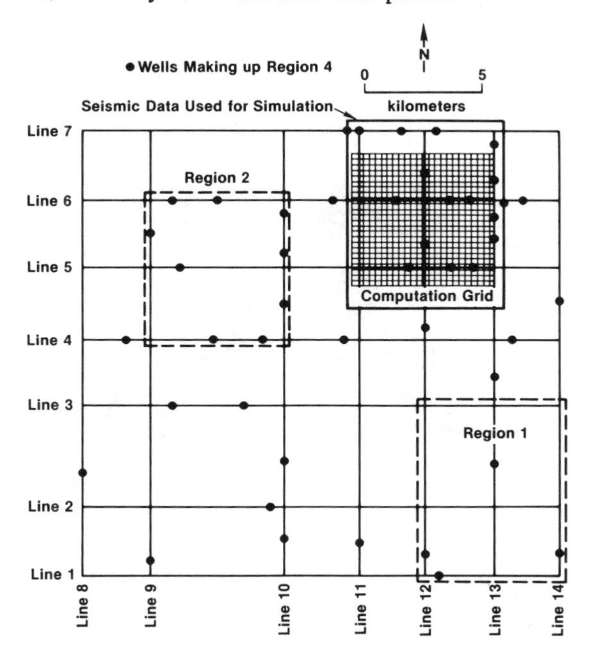

Figure 1

Basemap of seismic survey used in this study. Also shown are 3 of the 4 data subsets and the computation grid.

A. Geologic Interpretation

As can be seen from examples of a dip and a strike line (Figure 2), the study area is geologically complex. Because the stratigraphic packages (i.e., sequences) in this area were deposited in shelf to slope to basin floor environments, substantial lateral variation in lithology and thickness is observed. Any geostatistical procedure that interpolates seismic or well log attributes must include stratigraphic and structural constraints. Marker surfaces were interpreted as sequence stratigraphic boundaries using seismic stratigraphic techniques that analyze reflection terminations and configurations (Sheriff, 1980). Four of these markers, shown in Figure 2, were selected as constraints for the statistical methods to follow.

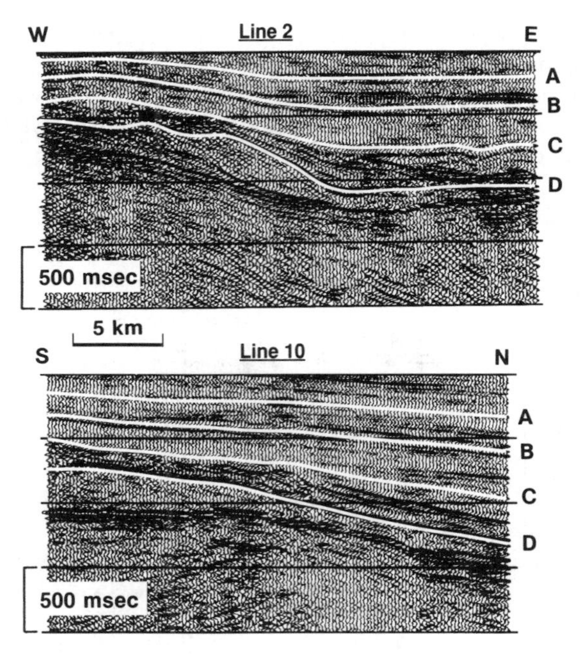

Figure 2

Examples of depositional dip (Line 2) and depositional strike (Line 10) seismic lines (see Figure 1 for line locations). Also shown are the four markers used to define the stratigraphic coordinate system.

B. Convolutional Model

A simplifying assumption is commonly made about seismic data that facilitates its usage in geologic interpretation. Namely, a seismic trace is the convolutional response of a waveform with the earth reflectivity for a single vertical profile. Figure 3 is a graphical representation of this process.

The acoustic impedance of a rock is defined as the product of its compressional velocity and bulk density. The velocity of a particular log interval is the reciprocal of the transit time recorded in the sonic log over that interval. The log transit times are accumulated (or integrated) to convert from depth to time. These time-depth pairs are used to convert other logs, such as density, to time. A reflectivity series is calculated from the impedance log (velocity times density) at each sample as the ratio of the difference over the sum in acoustic impedances immediately below and above the sample point. These reflection coefficients average about 0.05, but can be as large as 0.30.

Convolution is the superposition of a waveform "hung" on each reflection coefficient in the series. The waveform amplitudes are scaled by the magnitude and sign of the reflection coefficient. The resultant trace after all the summations is considered to be the seismic trace. This forward modelling scenario is often used where well logs tie seismic data directly for seismic identification of known well markers. Prominent reflections are traversed across a grid of seismic lines for structural mapping purposes. Time differences (isochrons) between two reflection events can yield information about thickness distributions for the intervening interval. Seismic amplitude variation may also be mapped as an indicator of stratigraphic or pore fluid variation.

Ideally for integration with other downhole measurements, seismic data should be "deconvolved" by removing the imbedded waveform and integrating the resulting reflectivity back to an impedance log. This process is commonly called seismic inversion (Lavergne and Willm, 1977; Lindseth, 1979; Bamberger et al., 1982).

C. Inverse Model and Waveform Estimation

The inverse model represented in Figure 4 is quite useful for outlining the process of creating pseudo impedance logs from seismic data and for demonstrating seismic temporal resolution (related to thickness). The starting point in most inversion schemes is the estimation of the inherent waveform within a specified time window.

The most common estimation algorithm generates a match filter (i.e., waveform) that maximizes the cross-correlation between the filtered, log-derived reflectivity and its corresponding seismic trace. This study used a match filter approach only to check waveforms estimated with a sharpness deconvolution process (Pusey, 1988). Waveforms estimated with this process are typically of broader bandwidth which translates to finer bed resolution. Sharpness deconvo-

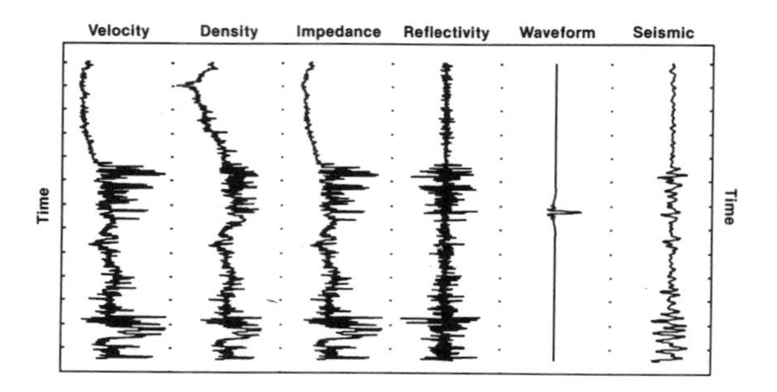

Figure 3

Graphical representation of the forward convolutional model.

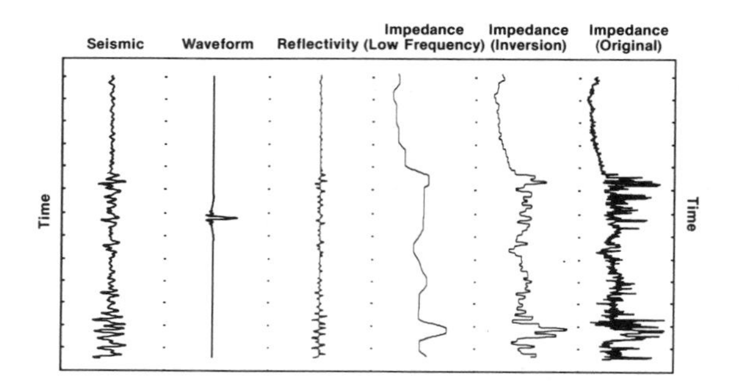

Figure 4

Graphical representation of the inverse convolutional model.

lution does not require an a priori knowledge of the reflectivity (i.e., well logs), so that the confidence level of prediction determined at the well location can be assumed elsewhere. This cannot be said for the match filter scheme. Figure 5 displays the waveform estimated from the case study dataset that was used to create the pseudo logs inverted from the seismic data. Also displayed is the amplitude spectrum of the waveform showing the spectral content (about 10-50 Hertz).

The application of a simple inverse produces a band-limited reflectivity trace which, if integrated, will not produce a realistic appearing log. The dominant period of the waveform persists in the integration and creates an undesirable modulation in the result. Variograms from these logs would indicate a single vertical correlation length of the dominant period throughout the data. This study included a spectral extrapolation step (Oldenburg et al., 1983; Walker and Ulrych, 1983), based on a "blocky" model, to sharpen the reflection coefficients to more of a spike (i.e., wider band).

D. Seismic Resolution

Note that the seismic-derived reflectivity in Figure 4 is sparser than the original log reflectivity in Figure 3. This sparseness translates to a reduction in bed thickness resolution because of the narrow bandwidth nature of the seismic

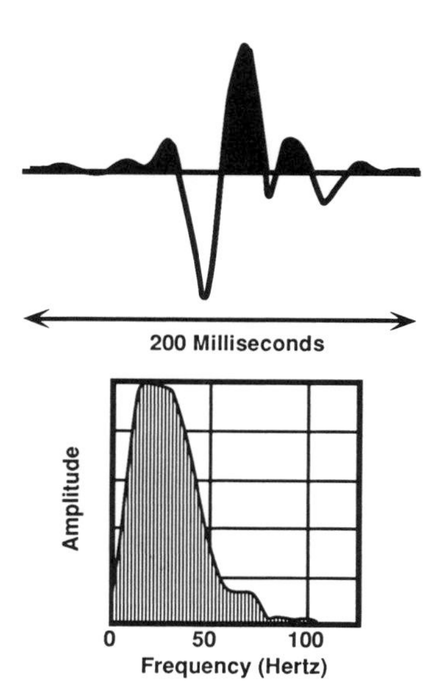

Figure 5

Estimated waveform and the amplitude spectrum used in the seismic inversion process.

waveform, and this reduced resolution can be seen by comparing the original and inverted impedance logs in Figure 4.

A general rule of thumb suggests that the expected minimum resolvable thickness from the seismic technique is one-quarter of a wavelength (Koefoed, 1981; Widess, 1983; Kallweit and Wood, 1982). Wavelength is related to frequency through the seismic velocity; that is, velocity is equal to the product of wavelength and frequency. Given our waveform with a high frequency end of 50 Hertz, clastic rock units with an average velocity of 2500 meters per second would be resolved down to thicknesses of about 12.5 meters. In comparison, this same interval would yield about 75 measurements in a typical well log.

E. Problems and Pitfalls

The integration of seismic data with other types of data in the characterization of a reservoir can be quite powerful. The spatial sampling far exceeds the relative spareness of well control and well-based data. However, the conditioning of seismic data for this integration is wrought with potential problems and pitfalls.

The first problem is that seismic data are recorded in the time domain, not depth. This means that the conversion from time to depth requires information about the velocity field which, depending upon the accuracy of depth conversion, may be quite complex. Unfortunately, even with check shot surveys or vertical seismic profiles acquired at well locations (and these are not common), velocity information is scarce and only approximate. Integrated sonic logs generally mistie the seismic data by 5-10 percent because of poor hole conditions, mud invasion, and possible seismic dispersion (different propagation velocities at different frequencies). Unless stacking velocities were generated using a model based approach where interval velocities from a well constrain their selection, they usually represent only velocities which best stack the data and may bear no relation to earth velocities.

Additionally, because the recorded bandwidth of seismic data does not include frequencies below 8 to 10 Hertz, the impedance information contained in these missing frequencies must be supplied from somewhere. Figure 4 shows the low-frequency contribution for the model well. If absolute numbers are required from the inverted traces for conversion to some rock property such as porosity, the determination of the low frequency curve for each seismic trace is paramount. Also, the processing of the seismic data must be done so that seismic amplitudes are directly related to the earth reflectivities as assumed in the convolutional model. Although this may seem trivial, in practice it is quite difficult to achieve, or is seldom regarded. Deterministic amplitude balancing procedures should always be favored over statistically based methods.

Another common assumption made in inversion algorithms is the stationarity of the waveform, both temporally and spatially. Few programs can handle varying waveforms. Again, the choice of processing algorithms may violate this need for stationarity. Trace based deconvolution programs that determine a separate operator for each trace are especially at fault for contributing to poor and inaccurate inversion results. Surface consistent schemes should be used on prestack data and multitrace, single operator programs for poststack data.

F. Synthetic Well Log Generation

For the reasons stated above, tremendous care must be exercised to properly condition the seismic data for integration with wellbore measurements. This conditioning takes time and must be based on sound geologic models. Unfortunately, because many of the concerns expressed above regarding seismic data processing were not properly addressed in our contractor-processed dataset, the correlation of the inverted pseudo logs to the available well logs was generally dismal. Fortunately, the goal of this paper was to demonstrate the utility of geostatistical procedures for multidata integration and not a specific reservoir characterization.

Synthetic well logs of seismic impedance were generated for all of the inverted seismic traces. In this way, problems associated with time or depth domain misties could be eliminated. The procedure used was quite simple. The inverted traces were transformed back to reflection coefficients, 10 percent random noise was added, a high frequency spectral extrapolation was performed to add additional fine scale resolution, and the resultant reflectivity was integrated back to impedance. In subsequent analysis, subsets of these synthetic traces will be referred to as the well log data. Future work will utilize a dataset in which rigorous controls have been enforced in the collection and processing of the seismic data for actual use in creating a reservoir model.

V. DATA ANALYSIS

The data were analyzed using an interactive workstation application which provides access to the data and geostatistical functions through an interactive graphical interface. The data are analyzed, and models are constructed, in a stratigraphic reference frame based on bed markers defined by a geoscientist familiar with all of the log, core, and seismic data. These bed markers are kriged to interwell locations using variograms computed for the markers. Within intervals, correlations are established and computation grids are defined to lie along surfaces which maintain a proportional spacing between the bounding bed markers. The areal locations of data traces are selected from a basemap, and depth intervals from the traces are selected by their relation to the bed markers. Variogram models are interactively fit to the data and saved for later use. Areal simulation grids are defined by digitizing an area on the base map and specifying the grid dimensions. Three-dimensional grids are defined by specifying the areal extent of the grid in a similar fashion and then specifying a vertical interval defined relative to the bed markers. The number of layers between markers are then specified. Arbitrary vertical cross-sections may be specified by digitizing a sequence of straight line segments on the basemap and specifying the vertical interval as described above. The data which informs the calculations for a grid are also specified interactively. Once the data and computation grid have been defined, the program user is then presented with a variety of methods for model construction, including kriging and a selection of conditional simulation methods. The resulting models are displayed graphically and may be written out to files for later use with reservoir simulation models.

Given the large amount of data available, several different subsets of the data were analyzed separately and compared. The locations of the four different data sets are shown on Figure 1. The first two consist of the contiguous vertical traces within the rectangles defined on the basemap. These represent the densest data sampling, but do not cover the entire map. The third is a uniform sampling of traces from the entire data array. The fourth is an irregular sampling of traces from the entire data array. These subsets of the data are referred to as Datasets 1-4, respectively.

A. Stratigraphic Markers

Four bed markers, denoted by the letters A-D, were identified in the seismic data, as described previously. The marker surfaces were first analyzed to determine their spatial structure and to define the stratigraphic reference frame. The results of a directional variogram computation for the top bed marker, Marker A, is shown in Figure 6. The data are fit with a fractional Brownian motion power law model and show a geometric anisotropy with the dominant trend in the N 154° E direction. Variograms for the other markers were also fit with anisotropic power law models. These variogram models were then saved and used in the construction of the stratigraphic reference frame.

B. Vertical Structure

Only one sonic log from the area under consideration was available for analysis. Rescaled range (R/S) analysis (Hewett, 1986) was performed on this log and indicated an intermittency exponent of $H = 0.76$. The trace was fit with a fractional Gaussian noise (fGn) variogram with the same value of H and a range of 26 ms., a sill of 1.02×10^6, and a relative nugget of 0.51. The variogram is shown in Figure 7. The histogram of data values was normally distributed as shown in Figure 8. Several of the synthetic well logs derived from the seismic traces were subjected to the same analysis. They showed values of H ranging from 0.80-0.84 and also were fit by fGn variogram models. This indicates that the synthetic logs have a statistical character similar to the mea-

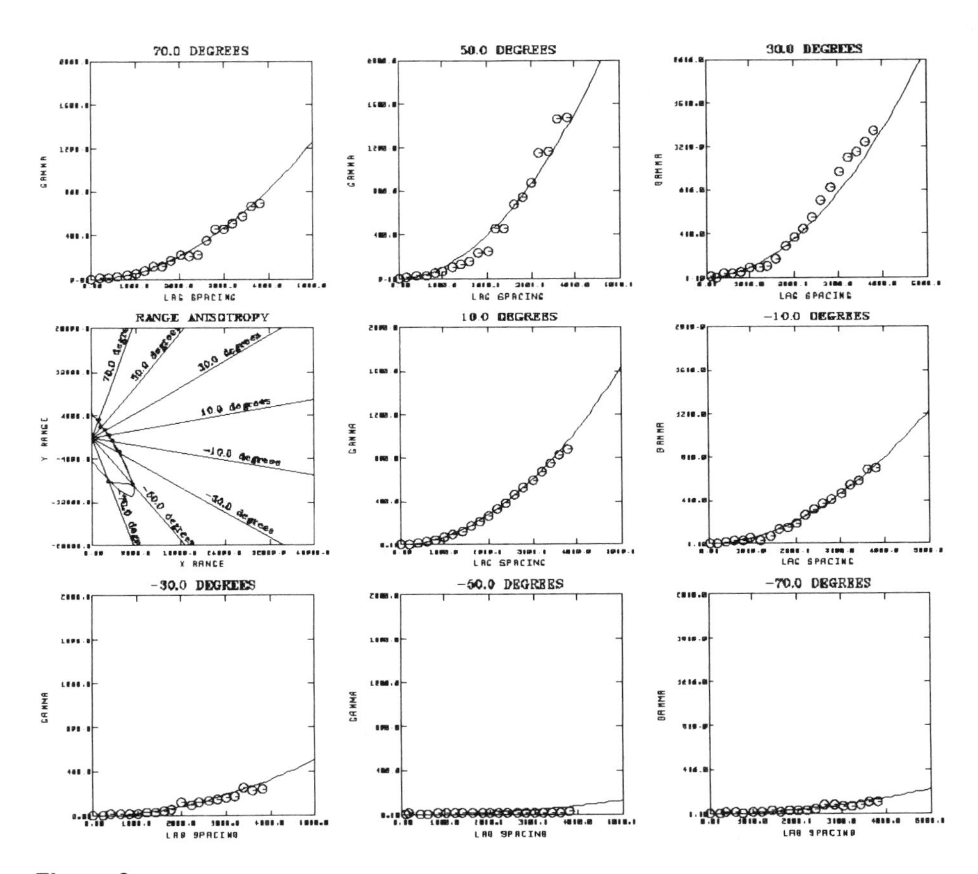

Figure 6

Directional variogram for the top bed marker (Marker A) fit with an anisotropic power law model.

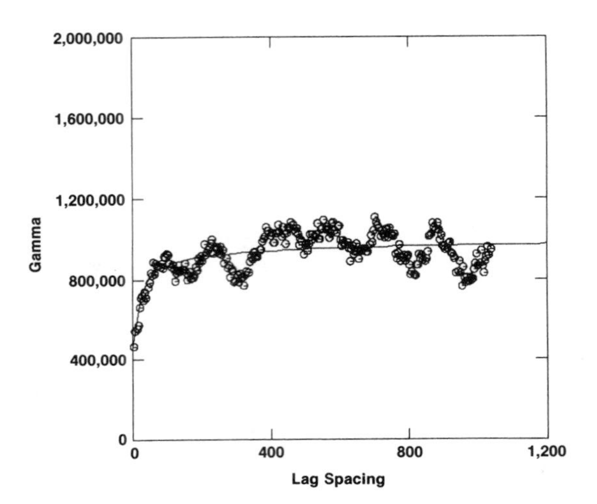

Figure 7

Vertical variogram of sonic velocity log fit with an fGn model and relative nugget.

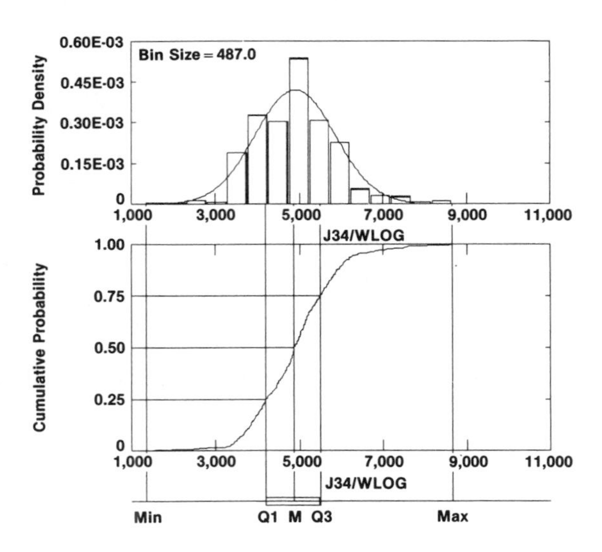

Figure 8

Histogram of sonic log values.

sured log from this area. The slightly higher values of H indicate somewhat more persistent vertical correlations in the synthetic logs than was observed in the measured log.

A comparison of the synthetic logs and the seismic inversions from which they were derived are shown for several wells along an E-W line in Figures 9 and 10. The low frequency character of the two types of traces is very similar, by construction. The synthetic well logs differ from the seismic traces primarily in the high frequencies. A cross-plot of the seismic and well log values along this section is shown in Figure 11. This is typical of the level of correlation between the two datasets across the study area. The correlation coefficient is r = 0.68.

C. Areal Structure

The well logs and seismic traces from the four subsets of the data were analyzed for areal structure by calculating areal variograms for the different stratigraphic intervals. Within each interval, the areal variograms were calculated for several (usually ten) stratigraphic layers and averaged together. This provided more statistical mass and reduced fluctuations observed in the experimental variograms calculated for a single layer. The averaged variograms all show a nearly isotropic exponential form with little or no nugget effect. The ranges and sills vary by stratigraphic interval and data subset, indicating that strict stationarity could not be assumed. The seismic data typically show correlation ranges 2-3 times that of the well log data. Similarly, the sills of the well log variograms are higher than those of the seismic data by as much as 20%. The differences between the variograms of the seismic and well log values can be attributed to the high-frequency fluctuations added to the well-log traces in their synthesis.

The variograms for the seismic and well log data from Region 1, Interval C-D, are shown in Figures 12 and 13. These are typical of the variograms from the densely sampled subsets (Datasets 1 & 2). The circles on these plots are the average variograms for the lag interval. The crosses represent the range of fluctuations observed for individual pairs of points making up the variogram averages.

Variograms for the less densely sampled, but more areally complete, data subsets (Datasets 3 & 4) show similar behavior. With the less dense sampling, the structure near the origin of the variogram is not as well captured. At large lags, a transition to a second structure, which was not modeled, was observed in some of the variograms. The areal variograms for the well log and seismic data from Dataset 3, Interval A-B, are typical of what was observed. These are shown in Figures 14 and 15. All of the variogram models were saved. In the subsequent model construction, the variograms from Dataset 4 were used. This is a global subset of the data with higher sampling in the upper right corner of the field near the simulation grid.

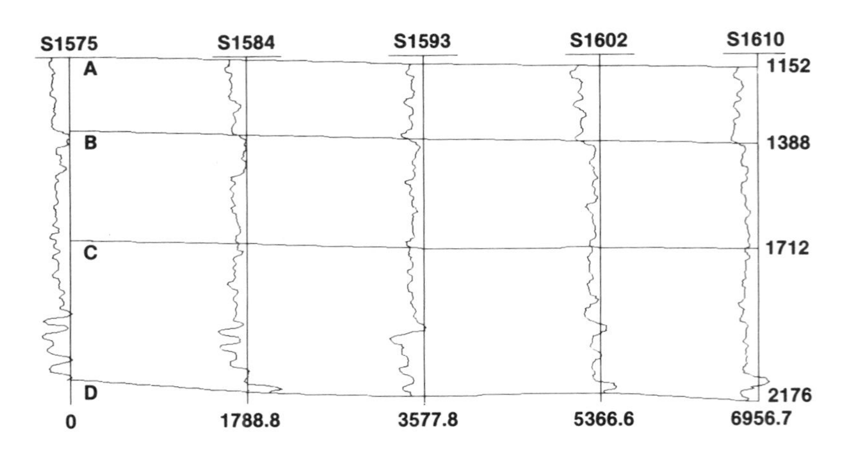

Figure 9

Typical well log data traces along an E-W line.

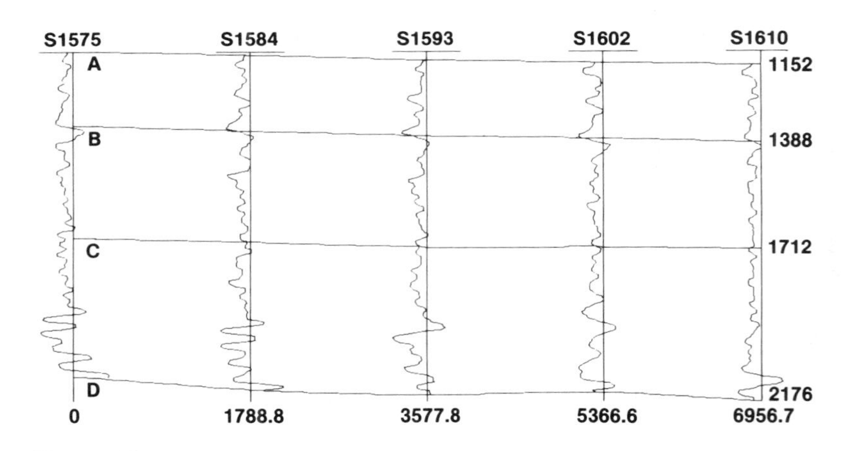

Figure 10

Typical seismic data traces along the same E-W line.

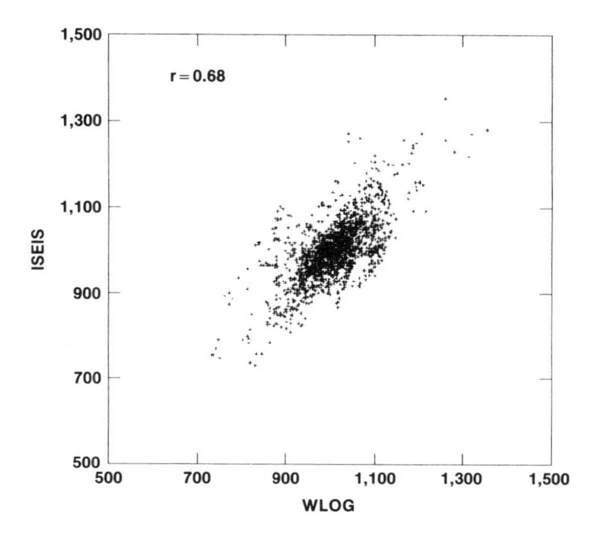

Figure 11

Scattergram of well log and seismic data values from Figures 9 and 10.

In addition to the variograms of the acoustic impedance values, variograms for nine classes of indicators were also calculated. The indicator cutoffs correspond to the decile boundaries of the data within an interval. The procedure of averaging the variograms of several layers within a stratigraphic interval, which worked well to smooth fluctuations in the experimental variograms of acoustic impedance values, introduced a bias when applied to the indicator variograms. These showed shorter ranges for all of the cutoffs than was observed in the variograms of the data values themselves.

This bias can be explained by considering what is known about indicator variograms of normally distributed data. For this case, the median indicator will show the most structure (i.e., largest range), with increasing destructuring being observed for cutoffs away from the median (Journel, 1989). In our data set, the histograms of the ten layers whose variograms were averaged are slightly different. This means that the decile cutoffs defined for the pooled data may not correspond to the same decile for each individual stratigraphic level. If a cutoff corresponding to the median value of the pooled data falls one or two deciles above or below the median for an individual level, the variogram for that cutoff will be less structured in that layer than it would be if the cutoff corresponded to the median of the layer. The averaging of variograms from different levels then may entail averaging variograms with less structure than would be observed in a

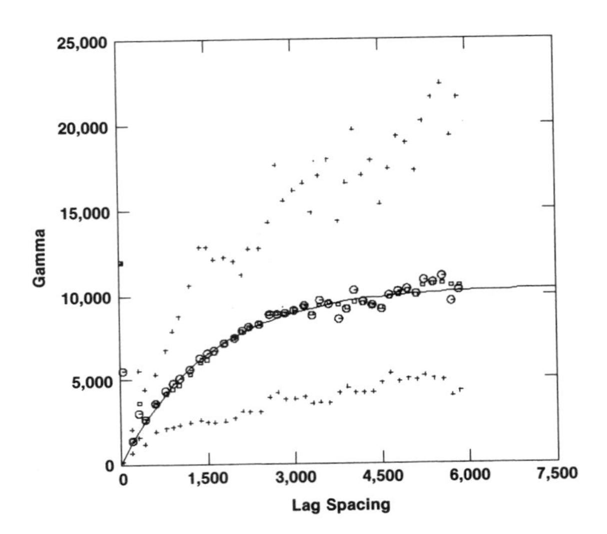

Figure 12

Isotropic areal variogram of seismic data from Dataset 1, Interval C-D.

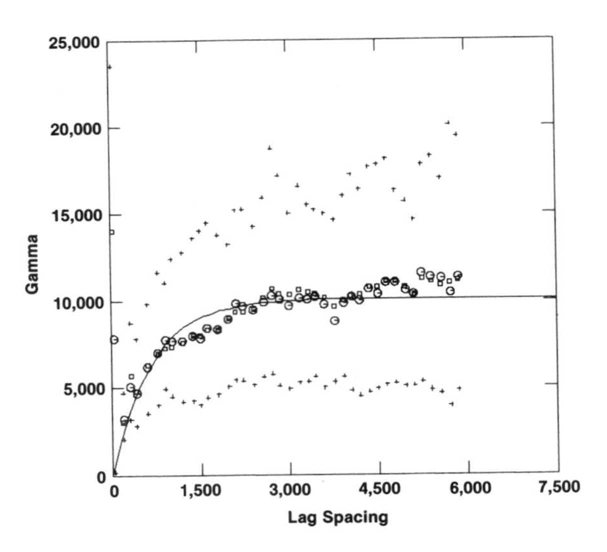

Figure 13

Isotropic areal variogram of well log data from Dataset 1, Interval C-D.

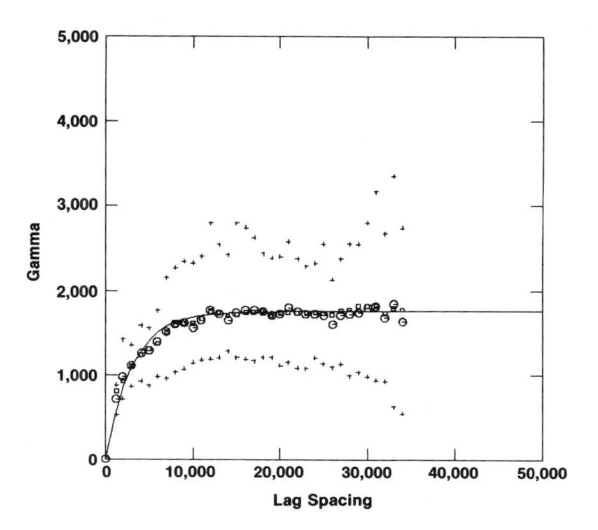

Figure 14

Isotropic areal variogram of seismic data from Dataset 3, Interval A-B.

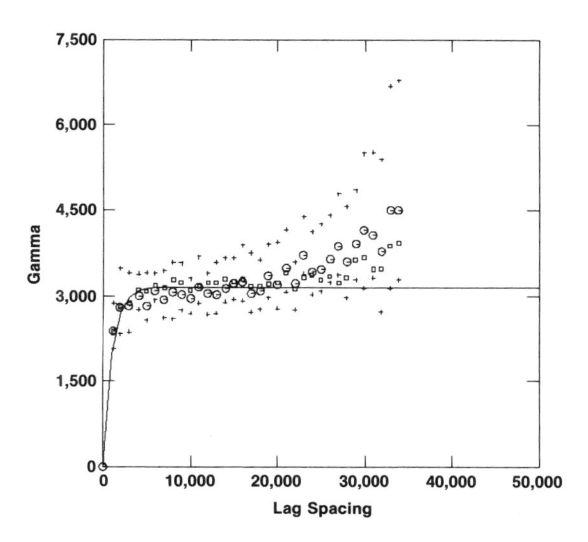

Figure 15

Isotropic areal variogram of well log data from Dataset 3, Interval A-B.

single layer with the same median as the pooled data. This means that in order to average indicator variograms from neighboring levels, they must be strictly stationary, i.e., they must have exactly the same histogram, not just the same mean.

As an example of this phenomenon, consider the variograms of the well log values from Dataset 2, Interval B-C, shown in Figures 16-19. The first shows the variogram calculated using the values from a single level within that interval. The model shown fit to that data has a range of 575 ft. Figure 17 shows the average of ten variograms calculated for ten levels from within the same interval. These may be fit with a model having the same range. However, the fluctuations in the experimental estimates of the variogram are considerably reduced when the average is considered rather than an individual level. Figure 18 shows the indicator variogram for Cutoff #5 from a single level in the same region. It shows fluctuations in the experimental variogram estimates, but they may be fit with a model having the same range as the variograms of the data values. When the average indicator variogram is calculated for ten levels, as shown in Figure 19, the resulting model fit shows a range of only 356 ft. Thus, in calculating indicator variograms, care must be taken in the trade-off between the desire for increased statistical mass to reduce estimation fluctuations, and the requirements of strict stationarity to prevent bias in the estimate of the ranges. The indicator variograms used in this study were based on individual layers.

D. Calibration of the Seismic and Well Log Data

Before the Markov-Bayes procedure can be used for inferring the cross-variogram needed for cokriging, a calibration of the seismic data against collocated well log data is required. This calibration provides the local prior distributions of well data based on the seismic data. For the data used in this study, the local prior distributions are summarized in Table 1 below. Local prior Z-cdf values are read row-wise from this table.

The calibration also provides several parameters which inform the system about the accuracy and precision of the seismic data in predicting the corresponding well-log values. The first is $m^{(1)}(z_k)$ which is the average of those seismic indicators corresponding to well data with indicators equal to unity. This is a measure of the accuracy of the seismic data in predicting well data values greater than the indicator cutoff. The second is $\sigma^{2(1)}(z_k)$, the variance of the seismic data corresponding to well data with indicators equal to unity. This is a measure of the precision of the seismic data in making that prediction. Similarly, $m^{(0)}(z_k)$ is the average of those seismic indicators corresponding to well data with indicators equal to zero. This is a measure of the accuracy of the seismic data in predicting well data values less than the indicator cutoff. The variance of these seismic indicators, $\sigma^{2(0)}(z_k)$, is a measure of the precision of

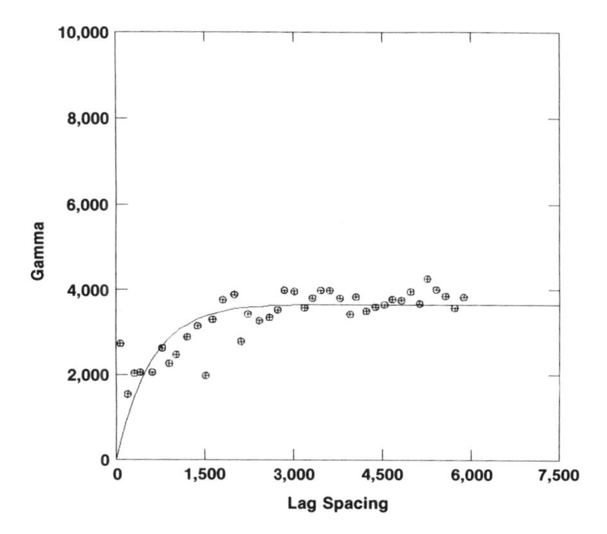

Figure 16

Isotropic areal variogram of well log data from Dataset 2, Interval B-C, using data from a single level.

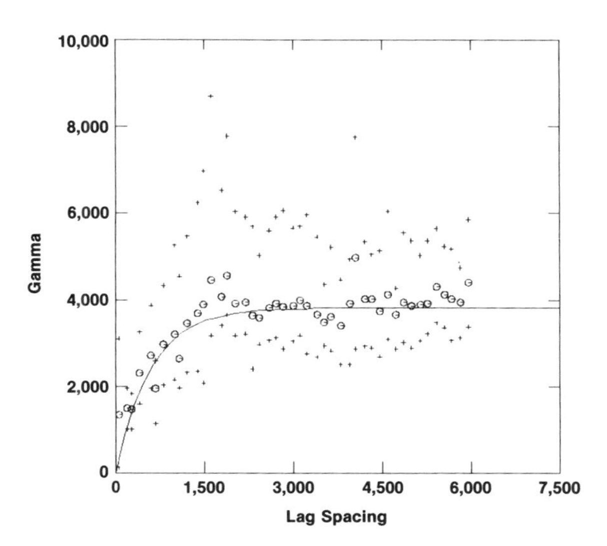

Figure 17

Isotropic areal variogram of well log data from Dataset 2, Interval B-C, averaging variograms from ten levels.

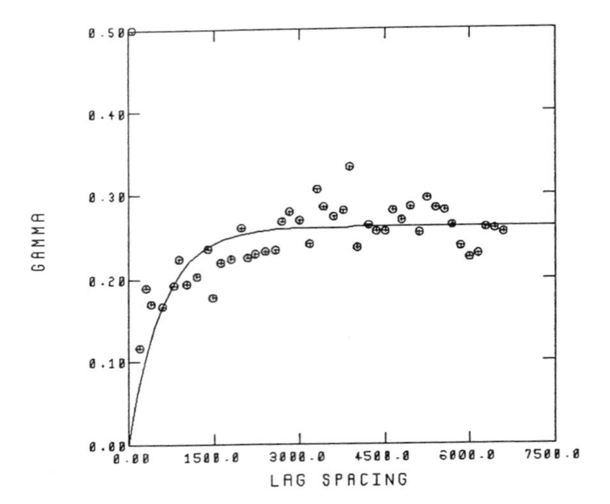

Figure 18

Isotropic areal indicator variogram of well log data from Dataset 2, Interval B-C, Cutoff 5, using data from a single level.

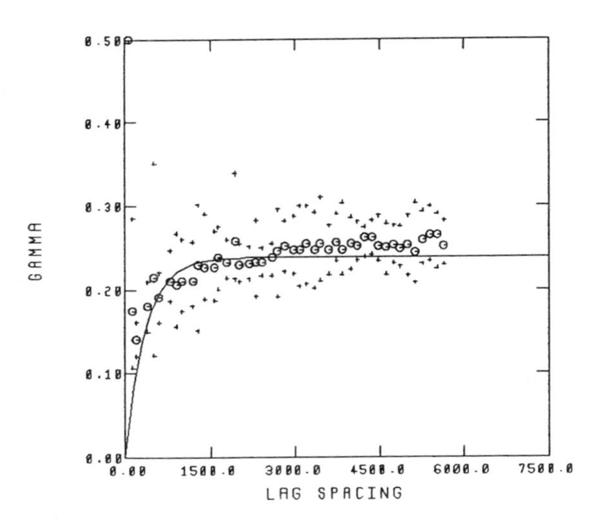

Figure 19

Isotropic areal indicator variogram of well log data from Dataset 2, Interval B-C, Cutoff 5, averaging variograms from ten levels.

Table I. Conditional distributions of well data, Z(x), for given classes of seismic data, S(x)

		z_k									
		933.2	954.1	969.6	981.7	992.7	1003.9	1016.1	1030.9	1054.8	z_{max}
	933.5	.48	.62	.723	.771	.813	.877	.911	.956	.982	1.000
	969.1	0.21	.36	.501	.622	.705	.774	.857	.908	.963	1.000
s_l	997.3	.11	.21	.319	.413	.519	.629	.749	.849	.932	1.000
	1020.1	.0657	.107	.171	.240	.333	.452	.609	.770	.886	1.000
	1054.3	0.0351	0.069	.104	.159	.214	.316	.411	.581	.768	1.000
	s_{max}	.0229	0.032	0.053	.0781	.105	.155	.196	.275	.436	1.000

that prediction. The difference between $m^{(1)}(z_k)$ and $m^{(0)}(z_k)$, denoted by $\mathbf{B}(z_k)$, provides an accuracy index for the seismic data in predicting well-log values (Journel and Hzu, 1990). For the data used here, the values of the calibration parameters are summarized in Table 2 below.

The values of $\mathbf{B}(z_k)$ determine the scaling factor for the cross-variograms of the seismic and well log data which are derived from the variograms of the well log data alone. They determine the degree of influence of the seismic data relative to the well log data when the cokriging of indicators is performed.

The values of $\mathbf{B}(z_k)$ along with the variances tabulated above also determine the shape of the variogram of the seismic data alone under the Markov-Bayes approximation. Under this hypothesis, the seismic data variogram will have a nugget effect determined by both $\mathbf{B}(z_k)$ and the variances, and the shape of the rest of the variogram will be scaled by $\mathbf{B}^2(z_k)$. For the low values of $\mathbf{B}(z_k)$ shown here, this implies a low level of correlation in the seismic data and a large nugget effect. A review of the measured variograms for the seismic data show that this conclusion from the Markov-Bayes hypothesis is not borne out. Fortunately, the role of the seismic data variograms is primarily to decluster the seismic data when determining the cokriging weights (Chu et al., 1991). The primary influences actually determining the cokriging weights are the well data variograms and the well and seismic data cross-variograms. A previous study using the Markov-Bayes approximation (Chu et al., 1991) also found that seismic data variogram inferred under the Markov-Bayes hypothesis indicated less correlation than the variograms determined from the actual seismic data. That study also showed that the cross-variograms were accurately inferred using

Table II. Calibration parameters

k	z_k	$m^{(1)}(z_k)$	$\sigma^{2(1)}(z_k)$	$m^{(0)}(z_k)$	$\sigma^{2(0)}(z_k)$	$B(z_k)$
1	933.2	0.322	0.0477	0.085	0.0142	0.237
2	954.1	0.414	0.0702	0.133	0.0216	0.281
3	969.6	0.472	0.0706	0.176	0.0286	0.295
4	981.7	0.511	0.0662	0.227	0.0338	0.283
5	992.7	0.558	0.0540	0.294	0.0417	0.263
6	1003.9	0.626	0.0451	0.357	0.0530	0.268
7	1016.1	0.698	0.0381	0.418	0.0645	0.280
8	1030.9	0.767	0.0323	0.504	0.0628	0.263
9	1054.8	0.856	0.0242	0.598	0.0542	0.258

the Markov-Bayes approach. A comparison of simulations using the actual variograms calculated from the data with those using the variograms inferred using the Markov-Bayes hypothesis showed barely detectable differences.

VI. RESERVOIR MODELS

All of the reservoir models were constructed on a 25 x 25 x 42 grid located in the upper right corner of the map as shown in Figure 1. The 42 layers conformed to the stratigraphic reference frame and spanned the A-D markers with 10 layers in the A-B interval, 15 layers in the B-C interval, and 17 layers in the C-D interval. The hard data which was used in the model construction consisted of 45 well-log traces with 11 of them inside the grid, and the remainder surrounding it in its immediate neighborhood. The seismic data consisted of 304 traces along 3 N-S lines and 3 E-W lines. Two of the lines in each direction were within the grid boundaries. The configuration of the well and seismic data is shown in relation to the computation grid in Figure 1.

Once constructed, each of the reservoir models can be displayed layer-by-layer, or as a series of vertical slices through the reservoir volume. Because of the large number of construction techniques employed, and the large number of possible views available for each model, we will focus on the results for a single stratigraphic layer, Layer 40. In viewing these figures, one artifact of the plotting procedure should be kept in mind. The figures originally were produced in color before photographic reduction to gray scale images. Unfortunately, the colorscale used had dark colors at the two extremes and light colors in the central range, so both high and low values appear dark gray. In the discussion of the figures, reference will be made to the location of different features, as well as their range of values, to aid in identifying them.

A. Kriging and Regression Analysis

A contour map of the kriged well log values and the locations of the data within the grid are shown in Figure 20. The sparse sampling of the data and the limited range of correlations are evident in the "bull's-eye" patterns in the resulting map. The kriged seismic values are shown in Figure 21. The data locations were omitted to avoid clutter in this figure. Both maps show low values in the center of the map, but the low values along a N-S line through the center of the seismic map appear more continuous than the single sample from this region in the well-log data indicates. This line corresponds to the seismic line in Figure 1. Both maps show two high regions along the right boundary and a band of mid-range to high values beginning in the center of the top of the map and running down the left-hand side. Since both the seismic data and the

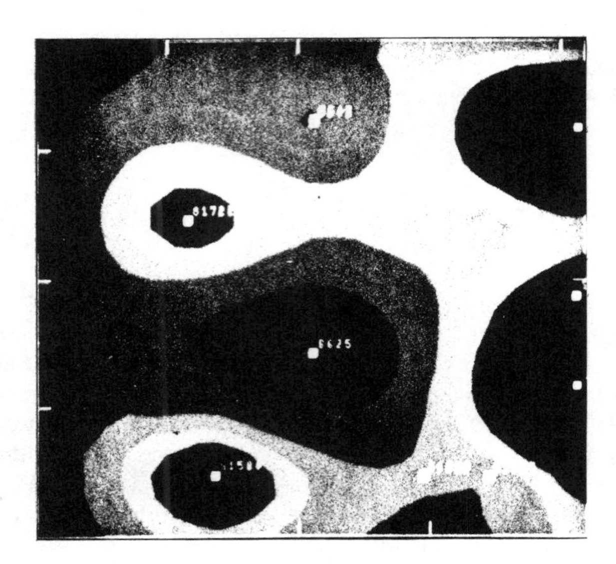

Figure 20

Contour map of Layer 40 from kriged well log data values.

Figure 21

Contour map of Layer 40 from kriged seismic data values.

well-log data are measured in the same units and have the same scale, a map constructed by regression of the seismic data appears identical to the kriged seismic map.

Both of these maps are smooth, a consequence of the kriging procedure used to construct them. Small scale variability can be introduced into them to provide more realistic models for flow simulation by several methods. Figures 22 and 23 show conditional simulations of the two data sets constructed using Gaussian Sequential Simulation. Figures 24 and 25 show conditional simulations based on the Mandelbrot-Weierstrass random fractal function conditioned by kriging (Voss, 1988; Hewett and Behrens, 1990). Each of these has additional small scale variability compared to the kriged maps, but each reflects only the large scale features of the data to which they are conditional.

B. Kriging with an External Drift

Figure 26 shows the results of kriging the well log data using the seismic data as an external drift. The inclusion of the seismic drift function serves to connect up the high values along the right side of the map and increases the N-S continuity of the low values in the center of the map compared to what was seen in the kriged well data alone. It has also increased the continuity of the isolated high value in the upper left quadrant of the well data to extend it toward the left boundary. Small scale variability can be added to the kriged map in this case as well. Figure 27 shows a conditional simulation based on Gaussian Sequential Simulation of the well log values using the external drift function from the seismic data in the kriging steps. In this particular realization, the region of high values in the upper left quadrant is expanded compared to either kriged map, but the continuity of low values along the central N-S line is preserved.

C. Cokriging with the Markov-Bayes Indicator Formalism

With the data coded as indicators and all of the indicator variograms modeled, Sequential Indicator Simulations (SIS) can be constructed on either data set alone, or they may be combined by cokriging using the Markov-Bayes formalism to infer the cross-variogram. SIS simulations of the well log data and seismic data considered separately are shown in Figures 28 and 29. Since the data are normally distributed, we do not expect much difference in the character of these simulations compared to the Gaussian Sequential Simulations shown earlier. The primary difference should be in the destructuring of values away from the median, as discussed previously. This is seen in both simulations, where the highs and lows show less continuity than the mid-range values compared to the Gaussian Sequential Simulations. However, caution is advised in drawing conclusions from the observation of individual realizations.

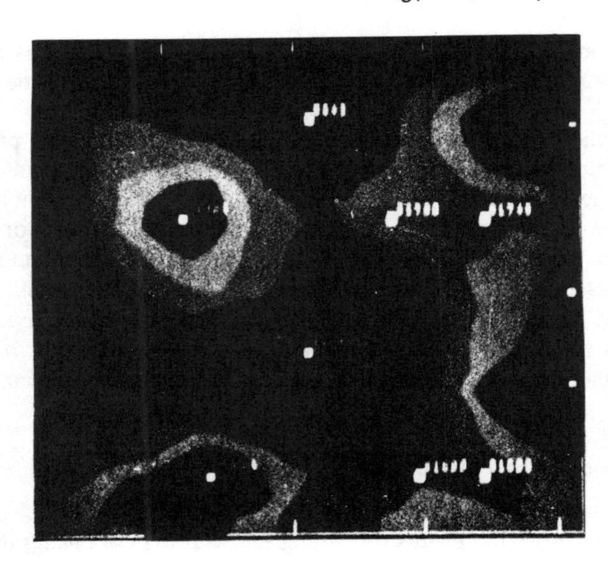

Figure 22

Contour map of Layer 40 from Gaussian Sequential Simulation of well log values.

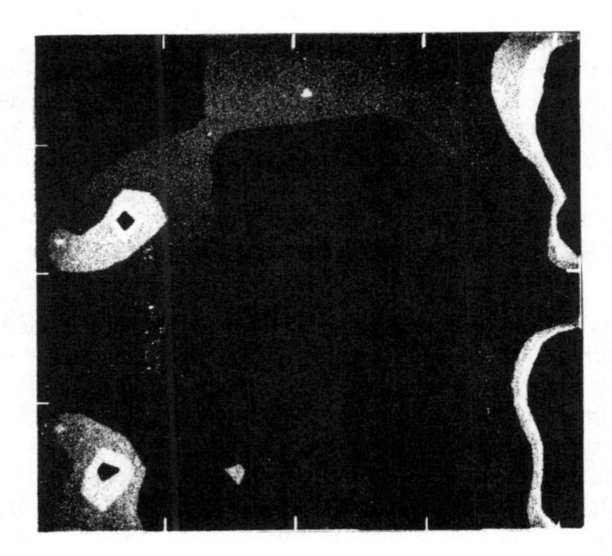

Figure 23

Contour map of Layer 40 from Gaussian Sequential Simulation of seismic data values.

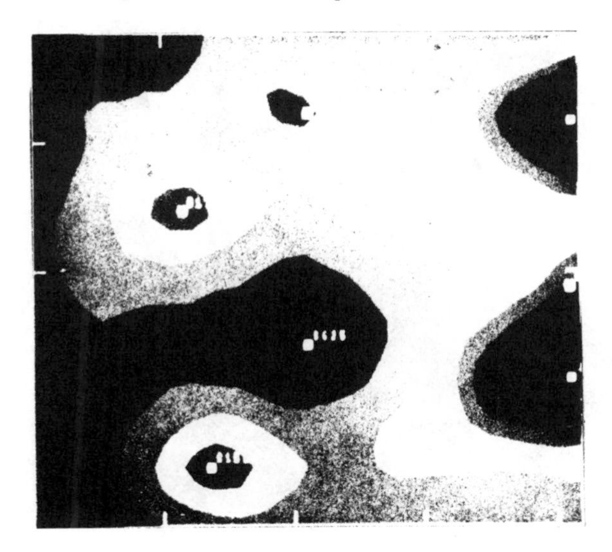

Figure 24

Contour map of Layer 40 from Mandelbrot-Weierstrass conditional simulation of well log values.

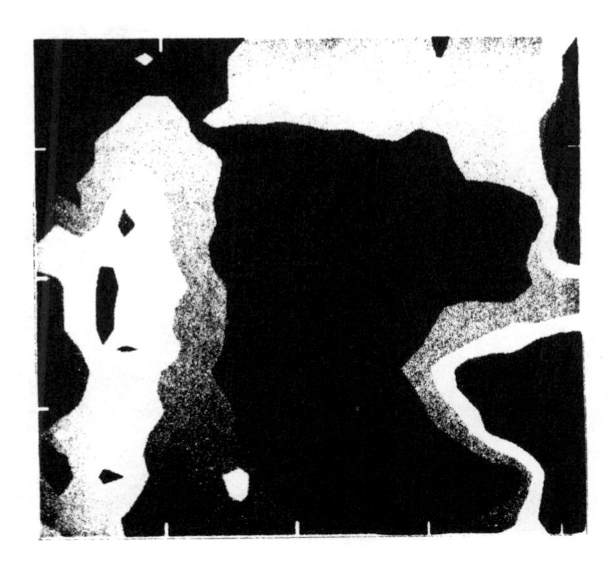

Figure 25

Contour map of Layer 40 from Mandelbrot-Weierstrass conditional simulation of seismic data values.

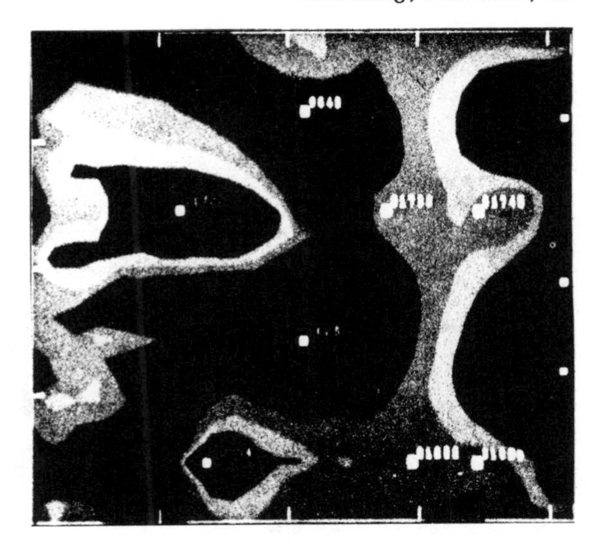

Figure 26

Contour map of Layer 40 from kriged well log values using the seismic data as an external drift.

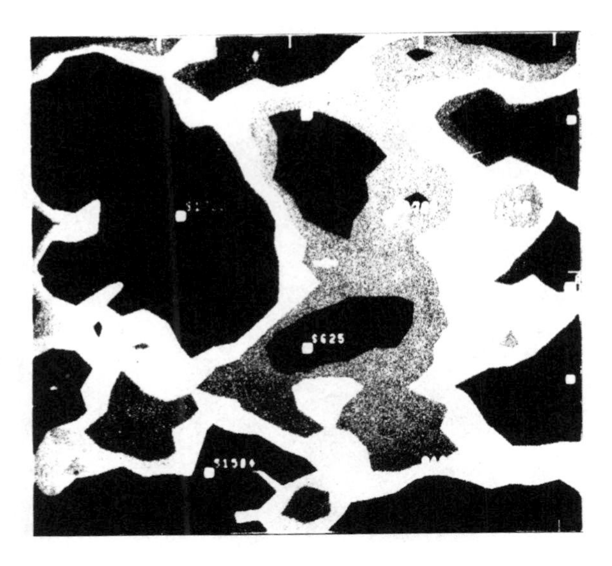

Figure 27

Contour map of Layer 40 from a Gaussian Sequential Simulation based on kriging the well log values using the seismic data as an external drift.

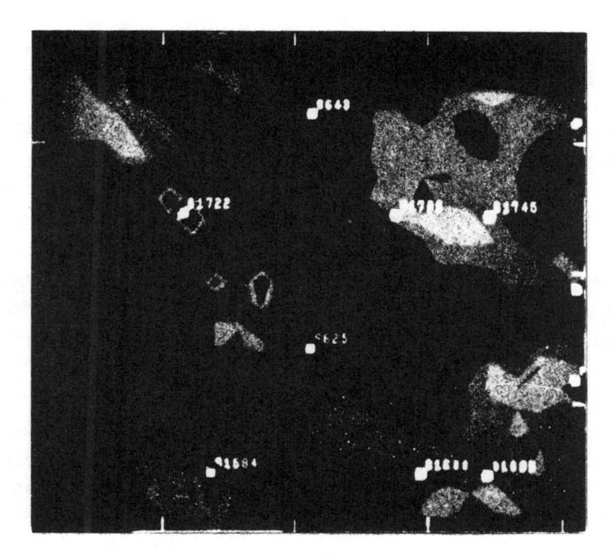

Figure 28

Contour map of Layer 40 from a Sequential Indicator Simulation of the well log data alone.

Figure 29

Contour map of Layer 40 from a Sequential Indicator Simulation of the seismic data alone.

A SIS simulation based on cokriging the seismic and well data indicators using the variograms derived with the Markov-Bayes calibration parameters is shown in Figure 30. The features of the two SIS simulations appear to be combined in this simulation. The destructuring of the extremes is also evident in this realization.

An alternative way of seeing the effects of combining the indicator data by cokriging is to compare maps showing the probability that individual cutoffs are exceeded using the two datasets alone and in combination through cokriging. Figure 31 shows the probability that each of the nine cutoffs is exceeded based on indicator kriging of the well values alone. The lowest cutoff is in the upper left hand corner and the highest cutoff is in the lower right hand corner, with the cutoffs increasing from left to right. Figure 32 shows a similar plot based on indicator kriging of the seismic data alone. Figure 33 shows the probability of exceedance based on indicator cokriging of both types of data, using the Markov-Bayes approximation. Looking at the well data alone, we see from the plot for the lowest cutoff, in the upper left corner, that the probability that the lowest cutoff is exceeded is relatively high everywhere. Based on the seismic data alone, we expect the lowest cutoff to be exceeded everywhere except in a strip running down the center. When the two types of data are combined by cokriging, the low probabilities of the central vertical strip in the seismic data

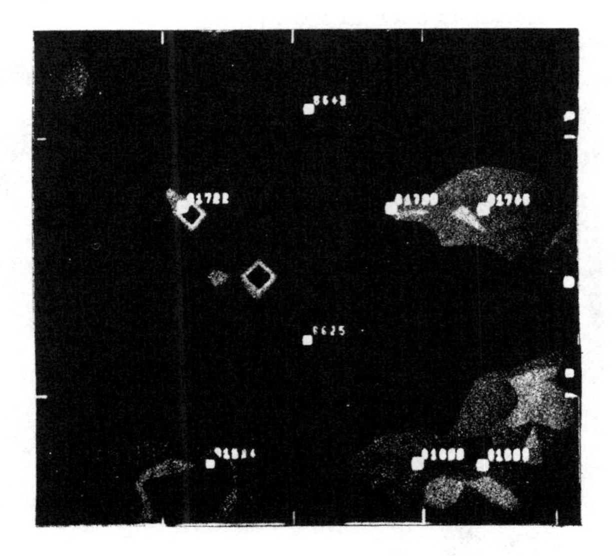

Figure 30

Contour map of Layer 40 from a Sequential Indicator Simulation based on cokriging the well log and seismic data using the Markov-Bayes approximation.

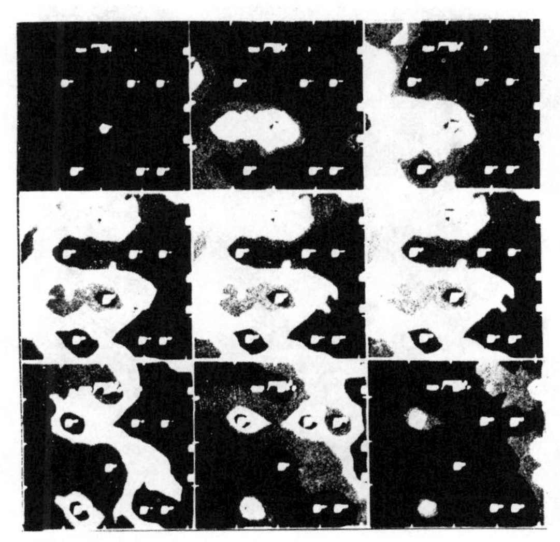

Figure 31

Maps of the probability of exceedance of the nine cutoffs based on indicator kriging of the well log data alone.

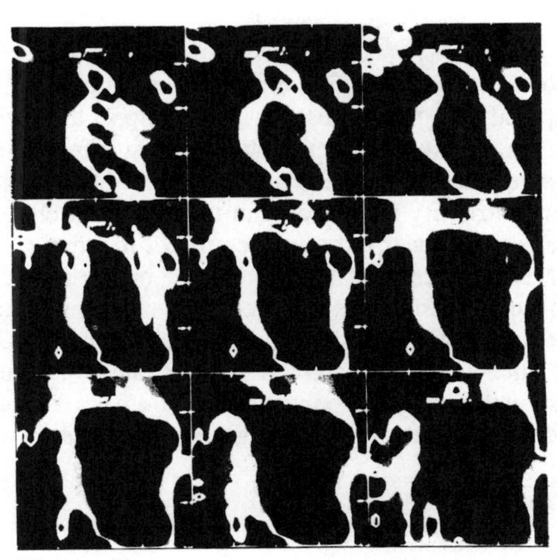

Figure 32

Maps of the probability of exceedance of the nine cutoffs based on indicator kriging of the seismic data alone.

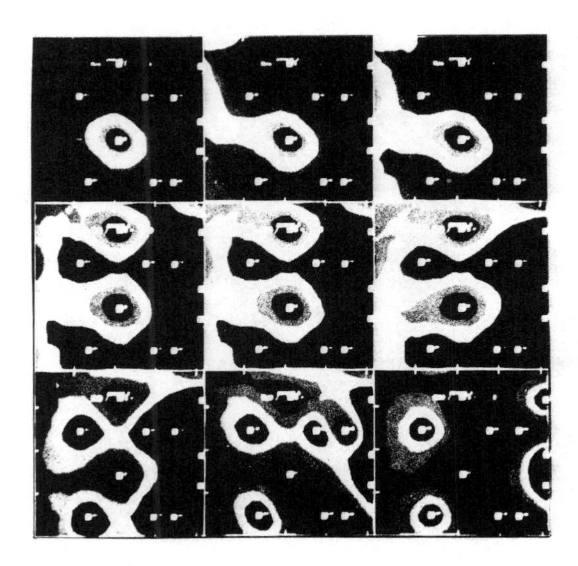

Figure 33

Maps of the probability of exceedance of the nine cutoffs based on indicator cokriging of the seismic and well log data using the Markov-Bayes approximation.

pull down the probabilities around first one of the wells along that strip in the lowest cutoff map and then both wells along that strip in the higher cutoff probability maps. In the probability plots for other cutoffs, the influence of the band of high seismic values along the left hand edge of the maps can be seen to expand the higher probability regions around the leftmost wells. The low values of the seismic data in the lower right hand corner of the map also tend to lower the probability of exceedance in that region for the higher cutoff values. It should be mentioned in passing that the irregular contours shown in the probability maps of the kriged well data indicators alone are an artifact of the shorter ranges of the indicator variograms and the search strategy used in selecting the data for use in estimating different points on the grid. For variograms with larger ranges, and data with denser sampling, these artifacts do not appear.

VII. DISCUSSION

A variety of methods for constructing reservoir models have been demonstrated and compared. One of the features seen in all of the models is the strong influence of the data configuration in this dataset on the resulting models. In all of the maps created using the well data alone, the sparse sampling and the relatively short range of the variograms compared to the data spacing created islands of similar values around individual wells. Although the various simulation techniques added small scale variability which tended to obscure these islands, the influence of the large scale structure dictated by the data configuration was still evident in all of them.

Similarly, in all of the maps created using the seismic data alone, the dense sampling of data along the seismic lines was evident. This is a product of the fact that the individual lines were shot and processed independently, so they could only provide information along the lines. This shortcoming could be overcome with a true three-dimensional seismic survey which would provide seismic data uniformly over the survey area. The dense sampling along orthogonal lines provided good estimates of variogram shapes for the seismic data, but the spacing of the lines relative to the resulting variogram ranges means that the linear nature of the data sampling is reflected in the resulting maps. Considering the high degree of correlation between the seismic and well log data ($r = 0.68$), the visual differences in the maps based on the two datasets considered independently are striking. Most of these differences are due to the fact that the low streak coinciding with the central N-S seismic line was only sampled by a single well.

The maps created using both datasets showed fewer artifacts from the data sampling, and combined features of the reservoir distributions seen in the individual datasets. Kriging of the well data with the seismic data treated as an external drift did introduce more continuity of the low values along the seismic line as expected. A similar effect was seen in the simulations based on indicator cokriging under the Markov-Bayes hypothesis, but the destructuring of the extreme cutoff values, which occurs in normally distributed data, tended to produce maps with a preponderance of near median values. In the absence of a knowledge of the true distribution, it is unclear how much closer to reality any of the combined maps really are.

VIII. CONCLUSION

1) The preparation of seismic data for integration with well log data in reservoir modeling requires careful processing, with this objective in mind from the beginning stages.

2) Areal correlations among synthetic acoustic impedance well logs based on a spectral extrapolation of inverted seismic traces show smaller ranges and higher sills than the corresponding seismic data when their correlation structure is measured using variograms.

3) Averaging of variograms from several levels within a stratigraphic interval can reduce the experimental variogram estimation fluctuations when applied to the measured data itself, but can bias the estimation of indicator

variograms for normally distributed data. The bias is toward a shorter range of correlations than is observed in the data values themselves or the median cutoff indicator for a single level.

4) Kriging sparsely sampled well data with seismic data treated as an external drift modifies the kriged values to include the shape of the seismic data. For the data configuration considered here, the extent of this modification was apparent, but not dramatic.

5) Cokriging of seismic and well log data incorporates the seismic data directly in the estimates of well data at unsampled locations. Sequential Indicator Simulations based on cokriging indicator variables using the Markov-Bayes hypothesis showed features of both datasets, but tended to have a lower continuity of extreme values, owing to the normality of the data used. Maps showing the probability of exceedance of individual cutoffs demonstrate the influence of the seismic data in modifying the probabilities determined from the well data alone.

6) Conditional simulations which include the small scale variability needed in realistic fluid flow simulations can be constructed using any of the data integration methods discussed.

REFERENCES

Araktingi, U. G., and Orr, F. M., "Viscous Fingering in Heterogeneous Porous Media", SPE 18095 presented at the 63rd Annual Technical Conference and Exhibition of SPE, Houston, TX, Oct 2-5, 1988.

Bamberger, A., Chavent, G., Hemon, C., and Lailly, P., "Inversion of normal incidence seismograms", Geophysics, v. 47, 1982, p. 757-770.

Barker, J. W., and Fayers, F. J., "Transport Coefficients for Compositional Simulation with Coarse Grids in Heterogeneous Media", presented at the 66th Annual Technical Conference of SPE, Dallas, Oct., 1991.

Chu, J., Wenlong, X., Zhu, H., and Journel, A. G., "The Amoco Case Study", Stanford Center for Reservoir Forecasting, June, 1991.

Dagbert, M., David, M., Crozel, D., and Desbarats, A., "Computing Variograms in Folded Strata-Controlled Deposits", in G. Verly et al. (Eds.), Geostatistics for Natural Resources Characterization, D. Reidel, Dordrecht, Holland, Part I, 198 4, pp. 71-89.

Deutsch, C. V., "The Relationship Between Universal Kriging, Kriging with an External Drift, and Cokriging", Stanford Center for Reservoir Forecasting Annual Report, May, 1991.

Deutsch, C. V., and Journel, A. G., GSLIB: Geostatistical Software Library User's Guide, Stanford Center for Reservoir Forecasting, Stanford, CA, 1991., pp. 55-59.

Doyen, P. M., "Porosity from Seismic Data: A Geostatistical Approach", Geophysics, 53, 10, October, 1988, pp. 1263-1275.

Fayers, F. J., Blunt, M. J., and Christie, M. A., "Accurate Calibration of Empirical Viscous Fingering Models", 2nd European Conference on the Mathematics of Oil Recovery, D. Guerillot and O. Guillon (Ed.), Editions Technip, Paris, 1990, pp. 45 -55.

Galli, A., and Meunier, G., "Study of a Gas Reservoir Using the External Drift Method", Geostatistical Case Studies, G. Matheron and M. Armstrong (Ed.), D. Reidel Publishing Co., 1987, pp. 105-109.

Haldorsen, H. H., and Damsleth, E., "Stochastic Modeling", JPT, April, 1990, pp. 404-412.

Hewett, T. A., "Fractal Distributions of Reservoir Heterogeneity and Their Influence on Fluid Transport", SPE 15386, presented at the 61st Annual Technical Conference of SPE, New Orleans, Oct 5-8, 1986 (also in Reservoir Characterization-2, SPE Reprint Series No. 27, 1989).

Hewett, T. A., and Behrens, R. A., "Conditional Simulation of Reservoir Heterogeneity with Fractals", SPE Formation Evaluation, September, 1990, pp. 217-225.

Hewett and Behrens, "Consideration Affecting the Scaling of Displacements in Heterogeneous Porous Media", SPE 20739, presented at the 65th Annual Technical Conference of SPE, New Orleans, Sept. 23-26, 1990.

Journel, A. G., Fundamentals of Geostatistics in Five Lessons, Short Course in Geology: Volume 8, AGU, 1989, p. 33.

Journel, A. G., and Alabert, F. G., "Focusing on the Spatial Connectivity of Extreme-Valued Attributes: Stochastic Indicator Models of Reservoir Heterogeneities", SPE 18324, presented at the 63rd Annual Technical Conference and Exhibition of SPE, Houston, TX, Oct 2-5, 1988.

Journel, A. G., and Zhu, H., "Integrating Soft Seismic Data: Markov-Bayes Updating, An Alternative to Cokriging and Traditional Regression", Stanford Center for Reservoir Forecasting Annual Report, May, 1990.

Kallweit, R. S., and Wood, L. C., "The limits of resolution of zero-phase wavelets", Geophysics, v. 47, 1982, p. 1035-1046.

Kyte, J. R., and Berry, D. W., "New Pseudo Functions to Control Numerical Dispersion", SPEJ, Aug., 1975, pp. 269-276.

Koefoed, O., "Aspects of vertical seismic resolution", Geophysical Prospecting, v. 29, 1981, p. 21-30.

Lavergne, M., and Willm, C., "Inversion of seismograms and pseudo velocity logs", Geophysical Prospecting, v. 11, 1977, p. 231-250.

Lindseth, R. O., "Synthetic sonic logs — A process for stratigraphic interpretation", Geophysics, v. 44, 1979, p. 3-26.

Luster, G. R., "Raw Materials for Portland Cement: Applications of Conditional Simulation of Coregionalization", Ph.D. Thesis, Stanford Univ., Branner Library, Stanford Univ., 1985.

Marechal, A., "Kriging Seismic Data in the Presence of Faults", in G. Verly et al. (Eds.), Geostatistics for Natural Resources Characterization, D. Reidel, Dordrecht, Holland, Part I, 1984, pp. 271-294.

Oldenburg, D. W., Scheuer, T., and Levy, S., "Recovery of the acoustic impedance from reflection seismograms", Geophysics, v. 48, 1983, pp. 1318-1337.

Omre, H., "Stochastic Models for Reservoir Characterization", Norwegian Computing Center, 1991.

Pusey, L. C., "Band-limited sharpness deconvolution", presented at Australian Society of Exploration Geophysics meeting, 1988.

Rendu, J. M., and Ready, L., "Geology and the Semi-Variogram — A Critical Relationship", 17th APCOM Symposium, A.I.M.E., New York, 1982, pp. 771-783.

Sheriff, R. E., Seismic Stratigraphy, IHRDC, Boston, 1980,.

Voss, R. F., "Fractals in nature: From characterization to simulation", in The Science of Fractal Images, Peitgen and Saupe (Eds.), Springer-Verlag, New York, 1988.

Waggoner, J. R., Castillo, J.L., and Lake, L.W., "Simulation of EOR Processes in Stochastically Generated Permeable Media", SPE 21237, presented at the 11th SPE Symposium on Simulation, Anaheim, Feb. 17-20, 1991.

Walker, C., and Ulrych, T. J., "Autoregressive recovery of the acoustic impedance", Geophysics, v. 48, 1983, pp. 1338-1350.

Widess, M. B., "Quantifying resolving power of seismic systems", Geophysics, v. 47, 1982, pp. 1160-1173.

SESSION 4

Optimization of Reservoir Management

Co-Chairmen

Mike Fowler, Exxon Company

Susan Jackson
National Institute for Petroleum and Energy Research

Assessment of Uncertainty in the Production Characteristics of a Sand Stone Reservoir

Henning Omre
Håkon Tjelmeland

Norwegian Computing Center
Oslo, Norway

Yuanchang Qi

IBM/EPAC
Stavanger, Norway

Leif Hinderaker

Norwegian Petroleum Directorate
Stavanger,Norway

1 Introduction

Decisions concerning development and depletion of petroleum reservoirs must be made under uncertainty in the decision supporting information. The incomplete knowledge of the reservoir characteristics contributes significantly to this uncertainty. The primary objective of reservoir evaluation is to reduce the uncertainty in the decision supporting information, the target frequently being predicted hydrocarbon production with time.

The efforts to reduce uncertainty will of course be subject to economical constraints.

Three actions can be imagined to reduce uncertainty:

- collect more reservoir specific observations, then more exact knowledge about the reservoir will be available. It may be obtained by more sampling in existing wells, new wells or other sampling techniques. Development of new data acquisition equipment will often be necessary, and this has been the trend in recent years. The cost associated with both development and operation are normally huge.

- collect the same amount of information more cleverly. A good sampling plan avoiding too much redundant information with the possibility to collect more areal covering information can result in reduced uncertainty. Statistical methodology can be used to improve the sampling design, and improvements are expected to be made for low costs.

- combine the existing information more optimally. By combining the data according to their precision and compensate for redundancies, the uncertainty in the decision supporting information will be reduced. Stochastic modeling and statistics are necessary methodology for obtaining this, and considerable improvements for relatively small investments are expected.

In order to trade-off among the three actions a reliable quantification of uncertainty in the decision supporting information has to be available. This paper aims at establishing a formalism for assessment of uncertainty in predictions of recovery from petroleum reservoirs.

The assessment of uncertainty has been mentioned briefly in many publications, usually in association with heterogeneity modeling, see Journel (1990), Hewett and Behrens (1988), Damsleth et al. (1990) and Høiberg et al. (1990). A serious discussion of the topic has not been found, however, probably since assessing the uncertainty realistically appears as much more complicated than reproducing heterogeneity.

In the paper, many aspects of statistics are touched. A sufficient collection of basic references is: Ripley (1987), Berger (1980) and Johnson and Kotz (1969) and (1970). Standard statistical terminology is used, in particular, random variables with upper case and constants with lower case letters.

2 Stochastic Formalism

Consider a particular petroleum reservoir subject to evaluation. The evaluation will typically be at the appraisal stage where a development plan is about to be prepared. Let the true hydrocarbon production characteristics as a function of time, t, under the recovery strategy, p, be denoted $q_p(t)$. These production characteristics will of course be unknown at the stage of appraisal. The characteristics can be measured by several variables: oil and gas rate per well or accumulated, pressure changes etc. The variable $q_p(t)$ includes all measures of interest hence it is vectorial. For reasons of convenience, the vector property is suppressed in the notation. The objective of this section is to establish a formalism for prediction of $q_p(t)$ with reliable uncertainty specification due to incomplete knowledge of the reservoir properties.

The use of reservoir simulators, or more specifically reservoir production simulators, is widespread in the petroleum industry. These simulators provide a prediction of the production characteristics given the reservoir properties and the recovery strategy. Let an asterix denote such a prediction, then:

$$q_p(t) = q^* \left[r_0(x), p(x,t)\right] + \delta_*(t) = q_p^*(t)|r_0(x) + \delta_*(t) \quad (1)$$

The true production characteristics are expressed as the sum of the prediction from the reservoir production simulator and a residual term. The production simulator is a function of $r_0(x)$, being the reservoir variables prior to production start and $p(x,t)$ being the recovery variables. The reservoir variables are a function of location, x, only, since the reservoir is assumed to be in approximate equilibrium before production starts. The recovery variables are functions of both location, x, and time, t. The reservoir production simulator, $q^*[\cdot,\cdot]$, is defined to include two components: the support-change or homogenization module and the fluid flow module. The former changes the scale of representation from a relatively detailed reservoir description to a coarse system of blocks as required by the fluid flow simulator. The latter is normally one of the widely used reservoir fluid flow simulators.

The initial reservoir variables, $r_0(x)$, represent the characteristics of the reservoir relevant for fluid flow. The representation is assumed to be sufficiently dense and accurate not to add uncertainty to the prediction. This includes variables like geometry, porosity distribution, permeability properties, initial phase saturations etc. The term, $r_0(x)$, is of course vectorial, but this is not represented in the notation. Exactly which variables to include will be determined by the characteristics of the reservoir, the recovery strategy and the requirements from the reservoir production simulator to be applied. Note that $r_0(x)$ represents the true reservoir characteristics prior to production, and that these of course are largely unknown at the appraisal stage. The challenge is to quantify the uncertainty in the prediction of the production characteristics due to this incomplete knowledge.

The recovery variables, $p(x,t)$, represent the development plan and the depletion strategy. This includes variables like location of injection/production wells, the perforation pattern, processing capacity on platforms, depletion procedure etc. The values of the recovery variables have to be such that they are robust with regard to the uncertainty in the reservoir characterization. The depletion procedure will be defined in a predictive setting and hence relate to the production volumes obtained from the individual wells. The term, $p(x,t)$, is vectorial, and exactly which variables to include is determined by the recovery strategy and the requirements from the reservoir production simulator to be applied. Note that $p(x,t)$ is related to development and depletion, hence subject to human control.

The residual, $\delta_*(t)$, is defined to fill the gap between the prediction based on the reservoir production simulator and the truth. In order to be able to evaluate the uncertainty, one has to conjecture the following: The variability in $\delta_*(t)$ is small relative to the variability in $q_p(t)$. Consequently, the existence of a reservoir production simulator, $q^*[\cdot,\cdot]$, for which this holds for the reservoir under study, has to be assumed. A thorough discussion of this is important, but not the topic of this article. The existence of a suitable production simulator is assumed for the rest of this article.

So far, no stochastic elements are defined in the formalism. In order to evaluate the uncertainty consistently, however, a stochastic approach is required.

2.1 Stochastic Model

The uncertainty due to incomplete knowledge of the initial reservoir characteristics represented by $r_0(x)$, is the main subject of the study. Hence it is convenient to represent $r_0(x)$ by a stochastic model:

$$R_0(x)|\theta \sim f_{R_0|\theta}(r(x)|\theta) \tag{2}$$

The $f_{R_0|\theta}(\cdot|\cdot)$ is the probability density function, pdf, for $R_0(x)|\theta$, assigning probabilities to all possible outcomes of the initial reservoir variables. The term $R_0(x)|\theta$ defines a multivariate random function with, x, reference in location. Furthermore, the probability structure of the random function is dependent on a set of model parameters, θ. The multivariability appears from reservoir geometry, porosity distribution, permeability properties, initial phase saturations etc. The reference in location is needed for representing spatial variability. The model parameters, θ, are typically expected values, variance and spatial dependence of each of the elements of the reservoir variable, interdependence among the elements etc.

The challenge of stochastic reservoir description is to obtain a representative and reliable pdf for the initial reservoir variables, $f_{R_0(x)|\theta}(\cdot|\cdot)$, for the reservoir under study. The pdf may appear as a probability weighted combination of pdf's representing different geological interpretations of the reservoir under study, if uncertainty exists on that level. For a particular geologic interpretation, the specification of the pdf must be based on general geological knowledge and experience. In order to keep the model simple, only the factors with major influence on the production characteristics should be included. The effects being represented by model parameters, θ, do usually correspond to reservoir properties which vary considerably from one reservoir to the other. The fact that mathematical tractability usually is a constraining factor should also be kept in mind. Haldorsen and Damsleth (1990) and Omre (1992) provide overviews over available stochastic model formulations. The stochastic modeling part, i.e. the specification of $f_{R_0(x)|\theta}(\cdot|\cdot)$, is a piece of art and should be the focus in reservoir description. Note that the actual values of the model parameters, θ, are not yet assigned.

Numerous studies on stochastic reservoir description are reported in the literature, see Mathews et al. (1989), Rudkiewicz et al. (1990), Alabert and Massonnat (1990) and Damsleth et al. (1990). A general impression is that frequently far too little effort has been made to obtain a representative stochastic model for the reservoir under study. There is a tendency to take the model formulation as granted, and only adjust the parameter values to the actual reservoir. Hence, the stochastic modeling aspects are underrated, and in most cases this results in a too simplistic model.

Returning to expression (1), and introducing the stochastic model for the reservoir variables provide:

$$Q_p(t) = q^* \left[(R_0(x)|\theta), p(x,t)\right] + \Delta_*(t)|\theta = Q^*_p(t)|\theta + \Delta_*(t)|\theta \qquad (3)$$

This entails that the true production characteristics are represented by a random function, $Q_p(t)$. Conceptually, realizations from it could be true production characteristics from different reservoirs having comparable geological origin, geometrical properties and are being depleted similarly. The predictor based on the reservoir production simulator, $Q_p^*(t)|\theta$, will be stochastic as well and conditioned on the values of the model parameters, θ. The $Q_p^*(t)|\theta$ will of course be highly dependent on the actual choice of $f_{R_0(x)|\theta}(\cdot|\cdot)$, although not explicitly expressed in the notation. The residual term, $\Delta_*(\cdot)|\cdot$, is also defined to be stochastic. The conjecture concerning expression (1) can now be formalized to: residual approximately centered at zero, i.e. $\mathrm{E}\{\Delta_*(t)|\theta\} \approx 0.0$ and the variance in the residual much less than the variance in the prediction, i.e. $\mathrm{Var}\{\Delta_*(t)|\theta\} \ll \mathrm{Var}\{Q_p^*(t)|\theta\}$.

So far, effects of appraisal and observations specific to the reservoir under study, have not been introduced. By assuming the reservoir to be in approximate equilibrium prior to production, all reservoir specific appraisal will be representative of $R_0(x)$. Note that the data collection will be far from complete, usually it will result in observations of the initial reservoir variables in a small number of appraisal wells only. Denote the available reservoir specific observations, o_r, and its stochastic counterpart O_r. The stochastic model for the initial reservoir characteristics should in the evaluation be conditioned on the available observations in order to use the information optimally. Hence:

$$R_0(x)|\theta, o_r \sim f_{R_0|\theta,O_r}(r(x)|\theta, o_r) \tag{4}$$

with $f_{R_0|\theta,O_r}(\cdot|\cdot,\cdot)$ being the corresponding pdf. It is obvious that the uncertainty in $(R_0(x)|\theta, o_r)$ will decrease with increasing o_r. In the limit, if the initial reservoir variables were completely observed, then $O_r \equiv o_r = r_0(x)$. Consequently $(R_0(x)|\theta, r_0(x)) \equiv r_0(x)$, the true initial state, and it would be so regardless of choice of pdf, $f_{R_0|\theta,O_r}(\cdot|\cdot,\cdot)$, and of values on θ. In reservoir description, data collection is the best cure against a poor stochastic model, but on the contrary, a good stochastic model can to some extent compensate for data collection and hence additional expenses.

Returning to expression (3), and introducing the conditioning in the predictor for the production characteristic, it follows:

$$Q_p^*(t)|\theta, o_r = q^*\left[(R_0(x)|\theta, o_r), p(x,t)\right] \tag{5}$$

Since $q^*[\cdot,\cdot]$ usually consists of a set of differential equations, the problem could be solved within a framework of stochastic differential equations, see Holden and Holden (1992). This framework is not yet available, however, and presently one have to rely on solutions based on stochastic simulation, or the Monte Carlo approach. Although stochastic differential equations is an interesting and prosperous field of research, this article is limited to consider simulation approaches. Expression (5) is important for understanding the various components of the uncertainty, hence it is further discussed:

- the predicted production characteristics, $Q_p^*(t)|\cdot,\cdot$, are interpretated as a multivariate stochastic function of time t, hence it is possible to perform a formal evaluation of uncertainty.

- the reservoir production simulator, $q^*[\cdot,\cdot]$, is a deterministic function for predicting fluid flow, and it is assumed to be representative of the reservoir under study.

- the stochastic model representing the initial reservoir characteristics, $R_0(x)|\theta, o_r$, is conditioned on the values of the model parameters, θ, and the reservoir specific observations available from appraisal, o_r. The θ is expected to have large influence on $Q_p^*(\cdot)|\cdot,\cdot$, since it represents reservoir global characteristics like average porosity, permeability and initial saturations. The θ is unknown, and assigning constants to it, hence ignoring the uncertainty in assessing it, is expected to introduce severe bias in the evaluation of uncertainty in $Q_p^*(\cdot)|\cdot,\cdot$. A Bayesian approach is recommended in the next paragraph in order to include this source of uncertainty in the evaluation. The reservoir specific observations, o_r, have impact on the model in two respects: firstly, it contributes to improved assessment of the model parameters θ. Secondly, it constrains the possible realization of $R_0(x)$ to reproduce o_r. When the reservoir specific observations, o_r, are scarce, as in this case, the contribution through θ is expected to have largest impact because it provides reservoir global influence, while the conditioning on the realizations are local only.

- the recovery variable, $p(\cdot,\cdot)$, will of course have considerable impact on $Q_p^*(\cdot)|\cdot,\cdot$. Hence the results must be considered as conditioned on the set of recovery variables assigned, as indicated by the notation. This does not create problems, since the recovery variables can be controlled by the reservoir management. A procedure for optimal selection of $p(\cdot,\cdot)$ in a stochastic setting can be developed

- the expression for $Q_p^*(\cdot)|\cdot,\cdot$ can be used to demonstrate why heterogeneity modeling of reservoir characteristics is important. The following relation is true,

$$\mathrm{E}\{Q_p^*(\cdot)|\cdot,\cdot\} = \mathrm{E}\{q^*[(R_0(\cdot)|\cdot,\cdot), p(\cdot,\cdot)]\} \tag{6}$$
$$\neq q^*[\mathrm{E}\{R_0(\cdot)|\cdot,\cdot\}, p(\cdot,\cdot)]$$

  for all reliable reservoir production simulators, because the reservoir variables are non-linearly involved in the formulation. The $\mathrm{E}\{R_0(\cdot)|\cdot,\cdot\}$ represents some sort of best guess on the reservoir variables, and by using this biased predictions will appear. Unbiased predictions can only be provided by stochastic simulation of $(R_0(\cdot)|\cdot,\cdot)$, reflecting the heterogeneity in the reservoir characteristics. This will be more thoroughly discussed later. In the literature on heterogeneity modeling, predictions of the production characteristics are frequently made based on one single realization of the reservoir variables. For this to be an improvement of traditional, deterministic predictions, it is necessary that the bias is relatively large compared to the variability of $Q_p^*(\cdot)|\cdot,\cdot$. There are no reason to believe that this is generally true, hence the predictions from heterogeneity modeling should be based on several realizations.

2.2 Assessment of Uncertainty

The total uncertainty can only be realistically assessed by taking the uncertainty in the assignment of values to the model parameters into consideration. In order to do so, the parameters will be considered as stochastic variables, Θ, according to the Bayesian paradigm. A qualified prior guess on the probability structure of Θ can be based on general geological knowledge, experience from similar types of reservoirs, observations of outcrops of comparable geology etc. Note that the reservoir specific observations, o_r, shall not be used. The qualified prior guess will be represented by:

$$\Theta \sim f_\Theta(\theta) \tag{7}$$

with $f_\Theta(\cdot)$ being a multivariate pdf.

Recall that the reservoir specific observations $O_r \equiv o_r$ actually are observations of parts of the reservoir variable $R_0(x)$. Hence from the pdf $f_{R_0|\Theta}(\cdot|\cdot)$ it is possible to obtain the pdf for the reservoir specific observations, O_r, given the parameters Θ, $f_{O_r|\Theta}(\cdot|\cdot)$. This is termed the likelihood function.

By applying Bayes relation, the impact of observing o_r on the pdf of Θ can be determined:

$$\Theta|o_r \sim f_{\Theta|O_r}(\theta|o_r) = c \cdot f_{O_r|\Theta}(o_r|\theta) f_\Theta(\theta) \tag{8}$$

with c being a normalizing constant. The $f_{\Theta|O_r}(\cdot|\cdot)$ is termed the posterior pdf for Θ given o_r.

At this stage of evaluation the stochastic model is specified, and qualified prior guesses on the model parameters are available. The only reservoir specific observations are o_r, however, and the prediction for the production characteristic conditioned on the reservoir specific observations, $Q_p^*(t)|o_r$, should be considered. The corresponding probability structure is defined by:

$$Q_p^*(t)|o_r \sim f_{Q_p^*|O_r}(q(t)|o_r) = \int_{\mathcal{D}} f_{Q_p^*|\Theta,O_r}(q(t)|\theta, o_r) \cdot f_{\Theta|O_r}(\theta|o_r)\mathrm{d}\theta \tag{9}$$

with $\mathcal{D}$ being the domain of all outcomes of Θ. The pdf $f_{Q_p^*|\Theta,O_r}(\cdot|\cdot,\cdot)$ is obtainable through the functional relationship between $Q_p^*(\cdot)|\cdot,\cdot$ and $(R_0(\cdot)|\cdot,\cdot)$ in expression (5) and $f_{R_0|\Theta,O_r}(\cdot|\cdot,\cdot)$ in relation (4). The pdf $f_{\Theta|O_r}(\cdot|\cdot)$ is defined in expression (8).

The pdf for $Q_p^*(t)|o_r$, which is the pdf of interest, cannot normally be determined analytically from expression (9). Stochastic simulation techniques have to be utilized. A procedure for assessing the uncertainty in $Q_p^*(t)|o_r$ is outlined in Figure 1.

The pdf's $f_{R_0|\Theta,O_r}(\cdot|\cdot,\cdot)$ and $f_{\Theta|O_r}(\cdot|\cdot)$ are obtained by procedures previously described. Use $f_{\Theta|O_r}(\cdot|\cdot)$ to generate S realizations of the model parameters. For each of these realizations of parameters, use $f_{R_0|\Theta,O_r}(\cdot|\cdot,\cdot)$ to generate T realizations of the initial reservoir variables conditioned on the reservoir specific observations. This provides $S \cdot T$ realizations, $(r_0(x)|o_r)_s; s = 1, 2, \ldots, S \cdot T$, each of them reproducing o_r. This set of realizations reflects the total uncertainty in the stochastic model for the initial reservoir characteristics. By activating the reservoir production simulator on each of these realizations, a set of realizations of the predicted production characteristics is obtained. This set, $(q_p^*(t)|o_r)_s; s = 1, 2, \ldots, S \cdot T$ reflects the total uncertainty in the predicted production characteristics. Recall that $q_p^*(\cdot)|\cdot$ is a vector of different measures of production. Graphical displays should be used to evaluate the results. For each vector element one may plot all the functions corresponding to the set of realizations, and this will represent the total uncertainty in that element of $Q_p^*(\cdot)|\cdot$. Dependence between the various elements can also be evaluated graphically.

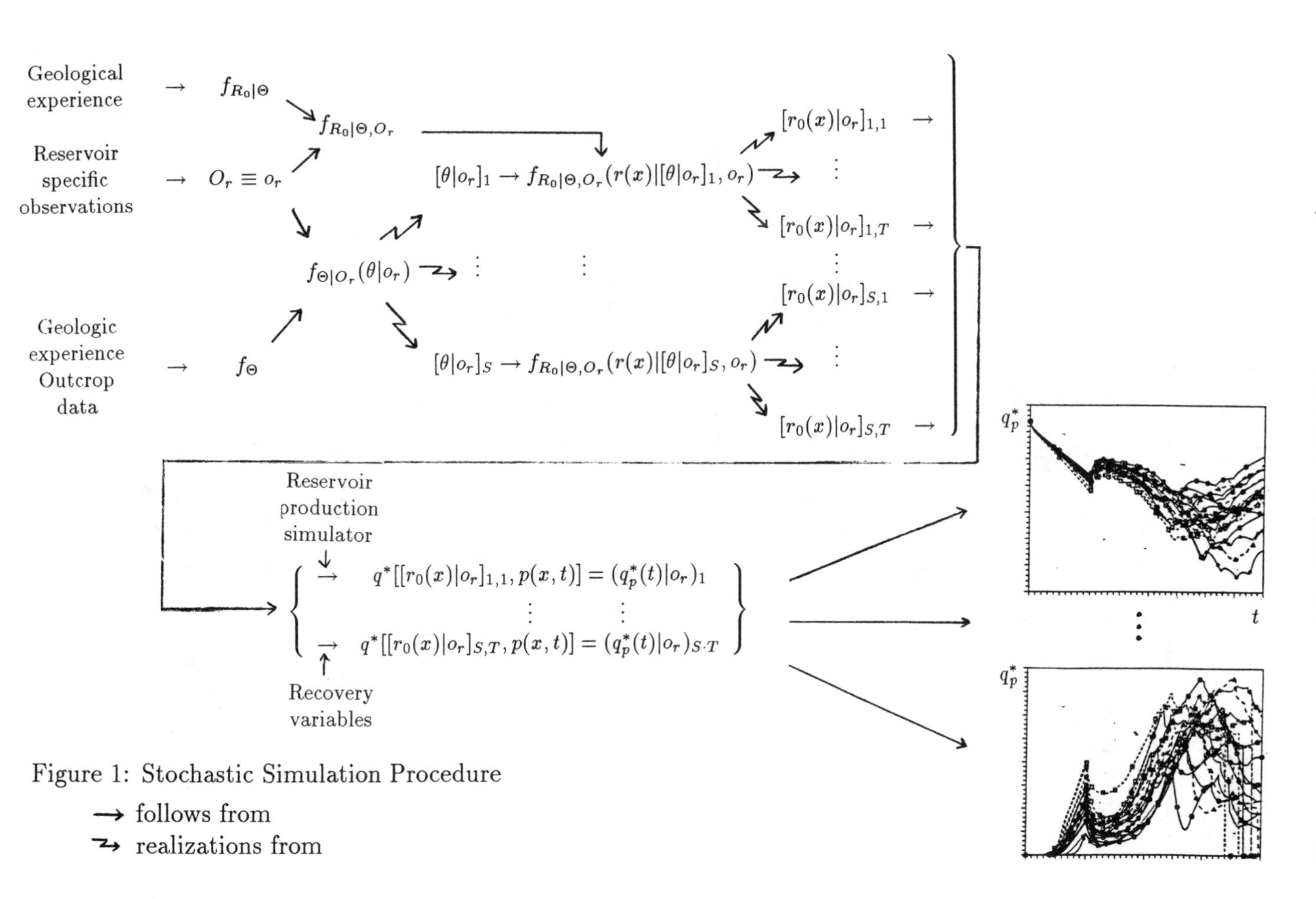

Figure 1: Stochastic Simulation Procedure

→ follows from

⇝ realizations from

The number of realizations required, $S \cdot T$, will vary according to the characteristic to be evaluated, the properties of the reservoir and the recovery strategy. The fact that most production characteristics are some average property over the reservoir help reduce the number and by applying importance sampling, see Ripley (1987), the number can be reduced even more.

From the set of realizations, $(q_p^*(t)|o_r)_s; s = 1, 2, \ldots, S \cdot T$, the best predictor for the production characteristics according to a specified criterion can be obtained. As an example consider the minimum variance predictor at an arbitrary time t'

$$\hat{\mathrm{E}}\{Q_p^*(t')|o_r\} = \frac{1}{S \cdot T} \sum_{s=1}^{S \cdot T} (q_p^*(t')|o_r)_s \tag{10}$$

with corresponding variance estimate:

$$\hat{\mathrm{Var}}\{Q_p^*(t')|o_r\} = \frac{1}{S \cdot T} \sum_{s=1}^{S \cdot T} \left[(q_p^*(t')|o_r)_s - \hat{\mathrm{E}}\{Q_p^*(t')|o_r\}\right]^2 \tag{11}$$

Note, however, that the true production characteristics is defined by:

$$Q_p(t) = Q_p^*(t)|o_r + \Delta_*(t)|o_r \tag{12}$$

A reliable predictor for the true production characteristics at t' is:

$$\mathrm{E}\{Q_p(t')\} \approx \hat{\mathrm{E}}\{Q_p^*(t')|o_r\} \tag{13}$$

since the residual term is assumed to be approximately centered. The corresponding variance estimate is:

$$\mathrm{Var}\{Q_p(t')\} \approx \mathrm{Var}\{Q_p^*(t')|o_r\} + \mathrm{Var}\{\Delta_*(t')|o_r\} \approx \hat{\mathrm{Var}}\{Q_p^*(t')|o_r\} \tag{14}$$

since the variance in the residual is assumed to be neglectible. Several other characteristics can also be predicted along these lines.

2.3 Parameter Sensitivity

The uncertainty in assessing the value of the model parameters is included in the total uncertainty. Frequently, one wishes to evaluate the sensitivity to certain parameters of the model through a user controlled test plan. Evaluation of parameter sensitivity can be formally expressed by splitting the stochastic parameter vector into two components,

$$\Theta = (\Theta_U, \Theta_S) \tag{15}$$

with Θ_U being the parameters to be randomized and Θ_S being the parameters on which sensitivity shall be evaluated.

The test plan for sensitivity evaluation can be defined by using experimental design techniques. Consider a testplan requiring for example N choices of Θ_S , i.e. $\Theta_S^i = \theta_S^i; i = 1, 2, \ldots, N$. Methodology for efficient choosing such designs is discussed in Damsleth et al (1991). The variable of interest is then the predicted production characteristics conditioned to the design values of θ_S and the available observations:

$$Q_p^*(t)|\theta_S, o_r \sim f_{Q_p^*|\Theta_S,O_r}(q(t)|\theta_S, o_r) \tag{16}$$

$$= \int_{\mathcal{D}} f_{Q_p^*|\Theta,O_r}(q(t)|\theta, o_r) \cdot f_{\Theta_U|\Theta_S,O_r}(\theta_U|\theta_S, o_r) \mathrm{d}\theta_U$$

The notation corresponds to the one used in expression (9). Frequently it is reasonable to assume independence between Θ_U and Θ_S. This will simplify the expression considerably. Note that in order to compute the total uncertainty, see expression (9), one has to assign a probability structure to the sensitivity parameters Θ_S.

The sensitivity evaluation should be based on estimates of

$$\mathrm{E}\{Q_p^*(t)|\theta_S^i, o_r\}; i = 1, 2, \ldots, N \tag{17}$$

$$\mathrm{Var}\{Q_p^*(t)|\theta_S^i, o_r\}; i = 1, 2, \ldots, N \tag{18}$$

which can be obtained by using a stochastic simulation approach corresponding to the one in Figure 1.

2.4 Partly Stochastic Approach

This approach to evaluation of uncertainty of production characteristics also requires a stochastic model. Usually the uncertainty in assessing the value of the model parameters is not included. The parameters are considered as constants without uncertainty and values are assigned based on experience and reservoir specific observations in an informal manner. Denote the parameter values obtained from this by $\hat{\theta}$.

In the formalism previously defined the partly stochastic approach would entail that the variable being studied is:

$$Q_p^*(t)|\hat{\theta}, o_r \sim f_{Q_p^*|\Theta,O_r}(q(t)|\hat{\theta}, o_r) \tag{19}$$

with $f_{Q_p^*|\Theta,O_r}(\cdot|\cdot,\cdot)$ being the corresponding pdf. From this it is obvious that the parameter uncertainty is ignored, since the production characteristics is conditioned on $\hat{\theta}$ being the correct values for the model parameters. One will expect this approach to underestimate the true uncertainty in the production characteristics. The partly stochastic approach is advocated in papers up to now, see Hewett and Behrens (1988), Journel (1990), Damsleth et al. (1990) and Høiberg et al. (1990).

3 Case Study - North Sea Sandstone Reservoir

The case study is based on the evaluation of a reservoir in the Norwegian sector of the North Sea. The values of the variables and parameters are changed to make them non-recognizable, but not beyond representativity. Parts of the same evaluation is reported in Høiberg et al. (1990 a and b). An outline of the stochastic reservoir description model is given in this article, while a more through presentation can be found in the references.

The reservoir consists of a complex sequence of sand with excellent quality and interbedded shales. It contains oil over an extensive water aquifer. At the stage of evaluation high-quality seismic data and observations in 17 wells are available. No production has been made, and it is the subject for prediction. The depletion will be performed by partly natural water drive and partly by gas injection.

The evaluation of uncertainty will be performed along the lines described in the previous section. Note, however, in that section a procedure for quantifying the total uncertainty with respect to the reservoir characteristics is defined. In this case study only contributions from the sedimentary characteristics are included. An extension to also include geometrical uncertainty can easily be done by applying the stochastic model for seismic depth conversion presented in Abrahamsen et al. (1991).

3.1 Stochastic Model for Production Characteristics

The formalism established in the previous section is used. The objective is to define the various components in the expression for the prediction of the reservoir characteristics, $Q_p^*(t)|o_r$, see expression (9). This also make sensitivity analysis and the partly stochastic approach of uncertainty easily available.

3.1.1 Production Characteristics - $q_p(t)$

The following measures for production are being evaluated:

- oil production rate - reservoir total and per well
- production GOR - reservoir total and per well
- water cuts - reservoir total and per well
- reservoir pressure - reservoir average

3.1.2 Reservoir Production Simulator - $q^*[\cdot,\cdot]$

The predictions of the production characteristics are based on a production simulator containing two components:

- support change or homogenization, applying arithmetic average for porosity and saturations, Haldorsen and Lake (1984) turtuosity procedure for absolute permeability and proportionality to flow area for transmissibility.
- flow simulator, using black oil version of ECLIPSE (1988).

3.1.3 Initial Reservoir Variables - $R_0(x)|\theta$

The model for the initial reservoir variables is stochastic, although with several deterministic components. The model parameters are termed θ. The initial reservoir variables are of three types: geometrical, sedimentary and fluid. The stochasticity is introduced through the sedimentary model in this study.

The reservoir characteristics can partly be inferred from general geological knowledge from the same region, partly from other reservoirs of similar origin and partly from comparable outcrop data. The main source of information is, however, reservoir specific seismic data and observations in wells.

Reservoir geometrical aspects, like top of reservoir and zone borders, are treated deterministically, and will therefore not contribute to the uncertainty. This choice is done partly out of convenience in order to limit the study, and partly because the geometrical characteristics are relatively well known. An extension to include geometrical uncertainty is considered to be simple but somewhat tedious.

The lateral extent of the reservoir is $(6300 \times 3370)m^2$ and the depth to the top of the reservoir is $2557m$. The reservoir is outlined by a

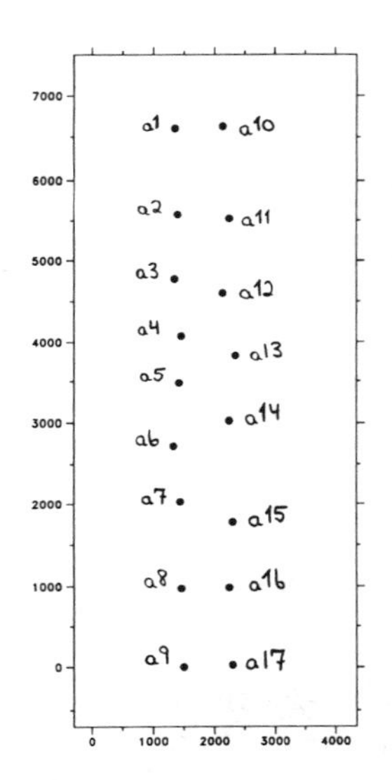

A.Well locations (17)

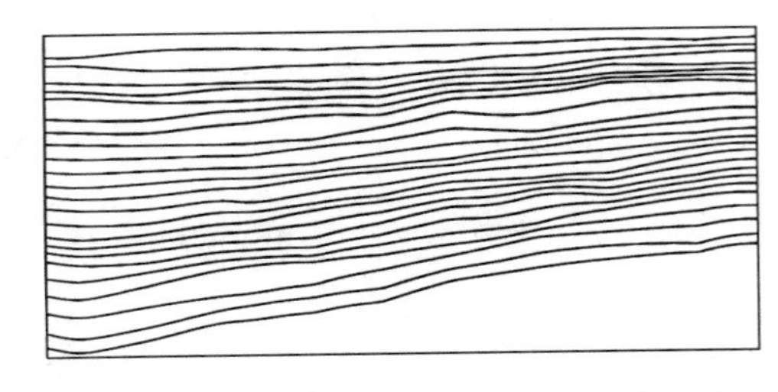

B.Cross section presenting zone borders

Figure 2: Geometric characteristics of reservoir

west dipping structure and a deformation zone to the east. Vertically, the reservoir is divided into 23 zones. Each zone is outlined by what is believed to be non-intersecting time horizons, frequently containing shale units of significant lateral extent. Internally, the zones appear to have fairly similar properties. A cross section indicating the zones is presented in Figure 2.B. The thickness of the reservoir is in the range $150 - 220m$. All these values are determined by geologists and geophysicists working on the field.

Reservoir sedimentological aspects are mostly treated as stochastic since the uncertainty concerning them is large and that is expected to have considerable impact on production uncertainty.

The geological interpretation is that the reservoir sand is reasonably homogeneous with excellent quality. Interbedded in the sand, several units of shale are identified. The shales are of two classes: Firstly, interchannel floodplains of large areal extent containing traces of soil production. These could in many instances be interpreted to correlate between wells being several kilometers apart. Secondly, abandoned channels, barform drapes and mudstone breccia facies of small areal extent, much less than the well spacing.

In the model, the reservoir is considered as a homogeneous sand matrix with constant porosity of 0.23 and isotropic absolute sand permeability in each reservoir zone. In the uppermost zone $2250mD$ is used, $750mD$ in the next 10 zones and $250mD$ in the 12 lowermost reservoir zones. In the matrix; zero-porosity, non-permeable shale units are interbedded. Their location and size being modeled stochastically. The shales are classified into two classes:

- Large-extent shales are being located in the 22 time horizons outlining the reservoir zones. These horizons are termed correlation surfaces. The shale units have considerable thickness which vary within each unit.

  The stochastic model for these shales is defined in $2D$ over each of the correlation surfaces and is based on Markov random field theory, see Besag (1974) and Geman and Geman (1984). Each correlation surface is divided into 67×118 pixels denoted $\mathcal{D}$. To each pixel $(i,j) \in \mathcal{D}$ there is associated a stochastic variable T_{ij} representing shale thickness taking a value from the set $\{0, 1, \ldots, 45\}$. The value 0 indicates sand. Each class in the set is $0.33m$, hence making the maximum thickness $15.0m$.

  For each of the correlation surfaces the following probability structure is defined:

$$\begin{aligned}
\text{Prob}\{T_{ij} = t | t_{kl}; (k,l) \in \mathcal{D}; (k,l) \neq (i,j); (\partial., \boldsymbol{\beta}, \boldsymbol{\Delta}), \text{cr}, (\mu, \sigma^2)\} \\
= c \cdot \exp\{- \sum_{(i,j)\in\mathcal{D}_{ij}} [\beta_{k-i,l-j} \cdot |t - t_{kl}| \\
+ \Delta_{k-i,l-j} \cdot \delta(t = 0 \neq t_{kl} \vee t \neq 0 = t_{kl})]\} \\
\cdot \exp\{-\tau_1 \cdot |\text{cr} - \hat{\text{cr}}(t_{kl}; (k,l) \in \mathcal{D})|^2\} \\
\cdot \exp\{-\tau_2 \cdot |f_T(t) - \hat{f}(t_{kl}; (k,l) \in \mathcal{D})|^2\}
\end{aligned} \tag{20}$$

  with $[(\partial., \boldsymbol{\beta}, \boldsymbol{\Delta}), \text{cr}, (\mu, \sigma^2)]$ being the parameters of the model. The c is a normalizing constant. The second factor governs the general spatial pattern of shales in the correlation surface. The $(\partial., \boldsymbol{\beta}, \boldsymbol{\Delta})$ are related to window of influence, smoothness of shale thickness and shape of intersection between shale and sand, respectively. The $\delta(\text{A})$ is an indicator taking the value 1 if A is true and 0 otherwise. The two last factors in the expression are constraining the realizations to reproduce the required coverage ratio of sand, cr, and required thickness distribution of shale, $f_T(t)$, respectively. The cr is a constant while $f_T(t)$ is a pdf with mean μ and variance σ^2. The $\hat{\text{cr}}(\cdot)$ is an estimate for the coverage ratio based on the realization of shale thicknesses. The $\hat{f}(\cdot)$ is the empirical pdf of thicknesses based on the same realization. The strength variables τ_1 and τ_2 are defined to be sufficiently large to ensure that the required coverage ratio of sand and required Gaussian distribution of shale thicknesses are reproduced almost exactly.

It can be shown that the expression above, with some weak conditions on the parameters, ensures the existence of the simultaneous probability,

$$\text{Prob}\{T_{ij} = t_{ij}; (i,j) \in \mathcal{D} | (\partial., \boldsymbol{\beta}, \boldsymbol{\Delta}), \text{cr}, (\mu, \sigma^2)\}. \tag{21}$$

The $(\partial., \boldsymbol{\beta}, \boldsymbol{\Delta})$ is assumed to be identical for all correlation surfaces in the reservoir, while cr and (μ, σ^2) will vary from one correlation surface to the other.

- Small-extent shales being located anywhere in the sand matrix. The stochastic model for these is based on marked point process theory, see Stoyan et al.(1987). Each shale unit is represented by a marked point $\mathcal{B} = b = (x, w, l, h)$, with x being the location reference, (w, l) being the lateral width and length and h being the height.

  For the N shale units, the following simultaneous probability structure is defined:

$$\begin{aligned} &\text{Prob}\{\mathcal{B}_i = b_i; i = 1, 2, \ldots, N | \alpha(\cdot, \cdot|\cdot), \mu_h, \sigma_h^2, \text{fq}\} \\ &= c \cdot \exp\left\{-\sum_{i=1}^{N}\left[\alpha(w_i, l_i|h_i) + \frac{1}{2\sigma_h^2}(h_i - \mu_h)^2\right]\right\} \\ &\quad \cdot \exp\left\{-\tau_1 |\text{fq} - \hat{\text{fq}}(b_i; i = 1, 2, \ldots, N)|^2\right\} \end{aligned} \tag{22}$$

  with $(\alpha(\cdot, \cdot|\cdot), \mu_h, \sigma_h^2, \text{fq})$ being the parameters of the model. The c is a normalizing constant. The second factor governs the relations between the marks. This can be seen by:

$$\begin{aligned} &\exp\left\{-\sum_{i=1}^{N}\left[\alpha(w_i, l_i|h_i) + \frac{1}{2\sigma_h^2}(h_i - \mu_h)^2\right]\right\} \\ &= \prod_{i=1}^{N} \exp\{-\alpha(w_i, l_i|h_i)\} \times \exp\{-\frac{1}{2\sigma_h^2}(h_i - \mu_h)^2\} \end{aligned} \tag{23}$$

$$\sim \prod_{i=1}^{N} f_{W,L|H}(w_i, l_i|h_i) \cdot f_H(h_i)$$

$$= \prod_{i=1}^{N} f_{WLH}(w_i, l_i, h_i)$$

It is clear from this that no dependence between shale units appear from this term. Moreover, the pdf for shale height is Gaussian with parameters (μ_h, σ_h^2). The $\alpha(\cdot, \cdot|\cdot)$ is related to the conditional pdf for width and length given the height. The last factor in the simultaneous probability expression is constraining the realizations to appear with the required frequency of shale units, fq. The fq is the frequency of shales along vertical transects penetrating the reservoir. The $\hat{\text{fq}}$ is an estimate of this frequency based on the realization of shale units. The strength variable τ_1 is defined to be sufficiently large to ensure that the frequency of shales is reproduced almost exactly.

Reservoir fluid properties are considered to be relatively well defined and are therefore treated deterministically.

The reservoir contains undersaturated oil. The initial OWC is at subsurface $2855m$. The initial pressure is 404.4 Bara at reference depth $2701m$. The initial GOR is $155m^3/m^3$, and the initial water saturation is about 0.20 on average above OWC. The other relevant fluid properties and relative permeability curves used are shown in Figure 3. Capillary pressure is assumed to be zero. They are implemented in the fluid flow simulator according to normal use in reservoir engineering. To discuss this in larger detail would be beyond the scope of this paper.

Several of the fluid variables will of course have influence on the production characteristics of the reservoir. Their effect is expected to be fairly independent of the effect caused by the sedimentary variables, however. Hence they can be studied separately.

This defines the probability structure for the initial reservoir variables, dependent on the following set of parameters:

$$\theta = [(\partial., \boldsymbol{\beta}, \boldsymbol{\Delta}), (\text{cr}_i, (\mu_i, \sigma_i^2); i = 1, 2, \ldots, 22), \alpha(\cdot, \cdot|\cdot), (\mu_h, \sigma_h^2), \text{fq}] \quad (24)$$

If other characteristics of the initial conditions of the reservoir are considered to have significant impact on production uncertainty, it is possible to include them as stochastic elements of the model.

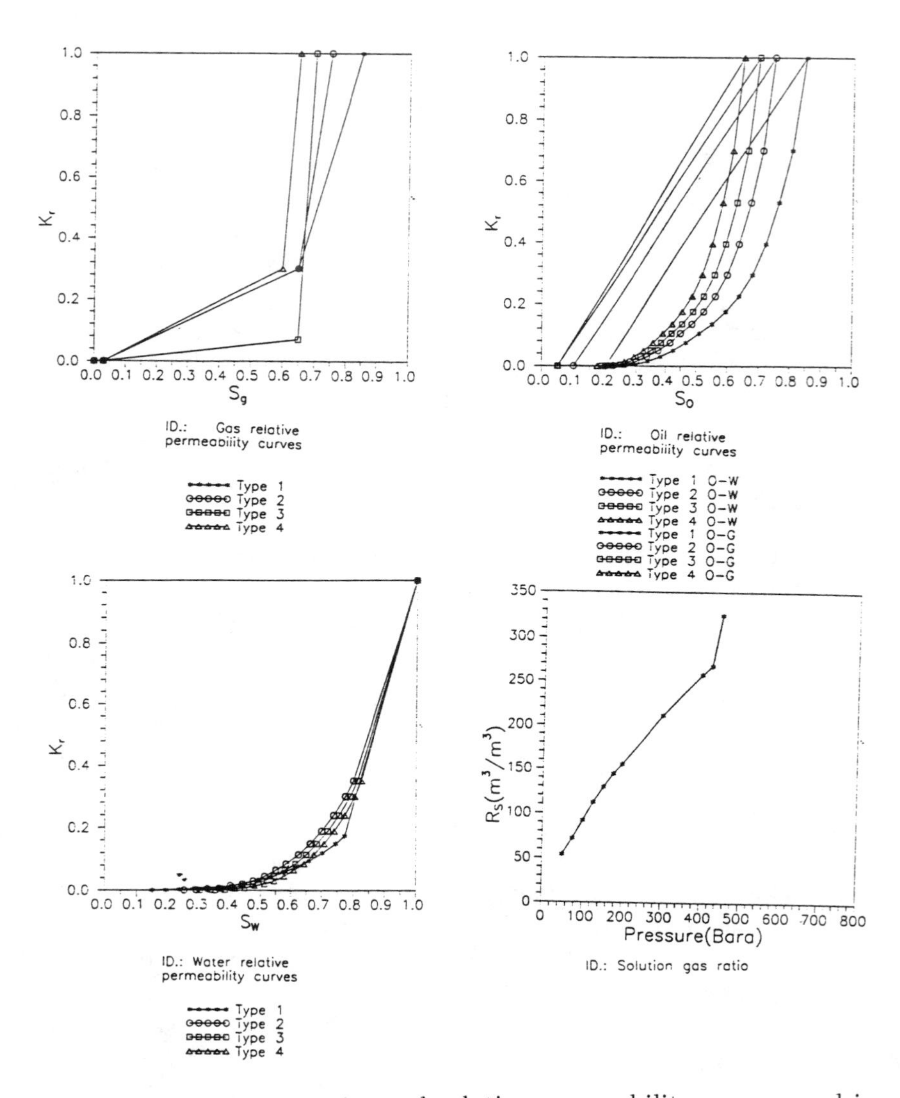

Figure 3: Fluid properties and relative permeability curves used in the simulations.

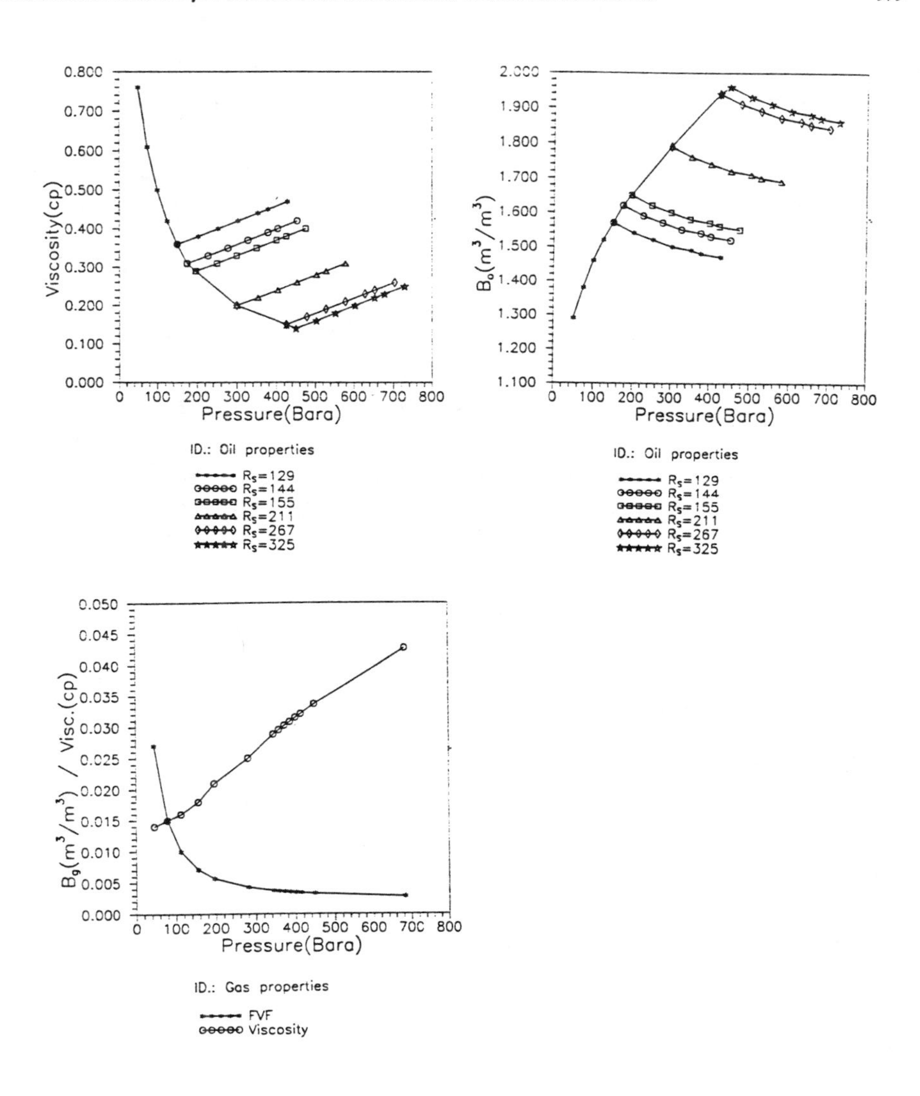

Figure 3 continued.

3.1.4 Prior Probabilities on Model Parameters - $f_\Theta(\theta)$

The Bayesian approach used here provides the possibility to assign prior probability distributions to the model parameters. These priors shall be based on general information and expectations, and not on the reservoir specific observations used in later conditioning. The experienced geologists play an important role in assigning these priors.

The prior qualified guesses on pdf's for θ, $f_\Theta(\theta)$, are specified in Figure 4. The priors on the parameters of the spatial shale pattern in the correlation surfaces, $(\partial., \boldsymbol{\beta}, \boldsymbol{\Delta})$, are complicated to obtain. The exact relation between parameter values and spatial shale pattern is not simple to understand. The priors are obtained by generating realizations based on models with a large variety of parameter values, and then let the geologist assign probabilities for each of the realizations representing the true one. The procedure resulted in assigning non-zero probability to four sets of parameter values, see Figure 4A. Simulations based on parameter set 1 contain small scattered shale units, while simulations on set 4 contain large homogeneous units and sets 2 and 3 provide shales of intermediate size. The coverage ratio of sand in the correlation surfaces, $\mathrm{cr}_i; i = 1, 2, \ldots, 22$, and the mean and variance in the thickness distribution for large-extent shales, $(\mu_i, \sigma_i^2); i = 1, 2, \ldots, 22$, are assigned identical prior distributions for all correlation surfaces. The prior on cr is difficult to determine from experience and a uniform distribution between zero and one, denoted a diffuse prior, is chosen. The prior distribution on (μ, σ^2) is obtained by inspecting all thickness observations of large-extent shales and by literature studies, see Figure 4B. The (width, length, height) distributions of small-extent shale are extensively studied in comparable geological environments, and data from these studies are used. The prior distribution on (width, length) given height is represented by $\alpha(\cdot, \cdot|\cdot)$ and can only be graphically represented, see Figure 4C. The prior distributions on the parameters in the Gaussian distribution representing the height of small-extent shales, (μ_h, σ_h^2) and the prior distribution for the frequency of small shales along transects, fq, are obtained from these studies as well, see Figure 4C and Figure 4D, respectively. The class of pdf's used as priors appears as a trade off between what is expected to be most representative and mathematical convenience.

At this stage all model parameters have been assigned a prior probability structure, and the stochastic model is operable. The reservoir specific observations in the 17 wells have not been used so far, however. All inference is based on fenomenological information available through geological experience and information collected in comparable geological

	Prob	Dir.	$\boldsymbol{\beta}$	$\boldsymbol{\Delta}$	$\partial.$
1	0.1	E-W	0.00	0.02	3×3
		N-S	0.00	0.02	
		NE-SW	0.00	0.02	
		SE-NW	0.00	0.02	
2	0.2	E-W	0.12	0.12	3×3
		N-S	0.12	0.12	
		NE-SW	0.12	0.12	
		SE-NW	0.12	0.12	
3	0.5	E-W	0.25	0.25	3×3
		N-S	0.50	0.50	
		NE-SW	0.25	0.25	
		SE-NW	0.25	0.25	
4	0.2	E-W	0.50	1.00	9×9
		N-S	5.00	4.00	
		NE-SW	0.50	1.00	
		SE-NW	0.50	1.00	

A. Prior probability structure for $(\partial., \boldsymbol{\beta}, \boldsymbol{\Delta})$

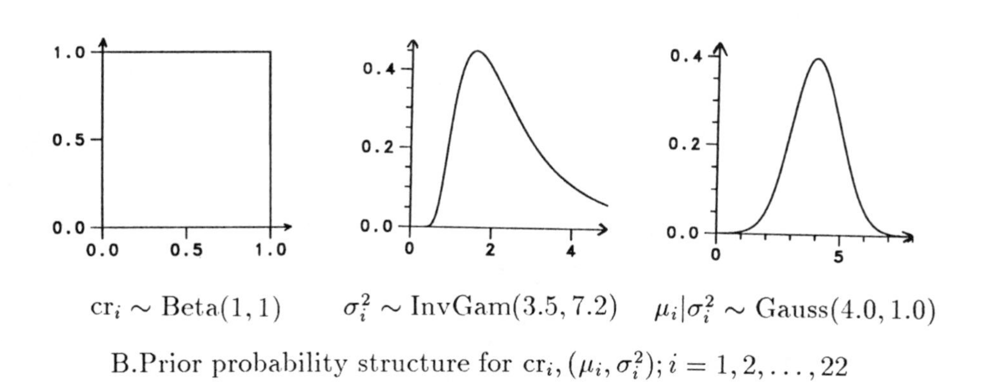

$\mathrm{cr}_i \sim \mathrm{Beta}(1,1)$ $\sigma_i^2 \sim \mathrm{InvGam}(3.5, 7.2)$ $\mu_i|\sigma_i^2 \sim \mathrm{Gauss}(4.0, 1.0)$

B.Prior probability structure for $\mathrm{cr}_i, (\mu_i, \sigma_i^2); i = 1, 2, \ldots, 22$

Figure 4: Prior probabilities structure on model parameters

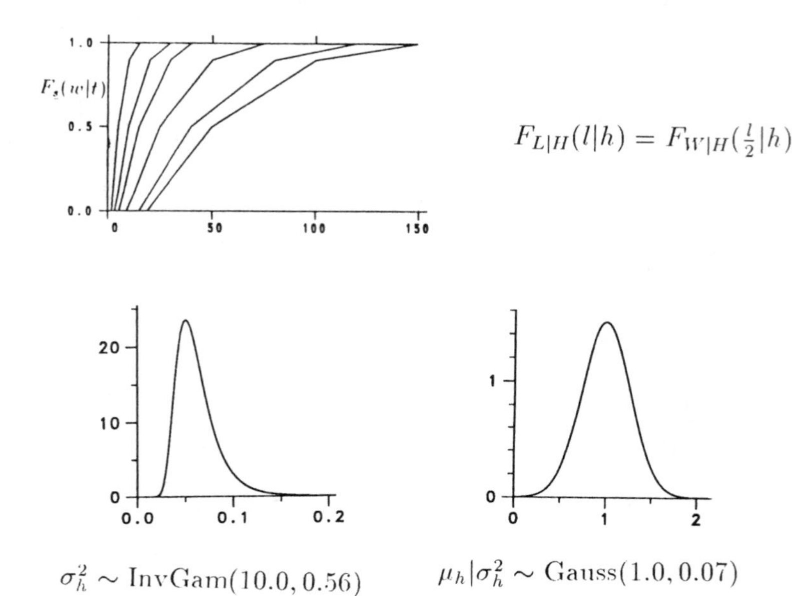

$$F_{L|H}(l|h) = F_{W|H}(\tfrac{l}{2}|h)$$

$\sigma_h^2 \sim \text{InvGam}(10.0, 0.56)$ $\qquad$ $\mu_h|\sigma_h^2 \sim \text{Gauss}(1.0, 0.07)$

C. Prior probability structure for $\alpha(\cdot,\cdot|\cdot)$ through $F_{W|H}(\cdot|\cdot)$ and $F_{L|H}(\cdot|\cdot)$, and μ_h and σ_h^2

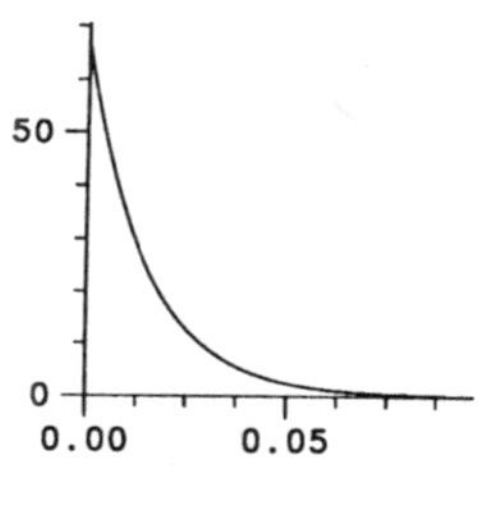

$\text{fq} \sim \text{Gam}(1.0, 66.7)$

D. Prior probability structure for fq

Figure 4 continued

environments.

3.1.5 Conditioning to Reservoir Specific Observations - $R_0(x)|\theta, o_r$ and $\Theta|o_r$

The reservoir specific observations, o_r, consists of the sand/shale sequence in the 17 available wells. The well locations are presented in Figure 2A, while the observations are displayed in Figure 5. The shale units are classified into large-extent and small-extent shales.

The stochastic model for initial reservoir characteristics, $f_{R_0|\Theta}(\cdot|\cdot)$, must be conditioned on o_r to ensure that all realizations generated from it reproduces the reservoir specific observations. The conditional stochastic model, $f_{R_0|\Theta,O_r}(\cdot|\cdot,\cdot)$, is defined by,

- large-extent shales:

$$\begin{aligned} &\text{Prob}\{T_{ij}; (i,j) \in \mathcal{D}|(\partial., \boldsymbol{\beta}, \boldsymbol{\Delta}), \text{cr}, (\mu, \sigma^2), o_r\} \\ &= c \cdot \text{Prob}\{T_{ij} = t_{ij}; (i,j) \in \mathcal{D}|(\partial., \boldsymbol{\beta}, \boldsymbol{\Delta}), \text{cr}, (\mu, \sigma^2)\} \\ &\quad \cdot \exp\{-\tau_0|o_r - \hat{o}_r(t_{ij}; (i,j) \in \mathcal{D})|^2\} \end{aligned} \tag{25}$$

- small-extent shales:

$$\begin{aligned} &\text{Prob}\{\mathcal{B}_i = b_i; i = 1,2,\ldots,N|\alpha(\cdot,\cdot|\cdot), (\mu_h, \sigma_h^2), \text{fq}, o_r\} \\ &= c \cdot \text{Prob}\{\mathcal{B}_i = b_i; i = 1,2,\ldots,N|\alpha(\cdot,\cdot|\cdot), (\mu_h, \sigma_h^2), \text{fq}\} \\ &\quad \cdot \exp\{-\tau_0|o_r - \hat{o}_r(b_i; i = 1,2,\ldots,N)|^2\} \end{aligned} \tag{26}$$

The first term in each of the expressions corresponds to the ones defined in expression (20) and (22) respectively. The second terms include $\hat{o}_r(t_{ij}; (i,j) \in \mathcal{D})$ and $\hat{o}_r(b_i; i = 1,2,\ldots,N)$ being estimates of the reservoir specific observations based on the realizations of large-extent and small-extent shales, respectively. The strength variable, τ_0, is defined to be sufficiently large to ensure that o_r is reproduced almost exactly.

The prior probability distributions for the model parameters, $f_\Theta(\cdot)$, must be conditioned on the reservoir specific observations to adapt the model as well as possible to the reservoir under study. The conditioning is performed by using Bayes rule as specified in the previous section. The basic assumptions and results are presented in Figure 6. The results are obtained under the assumption that the observations in each well are independent and that the following sub-sets of parameters, $(\partial., \boldsymbol{\beta}, \boldsymbol{\Delta})$,

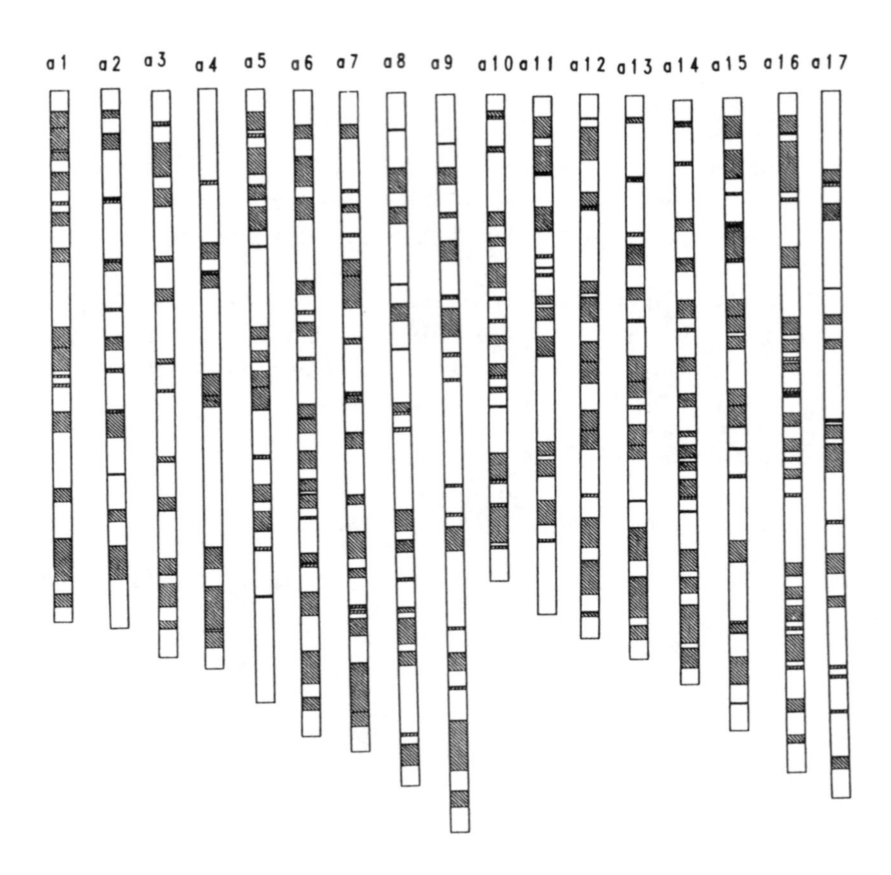

Figure 5: Reservoir specific observations

sand large extent shales small extent shales

Param type	Param θ	Param influence	Prior f_Θ	Observation variable O_r	Likelihood $f_{O_r\|\theta}$	Observ. $O_r = o_r$	Posterior $f_{\Theta\|O_r}$
E	$\partial.$ β Δ	Spatial pattern	Fig.3.A	Non-informative	—	—	Equal f_Θ
EE	cr_i	Coverage ratio	Beta(1,1)	n_i-no. of wells without shale	Bin(17,cr_i)	see sub-table	see sub-table
EE	σ_i^2 $\mu_i\|\sigma_i^2$	Gauss-par. in shale thickness distr.	InvGam(3.5,7.2) Gauss(4.0,σ_i^2)	n_i-no. of thick. obs. $\hat{\mu}_i$-average obs-thickness $\hat{\sigma}_i^2$-emp.var. obs-thickness	Gauss($\mu_i, \frac{\sigma_i^2}{n_i}$) $\chi_n^2 \cdot \sigma_i^2$	see sub-table	see sub-table
E	$\alpha(\cdot,\cdot\|\cdot)$	Prob.distr. (width,length) given height	Fig.3.C	Non-informative	—	—	Equal f_Θ
EE	σ_h^2 $\mu_h\|\sigma_h^2$	Gauss.par. in shale height distr.	InvGam(10,0.56) Gauss(1.0,σ_h^2)	n-no of height obs. $\hat{\mu}_h$-average obs. height $\hat{\sigma}_h^2$-emp.var. obs.height	Gauss($\mu_h, \frac{\sigma_h^2}{n}$) $\chi_h^2 \cdot \sigma_h^2$	$n = 89$ $\hat{\mu}_h = 0.75$ $\hat{\sigma}_h^2 = 0.27^2$	InvGam(54.5,3.83) Gauss(0.75,$\frac{\sigma_h^2}{n+1}$)
EE	fq	Frequency of shales	Gam(1,66.7)	l-length of well trace n-no. of shales	Po(fq $\cdot\, l$)	$l = 1580$ $n = 89$	Gam(90,1646.7)

Figure 6: Prior and Posterior probability distributions for model parameters

θ	$E = e_0$	$f_{\Theta\mid E}$
cr_1	$n_1 = 4$	Beta(5, 13)
σ_1^2	$n_1 = 13$	InvGam(10.0, 18.17)
$\mu_1\mid\sigma_1^2$	$\hat{\mu}_1 = 1.46$ $\hat{\sigma}_1^2 = 1.09^2$	Gauss(1.64, $\frac{\sigma_1^2}{14}$)
cr_2	$n_1 = 7$	Beta(8, 10)
σ_2^2	$n_1 = 10$	InvGam(8.5, 22.57)
$\mu_1\mid\sigma_2^2$	$\hat{\mu}_2 = 5.17$ $\hat{\sigma}_2^2 = 1.71^2$	Gauss(5.06, $\frac{\sigma_2^2}{11}$)
cr_3	$n_1 = 6$	Beta(7, 11)
σ_3^2	$n_1 = 11$	InvGam(9.0, 24.40)
$\mu_1\mid\sigma_3^2$	$\hat{\mu}_3 = 2.87$ $\hat{\sigma}_3^2 = 1.73^2$	Gauss(2.97, $\frac{\sigma_3^2}{12}$)
cr_4	$n_1 = 14$	Beta(15, 3)
σ_4^2	$n_1 = 3$	InvGam(5.0, 8.68)
$\mu_1\mid\sigma_4^2$	$\hat{\mu}_4 = 4.44$ $\hat{\sigma}_4^2 = 0.96^2$	Gauss(4.33, $\frac{\sigma_4^2}{4}$)
cr_5	$n_1 = 14$	Beta(15, 3)
σ_5^2	$n_1 = 3$	InvGam(5.0, 8.25)
$\mu_1\mid\sigma_5^2$	$\hat{\mu}_5 = 3.67$ $\hat{\sigma}_5^2 = 0.82^2$	Gauss(3.75, $\frac{\sigma_5^2}{4}$)
cr_6	$n_1 = 12$	Beta(13, 5)
σ_6^2	$n_1 = 5$	InvGam(6.0, 15.50)
$\mu_1\mid\sigma_6^2$	$\hat{\mu}_6 = 1.47$ $\hat{\sigma}_6^2 = 1.43^2$	Gauss(1.89, $\frac{\sigma_6^2}{6}$)
cr_7	$n_1 = 2$	Beta(3, 15)
σ_7^2	$n_1 = 13$	InvGam(10.0, 18.17)
$\mu_1\mid\sigma_7^2$	$\hat{\mu}_7 = 1.46$ $\hat{\sigma}_7^2 = 1.09^2$	Gauss(1.64, $\frac{\sigma_7^2}{14}$)
cr_8	$n_1 = 3$	Beta(4, 14)
σ_8^2	$n_1 = 14$	InvGam(10.5, 16.88)
$\mu_1\mid\sigma_8^2$	$\hat{\mu}_8 = 2.48$ $\hat{\sigma}_8^2 = 1.10^2$	Gauss(2.58, $\frac{\sigma_8^2}{15}$)
cr_9	$n_1 = 15$	Beta(16, 2)
σ_9^2	$n_1 = 2$	InvGam(4.5, 8.91)
$\mu_1\mid\sigma_9^2$	$\hat{\mu}_9 = 5.83$ $\hat{\sigma}_9^2 = 0.17^2$	Gauss(5.22, $\frac{\sigma_9^2}{3}$)
cr_{10}	$n_1 = 13$	Beta(14, 4)
σ_{10}^2	$n_1 = 4$	InvGam(5.5, 9.55)
$\mu_1\mid\sigma_{10}^2$	$\hat{\mu}_{10} = 2.25$ $\hat{\sigma}_{10}^2 = 0.64^2$	Gauss(2.60, $\frac{\sigma_{10}^2}{5}$)
cr_{11}	$n_1 = 7$	Beta(8, 10)
σ_{11}^2	$n_1 = 10$	InvGam(8.5, 13.37)
$\mu_1\mid\sigma_{11}^2$	$\hat{\mu}_{11} = 4.33$ $\hat{\sigma}_{11}^2 = 1.10^2$	Gauss(4.30, $\frac{\sigma_{11}^2}{11}$)

θ	$E = e_0$	$f_{\Theta\mid E}$
cr_{12}	$n_1 = 2$	Beta(3, 15)
σ_{12}^2	$n_1 = 15$	InvGam(11.0, 34.46)
$\mu_1\mid\sigma_{12}^2$	$\hat{\mu}_{12} = 3.98$ $\hat{\sigma}_{12}^2 = 1.91^2$	Gauss(3.98, $\frac{\sigma_{12}^2}{16}$)
cr_{13}	$n_1 = 14$	Beta(15, 3)
σ_{13}^2	$n_1 = 3$	InvGam(5.0, 9.96)
$\mu_1\mid\sigma_{13}^2$	$\hat{\mu}_{13} = 2.61$ $\hat{\sigma}_{13}^2 = 1.09^2$	Gauss(2.96, $\frac{\sigma_{13}^2}{4}$)
cr_{14}	$n_1 = 4$	Beta(5, 13)
σ_{14}^2	$n_1 = 13$	InvGam(10.0, 19.96)
$\mu_1\mid\sigma_{14}^2$	$\hat{\mu}_{14} = 3.59$ $\hat{\sigma}_{14}^2 = 1.39^2$	Gauss(3.62, $\frac{\sigma_{14}^2}{14}$)
cr_{15}	$n_1 = 6$	Beta(7, 11)
σ_{15}^2	$n_1 = 11$	InvGam(9.0, 18.05)
$\mu_1\mid\sigma_{15}^2$	$\hat{\mu}_{15} = 3.12$ $\hat{\sigma}_{15}^2 = 1.38^2$	Gauss(3.20, $\frac{\sigma_{15}^2}{12}$)
cr_{16}	$n_1 = 12$	Beta(13, 5)
σ_{16}^2	$n_1 = 5$	InvGam(6.0, 8.65)
$\mu_1\mid\sigma_{16}^2$	$\hat{\mu}_{16} = 4.67$ $\hat{\sigma}_{16}^2 = 0.70^2$	Gauss(4.55, $\frac{\sigma_{16}^2}{6}$)
cr_{17}	$n_1 = 10$	Beta(11, 7)
σ_{17}^2	$n_1 = 7$	InvGam(7.0, 14.28)
$\mu_1\mid\sigma_{17}^2$	$\hat{\mu}_{17} = 1.04$ $\hat{\sigma}_{17}^2 = 0.88^2$	Gauss(1.41, $\frac{\sigma_{17}^2}{8}$)
cr_{18}	$n_1 = 14$	Beta(15, 3)
σ_{18}^2	$n_1 = 3$	InvGam(5.0, 11.61)
$\mu_1\mid\sigma_{18}^2$	$\hat{\mu}_{18} = 1.40$ $\hat{\sigma}_{18}^2 = 0.83^2$	Gauss(2.05, $\frac{\sigma_{18}^2}{4}$)
cr_{19}	$n_1 = 2$	Beta(3, 15)
σ_{19}^2	$n_1 = 15$	InvGam(11.0, 25.75)
$\mu_1\mid\sigma_{19}^2$	$\hat{\mu}_{19} = 5.44$ $\hat{\sigma}_{19}^2 = 1.53^2$	Gauss(5.35, $\frac{\sigma_{19}^2}{16}$)
cr_{20}	$n_1 = 13$	Beta(14, 4)
σ_{20}^2	$n_1 = 4$	InvGam(5.5, 7.77)
$\mu_1\mid\sigma_{20}^2$	$\hat{\mu}_{20} = 3.58$ $\hat{\sigma}_{20}^2 = 0.49^2$	Gauss(3.66, $\frac{\sigma_{20}^2}{5}$)
cr_{21}	$n_1 = 2$	Beta(3, 15)
σ_{21}^2	$n_1 = 15$	InvGam(11.0, 49.58)
$\mu_1\mid\sigma_{21}^2$	$\hat{\mu}_{21} = 9.16$ $\hat{\sigma}_{21}^2 = 1.97^2$	Gauss(8.83, $\frac{\sigma_{21}^2}{16}$)
cr_{22}	$n_1 = 2$	Beta(3, 15)
σ_{22}^2	$n_1 = 15$	InvGam(11.0, 20.42)
$\mu_1\mid\sigma_{22}^2$	$\hat{\mu}_{22} = 3.65$ $\hat{\sigma}_{22}^2 = 1.32^2$	Gauss(3.67, $\frac{\sigma_{22}^2}{16}$)

Figure 6 continued

cr, (μ, σ^2), $\alpha(\cdot, \cdot|\cdot)$, (μ_h, σ_h^2), fq, are independent. This is expected to be approximately correct.

From Figure 6, it can be seen that the reservoir specific observations carry no information concerning the parameters $(\partial., \boldsymbol{\beta}, \boldsymbol{\Delta})$ and $\alpha(\cdot, \cdot|\cdot)$. These parameters are denoted E-parameters since their properties must be inferred from experience only. For the other parameters, the observations will carry information and the posterior pdf, $f_{\Theta|O_r}(\cdot|\cdot)$, will deviate from the prior one, $f_\Theta(\cdot)$. These parameters are denoted EE-parameters since they can be inferred from both experience and estimates from observations. The uncertainty in E-parameters will usually be larger and less reliably specified than for EE-parameters. This encourages use of sensitivity analysis with respect to E-parameters to evaluate their relative impact on the predicted production characteristics. It should be mentioned that the thickness distribution for large extent shales, $f_T(\cdot)$, is assumed to be Gaussian distributed with mean μ_i and variance σ_i^2 when using Bayes rule.

3.1.6 Recovery Variables - $p(x, t)$

The recovery variables define the development plan and depletion strategy. The oil is recovered by using 13 wells, six gas injectors and seven producers, see Figure 7. The injectors are located on the upper slope part and they are perforated from layer 1 to layer 4.

The depletion strategy is as follows:

- start by for each producer, produce oil by natural water drive at constant rate: 2600 or 3000 sm^3/d

- after two years, inject gas at constant rate for each injector: 1.5 or 2.0 $mmsm^3/d$.

- for each producer, reaching bottom hole pressure 80 Bara, switch from rate to pressure controlled mode

- for each producer, if GOR exceeds 1500 sm^3/sm^3 or water cut exceeds 0.9 - terminate the production.

The reason for using two production and two injection rates is to balance water and gas advances over the reservoir.

The grid size in the production simulation is (18 x 27 x 23) with the vertical grid sides coinciding with the correlation surfaces, see Figure 7. The production is simulated over eight years assuming black oil properties.

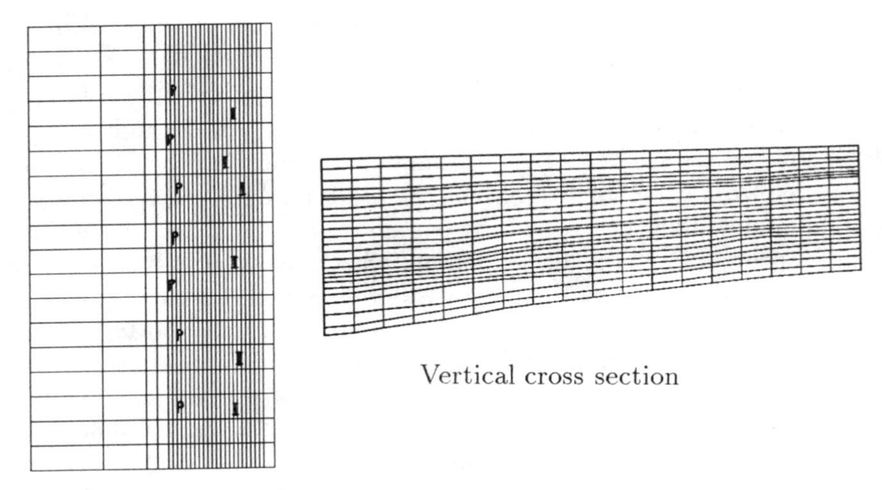

Horizontal cross section

Figure 7: Injection and production wells, and simulation grid
I—injectors (6) P—producers (7)

Note that all the 13 wells used in the recovery are identical to wells used for conditioning in the stochastic simulation of the initial reservoir characteristics, $r_0(x)$, i.e. all wells used in the recovery are already drilled. If wells not used for conditioning in the stochastic simulation, i.e. wells not yet drilled, were used in the recovery, increased uncertainty in the production characteristics is expected.

3.2 Stochastic Simulation of Production Characteristics

The probability structure representing the uncertainty in the predicted production characteristics,

$$Q_p^*(t)|o_r \sim f_{Q_p^*|O_r}(q(t)|o_r) \tag{27}$$

as defined in the previous section, cannot be determined analytically. It can only be assessed by stochastic simulation as outlined in Figure 1.

The specification of the posterior pdf of the model parameter, $f_{\Theta|O_r}(\cdot|\cdot)$, and the conditional pdf of the initial reservoir variables, $f_{R_0|\Theta,O_r}(\cdot|\cdot,\cdot)$, provide the base for the simulation. A set of realizations of model parameters is generated from $f_{\Theta|O_r}(\cdot|\cdot)$. The set, $[\theta|o_r]_s; s = 1, 2, \ldots, 20$, represents the uncertainty in the stochastic model for these parameters. It should be mentioned that the simulation procedure is stratified with respect to the parameters $(\partial., \boldsymbol{\beta}, \boldsymbol{\Delta})$, i.e. one has ensured that they appear exactly according to the probabilities specified in figure 4A. This makes the simulation procedure more efficient. Each realization of parameters is taken as conditioning value in $f_{R_0|\Theta,O_r}(\cdot|\cdot,\cdot)$ and a corresponding realization of the initial reservoir variables, $[r_0(x)|o_r]$, is generated. The conditioning on o_r ensures that the reservoir specific observations are reproduced. The simulation procedure actually used is based on the Metropolis algorithm and the Ripley-Kelly algorithm, and they are presented in more detail in Høiberg et al. (1990 a and b). In this case study, the number of repetitions for each realization of parameters is one, hence $T = 1$. Consequently, the set of realizations of initial reservoir variables, $[r_0(x)|o_r]_s; s = 1, 2, \ldots, 20$ is obtained. This set represents the uncertainty of the stochastic model for the initial reservoir variables. Note in particular that all realizations will reproduce the reservoir specific observations, o_r, due to the conditioning.

In Figure 8 through 11, four realizations of the initial reservoir variables are presented. The most spectacular differences are caused by the parameters for spatial pattern of large-extent shales in the correlation surfaces, $(\partial., \boldsymbol{\beta}, \boldsymbol{\Delta})$. This is not surprising since this is an E-parameter on which the reservoir specific observations has no influence, and hence it will be associated with large uncertainty. The realizations in Figure 8, Figure 9 and 10, and Figure 11 are based on set 1, 3 and 4 of $(\partial., \boldsymbol{\beta}, \boldsymbol{\Delta})$ respectively. The two realizations in Figure 9 and 10 are both from set 3 and are hence exposing the variability due to other sources of uncertainty.

Each of the realizations of the initial reservoir variables, $[r_0(x)|o_r]$, are then taken as input to the reservoir production simulator, $q^*[\cdot|\cdot]$, together with the recovery variables, $p(\cdot,\cdot)$. The first module in $q^*[\cdot,\cdot]$ performs change of support, or homogenization, to the simulation grid size defined for fluid flow. The exact procedure for this is specified in Høiberg et al. (1990 a and b). In Figure 12, the homogenized variables for one realization suitable as input to ECLIPSE, is presented. The second module, the fluid flow simulator ECLIPSE, is then activated for each of the homogenized realizations of the initial reservoir variables. The predictions for oil rate, production GOR, water cut and pressure, represented by $[q_p^*(t)|o_r]_s; s = 1, 2, \ldots, 20$, are obtained.

In Figure 13, the 20 realizations of the oil rate, production GOR, water cut and pressure for the reservoir total over the eight years being

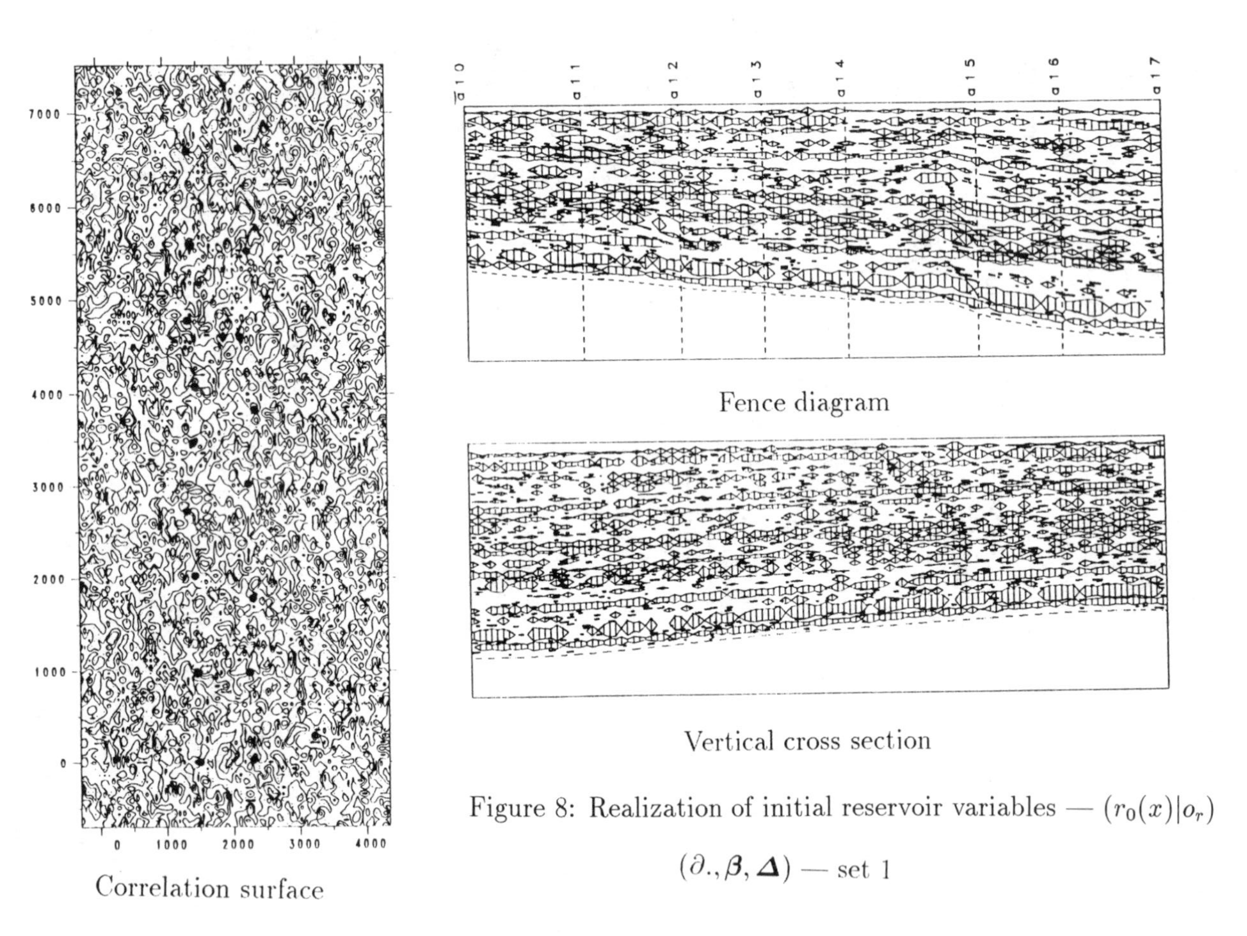

Figure 8: Realization of initial reservoir variables — $(r_0(x)|o_r)$

$(\partial., \boldsymbol{\beta}, \boldsymbol{\Delta})$ — set 1

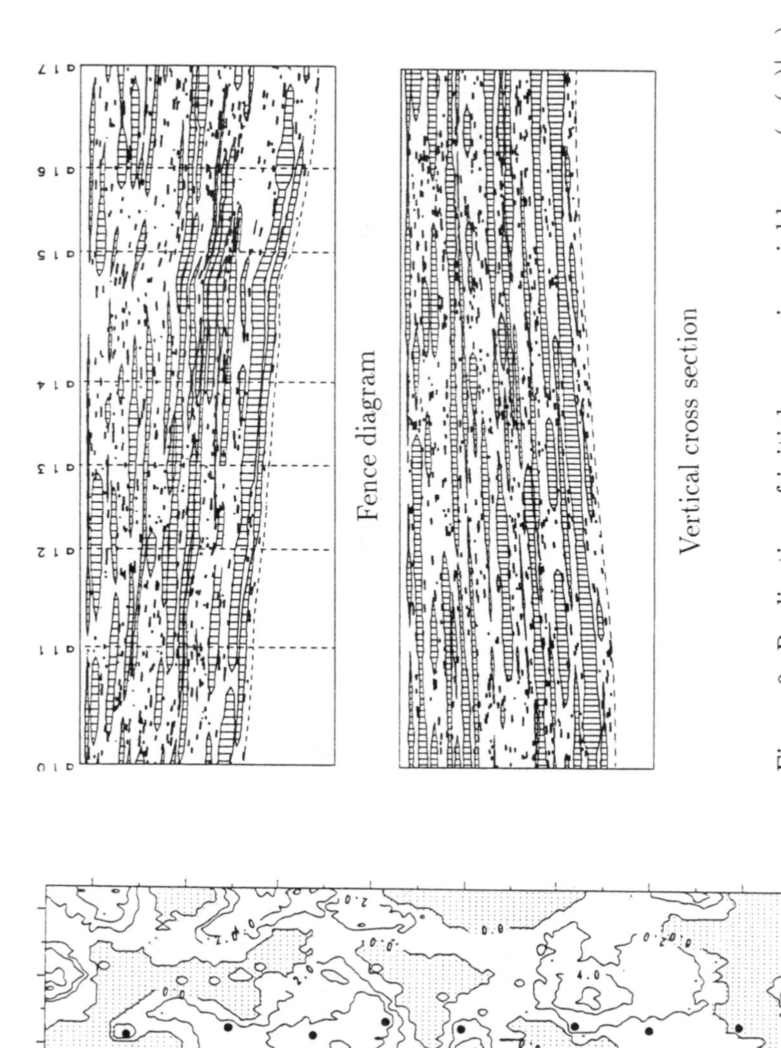

Figure 9: Realization of initial reservoir variables — $(r_0(x)|o_r)$

$(\partial_{\cdot}, \boldsymbol{\beta}, \boldsymbol{\Delta})$ — set 3

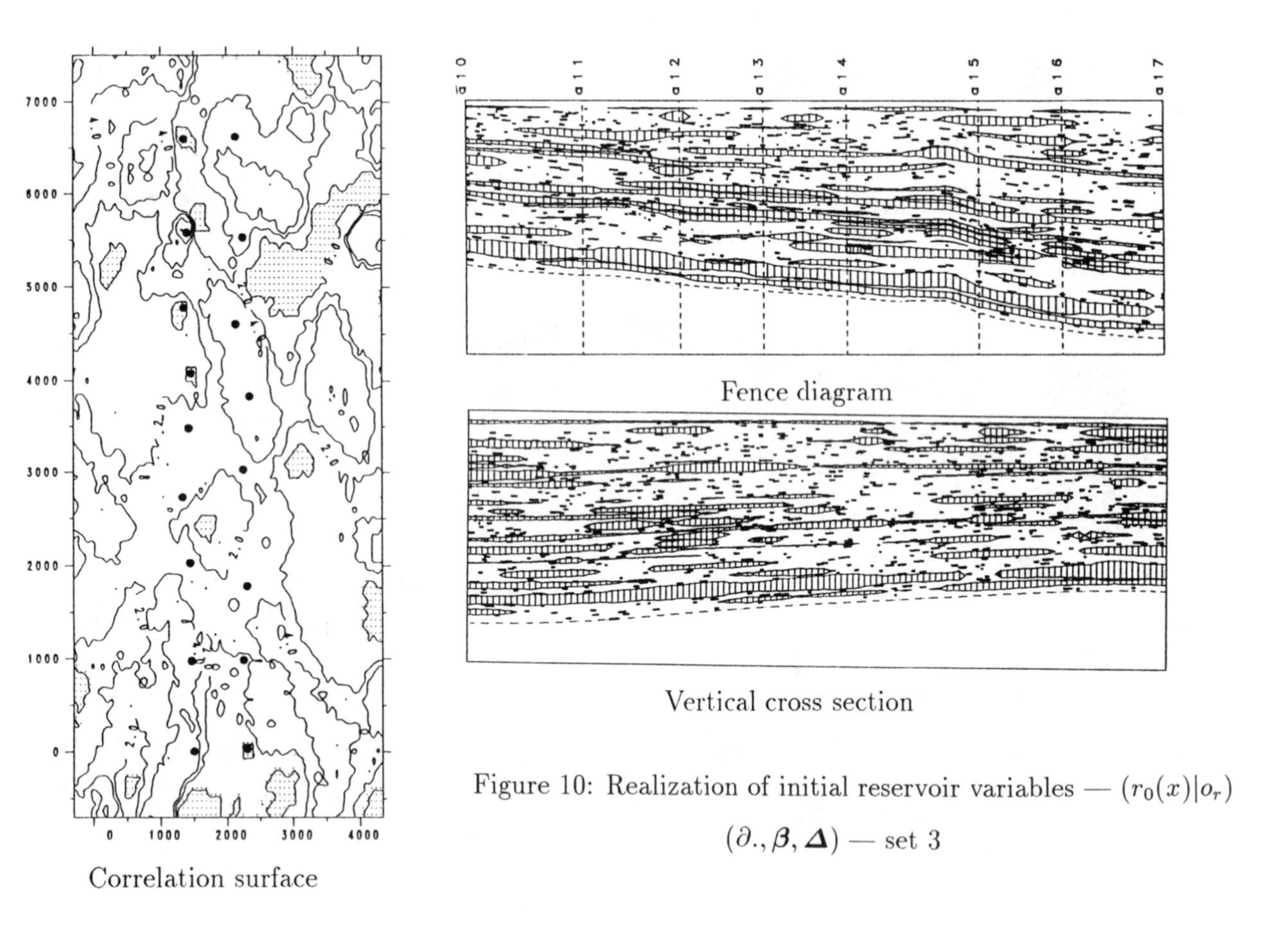

Figure 10: Realization of initial reservoir variables — $(r_0(x)|o_r)$

$(\partial., \boldsymbol{\beta}, \boldsymbol{\Delta})$ — set 3

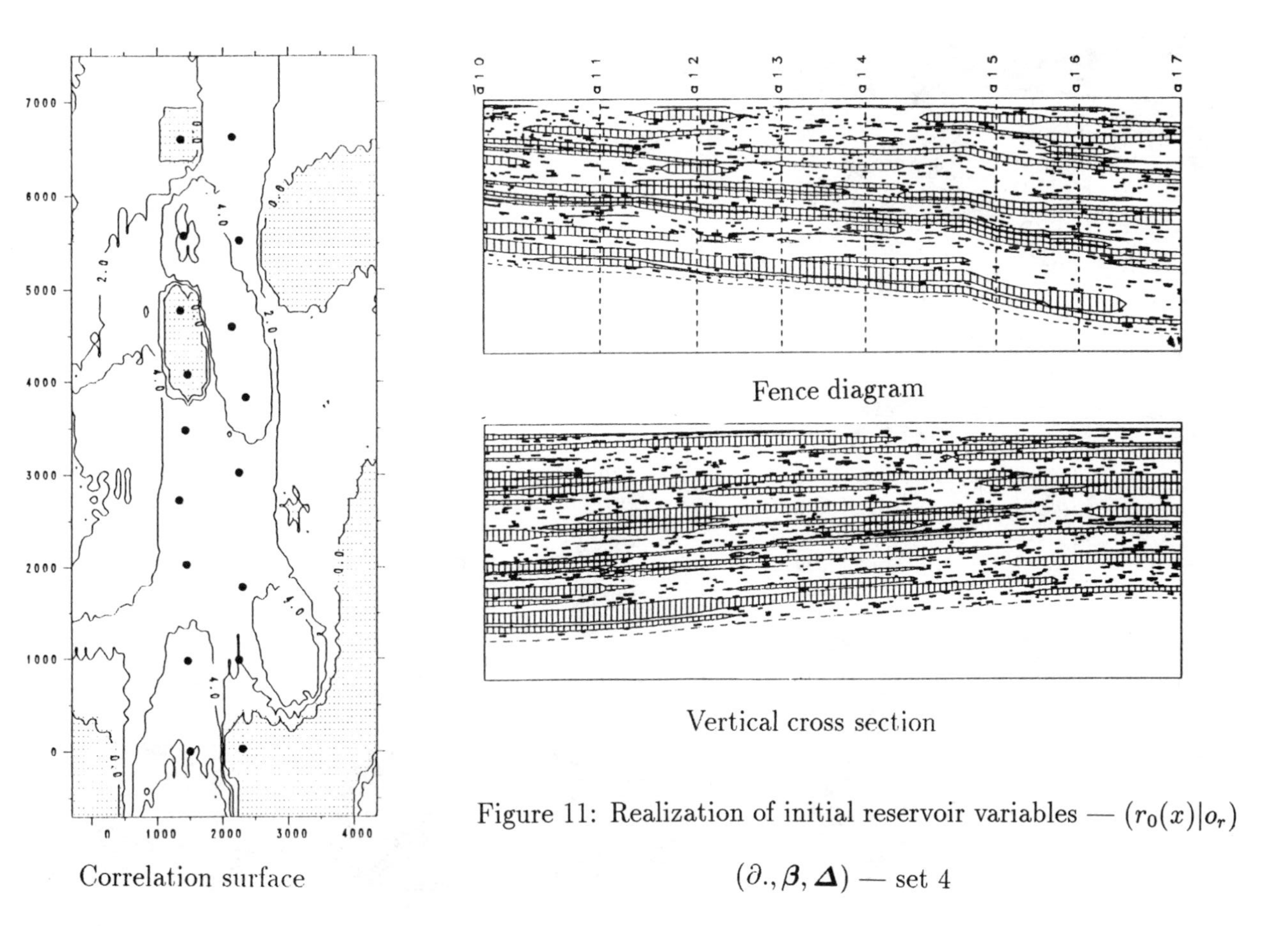

Figure 11: Realization of initial reservoir variables — $(r_0(x)|o_r)$

$(\partial_{\cdot}, \boldsymbol{\beta}, \boldsymbol{\Delta})$ — set 4

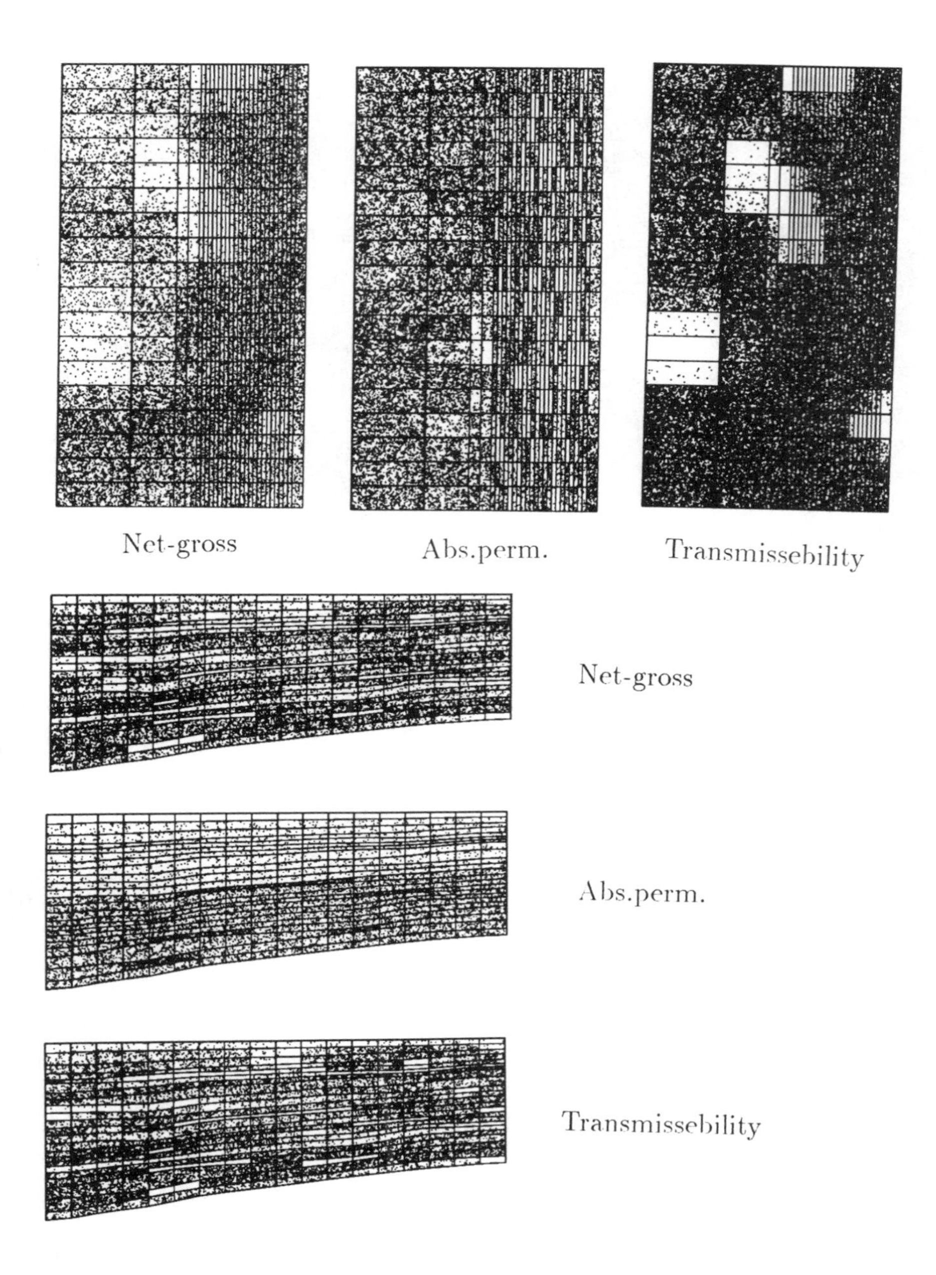

Figure 12: Realization of homogenized input to ECLIPSE

simulated, are displayed. These displays represents $f_{Q_p^*|O_r}(q(t)|o_r)$, and hence the uncertainty in $Q_p^*(t)|o_r$ for the reservoir total. In Figure 14, the realizations of oil rate, production GOR and water cut in one arbitrary production well, are presented. They represent the uncertainty in $Q_p^*(t)|o_r$ for this particular well.

The most spectacular differences in the realizations of the initial reservoir variables are due to $(\partial., \boldsymbol{\beta}, \boldsymbol{\Delta})$, see Figure 8 through 11. The same is expected to be true for the production characteristics, hence it is interesting to evaluate the sensitivity with respect to these parameters. To perform sensitivity analysis, split the parameter set $\theta = (\theta_U, \theta_S)$, with $\theta_U = (\mathrm{cr}_i, (\mu_i, \sigma_i^2); i = 1, 2, \ldots, 22), \alpha(\cdot, \cdot|\cdot), (\mu_h, \sigma_h^2), \mathrm{fq})$ being the parameters to be randomized, and $\theta_S = (\partial., \boldsymbol{\beta}, \boldsymbol{\Delta})$ being the parameters on which sensitivity are evaluated. The test design is based on $\theta_S^i; i = 1, 3, 4$, corresponding to parameter set 1, 3 and 4 in Figure 4A. The predicted production characteristics conditioned on the respective parameter values and reservoir specific observation, $Q_p^*(t)|\theta_S^i, o_r; i = 1, 3, 4$, are then to be evaluated. Realizations of these can be obtained by a procedure as outlined in Figure 1, except for θ_S being kept constant to the corresponding parameter values. The realizations are denoted $[q_p^*(t)|\theta_S^i, o_r]_s; i = 1, 3, 4; s = 1, 2, \ldots, 9$, with the number of realizations being nine.

In Figure 15 through 17, the nine realizations of the oil rate, production GOR, water cut and pressure for reservoir total for each of the parameter sets for $(\partial., \boldsymbol{\beta}, \boldsymbol{\Delta})$, are displayed. These displays represent $f_{Q_p^*|\Theta_S, O_r}(q(t)|\theta_S^i, o_r); i = 1, 3, 4$, respectively, hence they represents the uncertainty in $Q_p^*(t)|\theta_S^i, o_r; i = 1, 3, 4$, for the reservoir total.

An evaluation along the lines of the partly stochastic approach usually applied is also performed. This entails assigning the model parameters the most probable value, $\hat{\theta}$, without considering the uncertainty in assessing this value. The values for $\hat{\theta}$ are obtained from both general experience and the reservoir specific observations, and they are specified in Figure 18. The predicted production characteristics of interest is, $Q_p^*(t)|\hat{\theta}, o_r$, and realizations can be obtained along the lines previously defined, $[q_p^*(t)|\hat{\theta}, o_r]_s; s = 1, 2 \ldots, 9$. In Figure 19, the realizations of predicted production characteristics of the reservoir total based on the partly stochastic approach, are displayed. These displays represent $f_{Q_p^*|\Theta, O_r}(q(t)|\hat{\theta}, o_r)$, hence the uncertainty in $Q_p^*(t)|\hat{\theta}, o_r$, for the reservoir total.

The processing time for completing a study like this is considerable. The computer resources for obtaining one realization of the production

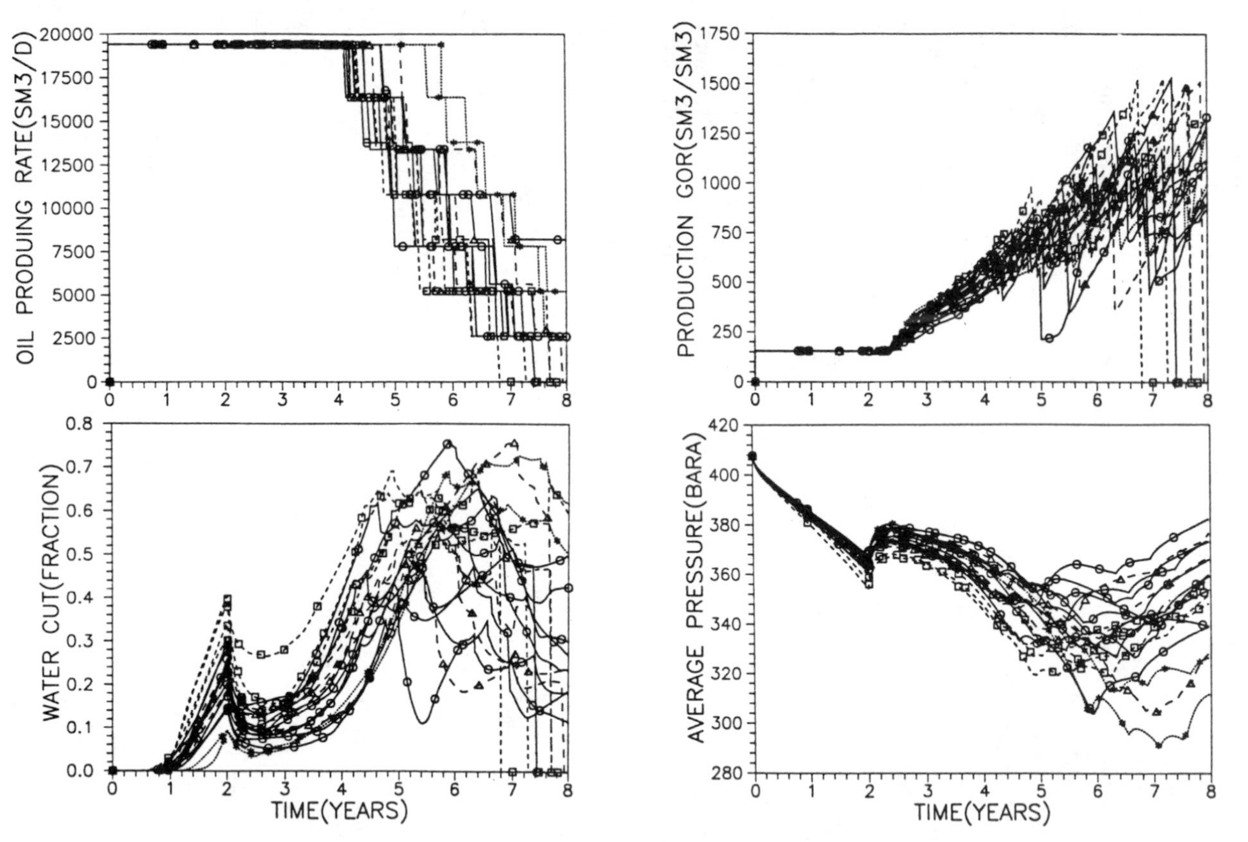

Figure 13: Predicted production characteristics — $Q_p^*(t|o_r)$ for reservoir total, representing total uncertainty

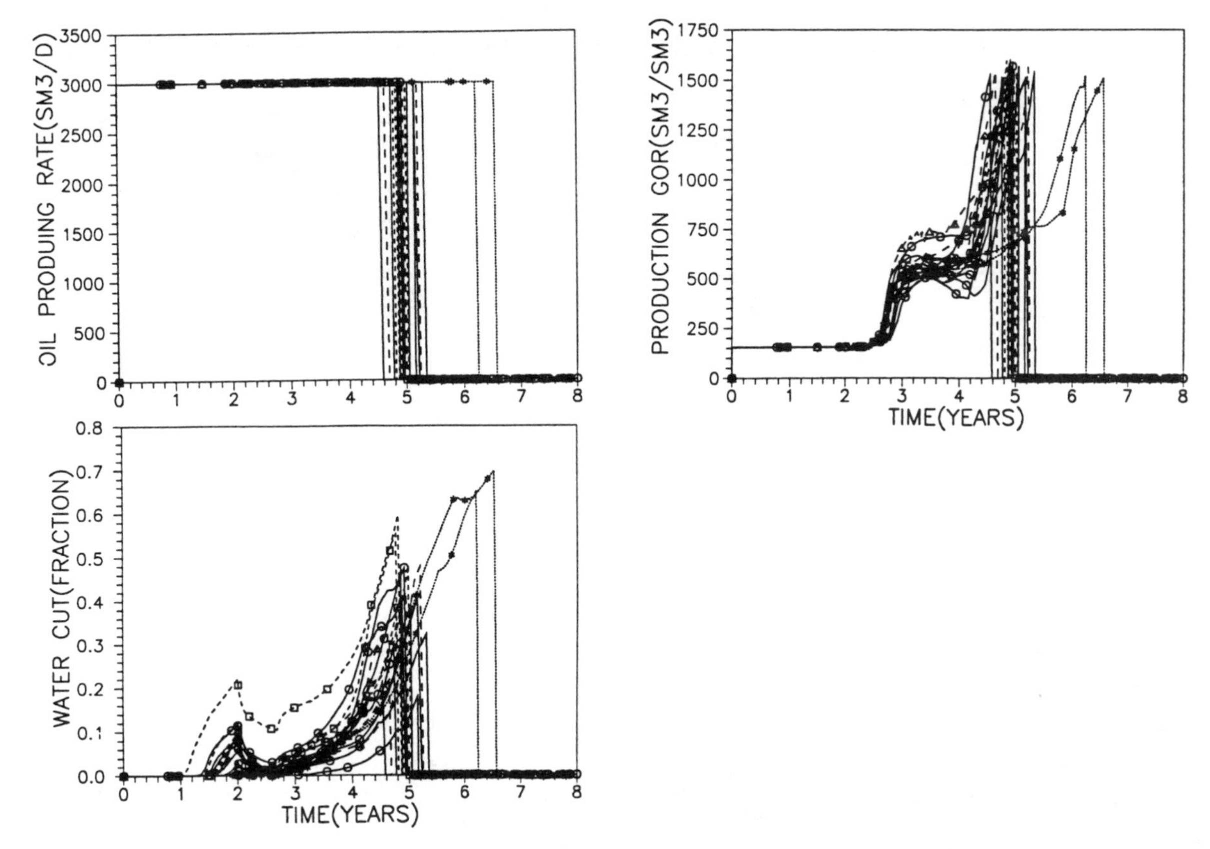

Figure 14: Predicted production characteristics — $Q_p^*(t|o_r)$ for well a5, representing total uncertainty

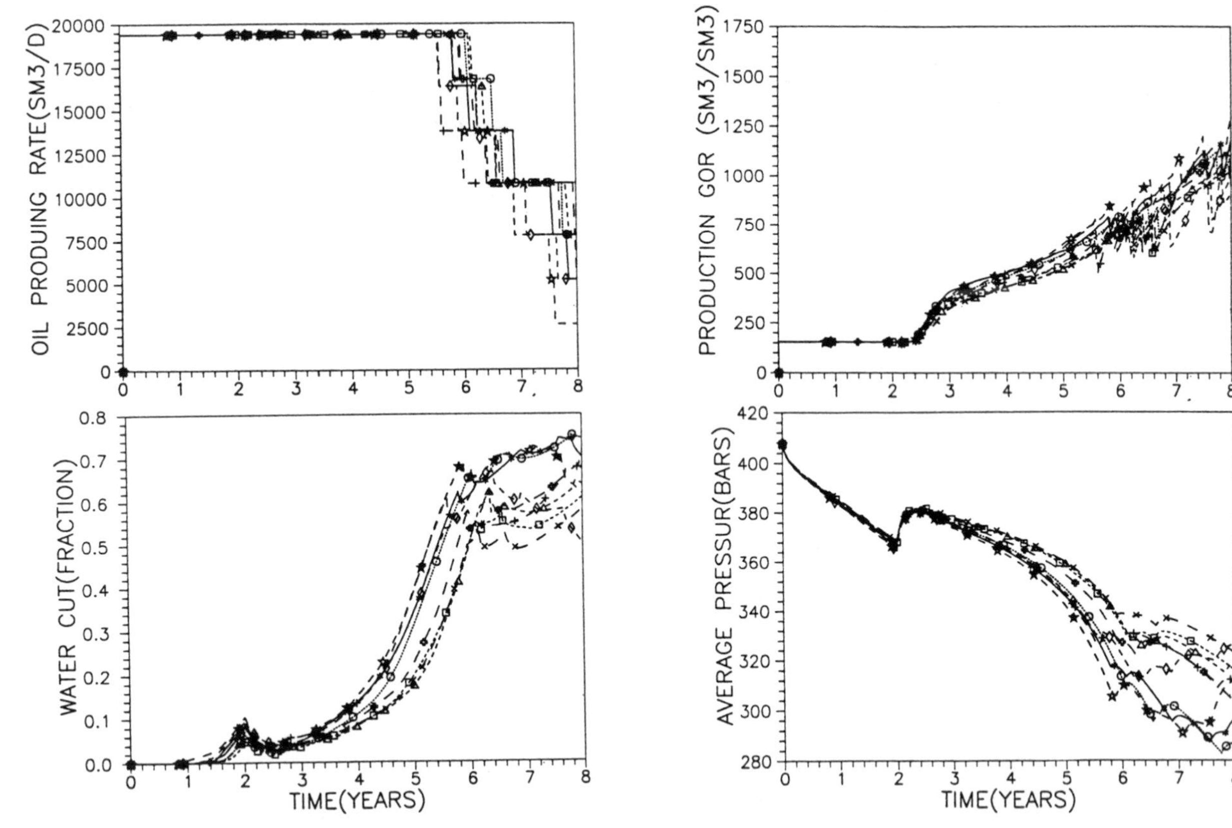

Figure 15: Predicted production characteristics for $(\partial., \boldsymbol{\beta}, \boldsymbol{\Delta})$ of set 1 — $Q_p^*(t|\theta_S^1, o_r)$ for reservoir total.

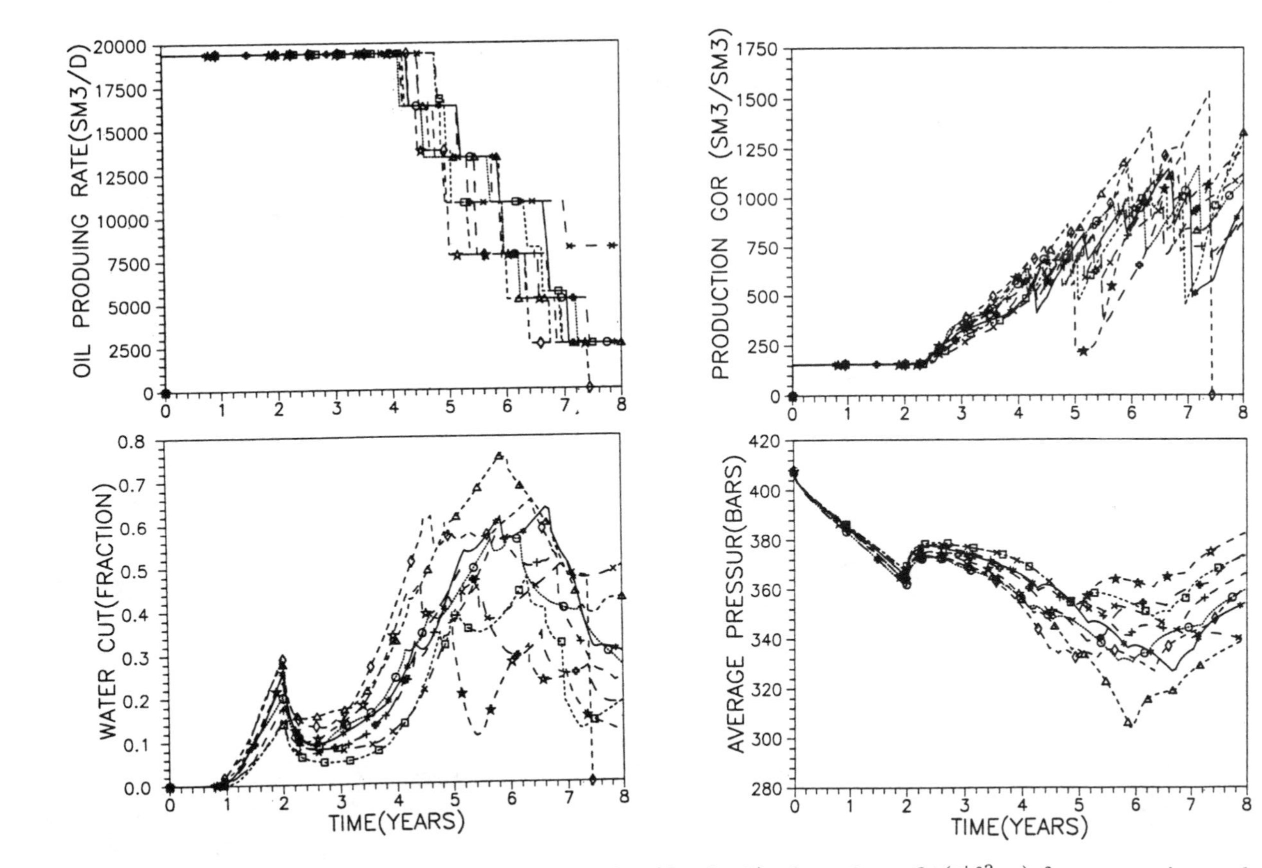

Figure 16: Predicted production characteristics for $(\partial., \boldsymbol{\beta}, \boldsymbol{\Delta})$ of set 3 — $Q_p^*(t|\theta_S^3 o_r)$ for reservoir total.

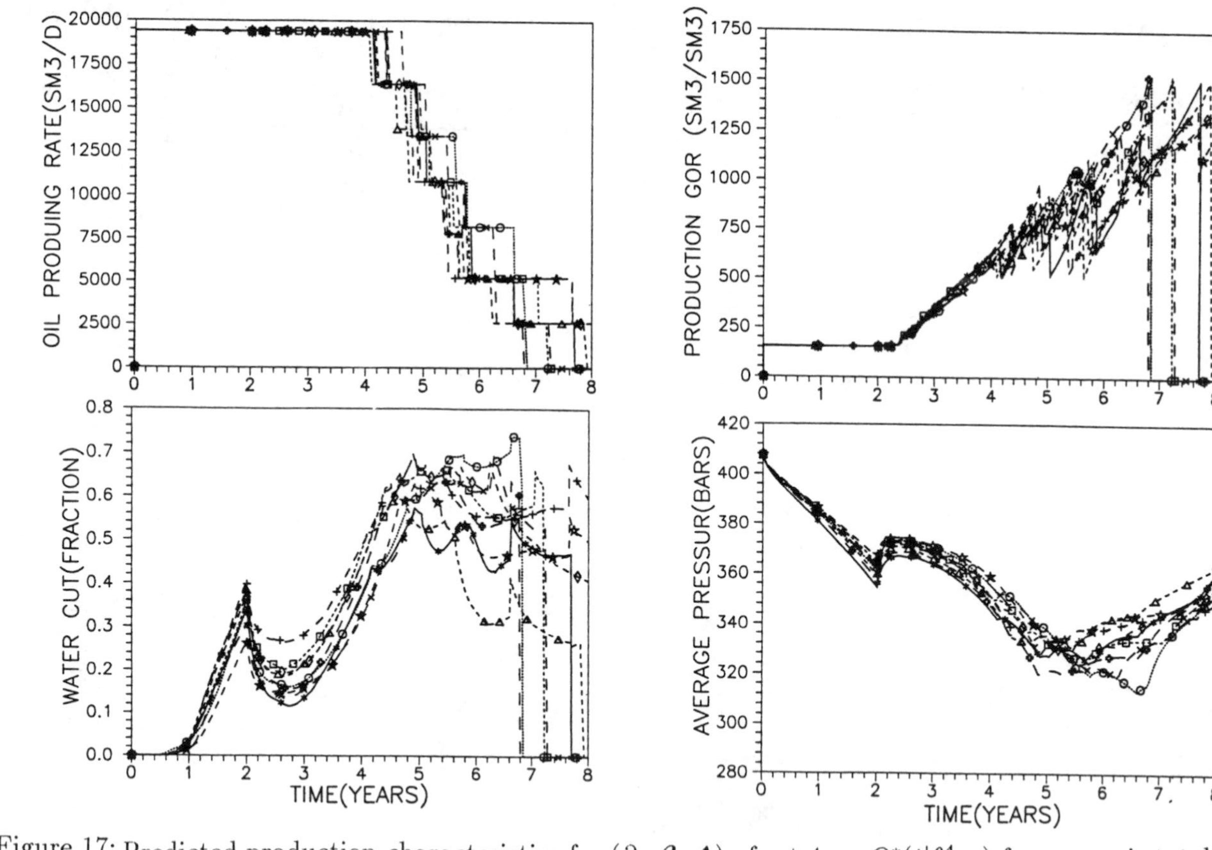

Figure 17: Predicted production characteristics for $(\partial., \boldsymbol{\beta}, \boldsymbol{\Delta})$ of set 4 — $Q_p^*(t|\theta_S^4 o_r)$ for reservoir total.

Param θ	$\hat{\theta}$	Param θ	$\hat{\theta}$	Param θ	$\hat{\theta}$
cr_1	0.12	cr_{10}	0.12	cr_{19}	0.18
cr_2	0.82	cr_{11}	0.12	cr_{20}	0.88
cr_3	0.24	cr_{12}	0.24	cr_{21}	0.76
cr_4	0.35	cr_{13}	0.41	cr_{22}	0.41
cr_5	0.71	cr_{14}	0.35	$(\partial., \beta, \Delta)$	set 3
cr_6	0.59	cr_{15}	0.82	μ_h	0.75
cr_7	0.82	cr_{16}	0.82	σ_h^2	0.27^2
cr_8	0.12	cr_{17}	0.71	fq	0.056
cr_9	0.76	cr_{18}	0.12		

Figure 18: Most probable values for model parameters. The thickness distribution for large-extent shales, $f_T(\cdot)$, is estimated from well observations non-parametrically for each correlation surface.

characteristics, $q_p^*(t)|o_r$, all the way from the generation of the model parameters, $[\theta|o_r]$, is typically 8.5 CPU-hours, varying from 5.5 to 13.5 CPU-hours, on an IBM 3090 J with vector facilities. The resources can be decomposed into: generation of model parameters - negligible; generation of initial reservoir characteristics - 0.25 CPU-hours; support change - 0.04 CPU-hours; fluid flow simulation - 8 CPU-hours; post-processing of results - 0.1 CPU-hours. These numbers refers to one realization, the number of realizations in the study is 39, hence the total computer resources are approximately 330 CPU- hours on IBM 3090 J with vector facilities. A study like this would be prohibited if any manual work was required in preparing the realizations of the initial reservoir characteristics for the fluid flow simulator. The processing in this study was done completely automatic all the way from generation of model parameters to extraction of interesting features from the results of the fluid flow simulator. Hence evaluation of uncertainty and sensitivity analysis can be done without human interference. Considerable manual work has of course been invested in specifying a stochastic model representative for the reservoir under study, implementing it and preparing the prior probability distributions for the model parameters.

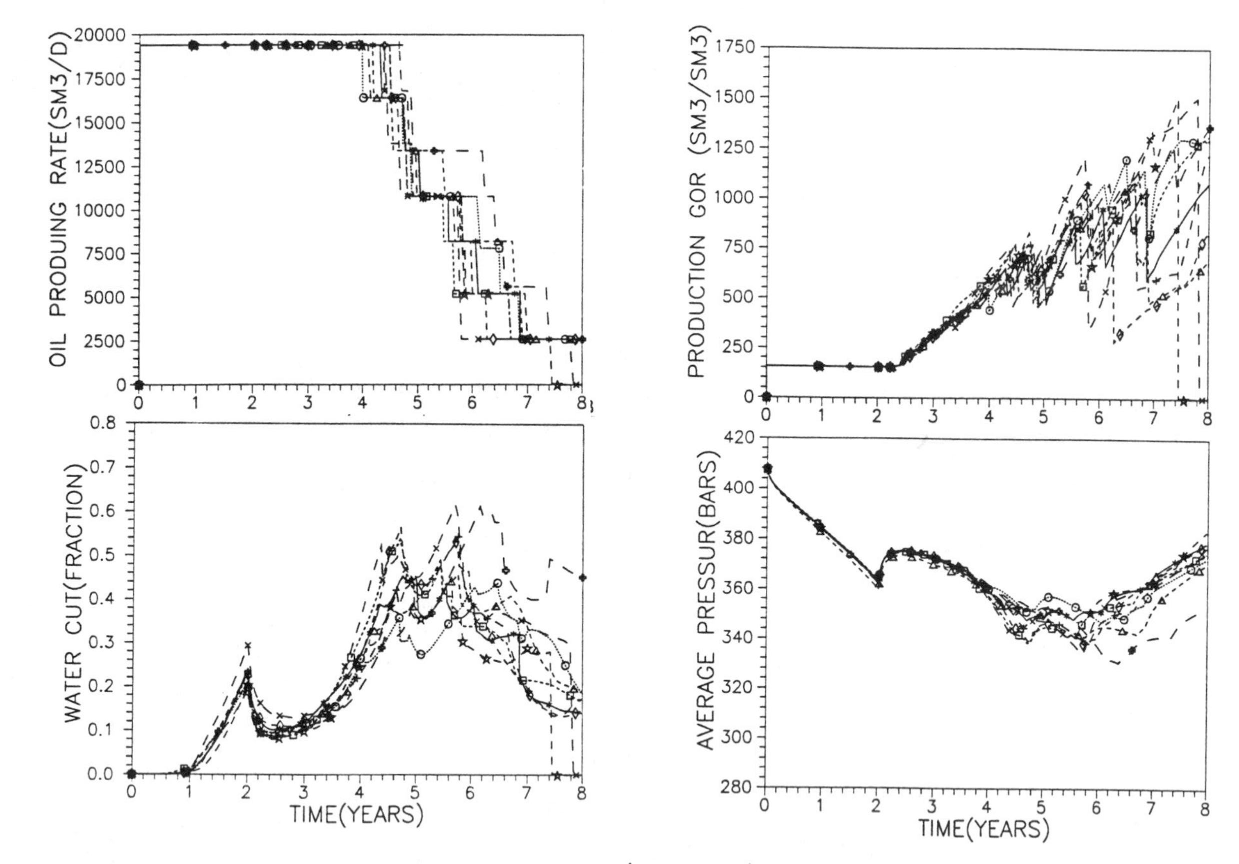

Figure 19: Predicted production characteristics for $\hat{\theta}$ — $Q_p^*(t|\hat{\theta}, o_r)$ for reservoir total

4 Discussion of Results

The uncertainty in the predicted production characteristics is due to incomplete knowledge of the sedimentary properties, the shale pattern in particular. In a complete uncertainty evaluation, geometrical, fluid and rock properties should be modeled stochastically as well. The interpretations from the results in this study are not expected to be influenced by such extension.

The total uncertainty for the predicted production characteristics is presented in Figure 13. During the two first years natural water drive is used, production is constant, and so is GOR, water cut increases slightly and pressure decreases. With gas injection, after year two, production remain constant for some time until about 4.5 years when the first well reaches GOR or water cut thresholds and are terminated. In the subsequent years more wells are terminated after reaching the thresholds, and for some realizations all production wells are terminated before the eight years long simulation period is completed. After year two, GOR is increasing and the threshold 1500 sm^3/sm^3 is reach for some wells. Water cut declines slightly after gas injection starts for later to increase considerably. For many wells the threshold 0.9 is reached, which causes termination. Pressure increases for a short period just after injection starts for then to decline until the wells are being terminated. Thereafter it increases again. In the two first years, the variability is reasonably small. After gas injection is initiated, the variability increases. When, after about 4.5 years, the termination of production wells are done in some realizations, the variability increases remarkably. The picture is approximately the same for all production variables since they are highly dependent.

The total uncertainty for one representative production well is presented in Figure 14. The time to termination varies from 4.5 years to 6.5 years. The overall picture is similar to the one for the reservoir total.

The sensitivity with respect to the areal extent of the large-extent shale units is presented in Figure 15 through 17. Small scattered shale units provide the largest production volume, see Figure 15. Most wells seem to be terminated due to the water cut threshold, hence water breaks through from the aquifer. The small units are not creating barriers and the reservoir is almost in equilibrium during production. This is reflected in the relatively small variability. Intermediate shale units provide the intermediate production volumes, see Figure 16. The GOR and water cut seem to be in good balance. The shale units create small barriers which vary between realizations, and this creates large variability. Large shale units provide the smallest production volumes, see Figure 17. The GOR and water cut seem to be in good balance. The large shale units create extensive barriers in the reservoir and tend to channel the water and gas to the wells. It is remarkable that the variability for this case is relatively small. It can probably be explained by the fact that the conditioning to the observations in the wells and the large shale units forces the shale to be located almost in the same locations in all realizations. Hence it will be small variation in production characteristics.

By comparing the results from the sensitivity evaluation, Figure 15 through 17, with the total uncertainty, Figure 13, it is easy to see that the major contribution to the latter comes from variation in the areal extent of large-extent shales.

The results from the partly stochastic approach to assessment of uncertainty are displayed in Figure 19. Recall that this entails ignoring the uncertainty in the model parameters. The graphs have, in average, the same appearance as for the total uncertainty in Figure 13. The variability is significantly smaller, however. There are reasons to believe that the uncertainty estimated from the partly stochastic approach is severely downward biased, since uncertainty in the model parameters is ignored. The partly stochastic approach should also be compared with the intermediate case in the sensitivity study of shale sizes, see Figure 16, since they both are based on the same shale size distributions. The variability is less in the partly stochastic approach since uncertainty in the parameters to be randomized is ignored.

5 Conclusions

Assessment of uncertainty in predicted production characteristics is frequently mentioned in the literature, particularly related to heterogeneity modeling. No thorough discussion of the topic seems to be available, however. The discussion and example in this article justifies the following conclusions:

- assessment of uncertainty must be founded on a clear statistical formalism — it is important to distinguish between heterogeneity modeling, as used today, and assessment of uncertainty. The latter being a more challenging task.

- realistic assessments of uncertainty require a reliable and consistent stochastic model for reservoir characteristics and the uncertainty in estimating the model parameters must be taken into consideration. Assessments based on general models and the best guess on parameter values will almost always provide dramatically overoptimistic uncertainty estimates.

- the partly stochastic approach, using a reliable stochastic model and the best guess on parameters, provides acceptable predictions for the expected production pattern — which is simpler to obtain than quantification of uncertainty. The use of an average of several realizations is recommended.

- in the type of reservoir studied in this article, the uncertainty in the predicted production characteristics is largest if the shale units have intermediate areal extent.

The conclusions can be summarized by:

	Stochastic Approach		Partly Stochastic Approach		Deterministic Approach
Best Prediction	$\mathrm{E}\{Q_p^*(t)\|o_r\}$	$\approx$	$\mathrm{E}\{Q_p^*(t)\|\hat{\theta}, o_r\}$	$\neq$	$q_p(t)$
Prediction Variance	$\mathrm{Var}\{Q_p^*(t)\|o_r\}$	$\gg$	$\mathrm{Var}\{Q_p^*(t)\|\hat{\theta}, o_r\}$		not available

The quantification of uncertainty in the predicted production characteristics is of great interest in itself. The fact that this quantification also makes fair comparison between different sampling strategies possible, is expected to have impact on future appraisal as well.

6 Acknowledgments

The authors are grateful to Eva Halland, NPD and Alister MacDonald, IFE for providing the geological interpretations to the case study, and Reidar Bratvold, IBM/EPAC for support on the reservoir engineering side. The founding is provided by Norwegian Petroleum Directorate, Norwegian Computing Center and IBM/EPAC.

7 References

Abrahamsen, P., Omre, H. and Lia, O.: "Stochastic Models for Seismic Depth Conversions of Geological Horizons",paper 23138 SPE, presented at Offshore Europe 1991, Aberdeen, sept. 3–6.

Alabert, F.G. and Massonnat, G.J.: "Heterogeneity in a Complex Turbiditic Reservoir: Stochastic Modeling of Facies and Petrophysical Variability",paper SPE 20604, presented at the 1990 SPE Annual Technical Conference and Exhibition, New Orleans, LA, sept. 23–26.

Berger, J.O.: *Statistical Decision Theory — Foundations, Concepts, and Methods.* Springer-Verlag, New York (1980).

Besag, J.:"Spatial interaction and the statistical analysis of lattice systems [with discussion]", *J. Royal Statist. Soc. B*, (1974), Vol. **36**, 192–236.

Damsleth,E., Tjølsen, C.B., Omre, K.H. and Haldorsen, H.H.:"A Two-Stage Stochastic Model Applied to a North Sea Reservoir", SPE JPT March 1992.

Damsleth, E., Hage, A. and Volden, R.: "Maximum Information at Minimum Cost",paper 23139 SPE, presented at Offshore Europe 1991, Aberdeen, sept. 3–6.

Geman, S. and Geman, D.:"Stochastic Relaxation, Gibbs Distribution, and the Bayesian Restoration of Images", *IEEE Trans.* (1984) PAMI 6 721–741.

Haldorsen, H.H. and Damsleth, E.: "Stochastic modeling",SPE JPT (Apr.1990) 404–412.

Haldorsen, H.H. and Lake, L.W.: "A new Approach to Shale Management in Field Scale Models",Soc. of Petr. Eng. Journel, 1984.

Hewett, T.A. and Behrens, R.A.: "Conditional Simulation of Reservoir Heterogeneity with Fractals", paper SPE 18326, presented at the 1988 SPE Annual Technical Conference and Exhibition, Houston, TX, oct. 2–5.

Høiberg, J., Omre, H. and Tjelmeland, H. (a): "Large-Scale Barriers in Extensively Drilled Reservoirs", *Proc.*, 2nd European Conference on the Mathematics of Oil Recovery, Arles (1990), 31–41.

Høiberg, J., Omre, H. and Tjelmeland, H. (b): "A Stochastic Model for Shale Distribution in Petroleum Reservoirs", Proceedings from the Second CODATA Conference on Geomathematics and Geostatistics, Leeds, Sept. 10–14, 1990; To appear in Science de la Terre.

Holden, H. and Holden L.: "Reservoir Evaluation by Stochastic Partial

Differential equations", Albeverio, S. and Merlini, D (eds.); World Scientific, In press.

Johnson, N.I. and Kotz, S.: *Discrete Distributions.* Houghton Mifflin Company, Boston (1969).

Johnson, N.I. and Kotz, S.: *Continuous Univariate Distributions.* Houghton Mifflin Company, Boston (1970).

Journel, A.G.: "Geostatistics for Reservoir Characterization", paper SPE 20750, presented at the 1990 SPE Annual Technical Conference and Exhibition, New Orleans, LA, sept. 23–26.

Mathews, J.L., Emanuel, A.S. and Edwards, K.A.: "Fractal Methods Improve Mitsue Miscible Predictions", SPE JPT (Nov.1989) pp 1136–1142.

Omre, H.: "Stochastic Models for Reservoir Characterization",in *Recent Advances in Improved Oil Recovery Methods for North Sea Sandstone Reservoirs*, Kleppe, J. and Skjæveland, S.M. (eds.); Norwegian Petroleum Directorate, Stavanger, Norway; 1992.

Ripley, B.D.: *Stochastic Simulation.* Wiley, New York (1987).

Rudkiewicz, J.L., Guerillot, D., Galli, A. and Group Heresim: "An Integrated Software for Stochastic Modelling of Reservoir Lithology and Property with an Example from the Yorkshire Middle Jurassic",*North Sea Oil and Gas Reservoirs - II*, A.T. Buller et al. (ed.); Graham & Trotman (1990) 399–406.

Stoyan, D., Kendall, W.S. and Mecke, J.:*Stochastic Geometry and Its Applications.* Akademie-Verlag, Berlin (1987).

ECLIPSE:*ECLIPSE 100 — Fully Implicit Black Oil Simulator, Reference Manual*, Exploration Consultants Ltd.,England.

SCREENING ENHANCED OIL RECOVERY METHODS WITH FUZZY LOGIC

W. J. Parkinson
K. H. Duerre
J. J. Osowski

Los Alamos National Laboratory
Los Alamos, New Mexico

G. F. Luger
Department of Computer Science
University of New Mexico
Albuquerque, New Mexico

R. E. Bretz
Department of Petroleum Engineering
New Mexrco Institute of Mining and Technology
Socorro, New Mexico

ABSTRACT

Three reasons many potential users argue against using expert systems for solving problems are (1) because of the relatively high cost of specialized LISP Machines and the large expert system shells written for them; (2) because some expert systems are used for jobs that the average professional could do with a relatively short literature search, a few hours of reading, and a few calculations; and (3) because some classical "crisp" rule-based expert systems are limited by their inflexible representation of human decision making, which is sometimes needed in problem solving. This paper demonstrates how a small, but useful expert system can be written with inexpensive shells that will run on inexpensive personal computers.

Rule-based expert assistants have been developed to help petroleum engineers screen possible enhanced oil recovery (EOR) candidate processes. Though the final candidate process is selected on the basis of an economic evaluation, the expert assistant greatly reduces the amount of work involved. Rather than requiring exhaustive economic calculations for all possible processes, the work is reduced to an economic comparison between two or three technically feasible candidates.

The expert system approach is compared with standard hand calculations that were performed using various graphs and charts. This manuscript also shows the advantages of the expert system method, solves several EOR screening problems using both the crisp expert system and the more flexible "fuzzy" expert system, and compares the two approaches.

INTRODUCTION

Reasons for studying enhanced oil recovery (EOR) techniques are summarized in a 1986 paper by Stosur [1]. When his paper was published, only 27% of all the oil discovered in the United States had been produced. Under current economic conditions, only about 6% more will be produced using existing technology. The remaining 67% is a target for EOR. Currently, about 6% of our daily oil production comes from EOR. Even in these times of reduced concern of an energy crisis, these numbers indicate that the study of EOR processes can be rewarding because of the potentially high payoffs.

Because, in general, EOR processes are expensive, engineers must choose the best recovery method for the reservoir in question to optimize profits or to make any profits at all. The screening methods are also expensive and, typically, involve many steps, one of which is to consult the technical screening guide. This screening step is the subject of this paper.

Screening guides contain tables or charts that list the rules of thumb for picking a proper EOR technique as a function of reservoir and crude oil properties. Once a candidate EOR technique is determined, further laboratory flow studies are often required. Data obtained from these studies are then used to demonstrate the viability of the selected technique. Throughout the screening process, economic evaluations are carried out .

In this paper, we present two expert systems for screening EOR processes. In the first, we developed a crisp, rule-based assistant, which replaces the previously published screening guides. These guides are based on tables and graphs designed for hand calculations. The expert assistant provides essentially the same information as the table and graph method, but is more comprehensive and easier to use than the screening guides. The second, fuzzy expert assistant was then developed to eliminate some of the weaknesses observed in the first expert system. These expert assistants provide users with a ranked list of potential techniques that would be difficult to compile using the tables. With both expert systems, the user must enter oil gravity, viscosity, composition, formation salinity, formation type, oil saturation, thickness, permeability, depth, temperature, and porosity. Although the final choice of technique will be based upon economics, the first screening step is quite important, first because the screening process itself is expensive and second because of the absolute necessity of choosing the most technically optimum EOR technique.

THE EOR SCREENING PROBLEM

For this study, EOR is defined as any technique that increases production beyond water flooding or gas recycling. This process usually involves the injection of an EOR fluid. Both of the expert systems discussed here are rule based and both rely mainly on the work of Taber and Martin [2] and Goodlet et al. [3,4] for their rules.

EOR techniques can be divided into four general categories: thermal, gas injection, chemical flooding, and microbial. Thermal techniques are subdivided into *in situ* combustion and steam flooding, which require reservoirs with fairly high permeability. Steam flooding has, traditionally, been the most used EOR method. It was previously applied only to relatively shallow reservoirs containing viscous oils. In this application, screening criteria are changing because the improved equipment allows economic operations on deeper formations. New studies show that, in addition to their effect on viscosity and density, steam temperatures also affect other reservoir rock and fluid properties. Thus, reservoirs previously not considered as candidates for steam flooding are being reevaluated. The expert system format is a good one to use here because we can easily change the program as the knowledge of a technology changes. Gas injection techniques, however, are at the opposite extreme from steam flooding. They are divided into hydrocarbon, nitrogen and flue gas, and carbon dioxide. These techniques tend to work best in deep reservoirs containing light oils. Chemical flooding techniques are divided into polymer, surfactant/polymer, and alkaline recovery techniques. Microbial techniques are new, and primarily experimental, at this time. The microbial category is not subdivided. Figure 1 shows all four of these categories and their associated EOR methods as the search tree for both expert assistants.

We often hear the comment "We have excellent papers on this subject with graphs and tables and information to help us solve the problem. Why do we need an expert system"? The response is that an expert system is not absolutely necessary, but the problem can be solved more quickly, and often better and with greater repeatability or consistency with the expert system. Table I, taken directly from Ref. 2, is a matrix of eight EOR techniques and nine EOR criteria.

Theoretically, if the values of the EOR criteria for the reservoir in question are known, engineers can pick some candidate processes from Table I, even without having much knowledge about EOR.

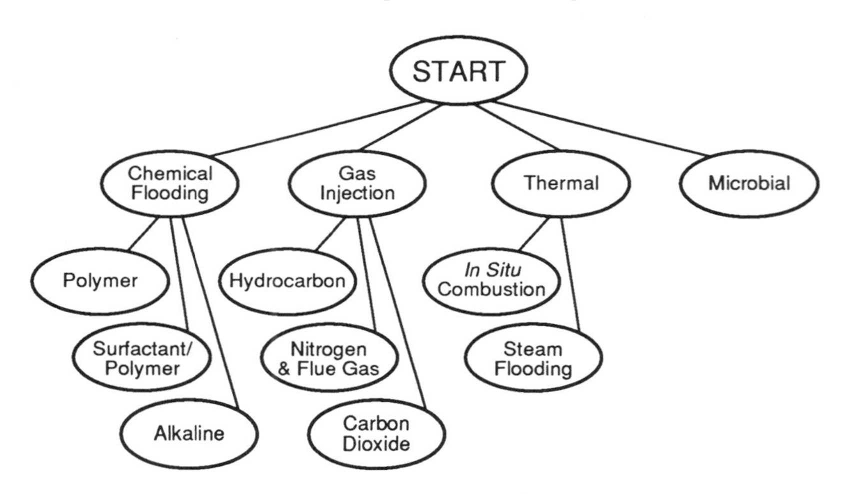

Figure 1. Search tree for the expert assistants.

The following simple examples show some of the problems with this argument. For Example 1, the following EOR criteria are used with Table I:

Example 1

(1) Gravity = 18 degrees API
(2) Viscosity = 500 cp
(3) Composition = high percent of C_4 - C_7
(4) Oil saturation = 50%
(6) Payzone thickness = 35 ft
(7) Average permeability = 1000 md

(8) Well depth = 2000 ft
(9) Temperature = 110°F

If we search the table, starting at the top, and move left-to-right before moving down a row, we are using a backward-chaining, or goal-driven, method. That is, we first assume a solution (e.g,. hydrocarbon gas-injection), then check the data either to verify or to disprove that assumption. On the other hand, a data-driven, or forward-chaining approach, would begin the search in the upper left-hand corner of the table and would move down, row by row, to the bottom before moving to the next column. That is, the search would start with the datum value for the oil gravity and would check that value against every EOR method before moving on to the other data. In this example, we use backward-chaining to find that steam flooding is the only good method to use for this example. The results of this search are shown in Fig. 2. *In situ* combustion techniques might also work. In Table I the meaning of the statement "greater than 150°F preferred" for the reservoir temperature is not perfectly clear. This is one example of how fuzzy logic can be useful, but we will discuss fuzzy logic further in a later paragraph.

The preceding situation, is not ideal because there is only one candidate for the next screening step, and this candidate could be eliminated, for other reasons, in a later screening step; then there would be no candidate recovery methods for this example. Having a reservoir that is not recommended for EOR is certainly legitimate, but we shouldn't eliminate the possibility of EOR because of too little knowledge. By changing the previous example slightly, we can have the opposite problem, as shown in Example 2, which has the following values for the EOR criteria:

Example 2

(1) Gravity = 35 degrees API
(2) Viscosity = 5 cp
(3) Composition high percent of C_4 - C_7 and some organic acids
(4) Oil saturation 50%
(5) Formation type = sandstone
(6) Payzone thickness = 10 ft
(7) Average permeability = 1000 md
(8) Well depth = 5000 ft
(9) Temperature = 150°F

TABLE I. Summary for screening criteria for enhanced recovery methods[c]

	Oil Properties				Reservoir Characteristics				
	Gravity °API	Viscosity (cp)	Composition	Oil Saturation	Formation Type	Net Thickness (ft)	Average Permeability (md)	Depth (ft)	Temperature (°F)
Gas Injection Methods									
Hydrocarbon	>35 >24	<10	High % of $C_2 - C_7$	>30% PV	Sandstone or Carbonate	Thin unless dipping	NC NC	<2000 (LPG) to <5000 HP gas	NC
Nitrogen & Flue Gas	>35 for N_2	<10	High % of $C_1 - C_7$	>30% PV	Sandstone or Carbonate	Thin unless dipping	NC	>4500	NC
Carbon Dioxide	>26	<15	High % of $C_5 - C_{12}$	>30% PV	Sandstone or Carbonate	Thin unless dipping	NC	>2000	NC
Chemical Flooding									
Surfactant/ Polymer	>25	<30	Light intermediates desired	>30% PV	Sandstone preferred	>10	>20	<8000	<175
Polymer	>25	<150	NC	>10% PV Mobile oil	Sandstone Preferred Carbonate possible	NC	>10 (normally)	<9000	<200
Alkaline	13-35	<200	Some organic acids	Above water-flood residual	Sandstone preferred	NC	>20	<9000	<200
Thermal									
Combustion	<40 (10-25 normally)	<1000	Some asphaltic components	>40-50% PV	Sand or Sandstone with high porosity	>10	>100[a]	>150 preferred	>150 preferred
Steam Flooding	<25	>20	NC	>40-50% PV	Sand or Sandstone with high porosity	>20	>200[b]	300-5000	NC

NC = not critical
[a]Transmissibility>20 md ft/cp
[b]Transmissibility>100 md ft/cp
[c]From reference 2

By searching Table I, again with a backward-chaining technique, we obtain the results shown in Fig. 3. This time only the steam flooding EOR method has been eliminated. This takes us to the second step with, possibly, too many candidates.

This is not a criticism of Ref. 2 or of tables like Table I. It is merely an effort to point out that in order to do a good first screening step, we will often need more information than is available in these tables. Much of this needed information is available in Refs. 2-4. References 3 and 4 include tables similar to Table I. Table II contains all of the material from Table I, as well as some of the information from the table in Ref. 4, including the microbial drive EOR method. The additional information improves the results of our search, but is still insufficient. We need information that will tell us what the impact of a reservoir temperature of 110°F will be when, as listed in Tables I and II, a temperature of greater than 150°F is preferred and information that will help us rank two or more methods when the methods fall within the acceptable range. That is, we need a ranked list of methods. A nonexpert can obtain a ranked list by reading the papers, and, possibly, by undertaking a short literature search, in addition to using Table I or II. But this screening step may require far more time than the few minutes it takes to search the tables. If the exercise must be repeated several times, or by several different nonexperts, then a small PC-based expert system can be easily justified for the job.

Figures 4-14 demonstrate the basis of the scoring system for the various EOR criteria and for the EOR methods used in a first attempt to solve this problem using a crisp rule-based expert system (see reference 5). Figures 5, 11, and 12 were taken from Ref. 2 and modified. The others were created by studying Ref. 2 through 4 and 6 through 8. Figures 4-14 are bar graphs showing the relative influence of each EOR criterion on each EOR method. The scoring system is empirical and was designed to add some judgement expertise to the expert system. Much of the information in Figs. 4-14 is based on experience and judgement, and it is influenced by the study of the more than 200 EOR projects listed in Ref. 8. The scoring system used in either expert system can easily be changed by someone with different experience or with new information.

Gas Injection Methods	Gravity	Viscosity	Composition	Oil Saturation	Formation Type	Net Thickness	Average Permeability	Depth	Temperature
Hydrocarbon	no	→	→	→	→	→	→	→	→
Nitrogen & Flue Gas	no	→	→	→	→	→	→	→	→
Carbon Dioxide	no	→	→	→	→	→	→	→	→
Chemical Flooding									
Surfactant/Polymer	no	→	→	→	→	→	→	→	→
Polymer	no	→	→	→	→	→	→	→	→
Alkaline	yes	no	→	→	→	→	→	→	→
Thermal									
Combustion	yes	yes	yes	yes	yes	yes	yes	yes	no
Steam Flooding	yes	yes	NC	yes	yes	yes	yes	yes	NC

NC = not critical

Fig. 2. Solution to Example Problem 1.

Gas Injection Methods	Gravity	Viscosity	Composition	Oil Saturation	Formation Type	Net Thickness	Average Permeability	Depth	Temperature
Hydrocarbon	yes	yes	ok	yes	yes	ok	NC	yes	NC
Nitrogen & Flue Gas	yes	yes	ok	yes	yes	ok	NC	yes	NC
Carbon Dioxide	yes	yes	ok	yes	yes	ok	NC	yes	NC
Chemical Flooding									
Surfactant/Polymer	yes	yes	ok	yes	yes	yes	yes	yes	yes
Polymer	yes	yes	NC	yes	yes	NC	yes	yes	yes
Alkaline	yes	yes	ok	yes	yes	NC	yes	yes	yes
Thermal									
Combustion	yes	yes	ok	yes	yes	yes	yes	yes	NC
Steam Flooding	no	→							
NC = not critical									

Fig. 3. Solution to Example Problem 2.

TABLE II. Summary for screening criteria for enhanced recovery methods[e]

	Oil Properties				Reservoir Characteristics						
	Gravity °API	Viscosity (cp)	Composition	Salinity (ppm)	Oil Saturation	Formation Type	Net Thickness (ft)	Average Permeability (md)	Depth (ft)	Temperature (°F)	Porosity (%)
Gas Injection Methods											
Hydrocarbon	>35	<10	High % of $C_2 - C_7$	NC	>30% PV	Sandstone or Carbonate	Thin unless dipping	NC	>2000 (LPG) to >5000 HP gas	NC	NC
Nitrogen & Flue Gas	>24 >35 for N_2	<10	High % of $C_1 - C_7$	NC	>30% PV	Sandstone or Carbonate	Thin unless dipping	NC	>4500	NC	NC
Carbon Dioxide	>26	<15	High % of $C_5 - C_{12}$	NC	>30% PV	Sandstone or Carbonate	Thin unless dipping	NC	>2000	NC	NC
Chemical Flooding											
Surfactant/ Polymer	>25	<30	Light intermediates desired	<140,000	>30% PV	Sandstone preferred	>10	>20	<8000	<175	≥20
Polymer	>25	<150	NC	<100,000	>10% PV Mobile oil	Sandstone Preferred Carbonate possible	NC	>10 (normally)	<9000	<200	≥20
Alkaline	13-35	<200	Some organic acids	<100,000	Above water-flood residual	Sandstone preferred	NC	>20	<9000	<200	≥20
Thermal											
Combustion	<40 (10-25 normally)	<1000	Some asphaltic components	NC	>40-50% PV	Sand or Sandstone with high porosity	>10	>100[a]	>500	>150 preferred	≥20[c]
Steam Flooding	<25	>20	NC	NC	>40-50% PV	Sand or Sandstone with high porosity	>20	>200[b]	300-5000	NC	≥20[d]
Microbial											
Microbial Drive	>15	*	Absence of toxic cone, of metals, No biocides present	<100,000	NC	Sandstone or Carbonate	NC	>150	<8000	<140	–

NC = not critical
[a]Transmissibility > 20md ft/cp
[b]Transmissibility >100 md ft/cp
[d]Ignore if saturation times porosity >0.1
[c]Ignore if saturation times porosity >0.08
[e]Modified from Refs. 2 and 4

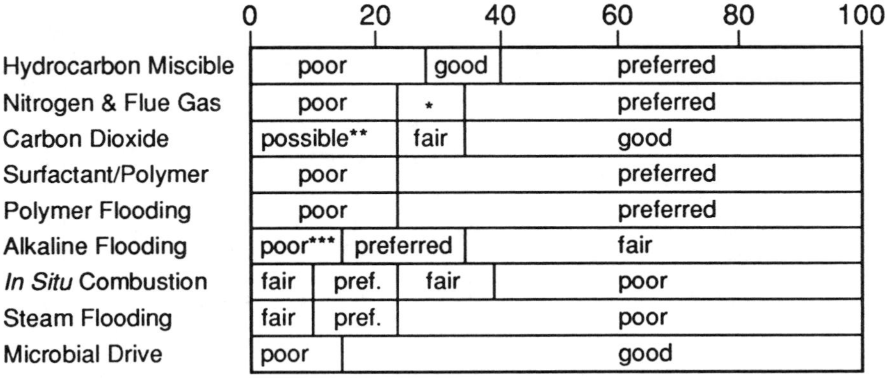

Scale: 0, 20, 40, 60, 80, 100

Hydrocarbon Miscible	poor	good	preferred
Nitrogen & Flue Gas	poor	*	preferred
Carbon Dioxide	possible**	fair	good
Surfactant/Polymer	poor	preferred	
Polymer Flooding	poor	preferred	
Alkaline Flooding	poor***	preferred	fair
In Situ Combustion	fair	pref.	fair	poor
Steam Flooding	fair	pref.	poor	
Microbial Drive	poor	good		

* Minimum preferred, 24 for flue gas and 35 for nitrogen.
** Possible immiscible gas displacement.
*** No organic acids are present at this gravity.

Fig. 4. Oil gravity screening data (°API).

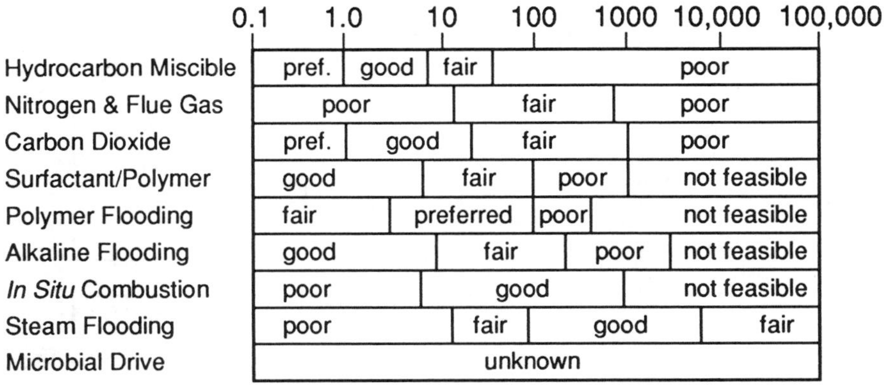

Scale: 0.1, 1.0, 10, 100, 1000, 10,000, 100,000

Hydrocarbon Miscible	pref.	good	fair	poor
Nitrogen & Flue Gas	poor	fair	poor	
Carbon Dioxide	pref.	good	fair	poor
Surfactant/Polymer	good	fair	poor	not feasible
Polymer Flooding	fair	preferred	poor	not feasible
Alkaline Flooding	good	fair	poor	not feasible
In Situ Combustion	poor	good	not feasible	
Steam Flooding	poor	fair	good	fair
Microbial Drive	unknown			

Fig. 5. Oil viscosity screening data (cp).

	High % C_2- C_7	High % C_1- C_7	High % C_5- C_{12}	Organic Acids	Asphaltic Components
Hydrocarbon Miscible	preferred	good	fair	NC	NC
Nitrogen & Flue Gas	good	preferred	fair	NC	NC
Carbon Dioxide	fair	fair	preferred	NC	NC
Surfactant/Polymer	fair	fair	preferred	NC	NC
Polymer Flooding	NC	NC	NC	NC	NC
Alkaline Flooding	NC	NC	NC	preferred	NC
In Situ Combustion	NC	NC	NC	NC	preferred
Steam Flooding	NC	NC	NC	NC	NC
Microbial Drive	NC	NC	NC	NC	NC

NC = not critical

Fig. 6. Oil composition screening data.

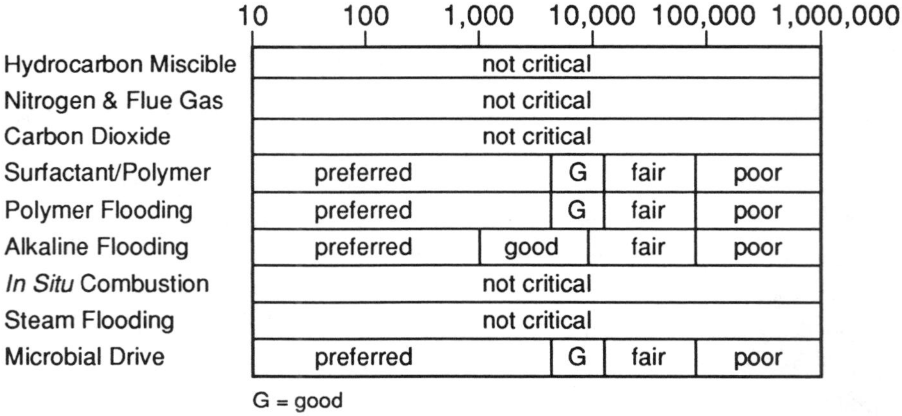

10 100 1,000 10,000 100,000 1,000,000

Hydrocarbon Miscible	not critical			
Nitrogen & Flue Gas	not critical			
Carbon Dioxide	not critical			
Surfactant/Polymer	preferred	G	fair	poor
Polymer Flooding	preferred	G	fair	poor
Alkaline Flooding	preferred	good	fair	poor
In Situ Combustion	not critical			
Steam Flooding	not critical			
Microbial Drive	preferred	G	fair	poor

G = good

Fig. 7. Formation salinity screening data (ppm).

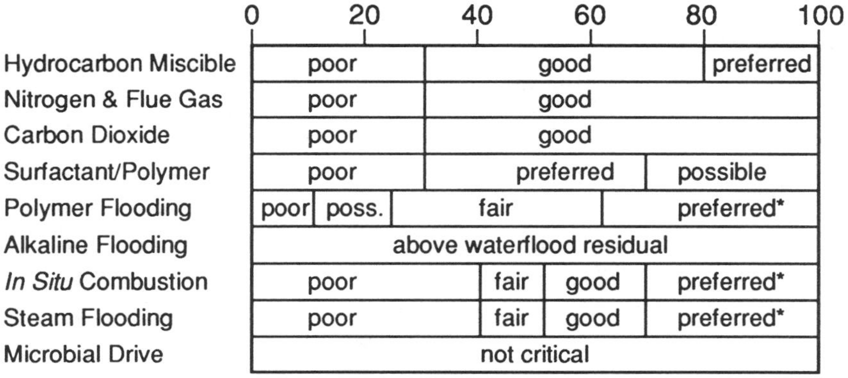

	0–20	20–40	40–60	60–80	80–100
Hydrocarbon Miscible	poor		good		preferred
Nitrogen & Flue Gas	poor		good		
Carbon Dioxide	poor		good		
Surfactant/Polymer	poor		preferred		possible
Polymer Flooding	poor | poss.		fair		preferred*
Alkaline Flooding	above waterflood residual				
In Situ Combustion	poor		fair | good		preferred*
Steam Flooding	poor		fair | good		preferred*
Microbial Drive	not critical				

* Preferred status is based on the starting residual oil saturations of successfully producing wells as documented by Ref. 8.

Fig. 8. Oil saturation screening data (%PV).

	Sand	Homogeneous Sandstone	Heterogeneous Sandstone	Homogeneous Carbonate	Heterogeneous Carbonate
Hydrocarbon Miscible	good	good	poor	good	poor
Nitrogen & Flue Gas	good	good	poor	good	poor
Carbon Dioxide	good	good	poor	good	poor
Surfactant/Polymer	preferred	preferred	poor	good	poor
Polymer Flooding	preferred	preferred	good	fair	poor
Alkaline Flooding	poor	preferred	fair	not feasible	not feasible
In Situ Combustion	good	good	good	good	fair
Steam Flooding	good	good	fair	good	fair
Microbial Drive	good	good	good	good	poor

Fig. 9. Formation type screening data.

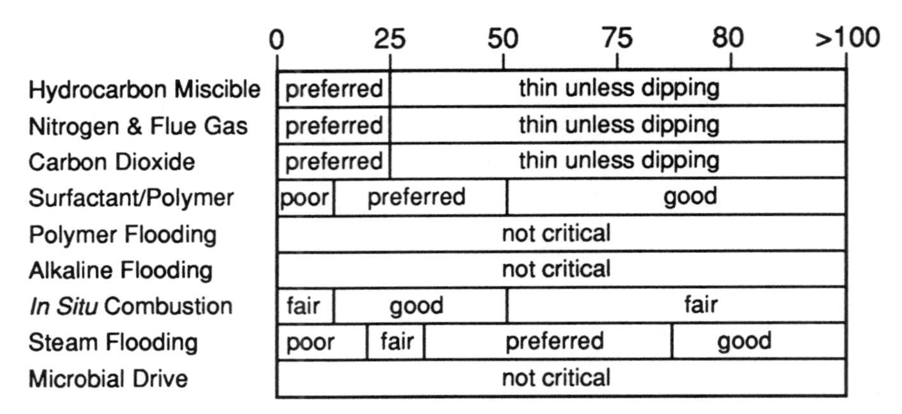

Fig. 10. Net thickness screening data (ft).

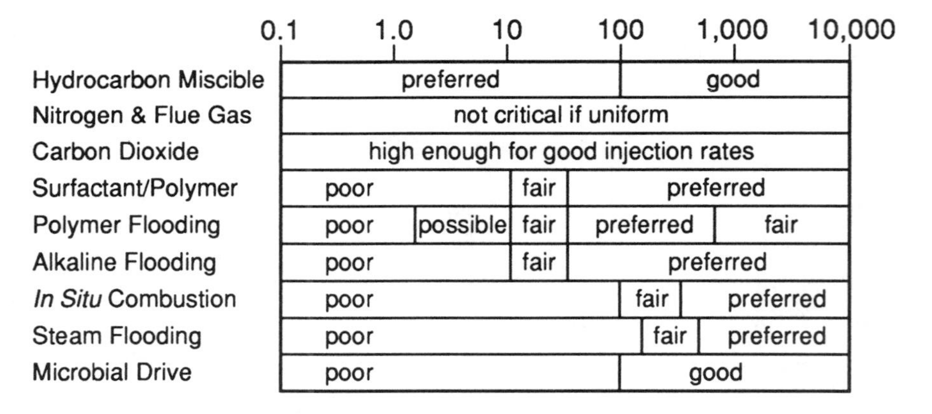

Fig. 11. Permeability screening data (md).

	0 – 2,000 – 4,000 – 6,000 – 8,000 – 10,000
Hydrocarbon Miscible	poor / fair / good
Nitrogen & Flue Gas	poor / fair / preferred
Carbon Dioxide	poor / possible / preferred
Surfactant/Polymer	preferred / poor
Polymer Flooding	preferred / poor
Alkaline Flooding	preferred / poor
In Situ Combustion	N / P / good
Steam Flooding	P / preferred / possible / poor
Microbial Drive	good / poor

P = possible N = not feasible

Fig. 12. Well-depth screening data (ft).

	0 – 100 – 200 – 300 – 400 – 500
Hydrocarbon Miscible	not critical
Nitrogen & Flue Gas	good / better
Carbon Dioxide	not critical
Surfactant/Polymer	preferred / good / poor / not feasible
Polymer Flooding	preferred / good / poor / not feasible
Alkaline Flooding	good / fair / poor
In Situ Combustion	poor / good / preferred
Steam Flooding	not critical
Microbial Drive	good / not feasible

Fig. 13. Formation temperature screening data (°F).

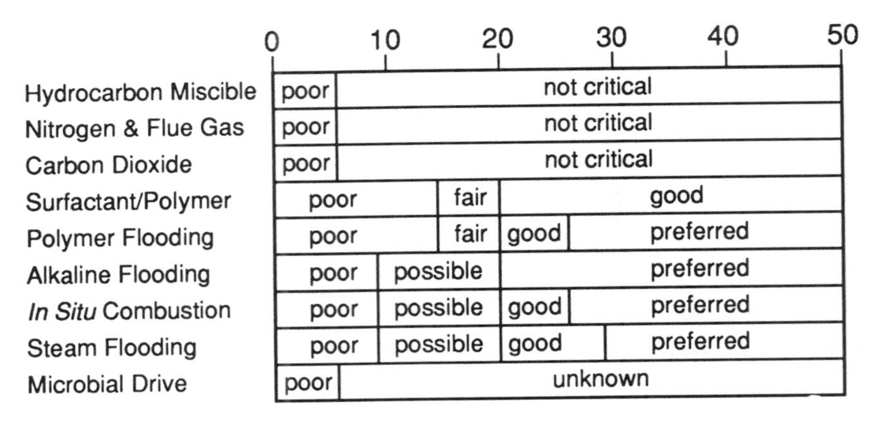

Fig. 14. Formation porosity screening data (%).

The crisp scoring system is based on the key words in Figs. 4-14, and works like this:

Not feasible	-50	Fair	6
Very poor	-20	Good	10
Poor	0	Not critical	12
Possible	4	Preferred	15

Note that "Not Critical" is a very good situation to have.

For the microbial drive method, the effect of viscosity, and, to a large extent, porosity, is unknown. Until more information is obtained, the viscosity and porosity are assigned a grade of 6 for an "Unknown," which is the same score as a "Fair."

For an example of the scoring system, observe Fig. 5 and consider an oil with a viscosity of about 500 centipoise. The hydrocarbon gas injection, surfactant-polymer, and alkaline chemical flood techniques are all "Poor," with scores of zero. The other two gas injection techniques, nitrogen and flue gas and carbon dioxide, are both "Fair," with scores of 6. The polymer flooding technique cannot be used with a viscosity this high, so it gets a score of -50. Each of the thermal techniques is "Good," and each gets a score of 10. The microbial drive method has an "Unknown," so it gets a score of 6.

The system is a significant improvement over the tables because each category is broken into many increments or sets. However, this system is still not adequate because the sets are crisp and they have a membership of either 0 or 1. This works fine for many problems but not for others. For example, in Fig. 13, the influence of the formation temperature on the microbial drive method is tremendous. With a change of one degree, the choice can go from "Good" to "Not Feasible." This is a change of 60 points. Although there is a temperature above which the microbes die, it is unlikely that the demarcation is that sharp.

Some EOR criteria carry more weight than others, and, in some cases, a given criteria may affect one method more than another, which explains why the maximum and minimum scores for each method vary within a given criterion (see Fig. 4). The variation in oil gravity allows the score of the hydrocarbon miscible gas injection method to range from "Poor" to "Preferred," a point spread of 0 to 15. The same gravity variation allows the score of the carbon dioxide gas injection method to range from "Possible" to "Good," a point spread of 4 to 10. This indicates that oil gravity has a larger influence on the hydrocarbon miscible method than on the carbon dioxide method.

The crisp expert system is a great improvement over the hand calculation method that utilizes graphs and charts. Considerable information has been added to the expert system, as can be seen in Figs. 4-14. A example of this improvement is our first example problem in which two conditions have been added from Table II. The salinity is 50,000 ppm and the porosity is 28%. Using this information with Table II one would get the same solution we obtained in our sample session, as shown in Fig. 2. This example, again, shows that the only method that can be used is steam flooding. The expert assistant, however, produces a ranked list of five different candidate processes. They are, in order, as follows:

Crisp Expert System Rankings for Example 1 Solution

	Score (%)
(1) Steam flooding	89
(2) In situ combustion	85
(3) Alkaline flooding	76
(4) Polymer flooding	73
(5) Microbial drive	72

The method for normalizing the scores is given later in this section. The expert system has provided the solutions to the two problems we had earlier, when using only Table I. It has given us a ranked list, instead of just one candidate or a large unranked list of candidates.

Methods such as *in situ* combustion can be ranked because it can also weigh problems such as "What does it mean to have a temperature of 110°F when the table says greater than 150°F preferred"? and it gives the method a relative score. This weighting is possible because of all the additional information provided in Figs. 4-14. As pointed out earlier, this expert system works very well on most real world cases. Although the scoring system described does quite well in most cases, some notable exceptions are described in the next two examples. Example 3 has the following values for the EOR criteria for two similar scenarios:

Example 3 - Scenario One

(1) Gravity = 23 degrees API
(2) Viscosity = 30 cp
(3) Composition = high percent C_5 -C_{12}
(4) Salinity = 101,000 ppm
(5) Oil saturation = 29%
(6) Formation type = sandstone (homogeneous)
(7) Payzone thickness = 26 ft
(8) Average permeability = 24 md
(9) Well depth = 1999 ft
(10) Temperature = 91°F
(11) Porosity = 19%

Example 3 - Scenario Two

(1) Gravity = 24 degrees API
(2) Viscosity = 22 cp
(3) Composition = high percent of C_5 - C_{12}
(4) Salinity = 99,000 ppm
(5) Oil Saturation = 31 %
(6) Formation type = sandstone (homogeneous)
(7) Payzone thickness = 24 ft
(8) Average permeability = 26 md
(9) Well depth = 2001 ft
(10) Temperature = 89°F
(11) Porosity = 21 %

The differences between these two scenarios are hardly measurable. Yet the crisp expert system gives them the following rankings and raw scores:

Crisp Expert System Solution for Example 3

Scenario One (Rankings)

1-	Polymer flooding	102 points
2-	Alkaline flooding	97 points
3-	*In situ* combustion	93 points
4-	Steam flooding	92 points
5-(tie)	Microbial drive	88 points
6-(tie)	Surfactant/polymer	88 points
7-	Carbon dioxide	85 points
8-	Hydrocarbon miscible	77 points
9-	Nitrogen and flue gas	72 points

Crisp Expert System Solution for Example 3

Scenario Two (Rankings)

1-	Surfactant/polymer	142 points
2-	Polymer flooding	136 points
3-	Alkaline flooding	127 points
4-	Carbon dioxide	116 points
5-	Nitrogen and flue gas	114 points
6-	Hydrocarbon miscible	104 points
7-	Microbial drive	94 points
8-	Steam flooding	83 points
9-	*In situ* combustion	80 points

It is easy to see that the rankings of these scenarios are completely different. The scores for the second scenario, except for *in situ* combustion and steam flooding, are much higher than those for the first scenario. (The relevance of the magnitude of these scores is discussed at the end of this section.) Figures 4-14 show that the scores for many of the EOR methods fall on one side of a crisp boundary in the first scenario and on the other side in the second scenario. The differences are increased because this occurs several times for each method as the expert system searches through the EOR criteria. This example is a worst case. It was set up so that the differences in scores would propagate, rather than cancel, from one criterion to another. But it is realistic in that most measurement techniques are not accurate enough to determine which side of a crisp boundary the data should really be on. The problem is exacerbated by the fact that a small change in the state of an EOR criterion can dramatically influence some EOR methods. For example, Fig. 4 shows that a small change in the API gravity of an oil can change the potential for surfactant/polymer and polymer flooding from "Poor" to "Preferred." Another example is the effect of viscosity on *in situ* combustion (see Fig. 5). A sharp change occurs, from "Poor" to "Good," as the viscosity increases. Another sharp change occurs, from "Good" to "Not Feasible," as the viscosity increases further. Even though these changes are relatively sharp, they are not as crisp as those shown in Figs. 4-14 or as used as those in the crisp expert system.

A fourth example demonstrates yet another problem related to use of a crisp expert system. If we add information about salinity and porosity to the scenario previously discussed in Example 1 so that we can use all of Figs. 4-14, and if we negligibly change the oil viscosity, gravity, and composition, we can demonstrate a drastic change in rating/score for *in situ* combustion and surfactant/polymer processes. The difference between these two scenarios is only 2 centipoise or 0.2% in viscosity.

Example 4 - Scenario One

(1) Gravity = 15 degrees API
(2) Viscosity = 999 cp
(3) Composition = high percent of C_5 - C_{12}
(4) Salinity = 50,000 ppm
(5) Oil saturation = 50%
(6) Formation type = sandstone (homogeneous)
(7) Payzone thickness = 35 ft
(8) Average permeability = 1000 md
(9) Well depth = 2000 ft
(10) Temperature = 110°F
(11) Porosity = 28%

Example 4 - Scenario Two

(1) Gravity = 15 degrees API
(2) Viscosity = 1001 cp
(3) Composition = high percent of C_5- C_{12}
(4) Salinity = 50,000 ppm
(5) Oil saturation = 50%
(6) Formation type = sandstone (homogeneous)
(7) Payzone thickness = 35 ft
(8) Average permeability = 1000 md
(9) Well depth = 2000 ft
(10) Temperature = 110°F
(11) Porosity = 28%

If we list the rankings of the top four methods computed from Scenario One, we find *in situ* combustion ranked second and surfactant/polymer ranked fourth.

Crisp Expert System Solution for Example 4

Scenario One (Rankings)

1- Steam flooding	132 points
2- *In situ* combustion	125 points
3- Alkaline flooding	117 points
4- Surfactant/polymer	116 points

Crisp Expert System Solution for Example 4

Scenario Two (Rankings)

1- Steam flooding	132 points
*- *In situ* combustion	65 points (Not Feasible)
2- Alkaline flooding	117 points
*- Surfactant/polymer	66 points (Not Feasible)

In scenario 2, the *in situ* combustion and surfactant/polymer techniques drop from the second and fourth ranked methods to methods that are Not Feasible. Even though a rather sharp drop in feasibility occurs, the small viscosity increase makes a drop this sharp seem unreasonable.

The preceding examples demonstrate the kinds of problems experienced by some expert systems' decision boundaries. Although there are several ways to reduce these problems, the problem of screening EOR methods is ideally suited to fuzzy logic. Fuzzy logic is like human logic at those boundaries. Instead of deciding which side to be on, a weighted average of each side is used. This makes the transition from one side of the boundary to the other much smoother.

An important task of the expert system is to give, the user meaningful advice about the individual EOR methods on the basis of the raw scores computed by the program. For these expert systems, the raw scores were normalized on the basis of a maximum possible best score of 100% for the best possible process, which is steam flooding. That is, if all methods were to receive the best possible score, steam flooding would get the highest score, with 148 points. This method also has the largest number of "Preferred" ratings of the methods shown in Figs. 4-14. The other EOR methods (except the microbial drive) are all rated quite close to the steam flooding method. The raw score of 148 corresponds to 100%. All raw scores are divided by 148 to produce a normalized score relative to the best score possible.

At the end of a session, the scores are tallied, providing the user with a ranked list of candidates to take to the next screening step as well as an idea of how good the candidates are relative to the best possible score. So far in these examples, both expert systems have given realistic results, except in those cases where the fuzzy decision was important. These expert systems have been run using much of the information given in Ref. 8 for actual EOR projects. In about 60% of the cases run, the method ranked highest by the expert system was the method that was actually selected and used in that project. In most of the other cases, the actual method used was ranked in the top three by the expert system. This is not unexpected because the actual test data influenced the scores used by the expert system. Expert systems are often built by comparing the results of the expert system with the results given by the experts, then modifying the system until it is as good as the experts. This approach gives us confidence in the accuracy of the results predicted by the expert systems.

EXPERT SYSTEMS AND FUZZY LOGIC

Most texts on artificial intelligence and expert systems [9,10] point out that almost all real expert systems have to deal with some kind of uncertainty. In building the first (crisp) expert system, considerable effort was expended examining the literature and working with raw data to reduce the uncertainty. If, for example, an EOR method gets a "Good" rating for an EOR criterion, that rating is assumed with

100% confidence, to be worth 10 points. Considerable effort went into defining the boundaries of the various ratings within each EOR criteria. For the crisp expert system, each rating block is considered to be a crisp set, that is, either the EOR method gets a particular rating or it doesn't. For instance (see Fig. 4), if the API gravity of the oil is greater than 40, the hydrocarbon miscible method gets a "Preferred" rating. If the gravity is not greater than 40, the method gets some other rating. This works fine as long as the gravity is not near the boundary (in this case 40). But if it is, then some uncertainty arises. For example, what if the gravity is 27, right about the boundary between "Good" and "Poor" for the hydrocarbon miscible method? Should the score be 0 for "Poor" or 10 for "Good"? The crisp expert system makes a decision and assigns a membership, to either "Poor" or "Good" for the hydrocarbon miscible gravity, and the score for the hydrocarbon miscible method is incremented appropriately.

The fuzzy expert system reduces the uncertainty caused by set boundaries by replacing the crisp sets with fuzzy sets [11]. Fuzzy logic is conventional logic, or inference rules, that is applied to fuzzy sets rather than crisp sets. Fuzzy sets are represented by membership functions. Unlike the crisp sets, the value for an EOR criterion for an EOR method can have membership in more than one set. Figures 15-23 are the membership functions, or fuzzy sets, that correspond to the crisp sets in Figs. 4-14.

There are no corresponding fuzzy sets for Fig. 6 (oil composition), or Fig. 9 (formation type). These two EOR criteria remained crisp for this study. Note that some of the abscissae of the fuzzy sets are different from those shown for the corresponding crisp sets. In these cases, simple transformations were used on the EOR criteria variables to better fit them to the fuzzy expert system shell.

As can be seen by observing Figs. 15-23, each of the values for each of the EOR criteria, for each EOR method has a membership in more than one fuzzy set. If we continue our example with the API gravity and observe Fig. 15a for the hydrocarbon miscible method, we see that for a gravity of 27 the hydrocarbon miscible method has a membership of about 0.3 in "Poor" and a membership of about 0.3 in "Good." These memberships are combined to produce a crisp score. Our example demonstrates how memberships are combined to produce a crisp score.

Since a gravity of 27 for the hydrocarbon miscible method has membership in two sets, two rules are fired, each with a "strength" relative to the set membership value (in this case 0.3 for each rule). The two rules are as follows:

1. If gravity_Hydrocarbon_Miscible is Poor
   Then Score = Poor
2. If gravity_Hydrocarbon_Miscible is Good
   Then Score = Good.

Figure 24 shows the membership functions for the output or the Score. From the rules above we can see that the score should be part "Good" and part "Poor" resulting in a crisp value somewhere between 0 and 10. There are several methods for combining memberships. The one used by our fuzzy expert system is called the Max-Min Inference Method. This method combines the "Good" and "Poor" scores by clipping the output membership function triangles at the height of the membership function value. (in this case the height is 0.3 for both "Good" and "Poor"). The crisp value for the score is the centroid of the combination of these two truncated triangles (in this case, it is the integer value 4). This procedure for returning a crisp value from a fuzzy calculation is known as the centroid defuzzification method. Figure 25 is a composite drawing of a portion of Fig. 15a and a portion of Fig. 24. It shows how the input and output membership functions are connected by the rules and how the crisp output score is computed based on the number of rules fired and the value of the membership function for the rule premises. (e.g., in this case the membership function value for each premise for each rule was 0.3.)

HOW THE EXPERT SYSTEMS WORK

If an engineer were to solve the EOR screening problem by hand, using the backward-chaining or goal-driven method he would pick a goal (for example, the hydrocarbon gas injection method from the left-hand side of Tables I and II). The engineer would then pick the subgoals that would have to be met before the original goal could be satisfied (for example, the gas injection category). This process of picking subgoals would continue as long as necessary, but in our case, it would stop here. The engineer would ask only those questions necessary to determine whether gas injection would be a feasible category. If the feasibility of the gas injection category were established, the engineer would ask only those questions necessary to determine whether the hydrocarbon method would be feasible. If it were not feasible, another goal would be picked. Then the problem would be solved, unless more than one solution was desired, in which case, another goal would be picked and the process continued.

With the forward-chaining, or data-driven, approach, the engineer

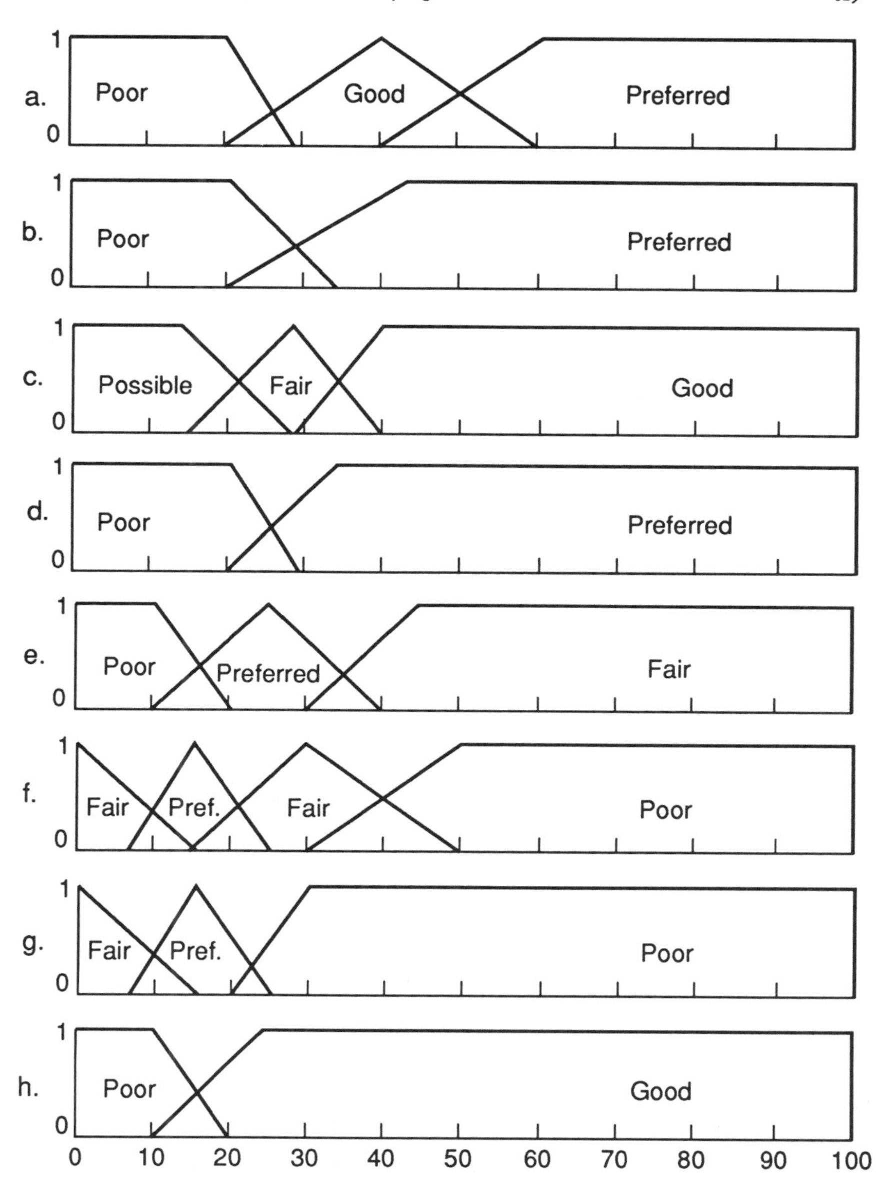

Fig. 15 is Membership functions for the gravity for: (a) Hydrocarbon Miscible, (b) Nitrogen and Flue Gas , (c) Carbon Dioxide, (d) Surfactant/Polymer and Polymer Flooding, (e) Alkaline Flooding, (f) *In Situ* Combustion, (g.) Steam Flooding, and (h) Microbial Drive Method.

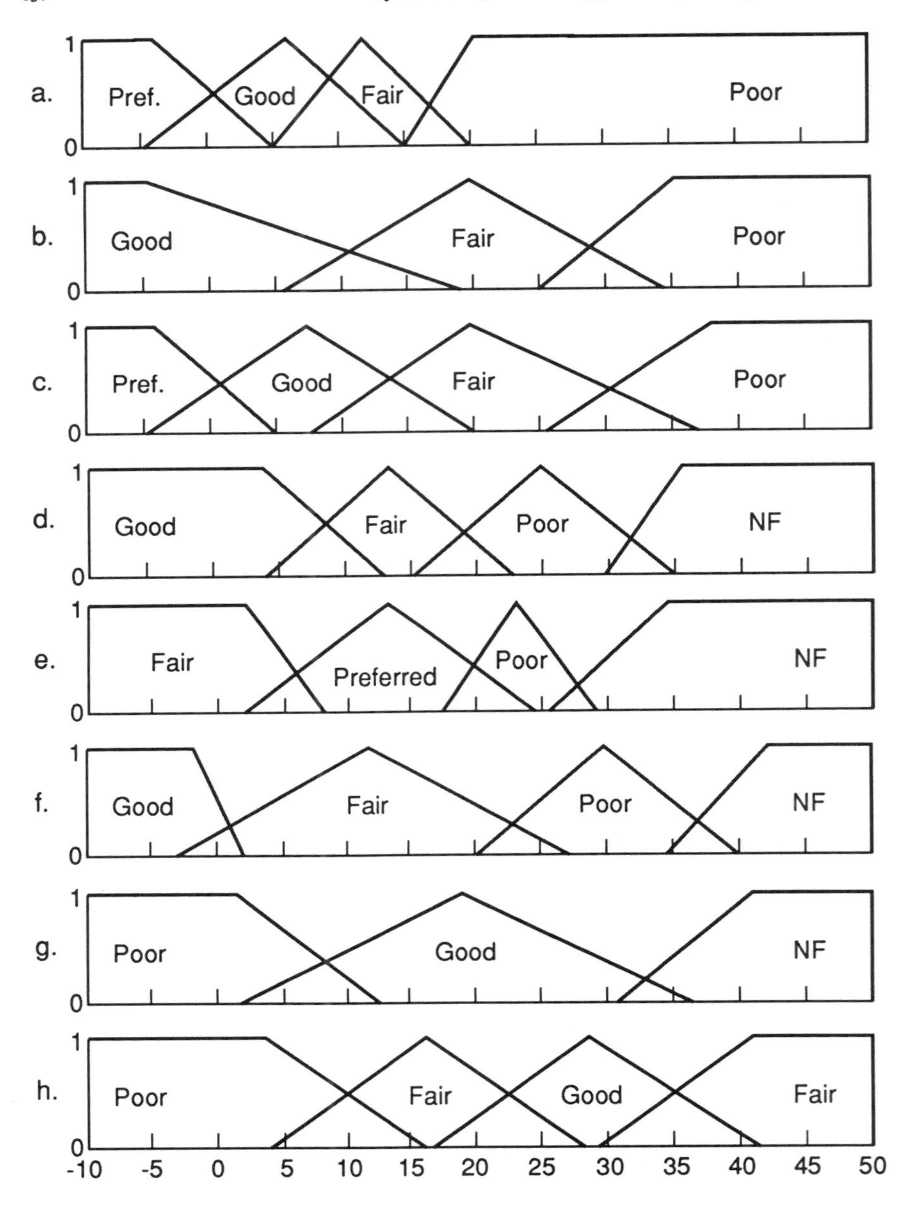

Fig. 16 is Membership functions for viscosity for: (a.) Hydrocarbon Miscible, (b.) Nitrogen and Flue Gas, (c.) Carbon Dioxide, (d.) Surfactant/Polymer, (e.) Polymer Flooding, (f.) Alkaline Flooding, (g.) *In Situ* Combustion, and (h.) Steam Flooding Method.

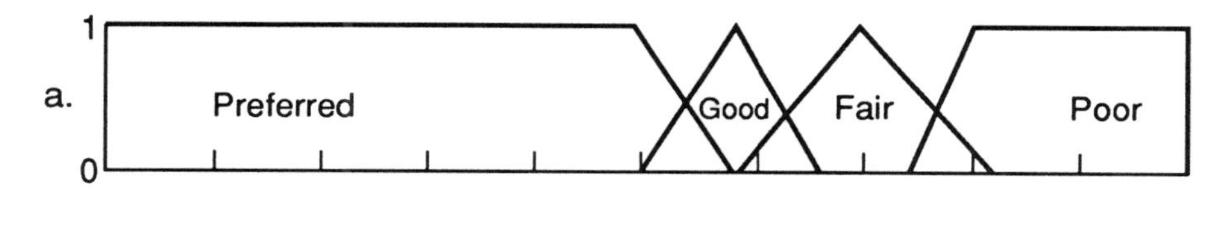

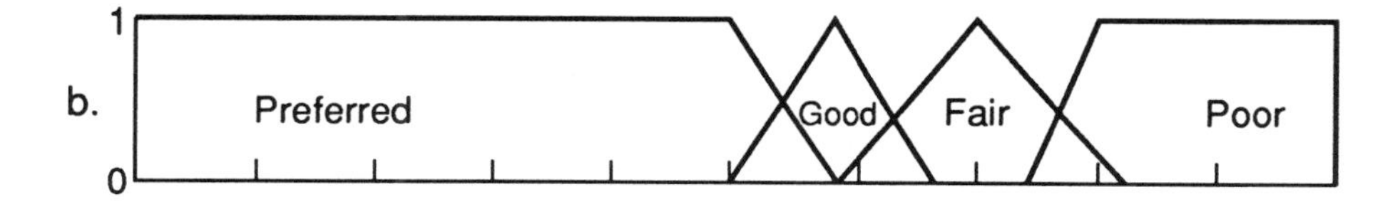

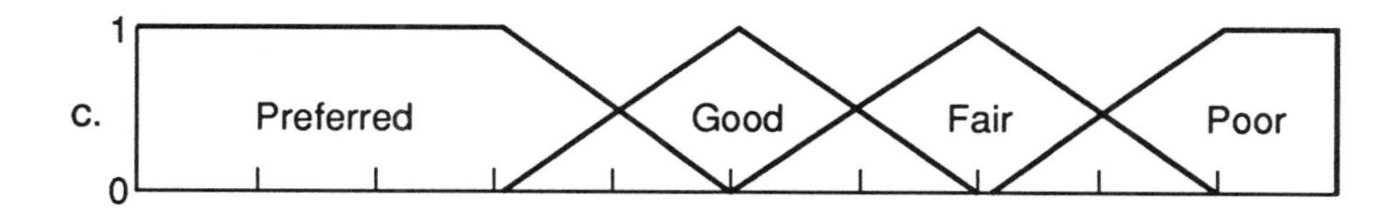

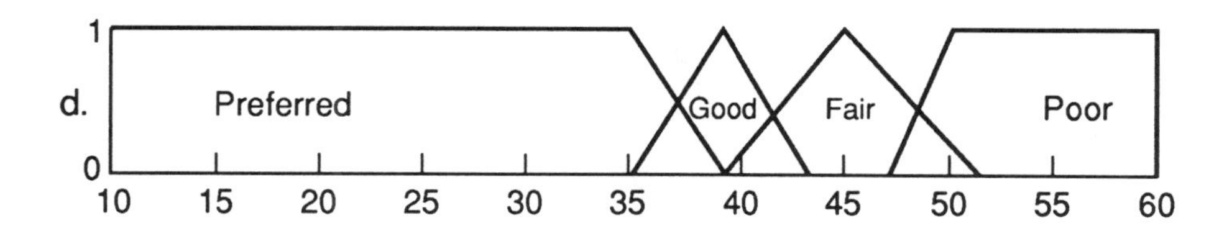

Fig. 17 is Membership functions for salinity for: (a.) Surfactant/Polymer, (b.) Polymer Flooding, (c) Alkaline Flooding, (d.) Microbial Drive method.

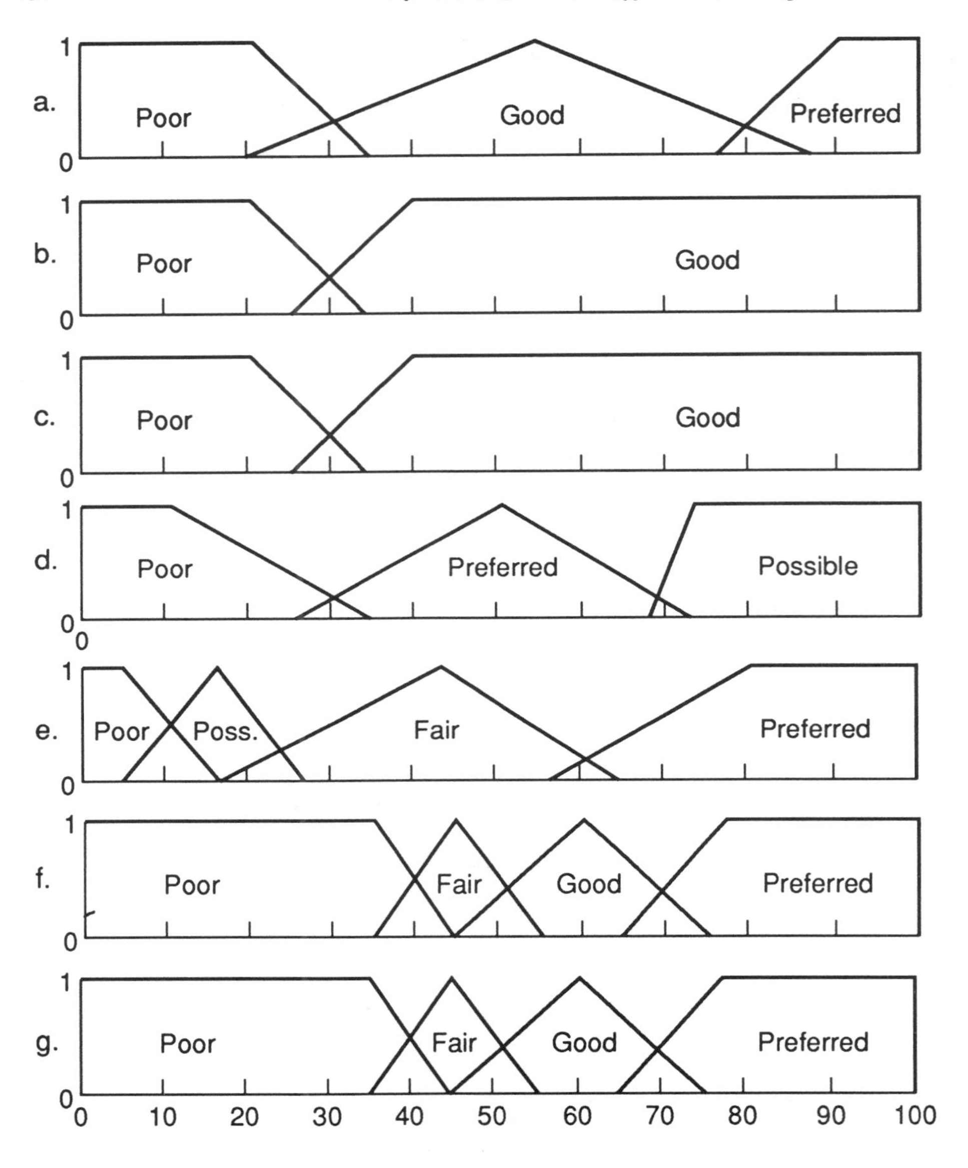

Fig. 18 is Membership functions for oil saturation for: (a.) Hydrocarbon Miscible, (b.) Nitrogen and Flue Gas, (c.) Carbon Dioxide, (d.) Surfactant/Polymer, (e.) Polymer Flooding, (f.) *In Situ* Combustion, and (g.) Steam Flooding Method.

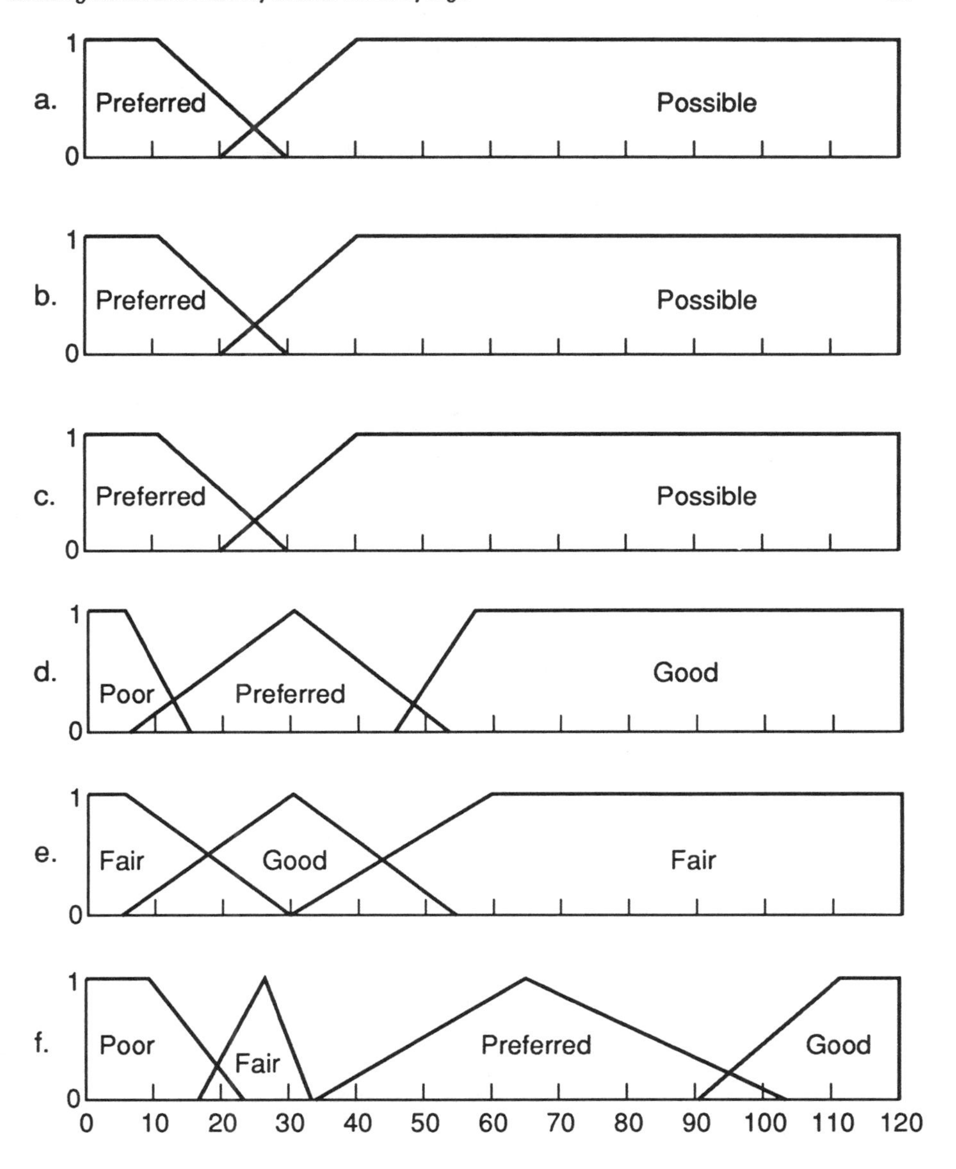

Fig. 19 is Membership functions for thickness for: (a.) Hydrocarbon Miscible, (b.) Nitrogen and Flue Gas, (c.) Carbon Dioxide, (d.) Surfactant/Polymer, (e.) *In Situ* Combustion, and (f.) Steam Flooding Method.

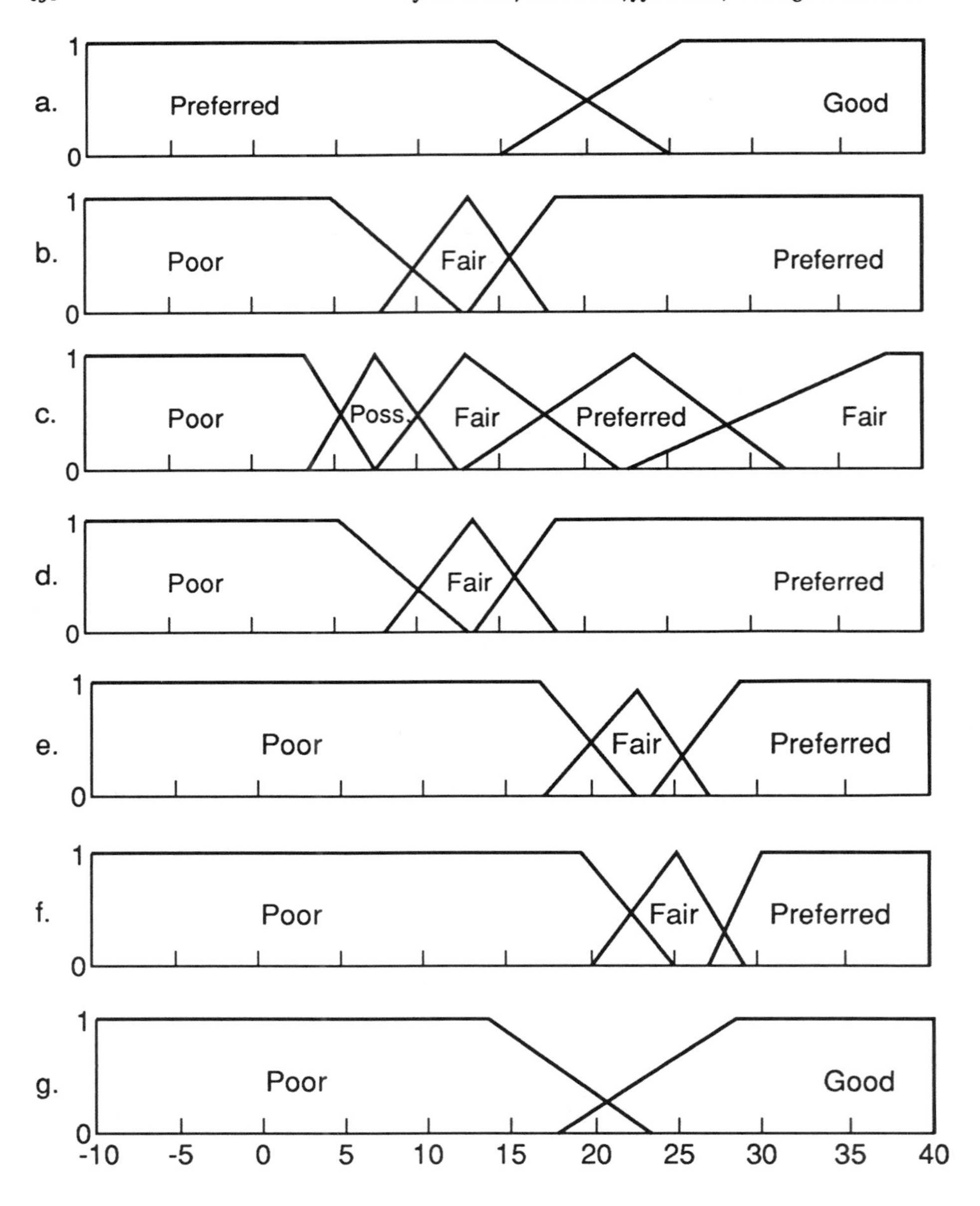

Fig. 20 is Membership functions for permeability for: (a.) Hydrocarbon Miscible, (b.) Surfactant/Polymer, (c.) Polymer Flooding, (d.) Alkaline Flooding, (e.) *In Situ* Combustion, and (g.) Steam Flooding Method, (g.) Microbial Drive Method.

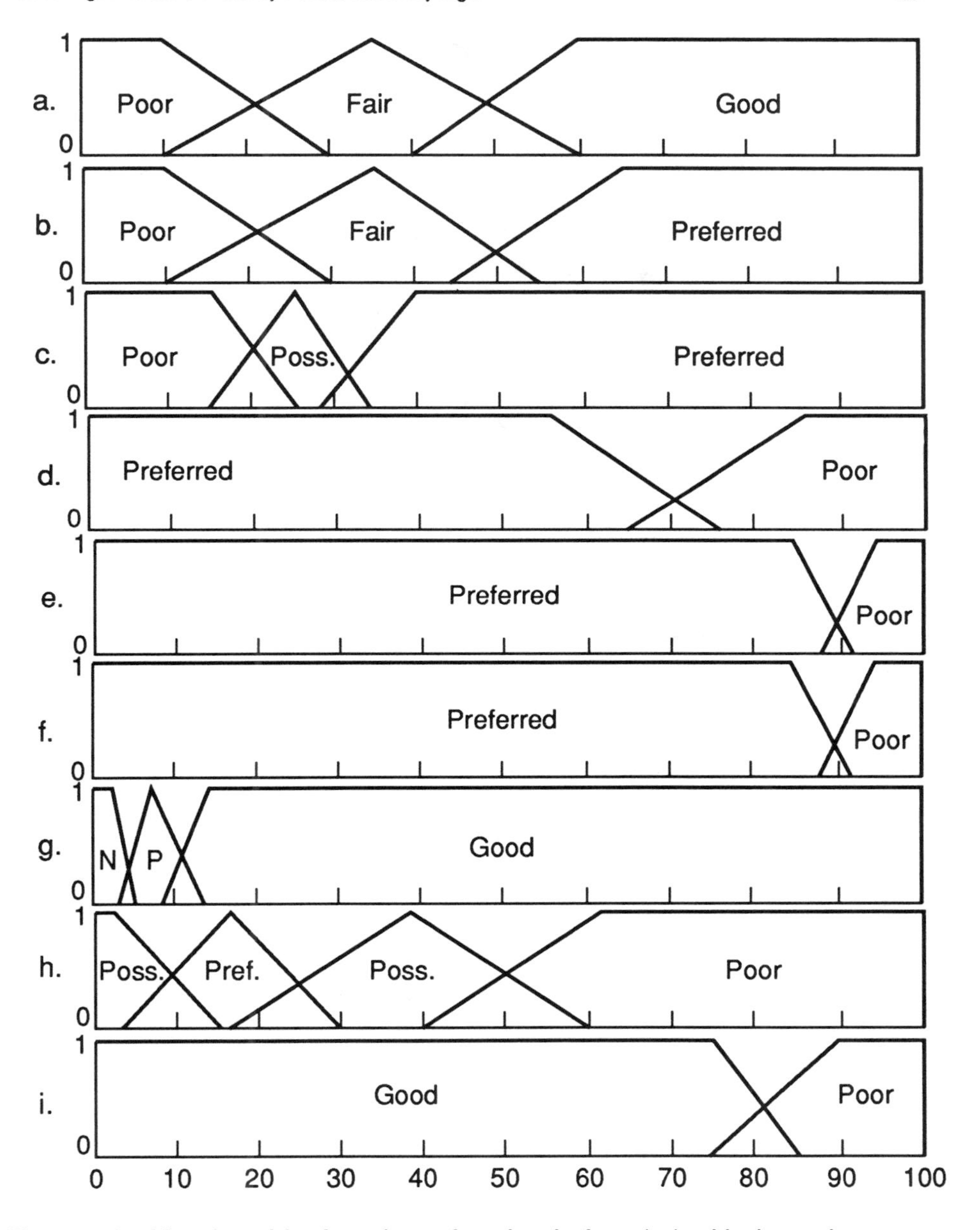

Fig. 21 is Membership functions for depth for: (a.) Hydrocarbon Miscible, (b.) Nitrogen and Flue Gas, (c.) Carbon Dioxide, (d.) Surfactant/Polymer, (e.) Polymer Flooding, (f.) Alkaline Flooding, (g.) *In Situ* Combustion, and (g.) Steam Flooding, (i.)Microbial Drive Method.

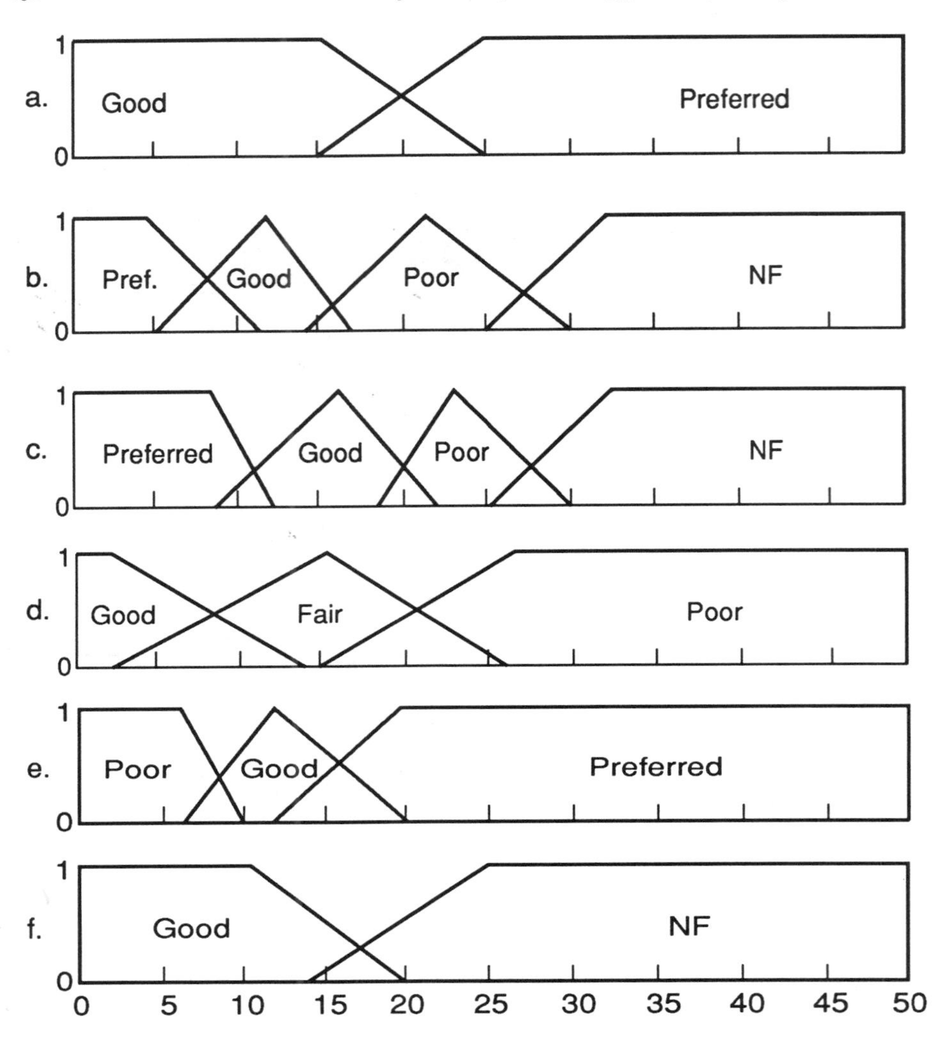

Fig. 22 is Membership functions for temperature for: (a.) Nitrogen and Flue Gas, (b.) Surfactant/Polymer, (c.) Polymer Flooding, (d.) Alkaline Flooding, (e.) *In Situ* Combustion, and (f.) Microbial Drive Method.

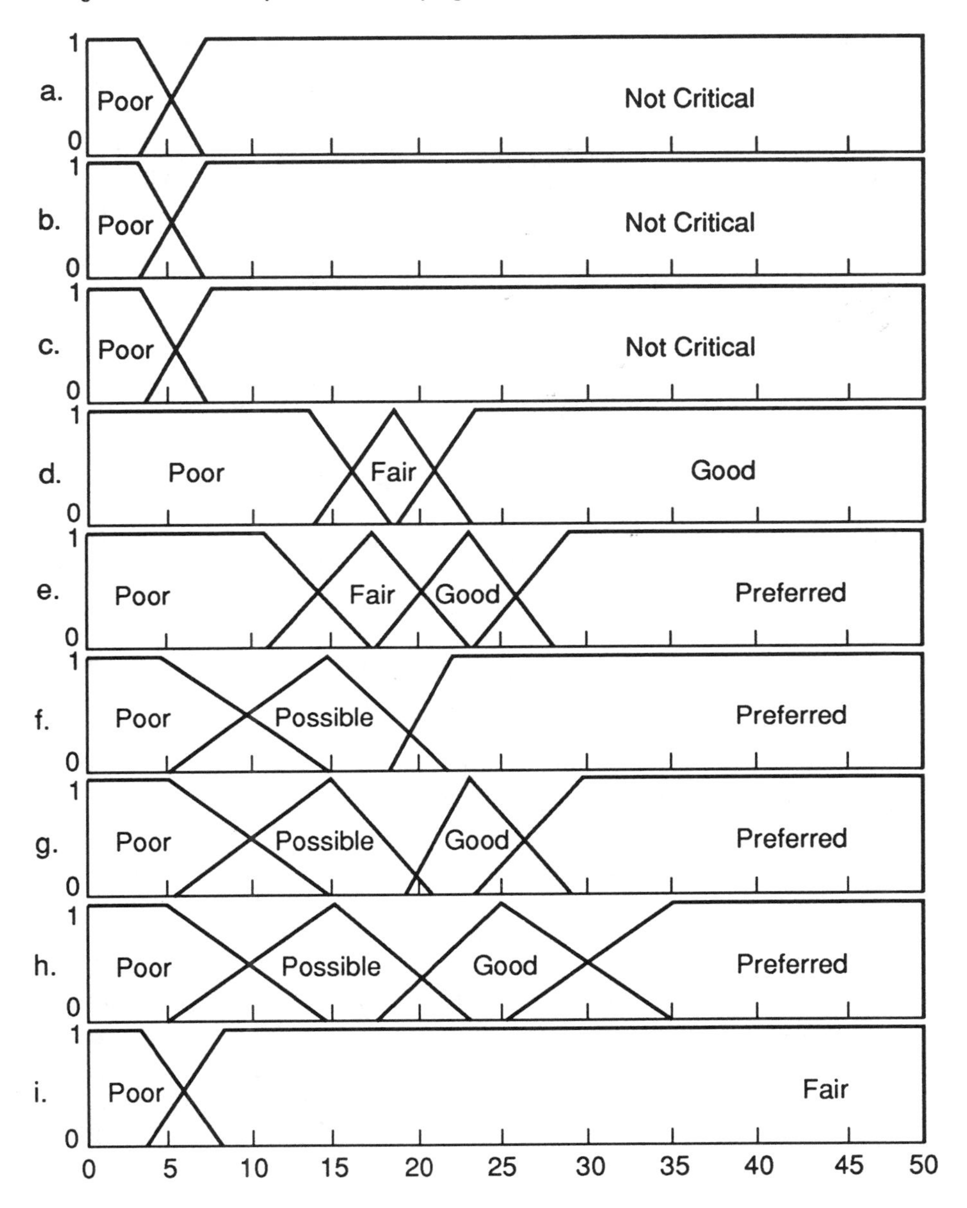

Fig. 23 is Membership functions for porosity for: (a.) Hydrocarbon Miscible, (b.) Nitrogen and Flue Gas, (c.) Carbon Dioxide, (d.) Surfactant/Polymer, (e.) Polymer Flooding, (f.) Alkaline Flooding, (g.) *In Situ* Combustion, and (h.) Steam Flooding, (i.) Microbial Drive Method.

lets the data help search through the search tree (the system keeps asking questions until it is clear which node to move to next).

The crisp expert system, the first one assembled, uses backward-chaining. With this system, the approach is to first assume that hydrocarbon injection is going to work. For hydrocarbon injection to work, the category of gas injection must be applicable. In order for gas injection to be applicable both the oil property data and the reservoir data shown in Figs. 4-14 must have scores greater than preprogrammed threshold values.

The program first tries to verify these subgoals by asking questions about gravity, viscosity, oil composition, etc, then continues until a final goal is met or until an assumption is rejected at some level. When an assumption is rejected, that branch of the search tree is pruned, and the program then moves to the next unpruned branch to the right and picks that EOR process as a goal. This process continues until a solution is found. Since we want a ranked list of candidate EOR methods, the program searches the tree until all possible solutions are found. When the search is finished, the solutions are printed, with a score for each qualifying method.

Figure 26 is a portion of an *and/or graph* for a portion of the search space for the crisp version of the expert assistant. The graph is so called because the branches connected by an arc are ***and*** branches (all of the leaves must be true, and in this case, must have a preprogrammed minimal score, before the branch is resolved). The unarced branches are ***or*** branches. They require only a single truth (minimal score) for resolution.

The fuzzy expert system was written next. It uses forward-chaining and, essentially, an exhaustive search. It starts with the API gravity of the oil in the reservoir (Fig. 15) and assigns a score to each EOR method. Then it moves on to viscosity (Fig. 16) and repeats the procedure. This procedure is repeated until all 11 EOR categories are checked. The fuzzy expert system actually uses some crisp rules, in combination with the fuzzy rules. Figure 6 shows oil composition screening data. Figure 9 shows the screening data for the reservoir rock formation type. In both of these cases there was not enough information to fuzzify the data. So for this study the rules for these two figures remained crisp.

Another area where the rules remain crisp is one in which an EOR criterion offers no options for an EOR method. Figure 7 shows screening data for formation,salinity. For five of the nine EOR methods formation salinity is not critical. This gives rise to five crisp rules.

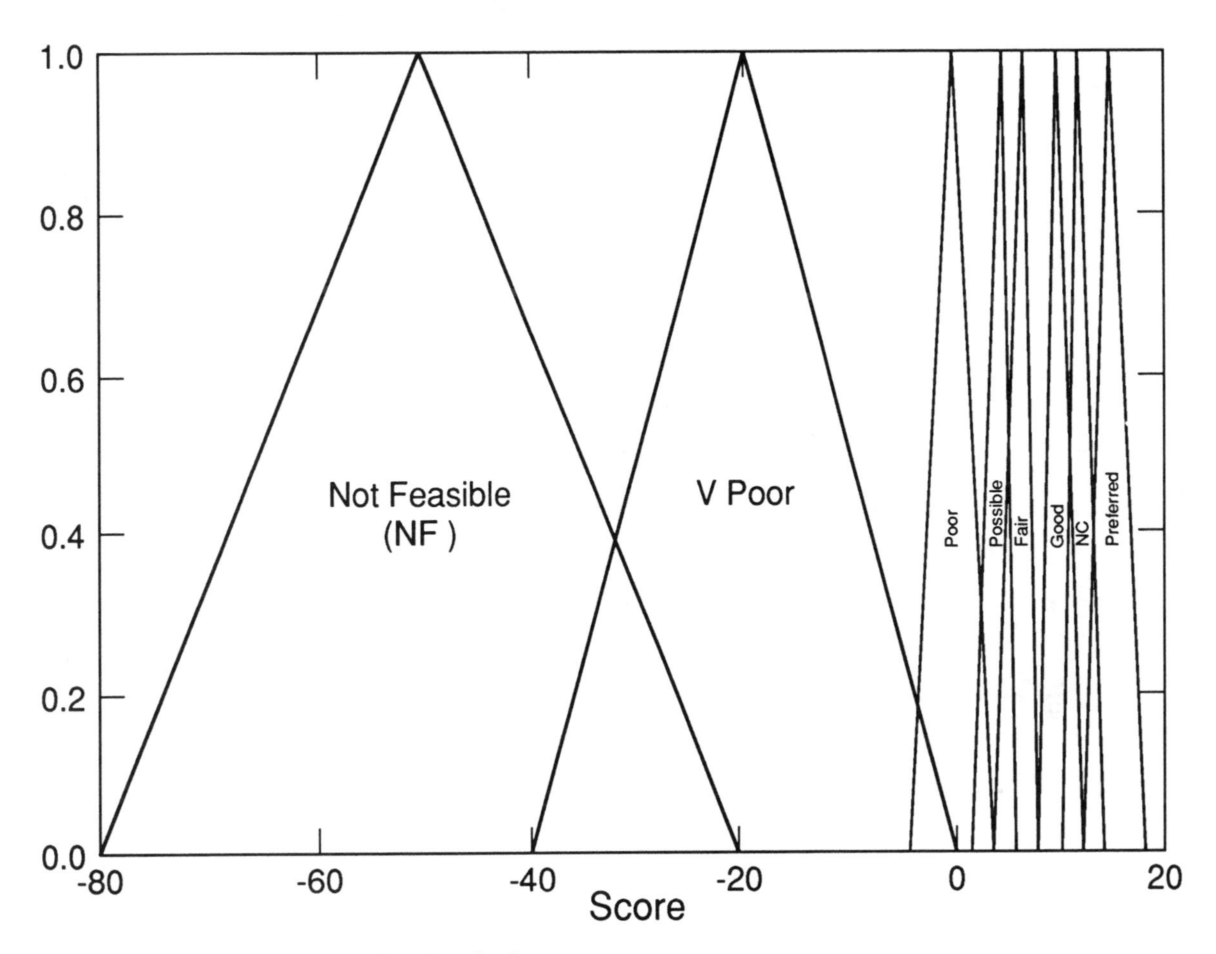

Fig. 24. Output membership functions for the score.

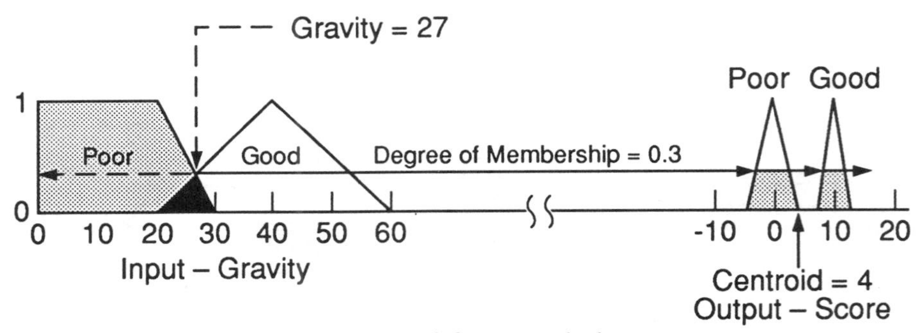

Fig. 25. Demonstration of the max-min inference method.

PROGRAM COMPARISONS AND SUMMARY

One difference between the two expert systems is the tool, or expert system shell, used. Each system uses a different shell. The crisp expert system was written with the expert system shell, CLIPS [12], developed by NASA. CLIPS is a forward-chaining shell written in

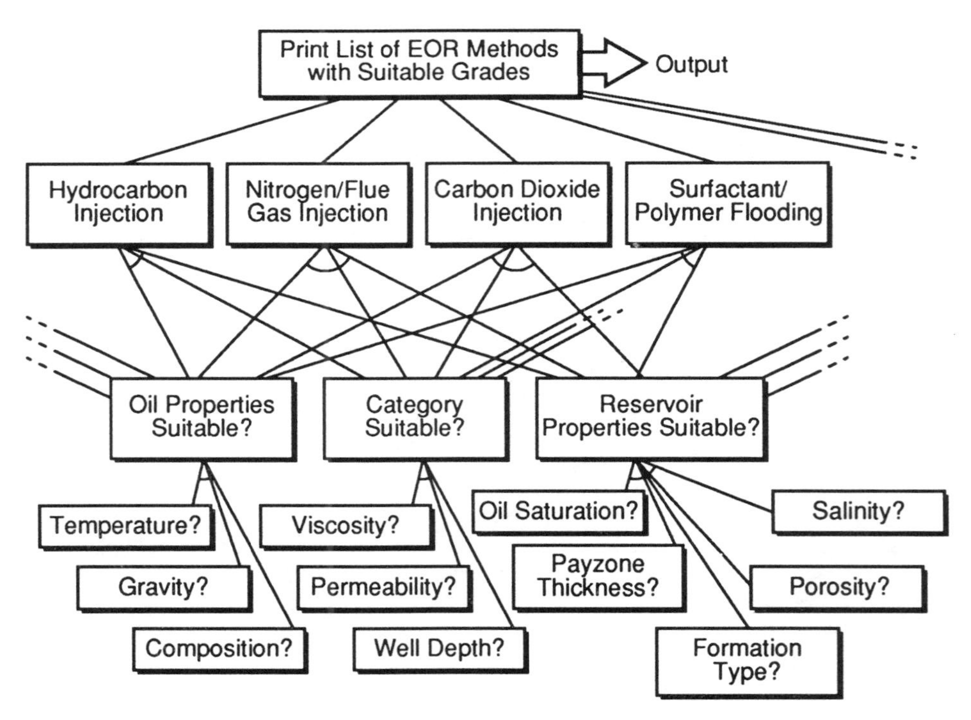

Fig. 26. *And/or* graph for a portion of the search space for the CLIPS backward-chaining version of the problem.

the C programming language. It is a very versatile and flexible shell, which can also be used to write expert systems in the backward-chaining mode, as was done for the crisp expert system (backward-chaining was used because it is more intuitive and, therefore, easier to prune search trees).

Examples 3 and 4 point out, however, that there is a definite potential for serious errors because of the sharp boundaries of the crisp sets shown in Figs. 4-14.

The fuzzy expert system was written to eliminate this potential problem and to add some human-like fuzzy reasoning to the otherwise rigid crisp expert system. This expert system was written with the Togai Fuzzy C development system (13). This system does a lot of work for the programmer; it makes it easy to enter membership functions, such as those shown in Figs. 15-23, and it computes the necessary centroids, as demonstrated in Fig. 25. This system shell is harder to use than CLIPS because the programmer must write a C language program to drive the Fuzzy C program. This means that the programmer has to write the search routines and other peripheral management software that is typically already supplied with shells like CLIPS. Although this allows more flexibility, a great deal of time is required to write search routines with the sophistication of those found in CLIPS. Because it was easiest to write, a forward-chaining exhaustive search was used on this expert system. Still, extensive coding was required.

This expert system does a much better job on problems such as those discussed in Examples 3 and 4. In Example 3, the crisp expert system causes dramatic changes between the two Scenarios, even though the input data for the two scenarios are very similar. The results shown in the ranked list presented with Example 3 are from the crisp expert system. The following results are from the fuzzy expert system.

Fuzzy Expert System Solution for Example 3

Scenario One (Rankings)

1-	Alkaline flooding	109 points
2-	Polymer flooding	107 points
3-	Surfactant/polymer	101 points
4-	Carbon dioxide	97 points
5-	Microbial drive	89 points
6-	Hydrocarbon miscible	86 points
7-	*In situ* combustion	83 points

8-	Nitrogen and flue gas	82 points
9-	Steam flooding	81 points

Fuzzy Expert System Solution for Example 3

Scenario Two (Rankings)

1-	Alkaline flooding	112 points
2- (tie)	Polymer flooding	109 points
3- (tie)	Surfactant/polymer	109 points
4-	Carbon dioxide	102 points
5- (tie)	Microbial drive	89 points
6- (tie)	Hydrocarbon miscible	89 points
7- (tie)	*In situ* combustion	87 points
8- (tie)	Nitrogen and flue gas	87 points
9-	Steam flooding	78 points

Only small changes occur between Scenarios One and Two when the fuzzy expert system is used. In fact the only changes are small changes in the total points awarded. The relative rankings are not really changed.

Example 3 is intended to be a realistic problem, but it is a worst case. The overall raw scores or points produced in the fuzzy version of Example 3 show little increase from Scenario One to Scenario Two. This means that the predicted viability of the EOR methods will not be unduly enhanced by small changes in the input data by the fuzzy expert system.

In Example 4 (Scenario One) the crisp expert system ranked *in situ* combustion as the second best method and surfactant/polymer as fourth best. In Scenario Two, the only change in the input data was an increase of 0.2% in the oil viscosity, hardly a measurable change. This change caused the *in situ* combustion and surfactant/polymer methods to be discarded because they were "Not Feasible." The fuzzy expert system keeps *in situ* combustion as the second best method and surfactant/polymer as the fourth best method in both scenarios, partly because the abscissae, shown in Fig. 5 and used in the crisp expert system, were converted to a logarithmic scale and plotted linearly in Fig. 16. This is also the way they are used in the fuzzy expert system.

The transformation equation is as follows: transformed-viscosity = (integer) ($10^*\log_{10}$ (viscosity) + 0.5). (The scale shown in Fig. 16 is linear data plotted on a logarithmic graph.) The transformation itself tends to fuzzify the set boundaries. The transformation was made because the fuzzy expert system shell doesn't handle very large numbers or long scales very well. The fuzzy membership functions also help fuzzify the set boundaries. But when any other output membership function is combined with the "Not Feasible" output membership function, with its centroid at -50, it's hard to make the result of the boundary change very gradual. The -50 score was designed to dramatically reduce the raw score of an EOR method that was thought to be "Not Feasible". This is a good idea if the criterion value is not near the set boundary. Even though a change in feasibility may be quite dramatic as the criterion value changes, it most likely is not a step function. Complete resolution of this problem will require a little more work.

The fuzzy expert system is much better at solving problems such as those in Examples 3 and 4 than the crisp expert system is. Although these "worst case" problems do not represent the majority of EOR screening problems, they are real, and some degree of the crisp set boundary problem is present in almost every EOR screening problem. Our crisp expert system works more like a classical expert system than the fuzzy expert system does. The crisp system works interactively with the user. It tries to prune the search tree and it offers a simple explanation facility. On the other hand, with the fuzzy expert system, users enter the data and wait for all of the scores to be computed. If the users want some explanation, they can request a dump and watch the progress of the score calculation.

Some of the differences between the two expert systems occur because fuzzy expert systems are designed to fire all the rules that apply to the problem, even those that have only a minor influence on the outcome. A conventional expert system, like the crisp expert assistant, does just the opposite, that is, it tries to prune the search tree by eliminating any consideration of rules that have little or no influence on the problem outcome.

The final issue we will discuss is the development of the membership functions for the fuzzy sets shown in Figs. 15-24. Reference 13 states that, "Determining the number, range, and shape of membership functions to be used for a particular variable is somewhat of a black art." It further states that trapezoids and triangles, such as those shown in Figs. 15-24, are a good starting point for membership functions. Trapezoids and triangles served as a starting point for membership functions for this project. The membership functions in Figs. 15-24 are still trapezoids and triangles, but many of them are different from those used as the starting points. Some effort was expended polishing the membership functions, and several changes were made. In many cases the changes made little difference in the final scores, but in some cases they made a great deal of difference. Ideally, we would expect the triangular membership functions to resemble bell-shaped curves and the trapezoids to resemble S-shaped curves. References 10 and 14-16 suggest methods for determining better membership functions. Example 4 shows that, in at least some cases, there is a need for improved membership functions. Improving the membership functions will require taking a harder look at the available data and will be the subject of another study. The idea of using neural nets, fuzzy pattern recognition, or genetic algorithms [16] to "teach" the membership functions to improve their shape is intriguing and should be considered for a future project.

REFERENCES

1. Stosur, J. G. ,"The Potential of Enhanced Oil Recovery," *International Journal of Energy Research, Vol.* 10, 357-370 (1986).

2. Taber J. J. and F. D. Martin, "Technical Screening Guides for Enhanced Recovery of Oil," paper presented at the 58th Annual Society of Petroleum Engineers Technical Conference, San Francisco, California, October 5-8, 1983 (SPE 12069).

3. Goodlett, G. O., H. M. Honarpour, H. B. Carroll, and P. S. Sarathi, "Lab Evaluation Requires Appropriate Techniques-Screening for EOR-I," *Oil and Gas Journal,* 47-54 (June 23, 1986).

4. Goodlet, G. O., H. M. Honarpour, H. B. Corroll, P. Sarathi, T. h. Chung, and D. K. Olsen, "Screening and Laboratory Flow Studies for Evaluating EOR Methods," Topical Report DE87001203, Bartiesville Project Office, USDOE, Bartlesville, Oklahoma, November 1986.

5. Parkinson, W. J., G. F. Luger, R. E. Bretz, and J. J. Osowski, "An Expert System for Screening Enhanced Oil Recovery Methods," paper presented at the 1990 Summer National Meeting of the American Institute of Chemical Engineers, San Diego, California, August 19-22, 1990.

6. Donaldson, E. C., G. V. Chilingarian, and T. F. Yen, (Editors), *Enchanced Oil Recovery, I Fundamentals and Analysis* (Elsevier, New York, 1985).

7. Poettmann, F. H., (Editor), *Improved Oil Recovery* (The Interstate Oil Compact Commission, Oklahoma City, Oklahoma, 1983).

8. "Enhanced Recovery Methods are Worldwide" (Petroleum Publishing Company, 1976). (Complied from issuses of *The Oil and Gas Jounal*).

9. Luger, G. F. and W. A. Stubblefield, *Artificial Intelligence and the Design of Expert Systems* (The Benjamen/ Cummings Publishing Company, Inc., Redwood City, California, 1989).

10. Giarratano, J. C. and G. Riley, *Expert Systems - Principles and Programming* (PWS - Kent Publishing Company, Boston Massachusetts, 1989).

11. Parkinson, W. J., K. H. Duerre, J. J. Osowski, G. F. Luger, and R. E. Bretz, "Screening Enchanced Oil Recovery Methods with Fuzzy Logic," paper presented at the Third International Reservoir Characterization Technical Conference, Tulsa Oklahoma, November 3-5, 1991.

12. Giarratano, J. C., *CLIPS User's Guide, Version 4.3*, Artificial Intelligence Section, Lyndon B. Johnson Space Center, June 1989.

13. Hill, G., E. Horstkotte, and J. Teichrow, *Fuzzy-C Development System User's Manual, - Release 2.1*, Togai Infralogic, Inc., June 1989.

14. Turksen, I. B., "Measurement of Membership Functions and Their Acquisition," *Fuzzy Sets and Systems* (40) 538 (Elsevier Science Publishers B. V., North-Holland, 1991).

15. Klir, G. J. and T. A. Folger, *Fuzzy Sets, Uncertainty, and Information* (Prentice Hall, Englewood Cliffs, New Jersey, 1988).

16. Karr, C., "Genetic Algorithms for Fuzzy Controllers," AI Expert, February 1991, pp. 2633.

AFTER THE FIRE IS OUT: A POST *IN-SITU* COMBUSTION AUDIT AND STEAMFLOOD OPERATING STRATEGY FOR A HEAVY OIL RESERVOIR, SAN JOAQUIN VALLEY, CA

Paul G. Soustek
James M. Eagan[1]
Mark A. Nozaki

Mobil Exploration & Producing U.S. Inc.
Bakersfield, CA 93389

Mary L. Barrett

Mobil Exploration & Producing U.S. Inc.
Denver, CO 80217

I. INTRODUCTION

The purposes of this paper are to: (1) document the mineralogical changes within *in-situ* combustion zones of the Webster reservoirs; (2) show that consistent change in log signatures permit mapping of the postburn intervals; and (3) demonstrate how this analysis was incorporated into a steamflood simulation to yield an operating policy for a dipping, partially burned reservoir. This detailed work resulted from the combined efforts of geologists, petrophysicists, and reservoir engineers.

Midway-Sunset Field is in the southwest corner of the San Joaquin valley in Kern County, California approximately 40 miles southwest of Bakersfield and 100 miles north of Los Angeles (Figure 1). Midway-Sunset is one of twelve giant oil fields in Kern County. In 1990 it was the second largest U.S. oil field with annual production exceeding 59 million barrels of oil. Cumulative field production exceeded 2

---

[1]Present address: Consultant, Denver, CO 80231

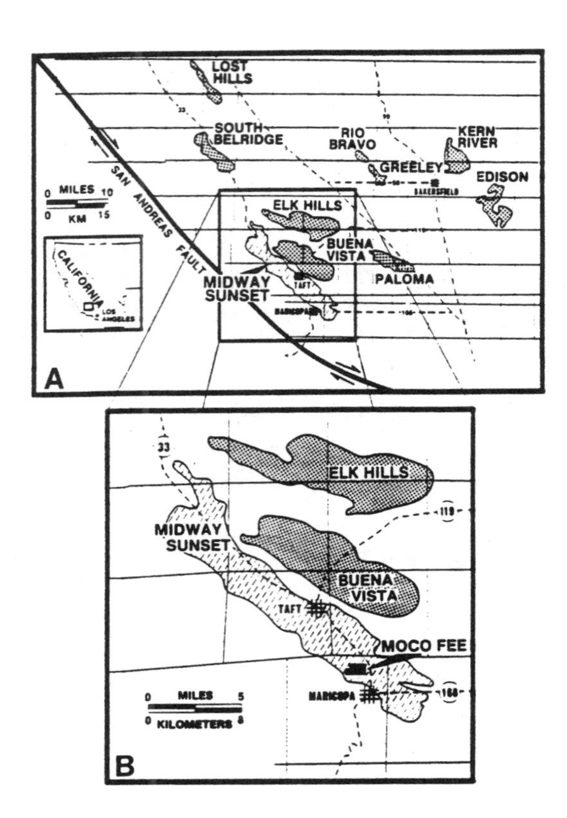

Fig. 1. Location maps showing (A) Midway-Sunset Field and other major oil fields of the southern San Joaquin Valley and (B) Mobil's MOCO fee property within Midway-Sunset Field reprinted from Link and Hall (1990).

billion barrels in early 1991. Midway-Sunset is approximately 25 miles (40 kilometers) long and 4 miles (6 kilometers) wide. MOCO is a one and three-quarter section Mobil fee property in the southern part of the field. This paper is limited to activities on the MOCO property in the Webster turbidite reservoirs.

Field production is from numerous unconsolidated sand reservoirs of Oligocene to Pleistocene age with local Miocene diatomaceous mudstone reservoirs. Sands within the upper Miocene Monterey Formation are the dominant reservoirs. The Webster sands occur within the Antelope Shale member of the Monterey Formation (Figure 2). Figure 3 shows a paleogeographic reconstruction of the southern San Joaquin

AGE	FORMATION	MEMBER	PRODUCING ZONE	DRILL DEPTH FEET (METERS)
PLEIST.	TULARE			200 (60)
PLIOCENE	SAN JOAQUIN		"O" SAND	
PLIOCENE	ETCH-EGOIN			
MIOCENE	MONTEREY	REEF RIDGE	LAKEVIEW	
MIOCENE	MONTEREY	ANTELOPE SHALE	MONARCH	
			WEBSTER INTERMEDIATE	1300 (390)
			WEBSTER MAIN	1600 (480)
			OBISPO SHALE	
			MOCO "T"	
			PACIFIC SHALE	3200 (970)
		McDONALD SHALE		

Fig. 2. Generalized stratigraphic column for the MOCO area, showing the positions of the Webster Intermediate and Webster Main reservoirs relative to other producing sands and their drill depths.

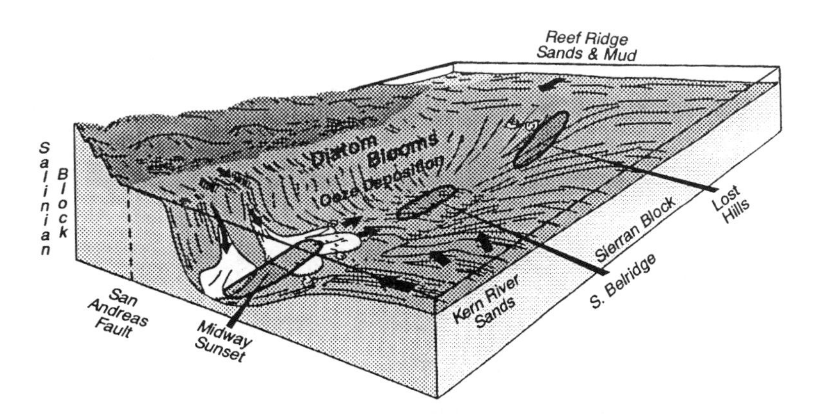

Fig. 3. Schematic paleoreconstruction of the southern San Joaquin Valley during late Miocene time reprinted from Schwartz (1988).

valley in late Miocene time. Sediments derived from highlands within the Salinian block flowed downslope as turbidites, being deposited in an intraslope basin separated from the Maricopa-Tejon sub-basin (Webb, 1981; Bartow, 1987; Ryder and Thomson, 1989). The narrow, tectonically active shelf of the Salinian block was adjacent to the San Andreas Fault. A series of en echelon folds evolved along the west flank of the San Joaquin basin beginning in middle Miocene time with folding continuing through the Pleistocene (Harding, 1976; Bartow, 1987).

Midway-Sunset field includes three east-west trending, east plunging anticlines within an overall area having eastern homoclinal dip along the east flank of the Temblor uplift with converging unconformities. The 35 anticline, present at MOCO, is one of these structures (Figure 4). Traps include 4-way anticlinal closures, truncation traps, updip tar seals on homoclinal dip and structural-stratigraphic traps. Oil was derived from organic-rich diatomaceous mudstones of the Monterey Formation, the main source rock sequence in the San Joaquin basin (Isaacs, 1987).

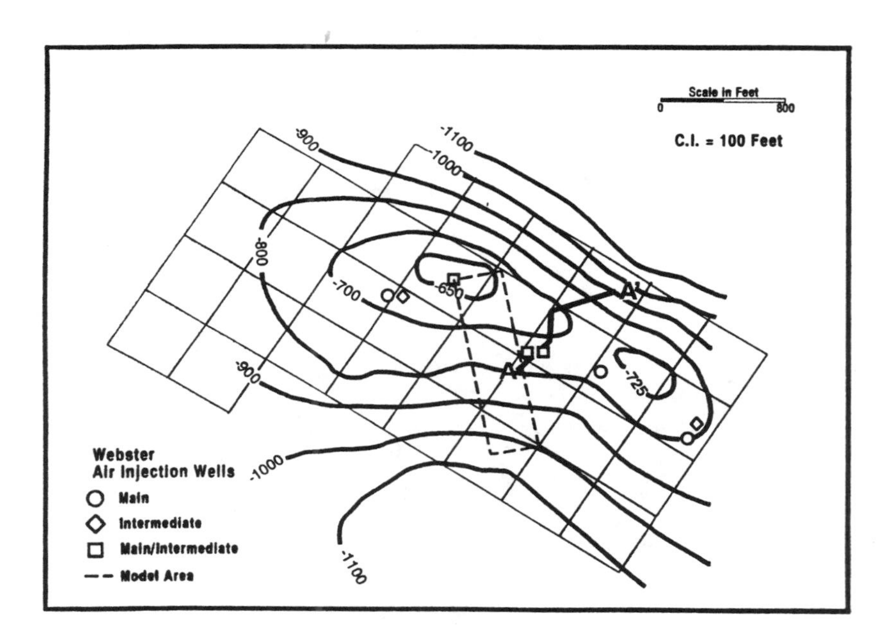

Fig. 4. Structure map contoured on the base of the Webster Main D sand within the steamflood development area. The location of stratigraphic cross-section A-A' is shown, as is the steamflood model study area.

Individual reservoir sands were first discovered in 1894 within present field limits. The Webster turbidite sands were first penetrated in 1910 with first production in 1913. Primary development surged in the late 1910s and the 1940s due to war. *In-situ* combustion began inadvertently in 1961 when air injected to promote the MOCO T *in-situ* combustion project leaked to the overlying Webster and ignited spontaneously. An *in-situ* combustion project in both the Webster Intermediate and Main ran between 1964 and 1976. Cyclic steam stimulation of Webster producers began in 1966 and continues. Mobil initiated a steamflood pattern development in 1989. This development is expected to continue through 1996. The Webster reservoirs on the MOCO property have yielded 13.9 million barrels of oil from 1913 to 1990.

In 1989, 66 Webster wells were drilled to fully develop 12 steamflood patterns. Thirty-three wells had extensive logging programs and three wells were continuously cored. In 1990, an additional 11 Webster wells were drilled. All had detailed logging programs and one was continuously cored. This intensive drilling program to develop 100 acres permits detailed correlation and analysis.

The forward dry combustion process involves air injection, ignition in the reservoir and propagation of a combustion front. Oil is displaced by hot gases moving ahead of the combustion front and by steam resulting from combustion and vaporization of connate water (Boberg, 1988). This process is depicted in Figure 5 which shows temperature distribution and location of various zones for an idealized one-dimensional case.

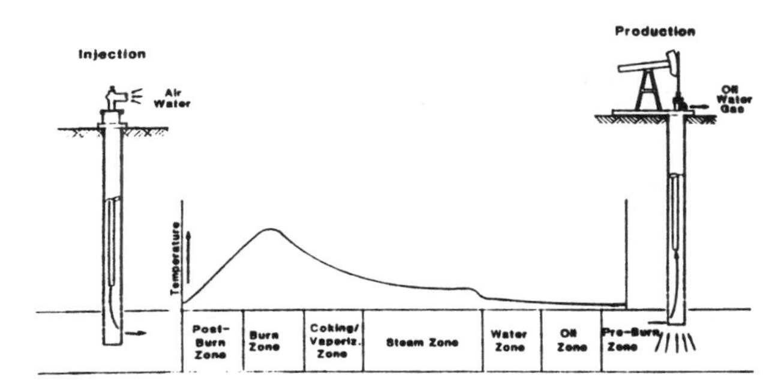

Fig. 5. The zonation of *in-situ* combustion, reprinted from Tilley and Gunter (1988). The first rock alteration probably begins in the hot water zone in front of steam. After completed coke combustion in the burn zone, injected air may saturate rocks, oxidize Fe-rich minerals and leave a reddish color.

Nearly 16.5 BCF of air was injected into the Webster reservoirs during this project with an average injection rate of 3.4 MMCFPD. This dry combustion project was initiated as a crestal drive in both the Webster Intermediate and Main. The burn was designed, in part, to heat the oil at the top of the structure and promote migration of oil toward the flanks of the anticline. New producers were not drilled for this project; air injectors were both drilled and converted.

II. WEBSTER STRATIGRAPHY

The Webster Main reservoirs within the steamflood development area were subdivided into 5 sand bodies (A-E), which are laterally continuous (Figure 6). The Webster Intermediate sand bodies are generally elongate, lenticular bodies which onlap one another becoming younger from northwest to southeast. Figure 7 shows the center of the Webster Main depositional system, thinning to the east and west. An isochore thin tracks the axis of the 35 anticline, thickening on both the north and south flanks. The individual (A-E) sand body isochores have very similar geometries.

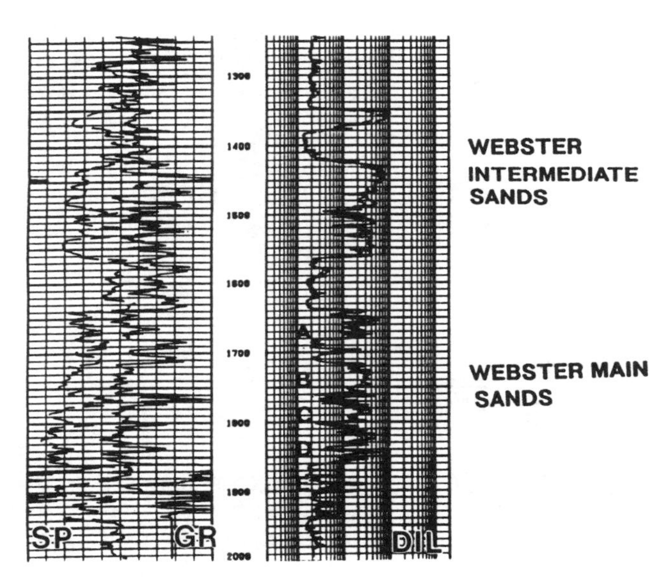

Fig. 6. Type log of the Webster reservoirs at the MOCO property showing the individual Webster Main sands (A-E).

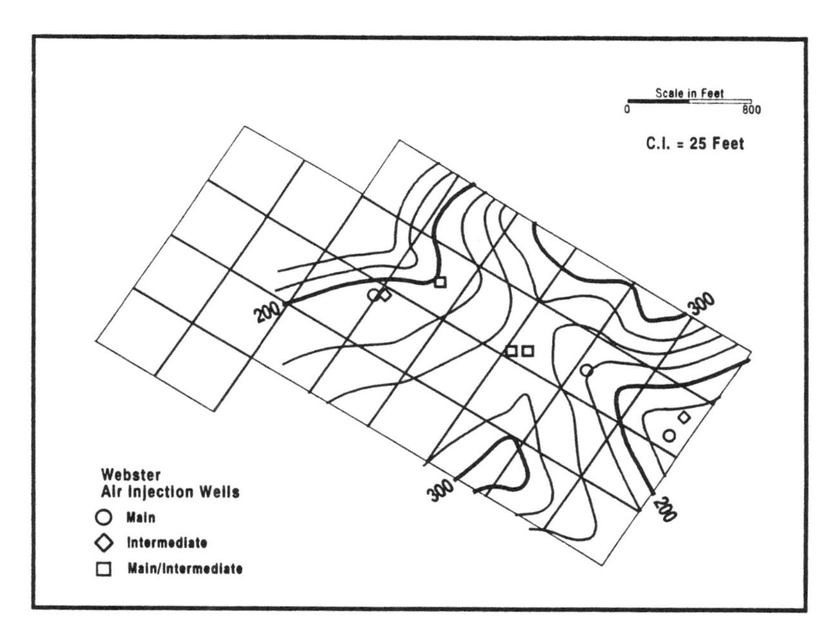

Fig. 7. Isochore map of the gross Webster Main (A-E) interval across the steamflood development area.

The Webster Main reservoirs are very fine to medium-grained, well sorted, arkosic sands. They are separated by laterally continuous, 5- to 30- foot intervals of diatomaceous mudstone. The sands have porosities ranging between 25 and 33% and air permeabilities between 0.5 and 2 darcies, while the mudstones have greater than 45% porosity with air permeabilities less than 50 millidarcies. The Webster Intermediate sands consist of very fine to very coarse, poorly sorted, arkosic sands with pebble to boulder conglomerates. The Webster Intermediate sand bodies are commonly in communication; muds deposited between periods of sand deposition were eroded so conglomerates are deposited directly on older sands.

Link and Hall (1990) proposed that Webster Main reservoirs were sheet-like depositional lobes and the overlying Webster Intermediate sand bodies were primarily more proximal turbidite channel-fill deposits. The additional 1989 and 1990 drilling confirms this depositional model.

III. WEBSTER MAIN PETROPHYSICAL PROPERTIES

The Webster reservoirs at the MOCO property have a basic problem typical of most, old producing reservoirs. Early petrophysical and reservoir data is lacking. The original reservoir pressure is unknown as is the original Rt, Rw, Sw, So and Sg. The first usable E-logs were from wells drilled in 1948, 35 years after initial production.

Old E-logs from wells drilled between 1948 and 1960 have shallow resistivities ranging between 25 and 100 ohm-meters in oil saturated Webster sands. Since reservoir porosity has remained constant, pre-fireflood oil saturations are estimated at 66%.

TABLE I. Reservoir Characteristics of Webster Main Sands

Gross thickness (feet)	150-300
Net pay thickness (feet)	120-240
Porosity (percent)	25-33
Permeability (darcies)	.5-2
Initial O.I.P./acre ft. (bbls.)	1673
Initial O.I.P., total (bbls.)	84×10^6
Viscosity at reservoir temperature (cp)	1630 at 110°F
Oil gravity (°API)	14

The 1989 steamflood development was installed on the crest of the 35 anticline thus overlying the prior Webster *in-situ* combustion project. The shallow resistivities recorded in the 1989-1990 wells were generally reduced to 15 to 40 ohm - meters over the intervals of pay. Large areas of gas saturation were encountered at many locations, indicated by neutron-density crossover exceeding 3 log divisions. The induction logs exhibited normal or higher than normal resistivities where the crossover occurred. This log character was correlated to cores and identified as responses to burned reservoir (Figure 8). Some other wells exhibited high neutron-density crossover but the resistivities were reduced. This log character was not attributed to burn but to the well's close proximity to a cyclic steam producer. Thus the crossover is attributed to steam and the lower resistivity is due to reduced oil saturation and an increase in water saturation.

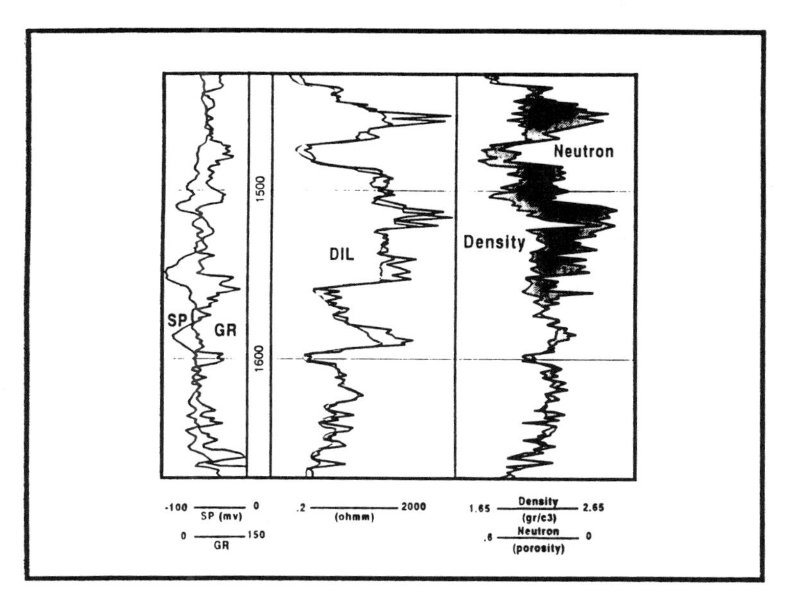

Fig. 8. Log character within a portion of the Webster Main interval. The high resistivities plus large neutron-density crossover from 1460 to 1560 indicates burn. The interval below 1600 feet shows reduced resistivities and minor neutron-density crossover due to oil production.

IV. WEBSTER MAPPING

All the 1989-1990 well logs were examined and all crossover that was present was counted. Isochore maps of crossover (gross gas saturation) were then created for the Webster Intermediate, Webster Main, and the five subzones (A-E) in the Main. All the logs that exhibited any crossover were then examined a second time. By using the resistivity response and the wells geographic location in relation to old air injectors or cyclic producers, burned reservoir footages were tabulated. These thicknesses were then plotted and contoured for all seven of the previous Webster maps (Figures 9 and 10). We realized the isochore maps represented not only burned reservoir sands but also burned diatomaceous mudstone barriers. These burned barriers had high crossover and high resistivities in many wells. The burned volume isochores represent gross altered interval, not just burned

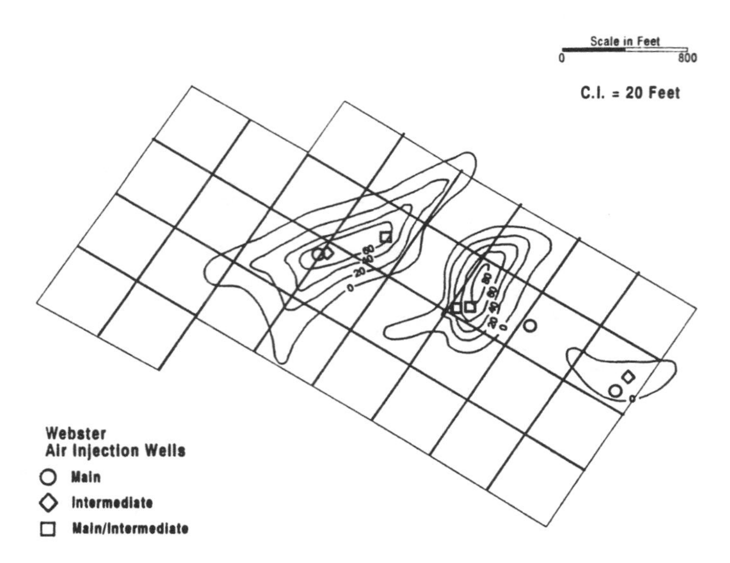

Fig. 9. Isochore map of the total burn within Webster Main (A-E) interval.

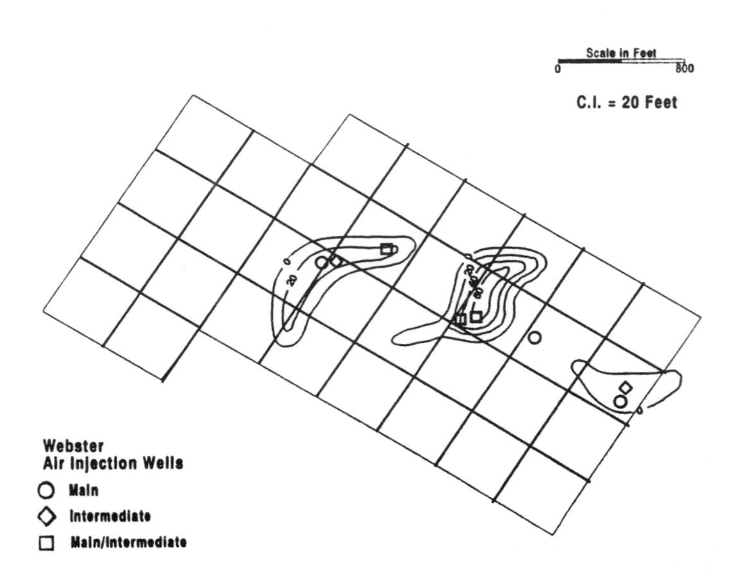

Fig. 10. Isochore map of the total burn in the Webster Main B subzone of the Webster Main interval.

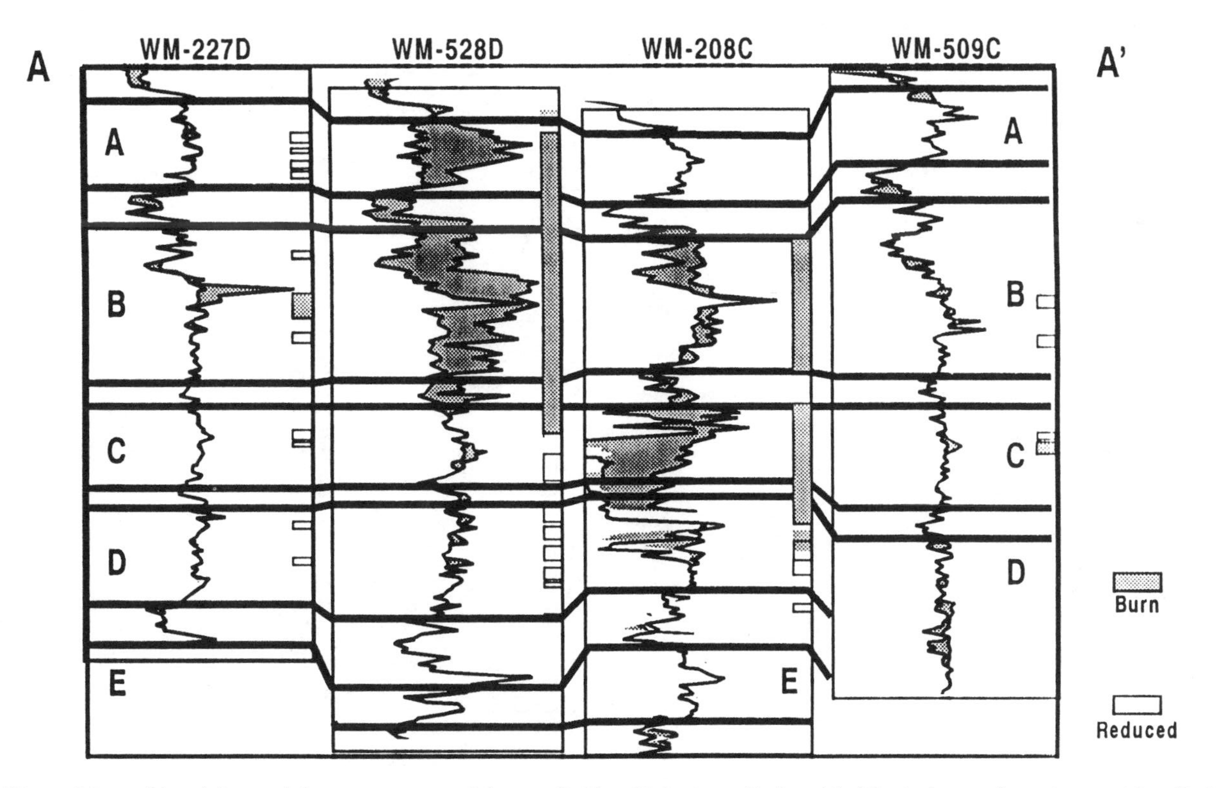

Fig. 11. Stratigraphic cross-section of the Webster Main (A-E) interval, shows the lateral and vertical discontinuities of burn in reservoir and diatomaceous mudstones. Cross-section is located on Figure 4, which shows the relation to structure and the proximity to air injectors.

reservoir. This log response was also correlated back to core data where highly oxidized mudstones were present. Figure 11 is a stratigraphic cross-section of four wells oriented south to north across the anticline. The location of this cross- section is depicted in Figure 4 which shows the well locations and their proximity to the old *in-situ* air injectors. The cross-section shows the burned reservoir to be both laterally and vertically discontinuous. Diatomaceous mudstone barriers also show discontinuity where they are altered and burned in one well but not in the next offsetting well. The overall pattern of the burned reservoir in the Webster Intermediate, Main, and the 5 subzones of the Main are not radial but are elongated across the axis of the anticline (Figures 9 and 10). This orientation is parallel to depositional strike and probably reflects directional permeability along channel axes.

V. SEDIMENT ALTERATION FROM COMBUSTION

A well was selected to be cored in an area near a Webster air injector. The MOCO 35 WIM 204B well was cored through the Webster Main and Intermediate intervals. Rock data was obtained to evaluate the affect of layering and sediment type on fireflood progression, plus evaluate sediment alteration associated with combustion. Interval depths studied were 1375'-1483' and 1505'-1705'.

Macroscopic alteration was first characterized and drawn against sedimentary layering. Core photography under ultraviolet lighting was used to note partially desaturated sands. Following macroscopic zonation of alteration types, sample analysis from x-ray diffraction, SEM, and thin sections was used to describe the detailed pore-size alteration. An extensive database on unaltered Webster sediments allowed further comparison of original and altered reservoir rock.

A. Macroscopic Alteration Zones

Distinct zones macroscopically recognized in the sediment reflect different temperature and oil conditions of a progressing fireflood front (Figure 5). The first evidence is seen as reduced oil saturation in sands (especially noticed under ultraviolet light). Next, a zone of combustion is recorded by the presence of black coke in the sediment pores. The final alteration after the total burn is a reddish oxidized sediment reflecting the injected air.

The different apparent fireflood zonations follow the bedding and layering characteristics of the Webster sands (Figures 12 and 13). Two major burn fronts, reflected by both coke-containing and oxidized clean sediments, are mapped in the core. Front progression was associated with the major sand body packages of the Webster Main and Intermediate. The slower-moving parts of the burn front are within the finer-grained thin- to medium-bedded sands interbedded with siliceous mudstones. A totally unaltered section between the Webster Main and Intermediate sand bodies occurs in thick siliceous mudstones interbedded with thin sand layers.

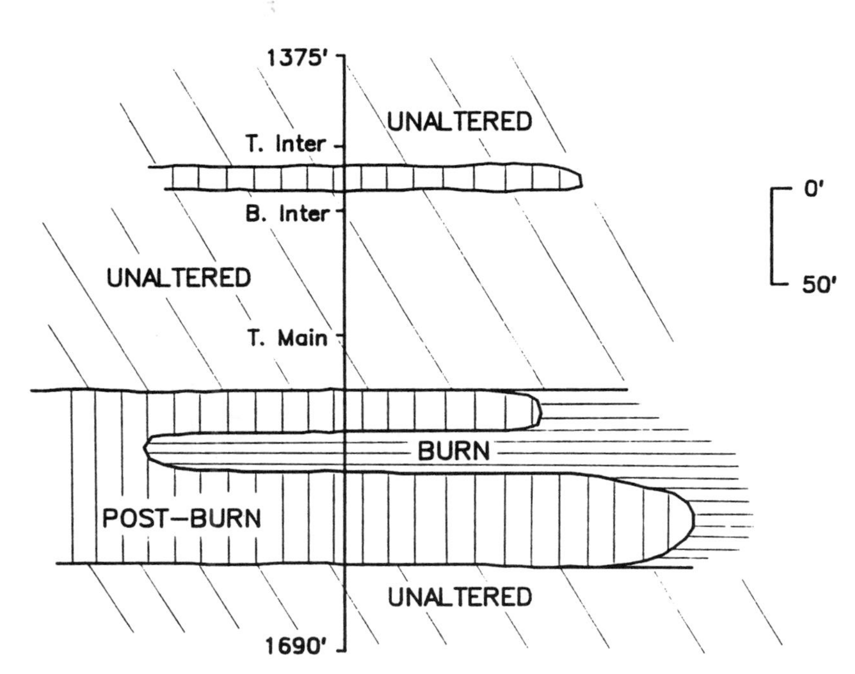

Fig. 12. Schematic drawing of the Webster Main and Intermediate burn. Zonations are based on Figure 13.

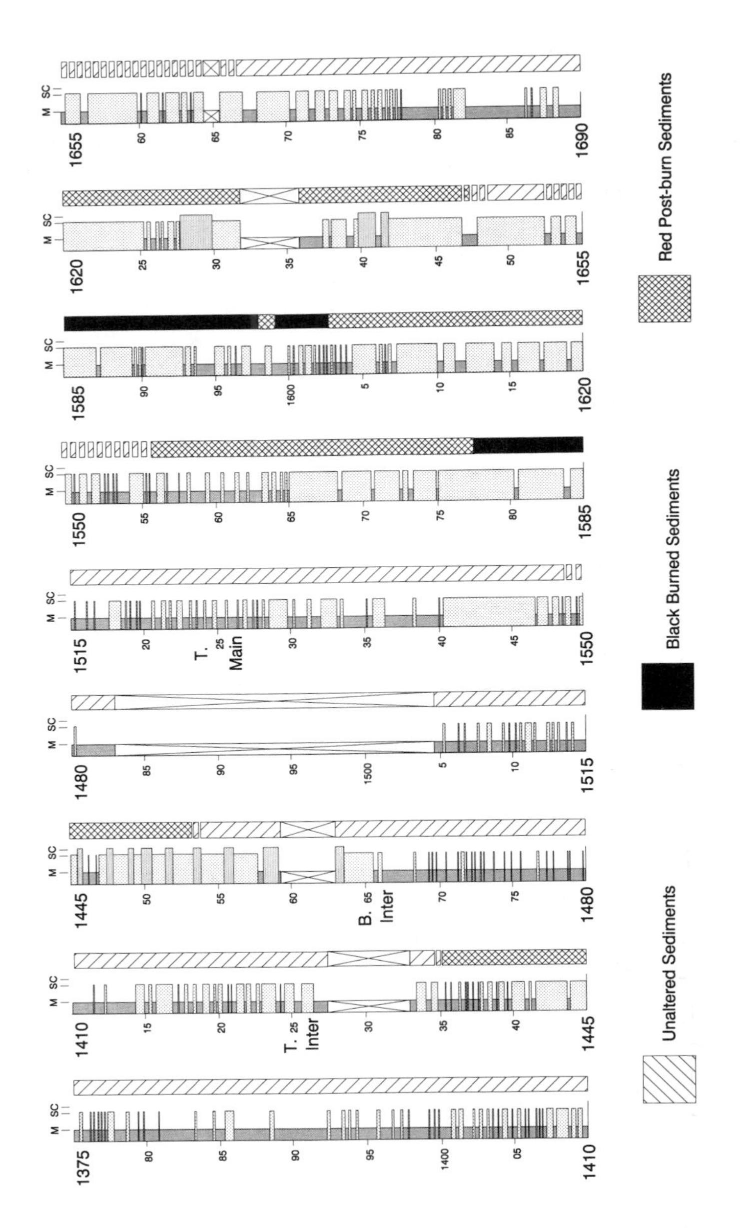

Fig. 13. Drawing of the MOCO 35 WIM 204-B core and related macroscopic alteration zones. Generalized grain sizes are M (mudstone), S (sand), and C (conglomerate).

B. Detailed Sediment Reactions

The unconsolidated nature of the original sands changed as the burn front progressed. Table II and Figure 14 summarize mineralogical changes thought to be associated with specific fireflood zones. The bulk of mineral alteration occurred within the finer-grained matrix material (mainly clays dominated by smectite and the poorly crystalline to amorphous siliceous matrix).

Earliest sediment alteration was due to the passage of hot water/steam in the progressing fireflood front. Sand-size reactions were limited to the dissolution of calcite grains (originally feldspar grains replaced by calcite during burial diagenesis). Calcite began to break down, releasing calcium ions in solution and carbon dioxide gases. Siliceous matrix and small amounts of kaolinite plus mica/illite dissolved. These reactants were reprecipitated as dominantly smectite and some zeolites.

The most intense temperatures occurred in the combustion zone itself (possibly 500+°C). Sediments now black in color are consolidated, with partial pore filling by coke (Figure 15-a). Calcite is either gone or altered. Reaction rims between calcite grains and coke are characterized by calcium sulphate crystals (Figure 15-b). A fraction of the smectite changed to illite/smectite, but most remained as expanding clay. Opal cements and zeolites are characterized by dissolution fabrics (Figure 15-c and 15-d).

Post-burn sediments are characterized by their clean, often reddish appearance (Figure 16). The reddish color is from the precipitation of hematite and possible amorphous iron oxide (Figure 16-d). The iron source may be from pyrite; however, altered sediments still have pyrite present. There is a visual variability in the amount of iron oxide related to grain size. The highest amount appears associated with both mudstones and matrix-rich sandstones. The final sediment product after the burn is a lightly consolidated sandstone.

Grain size is an obvious control of the lateral rate of movement in the fireflood. The mudstones burned at a slower rate than the sands when they were part of the combustion front. Non-oil organics are believed the source of "fuel" for the combustion. These fine-grained siliceous sediments were originally composed of opal-CT and dominantly smectitic clays. X-ray diffraction patterns of the opal-CT in unaltered to totally burned mudstones did not vary. In other words, the opal-CT did not appear to recrystallize during this time-frame of very high temperatures. Petrographic fabrics do not differ significantly from the original unaltered fabrics. Silica dissolution was noted under the SEM.

TABLE II. XRD Mineralogy of Original and Altered Webster

I. SANDS

	Qtz	Feld	Gyp	Pyr	Hem	Opal	Cal	Kaol	Smec	Ill/Smec	Ill/Mica
Original	41-50	39-50	ND	ND-1	ND	ND-19	ND-2	tr	2-6	ND	1
Altered	40-48	36-52	ND-3	ND-1	ND-1	ND-11	ND-1	ND-tr	3-7	ND-4	1
Significant	N	N	Y	N	Y	Y	Y	Y	Y	Y	N

II. MUDSTONES

	Qtz	Feld	Gyp	Pyr	Hem	Opal	Cal	Kaol	Smec	Ill/Smec	Ill/Mica
Original	10-18	10-25	ND	ND-3	ND	35-65	ND	tr	10-20	ND	ND-1
Altered	8-20	9-31	ND-12	ND-4	ND-10	20-70	ND	ND	9-20	ND-9	1-2
Significant	N	N	Y	?	Y	?	N	Y	?	Y	Y

NOTES --

Numbers in weight %

ND = not detected

"significant" (yes or no) means number variability indicates rock reactions rather than original compositional variability. From combined XRD, SEM, thin sections

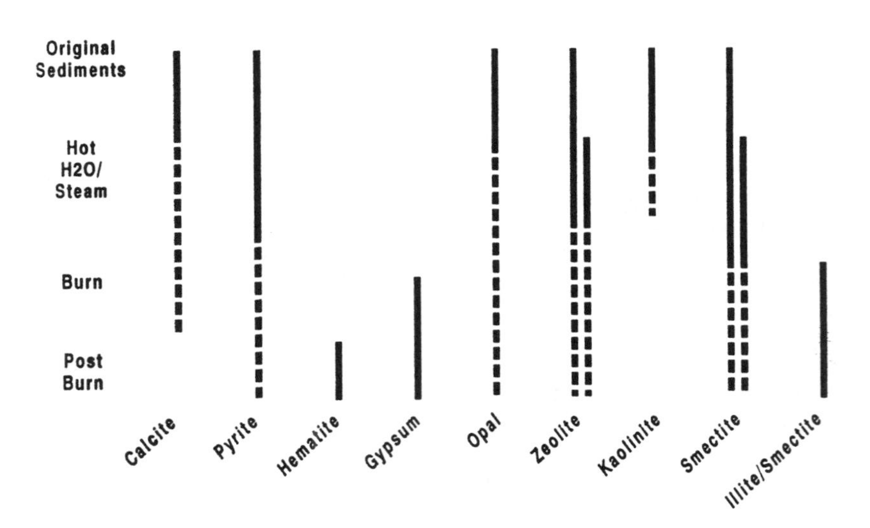

Fig. 14. Diagram showing the proposed diagenetic alteration associated with the moving combustion front. Dashed lines indicate dissolution; double lines indicate precipitation.

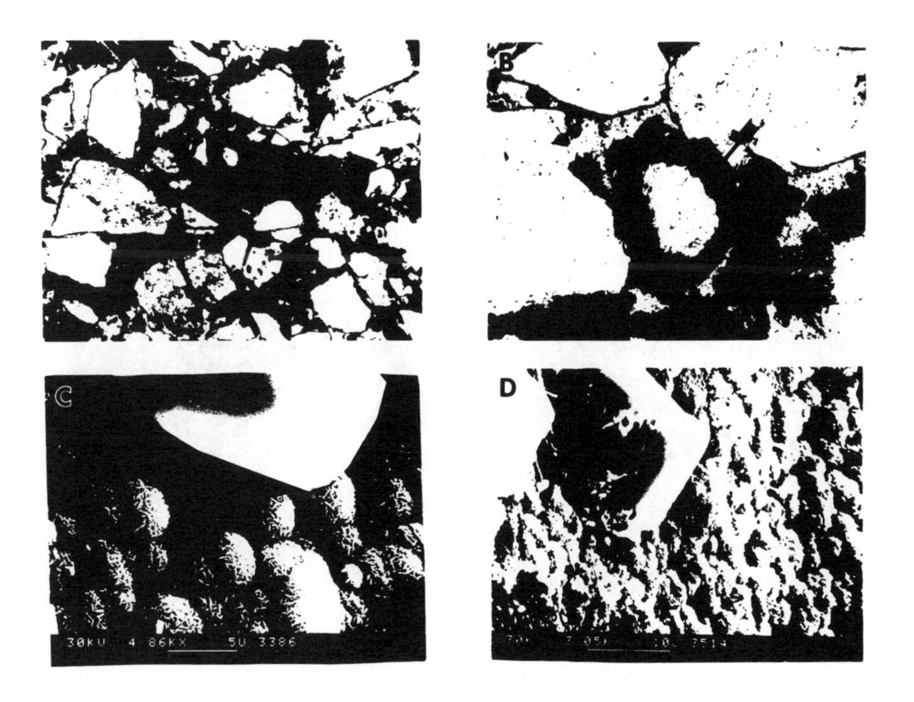

Fig. 15. Photomicrographs from a) coke-containing sand in thin-section (3.8mm field of view); b) thin section close-up of a coke/calcite grain reaction (0.4mm field of view), c) scanning electron view of unaltered opal-CT lepispheres and zeolite crystal, d) scanning electron view of burned, partially dissolved opal-CT and zeolite cements.

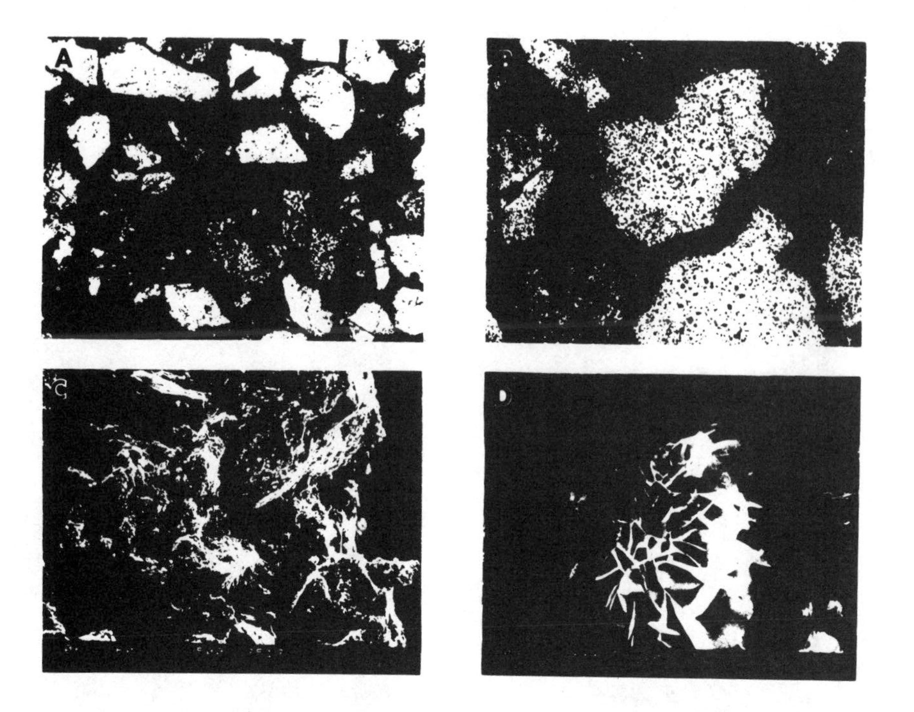

Fig. 16. Photomicrographs from a) post-burn lightly consolidated sand in thin section (3.8mm field of view), b) post-burn thin section close-up of remaining zeolite crystals (arrow) and clay lining pores (0.4mm field of view), c) scanning electron view of lightly consolidated post-burn sand, d) scanning electron view of hematite growth form present in reddish post-burn sands.

VI. RESERVOIR SIMULATION AND OPTIMIZED OPERATING STRATEGY

During 1989, the Webster reservoirs were developed for continuous steamflood. Inverted five spot patterns were drilled on five acre spacings. The Webster Main and Intermediate were developed separately except to the northwest where, due to reservoir thinning, the Main and Intermediate producers were combined in single wellbores. Continuous injection into four Webster Main patterns began during early 1990. Cyclic steaming of all Webster production wells began during this time as well. During 1990, oil production from the Webster zone increased from 250 BOPD to approximately 750 BOPD. However, little, if any, of the noted production increase resulted from continuous steam injection.

Prior to beginning continuous steam injection into the Webster Main, it was felt that the areal extent of burned reservoir could be quite extensive and behavior of steam in the burned portion of the reservoir unpredictable. Because of these concerns, initial steam injection patterns were separated from known areas of burned reservoir. It was also believed that contiguous patterns should be placed on continuous injection to concentrate reservoir heating in a specific area in order to evaluate the flow process and to optimize recovery efficiency. Initially, a low flux of 0.75 b/d/ac-ft. was used as a starting point for continuous injection. All injectors were designed for limited entry perforations with subzones A through E completed. In addition, a policy of low volume cyclic injection was implemented. Each production well received 10,000 barrels of steam per cycle. The injection cycle typically lasted 14 days. This was followed by a 7-day soak period and a production cycle lasting from 77 to 84 days.

As the vertical and lateral extent of the burn was mapped and quantified (Figure 9) it became apparent that not only was the volume of burned reservoir rock significant, but there was some doubt as to whether continuous steamflooding was the proper operating strategy for the Webster reservoir. In order to understand the reservoir recovery mechanism and develop an optimized operating strategy, it was decided to perform a reservoir simulation study of the Webster Main.

Three study objectives were identified. First, optimize recovery from the existing inverted five-spot patterns. Second, evaluate the effectiveness of updip and downdip line drive injection. Third, determine the effect of burned reservoir on performance.

A study area was selected to include a significant area of burned reservoir. Patterns with a history of continuous and cyclic steam injection and a sufficient number of producers and injectors was required to fully study the effects of converting from pattern to line drive injection. A model study area of 10.3 acres or approximately two patterns was selected. In order to capture the effect of gravity on oil recovery, the model area included the crestal portion of the reservoir and was extended down-structure to the lowest line of development. The model area was selected on the basis of reservoir structure. Ten wells were included in the model. Under existing conditions, six wells were producers and four were injectors. Each well was approximately 300-350 feet apart along dip while each strike pair was 340 feet apart. Each of the five major flow units was defined in the model with additional layering provided in each flow unit to more accurately establish flow behavior. The model grid configuration of the Webster Main Simulation Study is shown in Figure 17 while reservoir data is summarized in Table III.

The initial model assembled for this study was based on the existing operating policy of low continuous injection flux and frequent low volume cyclic injection. History matching was based on a 242-day injection/production period

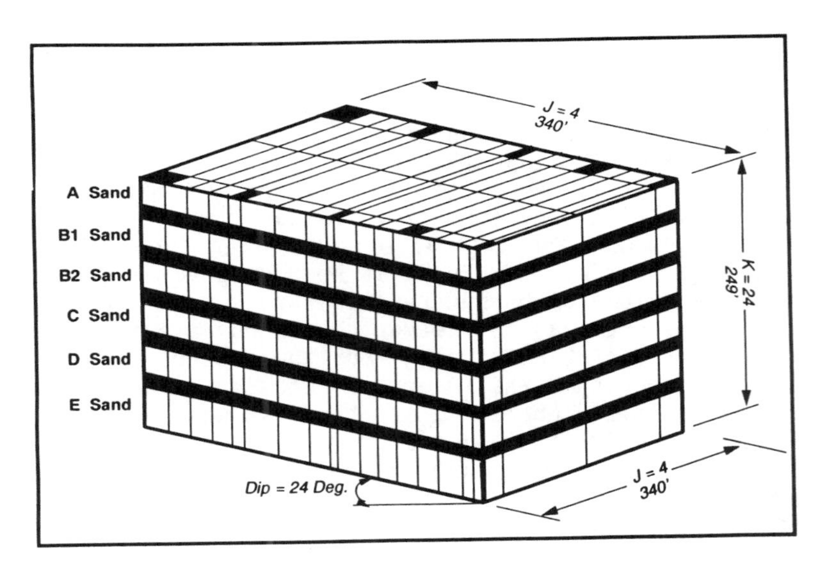

Fig. 17. Webster Main simulation study - model grid configuration.

TABLE III. Parameters of Webster Main Simulation Study

Model area (acres)	10.3
Gross thickness (feet)	249
Porosity (percent)	25-30
Horizontal permeability (millidarcies)	178-1399
Current oil saturation (percent)	2-72
Current O.I.P. (barrels)	2.4×10^6

beginning in early 1990. A 15-year predictive run based on this same operating policy was then performed. Results from the predictive run indicated very low recovery (approximately 20 percent of oil-in-place) during the 15 years. It was concluded that the Webster Main steamflood would not be a viable project under these operating conditions and alternative solutions were needed.

An initial optimization action taken was to increase injection flux from 0.75 to 1.0 b/d/ac-ft. No revision was made to the low volume cyclic policy. It was decided to discontinue cyclic operations once thermal communication was established between injector and producer. High vent zones were shut off as steam breakthrough occurred in individual flow units. These policies remained in effect throughout the entire 15-year predictive run.

A second optimized policy was to increase injection flux from 0.75 to 1.0 b/d/ac-ft. and also double the cyclic injection policy. Steam injection was increased from 10,000 to 20,000 barrels of steam per cycle per well. As in the first optimization case, cyclic activity was discontinued after thermal communication was established and steam breakthrough was shut off in high vent flow units.

Technical literature (Hong, 1988; Hong, 1990) and nearby competitor activity (Monghamian et al, 1982) suggested that a pattern steamflood might not be the optimum configuration for the Webster Main. These studies concluded that line drive steam injection was preferred. Two additional optimization cases were developed to examine these policies. In both the updip and downdip cases it was necessary to convert producers to injectors and vice versa in order to attain the proper configuration for line drive injection. As in the earlier cases, injection flux and cyclic volumes were increased. The same operating policies regarding thermal communication and steam breakthrough were implemented. A 15-year predictive run was performed for all cases.

A. Simulation Study Results

Analytical modeling was performed in addition to the numerical modeling study of the Webster Main. A history match of an offsetting steamflood project in the Webster Zone was obtained using Miller-Leung's model (Miller and Leung, 1985). Vogel's model (Vogel, 1984) was used to generate cumulative oil-steam ratios and recovery efficiencies for the Webster Main. Both of these analytical models were used for comparison with the numerical models in understanding the flow processes and recoveries of the reservoir.

Early predictive runs showed that heat would travel up structure regardless of injector location and thermal communication occurred in each flow unit at significantly different times due to a wide range of horizontal permeabilities. Each of these observations presented some problems in effectively managing the steamflood. Because heat from injected steam was moving up structure toward the burned portion of the reservoir, the two crestal wells (one injector and one producer) were shut-in throughout the life of the project in order to retain heat in this portion of the reservoir. Initial measurements taken prior to steam injection showed reservoir temperature due to the combustion project was still over 200°F in the crestal portion of the anticline. As long as heat was retained in this portion of the reservoir, it was believed that the burned reservoir would not cause serious problems to the rest of the steamflood.

Differential heating of the individual flow units presented another problem. While quick response was obtainable through massive cyclic steaming, the high permeability flow units would also steam out earlier. Steam breakthrough is a potentially serious threat to the success to any steamflood project. However, we believe that the Webster Main with its many flow units are effectively isolated by continuous barriers. These barriers should accommodate effective shut offs as steam breakthrough occurs.

Results generated from all four optimization case predictive runs showed substantial improvements in reservoir performance over the original base case. Figure 18 shows the oil rate versus time relationship between the two optimized pattern steamflood cases. The performance difference between the high volume cyclic case and the low volume cyclic case is remarkable. Not only does oil rate increase substantially once the increased flux and massive cyclic policies are implemented, but the rate remains higher through a major portion of the predictive run. The troughs shown in the figure are periods where steam breakthrough in a flow unit

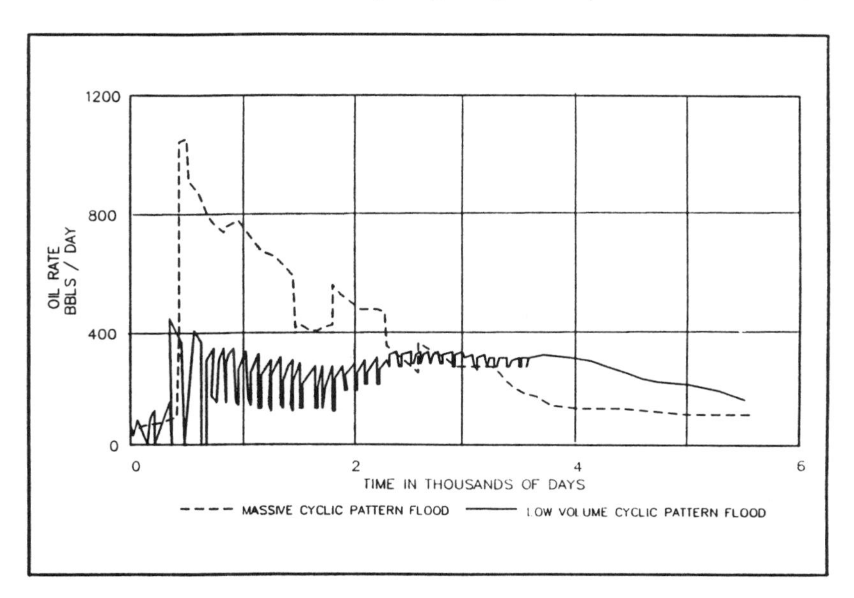

Fig. 18. Comparison of 2 pattern flood cases - oil rate versus time.

has occurred and the entire zone is shut off. Time is required for heat to be redirected to other flow units. Once this process occurs, the oil production rate returns to the previous level. In the high volume cyclic case, thermal communication was established after the first massive injection period and further cyclic activity was not necessary. Several predictive runs demonstrated continuous injection could be reduced through time without decreasing performance.

The results of the two line drive injection cases were fairly similar. As in the increased flux, high volume cyclic case, thermal communication was established after the first massive injection cycle. Steam breakthroughs over the life of the predictive run caused troughs in oil production rates. Predictive runs showed that injection flux could also be reduced over time in both of these cases.

In evaluating the merits of all four cases, it was decided that the most unbiased indicator was to compare net recovery efficiency versus time. Net recovery takes into account the amount of fuel gas required to generate the steam used in the predictive runs. By converting this fuel to equivalent barrels of oil it can be subtracted from the total

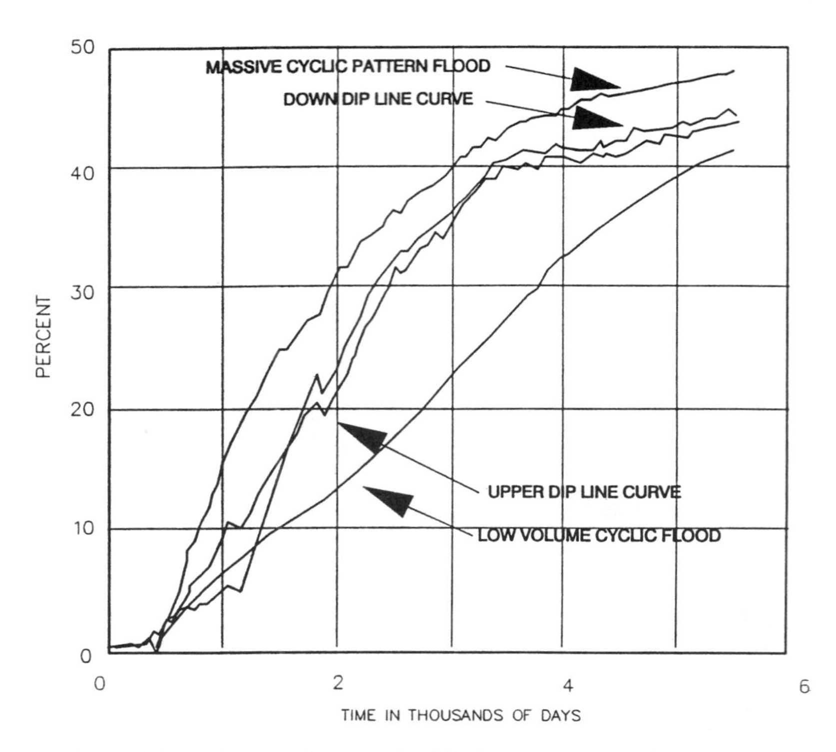

Fig. 19. Comparison of all 4 cases - net recovery efficiency versus time.

oil recovery over the entire predictive run. Hence, the net recoveries for each case may be compared on an equal basis and the optimum policy may be determined.

Figure 19 shows net recoveries versus time for all four optimization scenarios. For all points in time the increased flux, massive cyclic, pattern injection case has the best net recovery. The updip and downdip line drive cases had lower net recoveries over the 15-year predictive runs and the low volume cyclic pattern injection case had the lowest net recovery. Cumulative oil-steam ratios were also compared. Three of the four optimization cases had similar cumulative OSRs of approximately 0.21. The low volume cyclic pattern case had a substantially higher cumulative OSR, but a much lower net recovery. The cost of converting producers and injectors to accommodate the line drive injection cases was considered, as was the cost of shutting off high vent zones due to steam breakthrough. In this comparison, the increased flux, massive cyclic, pattern case was clearly superior on a net present value (NPV) basis. The other three cases had similar NPVs, ranging from 15 to 18 percent less than the optimum policy.

VII. OBSERVATIONS AND CONCLUSIONS

The areal extent of the Webster burns have been mapped by using consistent changes in the log signature of the resistivity, neutron and density tools. The burns are elongated across the axis of the 35 anticline and parallel to depositional strike. This probably reflects directional permeability along channel axes. The burns are both laterally and vertically discontinuous, including both reservoir sands and diatomaceous mudstone barriers. The burn front advanced more rapidly in high-permeability medium to coarse-grained sands than through finer-grained, thin-to medium-bedded sands interbedded with mudstones. The bulk of mineral alteration occurred within the fine matrix, particularly amorphous silica and clays.

The Webster Main Reservoir Simulation Study determined that the recovery process and optimized operating policy is strongly influenced by: 1) gravity dominance, 2) crestal burn, 3) high permeability contrast and 4) non-communicating flow units. In addition, the simulation study confirmed analytical model performance predictions.

The unique process dynamics of the Webster Main reservoir resulted in an atypical ranking of development alternatives. The increased flux, massive cyclic, pattern steamflood yields a higher net recovery and NPV than either of the line drive scenarios or the low volume cyclic pattern flood. Finally, although the simulation study identified an optimum operating policy, the ultimate success of this project will rely heavily on proper reservoir surveillance. Steam breakthroughs must be identified in a timely manner and then the high vent zones must be successfully shut off.

VIII. ACKNOWLEDGEMENTS

We would like to express our thanks to the management of Mobil Exploration and Producing U.S. Inc. for allowing us to publish this paper. We wish to thank individually our colleagues Robert Harrigal, John Turner, Kevin Crook, Amy

Sullivan, and Skip Mathias. We also want to acknowledge the help of Nagea Hunter and Ellen Heimbecher in the final preparation of this paper.

REFERENCES

Bartow, J.A., 1987, The Cenozoic Evolution of the San Joaquin Valley: U.S. Geological Survey Open-File Report 87-581, 74 p.

Boberg, T.C., 1988, Thermal methods of oil recovery, Wiley and Sons, New York.

Harding, T.P., 1976, Tectonic significance and hydrocarbon trapping consequences of sequential folding synchronous with San Andreas faulting, San Joaquin Valley, California: American Association of Petroleum Geologists Bulletin, v. 60, p. 356-378.

Hill, F.L. and Land, P.E., 1971, "Moco" combustion projects, Midway-Sunset Field: California Department of Oil and Gas Bulletin, v. 57, p. 51-59.

Hong, K.C., 1988, Steamflood strategies for a steeply dipping reservoir: Society of Petroleum Engineers Reservoir Engineering, v. 3, p. 431-439.

Hong, K.C., 1990, Effects of gas cap and edge water on oil recovery by steamflooding in a steeply dipping reservoir: Society of Petroleum Engineers paper 20021 presented at the 60th California Regional Meeting, Ventura, California.

Isaacs, C.M., 1987, Sources and deposition of organic matter in the Monterey Formation, south-central coastal basins of California, *in* Meyer, R.F. (ed.), Exploration for heavy crude and natural bitumen: American Association of Petroleum Geologists Studies in Geology 25, p. 193-205.

Link, M.H. and Hall, B.R., 1990, Architecture and sedimentology of the Miocene MOCO T and Webster turbidite reservoir, Midway-Sunset Field, California, *in* Kuespert, J.G. and Reid, S.A. (eds.), Structure, stratigraphy and hydrocarbon occurrences of the San Joaquin basin, California: Society of Economic Paleontologists and Mineralogists, Pacific section, Special Publication 64, p. 115-129.

Miller, M.A. and Leung, W.K., 1985, A simple gravity override model of steamdrive: Society of Petroleum Engineers paper 14241 presented at the 60th Annual Technical Conference and Exhibition, Las Vegas, Nevada.

Moughamian, J.M., Woo, P.T., Dakessian, B.A., and Fitzgerald, J.G., 1982, Simulation and design of a steam drive in a vertical reservoir: Journal of Petroleum Technology, v. 34, p. 1546-1554.

Ryder, R.T. and Thomson, A., 1989, Tectonically controlled fan delta and submarine sedimentation of late Miocene age, southern Temblor Range, California: U.S. Geological Survey Professional Paper 1442, 59 p.

Schwartz, D.E., 1988, Characterizing the lithology, petrophysical properties, and depositional setting of the Belridge Diatomite, South Belridge field, Kern County, California, *in* Graham, S.A. and Olson, H.C. (eds.), Studies of the geology of the San Joaquin basin, Society of Economic Paleontologists and Mineralogists, Pacific section, Special Publication 60, p. 281-301.

Tilley, B.J. and Gunter, W.D., 1988, Mineralogy and water chemistry of the burnt zone from a wet combustion pilot in Alberta: Bulletin of Canadian Petroleum Geology v. 36, p. 25-36.

Vogel, J.V., 1984, Simplified heat calculations for steamfloods: Journal of Petroleum Technology, v. 36, p. 1127-1136.

Webb, G.W., 1981, Stevens and earlier Miocene turbidite sandstones, southern San Joaquin Valley, California: American Association of Petroleum Geologists Bulletin, v. 65, p. 438-465.

THE DEPLETION OF THE RANNOCH-ETIVE SAND UNIT IN BRENT SANDS RESERVOIRS IN THE NORTH SEA

J M D Thomas
R Bibby

PEDSU, AEA Petroleum Services
Winfrith Technology Centre
Dorchester, Dorset DT2 8DH
ENGLAND

1. INTRODUCTION

This paper considers the depletion of the Etive-Rannoch Units in oil fields in the Brent Sand fields of the UK North Sea. These fields are the most prolific in the UK North Sea and lie within the East Shetland Basin off the Scottish coast (figure 1). The Brent Sand sequence consists of 5 Sand Units; the Broom, Rannoch, Etive, Ness and Tarbart, in ascending order. In this work we specifically consider the depletion of the Etive and Rannoch Sands in the Lower Brent, which have proved particularly problematic in reservoir engineering terms.

In some of the early field developments in the late 1970's such as Thistle and Dunlin, there were severe problems with premature water breakthrough of injected water into production wells. In some fields, breakthrough occurred within a few months of the start of production and wells had to be produced at sustained high water cuts (Barbe, 1981; Byat and Tehrani, 1985).

---

[1]Funded by the Petroleum Engineering Directorate of the United Kingdom Department of Energy

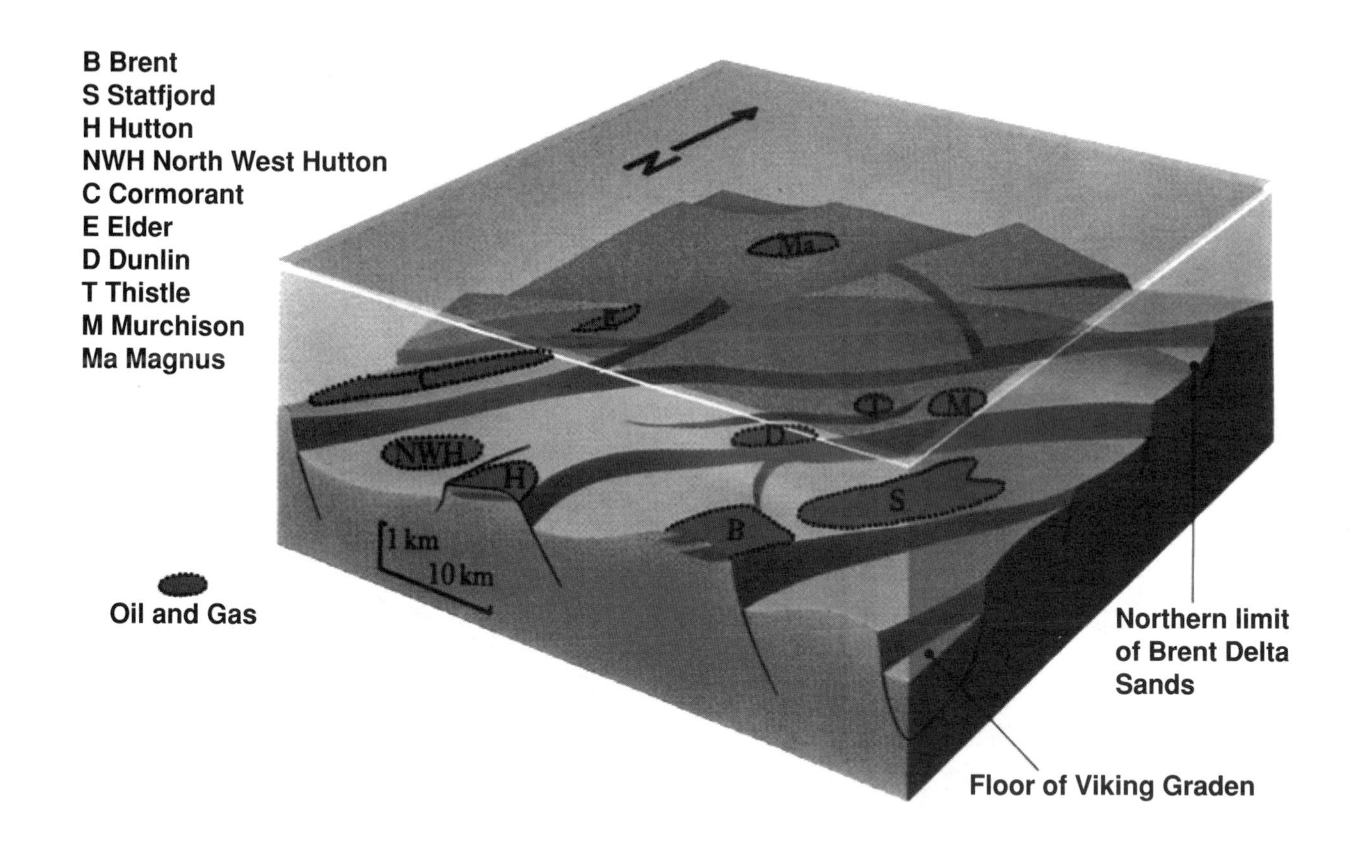

FIGURE 1 The Brent Province of the UK North Sea

Premature water breakthrough occurred basically because the permeability of the Etive layer is more than an order of magnitude greater than that of the Rannoch layer and there is, in many places, a low permeability 'tight' zone between them. The problem is illustrated schematically in figure 2, which shows that, even where perforations at the injection and production well are limited to the Rannoch, water still overrides in the high permeability Etive layer. The usual perforation policy which has been adopted is to complete wells initially on both the Etive and Rannoch and to plug back the perforations at the production wells to Rannoch-only when water breakthrough occurs. This action usually gives a brief period of dry oil production before water cones **downwards** (ie reverse coning) from the Etive. Figure 2 also illustrates two further recovery mechanisms that occur during Rannoch-Etive depletion: sudation (ie. gravity and capillary crossflow) at the boundary as water overriding in the Etive "soaks" downward into the Rannoch; and a "secondary" flood front moving along the highest permeability zones in the Rannoch.

In several oil fields, the premature water breakthrough was so serious and unexpected that target oil production rates could not be achieved and facilities for water injection and production were severely strained. It has also been common for the Rannoch reserves to be significantly downgraded. In the more mature fields the majority of the remaining oil reserves lie in the Rannoch Sands.

Three main strategies have been considered for improving recovery from the Rannoch. The first is to rely on sudation, the process whereby water moves downwards from the Etive, under the forces of gravity and capillarity, and displaces an equivalent amount of oil upwards from the Rannoch. The oil is then swept through the Etive to the production wells. The sudation process is an inevitable consequence of water override. Its effectiveness as a recovery process depends primarily on the rate at which it occurs, which is controlled mainly by the vertical permeability. The presence of a 'tight' zone at the top of the Rannoch in some fields is therefore critically important. Attempts have been made in several wells to estimate the sudation rate where the Etive has been flooded for several years. Frontal advance rates of

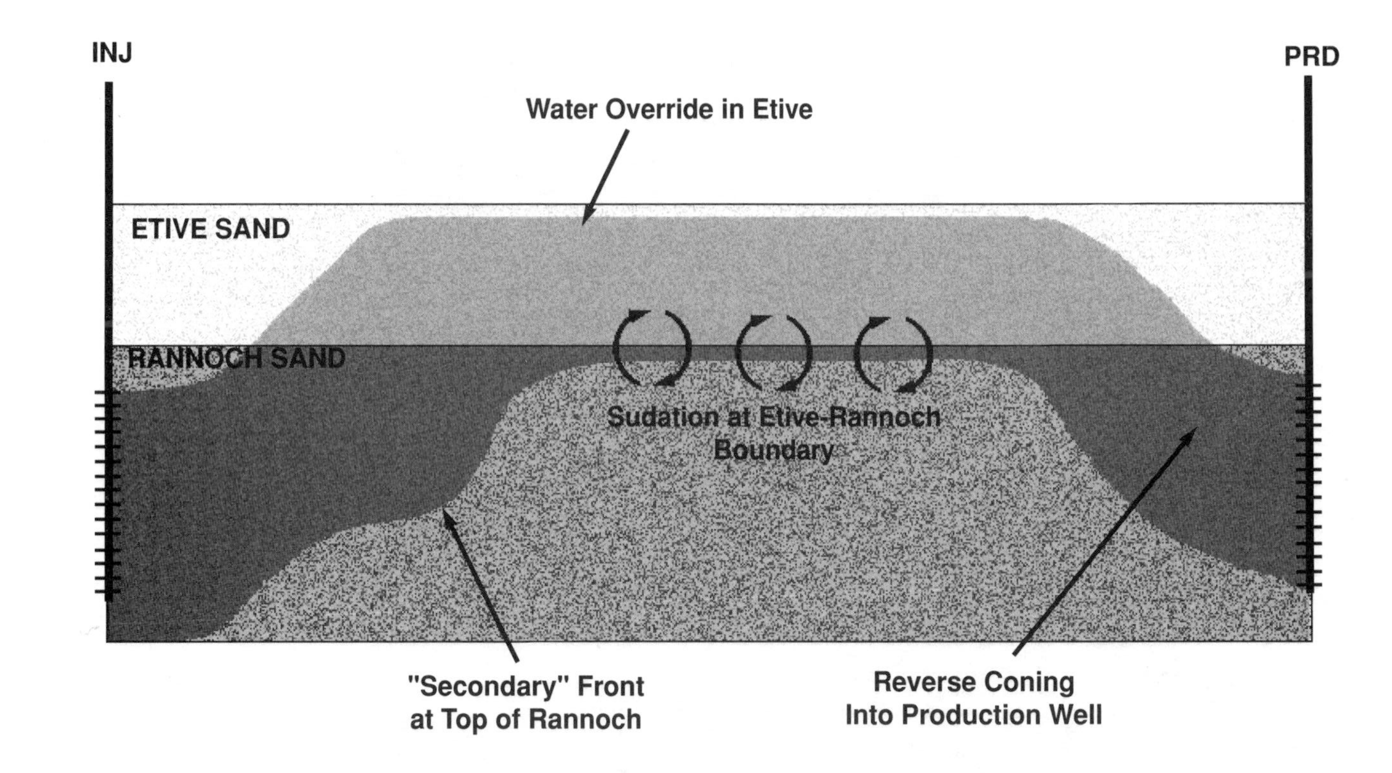

FIGURE 2 A Schematic of the Etive-Rannoch Depletion Mechanism

about 5 ft/year have been reported. However, we are not aware of any development relying solely on sudation as the mechanism for increasing recovery from the Rannoch.

The second method for increasing recovery from the Rannoch is to drill infill wells, either vertically or horizontally, into areas where water override has occurred. Production from such wells is usually followed fairly quickly by water breakthrough as water cones downwards from the Etive. The cost effectiveness of the infill wells depends mainly on the productivity of the Rannoch Sands and on how quickly the watercut develops. The latter is controlled mainly by the vertical permeability in the upper part of the Rannoch. Again, the presence of a 'tight' zone at the top of the Rannoch is critically important. Infill wells are probably the most commonly used approach to increasing Rannoch recovery.

The third strategy which has been considered to increase recovery from the Rannoch is by the use of selective perforation intervals at the injectors and producers. For example, Massie (1985) suggested that the Rannoch Sand recovery could be improved substantially by confining oil production to the Rannoch layer only. In some fields, such as Murchison, which have come onto stream after the problem was recognised, this kind of policy has been used with some success. In older fields, the tendency has been to perforate initially on both Etive and Rannoch and recomplete to Rannoch-only after water breakthrough has occurred.

This paper describes simulation calculations to investigate this third strategy. Our basic model consists of a producer-injector pair within the Etive-Rannoch Sands which we have used to examine the effects on recovery of overall reservoir quality and the importance of the Kh contrast between the Etive and Rannoch. We have investigated the benefit of various completion strategies for both producer and injector, varying from dedicated Rannoch perforations, to perforations over the entire Etive plus Rannoch interval. We have also investigated the benefit of recompletion of the production well when inevitable early water breakthrough occurs in the Etive Sand.

2. A BRIEF DESCRIPTION OF THE GEOLOGY OF THE ETIVE AND RANNOCH SANDS

The Brent Sands are of mid-Jurassic age and were deposited in a deltaic environment. The Rannoch and Etive Sands typically represent a prograding delta sequence which coarsens upwards from very fine sands and muds at the base to medium sands at the top. It is extremely rich in mica, which is present as very thin, closely spaced laminae within hummocky cross-beds. Calcareous doggers formed by early diagenetic cement are also common.

The Etive is generally a much cleaner, coarser massive sand. In some fields it was deposited as an upper shoreface or beach sand continuing the delta-front sedimentary sequence. In other fields it was deposited in distributary channels, or as barrier bars cut by tidal inlet channels.

There is evidence in several Brent sands reservoirs for a zone of low permeability, a 'tight' zone, at the top of the Rannoch. Where the zone is present it appears to be the result of calcite and siderite cementation. However, even where it is present, the Etive and Rannoch always appear to be in good pressure communication and act as a single reservoir.

Figure 3 shows a typical permeability profile through the Etive and Rannoch from a well in the Thistle field (Dake, 1982) in which the 'tight' zone is present. It may be seen that the Etive permeabilities range up to several Darcies and that Rannoch permeabilities vary from about 200 mD at the top to 50 mD at the base. There is uncertainty in the large scale effective vertical permeability of the Rannoch, because of the presence of the mica laminae, hummocky cross-beds and doggers.

3. A DESCRIPTION OF THE PRODUCER-INJECTOR PAIR MODEL

A. Introduction

This model comprises a producer/injector pair and is designed to enable Etive-Rannoch depletion strategies to be investigated.

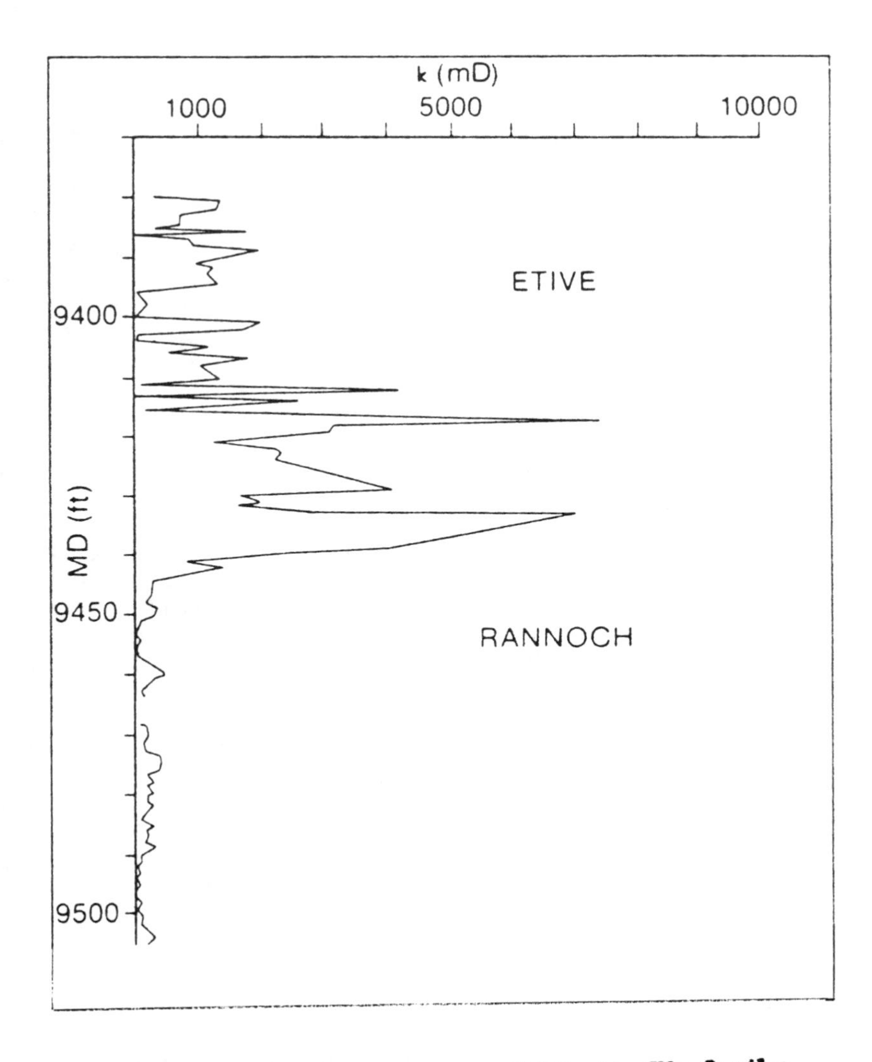

FIGURE 3 A Typical Permeability Profile for the Etive-Rannoch Sands (Thistle Field)

The geometry of this model is illustrated in figure 4. It comprises two "layer-cake" radial models which are joined at their outer boundary using the special connections facility within the "PORES" black oil reservoir simulation model. This particular geometry was chosen to provide an approximate representation of the central streamline of a line drive system, with the essential requirements of approximately linear flow away from the wells and radial flow close to the wells being met. A radial flow region close to the well is particularly important to represent the coning behaviour that dominates Rannoch performance. The well spacing may be varied to any chosen distance.

B. Model Description

The simulation model properties are based upon typical values for reservoirs in the Brent Province and are outlined in table 1 and figure 4.

The well spacing in the Base Case is 1900 ft and the angle subtended at each of the wells is 5 degrees. The dip angle is zero. The radial grid size increases logarithmically from a value of 5 ft close to the wellbore, to a maximum of 40 ft away from the well. Comparison with finer gridded models indicated that this grid was sufficiently fine to represent coning close to the well and frontal advance away from the well, without significant numerical dispersion.

The vertical gridding was chosen carefully to allow sufficient refinement (2 ft) at the interface between the Etive and Rannoch Sands to represent sudation mechanisms at the boundary. The vertical water saturation due to sudation for this model was compared to the predictions of an extremely fine grid (cell thicknesses down to 0.1 ft) 1D model and the results supported the adequacy of a 2 ft grid refinement at the Etive-Rannoch boundary. Most of the cells had a dimension of 10 ft in the vertical direction, although this increased to 20 ft in the very low permeability layer at the base of the Rannoch.

An important feature of the sudation process is the **counter-current** flow of oil and water across the Etive-Rannoch boundary. The literature on this topic (Bourbiaux and Kalaydjian, 1988) would suggest that countercurrent flow

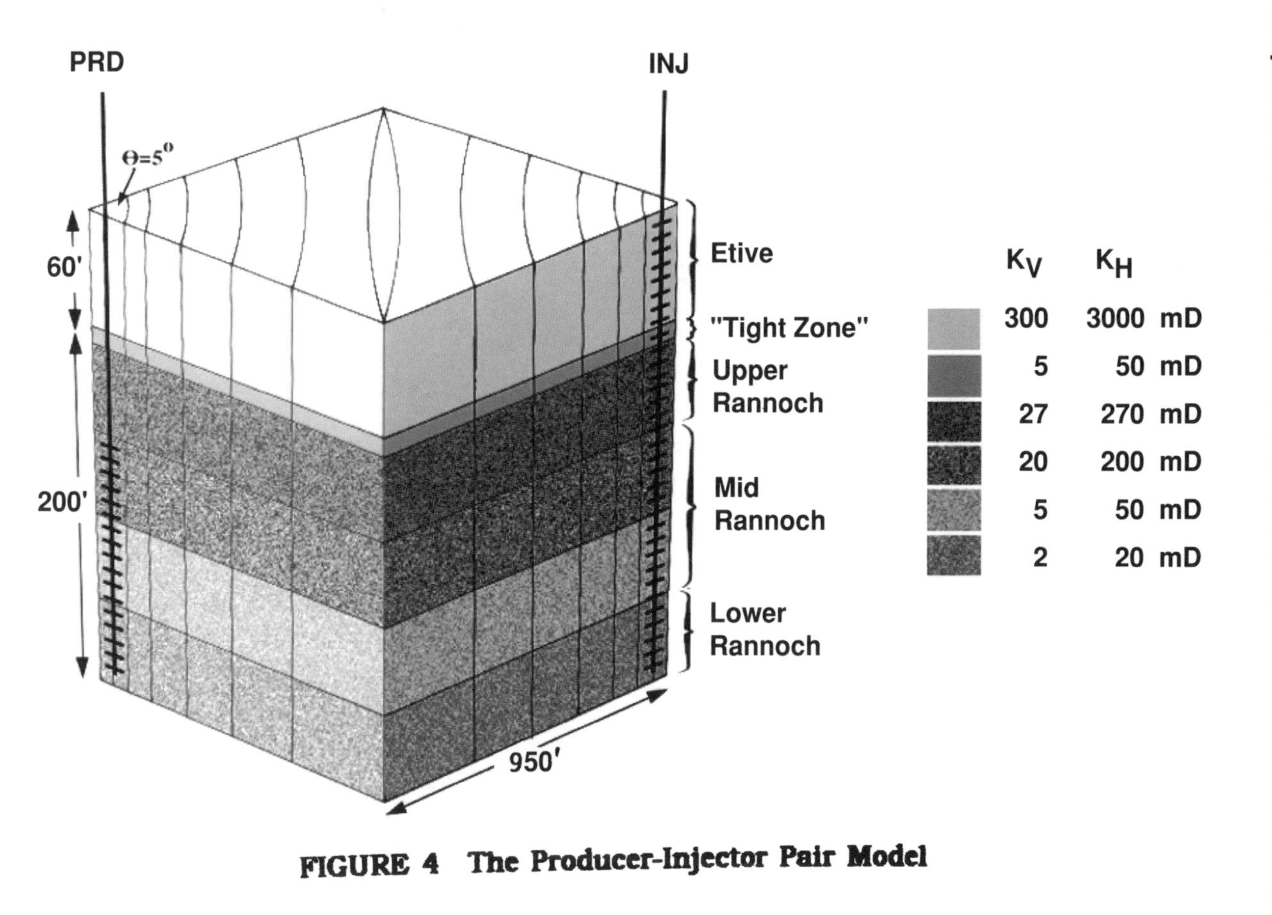

FIGURE 4 The Producer-Injector Pair Model

relative permeabilities are somewhat lower than those for concurrent flow, due to the interference effect caused by fluids flowing in the opposite, rather than the same, directions. Since the effect on relative permeabilities has not been quantified and in view of other major uncertainties, such as the imbibition capillary pressure, we have used the "rock" curve relative permeabilities to represent sudation. The injection and production wells are completed over any specified depth interval and operated with any of the usual simulator options, such as fixed production rate, minimum bottom-hole pressure and voidage replacement by water injection.

The Base Case model consists of 6 geological layers: a 60 ft thick zone of 3000 mD (K_h) representing the Etive Sand, a 10 ft zone of 50 mD permeability representing the "Tight Zone" at the top of the Rannoch, a 50 ft zone of 270 mD permeability, a 50 ft zone of 200 mD, a 50 ft zone of 50 mD permeability, and a basal layer 40 ft thick of 20 mD permeability. The (K_v/K_h) ratio was set at 0.1 throughout. The model properties were taken from the databases for several Brent Sands reservoirs which are supplied to the UK Department of Energy by Operating Oil Companies. The capillary pressure function for each of the 6 geological layers was scaled according to the Leverett J-function and the maximum capillary pressure was 25 psi.

The model was initialised with an oil-water contact at 100 ft below the base of the Rannoch Sand, in line with typical initial reservoir conditions.

The model has been run with the single producer completed throughout the Etive and Rannoch, and in the Rannoch Sand alone. The Etive and Rannoch "Kh" factors are 180,000 mD-ft and 27,300 mD-ft respectively and imply substantial well productivity from Rannoch completions alone.

Injection is controlled by voidage replacement of produced fluids with injection over the whole interval, or the Rannoch Sand alone.

The model described above has been used to investigate Etive-Rannoch depletion strategies in reservoirs with various levels of Etive and Rannoch permeability and permeability contrast, covering the range of observed North Sea values.

4. ETIVE PERMEABILITY ABOUT 1 DARCY, ETIVE-RANNOCH PERMEABILITY CONTRAST ABOUT 10:1

A. CASE01: Production from the Rannoch Only, Injection into both Etive and Rannoch (Maximum Rate = 20% STOIIP/yr)

In this case production is confined to Rannoch perforations (the well is completed from 60 ft below the top of the Etive Sand to the base of the Rannoch Sand) from initial conditions, with injection taking place into both the Etive and Rannoch layers (voidage replacement). The initial oil production rate corresponded to an offtake of approximately 20% STOIIP per year. This rate was achieved with a drawdown of about 1300 psi at the end of the simulation.

Figure 5(a) illustrates the water saturation profile at water breakthrough (after 2 years). Water has advanced rapidly along the Etive layer and has then coned across the "Tight Zone" at the top of the Rannoch into the producer.

The overall sweep after 7 years, when the watercut is 90%, is shown in figure 5(b). Note the "secondary" water front that develops in the top portion of the Rannoch layer.

Figure 6 illustrates the watercut and cumulative oil recovery profile for this case. The point of inflexion in the watercut profile at about 2600 days corresponds to the point when the secondary front reaches the production well.

B. CASE02: Production from both Etive and Rannoch, Injection into Etive and Rannoch (Maximum Rate = 20% STOIIP/yr)

In this case production takes place from both the Etive and the Rannoch layers throughout field life. The model is again limited to an offtake rate of 20% STOIIP per year, a rate achieved with a maximum drawdown of about 50 psi in this case. Approximately 85% of production occurs from the Etive and very early water breakthrough (within 1 year) occurs, as shown in figure 7 (a).

TABLE 1: RESERVOIR PROPERTIES FOR THE BASE CASE 'ETIVE-RANNOCH MODEL

Horizontal Permeability = 3000 mD (Etive, 60 ft thick)
50 mD ('Tight Zone', 10 ft thick)
270 mD (Top Rannoch, 50 ft thick)
200 mD (Upper Middle, 50 ft thick)
50 mD (Lower Middle, 50 ft thick)
20 mD (Base Rannoch, 40 ft thick)

(K_v/K_h) Ratio = 0.1

Oil Viscosity = 0.63 cp
Water Viscosity = 0.35 cp

Oil-Water Relative Permeabilities

Etive Formation

S_w	k_{rw}	k_{ro}
0.15	0.000	1.00
0.20	0.001	0.83
0.30	0.015	0.60
0.40	0.050	0.42
0.50	0.100	0.26
0.60	0.200	0.13
0.70	0.300	0.00

Rannoch Formation

S_w	k_{rw}	k_{ro}
0.20	0.000	1.00
0.30	0.005	0.75
0.40	0.025	0.53
0.50	0.065	0.36
0.60	0.130	0.22
0.70	0.220	0.09
0.77	0.30	0.00

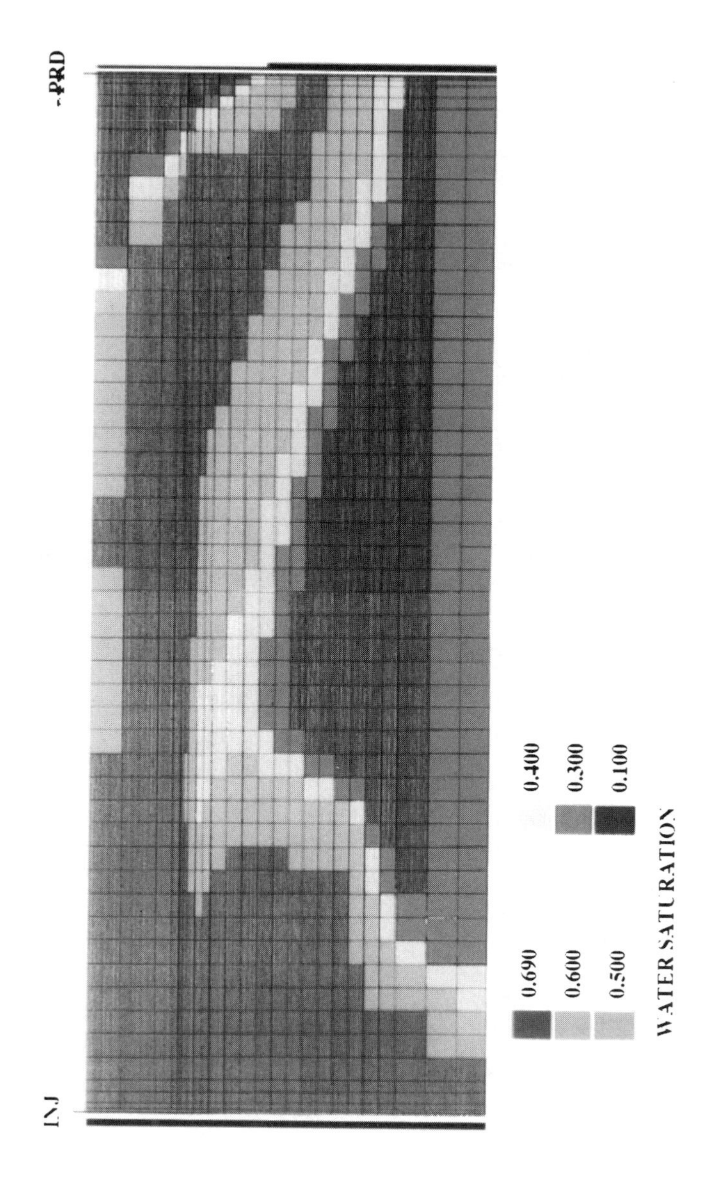

FIGURE 5(a) Saturation Profile at 2 years for CASE01

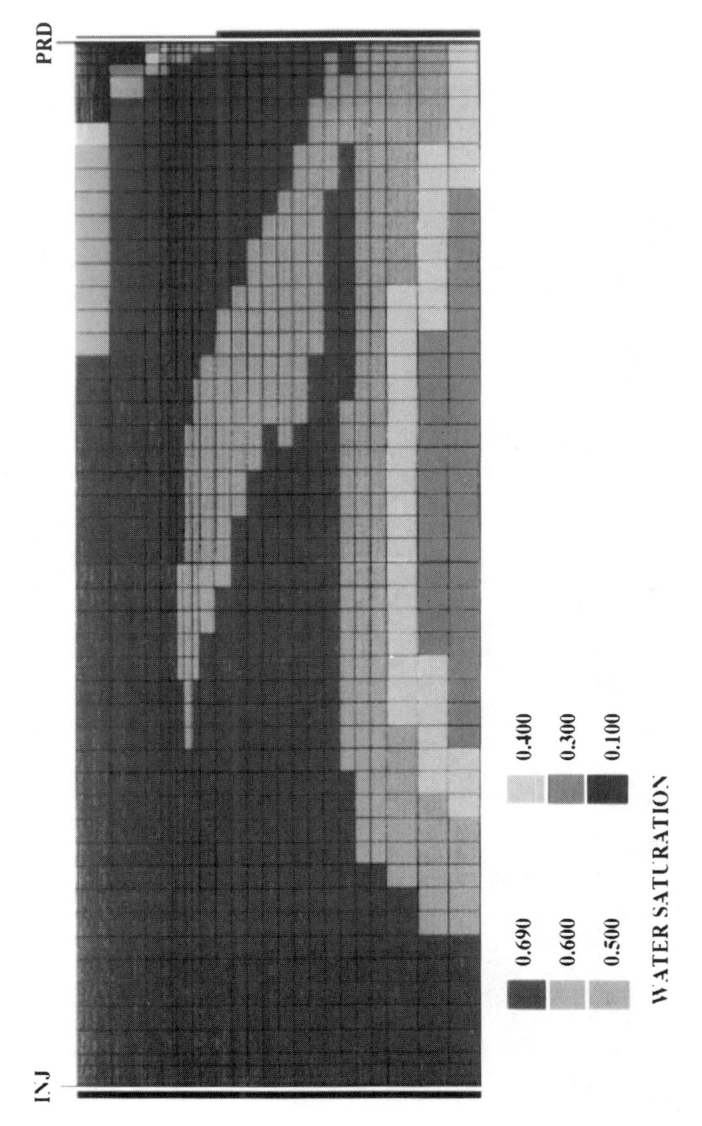

FIGURE 5(b) Saturation Profile at 7 years for CASE01

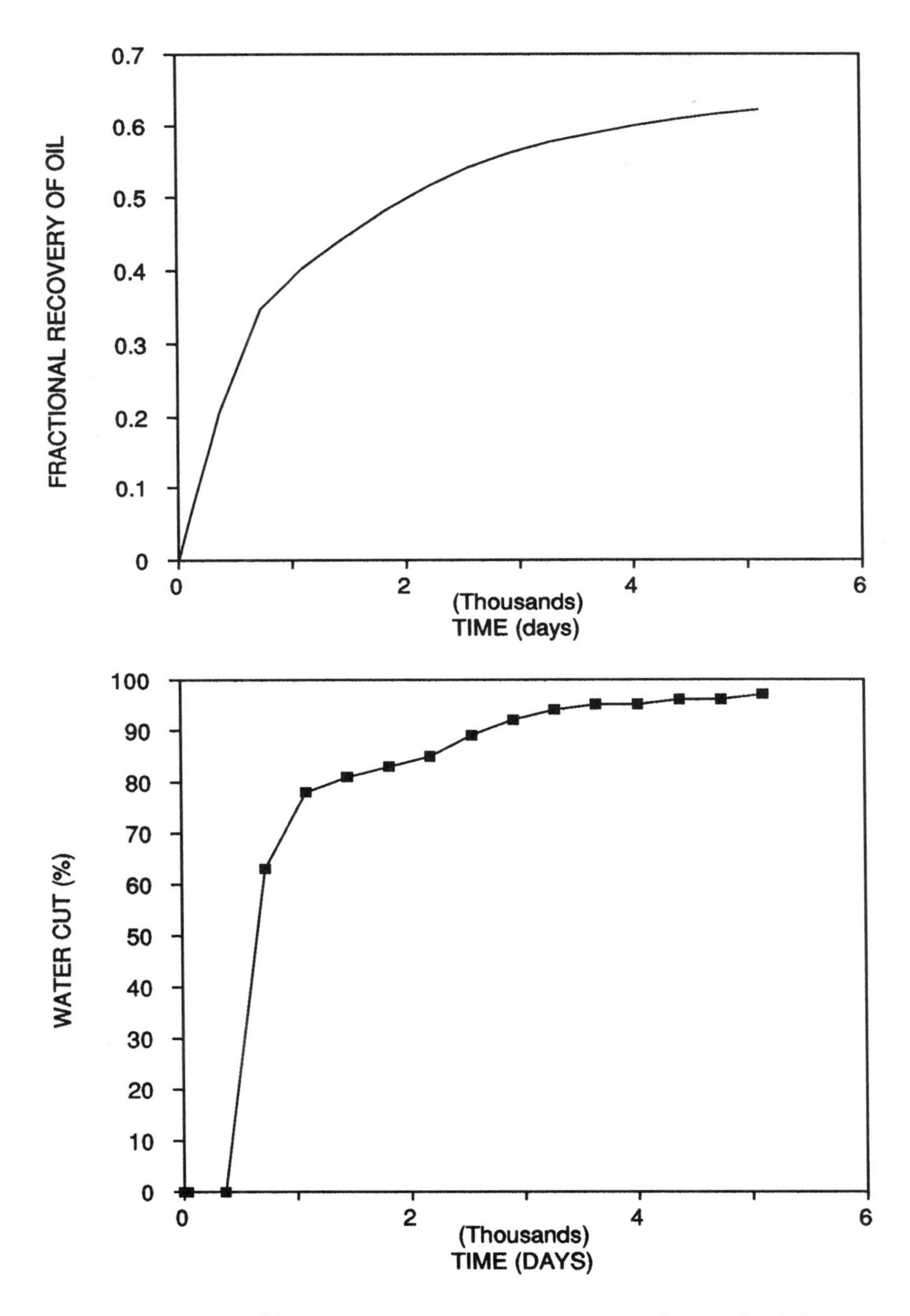

FIGURE 6 Oil and Watercut Profile for CASE01

The producing watercut very rapidly builds up to about 85% and then stabilises. The reason for this can be understood by considering figure 7(b) which illustrates the water saturation distribution after 7 years. After water breakthrough has occurred in the Etive layer, the watercut for the layer reaches over 90% very quickly, whilst the Rannoch perforations continue to produce dry oil.

As in CASE01, a secondary front develops in the Rannoch layer. The Rannoch perforations will continue to produce essentially dry oil until this secondary front reaches the producer, and the watercut will remain constant at about 85% until the secondary front arrives. This will, however, take a considerable time (about 10 years) and throughout this time the well will be producing very large amounts of water from the Etive layer. This behaviour is very clear from an inspection of figure 8 which illustrates the oil production and watercut profiles for CASE02. Long production periods at high, but relatively constant, watercuts are a characteristic feature in data from North Sea reservoirs in the Brent Province.

The process of sudation is also very evident from the saturation plots. The water penetration depth due to sudation is about 20 ft after 1000 days, a figure which agrees with more detailed (vertical resolution= 0.1 ft) 1D simulation calculations.

C. CASE03: Effect of Recompletion to Rannoch-Only Upon Water Break Through

This case examined the benefit of working over the producer upon water breakthrough, to Rannoch-only production. The aim of this policy is to cut out the high water production conduit provided by the Etive layer and is the policy that has generally been adopted in the North Sea. The results of this case are compared with the first two cases in figure 8.

It can be seen from figure 8 that the cumulative recovery for CASE03 lags slightly behind the Base Case (CASE01). This is a result of the earlier water breakthrough resulting from initial perforation on the Etive layer. Note that the initial perforation of the production well across the whole interval

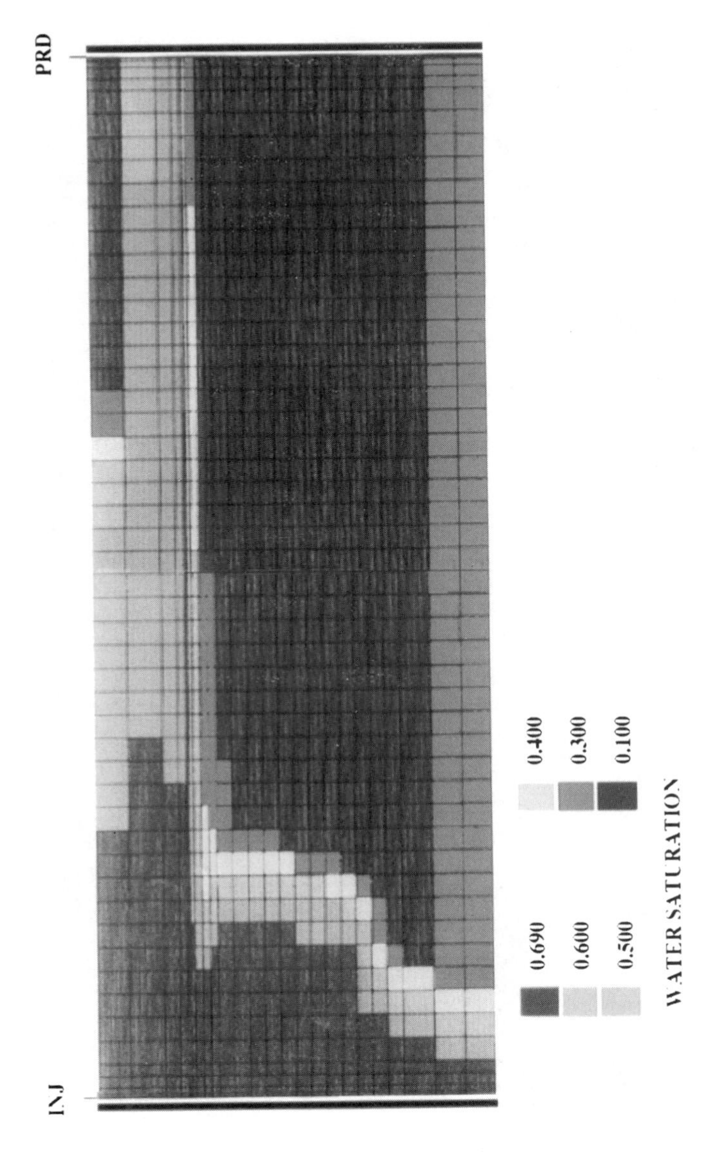

FIGURE 7(a) Saturation Profile at 1 year for CASE02

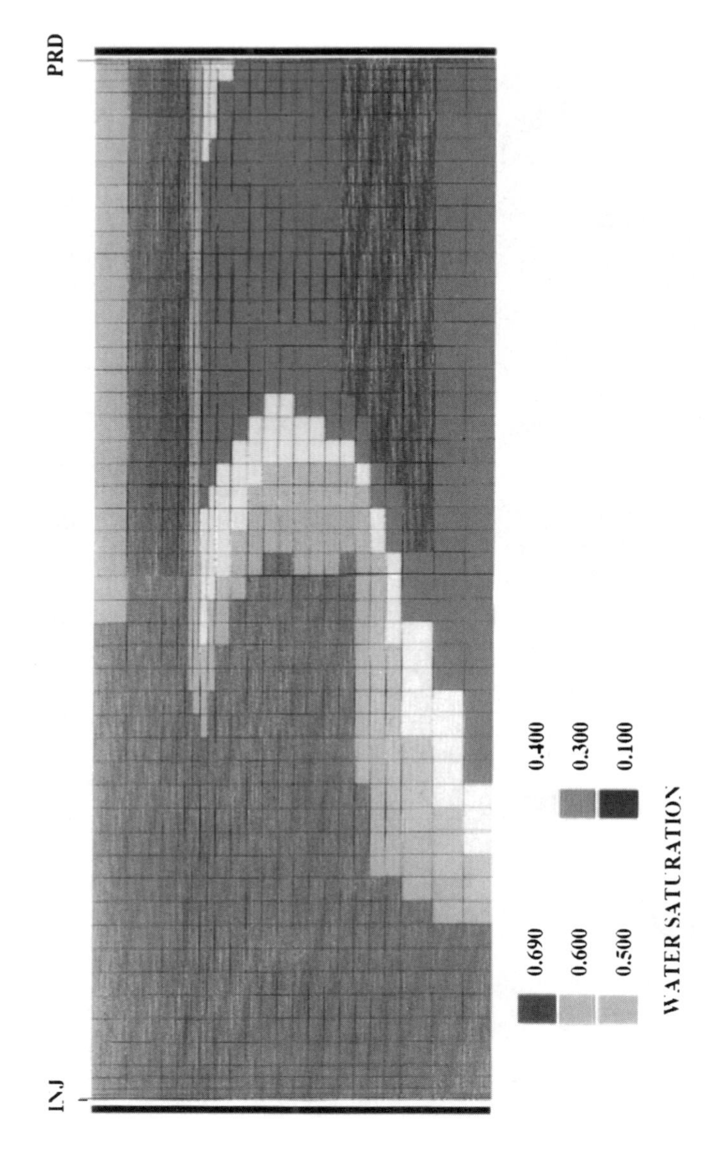

FIGURE 7(b) Saturation Profile at 7 years for CASE02

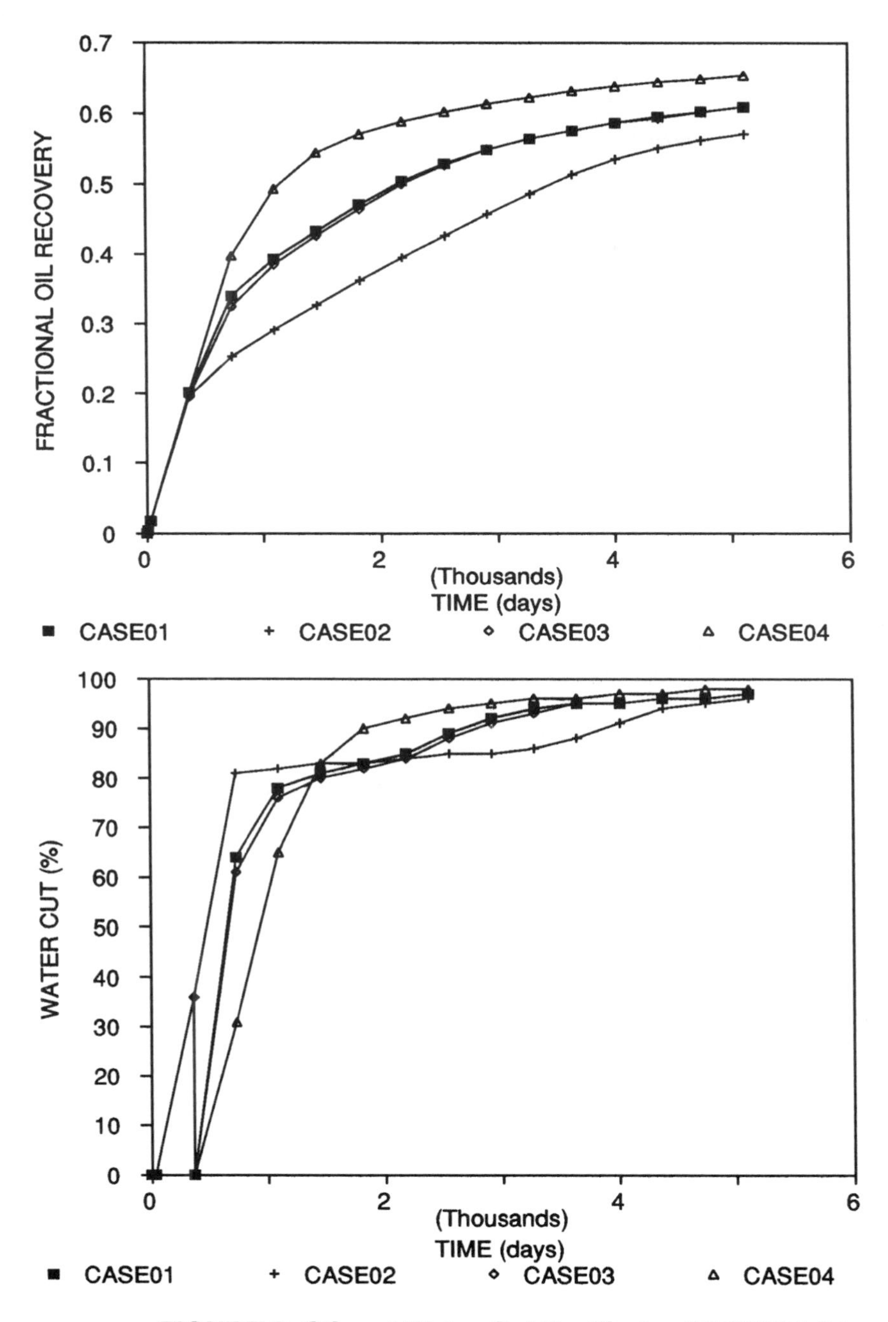

FIGURE 8 Oil and Water Cut Profile for CASES01-04

(Etive + Rannoch) has no benefit in terms of accelerated oil production since the production well is rate limited (at the liquid rate equivalent to 20% STOIIP/year) and Rannoch-only production is sufficient to achieve the desired liquid production rate.

D. CASE04: Effect of Rannoch-Only Injection

This case examined the benefit of confining water injection to the Rannoch Sand only, rather than injecting over the whole Etive and Rannoch interval. All other conditions were the same as CASE01.

Figure 8 compares the recovery and watercut profiles for this case with the Base Case. There is a substantial improvement in oil recovery due to a much better sweep of the base of the Rannoch Sands.

E. Discussion of CASES 01-04

The conclusion from these cases is that Rannoch-only production is preferable if the Rannoch productivity is high enough to supply a good fluid throughput at the well. In the current model the Rannoch productivity is rather high (Kh about 27000 mD-ft) and the producer would have to be choked back on Rannoch-only production to avoid excessively high offtake rates (ie. more than 20% STOIIP/year). Under these circumstances, it would always be preferable to confine production to the Rannoch perforations to delay water breakthrough, since the extra productivity supplied by Etive perforations has no benefit of accelerated production in the initial period. The case of re-perforating onto the Rannoch is never as successful as the Rannoch only production case, for similar reasons.

These preliminary conclusions may explain why the policy of Rannoch-only production in the Murchison reservoir in the UK North Sea has been so successful. The Rannoch productivity in this field is high (oil rates over 17,500 stbpd were achieved from dedicated Rannoch producers (Massie et al, 1985) while dual producers were choked back to 30,000 stbpd to avoid excessive production) and the extra production benefit supplied by the Etive completions has little real benefit if the wells are severely choked. In the Thistle and Dunlin fields, however, the production rate achieved by Rannoch-only completions is the order of a few thousand bpd (Bayat and Tehrani, 1985) and very significant production capacity is lost by excluding the Etive layer completions. In newly discovered fields that are operated by Rannoch-only production it may be advisable to ensure that artificial lift facilities are available from the beginning. This would increase the likelihood of economic production rates from the Rannoch completions alone.

It is also beneficial to confine water injection to the Rannoch Sands. This provides increased pressure support to the base of the Rannoch unit compared to injection over the whole Etive plus Rannoch interval and a resulting increase in sweep efficiency.

The effects of sudation (the combined effects of gravity segregation and capillary cross flow) at the Etive-Rannoch boundary have little influence on the behaviour of these models.

5. ETIVE PERMEABILITY ABOUT 1 DARCY, ETIVE-RANNOCH PERMEABILITY CONTRAST ABOUT 100:1

A. Introduction

In some Brent Sand reservoirs, the effective Rannoch permeability appears to be much lower than core measured values (reference 2), by over an order of magnitude in some cases. The reasons for this are not entirely clear, but may be related to the effects of the interfaces between the hummocky stratified cross-beds that are thought to make up the Rannoch Sand. The net effect is that initial oil rates from dedicated Rannoch producers are often of the order of only a few thousand stbpd rather than anticipated rates of over 20000 stbpd

We repeated the sensitivities described in section 3 with a revised model in which the Rannoch permeability was reduced by a factor of 20 (excluding the "Tight Zone" at the top of the Rannoch). The geological model was the same as the Base Case in all other respects. The resulting model represents fields where the oil production rate achievable from Rannoch only perforations is fairly modest and where substantial increases in production rate are achievable by using both Etive and Rannoch perforations. The effective Etive-Rannoch "Kh" contrast is about 100 in this type of reservoir.

B. CASE05: Production from Rannoch-Only Perforations (Minimum BHP = 1500 psi)

In this case the production well was unable to provide the desired initial oil offtake rate of 20% STOIIP/yr because of the low Rannoch Sand productivity. The production well switched to the subsidiary well control of a minimum BHP of 1500 psi, with an initial oil production rate of approximately 5% STOIIP/yr.

The results of this sensitivity are compared with CASE01 in figure 9. There is a very large reduction in the recovery factor with the reduction in Rannoch permeability and the watercut is considerably more adverse. The water saturation profile for this case at water breakthrough, and after 15 years production, is illustrated in figure 10. The poorer sweep compared to the Base Case (CASE01) is evident, and is a direct result of the order of magnitude increase in the Etive-Rannoch permeability contrast.

C. CASE06: Production from Rannoch + Etive, Injection into both Rannoch and Etive (Maximum Rate = 20% STOIIP/yr)

In this case the provision of initial Etive completions enabled the desired offtake rate of 20% STOIIP/yr to be achieved, with a modest drawdown of 70 psi. Water broke through at the production well after only 8 months, however, and there was an extremely rapid buildup in watercut, to over 90% after two years production. The watercut stabilised at about 95%.

The results of this case are compared to CASE05 in figure 11. The results can be interpreted as follows. The Etive layer contributes over 95% of the "Kh" product for the production well and waters out very rapidly. The watercut then stabilises at about 95%. The important point to note is that the Rannoch continues to produce essentially dry oil, albeit in extremely small quantities, until the secondary front in the Rannoch layer breaks through. This process will take over 50 years and is thus extremely inefficient.

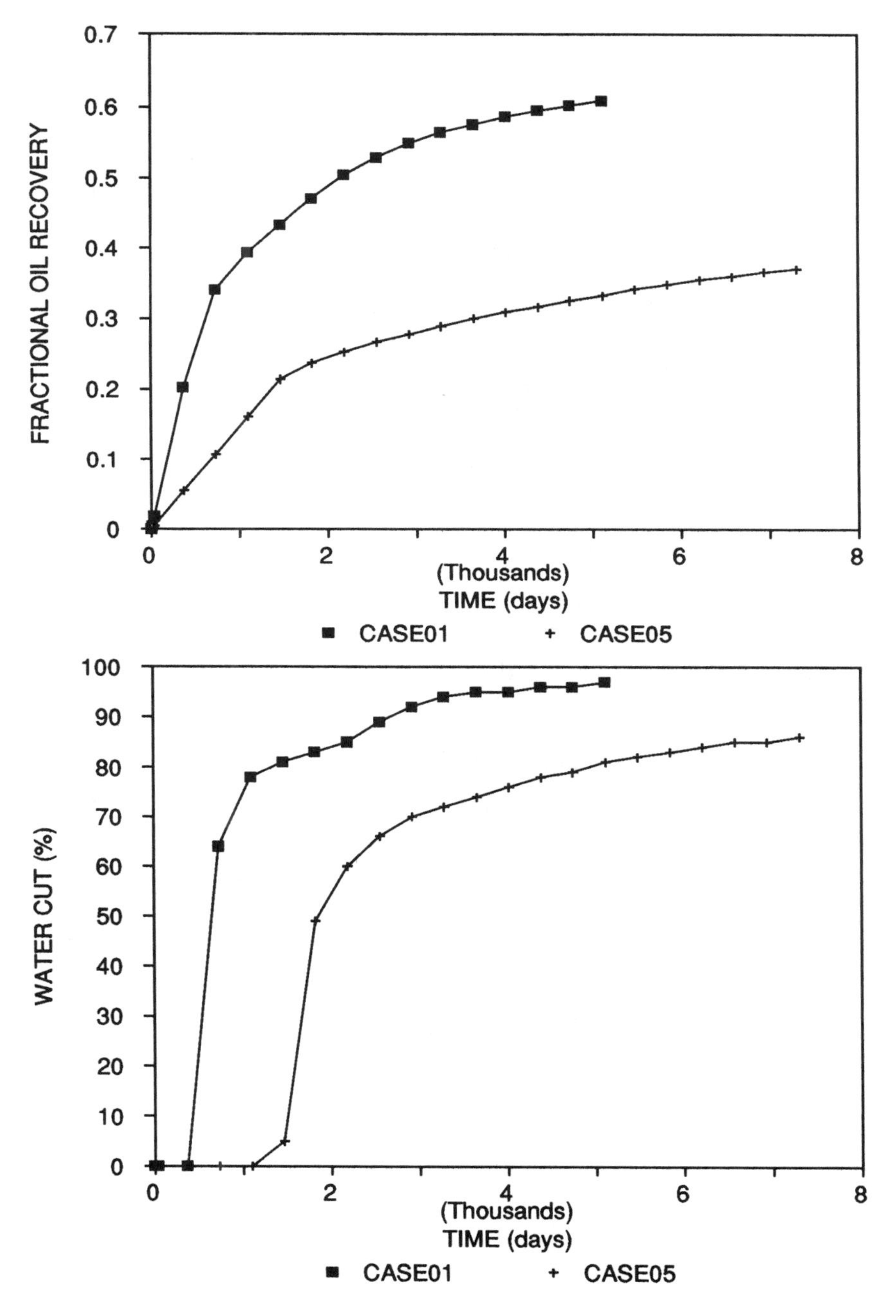

FIGURE 9 Comparison of Oil and Water Cut Profiles for CASE01 and CASE05

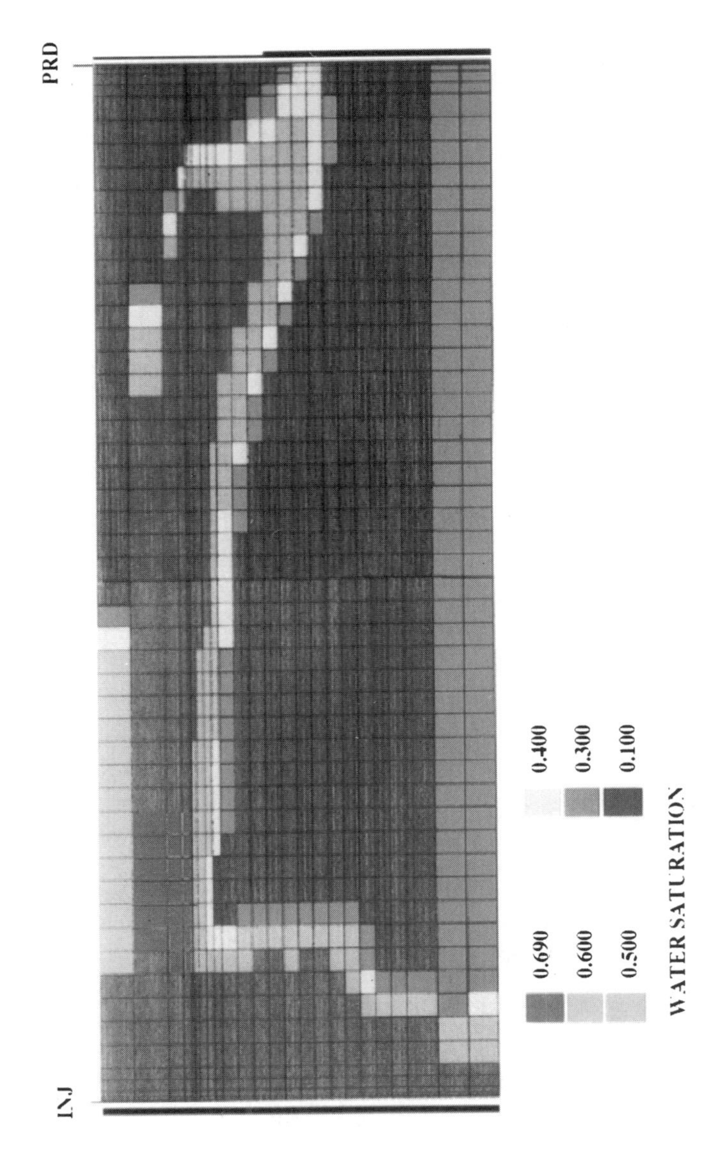

FIGURE 10(a) Saturation Profile at 1 year for CASE05

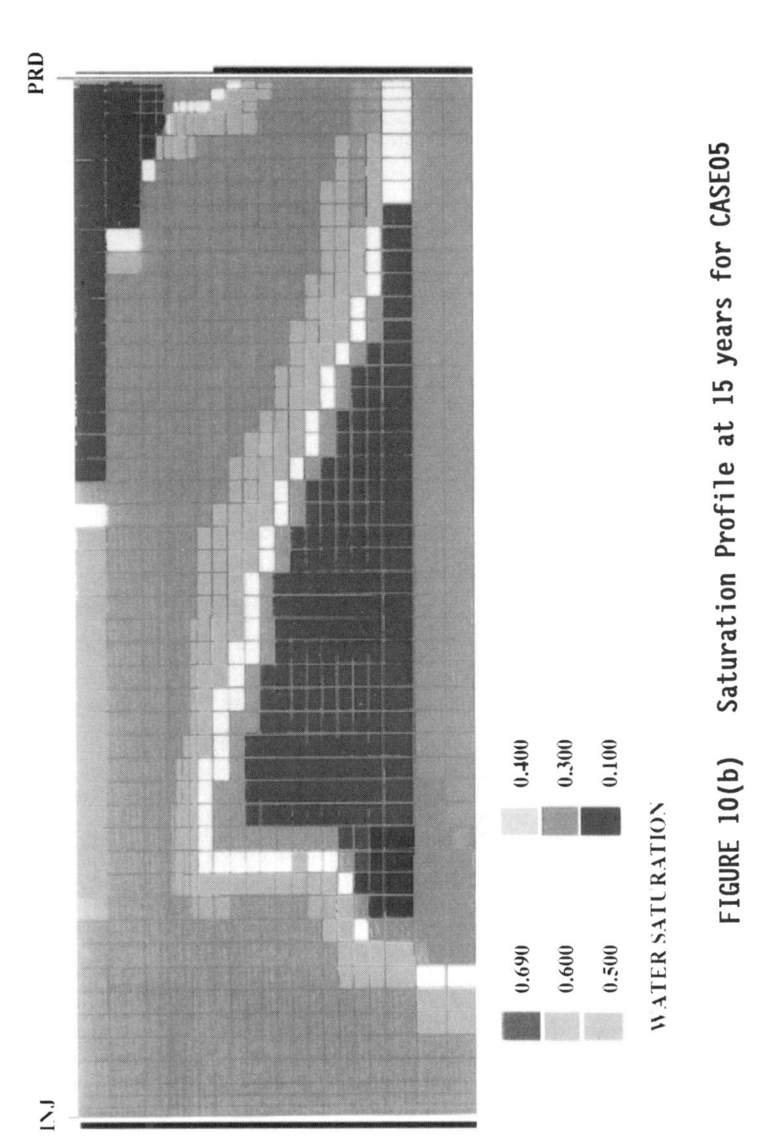

FIGURE 10(b) Saturation Profile at 15 years for CASE05

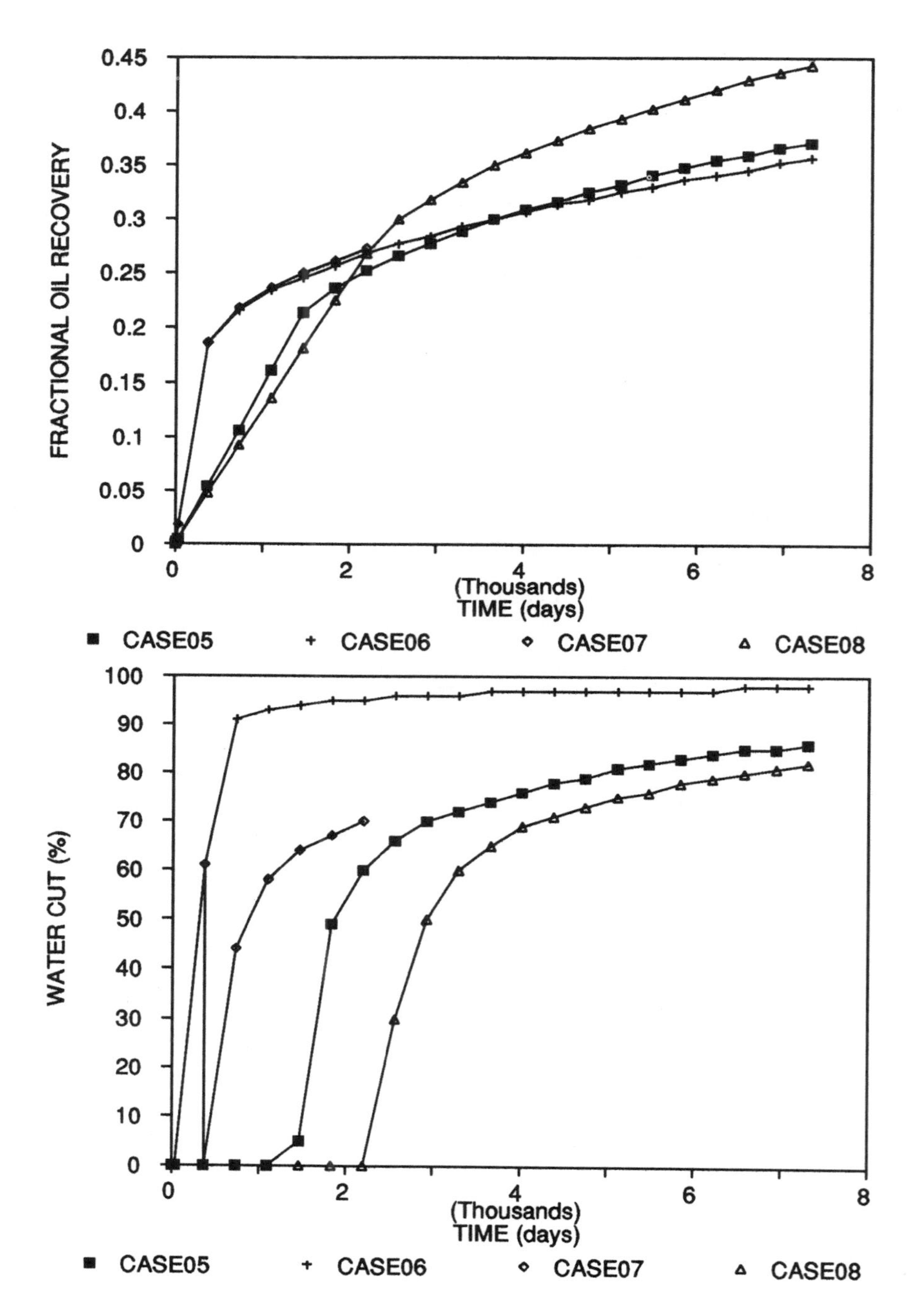

FIGURE 11 Oil and Water Cut Profile for CASES05-08

The Etive plus Rannoch completion strategy leads to a dramatic acceleration in early production, with the initial 20% of reserves being produced within 1.5 years (cf. 4 years for CASE05). The oil production from the Rannoch-only completion case (CASE05) takes 10 years to catch up with CASE06. The early oil production from CASE06 is at the expense of a far more severe watercut development, however. The watercut in CASE06 builds up to more than 90% within two years, whilst the watercut in CASE05 takes 15 years to build up to 80%.

The decision over which is the better strategy for reservoir depletion is not clear cut. The acceleration in oil production provided by Etive completion is very attractive in economic terms, but relies on extended production at very high watercuts if a reasonable recovery factor is to be achieved. The amounts of water produced under these circumstances may well exceed platform water handling constraints.

The Rannoch-only completion strategy provides a higher overall oil recovery and produces much less water. The maximum oil rate is very low (2-3000 stbpd) and may not be economic as an initial rate in reservoirs where the initial capital investment has been substantial.

D. CASE07: Effect of Recompletion to Rannoch-Only Production Upon Water Break Through

This case is identical to CASE06 for the first year of production. The production well is then recompleted to Rannoch-only production to limit the watercut buildup and to attempt to improve Rannoch sand productivity.

The results of this case are compared to CASES05-06 in figure 11. Due to simulator convergence difficulties we were only able to obtain results for the first 6 years of production. The main features of the case are apparent, however. The strategy appears to be very successful, combining the best features of CASE05 (the relatively slow buildup in watercut at late time) and CASE06 (accelerated early oil production). The rate of watercut buildup is very substantially reduced by the recompletion after 1 year to Rannoch-only production. The oil production rate only drops slightly upon well recompletion, by about 20%, and the oil production rate in subsequent years is always higher. The recompletion strategy has the dual benefit of both increasing the oil production rate and drastically reducing the water rate.

The results for CASE07 can be understood as follows. The fluid production rate is high during the period of dry oil production and the Etive layer is essentially fully swept during this initial period. Very little oil production takes place from the Rannoch Sand during this time. The recompletion of the production well after water breakthrough is then equivalent to drilling an infill well in a region of the reservoir where the Etive has been water flooded. The bulk of the remaining oil production is that displaced by the reverse coning of water from the Etive to the Rannoch layer. This is illustrated in figure 12 which shows the water saturation distribution for CASE07 after 6 years production. The increased Rannoch sweep efficiency due to reverse coning is evident.

E. CASE08: Effect of Rannoch-Only Injection

This case examined the benefits of confining water injection to the Rannoch sand only rather than injecting over the whole Etive plus Rannoch interval. All other conditions were the same as for CASE05.

Figure 11 compares the recovery and watercut profiles for this case with CASE05. There is a substantial improvement in the oil recovery at a given watercut with Rannoch-only injection, due to the better sweep at the base of the Rannoch sands. The production rate achievable is even lower than CASE05, however, and a bottom hole pressure of over 8000 psi was necessary at the injector to provide voidage replacement. If the Rannoch quality is poor a Rannoch-only injector is unlikely to provide sufficient injectivity, particularly if initial oil production is from a dual (Etive plus Rannoch) producer.

F. Discussion of CASES05-08

Cases 05-08 explored the Etive-Rannoch depletion strategy in a reservoir where the Rannoch quality is poor (the effective

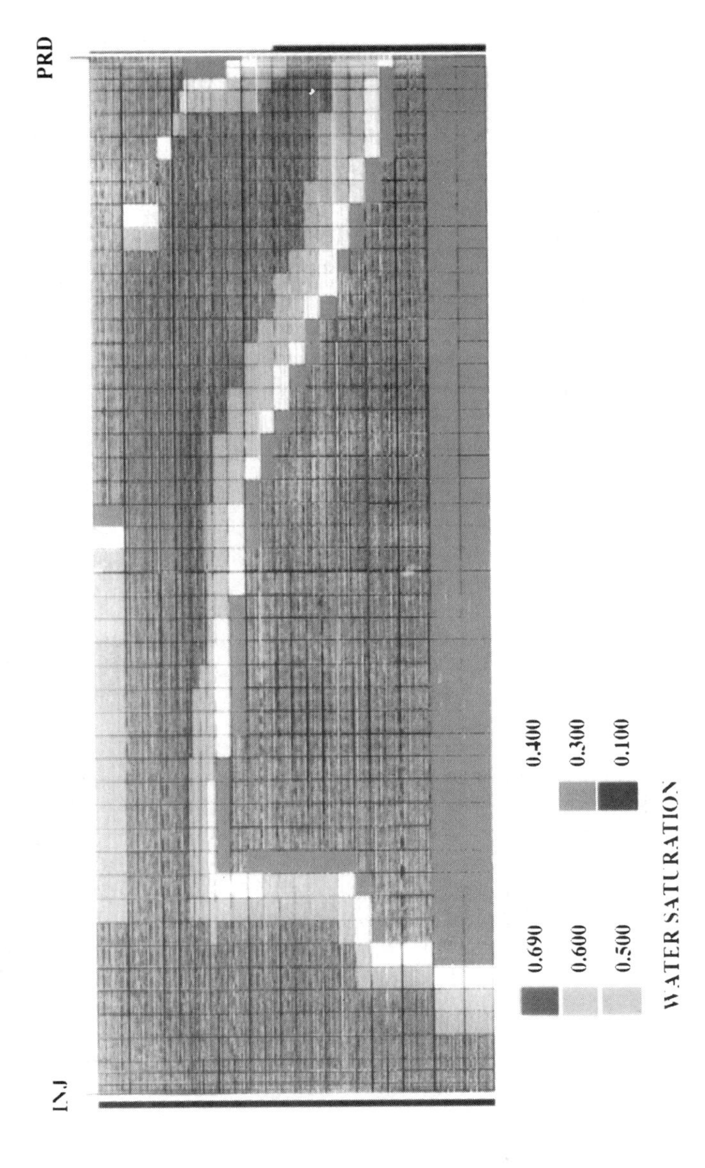

FIGURE 12 Saturation Profile at 15 years for CASE07

maximum horizontal permeability is of order 20 mD, and drops to order 1 mD at the base of the Rannoch) and where the Etive-Rannoch permeability contrast is very high (the ratio of "Kh" products for the Etive and Rannoch layers is about 100:1).

The situation is significantly different from that for CASES01-04 where the Rannoch permeability was high enough to provide individual production rates of 20% STOIIP/yr and a Rannoch-only production strategy was recommended. In CASES05-08 the Rannoch productivity is much lower (5% STOIIP/yr) and may well be uneconomic. Initial well perforation on both the Etive and Rannoch provides a significant boost to early production, but can lead to a very adverse watercut development (watercut rapidly rising above 90%) if Etive production is continued after water breakthrough.

The preferred strategy under these circumstances is to produce the well initially from a dual completion. This enables a substantial dry oil production to be achieved in the initial period where a high rate is economically more attractive. The well should be worked over upon water breakthrough to Rannoch only production. This step will lead to substantial reduction in watercut without substantial change in oil production rate (there is a small initial drop in oil rate upon recompletion, but the rate rises above that for CASE07 in subsequent years) and is thus very beneficial.

The Etive perforations are producing at well over 90% watercut when the recompletion takes place, whilst the Rannoch perforations continue to produce dry oil. The re-perforation concentrates the pressure drawdown in the Rannoch, where it is most needed, and future oil production is mainly from the drawdown cone that forms around the production well as water is drawn down from the swept out Etive to the Rannoch sand.

It may be advantageous to convert the producer to a horizontal well when the workover takes place. Horizontal wells have several important benefits, which are particularly suited to the problem of Rannoch production in situations where the Etive layer has already been swept. The first benefit is that a drawdown "crest" develops which has a much larger volume than a comparable drawdown "cone" for a conventional well. The amount of oil recovered from the Rannoch is highly dependent on the volume of water swept out in the drawdown cone. A second benefit is that a horizontal well may be placed significantly lower down in the Rannoch than a comparable vertical well. The Rannoch Sand coarsens upwards and a major problem with vertical well placement is the need to be in high permeability sand (ie. requires a perforation towards the top of the Rannoch) whilst avoiding rapid coning of water from the Etive (ie requires perforations as far below the Etive as possible). There is also the obvious benefit of the improved productivity index for a horizontal well.

Dedicated Rannoch infill wells are normally necessary to achieve adequate Rannoch recovery factors when the Rannoch permeability is poor. This is because the "secondary front" in the Rannoch moves so slowly when the Etive-Rannoch permeability contrast is of the order of 100:1.

Rannoch-only injection (CASE08) can improve ultimate oil recovery, but it is not considered feasible because of the very poor injectivity of the Rannoch Sand. An over-pressure of 4000 psi was necessary to provide voidage replacement, even when production was confined to the Rannoch.

6. ETIVE PERMEABILITY ABOUT 100 mD, ETIVE-RANNOCH PERMEABILITY CONTRAST ABOUT 10:1

A. CASES 09-11: Moderate Etive Permeability (of order 100 mD) Combined with Low Rannoch Permeability (of order 10 mD)

In this suite of cases we examined the situation where the Rannoch is again of low quality, but where the permeability contrast between the Etive and Rannoch remains about an order of magnitude. This permeability distribution was achieved by reducing the permeability in the Base Case model (CASE01) by a factor of 20 for all layers in the model (ie not just the Rannoch as in CASES 05-08).

Figure 13 illustrates the oil and watercut profiles for the three cases examined with this model (The injection well was perforated across the whole interval; whilst the producer had

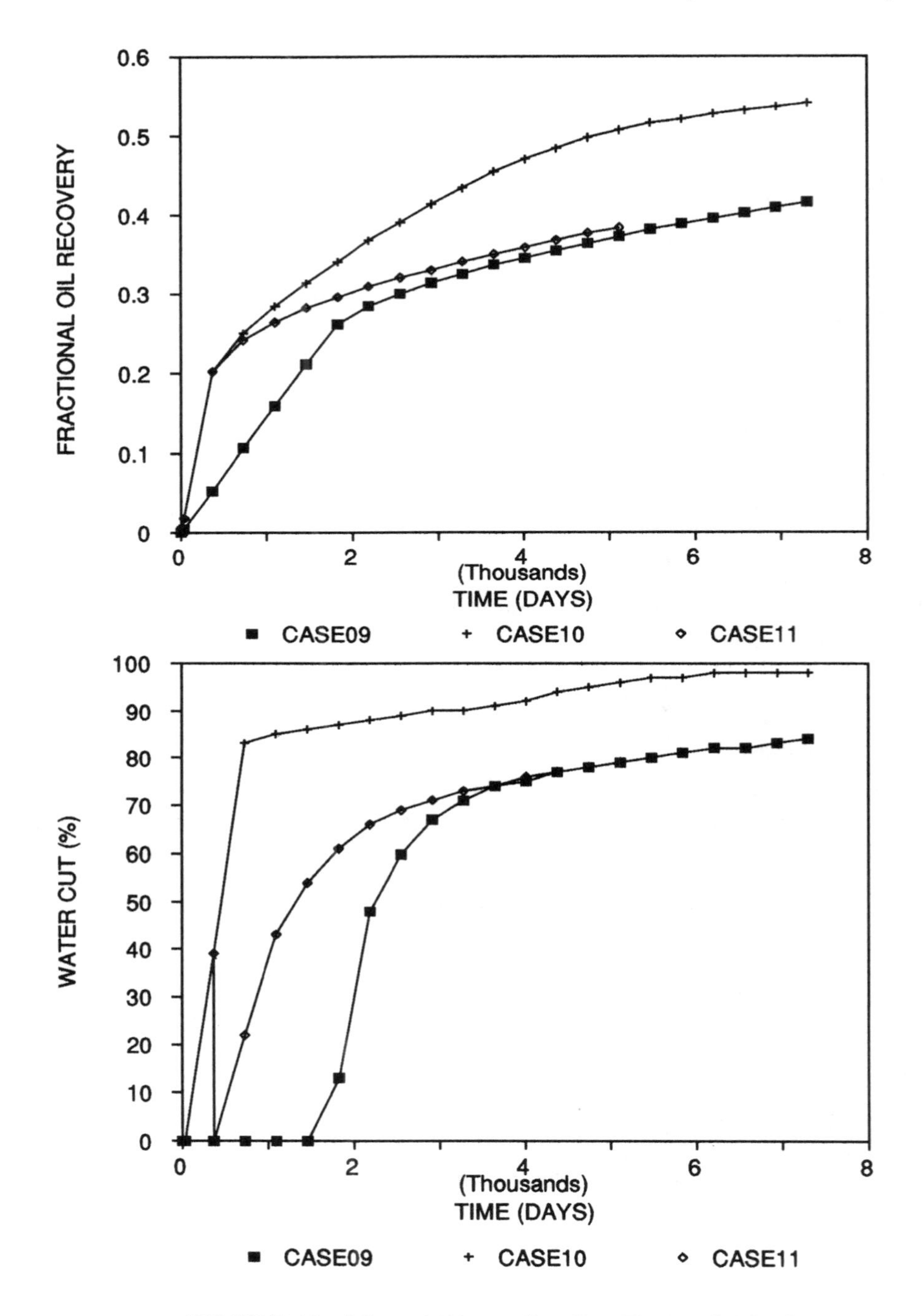

FIGURE 13 Oil and Water Cut Profile for CASES09-11

Rannoch-only perforation; Etive + Rannoch perforation; and initial perforation on Etive + Rannoch, with re-perforation to Rannoch-only after water breakthrough; respectively).

In the Rannoch-only perforation case (CASE09) the oil offtake rate at the limiting BHP was, as in CASE05, less than 5% STOIIP/yr. The oil recovery was substantially improved when the Etive layer was also perforated (CASE10). In this case the Etive watered out rapidly (within 1 year) and the watercut stabilised at about 85%. The important feature, however, was that the Rannoch layer continued to produce essentially dry oil until the secondary front in the Rannoch layer broke through and this 15% oil production was close to the total fluid production rate for CASE09. This is why the Rannoch-only perforation case failed to catch up with the cumulative production from the Etive+Rannoch producer.

Recompleting the producer to Rannoch-only production after water breakthrough led to a reduction in oil recovery. The reason for this is again that the dry oil production from the Rannoch perforations in the dual producer case (CASE10) exceeds the oil production rate from the Rannoch-only perforations after the workover. It is probable that the extra oil production from CASE10 would make a dual-perforation strategy more attractive than CASE11, if the high watercut associated with the dual producer can be tolerated.

B. Conclusions from CASES09-11

The conclusion to be drawn from CASES09-11 is that it can be beneficial to use dual producers in some circumstances, if the watercut associated with Etive completion can be tolerated. These circumstances are where the Rannoch permeability is poor and where the Etive-Rannoch permeability contrast is an order of magnitude or less. The low Rannoch productivity requires that Etive perforations are used, at least initially, to achieve an economic oil production rate. If the Etive-Rannoch permeability contrast is an order-of-magnitude or less, then the oil rate from the Rannoch Sand for a dual producer is likely to exceed that from dedicated Rannoch producer. The reason for this is that the Rannoch completions produce virtually dry oil in the case of a dual producer until the secondary flood front in the Rannoch Sand reaches the producer, whilst the Rannoch perforations produce at a significant watercut for a Rannoch-only producer.

The price to be paid for the extra oil production for the case of a dual producer is obviously the greatly increased water production. After 20 years of production the Rannoch-only producer has produced 41% STOIIP and approximately the same amount of water. An Etive plus Rannoch perforation produces an equivalent amount of oil in less than half the time, but produces three times as much water.

7. ETIVE PERMEABILITY ABOUT 10 DARCIES, ETIVE-RANNOCH PERMEABILITY CONTRAST ABOUT 10:1

A. High Etive Permeability (of order 10 Darcies) Combined with Very Good Rannoch Permeability (of order 1000 mD)

In the final suite of cases we examined reservoirs where both the Etive and Rannoch are of high quality. Our model was based upon the Statfjord reservoir where Etive permeabilities are of the order 10 Darcies, and Rannoch permeabilities are of order 1 Darcy (McMichael, 1978). There was also again a "Tight Zone" of reduced permeability at the top of the Rannoch, with a permeability of approximately 100 mD.

The results for this case showed that water-override in the Etive layer still occurred, despite the extremely high reservoir permeability. The degree of override was controlled by the order-of-magnitude permeability contrast between the Etive and Rannoch Sands, the fluid injection rate and the vertical permeability at the Etive-Rannoch interface.

The degree of water override and the fluid production profiles were insensitive to well completion strategy in this case. For example, there was an insignificant difference in the oil recovery profiles for the sensitivity cases where the production well were completed on the Rannoch Sand only throughout field life, and the case where production was allowed over the whole Etive plus Rannoch interval. The reason for this was that the high reservoir quality leads to conditions very close to hydrostatic equilibrium throughout field life.

8. CONCLUDING DISCUSSION

The results obtained from the fine-grid simulation model of a producer-injector pair in the Etive-Rannoch sands suggest that the correct depletion strategy depends on a number of factors. We have divided the possible Etive-Rannoch reservoir types into three main categories.

A. Reservoirs with Good Quality Rannoch Sands

Reservoirs falling into this category include Murchison, Statfjord and Brent (Massie et al, 1985; Haugen et al, 1988). In the very highest quality reservoirs such as Statfjord (Etive permeability about 10,000 mD, Rannoch permeability about 1000 mD) the completion strategy is unimportant since the bulk of the reservoir is under hydrostatic equilibrium at all times and the layer production rates are simply proportional to layer permeabilities. A conservative approach in these circumstances is to perforate both the injection well and the producer on the Rannoch Sands.

In more moderate quality reservoirs (Etive permeability about 1000 mD, Rannoch permeability about 200 mD) the best strategy is to perforate both production and injection wells over the Rannoch interval initially. If the Rannoch permeability is of order 100 mD or more, Rannoch completions are likely to have sufficient productivity to provide an economic rate (about 10,000 stbpd). A dual producer may have the potential to deliver about 100,000 stbpd in this case, but this would not be of much benefit as the well would have to be severely choked to confine production to less than about 20% STOIIP/yr. The point is that a dual producer provides no significant advantage in terms of well productivity, but leads to much earlier water breakthrough and a reduced ultimate recovery.

In this case there may be some advantage in recompleting wells at water breakthrough to perforate both the Etive and Rannoch in order to maintain dry oil production from the Rannoch until the secondary front arrives.

B. Reservoirs with Moderate to Poor Quality Rannoch Sands and High Permeability Contrast Between the Etive and Rannoch Sands

In this type of reservoir we suggest that the preferred strategy is to use dual producers until water breakthrough occurs and then plug back the wells to Rannoch-only production at this point. The low Rannoch productivity makes Rannoch-only production unattractive during the period of dry oil production, whilst the excessive water production makes a dual producer unattractive after water breakthrough has occurred.

Rannoch-only infill wells will be required since the secondary front will be moving too slowly to reach the principal producers in a reasonable period of time.

Dedicated Rannoch injection is not likely to be attractive in this type of reservoir due to the very poor Rannoch Sand injectivity.

C. Reservoirs with Moderate to Poor Quality Rannoch Sands and Moderate Permeability Contrast Between the Etive and Rannoch Sands

In cases where the Rannoch permeability is moderate to poor and where the Etive-Rannoch permeability contrast remains about an order-of-magnitude, the preferred depletion strategy is to use Etive plus Rannoch perforated producers throughout field life if the high watercut associated with a dual producer can be tolerated. In the case of a dual producer we obtain the expected early water breakthrough. After water breakthrough the Etive perforations produce at nearly 100% watercut, whilst the Rannoch perforations produce dry oil until the secondary front moving along the upper Rannoch reaches the producer. The oil production rate from the Rannoch Sand in the case of a dual producer significantly exceeds that from a dedicated Rannoch producer over most of field life and the cumulative oil production from a the dual producer is always higher (at least 10% STOIIP) than that from the dedicated Rannoch producer.

Infill wells will be required to drain the Rannoch effectively, because the secondary front in the Rannoch is moving very slowly.

Table 2 summarises the recommended policies for initial perforations, recompletions and infill drilling for various combinations of Etive and Rannoch permeabilities. In this table it is noted that horizontal wells will generally be more effective than Rannoch-only recompleted or infill wells.

9. ACKNOWLEDGEMENTS

This work was part of a more detailed study of Rannoch Sand depletion in UKCS Brent reservoirs commissioned by the United Kingdom Department of Energy. It is a pleasure to acknowledge our co-workers for this study, Laurie Dake (Reservoir Engineering) and Susan and Tony Corrigan (Geology).

10. REFERENCES

(1) Barbe, J.A., 1981, Reservoir Management at Dunlin, Presented at Offshore Europe 81 Conference, Aberdeen, UK

(2) Bayat, M.G. and Tehrani, D.H,, The Thistle Field - Analysis of its Past Performance and Optimisation of its Future Development, Presented at Offshore Europe 85 Conference, Aberdeen, UK

(3) Bourbiaux, B.J. and Kalaydjian, F.J, 1988, Experimental Study of Co-current and Countercurrent Flows in Porous Media, SPE 18282

(4) Dake, L.P., 1982, Application of the Repeat Formation Tester in Vertical and Horizontal Pulse Testing in the Middle Jurassic Brent Sands, EUR270, 1982 European Petroleum Conference

(5) Haugen, S.A, et al., 1988, Statfjord Field: Development Strategy and Reservoir Management, J, Pet. Technol.

(6) Massie, I, et al, 1985, Murchison: A Review of Reservoir Performance During the First Five Years, SPE 14343

Etive Kh (K)	Rannoch Kh (K)	Etive: Rannoch Kh Ratio	Initial Perforation Policy of Producers and Injectors	Recompletion Policy at Water Breakthrough	Infill Policy
Excellent (10 Darcies)	Very Good (1 Darcy)	10	**Any.** Rannoch-only perforations for conservative approach.	**Not necessary.**	**Not necessary.**
Excellent (10 Darcies)	Good (100mD)	100	**Rannoch-only.** Rannoch PI is high enough for good rate. Dual perforations would give a 99% watercut.	**No.**	**Rannoch-only.** May be needed if the "secondary" front in the Rannoch is slow moving.
Very Good (1 Darcy)	Good (100mD)	10	**Rannoch-only.** Rannoch PI is sufficiently high for a good oil rate.	**Dual.** Stabilised watercut will be about 90%, but oil rate can be increased by dual completion.	**Rannoch-only.** May be needed if the "secondary" front in the Rannoch is slow moving.
Very Good (1 Darcy)	Moderate (10mD)	100	**Dual.** PI on Rannoch too poor to give economic plateau rate.	**Rannoch-only.** Stabilised watercut for a dual producer will be 99%.	**Rannoch-only.** The front "secondary" front is moving very slowly.
Moderate (100mD)	Moderate (10mD)	10	**Dual.** PI on Rannoch too poor to give economic plateau rate.	**No.** Stabilised watercut is 90% and the wells will produce more oil than a dedicated Rannoch well.	**Dual Producer.** Stabilised watercut of 90%.
Moderate (100mD)	Poor (1mD)	100	**Dual.** PI on Rannoch too poor to give economic plateau rate.	**Uneconomic.**	**Uneconomic.**
Poor (10mD)	Poor (1mD)	10	**Dual.** PI on Rannoch too poor to give economic plateau rate. May be uneconomic.	**Uneconomic.**	**Uneconomic.**

TABLE 2: SUMMARY OF RESULTS

(7) McMichael, C.L., 1978, Use of Reservoir Simulation Models in the Development Planning of the Statfjord Field, EUR89, 1978 European Petroleum Conference

RESERVOIR DESCRIPTION AND MODELLING OF THE PEN FIELD, GRAHAM COUNTY, KANSAS

G. E. Gould[1]
D. W. Green
G. P. Willhite

Tertiary Oil Recovery Project
University of Kansas
Lawrence, Kansas

R. A. Phares[2]
A. W. Walton

Geology Department
University of Kansas
Lawrence, Kansas

The Pen Field lies in township 6S-22W in Graham County, Kansas, and produces from the Lansing and Kansas City (Upper Pennsylvanian) Groups. After discovery in May 1985, the field was fully developed on 40- acre spacing by March 1988, with seventeen producing wells and eight dry holes. Water injection began in November 1990 after four producers were converted to injectors.

The main producing zones are oolitic or skeletal grainstone lenses in the J and I zones that range up to 6 feet thick. Diagenesis immediately after deposition greatly reduced porosity in a few areas, making grainstones there unproductive. Subsequent diagenesis during exposure altered the upper part of each reservoir lens, creating non-permeable caprocks. Despite these effects and several later diagenetic events, most pores are primary with a small enhancement of vuggy or

[1]Present address: Exxon U.S.A., Midland, Texas.
[2]Present address: Consultant, Wichita, Kansas.

moldic pores. Pressure data indicate that the main reservoir in the J zone is continuous over most of the field.

The reservoir thickness was interpretatively contoured using standard geological methods. Well-log porosity values were assigned at grid blocks centering on wells and porosity was interpolated for inter-well blocks with adjustments based on geological observations. Permeabilities were assigned from correlations with porosity obtained from core analyses. Subsequently, geostatistical variogram analysis and punctual kriging were performed (using the EPA's GEO-EAS freeware) on the J Alpha grainstone properties to improve the estimated values of porosity and permeability. Primary production was history-matched using a black oil simulator. Waterflood plans were simulated to compare alternatives and to determine the plan likely to produce the most incremental oil.

In the Pen Field, timely data acquisition throughout the development and operation of a field was essential in developing the reservoir description. Reservoir modeling based on this description enabled improved estimates of additional oil recovery for various secondary recovery scenarios. This is essential in efficient recovery of our natural resources and in economic decision making.

1. INTRODUCTION

A. Background--Field History

The Pen Field lies in sections 17-20 of township 6S-22W in Graham County, Kansas, (Fig. 1) and produces from the I, J, K and L zones of the Lansing and Kansas City (Upper Pennsylvanian) Groups. PanCanadian Petroleum Company discovered the field in May 1985, and it was fully developed on 40 acre spacing by March 1988, with seventeen producing wells and eight dry holes (Fig. 2). Prior to unitization, the field consisted of six leases (Pennington, Shuck, White, Bethell, Griffey, and Demuth) as shown in Figure 3.

The reservoir was initially undersaturated, but pressure fell below the bubble point within one year. By 1989, reservoir pressure had declined to 150 psi and the field production rate was 60 barrels/day. Cumulative primary production was 538,000 barrels of oil. Very little water or gas was produced during this period. Production in each well was commingled in the wellbore. The field was unitized in October, 1990. Four producers were converted to injectors.

Water injection began in late November, 1990, and little waterflood response had been seen until June, 1991 when production began to rise. Injection profiles, run in June, 1991, indicated 95% of the water injected into Shuck #2 and 100% of the water injected into other injection wells was going into the J zone. The injection profiles suggested that permeability in the J zone was much larger than

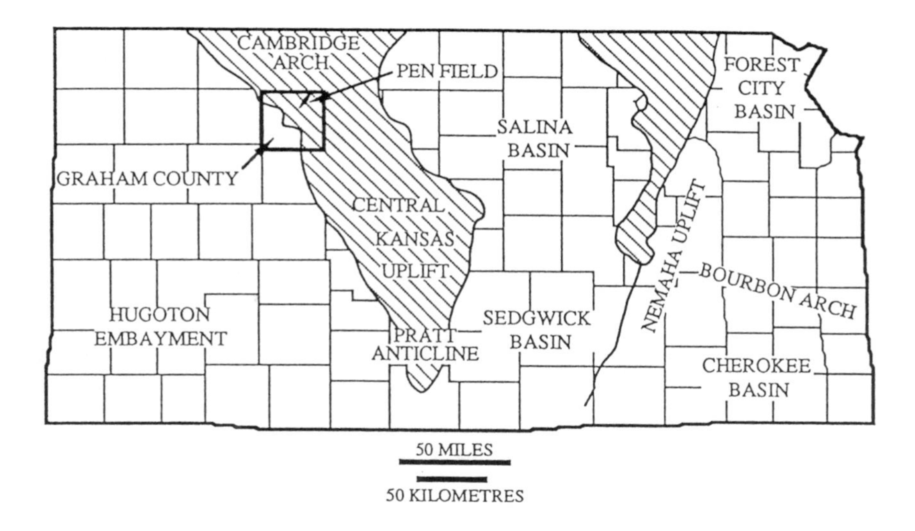

Fig. 1. Location map showing Pen Field in the Central Kansas Uplift.

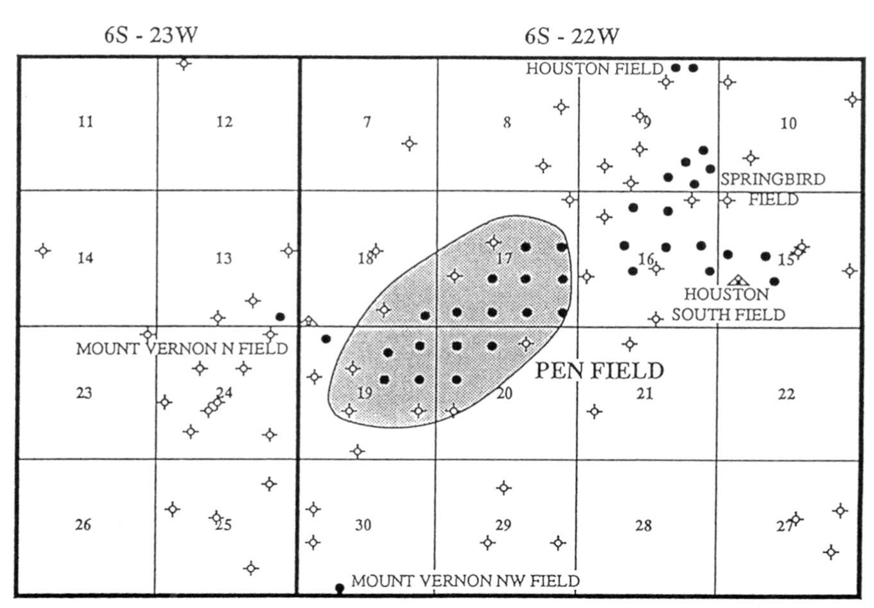

Fig. 2. Pen Field Region Map.

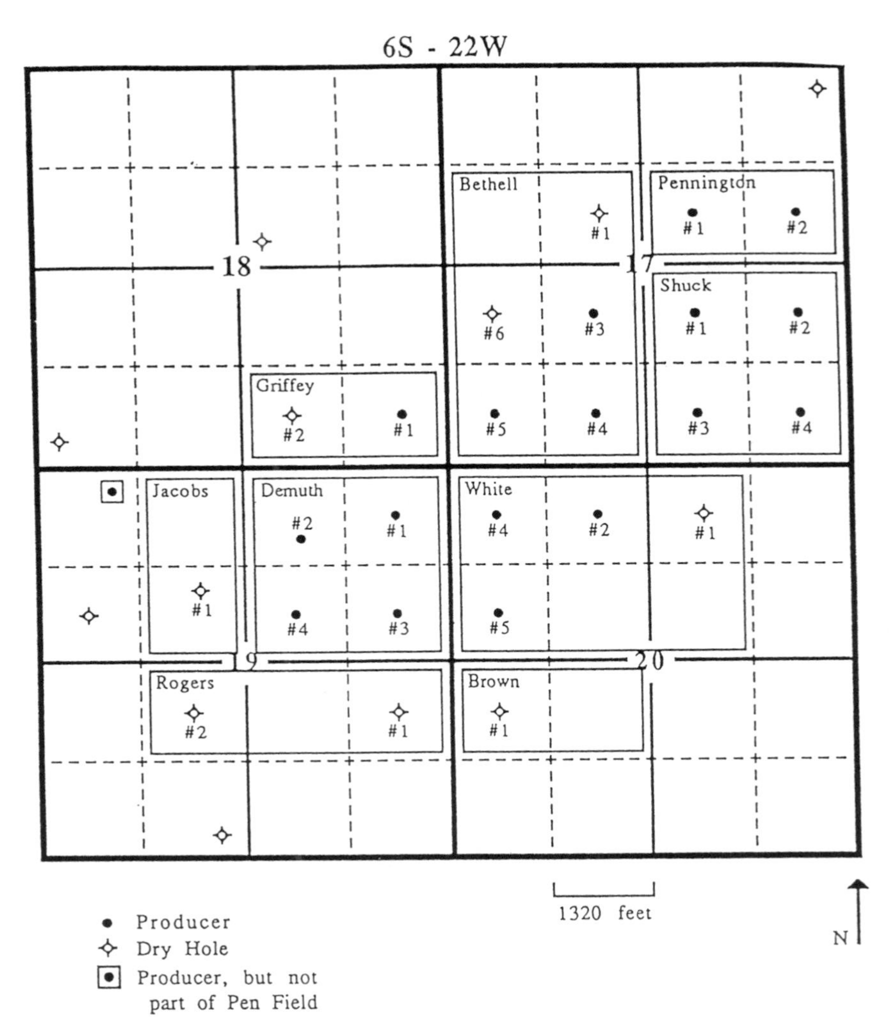

Fig. 3. Lease Map with Well Number Designation.

permeability in the I zone. Currently, production continues to rise and most of the waterflood response will occur in the future.

The Pen Field was selected for this study because a large amount of data was available for the field including well production records with monthly barrel tests for the first three years of field life, cores in eight producing wells and three dry holes (Fig. 4), modern log suites designed to evaluate carbonate lithologies, and drillstem tests of the productive interval in each well. Special core analyses were available for three core plug samples. Five pressure buildup tests were run in January, 1989. This is an unusual set of data for a Lansing-Kansas City reservoir because cores are seldom available.

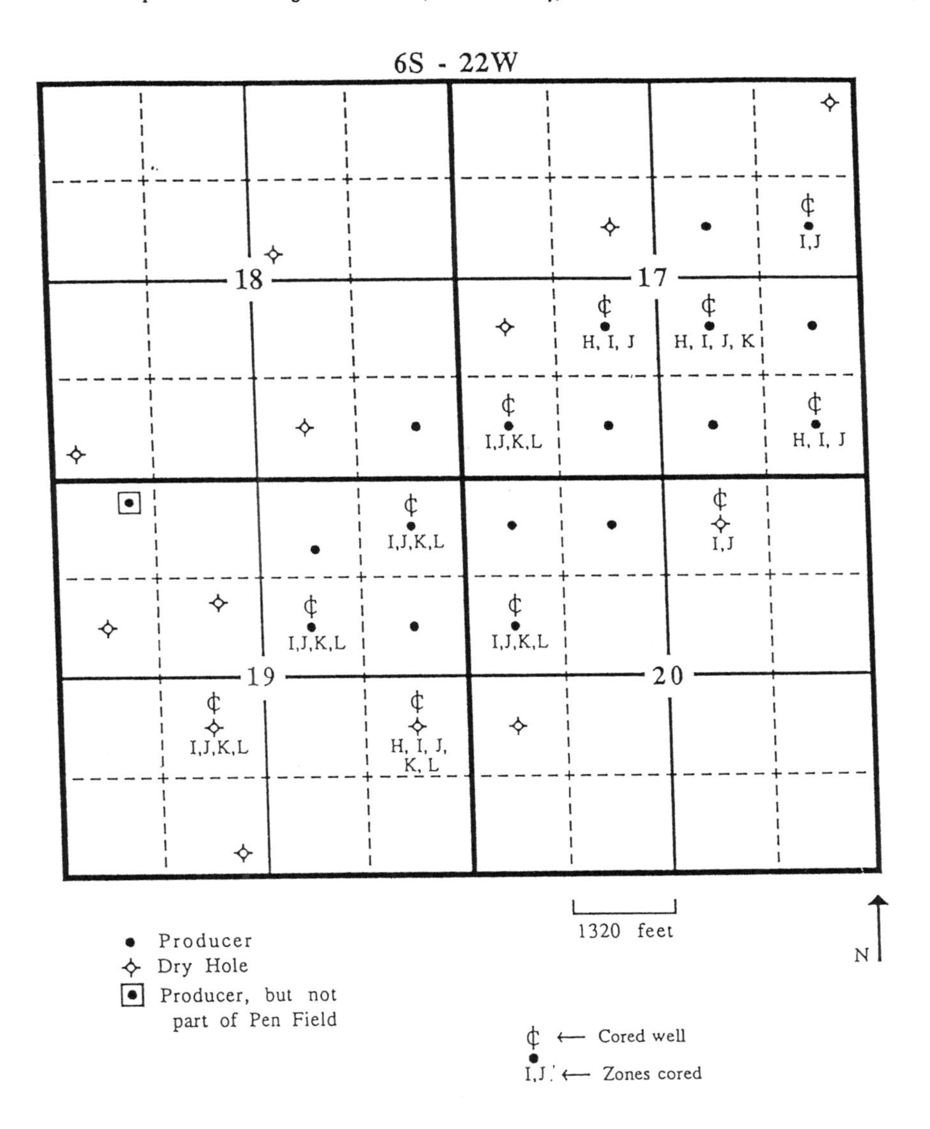

Fig. 4. Cored Wells in Pen Field.

This paper describes a reservoir management study in which geological description, core data, production data and reservoir simulation were integrated to develop a reservoir description of the Pen Field and to explore possible waterflooding plans. The overall plan of attack is outlined in the following section.

Lithofacies and paragenetic features of the main producing zones were described in cores and thin sections (1). These studies were combined with well logs, production and engineering data to estimate the continuity and heterogeneity of productive zones. Constraints on the analysis were to provide sufficient detail to permit identification of main producing units and to assign properties to grid blocks when the reservoir was subdivided in preparation for reservoir simulation.

A goal of this study was to compare two methods of assigning properties to individual grid blocks for reservoir simulation. The first method used linear interpolation with modification by geological insight between well locations to assign properties such as thickness and porosity to grid blocks. A second method based on geostatistics was used to estimate porosity for each grid block.

The reservoir descriptions were used in a reservoir simulator to match primary production history and to predict the performance of various waterflood plans. An iterative process was used in which the reservoir description was refined to fit the observed production response.

II. GEOLOGICAL DESCRIPTION

The Pen Field lies in the saddle between the Central Kansas uplift and the Cambridge arch of Nebraska and northern Kansas (Fig. 1). The Lansing-Kansas City interval (Missourian, Upper Pennsylvanian) of central and western Kansas consists primarily of limestone and shale interbedded at a scale of 3 to 33 ft. The Lansing-Kansas City interval is divided into a series of informal zones with letter designations (2). Alternating limestones and shales of the Missourian are commonly described as cyclothems, cyclic repetitions of lithologic units representing changes in environment of deposition (3,4).

The Pen Field is a stratigraphic trap, with production controlled by local development and preservation of porous carbonates (1). Although five zones are productive in parts of the field, the J zone is the only interval which produces throughout the field. The I zone is productive in parts of the field. The D, K, and L zones are productive in some wells but the zones are either small isolated reservoirs or have low permeability (<5 md). Detailed geological analysis of all zones is summarized by Phares (1). The I and J zones are the principal productive intervals and were the zones studied intensively.

A. The J Zone

The J zone in the Pen Field consists of two limestone beds (lower and upper carbonates) and two mudrock beds (lower and upper shales; Fig. 5) which are continuous across the field. Production in the Pen Field is primarily from a grainstone of the upper or Alpha unit of the upper carbonate, where remaining

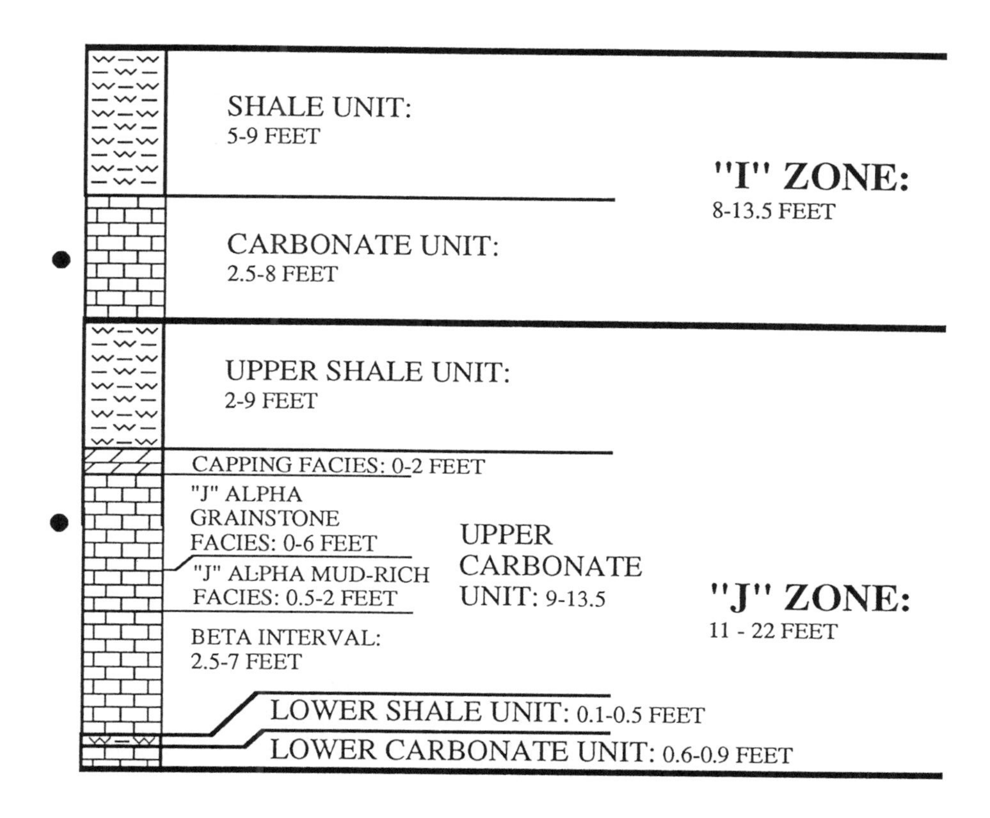

Fig. 5. I and J Zone Facies Identification.

interparticle pores have been augmented by molds and a few vugs. Minor production comes from molds and vugs in the cap unit at the top of the J upper carbonate and from Beta grainstones and wackestones in the lower part of the upper carbonate, where porosity is either interparticle or moldic.

Limestones of the J zone accumulated on a shallow marine shelf in generally oxygenated water. Micrite-rich lithologies accumulated in less agitated water than grainstones, especially the oolitic alpha grainstone. Microkarst, root molds, rhizoliths, and autoclastic breccia in the cap unit indicate subaerial exposure after accumulation of the Alpha grainstone but before deposition of the upper shale of the J zone.

The J zone superficially resembles the ideal cyclothem (3; Table 1): a thin transgressive limestone, a thin core shale, a thicker regressive limestone with an exposure surface, and then a shale formed at low sea levels separating the J zone from the basal transgressive limestone of the overlying I zone.

B. The I Zone

The I zone is thinner than the J zone, ranging from 7.8-13.4 ft thick, and contains only a carbonate unit and an overlying shale. The carbonate unit ranges from 2.6 to 7.87 ft thick and is cream, light gray, or light to medium brown. It contains a suite of oolites and normal marine fossils like that in the J zone, but with a greater abundance of Osagia-coated grains and coated grains. The grains are coarser in the lower part where the lithology ranges from carbonate mudstone to grainstone. The upper part is normally a muddy lithology – carbonate mudstone or wackestone and rarely a mixed wackestone-packstone. I zone grainstone is the secondary productive reservoir, contributing 10% of the Pen Field's production. The top of the I carbonate and the overlying shale displays extensive paleosal features.

C. Diagenesis and Porosity History in Grainstones

Diagenesis in the rocks of the Pen Field includes two broad episodes: a period of dissolution, cementation, recrystallization, and paleosol formation during subaerial exposure in marine, brackish, or fresh water, and a late period of fracturing, cementation, and dissolution that post-dated exposure. Diagenesis is described completely in Phares (1).

The earliest episodes of cementation differ from bed to bed, but took place in marine or mixing zone environments. One of the early cements, a non-ferroan, isopachous, cloudy, finely-bladed marine calcite cement occurs in the J Alpha grainstone. This cement filled the primary pores and rendered the Alpha grainstone impermeable along the southern margin of the field. It is the only cement that demonstrably converts this grainstone to non-reservoir rock. In the rest of the field, most interparticle pores remained open at the end of this initial phase of diagenesis.

Exposure of carbonate beds created paleosols. These effects are most pronounced at the top of grainstones or overlying them, forming the capping layers with drastically different reservoir properties than the underlying grainstones. Generally, paleosols are non-porous, impermeable, and not saturated with oil; they form part of the seal. The effects of paleosol formation were greater in the I zone than in the J zone; I zone effects reach through the entire I zone to the top of the J upper shale in one core. However, most primary pores in the grainstones and secondary pores in a few areas in cap-facies rocks remained open after the episode of exposure. In a few wells, the cap of the J upper carbonate has oil in vugs and molds; these may add minor production.

After exposure and subsequent burial, diagenetic events took place that were either parochial to one layer or widespread throughout the vertical section. The cements, fractures, and pressure solution features seem to have left many primary and secondary pores open after their formation, and the late diagenesis had little effect on production of oil.

D. Continuity in the Pen Field

The Pen Field was developed between 1985-1988. During this time, reservoir pressure fell rapidly as the reservoir energy from expansion of the undersaturated oil was exhausted. Since the reservoir is small and wells were developed over a period of pressure decline, drillstem test data provided insight into reservoir continuity. The pressure data available were DST pressures for each well and five pressure buildup tests run in January, 1989 (5).

Both the I and J zones are within the tested intervals of the DST's. Some tests showed two distinct buildup pressure curves during both the initial and final shut-in periods. Data from such tests are questionable due to cross-flow between the two zones, and the two zones should have been tested individually (6).

In a DST, the extrapolated initial reservoir pressure from the final shut-in period should be close to that from the initial shut-in period. A significant decrease in the extrapolated initial reservoir pressure indicates either a very small reservoir or a bad test (7,8). The Shuck #2, Shuck #4, Bethell #3, White #4, and Demuth #1 wells showed decreases of greater than 5%, so their DST pressure data provide qualitative support for reservoir continuity. These DST data were ignored in the history-match of primary production performance as discussed later. Matthews and Russell commented that if a decrease of more than 5% occurs, the test should be repeated with a longer final flow period (8).

1. J Zone

Drillstem tests sampling the J alpha grainstone interval were run on most of the wells in the Pen Field. Reservoir pressure within the drainage volume probed by the DST was determined by interpretation of pressure data. The graph of reservoir pressure versus cumulative production as of the date of the test is shown in Figure 6. The first four tests show a rapid decline from an initial pressure of 1342 psi to about 285 psi after only 63,000 barrels of production by fluid expansion drive. Four subsequent tests show a slow decline of pressure as production rose to 400,000 barrels by solution gas drive after the bubble point (ca. 285 psi) was reached. Reservoir pressures estimated from DSTs indicate a high degree of communication throughout the grainstone interval. These results suggest that the

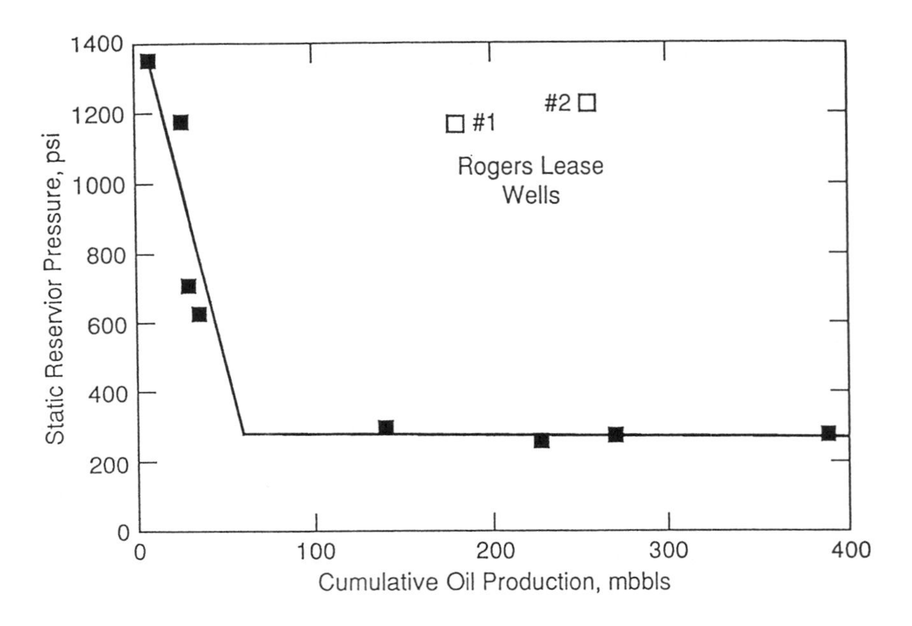

Fig. 6. Reservoir Pressure Versus Cumulative Production.

J zone reservoir has no significant internal barriers to flow and can be treated as a single unit in waterflood operations.

While most drillstem tests show a clear relationship between reservoir pressure and cumulative production, two tests do not (Fig. 6). The two wells in question, the Rogers #1 and #2, are in the southwestern part of the field. The J zone Alpha grainstone in these wells was porous but it was impermeable and the wells were not commercial. Their high pressure at a late date indicates that they are not connected to the reservoir in the Alpha grainstone. In cores from these wells, interparticle spaces in the Alpha grainstone are almost completely filled with isopachous marine calcite cement, rendering the rock impermeable, although subsequently-formed molds may give the rock up to 8% core and log porosity.

2. *I-Zone*

Production from the I zone comes from four separate areas (Fig. 8). A single, isolated well, the White #5 in the southern part of the field, produces from a porous wackestone; the other three areas are in two separate lobes of grainstone. The easternmost of these lobes is divided into two parts by tightly cemented, impermeable rock. These four compartments were mapped from the available log and core data, but it is possible they are even further subdivided. The porosities and permeabilities in the I zone are much less than those in the J zone.

An anomalously high DST pressure of 1340 psi (original pressure equal to Shuck #1 original pressure) occurred for the Griffey #1 after 276 days of field production. Since J zone pressures at this time were already below 300 psi for most of the field, it appears that the 1340 psi was indicative of the I zone.

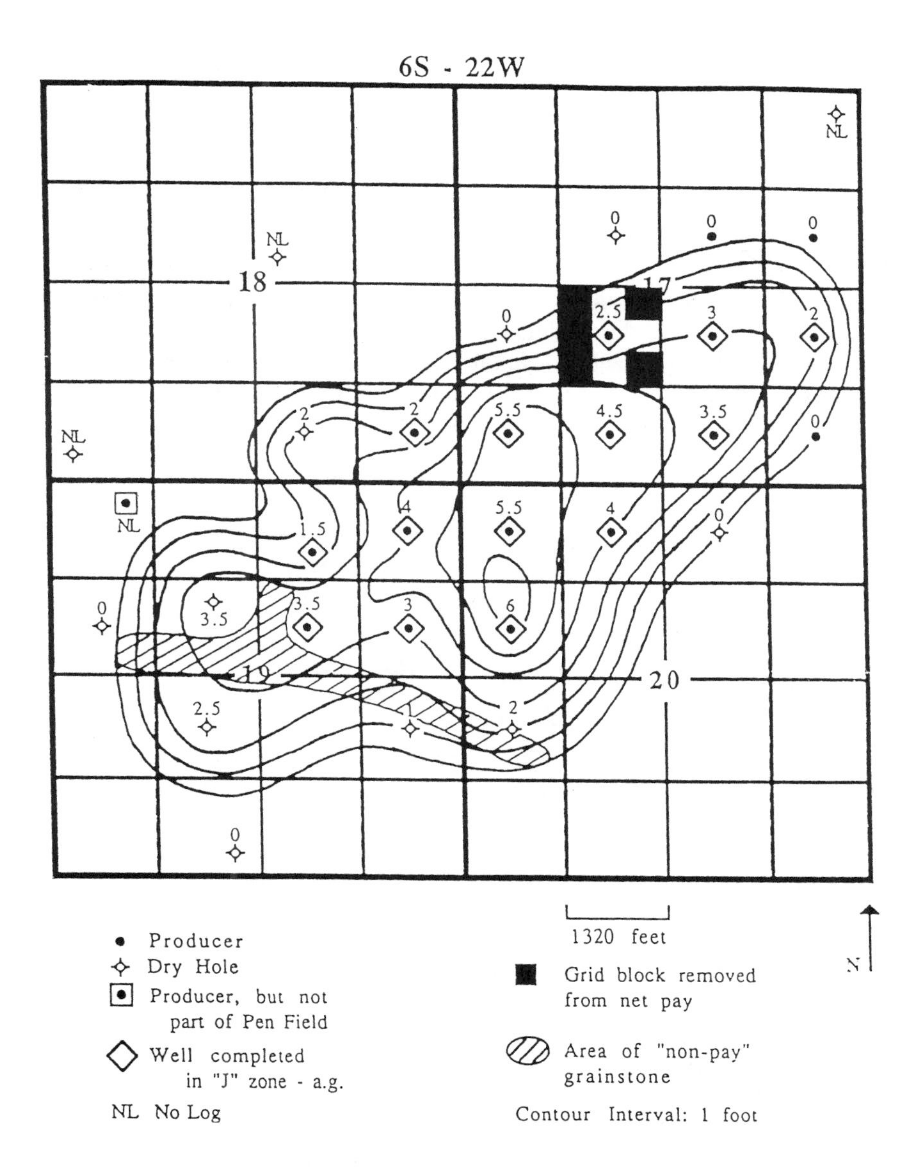

Fig. 7. Net Pay Thickness for J Zone Alpha Grainstone.

III. RESERVOIR MANAGEMENT

The reservoir management plan was based on using the reservoir description developed in the previous section to estimate properties required for simulation of reservoir performance using a 3D black oil simulator.

A. Net Pay Maps

Net-pay maps for the Alpha grainstone and other reservoir intervals were constructed using core or log thicknesses for porous intervals and contoured using geological insight (Fig. 7 and 8). In the J zone, net pay was defined as pay with porosity greater than 7%, while a 5% porosity cutoff was used in the I zone. The J zone porosity cutoff was based on a set of samples from Rogers #1 and #2 in which all samples with porosity less than 7% had permeabilities on the order of 1 md. The porosity in these samples was moldic rather than interparticle as was found in the other areas of the field. Permeability for Alpha grainstone with interparticle porosity was correlated with porosity using Equation 1.

$$k = 6.21 \times 10^{0.069\phi} \tag{1}$$

where:

ϕ = porosity, %

k = permeability, md

At 7% porosity, the permeability from the correlation is about 18.9 md. Since moldic porosity cannot be distinguished from interparticle porosity on the well logs, the net pay estimates are conservative.

Porosity values calculated from neutron and density logs were used in wells where cores were not available. Porosity derived from the well logs and corrected for dolomite content was generally in good agreement with core data. Conditioning the core data using a three foot running average to compensate for the vertical resolution of the logs significantly improved the correlation (1).

The log suites available for the Pen Field were useful in recognizing and quantifying porosity, calculating water saturation and identifying lithologic units to trace across the field. They do not differentiate non-porous wackestone or mudstone from non-porous grainstone and packstone, or porous grainstone from porous wackestone or mudstone. Consequently, where cores were absent, identification of subtypes of limestone was not possible.

The net pay interval (Fig. 7 and 8) in wells without cores may include some pay where numerous vugs and molds form pores in fine-grained limestone that are not part of the productive grainstone and may not be connected.

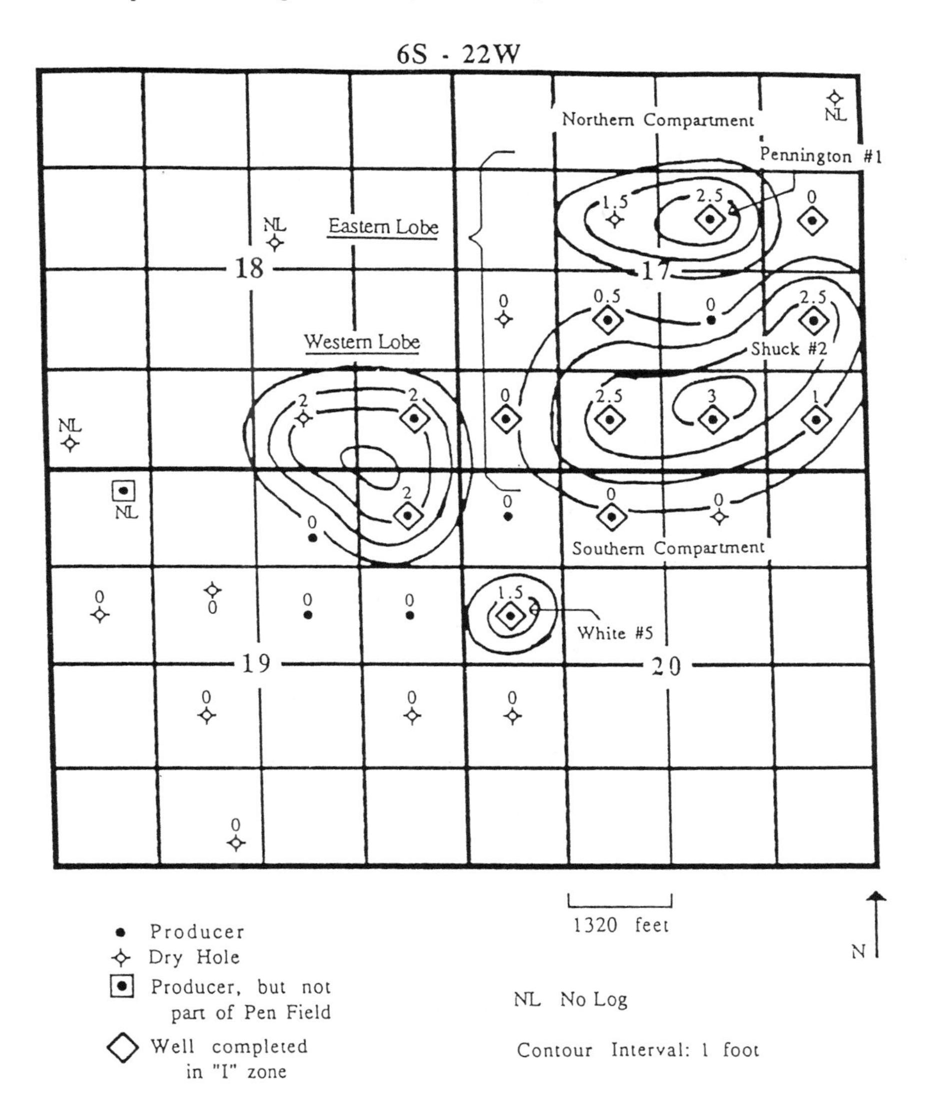

Fig. 8. Net Pay Thickness for I Zone Alpha Grainstone.

B. Reservoir Simulation

The simulator used in this study was Integrated Technologies VIP CORE simulator which was made available to the University of Kansas through a licensing agreement. Data input to a reservoir simulator requires representation of reservoir properties using a grid system such as shown in Figure 9. Properties for each node of the reservoir grid were estimated from the available data.

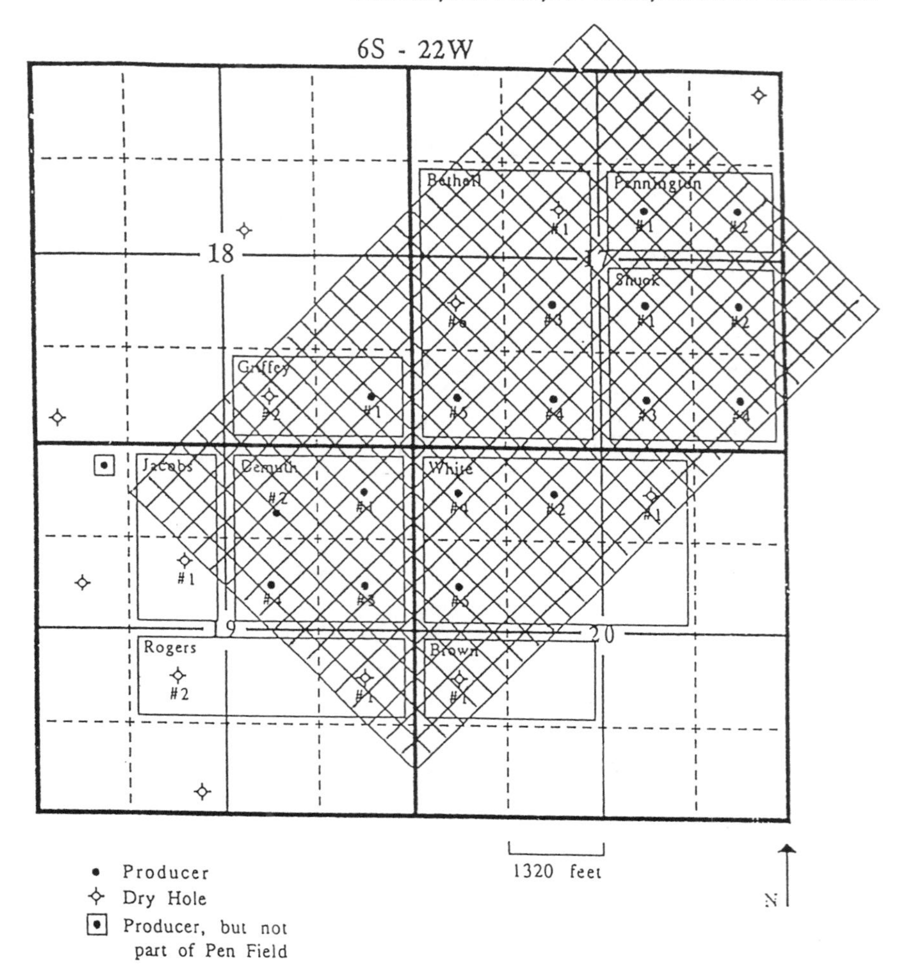

Fig. 9. Grid for Reservoir Simulation.

The square grid pattern shown in Figure 9 aligned with the NE-SW directional trend of the Pen Field was selected for the simulation to minimize the number of grid blocks required in the model. This allowed for directional permeability along the depositional trend, placement of wells in the centers of grid blocks, and at least one empty grid block between all grid blocks containing wells. The final history-match and most of the sensitivity runs were for a grid pattern of 18 by 30 blocks, with a grid size of 2.22 acres. One zone was used to represent the "J" alpha grainstone and one zone was used to represent the "I" zone. Therefore the total number of grid blocks was 1080. The porosity at each grid node was the thickness averaged porosity for that location. Permeability at each grid node was determined from the correlation of permeability with porosity for the particular zone.

1. Porosity and Permeability

Two methods of assigning porosity to individual grid nodes were investigated. In the first method, values of porosity were assigned to each foot of pay in the Alpha grainstone and other reservoirs in each grid block, based upon linear interpolation between wells. The initial grid was oriented N-S with 4.4 acre blocks (1). Permeability was assigned by a correlation of core data (regression of log permeability versus porosity given as Equation 1) from productive wells for each zone. These grid values, the thickness of productive intervals and the extent of drainage areas were modified by information on the productivity of individual wells as compared to the amount of oil that was estimated to be present based on porosity and net pay thickness (1).

The second method for assigning grid properties was based on geostatistics. Geostatistics provides a set of probabilistic techniques that can be used to estimate values of a "regionalized variable" at discrete points within the reservoir boundary. In reservoir engineering applications, the regionalized variable is a reservoir property such as the porosity.

The basic measure of geostatistics is the "semivariance," which describes the rate of change of the regionalized variable with distance along a specific orientation (9). In application, semivariance is plotted as a function of distance on what is termed a "semivariogram." Then, the semivariance points are fitted with a continuous curve, or semivariogram model. Finally, the semivariogram model is used as input for a technique called kriging to estimate the thickness averaged porosity at each grid point. Detailed geostatistical theory is described in the literature (9-13). Geostatistical analysis was used to estimate porosity for the J zone porosity grid. The discontinuous nature of the I zone prohibited geostatistical analysis. That is, the I zone did not have enough data points in each individual pod to generate a variogram. Two-dimensional geostatistical analyses were done using Geo-EAS (14) (Geostatistical Environmental Assessment Software), a collection of interactive freeware tools written for the U.S. Environmental Protection Agency.

Geo-EAS was used to plot and model the porosity semivariogram. The J zone porosity histogram in Figure 10 shows the distribution of the porosity data. The isotropic semivariogram model selected for this simulation was a spherical model having a sill of 18.5 and a range of 4,500 feet. This model matched data for a lag interval of 500 feet very well (Fig. 11), but was also reasonable for lag intervals of 1000 and 1500 feet. The points that vary significantly from the model are averages of fewer data points that those that correlate better. With the sill fixed at a value of 18.5, analysis of directional semivariograms (by specifying an angle with a tolerance of +/- 22.5°) showed the anisotropic semivariogram to have a major range of 8,000 feet along the NE-SW apparent depositional trend (Fig. 12) and a minor range of 3,300 feet in the NW-SE direction (Fig. 13). This anisotropic semivariogram model was used for kriging J zone porosity.

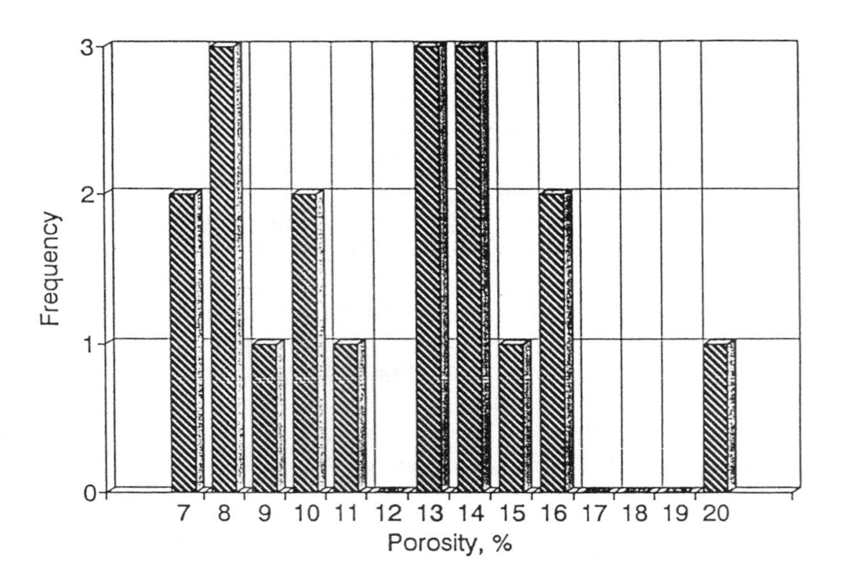

Fig. 10. Porosity Histogram for J Zone Alpha Grainstone.

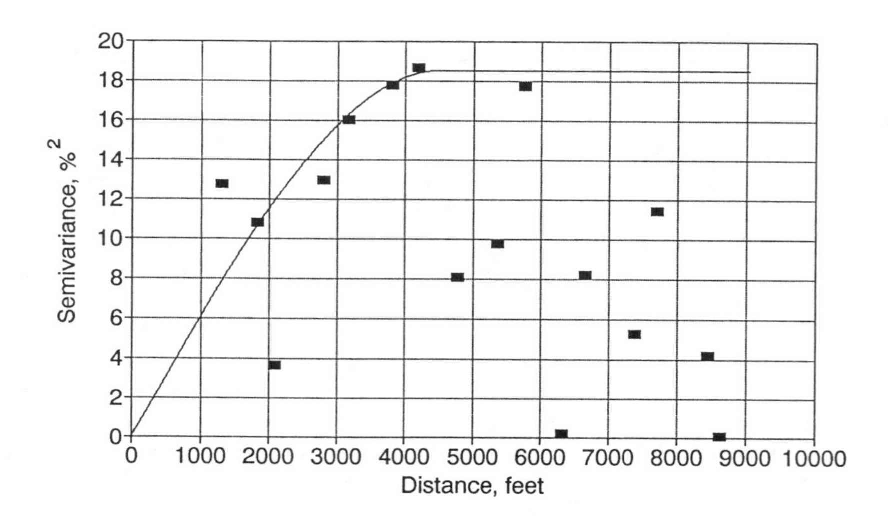

Fig. 11. Isotropic Semivariogram for J Zone Alpha Grainstone.

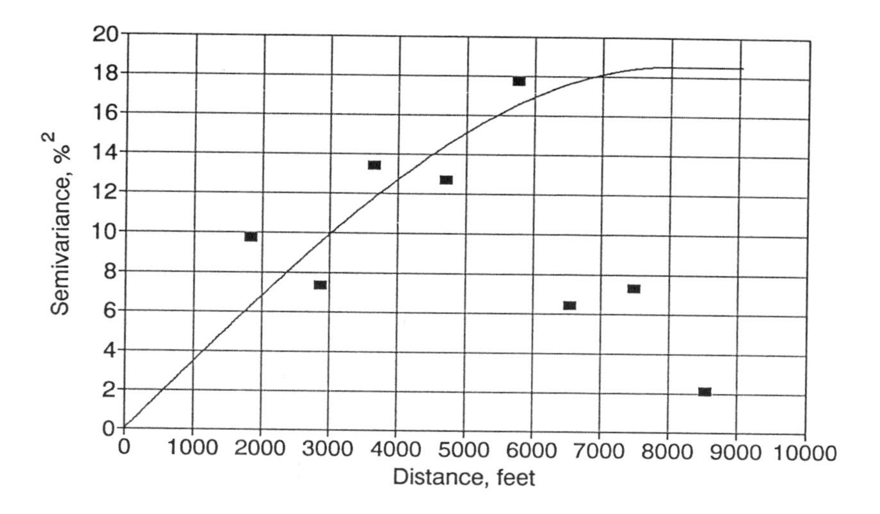

Fig. 12. Anisotropic Semivariogram for J Zone Alpha Grainstone-Major Axis.

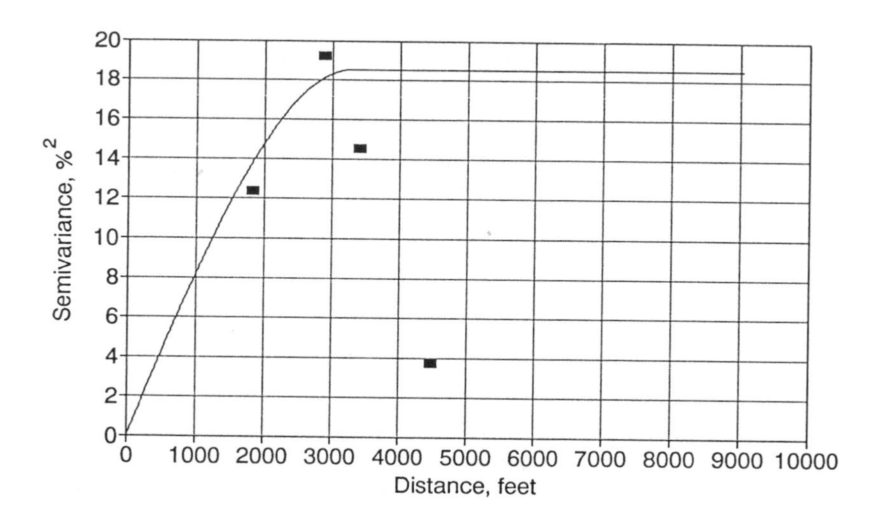

Fig. 13. Anisotropic Semivariogram for J Zone Alpha Grainstone-Minor Axis.

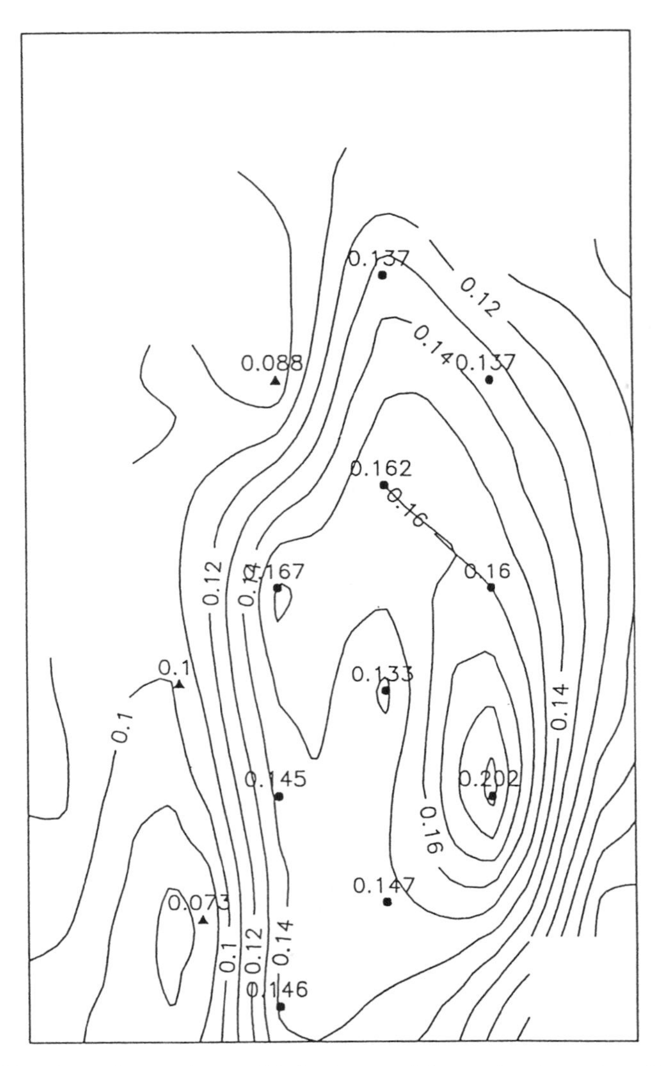

Fig. 14. Comparison of kriged porosities and values at wells.

Several data points lie below the sill at distances greater than the range. This occurs when applying geostatistics to an entire reservoir as opposed to only a part of a large reservoir. Data values along opposite edges of a reservoir are separated by a large distance but have similar (usually low) values, resulting in low semivariance values. Since these values are at distances greater than the range, they have no effect on the calculation of values at grid points. In addition, the accuracy of a semivariogram model decreases as distance increases.

Punctual kriging was used to assign porosity values to grid nodes using the semivariogram determined from the J zone porosity data. A contour map of kriged porosities developed using GEO-EAS is presented in Figure 14 with known porosity values from cored wells indicated. The agreement between kriged and known porosity values is acceptable.

Cross-validation (removing a known value at a well location from the data set, kriging neighborhood values to estimate a value at that location, and then comparing the estimated value with the known value) was performed. Comparison of these estimates to known well values can be helpful in testing semivariogram models for accuracy. A scatter plot of the kriged porosity versus actual well values is displayed in Figure 15. Although the plot indicates no correlation along the unit slope line, actual kriged results (Fig. 14) are more accurate than the scatter plot suggests, because the actual well values *are* included for kriging at nearby grid locations, and actual well values are excluded for cross-validation.

The high degree of variability along the unit slope line is attributed to the large semivariance values for the first data point on the semivariogram (Fig. 12). The first semivariance value is above 10.5 with the sill occurring at a value of 18.5, so most of the variance of the data occurs in a distance less than the closest well spacing (about 1,320 feet). Thus, prediction of porosity values at this distance is difficult. Also, it should be noted that the Pen Field contained only nineteen wells with J zone porosity, many less than the minimum of 50 wells which was recommended as a minimum number by Jones (13). With so few wells and such variability over the smallest well spacing, the geostatistical results are questionable.

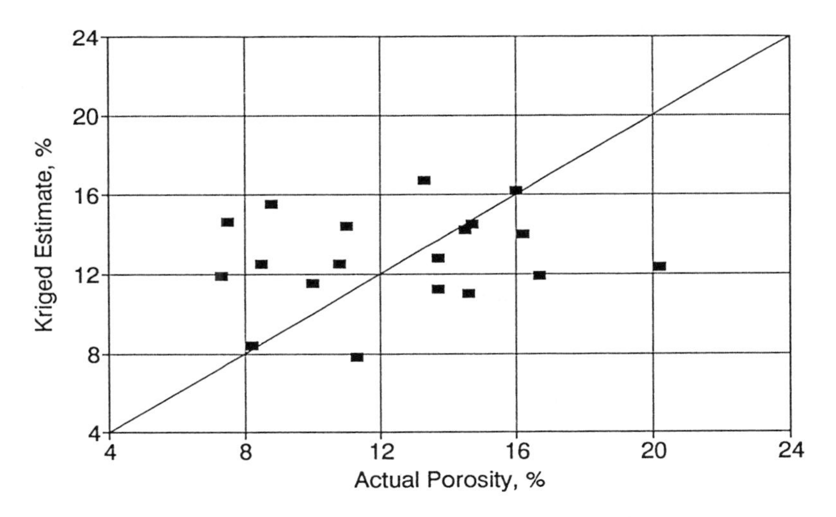

Fig. 15. Comparison of kriged porosities and Values at Wells.Cross-validation of kriged porosities.

2. *Primary Production History Match.*

Oil production data were available by lease until unitization in October, 1990. Barrel tests were run monthly from May, 1985 (field discovery date) until September, 1988, so well production data were excellent for this period. For the period of October, 1988 to September, 1990, lease data were allocated to wells based on exponential decline analysis of each well. The field did not produce much water or gas. These oil-rate data were input into the simulation, and the primary history match was based on pressure data interpreted from DST tests and pressure buildup tests.

Validity of DST pressures was based on less than a 5% decrease from the initial extrapolated reservoir pressures to the final extrapolated reservoir pressure. Nine DST pressures and five pressures from pressure buildup tests were used in the history match. A transmissibility-weighted average of I and J zone pressures at the well's grid cell was selected to be the simulation pressure to match with each field test pressure.

The base case for matching the primary production history was developed from the I zone property grids and the J zone thickness grid developed from linear interpolation between wells and the J zone porosity grid developed using geostatistics. The gas-oil relative permeability curve was obtained from an average relative permeability ratio, k_{rg}/k_{ro}, curve for limestones determined from the literature (15).

The primary production history match was run using the 18 x 30 grid with 2.22-acre spacing. Simulation results for the original data (Fig. 16) did not provide a good history match because simulated pressures were too low. The simulated reservoir pressure at the end of the history match period was 25 psi while the observed pressure was about 150 psi. Adjustments were made in the reservoir description to obtain an acceptable match. The best match (Fig. 16) was obtained by: 1) adding about 50% to the volumetric original oil in place (OOIP) (by multiplying thickness and porosity by 1.25), 2) placing the Shuck #2 in a separate J zone pod, 3) establishing directional permeability of $k_y = 5\ k_x = 5\ k_{core}$ along the NE-SW trend of the J zone, 4) multiplying the average permeabilities of the J zone by about 2.24, 5) shifting the average gas-oil relative permeability curve by +3.5%, and 6) adjusting skin factors. Justification of these changes is presented in the following paragraphs.

Adding 50% to the volumetric OOIP was required to raise simulation pressures to field test pressures. It was performed by multiplying thickness and porosity by 1.25. It was supported by 1) the J zone material balance and 2) the volume of primary recovery.

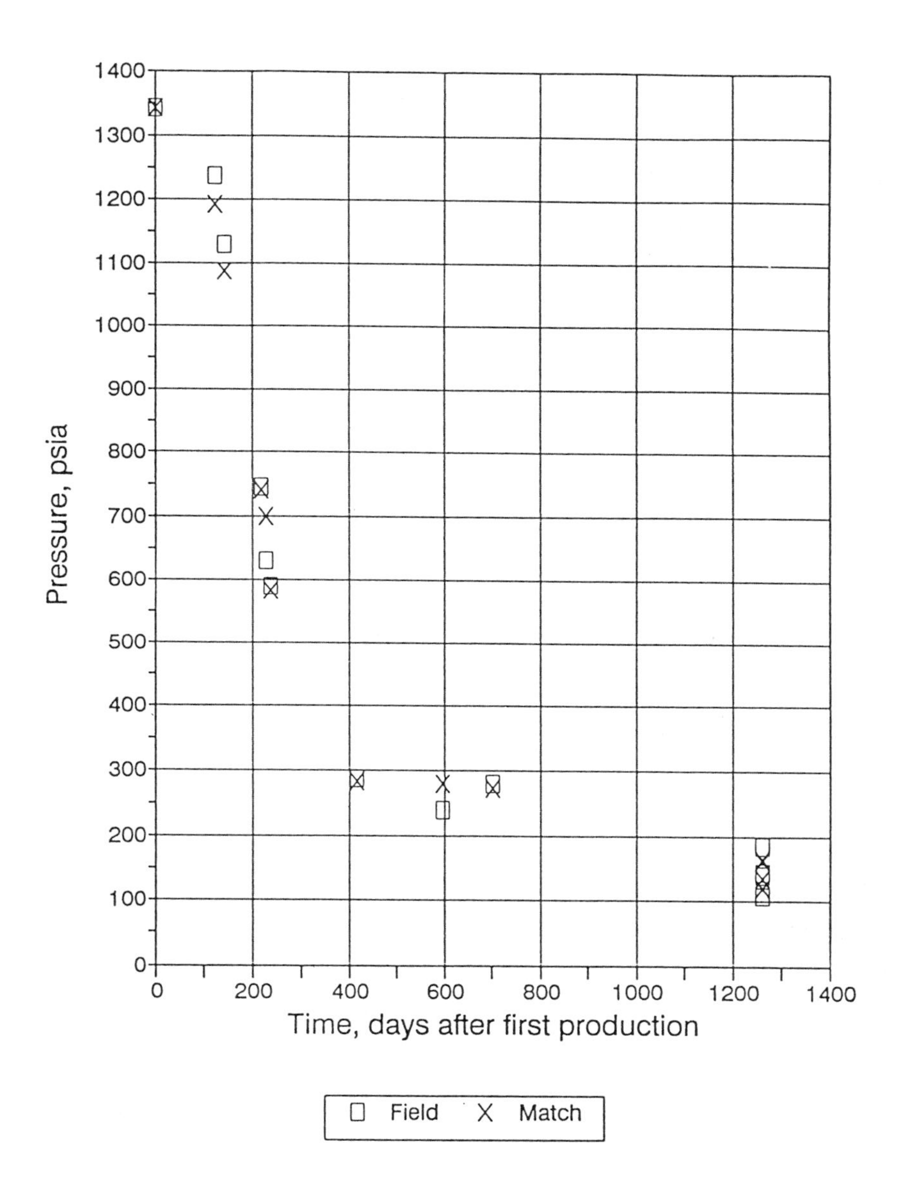

Fig. 16. History Match of Pen Field Reservoir Pressure During Primary Production with J Zone Porosities Estimated by Geostatistics.

The following equation describes the material balance for an undersaturated reservoir:

$$N_p B_o = NB_{oi} \left[\frac{B_o - B_{oi}}{B_{oi}} + \frac{c_w S_{wc} + c_f}{1 - S_{wc}} \right] \Delta P \tag{2}$$

where:

N = original oil in place (OOIP), STB
N_p = production, STB
B_o = formation volume factor, bbl/STB
B_{oi} = initial formation volume factor, bbl/STB
c_w = water compressibility, psi^{-1}
c_f = formation compressibility, psi^{-1}
S_{wc} = connate water saturation, fraction
ΔP = change in reservoir pressure, psia

Regression of $Y = N_pB_o$ versus $X = B_{oi}((B_o\text{-}B_{oi})/B_{oi} + ((c_wS_{wc}+c_f)/(1\text{-}S_{wc}))\Delta P)$ results in a slope equal to N, the OOIP. Such regression can be performed once values of the parameters on the right hand side of Equation 2 are known. Fluid properties (B_o, B_{oi}, c_w and c_f) were estimated from correlations and measurements (1).

Regression for data from wells completed only in the J zone resulted in an OOIP of 2,853,000 STB, approximately 50% higher than the OOIP of 1,920,000 STB estimated from a volumetric calculation. The material balance did not include production from the Pennington Lease or the Shuck #2 Well. It was assumed that 90% of the remaining field production came from the J zone. With a total field OOIP of 3,669,000 STB, as was used for the history match, the estimated ultimate primary recovery of 550,000 STB was 15% of OOIP, a reasonable amount for a Lansing-Kansas City reservoir with such a low bubble point pressure (285 psi) and low solution gas-oil ratio at the bubble point (60 scf/STB). Solution-gas drive reservoirs typically recover 10-15% OOIP during primary production (16).

Data from the ongoing waterflood were used to select parameters for the history match. When water injection began in November 1990, water injection for Shuck #2 averaged 180 barrels/day. In April, 1991, the wellhead tubing pressure for the Shuck #2 climbed from 0 psi to 800 psi. The tubing pressure was thereafter held constant at 800 psi to prevent fracturing the reservoir, and the injection rate fell to 130 barrels/day by June. This pressure response suggested that the Shuck #2 is completed in a separate J zone pod, probably extending to the east and discontinuous from the main reservoir.

It was necessary to increase the average permeability (obtained from a log permeability versus porosity correlation of core data) of the J zone by a factor of 2.24 for production in the simulation to match production in the field without bottom-hole pressures falling below 25 psia. This factor is consistent with the observation that permeabilities calculated from initial potentials (IP's) of the wells were larger than permeabilities measured from cores by about a factor of 3.

Thus,

$$(k_{IP})(h_{IP}) = 3(k_{core})(h_{core}) \tag{3}$$

However, based on the material balance calculation previously described, the thickness should be increased by a factor of 1.25. Uncertainty in the net thickness is due to two components. First, the vertical resolution of the porosity logs is 0.5 ft at best. This is significant in thin reservoirs. Secondly, the well logs cannot distinguish between interparticle porosity and moldic porosity. There may be net pay in the Alpha grainstone with interparticle porosity below the 7% cutoff. When the net thickness is increased by a factor of 1.25,

$$(k_{IP})(1.25h_{core}) = 3(k_{core})(h_{core}) \tag{4}$$

which reduces to:

$$k_{IP} = 2.4k_{core} \tag{5}$$

This was approximately equal to the average permeability which resulted in the best history match.

Directional permeability of $k_y = 5\ k_x$ assuming $k_x = k_{core}$, provided a better history match than isotropic permeability. The directional permeability was in a NE-SW direction which was along the long axis of the reservoir and therefore along the apparent trend of the field. This directional permeability results in the following average permeability determined from initial potential calculations:

$$k_{avg} = \sqrt{k_y k_x} = \sqrt{5k_{core}k_{core}} = 2.24k_{core} \tag{6}$$

$$k_{avg} = 2.24k_{core} \cong 2.4k_{core} = k_{IP} \tag{7}$$

The shifting of the gas-oil relative permeability curve by +3.5% was required to match the pressure buildup pressures below the bubble point. The curve was not obtained from a Pen Field core analysis, as no such data were available. Relative permeability curves are often shifted for a history match (17).

Wells were acidized upon completion. Skin factors measured for the five pressure buildups taken in January, 1989, ranged from -4.6 to -5.2. Similar to the higher permeabilities, skin factors ranging from 0 to -5 were required for production in the simulation to match production in the field. Values higher than -5 were not reasonable since as wells produce, they tend to become less stimulated.

a. Comparison of Linear Interpolation Grid with Geostatistical Method. A simulation was run using a J zone porosity grid determined by linear interpolation to compare results with those attained with the geostatistical grid. There was little difference in the computed pressure history for the two methods of assigning porosity to the J zone grid system for this particular reservoir.

2. Waterflood Recovery Forecasts

The reservoir management plan included predicting waterflood recovery for possible waterflooding plans. Reservoir properties were fixed at values determined from the history match of the primary production. Waterflood simulations were confined to the Alpha grainstone unit of the J zone pod which has contributed an estimated 90% of the field production. The first simulation was of the current waterflood. Next, since directional permeability ($k_y = 5\ k_x$) of the model based on history matching could have a strong effect on waterflood performance, forecast results for isotropic permeability ($k_y = k_x$) are presented. Finally, an alternative waterflood plan was investigated in which injection wells were centered along the NE-SW axis of the field.

For consistency, each waterflood simulation was run with 180 barrels/day injected into the Shuck #2 and 190 barrels/day injected into three injection wells completed in the main J zone pod. This is consistent with the current field operation of the waterflood. A maximum bottom-hole pressure and fracture pressure was assumed to be 2,560 psia. Production wells were shut-in when they reached a 98% water cut.

a. Current Injection Pattern. The existing waterflood plan is a line drive from the Northwest edge of the field. The injection wells are Shuck #2, Bethell #3, Griffey #1, and Demuth #2. Simulation runs predict that oil production will increase from 60 barrels per day in 1991 to 200 barrels per day in late 1992 and then decline to 56 barrels/day by 2010. (Fig. 17). Between 2010 and 2015, Bethell #5 and Demuth #4 reached water cuts of 98% and were shut-in, allowing water from injection wells to sweep by them and contact oil at other reservoir locations, thus resulting in increased oil production. Cumulative oil production will be 1,505 MSTB by 2020. With an estimated ultimate primary production of 550 MSTB, waterflood recovery would be 26% of OOIP, or 1.7 times that of primary recovery. There is some concern about the total recovery because single zone Lansing-Kansas City reservoirs seldom yield secondary recoveries in excess of 1.0 times primary. Simulation runs predicted that water production (Fig. 18) would increase from 0 barrels/day in 1991 to 70 barrels/day at the beginning of 1992 to 500 barrels/day in 2000. The drop in water production between 2010 and 2015 is due to shutting-in of Bethell #5 and Demuth #4.

b. Current Injection Pattern—$k_y = k_x$ Sensitivity. A waterflood simulation was run to predict performance assuming isotropic permeability, $k_y = k_x$. Comparisons of oil and water production between the isotropic and directional permeability cases are presented in Reference 5. In the isotropic permeability case, oil production does not peak as high or as soon as for the directional permeability case, but more oil is produced over the life of the waterflood. Cumulative production by 2020 was forecast to be 1,552 MSTB, 47 MSTB higher than if

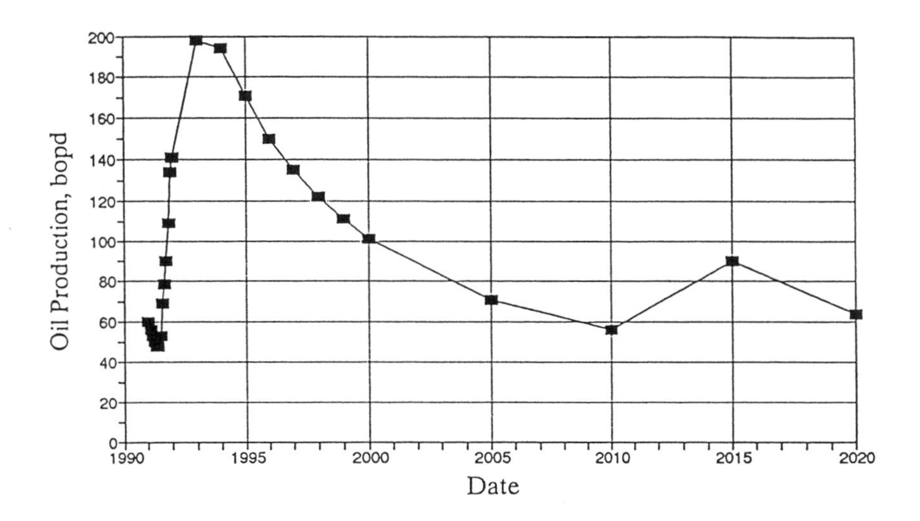

Fig. 17. Estimated Oil Production from Simulation of Existing Waterflood Plan.

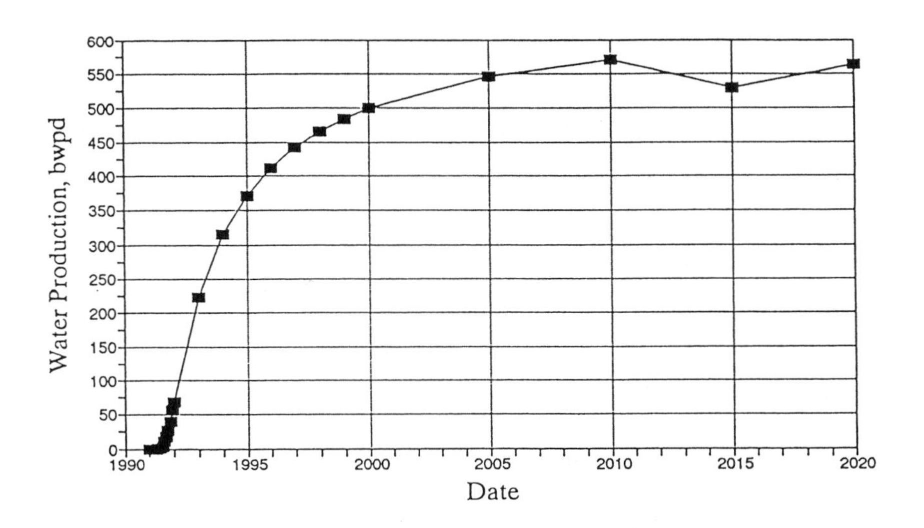

Fig. 18. Estimated Water Production from Simulation of Existing Waterflood Plan.

directional permeability of $k_y=5k_x$ exists. Water production is similar in the two cases. Considering the uncertainties in input data, there is no significant difference between an isotropic reservoir and an anisotropic reservoir.

c. Alternate "Axis 1" Waterflood Plan. An alternate waterflood plan, with injection wells lying along the axis of the J zone, was simulated to see if it would be more effective in recovering oil from the reservoir. The "axis 1" injection wells consisted of Shuck #2, Bethell #4, White #4, and Demuth #3. Oil production from this plan (Fig. 19) peaked later but higher, and remained higher until 2010. Oil production peaked at 300 barrels/day in 1993. By 2020, cumulative production was 1,657 MSTB, 152 MSTB (4% of OOIP) higher than the production forecast for the current pattern. Water production (Fig. 20) was generally lower than for the current pattern. Pressure maps and the water saturation maps show that the higher net pay and porosity near these injection wells cause a later waterflood response than in the current pattern (5). This occurs because pressure and saturation changes take longer to reach offset producers. However, as the waterflood progresses, the pressure and saturation changes cover larger areas of the reservoir than with the current pattern. Thus an additional 152 MSTB of oil is produced.

IV. DISCUSSION OF DATA

Data available for the Pen Field included wireline logs, cores, drillstem test results, measurements of porosity and permeability, and production records. These data represent information from different scales of measurement ranging from core plugs which sample a small volume of the reservoir rock to pressure buildup tests which provide information on the drainage volume sampled by the well to primary production-pressure data which is averaged over the entire reservoir volume. **In characterizing the reservoirs, each source provided useful information, and the final results were reached only through integration of all data.**

Cores provided stratigraphic, lithologic and petrographic information so that the nature of the reservoir rock was well known. Whole core measurements of porosity and permeability provided the baseline for interpreting permeability from porosity calculated from wireline logs and water saturations for thin reservoir beds and for those portions of reservoir intervals where the logs averaged dissimilar underlying or overlying beds. This combination was necessary to provide the comprehensive assignment of values to the data grid for the Pen Field.

The drillstem test results provided pressures and demonstrated the interconnection of the Alpha grainstone across the field. The drillstem tests would have been

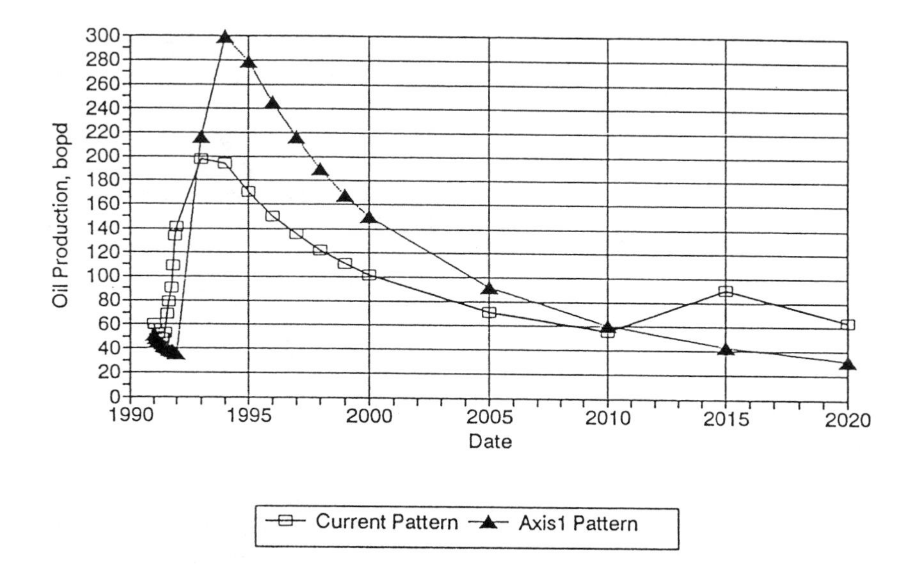

Fig. 19. Comparison of Estimated Oil Production from Simulation of Alternate "Axis1" Waterflood Plan with Existing Waterflood.

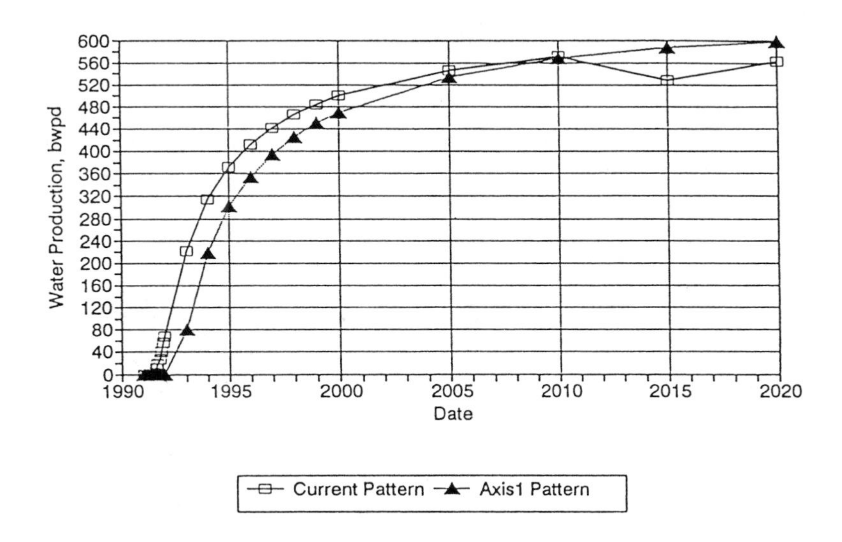

Fig. 20. Comparison of Estimated Water Production from Simulation of Alternate "Axis1" Waterflood Plan with Existing Waterflood.

even more useful for this purpose if each had been restricted to a single zone in the tests covering the minor pay intervals and were run with a longer final flow period. Monthly, well-by-well production data provided the basis for history matching and predicting future recovery.

Cores were the only source of data to develop a permeability-porosity correlation which could be used to estimate permeability in wells where no core was available. Unless a reliable relationship between porosity and permeability is discovered and found applicable to all reservoirs of a certain type, some cores are a necessary part of any complete characterization of any oil field.

V. CONCLUSIONS

Our object in studying the Pen Field was to develop a reservoir description for use in history matching and predictive modeling of production from the field. An integrated approach was necessary using data covering a range of scales from cores to pressure decline during primary production. The results provide estimates of thickness, thickness averaged porosity, permeability and fluid saturation for all of the 2.22-acre grid blocks.

The understanding of the stratigraphy and lithology of the Pen Field came from studies of cores and interpretation of wireline logs. Most reservoir rock is oolitic carbonate grainstone with primary interparticle porosity preserved. Some oil is produced from vuggy and moldic porosity in matrix-rich limestones. The main reservoir, the Alpha grainstone, is one of three horizons in the J zone that produce, although the other two are minor.

Pressure data from drillstem tests show that the Alpha grainstone in the J zone is a single, continuous reservoir roughly coextensive with grainstone development. Early marine calcite cement destroyed porosity and permeability in parts of the Alpha grainstone. Grainstones of the I zone formed in three separate pods in the field area, and one of the pods was divided by an impermeable area, possibly the result of diagenetic cementation, into two separate reservoirs. Paleosol processes created impermeable carbonate cap rocks on the top of the I and J zones.

A satisfactory history match of the reservoir pressure decline during primary production was obtained by altering reservoir properties estimated from the geological analysis. The analysis would have been improved if zone-by-zone drillstem tests, rather than the commingled ones, were used with sufficient flow time to interpret reservoir pressure accurately.

The current edge line drive waterflood pattern is predicted to recover 955 MSTB (26% of OOIP), 1.7 times that of primary production, by the year 2020. An alternative axis line drive waterflood plan was predicted to recover 150 MSTB (4% of OOIP) more oil than the current edge line drive pattern. One well, Shuck

#2, is completed in a separate J zone pod and therefore will be an ineffective injection well.

Although geostatistics provided an effective description of the thickness averaged porosity for the J zone, similar reservoir performance was predicted using thickness averaged porosities estimated by linear interpolation between adjacent wells.

ACKNOWLEDGMENTS

This research was supported by the Tertiary Oil Recovery Project (TORP) of The University of Kansas. The field was brought to our attention by Lynn Watney of the Kansas Geological Survey. PanCanadian Petroleum Corporation provided cores, logs, and data on completion and production for the wells. David Pauley, consultant to the Standard Operating Company, continued this support after he assumed responsibility for operation of the field. Randy Koudele discussed various aspects of the operation of the field. The VIP reservoir simulator was provided through the cooperation of Integrated Technologies.

REFERENCES

1. Phares, Roderick A, 1991, "Characterization and Reservoir Performance of the Lansing-Kansas City I and J Zones (Upper Pennsylvanian) in the Pen Oil Field, Graham County, Kansas", unpub. MS thesis, University of Kansas.
2. Morgan, J.V., 1952, "Correlation of Radioactive Logs of the Lansing and Kansas City Groups in Central Kansas, ***J. Pet. Tech.***, v. 4, 111-118.
3. Heckel, Phillip H., 1977, "Origin of Phosphatic Black Shale Facies in the Pennsylvanian Cyclothems of Midcontinent North America", ***Am. Assoc. Pet. Geologists Bull.***, v. 61, 1045-1068.
4. Moore, Raymond C., 1939, "Stratigraphic Classification of Pennsylvanian System in Kansas", ***KS Geol. Survey Bull. 83***, 203 pp.
5. Gould, Gary E., 1991, "Geostatistical Study and Reservoir Simulation of the Pen Field, Graham County, Kansas", unpub. M.S. Thesis, University of Kansas.
6. Edwards, A. G. and R. H. Winn, 1974, "A Summary of Modern Tools and Techniques Used in Drill Stem Testing", Duncan Oklahoma: Halliburton Services.

7. Earlougher, Robert C., Jr., 1977, ***Advances in Well Test Analysis***. New York: Society of Petroleum Engineers.
8. Matthews, C. S. and D. G. Russell, 1967, ***Pressure Buildup and Flow Tests in Wells***. New York: Society of Petroleum Engineers.
9. Davis, John, 1986, ***Statistics and Data Analysis in Geology***. 2nd ed. New York: John Wiley and Sons.
10. Henley, Stephen, 1981, ***Nonparametric Geostatistics***. London: Applied Science Publishers.
11. Journel, A. G. and Ch. J. Hijbregts, 1978, ***Mining Geostatistics***. London: Academic Press.
12. Isaaks, Edward H. and R. Mohan Srivastrava, 1989, ***Applied Geostatistics***. Oxford: Oxford University Press, Inc.
13. Jones, Thomas A., 1983, "Problems in Using Geostatistics for Petroleum Applications", ***Geostatistics for Natural Resources Characterization***, Part 2, ed. by Verly, G., M. David, A. G. Journel, and A. Marechal. Boston: D. Reidel Publishing Company; Proceedings of the NATO Advanced Study Institute on Geostatistics for Natural Resources Characterization, South Lake Tahoe, California, September 6-17.
14. Englund, Evan, 1988, "Geo-EAS (Geostatistical Environmental Assessment Software) User's Guide". Las Vegas, Nevada: U.S. Environmental Protection Agency.
15. Honarpour, Mehdi, Leonard Koederitz and A. Herbert Harvey, 1986, ***Relative Permeability of Petroleum Reservoirs***. Boca Raton, Florida: CRC Press, Inc.
16. Willhite, G. Paul, 1986, ***Waterflooding***. Richardson, Texas: Society of Petroleum Engineers.
17. Mattax, Calvin C. and Robert L. Dalton, 1990, ***Reservoir Simulation***, S.P.E. Monograph, Vol. 13. Richardson, Texas: Society of Petroleum Engineers.

USING PRODUCTION DECLINE TYPE CURVES TO CHARACTERIZE VERTICAL AND HORIZONTAL AUSTIN CHALK WELLS[1]

S. W. Poston
H. Y. Chen
A. Aly

Department of Petroleum Engineering
Texas A&M University
College Station, Texas

I. ABSTRACT

Production decline type curves representing the expected flow characteristics from a dual fracture-matrix flow system have been developed. The validity of the generalized geological assumptions forming the basis for the mathematical model are based on characterizing studies of the Austin Chalk formation at the outcrop.

Type curves representing the expected producing characteristics for a reservoir possessing at least two different permeability fracture systems and a matrix block system are presented. These curves represent a combination of flow through a major fracture system with infinite conductivity, linear flow through a set of lesser, subsidiary micro-fractures, and flow from the matrix block system into the micro-fracture system.

---

[1] Supported by Award Number: DE-FG07-89BC14444, Project No. [illegible], Relating to Fossil Energy Resource Characterization, Research, Technology Development and Technology Transfer - ANNEX IV

Field case studies of production records from Austin Chalk, vertical and horizontal wells located in both the Pearsall and Giddings Fields illustrate the utility of the new, dual fracture-matrix concept to determine the variation of reservoir character, the degree of connectivity of offsetting wells as well as the ability to generate area wide "field" type curves. Additionally, the interpretation of these curves also indicates the generally dissimilar production characteristics of the two different production regions within the Austin Chalk producing trend.

II. INTRODUCTION

Figure 1 shows the Austin Chalk formation is located in the Gulf Coast basin and trends approximately parallel to the Gulf of Mexico in a general southwest-northeast direction. Faults and major fracture systems found within the Austin generally parallel and are located immediately downdip of the trends of the Luling, Mexia, and Talco fault zones. The majority of the production from the naturally fractured-low matrix permeability formation has been divided by the Texas Railroad Commission into the Pearsall Field located in south Texas and the Giddings Field located in central Texas.

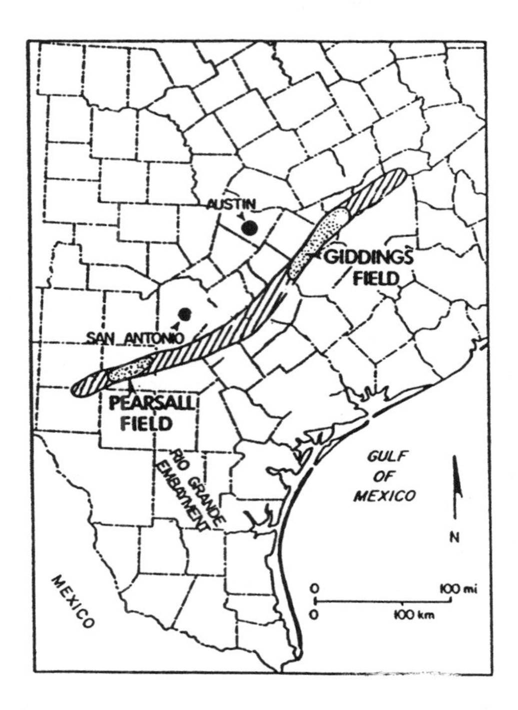

Fig. 1 - Geographic Reference Map.

Austin Chalk wells present a particular problem because of the extreme heterogeneity of the producing system. The permeability of the matrix rock is on the order of 0.01 to 0.0001 md, while the fracture permeabilities usually range from 10 to 100 times greater.[1] This type of producing characteristic is usually classed as a "so called" dual porosity, naturally fractured producing system even though the configuration of the flow system is known to be more complicated. Porosities range from 30% at the outcrop near Dallas decreasing to the southwest to 9% near Langtry.[2]

Most Austin Chalk wells are thought to produce from this type of a naturally fractured, dual-porosity system. Additionally, most Austin Chalk wells are hydraulically fractured. Flow from this type of reservoir would be expected to be predominantly linear in nature, not radial. The productivity of an individual Austin Chalk well depends to a very great extent on the intensity of encountered fracture system, the extent of the natural fractures and the permeability of matrix material. Estimation of remaining reserves and other reservoir characteristics is particularly difficult because of these uncertainties.

Well test and production records are usually the only information available for analyses. Decline curve analysis is the usual method used to predict future production characteristics of Austin Chalk wells. The most widely known and most convenient method of production decline analysis is the method of Arps[3] and Fetkovich.[4]

No one has attempted to quantify why the decline curve characteristics from the Pearsall and Giddings fields within the Austin Chalk producing trend are different. The following paper discusses the results of using type curves expressing dual fracture-matrix permeability flow to describe the drainage characteristics of Austin Chalk wells.

A. Outcrop Studies and the Geological Model

The lowermost portion of the Austin Chalk Formation supplies the majority of the production within the entire Austin Chalk interval. The rhythmically bedded chalk-marl sequences[1] generally range from 1 to 5 feet thick and are usually separated by marls ranging from 1 to 6 inches thick.

Outcrop studies have shown the differing degrees of "brittleness" between the limestone and the marls and shales tend to make the fractures "bed contained". Only a very small number of the fractures extend through the interbeded shales and marls. The bed contained fractures coupled with the rather close spacing of the fractures within the fracture zones produces in most cases block sizes on the order of 1 to 3 feet on a side. Extrapolation of these block sizes to the subsurface is uncertain, however, the presence of natural fractures divides the producing reservoir into a large number of moderate sized matrix blocks.

Geological mapping of Austin Chalk outcrops has disclosed a number of interesting characteristics of the fracture patterns.[5-7] Fig. 2 is a typical example of a mapped outcrop, while Fig. 3 is a typical plot showing the relation of frequency vs. fracture spacing. Note the regularity of the patterns and fracturing system. Subsidiary, relief(micro)-fractures with dip angles of approximately 60 to 75° connect the major(macro)-fractures. These micro-fractures extend in nearly straight lines. The dip of these micro- fractures varies from horizontal to vertical. The productivity of a typical Austin Chalk vertical well is probably a function of the number of these subsidiary fractures penetrated by the wellbore. The subsidiary micro-fractures act as conduits to carry oil and gas supplied by the matrix blocks to the wellbore. There is a pronounced lessening of the fracture intensity away from the center of the flexure. The geological model explains why horizontal drilling has been such a success in the Austin Chalk.

Fig. 2 - A Typical Fracture Pattern from the Outcrop.

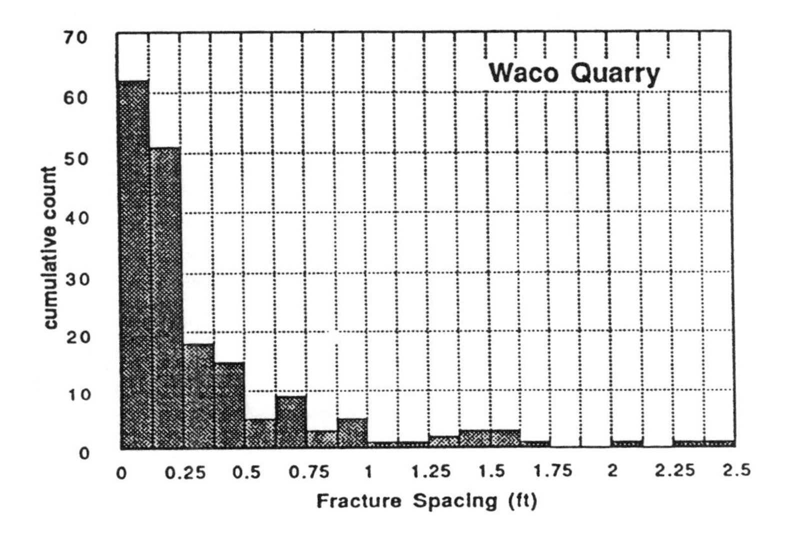

Fig. 3 - A Typical Frequency vs. Fracture Spacing Plot.

Concepts derived from the geological study may be summarized as follows:

- Master (macro) fractures have a periodic spacing with a common orientation. The permeability of these "master" fractures is significantly greater than the "minor" fractures.
- There is a set of "minor" fractures roughly aligned normal to the "master" fractures which connect the "master" fractures to each other.
- There is at least one other set of fractures which are in more or less random in nature.

Figs. 4 and 5 are idealized representations of a typical Austin Chalk flexure fracture system. Refs. 8, 9, 10 and 11 present a more extensive discussion on the Austin Chalk fracture-matrix relationships.

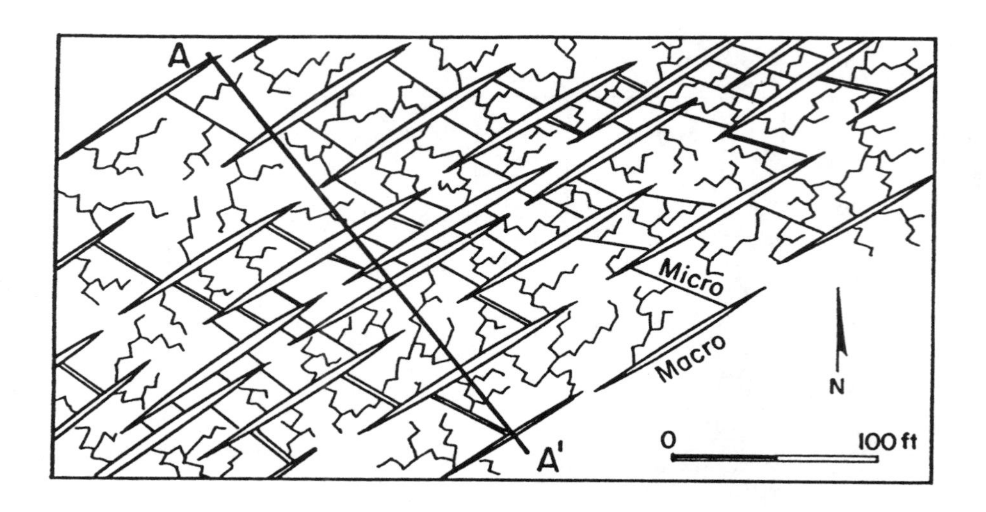

Fig. 4 - Idealized Macro, Micro-Fracture and Matrix System - Plan View.

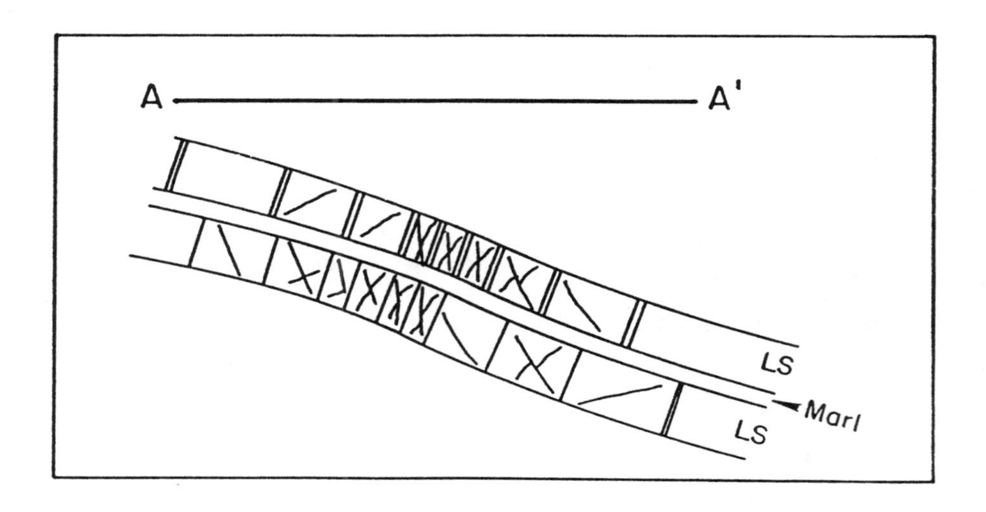

Fig. 5 - Idealized Macro, Micro-Fracture and Matrix System - Cross Section.

B. Mathematical Model Considerations

The geological discussion indicates the fracture network is comprised of a dominant array of near-vertical macro-fractures with subsidiary sets of micro-(hairline)fractures supplying varying degrees of interconnection between the macro-fractures. The macro- and micro- expressions are relative terms with the macro-fracture having larger dimensions, openings and extensions relative to the micro-fractures.

The flow pattern near a producing well will be governed more by the fracture geometry of the fracture/s intersecting the wellbore than by wellbore effects. The flow pattern out-in the reservoir is governed by the connectivity and distribution of the natural fracture system to the rock matrix blocks and the permeability ratios of the various flow regimes in the system.

The previous discussion shows why there is oftentimes extreme variability of the productive potential of Austin Chalk wells within a very small area. "Good" vertical wells must be drilled close to the center of macro-fracture zone to maximize the probability of encountering macro-fractures. Wells drilled only a very small distance from the center will produce substantially lesser quantities of oil since they will probably encounter substantially fewer macro- and micro-fractures. Additionally, a slight increase in the rock matrix permeability may profoundly affect the shape of the production decline curve.

A realistic mathematical model of the Austin Chalk must consider the spatially dependent fracture orientation, connectivity, distribution and intensity. Specifically, there is a necessity to distinguish macro- from micro-fractures when describing the producing characteristics of an Austin Chalk well.

One way to model the combined system of macro- and micro-fractures is by coupling a single-fracture type model to a dual-porosity type model. Refs. 10 to 12 describe the mathematical flow equations replicating the expected geological model.

The following assumptions concerning the expected fracture characteristics and producing mechanisms are used to simplify the mathematical treatment.

- The macro-fractures deplete under essentially steady state conditions and are connected to the micro-fractures.
- The micro-fractures are more or less connected and are considered as a continuous network. Flow from the microfractures may be at transient or pseudosteady state conditions which subsequently feeds into the macro-fracture system. The matrix blocks are enclosed by the continuous fractures.
- Transient or pseudosteady state production from the matrix is fed into the micro-fractures. The matrix blocks act as supporting sources to feed the fractures with fluid.

In summary, the natural fracture network and matrix blocks are conceptualized as distinct, yet overlapping continua. The macroscopic flow pattern in the reservoir is represented by the flow in the natural fracture network superimposed by the flow system within the matrix blocks. Flow is from the fractured reservoir, into the wellbore intercepted fracture and then into the well.

Fig. 6 is an idealized comparison of the macro- and micro-fractures and the supporting matrix. The initial decline is controlled by flow from the micro-fracture system, while late time flow is principally a function of the matrix properties.

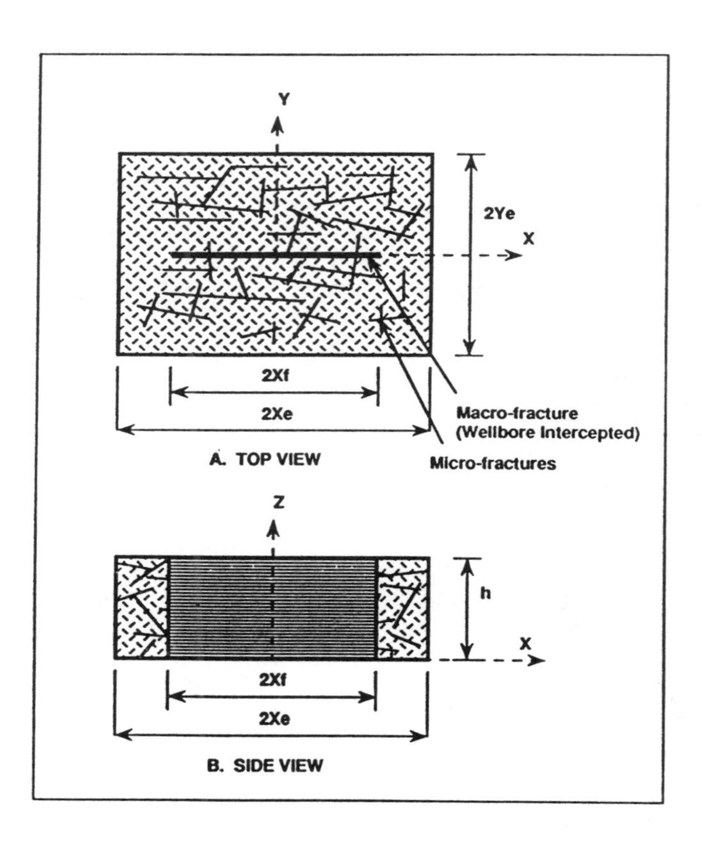

Fig. 6 - Model Elements.

C. Type Curves

The type curves shown in Fig. 7 were constructed in a manner similar to the Fetkovich decline-curve type curve.[4] A more extended discussion of the formulation of the type curves may be found in Refs. 9 to 12. The dimensionless parameters characterizing the dual fracture-matrix flow system are a storage-compressibility term, ω, and a fracture intensity term, γ, defined as:

$$\omega = \frac{(\phi c_t)_f}{(\phi c_t)_f + (\phi c_t)_m} = \frac{(\phi c_t)_f}{(\phi c_t)_t}, \tag{1}$$

$$\gamma = (\mathrm{FI})^2 x_f^2 \frac{k_m}{k_f} \frac{(\phi c_t)_m}{(\phi c_t)_f} \beta_1 \beta_2 = \left(\frac{6}{\pi} \frac{y_e}{l_c}\right)^2 \frac{k_m}{k_f} \frac{(\phi c_t)_m}{(\phi c_t)_f} \tag{2}$$

D. The Meaning of the ω and γ Terms

ω, Eq. 1, is defined as the ratio of the storage-expansion values for the fracture system compared to that of the total system.

Example 1 - The fracture and matrix compressibility may be assumed to be about the same for a slightly compressible fluid even though we would intuitively expect the fracture compressibility to be greater than the matrix compressibility. An average value of ϕ_f of Austin Chalk is about 0.0005.[10] Therefore, the ω value for a typical fractured reservoir would be approximately in the order of:

$$\omega \cong \frac{\phi_f}{\phi_f + \phi_m} \cong \frac{\phi_f}{\phi_m} \cong 10^{-3}$$

Example 2 - The solution-gas drive effect appears to usually be present in Austin Chalk producing situations. If the gas compressibility effect in the fracture system is assumed to be approximately 100 times greater than oil or the formation, then, $\phi_f c_f = \phi_m c_m$ (e.g., c_f is replaced by c_g). Therefore, the "apparent" ω may be as high as 10^{-1}. In any event, one would hope to see the value of ω increase over the life of the well or reservoir.

A high value of ω and homogeneous reservoir like production behavior, i.e., poorly developed natural fractures, may be an indication of poor fracture connectivity with relatively large matrix block sizes.

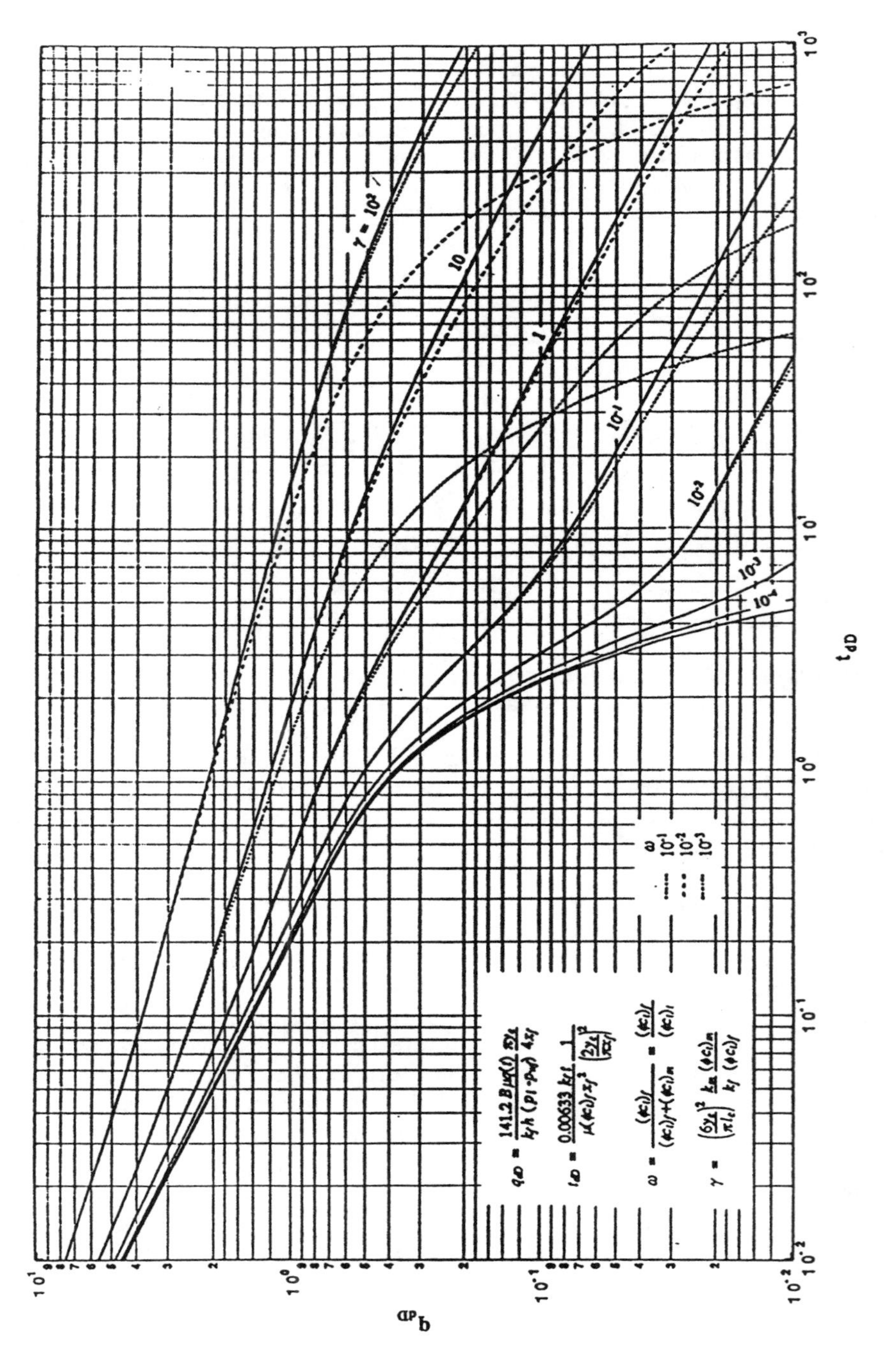

Fig. 7 - Type Curves.

The fracture intensity term is defined by Eq. 2 where FI denotes the "Fracture Intensity" in terms of "number of fractures per foot" and l_c is the characteristic length of the matrix block. β_1 and β_2 are normalizing factors and are defined in Ref. 10.

The unique feature for the dual fracture-matrix type curve shown in Fig. 7 is the extended production tail. This production extension is a direct consequence of the matrix contribution in the dual-porosity formulation. Naturally fractured reservoirs with the permeability of the matrix block being to small to permit significant fluid feed-in to the micro-fractures should display a pronounced "fall off" later in the life of the well. The extended "tail" with a much lesser difference between the micro-fracture and matrix dominated curves will be more pronounced if the permeability displayed by the matrix blocks is large enough to permit significant fluid feed in. An additional energy term, i.e., gas expansion, must be introduced into the reservoir in order to maintain the production rate above that expected from the compressibility effects which could be caused from single phase flow.

In conclusion and for a *given* production rate:

- A large value of γ tends to be associated with small values of ω because fluid flow to the well is being maintained by the extended fracture system. Conversely,
- A small value of γ would have to be associated with large values of ω because the gas compressibility effect would have to be significant to maintain a comparable production rate. The implication of the foregoing analysis is: although the type-curves were generated with a single liquid solution, qualitative interpretation of more than one phase producing situations is realistic with these type curves.

Fig. 8 is an idealized representation of the storage-compressibility term ω. Note, there is less deflection in the extended production decline tail as the value of ω increases. One would expect production levels to be maintained if the gas compressibility term becomes significant.

Fig. 9 indicates the fracture intensity term γ maintains production levels if the value γ is large. A high production level will be maintained for a longer period of time if a particular well intersects more micro-fractures than offsetting wells.

Relatively similar values of micro-fracture and matrix permeability will cause a smooth decline in production decline curve. Greatly dissimilar values will cause a significant deflection in the production decline curve.

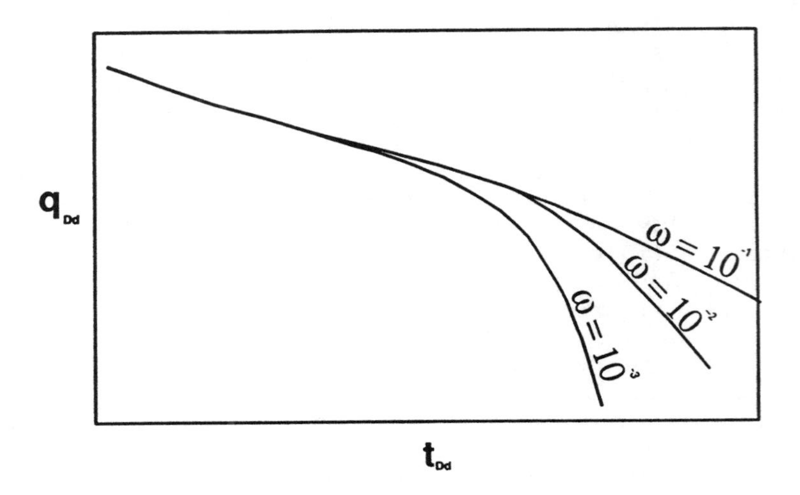

Fig. 8 - Idealized Change in Fracture Storage-Compressibility Term, ω.

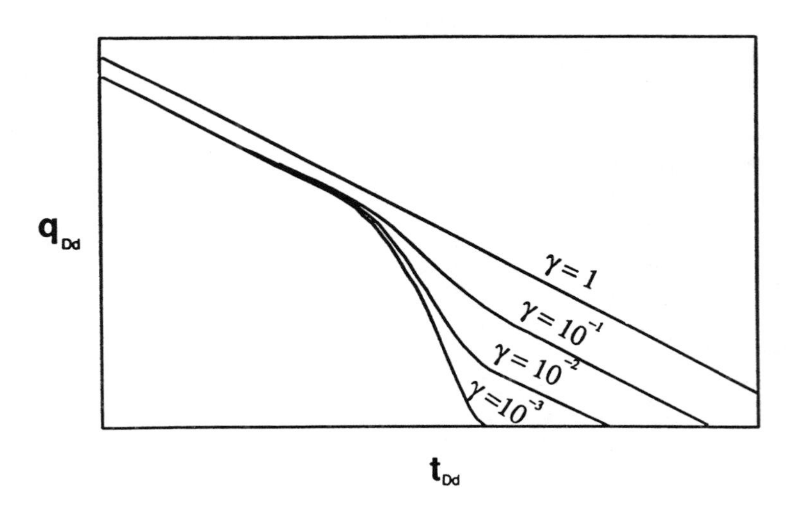

Fig. 9 - Idealized Change in Fracture Intensity Term, γ.

III. COMPARISON WITH FIELD DATA

A. The Pearsall Area - The Bagget #8 Vertical Well[9]

The initial production test for the well was 67 BOPD on 01/79. The interpreted match characteristics in Fig. 10 were found to be, $\omega \leq 0.01$, $\gamma = 0.05$. The ω and γ terms are the same as was found in the immediately adjacent #7 and #10 wells. Well #7 had been completed one year previous to #8 while #10 had been completed at approximately the same time as #8.

The fracture portion of the decline curve was extended to the minimum economic limit of 10 BOPD. Refer to Fig. 10. Average rate values were then used to calculate an ultimate recovery of 153,000 stb. The rate differences between fracture and matrix curves, i.e., the shaded area in Fig. 10, was determined. Cumulated oil production from this "matrix" portion of the curve was calculated to be 3,200 stb. Note, the significant difference in the levels between micro-fracture flow and matrix production contributions. All of the wells evaluated in this area showed this large difference in the two levels. There appears to be a significant difference between the micro-fracture and the matrix permeabilities.

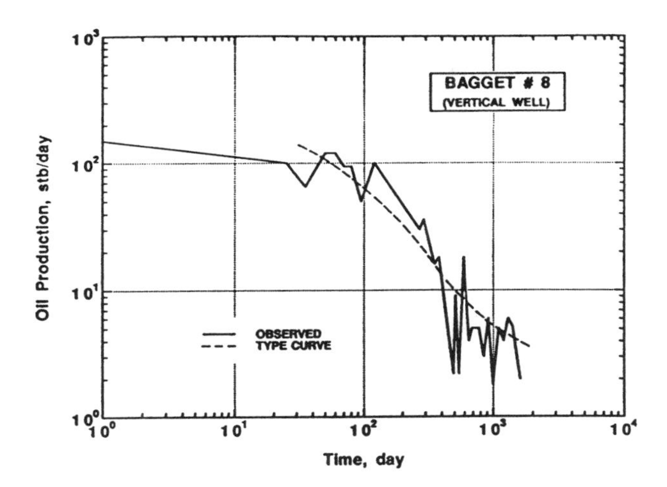

Fig. 10 - Initialized Decline Curve - Bagget No. 8 Well.

B. The Giddings Area - Vertical Wells

The Knesek A-1 was completed in late 1981. The initial production test on 10/81 was 122 BOPD and the cumulative production on 11/89 was 65,000 stb.

The match characteristics in Fig. 11 are, $\omega = 0.1$, $\gamma = 1$.

Note the decline from micro-fracture flow to the matrix system appears to be essentially linear. This type of decline indicates the permeability of the micro-fracture and the matrix system are reasonably close to each other and explains the absence of the expected tail due to matrix effect.

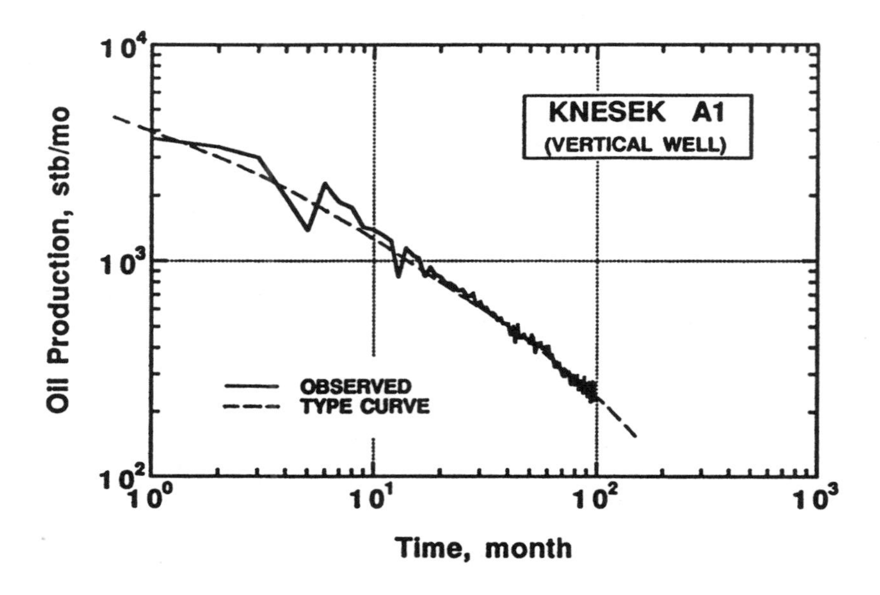

Fig. 11 - Initialized Decline Curve - Knesek A-1.

C. The Giddings Area - Field Curve

Figure 12 presents field curve matching of the Knesek-A1, Krobot-FJ1, and Norman-M2 wells located within a 640 acre area. The ω terms are the same for the three wells, which shows the wells are depleting by the same mechanism.

The match points derived from the type curve match are:

Well	ω	γ	N_p (stb)
Knesek-A1	0.1	1	65,000
Krobot-FJ1	0.1	10	48,000
Norman-M2	0.1	1	140,000

Note, the similarity of the final decline rates. The difference in ultimate recovery between the individual wells is a function of the time each well experiences macro-fracture flow. The Krobot well shows a greater fracture intensity term than the other two wells even though it was completed 6 months afterwards and in actuality has less ultimate recovery.

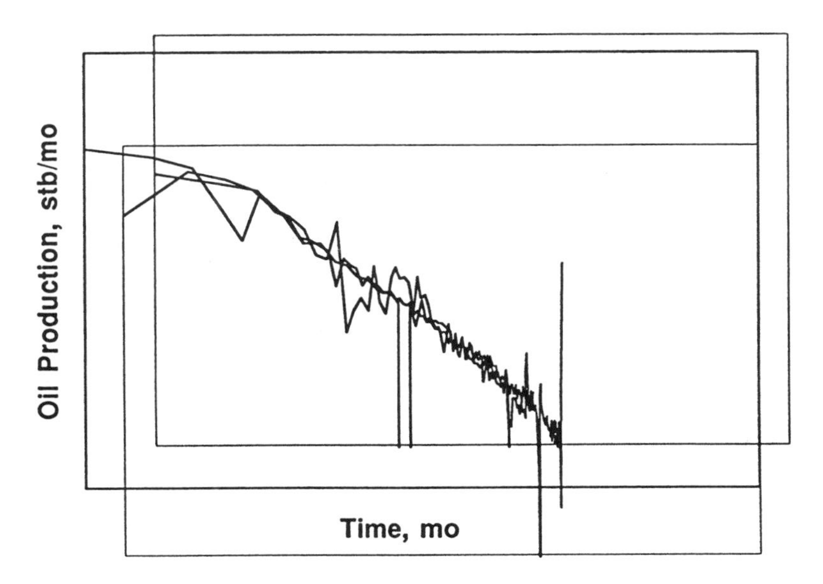

Fig. 12 - Family of Curves of Vertical Wells in the Giddings Area.

D. The Giddings Area - A Horizontal Well

Well A was drilled and completed in 1990. The initial production test on 11/90 was 833 BOPD and the cumulative production on 07/91 was 112,000 stb. Match characteristics from Fig. 13 are, $\omega = 0.1$, $\gamma = 1$.

Once again we see a smooth transition between micro-fracture and matrix flow which indicates a close similarity between the two permeabilities.

E. The Giddings Area - Family of Curves - Three Horizontal Wells

Wells A, B and C were drilled and completed in 1990. The wells are located in close proximity to each other. The ω term was the same for the three wells, which shows that they are producing from the same fracture system. Match points derived from Fig. 14 are:

Well	ω	γ	N_p (stb)
A	0.1	1	112,000
B	0.1	10	79,000
C	0.1	10	96,000

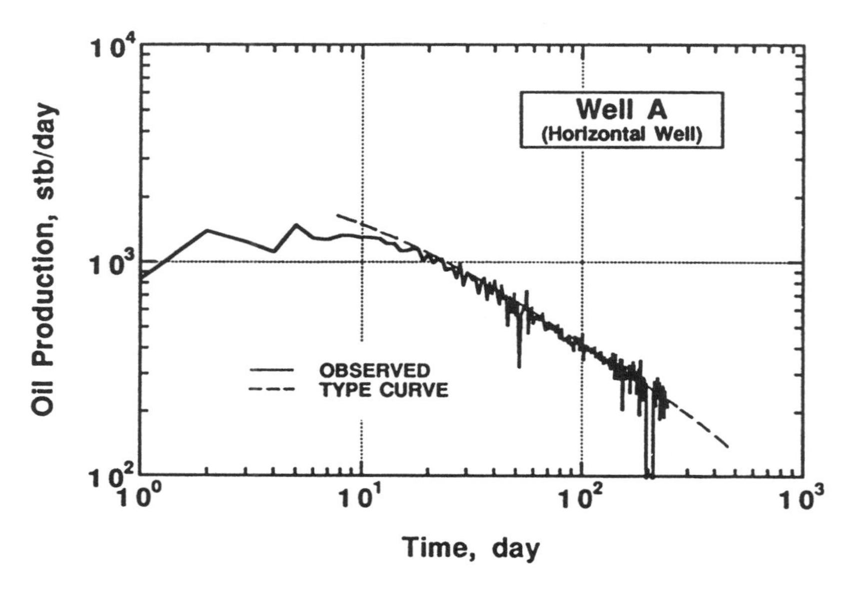

Fig. 13 - Initialized Decline Curve - Horizontal Well "E."

Note, the extended time interval before the decline was initiated. One would intuitively expected a horizontal well to intersect a greater number of fractures and have a different characteristic decline curve than vertical wells. On the contrary, the horizontal well's family of curves match those of vertical wells except for early data which is due to the transient period being masked in horizontal wells. This match of horizontal and vertical decline curves may be due to drilling of vertical wells in the center of the flexure of the fractures which yields production decline curves similar to those of horizontal wells. The ω term was the same for the Pearsall area vertical wells and the Giddings area horizontal wells, which may lead to the assumption that both fields are producing from similar fracture system which is characteristic of the Austin chalk in this area.

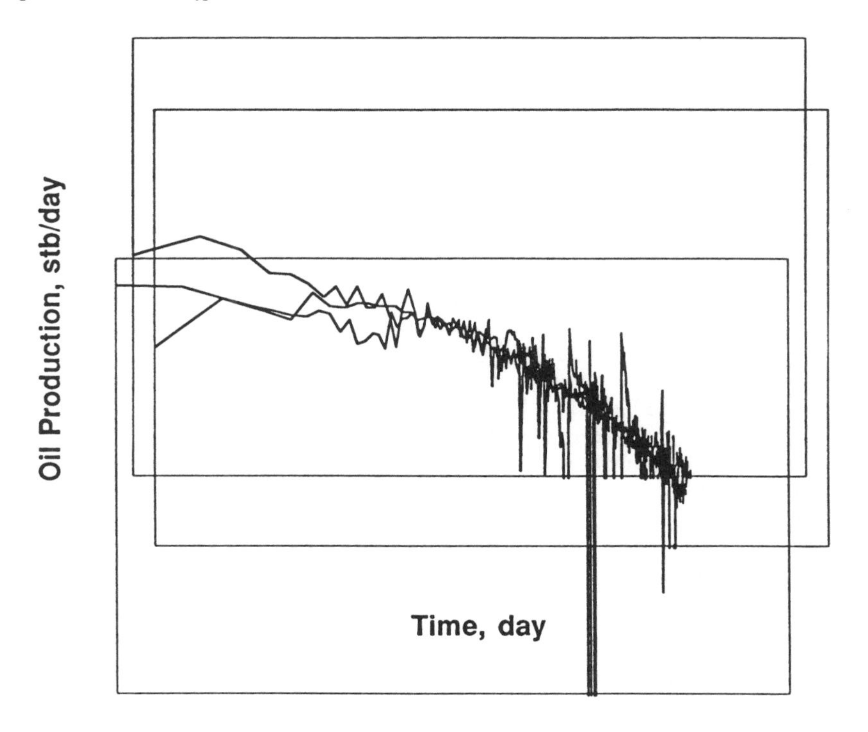

Fig. 14 - Family of Curves - Horizontal Wells.

IV. CONCLUSIONS

1. Type curves representing flow equations describing a generalized geological model describing the expected fracture network system within the Austin Chalk are presented.
2. The utility of the method to differentiate between dissimilar producing characteristics has been illustrated by comparing decline curves from two different areas in the Austin Chalk producing trend.
3. Decline curves within a particular area should follow similar production declines.
4. Pronounced curvature between the micro-fracture flow region and the matrix flow region result in a qualitative estimate in the difference in the permeabilities between the two flow regimes.

V. NOMENCLATURE

B = formation volume factor, b/stb
c_t = total compressibility, 1/psi
FI = fracture intensity
h = producing thickness, ft
k_f = fracture permeability, md
k_m = matrix permeability, md
l_c = characteristic dimension of matrix block, ft
p_i = initial pressure, psi
p_{wf} = flowing pressure, psi
q = flow rate, stb/day
t = time, day
x_e = drainage half-length in x-direction, ft
x_f = characteristic macro-fracture half-length, ft
y_e = drainage half-length in y-direction, ft
μ = viscosity, cp
ϕ = porosity, fraction
ω = storage ratio (Eq. 1), dimensionless
γ = fracture intensity parameter (Eq. 2), dimensionless

Subscripts
dD = decline curve dimensionless
f = fracture system (macroscopic)
m = matrix system (macroscopic)
t = total

VI. ACKNOWLEDGEMENTS

We wish to thank the Department of Energy and the Petroleum Engineering Department at TAMU for supplying funds to conduct this work. Special thanks to Oryx Energy company and Union Pacific Resources for suppling the information for analysis and Dr. Bob Berg of the Geology Department at TAMU for his illuminating discussions.

VII. REFERENCES

1. Stapp, W.L.: "Comments on the Geology of the Fractured Austin and Buda Formations in the Subsurface of South Texas," *Bull.* South Texas Geol. Soc. (1977) **17**, 13-47.
2. Cloud, K.W.: "The Diagenesis of the Austin Chalk," M.S. Thesis, University of Texas at Dallas, Dallas, TX (1975).
3. Arps, J.J.: "Analysis of Decline Curves," *Trans.* AIME (1945) **160**, 228-247.
4. Fetkovich, M.J.: "Decline Curve Analysis Using Type Curves" *J. Pet. Tech* (June, 1980) 1065-1077.
5. Corbett, K., Friedman, M., Wiltschko, D., and Hung, J.: "Controls on Fracture Development, Spacing, and Geometry in the Austin Chalk," 1991 AAPG Annual Convention, Field Trip #4.
6. Corbett, K., Friedman, M., and Spang, J.: "Fracture Development and Mechanical Stratigraphy of Austin Chalk, Texas," *AAPG Bulletin*, (1987) **71**, 17-28.
7. Friedman, M.: "Interpreting and Predicting Natural Fractures - Subtask 2.1.1," 1989-1990 Annual Report, ANNEX IV, Grant No. DE-FG22-89BC14444: Oil Recovery Enhancement from Fractured, Low Permeability Reservoirs, available at U.S. Department of Energy.
8. Chen, H. Y., Raghavan, R., and Poston, S. W.: "The Well Response in a Naturally Fractured Reservoir: Arbitrary Fracture Connectivity and Unsteady Fluid Transfer," paper SPE 20566 presented at the 1990 SPE Annual Technical Conference and Exhibition, New Orleans, LA, Sept. 23-26.
9. Poston, S.W., Chen, H.Y., and Sandford, J.R.: "Fitting Type Curves to Austin Chalk Wells," paper SPE 21653 presented at the 1991 Production Operations Symposium, Oklahoma City, OK, April 7-9.
10. Chen, H.Y.: "Relating Recovery to Well Log Signatures - Subtask 3.2.2," 1989-1990 Annual Report, ANNEX IV, Grant No. DE-FG22-89BC14444: Oil Recovery Enhancement from Fractured, Low Permeability Reservoirs, available at U.S. Department of Energy.
11. Chen, H.Y., Poston, S.W., and Raghavan, R.: "An Application of the Product Solution Principle for Instantaneous Source and Green's Function in the Laplace Domain," *Soc. Pet. Engn. Formation Evaluation* (June 1991) 161-168.
12. Chen, H.Y., Poston, S.W. and Raghavan, R.: "Mathematical Development of Austin Chalk Type Curves," paper SPE 23527, submitted for publication, Dec., 1991.

SESSION 5

Poster Presentations

Chairman

Dwight Dauben, K&A Energy Consultants

A TECHNIQUE OF THE THIN LAYER SANDBODY INTERPRETATION IN EXPLORATION AND DEVELOPMENT STAGE

Liu Zerong Hou Jiagen and Xin Quanlin

Department of Exploration
University of Petroleum
Dongying, Shandong, P. R. China

I. INTRODUCTION

The NZ oilfield is mainly in primary lithological primary lithological pools in sandstone/mudstone seqe-nces of the Tertiary in east of China. There had been no significant achievement in oil production for a long time, because the subsurface geology is quite complic-ated and that the reservoir rocks, the deep water tur-bidite sandstones, are small in size, thin layer, deep burial. Up to now we have discovered quntitativly 55 sandbodies by means of the technique of the thin layer sandbidy interpretation. The Z20 fault blick in which three sandbodies are interpretation quntitatively is an experiemental developing area in this oilfield. Five evaluation wells are designed, The drilling program ve rified our interpretation quite exact. In about 3000m deep target intevals, the predicted depth error is less than 10m, the predictable minimum thickness is about 10m. These evaluation wells also produce commercial oil from three sandbodies.

II. THE THIN LAYER SANDBODY INTERPRETATION TECHNIQUE

The characteristics of the technique is combind usage of the seismic, geological and well-logging inf-ormation. It works on following steps:

A. Reservoir Calibrating

The seismic interpretation of reservoir sandbodies is generally starting with available well data. For this reason, it is necessery to know the locations of the seismic records corresponding to the actual sandb-odies. This process is known as reservoir calibrating. We have used three methods of calibrating: the depth-time conversion, VSP and the synthtic seismogram.

B. Horizontal Tracing

Considing the geological information of the studied ares and analysing one-dimension or two-dimension seismic models, we have understanded the reflection characters of different sandbodies with different shapes, thickness and combinations with mudstones. In conclusion; we have established four sandbody eflection patterns which are the basement on tracing and predicting the samdbodies in this area.

1. Lens single wave, correspond to the sanbody less than $\lambda/4$ in thickness(λ is wavelet length).
2. Lens compound wave, correspond to the $\lambda/4-\lambda/2$ thick sandbody.
3. Parrallel double track wave, correspond to the seismic responses of the $\lambda/2-\lambda$ thick sandbody.
4. Apparent high frequence multi-wave, correspond to the sandstone groups over λ thick which are the river-month sandbars.

C. Model Checking

The model checking will improve our recognizations of the reservoir relection dynamics. As a lot of experiences obtained, we can summarize or set up reservoir reflcetion models.

D. Integral Interpretation

After horizotal tracing, we can draw a series of maps for the interpreted sandbodies such as the sandbody distribution map, isotime (TO) map or isopach of the sandbody's top deepth, isopach of the sandbody's thickness. All of the maps mentioned above is on the base of interpretation of the waveform and amplitudes of the seismic reflection. In order to be accurate and reliable, one must make use as much information as possible in the integral interpretations i.e. synthetic acoustic logging section, freqeuncy spectrum, test well date, and so on,

E. Effect Testing

The thin sandbody interpretion can only be tested by well drilling.

III. DISCUSSION

We have detailly analysed the key factors that the technique are applied successfully in the NZ oilfield and summarized some points of views as follows:

A. Special Geological Conditions

The NZ oilfield locates in the deep lacuna. The structure is simple or not complicated by many faults.

B. Favourable Stratum column

In this ares, the target intervals are mainly turbidites sandstones frequently alternating with the deep lacuna. The structure is simpe or not complicated by many faults.

C. The Study of Velocity as a Foundation of Sandbody Interpretation

D. High quality seismic data

The main fators of high quality seismic data is that the area has been explorated by high resolusion seismics for many years and there are the seismic experiments in the field and in the liboratory. The seismic frequency is high and wide. So the high-fidelity waveform and amplitude are obtained.

E. Application of Integral Information

It is necessary to apply many kinds of seismic methods, including waveform, amplitude, frequency spectrum, impendence and theoritical model etc., to compensate and verify each other.

F. A Special Reservoir Forming Condition

In the oil source lacuna, the turbidite sandbodies are surrounded by source rocks and form primary lithological oil pools. Therefore the problem of surveying complex lithological oil pools is simplified to the recognizing reservoir bodies.

HOW TO ESTIMATE THE RELIABILITY OF RESERVES? A QUICK AND PRACTICAL SOLUTION

Tibor Kuhn
Zsolt Komlosi

MOL Rt. - SZKFi
Hungarian Hydrocarbon Institute
Budapest, Hungary

Summary: The method developed for estimating of confidentiality of a determined reserve is based on analysis of important factors. Natural conditions and technical interventions form a complete system. Sources of objective and subjective errors in input data and reserve calculation process have been analyzed. At the end a special weighting factor system has been established arbitrarily for quantification of conditions (e.g., rock type of formation, reservoir shape complexness, producing status) influencing accuracy of determined reserve significantly. The result of the method is a ratio characterizing uncertainty of the determined reserve. An application of the method is shown in case of a Hungarian field.

The aim of this work is to present a method to find calculation quality of the original and recoverable reserves, which have utmost importance in making decision on hydrocarbon production. The main criteria for the method are as follows:

- It should be universal regarding applicability on any kind of reservoir;
- It should be applicable to any method of reserve estimation;
- It should fit to the present practice of reserve estimation;
- No significant extra work should be necessary for its use.

1. BACKGROUND - SYSTEMATIZATION

The method is based on analysis of knowledge and information necessary for determination of reserves generally available to the analyst. A simple composition model of recoverable reserve is shown in Fig.1. The lower part of Fig.1 contains various 'disciplines', i.e. sources of information, applied for estimating of unknown parameters of reser-

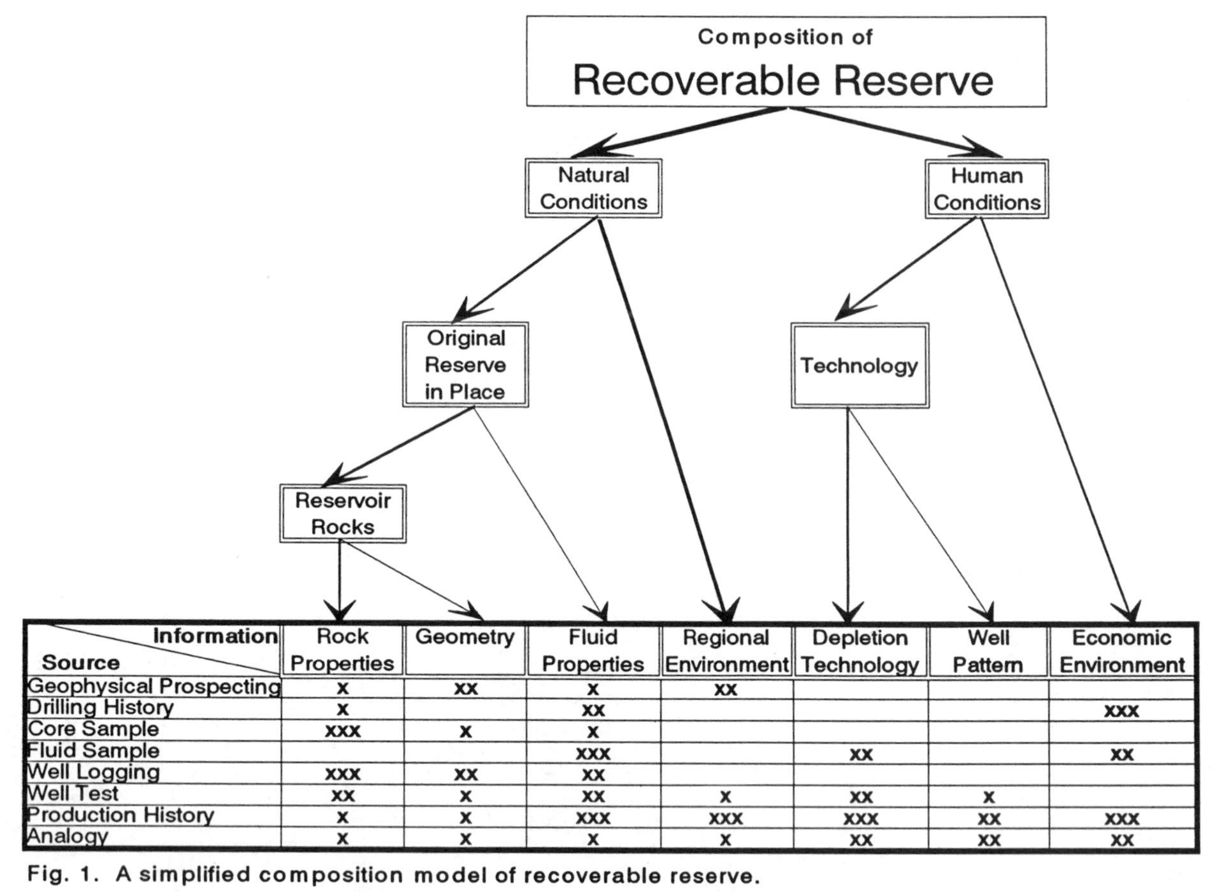

Source \ Information	Rock Properties	Geometry	Fluid Properties	Regional Environment	Depletion Technology	Well Pattern	Economic Environment
Geophysical Prospecting	x	xx	x	xx			
Drilling History	x		xx				xxx
Core Sample	xxx	x	x				
Fluid Sample			xxx		xx		xx
Well Logging	xxx	xx	xx				
Well Test	xx	x	xx	x	xx	x	
Production History	x	x	xxx	xxx	xxx	xx	xxx
Analogy	x	x	x	x	xx	xx	xx

Fig. 1. A simplified composition model of recoverable reserve.

Legend: Role of parameter is ... x = indicator, xx = qualitative, xxx = quantitative ves and their capability in the process. It is important to emphasize, while original reserve is determined by natural conditions, recoverable reserve is influenced by human conditions and technical interventions as well.

Results of analysis of information sources are shown in Fig. 2, as a summary, containing the main parameters obtained and 'critics' of the sources. To reach the best solution, the different sources are to be analyzed simultaneously integrating with professional experience and creativity of the analyst.

2. DEFINITIONS OF ERRORS USED

At-well errors characterize uncertainties of parameters obtained in separate wells; Their absolute or relative magnitude is independent of the number of the wells. They can be divided into two groups by the origin:

- Measurement and data processing errors are caused by the measuring instrument, the method used and, eventually, circumstances of measurement. Their magnitude can be quantified by modeling of measurement process or analyzing results of measurement.
- Interpretation or methodological errors arise due to the parameter is not measured directly, but its value is concluded from the results of other measurements using some interpretation method.

Inter-well errors originate from the process of spatial extension of the parameter values available only at separate wells. Their magnitude depends upon the well number i.e. the more dense well pattern the less error value occurs. The inter-well errors can be divided into two groups:

- Errors that can be estimated numerically, e.g., mapping and volume calculation errors.
- Qualitative errors originate from qualitative conditions influencing accuracy of estimated reserve, e.g., genetics, rock type, structure complexity.

Modeling errors characterize difference between the geological, petrophysical or depletion (etc.) model used and the unknown true reservoir conditions. They can occur as either at-well or inter-well error. Their magnitude cannot be quantified analytically usually, but selection of the proper model is an essential problem of reserve estimation.

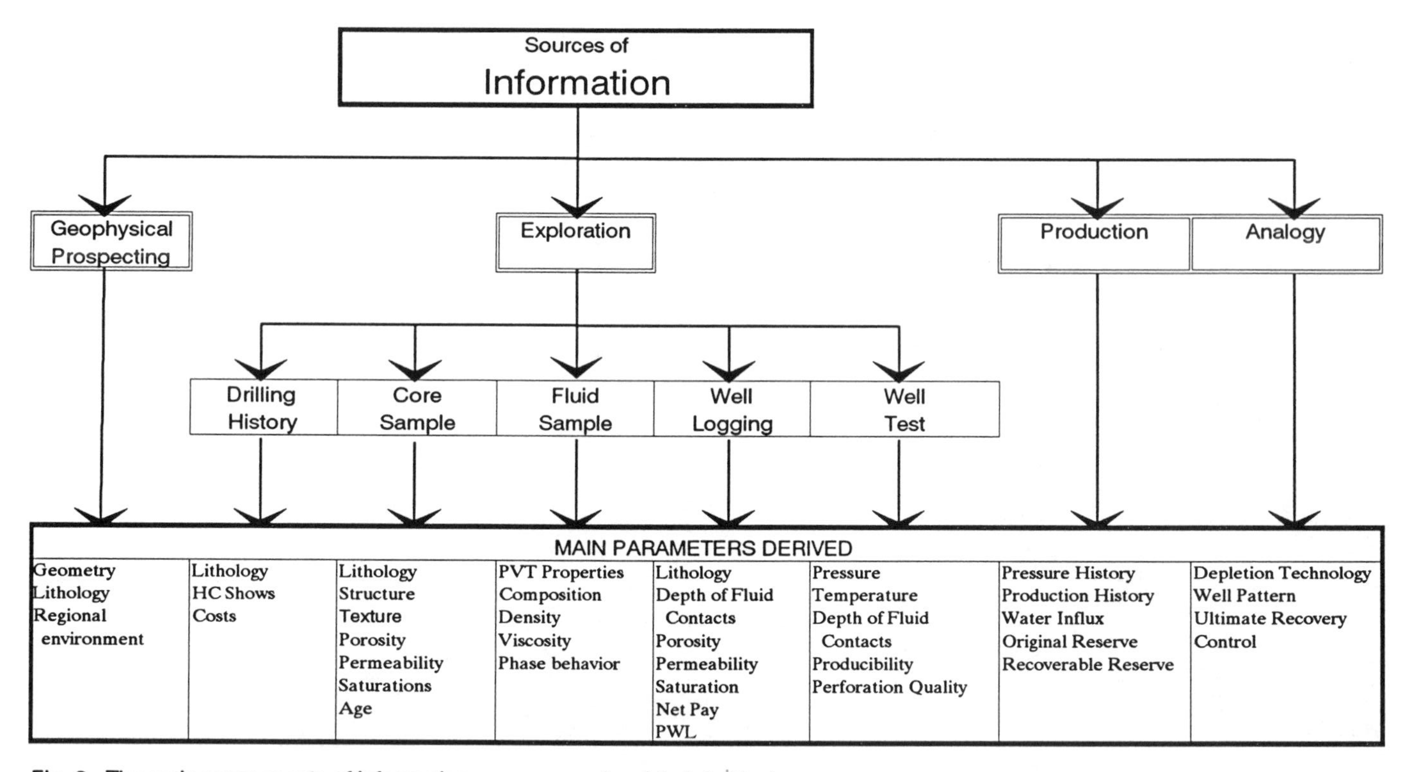

Fig. 2. The main components of information sources used and their 'critics'.

	Geophysical Prospecting	Drilling History	Core Sample	Fluid Sample	Well Logging	Well Test	Production	Analogy
ADVANTAGES	Cheap, since no drilling is needed. Information derived via cross section even 3D Good structure indication	Cut is a direct material of formation. Additional infortmation to geological exploratio (except MWD) Data are derived immediately	The only way to have direct information of formation. Useful at calibration of well log data. Good vertical resolution.	Direct information from formation fluid content.	Continuous vertical information. Relative cheap. Quick. A relative big rock volume yields information.	The best information of formation productivity. Inter-well volumes are showed as well.	No additonal equipment is necessary. The best for prediction Good control for project management.	Practicaly free of charge Substitute lack of information ("bridge" over it)
DISADVANTAGES	Indirect method Limited resolution High uncertanity	Indirect information Depth accuracy is limited MWD is expensive	Expensive Processing is fairly time-consumer Depth accuracy can be unreliable Rock material is disturbed Small (volume) representativity	Its representativity is highly influenced by sampling conditions	Indirect method Vertical resolution is worse than core sample's one	It depends on drilling and perforating technique applied. Technical problems can be caused	Its confidentality is growing by producing time Recovery factor is not defined Subsequent character	Well educated experienced specialists are needed Only estimation

Fig. 2. (Continued)

3. PROPOSED METHOD FOR UNCERTAINTY DETERMINATION

Attention was focused on the key parameters while the method was developed assuming that obvious false information was previously omitted. The economic aspects of the question have not been studied.

3.1. Error in original reserve in place

Error $\mathbf{m_Q}$ in reserve is assumed as a composition of at-well error ($\mathbf{m_a}$) and inter-well error ($\mathbf{m_i}$) weighted with a modeling error factor ($\mathbf{W_M}$).

$$m_Q = W_M[m_a^2 + m_i^2]^{1/2}$$

The error $\mathbf{m_Q}$ in calculated reserve is interpreted as an over estimating factor of the true reserve to avoid the problem of an $\mathbf{m_Q}$ higher than 100%. Therefore, relative error $\mathbf{m_{Qo}}$ relating to reserve estimated can be calculated as:

$$m_{Qo} = \pm m_Q/(1+m_Q)$$

The components of the error $\mathbf{m_Q}$ can be calculated as follows:

3.1.1. **The at-well error** is characteristic of parameters derived in separate wells. It is composed from errors of parameters such as porosity ($\mathbf{m_p}$), water saturation ($\mathbf{m_s}$) and net pay ($\mathbf{m_h}$), so called basic error (in the parenthesis) weighted with a lithological factor ($\mathbf{W_L}$) at each well.

$$m_a = [(m_p^2 + m_s^2 + m_h^2)W_L]^{1/2}$$

The absolute error of effective porosity (considering publications and our own experience [1, 2, 3]) is assumed as much as one per cent, therefore relative error in porosity $\mathbf{m_p}$ is

$$m_p = 100/PHI \qquad \text{if PHI} > 2\%$$

In low porosity reservoirs (where **PHI<2%**) the relative porosity error $\mathbf{m_p}$ can be considered as much as 50%.

The error in original hydrocarbon saturation ($\mathbf{S_w}$) is derived from analysis of the Archie-equation used for clean sandstones. Absolute error in water saturation vs. porosity and water saturation is shown in Fig. 3. The relative error $\mathbf{m_s}$ in water saturation is

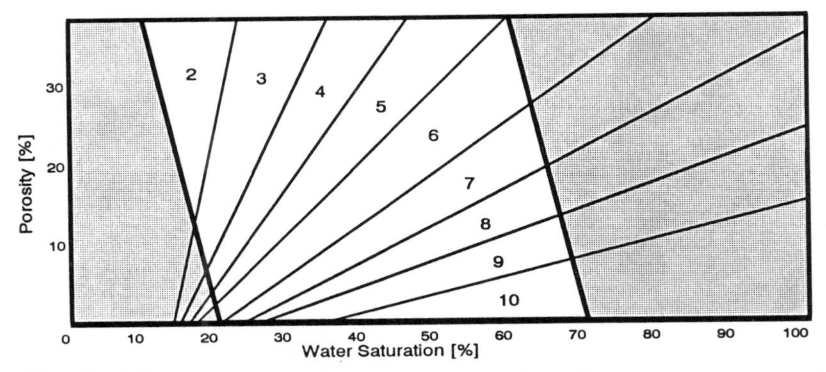

Fig. 3. Absolute error in water saturation (curve parameters) vs. porosity and water saturation. (When the point is falling onto either shaded areas some special treatment is required).

$$m_s = 100\{\text{value taken from Fig.3}\}/(100\text{-}S_w)$$

The error $\mathbf{m_h}$ in net pay is composed of the mean error of depth measurement considered to be $\mathbf{e}=1\text{m}$[1] [3] and the ratio of the total thickness $\mathbf{h_{to}}$ and net pay **h**.

$$m_h = 100e/h + (h_{to}/h)^2$$

The at-well error calculation for the entire reservoir needs mean values of errors derived at each well. They can be averaged either by weighting the area (i.e. mapping the values), the net pay, or not weighting. In the worst case, an average error can be calculated using the mean parameter values of the reservoir.

The errors determined above are valid in a clean (i.e. clay and silt free) sandstone reservoir. To handle reservoirs of other lithology, an empirical weighting factor value $\mathbf{W_L}$ can be taken from Table I which has been invented by the authors arbitrarily.

3.1.2. **The inter-well error $\mathbf{m_i}$** is characteristic of the space between wells including uncertainty of map construction. It is assumed to be determined by the shape of the reservoir and complexity of tectonics. The proper error value $\mathbf{m_i}$ can be selected from Table II by experience of the analyst.

[1] If the dimension of net pay is other than a meter, e should be redimensioned.

TABLE I. The weighting factor value W_L allows handling reservoirs of other lithology than a clean sandstone reservoir.

Clay (silt) content / Matrix material	Negligible (<5 %)	Corrective (5-20 %)	Significant (>20%)
	Lithological Factors W_L		
Sandstone	1.0	1.5	2.0
Conglomerate or breccia	1.5	1.8	2.1
Carbonate of double porosity	1.8	2.0	2.3
Carbonate of secondary porosity	2.2	2.3	2.5
Metamorphic or basement rock	2.5	2.7	3.0

3.1.3 **The modeling error** factor W_M characterizing the difference between the models used and the true reservoir conditions is determined by product of three weighting factors. The first of them (W_{iq}) characterizes the quality and quantity of the available core, well log and well test data. The second one (W_{ph}) shows the effect of production history if it has been used at reserve calculation. Thirdly, (W_{ia}) depends on the 'accordance' among different information judged by the analyst subjectively.

$$W_M=(W_{iq}W_{ph})W_{ia}$$

The factor values (see Table III) have been arbitrarily derived by the authors. The factor W_{iq} can be calculated by multiplication of partial factors yielded by cores, well logs and well tests available respectively depending on percentages of wells where they are available.

TABLE II. The inter-well error m_i is decided by the shape of the reservoir and complexity of tectonics.

Tectonics / Type of Reservoir	No	Simple	Disturbed
	Inter-Well Error m_i [%]		
Plain Sandstone	5	7	10
Anticline with more maximum	7	10	13
Turbidite, Channel Sediment	10	12	17
Lithological Trap	9	11	16
Blocked Structure	--	12	20

TABLE III. The modeling error factor $\mathbf{W_M}$ characterizes the difference between the models used and the true reservoir conditions.

W_{iq}	Well Information	Percentage of Wells <30	30-70	70<
		Factor		
Core	a. A few, not processed	1.15	1.10	1.05
	b. Standard process	1.10	1.05	1
	c. Key well, processed in detail	1.05	1	0.95
Well Log	a. Qualitative logs	1.30	1.20	1.10
	b. Quantitative logs	1.20	1.10	1
	c. Borehole compensated logs	1.10	1	0.90
Well Test	a. DST	1.15	1.10	1.05
	b. Standard well test	1.10	1.05	1
	c. Well-flow test	1.05	1	0.95
W_{ph}	Production History	NO	There is	
	a. OIL produced		<5%	>5%
	b. GAS produced		<10%	>10%
	Factor	1	0.8	0.6
W_{ia}	Information Accordance		Factor	
	a. They are in accordance		0.8	
	b. Contradictions may be solved		1	
	c. Contradictions are significant		1.2	
	d. Unresolvable contradictions		1.4	

3.2. **Error in recoverable reserve**

At the beginning of production the error in recoverable reserve must be higher than that in original reserve. After a longer production period the recoverable reserve predicted by the production history becomes more independent of the original reserve, moreover, its error is negligible at the normal technical abandonment of production [4]. Therefore the error $\mathbf{m_{Qr}}$ of recoverable reserve depends on cumulative production ($\mathbf{q}$) and final cumulative production estimated until ultimate recovery ($\mathbf{q_f}$).

$$m_{Qr} = [1-q/q_f][m_{Qo}^2 + W_R m_k^2]^{1/2}$$

TABLE IV. The weighting factor $\mathbf{W_R}$ is composed from partial factors.

W_{lt}	Production Lifetime	Recovery [%]				
	OIL	0	5	10	20	30
	GAS	0	10	20	40	60
	Factor	1	0.8	0.6	0.4	0.2
W_{pt}	Production Technology			Primer	Secunder	Tertier
	Factor			1	1.2	1.5
W_{nw}	Natural Waterdrive		No	Some	Active	Uncertain
	Factor		1	1.05	1.1	1.2
W_{pp}	Production Project		Taylored	Simulation	Mat. Bal.	Analogy
	Factor		1	1.1	1.2	1.3
W_{ea}	Quality of Economic Analysis			Good	Reliable	Uncertain
	Factor			1	1.1	1.2

The uncertainty of reservoir permeability $\mathbf{m_k}$ is defined with porosity and saturation errors (see item 3.1.1).

$$m_k = 1.5[W_L(m_p{}^2 + m_s{}^2)]^{1/2}$$

The weighting factor $\mathbf{W_R}$ is composed of partial factors (see Table IV) showing the effect of the date of calculation (i.e. maturity of the field studied) $\mathbf{W_{lt}}$, the production technology applied $\mathbf{W_{pt}}$, natural water inflow expectation $\mathbf{W_{nw}}$, quality of the production project used $\mathbf{W_{pp}}$, and quality of economic analysis done $\mathbf{W_{ea}}$.

$$W_R = W_{lt} W_{pt} W_{nw} W_{pp} W_{ea}$$

4. APPLICATION OF THE METHOD

Use of the method is illustrated by the example of the deep horizon 'X', which is a massive type reservoir, consisting of Paleozoic metamorphic basement and breccia. The nonhomogeneous rock of the reservoir having abundant secondary porosity is attributable to more initial rocks and the degrees of the metamorphism. The most frequent rock types are gneiss and mica schist. The area is tectonized by lines of strike NE-SW mainly. Storage and flowing conditions of the field are controlled by faults.

There were 50 wells used for reserve calculation. 20 of the 50 wells were cored including a key well with a 110m long cored section. The 31 wells drilled later have quantitative logs run by borehole compensated

TABLE V. The main parameter values of the field 'X'.

Parameter	Symbol	Unit	Value
Porosity	PHI	%	3.1
Water Saturation	S_w	%	28
Reservoir Thickness	h_{to}	m	140
Net Pay	h	m	63
Reserve	OOIP	MMt	4.3
Recovery	RF	%	37

tools rendering up-to-date quantitative log interpretation. Well logs run in 11 older wells were processed with a special method approaching quantitative results.

The main parameter values of the field (see Table V) allow learning errors in porosity, water saturation and net pay. Other errors and weighting factors chosen from Tables are as follows: $W_L=2.7$, $m_i=20\%$, $W_M=1.1*0.8*1=0.88$, $W_R=0.4*1*1.05*1.2*1.1=0.55$ and $q/q_f=0.2$.

The calculated errors and their distribution are shown in Table VI. The calculation errors in original and recoverable reserve are as much as 1.5 MMt and 0.9 MMt respectively.

TABLE VI. The weighted mean errors of the two reservoir parts and their distribution.

Source of Error	Error	Uncertanity of Reserve [%]	Uncertanity of Reserve Distribution
Recoverable Reserve	m_{Or}	57.0	100.0
Permeability	m_K	83.4	70.7
Original reserve	m_{Oo}	34.9	29.3
Original Reserve	m_{Oo}	34.9	100.0
Inter-Well Parameters	m_i	20.0	26.1
At-Well Parameters	m_a	56.6	73.9
Porosity	m_p	32.3	48.7
HC Saturation	m_s	10.2	15.4
Net Pay	m_h	6.5	9.8

CONCLUSION

Evaluation of confidentiality of reserve is a very complex task. The method developed by the authors focus attention on geological and technical aspects of reserve projection. It uses artificial parameters defined by subjective judgments based on experience and knowledge of the analyst. The method is very simple one and easy to do. It lets the evaluator and management to compare different fields and different conditions.

During application the method can be adjusted (e.g., selection of better coefficients, assuming effect of well number) further. The authors are open to receive any suggestion and contribution.

ACKNOWLEDGEMENTS

We are grateful to the Lowland Hydrocarbon Production Company (Hungary) for funding the study required preparing this paper. Thanks are due Hungarian Oil and Gas Corporation for permission to publish this paper. We thank many people inside and outside Hungarian Hydrocarbon Institute, who have listened to these arguments and given constructive criticism.

REFERENCES

[1] Surguchev, M.L., Fursov, A.Ya. and Taldikin, K.S. (Dec. 1979) "Methodology for establishing requirements for parameters used for projecting & exploiting reservoirs" Neftyanoe Khozyastvo p. 23-28 (in Russian)

[2] Komlosi Zs. (Oct. 1987) "Utilization of corelab data in well log interpretation" BKE Kőolaj- és Földgáz p. 297-317 (in Hungarian)

[3] Kuhn, T. and Gombos, Z. (July 1986) "Information system for production project preparation" BKE Kőolaj- és Földgáz p. 193-207 (in Hungarian)

[4] Garb, F.A. (June 1988) "Assessing Risk in Estimating Hydrocarbon Reserves and in Evaluating Hydrocarbon-Producing Properties" Journal of Petr. Tech. p. 765-778.

AN OPTIMAL METHOD FOR AVERAGING THE ABSOLUTE PERMEABILITY[1]

Thierry Gallouët

Département de Mathématiques
Université de Savoie,
73011 Chambéry, France

Dominique Guérillot[2]

Division Gisements
Institut Français du Pétrole,
92506 Rueil Malmaison, France

I. INTRODUCTION

Improving the recovery in hydrocarbon reservoirs demands a very detailed knowledge of the subsurface porous media and, in particular, of the spatial variations in their properties. The means available for analysing and modeling these often complex variations are increasingly varied. However, accounting for this complexity in the numerical simulations of the fluid flows in the porous

---

[1] Supported by the "Fonds de Soutien aux Hydrocarbons - COPREP", joint project IFP-ELF

[2] now seconded to ELF-U.K. Geoscience Research Center, 114A Cromwell Road, London SW7

media still remains. Discretized models cannot take into account heterogeneities of intermediate scale (decimetre to decametre) for reasons of memory size and calculation time of present-day computers. One of the practical problems faced by the user of a reservoir model is the following:

"How to average the physical quantities available so that the 'simulator' best reconstructs the real behaviour of the reservoir ?"

Note that this problem is different from the problem entitled *equivalent homogeneous medium* in so far as the attempt is not to model the structure of the reservoir by a continuous medium, but by a discrete structure, directly usable by the 'simulators'.

This general problem is a vast one. Among the physical quantities that have to averaged, we shall focus here only on the absolute permeability.

II. PROPOSED METHOD

As proposed by Marle: "Let us consider a heterogeneous porous medium H (see next figure) in which an incompressible fluid flow takes place and replace a part Ω of this medium by a homogeneous medium with characteristics such that the overall flow is disturbed as little as possible".

It appears that the very definition of an equivalent permeability demands assumptions on the type of flow to be reconstructed to return to a local problem over a region R containing Ω, and requires the introduction of a criterion defining the information which is to be preserved.

We propose the following method:

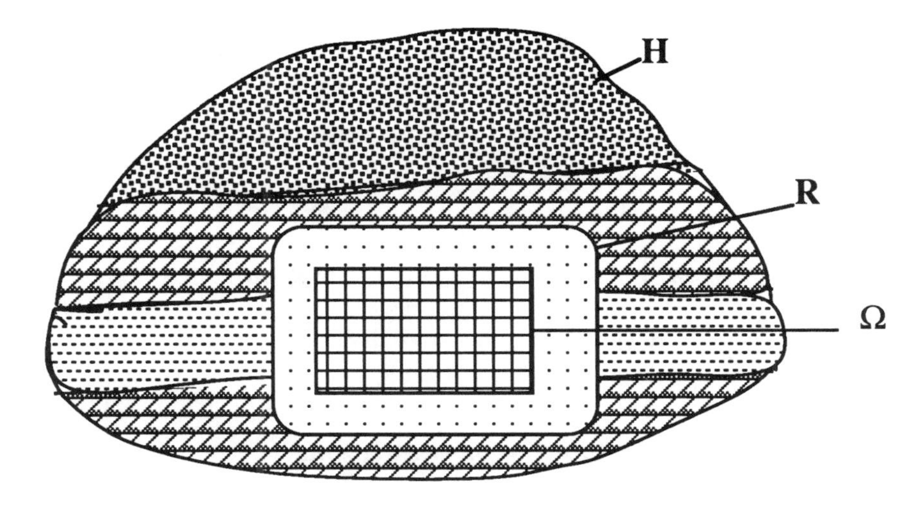

(1) To define an average permeability which is independent of the solution over the whole domain H, a family of pressure fields P is considered, whose gradient is constant over H:

$$P(x) = \lambda . x \text{ with } x=(x_1,x_2,x_3)^t \text{ and } \lambda=(\lambda_1,\lambda_2,\lambda_3)^t$$

This hypothesis implicitly presumes a large distance from any well. Other families of pressure fields could be selected for flows in their neighbourhood.

(2) A coupling is assumed on the edge of R between this *average* pressure P and the *local* pressure in R.

These hypotheses serve to return to the resolution of the following local problem (depending on parameter λ): "Find p and q such that:

$$q = - k/\mu \text{ (grad } p - \rho g) \text{ and div } q = 0, \text{ on } R$$
$$p = \lambda . x, \text{ on } \delta R,$$

where the permeability **tensor** k is the only parameter here characterizing the porous medium, and the viscosity μ and the density ρ characterizing the fluid "

Since this problem is linear, it is necessary, in practice to solve d (d=1, 2 or 3) local problems to calculate the average matrix K_m by using the following criterion:

K_m is such that $J(K_m) \leq J(K)$,

$$\text{with: } \mathbf{J(K)} = \int_D \{ \int_{\delta\Omega} [(\mathbf{Q}-\mathbf{q}).\mathbf{n}]^2 \, d\sigma \} d\lambda,$$

where $D=\{\lambda \text{ such that } \lambda_1^2 + \lambda_2^2+\lambda_3^2=1\}$, $K= k$ on $R \backslash \Omega$,

$$Q= - K/\mu \ (\text{grad } P - \rho g) \text{ and div } Q = 0, \text{ on } R$$
$$P= \lambda \, . \, x, \text{ on } \delta R,$$

and $n=(n_1,n_2,n_3)^t$ is the vector normal outside to the boundary.

A **physical interpretation:** try to minimize the square of the differences between local q and average Q filtration velocities, by averaging the differences with respect to all directions λ of the average flow.

Note that it is not assumed *a priori* that the average matrix K_m is symmetrical.

Solving this minimisation problem analytically gives the expression of the coefficients of the matrix K_m:

$$\mathbf{K_m}^{ij} = - \frac{1}{\text{mes } \Omega} \int_{\delta\Omega} \mathbf{q_{ji}}^2 \, . \, \mathbf{n_i}^2 \, d\sigma$$

where $q_1=(q_{11},q_{12},q_{13})^t$ is the filtration speed solution of the local problem with $\lambda=(1,0,0)^t$, q_2 with $\lambda=(0,1,0)^t$ etc.

III. APPLICATION

Applying this method to an example published by White and Horne, 1987, and also used by Corre, and Samier, 1990 (here d=2 and R=Ω) (see Figure below) gives the results on the following table:

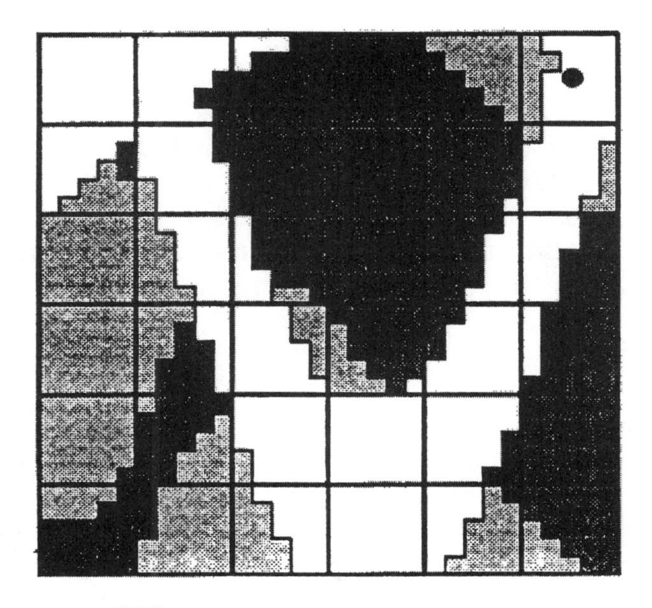

k = 0.10 D

k = 1.0 D

k = 10.0 D

<table>
<tr><td>10. 0.
0. 10.</td><td>6.97 -1.08
-0.71 7.10</td><td>1.32 1.20
1.13 1.25</td><td>0.1 0.
0. 0.1</td><td>0.5 -0.04
-0.04 0.54</td><td>5.34 0.21
6.51 7.43</td></tr>
<tr><td>5.05 0.56
0.54 5.28</td><td>6.69 2.16
1.37 8.15</td><td>0.1 0.
0. 0.1</td><td>0.1 0.
0. 0.1</td><td>0.76 0.66
0.66 0.76</td><td>5.67 -0.97
-1.2 7.35</td></tr>
<tr><td>1. 0.
0. 1.</td><td>4.60 -0.64
-0.30 5.88</td><td>2.27 -0.69
0.60 2.71</td><td>0.1 0.
0. 0.1</td><td>2.58 0.99
0.90 3.25</td><td>3.4 0.56
0.34 4.52</td></tr>
<tr><td>1. 0.
0. 1.</td><td>2.55 -0.60
-0.53 2.97</td><td>6.25 0.34
0.44 7.77</td><td>0.93 0.52
0.52 0.93</td><td>6.96 -0.72
-0.57 8.08</td><td>0.86 0.72
0.98 1.1</td></tr>
<tr><td>0.92 -0.06
-0.06 0.92</td><td>0.37 0.22
0.23 0.37</td><td>8.44 1.01
0.65 9.13</td><td>10. 0.
0. 10.</td><td>6.96 -0.72
-0.57 8.08</td><td>0.1 0.
0. 0.1</td></tr>
<tr><td>0.27 0.14
0.11 0.26</td><td>0.75 -0.10
0.06 0.90</td><td>4.51 -0.65
-0.34 5.53</td><td>10. 0.
0. 10.</td><td>3.72 0.99
0.90 4.20</td><td>0.26 -0.14
-0.14 0.26</td></tr>
</table>

Each cell of this table corresponds to the averaged matrix obtained on the coarse mesh. If one wants a symmetrical matrix, it averages the 2 extra-diagonal coefficients. A 9-point finite volume scheme allows to take into account these unsymmetrical extra-diagonal coefficients.

Other tests have been done to (1) study the influence of choosing different overlapping R domain around the macro-cells, (2) compare the pressure and saturation contours from the microlevel and macrolevel simulations using different maps of heterogeneous permeability distribution.

IV. CONCLUSIONS

A new method for averaging the absolute permeability in 3 D is presented briefly. This method:

(1) gives a rigourous formalism for the practical problem faced by the user of a fluid flow simulator who wants to loose as little as possible his available information on the reservoir heterogeneities. Here, the error on the filtration speed is minimized (and even, it is possible to make *a posteriori* checks of this error),

(2) allows to consider 2 different supports: R which size should be linked to the correlation lengths of the absolute permeability and Ω which is often chosen as the reservoir grid cell;

(3) can be extended to take into account the neighbourhood of wells.

V. REFERENCES (short list)

Samier, P., 1990, A Finite Element Method for Calculating Transmissibilities, Second European Conference on the Mathematics of Oil Recovery, Editions Technip, Arles, France.

White, C.D. and Horne, R.N., 1987, Computing Absolute Transmissibility in the Presence of Fine-Scale Heterogeneity, SPE 16011, presented at the Ninth Symposium on Reservoir Simulation, San Antonio, Texas.

FRACTAL ANALYSIS OF CORE PHOTOGRAPHS WITH APPLICATION TO RESERVOIR CHARACTERIZATION

H. H. Hardy

Conoco
Ponca City, OK

I. INTRODUCTION

This article describes the one- and two-dimensional statistical character of rock property distributions. This character is used to produce statistical distributions of reservoir properties between wells at the reservoir scale.

Interwell data is difficult to obtain at the scale needed to create reservoir property distributions for reservoir simulations. We must, therefore, try to deduce this information indirectly. Analyzing data from several sources gives us more confidence in our conclusions than any single data set. For this reason, this paper addresses the analyses of data from core photos, outcrop photos, and horizontal well logs.

Hewett[1] demonstrated that vertical well logs have a fractal structure called fractional Gaussian noise (fGn). He inferred the horizontal fractal character is different based on miscible flow data and seismic data. The horizontal character he assumed is called fractional Brownian motion (fBm). The present study indicates the statistical structure is not fBm, but fGn in both vertical and horizontal directions.

II. DATA ANALYSES

Core photos are analyzed to define "models" of reservoir property distributions. Core photos only capture light reflectance. Light reflectance is not a direct measure of the reservoir properties needed for reservoir simulation (e.g., porosity or permeability). Care must be taken in lighting conditions, camera location, film speed, and the like. For outcrop photos, vegetation, shadows, and weathering further complicate photo analysis. Nevertheless, geologists have long used the visual descriptions (light reflectance) of outcrops and core photos to help define geologic trends and depositional units.

Cores were slabbed and polished smooth. The cores were oriented so that bedding planes were horizontal, and core photos were taken. The core photos (e.g., Figures 1a, 2a, and 3a) were digitized. Digitization produces an array of numbers which capture the light intensity on the photo at each "point" on the photo. Once light reflectance is reduced to a set of numbers, many statistical tests can be made. One- and two-dimensional statistical tests were made for this study.

The one-dimensional tests were the same as those used by Hewett[1] for vertical well logs. These were a variogram, Fourier transform, and RS analyses. These were applied to traces parallel, perpendicular, and at 45 degrees to bedding planes in the core photos. Examples of these techniques applied to a horizontal trace are shown in Figure 4.

These tests indicated fGn in for all three directions. The major difference in the vertical and horizontal directions was the total variance of the sample values rather than the fractal character. For example, in Figure 1a, both vertical and horizontal traces are fGn with H=0.8. The standard deviation of traces in the vertical direction is twice that of traces in the horizontal direction.

In addition to the one-dimensional tests, two-dimensional Fourier transforms were taken of the photos. Two-dimensional amplitude and phase arrays result. The amplitude arrays are used to calculate the spectral density array for a photo. In order to try to capture the full two-dimensional description, an equation was fit to this two-dimensional spectral density. The result was the following:

$$SD(\omega_x, \omega_y) = A\left[\frac{\delta(\omega_x)}{|\omega_y|^{\alpha}} + \frac{B}{|\omega_x^2 + \omega_y^2|^{\beta}}\right] \tag{1}$$

$$\alpha = 2H - 1 \ , \quad \beta = H \ , \quad and \quad 0.5 < H < 1.0$$

$$\delta(\omega_x) = [1 \ if \ \omega_x = 0, \ zero \ otherwise]$$

If this equation is a good description of core photos, varying the parameters in this equation should produce other core photos. Indeed, this was found to be the case.

Figures 1b, 2b, and 3b were generated using Equation 1. Comparing these to core photos, Figures 1a, 2a, and 3a show that the computer-generated photos are indeed similar to many core photos.

The details of these analyses and the reconstruction of the core photos can be found in Hardy[2].

The one-dimensional analysis was also applied to the outcrop photo along the traces highlighted in Figure 5. The results of this study was also fGn in the vertical and horizontal directions.

The one-dimensional analysis applied to a neutron porosity log from a horizontal well also gives fGn. Crane and Tubman[3] have published similar results for other horizontal wells. Figure 6 shows the similarity of horizontal well data with fGn rather than fBm. In this figure, fGn and fBm of $H=.8$ is plotted for comparison with the actual horizontal well data. The "smoothness" of fBm is not apparent in the horizontal well trace. Figure 7 shows the same statistical measures used for the horizontal trace in Figure 4, applied to about 2,000 feet of a horizontal log.

III. GENERATING WELL TO WELL CROSS SECTIONS

The spectral density analysis can be used to produce well-to-well cross sections using porosity log data. The technique is based on the observation that the spectral density equation, Equation 1, is made up of two parts. The first part produces "layers;" the second produces "splotches." For example, Figure 2b has $B \ll 1$ so that the first term in Equation 1 dominates. Figure 3b has $B \gg 1$ so that the second term dominates. Figure 2b is mostly "layers" whereas Figure 3b is mostly "splotches."

Well-to-well cross sections are generated by using linear interpolation between the wells to generate the layers. "Splotches" (noise) are added using Equation 1 with $B \gg 1$. H is set to the value found from analyzing the vertical well data. The amount of noise added is determined by the difference between the two well log readings. The details and result of this procedure are described by Beier and Hardy[4]. Crane and Tubman[3] have generated cross sections with fGn noise but conditioned the cross section to the well data by the same technique used by Hewett[1] for fBm. The procedure described by Beier and Hardy[4] is different and is based on our success of producing computer forgeries of core photos.

REFERENCES

1. Hewett, T. A.: "Fractal Distributions of Reservoir Heterogeneity and Their Influence on Fluid Transport," SPE paper 15386 presented at the 1986 SPE Annual Technical Conference and Exhibition, New Orleans, Oct. 5-8.
2. Hardy, H. H.: "The Fractal Character of Photos of Slabbed Cores," to appear in Mathematical Geology.
3. Beier, R. A. and Hardy, H. H.: "Comparison of Different Horizontal Fractal Distributions in Reservoir Simulations."
4. Crane, S. D. and Tubman, K. M.: "Reservoir Variability and Modeling With Fractals," paper SPE 20606 presented at the 1990 SPE Annual Technical Conference and Exhibition, New Orleans, LA, September 23-26.

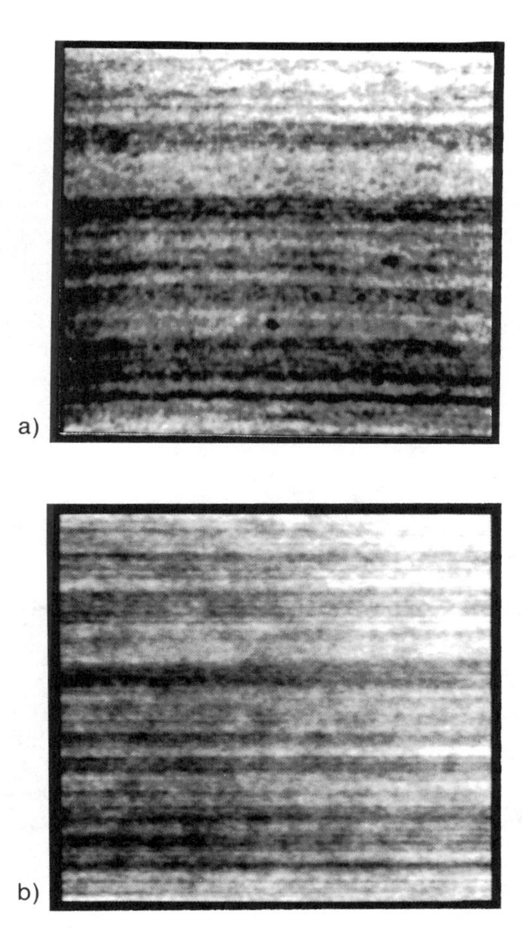

Figure 1. a) Photo of a slabbed sandstone core about 2"x2". b) Computer generated core photo.

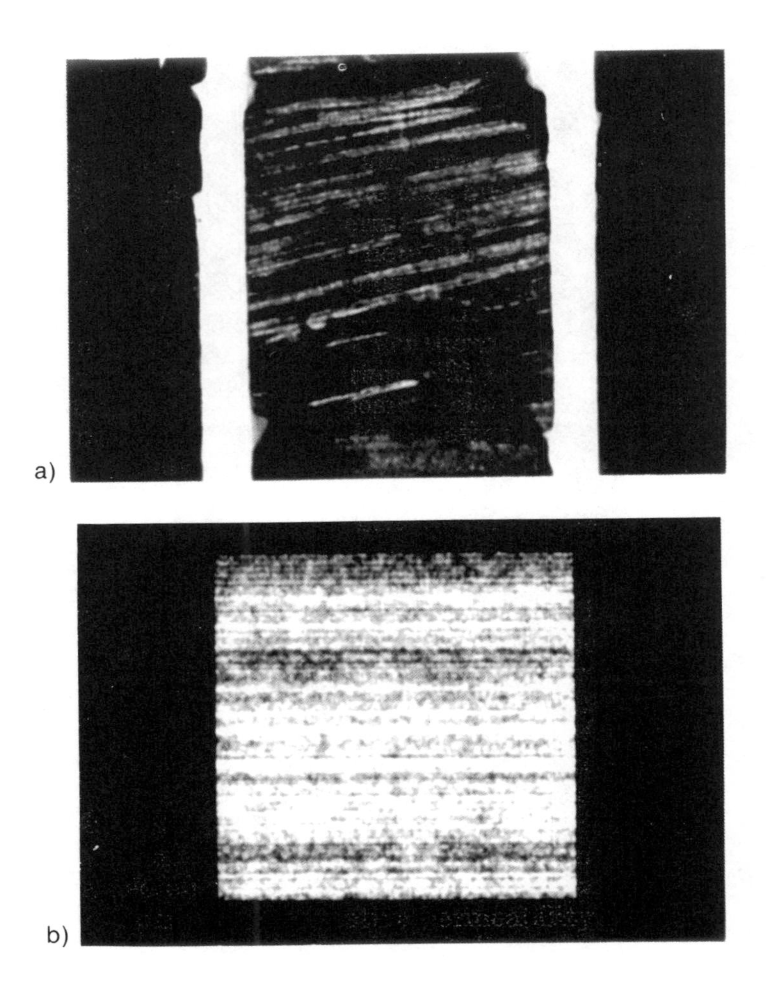

Figure 2. a) Slabbed core photo using UV light of an oil-stained, cross-bedded sandstone about 2"x2". b) Computer generated core photo.

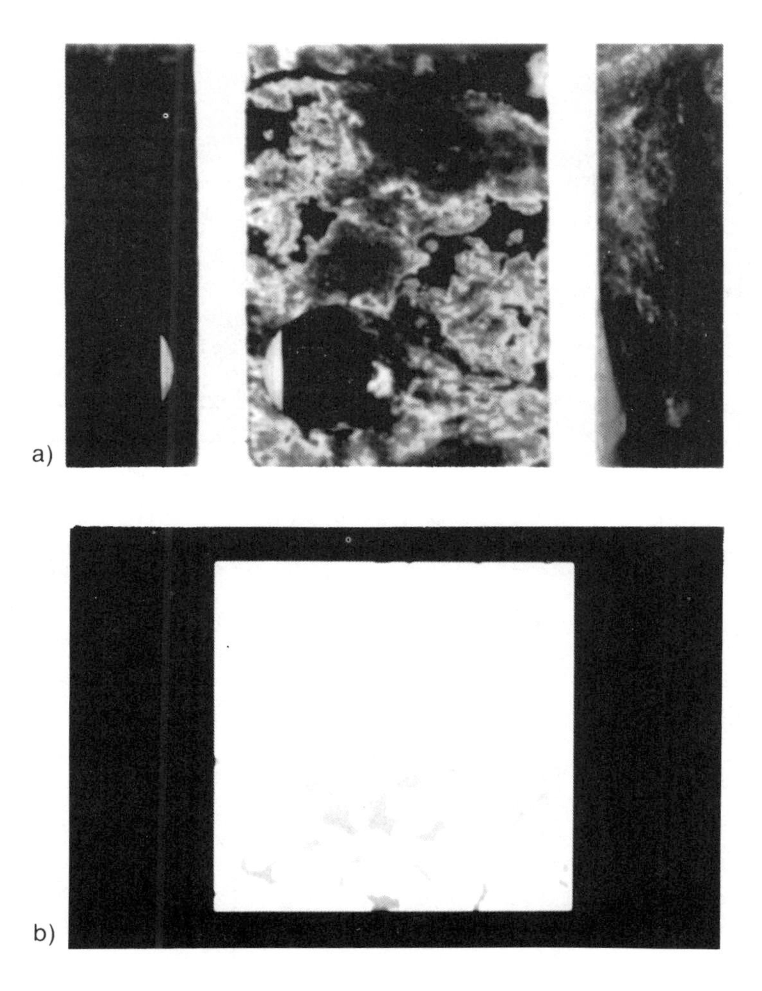

Figure 3. a) Slabbed core photo showing unevenly oil-stained sandstone core about 2"x2". b) Computer generated core photo.

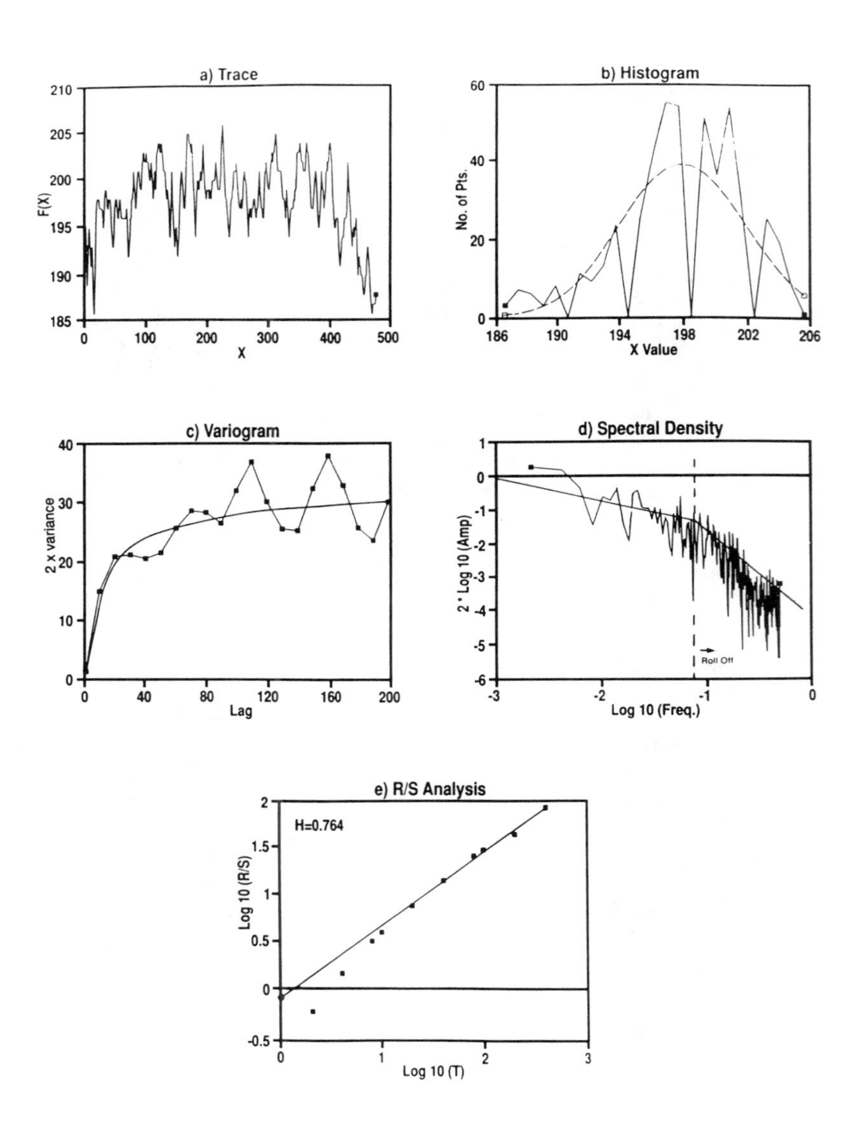

Figure 4. One dimensional statistical analysis applied to a horizontal trace: a) trace, b) histogram, c) variogram, d) spectral density, e) RS analysis.

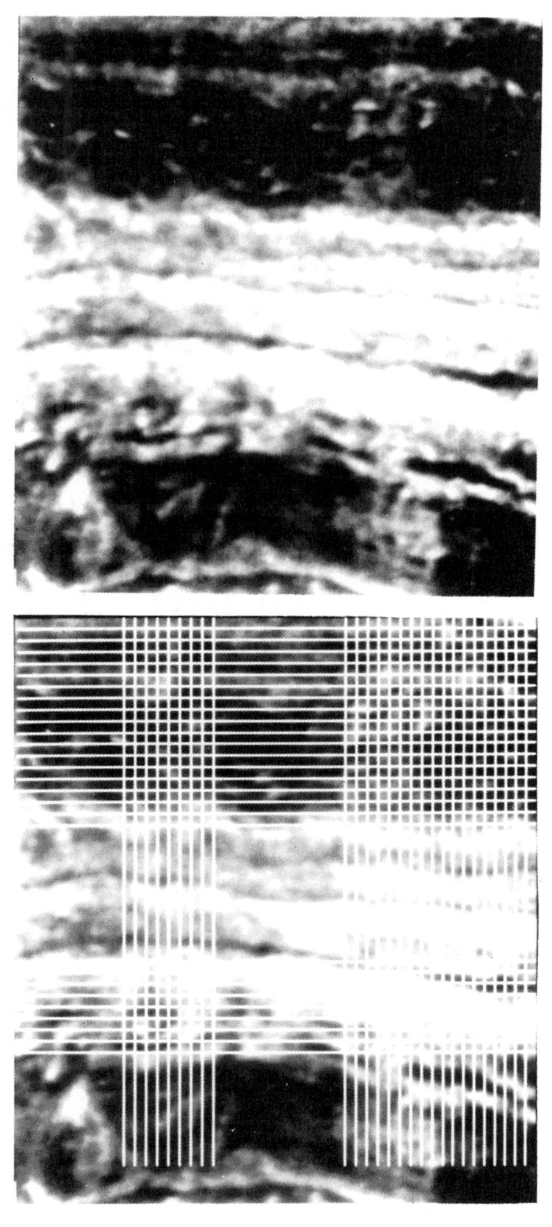

Figure 5. a) Photo of a carbonate outcrop in the Canadian Rockies about 100'x100'. b) Traces analyzed from the outcrop photo.

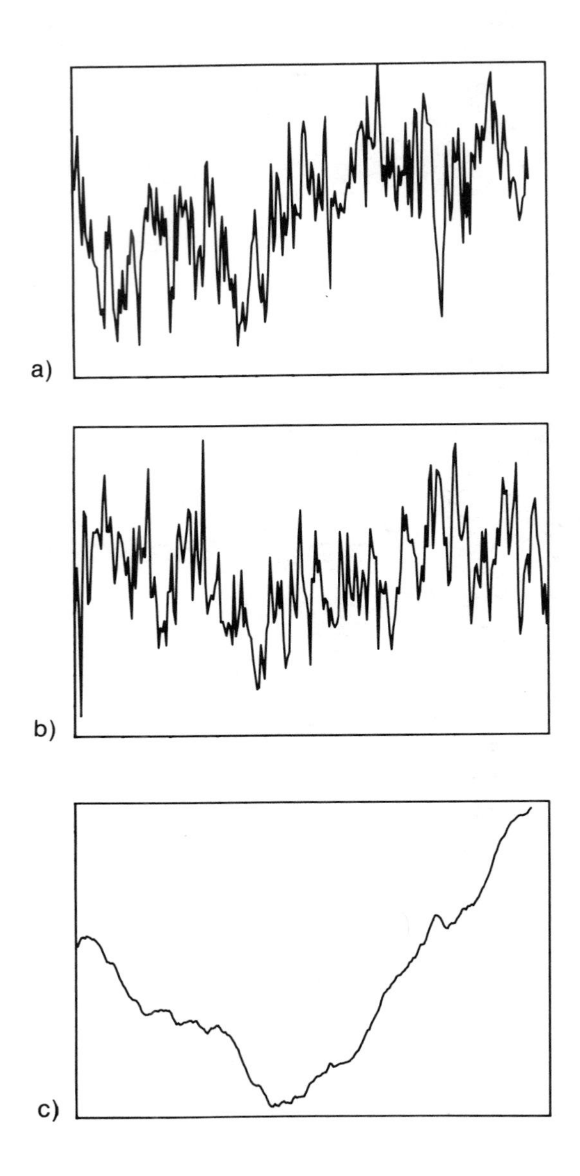

Figure 6. a) fGn, b) Neutron porosity log (7 to 17 porosity units) from a horizontal well (1000 foot interval) through an aeolian sandstone, c) fBm.

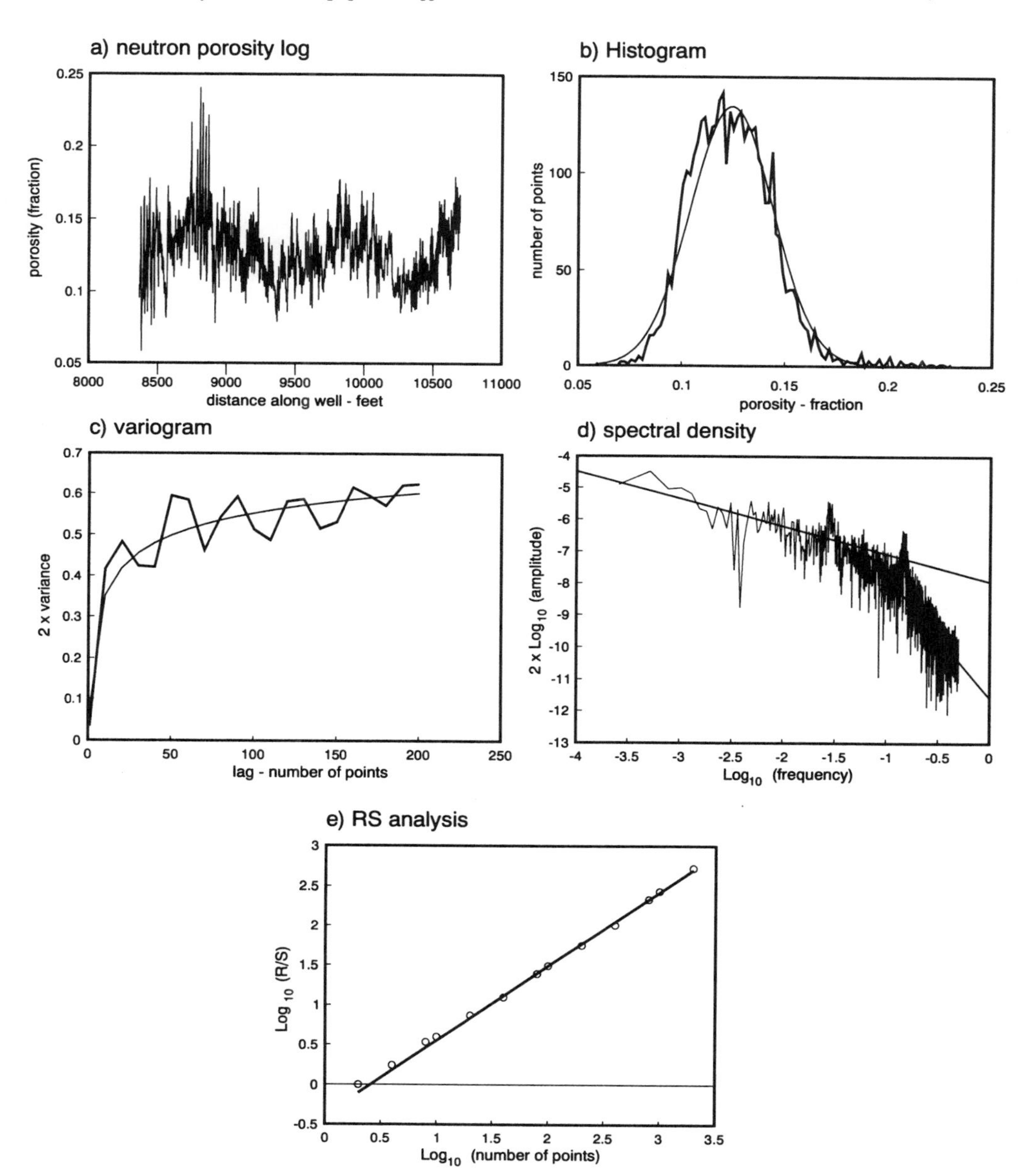

Figure 7. One dimensional statistical analysis of a horizontal log through an aeolian sandstone. a) neutron porosity log, b) histogram, c) variogram, d) spectral density, e) RS analysis.

EARTH STRESS ORIENTATION - A CONTROL ON, AND GUIDE TO, FLOODING DIRECTIONALITY IN A MAJORITY OF RESERVOIRS

K J Heffer

BP Research, Sunbury, England

J C Lean

Imperial College of Science
Technology and Medicine
London

I. INTRODUCTION

This poster describes analysis of empirical field data relating to lateral directionality (anisotropy) in flooding of oil reservoirs. It tests the suggestion that anisotropy caused by the presence of natural fractures and faults in the reservoir is widespread. Although this has often been realised during the life of a flood, the influence of natural fractures is not always obvious nor anticipated.

In a network of existing natural fracture sets, any anisotropy in the present earth stresses will influence the distribution and orientation of fracture widths, either with or without injection; hence stresses can govern the anisotropy in fluid conductivity. Perhaps the best direct demonstration of this has been obtained from observations of microseismicity during water injection into jointed granitic geothermal reservoirs (Pine and Batchelor, 1984; Fehler,

1987). If fractures, natural or induced, are conductive over extensive areas of a reservoir, in which the local minimum principal stress axis is (sub) horizontal, then we anticipate that the preferred flooding directions would tend to be at small angles to the local present major horizontal principal stress axis (SHMAX).

The small inclination to SHMAX would be symptomatic of the involvement of extension, shear or hybrid shear fractures, a mixture of these fracture types or local variation of stress directions. The anticipated correlation between the anisotropies of flooding and horizontal stresses was indeed found for 36 field cases in Heffer & Dowokpor (1990). The database for that study has now been more than doubled and we report below the confirmation and further analysis of that correlation.

Other sources of flooding directionality, such as variations in interwell distances or offtake rates, large-scale pressure gradients or formation dip could distort anisotropy due to fractures. For the purposes of this study those other influences were again put aside in order to test the hypothesis of the influence of fracture. The existence of a correlation amidst these potentially conflicting influences on directionality could only strengthen the inference that stress orientation is a prime indicator for flooding directionality.

II. ANALYSIS

The relationship between the flood and stress anisotropies was tested for some 80 fields involving water floods, surfactant/polymer floods and gas floods across North America, the North Sea, continental Europe, the Middle East and China, using data from both BP operations and the literature. As previously, preferred flood directions were determined by a consistent method, designed to be as objective as possible and utilising data on injected fluid breakthrough, tracer production, oil production response and multiwell pressure analysis. The methodology is to fit, by maximum likelihood, an ellipse representing variation of probability of breakthrough with azimuth to the observed data. It is described in detail in Heffer & Dowokpor, 1990, but a short summary follows:

In any one case, "available" directions for breakthrough were generally defined as those directions from injector wells to all nearest neighbour producer wells. Each of those directions was then assigned a weighting factor which was either 1 or 0 according to whether injected fluid respectively had, or had not, broken through between the particular injector - producer pair of wells. The weighting was more general if quantitative data were available which indicated the degree of breakthrough (e.g. relative time delays, percentage of tracer produced etc.) or if the principal axes of the directional permeability tensor could be assessed from interference testing. Directions with non-zero weights are referred to as "taken" directions. A model for breakthrough probability variation with orientation was defined as the ellipse:

$$P(\theta;\Phi,a,b) = [\cos^2(\theta-\Phi)/a^2 + \sin^2(\theta-\Phi)/b^2]^{-\frac{1}{2}}$$

in which Φ is the orientation of the major axis relative to the local azimuth of SHMAX
a,b are respectively the major, minor semi-axes of the ellipse
and θ is the orientation of a potential breakthrough path relative to the azimuth of SHMAX.

A likelihood function for the parameters of this model with a particular set of observations was defined as:

$$L(\Phi,a,b;B,N) = \prod_{i\varepsilon B}[P(\theta_i;\Phi,a,b)]^{w_i} \times \prod_{j\varepsilon N}[1-P(\theta_j;\Phi,a,b)]$$

where B is the set of observations of breakthrough and N is the set of observations of non-breakthrough.

The parameters of the ellipse were adjusted to maximise the likelihood function. The preferred flooding direction was taken as the major axis of the fitted ellipse. Stress data were taken mainly from published maps, particularly Zoback *et al.* 1989, supplemented of course with any locally available data. In general there is an uncertainty of 20 to 30 degrees in stress azimuth measurements, except where more concentrated data has allowed this uncertainty to be reduced to 15°. Interpolation between published data points and local variability of stress direction within structures will have further degraded the accuracy.

Figure 1 gives an example of the type of data. Water breakthrough in this group of neighbouring fields showed a strong NW-SE trend, in line with the local orientation of SHMAX.

The 80 field cases studied covered wide variations in reservoir characteristics, such as permeabilities, depths, thickness, lithologies, and tectonic setting. The large majority of cases involved flooding with water or water-based fluids or WAG schemes. Some cases of just gas injection were also included. It is emphasised that no screening of cases was made other than for availability of data on both flood behaviour and local stress orientation.

III. RESULTS

The overall correlation between principal flooding and stress directions for all 80 cases is shown in Figure 2. A very strong trend of flooding in the direction of the major horizontal stress axis is seen; 80% of cases (64) have a deviation between these axes of less than 45 degrees. The data involved more than 1700 individual injector-producer pairs between which flood breakthrough was possible. Figure 3 shows the circular frequency of both "available" flooding directions (outer rose diagram) as well as those directions "taken" by the flood (inner rose diagram). Figure 4 shows directional variation in the proportion of those "available" paths along which breakthrough had already occurred. It is seen that a significantly higher proportion of breakthrough has occurred along directions close to SHMAX. The probability of breakthrough parallel to SHMAX is interpreted as being approximately 50% greater than the probability in the orthogonal direction. Figure 5 is the "aggregate" ellipse which best fits all of the data: it confirms the trend of Figure 4.

The success of the correlation despite variability in local stress orientations is attributed to the fact that <u>averages</u> of both stress and flooding directions are being compared for a given field.

The greater quantity of data in this study has allowed screening of different characteristics of reservoirs. In particular we have divided the cases into those which are deemed "naturally fractured" and those for which natural fractures are not a prominent feature of the reservoir description. Figure 6 shows the deviations of the major flood axes for the "naturally fractured" set of reservoirs: there is a peak at 0 degrees with side peaks at ±15°. Figure 7 shows the aggregate ellipse of this set of data. Figures 8 & 9 record the results for reservoirs deemed to be "not naturally fractured".

There is little difference in the strength of the trend in either case. Work is currently in progress to determine whether the relative low in frequencies at precisely zero degrees in Figure 8 is meaningful.

Additionally, screening of other reservoir characteristics has indicated:

- floods below 2000ft deep show high anisotropy whilst those shallower than 2000ft show hardly any anisotropy. This would be in accordance with expectations of fractures being responsible for the anisotropy if at greater depths the fractures are sub-vertical whilst at shallower depths they are sub-horizontal or shallow dipping. (Figures 10 to 13).

- floods in extensional tectonic regimes show much higher anisotropies than do those in compressional regimes (defined by maximum principal stress being sub-horizontal, i.e. strike-slip or thrust conditions). This is again in accordance with the involvement of fractures: extensional regimes are favourable for subvertical fractures parallel to SHMAX whilst compressional regimes tend to give vertical fractures striking approximately ± 30° to SHMAX or thrust fractures with little lateral anisotropy. (Figures 14 to 19).

Two further results are contrary to the possibility that hydraulic fractures induced by injection are responsible for the correlation:

- the more permeable reservoirs show tighter clustering in flood directionality around SHMAX (Figures 20 and 21) whilst less permeable reservoirs show greater directionality towards ±30° from SHMAX (Figures 22 and 23).

- the strength of the correlation is much more marked for those cases in which well spacing is greater than 1000 feet. (Figures 24 to 27).

The latter result also weighs against the possibility that the correlation is produced by sedimentary controls; conversely, directionality inherent in the depositional trends contrary to stress directions may destroy the correlation at smaller well spacings.

- flooding with gases appears to give rise to a similar degree of anisotropy as for liquid floods (Figures 28 to 31). Two counteracting factors which need to be taken into account in understanding this observation are: (i) gas injection provides less cooling than liquid and therefore less thermal stress to promote fracture dilation and (ii) the higher mobility ratios of gas floods imply that gases will "exploit" any fracture conductivity to a greater extent.

- Figure 32 is an histogram of the ratios of major to minor axes of the fitted ellipses for the 80 cases. It is truncated at 7:1, which was the highest degree of eccentricity allowed in the fitting process.

IV. CONCLUSIONS

Whilst full explanation of these results will require coupled numerical modelling of hydraulic and geomechanical processes, they are strongly indicative of the influence of natural fractures in flooding directionality. Whatever the explanation the study has strengthened the implication that local stress orientation measurements are immediate guides to the likely preferred directions of flooding in a reservoir, to which well patterns and development plans may be adjusted. It is also a pointer towards better understanding, and therefore better prediction of flooding processes.

REFERENCES

Fehler, M., House, L. and Kaieda, H. (1987). Determining planes along which earthquakes occur: method and application to earthquakes accompanying hydraulic fracturing. J. Geophys. Res.92 (B9), 9407-9414.

Heffer, K.J. and Dowokpor, A.B. (1990). Relationship between Azimuths of Flood Anisotropy and Local Earth Stresses in Oil Reservoirs, North Sea Oil and Gas Reservoirs II, Graham and Trotman.

Pine, R.J. and Batchelor, A.S. (1984). Downward Migration of Shearing in Jointed Rock During Hydraulic Injections, Int. J. Rock Mech. Min. Sci. & Geomech. Abstr. Vol.21, No.5, pp.249-263.

Zoback, M.L. *et al.* (1989). Global Pattern of Tectonic Stress, Nature, 341 (6240), 291-298.

Fig 1. ORIENTATION OF WATER BREAKTHROUGH PATHS AND PREFERRED PERMEABILITY DIRECTIONS

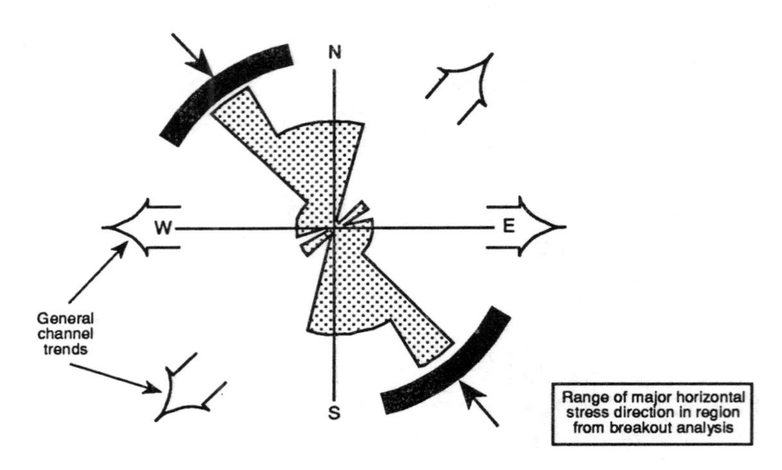

Please Note:
On all subsequent plots directions are **RELATIVE** to the local horizontal stress axes, for which the arrangement on the page is as follows:

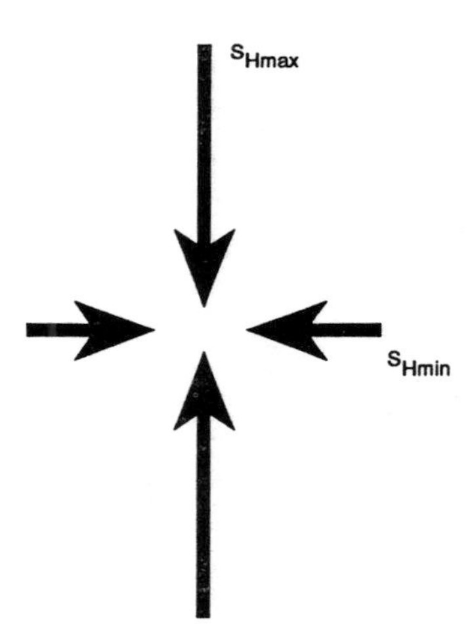

Fig 2. DEVIATION OF MAJOR FLOOD AXES FROM S_{Hmax} - ALL CASES

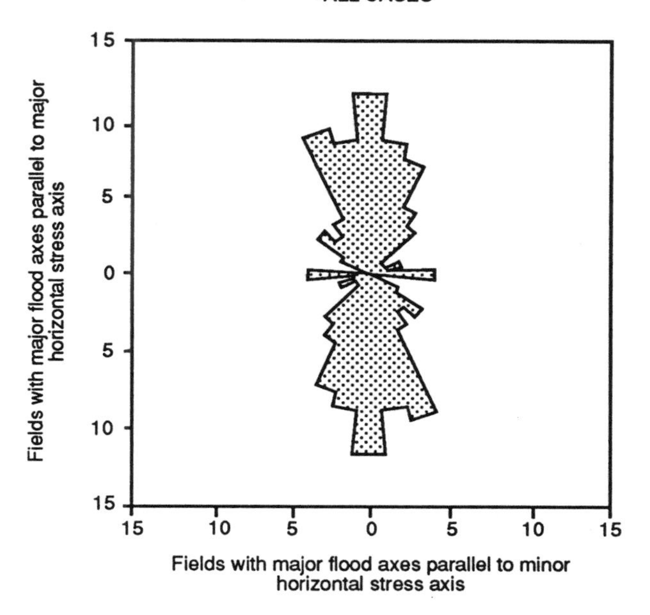

Fig 3. TAKEN AND AVAILABLE DIRECTIONS - ALL CASES

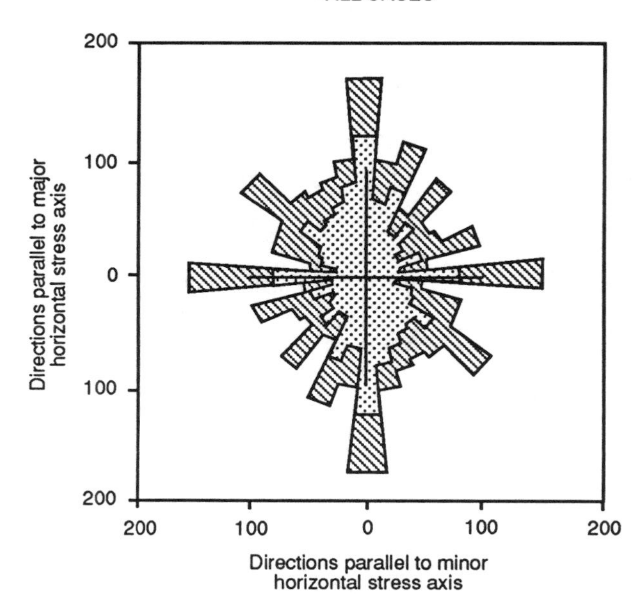

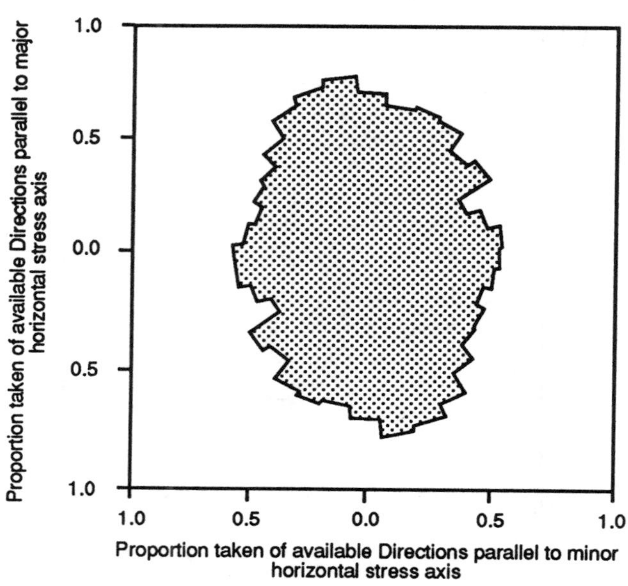
Fig 4. PROPORTIONAL BREAKTHROUGH
- ALL CASES
1.0
0.5
0.0
0.5
1.0
1.0
0.5
0.0
0.5
1.0
Proportion taken of available Directions parallel to major horizontal stress axis
Proportion taken of available Directions parallel to minor horizontal stress axis

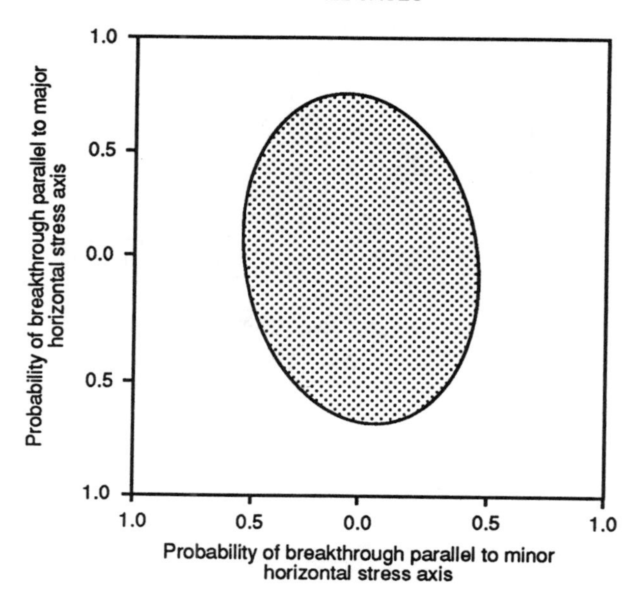
Fig 5. AGGREGATE BREAKTHROUGH PROBABILITY ELLIPSE
- ALL CASES
1.0
0.5
0.0
0.5
1.0
1.0
0.5
0.0
0.5
1.0
Probability of breakthrough parallel to major horizontal stress axis
Probability of breakthrough parallel to minor horizontal stress axis

Fig 6. DEVIATION OF MAJOR FLOOD AXES FROM S_{Hmax} - NATURALLY FRACTURED

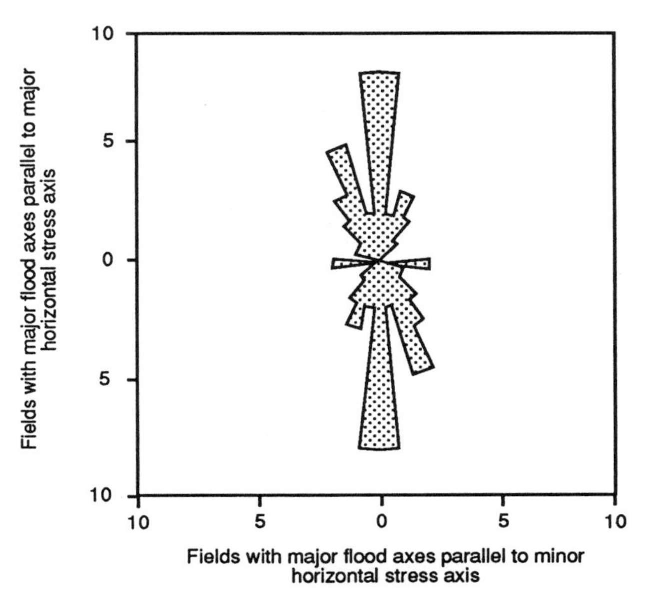

Fig 7. AGGREGATE BREAKTHROUGH PROBABILITY ELLIPSE - NATURALLY FRACTURED

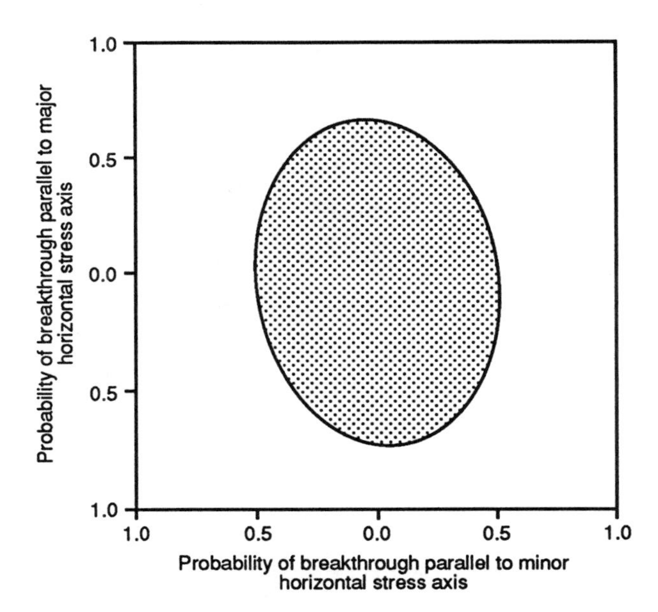

Fig 8. DEVIATION OF MAJOR FLOOD AXES FROM S_{Hmax} - NOT FRACTURED

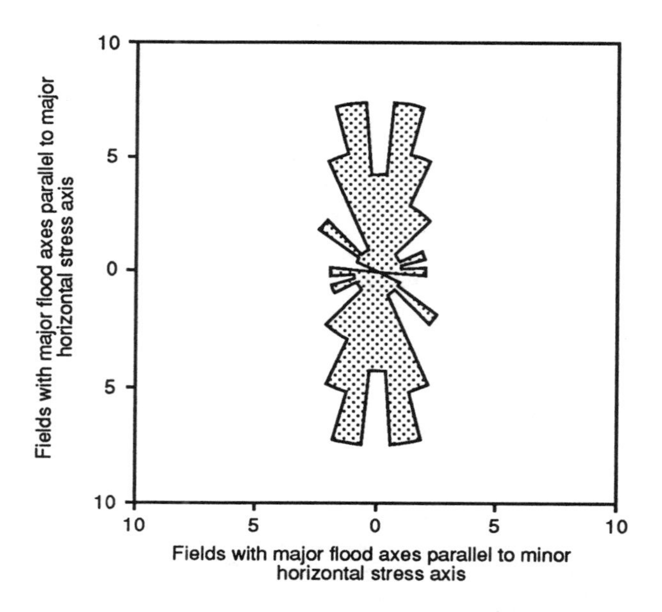

Fig 9. AGGREGATE BREAKTHROUGH PROBABILITY ELLIPSE - NOT FRACTURED

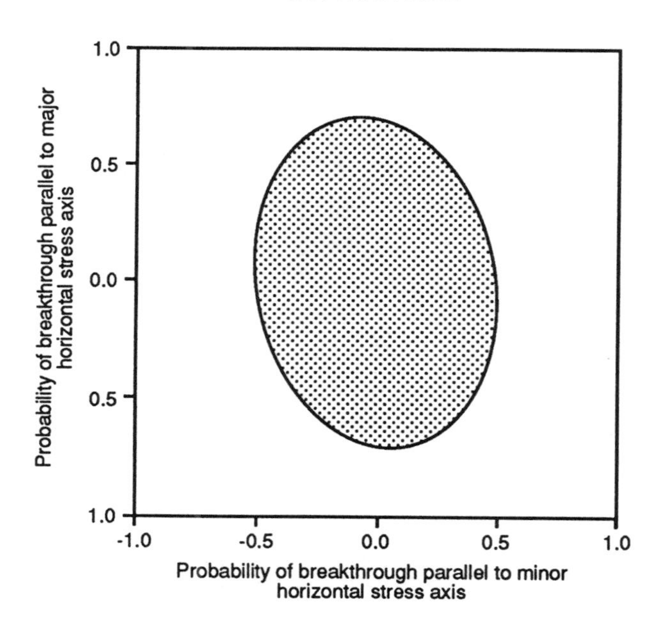

Fig 10. DEVIATION OF MAJOR FLOOD AXES FROM S_{Hmax} - FLOODS BELOW 2000 ft

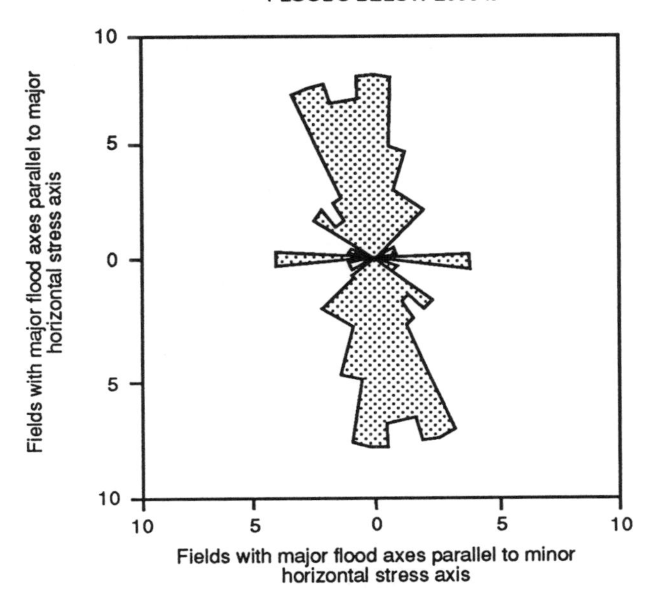

Fig 11. AGGREGATE BREAKTHROUGH PROBABILITY ELLIPSE - FLOODS BELOW 2000 ft

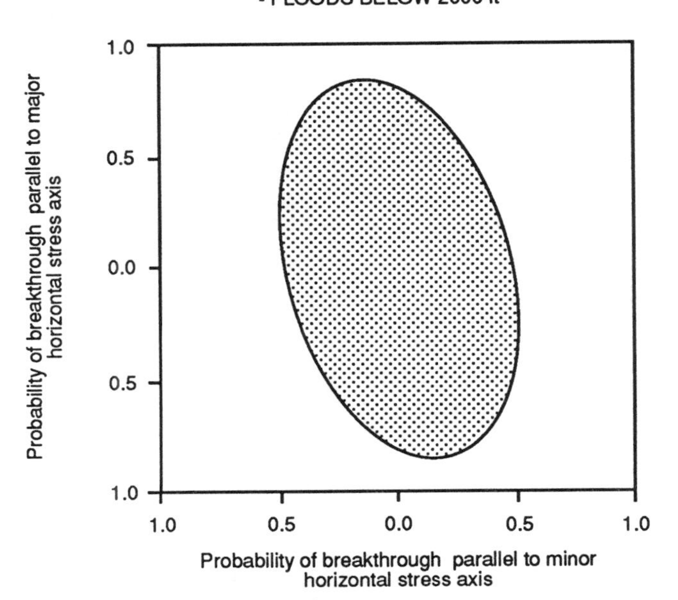

Fig 12. DEVIATION OF MAJOR FLOOD AXES FROM S_{Hmax} - FLOODS ABOVE 2000 ft

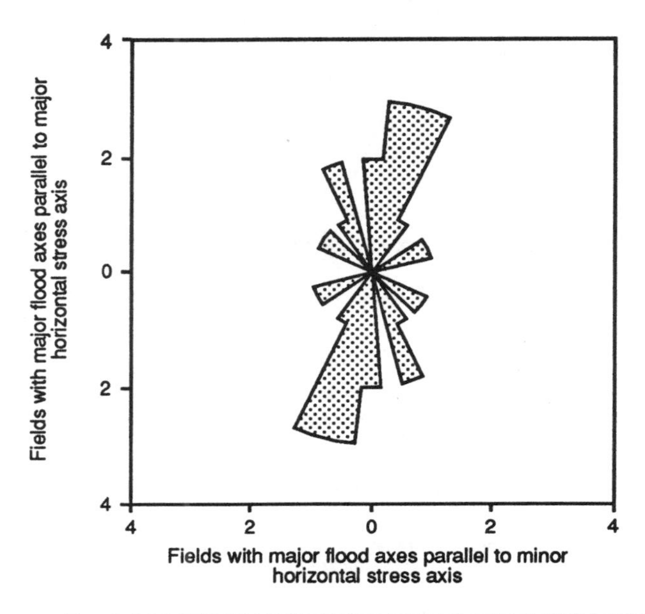

Fig 13. AGGREGATE BREAKTHROUGH PROBABILITY ELLIPSE - FLOODS ABOVE 2000ft

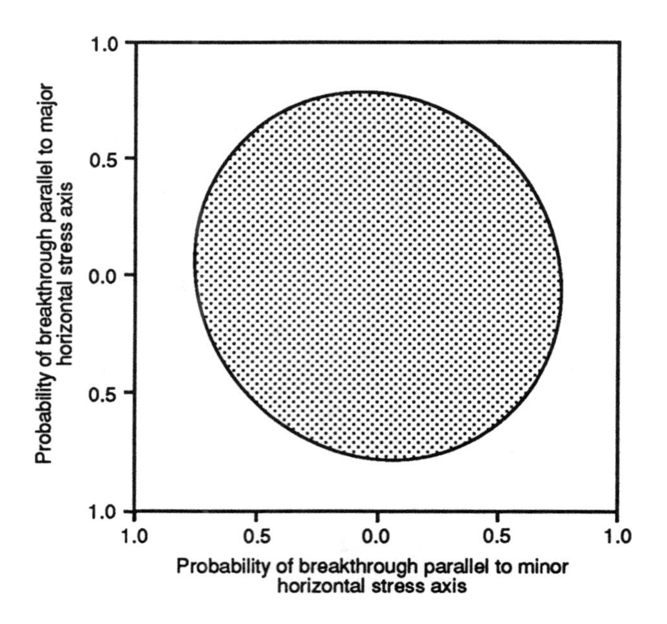

Fig 14. DEVIATION OF MAJOR FLOOD AXES FROM S_{Hmax}
- EXTENSIONAL REGIME

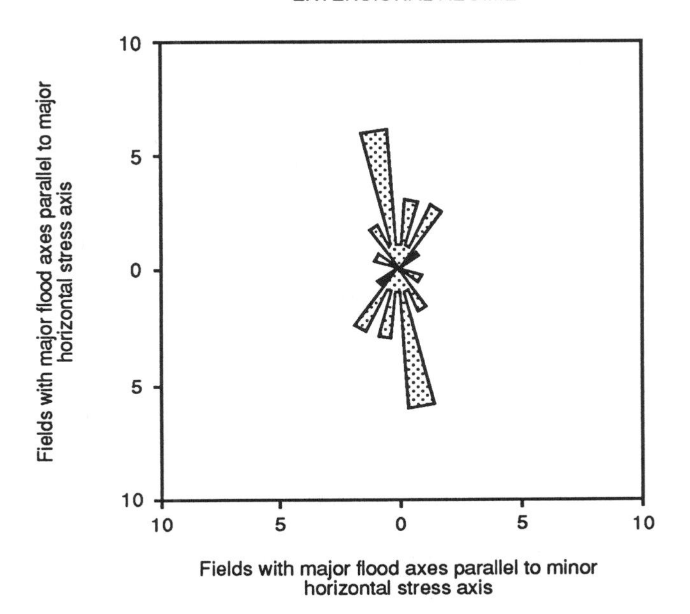

Fig 15. AGGREGATE BREAKTHROUGH PROBABILITY ELLIPSE
- EXTENSIONAL REGIME

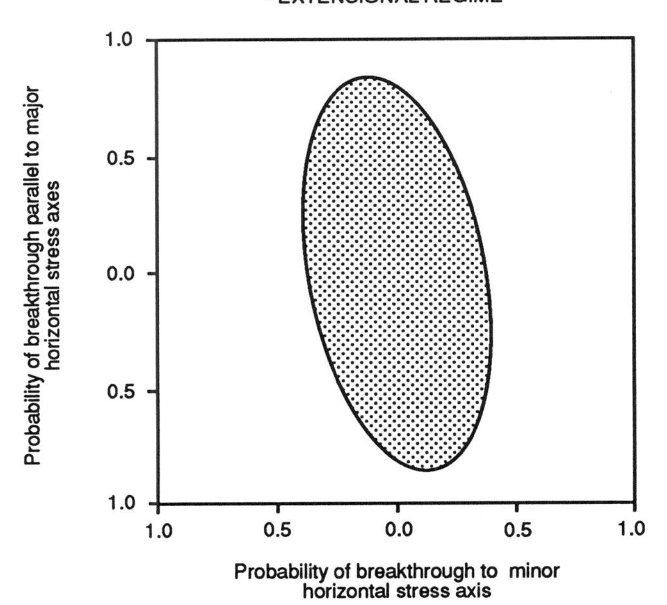

Fig 16. DEVIATION OF MAJOR FLOOD AXES FROM S_{Hmax} - COMPRESSIONAL REGIME

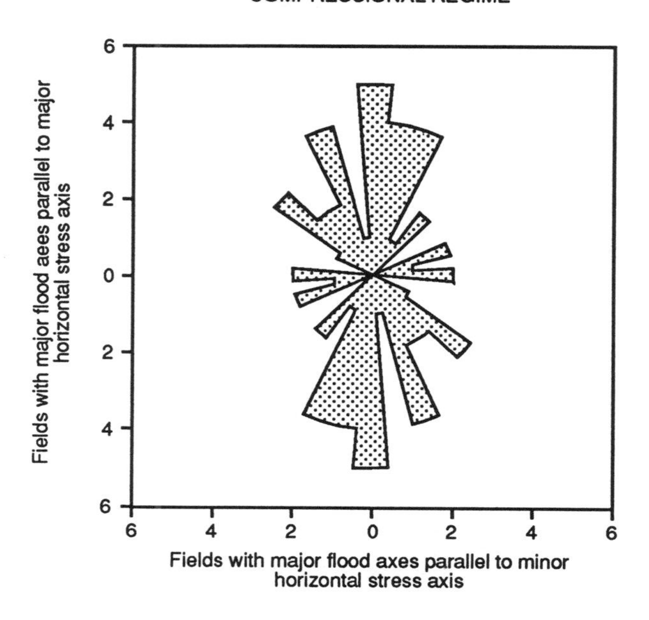

Fig 17. AGGREGATE BREAKTHROUGH PROBABILITY ELLIPSE -COMPRESSIONAL REGIME

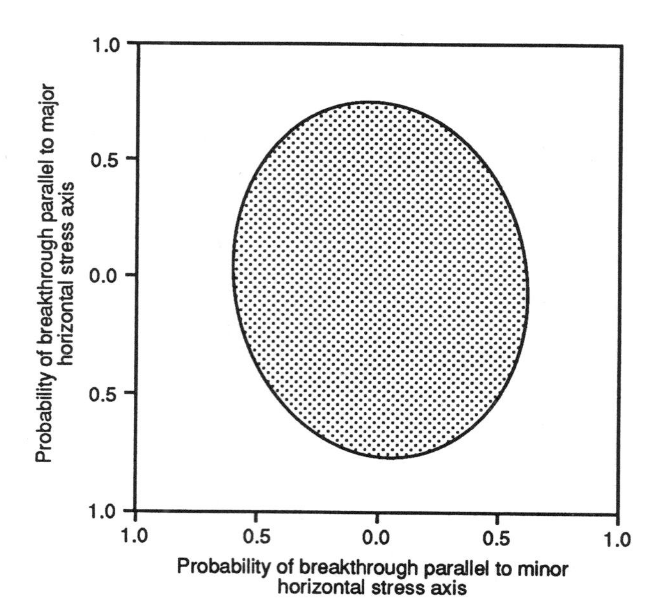

Fig 18. "EXTENSIONAL" REGIME

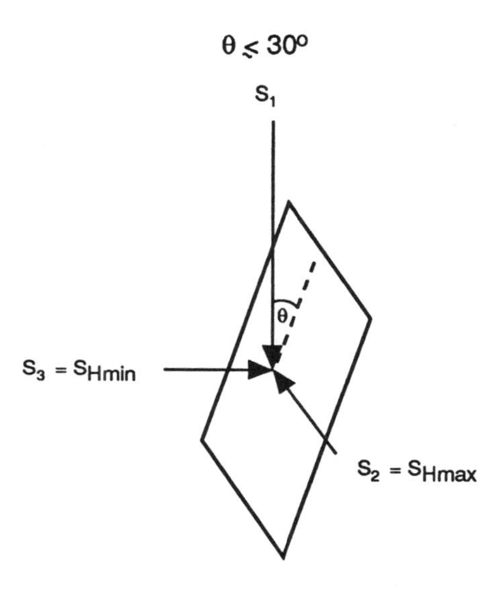

Fig 19. "COMPRESSIONAL" REGIME

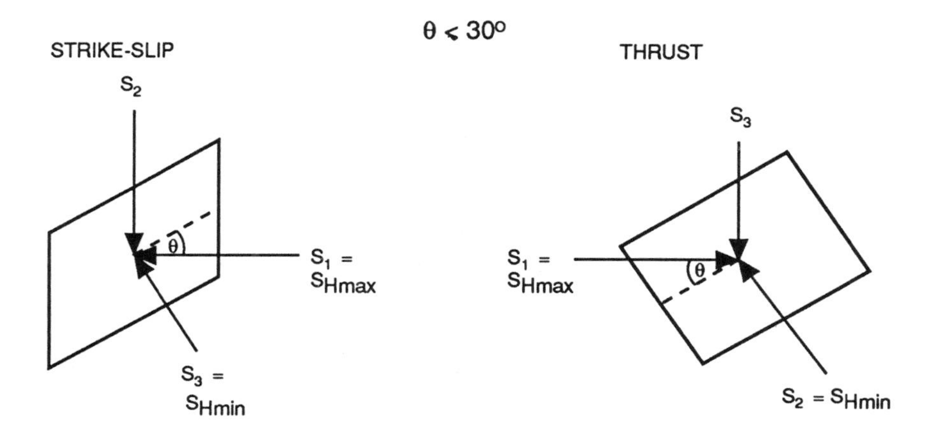

Fig 20. DEVIATION OF MAJOR FLOOD AXES FROM S_{Hmax} - PERMEABILITIES > 100mD

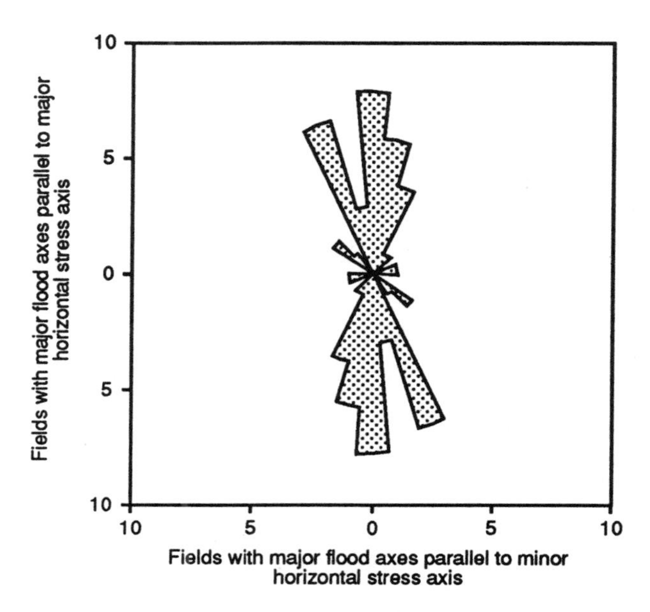

Fig 21. AGGREGATE BREAKTHROUGH PROBABILITY ELLIPSE - PERMEABILITIES > 100 mD

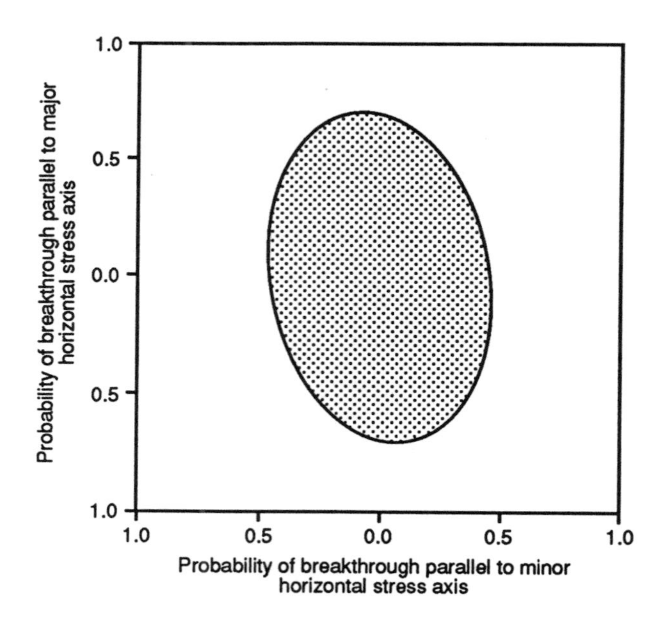

Fig 22. DEVIATION OF MAJOR FLOOD AXES FROM S_{Hmax} - PERMEABILITIES < 100 mD

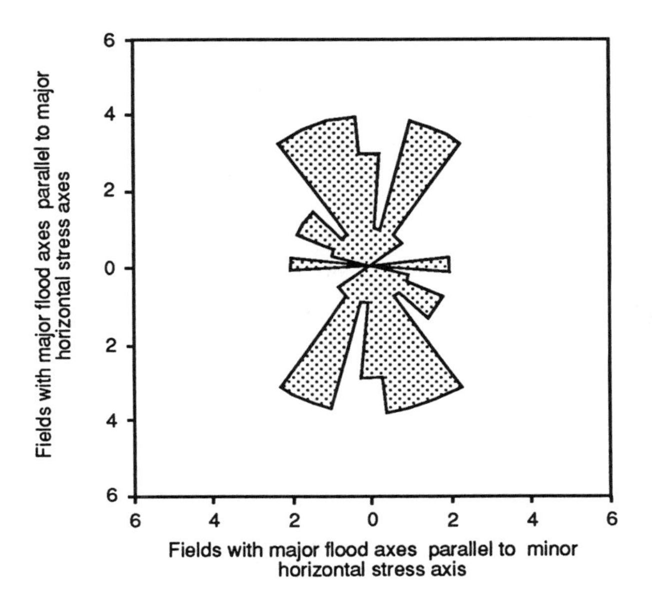

Fig 23. AGGREGATE BREAKTHROUGH PROBABILITY ELLIPSE - PERMEABILITIES < 100 mD

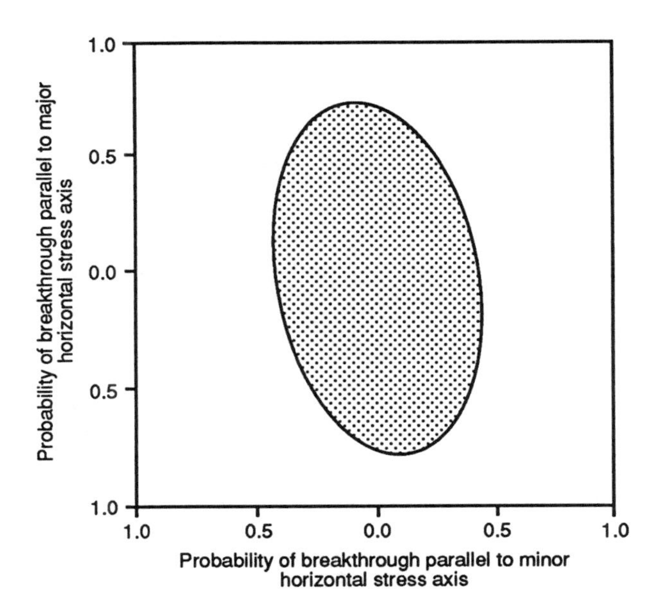

Fig 24. DEVIATION OF MAJOR FLOOD AXES FROM S_{Hmax} - SPACINGS > 1000 ft

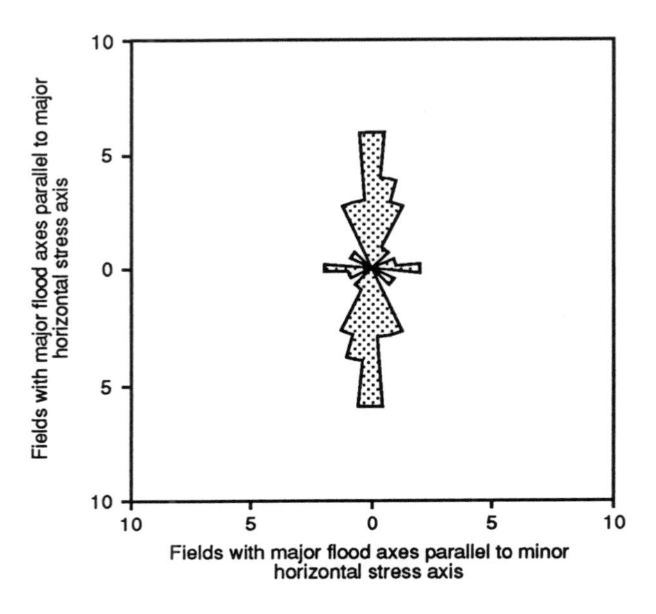

Fig 25. AGGREGATE BREAKTHROUGH PROBABILITY ELLIPSE - SPACINGS > 1000 ft

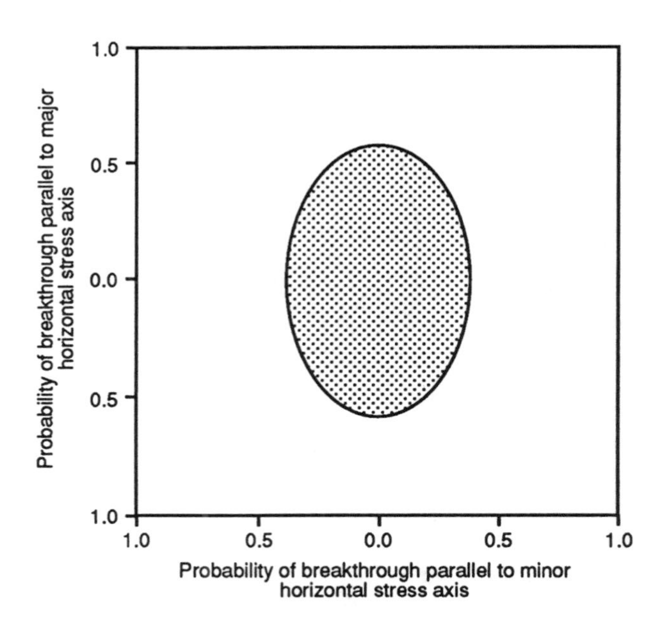

Fig 26. DEVIATION OF MAJOR FLOOD AXES FROM S_{Hmax} - SPACINGS < 1000ft

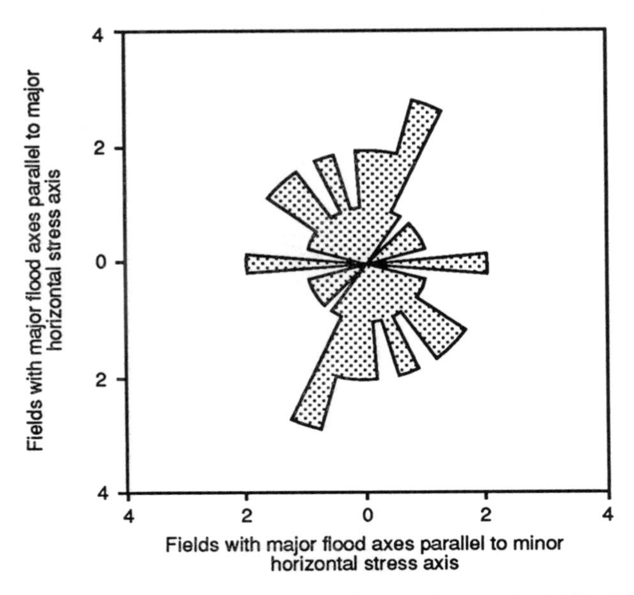

Fig 27. AGGREGATE BREAKTHROUGH PROBABILITY ELLIPSE - SPACINGS < 1000ft

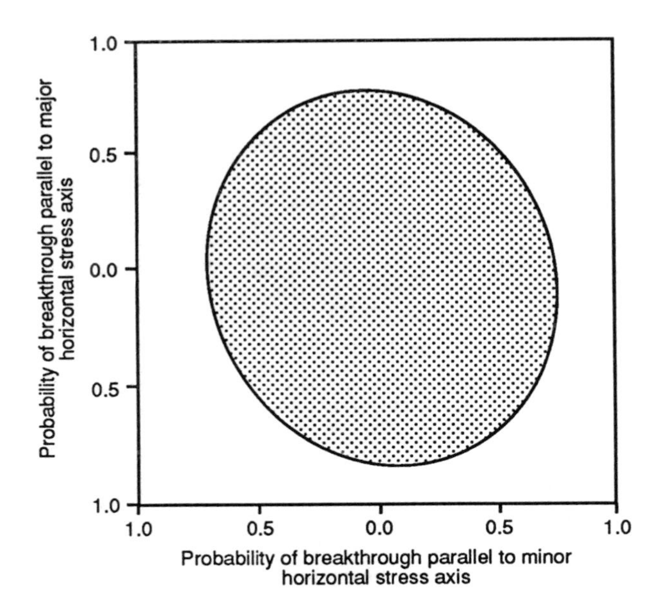

Fig 28. DEVIATION OF MAJOR FLOOD AXES FROM S_{Hmax} - NON GAS FLOODS

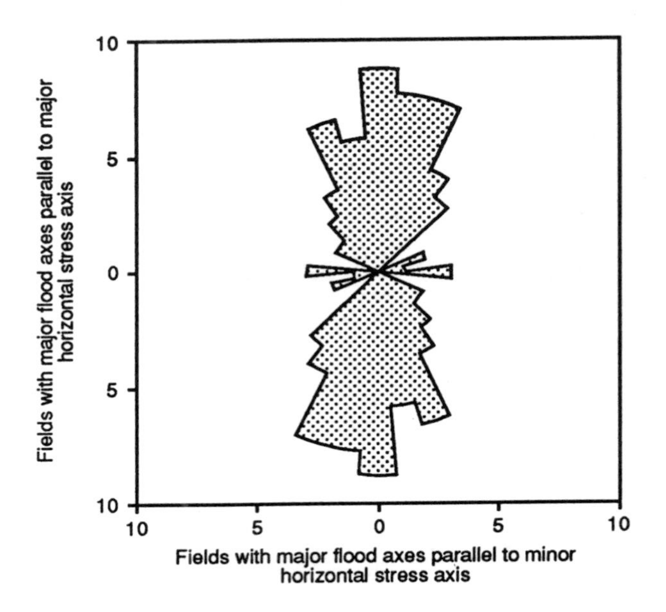

Fig 29. AGGREGATE BREAKTHROUGH PROBABILITY ELLIPSE - NON GAS FLOODS

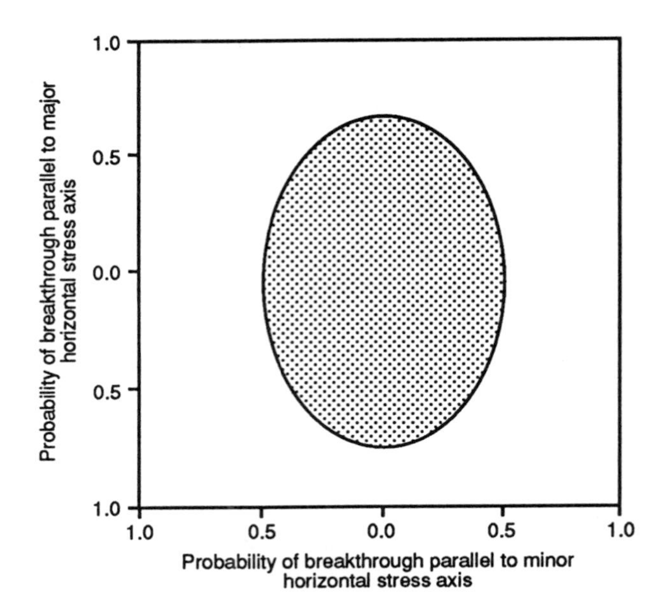

Fig 30. DEVIATION OF MAJOR FLOOD AXES FROM S_{Hmax} - GAS FLOODS

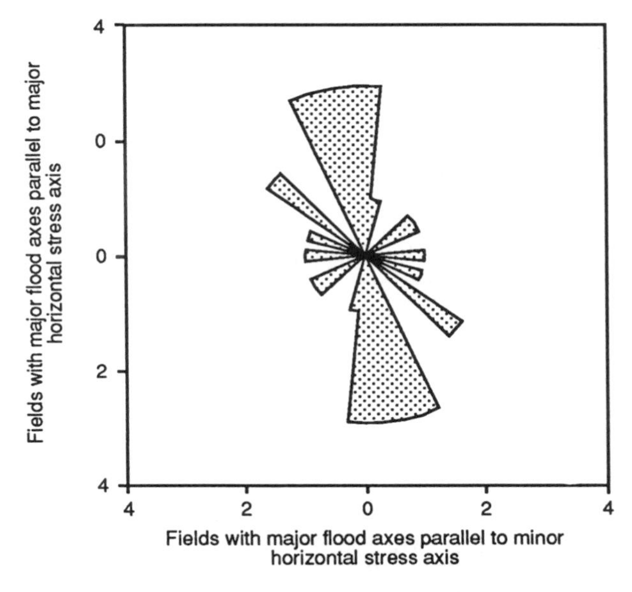

Fig 31. AGGREGATE BREAKTHROUGH PROBABILITY ELLIPSE - GAS FLOODS

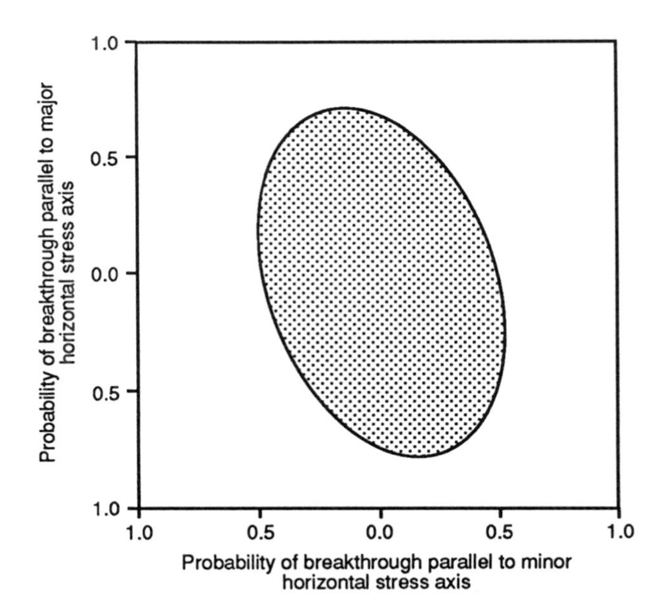

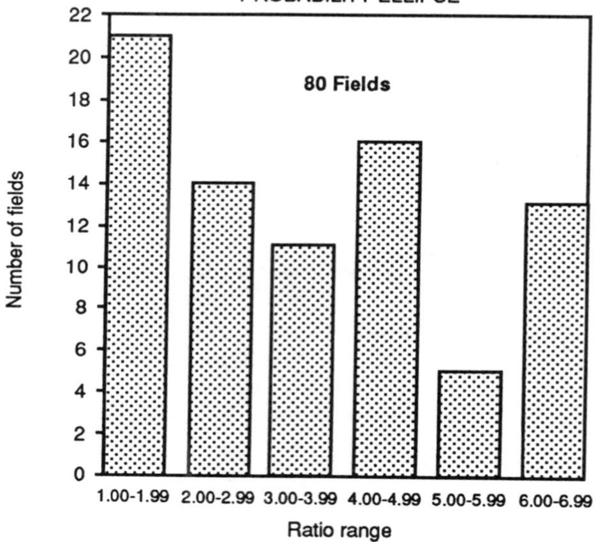
Fig 32. BREAKTHROUGH ANISOTROPY
RATIO OF MAJOR TO MINOR AXIS OF BREAKTHROUGH PROBABILITY ELLIPSE
80 Fields
Number of fields
0
2
4
6
8
10
12
14
16
18
20
22
1.00-1.99
2.00-2.99
3.00-3.99
4.00-4.99
5.00-5.99
6.00-6.99
Ratio range

RENORMALIZATION: A MULTILEVEL METHODOLOGY FOR UPSCALING.

Alistair Jones, Peter King, Colin McGill and John Williams

BP Research,
Sunbury Research Centre, Chertsey Road,
Sunbury-on-Thames,
Middlesex, TW16 7LN, U.K.

1. Introduction.

Reservoir descriptions are invariably generated on a fine scale, in part reflecting the scale of the input information. Reservoir performance simulators require input on a much coarser scale for manageable computation. The averaging process for changing scale — upscaling — is complicated, particularly for a property like permeability which is non-additive. The purpose of our poster is to describe the renormalization approach to upscaling and to give some impression of the wide range of problems that can be tackled using such an approach. In this note we outline the renormalization strategy and describe a simple implementation of the rescaling procedure. This is followed by some illustrative examples and a brief look towards the future.

Our principal aim in reservoir characterization is to determine the large-length-scale flow behaviour of the reservoir (i.e. at the scale of a simulator grid-block). For clarity, we focus on the determination of effective permeability. The main difficulty lies in the interdependent influences of permeability heterogeneities on many length scales. In essence the renormalization strategy tackles such problems via a multilevel or hierarchical approach. At each stage the current grid is divided up into cells, each consisting of a number of grid-blocks, and the effective permeability of each cell determined. These cells then become the grid-blocks of the next coarser grid. In effect, repeated application of the rescaling operation captures successively the effects of increasingly larger-scale heterogeneities on the large-scale flow behaviour of concern: as explicit reference to permeability heterogeneities on a given scale is eliminated by rescaling; their influence is carried forward implicitly in the coarse grid parameters (i.e. renormalized permeabilities).

The renormalization approach is remarkably accurate, efficient and robust — by robust we mean that it can be successfully applied to a wide variety of heterogeneous problems. Its power comes from choosing an appropriate rescaling prescription.

Renormalization: a Multilevel Methodology for Upscaling

2. Rescaling Prescriptions.

Whenever possible, it is advantageous to maintain a 1:2 ratio in length scales between successive grids: this entails using $2 \times 2 \times 2$ unit cells as shown in Figure (1). In order to calculate the cell effective permeability in a given direction we set constant pressures over the inlet and outlet cell boundaries and no flow through the other boundaries, and then replace the cell by a network of transmissibilities or equivalent resistors (using the analogy between Darcy's Law and Ohm's Law). Using simple resistor transformations we can find the equivalent network resistance and hence the effective permeability [1,2]. We can then 'rotate' the boundary conditions and repeat the procedure to obtain the cell effective permeabilities in the other directions. The main elements are outlined schematically in Figure (1) for both uniform and radial flow; standard transformations are used for the latter to map from a radial sector grid to a cartesian computational grid (non–uniform grid spacings are easily handled).

The choice of small cells renders the calculations extremely efficient with, under most circumstances, remarkably little loss in accuracy. In comparison with full numerical simulation 100 to 1000–fold speed–ups are not atypical, with errors usually less than 10 %.

3. Representative Examples.

In Figure (2) we indicate how renormalization can be used for upscaling. The problem is to reconcile core data with results from a well–test via a plausible reservoir description. The approach is to generate stochastic realizations of the large scale reservoir architecture (lithofacies) using SIS (Sequential Indicator Simulation) [3], taking facies abundance from well data and estimating dimensions/spatial correlations from analogue data. The permeability distribution within each lithofacies is simulated by using SGS (Sequential Gaussian Simulation) [4], using core data as input together with estimated spatial correlations. These permeability distributions are upscaled using renormalization to the scale of the grid–blocks used in the SIS simulation. This yields a full permeability grid at that scale, which is then renormalized to obtain the final effective permeability values for comparison with the well–test. The final agreement is good given the various assumptions made in producing the reservoir description.

In Figure (3) we give an example of an application to an idealised 2D fracture network. The model consists of a set of orthogonal fractures resolved onto a 200 by 200 grid. The fractures are assigned different conductivities in the horizontal and vertical directions and the macroscopic flow direction is assumed to be horizontal. The fracture network can be treated as an analogue resistor network and a large cell renormalization technique [2,5] used to determine the effective permeability. In 2D it is possible to obtain essentially exact results as confirmed by the agreement between the results obtained using either the full network or the backbone (those fractures carrying the flow). These results can be obtained at a fraction of the cost of a conventional numerical solution.

4. Concluding Remarks.

The renormalization technique provides a powerful approach to the general problem of upscaling. In many instances simple small cell schemes yield very accurate results at minimal computational expense. In those cases where the small cell schemes are less satisfactory (usually when there is significant macroscopic flow tortuosity) other options are available. One option is to use a large cell approach (as in the fracture example above); another is to improve the small cell approach by combining it with a multi–grid aproach [6]. The latter has so far received little attention in the literature.

We believe that the most promising path to the solution of many difficult upscaling problems lies in further development of renormalization and multi–level methods.

Acknowledgements.

The authors would like to thank the British Petroleum Company plc. for permission to publish this note.

References.

[1] King P R, Transport in Porous Media, **4**, p37 (1989)

[2] Williams J K, in 'Mathematics of Oil Recovery', Oxford UP (1991)

[3] Journel A G and Alabert F G, J.P.T. p212 (**Feb** 1990)

[4] Verly G, presented at EAPG Conf. (Florence) (1991)

[5] Frank D J and Lobb C J, Phys. Rev. B **37**, p302 (1988)

[6] Brandt A, in 'Multigrid Methods', Marcel Dekker (1988)

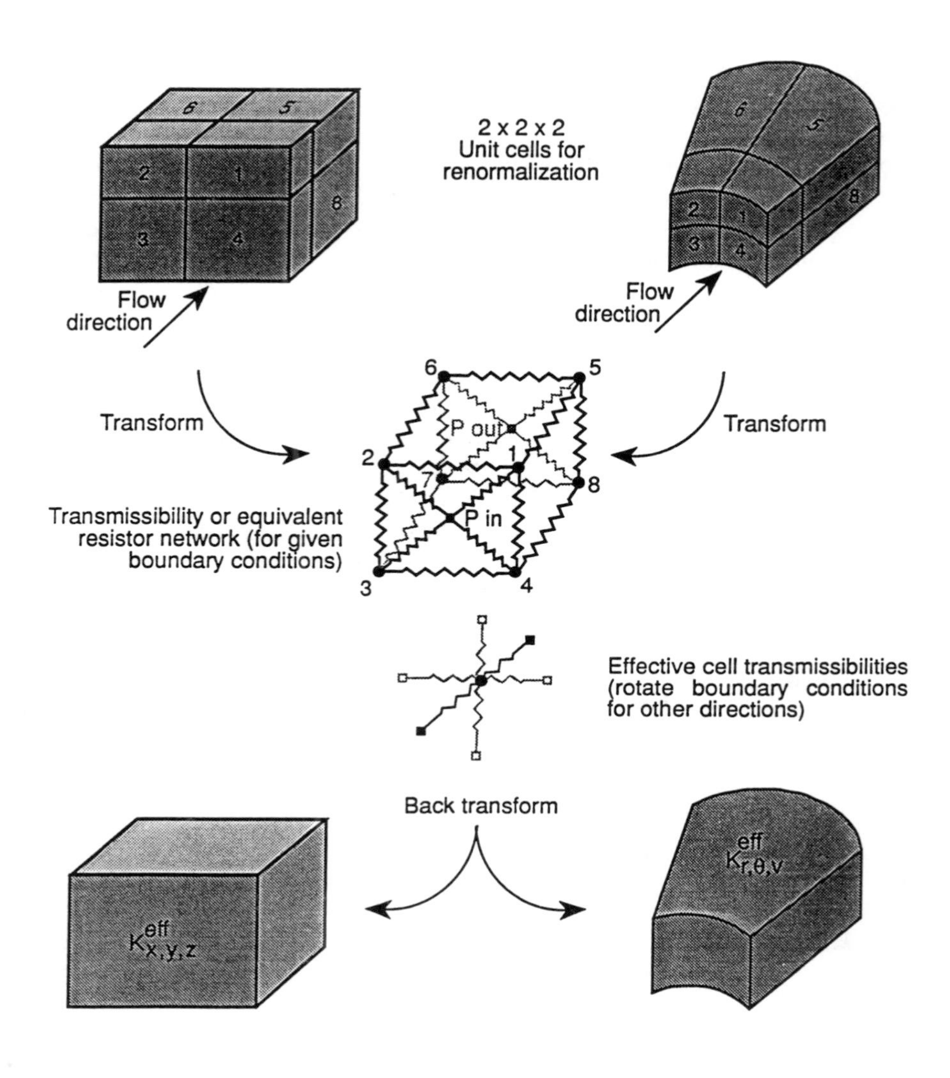

Fig 1. SMALL CELL RESCALING

RCS 32517

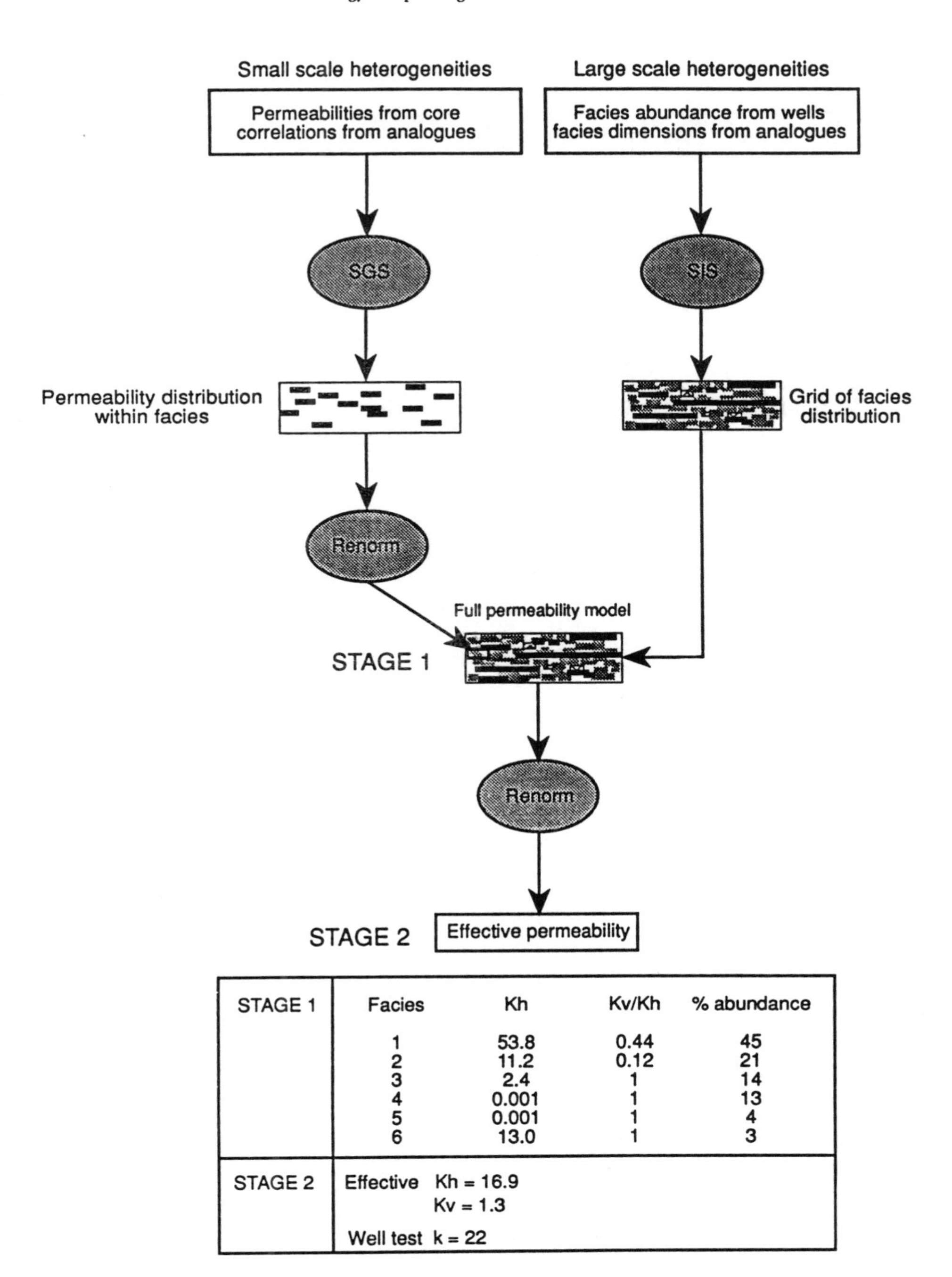

	Facies	Kh	Kv/Kh	% abundance
STAGE 1	1	53.8	0.44	45
	2	11.2	0.12	21
	3	2.4	1	14
	4	0.001	1	13
	5	0.001	1	4
	6	13.0	1	3
STAGE 2	Effective Kh = 16.9 Kv = 1.3 Well test k = 22			

Fig 2 EXAMPLE OF USE OF UPSCALING
RCS 32518

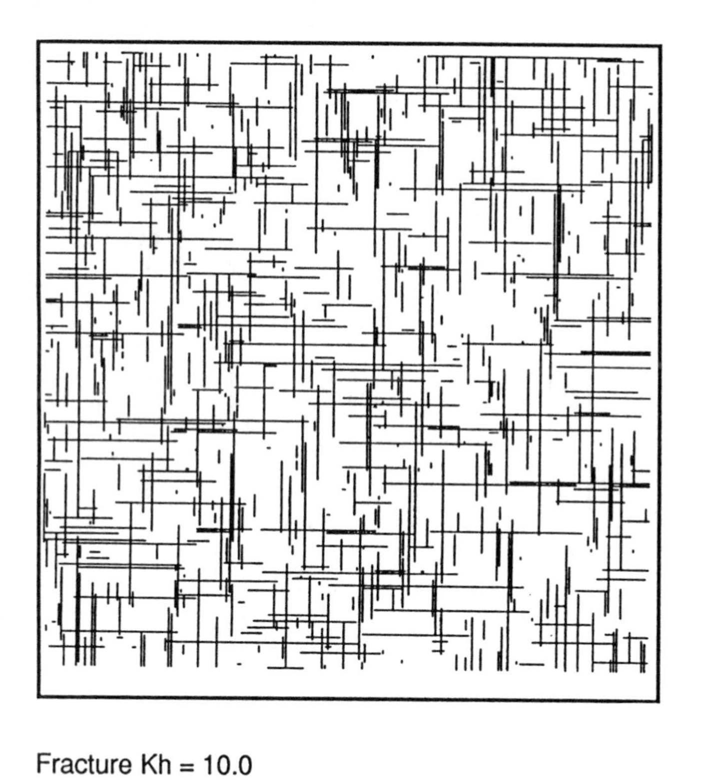

Fracture Kh = 10.0
Fracture Kv = 1.0
Effective Kh = 0.1005

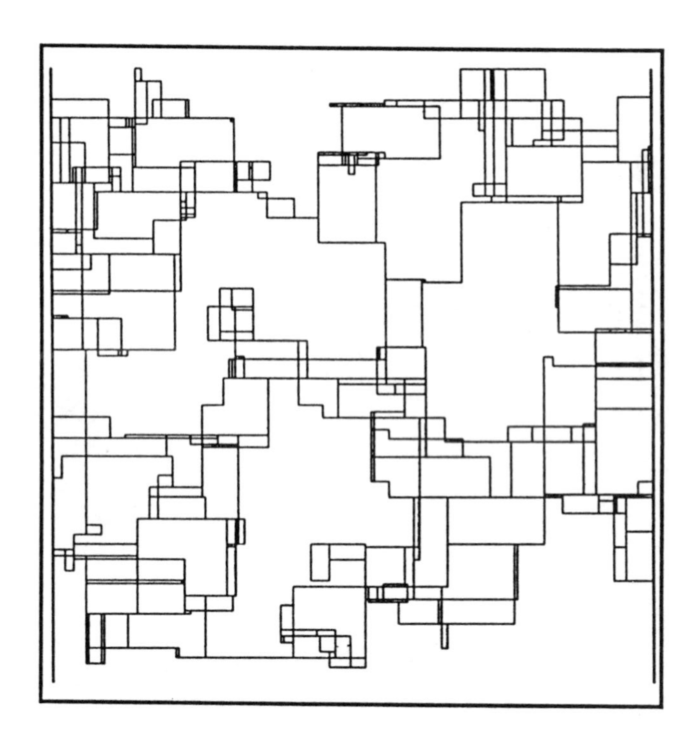

Permeable Backbone
Effective Kh = 0.1005

Estimating Effective Permeability: A Comparison of Techniques

Colin McGill, Peter King & John Williams

BP Research,
Sunbury Research Centre, Chertsey Road,
Sunbury-on-Thames, Middlesex, TW16 7LN, U.K.

1 Introduction

Effective permeability is an important control on the behaviour of reservoir models. With the advent of multi-million block stochastic models, calculating effective permeability assumes an even greater importance. A number of different techniques for evaluating this quantity have been developed by various workers. This paper presents a quantitative comparison of these methods. We begin by providing a brief description of the methods to be studied.

2 Simple Methods

It is useful to divide the methods for estimating effective permeability, k_{eff}, into two groups: those that consider only the *frequency* distribution of permeabilities, and those that also allow for the *spatial* distribution. The techniques described in this section belong to the first category.

2.1 Averages

It is common practice to use averages as estimates of k_{eff}. The most general form is the *power average* advocated by Deutsch (1986):

$$k_{\text{eff}} = \left\{ \int p(k) k^{\omega} \, dk \right\}^{1/\omega} \tag{2.1}$$

where $p(k)$ is the frequency distribution of permeabilities. In this formulation, the arithmetic, geometric and harmonic means may be represented by choosing ω to be 1, 0, and -1, respectively. The difficulty with using the power average is that the choice of ω depends upon the spatial distribution of permeabilities. Only the arithmetic, harmonic and geometric means will be used in this study. A full commentary on the power average can be found in McGill, Williams & King (1991)

2.2 Perturbation Theory

Perturbation theory assumes that the permeability is roughly constant with only small deviations from the mean. Standard theory performs a Taylor series expansion in the parameter $\delta k \equiv (k - \bar{k})/\bar{k}$ up to order δk^2, where $\bar{k}$ is the mean permeability. This yields

$$k_{\text{eff}} = \bar{k}(1 - \overline{\delta k^2}/D), \tag{2.2}$$

where D is the number of dimensions (2 or 3). This has been extended by King (1987) to include more terms in the Taylor expansion:

$$k_{\text{eff}} = k_g \exp\left(\sigma_g^2 \left[1/2 - 1/D\right]\right) \tag{2.3}$$

where k_g is the geometric mean of the permeabilities and σ_g^2 is the variance of the logarithm of the permeabilities. Recently, Noetinger (1990) has shown that King's results are exact for an uncorrelated medium with a log-normal distribution of permeabilities.

2.3 Effective Medium Theory

This is an example of a mean field theory. It assumes that any region of permeability behaves as if embedded within the average medium. This problem may be solved analytically. By demanding that the sum of the permeability regions reproduces the average medium a self-consistent solution may be found. This constraint produces the integral equation (Kirkpatrick 1973)

$$\int \frac{p(k)(k - k_{\text{eff}})}{(D-1)k_{\text{eff}} + k}\, dk = 0. \tag{2.4}$$

3 Better Estimates

The techniques discussed above have two main restrictions. They do not take the spatial distribution of permeabilities into account and they can deal only with scalar (isotropic) permeabilities. The techniques discussed in this section can address both these considerations, although the formulations considered here are restricted to permeability tensors which have diagonal components only. While this is a common assumption in reservoir engineering studies, some authors (e.g., White & Horne 1987) claim that the full tensor nature of the permeability must be taken into account.

3.1 Layer and Column Estimates

These techniques are quick and allow bounds to be placed on each component of effective permeability. Note that even if the permeability is locally isotropic, the spatial distribution of permeabilities will in general mean that k_{eff} will be anisotropic.

The column technique divides the region into one-dimensional columns in the direction of flow. By assuming that no flow is allowed between columns, an estimate of effective permeability is readily constructed. This is an underestimate of the permeability since inter-column flow is restricted. In contrast, the layer method breaks the region down into a series of layers perpendicular to the flow direction. The effective permeability of each layer is given by the arithmetic mean of the permeabilities in the layer. The layer permeabilities can then be combined by taking their harmonic mean. This produces an overestimate of effective permeability since inter-column flow is unimpeded.

3.2 Renormalisation

Renormalisation is a quick yet sophisticated method for calculating effective permeabilities (King 1989). The technique works by dividing the grid into $2 \times 2 \times 2$ sub-blocks. The effective permeability of each sub-block can be calculated analytically, yielding a new grid with 8 times fewer blocks. The process can be repeated again and again until only one block remains. This block has the effective permeability of the original grid. Renormalisation is discussed in detail by Jones, King, McGill & Williams elsewhere in this volume.

3.3 NUMERICAL SOLUTION

It is possible to obtain the 'exact' effective permeability by solving the flow equations (Darcy's law plus conservation of mass) numerically. This technique is much slower than the others discussed here but it does provide the yardstick by which others may be judged. Full details may be found in Begg & King (1985).

3.4 RESISTOR NETWORK METHODS

It is an interesting and useful observation that the discretised form of the flow equations can be cast in the form of a resistor network. In two dimensions, the resistor network can be manipulated using the star-triangle transformation. The result is a single resistor which can be interpreted as an effective permeability. This is an *exact* solution for effective permeability, in full agreement with the numerical answer. However, the technique is much faster than the conventional numerical solution. Full details can be found in Williams (1991).

4 Comparison of methods

The methods described above were used to evaluate the effective permeabilities for several permeability data sets, all generated using the Sequential Gaussian Simulation (SGS) technique (Verly 1991). SGS produces output with normal scores (normally distributed with a mean of zero and variance of one). This may be mapped into a log-normal distribution of permeabilities by using a transformation of the form:

$$k = k_g \exp(\sigma_g \phi), \tag{4.1}$$

where ϕ is the normal score value, k_g the desired geometric mean of the permeability distribution and σ_g is the desired variance in logarithm of the permeabilities.

For the first test, an uncorrelated field was generated and converted to permeabilities using σ_g values of 0.1, 0.5, 1.0, 2.0 and 3.0. k_g was taken to be 100 throughout. The results of applying the different methods are shown in Table 1.

Method	k_{eff} $\sigma_g = 0.1$	k_{eff} $\sigma_g = 0.5$	k_{eff} $\sigma_g = 1.0$	k_{eff} $\sigma_g = 2.0$	k_{eff} $\sigma_g = 3.0$	CPU s
Arithmetic Mean	100.52	113.39	164.68	727.45	7928.22	0.13
Geometric Mean	100.02	100.12	100.24	100.47	100.71	0.13
Harmonic Mean	99.52	88.25	60.19	12.77	1.10	0.13
Standard P.T.	100.02	97.47	27.05	—	—	0.17
Extended P.T	100.02	100.12	100.24	100.47	100.71	0.21
E.M.T.	100.02	100.15	100.41	100.95	101.31	1–4
Column Estimate	99.53	88.50	61.53	16.33	3.21	0.14
Layer Estimate	100.51	112.73	161.23	573.27	3234.43	0.16
Renormalisation	99.94	98.12	93.34	80.15	66.37	0.65
KEFF	99.84	96.21	87.96	71.40	56.68	50–180
Resistor Network	99.84	96.21	87.96	71.40	56.68	4.91

Table 1. The results of applying the methods for calculating effective permeabilities discussed above to datasets with increasing variability. It should be noted that for small σ_g virtually any method will give an acceptable answer. However, as σ_g increases, only the renormalisation method gives an acceptable answer. The last two methods are assumed to represent the correct answer. — indicates a negative value.

As can be seen from Table 1, for small σ_g virtually any method will give an acceptable answer. As the variability increases, however, renormalisation alone of all the approximate methods yields an acceptable answer. Note that renormalisation is about 200 times faster than KEFF for this grid of 250×40 points.

The second test was designed to show the influence of lateral correlations on the resulting effective permeabilities. Data sets with a vertical correlation length of 4 grid points and horizontal correlation lengths of 0, 5, 25, 125 and 625 grid points were generated. They were converted to permeabilities using $k_g = 100$ and $\sigma_g = 2$. The results are shown in Table 2. Note that for the simple methods only a value of the isotropic k_{eff} can be obtained whilst the other

Method	k_{eff} $l_c = 0$	k_{eff} $l_c = 5$	k_{eff} $l_c = 25$	k_{eff} $l_c = 125$	k_{eff} $l_c = 625$	CPU s
Arithmetic Mean	724.47	732.93	654.04	916.61	438.12	0.10
Geometric Mean	100.47	97.81	109.45	143.50	97.49	0.16
Harmonic Mean	12.77	12.22	14.88	25.91	13.05	0.13
Standard P.T.	—	—	—	—	—	0.17
Extended P.T	100.47	97.81	109.45	143.50	97.49	0.21
E.M.T.	100.95	98.68	111.06	142.69	105.02	2

Method	k_h $l_c = 0$	k_h $l_c = 5$	k_h $l_c = 25$	k_h $l_c = 125$	k_h $l_c = 625$	CPU s
Column Estimate	14.23	14.75	24.20	66.73	213.70	0.14
Layer Estimate	504.60	415.42	386.05	749.60	414.28	0.16
Renormalisation	78.95	87.15	193.36	473.87	323.05	0.65
Resistor Network	69.40	86.52	177.10	416.45	321.97	4.91
KEFF	69.40	86.51	177.07	416.40	322.00	~ 100

Method	k_v $l_c = 0$	k_v $l_c = 5$	k_v $l_c = 25$	k_v $l_c = 125$	k_v $l_c = 625$	CPU s
Column Estimate	19.18	22.77	29.42	35.03	15.15	0.14
Layer Estimate	652.77	607.29	450.01	269.33	34.29	0.16
Renormalisation	81.17	80.81	50.78	48.11	15.55	0.65
Resistor Network	72.34	80.98	57.82	55.53	15.69	4.91
KEFF	72.34	80.98	57.82	55.54	15.70	~ 100

Table 2. The results of applying the different methods for obtaining k_{eff} to datasets with increasing correlation lengths. Note again that only renormalisation produces results in accord with the exact results.

methods can provide values for both k_h and k_v. Since the correlation lengths are anisotropic it is expected that these should now be different.

Before comparing the methods, it is instructive to note that the bounds on permeability given by the column and layer estimates converge to the correct solution as the correlation length increases. This is because for infinite correlation length the system is layered; in this limit the column and layer estimates become *exact*. Table 2 also indicates that correlation has little effect when the correlation lengths are much smaller than the size of the system. In this limit, the system behaves as though uncorrelated. Thus, the more approximate estimates can work well in the limits of small or infinite correlation. However, in the intermediate regime, more sophisticated methods are needed.

Again the figures indicate that renormalisation gives quite impressive results. Typically, renormalisation gives results which are correct to within 10–20% whilst taking only five times as much CPU as a straight average. The time taken is at least two orders of magnitude less than the numerical solution. Although we have not performed experiments on larger systems here, this impressive speed ratio can be expected to improve with system size.

The impressive performance of renormalisation has also been noted by other authors (e.g.,Grindheim & Aasen 1991). A more complete comparison can be found in McGill, Williams & King (1991).

Acknowledgements

The authors would like to thank the British Petroleum Company plc for permission to publish this paper.

References

Begg, S.H. & King, P.R. (1985) *SPE* 13529

Deutsch, C. (1986) *SPE* 15991

Grindheim, A.O. & Aasen, J.O. (1991) Presented at *Lerkendal Petroleum Engineering Workshop,* Trondheim.

King, P.R. (1987) *J.Phys.A*, **20**, 3935.

King, P.R. (1989) *Transport in Porous Media*, **4**, 37.

Kirkpatrick, S. (1973) *Rev. Mod. Phys.*, **45**, 574.

Jones, A.D.W., McGill, C.A., King, P.R. & Williams, J.K. (1991) This volume.

McGill, C.A., Williams. J. K. & King, P.R. (1991) In preparation.

Noetinger, B. (1990) Submitted to *Transport in Porous Media.*

Verly, G.W. (1991) Presented at *EAPG Conference,* Florence.

White, C.D & Horne. R.N. (1987) *SPE* 16011

Williams, J.K. (1991) in *Mathematics of Oil Recovery,*Oxford University Press

DEFINING WATER-DRIVE GAS RESERVOIRS: PROBLEMS IN DATA ACQUISITION AND STRATEGIES IN EVALUATION.

William M. Colleary
John R. Hulme
Saad M. Al-Haddad

Colorado School of Mines
Golden, Colorado

Gas Research Institute
Chicago, Illinois

This paper presents the steps which must be taken, and discusses the data necessary, to identify water-drive gas reservoirs, and addresses the importance of making this determination. This work presents the interim results of a research project presently underway at the Colorado School of Mines, and funded through the Gas Research Institute (GRI).

A comprehensive search for water-drive reservoirs was conducted throughout the continental United States, with emphasis on the Rocky Mountain and Mid-Continent regions. This search included a review of geologic and engineering literature, contacts with operating companies, and search through the Petroleum Information (TM) Production and Historical Databases.

Early identification of water-drive commences with detailed geologic appraisal of a reservoir, and provides the data necessary and time scale to develop and evaluate alternate programs for maximizing recovery of reserves through field development, completion and operations, and management of produced water. One traditional production strategy calls for high production rates to "out-run" encroaching water, often leaves a significant amount of unrecoverable gas within the reservoirs. Alternative strategies are proposed.

The identification of pressure support within a reservoir is the most reliable engineering method to determine the presence of water-drive. To verify this, a complete pressure history, along with gas composition, is needed to construct P/Z plots. An example of a classic strong water-drive is shown in Figure 1, from a Delaware Basin Field in West Texas. Pressure data should be collected for each individual well, since the water-drive often only appears in certain wells in the field. As an example, Figures 2 and 3 were taken from wells less than one mile apart in another West Texas Field. Figure 2 shows a volumetric depletion, whereas Figure 3 shows water-drive pressure support.

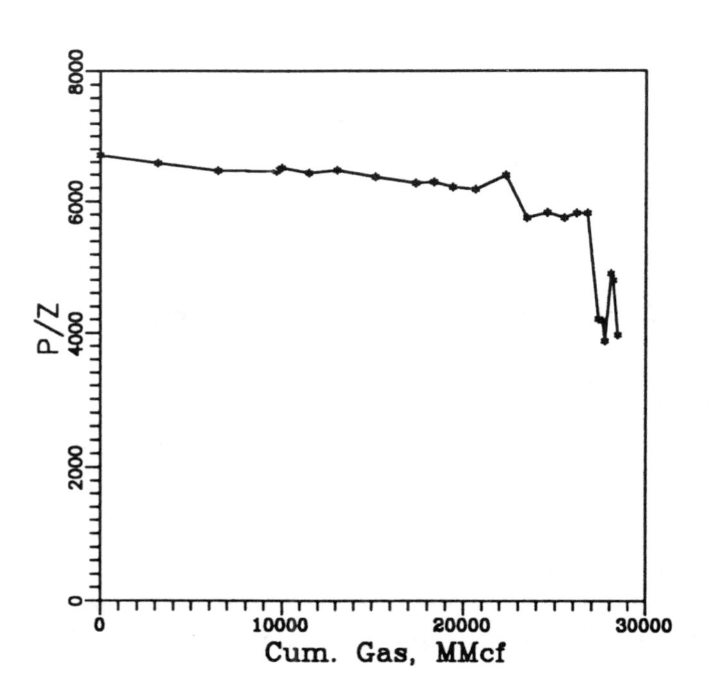

Figure 1. P/Z plot of deep Delaware Basin Reservoir illustrating strong water-drive support.

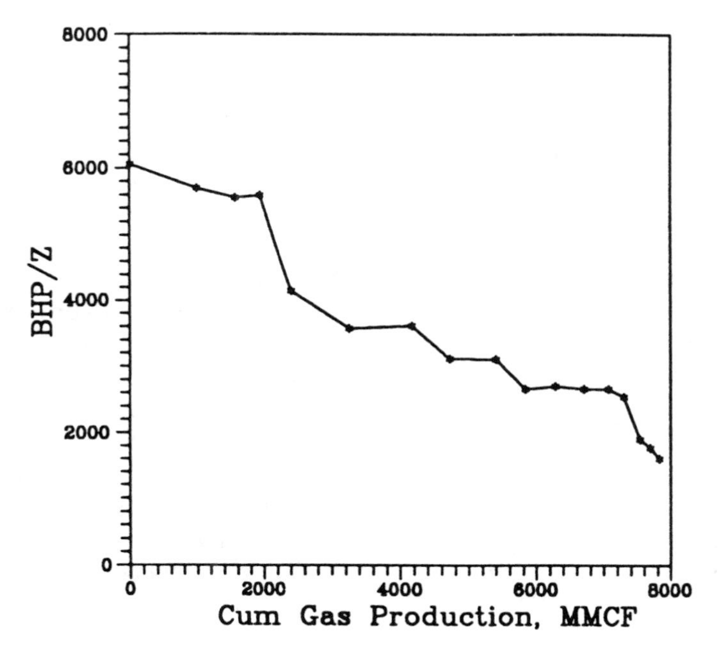

Figure 2. P/Z plot of West Texas Well showing volumetric depletion.

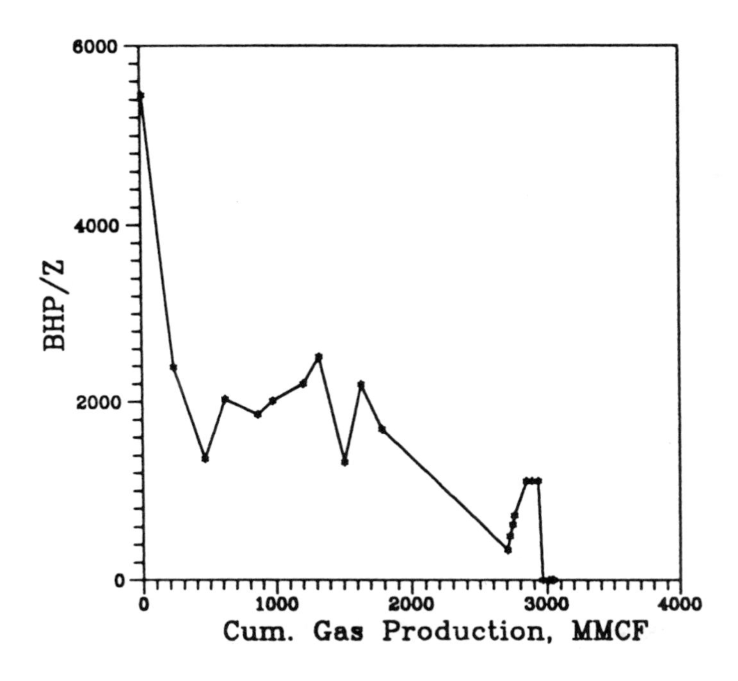

Figure 3. P/Z plot of West Texas Well showing pressure support.

Comprehensive production histories are also important in investigating future reservoir management strategies through reservoir simulation. These data should include water, gas, and condensate production, and water injection volumes, if any. Complete workover and completion records, with any production logging surveys are also available.

Important clues to the internal architecture of the reservoir can only be obtained through the analysis of well-by-well production histories. The existence of permeability barriers, such as small-scale sealing faults, or facies changes, which affect reservoir behavior, can be identified, and the geologic interpretation may be confirmed. Pressure data should also be taken throughout the life of the well, since a straight line depletion does not assure a closed system. "Many curves show a rapid decline in the early stages of production after which they flatten out." (Bruns, Fetkovich, Meitzen, SPE #898, March 1965). An example of this is shown in Figure 4, from a Gulf Coast Region reservoir in Texas.

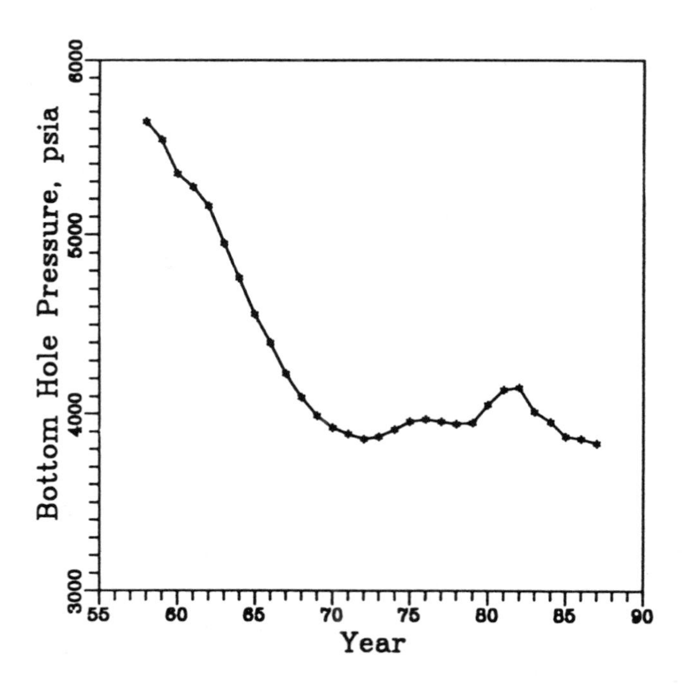

Figure 4. Pressure versus time plot of Texas Gulf Coast Field.

The primary problem encountered in any analysis is the incomplete or total lack of pressure data. In various regions, reporting of reservoir pressures is not required by regulatory agencies, and many operators do not recognize the economic value of such data. In such cases, the possibility of water-drive would not be considered until wells show an increase in water production. This would occur, generally, at a later stage in the life of the reservoir, when it is too late to carry out the most effective management strategies. As water production begins, it may rise sharply, causing problems if there has been no preparation for water disposal. An example of such a sharp increase is shown in Figure 5, from a field in the San Juan Basin of New Mexico.

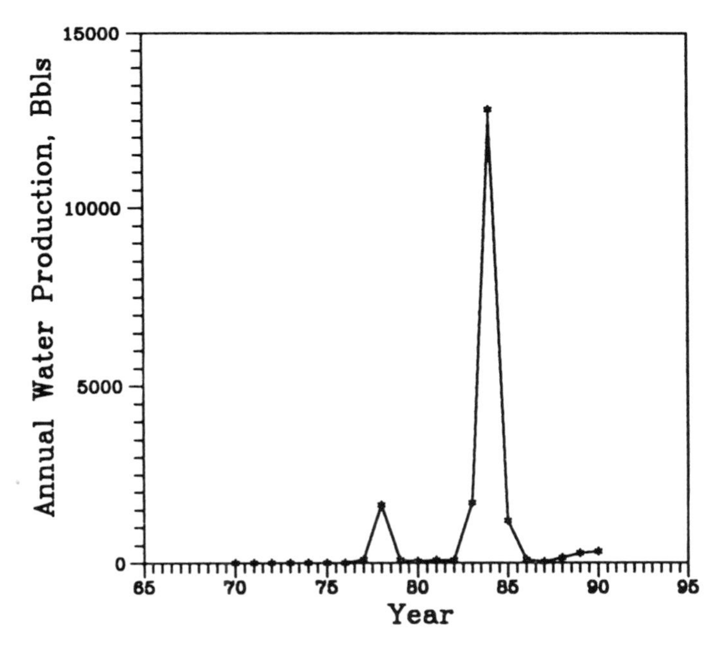

Figure 5. Plot of water production of a San Juan Basin Field, NM.

In conclusion,the early recognition of water-drive in gas reservoirs is critical to maximizing recovery of natural gas reserves from the reservoir. This recognition can be achieved when the detailed reservoir geologic description and classic reservoir engineering methods are integrated to provide the data necessary for the development of an effective management strategy. Detailed pressure data, fluids production histories, compositional analyses, open hole and production logging, and completion data are essential for such early recognition.

SCALE-UP OF TENSORIAL EFFECTIVE PERMEABILITY TO DETECT ANISOTROPY THROUGH MFP MEASUREMENTS

Yngve Aasum
Ekrem Kasap
Mohan Kelkar

The University of Tulsa

I. INTRODUCTION

In optimizing recovery and production of hydrocarbons, numerical simulations are more cost effective than lab experiments for field scale studies. Field development and reservoir management practices require large scale numerical simulations. Although the development of more powerful computers and simulation techniques allows us to use very fine grids and include rather detailed features, the size of a simulation grid-block in a field scale numerical simulation is on the order of several hundreds of feet, which is much larger than the scale of inter-well scale laminations and cross-bedding structures. For these simulation blocks, the heterogeneities can only be accounted for by calculating an effective permeability. An *effective* permeability is the one that preserves the fluid flux-potential drop quotient between a heterogeneous block and an equivalent (same size, geometry and fluid viscosity) homogeneous block.[1] An analytical method to calculate the effective permeability is required, because of its generality and ease in inserting into numerical simulators.

Simulation results indicate that not only the magnitudes of permeability heterogeneities but also their locations and positions will alter the calculated effective permeability. The positions and locations will cause fluid flow direction to diverge from the direction of applied pressure gradient. This phenomenon should be included in effective permeability calculations by calculating the effective permeability in tensorial form.

In this study we developed an analytic method to calculate effective block permeability values as a full tensor based on geometry, block size, local permeability values and the location of heterogeneities within the simulation block. This method can be applied to Mini-Field-Permeameter (MFP) measurements to detect the anisotropy in outcrop formations, thereby improving their descriptions.

II. DISCUSSION

Mini-Field-Permeameters (MFP) have been intensively used to characterize inter-well scale heterogeneities of analog outcrop formations. Permeability anisotropy, however, cannot be detected with MFP, because a MFP often measures the permeability on a scale that is much smaller than the scales of lamination and cross-bedding which are considered to be prime cause for anisotropy. Therefore, a method to detect the cross-bedding and the permeability anisotropy from MFP measurements is needed for better description of outcrop analogs.

Non-uniform distributions of heterogeneities cause a transverse pressure gradient that results in a cross-flow or redirection of flow within the block (Fig. 1). The numerical simulation of flow in such a block also indicates flow redirectioning, although the applied pressure gradient is unidirectional (Fig. 2). This flow redirectioning can be accounted for by calculating the effective permeability as a tensor, because non-zero off-diagonal elements of the permeability tensor show similar effects on velocity field (mean flow direction is different from the applied pressure gradient).

An analytical effective permeability calculation method has been developed to account for the cross-flow caused by the transverse pressure gradient. Two permeability values, one obtained by assuming no communication in the transverse direction (normal to the applied pressure gradient) and the other one by assuming perfect vertical communication in the same direction, are calculated.[1] The transverse pressure gradient is calculated based on an anisotropy-weighted average of those two permeability values and off-diagonal contributions to the effective permeability tensor are based on those transverse pressure gradients. Because of the lengthy derivations of the analytical equations, we refer the reader to the research reports of TUPREP at the University of Tulsa[2] for further details.

A series of numerical simulations was also conducted for various cases of local permeability values. The effective permeability tensors calculated with the analytical method and from the numerical simulations are compared in Figs. 3-6. Also shown in Figs. 3-6 are the local permeability values used in the calculations. The comparison indicates an excellent agreement between the numerical simulation results and the analytical method calculations.

The analytical method has been applied to MFP measurements from San Andres strata of the Algerita outcrop to detect permeability anisotropy and existing cross-bedding. The analytical method employs the measured point permeability values and their locations to calculate the effective permeability tensor for consecutively expanding scales. Figure 7 shows a grainstone permeability field obtained by 16x16 ft measurements.[3] Figure 8 shows the calculated permeability values during scale-up process. The final permeability tensor indicates that the principal directions of permeability are very close to the horizontal and the vertical directions and there exists an anisotropy ratio of approximately 4 between horizontal and vertical permeability values.

The method has been applied to a generic cross-bedded permeable medium, shown in Fig. 9, in order to test the method's ability to detect the cross-bedding. Only two permeability values are used in the generic cross-bedded flow field, dark cells =20 md and light cells=1000 md. The calculated effective permeability tensor is:

$$k = \begin{bmatrix} 304 & -209 \\ -209 & 304 \end{bmatrix}$$

which compares well with the expected effective permeability tensor obtained by calculating the principal direction permeability values[4] in the parallel and perpendicular directions to the cross-bedding and then applying a coordinate rotation procedure:

$$k = \begin{bmatrix} 275 & -235 \\ -235 & 275 \end{bmatrix}$$

The analytical method calculates the cross-bedding angle correctly (-45°), because isotropic diagonal and negative off-diagonal tensorial elements for a cross-bedded medium will only come from -45° cross-bedding angle. The method calculates permeability values in the principal directions as

$$k = \begin{bmatrix} k_p=513 & 0 \\ 0 & k_s=95 \end{bmatrix}$$

which compare reasonably well with the results of simple calculations of parallel and serial beds model as

$$k = \begin{bmatrix} k_p=510 & 0 \\ 0 & k_s=39 \end{bmatrix}$$

where k_s and k_p refer to permeability values in the serial and parallel directions of cross-bedding and all permeability values are in md.

III. CONCLUSIONS

1- By using transverse pressure gradient, an analytic method has been developed to account for the effects of asymmetrically located heterogeneities.

2- The effective permeability values from the analytic method and detailed numerical simulations are in excellent agreement.

3- By using scale-up procedures, the analytic method is able to capture the anisotropy in San Andres grainstone strata. For the scale studied, the anisotropy ratio was approximately 4.

4- Additional work is needed for cross-bedded media since the preliminary work reported here is promising. Although the permeability in serial principal direction was off, the parallel direction permeability was almost exact and the cross-bedding angle was captured correctly (-45°).

VI. REFERENCES

1. Kasap E. and Lake W. L.: "Calculating the Effective Permeability Tensor of a Gridblock," SPE Formation Evaluation, June 1990, 192-200.

2. Aasum, Y., Kasap, E., Kelkar, M.: "Effective Properties of Reservoir Simulator Grid Blocks," Status Report TUPREP, Research Report 5, May 14, 1991.

3. Lucia, F. J., Senger, R. K., Fogg, G. E., Kerans, C., Kasap, E.: "Scales of Heterogeneity and Fluid Flow Response in Carbonate Ramp Reservoirs: San Andres Outcrop, Algerita Escarpment, New Mexico," paper SPE 22744 presented in the 1991 SPE Annual Technical Conference and Exhibition, October 6-9, Dallas, Texas.

4. Kasap E. and Lake W. L.: "Dynamic Effective Relative Permeabilities for Cross-Bedded Flow Units," paper SPE 20179, Proceedings of SPE/DOE Seventh Symposium on Enhanced Oil Recovery, Tulsa, OK, April 22-25, 1990.

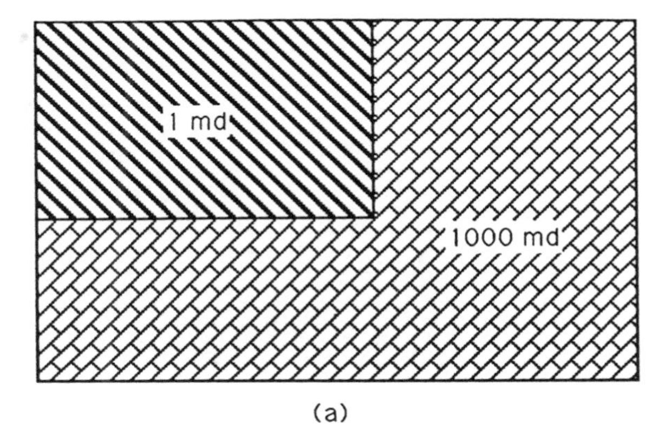

(a)

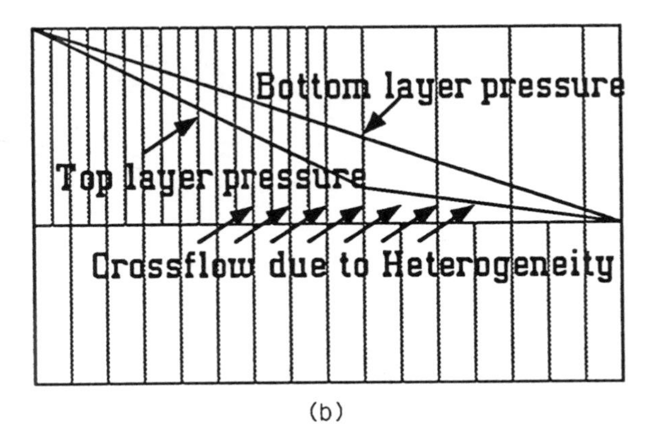

(b)

Figure 1. a) A Simulation Grid-Block with Megascopic Heterogeneities b) Schematic Pressure Distribution In the Block.

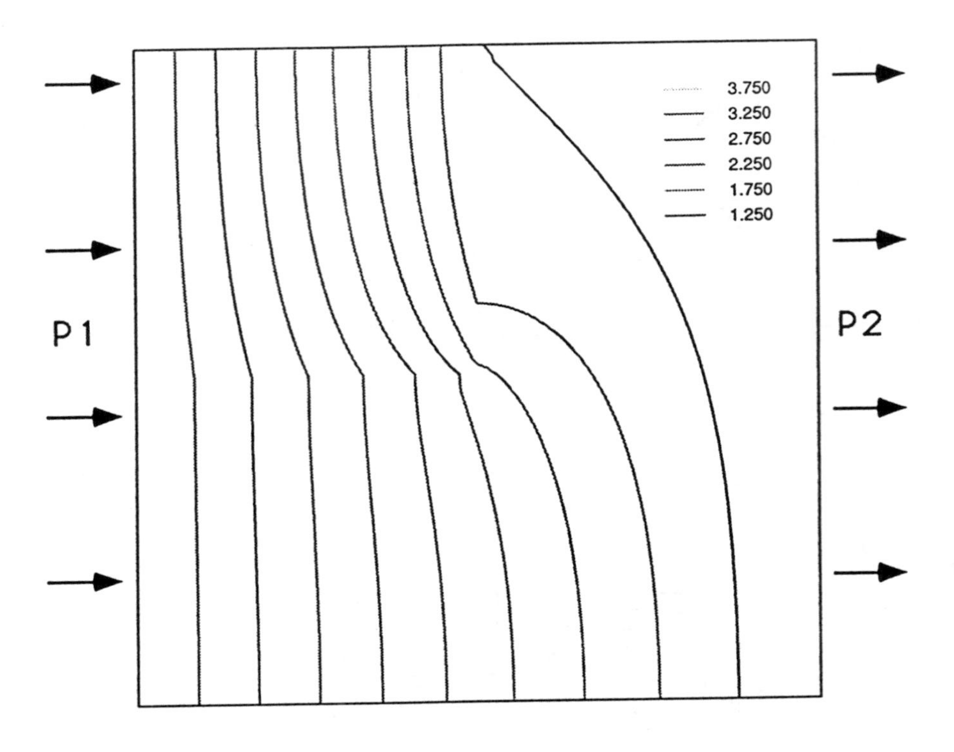

Figure 2. Pressure Contours from the Numerical Simulation of Flow in the Permeability Field Shown in Fig. 1-a. Pressures Are in atm. and Horizontal Boundaries are Sealed, P1>P2

1	0.	1000	0.
0.	1	0.	1000
1000	0.	1000	0.
0.	1000	0.	1000

(a)

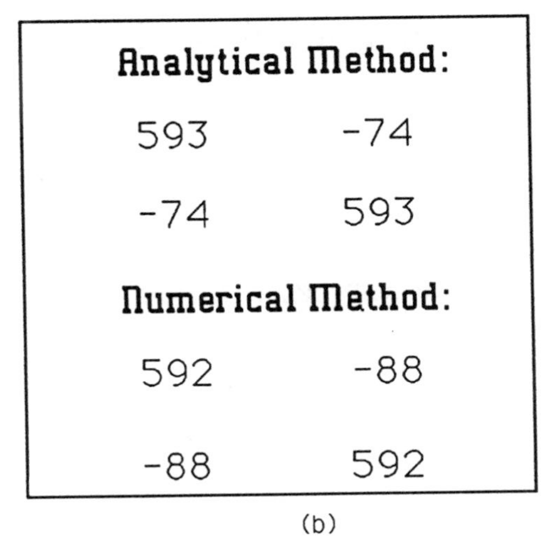

(b)

Figure 3. Comparison of Effective Permeability Tensors Calculated by Analytical and Numerical Methods. a) Isotropic Local Permeabilities (md) b) Calculated Effective Permeability Tensors (md)

1	0.	1000	0.
0.	.1	0.	100
1000	0.	1000	0.
0.	100	0.	100

(a)

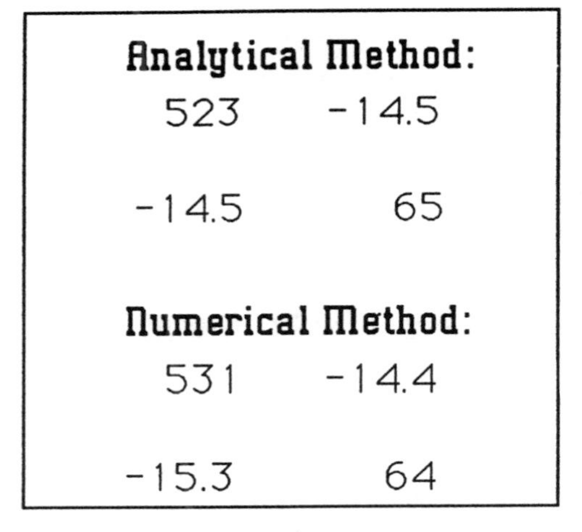

(b)

Figure 4. Comparison of Effective Permeability Tensors Calculated by Analytical and Numerical Methods. a) Anisotropic Local Permeabilities (md) b) Calculated Effective Permeability Tensors (md)

1	.1	1000	100
.1	1	100	1000
1000	100	1000	100
100	1000	100	1000

(a)

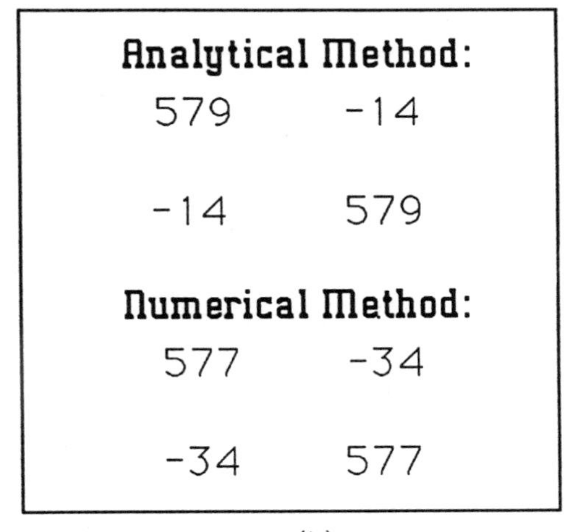

(b)

Figure 5. Comparison of Effective Permeability Tensors Calculated by Analytical and Numerical Methods. a) Isotropic Local Tensorial Permeabilities (md) b) Calculated Effective Permeability Tensors (md)

1	.1	1000	100
.1	.1	100	100
1000	100	1000	100
100	100	100	100

(a)

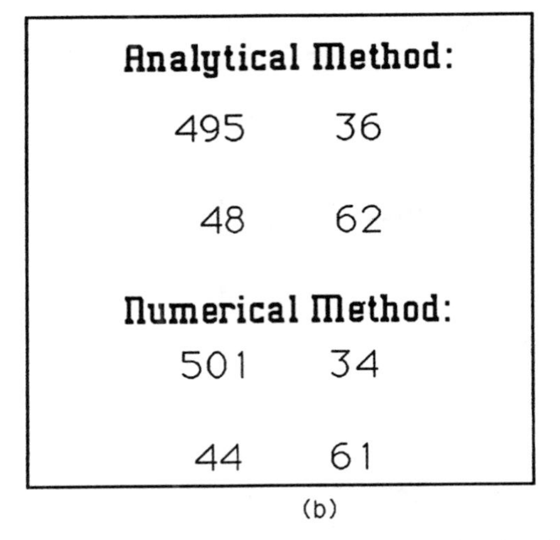

(b)

Figure 6. Comparison of Effective Permeability Tensors Calculated by Analytical and Numerical Methods. a) Anisotropic Local Tensorial Permeabilities (md) b) Calculated Effective Permeability Tensors (md)

Figure 7. Grainstone Permeability Field, 16X16 MFP Measurements on San Andres Carbonate Outcrop, Algerita, New Mexico.

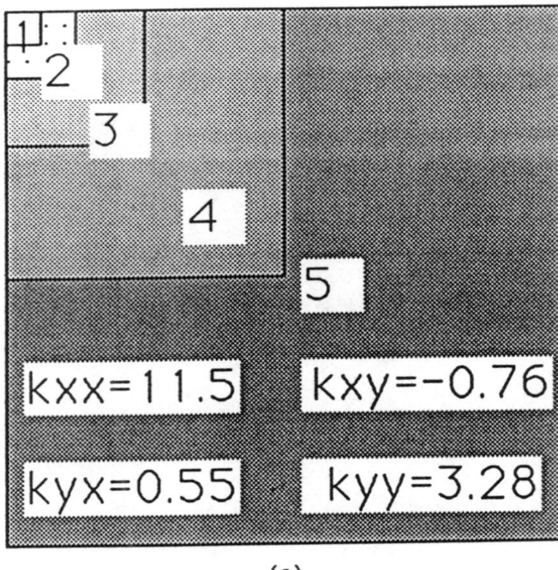

(a)

14.7 0. (1) 0. 14.7	19.73 0.61 (2) 0.56. 19.16
19.1 -1.17 (3) -0.11. 13.45	16.64 -1.04 (4) 0.16. 9.71

(b)

Figure 8. Scaling-Up of a Grainstone Permeability Field (16x16) Shown in Fig. 7 from San Andres Carbonate Outcrop, Algerita, New Mexico a) Five Scales and the Final Permeability Tensor. b) Permeability Tensors at the Upper Left Corner for Other Scales.

Figure 9. Generic Cross-Bedded Permeable Medium 64x64. Only Two Permeability Types, Dark Cells= 20 md, Light= 1000 md.

RESERVOIR SIMULATION OF FLUVIAL RESERVOIRS: TWO PROGNOSTIC MODELS

Kelly Tyler
Adolfo Henriquez

Statoil
Stavanger, Norway

I. INTRODUCTION

Fluvial reservoirs are specially difficult to model in an offshore situation and in a predevelopment phase, as the sandbodies are rarely correlatable when the distance between wells may be kilometers. The main challenge has been to simulate hydrocarbon bearing isolated channels imbedded in an impermeable matrix which honors (i) the known data at the wells, (ii) the presence or absence of channels, their direction and thickness, (iii) outcrop data (thickness to width ratio) and (iv) seismic data (top and bottom of reservoir or other intrareservoir reflectors).

The sophisticated programs to model these types of reservoirs represent a large step forward in modelling medium-scale heterogeneities and away from the layered or zoned concepts. However, there exist several drawbacks (1):

1. The input data is difficult to obtain
   a. The probability distribution of channel orientation is generally deduced from dipmeter data. These measurements are local, and as the channels meander or are 'braided', the main direction is lost.
   b. The thickness from log data is not unambiguous, as the channels are stacked or eroded.
   c. The width to thickness ratio, generally from analog outcrop data or other more densely drilled fields, can vary for different interpreters of the data. In addition, studies in the literature do not (or cannot) record parameters like compaction, avulsion or subsidence rates, which are essential to the stacking or spreading of the channels (5).
2. The geological model loses many of its details when being transferred to the coarser grids of the reservoir simulation model.
3. The geological simulation is generally computer-intensive.

4. It is difficult to integrate sedimentological understanding ('geological intelligence') developed by the geoscientists through their experience.

In spite of all of these challenges, which are being tackled, the status of modelling fluvial reservoirs is much more advanced now than just a few years ago. There is, however, a need for simple models, which capture some of the important features of fluvial reservoirs and which can be simulated quickly. Two such models are presented here. The net/gross ratio (NG) (percent sand) is one of the few geological parameters which can be trusted when few wells are drilled. Therefore, these models rely mainly on this parameter. The stochastic simulation model described in Augedal et al. (2) and Clementsen et al. (3) has been used to verify the two proposed models.

II. HOMOGENEOUS MODEL

A homogeneous model, with all cells having an average net/gross ratio and transmissibility can be used to approximate the upper limits for recovery and water breakthrough. The transmissibility multipliers (TM) should be a function of the area of the common face of the two blocks occupied by sand. This common area depends on the ratio of sand and shale in each block, and it can be shown that the average is the average net/gross ratio (NG) for the region. This assumption has been tested with the stochastic simulation model, using several net/gross ratios, channel widths, and varying grid size, and has been verified within 2%.

In testing the homogeneous model with a standard reservoir simulator, TM in the vertical direction was the average generated by the stochastic model. The results are shown in Table I.

The following arguments suggest methods to enhance the homogeneous model. Simple arguments show that the effect of channel orientation can be implemented in the approximation of TM in the following manner.

$$TMX(\Theta) = NG \times (\alpha - (1 - \alpha) \sin \Theta) \quad (1)$$

In the y-direction, $\sin \Theta$ is replaced with $\cos \Theta$, where Θ is the angle between the channel and the x-axis. α is a measure of the channel connectivity perpendicular to the channel direction and is of the type as the 'interconnectedness ratio' and 'average fractional contact - K' described in references 4 & 5.

In a geometric model with regular packing of channels (4), K may be approximated by:

$$K = 0 \text{ for } NG < 0.5 \quad \text{AND} \quad K = 2 \times NG - 1 \text{ for } N/G > 0.5 \quad (2)$$

In reservoirs generated with the stochastic model, K is approximately equal to NG. A realistic value of connectivity may then be K(regular packing) $< \alpha <$ K(stochastic model) . Connectivity in a fluvial reservoir generated by Bridge and Leeder (4) is included in figure 1.

Figure 2 shows TM in the x-direction from equation 1 and from the stochastic model for varying NG and channel orientation.

This simplified model is suited for sensitivity studies, such as testing fluid and relative permeability data, but does not reflect the geological detail of the reservoir. A heterogeneous model is proposed for this purpose.

III. HETEROGENEOUS MODEL

This model distributes NG for the individual grid blocks and then TM can be calculated as a function of NG.

It is expected that a fine grid system with NB number of grid blocks, $NG \times NB$ blocks will be filled with sand, and $(1 - NG) \times NB$ blocks will be full of shale (no sand) where NG is the average net/gross of the reservoir. The slope of such a frequency distribution of sand content in the grid blocks is $m = 2 \times NG \times NB - NB$.

With a coarse grid, the distribution is flat with a slope of 0; all grid blocks have the same NG. With a realistic grid system, the slope of the distribution lies between these extremes.

$$m = NG \times NB - 0.5 \times NB \tag{3}$$

Several reservoirs were simulated using the stochastic reservoir simulator with varying NG. The frequency of NG in the the grid blocks was fitted using a straight line from linear regression with slope $m = 0.93 \times NG \times NB - 0.46 \times NB$ (fig. 3). Thus the approximation made with equation 3 is suitable.

The y-axis intercept, y_o, is found by considering that the integral of the frequency distribution is NB. A simple calculation shows that the intercept is

$$y_0 = -0.5 \times NB \times NG + 0.27 \times NB \tag{4}$$

The simulations used before are fitted with the equation $y_0 = -0.43 \times NB \times NG + 0.30 \times NB$ (fig. 3).

$$TMX(i,i+1) = 2 \times NG_i \times NG_{i+1} / NG_i + NG_{i+1} \tag{5}$$

Here, i is the grid block number.

Equation 5 is used to calculate TMX and similarly TMY and TMZ. This equation was verified with a fit for the transmissiblity multipliers. The regression coefficients are 0.95, 0.95 and 0.64 for the x-, y-, and z-directions respectively (fig. 4). The fit is not as good in the z-direction due to the excessive amount of sand bodies generated in the stochastic model (1). A program using the package SAS was developed to generate grid block properties using equation 5.

The reservoirs were simulated with a black oil simulator and compared with reservoir realizations generated with the stochastic simulator. The results are shown in Table I and show good agreement.

IV. RESERVOIR SIMULATION RESULTS AND VALIDATION

Both the homogeneous and the heterogeneous reservoir models were simulated with a commercial black oil simulator and compared with the realization generated with the stochastic simulator of reference 2. A waterflood was performed with a line drive pattern for injection and production wells. The input data for all the geological and the reservoir simulation models was the same. The stochastic model needed more data, but this was consistent with the data of the simpler models.

In Table I the results of the reservoir simulation are shown for the models: the oil recovery fraction at the time when watercut was 90% (time to production stop) are similar.

The heterogeneous model, even if very simple, gives better results than the homogeneous model, as expected.

V. CONCLUSIONS

Two very simple models have been used for fluvial reservoirs in a standard reservoir simulator and have been verified using an advanced stochastic simulator. The recovery factors and breakthrough times were similar. Proper representation should be done with sophisticated geostatistical programs developed to tackle this challenge, but at a predevelopment phase, simple, naive models are useful.

REFERENCES

1. Henriquez, A., Tyler, K., and Hurst, A.: "Characterization of Fluvial Sedimentology for Reservoir Simulation Modeling", *Formation Evaluation* (1990), Vol 5, 211-216.
2. Augedal, H.O., Stanley, K.O. and Omre, H.: "SISABOSA, a program for stochastic modelling and evaluation of reservoir geology", presented at the Conference on Reservoir Description and Simulation with Emphasis on EOR, Oslo, September 1986.
3. Clementsen, R., Hurst, A.R., Knarud, R., and Omre, H., "A Computer Program for Evaluation of Fluvial Reservoirs", presented at the IInd International Conference on North Sea Oil and Gas Reservoirs, Trondheim, 1989 and references there in.
4. Bridge, J.S. and Leeder, M.R. : "A simulation model of alluvial stratigraphy", *Sedimentology* (1979), Vol 26, 617-644.
5. Allen, J.R.L. : "Studies in fluviatile sedimentation: an elementary geometrical model for the connectedness of avulsion-related channel sand bodies", *Sedimentary Geology* (1979), Vol 24, 253-267.

COMPARISON OF THE MODELS		
MODEL	TIME TO PROD. STOP	RECOVERY %
STOCHASTIC	9.80	44
HOMOGENEOUS	8.40	48
HETEROGENEOUS	9.60	43

Table I. Comparison of results using a numerical reservoir simulator with input from the stochastic, homogeneous, and heterogeneous models.

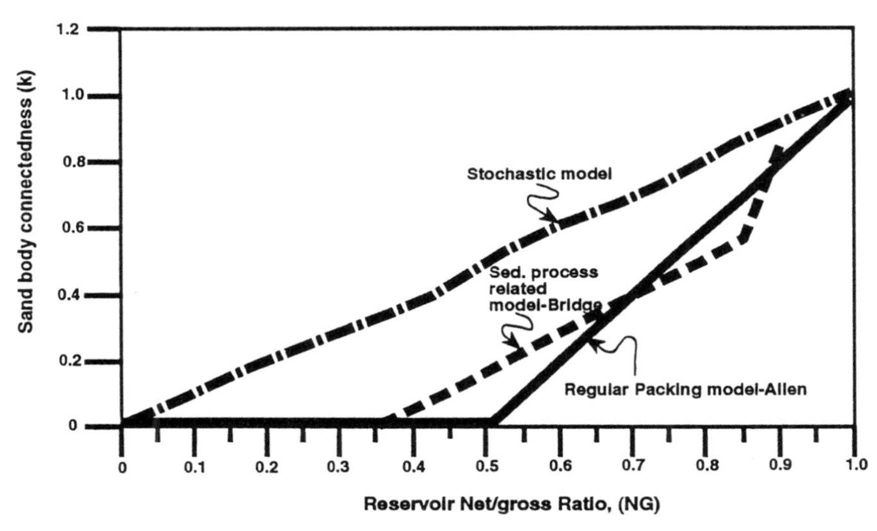

Fig. 1. Sand-body connectedness as calculated by different models. Taken from Henriquez et al. (ref. 1).

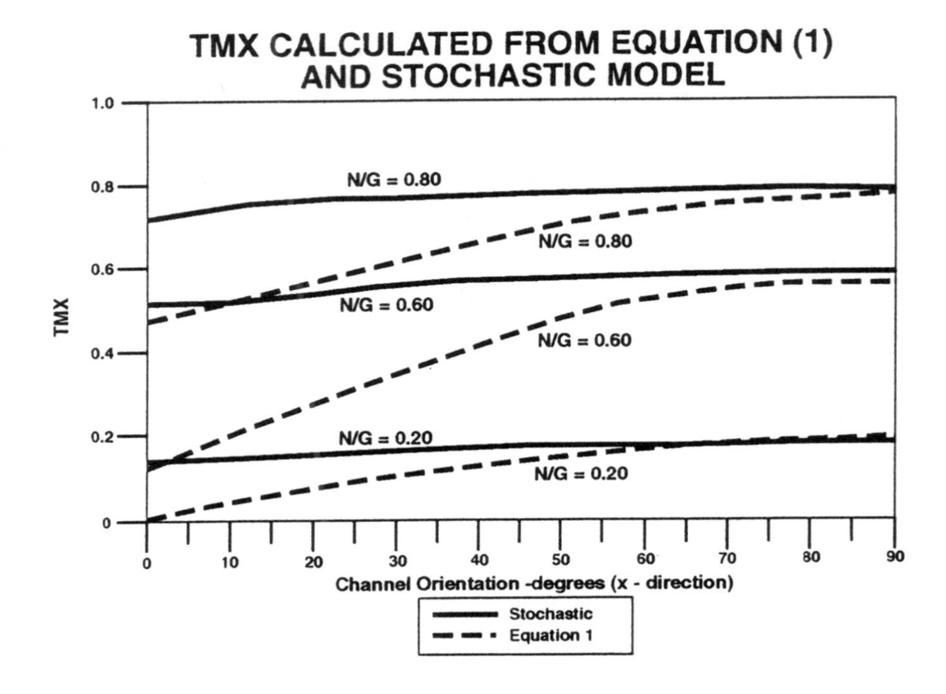

Fig. 2. TMX versus channel orientation, Θ, calculated from equation 1 and the stochastic model. α in equation 1 is derived from equation 2. When the channels are oriented along the y-axis (0°), the stochastic model is unable to calculate the TMX = 0. Generally, the stochastic model overestimates transmissibility.

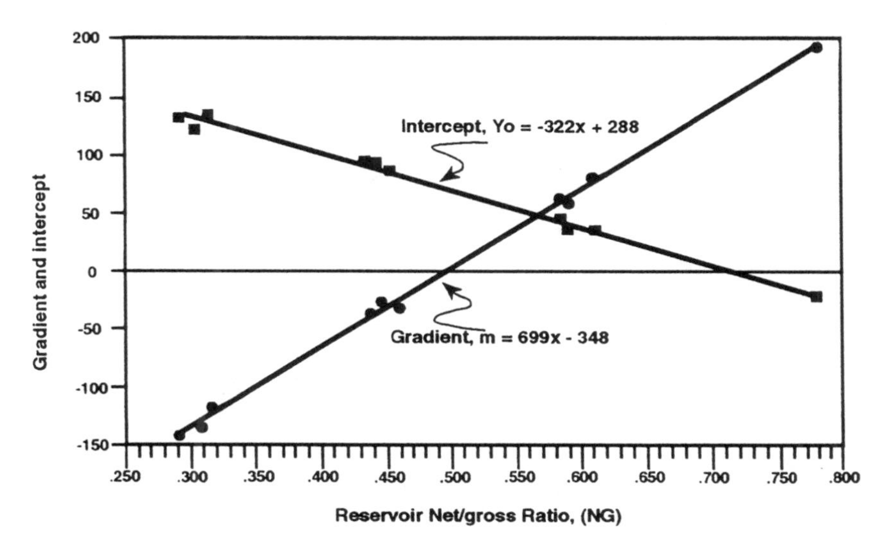

Fig. 3. Calculation of the parameters m and y_0, as a function of reservoar net/gross ratio, which are used in the distribution of NG in grid blocks.

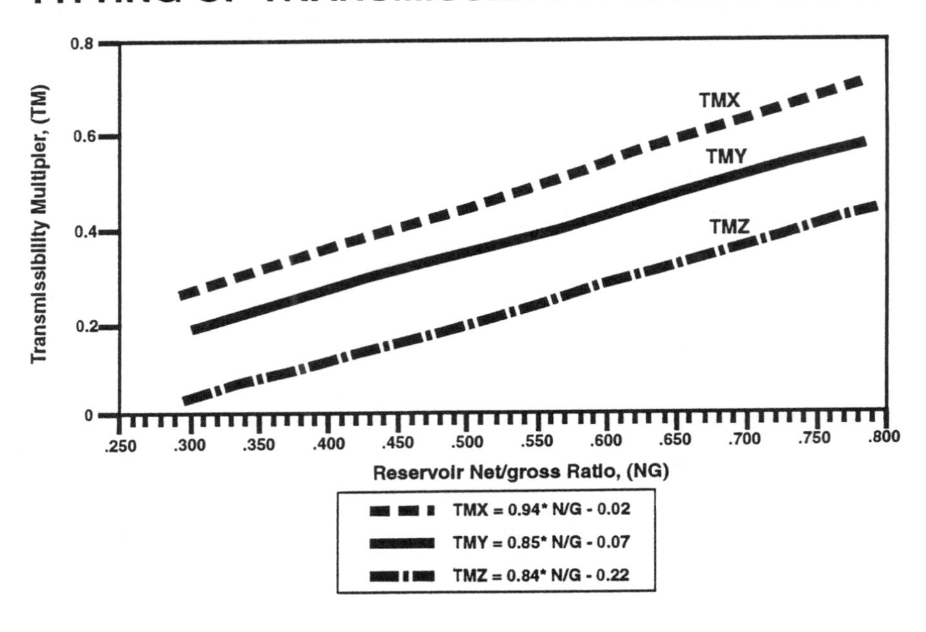

Fig. 4. Transmissibility multipliers from the stochastic model plotted as a function of reservoar NG. The regression coefficients are 0.95, 0.95 and 0.64 for TMX, TMY and TMZ, respectively.

ELECTROMAGNETIC OIL FIELD MAPPING FOR IMPROVED PROCESS MONITORING AND RESERVOIR CHARACTERIZATION[1]

J. R. Waggoner
A. J. Mansure

Division 6253
Sandia National Laboratories
Albuquerque, NM

Electromagnetic (EM) techniques are being developed to monitor oil recovery processes and improve overall process performance. Just as seismic surveys generate acoustic waves to detect changes in sonic velocity, and then infer rock properties, EM surveys generate electromagnetic fields to detect changes in electrical resistivity, and then infer fluid properties. The potential impact of the EM survey is very significant, primarily in the areas of locating oil, identifying oil inside and outside the pattern, characterizing flow units, and pseudo-real time process control to optimize process performance and efficiency.

Since a map of resistivity is of no direct use to a reservoir engineer, an essential part of the EM technique is understanding the relationship between the process and the formation resistivity at all scales, and integrating this understanding into reservoir simulation. This abstract discusses the process we have developed for integrating EM measurements into reservoir characterization and simulation.

1. CORE SCALE PETROPHYSICS

To interpret steamflood resistivity, it is necessary to distinguish between the various fluid zones (steam zone, hot condensate zone, cold condensate zone, etc.) of a steamflood (Mansure and Meldau, 1990). Laboratory and field examples have demonstrated that for clean sands, Archie's law, and for shaly sands, Waxman-Smits and dual-water models, can be used to understand and interpret steamflood resistivity (Mansure, et al., 1990). There are of course physical effects, such as chemical reactions, changes in wettability and formation

[1] Work performed at Sandia National Laboratories under DOE contract number DE-AC04-76DP00789.

factor, etc., that are not included in these petrophysical models. However, these effects appear to be less important than establishing proper estimates for temperature, saturation, salinity, and clay parameters. For a typical California heavy-oil steamflood like Kern River, steamflooding has been observed to lower formation resistivity a factor of about 3. As little as 3% clay can have a significant effect because of the low salinities frequently encountered in steamfloods.

2. COUPLING PETROPHYSICAL KNOWLEDGE AND RESERVOIR SIMULATION

To utilize the petrophysical insights described above, including Archie's Law, clay effects, and thermal effects, a post-processor named RESIST was written which calculates the electrical resistivity of each simulation grid block. The inputs to RESIST are both simulation-derived variables, such as saturation, porosity, and temperature, and user-defined constants, such as saturation and porosity exponents and clay effect parameters. The post-processing approach requires that the clay effects do not affect fluid flow, which is a reasonable assumption when the percentage of clay in the reservoir is low.

One additional RESIST input parameter is the water salinity, which is used along with temperature to estimate the water resistivity from published correlations (Arps, 1953). Since water salinity is not typically monitored in non-compositional simulators, a post-processor named TRACK was written to track multiple components within the aqueous phase. Again, it is required that the component concentrations not affect the aqueous phase flow properties. Knowing component concentrations and salinity, it is possible to mix the components and estimate the salinity of the mixture.

The post-processing approach briefly described allows electrical resistivity to be estimated for each grid block without substantial modification to the reservoir simulator. Since both TRACK and RESIST are stand-alone programs, they can be run multiple times with the same simulation data as input to study sensitivities to clay and multicomponent effects. Reducing the number of simulations required and interfacing with existing simulators are two of the major advantages of the post-processing approach. This triad of programs is named ORRSIM, the Oil Recovery/Resistivity Simulation System.

3. EFFECT OF HETEROGENEITY AND ANISOTROPY ON EM TECHNIQUE

The geology of a reservoir is not homogeneous at the scale of resolution of EM measurements or the grid scale of a reservoir simulator. It is thus necessary to understand how the resistivity of rocks average to give bulk or grid-scale resistivity (Mansure, 1990). The bulk-averaged resistivity of a heterogeneous material is a function of the direction of current flow relative to any structure of the material, resulting in a bulk material that may be electrically anisotropic. Failure to recognize the importance of the direction of current flow can lead to significant errors in interpreting EM measurements, because the response of the bulk rock may depend upon the direction the EM measurement probes the rock. The bulk resistivity of rock also depends upon both the average and distribution of rock properties. Merely plugging average saturation into Archie's law can give resistivities that are off by 50% or more.

4. APPLICATION OF ORRSIM TO EM DATA INTERPRETATION

ORRSIM is a forward model following the path from reservoir description to reservoir simulation to TRACK and to RESIST to produce a resistivity map of the reservoir (path A on Fig. 1). Since the intent of the program is to interpret EM-derived resistivity in terms of reservoir parameters, ORRSIM can also be used in an iterative mode. After generating a good geologically-based reservoir description, ORRSIM will predict a process-derived resistivity map. Qualitative, or perhaps even quantitative, comparison with the EM-derived resistivity map may suggest ways to alter the reservoir description, the process simulation parameters, or TRACK and RESIST input variables. After making the changes, the map generated by a new ORRSIM run should yield a better match to the EM-derived map. This iteration can continue until a desired match is achieved.

REFERENCES

Arps, J.J., (1953)."The Effect of Temperature on the Density and Electrical Resistivity of Sodium Chloride Solutions," Journal of Petroleum Technology, Section 1 pp 17-22, October; Trans. AIME, vol. 198.

Mansure, A.J., to be published, Parameterizing and Block-Averaging Electrical Characteristics of a Reservoir: an Essential part of Electrical/Electromagnetic Evaluation of Production Processes, Trans. 1st Annual Archie Conf.,1990,AAPG.

Mansure, A.J., and R.F. Meldau, 1990, Steam-Zone Electrical Characteristics for Geodiagnostic Evaluation of Steamflood Performance, SPE Formation Evaluation, 5, 3, 241-247.

Mansure, A.J., R.F. Meldau, and H.V. Weyland, 1990, Field Examples of Electrical Resistivity Changes During Steamflooding, 65th Annual SPE Meeting, New Orleans, SPE 20539.

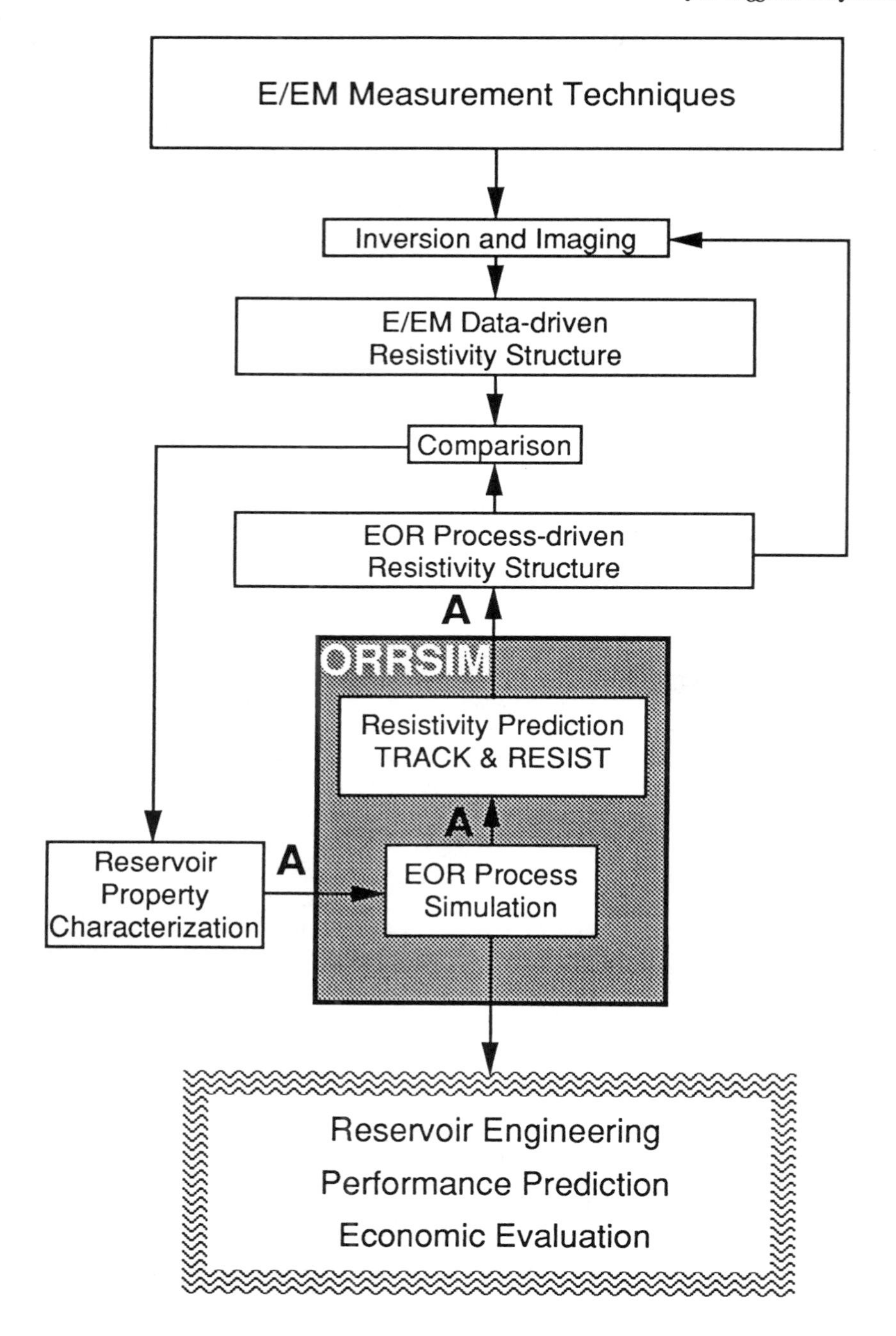

Figure 1: Electromagnetic (EM) Project Flowchart

THERMAL SIMULATION OF SHALLOW OIL ZONE LIGHT OIL STEAMFLOOD PILOT IN THE ELK HILLS OIL FIELD

Alan A. Burzlaff

ECL-Bergeson Petroleum Technologies, Inc.
Denver, Colorado

Bob R. Harris, Jr.

U. S. Department of Energy
Tupman, California

I. INTRODUCTION

A light oil steamflood (LOSF) pilot project has been operating since July 1987 in the Shallow Oil Zone (SOZ) at the Elk Hills oil field, Naval Petroleum Reserve No. 1, Kern County, California (Gangle *et al.*, 1990). The SOZ is still largely under primary recovery and represents a potential 100 million barrel enhanced oil recovery target. The performance of the LOSF pilot project was duplicated using a single distillable component thermal simulator. The principle objectives of the simulation study were to develop a useful geologic and engineering model to screen other prospective SOZ steamflood areas, and to evaluate expansion of the current project.

II. RESERVOIR CHARACTERIZATION

The LOSF pilot project (Fig. 1) is located on the southeastern flank of the Elk Hills anticline. Steam is injected into the SS-1 sands of the Eastern SOZ at an average depth of 3000 feet. The beds dip at three to five degrees in the pilot area. Normal faulting is extensive throughout the LOSF pilot area (Fig. 2) and there is a confining fault (S fault) to the south whose 70 to 90 foot of displacement is believed to be an effective barrier to fluid movement. Other minor faults with five to thirty feet of displacement act as partial barriers or conduits depending on the degree of displacement and juxtaposition.

Seven discrete SS-1 sand members, designated the A, A1, B, C1, C2, D and E subzones (Fig. 3) have been identified in the steamflood area. During the course of this study, the subzones were further divided into 16 layers to incorporate the permeability variation of the SS-1 sands into the simulation models. The total interval is about 90 feet thick, with net sand thicknesses ranging from 50 to 70 feet. The porosity varies between 27 to 35 percent, and air permeabilities range from 100 to 2000 millidarcies. The average oil gravity in the project area is 27 degree API.

Petrophysical relationships developed in a previous study (Bergeson, 1989) were applied to the log and core data in the LOSF study area to calculate net thicknesses, porosities, initial water saturations and permeabilities for each of the 16 layers. Vertical permeability as a function of horizontal permeability, and relative permeability relationships were generated using conventional and special core analysis studies from wells in the surrounding area.

A 20 x 15 x 16 conventional black oil model was constructed which covered the LOSF pilot area as well as a significant area surrounding the patterns. The conventional model[1] was used to duplicate 11 years of pre-pilot primary production performance.

A significant improvement in the reservoir description was obtained as a result of this history match. The improved reservoir description included adjusting the pore volume to match observed production and pressure data, identifying the effectiveness of faulting in preventing fluid migration, and quantifying the distribution of oil, water and gas saturations prior to the initiation of enhanced recovery operations. This increased the confidence in the performance projections and provided useful operational information, such as identifying trapping of oil against faults.

III. LOSF PILOT PERFORMANCE

Pilot steam injection began in July 1987 with the commencement of steam injection into the D and E subzones of well 18N-3G. In November 1987, three additional steam injection wells were added to the pilot project. Between November 1987 and May 1988, steam injection was limited to the D and E subzones of the SS-1. In June 1988, the A, B, C1 and C2 subzones were completed with limited entry perforations to regulate steam injection into each subzone. After June 1988, all subzones were open to steam injection except for the A1 subzone which was felt to be gas saturated.

The pilot area oil production doubled within 6 months after the start of steam injection increasing from 400 BOPD to a peak rate of 850 BOPD. Oil production also increased significantly in producers outside the pilot area as a result of repressuring of the reservoir with steam, in situ generated carbon dioxide, and hot water. Oil production has subsequently declined in the pilot area as a result of operational problems, increasing watercuts, and severe formation damage.

---

[1]ECLIPSE Black Oil Simulator

IV. THERMAL SIMULATION

A thermal simulation program[2], which models the flow of four components plus the energy associated with the system, was used to perform the steamflood simulation work. The four components are "heavy" oil, "light" oil, water and non-condensible gas. The model's light oil component is capable of being vaporized at increased temperature. Since the steamflood has not historically experienced particularly high temperatures, the use of a simulator with a single distillable component was suitable for this study.

A steamflood model of 2145 cells (11 x 15 x 13 layers) was constructed for a sector of the LOSF project area. This sector, shown in Fig. 4, was selected such that all four pilot injection wells would be included in the model area. In addition, a substantial portion of the responding area northeast of the pilot was included in the sector.

A satisfactory history match of the performance of this sector of the LOSF pilot was achieved with the steamflood model (Figures 5-6). Temperature profile surveys, run in several observation wells located in the pilot area, were also matched by the steamflood model (Fig. 7). A prediction of future pilot performance was made assuming a continued steaming operations scenario.

Steamflooding was compared to two alternative depletion scenarios, waterflooding and continued primary depletion. The waterflood and primary depletion scenarios were predicted using the history matched conventional model (Fig. 8). The major conclusions from the results of the steamflood model history match and the various prediction cases were:

1. The majority of the oil response has been due to waterflooding by the condensed steam. Steam distillation and distillate drive recovery mechanisms have been of secondary importance due to uneven thermal energy distribution and limited sustained steam zone temperatures in the pilot area (Fig 9).

2. The 'P' fault in the pilot area significantly affected the movement of steam and condensed hot water by restricting steam and heat flow to offsetting downdip producers.

3. Oil recovery from the LOSF pilot area, through September 1990, is estimated to be about 41 percent of the oil-in-place (OIP) at the beginning of steam injection (Fig. 10). Of this 41 percent, an amount of oil equal to approximately 14 percent of the OIP has been displaced outside of the pattern area. This amount of oil recovery is only about 5 percent greater than that predicted for a waterflood depletion scenario employing the same injection rates.

4. Based on the performance of the LOSF pilot, the predicted future performance from the steamflood model, and the price and cost projections provided by the

---

[2]Dynamic Reservoir Systems - Thermal (DRSTH), a product of Simtech Consulting Services, Inc.

Department of Energy, light oil steamflooding will not maximize economic oil recovery in the SOZ.

ACKNOWLEDGEMENTS:

We would like to acknowledge our appreciation of the Unit Management at the Naval Petroleum Reserve No. 1 for allowing publication of this paper. We also acknowledge the assistance provided by Vernon Breit and Joe Dosso of Simtech Consulting Services, Inc.

REFERENCES:

Bergeson & Associates, 1989, "Eastern Shallow Oil Zone Study - Fault Blocks 10, 11, 12, Phase IVA Report".

Gangle, F. J., Weyland, G. V., Lassiter, J. P., and Veith, E. J., 1990, "Light Oil Steamdrive Pilot Test at NPR-1, Elk Hills, California", SPE Paper #20032.

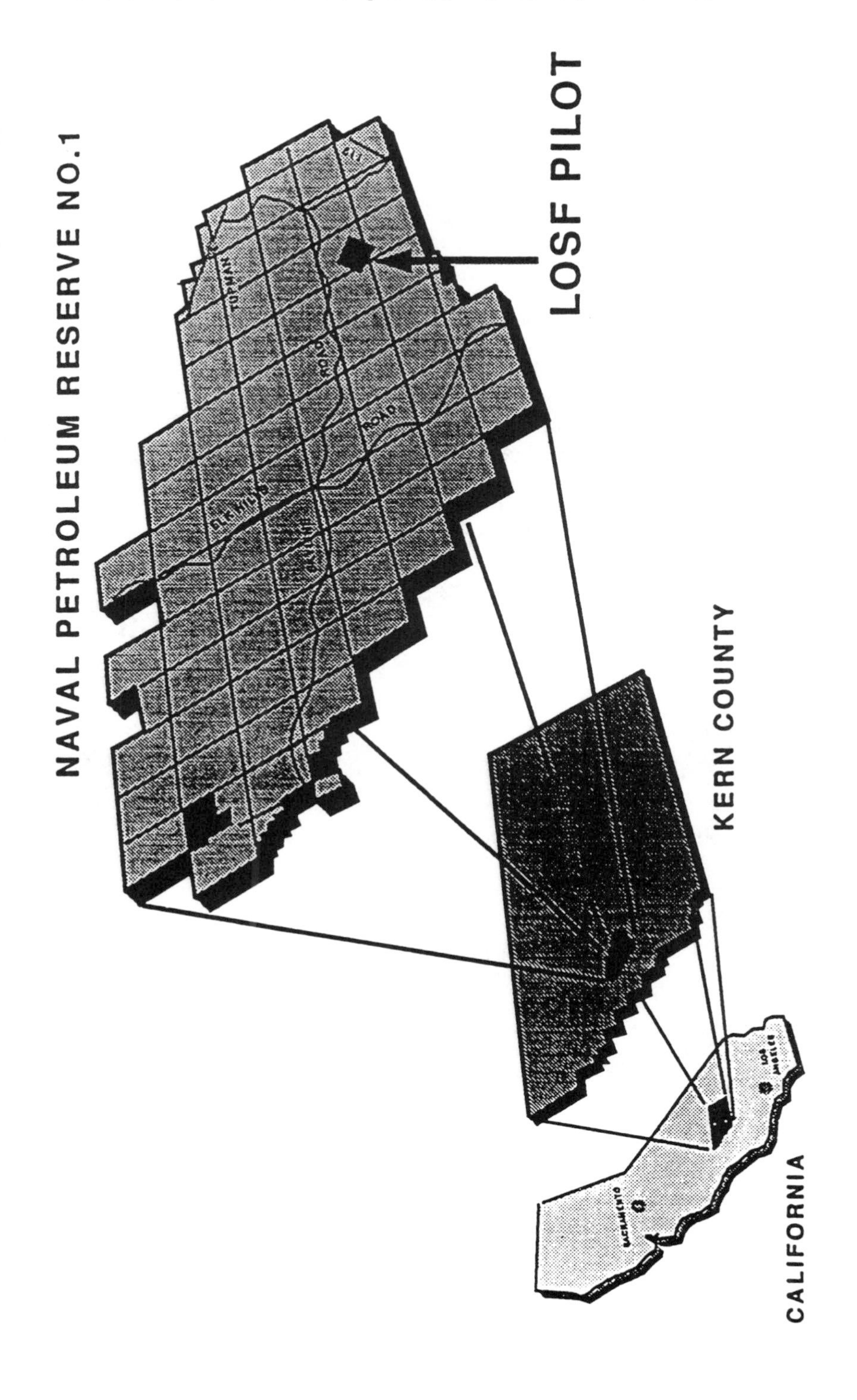

Fig. 1 - LOSF Pilot Area

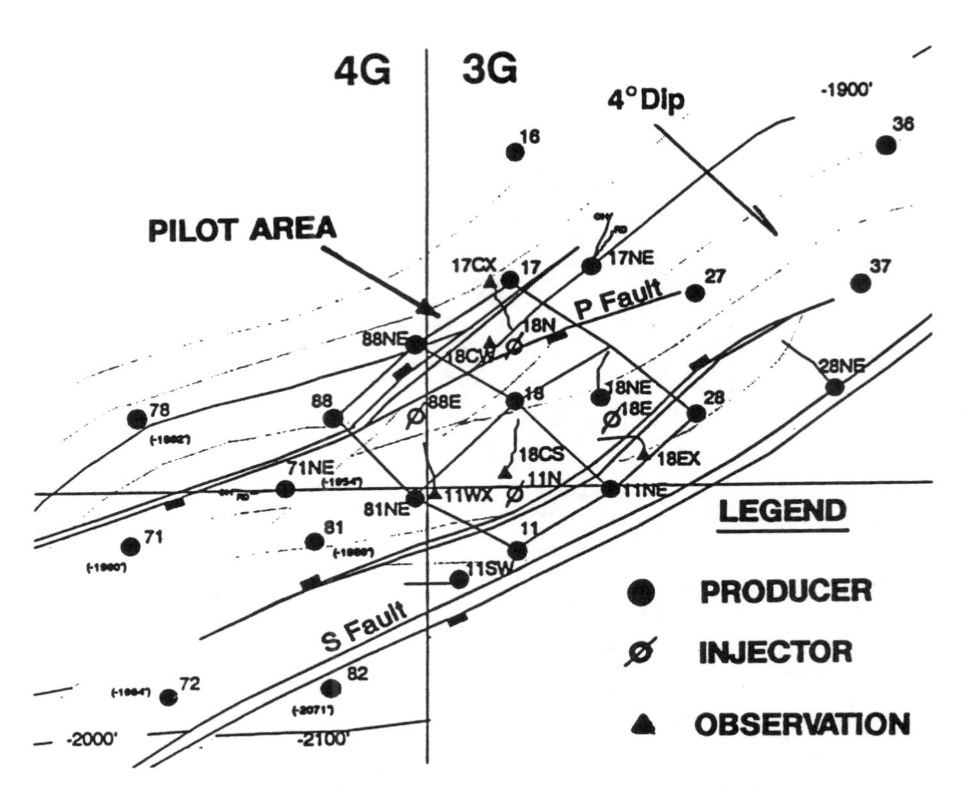

Fig. 2– Steamflood pilot area structure map and well configuration.

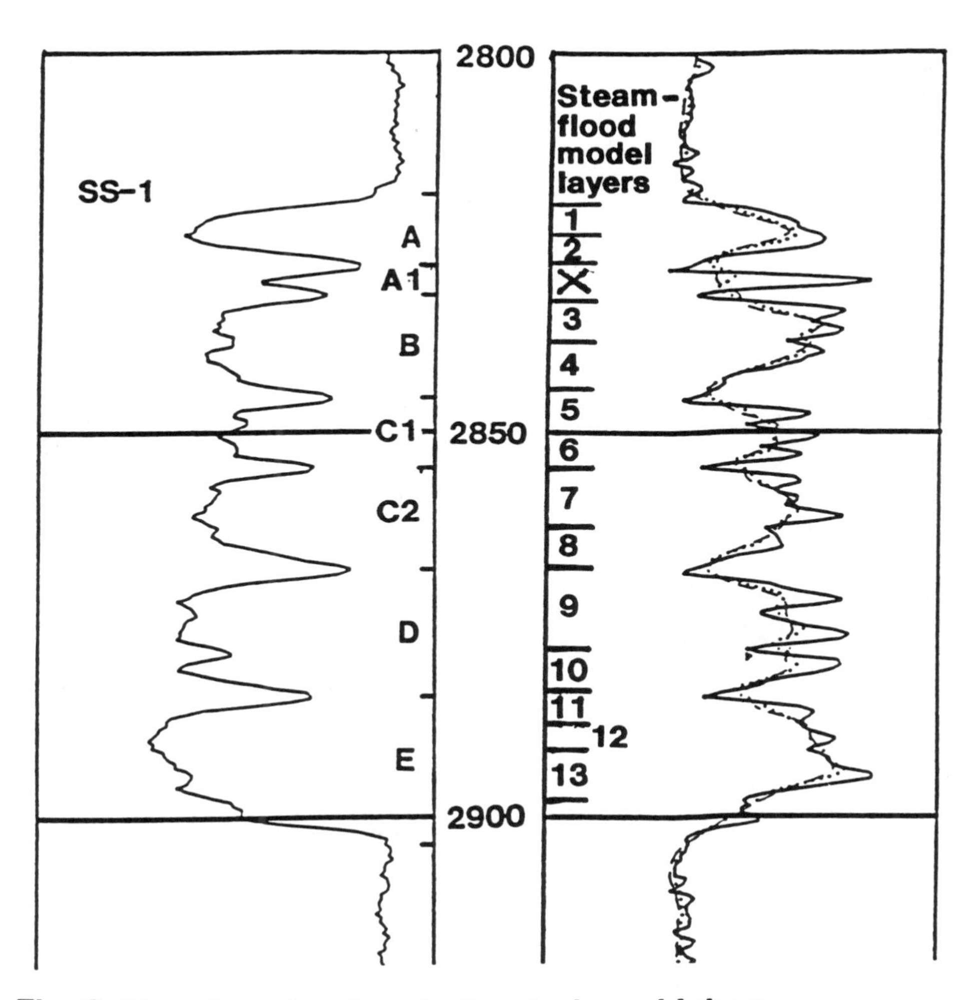

Fig. 3–Type log showing designated sand lobes, A through E.

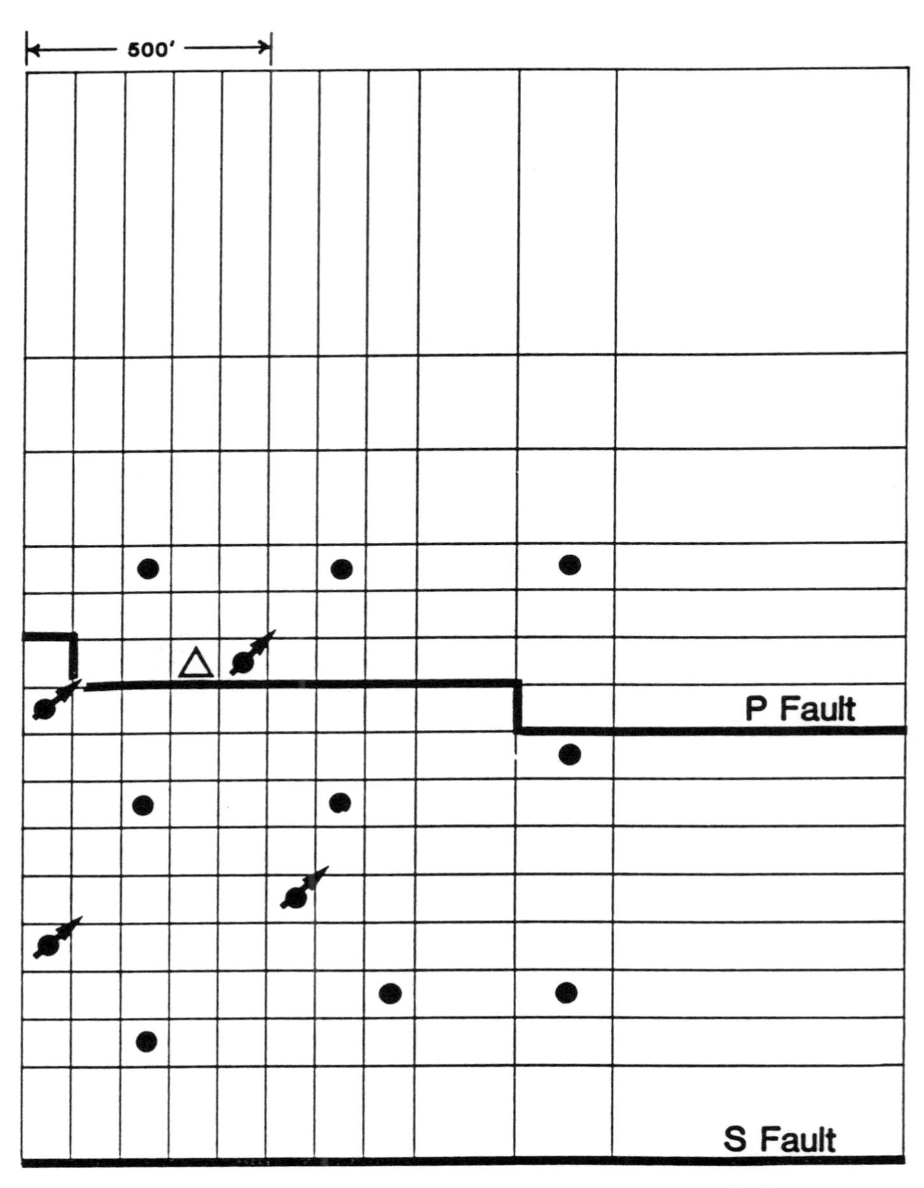

Fig. 4–Areal grid of steamflood model.

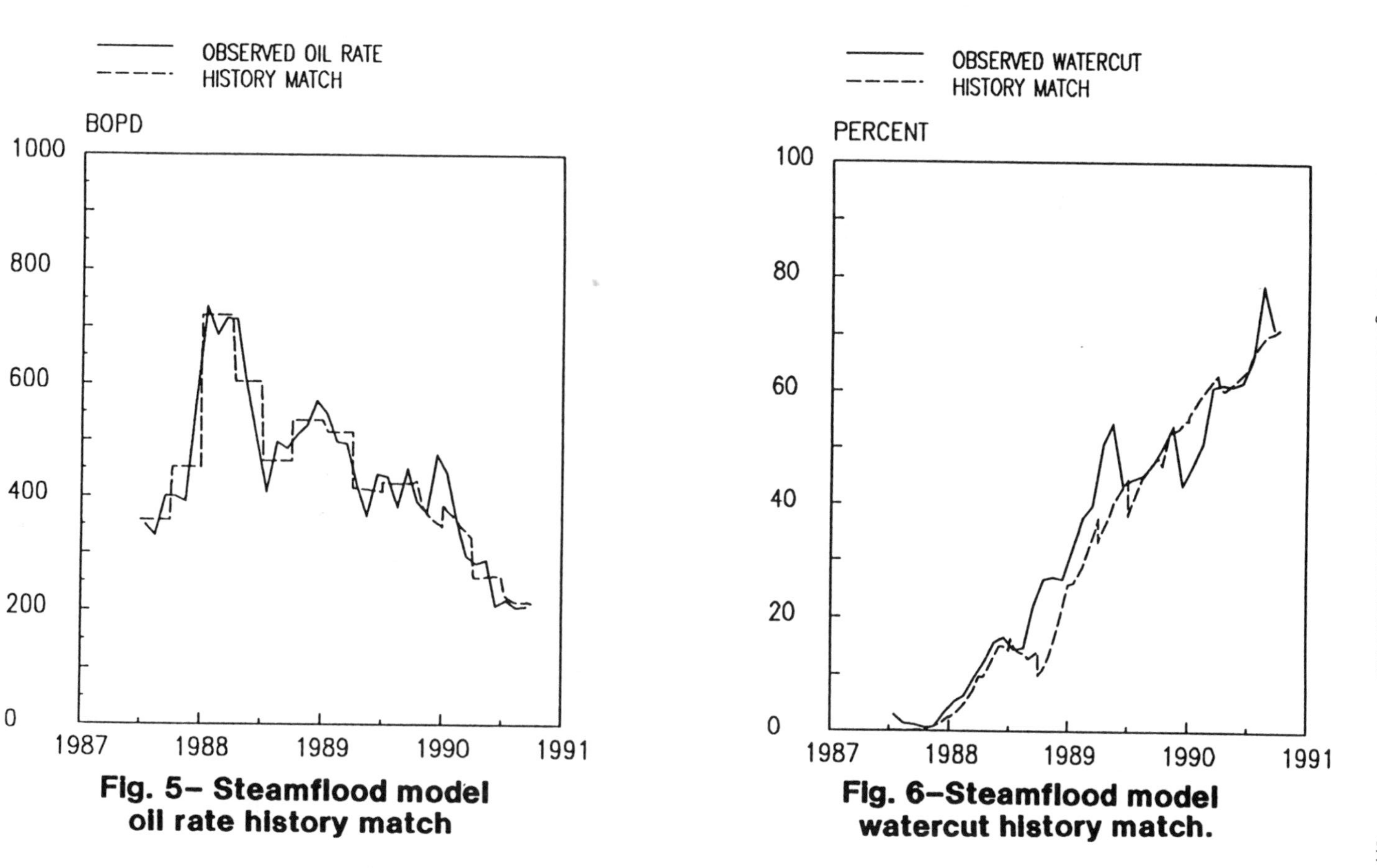

Fig. 5– Steamflood model oil rate history match

Fig. 6–Steamflood model watercut history match.

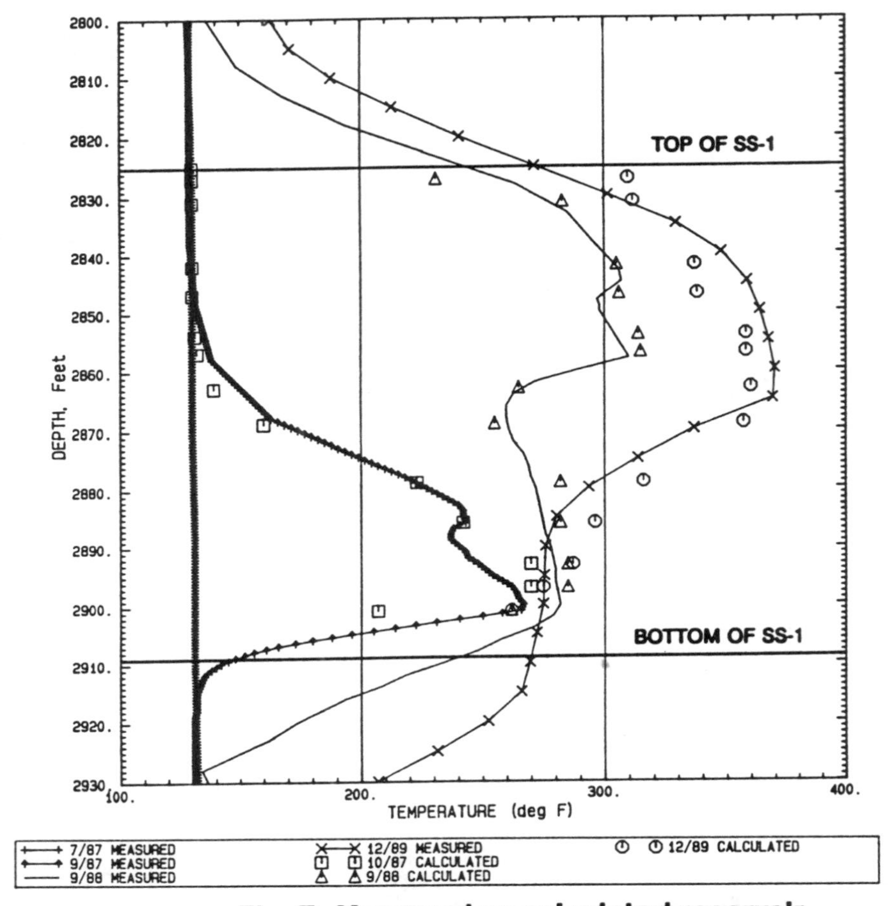

Fig. 7–Measured vs calculated reservoir temperatures–observation well 18CW.

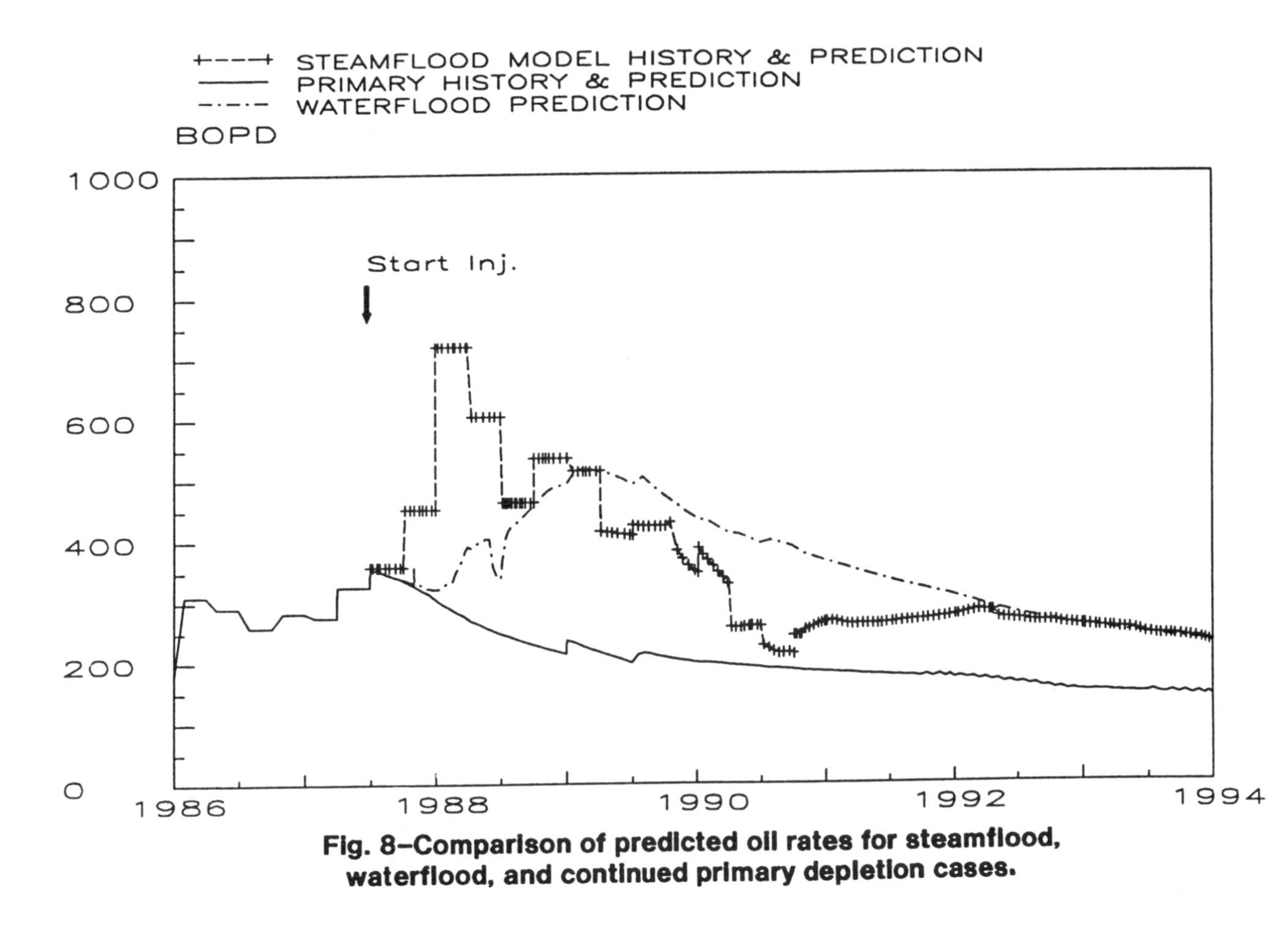

Fig. 8–Comparison of predicted oil rates for steamflood, waterflood, and continued primary depletion cases.

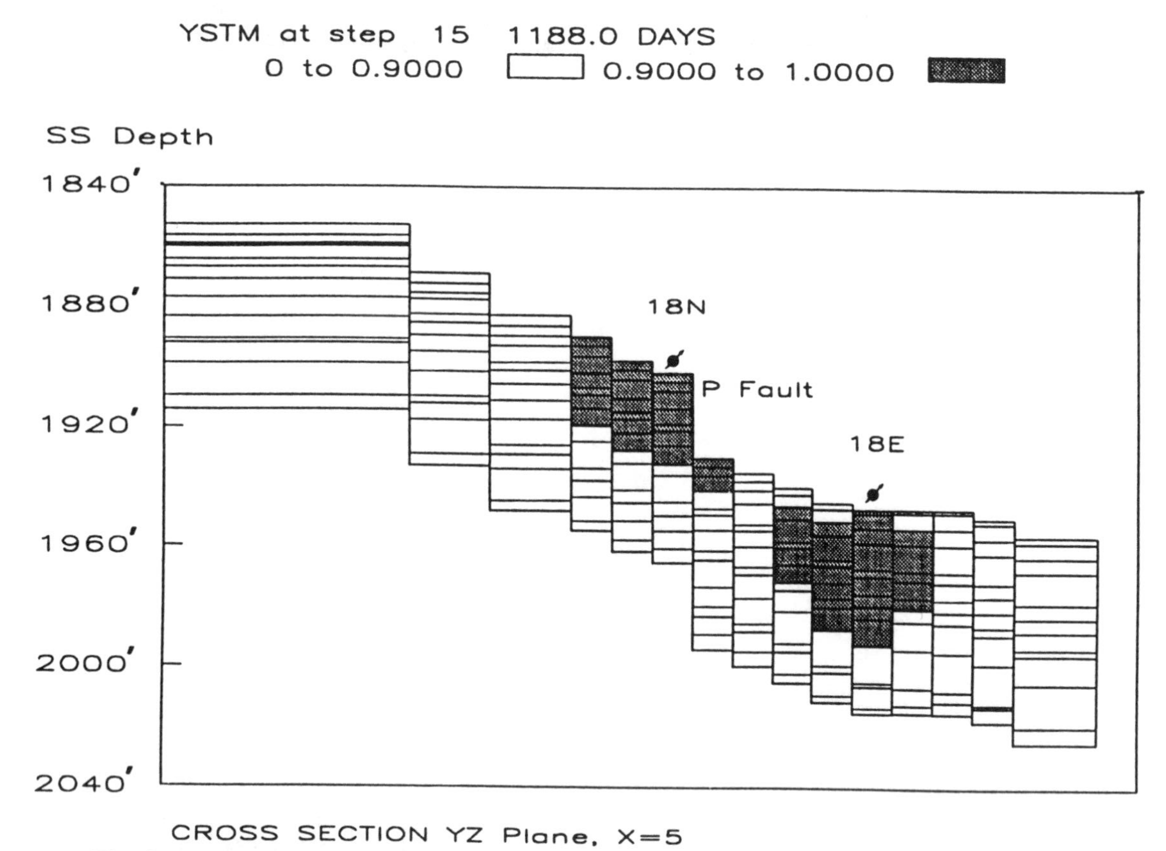

Fig. 9–Distribution of steam mole fraction in the gas phase at end of history match.

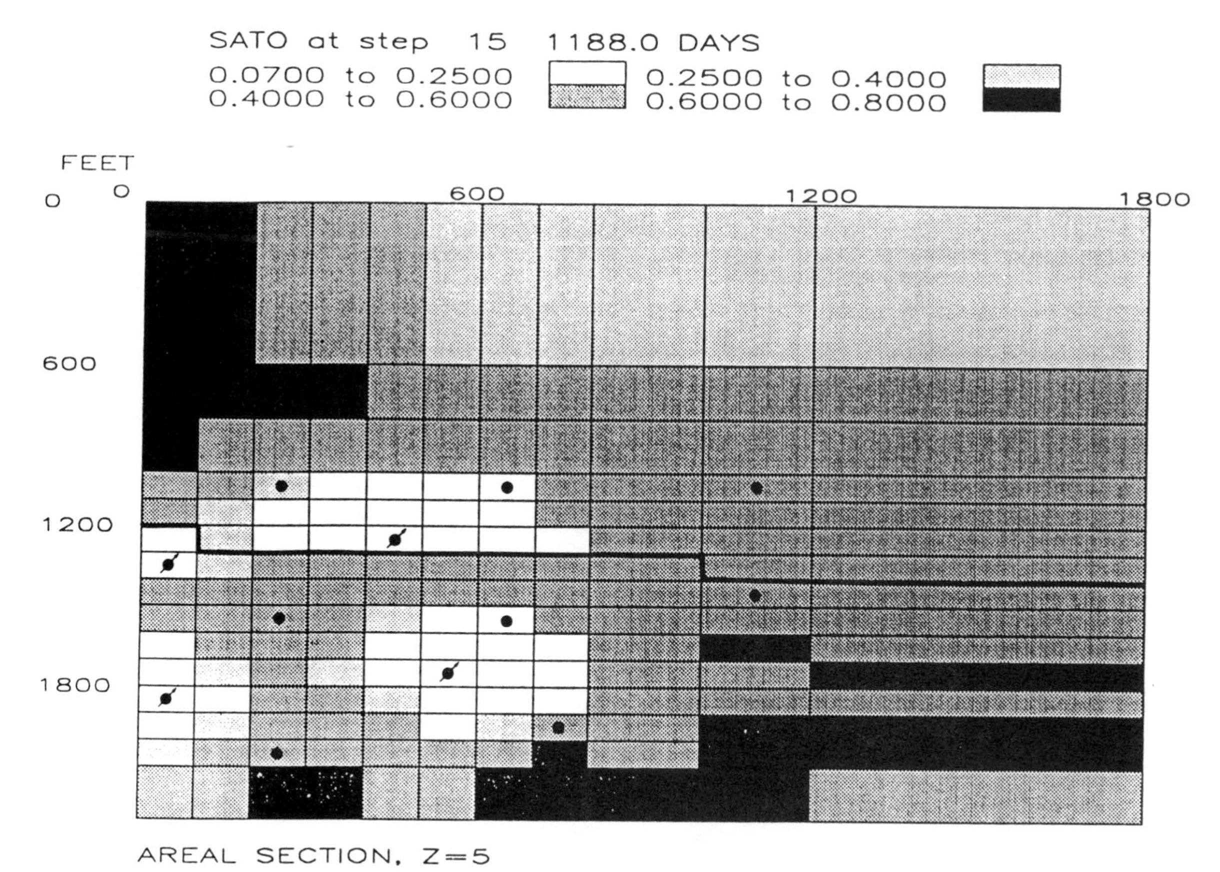

Fig. 10–Oil saturation at end of history match, layer 5 (C1 subzone).

CHARACTERIZATION OF RESERVOIR HETEROGENEITY, FULLER RESERVOIR AND HAYBARN FIELDS, PALEOCENE FORT UNION FORMATION, FREMONT COUNTY, WYOMING

C. Wm. Keighin
R. M. Flores

U.S. Geological Survey
Denver, CO 80225.

A major concern in petroleum exploration and development is the detailed architecture (internal and hierarchical) of sandstone reservoirs in a field or basin-wide setting. This concern is due, in part, to the effects of architecture on fluid flow (Tyler and others, 1992). Unifying models of reservoir architecture are seldom available because of the temporal and three-dimensional complexity of depositional systems (Keefer, 1965; 1990; Flores and others, 1990). Local and basinal scales of sandstone-reservoir heterogeneities can be recognized and classified in outcrops and subcrops. Outcrop studies provide information on grain size, internal structure, geometry, hierarchical arrangements, and continuity of reservoir sandstones. Subcrop investigations, including studies of core and geophysical logs, help extend mappability (lateral variability) and predictability of the geometry and architecture of reservoir sandstones in a basinal setting.

Fluvio-deltaic sandstones in the Paleocene Fort Union Formation, Fremont Co., WY, form shallow (approximately 2,500-5,000 ft in depth) discontinuous reservoirs for oil and gas in the Fuller Reservoir and Haybarn fields, Wind River Basin, Wyoming (Pirner, 1978; Robertson, 1984; Specht, 1989). These fields have yielded 1.79 million bbl of oil and 12.9 bcf of natural gas. An integrated investigation of these sandstones, utilizing facies descriptions from outcrops, wireline logs, cores, and petrographic examination of core samples, demonstrates a relationship between reservoir quality and facies architecture. These data also show that sandstone facies within the upper Fort Union Formation (Shotgun and Waltman Shale Members) are discontinuous across the Fuller Reservoir and Haybarn fields (fig. 1).

Fining-upward fluvial channel sandstones are the best reservoirs. These sandstones rarely exceed 20 ft in aggregate thickness; porosity is about 20%, and permeability may be as high as 100-150 md. Coarsening-upward delta-front sandstones, which may be as thick as, and only slightly less porous (15-18%) than, the fluvial channel sandstones, have measured ambient permeabilities typically less than 50 md. Porosity, even when good (10-20%), may be partly to extensively occluded by pore-filling clays, especially kaolinite. Locally, 1-2 ft-thick delta-front sandstone units tightly cemented by Fe-bearing carbonate cement may serve as permeability barriers. Framework architecture of the fluvial-channel sandstones reveals laterally stacked, partly interconnected reservoirs 0.1-0.6 miles in lateral extent. Delta-front sandstones are vertically stacked and form isolated reservoir compartments that may be 1.5-4.5 miles wide.

Our study indicates that reservoir characterization of Fort Union sandstones reflects heterogeneity on the scale from outcrop to thin section. Although potential reservoir facies include sandstones deposited in fluvial channels, crevasse channels, crevasse channel/splay deposits, or as delta fronts deposits, plots of permeability/porosity/lithology-facies data indicate that sandstones from fluvial channels are most porous and permeable (fig. 2a,b). Thus, reservoir architectural complexities due to depositional environments, as well as petrophysical differences, are vital aspects to developmental strategies.

REFERENCES

Flores, R.M., Keighin, C.Wm., and Keefer, W.R., 1990, Reservoir-sandstone paradigms, Paleocene Fort Union Formation, Wind River basin, Wyoming (abs.), Am. Assoc. Petroleum Geologists Bull., v. 74/8, p. 1323.

Keefer, W.R., 1965, Stratigraphy and geologic history of the Uppermost Cretaceous, Paleocene, and lower Eocene rocks in the Wind River basin, Wyoming, U.S. Geol. Survey Professional Paper 495-A, 77p.

Keefer, W.R., 1990, Cretaceous rocks in the Wind River basin, Central Wyoming (abs.), Am. Assoc. Petroleum Geologists Bull., v. 74/8, p. 1331.

Pirner, C.F., 1978, Geology of the Fuller Reservoir Unit, Fremont County, Wyoming, in Wyoming Geological Assoc. 30th Ann. Field Conf. Guidebook "Resources of the Wind River basin", Casper, WY, p. 281-288.

Robertson, R.D., 1984, Haybarn field, Fremont County, Wyoming: An Upper Fort Union (Paleocene) stratigraphic trap, Mountain Geologist, v.21, p. 47-56.

Specht, R.W., 1989, Fuller Reservoir, in Wyoming Geological Assoc. "Wyoming oil and gas fields symposium, Bighorn and Wind River basins", p. 178-180.

Tyler, N., Barton, M.D., Fisher, R.S., and Gardner, M.H., 1992, Architecture and permeability structure of fluvial-deltaic sandstones: A field guide to selected outcrops of the Ferron Sandstone, east-central Utah: Mesozoic Field Guidebook, Rocky Mountain Section SEPM.

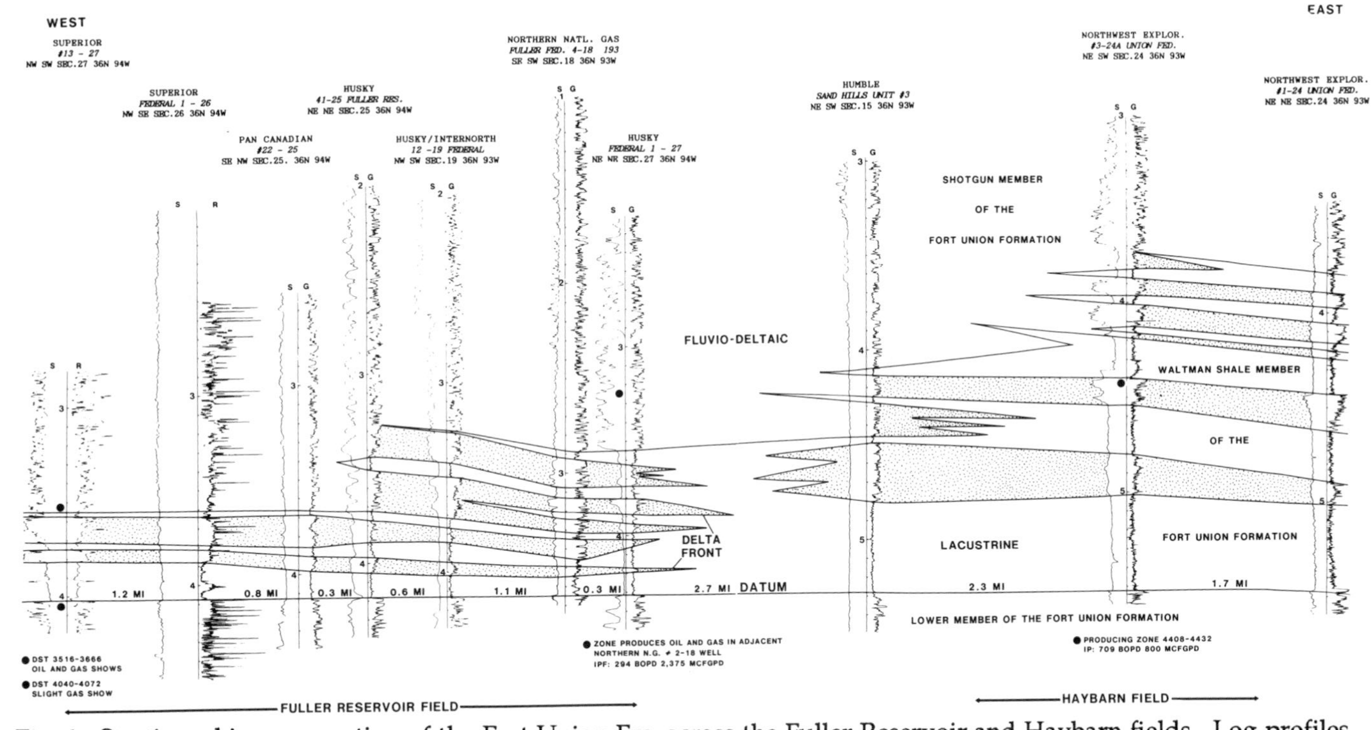

Fig. 1. Stratigraphic cross section of the Fort Union Fm. across the Fuller Reservoir and Haybarn fields. Log profiles are SP (S) and gamma (G), or SP and resistivity (R). Stippled pattern indicates deltaic-lacustrine sandstone reservoir facies I (delta front) pinching out and interfingering eastward with the lacustrine Waltman Shale Member.

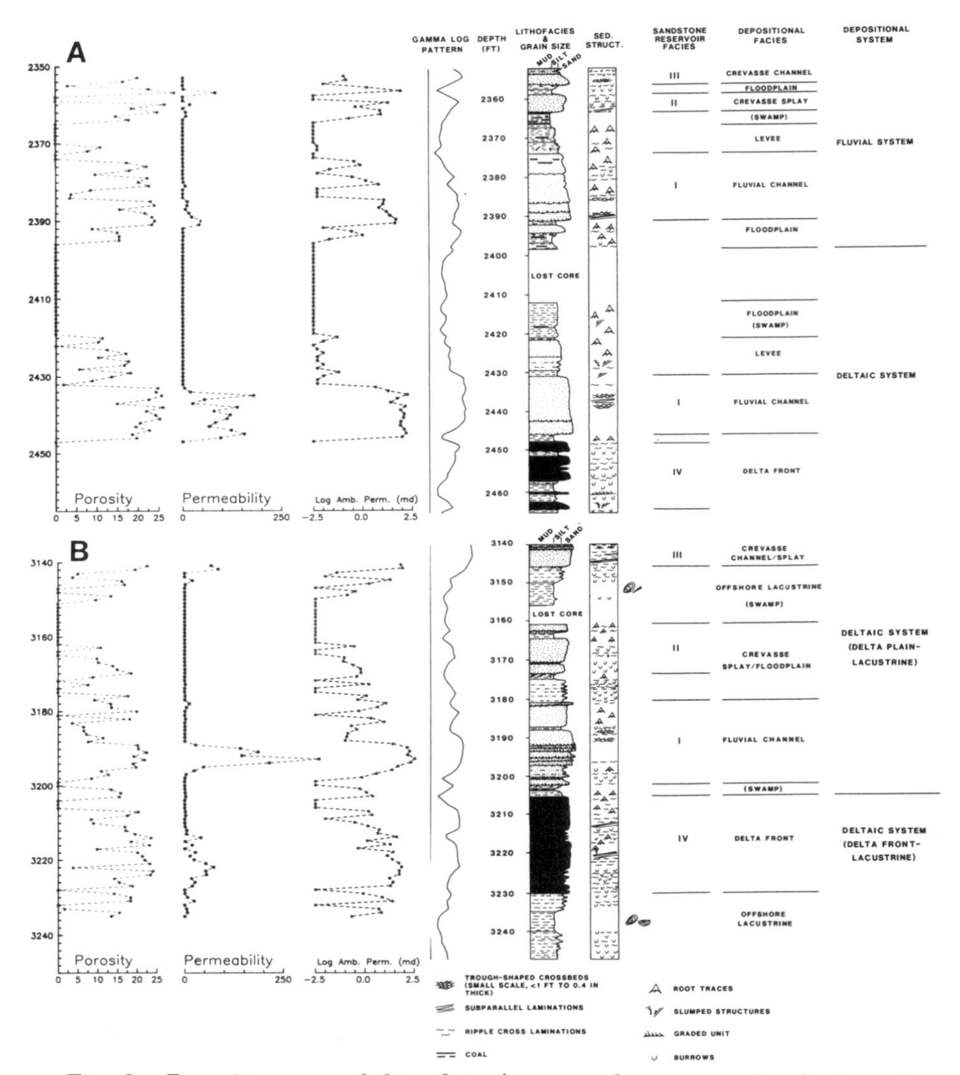

Fig. 2. Porosity-permeability data (measured on core plugs) plotted against various parameters, but especially depth (ft below surface), lithofacies, and depositional facies. Porosity/permeability values are closely related to the environment of deposition. (a) Northern Natural Gas #7-18 Federal 886 (SE NE SEC.18 36N 93W). Porosity values range between 3 and 25%; permeability (md) ranges between <0.01 and 179, but rarely exceeds 50 md. (b) Husky Oil Co. #8-19 Fuller Federal (SE NE SEC.19 36N 93W). Porosity values range between 3.6 and 24%; permeability (md) ranges between 0.01 and 339. Permeability above 100 md appears restricted to fluvial channel environments.

MICROSCOPIC TO FIELD-WIDE DIAGENETIC HETEROGENEITY IN DELTAIC RESERVOIRS FROM BRAZIL: DESCRIPTION, QUANTIFICATION AND INTEGRATION WITH OTHER RESERVOIR PROPERTIES

Marco A. S. Moraes
PETROBRAS/CENPES/DIGER
Ilha do Fundão, Cid. Universitária
Rio de Janeiro RJ 21910 Brazil

Ronald C. Surdam
Department of Geology and Geophysics
University of Wyoming
Laramie WY 82071 U. S. A.

I. INTRODUCTION

The description and quantification of reservoir heterogeneity has achieved significant progress in recent years. Most of the new developments have been in the application of depositional models and stochastic models to reservoir characterization (Haldorsen, 1983; Weber, 1986; Fogg and Lucia, 1990; Haldorsen and Damsleth, 1990; Weber and van Genus, 1990). In most cases, the main objective has been to model depositional features. Diagenetic studies commonly do not include an evaluation of the heterogeneity resulting from diagenetic processes. Such heterogeneity, however, is typically observed at different reservoir scales and its detailed description is essential to achieve a complete and adequate spatial representation of the reservoirs.

Deltaic lacustrine sandstones produce hydrocarbons in more than a hundred accumulations in the Cretaceous rift basins of northeastern Brazil. These sandstones show a complex diagenetic evolution that has been extensively documented by core, thin section and SEM studies in many different oil fields. Similarities observed in the depositional and diagenetic properties characterizing these reservoirs permitted the development of diagenetic models that can be used to improve reservoir quality evaluation and performance prediction.

II. GEOLOGIC SETTING

Deltaic reservoirs have been studied in the Potiguar and Recôncavo basins of northeastern Brazil (Fig. 1). The basins are onshore failed rift arms formed in Early Cretaceous and associated with the opening of the South Atlantic (Bertani et al., 1987; Milani and Davinson, 1988). The structural framework of the basins is characterized by deep assymetrical grabens separated by internal highs (Fig. 2). Deltaic reservoirs commonly consist of a series of stacked delta-front sandstone bodies deposited in lows associated with growth faults (see Fig. 2). The sandstone bodies typically pinch-out away from the faults. Extensive and fairly continuous lacustrine shales commonly separate the reservoir beds. The sandstones are predominantly fine-grained, with medium- to coarse-grained sandstones appearing locally.

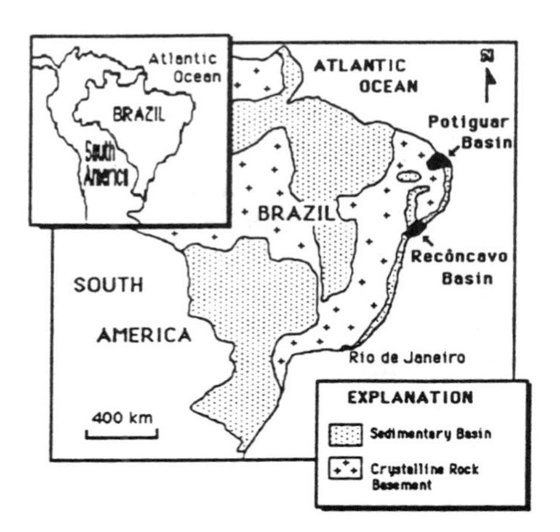

Fig. 1- Location of Potiguar and Recôncavo basins.

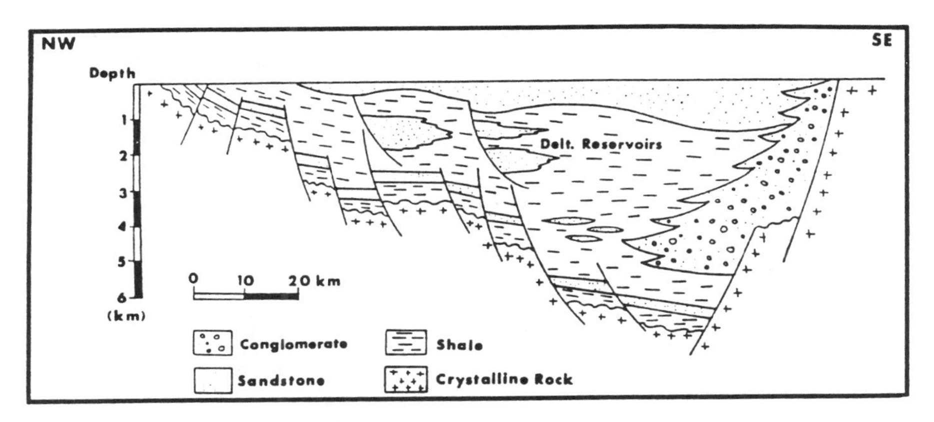

Fig. 2- Schematic geologic section of Recôncavo Basin showing main geologic features and the typical setting of deltaic reservoirs.

The petrographic characteristics and the diagenetic evolution of these rocks have been summarized by Becker (1984), Alves (1985), Anjos and Carozzi (1988) and Moraes (1991). Main diagenetic phases include, in an inferred temporal succession: 1. mechanical compaction, 2. quartz and feldspar overgrowths, 3. calcite precipitation, 4. dissolution of framework grains and calcite cement, 5. dolomite precipitation, and 6. chlorite formation.

III. MICROSCOPIC HETEROGENEITY

Burial and consequent diagenesis of these lacustrine sandstones produced strong pore modifications, not only in the geometry of individual pores, but also in the whole porosity structure of the sandstones. The evaluation of pore structure leads to a better understanding of fluid distribution, mobility and recovery at the microscopic level, where fluid properties are essentially controlled by capillary forces (Wardlaw and Cassan, 1979; Wardlaw, 1980).

Present porosity values in the main reservoir zones average 21% from petrophysical analyses, and 19% from thin-section determinations. At the microscopic scale, sandstone pore structure (Fig. 3) is characterized by discontinuous patches (islands) of small and badly connected reduced primary (RP)

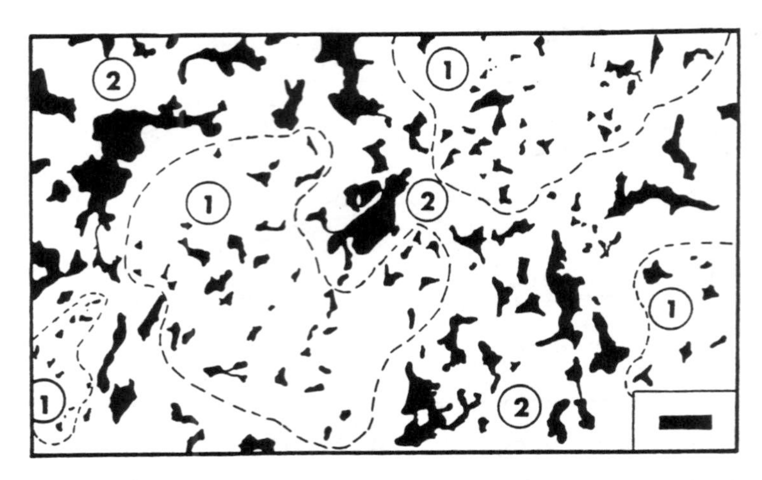

Fig. 3- Map obtained by tracing over a photomicrograph showing the typical pore geometry observed within a deltaic reservoir. 1- RP pores; 2- CD pores.

pores surrounded by a more continuous network of larger and better connected carbonate dissolution (CD) pores.
This type of pore structure is thought to strongly affect microscopic reservoir properties, particularly irreducible water saturation (Swi), which is typically high (>30%) in the reservoirs (Fig. 4).

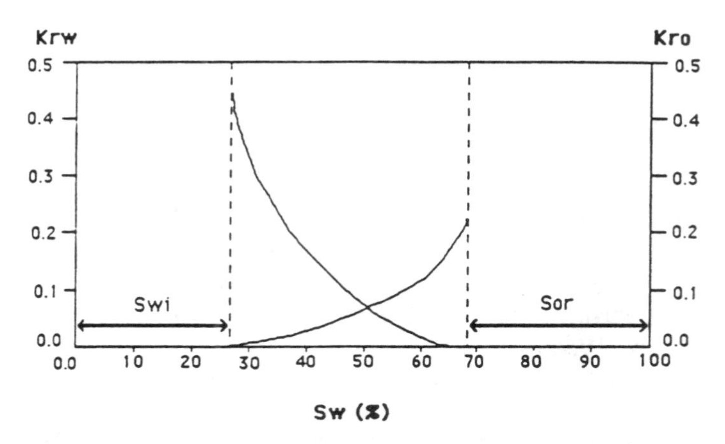

Fig. 4. Typical relative-permeability curve of deltaic reservoirs showing irreducible water saturation (Swi) and residual oil saturation (Sor).

Part of the Swi can be attributed to the microporosity developed within authigenic clay (chlorite) aggregates, or intragranular pores formed by the partial dissolution of framework grains. However, high saturations are found also in sandstones with low clay content and relatively low intragranular porosity. Retention of water as a discontinuous phase within the islands of RP porosity whereas the oil fills the main (CD porosity dominated) pore system is inferred to be another important factor producing high irreducible water saturation (Table 1).

Factors Controlling the Ocurrence of High (> 20%) Irreducible Water Saturations in the Sandstones Reservoirs

Element	Critical Amount
Authigenic Clays (chlorite)	> 5% [a]
Intragranular and Moldic Pores	> 30% [b]
RP porosity	> 30% [b]

[a] Percent of bulk rock volume (BRV)
[b] Percent of total thin section porosity

Table 1. Factors affecting irreducible water saturation (Swi) in the sandstones.

IV. INTERWELL HETEROGENEITY

The relationships between diagenetic properties and depositional facies strongly influence reservoir performance at the interwell level. Diagenetic transformations can enhance, decrease or simply modify the nature of the contrasts inherited from the depositional environment. In the sandstones, dolomite and chlorite are the diagenetic elements which have a significant impact at the interwell scale.

Dolomite cement is observed concentrated along discrete stratification planes, a feature that increases permeability anisotropy and lowers the effective horizontal permeability (Keff) of cross-bedded zones (Fig. 5).

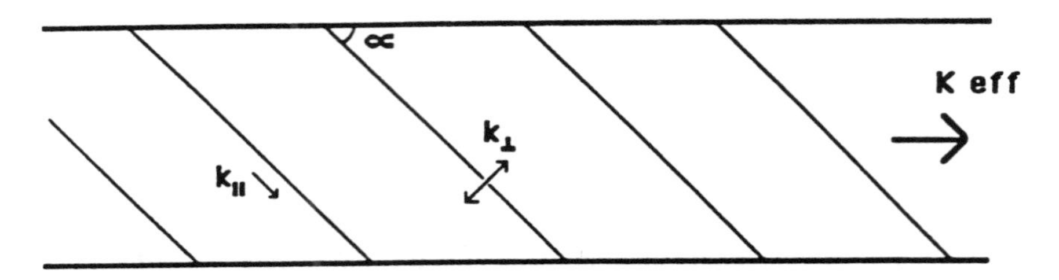

Fig. 5. Elements used for the calculation of effective directional permeability (Keff) in a cross-bedded reservoir.

The dolomite effect was evaluated with data from the Serraria field, Potiguar Basin, using the following relationship (Weber, 1982):

$$\frac{1}{k_{eff}} = \frac{\cos^2 \alpha}{k_{\parallel}} + \frac{\sin^2 \alpha}{k_{\perp}}$$

Where:

α = foreset dip angle

$k_{\parallel}$ = permeability parallel to the foresets

$k_{\perp}$ = permeability normal to the foresets

The calculations indicate that a reservoir layer with 125 md of averaged plug horizontal permeability would have only 8.17 md of effective horizontal permeability if dolomite (> 5% of the bulk rock volume) and high-angle (α>15°) cross-bedding are present.

Chlorite rims on framework grains were observed to reduce permeability in fine-grained sandstones, but had little influence on coarser grained sandstones (Fig. 6). This grain size dependent effect may change the permeability structure of the reservoirs because although deltaic reservoir units consist dominantly of fine-grained sandstones, narrow elongated zones composed of medium- to coarse-grained sandstones appear locally. Such coarser zones, which represent channel-mouth bar deposits, occur associated with the points where distributary channels fed into the delta-front (Alves, 1985; Moraes, 1991). If chlorite is present in both zones, it will have little influence on the coarser zones but will cause a significant permeability reduction in the fine-grained portion of the reservoir.

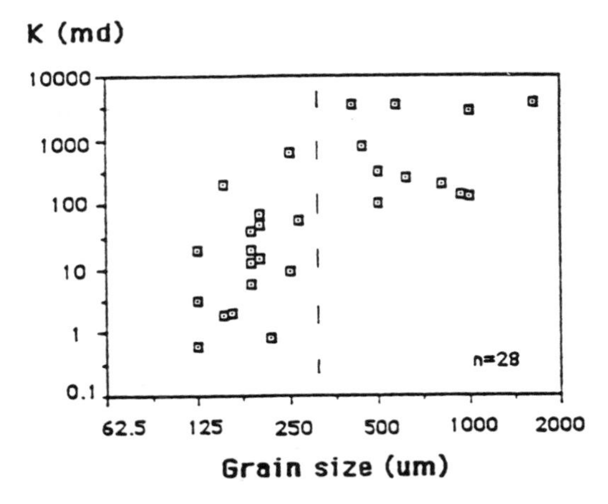

Fig. 6. Plot relating permeability to grain size in deltaic sandstone reservoirs.

Figure 7 illustrates the chlorite effect at the interwell level using an example from the Serraria field. The presence of such narrow high permeability zones encased in a moderate to low permeability region may cause strong irregularities (e. g., fingering phenomena) to develop at the fluid fronts.

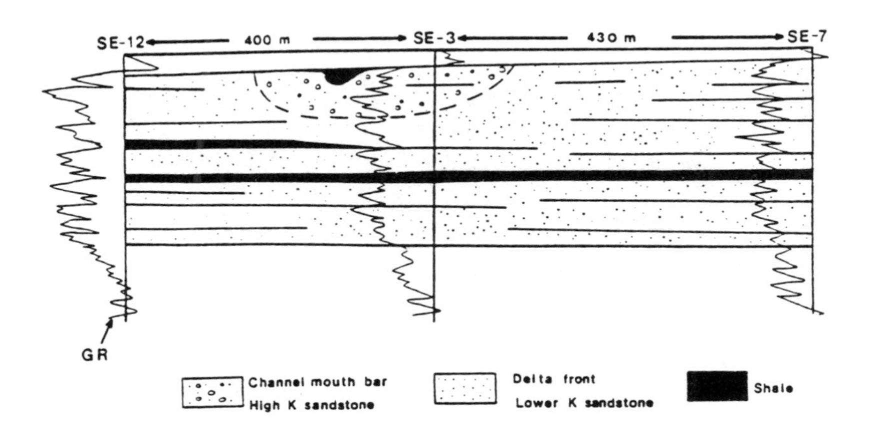

Fig. 7. Geologic section in the deltaic reservoirs of the Serraria field, Potiguar Basin, showing the location of a high permeability zone.

V. FIELD-WIDE HETEROGENEITY

Distribution of diagenetic elements at the field-wide scale is commonly related to both the depositional geometry and to the geological context of hydrocarbon trapping in the different reservoir types. Knowledge of reservoir heterogeneity at this scale serves as a guide for location of development wells and definition of production strategies. Field-wide water flooding and enhanced oil recovery (EOR) projects also will greatly rely on that kind of information.

The most significant diagenetic heterogeneity observed at the field-wide scale is the increase of carbonate cementation toward the outer boundaries of the sandstone accumulations (Anjos and Carozzi, 1988; Moraes, 1991). Such a carbonate cement distribution typically causes the occurrence of poorest reservoir quality in the outer portions of deltaic reservoirs (Fig. 8), and better reservoir quality toward the inner portions of the reservoir. In addition to the carbonate cement effect, the occurrence of narrow high permeability zones at the inner portion of the reservoir shown in Figure 8 is related to the lack of permeability reduction by chlorite in regions composed of coarser grained sandstones. The existence of such narrow high permeability zones at central portions in the reservoir will likely have a strong impact on fluid mobility at the field-wide scale.

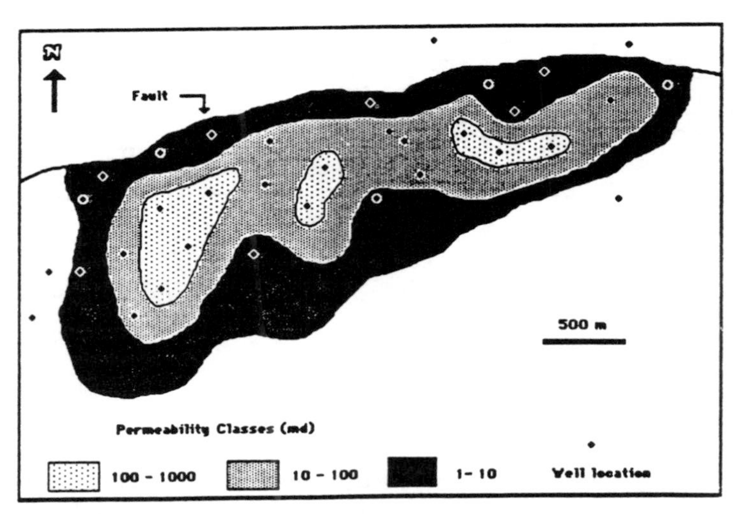

Fig. 8. Reservoir quality map of the deltaic reservoirs of Serraria field, Potiguar Basin.

VI. CONCLUSIONS

Diagenetic models can help to determine the critical aspects controlling fluid distribution and mobility at different reservoir scale levels. These diagenetic models are specially significant when viewed in a stratigraphic/sedimentologic framework. Integrating depositional and diagenetic characteristics at various scales permits a more accurate determination of effective properties used for performance prediction and numerical simulation of hydrocarbon reservoirs. Most of the patterns described herein were found to be repetitive in the lacustrine sandstone reservoirs of northeastern Brazil. Thus, the diagenetic models generated here are thought to have general applicability in the studied basins.

VII. REFERENCES

Alves, A. C., 1985, Petrografia e diagênese dos arenitos reservatório da Formação Pendência (Cretáceo Inferior) na Campo de Serraria, Bacia Potiguar [unpubl. M.S. thesis]: Universidade Federal de Ouro Preto, Ouro Preto, Brazil, 143 p.

Anjos, S. M. C. and Carozzi, A. V., 1988, Depositional and diagenetic factors in the generation of the Santiago arenite reservoirs (Lower Cretaceous): Araças oil field, Recôncavo Basin, Brazil: Jour. South American Earth Sciences, v. 1, p. 3-19.

Becker, M. R., 1984, Petrologia e geologia de reservatório da unidade Catu 5, Grupo Ilhas, campo de Miranga, Bacia do Recôncavo [unpubl. M. S. thesis]: Universidade Federal de Ouro Preto, Ouro Preto, Brazil, 207 p.

Bertani, R. T.; A. F. Apoluceno Netto and R. M. D. Matos, 1987, Habitat do petróleo e perspectivas exploratórias da Bacia Potiguar: Bol. Geoc. Petrobras, v. 1, p. 41-49.

Fogg, G. E. and F. J. Lucia, 1990, Reservoir modeling of restricted plataform carbonates: geologic/geostatistical characterization of interwell-scale reservoir heterogeneity, Dune field, Crane County, Texas: The Bureau of Economic Geology Report of Investigations no. 190, 66 p.

Haldorsen, H. H., 1983, Reservoir characterization procedures for numerical simulation [unpubl. Ph.D. dissertation]: University of Texas, Austin, 147 p.

Haldorsen, H. H. and E. Damsleth, 1990, Stochastic modeling: Jour. Pet. Technology, v. 42, p. 404-412.

Milani, E.D. and I. Davison,1988, Basement control and tectonic transfer in the Recôncavo-Tucano-Jatobá Rift, Brazil: Tectonophysics, v.154, p. 41-70.

Moraes, M. A. S., 1991, Multiscale diagenetic heterogeneity and its influence on the reservoir properties of fluvial, deltaic and turbiditic reservoirs, Potiguar and Recôncavo rift basins, Brazil [unpubl. Ph. D. dissertation]: University of Wyoming, Laramie, 176 p.

Wardlaw, N. C., 1980, The effects of pore structure on displacement efficiency in reservoir rocks and in glass micromodels: Society of Petroleum Engineers of AIME Preprint SPE 8843, p. 345-352.

Wardlaw, N. C. and J. P. Cassan, 1979, Oil recovery efficiency and the rock-pore properties of some sandstone reservoirs: Bull. Can. Pet. Geology, v. 27, p. 117-138.

Weber, K. S., 1982, Influence of common sedimentary structures on fluid flow in reservoir models: Jour. Pet. Technology, v. 34, p. 665-672.

Weber, K. J., 1986, How heterogeneity affects oil recovery, *in* Lake, L. W. and Carrol Jr., H. B. (eds.), Reservoir characterization, Orlando, Florida, Academic Press, p. 487-544.

Weber, K. S. and L. C. van Genus, 1990, Framework for constructing clastic reservoir simulation models: Jour. Pet. Technology, v. 42, p. 1248-1253/1296-1297.

VARIABILITY IN CARBONATE RESERVOIR HETEROGENEITY AMONG JURASSIC SMACKOVER OIL PLAYS OF SOUTHWEST ALABAMA

Ernest A. Mancini
Robert M. Mink
Berry H. Tew
David C. Kopaska-Merkel
Steven D. Mann

Geological Survey of Alabama
Tuscaloosa, Alabama 35486-9780

The Upper Jurassic Smackover Formation is the most prolific petroleum reservoir in southwest Alabama (figs. 1 and 2). To date, 70 fields in this area have produced 258 million barrels of oil and condensate (fig. 3 and Table 1). These fields can be grouped into five distinct oil plays (fig. 4): the basement ridge play, the regional peripheral fault trend play, the Mississippi interior salt basin play, the Mobile graben fault system play, and the Wiggins arch complex play. Plays are recognized by basinal position, relationships to regional structural features, and characteristic petroleum traps. Within the basement ridge play, two subplays, the Choctaw ridge complex subplay and the Conecuh and Pensacola-Decatur ridge complexes subplay, can be distinguished based on oil gravities and reservoir characteristics. Similarly, two subplays, the Pickens, Gilbertown, and West Bend fault systems subplay and the Pollard and Foshee fault systems subplay can be recognized in the regional peripheral fault trend play. The plays are classified into three groups that differ in the scale, type, and range of reservoir heterogeneity (Table 2).

Lithologically, Smackover reservoirs include primarily grainstone, dolograinstone and crystalline dolostone. Pore systems in these reservoirs are dominated volumetrically by particle molds and intercrystalline pores. Moldic, secondary intraparticle, and interparticle pores are a product of depositional fabric modified by diagenesis, whereas intercrystalline pores are created by fabric-destructive, nonselective dolomitization. The plays differ in reservoir characteristics and in the nature of heterogeneity as a result of depositional, diagenetic, and halokinetic processes. Reservoirs in the Choctaw ridge complex subplay are peritidal, nondolomitized to completely dolomitized, oolitic, peloidal, and oncoidal grainstone. Reservoir pore systems are dominated by molds of nonskeletal particles and by secondary intraparticle pores (partial molds). Heterogeneity is low, primarily because hydrocarbon accumulations are small and because the dominant reservoir type, grainstone with moldic and secondary intraparticle pore systems, forms large and homogeneous flow units. Reservoirs in the Conecuh and Pensacola-Decatur ridge complexes subplay are subtidal to supratidal, oolitic, oncoidal, intraclastic, and peloidal dolograinstone and dolopackstone, fenestral dolostone, quartz sandstone, and doloboundstone. The pore systems have substantial amounts of intercrystalline porosity; moldic and secondary intraparticle pores are also common. The variety of lithofacies and pore systems associated with the relatively small hydrocarbon accumulations result in moderate to high reservoir heterogeneity. Multiple reservoir zones are common. Reservoirs in the Pickens, Gilbertown, and West Bend fault systems subplay are peritidal, nondolomitic to completely dolomitized, oolitic, oncoidal, and peloidal grainstone. Reservoir pore systems are dominated by moldic and secondary intraparticle pores. Grainstone lithofacies are areally extensive and this, in combination with relatively well-developed moldic and secondary intraparticle pore systems, produces low reservoir heterogeneity. Reservoirs in the Pollard and Foshee fault systems subplay are subtidal to supratidal, partially to completely dolomitized, peloidal grainstone to wackestone and dolomitized boundstone. Reservoir pore systems are dominated by intercrystalline pores with a high percentage of moldic and secondary intraparticle pores. This subplay is characterized by a variety of reservoir lithofacies and pore systems and by multiple reservoir zones; therefore, heterogeneity is moderate to high. Due, however, to the generally higher reservoir pressures and relatively higher gravities of the liquids produced, hydrocarbon recovery is good. Reservoirs in the Mississippi interior salt basin play are peritidal, nondolomitic to completely dolomitized, peloidal and oolitic grainstone and packstone. In the northern part of this play, reservoir pore systems are dominated by oomolds and secondary intraooid pores, and in the southern part, pelmold and intrapeloid pores are the most important. Heterogeneity is low because the particle-supported reservoir strata are widespread and have well-developed moldic pore systems. Reservoirs in the Mobile graben fault system play are peritidal, peloidal and oolitic dolograinstone to dolowackestone and crystalline dolostone. Reservoir pore systems are dominated by intercrystalline pores. However, in the northern part of the play, moldic and interparticle pores are common. Heterogeneity is variable and ranges from low in the northern part of the play to high in the southern part. Reservoirs in the Wiggins arch complex play are subtidal to supratidal, peloidal, oolitic, oncoidal dolograinstone and dolopackstone and crystalline dolostone. Intercrystalline pores dominate reservoir pore systems. Heterogeneity is high in this play due to the variety of reservoir lithofacies and because of the highly variable permeability distribution associated with the intercrystalline pore system which has a significant effect on producibility.

The distribution of Smackover depositional facies exerts the primary control on reservoir heterogeneity. In general, upper Smackover strata, which include the principal reservoirs, consist of stacked, cyclical, upward-shoaling, carbonate mudstone to grainstone, highstand parasequences that accumulated in a keep-up, shallow carbonate setting during progressive overall relative sea level fall (fig. 5). Relative sea level, as used here, approximates water depth. Depositional facies distribution reflects basin paleogeography and the sedimentary processes associated with the geologic setting; these factors, in turn, relate to reservoir heterogeneity. For example, the basement ridge play occurs in an updip paleogeographic setting associated with basement paleotopography around the margin of the depositional basin. Within this play, the Conecuh and Pensacola-Decatur ridge complexes subplay is characterized by a variety of reservoir lithofacies and pore systems that, in part, result from the sedimentological response to small-scale sea-level fluctuations superimposed on the overall upper Smackover regressive trend. Heterogeneity is low within reservoir strata in the Choctaw ridge complex subplay. However, these grainstone reservoirs are generally thin and are intercalated with a variety of nonreservoir strata that result from the depositional conditions in the updip paleogeographic setting. Thus, rocks in the basement ridge play exhibit the effects of small-scale perturbations on depositional environments in updip areas around the basin periphery. Conversely, reservoirs of the Pickens, Gilbertown, and West Bend fault systems subplay and the Mississippi interior salt basin play, which include strata that were deposited in a more seaward position, are characterized by relatively homogenous, relatively thick, widespread lithofacies and pore systems that reflect deposition in a stable marine shelf setting which was not dramatically affected by minor sea-level changes.

Smackover reservoir heterogeneity is a product of primary depositional patterns modified by diagenesis, mainly dolomitization and leaching of metastable particles. Upper Smackover strata accumulated as progradational, highstand deposits within a unconformity-bounded depositional sequence (fig. 5). The upper sequence boundary is above the Buckner Anhydrite Member of the Haynesville Formation. In southwest Alabama, the lower Buckner is dominated by sabkha and salina deposits on the basin margins and on the flanks of paleohighs, and by impermeable anhydritic subaqueous saltern deposits in the basin centers. During Buckner time, refluxed hypersaline fluids from the Buckner dolomitized much of the uppermost Smackover. After deposition of the Buckner and the lower Haynesville, the climate became less arid and much of the region was subaerially exposed along the type 2 sequence boundary. At this time, meteoric waters entered the Smackover through exposed permeable strata on the basin margins and from isolated emergent paleohighs such as the Wiggins arch and the Conecuh ridge complex. Meteoric water mixed with marine water and/or with Buckner-derived brines in the subsurface, dissolving metastable particles and forming dolomite in the upper Smackover. Some later stage dolomite not associated with the sequence boundary also appears to be important in the Smackover.

Smackover oil plays and subplays in southwest Alabama can be classified into three groups based on relative heterogeneity ranking (fig. 6). Low heterogeneity typifies reservoirs in the Choctaw ridge complex subplay, the Pickens, Gilbertown, and West Bend fault systems subplay and the Mississippi interior salt basin play and results from the restriction of reservoirs to thin, homogeneous intervals and the presence of areally extensive grainstone lithofacies with associated homogeneous pore systems. Reservoirs in the Conecuh and Pensacola-Decatur ridge complexes and Pollard and Foshee fault systems subplays exhibit moderate to high heterogeneity which can be attributed to the variety of reservoir lithofacies and pore systems present. Heterogeneity in the Mobile graben fault system is variable, ranging from low to high. High reservoir heterogeneity in the Wiggins arch complex play results from the variety of lithofacies present and the highly variable nature of the intercrystalline reservoir pore systems.

This research was partially funded by the United States Department of Energy under Contract No. DE-FG22-89BC14425.

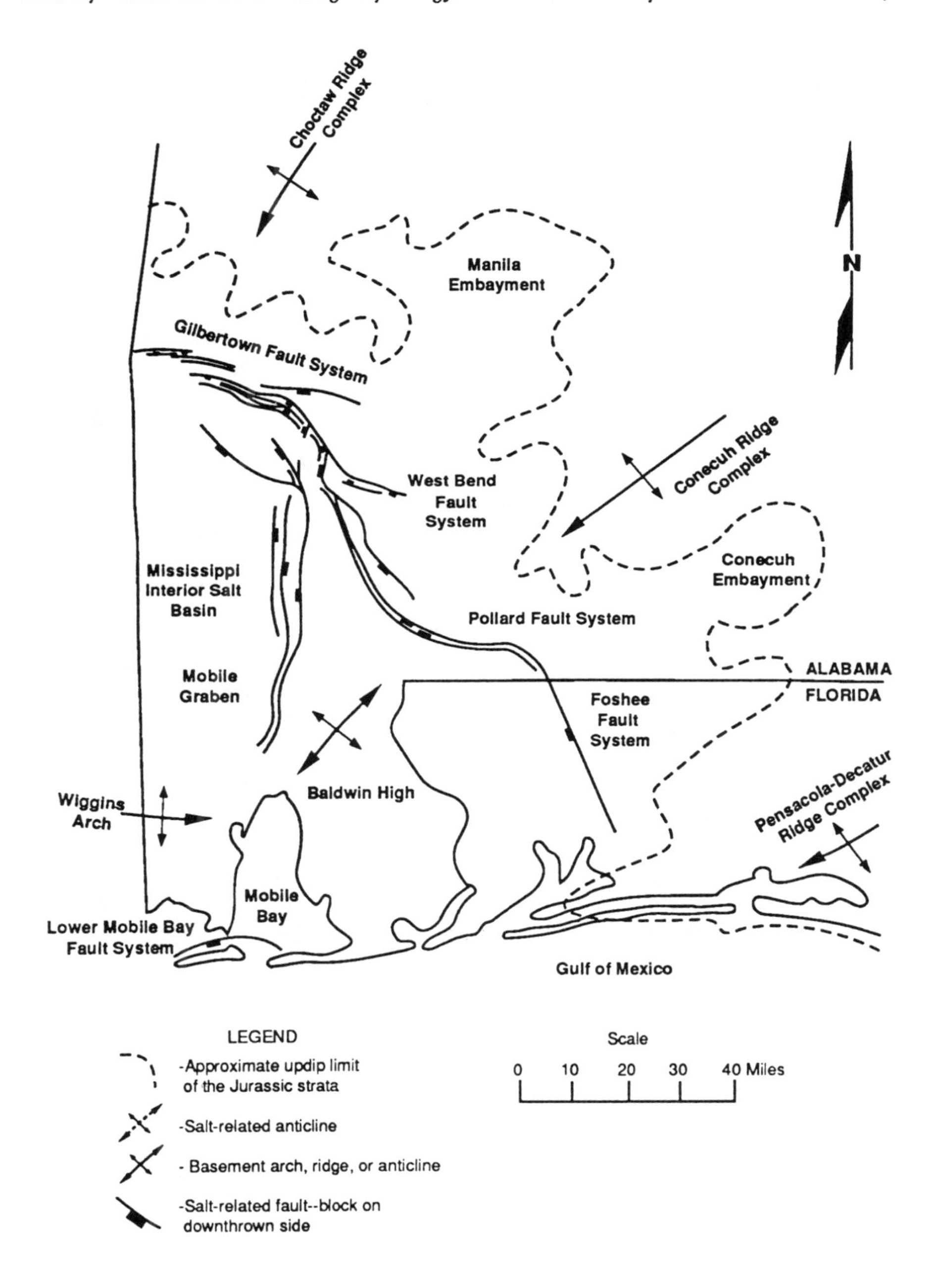

Figure 1. Map of southwest Alabama showing major structural features.

SERIES	STAGE	ROCK UNIT
Lower Cretaceous	Berriasian	Cotton Valley Group
Upper Jurassic	Tithonian	
	Kimmeridgian	Haynesville Formation
		Buckner Anhydrite Member
	Oxfordian	Smackover Formation
		Norphlet Formation
		Pine Hill Anhydrite Member
Middle Jurassic	Callovian	Louann Salt
		Werner Formation
Lower Jurassic/ Upper Triassic		Eagle Mills Formation
		Undifferentiated Paleozoic and Proterozoic "basement" rocks

Figure 2. Jurassic stratigraphy of southwest Alabama.

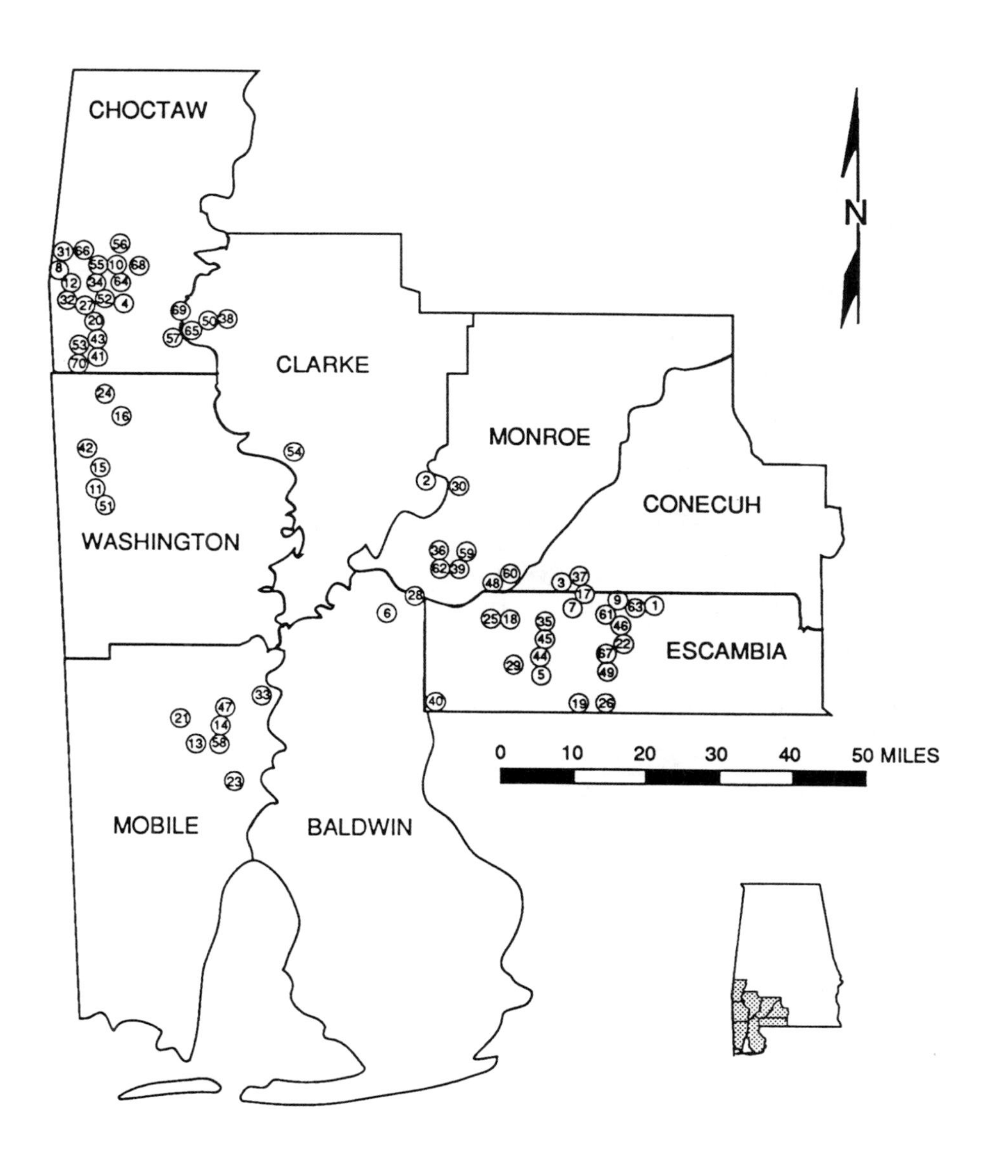

Figure 3. Location of Smackover fields in southwest Alabama.

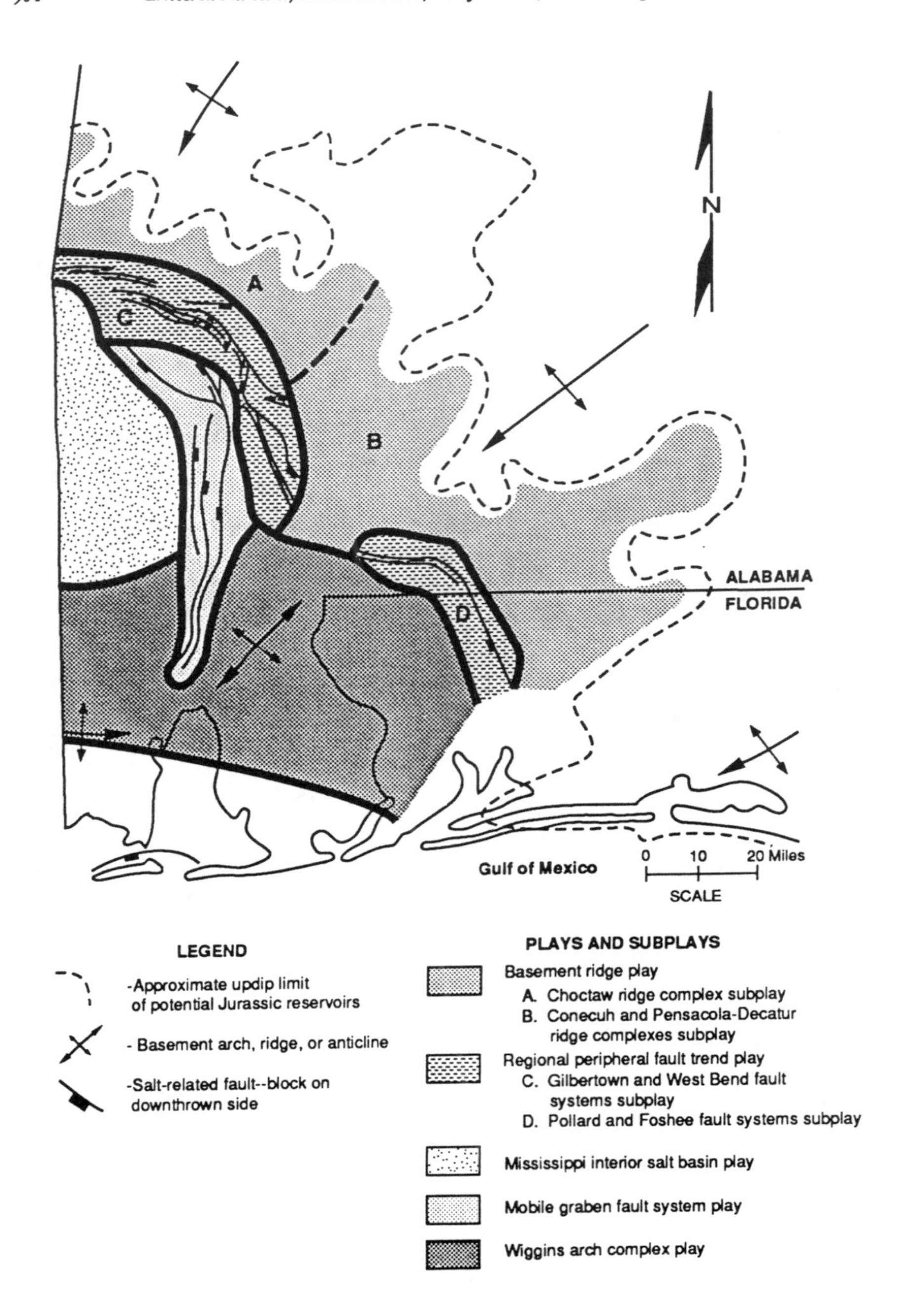

Figure 4. Smackover oil plays in southwest Alabama.

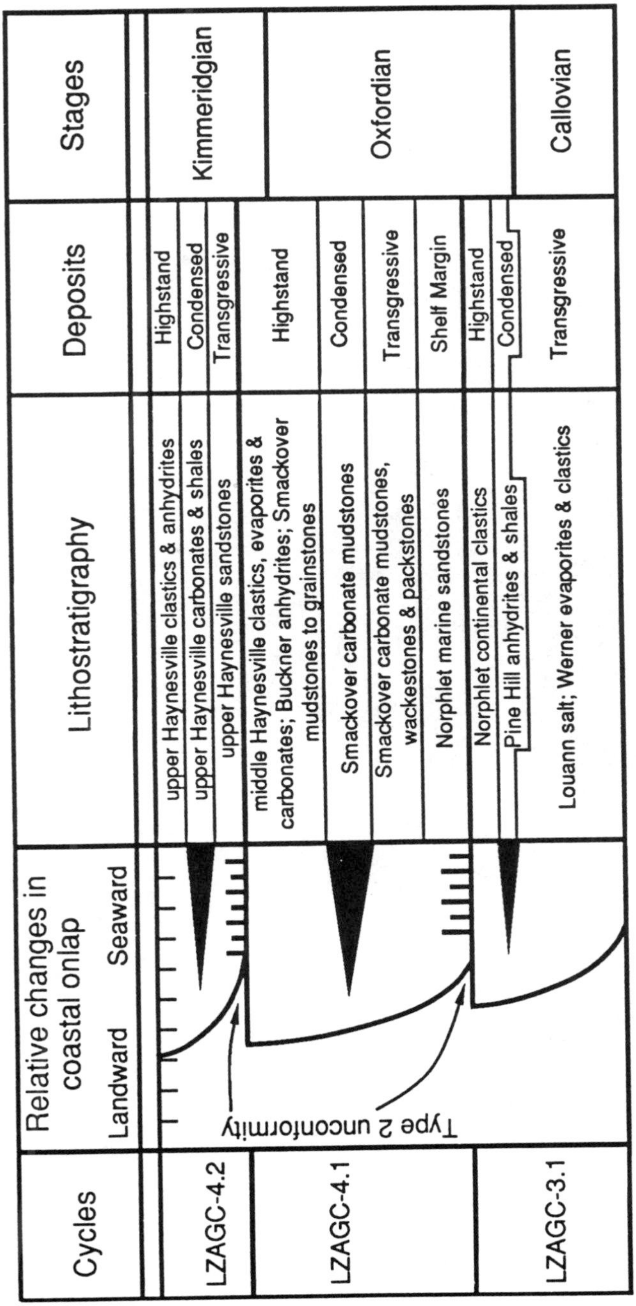

Cycles	Relative changes in coastal onlap (Landward – Seaward)	Lithostratigraphy	Deposits	Stages
LZAGC-4.2		upper Haynesville clastics & anhydrites	Highstand	Kimmeridgian
		upper Haynesville carbonates & shales	Condensed	
		upper Haynesville sandstones	Transgressive	
LZAGC-4.1	Type 2 unconformity	middle Haynesville clastics, evaporites & carbonates; Buckner anhydrites; Smackover mudstones to grainstones	Highstand	Oxfordian
		Smackover carbonate mudstones	Condensed	
		Smackover carbonate mudstones, wackestones & packstones	Transgressive	
		Norphlet marine sandstones	Shelf Margin	
LZAGC-3.1		Norphlet continental clastics	Highstand	
		Pine Hill anhydrites & shales	Condensed	
		Louann salt; Werner evaporites & clastics	Transgressive	Callovian

Figure 5. Jurassic sequence stratigraphy in southwest Alabama.

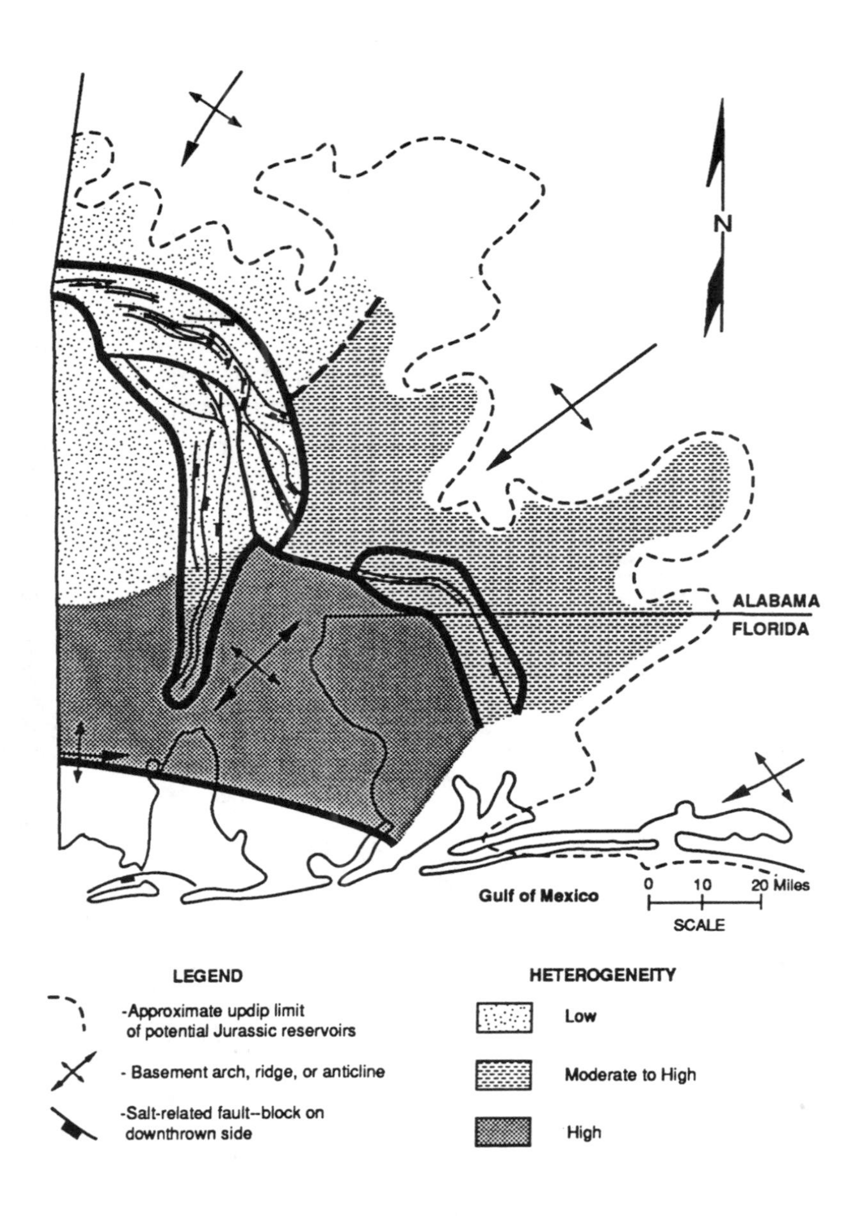

Figure 6. Relative levels of heterogeneity of Smackover oil plays and subplays in southwest Alabama.

Table 1. List of Smackover fields in southwest Alabama.

Field No.	Field Name	County	Hydrocarbon Type
1	Appleton	Escambia	Oil
2	Barlow Bend	Clarke and Monroe	Oil
3	Barnett	Conecuh and Escambia	Oil
4	Barrytown	Choctaw	Oil
5	Big Escambia Creek	Escambia	Condensate
6	Blacksher	Baldwin	Oil
7	Broken Leg Creek	Escambia	Oil
8	Bucatunna Creek	Choctaw	Oil
9	Burnt Corn Creek	Escambia	Oil
10	Chappell Hill	Choctaw	Oil
11	Chatom	Washington	Condensate
12	Choctaw Ridge	Choctaw	Oil
13	Chunchula	Mobile	Oil/Condensate
14	Cold Creek	Mobile	Oil
15	Copeland Gas	Washington	Condensate
16	Crosbys Creek	Washington	Condensate
17	East Barnett	Conecuh and Escambia	Oil
18	East Huxford	Escambia	Oil
19	Fanny Church	Escambia	Oil
20	Gin Creek	Choctaw	Oil
21	Gulf Crest	Mobile	Oil
22	Hanberry Church	Escambia	Oil
23	Hatter's Pond	Mobile	Condensate
24	Healing Springs	Washington	Condensate
25	Huxford	Escambia	Oil
26	Little Escambia Creek	Escambia	Oil
27	Little Mill Creek	Choctaw	Oil
28	Little River	Baldwin and Monroe	Oil
29	Liitle Rock	Escambia	Condensate
30	Lovetts Creek	Monroe	Oil
31	Melvin	Choctaw	Oil
32	Mill Creek	Choctaw	Oil
33	Movico	Baldwin and Mobile	Oil
34	North Choctaw Ridge	Choctaw	Oil
35	North Smiths Church	Escambia	Oil

Table 1. List of Smackover fields in southwest Alabama (continued).

Field No.	Field Name	County	Hydrocarbon Type
36	North Wallers Creek	Monroe	Oil
37	Northeast Barnett	Conecuh	Oil
38	Pace Creek	Clarke	Oil
39	Palmers Crossroads	Monroe	Oil
40	Perdido	Baldwin and Escambia	Oil
41	Puss Cuss Creek	Choctaw	Oil
42	Red Creek	Washington	Condensate
43	Silas	Choctaw	Oil
44	Sizemore Creek Gas	Escambia	Condensate
45	Smiths Church	Escambia	Condensate
46	South Burnt Corn Creek	Escambia	Oil
47	South Cold Creek	Mobile	Oil
48	South Vocation	Monroe	Oil
49	South Wild Fork Creek	Escambia	Condensate
50	South Womack Hill	Choctaw and Clarke	Oil
51	Southeast Chatom	Washington	Condensate
52	Southwest Barrytown	Choctaw	Oil
53	Souwilpa	Choctaw	Condensate
54	Stave Creek	Clarke	Oil
55	Sugar Ridge	Choctaw	Oil
56	Toxey	Choctaw	Oil
57	Turkey Creek	Choctaw and Clarke	Oil
58	Turnerville	Mobile	Oil
50	Uriah	Monroe	Oil
60	Vocation	Monroe	Oil
61	Wallace	Escambia	Oil
62	Wallers Creek	Monroe	Oil
63	West Appleton	Escambia	Oil
64	West Barrytown	Choctaw	Oil
65	West Bend	Choctaw and Clarke	Oil
66	West Okatuppa Creek	Choctaw	Oil
67	Wild Fork Creek	Escambia	Oil
68	Wimberly	Choctaw	Oil
69	Womack Hill	Choctaw and Clarke	Oil
70	Zion Chapel	Choctaw	OIl

Table 2. Summary of resevoir characteristics and levels of heterogeneity for Smackover oil plays and subplays in southwest Alabama.

Play/Subplay	Reservoir	Pore System	Heterogeneity
Basement Ridge Play			
Choctaw Ridge Complex Subplay	Grainstone Dolograinstone	Moldic Intraparticle	Low
Conecuh and Pensacola-Decatur Ridge Complexes Subplay	Dolograinstone Dolopackstone Doloboundstone Dolostone Sandstone	Intercrystalline Moldic Intraparticle	Moderate to High
Regional Peripheral Fault Trend Play			
Pickens, Gilbertown, and West Bend Fault Systems Subplay	Grainstone Dolograinstone	Moldic Intraparticle	Low
Pollard and Foshee Fault Systems Subplay	Grainstone Packstone Wackestone Boundstone Dolograinstone Dolopackstone Dolowackestone Doloboundstone	Intercrystalline Moldic Intraparticle	Moderate to High
Mississippi Interior Salt Basin Play	Grainstone Packstone Dolograinstone Dolopackstone	Moldic Intraparticle	Low
Mobile Graben Fault System Play	Dolograinstone Dolopackstone Dolowackestone Dolostone	Intercrystalline Moldic Intraparticle Interparticle	Low to High
Wiggins Arch Complex Play	Dolograinstone Dolopackstone Dolostone	Intercrystalline	High

Minipermeameter Study of Fluvial Deposits of the Frio Formation (Oligocene), South Texas: Implications for Gas Reservoir Compartments[1]

Dennis R. Kerr[2], Andrew R. Scott
Jeffry D. Grigsby, and Raymond A. Levey
Bureau of Economic Geology
University Station, Box X
Austin, Texas 78713-7508

I. INTRODUCTION

Fluvial deposits of the Oligocene Frio Formation in South Texas contain significant natural gas reserves (1). Production history and completion pressure reports offer evidence for the existence of gas reservoir compartments within mature reservoirs and fields. A challenging question to the Secondary Natural Gas Recovery Joint Venture Program (cosponsored by the Gas Research Institute, the U.S. Department of Energy, and the State of Texas) is whether incompletely drained compartments exist in gas reservoirs in which sandstone-on-sandstone contacts are common. To address this issue, a minipermeameter study was undertaken on cores collected from Stratton field (Fig. 1).

Fluvial deposits at Stratton field are included in the Gueydan depositional system of the Texas Gulf Coast Basin (2). Four facies comprise the fluvial Frio (3): channel-fill, splay, levee, and floodplain. The channel fill facies is divided into three subfacies (Figs. 2, 3, and 4): (1) lower channel fill consisting of 0.5- to 1.5-ft-thick sets of trough cross-stratified medium sandstone having intraclast lags, (2) middle channel fill consisting of low-angle parallel and ripple-stratified very fine to fine sandstone having chute channel-fill (Fig. 4) and lateral accretion

---

[1] Publication authorized by the Director, Bureau of Economic Geology, The University of Texas at Austin.

[2] Current address: Department of Geosciences, University of Tulsa, Tulsa, Oklahoma.

surfaces, and (3) upper channel fill consisting of structureless to poorly developed parallel and ripple-stratified, muddy, very fine sandstone to sandy mudstone. Commonly associated with the upper channel fill are pedogenic carbonate and clay cements, and infiltrated fines lining intergranular pores, compacted fractures, and root molds. Architectural stacking patterns of discrete genetic intervals (i.e., deposits of a discrete channel system with attendant splays) range from *laterally stacked*, in which sandstone bodies contact one another or are separated by only a few feet of floodplain mudstones, to *vertically stacked*, in which sandstone bodies are separated by several feet of floodplain mudstones. Channel-on-channel contacts are prevalent in the laterally stacked architectural style. The question of whether enough permeability contrast exists among the *sandstones of the channel-fill subfacies* to produce a barrier or baffle to gas flow is key to this study.

Two distinct reservoirs (types I and II) are differentiated on the basis of framework mineralogy, diagenetic history, and reservoir quality in fluvial Frio sandstones (4; Fig. 5). Type I sandstones are poorly to moderately sorted feldspathic litharenites to litharenites; lithic fragments are predominantly volcanic and carbonate rock fragments. Major authigenic minerals are calcite, chlorite, and kaolinite; minor amounts of quartz and feldspar overgrowths and pyrite are present. Type I reservoirs have a strong positive correlation between porosity and permeability, with permeability reaching the 1,000's of millidarcies (Fig. 5). This relationship indicates a well-connected intergranular pore system. By contrast, type II sandstones are poorly to moderately sorted lithic arkoses to feldspathic litharenites; lithic fragments are predominantly volcanic, with carbonate rock fragments constituting less than 1 % of the rock volume. Volcanic glass detritus occurs exclusively in type II sandstones. Depositional matrix averages 20 %. Major authigenic minerals are calcite, analcime, and illite-smectite; minor minerals include pyrite and very finely crystalline quartz, feldspar, and chlorite. Type II reservoirs have a weak positive correlation between porosity and permeability, with permeability being limited to less than 100 millidarcies (Fig. 5). This relationship indicates that the pore networks are poorly connected even in rocks that have high porosity.

II. MINIPERMEAMETER STUDY

More than 1,000 permeability measurements were made on three core segments (Figs. 2, 3, and 4) using a minipermeameter (mechanical field permeameter). To evaluate the permeability distribution within major channel-fill subfacies of types I and II reservoirs, two parallel transects of closely spaced (0.1-ft) intervals were made on slabbed, air-dried cores. The depth and effective radius of investigation of the minipermeameter are 0.35 and 0.18 in, respectively. Abnormally high permeability values, resulting from the desiccation of clays in clayey intervals or hairline surface fractures on the core face, were eliminated from the data base.

A very good correlation exists between the average permeability determined from minipermeameter measurements and the permeability relative to air determined by conventional methods (Fig. 6), suggesting that the minipermeameter can be used reliably to evaluate permeability trends. However, at low values of permeability the minipermeameter is apparently less reliable (Fig. 6). This, plus the detection limit of the minipermeameter (approximately 0.1 md), contributes to the distortion of the distribution of minipermeameter data (Fig. 6).

Semivariograms were generated to assess the spatial variation of permeability in types I and II reservoirs. The permeability data required a square-root transformation to achieve an approximation of a normal distribution required for the generation of semivariograms (Fig. 7). Square-root transformation of permeability data for statistical analysis of eolian sandstones has also been reported by Goggin *et al.* (5).

The upper channel-fill subfacies of both types I and II reservoirs show no spatial correlation of permeability, indicating that higher permeability zones are localized and have limited vertical extent (Fig. 8). Porosity and permeability are limited because of a clay matrix and calcite (micrite) cement. Semivariograms of the middle and lower channel-fill subfacies from both types of reservoirs indicate a spatial correlation of permeability (Fig. 8). Damped oscillations about the sill, called a hole effect, reflect the distribution of permeability. These quasi-periodic components are related to the thickness of cross-strata sets and stratification; at distances on the order of tens of feet, the larger oscillations are related to channel thicknesses (e.g., Figs. 2, 3, and 4). Both types of reservoirs have a nugget effect that is approximately 13 % of total variance, suggesting permeability heterogeneity at the finest scales. The middle and lower channel fill subfacies of type I reservoirs have a range larger (more than 1 ft) than that of type II reservoirs, which have ranges of less than 1 ft (Fig. 8), indicating that permeability can be correlated over relatively greater distances in type I reservoirs. The abundant detrital matrix material; pore-filling authigenic clays, calcite, and analcime; and grain-moldic porosity reduce the vertical continuity of permeability in type II reservoirs.

III. IMPLICATIONS

Thin-section petrography and semivariograms of upper channel-fill subfacies in both types I and II reservoirs indicate that the vertical distribution of permeability is limited because of abundant detrital clays and calcite cement. Therefore, the upper-channel-fill subfacies are likely baffles to gas flow, and understanding the distribution of this subfacies is important for evaluating reservoir compartmentalization within mature reservoirs and fields.

Semivariograms of middle and lower-channel-fill subfacies for types I and II reservoirs indicate vertical continuity of permeability. However, permeability continuity is more restricted in type II reservoirs because of extensive diagenesis.

ACKNOWLEDGMENTS

This study was funded by the Gas Research Institute under contract number 5088-212-1718, the U.S. Department of Energy under contract number DE-FG21-88MC25031, and the State of Texas through the Bureau of Economic Geology. The cooperation of Union Pacific Resources is gratefully acknowledged. We thank G.W. Davies of BP Research and Tucker F. Hentz of the Bureau of Economic Geology for their careful review of the manuscript. Autumn L. Laughrun assisted in data entry and permeability calculations. Drafting was done by Joel L. Lardon and Kerza A. Prewitt under the direction of Richard L. Dillon, chief carographer, Bureau of Economic Geology.

REFERENCES

1. Kosters, E.C., Bebout, D.G., Seni, S.J., Garrett, C.M., Jr., Brown, L.F., Jr., Hamlin, H.S., Dutton, S.P., Ruppel, S.C., Finley, R.J., and Tyler, Noel, 1989, Atlas of major Texas gas reservoirs: The University of Texas at Austin, Bureau of Economic Geology, 161 p.

2. Galloway, W. E., Hobday, D. K, and Magara, K., 1982, Frio Formation of the Texas Gulf Coast Basin--depositional systems, structural framework, and hydrocarbon origin, migration, distribution, and exploration potential: The University of Texas at Austin, Bureau of Economic Geology Report of Investigations No. 122, 78 p.

3. Kerr, D. R., and Jirik, L. A., 1990, Fluvial architecture and reservoir comparmentalization in the Oligocene middle Frio Formation, South Texas: Gulf Coast Association of Geological Societies v. 40, p. 373-380.

4. Grigsby, J. D., and Kerr, D. R., 1991, Diagenetic variability in middle Frio Formation gas reservoirs (Oligocene), Seeligson and Stratton fields, South Texas: Gulf Coast Association of Geological Societies Transactions, v. 41 (in press).

5. Goggin, D. J., Chandler, M. A., Kocurek, G., and Lake, L. W., 1989, Permeability transects in eolian sands and their use in generating random permeability fields: Society of Petroleum Engineers paper 19586, p.149-164.

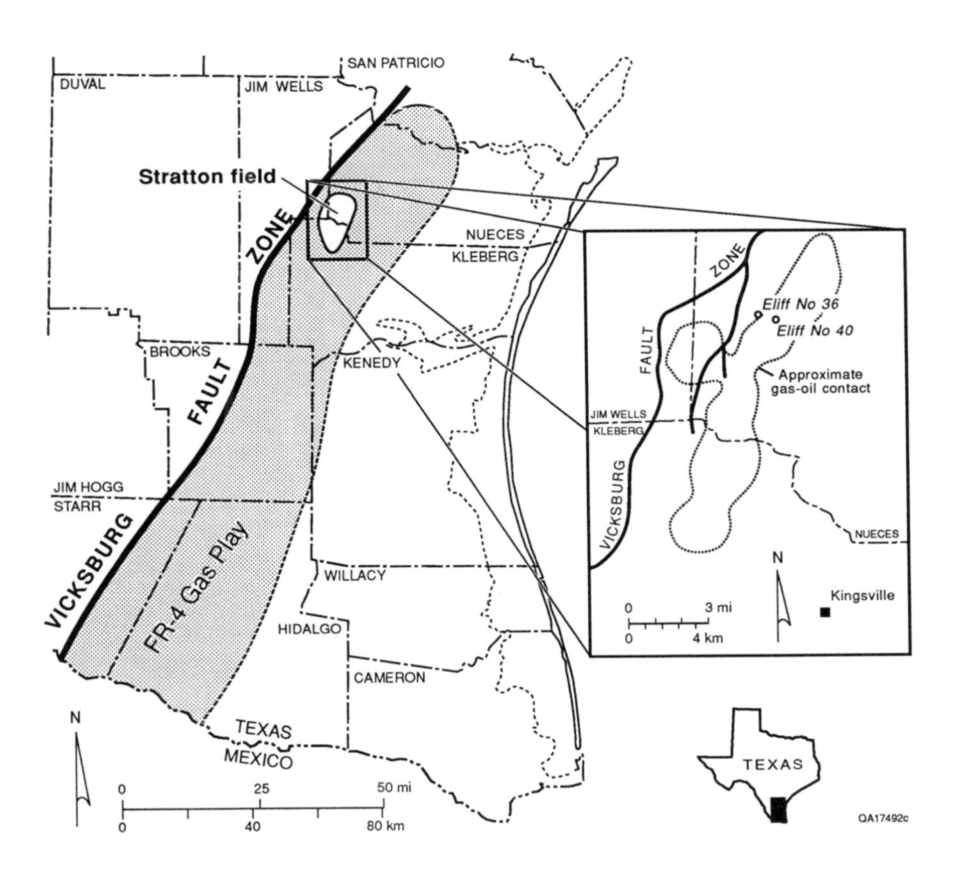

Fig. 1. Index maps showing the location of Stratton field within the FR-4 gas play (1) and the location of Elliff Nos. 36 and 40 wells within Stratton field.

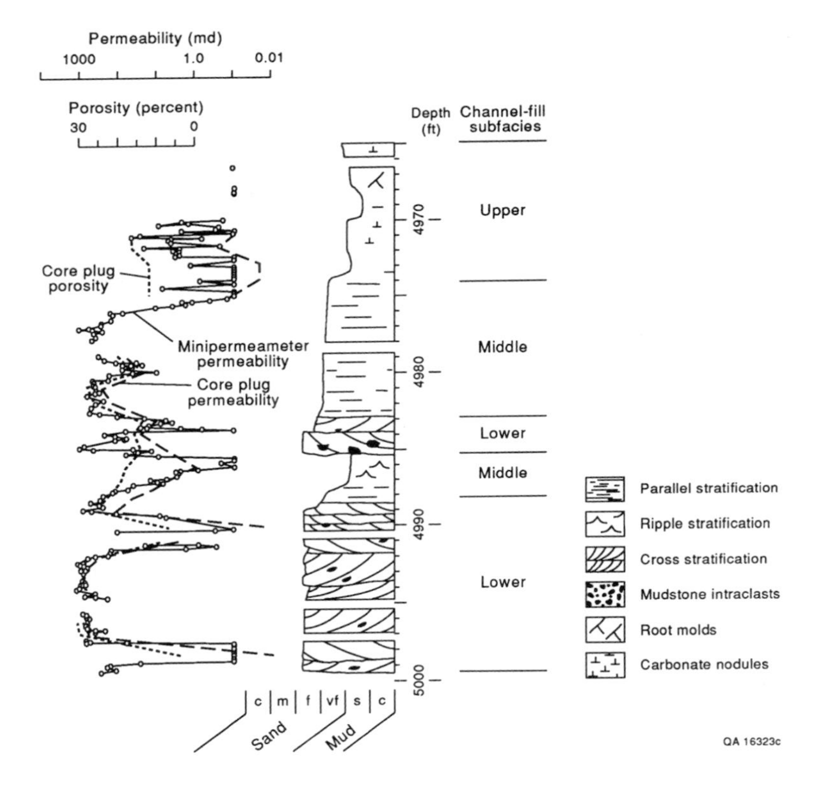

Fig. 2. Core graphic of Elliff No. 40 core 1 (C18 reservoir) showing minipermeameter, core plug permeability, and porosity profiles for a type I reservoir. Minipermeameter values are averaged from two closely spaced vertical transects.

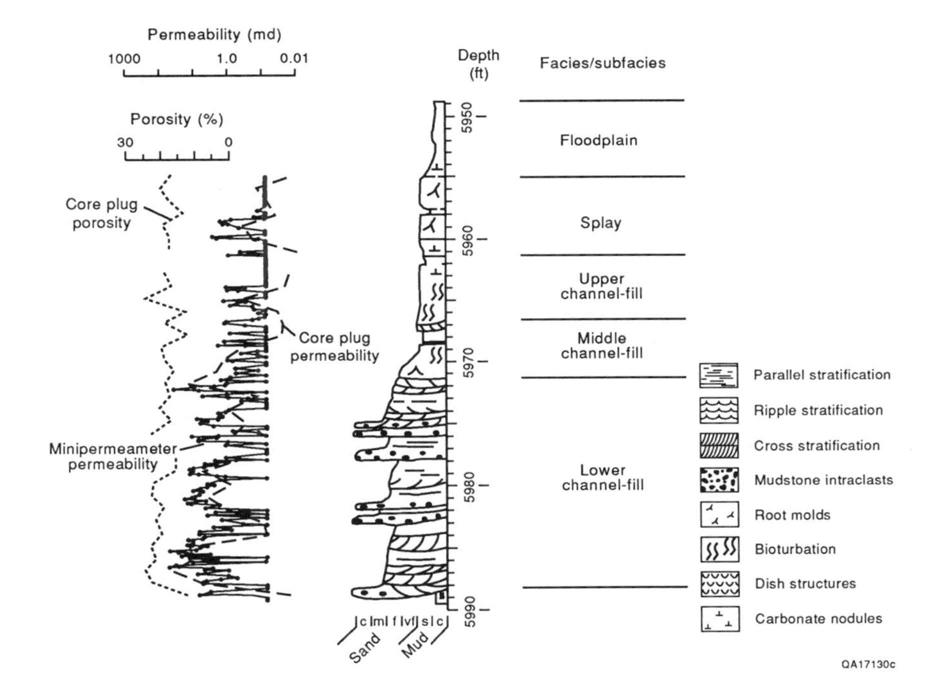

Fig. 3. Core graphic of Elliff No. 40 core 3 (E13 reservoir) showing minipermeameter, core plug permeability, and porosity profiles for a type II reservoir. Minipermeameter values are averaged from two closely spaced vertical transects. Permeability is significantly lower in type II reservoirs than in type I reservoirs (Fig. 2).

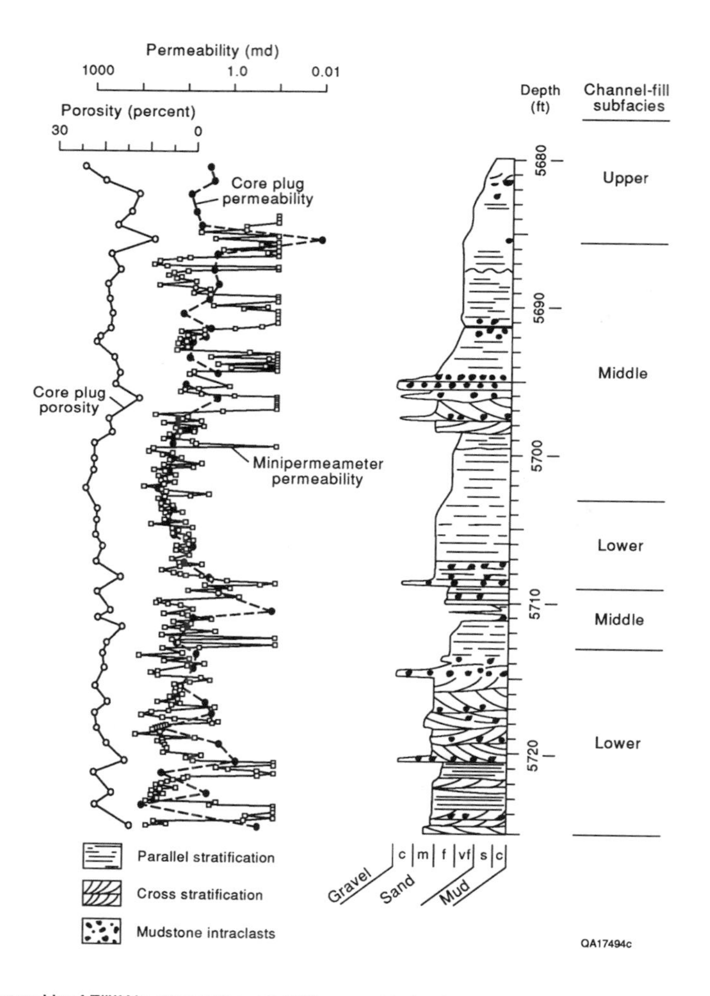

Fig. 4. Core graphic of Elliff No. 36 core 2 and 3 (D35 reservoir) showing minipermeameter, core plug permeability, and porosity profiles for a type II reservoir. Minipermeameter values are averaged from two closely spaced vertical transects. Chute channel-fill deposits (5691.0 to 5698.5) are interpreted from detailed cross-section correlation and facies mapping.

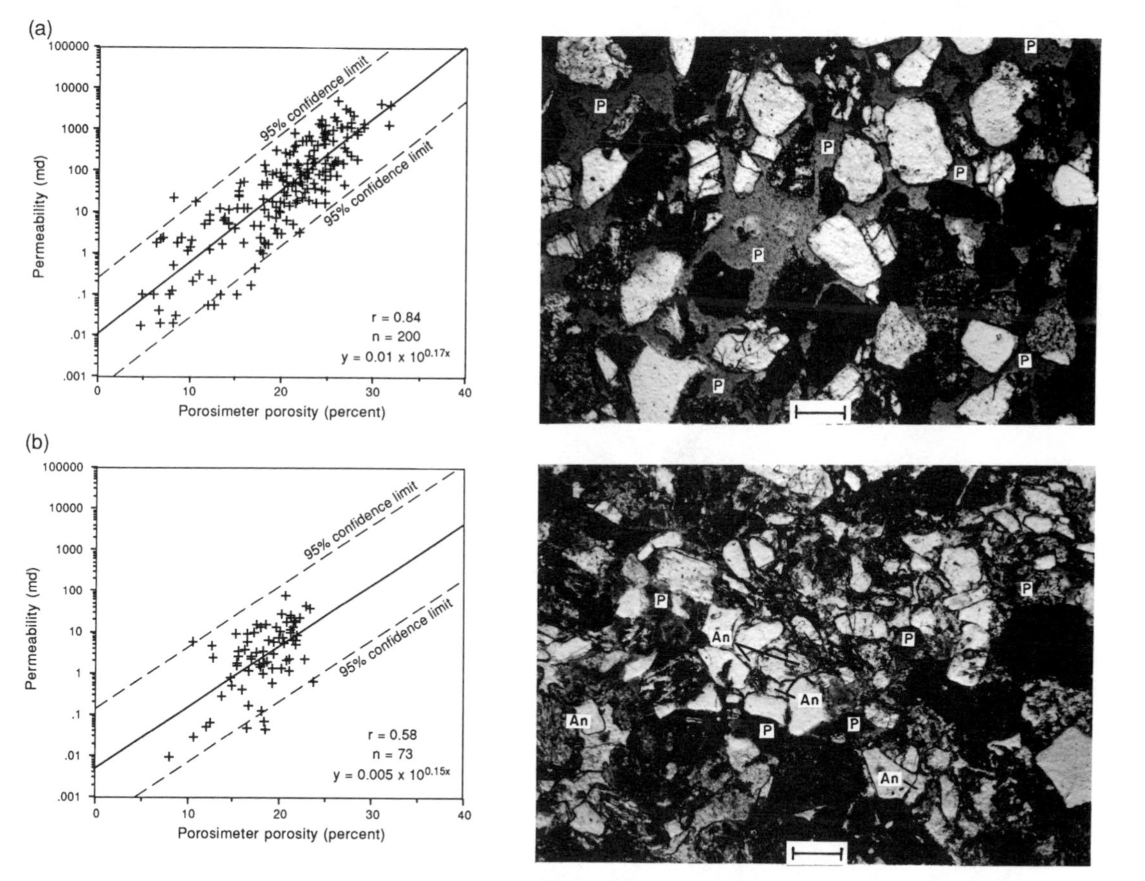

Fig. 5. Porosity-permeability cross plots for (a) type I and (b) type II reservoirs. The strong correlation between porosity and core-plug permeability in type I reservoirs indicates well-developed intergranular porosity and good permeability distribution. The weaker correlation in type II reservoirs demonstrates the effect of poorly developed intergranular porosity and lower permeability. Photomicrographs illustrate the differences in porosity development between types I and II reservoirs. Note poorly connected pores in type II reservoirs. An=analcime; P=pore. Length of bar is 250 microns.

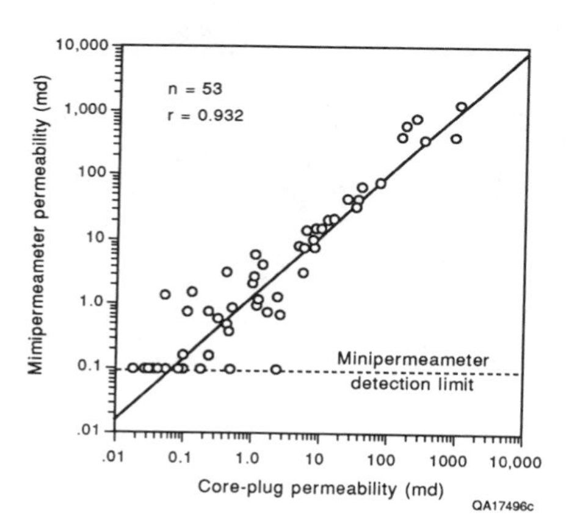

Fig. 6. Permeability-permeability cross plot showing the relation between minipermeameter and conventional core plug permeability measurements. Minipermeameter measurements were taken from the ends of the core plugs after the plug permeability measurements were made. The high correlation coefficient supports the reliability of minipermeameter measurements.

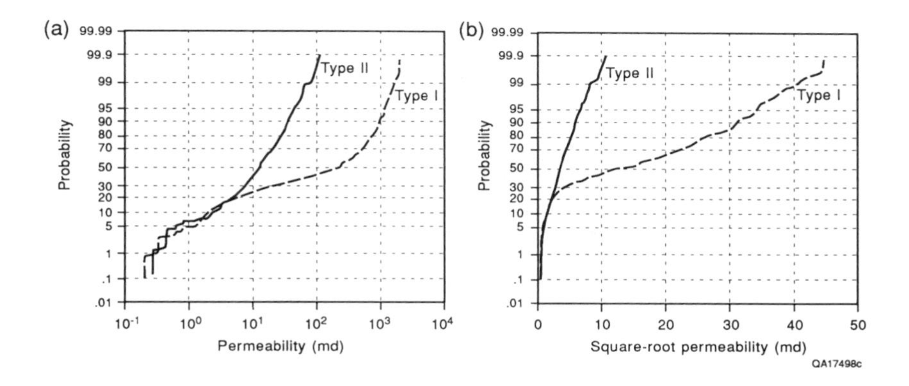

Fig. 7. Cumulative relative frequency plots of minipermeameter data from types I and II reservoirs: (a) log transformation and (b) square-root transformation. Note how the square-root transformation provides a better approximation of a normal distribution (except at low permeability values) than the log transformation.

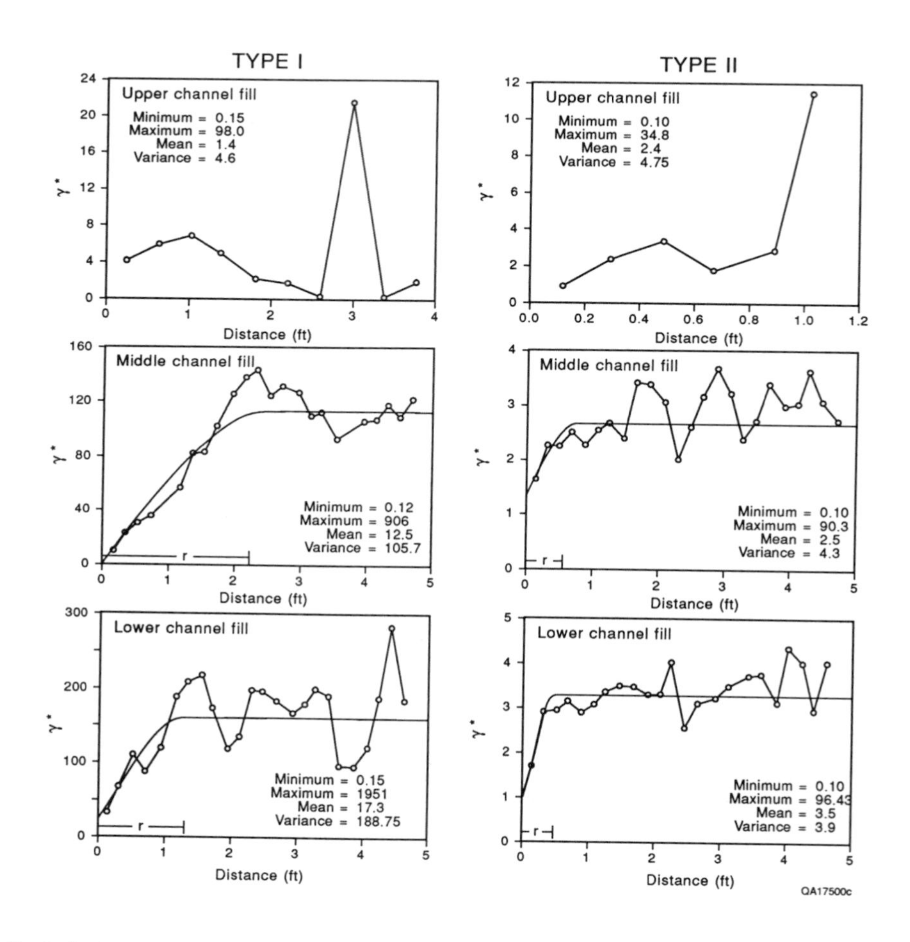

Fig. 8. Semivariograms for upper, middle, and lower channel-fill subfacies for types I and II reservoirs (Figs. 2, 3, and 4). Semivariograms for the upper channel-fill subfacies of both types of reservoirs indicate no spatial correlation of permeability, suggesting that higher permeability zones within this subfacies are localized. Structural analysis of the semivariograms was done using spherical modeling (γ*= gamma function; r=range). The mean and variance are given in square-root units.

HIGH RESOLUTION SEISMIC IMAGING OF FRACTURED ROCK

Ernest Majer
Larry Myer
John Peterson

Earth Sciences Division, Lawrence Berkeley Laboratory
University of California, Berkeley, California 94720

Peter Blumling

Swiss National Cooperative for the Storage of Nuclear Waste, NAGRA, Hardstrasse 73, CH-543o Wettington, Switzerland

I. INTRODUCTION

From 1987 through 1989 the U. S. Department of Energy (DOE) participated in an agreement with the Swiss National Cooperative for the Storage of Radioactive Waste (Nagra) to perform joint research on various topics related to geologic storage of nuclear waste. As part of this Nagra-DOE Cooperative (NDC-I) project Lawrence Berkeley Laboratory (LBL) participated in several projects at an underground research facility in fractured granite in Switzerland (Grimsel) which were directed towards improving the understanding of the role of fractures in the isolation of nuclear waste. Described here are the results of a series of experiments at Grimsel called the Fracture Research Investigation (FRI). The FRI project was designed to address the effects of fractures on the propagation of seismic waves and the relationship of these effects to the hydrologic behavior. The fundamental design of the project was to find a simple, well defined, accessible fracture zone surrounded by relatively unfractured rock and use this zone as a point of comparison for seismic, hydrologic and mechanical behavior. The FRI site was designated in the Grimsel Rock Laboratory for this purpose and field work was carried out during each year of the three year NDC-I project (1987-1989). The FRI project has involved separate and simultaneous, detailed geologic studies, field measurements of seismic wave propagation, geomechanical measurements of fracture properties, and the hydrologic response of the fractured rock.

Laboratory measurements of core have also been carried out to address the fundamental nature of seismic wave propagation in fractured rock. Because all these studies were focussed on the same fracture zone, the study provided insight into new theories of seismic wave propagation through fractures, how changes in the fracture properties affect seismic wave propagation, interpretation of seismic tomography to identify hydrologic features, and integration of seismic data into a hydrologic testing plan.

II. SEISMIC IMAGING EXPERIMENTS

The FRI experiment offered an excellent opportunity to perform calibrated experiments in a rock mass where the fracture locations and characteristics are relatively well known. An advantage of the FRI site was that there is access to the fracture zone from all four sides which allowed a comparison of techniques between two-, three-, and four-sided tomography. The greatest attraction, however, was the opportunity to evaluate theories of wave propagation in fractured media and to evaluate these theories at several different scales. For example, Pyrak *et al.*, (1990a) have performed laboratory experiments which have confirmed the effect of fracture stiffness at small scales. The scaling of this phenomenon to larger distances is yet unknown. Therefore, one of the main objectives of the FRI experiments was to observe the effect of individual fractures as well as a fracture zone on the propagation of seismic waves. A second objective was to relate the seismic response to the hydrologic behavior of the fractures, i.e., do all fractures effect the seismic wave, or do just fluid filled, or partially saturated fractures effect the seismic waves in a measurable amount. The final objective of the study was to assess the amount of seismic data necessary to provide useful information, and how one would process data for the maximum information in a routine fashion. These are important questions when one progresses to the point of applying these techniques to actual field sites.

A. FRI Zone Experimental Procedure

Shown in Figure 1 is a map showing the FRI zone in the plane of boreholes BOFR 87.001 and BOFR 87.002. As shown in this figure, there is a common shear feature crossing the FRI zone. The boreholes were drilled to intersect this fracture zone as shown in Figure 1. Three

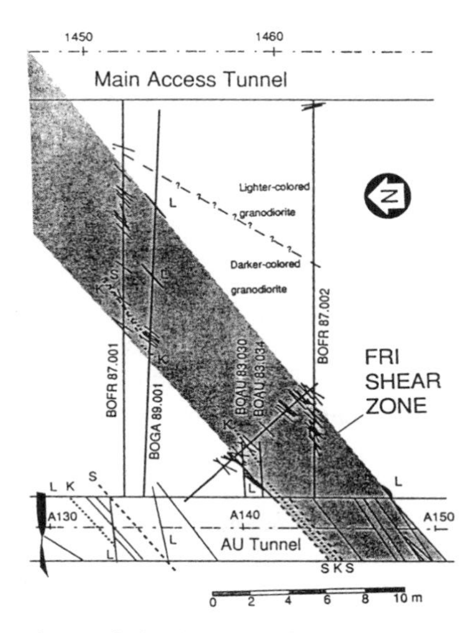

Figure 1. A plan view of the layout of the experimental area used in this study. The area is 10 meters by 21 meters.

long boreholes were drilled through the FRI zone for seismic investigations. Boreholes BOFR 87.001 and BOFR 87.002 are 86 mm diameter, 21.5 m long holes that were drilled from the AU tunnel to the access tunnel to provide a means of performing crosshole seismic work, core of the fracture zone, measuring fracture response in-situ, and for carrying out hydrologic experiments. Borehole BOFR 87.003 is a 127 mm diameter, 9 m long hole drilled through the fracture zone for obtaining large core for laboratory analysis and also for hydrologic testing. In addition to these holes, 76, 74 mm diameter, 50 cm long holes were drilled into the AU and access tunnel walls between boreholes BOFR 87.001 and BOFR 87.002 at 0.25 meter spacing to allow the placement of the seismic sources and receivers.

For the collection of the data piezoelectric seismic sources were placed in the holes (boreholes BOFR 87.001, BOFR 87.002, and the shallow holes in the sides of the tunnel) and activated. The data from a clamped three component accelerometer package was recorded at 0.5 meter spacing in boreholes BOFR 87.001 and BOFR 87.002. The receiver package was also placed in the shallow (.25 m spacing) holes to give complete four sided coverage. In the 1987 studies the source and three component receiver package were clamped to the bore hole wall to provide good seismic coupling. For the acquisition of 1987 cross well data, the source was in a dry hole and the receiver was in a fluid filled hole. In the 1988 studies, both boreholes were filled with fluid. Thus, the fluid in the borehole provided improved coupling between the source and the rock. We found that fluid coupling was more efficient than mechanical coupling and also allowed for faster data acquisition. The data were recorded on an in-field PC-based acquisition system. Four channels of data were acquired, the x, y, and z receivers and the "trigger" signal. The sample rate was 50,000 samples/sec on each channel with 20 milliseconds of data being recorded for each channel in the 1987 experiments and 250,000 samples/second in the 1988 data. Typical travel times were less than 5 milliseconds for the P-wave and 10 milliseconds for the shear wave. Seismograms were acquired from nearly 60,000 ray paths (X, Y, and Z components) in the FRI zone, at distances from 1/2 meter to nearly 23 m. The peak energy transmitted in the rock was at frequencies from 5,000 to 10,000 Hz, yielding a wavelength of approximately 1 to 0.5 meter in the 5.0 km/sec velocity rock.

B. Data Processing Sequence

The travel times were picked manually using an interactive picking routine developed at LBL. The times were picked on the radial component (component 3), which is in the direction of strongest P-wave motion. This was confirmed by rotating the data into the P-, SV- and SH-directions. Due to field equipment modifications the data quality was improved from 1987 to 1988. After the travel times have been picked, they were initially checked by plotting a time-distance curve, a velocity-distance curve, and a velocity vs. incidence angle curve. Any large variation from the general trend would indicate picking errors or acquisition problems. The most obvious deviating sweeps were removed entirely from the data used to perform the inversion. The removal of these data had little effect on the major features in resultant image. No additional travel time corrections were necessary in either the 1987 or 1988 data, and few of the signals were too noisy or did not have sufficient amplitude to pick times accurately.

The travel times were inverted using a simple algebraic reconstruction technique (ART) (Peterson, 1986). A 44 x 88 array of pixels was chosen for the tomographic inversion. This produces a pixel size of 0.25 m which is the size of the smallest anomaly we can expect to see given the wavelength of 0.7 m and station spacing of 0.5 m. Our previous experience has shown that for this geometry a pixel size of half the station spacing gives the optimal combination of resolution and inversion stability.

An image was produced using the entire 1987 data set, all four sides.

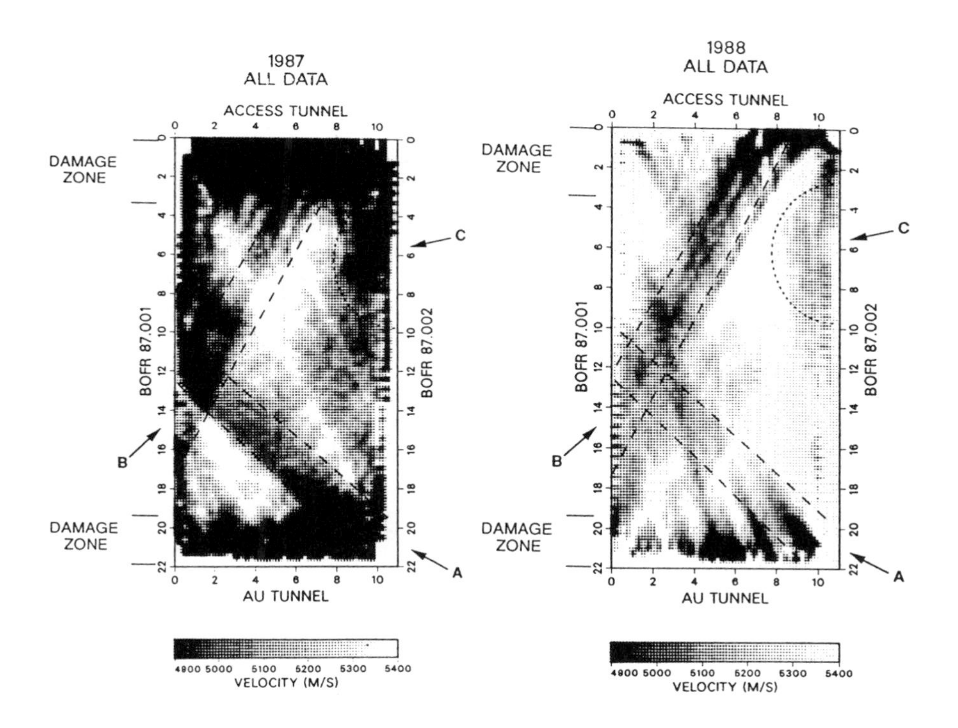

Figure 2. The final result of inverting all the good data from the 1987 tests.

Figure 3. The final result of inverting all of the good data from the 1988 tests. No anisotropic corrections, all poor data ray paths deleted.

This image can be compared to the 1988 results (Figure 3) using the same velocity intervals and same number of rays (4.9 km/s to 5.4 km/s). The main features identified in the 1987 results (Figure 2) are the low velocity zones adjacent to the tunnels, assumed to be damaged zones, and a low velocity zone (Feature A) extending from the middle of borehole BOFR 87.001 to the AU tunnel/borehole BOFR 87.002 intersection, and two other low velocity zones (B and C in Figure 2).

The 1988 image has many differences from the 1987 results. The main differences are:

1. There is little evidence of the extensive 1987 damaged zones near the tunnels in the 1988 results. Also, the average velocity values in the entire 1988 field (5.2 km/s) are higher than the average velocities in the 1987 field (4.9 km/s).
2. The prominent feature (Feature A, Figure 2) is observed in 1987 results as a single strong low velocity zone about 2 m wide. The corresponding zone in the 1988 results, Figure 3, consists of two or three very thin (< 0.5 m thick) zones which become discontinuous at about 4 or 5 m from the laboratory tunnel and are located in a different orientation and place. Also, Feature C is not as pronounced in the 1988 results as in the 1987 results.
3. There appears to be an additional structure, Feature B, in the 1988 image which was masked by the low velocity zone on the center-north edges of the 1987 result. This feature in the 1988 results extends from near the access tunnel/BOFR 87.002 intersection, to the middle of BOFR 87.001, and is, in fact, the dominant feature of the 1988 results. Evidence of this feature exists in the 1987 data, but is obscured by a low velocity feature on the east side of the tomogram.

The discrepancies between the two images seem severe and will be discussed in detail later. These differences between 1987 and 1988 results were very significant. The 1988 hydrologic testing program was based on the results of the 1987 tomogram. Features A and and C were the target of these tests and Feature B had not yet been identified as a major feature. Therefore we were very interested in the appearance of Feature B and its hydrologic significance.

C. Anisotropy Corrections

There is a strong foliation in the Grimsel granodiorite which suggests that the rock may be highly anisotropic with respect to wave propagation. An obvious step was to correct for this anisotropy in order to improve the image.

In general, the P wave anisotropy may be approximately represented as

$$V_P^2 = A + B\sin(2\phi) + C\cos(2\phi) + D\sin(4\phi) + E\cos(4\phi) \qquad (1)$$

where ϕ is the angle of direction of propagation. A function of this form is fitted to the data. The coefficients A, B, C, D and E represent the strength of the anisotropy. These values may be determined in the laboratory or in the field. The laboratory values are difficult to determine and may not adequately represent the in-situ anisotropy. In the field, the same travel times gathered for the tomographic survey may be used to determine the coefficients or a separate test may be set up in a more homogeneous (though not differing in anisotropy) area.

The anisotropy coefficients calculated from the 1987 and 1988 cross well data show rock matrix anisotropy in the direction of the foliation. Table 1 shows these values are slightly different.

Table 1. Anisotropy coefficients

coef	A	B	C	D	E
1987	26.211	0.544	-1.122	-0.331	-0.185
1988	27.942	1.375	-0.633	-0.309	-0.196

Although it is not likely that the background anisotropy changed from 1987 to 1988, the anisotropy was removed from each data set using their respective correction coefficients. In each case the contribution of the anisotropy was calculated and removed from the observed travel times. This was done by calculating the difference between the travel times calculated with coefficients A-E and the travel time calculated with only coefficient A, and then subtracting this value from the measured travel time.

The travel times corrected for anisotropy were inverted in the same fashion as the uncorrected data. The effect of the anisotropy corrections on the 1987 results (Figure 4) change the magnitude of the anomaly corresponding to the shear zone. The corrections also cause the amplitude of the anomaly to vary along the strike of the main fracture zone. Also, smaller zones within the large low velocity feature adjacent to borehole BOFR 87.001 are more resolved and coincide with similar features in the 1988 result.

The result of applying the correction to the 1988 data is shown in Figure 5. The uncorrected image has been smoothed with the low velocity zones more distinct. The inversion also appears to remove some artifacts

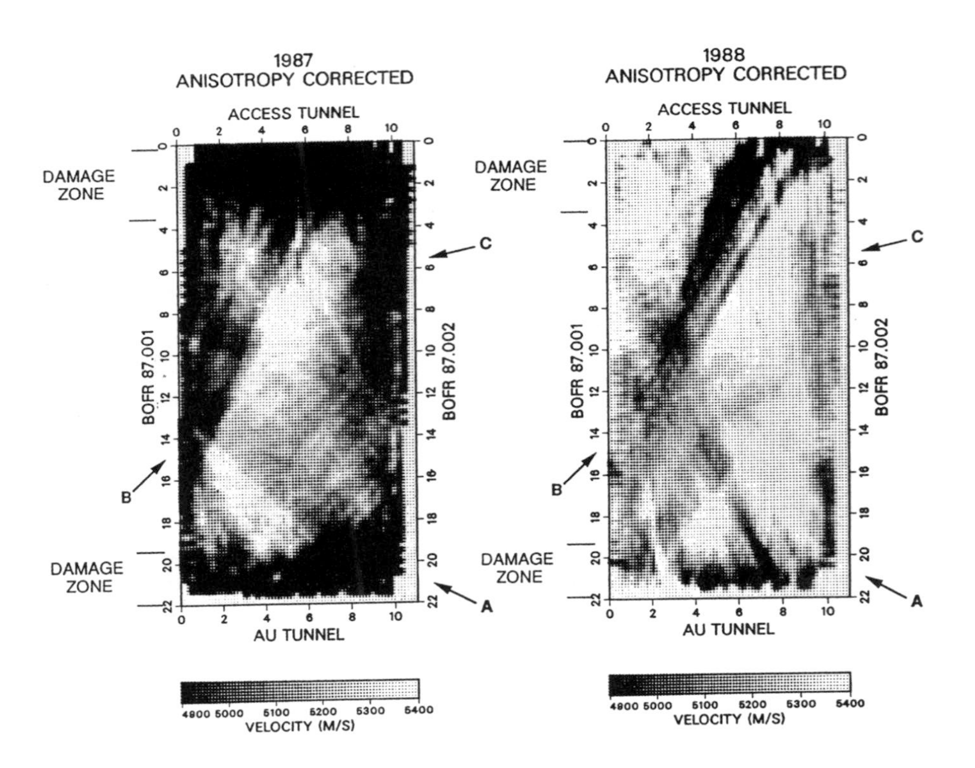

Figure 4. The final 1987 inversion after correcting for anisotropy.

Figure 5. The final 1988 inversion after correcting for anisotropy. The "unusual" rayshave been deleted.

that are produced in the original inversion. The smearing seen to the upper left of Figure 3 is reduced, as is the effect of a strong, thin low velocity feature extending from the middle of the laboratory tunnel to the center of BOFR 87.001. The "secondary" features which parallel the main low velocity zone (Feature A) in Figure 3 are also greatly reduced leaving a single zone whose intersection with the laboratory tunnel coincides with the large fracture observed on the wall of the AU tunnel.

D. Discussion of Results

Although not immediately obvious, after anisotropy corrections, the 1987 and the 1988 results show essentially the same features. We will first discuss how the two results compare, then analyze the best image in terms of what is known of the geology in the FRI site.

An obvious difference between the two results is the disappearance of the low velocity features near the tunnels in 1988 results (Figures 4 and 5). As previously stated, the data quality was much better in 1988 than 1987 for several reasons. The source was more powerful and its repeatability improved. Also, when the source and/or receiver were in BOFR 87.001 and BOFR 87.002 the coupling was improved over 1988 because both holes were water filled. We also stacked from 2 to 9 traces for each source receiver pair. These improvements increased the signal to noise ratio providing much more accurate travel time picks on higher frequency first arrivals. In effect, the reduced data quality in 1987 prevented the "proper" first arrival travel times to be picked in 1987. In some cases the value picked was a pulse or two later than the time picked in 1988. This is especially true where the attenuation is greater, e.g., the damaged zones adjacent to the tunnel and in the main shear zone. The entire 1987 travel time data for sources or receivers along the tunnels are probably picked consistently late, producing a velocity reconstruction which shows consistently lower velocities near the tunnels and resulting in a lower average velocity. This means that the 1987 tomogram was essentially a mixed velocity-attenuation tomogram.

Another difference in the results is that the shear zone (Feature A, Figures 4 and 5) becomes discontinuous and less dominant in 1988. This result is of great interest because this is the zone that we were initially trying to image. Also, Feature C is less obvious in the 1988 results, again a target of the hydrologic tests. In 1987, we had assumed that we had imaged Feature A satisfactorily as a several meter wide low velocity zone. However, the 1988 inversion does not show such an extensive feature, but a thinner zone which extends to about 4 m from the laboratory tunnel. The zone dies out for a meter, then reoccurs as a more massive feature with variable velocity. To show the actual difference in the results, the 1988 image is subtracted from the 1987 image pixel by pixel (Figure 6). (An inversion using the differences in travel times could not be performed because slightly different stations were used for a few of the sweeps). As can be seen from Figure 6, there is little difference between the two images except at the tunnels, suggesting that the 1987 low velocity zone in the region of the shear zone and Feature C exists in the 1988 result, but has a slightly different form and magnitude.

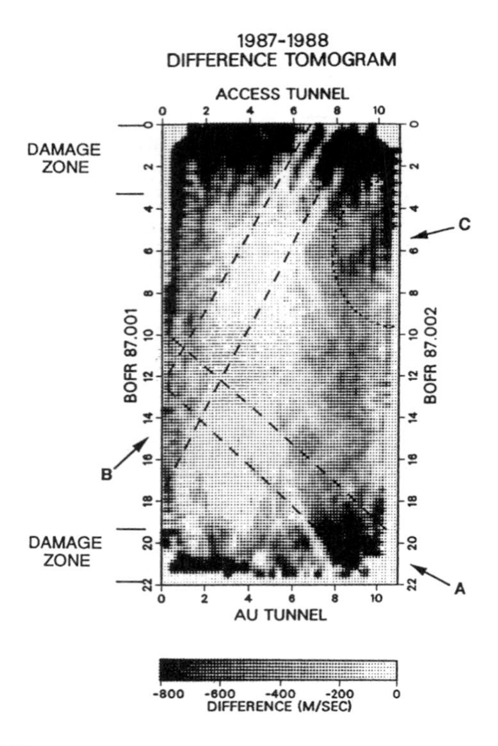

Figure 6. Difference between the 1988 and 1987 tomograms after anisotropy corrections.

Both 1987 and 1988 images show a lower velocity near the intersection of the AU tunnel and the shear zone where the excavation of the tunnel may have "loosened" the fractures. The large low velocity features toward borehole BOFR 87.001 on strike with the shear zone are also comparably imaged in the 1987 and 1988 results.

The most unexpected result from the 1988 inversion is the dominance of the low velocity feature (Feature B), which extends from the intersection of the access tunnel and BOFR 87.002 intersection to the large low velocity feature near BOFR 87.001. As mentioned earlier extensive efforts through careful examination of the data were made to determine whether this is an actual zone of low velocity material or an artifact of the inversion process or some kind of error. The 1987 result does show a hint of this feature protruding from the large low velocity zone adjacent to the access tunnel. However, in the 1987 results it is not a dominant continuous feature and is obscured by the extensive damaged zone.

Checking the plots of the difference between the 1987 and 1988 data, Figure 6, we see again that the difference is not significant. This indicates that the anomaly actually exists in the 1987 results, but it is overshadowed by the effect of the damaged zones. There is always the possibility that errors occurred in both years for several of the sweeps whose source was in this region, but this is unlikely. However, if these sweeps were removed from the data, the region of interest would not be fully sampled and the anomaly would not be adequately resolved.

E. Geologic Interpretation of the Results

The geologic map of the structures at FRI is shown next to the "final" 1988 tomographic image to assist in the interpretation (Figure 7). To be useful the interpretation of the tomographic images must include the geologic structure that is associated with each of the main features that are imaged and an explanation for the differences between the results of the 1987 and 1988 experiments. These include:

1. The large low velocity anomalies observed along the tunnels in the 1987 image which do not exist in the 1988 result.

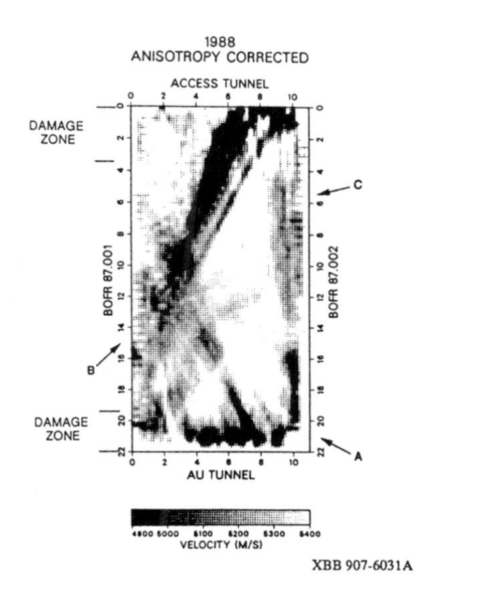

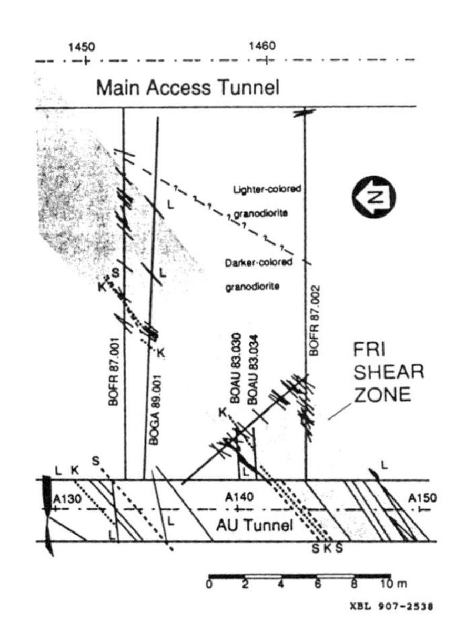

Figure 7. (a) The velocity tomogram for 1988 compared to (b) the geologic map of the FRI site.

2. The shear zone (Feature A) is observed in 1987 as a single large low velocity zone about two meters wide. The corresponding zone in the 1988 results consists of a very thin (< 0.5 m thick) zone which become discontinuous.
3. Feature B, which extends from near the access tunnel BOFR 87.002 intersection, to the middle of BOFR 87.001, and is, in fact, the dominant feature of the 1988 results.
4. Two strong low velocity features at the intersection of Feature B and the shear zone.

We have already mentioned that the low velocity zones associated with the tunnels in the 1987 results may be due to the initial P-wave pulse being highly attenuated. This was primarily due to a weaker source being used in the 1987 experiment. However, this does not explain why the 1988 velocity data did not resolve the damage zones, i.e., if there are damage zones with high fracture content, why did we not detect them in the 1988 velocity data. It is true that one explanation may be that the 1987 result only detected the damage zone by picking later arrivals because the initial pulse was attenuated, not slowed, and thus an artificially low velocity result was obtained. The attenuation data from 1988, however, did detect the damage zones near the tunnels. This suggests that at the frequencies we used, 5 to 10 Khz, the effect of these thin fractures on the velocity was much less than that on attenuation. This is in fact what the ''stiffness'' theory predicts. As frequency increases, for a constant stiffness, the velocity or delay becomes less relative to the attenuation effect. Apparently we were at frequencies where for the stiffnesses involved, attenuation is important and delay is less important.

In the final 1988 tomogram, the shear zone appears to produce a relatively weak velocity anomaly. The zone appears as expected from the 1987 results, but its form is altered in 1988. Although there is a visual difference, the actual differences are not great and may be due to the better resolution obtained in 1988. The 1988 results indicate that it is likely that the zone is not a simple single planar feature and thus the permeability along the zone may also be variable rather than being a single well connected feature. Figure 7 indicates that the shear zone produces a large velocity anomaly near the AU tunnel wall, until the point where this anomaly intersects a lamprophyre at about 4 m from the AU tunnel wall along the strike of the shear zone. The intersections of lamprophyres and shear zones are areas of more intense fracturing, probably causing larger velocity anomalies. This lamprophyre is probably discontinuous, being stretched along the shear zone during deformation. After this velocity anomaly dies out, another small low velocity anomaly is encountered at about

2 m further along strike of the shear zone. This anomaly may be another piece of lamprophyre or a region of high fracturing.

The most dominant feature in Figure 7 is Feature B, which extends from a highly fractured area in the access tunnel to the shear zone. It is unlikely that the anomaly is totally an artifact of the inversion or due to data errors since it occurs in the results from both years. The anomaly may not actually extend to the shear zone, but may be smeared somewhat in this direction. The visible fractured area at the access tunnel where the anomaly begins consists of subhorizontal fractures and a tension fissure.

From geologic considerations, it is most likely that this feature is associated with a lamprophyre or an especially large tension fissure. The strike is different from the lamprophyres in the immediate area, but as noted in the geology section, lamprophyres are not consistent in their behavior, especially when associated with shear zones. Since the geologic information about this feature is sparse, the only way to validate its presence is to drill into it. A subsequent borehole (BOGA 89.001) was drilled parallel to BOFR 87.001, but unfortunately it was several meters away from the anomaly and could not validate the prediction.

Where Feature B intersects the shear zone, Feature A, two large anomalies are also observed. These anomalies may be areas of intense fracturing, most likely due to lamprophyres intersecting the shear zone. A small lamprophyre was logged in borehole BOGA 89.001 which coincides exactly with one of these anomalies. The other anomaly coincides with a kakirite zone which also indicates a region of increased fracturing. These anomalies also suggest that there may be hydrologic communication across the shear zone in this region.

Except for Feature B, all the anomalous velocity zones are coincident with geologic structures. Feature C is still not verified, but the core suggests a different rock type, lighter colored granite, rather than fracturing may be the cause of this feature. Although the core from this region were not tested, the testing of the other core from the FRI zone suggested that the lighter colored granite has lower velocity than the more altered darker colored granite. The two low velocity anomalies near borehole BOFR 87.001 were interpreted as zones of intense fracturing likely due to the presence of lamprophyres. Borehole BOGA 89.001 was drilled, and validated this interpretation. Though the geologic information determined the possibility of such fractured regions, these anomalies could not be located by geologic data alone. It is always possible that feature B could be an artifact caused by some data error. However, there is no basis on which to reject its existence since it is observed in both the 1987 and 1988 results. Direct examination leaves little doubt that there is some anomalous zone that exists near the tunnel wall, probably a tension fissure.

III. IN-SITU GEOMECHANICAL MEASUREMENTS

In addition to the tomographic measurements, another test referred to as the inflation test was performed. The objectives of the inflation test were to determine the mechanical stiffness of the kakarite fracture and one to evaluate the effects of hydraulically pressurizing the fracture on its seismic response. To perform this test instrumentation was configured as shown in Figure 8. Bofex instrumentation was used to measure fracture displacements during pressurization in boreholes BO87.001 and BO87.003 across the main fracture zone. The linkage between the anchors is decoupled from the packers to prevent contamination of displacement measurements by deformation of the packer system. Displacement resolution was 0.6μ.

The locations of the seismic transmitter and receiver are indicated by T and R, respectively in Figure 8. During the first 45 hours of the testing the transmitter and receiver were held stationary at one location (R_1 and T_1 in Figure 8). The transmitter and receiver were then moved to locations R_2 and T_2 for the remainder of the testing.

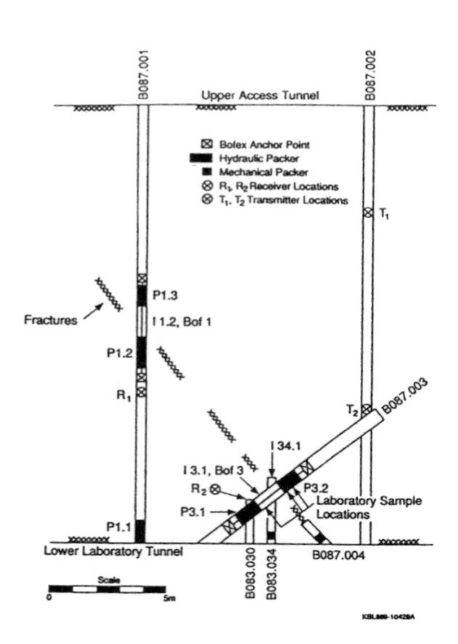

Figure 8. Plan view showing instrumentation locations for inflation test.

Measurements were made over the course of six days. For the first 45 hours the fracture was pressurized in interval I1.2 in BO87.001 while the water pressure was monitored in interval I3.1. Displacements across the fracture were monitored continuously during this time and seismic measurements were made periodically at locations R_1 and T_1. At 48 hours, in order to pressurize a larger portion of the area of the fracture, the pressure was made equal in both I1.2 and I3.1. Periodic seismic measurements were made with the receiver and transmitter in positions R_2 and T_2 respectively. After both intervals were depressurized displacements were monitored and seismic measurements were periodically made until the testing was concluded after 115 hours.

A. Analysis of Deformation Measurements

The displacement of the kakarite fracture in response to fluid pressurization was measured in order to determine the stiffness of the fracture. Fracture stiffness is given by the ratio of incremental stress to incremental displacement, when in concept, the displacement can be equated to the change in volume of voids in the plane of the fracture. Since the hydrologic properties of a fracture are closely related to the void geometry, fracture stiffness is an important parameter in the evaluation of the fracture fluid flow properties (Witherspoon *et al.*, 1980). Studies e.g. Majer *et al.*, 1988, have also shown that fracture stiffness affects both the velocity and attenuation of propagating seismic waves. Schoenberg, 1980 and Pyrak-Nolte *et al.*, 1990b developed explicit relationships between the stiffness of a single fracture and the group time delay and amplitude of a wave transmitted across it.

Displacements measured in BO87.001 increased steadily with time after interval I1.2 was initially pressurized to 19 bars, and more rapidly when the interval was pressurized to 36 bars. During this time negligable displacement was measured in BO87.003. The time dependent nature of the displacements meant that the fluid pressure in the fracture was increasing over an ever increasing area of the fracture.

Since fluid pressures were not uniform, it was necessary to assume a model for the fracture void geometry in order to calculate stiffness. As a first approximation the fracture was represented by a row of co-planar cracks of equal length, 2c, and uniform spacing, 2b, subjected to a pressure distribution as shown in Figure 9. In this model the ligaments of material between cracks represent the areas of contact in the fracture and the cracks represent the void space. The pressure distribution was approximated by a linear distribution with a maximum value of 19 bars at a

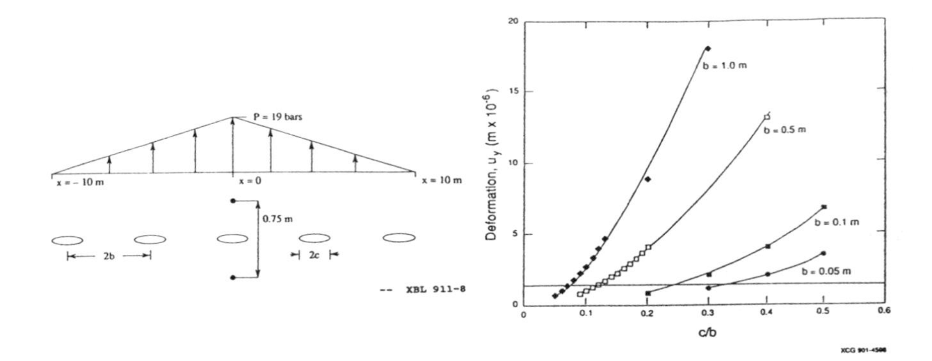

Figure 9. Model of fracture as co-planar parallel cracks. Pressure varying linearly with distance was applied internally to cracks.

Figure 10. Displacements between two points located 0.75 m either side of the midpoint of the center crack of a row of pressurized co-planar cracks. Horizontal line corresponds to in-situ measurement (after Hesler *et al.,* 1990).

location x = 0 and a value of 1.9 bars at x = ± 10 m. Calculations were carried out for different values of crack lengths and spacing to match the observed maximum fracture displacement measured in BOFR 87.001 under 19 bars of pressure (Hesler *et al.,* 1990). Results are shown in Figure 10. Displacements were calculated at a location of 0.75 m on either side of the row of cracks to coincide with the location of the Bofex anchors relative to the fracture. Figure 10 shows that for any given crack size large spacings are required to match the observed displacement of 1.4 x 10^{-6} m. These results suggested a contact area in the fracture on the order of 60%.

Having determined a range of crack sizes and spacings for the idealized model, the stiffness of the fracture could be estimated analytically. Using solutions form fracture mechanics, for a row of co-planar cracks the average displacement, $\overline{\delta}$, is given by:

$$\overline{\delta} = \frac{-8\sigma b(1-\nu^2)\ [\ln \cos(\pi c/2b)]}{\pi E} \tag{2}$$

where σ is the far field stress and ν and E are Poisson's Ratio and Young's modulus, respectively. Since, by definition, stiffness, κ, is related to $\overline{\delta}$ by:

$$1/\kappa = \overline{\delta}/\sigma,$$

the stiffness of a row of co-planar cracks is given by:

$$1/\kappa = \frac{-8b(1-\nu^2)[\ln \cos(\pi c/2b)]}{\pi E} \tag{3}$$

Using equation 3 and values of c and b from Figure 10, values of κ were calculated which ranged from 2 x 10^{11} Pa/m to 3 x 10^{11} Pa/m. These values are about one half to one order of magnitude less than those obtained in laboratory tests on a natural fracture from the FRI study area, but direct comparison is not possible since the laboratory specimen was not from the kakirite fracture and did not contain gouge material.

The effect of a single fracture with a stiffness of 3 x 10^{11} Pa/m on seismic wave propagation van be evaluated using the seismic displacement discontinuity model described by Pyrak-Nolte *et al.*, 1990b. Figure 11 is a plot of the magnitude of the transmission coefficient, $|T|$, and normalized group time delay, t_g/t_{go}, for a wave normally incident upon a fracture. Values of $[\omega/\kappa/Z)]$ were calculated for κ = 3 x 10^{11} Pa/m, z = 1.4 x 10^7 kg/m^{2}s, and ω corresponding to the center frequency of the P-waves in the tomographic survey. For the 1987 survey this frequency was about 6 kHz while for the 1988 survey it was about 10 kHz. In Figure 11 the corresponding values of $|T|$ and t_g/t_{go} are plotted as circles for the 1987 survey and squares for the 1988 survey.

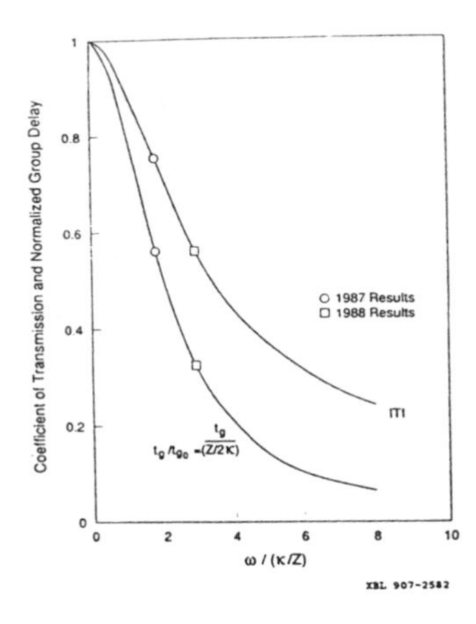

Figure 11. Magnitude of the transmission coefficient and normalized group delay for a seismic wave normally incident upon a displacement discontinuity as a function of normalized frequency.

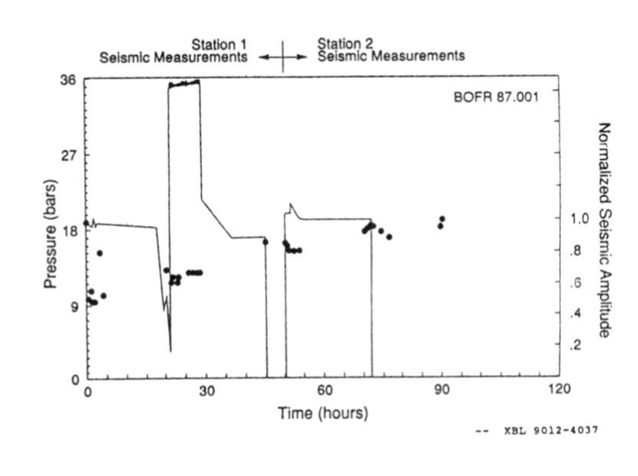

Figure 12. Normalized seismic amplitudes obtained at two stations superimposed on pressure history of interval I1.2.

For a 6 kHz wave normally incident on a fracture with stiffness of 3 x 10^{11} Pa/m the value of $|T|$ is seen to be about 0.75, while for a 10 kHz wave it is about 0.56. A value of $|T| = 0.75$ means that, a wave propagated across the fracture would have an amplitude about 25% lower than one propagating over the same path length of intact rock. Because of the higher frequency used in the 1988 survey, the amplitude reduction (i.e. $|T| = .56$) is predicted to be greater. From Figure 11 it is seen that the value of t_g/t_{go} for the 1987 survey was about 0.55 while for the 1988 survey it was about 0.31. This means that the delay in measured travel time caused by the fracture is predicted to be less for the 1988 survey run with the higher frequency source. This may explain in part why the fracture was less distinct in the 1988 tomogram.

Finally Figure 12 presents the results of the seismic measurements made periodically during the test. Amplitudes were normalized to the maximum P-wave amplitude measurement obtained while the source and receiver were held stationary. For position 1 (T_1 and R_1 in Figure 8) the maximum amplitude was just prior to pressurization of the interval I1.2. Almost immediately following pressurization the amplitudes dropped by over 40%. Thereafter, amplitudes generally increased over the course of the experiment, without any obvious correlation to the pressure history observed in intervals I1.2 and I3.1. At position 2 the smallest amplitudes were again at the time of pressurization. Amplitudes then increased with the maximum occurring after depressurization of the test intervals. If all other factors are constant and assuming elastic behavior, increasing the fluid pressure in a fracture results in a decrease in contact area and hence a reduction in stiffness and a concomitant decrease in the amplitude of the transmitted wave. This could explain the decrease in amplitude at stations 1 and 2 after pressurization. The fracture displacement which would produce this change in stiffness was not observed. This may be because the seismic wave received at either station sampled only a small portion of the fracture plane, so the observed changes do not reflect the average deformation of the entire area pressurized. The increase in amplitudes following pressurization can be explained if the fracture and/or the rock adjacent to the fracture were not fully saturated prior to pressurization. If a fracture is envisioned as a collection of co-planar thin cracks, the stiffness of the fracture will increase as more of the cracks are filled with water. This increase in stiffness, according to the displacement discontinuity model will lead to an increase in transmitted P-wave amplitudes. Such an increase in amplitudes upon saturation has been observed by Pyrak-Nolte I. et al., 1990b in laboratory tests on granite specimens containing a single natural fracture. Saturation of intact specimens of the schistose rock from the FRI area also resulted in increased amplitudes of the transmitted P-waves (Majer *et al.*, 1990).

IV. CONCLUSIONS

Overall, the tomographic inversions seem to successfully image the major structures in the test region. At the frequencies used, the tomograms do not successfully image the minor structures such as individual fractures. The main conclusions from these tests are:

1. The velocity anomalies observed associated with fracturing were not due to single fractures but groups of fractures.
2. The seismically important features associated with fracturing were fracture intersections and fracture-lamprophyre intersections. These are also the structures that should be hydrologically important.
3. An interpretation of the results can not be done adequately without knowledge of the geology.
4. Structures were resolved that could not be anticipated from the borehole and tunnel data.
5. Although shear wave data were sparse, it is obvious that given the proper source, S-wave data would greatly aid in the interpretation of the geologic features.
6. A method for obtaining in-situ fracture displacements for estimating fracture stiffness was successfully demonstrated.
7. The seismic visibility of the features from the amplitude and velocity tomograms seem to support the displacement discontinuity theory which relates these seismic properties to fracture stiffness.

ACKNOWLEDGEMENTS

This work was supported through U.S. Department of Energy Contract No. DE-AC03-76SF00098 by the DOE Office of Civilian Radioactive Waste Management, Office of Geologic Repositories. We want to thank the personnel at NAGRA and the Grimsel Rock Laboratory for their help and support, in particular Peter Blumling for his extensive support and scientific insight, Piet Zuidema and Gerdt Sattel. We would also like to thank Eric Wyss of SOLEXPERTS AG.

REFERENCES

Hesler, G.J., Zheng, Z. and Myer, L.R. (1990). "In-situ fracture stiffness determination," *in Rock Mechanics Contributions and Challenges,* Proceedings of 31st U.S. U.S. Rock Mechanics Symposium, A.A. Balkema 405-412.

Majer, E.L., McEvilly, T.V., Eastwood, F.S. and Myer, L.R. (1988). "Fracture detections using P- and S-wave vertical seismic profiling at the Geysers," *Geophysics 53 No.1,* 76-84.

Majer, E.L., Myer, L.R., Peterson, J., Karasaki, K., Long, J.C., Martel, S., Blumling, P., and Vomvoris, S. (1990). *Joint Seismic, Hydrogeogical, and Geomechanical Investigations of a Fracture Zone in the Grimsel Rock Laboratory, Switzerland, NDC-14, LBL-27913,* Lawrence Berkeley Laboratory, Berkeley, California, 173 pgs.

Pyrak-Nolte, L.J., Myer, L.R., and Cook, N.G.W. (1990a). "Anisotropy in seismic velocities and amplitudes from multiple parallel fractures," *Journal of Geophysical Research 95 No. B7,* 11345-11358.

Pyrak-Nolte, L.J., Myer, L.R., and Cook, N.G.W. (1990b). "Transmission of seismic waves across single fractures," *Journal of Geophysical Research 95 No. B6,* 8617-8638.

Schoenberg, M. (1980). "Elastic wave behavior across linear slip interfaces" *Journal of the Acoustical Society of America 68 No. 5,* 1516-1521.

Witherspoon, P.A., Wang, J.S.Y., Iwai, K. and Gale, J.E. (1980). "Validity of cubic law for fluid flow in a deformable rock fracture," *Water Resources 16,* 1016.

Natural Gas Reserve Replacement through Infield Reserve Growth: An Example from Stratton Field, Onshore Texas Gulf Coast Basin[1]

Raymond A. Levey[2]
Mark A. Sippel[3]
Richard P. Langford[2]
Robert J. Finley[2]

[2] Bureau of Economic Geology, The University of Texas at Austin
University Station, Box X, Austin, Texas 78713-7508
[3] Research Engineering Consultants, 7600 E. Orchard Street, Suite 106S
Englewood, Colorado 80111

The major objective of a joint venture sponsored by the Gas Research Institute (GRI), the U.S. Department of Energy (DOE), and the State of Texas is to assess the distribution of incremental natural gas resources in conventional reservoirs within mature fields. Analysis of publicly available production data for a 50-year-old gas field indicates that reserves may be effectively replaced with additional infield wells and recompletions within a mature gas field containing reservoirs that have conventional porosity (>15%) and permeability (>10 md). It is both technically and economically viable to undertake historical evaluation of decline curves and cumulative production from 1979 through 1989 to quantify reserve growth within a field that has undergone a successful infield drilling program resulting in replacement of gas reserves approaching 50 Bcf. The spacing of wellbore penetrations in the study area was less than 25 acres per wellbore across all reservoirs to an average completion depth of approximately 6,400 ft at the end of 1989.

---

[1] Funded by the Gas Research Institute contract no. 5088-212-1718, the U.S. Department of Energy contract no. DE-FG21-88MC25031, and the State of Texas through the Bureau of Economic Geology. The cooperation of Union Pacific Resources Company is gratefully acknowledged. We thank T. F. Hentz and Bill Keighin for technical review. Editing was by Amanda R. Masterson. Word processing was by Susan Lloyd, and pasteup was by Margaret L. Evans. Figures were drafted under the supervision of R. L. Dillon. Publication authorized by the Director, Bureau of Economic Geology, The University of Texas at Austin.

Reservoir geometry, controlled by the depositional system, is a major factor influencing the magnitude of incremental gas resource that may be accessed through additional infield completions. Analysis of produced and projected gas resources at less than 7,000 ft deep from a lease block within Stratton field of the Frio fluvial-deltaic (FR-4) play along the Vicksburg fault zone illustrates the impact of new infield drilling and recompletions within established wellbores in the onshore Texas Gulf Coast Basin. Gas reserve additions between 1979 and 1990 exceeded 13% of the developed extrapolated ultimate recovery at the end of 1989.

Stratton field (fig. 1), discovered in 1937, is 1 of 29 multireservoir fields containing major gas resources in the Frio fluvial-deltaic play associated with the Vicksburg fault zone (Kosters et al., 1984). Stratton field is an excellent candidate for detailed analysis because (1) it has a long (>50 yr) development and production history, (2) an aggressive infield drilling and recompletion program in this field during the late 1980's provides a strong basis for using a historical approach to resource assessment, and (3) the resulting development of the gas cap has provided a densely drilled area that is not commonly available in other Frio fluvial-deltaic fields.

Gas production data from the annual reports of the Railroad Commission of Texas indicate that more than 300 gas reservoirs have been defined in Stratton field from the initial field discovery in 1937 through continued development and production in 1989. Detailed well log stratigraphic correlations across Stratton field indicate that this number of reservoirs is relatively high because of difficulties in correlation and variations in reporting practices for reservoir designations. Figure 2 shows a plot of depth versus discovery year from 1940 to 1989. Fieldwide cumulative production analysis is based on records of annual production for Stratton field from 1950 through 1989 (fig. 3). The primary development of middle Frio fluvial reservoirs took place from 1950 to 1968, and during this period reservoir pressure was partly supported by gas cycling to produce associated oil. Annual production in Stratton field peaked at more than 95 Bcf in the early 1970's with major gas production from the prolific E41 and F11 (fig. 4) gas reservoirs. From 1974 to 1986 production declined substantially. The reversal in production decline from 1986 through 1989 is a function of recent infield drilling and recompletions of previous wells in the field.

This analysis examines the number of completions and reserve estimates at two times in the history of Stratton field. Reserves are addressed in two depth categories: reserves shallower than 7,000 ft and those deeper than 7,000 ft. The timeframe utilized in this analysis compares completion history and gas reserves based on all completions in January 1979 to the same completions and additional completions from infield drilling after 1978 but before January 1990.

An evaluation of secondary gas (incremental) reserve growth is indicated by assessment of the historical changes in gas reserves for a contiguous part of Stratton field (~7,400 acres) referred to as the study area (fig. 4). In January 1979 a total of 53 active completions existed in the study area (fig. 5). The average completion spacing within the study area was approximately 132 acres in January 1979. Of these original 53 completions, 22 were still active in January 1990 (fig. 5). A total of 149 new completions were made between 1979 and 1990. In January 1990 the cumulative completion density for both active and inactive completions was approximately 90 acres for the Frio reservoirs. Of the 149 new completions, 82 were still active in January 1990.

Reserve additions are derived from three sources: (1) new reservoirs that are deeper pool than the current production, (2) reservoirs already contacted in wellbores but not effectively drained by the current completion spacing, and (3) reservoirs that were previously bypassed or new reservoirs encountered by infield drilling as untapped reservoir compartments. A histogram of gas volume for the two timeframes illustrates the degree of reserve growth attributed to incremental drilling and completions (fig. 6). In January 1979 a recovery of 52 Bcf from 53 completions in 43 wells was projected. From 1979 to 1990 a total of 38 Bcf was produced, leaving 10 Bcf as the reserves remaining from the 52 Bcf projected in 1979. By comparison, 149 completions made in 84 wells between January 1979 and January 1990 added 48 Bcf of developed reserves to the projections made for completions active in January 1979. From these new completions a total of 30 Bcf was actually produced between 1979 and 1990, and an additional 18 Bcf was projected as remaining to be produced after January 1990 to an operational ending rate of 30 Mcf/d per completion. The additional completions from January 1979 to January 1990 provided more than 90% reserve replacement beyond the projected remaining developed reserves for all completions active in 1979.

Comparison of reserve estimates using a depth cutoff of 7,000 ft demonstrates the overprint of depositional system on incremental gas reserves. Because of the simple low-relief structure and consistent elevation across the study area, the 7,000-ft depth value approximates the transition from the middle Frio (<7,000 ft) to the lower Frio and Vicksburg Formations (>7,000 ft). For the completions less than 7,000 ft deep as of January 1979, a total of 39 Bcf is projected as the remaining recovery for the gas reserves, compared with a total of 70 Bcf as actually produced and estimated to be remaining as of January 1990 (fig. 7). Actual production and projected reserves in January 1990 indicate that of the projected 39 Bcf, 27 Bcf was produced, and 10 Bcf is projected as remaining for the completions made before January 1979.

For the 11 completions deeper than 7,000 ft made before January 1979 (fig. 8), 13 Bcf is indicated. Analysis of the completions deeper than 7,000 ft between January 1979 and January 1990, indicates an additional incremental 15 Bcf was estimated for 59 post-1978 completions within 35 wells. Of that 15 Bcf, 9 Bcf was actually produced, and an additional 6 Bcf is projected as the ultimate estimated recovery remaining for these completions.

The contrast in gas production between reserves shallower than 7,000 ft and those deeper than 7,000 ft is thought to reflect a major change in depositional setting. The middle Frio Formation reservoirs (<7,000 ft) are fluvial; in contrast, the lower Frio and Vicksburg reservoirs (>7,000 ft) are in a predominantly deltaic setting.

In summary, an aggressive infield drilling and recompletion program has demonstrated an incremental gas resource within a mature field. Depositional system, which strongly affects the magnitude of incremental gas resources, is also a key control on secondary (incremental) gas reserve growth.

REFERENCE

Kosters, E. C., Bebout, D. G., Seni, S. J., Garrett, C. M., Jr., Brown, L. F., Jr., Hamlin, H. S., Dutton, S. P., Ruppel, S. C., Finley, R. J., and Tyler, Noel, 1989, Atlas of major Texas gas reservoirs: The University of Texas at Austin, Bureau of Economic Geology, 161 p.

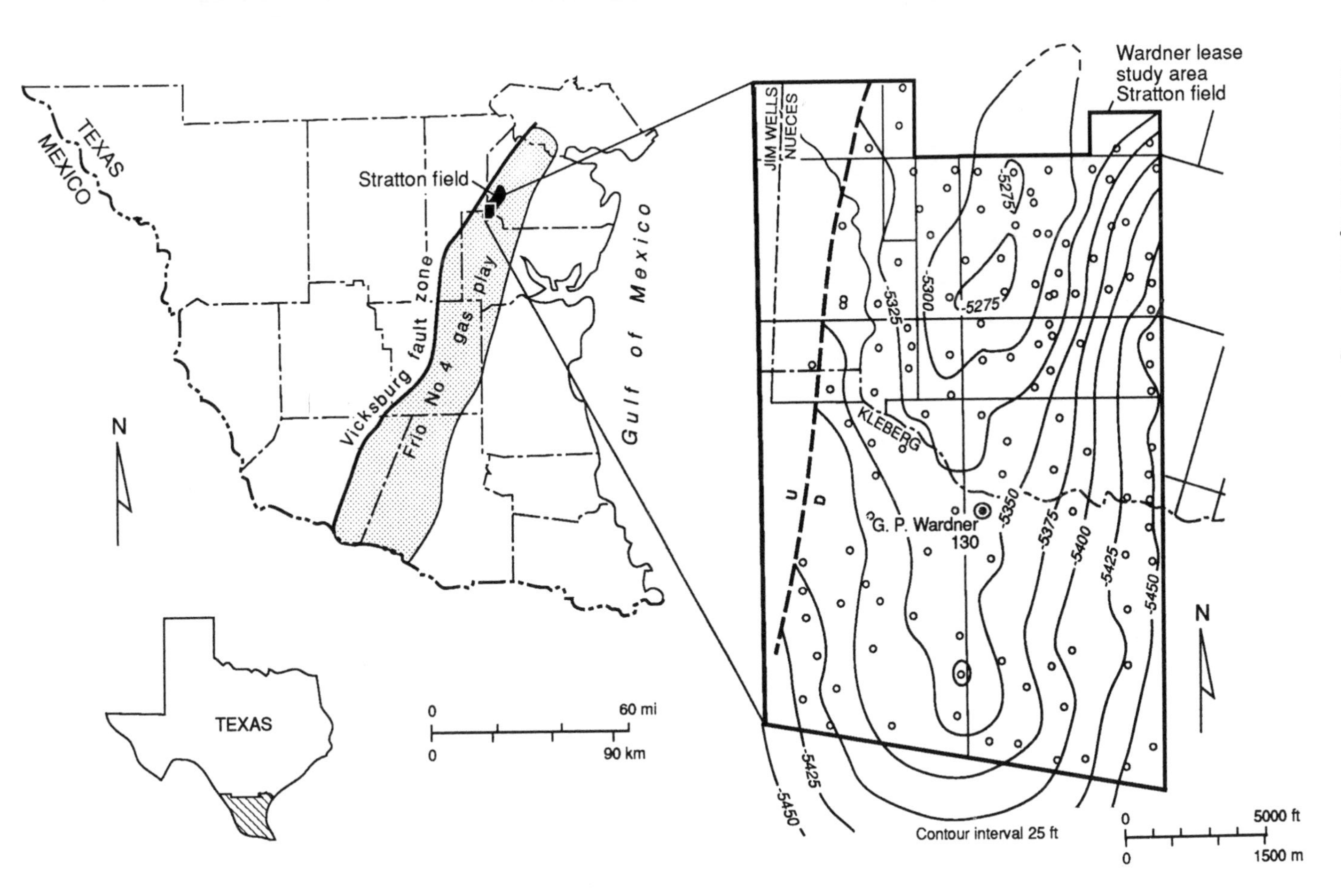

Fig. 1. Location map of Stratton field within the Frio fluvial-deltaic (FR-4) gas play and map of the Wardner lease study area.

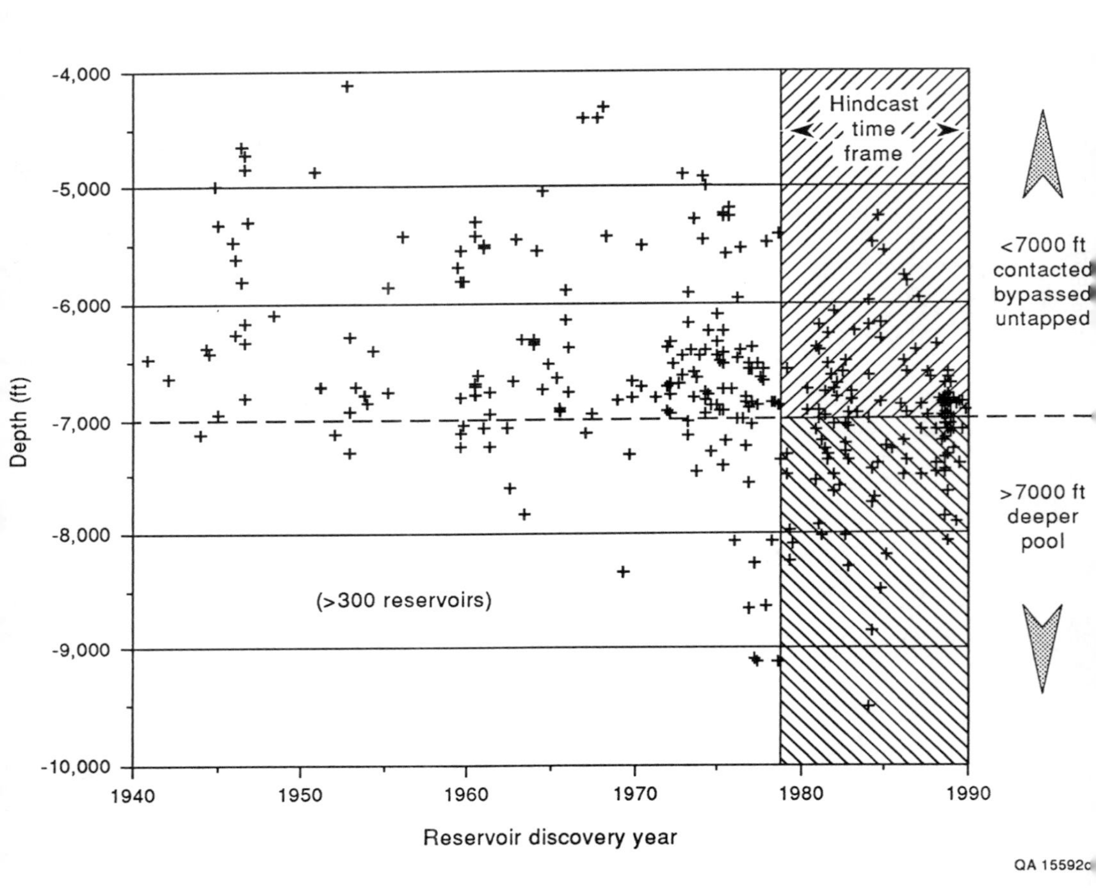

Fig. 2. Plot of discovery year of reservoir versus depth of the reservoir in Stratton field.

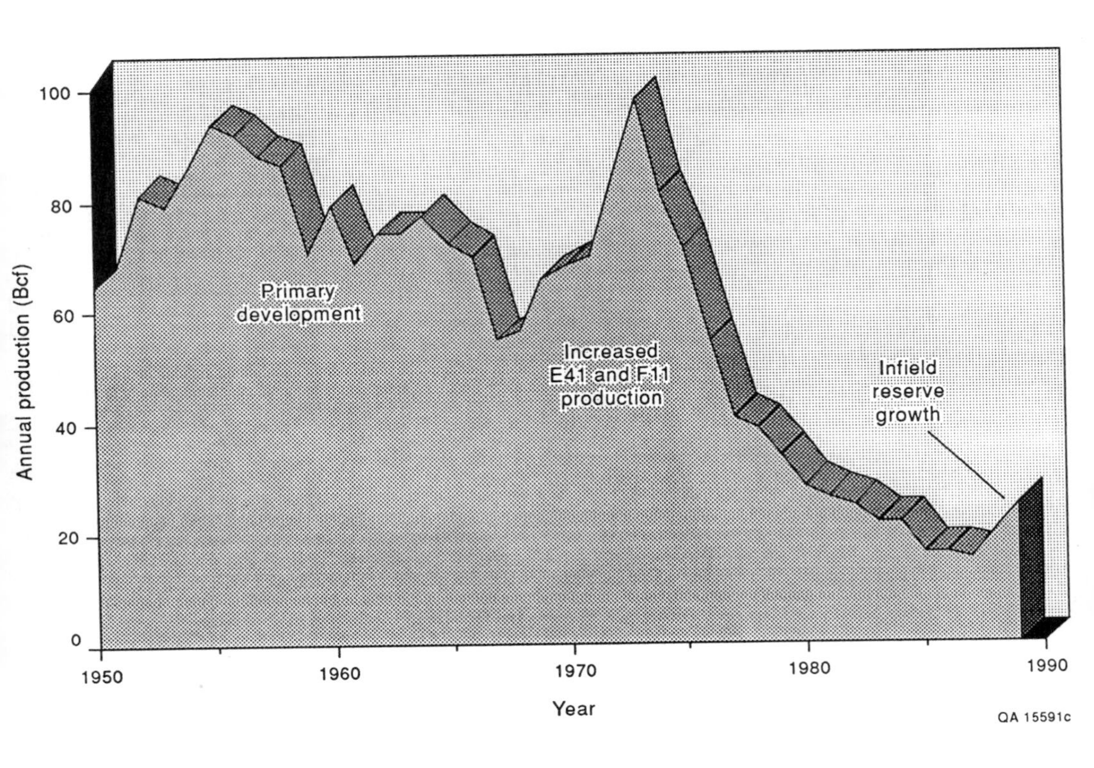

Fig. 3. Annual gas production of Stratton field from 1950 to 1990.

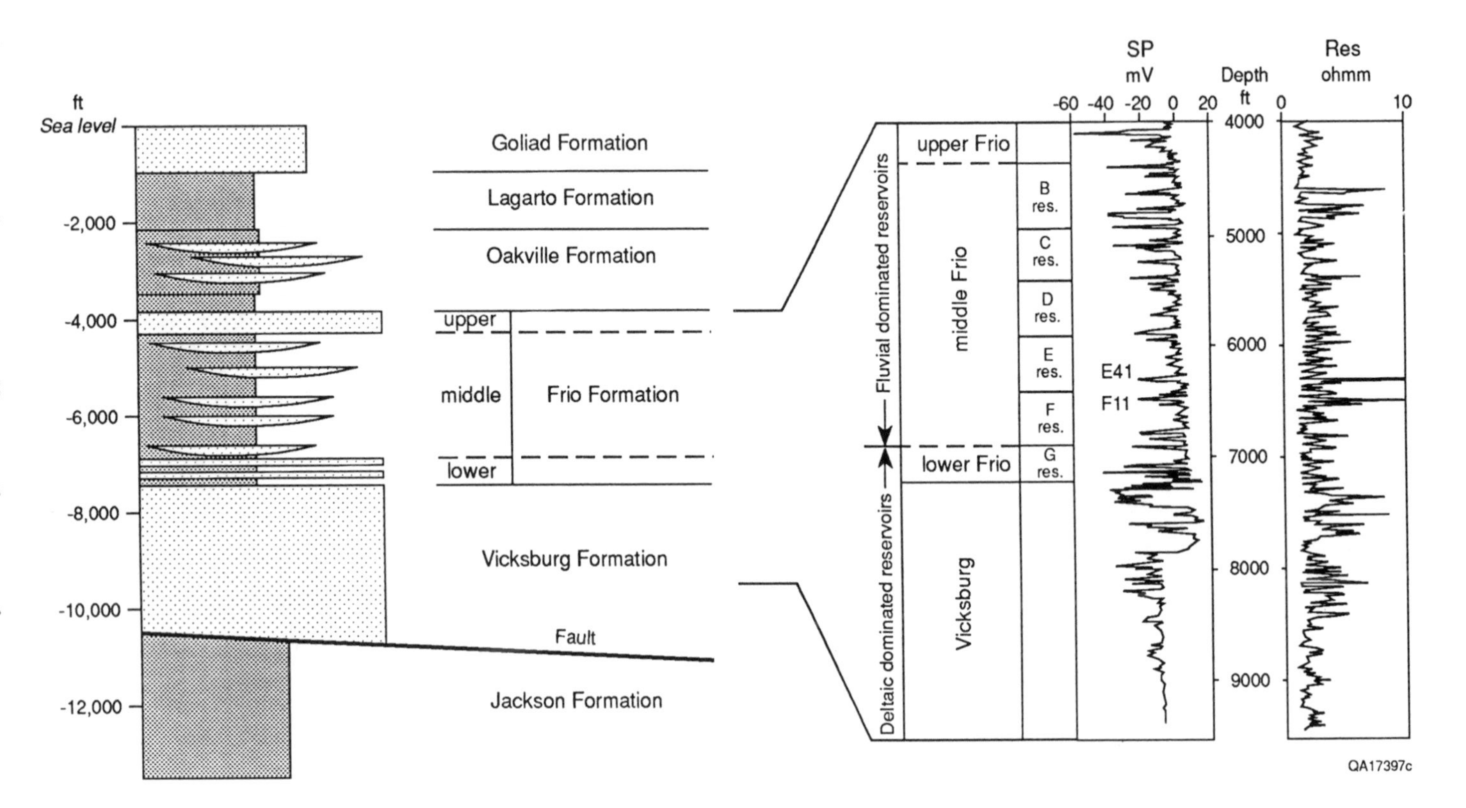

Fig. 4. Type well log from Wardner lease showing Frio and Vicksburg reservoir nomenclature.

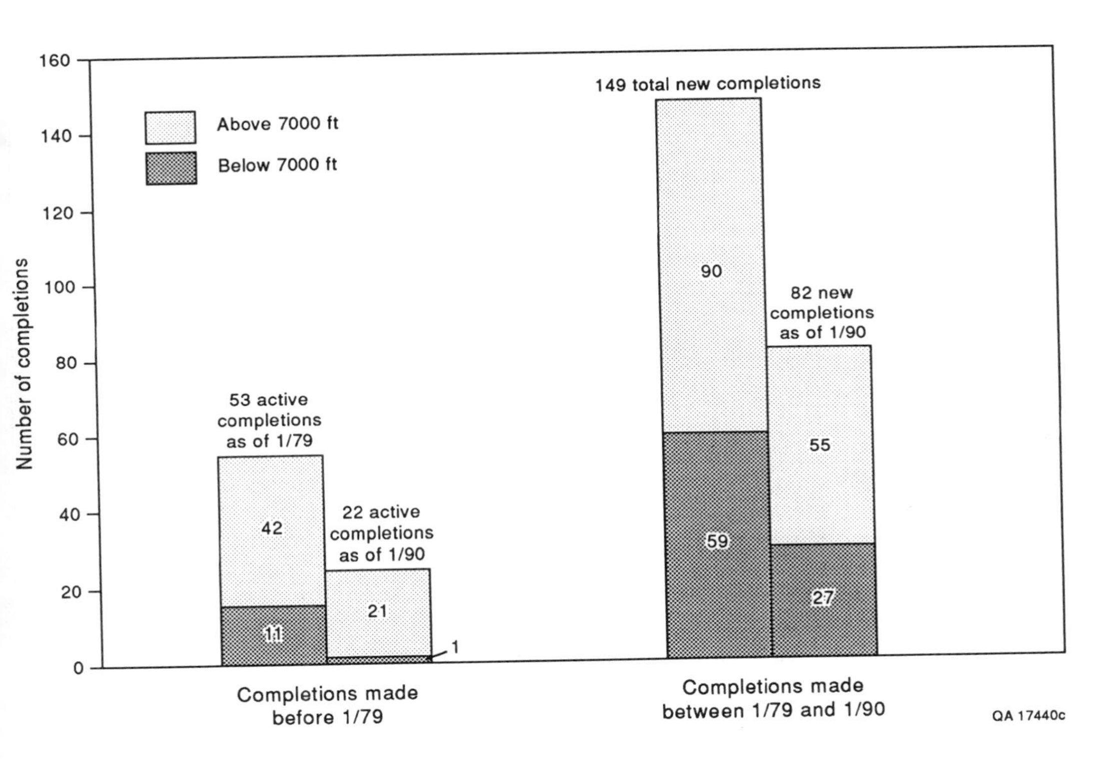

Fig. 5. Histogram of number of completions in the study area.

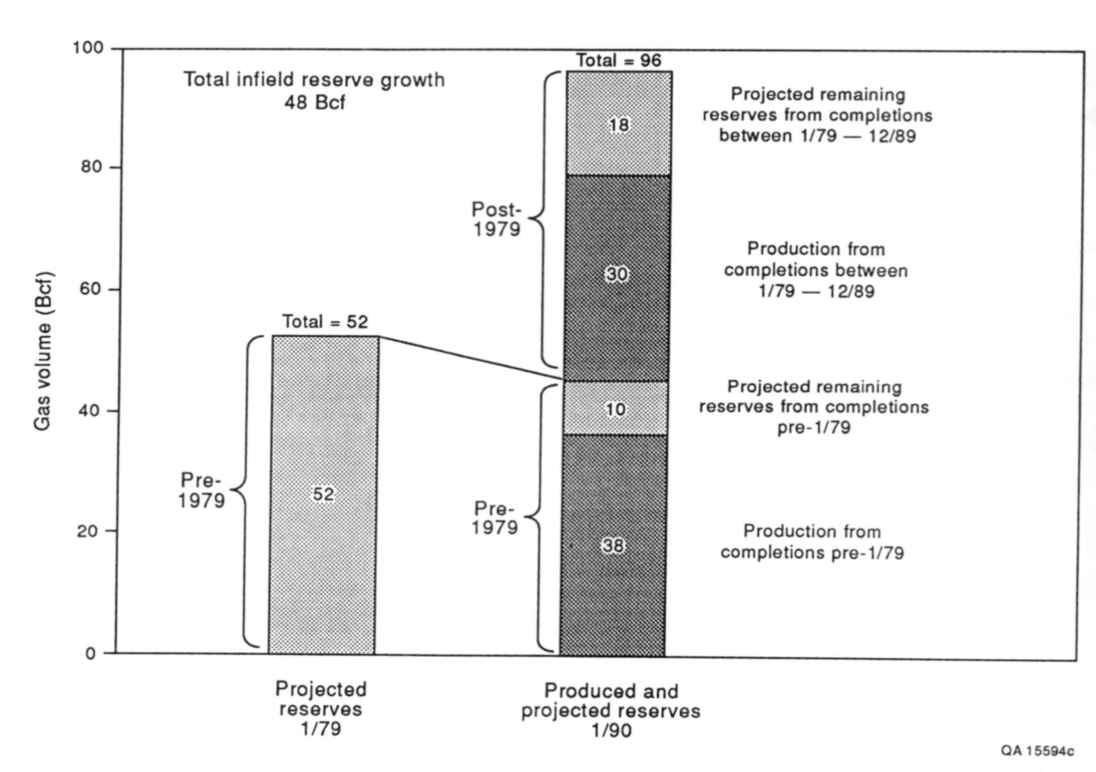

Fig. 6. Histogram of total gas volume for all completions.

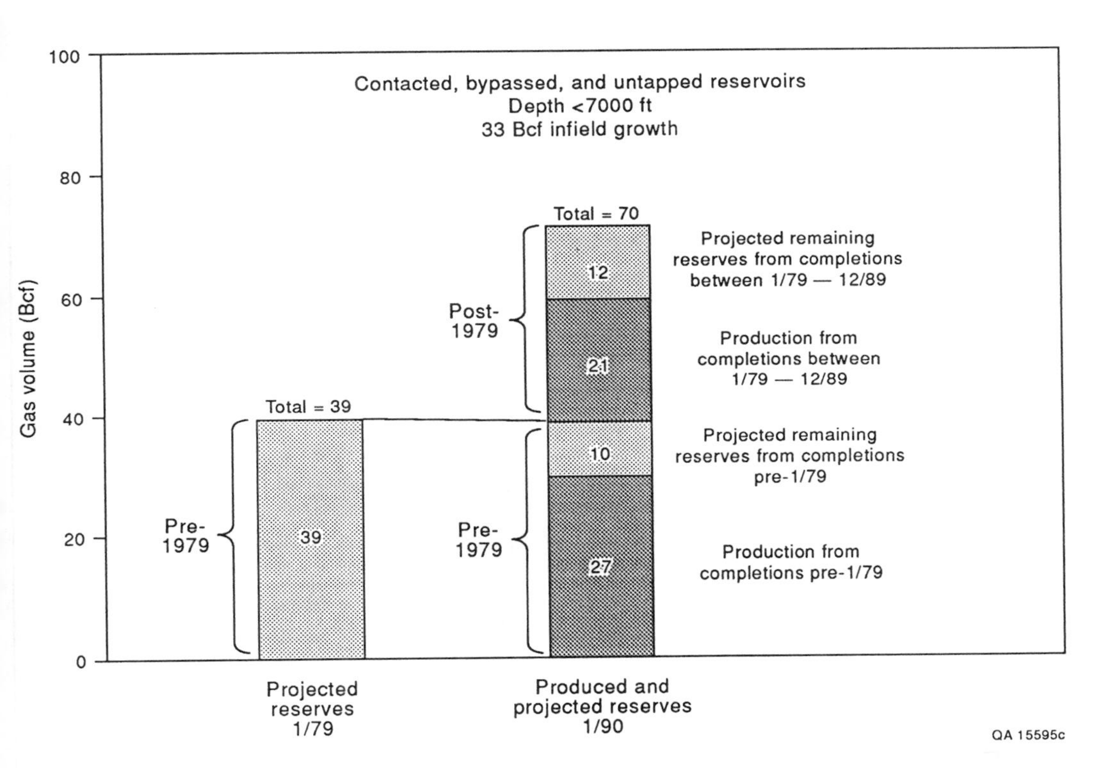

Fig. 7. Histogram of gas volume for completions <7,000 ft as of January 1979.

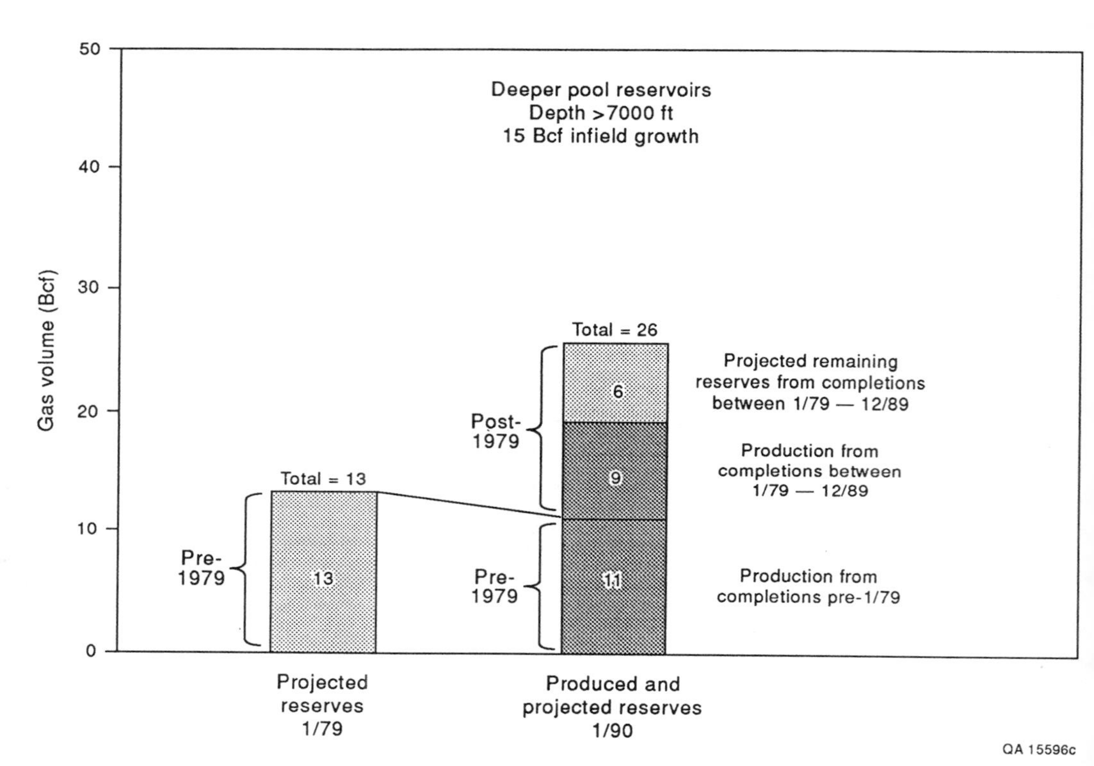

Fig. 8. Histogram of gas volume for completions >7,000 ft.

AN EXPERT SYSTEM FOR THE NETWORK MODELLING OF PORE STRUCTURE AND TRANSPORT PROPERTIES OF POROUS MEDIA

Ioannis Chatzis
Marios A. Ioannidis

Department of Chemical Engineering
University of Waterloo
Waterloo, Ontario, N2L 3G1

Apostolos Kantzas

Nova Husky Research Corp.
Calgary, Alberta

I. INTRODUCTION

The determination of transport and capillary equilibrium properties of sedimentary rocks, such as absolute and relative permeabilities, formation resistivity factor and capillary pressure curves, is of fundamental importance for the successful estimation of hydrocarbon reserves and for the design and application of production schemes. Despite the fact that much effort has been expended to theoretically simulate the aforementioned properties (1), the development of comprehensive predictive tools using microscopic pore structure parameters is still at its early stages. Consequently, the estimation of transport and equilibrium properties of reservoir rocks relies heavily on experimentation and on the use of empirical equations.

The purpose of this paper is to report on the development of an expert computer model, WATNEMO (WATerloo NEtwork MOdel), which can be used to predict the transport and equilibrium properties of sandstones from readily available pore structure parameters. The performance of WATNEMO is illustrated by means of comparisons between the predicted and measured properties for a variety of sandstones. The most important features of the model are its rigorous formulation based on bond-correlated site percolation theory (2) and its ability to incorporate all the available pore structure information in a consistent manner (porosity, breakthrough capillary pressure/ absolute permeability, and photomicrographic information on pore structure).

II. SYSTEM DESCRIPTION

The pore structure of a reservoir rock is modelled as a cubic network consisting of pore throats and pore bodies of rectangular cross-section following respective pore size distributions. It is generally acknowledged that, the differences between the pore size distributions of various samples is the major cause of different macroscopic behavior(3). The problem of estimating the pore throat and pore body size distributions of different media is handled in WATNEMO by employing a set of deterministic and heuristic criteria. For each porous medium, the WATNEMO constructs a geometrically defined porous medium model according to the following rules:

- The number based pore throat size distribution is selected such that the pore throat size corresponding to the percolation threshold of the site percolation-bond correlated-cubic network leads to the same breakthrough capillary pressure as that measured in the real porous medium(2-5).
- The pore body size distribution is selected in such a manner, so that the average pore body size and the range of pore body sizes observed in pore casts of the real medium are also present in the simulated medium.
- The porosities of the real and simulated medium are identical.

The prediction of drainage capillary pressure curves for Hg-air or oil-water immiscible displacements is accomplished by using the generalized results of pore accessibility studies (2-4). Details of methodology used can be found elsewhere(2-4).

The prediction of absolute permeability and formation factor is performed by solving the linear electrical-resistor network problem for a fully saturated medium (5). Similarly, the drainage relative permeability curves are obtained by solving the analogous problem for a partially saturated medium, where the fluid distributions are dictated by the pore space accessibility results for oil-water displacement (6). In these computations, the individual conductances of pore throats and pore bodies are calculated using solutions of the appropriate momentum/ electric current transport equations. A simplified flow diagram, indicating the basic components of WATNEMO, is given in Fig. 1.

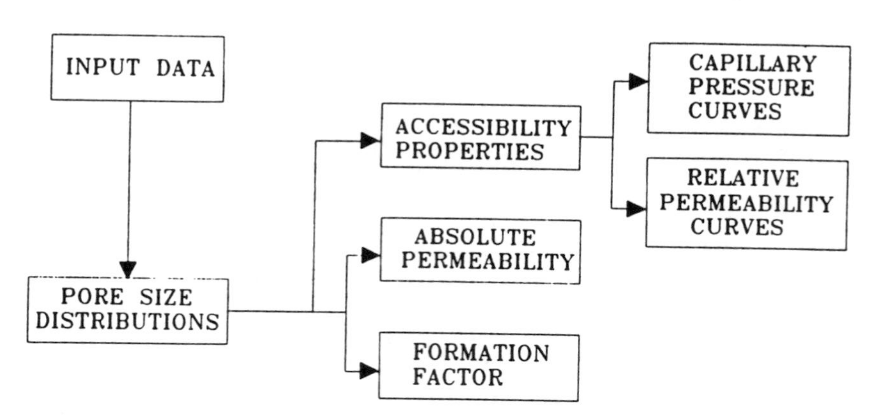

Fig. 1. Basic functions performed by the WATerloo NEtwork MOdel.

III. RESULTS AND DISCUSSION

The potential value of WATNEMO in Computer Aided Core Analysis is best illustrated by comparing the model predictions to experimental results for various sandstones. The only input data required to predict transport and equilibrium properties of different samples using WATNEMO are: 1) the sample porosity, 2) the breakthrough capillary pressure, 3) the volume average pore body size and 4) the range of pore body sizes seen in photomicrographs of pore casts. The breakthrough capillary pressure and porosity can be obtained from conventional mercury porosimetry experiments.

The predicted drainage capillary pressure curves for Hg-air and oil-water displacements in a Berea sandstone are compared to the experimental ones in Fig. 2. The agreement for both cases is very good. The successful prediction of mercury porosimetry curves of sandstones using realistic pore structure information has important implications with regards to the characterization of porous media using mercury porosimetry experiments. In Fig. 3, the partitioning of pore volume of a simulated Boise sandstone into pore throats and pore bodies is compared to the "pore size distribution" that is obtained from the classical method of analysis(3). The inadequacy of the bundle-of-tubes model to describe the pore structure of sandstones is obvious. However, it is more important to note that WATNEMO may be useful in deconvolving the capillary pressure curves into information about the pore throat and pore body size distributions.

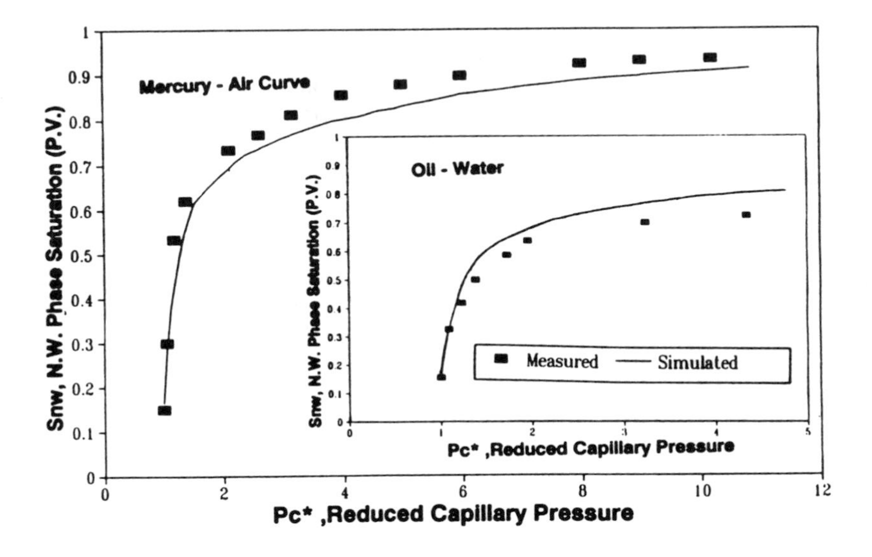

Fig. 2. Predicted vs. measured drainage capillary pressure curves of a Berea sandstone: (a) Mercury porosimetry; (b) Oil-water displacement.

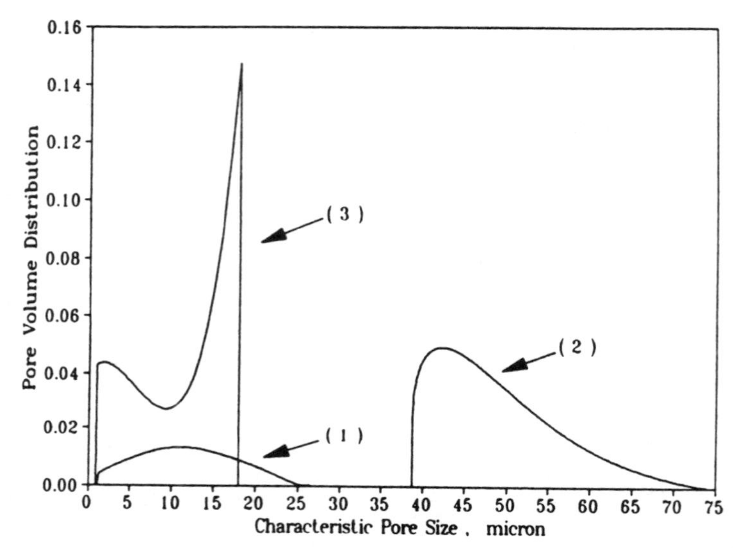

Fig. 3. Decomposition of model pore volume of Boise sandstone into its constituents: (1) Pore throat volume (20%); (2) Pore body volume (80%). Curve (3) is the result of differentiating the simulated intrusion mercury porosimetry curve.

The measured and the predicted drainage relative permeability curves of the same Berea sandstone sample are compared in Fig. 4. The successful prediction of both drainage relative permeability curves and drainage capillary pressure curves for oil-water displacement is strongly indicative of the appropriateness of the use of bond-correlated site percolation theory in combination with realistic estimates of the pore size distributions.

The measured and the predicted absolute permeabilities of 13 very different sandstones are compared in Fig. 5. The agreement is better than what could be expected from using empirical correlations(7). The prediction of the formation resistivity factor is accurate to within 20% of the measured values for sandstones with permeability higher than 200 mD. For lower permeability samples WATNEMO consistently overestimates the formation factor, probably because the conductance of clayey material is not taken into account in the model.

IV. CONCLUSIONS

An expert system (WATNEMO) was developed that is capable of approximating the pore structure of different sandstones from a minimum of information (porosity, breakthrough capillary pressure and photomicrographic data). Using this information, WATNEMO can successfully predict the absolute permeability, formation factor, drainage capillary pressure curves and drainage relative permeability curves of many different sandstones in a consistent manner. The results obtained provide strong indication that, upon further development, WATNEMO may become a useful tool for <u>Computer Aided Core Analysis.</u>

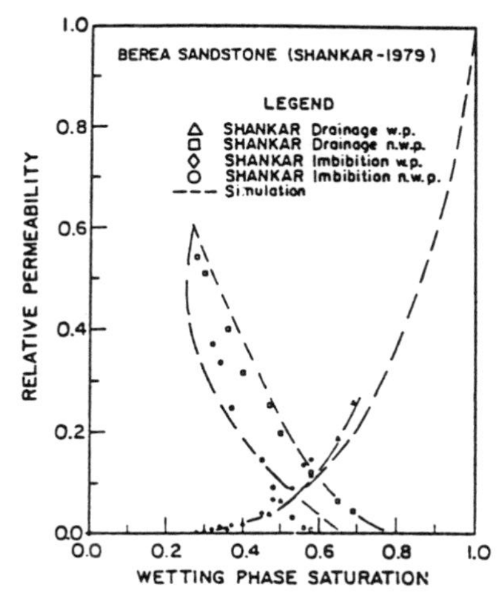

Fig. 4. Predicted vs. measured drainage relative permeability curves of Berea.

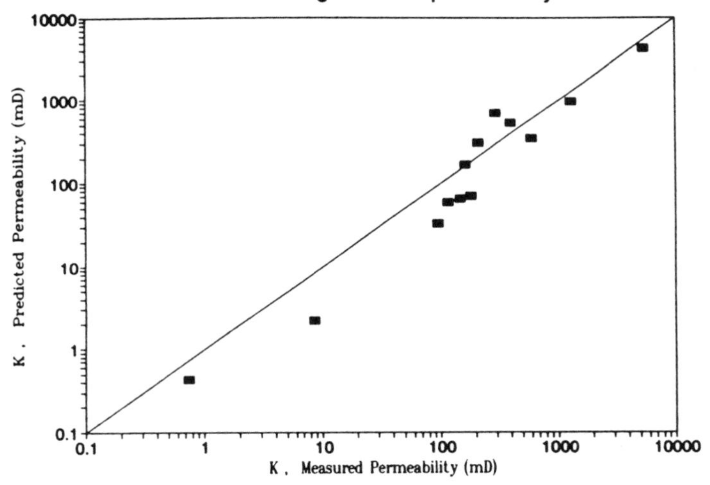

Fig. 5. Predicted vs. measured absolute permeabilities of 13 different sandstones.

REFERENCES

1. Adler, P.M., and Brenner, H. (1988). *Ann. Rev. Fluid Mech.* **20,** 35.
2. Chatzis, I., and Dullien, F.A.L. (1985). *Int. Chem. Eng.* **25,** 47.
3. Dullien, F.A.L., Porous Media: Fluid Transport and Pore Structure, 2nd Ed., *Academic Press Inc.* (1991)
4. Diaz, C.E., Chatzis, I., and Dullien, F.A.L. (1987). *Transp. Porous Media* **2,** 215.
5. Ioannidis, M.A., and Chatzis, I. (1991). Submitted to *Chem. Eng. Sci.*
6. Kantzas, A., and Chatzis, I. (1988). *Chem. Eng. Commun.* **69,** 169.
7. Kamath, J. (1988). *SPE paper No.18181,* 63th Ann. Fall Tech. Con., Oct.2-5, Houston.

FORMATION EVALUATION FOR IDENTIFYING SECONDARY GAS RESOURCES: EXAMPLES FROM THE MIDDLE FRIO, ONSHORE TEXAS GULF COAST BASIN[1]

Jose' M. Vidal
William E. Howard

ResTech Houston, Inc.
Houston, Texas

Raymond A. Levey

Bureau of Economic Geology
The University of Texas at Austin

Effective formation evaluation is critical for evaluating producing and bypassed gas reservoirs in existing gas fields. As a part of the Infield Natural Gas Reserve Growth Project sponsored by the Gas Research Institute, The U.S. Department of Energy and the State of Texas, an integrated analysis of open hole and cased hole well logs, wireline pressure tests, combined with production tests results and core data analysis were used to evaluate Frio fluvial-deltaic sandstone gas reservoirs. Results for the study indicate that gas producing intervals can be identified using the methods described below.

Well logs were evaluated from older wells with limited logging suites and in more recent wells with complete logging

[1] Funded by the Gas Research Institute contract No. 5088-212-1718, the U.S. Department of Energy contract no. DE-FG-21-88MC25031 and the State of Texas throught the Bureau of Economic Geology. The cooperation of Union Pacific Resources Company and Mobil Exploration and Producing U.S. Inc., and Oryx Energy Company is gratefully acknowledged.

suites. A shaly sand model (dual water method) was used in the formation evaluation.[2] Computed results including shale volume, porosity, water saturation and empirical permeability were obtained from wells logged by different service companies. Figure 1 is an example of open hole evaluation for a typical middle Frio penetration. The perforated zone has a cumulative production of 48,144 mcfg.

Because log responses differ among service companies, individual logs were corrected for environmental and borehole effects and standardized to the average field response. Figure 2 shows the standardization of a density log. Corrections[3] for borehole washout were applied to the affected logs (density and neutron) using a model based on responses of other porosity dependent measurements. Figure 3 shows before and after results of the environmental and borehole washout corrections.

The values of formation water resistivity vary from .32 to .17 ohm-m at 75 F°. These values were determined from log data and from measurements of resistivity and salinity on 14 collected formation water samples. The measured formation water salinities were also corrected for the dilution problem associated with gas production.

The results of available core data analysis were used to calibrate log derived values.[4] Figure 4 shows the agreement between the log derived porosity and shale volume vs. the core analysis results as well as displaying the irreducible water saturation derived by capillary pressure measurements. From core data a relationship was also obtained to predict permeability. Core permeability measurements revealed both high (>200 md) and low (<15 md) permeability sandstone reservoirs (see Figure 5). These two trends of permeability were incorporated into the computed log interpretation model.

Because many Frio reservoirs are gas depletion drive, identification of depleted zones is critical. Wireline pressure tests in open hole wells and short term production tests in cased hole wells are common techniques used to determine reservoir pressure. Some wireline formation pressures can be misleading due to tool defects, miscalibrations or incomplete pressure test sets. The validation of wireline pressures and close control of the operation at the wellsite is indispensable before selecting zones of interest.

One problem encountered obtaining valid formation pressure data is when insufficient setting time is allowed in order to measure stabilized shut-in wireline pressures. For example in Figure 6A, a 60 second wireline set leads a low pressure of 139 psi. By contrast, in Figure 6B, the tool was reset l/2 foot higher in the wellbore and allowed to set for 251 seconds, yielding a more representative though not quite stabilized pressure of 3556 psi. A valid test is shown in Figure 7; here the tool was set for an adequate amount of time so that the shut in pressure has stabilized and can be used with confidence. Another problem is depletion of the reservoir from the time that the open hole pressure was taken to the time that the zone was perforated and production tested.

For cased hole wells with incomplete logging suites a logging program was designed to complement the data already available.[5] The cased hole interpretation model was applied to a 39 year old well that penetrated to the base of the middle Frio formation in Stratton Field. The original open hole logs consisted only of an electrical survey and a few sidewall core samples. A cased hole logging program was designed to fully evaluate the well. Pulsed neutron, gamma ray, dipole sonic, temperature, noise and cement bond logs were acquired using currently available technologies.

Porosities were derived from the combination of the compressional sonic porosity and the pulsed neutron porosity. These logs respond in a similar manner in oil zones as well as in zones with high water saturations. Gas presence is easily detected by observing the characteristic crossover of porosities (sonic > neutron porosity). Gas crossover is also observed in the normalized far and near pulsed neutron count rates. Combining that information with the sigma derived cased hole water saturation, and comparing it to the original open hole water saturation , it was possible to identify potential gas productive zones. Figure 8 shows the results in two gas zones, E41L was perforated and produced 316 mcfd gas with 412 psi.

In conclusion: An interpretation model that identifies gas bearing intervals has been developed for open hole and cased hole wells in Frio sandstone gas reservoirs. This method can be applied to other sandstones with similar depositional environments. The method integrates logs with core analysis results and utilizes wireline pressures and production test results.

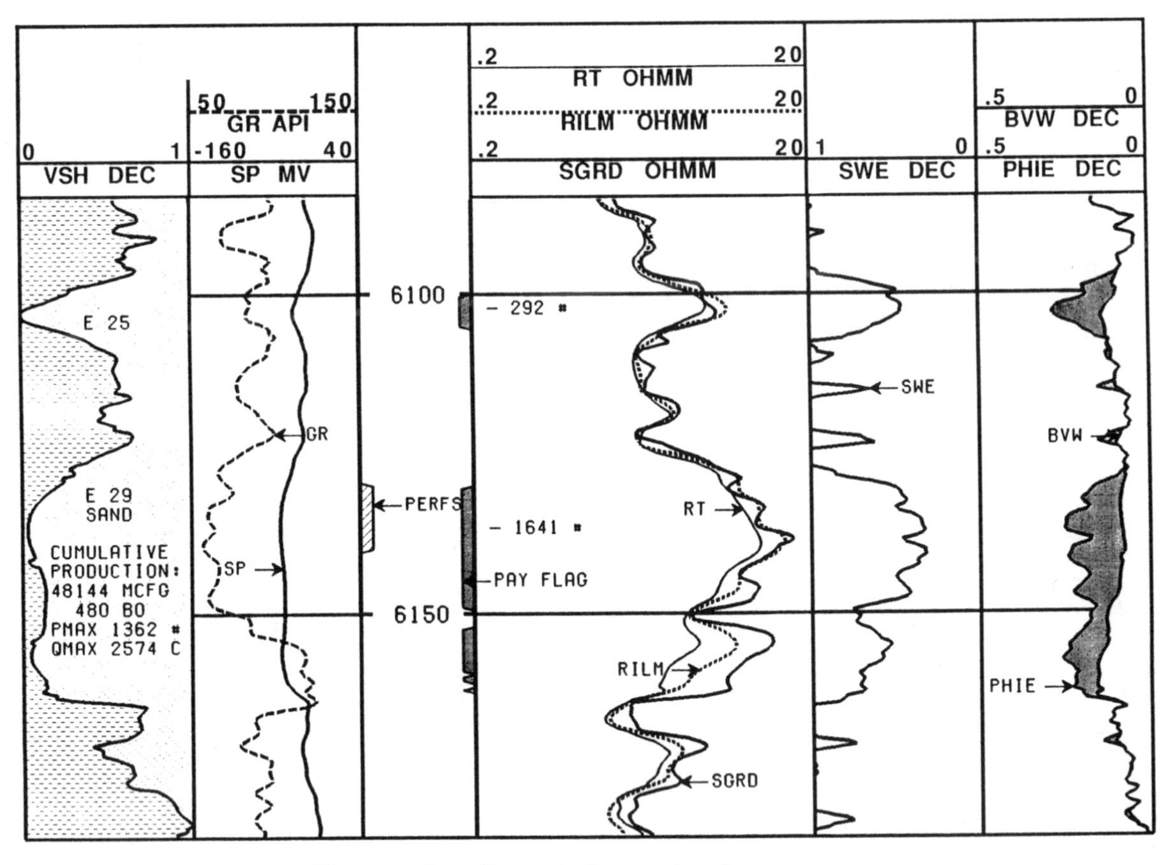

Figure 1. Open hole evaluation.

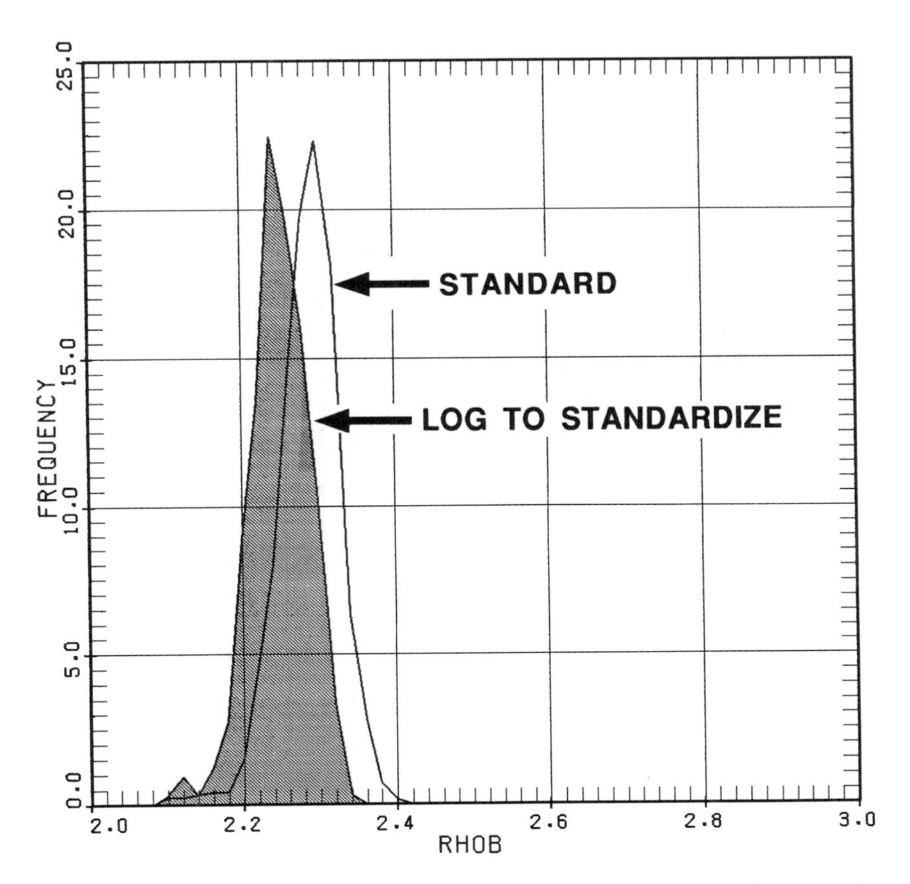

Figure 2. Standardization of density log to fieldwide standard.

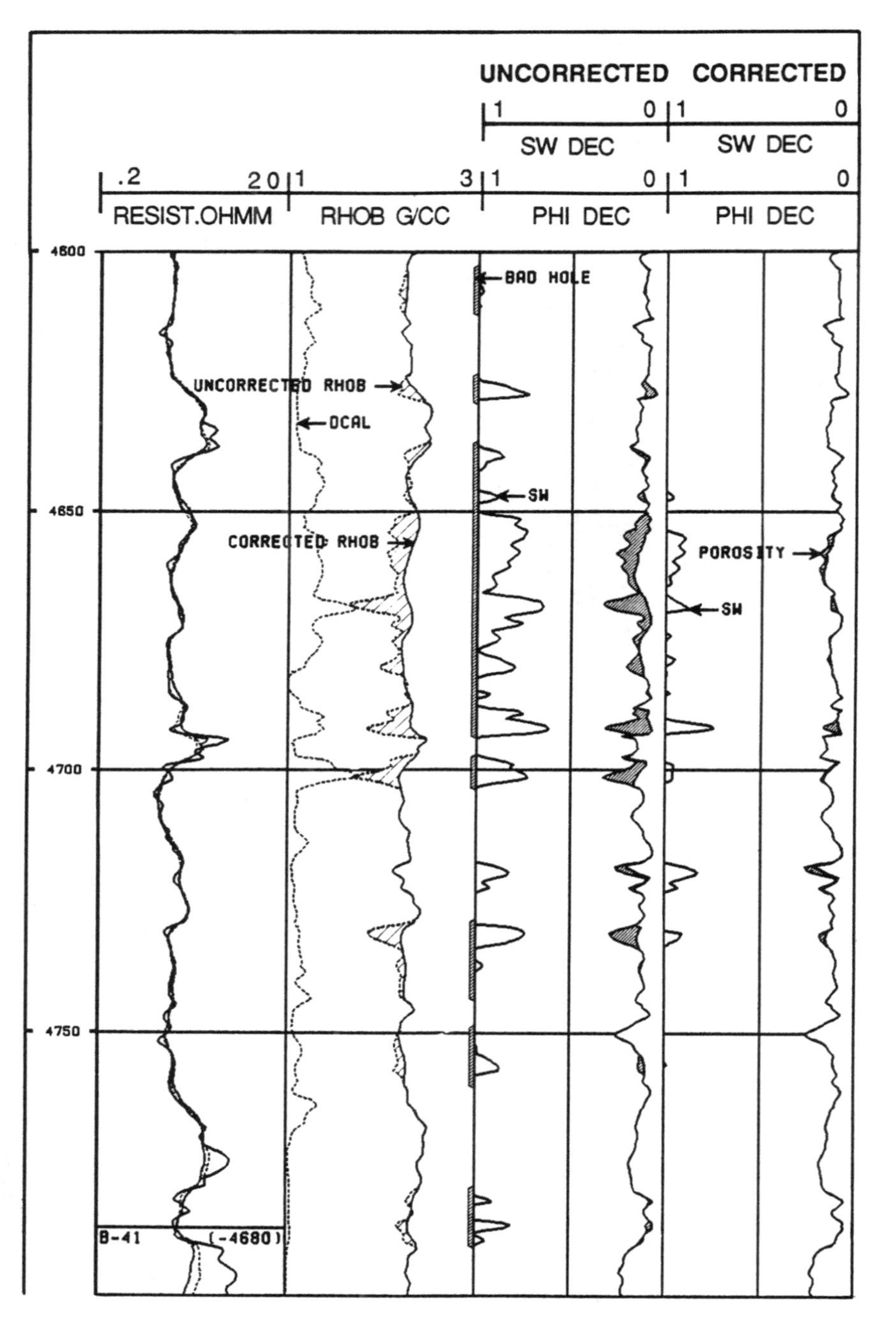

Figure 3. Log corrections in washouts.

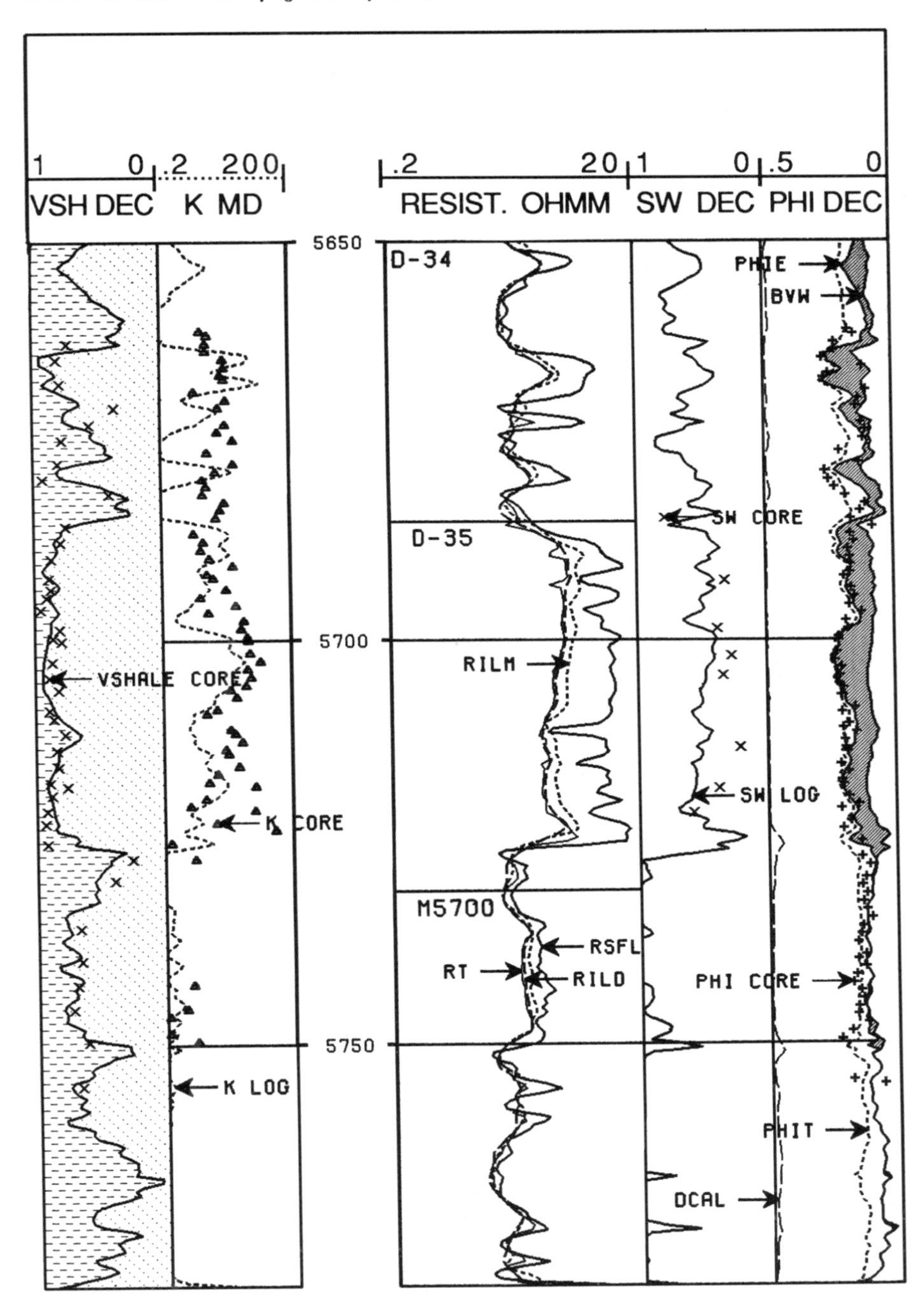

Figure 4. Comparison of results with core analysis.

Porosity and Permeability of Middle Frio Reservoirs Stratton Field, Kleberg and Nueces Counties

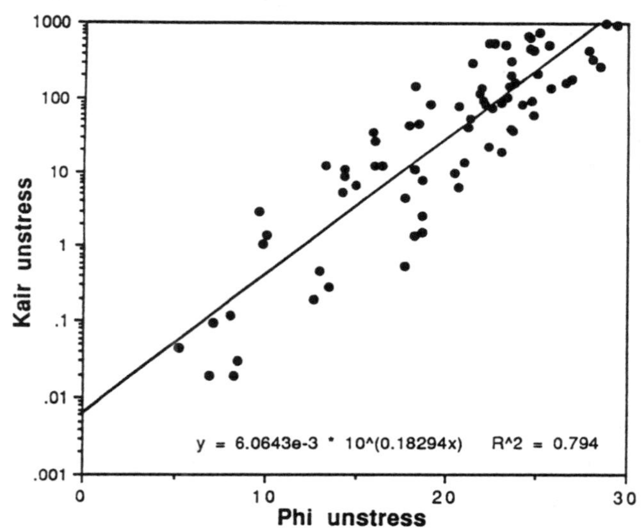

Lower Permeability Reservoirs

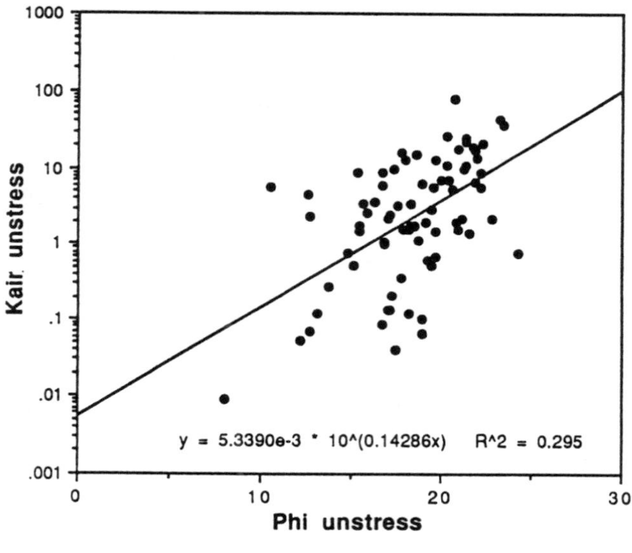

Figure 5.

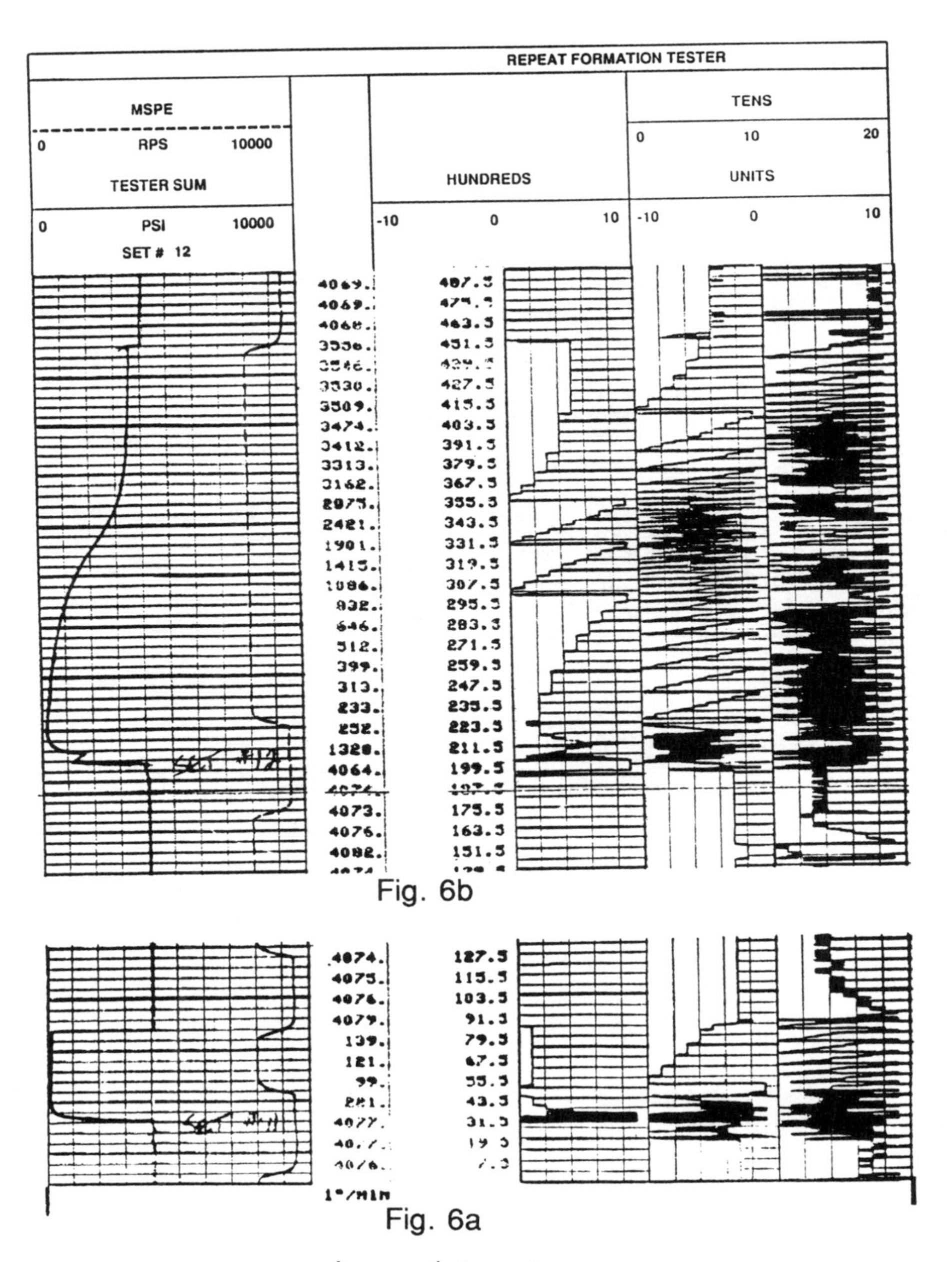

Fig. 6b

Fig. 6a

Incomplete sets.

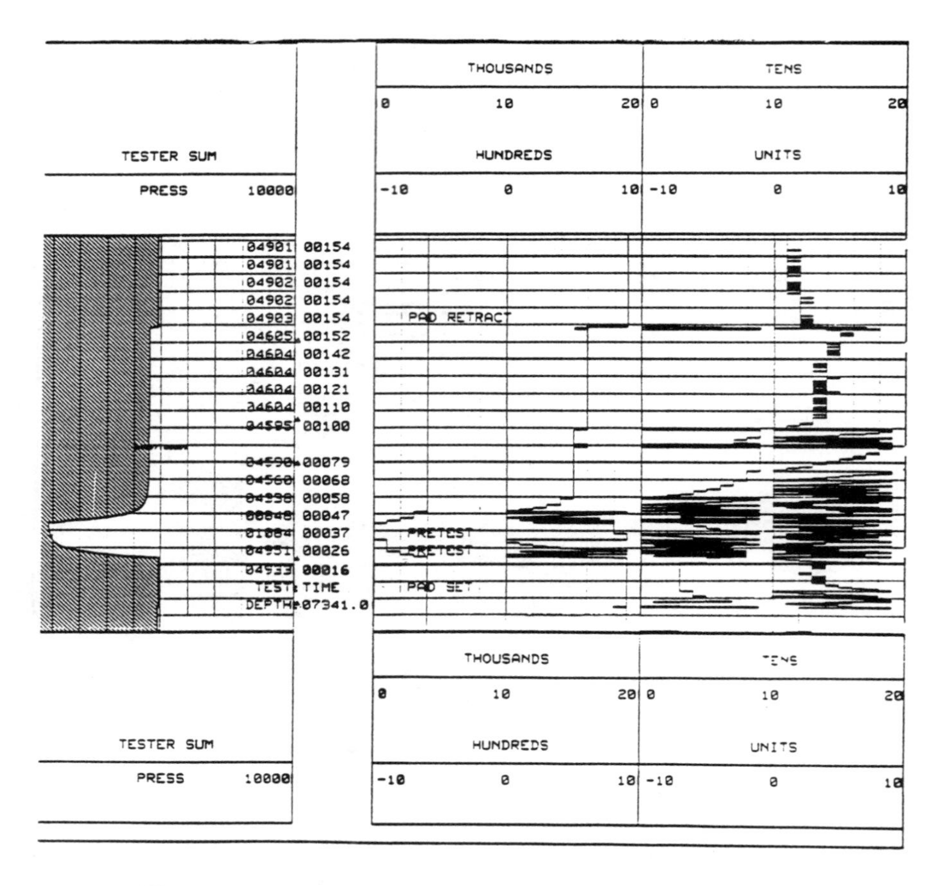

Figure 7. Valid wireline formation set.

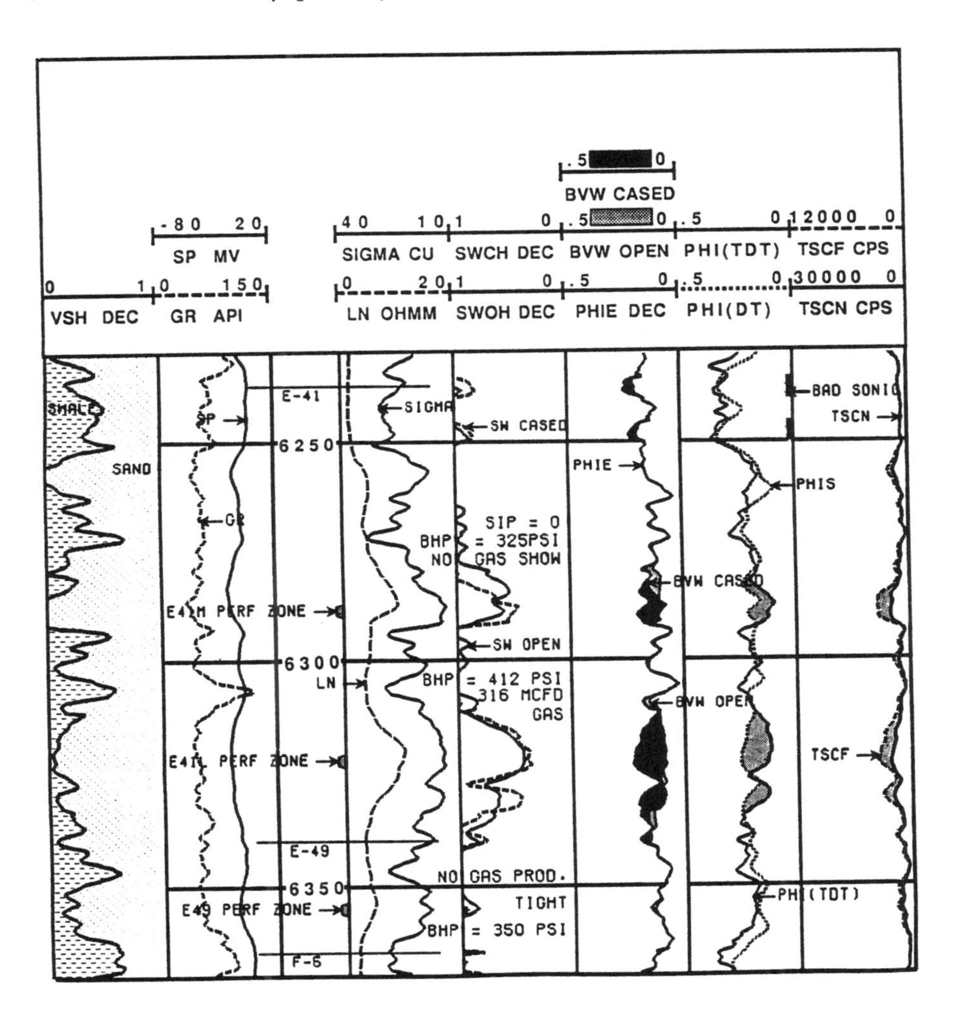

Figure 8. Cased hole evaluation.

References

2. Clavier C., Coates, G., and Dumanoir, J. 1977. " The Theoretical and Experimental Basis for the Dual Water Model for the Interpretation of Shaly Sands," SPE Paper 6859 presented at the 1977 SPE Annual Technical Conference, Denver, Colorado, Oct. 9-12.

3. Howard W. E., Hunt E. R. 1986. "Travis Peak: An Integrated Approach to Formation Evaluation," SPE Paper 15208 presented at the Unconventional Gas Technology Symposium of the Society of Petroleum Engineers held in Louisville, KY, May 18-21, 1986, pp 8-9.

4. Truman R. B., Davies D. K., Howard W. E. and Vessell R. K. "Utilization of Rock Characterization Data to Improve Well Log Interpretation." Paper V. SPWLA 27th Annual Logging Symposium June 9-13, 1986.

5. Jirik L.A., Howard W. E. and Sadler D. L. "Identification of Bypassed Gas Reserves through Integrated Geological and Petrophysical Techniques: A Case Study in Seeligson Field, Jim Wells County, South Texas." SPE Paper 21483 presented at the SPE Gas Technology Symposium held in Houston, Texas, January 23-25, 1991.

RESERVOIR CHARACTERIZATION AND GEOSTATISTICAL MODELING - THE INTEGRATION OF GEOLOGY INTO RESERVOIR SIMULATION: EAST PAINTER RESERVOIR FIELD, WYOMING

D. Scott Singdahlsen

CHEVRON USA, Inc.
Western Exploration Business Unit
Frontier Division Development Geology
Denver, Colorado USA

I. INTRODUCTION

Geologic heterogeneity is known to be an important control on fluid flow in petroleum reservoirs. Heterogeneity is present at all scales within the reservoir, ranging from gigascopic (km) to microscopic (μm). The task confronting reservoir engineers and geologists is to describe and characterize the varying scales of reservoir heterogeneity in a form that is usable in the daily application of reservoir flow simulation to field management. The proper representation of reservoir heterogeneity is the key to successful modeling of past and future production performance. Detailed geologic models, based on reservoir description and depositional environment, must be integrated into reservoir flow simulations to assure realistic results. The result of this effort is the application of reservoir flow simulation as a project oriented risk assessment tool used in daily reservoir management.

II. GENERAL GEOLOGIC SETTING

East Painter Reservoir field is one of a number

of structural traps located within the Overthrust Belt of southwestern Wyoming. Hydrocarbon production is from the Jurassic Nugget Sandstone, the primary reservoir for several fields along trend. The East Painter structure is a doubly plunging, westward verging, asymmetric anticline formed on the hanging wall of a back thrust imbricate near the leading edge of the Absaroka thrust. Structural heterogeneity is characterized from 3-D seismic interpretation and dipmeter analysis.

The approximately 900-foot thick Nugget Sandstone is a stratigraphically complex and heterogeneous unit deposited primarily by eolian processes in a complex erg setting. Previous workers (Lindquist, 1988; Tillman, 1989) have documented the stratigraphic variability and reservoir characteristics of the Nugget.

III. RESERVOIR DESCRIPTION

The Nugget is comprised of dune, transitional toeset, and interdune/sand sheet facies, each exhibiting different petrophysical distributions. The high degree of heterogeneity within the Nugget results from variations in grain size, sorting, mineralogy, and degree and distribution of lamination. Resultant reservoir anisotropy is inherited directly from the primary eolian depositional fabric. Porosity ranges from 2.5% to 21.5%, while air permeabilities, measured in core samples, span five orders of magnitude from hundredths to hundreds of millidarcies. Facies architecture results in both vertical and lateral stratification of the reservoir. Eolian depositional architecture within the Nugget, particularly the spatial orientation of dune sets, stratification types, and the bounding surfaces that truncate them, controls permeability patterns in the reservoir. Facies discontinuities, fractures, and faults offset permeability trends in portions of the field.

The Nugget is divided into three primary reservoir facies based on differences in stratification types and petrophysical properties identified from detailed core analysis. The best reservoir rock is found in the dune foreset facies. This facies consists of cross-stratified sandstones resulting from grainflow and wind-ripple processes (Hunter, 1977, 1981). Reworking of grainflow avalanche deposits along strike results in rippleform and translatent wind ripples, and causes slight to moderate degradation of reservoir quality. The dune foreset facies commonly has permeabilities from tens to hundreds of millidarcies with moderate anisotrophy (Kh/Kv ranging from 1/1 to 4/1). Transitional toeset facies consists of grainflow avalanche toesets that are extensively wind rippled. This facies exhibits intermediate reservoir quality and increased permeability anisotropy (Kh/Kv ranging from 2/1 to 7/1) when compared to the dune facies. The third reservoir facies consists of interdune and sand sheet deposits. These deposits are characterized by the predominance of low-angle to horizontal wind ripple stratification with zones of extensive bioturbation and wavy lamination (Ahlbrandt and Fryberger, 1981; Kocurek, 1981a). Both dry and wet interdune deposits are observed in the Nugget cores. Interdune rocks exhibit poor reservoir characteristics due to fine grain sizes, poor sorting, and thinner lamination. Porosities commonly are less than 10% and permeabilities range from hundredths to tenths of millidarcies. Wet interdune deposits exhibit the poorest reservoir quality due to the presence of interstitial fines, and early diagenetic cementation. From outcrop studies, the wet interdune deposits appear to be laterally continuous over field size areas and may result in reservoir scale vertical flow compartmentalization. Translatent wind ripple strata of the interdune/sand sheet facies have the largest permeability anisotropy with high horizontal to vertical permeability ratios (Kh/Kv ranging from 3/1 to 20/1) due to thin and texturally variable stratification.

Reservoir quality improves upward from the mixed eolian and non-eolian lower Nugget to the entirely eolian upper Nugget. The lower third of the Nugget was deposited in a low relief sand sheet environment with isolated small barchanoid dunes. Minor erg margin deposition characterized by fluvial, lacustrine and sabkha facies is also common. The upper two thirds of the Nugget is comprised of migrating large crescentic transverse dunes with minor dry interdune regions. The vertical segmentation of the Nugget at East Painter represents regional erg migration as conceptualized by Porter (1986). Thus, the lower Nugget represents the fore-erg sequence, dominated by sand sheet and erg margin deposition, while the upper Nugget represents a central-erg setting reflecting prolonged eolian deposition of large dune complexes. The back-erg sequence is absent at East Painter due to erosional removal at the post-Nugget unconformity.

IV. GEOLOGIC RESERVOIR MODEL

Once the depositional environment is deciphered from core and outcrop analog study, the depositional model can be placed into a reservoir flow context. Detailed relationships of petrophysical properties must be established to form the basis for flow simulation. Permeability, which plays the dominant role in flow simulation, is the variable with the least amount of reliability, and is commonly not extensively collected during the development phase of most fields. Permeability data can usually only be obtained from core analysis and well testing, but a general sparsity of core data generally requires a log/core porosity-permeability transformation to be utilized. At East Painter, high permeability zones corresponding to dune foreset facies, have historically been underestimated. Historic production data reveals that these high permeability zones account for the majority of reservoir processing and resultant field production. Specific high permeability core-based transforms, integrating gamma ray character, are used in an attempt to better reflect the distribution of these dominant flow zones.

Permeability anisotropy is determined from directional permeability core measurements, and appears to be related to primary depositional fabric. Tri-axial directional permeability measurements of cubic core plugs from East Painter core indicate that the maximum permeability direction is oriented along the strike of dune foreset cross-stratification. Intermediate permeability is oriented parallel to dune migration direction (paleowind direction). The minimum permeability direction is oriented perpendicular to laminations, or nearly vertical in horizontally laminated wind ripple deposits. Analysis of stratigraphic dipmeter data provides field-wide eolian bedform morphology and orientation for East Painter. Dipmeter analysis of the high angle stratification in the upper Nugget shows a strongly unimodal dip orientation with an azimuth mode of 230° indicating a general bedform orientation of 140° (NW-SE). The moderate spread of dip azimuths about the mode suggests that the crescentic dunes were sinuous rather than straight-crested. Calculated paleowind directions range from 215° to 245° which correlates well with the predicted paleowind directions of Parrish and Curtis (1982) and Parrish and Peterson (1988). This data is directly integrated with core based permeability patterns to yield reservoir-wide directional permeability trends.

Sparse core data at East Painter precludes direct facies correlation across the field. Petrophysical characteristics, stratigraphic dipmeter analysis, and borehole resistivity image (FMS) data are integrated and related to core facies descriptions to develop facies correlations between wells (Luthi and Banavar, 1988). Well to well correlations of stratification type and associated eolian bounding surfaces are applied to develop a reservoir facies model. Petrophysical data is also integrated into the reservoir model based on relationships developed from core and log analysis. The geologic model can then be used to determine stratigraphic controls on fluid flow within the reservoir (Chandler and others, 1989). This first-order stratigraphic heterogeneity is integrated into flow simulation models in the form of distributed flow layering.

V. GEOSTATISTICAL MODELING

Geostatistics are increasingly being used as a probabilistic method of describing the spatial relationships of reservoir properties including the uncertainties related to the incomplete sampling of the reservoir property distribution (Journel, 1990). Geostatistical modeling provides a method for modeling the spatial reservoir property distribution at East Painter based on the geologic model developed for the Nugget.

Due to sparse well data, additional reservoir geometric data was required to accomplish detailed variogram analysis of the relative lengths, widths, and heights of dune set packages in the Nugget at East Painter. Unfortunately, the Nugget crops out in very poor and fragmented exposures near East Painter, and is not suitable for direct, detailed field study and analysis of dune geometries and distribution. Field studies of outcrop analogs such as the Entrada Sandstone and Navajo Sandstone in Utah were used in combination with published research to determine probable facies distribution and geometry for the Nugget (Breed and Grow, 1979; Knight, 1986; Kocurek, 1981b; and Weber, 1987). Results of detailed core, log, and outcrop analysis of dune set orientation, geometry, size, and directional permeability trends are imposed on directional variogram models resulting in both vertical and lateral correlation structures of petrophysical variables (porosity and permeability) for the reservoir.

Vertical and lateral reservoir variability is integrated to characterize the spatial distribution of reservoir properties. Conditional simulation methodology proposed by Hewitt (1986) and Hewitt and Behrens (1988) was employed to generate equally probable realizations that combine large-scale reservoir heterogeneity with fractally generated stochastic small-scale variability. Conditional realizations, based on the calculated variogram models, are used to determine the geological and petrophysical variability input into reservoir flow simulation.

VI. RESERVOIR FLOW SIMULATION

Geostatistical modeling output of reservoir property variability and distribution are directly integrated into reservoir flow simulation grids to perform project oriented fluid flow modeling of the Nugget at East Painter. In general, the application of detailed geologic models to flow simulation indicate that reservoir heterogeneity plays the dominant role in controlling fluid flow. Simulation results commonly show increased stratigraphic channeling of fluids, more rapid breakthrough of injectants, less complete processing of the reservoir, and generally lower ultimate recoveries.

Direct application of results from geostatistical modeling to the simulator grid allows for multiple simulation runs to study and assess both geologic and production risks associated with projects. Multiple simulation runs on multiple reservoir realizations provide a probability distribution of results that can be applied to risk analysis and the daily management of reservoirs.

VII. SUMMARY AND CONCLUSIONS

With increasing emphasis on daily management and increased recovery in existing reservoirs, ever greater numbers of earth scientists are becoming involved in the detailed description and characterization of reservoir geology. It is incumbent upon earth scientists to become familiar with, and integrally involved in project oriented reservoir simulation, project risk assessment, and the daily management of developed fields. It is equally important that reservoir engineers gain an understanding of the geologic heterogeneities that control fluid flow in reservoirs. Only through an integrated team approach can significant results in reservoir characterization and management be achieved.

An integrated team approach to reservoir characterization and simulation was applied at East Painter Reservoir field to study the benefits and risks associated with continued field development and injection realignment. The results of this team approach were the construction of a detailed comprehensive geologic model of the Nugget Sandstone, the use of geostatistics to model the interwell geologic variability and uncertainty, and the application of reservoir flow simulation to risk assessment, decision analysis, and field management.

VIII. ACKNOWLEDGEMENTS

Fellow team members on the East Painter Simulation Study:

Doris Lambertz, WPBU Reservoir Engineer
Steven Prelipp, RMPBU Production Engineer
Jim Ricotta, RMPBU Reservoir Engineer

IX. REFERENCES

Ahlbrant, T.S. and Fryberger, S.G., 1981, Sedimentary features and significance of interdune deposits, in, F.G. Ethridge and R.M. Flores, eds., Recent and ancient nonmarine depositional environments: Society of Economic Paleontologists and Mineralogists Special Publication 31, p. 293-314.

Breed, C.S. and Grow, T., 1979, Morphology and distribution of dunes in sand seas observed by remote sensing, in, E.D. McKee, ed., A study of global sand seas: USGS Professional Paper 1052, p. 253-302.

Chandler, M.A., Kocurek, G., Goggin, D.J., and Lake, L.W., 1989, Effects of stratigraphic heterogeneity on permeability in eolian sandstone sequence, Page Sandstone, northern Arizona: American Association of Petroleum Geologists Bulletin, v. 73, p. 658-668.

Hewett, T.A., 1986, Fractal distributions of reservoir heterogeneity and their influence of fluid transport: SPE paper 15386 presented at the 1986 SPE Annual Technical Conference, New Orleans.

----- and Behrens, R.A., 1988, Conditional Simulation of reservoir heterogeneity with fractals: SPE paper 18326 presented at the 1988 SPE Annual Technical Conference, Houston, p. 645-660.

Hunter, R.E., 1977, Basic types of stratification in small eolian dunes: Sedimentology, v. 24, p. 361-387.

----- 1981, Stratification styles in eolian sandstones: some Pennsylvanian to Jurassic examples from the western interior USA, in, F.G. Ethridge and R.M. Flores, eds., Recent and ancient nonmarine depositional environments: Society of Economic Paleontologists and Mineralogists Special Publication 31, p. 315-329.

Journel, A.G., 1990, Geostatistics for reservoir characterization: SPE paper 20750 presented at the 1990 SPE Annual Technical Conference, New Orleans, p. 353-358.

Knight, J.B., 1986, Eolian bedform reconstruction: a case study from the Page Sandstone, northern Arizona: unpublished MS Thesis, University of Texas, Austin, 100 p.

Kocurek, G., 1981a, Significance of interdune deposits and bounding surfaces in aeolian sand dunes: Sedimentology, v. 28, p. 753-780.

-----, 1981b, Erg reconstruction: the Entrada Sandstone (Jurassic) of northern Utah and Colorado: Palaeogeography, Palaeoclimatology, Palaeoecology, v.36, p. 125-153.

Lindquist, S.J., 1988, Practical characterization of eolian reservoirs for development: Nugget Sandstone, Utah-Wyoming thrust belt: Sedimentary Geology, v. 56, p. 315-339.

Luthi, S.M. and Banavar, J.R., 1988, Application of borehole images to three-dimensional geometric modeling of eolian sandstone reservoirs, Permian Rotliegende, North Sea: American Association of Petroleum Geologist Bulletin, v. 72, no. 9, p. 1074-1089.

Parrish, J.T., and Curtis, R.L., 1982, Atmospheric circulation, upwelling, and organic-rich rocks in the Mesozoic and Cenozoic Eras:

Palaeogeography, Palaeoclimatology, and Palaeoecology, v. 40, p. 31-66.

-----, and Peterson, F., 1988, Wind directions from global circulation models and wind directions determined from eolian sandstones of the western United States - a comparison: Sedimentary Geology, v. 56, p. 261-282.

Porter, M.L., 1986, Sedimentary record of erg migration: Geology, v. 14, p. 497-500.

Tillman, L.E., 1989, Sedimentary facies and reservoir characteristics of the Nugget Sandstone (Jurassic), Painter Reservoir Field, Unita County, Wyoming, in, E.B. Coalson, ed., Petrogenesis and petrophysics of selected sandstone reservoirs of the Rocky Mountain region: Rocky Mountain Association of Petroleum Geologists, p. 97-108.

Weber,K.J., 1987, Computation of initial well productivities in aeolian sandstone on the basis of a geologic model, Leman Gas Field, U.K., in, R.W. Tillman and K.J. Weber, eds., Reservoir Sedimentology: Society of Economic Paleontologists and Mineralogists Special Publication 40, p. 333-354.

INTEGRATED FIELD, ANALOG, AND SHELF-SCALE GEOLOGIC MODELING OF OOLITIC GRAINSTONE RESERVOIRS IN THE UPPER PENNSYLVANIAN KANSAS CITY GROUP IN KANSAS (USA)[1]

John A. French
W. Lynn Watney

Kansas Geological Survey and the Department of Geology
University of Kansas
Lawrence, Kansas

I. INTRODUCTION

Large volumes of unswept mobile oil remain in cyclic Pennsylvanian reservoirs in the midcontinent region of the United States. To effectively increase recovery of this oil, geologic variables including relative sea-level history, topography, and depositional and diagenetic conditions that control heterogeneity within these reservoirs must be well understood. To accomplish this, reservoir-bearing carbonate units and associated strata are being studied at a variety of scales in surface exposures and in cores and wireline logs including an extensive set of existing wells in the adjacent shallow subsurface. The site described here serves as an analog to reservoirs that occur in the deeper subsurface. Regional mapping and study of oil fields are additional components in the broader study of this work.

At and near the outcrop belt in southeastern Kansas, the lower Missourian Bethany Falls Limestone (equivalent to the K zone in the subsurface) contains oolitic grainstones similar to the lithofacies that produce from the unit farther west. Depositional sequence analysis of the lower Missourian strata suggests that 1) this stratigraphic interval is comprised of multiple depositional sequences exhibiting a complex three-dimensional stacking geometry, 2) the Bethany Falls Limestone is a complex, shallowing-upward unit that is part of a major, shelf-wide, unconformity-bounded depositional sequence, 3) at least two cycles of lessor relative sea-level change controlled the vertical distribution of porosity zones within this major sequence, 4) the character and distribution of reservoir-scale grainstone buildups within this

---

[1]Supported by DOE Grant No. DE-FG22-90BC14434

depositional sequence were strongly influenced by pre-existing depositional topography, 5) a series of dip-oriented oolitic lobes occur laterally at similar elevations along a prominent paleoslope suggesting longer term residence of sea level at this position, and 6) lithologic and well log profiles of the Bethany Falls Limestone are closely comparable along a strike position of the paleoslope.

II. STRATIGRAPHY AND OCCURRENCE OF RESERVOIR-TYPE UNITS

A. Description and Depositional Sequence Analysis of the Study Interval

The focus of this study is the Bethany Falls Limestone Member of the lower Missourian Swope Limestone in southeastern Kansas, U.S.A. (FIGURES 1 and 2). The Bethany Falls Limestone is part of a cyclic transgressive-regressive unit of the type that typifies the lower Missourian succession in much of eastern Kansas and surrounding regions of the northern Midcontinent. Each carbonate-dominated cycle is an individual unconformity-bounded depositional sequence that ranges from 25 ft (7.6 m) to over 110 ft (33.5 m) in thickness in the study area. They typically consist of a basal transgressive shale-limestone couplet less than 10 ft (3 m) thick that is overlain by a deep-water black shale normally less than 3 ft (1 m) thick, which is in turn overlain by an upper, generally regressive limestone that is up to 90 ft (27.4 m) thick. In the midcontinent, many of these upper limestones contain reservoir-type oolitic grainstones that in most cases were deposited near the end of a period of overall relative sea-level fall. This is evidenced by a consistent upward transition from relatively deep-water deposits such as the black shale through open-shelf wackestones into the high-energy, shallow-water oolitic lithofacies, with subaerial exposure features common at and near the top of the oolite or the peritidal lithofacies that caps the oolite in some cycles. The regressive limestone ranges from a single shallowing-upward unit to a complex set of thinner, small-scale (minor) unconformity-bounded cycles. Cycle architecture is very similar at similar shelf locations while predictable changes in cycle architecture are noted in individual regressive limestones at different shelf locations; for example, a regressive limestone that is a simple shallowing-upward unit at high-shelf locations may exhibit several small-scale deepening-shallowing cycles lower on the shelf, presumably as more accommodation space is available to preserve more complex elements of the sea-level history.

The Bethany Falls Limestone is the upper, overall regressive limestone within the Swope depositional sequence (FIGURE 3). It ranges from about 22 ft (6.7 m) to 42 ft (12.8 m) in thickness in the study area, and consists of two basic lithofacies. The lower Bethany Falls consists mainly of nonporous, diversely fossiliferous bioclast wackestone, with fossils such as echinoderms, brachiopods, bryozoans, and phylloid algae constituting the bulk of the biota. The upper portion is more variable in thickness and consists of one or two units of generally porous, cross-bedded, ooid-bioclast

120 — 400
m — ft
0 — 0
Approximate scale (limestones expanded at expense of shales)

Interval studied: Mound Valley – Swope

Formation	Group	Stage	Series
Howard	Wabaunsee	Virgilian	Upper Pennsylvanian
Topeka	Shawnee		
Deer Creek			
Lecompton			
Oread			
Lawrence	Douglas		
Stranger			
Stanton	Lansing	Missourian	
Plattsburg			
Wyandotte	Kansas City		
Iola			
Drum			
Lwr Cherryvale			
Dennis			
Mound Valley			
Swope			
Hertha			
	Pleasanton		
Lenepah	Marmaton	Desmoinesian	Middle Penn.
Altamont			
Pawnee			
Fort Scott			
Excello	Cherokee		

Modified from Heckel 1977

Figure 1. Schematic section of a portion of the Pennsylvanian System in Kansas showing the interval discussed in this paper.

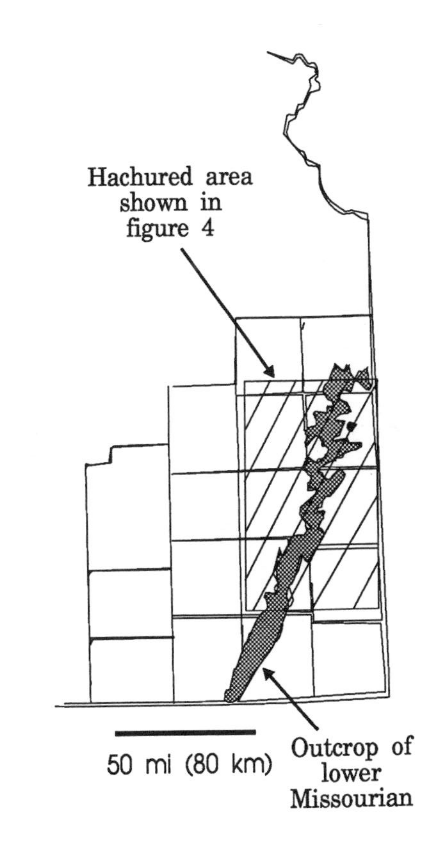

Figure 2. Portion of eastern Kansas showing study area and outcrop of lower Missourian strata.

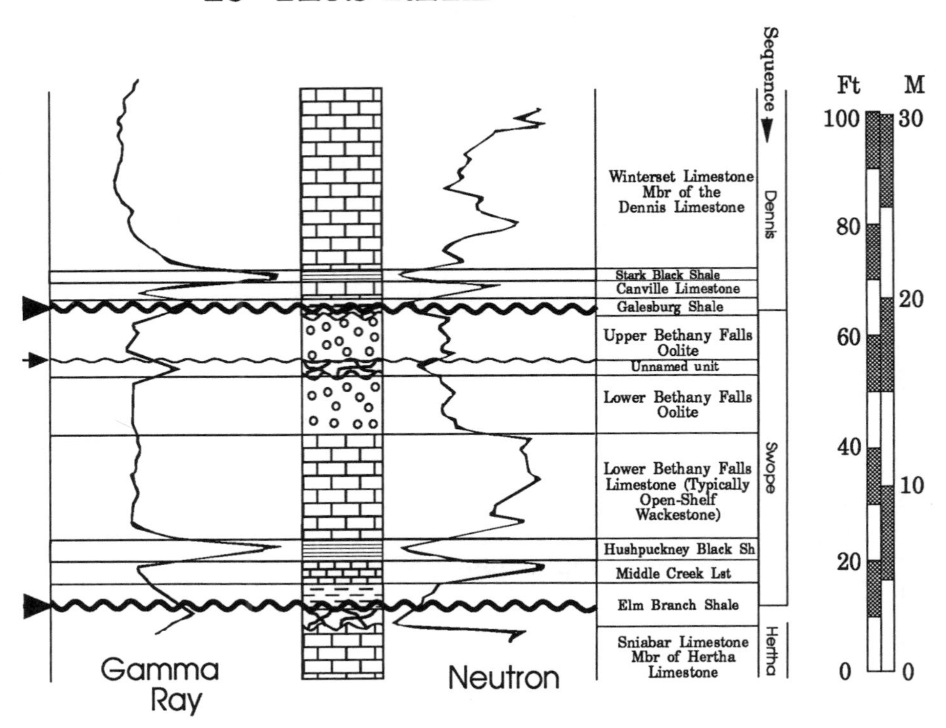

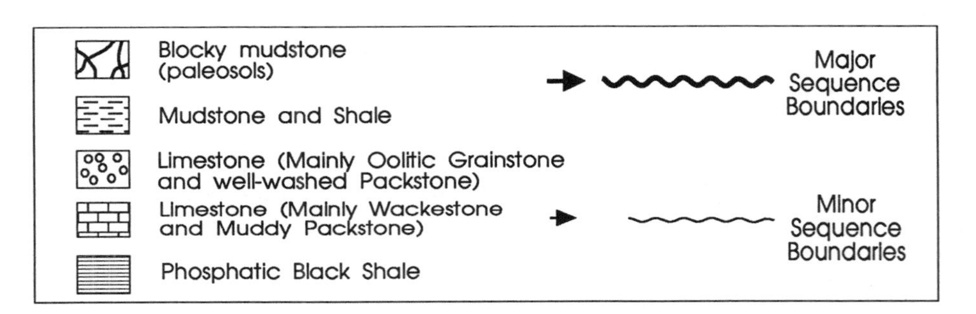

Figure 3. Representative gamma ray-neutron log and stratigraphic column of study interval in west-central Bourbon Co., Kansas. The grainstones and packstones of the Bethany Falls Limestone are the focus of this paper. See text for details.

grainstone and well-washed packstone. Within the upper few feet (about one meter) of each oolite there is commonly evidence of subaerial exposure, such as rhizoliths, dissolution features, and laminated crusts. In the study area the lower oolite is locally overlain by a mudstone unit that is interpreted to be a paleosol, at least in its lower part. In addition to petrographic evidence there are negative carbon isotope excursions at the top of each oolite, which also suggests that intervening episodes of subaerial exposure occurred.

B. Nature and Distribution of Reservoir-Type Lithofacies

The oolitic grainstones and packstones that comprise the upper portion of the Bethany Falls Limestone in the study area are lithologically very similar in thickness, lithology, and stratal succession to many producing zones in the subsurface of central and western Kansas. The localization and character of these reservoir analogs are similarly believed to have been controlled largely by a combination of favorable depositional topography acting in concert with small-scale sea-level fluctuations.

Within the study area, the Pleasanton shales and sandstones that underlie the lowermost Missourian carbonate units (see FIGURE 1) thin from over 180 ft (55 m) in east-central Allen Co. to less than 90 ft (27+ m) to the east in western Bourbon Co. (FIGURE 4). Selective isopach mapping of this underlying Pleasanton interval is a critical step to defining the depositional topography that influenced the accumulation of the oolites in the Bethany Falls Limestone. Although much of the thinning of the Pleasanton is compensated for by the immediately overlying Critzer and Hertha limestones, there was still sufficient slope to allow high-energy conditions suited for production of oolites. In downslope positions, additional accommodation space during deposition of the upper Bethany Falls allowed for the development of relatively thick oolites (FIGURE 5). These oolites occur as a series of elongate, approximately 1- to 2-mi- (1.6- to 3.2-km) wide accumulations of indeterminate length that are oriented roughly normal to the paleoslope of the Pleasanton Shale (FIGURE 4). These lobes strongly resemble oolitic tidal bars that occur in association with slope breaks in the modern Great Bahamas Bank area, and may well have originated in a similar manner.

In order to better understand the details of the internal geometry of these porous oolitic intervals, a series of six cores and associated wireline logs was acquired across the edge of one of the oolite lobes (FIGURE 6). The distance between cores ranged from 250 to 900 ft (76 to 274 m), which is less than the typical 40-acre well spacing in many oil fields in western Kansas. Within this transect, two porous oolites occur in the Bethany Falls Limestone. The lower oolite is up to 16 ft (4.9 m) thick and the upper oolite is up to 8 ft (2.4+ m) thick. These oolitic zones in the Bethany Falls are coarsest near the center of the inferred tidal bar, and thin and virtually pinch out over distances as little as 1/2 mile (0.8 km), with the lower oolite extending farther from the central axis of the bar (FIGURE 6). As previously stated, these two oolites are separated by either a nonfossiliferous,

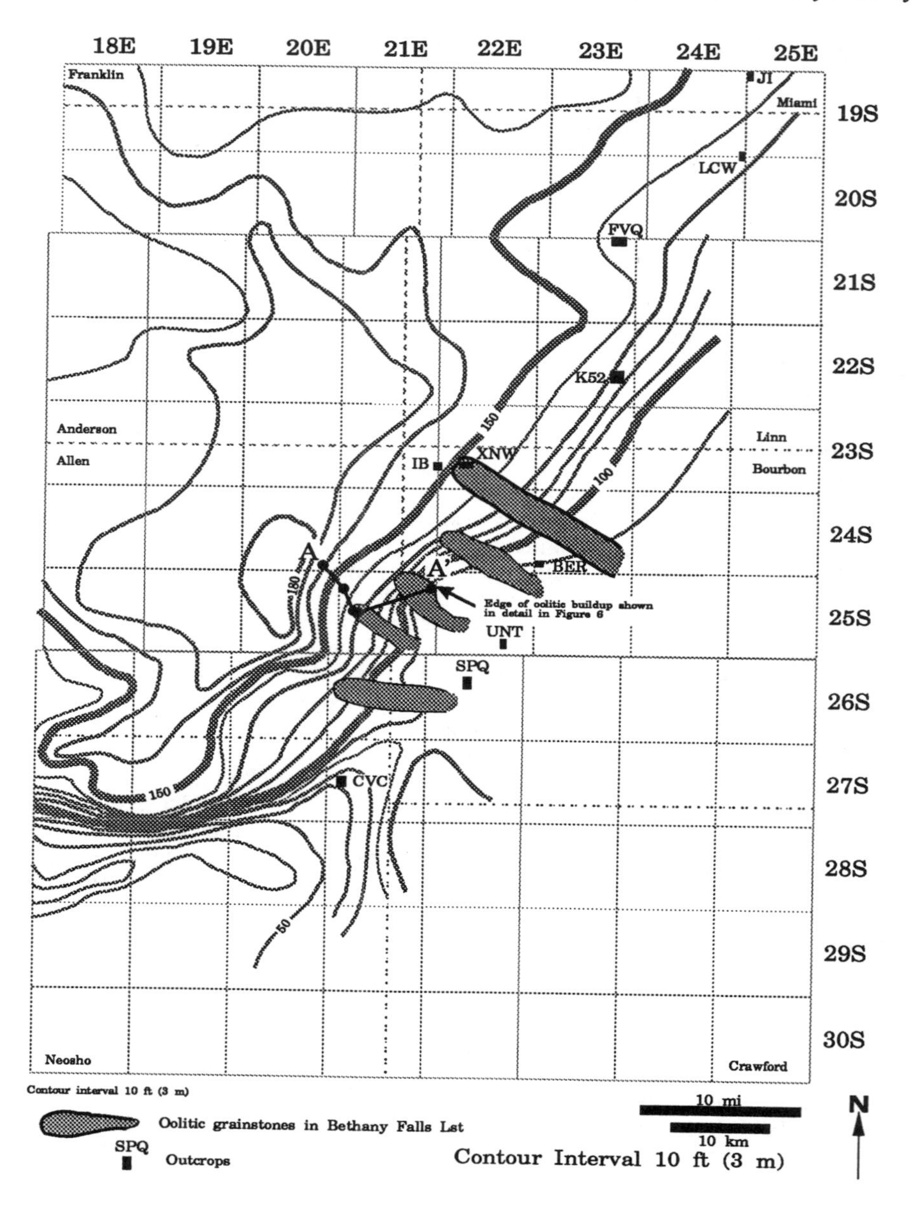

Figure 4. Isopach map of interval from base Nuyaka Creek Shale to top Pleasanton (two cycles below Bethany Falls Limestone). Stippled pattern shows approximate areas where the lower Bethany Falls Limestone-lower Bethany Falls oolite succession is > 25 ft (7.6 m) thick, which in most cases corresponds to areas of thick oolitic grainstone. Also shown is approximate location of detailed core transect at edge of one of the buildups (Figure 5). Also shown is the location of cross

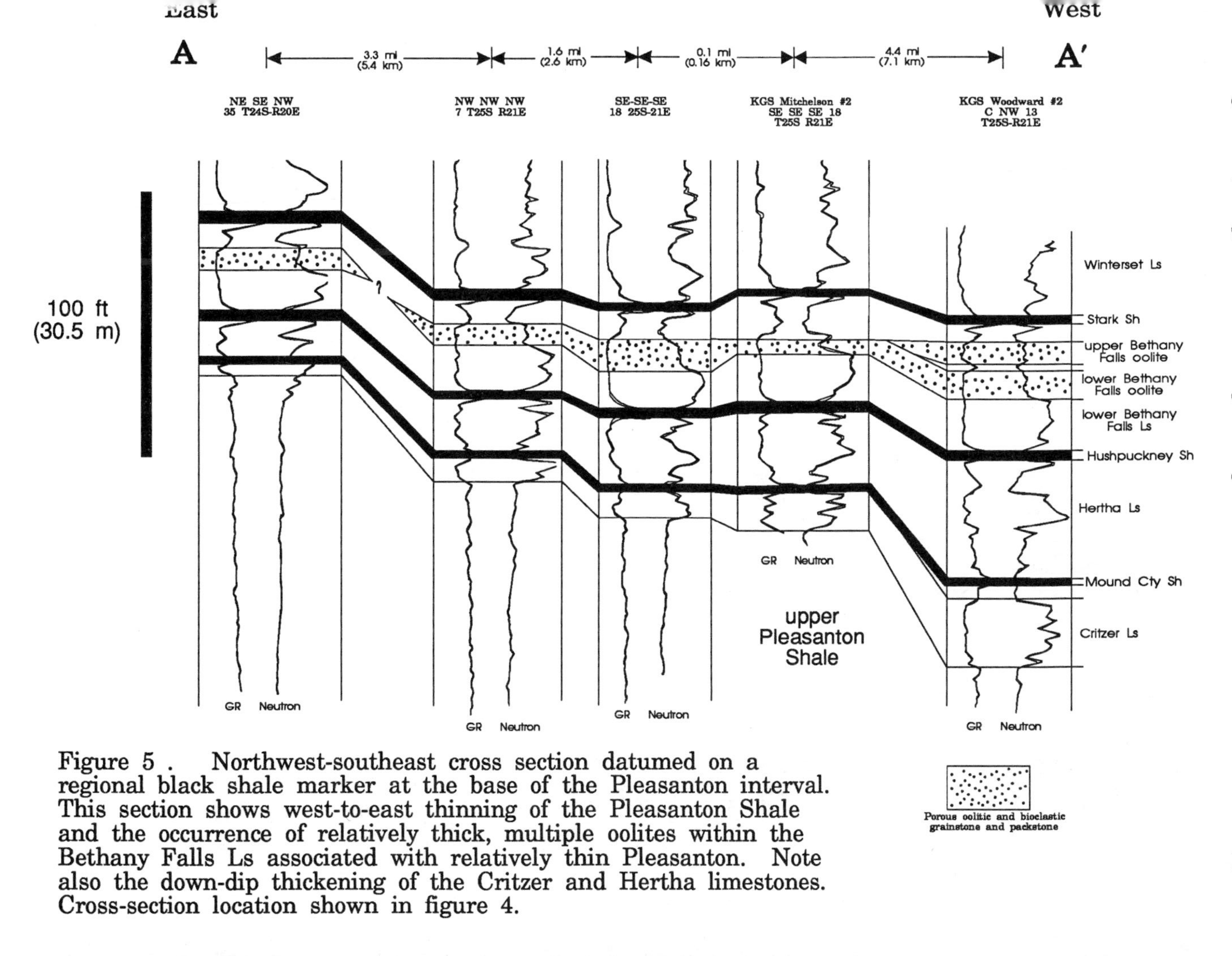

Figure 5 . Northwest-southeast cross section datumed on a regional black shale marker at the base of the Pleasanton interval. This section shows west-to-east thinning of the Pleasanton Shale and the occurrence of relatively thick, multiple oolites within the Bethany Falls Ls associated with relatively thin Pleasanton. Note also the down-dip thickening of the Critzer and Hertha limestones. Cross-section location shown in figure 4.

blocky siliciclastic mudstone roughly 2 ft (0.6 m) thick, (e.g in core 2 in FIGURE 6), or by a nonporous, skeletal-peloidal lime packstone to wackestone that is 2 ft to 6 ft (0.6 m to 1.8 m) thick (cores 3 and 4 in FIGURE 2). The blocky mudstone is interpreted to be at least in part a paleosol, based on the occurrence of rhizoliths and soil peds. The presence of a paleosol between the oolites almost certainly indicates that there was a significant period of subaerial exposure after deposition of the lower oolite. Although all evidence of this paleosol is missing only 900 ft (274 m) to the north in core 3 (FIGURE 6), there is a negative carbon isotope excursion at the top of the lower oolite in core 3, which also suggests that subaerial exposure occurred after deposition of the lower oolite at that location. Similar blocky mudstones and negative carbon isotope excursions are associated with the upper part of the upper oolite as well; these are associated with the major sequence boundary at the top of the Swope. Both of the surfaces of subaerial exposure appear to be regionally significant.

The potential reservoir quality of the two Bethany Falls oolites differs significantly, with the lower oolite tending to have higher porosities and permeabilities (FIGURE 7). These differences are due to variations in factors such as percent ooids, mud content, and early and late diagenesis. Although important, these differences are at best subtly recorded on wireline logs. Commingled production from these oolites could result in selective fluid flow through the more permeable lower zone. Furthermore, the two zones could be hydraulically isolated where the thin, relatively impermeable mudstone or muddy limestone is developed between the oolites. The recognition of such relationships in an oil field could aid in the recovery of additional oil via completion and treatment methods tailored to specific reservoirs.

The upper and lower oolites are part of time-distinct periods of deposition attributed to changes in relative sea level that can be traced along this slope in outcrop and core. As suggested in Figure 4 the lower subaerial surface merges with the upper subaerial surface and sequence boundary in higher-shelf positions. Accommodation space for these oolites diminishes westward, higher on the paleoslope, and accordingly their upper bounding subaerial surfaces converge. The upper oolite is believed to be the first to lap out updip, based on lack of evidence for it in an updip cored and logged well (FIGURE 5). This onlap at a lower elevation suggests a small sea level fluctuation during a slightly lower sea level than associated with the lower oolite. The lower oolite itself probably laps out farther updip (FIGURE 5) as suggested by the presence of several sets of strike-oriented oolite trends based on regional mapping.

It appears that depositional topography impacts the continuity of the oolitic grainstone in the Bethany Falls at a field-scale level. Such continuity or lack thereof could be verified in an oil field by examining pressure and fluid history or running flow and pressure tests from strategically selected wells in conjunction with the mapping of the depositional sequences.

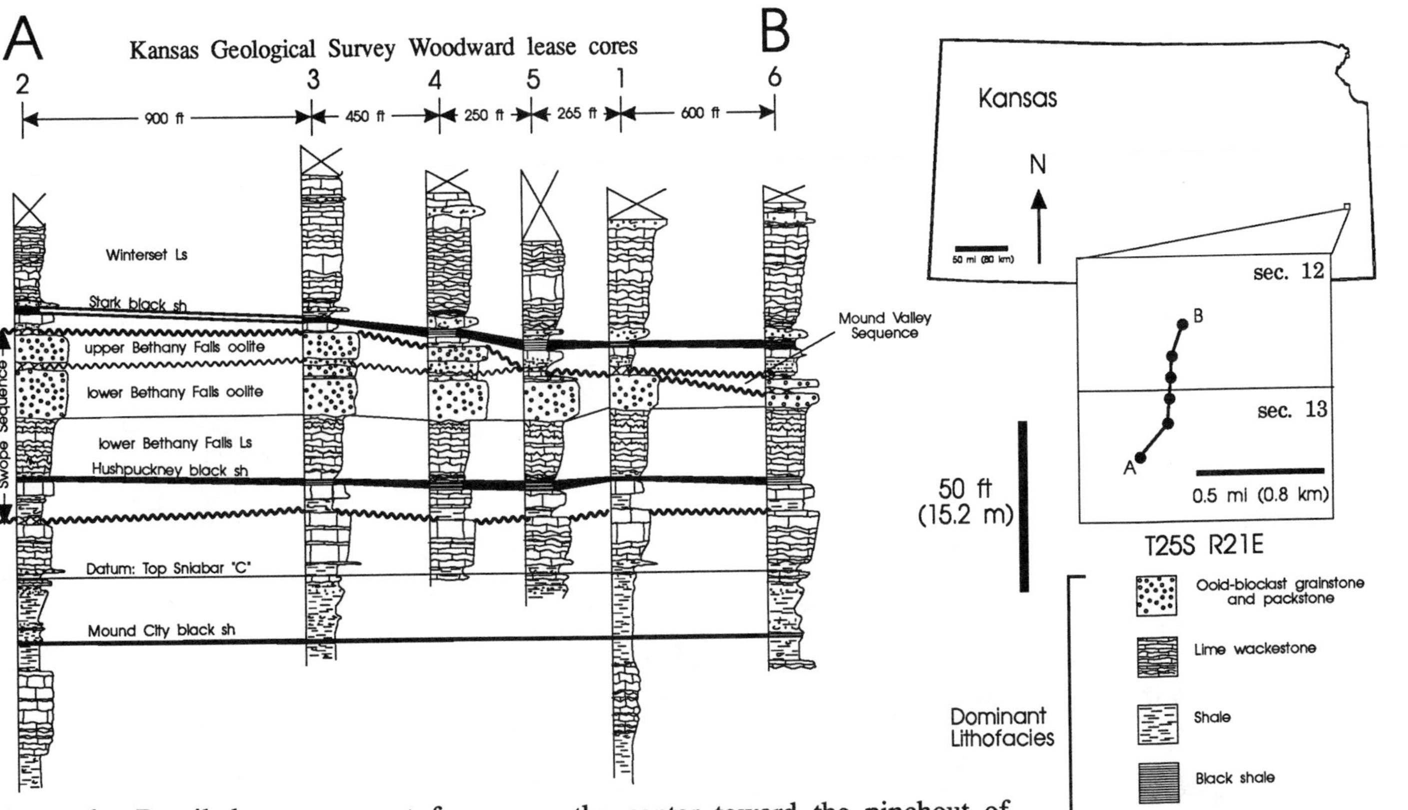

Figure 6. Detailed core transect from near the center toward the pinchout of an oolitic buildup in the Swope depositional sequence. Porosity in these oolites varies both laterally and vertically, being controlled in part by the distribution of paleosols and relatively fine-grained transgressive lithofacies. Cross-section location is shown in fig. 4. See text for additional details.

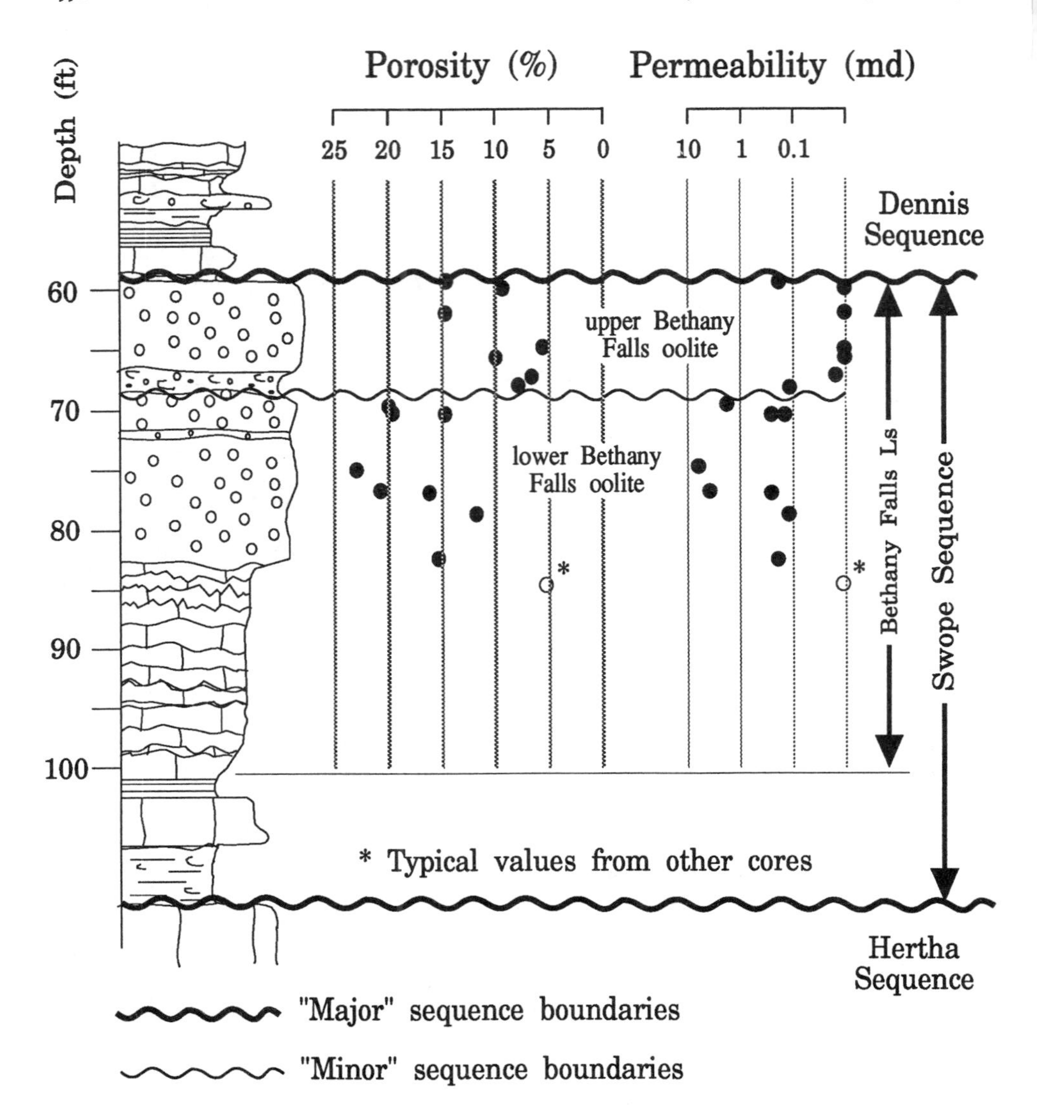

Figure 7.-- Porosity and permeability plots of core plugs from the Kansas Geological Survey Woodward #3 core. Porosity and permeability values in the Bethany Falls are generally higher in the lower oolite as compared to the upper oolite.

III. CONCLUSIONS

Carbonate grainstones similar to those that form important Missourian oil and gas reservoir facies are present in equivalent units at and near the outcrop belt in southeastern Kansas. At least two porous, oolitic zones, separated vertically and laterally by non-porous lithologies, are developed within the Bethany Falls Limestone at a lower shelf position, similar to oil field locations in western Kansas. The deposition, geometries, and distribution of the oolites were probably controlled by fluctuations of relative sea-level acting in concert with favorable depositional topography. These relationships provide evidence for predicting reservoir attributes in equivalent depositional sequences and shelf position in the deeper subsurface.

Detection and mapping changes in elevation of the depositional surface in this near-surface analog are shown to significantly aid in reconstructing and interpreting the spatial characteristics of reservoir rock at the field- and interwell-scale. The progressive changes in topography are estimated by delimiting genetic units within time-distinct, unconformity-bounded depositional sequences.

This paper represents a conceptual departure from conventional facies mapping of a reservoir to assess heterogeneity. The likelihood of changes in relative sea level interacting with depositional topography to produce field-scale and inter-well scale reservoir heterogeneity challenges us to redefine the appropriate scales that are needed in geologic mapping to fully constrain and model observed reservoir geometries in an oil field.

Finally, reservoir-bearing strata extending from the late Mississippian through early Permian are believed to have been influenced by significant excursions in sea level due to the influence of glacial eustasy. Similar approaches could be applied to these reservoirs. The midcontinent U.S., like many other developed provinces, provides an excellent, very accessible subsurface data base from which to make these multi-scale assessments for aid in extracting the remaining oil.

IV. ACKNOWLEDGEMENTS

Thanks is given to Paul Enos and Robert Goldstein for providing both logistical and conceptual ideas. This paper was prepared with the support of the U. S. Department of Energy (DOE Grant No. DE-FG22-90BC14434). However, any opinions, findings, conclusions, or recommendations expressed herein are those of the author and do not necessarily reflect the views of the DOE. This paper was reviewed by Paul Enos who is thanked for his suggestions which significantly improved the manuscript.

CHARACTERIZATION OF RESERVOIR HETEROGENEITIES FROM OUTCROP AND WELLS (UPPER TRIASSIC FLUVIATILE SANDSTONE, CENTRAL SPAIN).

Yves Mathieu
and
HERESIM Group

Institut Français du Pétrole
BP 311
92506 - Rueil Malmaison - France

François Verdier

Gaz de France DETN
93211 - La Plaine Saint Denis - France

I. INTRODUCTION

This paper deals with the initial phase of a project concerning reservoir heterogeneities, named PHEDRE (an IFP/GDF project). It includes an outcrop study and a well survey. All the results obtained on this field example constitute one of the available data banks used by HERESIM group (HEterogeneities REservoir SIMulations) for geostatistical studies.

II. OUTCROP STUDY

A 9 km-long cliff, located in central Spain (Ciudad Real province) has been analyzed. The 35m-thick series belong to the Upper Triassic continental red beds, deposited along the easter border of the Paleozoic Iberian

Meseta. Several geologic and geometric criteria determined the choice of this site:

-analogies with the subsurface case study (age, environment, sand body geometry),

-reliable key beds,

-continuous outcrops along a very long, sinuous, lowly tectonized and easily accessible cliff,

-presence of an open plateau just behind this cuesta, where wells could be drilled.

This part of the study was subdivided into three phases: data acquisition, interpretation and processing.

A. Data acquisition

A detailed lithological interpretation was made of the whole cliff, based on an atlas of 160 high-definition photographs. It was associated with a systematic topographic survey.

A practical facies code based on sandstone clay content (Facies 1 to 3), shale deposit subenvironment (Facies 4 to 6), carbonate presence (Facies 7) was used for the lithological interpretation. About 1000 measurements were performed during the topographic survey.

B. Data interpretation

The geologic series is subdivided into five lithostratigraphic units. On one hand, four fluviatile units, are composed of various channelized reservoir sand bodies and their associated floodplain deposits. On the other hand, a median lacustrine unit is made of purple shales and carbonates, including the main key bed. The sedimentology is typical of Triassic red beds, and the sand body geometry results from the evolution of fluviatile patterns correlated with base level variations (meandering, braiding).

C. Data processing

The cliff lithological draft was rectified by using various topographic methods. The obtained document was discretized along 942 digitized 10m-spaced vertical lines, considered as virtual wells. Hence, these were directly usable for geostatistical studies.

III. WELL SURVEY

A.Well data

Eleven wells (eight cored and three bored), 60 m-deep, were spudded on the plateau according to geologic and geostatistical criteria (less than 300 m spacing). The well grid gave a subsurface equivalent of the cliff section and allowed to extrapolate the observations. A complete data set is available for each well (composite logs, cores, petrophysical measurements, and core photographs)

B.Well data interpretation

It was necessary to establish a link between cliff lithofacies and electrofacies. For example, on logs it is important to discriminate between floodplain and plug shales because the size of the heterogeneities induced by these various argillaceous deposits, is quite different. To obtain this link, the following procedure was adopted:
- creation of a lithologic column for each cored well,
- attribution of a cliff lithofacies code number to each lithologic interval,
- calibration of the logging responses with these intervals,
- creation of a lithofacies column for the three bored wells.

C. Comparison between outcrop and subsurface cross-sections

Without the cliff section interpretated, it was difficult to obtain a reliable and detailed representation of sand body architecture. In consequence, all the subsurface cross sections related to the cliff were strongly influenced by the cliff observations. In this context, it appears that many precautions are necessary to have a satisfactory drawing of sand bodies on an hectometric scale.

IV.CONCLUSION

The result was the acquisition of a large set of data, from an outcrop equivalent to a 5 km-long oilfield in fluviatile series. This is the compulsory first step to make geostatistical calculations for low and medium heterogeneities.

THE GEOSCIENCE-ENGINEERING DATABASE NETWORK

Jim Myers

Cullen College of Engineering, University of Houston, Texas
Institute for Improved Oil Recovery -- Well Logging Lab

I. ABSTRACT

A second generation of conceptual and technological advance following the revolution of plate tectonics theory is beginning. The volume of geologic data and evaluations is expanding and being verified. Engineers use this knowledge on earth formations to guide and parameterize their work. They require in-depth access to this growing volume.

Access will become practical when engineers and geoscientists begin submitting their data, hypotheses, and questions to an international network of geoscience and engineering databases organized according to scientific schema, the geoscience-engineering database network, or GEDN. The breadth of questions and hypotheses to be posed and answers to be sought defies most other approaches.

Function for the engineer exploiting natural resources is most efficient with commitment to geoscientists, with their traditions and cultures of examination, description, and imagination, as colleagues. The engineer's quantitative results and evaluations are geologic data, but often are not analyzed and archived as such.

The hierarchical systems of geologic classification are scientific relationships. They hold great promise as keys for creating and using databases to organize engineering data and related textual information for easy retrieval. The development and maturity of engineering and scientific concepts will be accelerated by use of these structures.

Libraries of spatial data structures for geographic aspects, geographic information systems (GIS's), have been introduced. International electronic mail networks and structured query languages (SQL's) are likely structures for entry, retrieval, review, and query formats and utilities to support various industrial and academic activities. All these structures are available now.

Thus the activities and results of engineers and geoscientists can complement each other to allow analysis beyond local trends, rules of thumb, and current understanding. This will maximize our opportunities for identification and transfer of data and concepts between distant reservoirs with geologic similarities. Concepts and methods for marginally similar reservoirs, processes, traps, etc. can also be researched and considered for practical application.

II. INTRODUCTION: ECONOMIC STRATEGY JUSTIFIES GEDN

The international community can be classified, according to the economics of their petroleum consumption, into 3 groups:

1. nations glutted with reserves, like Saudi Arabia and Venezuela,
2. nations importing large petroleum volumes, and
3. undeveloped nations without significant production.

The world of petroleum technology can also be analyzed; significant expertise is highly localized. The nations of Western Europe have joined the U.S. as major sources of engineering expertise and geologic insight. They also share U.S. long-term expectations for petroleum imports. Their unique resource exploitable to retain and increase access, consideration, influence, investment, and confidence is superior access to data and professional excellence.

The value of knowledge-intense resources is directly related their depth and degree of organization. The scope of GEDN is outlined in Table 1. A dominantly positive factor encouraging rapid GEDN implementation is the recent warming of the Cold War. Peace threatens large industrial forces worldwide; necessary reordering of industry to adapt to safer international politics may allow emphasis on correct exploitation and management Earth resources.

Giant market sectors are opening to Western producers. More importantly, access is available to new geologic activities. Each oil company is potentially a think-tank. These companies support massive data libraries. Thus the GEDN has potentially vast geo-political potential. Finding and maximizing production is the first application anticipated. This list summarizes that applicaton; GEDN promises to return value to supporting entities in the following manners:

1. Allow the operator to locate data in residence otherwise inconvenient.
2. Allow the operator to broker his proprietary data with enhanced efficiency.
3. Offer the independent operator more efficient access to the proprietary data of major operators, providing a fee.
4. Support the free exchange of nonproprietary data.
5. Promote assembly of disjoint data into large, coherent structures.

The GEDN will allow traders, refiners, tankers, and pipelines to plan their facilities with the advantage of comprehensive graphic overviews. It will allow petrophysicists and geoscientists to access the porosity, permeability, wettability, and mineral associations they need to guide field development and completion design. Rig designers and service units will have additional input to improve their machines and procedures.

V. GEDN SCOPE

The database network will contain a wide variety of technical data and text within a framework unified by geology and application elements. Its scope is summarized in outline form in Table 1.

Table 1. The scope of GEDN user disciplines and data examples.

technical discipline	GEDN data examples
A. geography	general
B. geomorphology and remote sensing	surface reflectances
C. sedimentology	provenance, climate
D. geochemistry	
1. coal, petroleum, and oil shale	compositions
2. water	solute ions
3. atmosphere	global temperature
4. rock	wettabilities
E. mineralogy and petrology	mineral and rock types
F. historical geology	micropaleo. interpretations
G. paleontology	
1. micropaleontology	micrograph summaries
2. vertibrate paleontology	general summaries
H. structural geology	formation tops
I. stratigraphy	Markov Chain factors
J. geophysics	parameters, events
K. geostatistics	variogram equations
L. atmospheric studies	
1. climatology	long-term weather
2. meteorology	short-term weather
M. hydrology	runoff estimates
N. soil science	soil analysis
O. botany and zoology	general
P. architecture and urban studies	climate, land use
Q. chemical engineering	catalyst precursors
R. environmental engineering	
a. air quality	contaminant sources
b. groundwater	retardation factor
c. solid waste disposal	soil analysis
S. geotechnical engineering	soil analysis
T. petroleum engineering	
1. reservoir engineering	
a. oil reservoirs	
1. primary recovery	volumes, drive type
2. secondary recovery	rock wettabilities
3. enhanced oil recovery	$k_{rog}(S_g)$, $k_{row}(S_w)$
b. gas reservoirs	
1. conventional	volume, pVT data
2. unconventional	coal dewatering processes
2. formation evaluation	
a. petrophysics	$k(\phi)$, mineral associations
b. shaly sands	CEC, $V_{sh}(I_{\gamma-ray})$, ρ_b
c. fractured and vuggy reservoirs	m, n, core description
d. testing and sampling	mechanical properties
3. drilling engineering	abnormal pressure clues
4. production engineering	formation sensitivities

Near the top of the conceptual hierarchy of GEDN scope (not shown) are the systems of basin classification, organizing all data according to basin type (St. John *et al*, 1984). For brevity and clarity, only Klemme's classification scheme is summarized here in Table 2.; the reader is refered to the same reference for the scheme of Bally and Snelson. Other elevâted elements of GEDN hierarchy might be current climatic classification, demographic parameters, industrial parameters, parameters related to environmental concerns, etc, or any of the myriad specialized geologic systems.

Table 2. Klemme basin classification.

basin type:	regional stress	areal shape	basement profile	ratio: volume -area[1]	sequential architectures
I. craton interior basin type:					
	exten / sag	circular-elongate	symmetric	.95	sag
II. continental multicycle basin types:					
A. craton margin	comp / exten	elongate-circular	asymmetric	1.95	fordeep, sag, platform
B. craton/ac-creted margin	sag / exten	random	symmetrically irregular	1.6	sag or rift
C. crustal collision zone	comp / exten	elongate, large	assymetric	>2.5	fordeep, sag, platform
III. continental rifted basin types:					
A. craton/ac-reted zone	exten / wrench	elongate	irregular	2.35	rift or sag
B. rifted conver-gent margin	exten + wrench + comp	elongate	irregular	1.8	rift /wrench rift /sag
C. divergent margin	exten.	elongate	assymetrically irregular	2.0	rift / drift rift - 1/2 sag
IV. delta-type basin:					
	exten/sag	circ-elong	depocenter	3.5	modified sag
V. convergent margin (forearc):					
	comp/ext	elongate	assymetric	?	subduction

[1] Unit of basin volume-area ratio is cubic meter per square meter.

VI. GEDN APPLICATION

A wide range of industrial, academic, and governmental activities, humanitarian and economic are accessible using GEDN. All natural resources have resulted from the complex interplay and overprint of terrestrial physics and chemistry. Plate tectonics, global climate, and evolution of life are processes by which solar energy and geothermal energy have combined to shape the Earth's crust.

Patterns and cycles of sea level rise and glaciation, mountain building, vulcanism, and diapirism are spectacular results. These control deposition, burial, compaction and diagenesis, uplift, faulting, and metamorphosis of subsurface sediments. The geoscientist and engineer will use the GEDN differently. The geologist will dominate in qualitiative aspects, entering his interpretations, evaluating and enhancing other interpretations in his personal use. He is the dominant generator, referee, and presenter of the GEDN.

The engineer generates, enhances, and/or evaluates certain types of quantitative data. He presents data from GEDN to support his results or to facilitate evaluation of data and results. He uses geologic interpretations as thought-models for his evaluations. His input as a referee will be vital.

Management personnel will use GEDN. Preliminary to any transfer of proprietary information, GEDN elements allow systematic inventory of resources. Joint venture partners can use GEDN to transmit, capture, review, and critique elements of each interpretation. This process will assist in holding information proprietary when necessary, and offers the practical and attractive capacity to set far in advance the date such information is released.

A. evaluation of phenomena

The phenomena which result in petroleum accumulation include structure and stratigraphy, petroleum generation and migration, petroleum properties, capillary and wetting forces, and hydrodynamics. They are studied by collection of data such as porosity, permeability, composition, fluid saturations, etc. All these are within GEDN scope (Table 1.).

These phenomena are summarized for petroleum exploration by the classic elements "source", "trap", "seal", and "reservoir". The performance of a producing property is described by the property's recoverable reserves, the nature of the product underground and at the surface, and by the available production rate.

An example is now attempted; estimating before development the performance of a developed field is the classic exploitation problem. Oil displacement modeling depends upon relative permeability data, which is subject to considerable alteration by rescaling from the saturation dependence measured from core plugs in the laboratory. A different scaling problem exists with data from sections of whole core. A similar scaling problem is encountered in evaluation of other transport parameters, like diffusivity or dispersion constant. Documentation of scaling methods in the GEDN may assist some users.

Similarly, comparison of wireline log data to data from core plugs or whole cores is problematic. It is likely that the GEDN team responsible will not attempt to database such high-density data as complete well logs; wherever core data is available, however, its value would be multiplied by including documented log data within the tab file. This petrophysical file would then be available to professionals studying formations similar generically or genetically.

B. evaluation of processes

A process is a phenomenon which has been studied sufficiently for description as a formal event. All natural resources result from the geologic processes addressed by the GEDN.

The GEDN has potential applications far outside the petroleum industry, however. Preservation of human quality of life and even life of Earth itself may eventually be controlled by human manipulation of natural resources. For example, the enzymes in migratory birds allow their paths to be retraced regardless of capture location. Zoologists in one location to evaluate toxin buildup in birds; the birds blood enzymes then allow the birds toxic buildup to be assigned to specific sedimentary basins continents away. This technology offers environmental early warning potential.

Processes such as aquifer recharge, groundwater contaminant transport, air and surface water pollution will be studied. Agricultural studies of soil, air, water, and plant, animal, and human life will be conducted. All these phenomena, processes and subprocesses can be studied with GEDN resources. The large-scale features of Earth will be accessible for such studies as these and for studies which current investigators cannot envision.

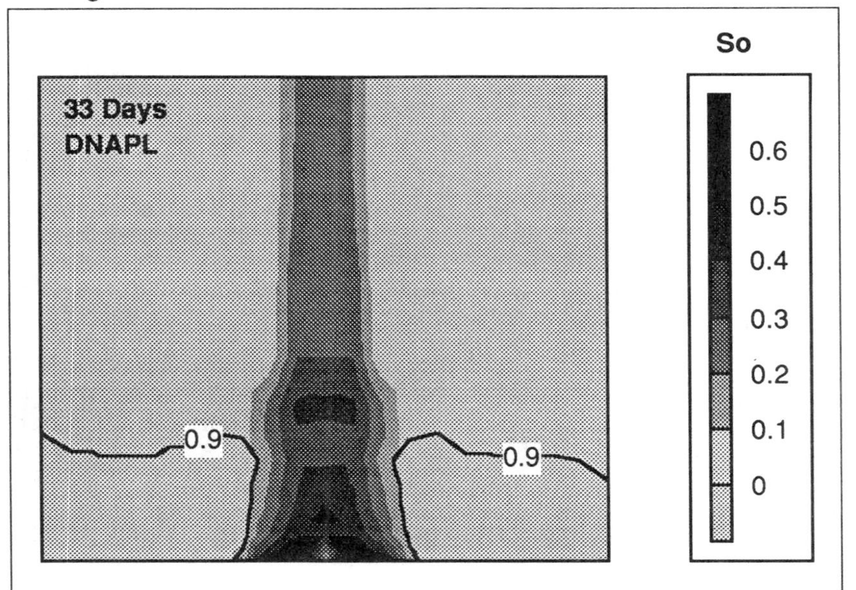

Figure 1. Contaminant saturation: application of E&P simulator to shallow soil contaminated by dense solvent, fuel, or lubricant, which falls through vadose ($S_W < 100$) zone (Hinckley and Killough, 1992).

A final example outside exploration and production is now considered. Figure 1. shows the scenario encountered during evaluation or remediation of the site where a dense non-acqueous liquid (DNAPL) has been spilled on a porous surface, gradually entering the porous medium and migrating under the influences of gravity and hydrodynamics.

Similar events occur frequently worldwide. The key parameters for evaluation include fluid properties, properties of soil, aquifer, and the vadose transition between soil and aquifer, soil-fluid interations, and geometric considerations. These are presented in Table 3. along with identification and communication information to facilitate interdisplinary contact. GEDN could be helpful in transfering valuable concepts and data between groups with backgrounds as dissimilar as those of groundwater contamination specialists and enhanced oil recovery specialists.

Table 3. Example database entry to facilitate communication between groups using different resources to address similar problems.

application:	groundwater contamination and transport, hydrodynamic velocity= 0.1 ft/d
contaminant type:	dense non-acqueous-phase liquid (DNAPL), pseudo-TCE, ρ_{DNAPL}=1.5 g/cc
study approach:	2D finite-difference simulation, semi-compositional, k-value, resolution 20 by 21
group:	Institute for Improved Oil Recovery, University of Houston, Drs. Kovarik, Killough, and Mohanty
electronic mail to:	johnk @lnic2.hprc.uh.edu

This application appears to fall in the domain of environmental engineering; the reservoir simulators developed for petroleum studies, however, have greatly superior capacity to model many aspects of the contaminant transport and remediation processes. The crossectional displays of DNAPL saturation in Figure 1. result from the application of a modern multiphase reservoir simulator to the DNAPL problem.

Here the value of an expensive product, already developed, is enhanced by extending its utility from petroleum studies to environmental studies. At a time when vendors of reservoir simulation software have saturated and limited markets, these new applications hold vital significance as sources of software development funds. Improved support of the software vendor benefits not only the vendor but all supporters.

While this saturation data and image files may be too detailed for GEDN entry, the entry of the names and types of simulation models and output files used for various scientific and engineering tasks in various basins will greatly facilitate this sort of interdisciplinary crosspollination. New compression algorithms and formats are so efficient to suggest limited online presentation, sale, and barter of compressed images.

IV. SUMMARY

The modular components and functions of GEDN are summarized in Appendix 1. A plan for developing the modules of GEDN is presented in Appendix 2.

Establishment of the GEDN will be a landmark events in the evolution of geoscience. The GEDN offers utility which can enhance the capacity of many geoscientists and engineers to make the breakthroughs society must anticipate for our societies to continue to thrive and advance. It has the potential to extend the value of each investigator's contribution farther and faster. It is economically attractive by virtue of its capacity to assist investigating operators to broker the data they own and to acquire data owned by other operators.

V. ACKNOWLEDGEMENTS

Dr. Marcelo Ramé has assisted considerably in assembling this paper and its poster presentation. Dr. Liang Shen and the Well Logging Lab and Dr. John Killough and the Institute for Improved Oil Recovery have provided support and permission.

This paper is dedicated to the beloved memory of the late Dr. Wilmer A. Hoyer, my mentor and academic advisor in 1987-88.

IX. REFERENCES

Bally, A.W., 1975, A geodynamic scenario for hydrocarbon occurrences, in Proceedings of the 9th World Petroleum Congress, vol. 2 (geology): Essex, England, Applied Sci. Pub., Ltd., p. 33-44.

Hinkley, R. and Killough, J., Simulation of Multi-Phase Contaminant Transport in the Vadose Zone: Model Case Studies and an Exploration of Air-Phase Dynamics, Proceeding of Computational Methods in Water Resources IX, Volume II: Numerical Methods in Water Resources, Elsevier, 1992.

Lake., L., Caroll, H.B., and Wesson, T.C., 1989, Reservoir Characterization II: London, England, Academic Press.

Samet, H., Applications of spatial data structures: computer graphics, image processing, and GIS, Addison-Wesley, 1990.

St. John, Bill, Bally, A.W., Klemme, H.D., Sedimentary Provinces of the World -- Hydrocarbon Productive and Nonproductive, AAPG, Tulsa, Oklahoma, 1984.

Appendix 1. NETWORK COMPONENTS AND FUNCTIONS

Functionality for the geoscience database network (GEDN) requires the elements outlined in Outline A1. Functionality is considered at two levels. The basic level suitable perhaps for "beta" testing of basis software is described by outline entries without asterisks. Entries with asterisks refer to the fully functional system with graphics and file transfer optimized to the results of "beta" tests.

In addition to the intrinsic advantages GEDN presents as a database network, each module within the structure is designed to support user graphics. While uncompressed images will not be stored, the data required to create images will be available so as to allow creation of unique original cross-sections, stratigraphic columns, and diagrams.

Outline A1. Functional features of GEDN are supported by these hardware and software elements.

1. local user facilities:
   a. smart terminal and modem
   b. graphic monitor and processor [1]
   c. convenient hard copy support [1]
2. international communication facilities:
   a. electronic mail networks
   b. file transfer utilities [1]
3. virtual and physical devices hold the databases, requiring --
   a. long-term data security system
   b. sufficient mass storage allocation.
4. GIS elements (Samet, 1990):
   a. spatial data structures
   b. conversion of local coordinates to world coordinates
5. Standard data formats, approved and respected internationally.
   a. allow effective archival
   b. eliminate compromise to data security
   c. facilitate international communication [1]
   d. support efficient object-oriented graphics [1]

[1] Efficient functionality for the GEDN would require these additional elements. Extension of the GEDN design to graphics allows the user to quickly create maps, cross-sections, engineering and scientific diagrams, and abstract feature illustrations for pattern recognition activities.

The system retains speed, using compressed image formats and sparse data sets in object-oriented graphic routines, not bit maps. These graphic presentations will assist users in detecting similarities and contrasts between geologic systems.

Appendix 2. DEVELOPMENT OF THE GEDN

In specification and development of GEDN, the vital qualities are the structured, modular, and portable nature of the design. The GEDN can only be developed through cooperative interdisciplinary work. Here are suggested steps for the project, requiring about 5 years to end of "beta" phase:

1. initial approval by industry and professional organizations
2. formation of grassroots user specification groups
   a. local groups for individual basins
   b. narrow groups for specialized databases
3. interdisciplinary critique of specifications
4. develop top-down scheme for implementation of features in
   a. Outline A1.
   b. grassroots user specifications
4. prototype implementation of selected GEDN modules
5. interdisciplinary critique of implementation
6. respecification with additional modularity and portability
7. implementation of "beta test" system(s)
8. respecification and final implementation.

Grassroots user specification groups will be formed by professional organizations like SEG, AAPG, SPE, SEPM, SPWLA, etc. Table A2. presents a prototype tmembership for a group to consider the data structures for an individual sedimentary basin (Anadarko, Permian, etc).

The term "knowledge engineer" is borrowed from the intelligent systems community. The knowledge engineer is an individual trained in elements of intelligent systems, computer graphics, and software engineering, with intimate familiarity and experience with geoscience and engineering principles and activities. Successful recruiting for this position is critical, as this individual is responsible for "the big picture" encompassing geoscience, engineering, and data processing.

Table A2. GEDN basin module development team description. The practical size for a team is between 5 and 10 members. The members' expertise must encompass virtually the entire scope of GEDN, in contrast to the members required to work on a specialized database.

member	data focus[2]	background
knowledge engineer	overall	comprehensive
global data specialist.	basin	geomorph., remote sensing
basin studies specialist	sub-basin	geomorph., sediment.
reservoir data specialist	field	reservoir engr. or geol.,
well data specialist	wellbore	petrophys., well tester, geol.
laboratory data specialist	sample analysis	mineral., petrol., petrophys.
geophysicist	field	geophys., petrophys.,
geostatistician	interwell	reservoir engr. or geol.
applications engineer	software	computer engr. or science

[2] These data scale descriptions are sometimes stated more generally: applicable terms are "microscopic", "meso-", "macro-", and "megascopic". (Lake *et al*, 1989)